MW00650498

# 5 Reasons to Use MyMathLab® for Applied Calculus

**1** **Thousands of high-quality exercises.** Algorithmic exercises of all types and difficulty levels are available to meet the needs of students with diverse mathematical backgrounds. We've also added even more conceptual exercises to the already abundant skill and application exercises.

**2** **Helps students help themselves.** Homework isn't effective if students don't do it. MyMathLab not only grades homework, but it also does the more subtle task of providing specific feedback and guidance along the way. As an instructor, you can control the amount of guidance students receive.

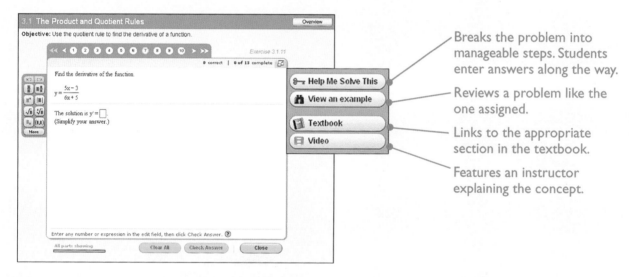

Breaks the problem into manageable steps. Students enter answers along the way.

Reviews a problem like the one assigned.

Links to the appropriate section in the textbook.

Features an instructor explaining the concept.

**3** **Addresses gaps in prerequisite skills.** Our *Getting Ready for Applied Calculus* content addresses gaps in prerequisite skills that can impede student success. MyMathLab identifies precise areas of weakness, then automatically provides remediation for those skills.

**4** **Adaptive Study Plan.** MyMathLab's Adaptive Study Plan makes studying more efficient and effective. Each student's work and activity are assessed continually in real time. The data and analytics are used to provide personalized content to remediate any gaps in understanding.

**5** **Ready-to-Go Courses.** To make it even easier for first-time users to start using MyMathLab, we have enlisted experienced instructors to create premade assignments for the Ready-to-Go Courses. You can alter these assignments at any time, but they provide a terrific starting point, right out of the box.

Since 2001, more than 15 million students at more than 1,950 colleges have used MyMathLab. Users have reported significant increases in pass rates and retention. Why? Students do more work and get targeted help when they need it. See www.mymathlab.com/success_report.html for the latest information on how schools are successfully using MyMathLab.

## Learn more at www.mymathlab.com

# Calculus for the Life Sciences

## SECOND EDITION

Raymond N. Greenwell
Hofstra University

Nathan P. Ritchey
Edinboro University

Margaret L. Lial
American River College

**PEARSON**

Boston   Columbus   Indianapolis   New York   San Francisco   Upper Saddle River
Amsterdam   Cape Town   Dubai   London   Madrid   Milan   Munich   Paris   Montréal   Toronto
Delhi   Mexico City   São Paulo   Sydney   Hong Kong   Seoul   Singapore   Taipei   Tokyo

Editor in Chief: Deirdre Lynch
Executive Editor: Jennifer Crum
Content Editors: Christine O'Brien and Katherine Roz
Editorial Assistant: Joanne Wendelken
Senior Managing Editor: Karen Wernholm
Production Project Manager: Sherry Berg
Associate Director of Design, USHE EMSS/HSC/EDU: Andrea Nix
EMSS/Program Design Lead: Heather Scott
Digital Assets Manager: Marianne Groth
Media Producer: Jean Choe
Software Development: Kristina Evans
Executive Marketing Manager: Jeff Weidenaar
Marketing Assistant: Brooke Smith
Senior Author Support/Technology Specialist: Joe Vetere
Rights and Permissions Advisor: Joseph Croscup
Image Manager: Rachel Youdelman
Senior Procurement Specialist: Carol Melville
Production Coordination, Composition, and Illustrations: Cenveo® Publisher Services
Cover Design: Heather Scott
Cover Image: Nagel Photography/Shutterstock

Credits appear on pages C-1 and C-2, which constitute a continuation of the copyright page.

Many of the designations used by manufacturers and sellers to distinguish their products are claimed as trademarks. Where those designations appear in this book, and Pearson was aware of a trademark claim, the designations have been printed in initial caps or all caps.

**Library of Congress Cataloging-in-Publication Data**
Greenwell, Raymond N., author.
  Calculus for the life sciences / Raymond N. Greenwell, Hofstra University,
Nathan P. Ritchey, Edinboro University of PA, Margaret L. Lial, American River
College.—Second edition.
     pages cm
  ISBN-13: 978-0-321-96403-8 (hardcover)
  ISBN-10: 0-321-96403-9 (hardcover)
  1. Calculus—Textbooks.   2. Life sciences—Mathematics—Textbooks.   I. Ritchey,
Nathan P., author.   II. Lial, Margaret L., author.   III. Title.
  QA303.2.G74 2014
  570.1'515—dc23
                          2013035845

19 2022

www.pearsonhighered.com

ISBN-10: 0-321-96403-9
ISBN-13: 978-0-321-96403-8

# Contents

Preface    vii

Prerequisite Skills Diagnostic Test    xvii

**CHAPTER R**

## Algebra Reference    R-1

R.1    Polynomials    R-2

R.2    Factoring    R-5

R.3    Rational Expressions    R-8

R.4    Equations    R-11

R.5    Inequalities    R-16

R.6    Exponents    R-21

R.7    Radicals    R-25

**CHAPTER 1**

## Functions    1

1.1    Lines and Linear Functions    2

1.2    The Least Squares Line    17

1.3    Properties of Functions    29

1.4    Quadratic Functions; Translation and Reflection    41

1.5    Polynomial and Rational Functions    51

CHAPTER 1 REVIEW    63

EXTENDED APPLICATION    Using Extrapolation to Predict Life Expectancy    70

**CHAPTER 2**

## Exponential, Logarithmic, and Trigonometric Functions    72

2.1    Exponential Functions    73

2.2    Logarithmic Functions    84

2.3    Applications: Growth and Decay    96

2.4    Trigonometric Functions    102

CHAPTER 2 REVIEW    117

EXTENDED APPLICATION    Power Functions    124

**CHAPTER 3**

## The Derivative    127

3.1    Limits    128

3.2    Continuity    146

3.3    Rates of Change    155

3.4    Definition of the Derivative    167

3.5    Graphical Differentiation    185

CHAPTER 3 REVIEW    192

EXTENDED APPLICATION    A Model for Drugs Administered Intravenously    197

**CHAPTER 4**

## Calculating the Derivative    201

4.1    Techniques for Finding Derivatives    202

4.2    Derivatives of Products and Quotients    214

4.3    The Chain Rule    220

4.4    Derivatives of Exponential Functions    230

4.5    Derivatives of Logarithmic Functions    239

4.6    Derivatives of Trigonometric Functions    247

CHAPTER 4 REVIEW    254
EXTENDED APPLICATION    Managing Renewable Resources    258

**CHAPTER 5**

## Graphs and the Derivative    261

5.1    Increasing and Decreasing Functions    262

5.2    Relative Extrema    273

5.3    Higher Derivatives, Concavity, and the Second Derivative Test    284

5.4    Curve Sketching    298

CHAPTER 5 REVIEW    308
EXTENDED APPLICATION    A Drug Concentration Model for Orally Administered Medications    314

**CHAPTER 6**

## Applications of the Derivative    316

6.1    Absolute Extrema    317

6.2    Applications of Extrema    326

6.3    Implicit Differentiation    337

6.4    Related Rates    343

6.5    Differentials: Linear Approximation    350

CHAPTER 6 REVIEW    356
EXTENDED APPLICATION    A Total Cost Model for a Training Program    360

**CHAPTER 7**

## Integration    362

7.1    Antiderivatives    363

7.2    Substitution    375

7.3    Area and the Definite Integral    385

7.4    The Fundamental Theorem of Calculus    398

7.5    The Area Between Two Curves    412

CHAPTER 7 REVIEW    418
EXTENDED APPLICATION    Estimating Depletion Dates for Minerals    424

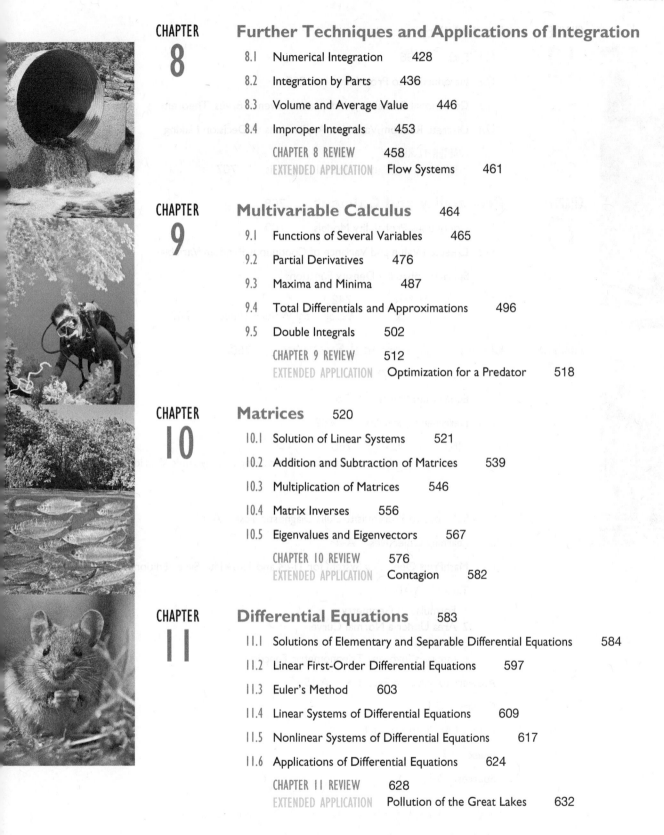

CHAPTER **8**

## Further Techniques and Applications of Integration    427

8.1 Numerical Integration    428

8.2 Integration by Parts    436

8.3 Volume and Average Value    446

8.4 Improper Integrals    453

CHAPTER 8 REVIEW    458

EXTENDED APPLICATION    Flow Systems    461


CHAPTER **9**

## Multivariable Calculus    464

9.1 Functions of Several Variables    465

9.2 Partial Derivatives    476

9.3 Maxima and Minima    487

9.4 Total Differentials and Approximations    496

9.5 Double Integrals    502

CHAPTER 9 REVIEW    512

EXTENDED APPLICATION    Optimization for a Predator    518


CHAPTER **10**

## Matrices    520

10.1 Solution of Linear Systems    521

10.2 Addition and Subtraction of Matrices    539

10.3 Multiplication of Matrices    546

10.4 Matrix Inverses    556

10.5 Eigenvalues and Eigenvectors    567

CHAPTER 10 REVIEW    576

EXTENDED APPLICATION    Contagion    582


CHAPTER **11**

## Differential Equations    583

11.1 Solutions of Elementary and Separable Differential Equations    584

11.2 Linear First-Order Differential Equations    597

11.3 Euler's Method    603

11.4 Linear Systems of Differential Equations    609

11.5 Nonlinear Systems of Differential Equations    617

11.6 Applications of Differential Equations    624

CHAPTER 11 REVIEW    628

EXTENDED APPLICATION    Pollution of the Great Lakes    632

**CHAPTER 12**

## Probability 635

12.1 Sets 636

12.2 Introduction to Probability 650

12.3 Conditional Probability; Independent Events; Bayes' Theorem 668

12.4 Discrete Random Variables; Applications to Decision Making 687

CHAPTER 12 REVIEW 699

EXTENDED APPLICATION Medical Diagnosis 707

**CHAPTER 13**

## Probability and Calculus 708

13.1 Continuous Probability Models 709

13.2 Expected Value and Variance of Continuous Random Variables 719

13.3 Special Probability Density Functions 728

CHAPTER 13 REVIEW 742

EXTENDED APPLICATION Exponential Waiting Times 747

**CHAPTER 14**

## Discrete Dynamical Systems 750

14.1 Sequences 751

14.2 Equilibrium Points 756

14.3 Determining Stability 762

CHAPTER 14 REVIEW 766

EXTENDED APPLICATION Mathematical Modeling in a Dynamic World 768

Appendix

A Solutions to Prerequisite Skills Diagnostic Test A-1

B Learning Objectives A-4

C MathPrint Operating System for TI-84 and TI-84 Plus Silver Editions A-8

D Tables A-10
   1 Formulas of Geometry
   2 Area Under a Normal Curve
   3 Integrals
   4 Integrals Involving Trigonometric Functions

Answers to Selected Exercises A-15

Credits C-1

Index of Applications I-1

Index I-5

Sources S-1

Special Topics Available in MyMathLab to Accompany *Calculus for the Life Sciences*

Markov Chains (online)

   Basic Properties of Markov Chains
   Regular Markov Chains
   Absorbing Markov Chains

# Preface

*Calculus for the Life Sciences* emphasizes those aspects of calculus most relevant to the life sciences. The application examples and exercises are drawn predominately from biology, medicine, ecology, and other life sciences. A prerequisite of two years of high school algebra is assumed.

## Our Approach

Our main goal is to present applied calculus in a concise and meaningful way so that students can understand the full picture of the concepts they are learning and apply it to real-life situations. This is done through a variety of ways.

**Focus on Applications** Making this course meaningful to students is critical to their success. Applications of the mathematics are integrated throughout the text in the exposition, the examples, the exercise sets, and the supplementary resources. *Calculus for the Life Sciences* presents students with myriad opportunities to relate what they're learning to life science and career situations through the *Apply It* questions, the applied examples, and the *Extended Applications*. To get a sense of the breadth of applications presented, look at the Index of Applications in the back of the book or the extended list of sources of real-world data at the back of this text and on www.pearsonhighered.com/mathstatsresources.

**Pedagogy to Support Students** Students need careful explanations of the mathematics along with examples presented in a clear and consistent manner. Additionally, students and instructors should have a means to assess the basic prerequisite skills. This can now be done with the *Prerequisite Skills Diagnostic Test* located just before Chapter R. In addition, the students need a mechanism to check their understanding as they go and resources to help them remediate if necessary. *Calculus for the Life Sciences* has this support built into the pedagogy of the text through fully developed and annotated examples, *Your Turn* exercises, *For Review* references, and supplementary material.

**Beyond the Textbook** Students today take advantage of a variety of resources and delivery methods for instruction. As such, we have developed a robust MyMathLab® course for *Calculus for the Life Sciences*. MyMathLab has a well-established and well-documented track record of helping students succeed in mathematics. The MyMathLab online course for *Calculus for the Life Sciences* contains exercises to challenge students and provides help when they need it. Students who learn best by seeing and hearing can view section- and example-level videos within MyMathLab. These and other resources are available to students as a unified and reliable tool for their success.

## New to This Edition

Based on the authors' experience in the classroom along with feedback from many instructors across the country, the focus of this revision is to improve the clarity of the presentation and provide students with more opportunities to learn, practice, and apply what they've learned on their own. This is done both in the presentation of the content and in new features added to the text.

### New Textbook Features

- **Exercises and examples have been updated** to reflect the latest data and to incorporate feedback received on the previous version of the text.

- **"Your Turn"** exercises following selected examples provide students with an easy way to stop and check their understanding of the skill or concept being presented. Answers are provided at the end of the section's exercises.

- The answers for each section (at the back of the book) now include a table listing the examples within the section that are most similar to each exercise.

- **Updated Chapter Reviews**
  - A list of important formulas and definitions has been added.
  - Review exercises now begin with Concept Check exercises—a series of true/false exercises designed to assess understanding of key ideas.
  - The answers for review exercises (at the back of the book) now include a table listing which section has examples most similar to each review exercise.

- The **Prerequisite Skills Diagnostic Test**, just prior to Chapter R, gives students and instructors an opportunity to assess students' skills on topics that are critical to success in this course. Answers reference specific review material in the text for **targeted remediation**.

- The updated design makes it **easier to identify graphing calculator and Excel® spreadsheet technology coverage** so instructors can more easily highlight (or skip) the material.

## New MyMathLab® Course

- **Extensive exercise coverage**—choose from hundreds of assignable exercises to help you craft just the right assignments.
- **Diagnostic quizzes and personalized homework** provide individualized remediation for any gaps in prerequisite skills.
- **Video program** features example and section coverage.
- **Graphing calculator and Excel® spreadsheet** guidance for students.
- **Application exercises** within MyMathLab are now labeled by type of application.

## New and Revised Content

**Chapter 1**    In Section 1.1, a new example has been added that models the prevalence of cigarette smoking in the United States, and then uses the model to make predictions. Section 1.2 has been revised, giving the formulas for the least squares line explicitly and making them more consistent with the formula for the correlation coefficient. In Section 1.4, an example on vaccination coverage, which illustrates how to derive a quadratic model, both by hand and with technology, has been added. In Section 1.5, material on identifying the degree of a polynomial has been rewritten as an example to better highlight the concept. An example on tuberculosis in the United States, which illustrates how to derive a cubic model, has also been added. Throughout the chapter, real-life exercises have been updated, and new exercises on topics such as cancer, diabetes, gender ratio, energy consumption, meat consumption, the demand for nurses, organic farming, and ideal partner height have been added.

**Chapter 2**    In Section 2.1, a new example on the surplus of food, which illustrates how to derive an exponential model, both by hand and with technology, has been added. In Section 2.4, an example using a trigonometric function to model the pressure on the eardrum has been included. A new Extended Application on power functions has been added. Throughout the chapter, real-life exercises have been updated and new exercises on topics such as the bald eagle population, minority population growth, carbon monoxide emissions, wind energy, metabolic rate, physician demand, and music have been added.

**Chapter 3**    In Section 3.1, the introduction of limits was completely revised. The opening discussion and example were transformed into a series of examples that progress through different limit scenarios: a function defined at the limit, a function undefined at that limit (a "hole" in the graph), a function defined at the limit but with a different value than the limit (a piecewise function), and then, finally, attempting to find a limit when one does not exist. New figures were added to illustrate the different scenarios. In Section 3.2, the definition and examples of continuity have been revised using a simple process to test for continuity. A medical device cost analysis has been added as an example. In Section 3.3, an example calculating the rate of change of the number of households with landlines has been added. The opening discussion of Section 3.5, showing how to sketch the graph of the derivative given the graph of the original function, was rewritten as an example. An Extended Application on the modeling of drugs administered intravenously has been added. Throughout the chapter, real-life exercises have been updated, and new exercises on topics such as Alzheimer's disease, body mass index, and immigration have been added.

**Chapter 4**   The introduction to the chain rule was rewritten as an example in Section 4.3. In Section 4.4, a new example illustrates the use of a logistic function to develop a model for the weight of cactus wrens. An Extended Application on managing renewable resources has been added. Throughout the chapter, real-life exercises have been updated, and new exercises on topics such as tree growth, genetics, insect competition, whooping cranes, cholesterol, involutional psychosis, radioactive iron, radioactive albumin, heat index, Jukes-Cantor distance, eardrum pressure, online learning, and minority populations have been added.

**Chapter 5**   Twenty-six new exercises were added throughout Chapter 5, nine of them applications based on scientific sources, such as three on foraging and two on cohesiveness, which involves the spacing between fish in a school or birds in a flock. There are now more trigonometric exercises in the chapter, along with an additional example involving trigonometry.

**Chapter 6**   Throughout Chapter 6, thirteen new exercises and three examples were added. Five of the new applications are based on scientific sources, such as the one on cancer cells and the one on honeycombs in Section 6.2 on Applications of the Derivative. In addition, there are now more trigonometric exercises in the chapter.

**Chapter 7**   The antiderivatives of the trigonometric functions have been moved to the first section so they can now be applied throughout the chapter. In addition, six exercises were added, some involving trigonometry and others involving integration using real data (such as those related to energy usage). The examples were improved by adding two new examples involving trigonometry and updating six involving real data.

**Chapter 8**   An example using tables to find trigonometric integrals has been included in Section 8.2. New graphics help illustrate the calculation of the volume of a solid of revolution in Section 8.3. Throughout the chapter, real-life exercises have been updated.

**Chapter 9**   In Chapter 9, seven of the new applications are based on scientific sources and fifteen new exercises were added, such as those on zooplankton growth, hypoxia (when the lungs receive inadequate oxygen), and Body Shape Index. In addition, many more trigonometric exercises are found throughout the chapter. The Extended Application has been revised to include an additional question involving technology as well as a group project option. Finally, material was added to justify the second derivative test for functions of two variables.

**Chapter 10**   In Section 10.1, a brief summary of solutions of two equations in two variables has been added. An alternate method for fraction-free Gauss-Jordan has been added to the first example. Section 10.5 has been revised, with the beginning discussion rewritten as examples on insect population growth, determinant calculations, and eigenvalue and corresponding eigenvector calculations. New exercises on finding the determinant of a matrix have also been included in this section. Throughout the chapter, real-life exercises have been updated, and new exercises on topics such as cancer, motorcycle helmets, calorie expenditure, basketball, and baseball have been added.

**Chapter 11**   The chapter opener was changed to relate to an exercise involving diseases in mouse populations. Material on equilibrium points and stability was added to Section 11.1. In addition, thirteen new exercises were added, six of them applications based on scientific sources. One of the new exercises compares the three population growth models discussed in the chapter: exponential, limited growth, and logistic. There are now more trigonometric exercises throughout the chapter. Finally, exercises were updated and an example, a Your Turn exercise, a For Review box, and a Technology Note were added.

**Chapter 12**   In Section 12.1, an example on endangered species has been added to illustrate set operations. In Section 12.2, an exam on student health has been included to show how Venn diagrams can be used to determine probabilities. In Section 12.3, the beginning discussion has been rewritten as an example of conditional probability. Examples on environmental inspections and on playing cards, both illustrating the product rule, have been added. The solution to the example on sensitivity and specificity has been rewritten for clarity. Challenging problems from actuarial exams have been included as exercises throughout this chapter as well.

**Chapter 13**   New material was added on the Poisson random variable and its connection to the exponential random variable. In addition, new material was inserted on a random sample, its mean, and the Central Limit Theorem. To support the new material as well as to better align with the existing content, four new examples were included, and thirty new exercises were added, of which four are applications based on scientific sources, such as the ones on plant growth and on the Hodgkin-Huxley model for excitable nerve cells.

**Chapter 14**   Chapter 14 in the last edition was available online for those who chose to cover discrete dynamical systems. With increased attention on this topic, the chapter has been heavily revised and supplemented to make for a robust topic in this course. Along with a new chapter opener, twelve new exercises were added to Section 14.1, six to Section 14.2, and thirteen to Section 14.3. A Chapter Review was added, with thirty exercises, two of them applications based on scientific sources. Finally, a new Extended Application was included, based on modeling the growth and decline of polio, as well as other life science situations that can be modeled by discrete dynamical systems.

## Prerequisite Skills Diagnostic Test

The Prerequisite Skills Diagnostic Test gives students and instructors a means to assess the basic prerequisite skills needed to be successful in this course. In addition, the answers to the test include references to specific content in Chapter R as applicable so students can zero in on where they need improvement. Full solutions to the questions in this test are in Appendix A.

## Your Turn Exercises

The **Your Turn** exercises, following selected examples, provide students with an easy way to quickly stop and check their understanding of the skill or concept being presented. Answers are provided at the end of the section's exercises.

## More Applications and Exercises

This text is used in large part because of the enormous amounts of real data contained in examples and exercises throughout. This second edition will not disappoint in this area. Every section of this text has new and updated examples, exercises and applications. The data has been updated in the applications whenever possible and new examples have been added altogether to provide additional guidance to students or new applications of the mathematics.

## Coverage of Technology

**Graphing calculator** or **Microsoft Excel®** material is now set off to make it easier for instructors to use. Figures depicting graphing calculator screens have been redrawn to create a more accurate depiction of the math. In addition, this edition provides students with a transition to the new MathPrint™ operating system of the TI-84+C through the Technology Notes, a new appendix, and the *Graphing Calculator Manual*. An *Excel® Spreadsheet Manual* is also available.

## MyMathLab®

Available now with *Calculus for the Life Sciences* are the following resources within MyMathLab that will benefit students in this course.

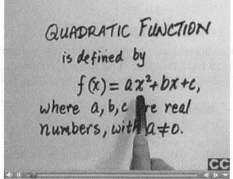

- "Getting Ready for Applied Calculus" chapter covers basic prerequisite skills.
- Personalized Homework allows you to create homework assignments based on the results of student assessments.
- Premade "Getting Ready for Chapter X" quizzes are available for instructors to assign as a way to assess student preparedness for upcoming topics.
- Videos with example and section coverage.
- Hundreds of assignable exercises.
- Application labels within exercise sets (e.g., "Life Sciences") make it easy to find types of applications appropriate for students with different interests.
- Additional graphing calculator and Excel spreadsheet help.

A detailed description of the overall capabilities of MyMathLab is provided on pages xiii and xiv.

## Source Lines

Sources for the exercises are now written in an abbreviated format within the actual exercise so that students immediately see that the problem comes from, or pulls data from, actual research or industry. The complete references are available at www.pearsonhighered.com/mathstatsresources as well as on page S-1.

# Continued Pedagogical Support

This second edition has undergone a significant revision to address the needs of today's students. However, the signature writing style and features that make this text accessible to the average student have been maintained. Some of those signature features are the following:

## Exposition and Examples

Nothing in a text is more important to a student than the examples and surrounding explanations. This text is known for its careful explanation of the mathematics and the fully developed examples. The examples often include explanatory comments in color, to the side of the worked-out examples, to help guide students through the steps. In addition, the authors frequently summarize key rules and formulas throughout the text to help students recognize the advantages of, or differences between, multiple techniques for solving a problem.

## Multiple Methods of Solution

An emphasis remains on multiple representations of a topic, whenever possible, by examining each topic symbolically, numerically, graphically, and verbally, and by providing multiple methods of solution to various examples. Some of these alternative methods involve technology, while others represent a different way of performing a computation. These multiple representations give instructors the flexibility to emphasize the method they prefer and will ultimately help students better understand a topic.

## Apply It

An **Apply It** question (previously named "Think About It"), typically at the start of a section, asks students to consider how to solve a real-life situation related to the math they are about to learn. The **Apply It** question is solved in an application within the section or the exercise set so students see it in context with the mathematics.

### Extended Applications

In-depth applied exercises called "Extended Applications" are included at the end of every chapter to stimulate student interest and for use as group projects. These applications are also ideal for use in honors sections of the course.

### For Review

**For Review** boxes are provided in the margin as appropriate, giving students just-in-time help with skills they should already know but may have forgotten. **For Review** comments sometimes include an explanation, while others refer students to earlier parts of the book for a more thorough review.

### Caution and Notes

Common student difficulties and errors are highlighted under the heading "Caution" to keep students on track. In addition, important hints and asides are highlighted with the heading "Notes."

### Full-Color Graphics and Photography

To make this book lively and fun to read, full-color graphics have been included. As needed, color is used to enhance the exposition. More often, color is used to clarify the different parts of a figure or to highlight important aspects of an illustration.

## Supplements

| STUDENT RESOURCES | INSTRUCTOR RESOURCES |
|---|---|
| **Student's Solutions Manual** <br> • Provides students with detailed solutions to odd-numbered section-level exercises, and all review exercises. <br> • Authored by Beverly Fusfield. <br> • ISBN 0321963830 / 9780321963833 | **Instructor's Edition** <br> • Includes all the answers, for quick reference in back of text. <br> • ISBN 0321963822 / 9780321963826 |
| **Graphing Calculator Manual (download only)** <br> • By Victoria Baker, Nicholls State University <br> • Contains detailed instruction for using the TI-83/TI-83+/TI-84+ C. <br> • Instructions are organized by topic. <br> • Available in MyMathLab. | **Instructor's Solutions Manual (download only)** <br> • Provides instructors with detailed solutions to all section-level exercises, review exercises, and Extended Applications. <br> • Authored by Beverly Fusfield. <br> • Available to qualified instructors within MyMathLab or through the Pearson Instructor Resource Center, www.pearsonhighered.com/irc |
| **Excel® Spreadsheet Manual (download only)** <br> • By Stela Pudar-Hozo, Indiana University—Northwest <br> • Contains detailed instructions for using Excel 2013. <br> • Instructions are organized by topic. <br> • Available in MyMathLab. | **PowerPoint® Lecture Presentation** <br> • Includes lecture content and key graphics from the book. <br> • Available to qualified instructors within MyMathLab or through the Pearson Instructor Resource Center, www.pearsonhighered.com/irc |
| **Video Lectures with Optional Captioning (download only)** <br> • Videos with extensive example and section coverage, for student use at home or on campus. <br> • Ideal for distance learning or supplemental instruction. <br> • Available in MyMathLab. | |

# Media Resources

## MyMathLab® Online Course (access code required)

MyMathLab delivers **proven results** in helping individual students succeed.

- MyMathLab has a consistently positive impact on the quality of learning in higher education math instruction. MyMathLab can be successfully implemented in any environment—lab-based, hybrid, fully online, traditional—and demonstrates the quantifiable difference that integrated usage has on student retention, subsequent success, and overall achievement.

- MyMathLab's comprehensive online gradebook automatically tracks your students' results on tests, quizzes, homework, and in the study plan. You can use the gradebook to quickly intervene if your students have trouble, or to provide positive feedback on a job well done. The data within MyMathLab is easily exported to a variety of spread-sheet programs, such as Microsoft Excel. You can determine which points of data you want to export, and then analyze the results to determine success.

MyMathLab provides **engaging experiences** that personalize, stimulate, and measure learning for each student.

- **Personalized Learning:** MyMathLab offers features that support adaptive learning: personalized homework and the adaptive study plan. These features allow your students to work on what they need to learn when it makes the most sense, maximizing their potential for understanding and success.

- **Exercises:** The homework and practice exercises in MyMathLab correlate to the exercises in the textbook, and they regenerate algorithmically to give students unlimited opportunity for practice and mastery. The software offers immediate, helpful feedback when students enter incorrect answers.

- **Chapter-Level, Just-in-Time Remediation:** The MyMathLab course for this text includes a short diagnostic, called "Getting Ready," prior to each chapter to assess students' prerequisite knowledge. This diagnostic can then be tied to Personalized Homework so that each individual student receives a homework assignment specific to his or her prerequisite skill needs.

- **Multimedia Learning Aids:** Exercises include guided solutions, sample problems, animations, videos, and eText access for extra help at point-of-use.

And, MyMathLab comes from an **experienced partner** with educational expertise and an eye on the future.

- Knowing that you are using a Pearson product means knowing that you are using quality content. This means that our eTexts are accurate and our assessment tools work. It also means we are committed to making MyMathLab as accessible as possible. MyMathLab is compatible with the JAWS 12/13 screen reader, and enables multiple-choice and free-response problem types to be read and interacted with via keyboard controls and math notation input. More information on this functionality is available at http://mymathlab.com/accessibility.

- Whether you are just getting started with MyMathLab, or have a question along the way, we're here to help you learn about our technologies and how to incorporate them into your course.

- To learn more about how MyMathLab combines proven learning applications with powerful assessment and continuously adaptive capabilities, visit **www.mymathlab .com** or contact your Pearson representative.

## MyMathLab® Ready to Go Course (access code required)

These new "Ready to Go" courses provide students with all the same great MyMathLab features, but make it easier for instructors to get started. Each course includes preassigned homework and quizzes to make creating a course even simpler. In addition, these prebuilt courses include a course-level "Getting Ready" diagnostic that helps pinpoint students' weaknesses in prerequisite skills. Ask your Pearson representative about the details for this particular course or to see a copy of it.

## MyLabsPlus®

MyLabsPlus combines proven results and engaging experiences from MyMathLab® and MyStatLab™ with convenient management tools and a dedicated services team. Designed to support growing math and statistics programs, it includes additional features such as:

- **Batch Enrollment:** Your school can create the login name and password for every student and instructor, so everyone can be ready to start class on the first day. Automation of this process is also possible through integration with your school's Student Information System.

- **Login from your campus portal:** You and your students can link directly from your campus portal into your MyLabsPlus courses. A Pearson service team works with your institution to create a single sign-on experience for instructors and students.

- **Advanced Reporting:** MyLabsPlus's advanced reporting allows instructors to review and analyze students' strengths and weaknesses by tracking their performance on tests, assignments, and tutorials. Administrators can review grades and assignments across all courses on your MyLabsPlus campus for a broad overview of program performance.

- **24/7 Support:** Students and instructors receive 24/7 support, 365 days a year, by email or online chat.

MyLabsPlus is available to qualified adopters. For more information, visit our website at www.mylabsplus.com or contact your Pearson representative.

## MathXL® Online Course (access code required)

**MathXL®** is the homework and assessment engine that runs MyMathLab. (MyMathLab is MathXL plus a learning management system.)

With MathXL, instructors can:

- Create, edit, and assign online homework and tests using algorithmically generated exercises correlated at the objective level to the textbook.

- Create and assign their own online exercises and import TestGen tests for added flexibility.

- Maintain records of all student work tracked in MathXL's online gradebook.

With MathXL, students can:

- Take chapter tests in MathXL and receive personalized study plans and/or Personalized Homework assignments based on their test results.

- Use the study plan and/or the homework to link directly to tutorial exercises for the objectives they need to study.

- Access supplemental animations and video clips directly from selected exercises.

MathXL is available to qualified adopters. For more information, visit our website at www.mathxl.com or contact your Pearson representative.

## TestGen®
## www.pearsoned.com/testgen

TestGen® enables instructors to build, edit, print, and administer tests using a computerized bank of questions developed to cover all the objectives of the text. TestGen is algorithmically based, allowing instructors to create multiple but equivalent versions of the same question or test with the click of a button. Instructors can also modify test bank questions or add new questions. The software and testbank are available for download from Pearson Education's online catalog at www.pearsonhighered.com.

## Acknowledgments

We wish to thank the following professors for their contributions in reviewing portions of this text:

Sviatoslav Archava, *East Carolina University*
Allen Back, *Cornell University*
Ian Besse, *Pacific University*
John Burke, *American River College*
Rodica Curtu, *University of Iowa*
Huimei Delgao, *Purdue University*
Alan Genz, *Washington State University*
Jeffrey L. Meyer, *Syracuse University*
Bharath Narayanan, *Pennsylvania State University*
Fran Pahlevani, *Pennsylvania State University, Abington*
Timothy Pilachowski, *University of Maryland*
Michael Price, *University of Oregon*
Brooke P. Quinlan, *Hillsborough Community College*
Jim Rogers, *University of Nebraska–Omaha*
Leonid Rubchinsky, *Indiana University–Purdue University Indianapolis*
David Snyder, *Texas State University*

We also thank Beverly Fusfield for doing an excellent job updating the *Student's Solutions Manual* and *Instructor's Solutions Manual*, an enormous and time-consuming task. Further thanks go to our accuracy checkers Damon Demas, Rhea Meyerholtz, and Lauri Semarne. We are very thankful for the work of William H. Kazez, Theresa Laurent, and Richard McCall, in writing Extended Applications for the book. We are grateful to Karla Harby and Mary Ann Ritchey for their editorial assistance. We especially appreciate the staff at Pearson, whose contributions have been very important in bringing this project to a successful conclusion.

*Raymond N. Greenwell*
*Nathan P. Ritchey*

# Dear Student,

Hello! The fact that you're reading this preface is good news. One of the keys to success in a math class is to read the book. Another is to answer all the questions correctly on your professor's tests. You've already started doing the first; doing the second may be more of a challenge, but by reading this book and working out the exercises, you'll be in a much stronger position to ace the tests. One last essential key to success is to go to class and actively participate.

You'll be happy to discover that we've provided the answers to the odd-numbered exercises in the back of the book. As you begin the exercises, you may be tempted to immediately look up the answer in the back of the book, and then figure out how to get that answer. It is an easy solution that has a consequence—you won't learn to do the exercises without that extra hint. Then, when you take a test, you will be forced to answer the questions without knowing what the answer is. Believe us, this is a lot harder! The learning comes from figuring out the exercises. Once you have an answer, look in the back and see if your answer agrees with ours. If it does, you're on the right path. If it doesn't, try to figure out what you did wrong. Once you've discovered your error, continue to work out more exercises to master the concept and skill.

Equations are a mathematician's way of expressing ideas in concise shorthand. The problem in reading mathematics is unpacking the shorthand. One useful technique is to read with paper and pencil in hand so you can work out calculations as you go along. When you are baffled, and you wonder, "How did they get that result?" try doing the calculation yourself and see what you get. You'll be amazed (or at least mildly satisfied) at how often that answers your question. Remember, math is not a spectator sport. You don't learn math by passively reading it or watching your professor. You learn mathematics by doing mathematics.

Finally, if there is anything you would like to see changed in the book, feel free to write to us at matrng@hofstra.edu or nritchey@edinboro.edu. We're constantly trying to make this book even better. If you'd like to know more about us, we have Web sites that we invite you to visit: http://people.hofstra.edu/rgreenwell and http://users.edinboro.edu/nritchey.

*Ray Greenwell*
*Nate Ritchey*

# Prerequisite Skills Diagnostic Test

Below is a very brief test to help you recognize which, if any, prerequisite skills you may need to remediate in order to be successful in this course. After completing the test, check your answers in the back of the book. In addition to the answers, we have also provided the solutions to these problems in Appendix A. These solutions should help remind you how to solve the problems. For problems 5–26, the answers are followed by references to sections within Chapter R where you can find guidance on how to solve the problem and/or additional instruction. Addressing any weak prerequisite skills now will make a positive impact on your success as you progress through this course.

1. What percent of 50 is 10?

2. Simplify $\frac{13}{7} - \frac{2}{5}$.

3. Let $x$ be the number of apples and $y$ be the number of oranges. Write the following statement as an algebraic equation: "The total number of apples and oranges is 75."

4. Let $s$ be the number of students and $p$ be the number of professors. Write the following statement as an algebraic equation: "There are at least four times as many students as professors."

5. Solve for $k$: $7k + 8 = -4(3 - k)$.

6. Solve for $x$: $\frac{5}{8}x + \frac{1}{16}x = \frac{11}{16} + x$.

7. Write in interval notation: $-2 < x \le 5$.

8. Using the variable $x$, write the following interval as an inequality: $(-\infty, -3]$.

9. Solve for $y$: $5(y - 2) + 1 \le 7y + 8$.

10. Solve for $p$: $\frac{2}{3}(5p - 3) > \frac{3}{4}(2p + 1)$.

11. Carry out the operations and simplify: $(5y^2 - 6y - 4) - 2(3y^2 - 5y + 1)$.

12. Multiply out and simplify $(x^2 - 2x + 3)(x + 1)$.

13. Multiply out and simplify $(a - 2b)^2$.

14. Factor $3pq + 6p^2q + 9pq^2$.

15. Factor $3x^2 - x - 10$.

16. Perform the operation and simplify: $\frac{a^2 - 6a}{a^2 - 4} \cdot \frac{a - 2}{a}$.

**17.** Perform the operation and simplify: $\dfrac{x+3}{x^2-1} + \dfrac{2}{x^2+x}$.

**18.** Solve for $x$: $3x^2 + 4x = 1$.

**19.** Solve for $z$: $\dfrac{8z}{z+3} \le 2$.

**20.** Simplify $\dfrac{4^{-1}(x^2y^3)^2}{x^{-2}y^5}$.

**21.** Simplify $\dfrac{4^{1/4}(p^{2/3}q^{-1/3})^{-1}}{4^{-1/4}\,p^{4/3}q^{4/3}}$.

**22.** Simplify as a single term without negative exponents: $k^{-1} - m^{-1}$.

**23.** Factor $(x^2+1)^{-1/2}(x+2) + 3(x^2+1)^{1/2}$.

**24.** Simplify $\sqrt[3]{64b^6}$.

**25.** Rationalize the denominator: $\dfrac{2}{4-\sqrt{10}}$.

**26.** Simplify $\sqrt{y^2 - 10y + 25}$.

# R Algebra Reference

R.1    Polynomials

R.2    Factoring

R.3    Rational Expressions

R.4    Equations

R.5    Inequalities

R.6    Exponents

R.7    Radicals

In this chapter, we will review the most important topics in algebra. Knowing algebra is a fundamental prerequisite to success in higher mathematics. This algebra reference is designed for self-study; study it all at once or refer to it when needed throughout the course. Since this is a review, answers to all exercises are given in the answer section at the back of the book.

# R.1 Polynomials

An expression such as $9p^4$ is a **term**; the number 9 is the **coefficient**, $p$ is the **variable**, and 4 is the **exponent**. The expression $p^4$ means $p \cdot p \cdot p \cdot p$, while $p^2$ means $p \cdot p$, and so on. Terms having the same variable and the same exponent, such as $9x^4$ and $-3x^4$, are **like terms**. Terms that do not have both the same variable and the same exponent, such as $m^2$ and $m^4$, are **unlike terms**.

A **polynomial** is a term or a finite sum of terms in which all variables have whole number exponents, and no variables appear in denominators. Examples of polynomials include

$$5x^4 + 2x^3 + 6x, \qquad 8m^3 + 9m^2n - 6mn^2 + 3n^3, \qquad 10p, \qquad \text{and} \qquad -9.$$

## Order of Operations
Algebra is a language, and you must be familiar with its rules to correctly interpret algebraic statements. The following order of operations have been agreed upon through centuries of usage.

- Expressions in **parentheses** are calculated first, working from the inside out. The numerator and denominator of a fraction are treated as expressions in parentheses.
- **Powers** are performed next, going from left to right.
- **Multiplication** and **division** are performed next, going from left to right.
- **Addition** and **subtraction** are performed last, going from left to right.

For example, in the expression $[6(x + 1)^2 + 3x - 22]^2$, suppose $x$ has the value of 2. We would evaluate this as follows:

$$[6(2 + 1)^2 + 3(2) - 22]^2 = [6(3)^2 + 3(2) - 22]^2 \quad \text{Evaluate the expression in the innermost parentheses.}$$
$$= [6(9) + 3(2) - 22]^2 \quad \text{Evaluate 3 raised to a power.}$$
$$= (54 + 6 - 22)^2 \quad \text{Perform the multiplication.}$$
$$= (38)^2 \quad \text{Perform the addition and subtraction from left to right.}$$
$$= 1444 \quad \text{Evaluate the power.}$$

In the expression $\dfrac{x^2 + 3x + 6}{x + 6}$, suppose $x$ has the value of 2. We would evaluate this as follows:

$$\frac{2^2 + 3(2) + 6}{2 + 6} = \frac{16}{8} \quad \text{Evaluate the numerator and the denominator.}$$
$$= 2 \quad \text{Simplify the fraction.}$$

## Adding and Subtracting Polynomials
The following properties of real numbers are useful for performing operations on polynomials.

### Properties of Real Numbers
For all real numbers $a$, $b$, and $c$:

1. $a + b = b + a$;
   $ab = ba$;  **Commutative properties**

2. $(a + b) + c = a + (b + c)$;  **Associative properties**
   $(ab)c = a(bc)$;

3. $a(b + c) = ab + ac$.  **Distributive property**

### EXAMPLE 1  Properties of Real Numbers

**(a)** $2 + x = x + 2$      Commutative property of addition

**(b)** $x \cdot 3 = 3x$      Commutative property of multiplication

**(c)** $(7x)x = 7(x \cdot x) = 7x^2$      Associative property of multiplication

**(d)** $3(x + 4) = 3x + 12$      Distributive property

One use of the distributive property is to add or subtract polynomials. Only like terms may be added or subtracted. For example,

$$12y^4 + 6y^4 = (12 + 6)y^4 = 18y^4,$$

and

$$-2m^2 + 8m^2 = (-2 + 8)m^2 = 6m^2,$$

but the polynomial $8y^4 + 2y^5$ cannot be further simplified. To subtract polynomials, we use the facts that $-(a + b) = -a - b$ and $-(a - b) = -a + b$. In the next example, we show how to add and subtract polynomials.

### EXAMPLE 2  Adding and Subtracting Polynomials

Add or subtract as indicated.

**(a)** $(8x^3 - 4x^2 + 6x) + (3x^3 + 5x^2 - 9x + 8)$

    **SOLUTION**  Combine like terms.

$$(8x^3 - 4x^2 + 6x) + (3x^3 + 5x^2 - 9x + 8)$$
$$= (8x^3 + 3x^3) + (-4x^2 + 5x^2) + (6x - 9x) + 8$$
$$= 11x^3 + x^2 - 3x + 8$$

**(b)** $2(-4x^4 + 6x^3 - 9x^2 - 12) + 3(-3x^3 + 8x^2 - 11x + 7)$

    **SOLUTION**  Multiply each polynomial by the coefficient in front of the polynomial, and then combine terms as before.

$$2(-4x^4 + 6x^3 - 9x^2 - 12) + 3(-3x^3 + 8x^2 - 11x + 7)$$
$$= -8x^4 + 12x^3 - 18x^2 - 24 - 9x^3 + 24x^2 - 33x + 21$$
$$= -8x^4 + 3x^3 + 6x^2 - 33x - 3$$

**(c)** $(2x^2 - 11x + 8) - (7x^2 - 6x + 2)$

    **SOLUTION**  Distributing the minus sign and combining like terms yields

$$(2x^2 - 11x + 8) + (-7x^2 + 6x - 2)$$
$$= -5x^2 - 5x + 6.$$      TRY YOUR TURN 1

**YOUR TURN 1** Perform the operation $3(x^2 - 4x - 5) - 4(3x^2 - 5x - 7)$.

## Multiplying Polynomials

The distributive property is also used to multiply polynomials, along with the fact that $a^m \cdot a^n = a^{m+n}$. For example,

$$x \cdot x = x^1 \cdot x^1 = x^{1+1} = x^2 \quad \text{and} \quad x^2 \cdot x^5 = x^{2+5} = x^7.$$

### EXAMPLE 3  Multiplying Polynomials

Multiply.

**(a)** $8x(6x - 4)$

    **SOLUTION**  Using the distributive property yields

$$8x(6x - 4) = 8x(6x) - 8x(4)$$
$$= 48x^2 - 32x.$$

**(b)** $(3p - 2)(p^2 + 5p - 1)$

**SOLUTION** Using the distributive property yields

$$(3p - 2)(p^2 + 5p - 1)$$
$$= 3p(p^2 + 5p - 1) - 2(p^2 + 5p - 1)$$
$$= 3p(p^2) + 3p(5p) + 3p(-1) - 2(p^2) - 2(5p) - 2(-1)$$
$$= 3p^3 + 15p^2 - 3p - 2p^2 - 10p + 2$$
$$= 3p^3 + 13p^2 - 13p + 2.$$

**(c)** $(x + 2)(x + 3)(x - 4)$

**SOLUTION** Multiplying the first two polynomials and then multiplying their product by the third polynomial yields

$$(x + 2)(x + 3)(x - 4)$$
$$= [(x + 2)(x + 3)](x - 4)$$
$$= (x^2 + 2x + 3x + 6)(x - 4)$$
$$= (x^2 + 5x + 6)(x - 4)$$
$$= x^3 + 5x^2 + 6x - 4x^2 - 20x - 24$$
$$= x^3 + x^2 - 14x - 24.$$ TRY YOUR TURN 2

**YOUR TURN 2** Perform the operation $(3y + 2)(4y^2 - 2y - 5)$.

A **binomial** is a polynomial with exactly two terms, such as $2x + 1$ or $m + n$. When two binomials are multiplied, the FOIL method (First, Outer, Inner, Last) is used as a memory aid.

---

### EXAMPLE 4 Multiplying Polynomials

Find $(2m - 5)(m + 4)$ using the FOIL method.

**SOLUTION**

$$(2m - 5)(m + 4) = \overset{F}{(2m)(m)} + \overset{O}{(2m)(4)} + \overset{I}{(-5)(m)} + \overset{L}{(-5)(4)}$$
$$= 2m^2 + 8m - 5m - 20$$
$$= 2m^2 + 3m - 20$$

---

### EXAMPLE 5 Multiplying Polynomials

Find $(2k - 5m)^3$.

**SOLUTION** Write $(2k - 5m)^3$ as $(2k - 5m)(2k - 5m)(2k - 5m)$. Then multiply the first two factors using FOIL.

$$(2k - 5m)(2k - 5m) = 4k^2 - 10km - 10km + 25m^2$$
$$= 4k^2 - 20km + 25m^2$$

Now multiply this last result by $(2k - 5m)$ using the distributive property, as in Example 3(b).

$$(4k^2 - 20km + 25m^2)(2k - 5m)$$
$$= 4k^2(2k - 5m) - 20km(2k - 5m) + 25m^2(2k - 5m)$$
$$= 8k^3 - 20k^2m - 40k^2m + 100km^2 + 50km^2 - 125m^3$$
$$= 8k^3 - 60k^2m + 150km^2 - 125m^3 \quad \text{Combine like terms.}$$

Notice in the first part of Example 5, when we multiplied $(2k - 5m)$ by itself, that the product of the square of a binomial is the square of the first term, $(2k)^2$, plus twice the product of the two terms, $(2)(2k)(-5m)$, plus the square of the last term, $(-5k)^2$.

> **CAUTION** Avoid the common error of writing $(x + y)^2 = x^2 + y^2$. As the first step of Example 5 shows, the square of a binomial has three terms, so
>
> $$(x + y)^2 = x^2 + 2xy + y^2.$$
>
> Furthermore, higher powers of a binomial also result in more than two terms. For example, verify by multiplication that
>
> $$(x + y)^3 = x^3 + 3x^2y + 3xy^2 + y^3.$$
>
> Remember, for any value of $n \neq 1$,
>
> $$(x + y)^n \neq x^n + y^n.$$

# R.1  EXERCISES

**Perform the indicated operations.**

1. $(2x^2 - 6x + 11) + (-3x^2 + 7x - 2)$

2. $(-4y^2 - 3y + 8) - (2y^2 - 6y - 2)$

3. $-6(2q^2 + 4q - 3) + 4(-q^2 + 7q - 3)$

4. $2(3r^2 + 4r + 2) - 3(-r^2 + 4r - 5)$

5. $(0.613x^2 - 4.215x + 0.892) - 0.47(2x^2 - 3x + 5)$

6. $0.5(5r^2 + 3.2r - 6) - (1.7r^2 - 2r - 1.5)$

7. $-9m(2m^2 + 3m - 1)$

8. $6x(-2x^3 + 5x + 6)$

9. $(3t - 2y)(3t + 5y)$

10. $(9k + q)(2k - q)$

11. $(2 - 3x)(2 + 3x)$

12. $(6m + 5)(6m - 5)$

13. $\left(\dfrac{2}{5}y + \dfrac{1}{8}z\right)\left(\dfrac{3}{5}y + \dfrac{1}{2}z\right)$

14. $\left(\dfrac{3}{4}r - \dfrac{2}{3}s\right)\left(\dfrac{5}{4}r + \dfrac{1}{3}s\right)$

15. $(3p - 1)(9p^2 + 3p + 1)$

16. $(3p + 2)(5p^2 + p - 4)$

17. $(2m + 1)(4m^2 - 2m + 1)$

18. $(k + 2)(12k^3 - 3k^2 + k + 1)$

19. $(x + y + z)(3x - 2y - z)$

20. $(r + 2s - 3t)(2r - 2s + t)$

21. $(x + 1)(x + 2)(x + 3)$

22. $(x - 1)(x + 2)(x - 3)$

23. $(x + 2)^2$

24. $(2a - 4b)^2$

25. $(x - 2y)^3$

26. $(3x + y)^3$

**▬▬ YOUR TURN ANSWERS**

1. $-9x^2 + 8x + 13$

2. $12y^3 + 2y^2 - 19y - 10$

# R.2  Factoring

Multiplication of polynomials relies on the distributive property. The reverse process, where a polynomial is written as a product of other polynomials, is called **factoring**. For example, one way to factor the number 18 is to write it as the product $9 \cdot 2$; both 9 and 2 are **factors** of 18. Usually, only integers are used as factors of integers. The number 18 can also be written with three integer factors as $2 \cdot 3 \cdot 3$.

## The Greatest Common Factor
To factor the algebraic expression $15m + 45$, first note that both $15m$ and 45 are divisible by 15; $15m = 15 \cdot m$ and $45 = 15 \cdot 3$. By the distributive property,

$$15m + 45 = 15 \cdot m + 15 \cdot 3 = 15(m + 3).$$

Both 15 and $m + 3$ are factors of $15m + 45$. Since 15 divides into both terms of $15m + 45$ (and is the largest number that will do so), 15 is the **greatest common factor** for the polynomial $15m + 45$. The process of writing $15m + 45$ as $15(m + 3)$ is often called **factoring out** the greatest common factor.

## EXAMPLE 1 Factoring

Factor out the greatest common factor.

**(a)** $12p - 18q$

**SOLUTION** Both $12p$ and $18q$ are divisible by 6. Therefore,

$$12p - 18q = 6 \cdot 2p - 6 \cdot 3q = 6(2p - 3q).$$

**(b)** $8x^3 - 9x^2 + 15x$

**SOLUTION** Each of these terms is divisible by $x$.

$$8x^3 - 9x^2 + 15x = (8x^2) \cdot x - (9x) \cdot x + 15 \cdot x$$
$$= x(8x^2 - 9x + 15) \quad \text{or} \quad (8x^2 - 9x + 15)x$$

**YOUR TURN 1** Factor $4z^4 + 4z^3 + 18z^2$.

TRY YOUR TURN 1

One can always check factorization by finding the product of the factors and comparing it to the original expression.

> CAUTION | When factoring out the greatest common factor in an expression like $2x^2 + x$, be careful to remember the 1 in the second term.
> $$2x^2 + x = 2x^2 + 1x = x(2x + 1), \text{ not } x(2x).$$

## Factoring Trinomials

A polynomial that has no greatest common factor (other than 1) may still be factorable. For example, the polynomial $x^2 + 5x + 6$ can be factored as $(x + 2)(x + 3)$. To see that this is correct, find the product $(x + 2)(x + 3)$; you should get $x^2 + 5x + 6$. A polynomial such as this with three terms is called a **trinomial**. To factor the trinomial $x^2 + 5x + 6$, where the coefficient of $x^2$ is 1, we use FOIL backwards.

## EXAMPLE 2 Factoring a Trinomial

Factor $y^2 + 8y + 15$.

**SOLUTION** Since the coefficient of $y^2$ is 1, factor by finding two numbers whose *product* is 15 and whose *sum* is 8. Since the constant and the middle term are positive, the numbers must both be positive. Begin by listing all pairs of positive integers having a product of 15. As you do this, also form the sum of each pair of numbers.

|        Products        |        Sums        |
|------------------------|--------------------|
| $15 \cdot 1 = 15$      | $15 + 1 = 16$      |
| $5 \cdot 3 = 15$       | $5 + 3 = 8$        |

The numbers 5 and 3 have a product of 15 and a sum of 8. Thus, $y^2 + 8y + 15$ factors as

$$y^2 + 8y + 15 = (y + 5)(y + 3).$$

The answer can also be written as $(y + 3)(y + 5)$.

If the coefficient of the squared term is *not* 1, work as shown below.

## EXAMPLE 3 Factoring a Trinomial

Factor $4x^2 + 8xy - 5y^2$.

**SOLUTION** The possible factors of $4x^2$ are $4x$ and $x$ or $2x$ and $2x$; the possible factors of $-5y^2$ are $-5y$ and $y$ or $5y$ and $-y$. Try various combinations of these factors until one works (if, indeed, any work). For example, try the product $(x + 5y)(4x - y)$.

$$(x + 5y)(4x - y) = 4x^2 - xy + 20xy - 5y^2$$
$$= 4x^2 + 19xy - 5y^2$$

This product is not correct, so try another combination.

$$(2x - y)(2x + 5y) = 4x^2 + 10xy - 2xy - 5y^2$$
$$= 4x^2 + 8xy - 5y^2$$

Since this combination gives the correct polynomial,

$$4x^2 + 8xy - 5y^2 = (2x - y)(2x + 5y).$$

**YOUR TURN 2** Factor
$6a^2 + 5ab - 4b^2$.

**TRY YOUR TURN 2**

## Special Factorizations  Four special factorizations occur so often that they are listed here for future reference.

**Special Factorizations**

| | |
|---|---|
| $x^2 - y^2 = (x + y)(x - y)$ | **Difference of two squares** |
| $x^2 + 2xy + y^2 = (x + y)^2$ | **Perfect square** |
| $x^3 - y^3 = (x - y)(x^2 + xy + y^2)$ | **Difference of two cubes** |
| $x^3 + y^3 = (x + y)(x^2 - xy + y^2)$ | **Sum of two cubes** |

A polynomial that cannot be factored is called a **prime polynomial**.

**EXAMPLE 4**  **Factoring Polynomials**

Factor each polynomial, if possible.

**(a)**  $64p^2 - 49q^2 = (8p)^2 - (7q)^2 = (8p + 7q)(8p - 7q)$    Difference of two squares

**(b)**  $x^2 + 36$ is a prime polynomial.

**(c)**  $x^2 + 12x + 36 = (x + 6)^2$    Perfect square

**(d)**  $9y^2 - 24yz + 16z^2 = (3y - 4z)^2$    Perfect square

**(e)**  $y^3 - 8 = y^3 - 2^3 = (y - 2)(y^2 + 2y + 4)$    Difference of two cubes

**(f)**  $m^3 + 125 = m^3 + 5^3 = (m + 5)(m^2 - 5m + 25)$    Sum of two cubes

**(g)**  $8k^3 - 27z^3 = (2k)^3 - (3z)^3 = (2k - 3z)(4k^2 + 6kz + 9z^2)$

    Difference of two cubes

**(h)**  $p^4 - 1 = (p^2 + 1)(p^2 - 1) = (p^2 + 1)(p + 1)(p - 1)$    Difference of two squares

**CAUTION**  In factoring, always look for a common factor first. Since $36x^2 - 4y^2$ has a common factor of 4,

$$36x^2 - 4y^2 = 4(9x^2 - y^2) = 4(3x + y)(3x - y).$$

It would be incomplete to factor it as

$$36x^2 - 4y^2 = (6x + 2y)(6x - 2y),$$

since each factor can be factored still further. To *factor* means to factor completely, so that each polynomial factor is prime.

# R.2  EXERCISES

Factor each polynomial. If a polynomial cannot be factored, write *prime*. Factor out the greatest common factor as necessary.

**1.** $7a^3 + 14a^2$

**2.** $3y^3 + 24y^2 + 9y$

**3.** $13p^4q^2 - 39p^3q + 26p^2q^2$

**4.** $60m^4 - 120m^3n + 50m^2n^2$

**5.** $m^2 - 5m - 14$

**6.** $x^2 + 4x - 5$

**7.** $z^2 + 9z + 20$

**8.** $b^2 - 8b + 7$

**9.** $a^2 - 6ab + 5b^2$

**10.** $s^2 + 2st - 35t^2$

**11.** $y^2 - 4yz - 21z^2$

**12.** $3x^2 + 4x - 7$

**13.** $3a^2 + 10a + 7$

**14.** $15y^2 + y - 2$

**15.** $21m^2 + 13mn + 2n^2$

**16.** $6a^2 - 48a - 120$

**17.** $3m^3 + 12m^2 + 9m$

**18.** $4a^2 + 10a + 6$

**19.** $24a^4 + 10a^3b - 4a^2b^2$

**20.** $24x^4 + 36x^3y - 60x^2y^2$

**21.** $x^2 - 64$        **22.** $9m^2 - 25$

**23.** $10x^2 - 160$      **24.** $9x^2 + 64$

**25.** $z^2 + 14zy + 49y^2$   **26.** $s^2 - 10st + 25t^2$

**27.** $9p^2 - 24p + 16$    **28.** $a^3 - 216$

**29.** $27r^3 - 64s^3$      **30.** $3m^3 + 375$

**31.** $x^4 - y^4$        **32.** $16a^4 - 81b^4$

▬▬ YOUR TURN ANSWERS

**1.** $2z^2(2z^2 + 2z + 9)$       **2.** $(2a - b)(3a + 4b)$

# R.3 Rational Expressions

Many algebraic fractions are **rational expressions**, which are quotients of polynomials with nonzero denominators. Examples include

$$\frac{8}{x - 1}, \quad \frac{3x^2 + 4x}{5x - 6}, \quad \text{and} \quad \frac{2y + 1}{y^2}.$$

Next, we summarize properties for working with rational expressions.

## Properties of Rational Expressions

For all mathematical expressions $P$, $Q$, $R$, and $S$, with $Q \neq 0$ and $S \neq 0$:

$$\frac{P}{Q} = \frac{PS}{QS} \qquad \textbf{Fundamental property}$$

$$\frac{P}{Q} + \frac{R}{Q} = \frac{P + R}{Q} \qquad \textbf{Addition}$$

$$\frac{P}{Q} - \frac{R}{Q} = \frac{P - R}{Q} \qquad \textbf{Subtraction}$$

$$\frac{P}{Q} \cdot \frac{R}{S} = \frac{PR}{QS} \qquad \textbf{Multiplication}$$

$$\frac{P}{Q} \div \frac{R}{S} = \frac{P}{Q} \cdot \frac{S}{R} \ (R \neq 0) \qquad \textbf{Division}$$

When writing a rational expression in lowest terms, we may need to use the fact that $\frac{a^m}{a^n} = a^{m-n}$. For example,

$$\frac{x^4}{3x} = \frac{1x^4}{3x} = \frac{1}{3} \cdot \frac{x^4}{x} = \frac{1}{3} \cdot x^{4-1} = \frac{1}{3}x^3 = \frac{x^3}{3}.$$

### EXAMPLE 1   Reducing Rational Expressions

Write each rational expression in lowest terms, that is, reduce the expression as much as possible.

**(a)** $\dfrac{8x + 16}{4} = \dfrac{8(x + 2)}{4} = \dfrac{4 \cdot 2(x + 2)}{4} = 2(x + 2)$

Factor both the numerator and denominator in order to identify any common factors, which have a quotient of 1. The answer could also be written as $2x + 4$.

**YOUR TURN 1** Write in lowest terms $\dfrac{z^2 + 5z + 6}{2z^2 + 7z + 3}$.

**(b)** $\dfrac{k^2 + 7k + 12}{k^2 + 2k - 3} = \dfrac{(k + 4)(k + 3)}{(k - 1)(k + 3)} = \dfrac{k + 4}{k - 1}$

The answer cannot be further reduced.

TRY YOUR TURN 1

> **CAUTION** One of the most common errors in algebra involves incorrect use of the fundamental property of rational expressions. Only common *factors* may be divided or "canceled." It is essential to factor rational expressions before writing them in lowest terms. In Example 1(b), for instance, it is not correct to "cancel" $k^2$ (or cancel $k$, or divide 12 by $-3$) because the additions and subtraction must be performed first. Here they cannot be performed, so it is not possible to divide. After factoring, however, the fundamental property can be used to write the expression in lowest terms.

**EXAMPLE 2**    **Combining Rational Expressions**

Perform each operation.

**(a)** $\dfrac{3y + 9}{6} \cdot \dfrac{18}{5y + 15}$

**SOLUTION**    Factor where possible, then multiply numerators and denominators and reduce to lowest terms.

$$\frac{3y + 9}{6} \cdot \frac{18}{5y + 15} = \frac{3(y + 3)}{6} \cdot \frac{18}{5(y + 3)}$$

$$= \frac{3 \cdot 18(y + 3)}{6 \cdot 5(y + 3)}$$

$$= \frac{3 \cdot \cancel{6} \cdot 3\cancel{(y + 3)}}{\cancel{6} \cdot 5\cancel{(y + 3)}} = \frac{3 \cdot 3}{5} = \frac{9}{5}$$

**(b)** $\dfrac{m^2 + 5m + 6}{m + 3} \cdot \dfrac{m}{m^2 + 3m + 2}$

**SOLUTION**    Factor where possible.

$$\frac{(m + 2)(m + 3)}{m + 3} \cdot \frac{m}{(m + 2)(m + 1)}$$

$$= \frac{m\cancel{(m + 2)}\cancel{(m + 3)}}{\cancel{(m + 3)}\cancel{(m + 2)}(m + 1)} = \frac{m}{m + 1}$$

**(c)** $\dfrac{9p - 36}{12} \div \dfrac{5(p - 4)}{18}$

**SOLUTION**    Use the division property of rational expressions.

$$\frac{9p - 36}{12} \cdot \frac{18}{5(p - 4)} \qquad \text{Invert and multiply.}$$

$$= \frac{9\cancel{(p - 4)}}{\cancel{6} \cdot 2} \cdot \frac{\cancel{6} \cdot 3}{5\cancel{(p - 4)}} = \frac{27}{10} \qquad \text{Factor and reduce to lowest terms.}$$

**(d)** $\dfrac{4}{5k} - \dfrac{11}{5k}$

**SOLUTION**    As shown in the list of properties, to subtract two rational expressions that have the same denominators, subtract the numerators while keeping the same denominator.

$$\frac{4}{5k} - \frac{11}{5k} = \frac{4 - 11}{5k} = -\frac{7}{5k}$$

**(e)** $\dfrac{7}{p} + \dfrac{9}{2p} + \dfrac{1}{3p}$

**SOLUTION** These three fractions cannot be added until their denominators are the same. A **common denominator** into which $p$, $2p$, and $3p$ all divide is $6p$. Note that $12p$ is also a common denominator, but $6p$ is the **least common denominator**. Use the fundamental property to rewrite each rational expression with a denominator of $6p$.

$$\frac{7}{p} + \frac{9}{2p} + \frac{1}{3p} = \frac{6 \cdot 7}{6 \cdot p} + \frac{3 \cdot 9}{3 \cdot 2p} + \frac{2 \cdot 1}{2 \cdot 3p}$$

$$= \frac{42}{6p} + \frac{27}{6p} + \frac{2}{6p}$$

$$= \frac{42 + 27 + 2}{6p}$$

$$= \frac{71}{6p}$$

**(f)** $\dfrac{x + 1}{x^2 + 5x + 6} - \dfrac{5x - 1}{x^2 - x - 12}$

**SOLUTION** To find the least common denominator, we first factor each denominator. Then we change each fraction so they all have the same denominator, being careful to multiply only by quotients that equal 1.

$$\frac{x + 1}{x^2 + 5x + 6} - \frac{5x - 1}{x^2 - x - 12}$$

$$= \frac{x + 1}{(x + 2)(x + 3)} - \frac{5x - 1}{(x + 3)(x - 4)}$$

$$= \frac{x + 1}{(x + 2)(x + 3)} \cdot \frac{(x - 4)}{(x - 4)} - \frac{5x - 1}{(x + 3)(x - 4)} \cdot \frac{(x + 2)}{(x + 2)}$$

$$= \frac{(x^2 - 3x - 4) - (5x^2 + 9x - 2)}{(x + 2)(x + 3)(x - 4)}$$

$$= \frac{-4x^2 - 12x - 2}{(x + 2)(x + 3)(x - 4)}$$

$$= \frac{-2(2x^2 + 6x + 1)}{(x + 2)(x + 3)(x - 4)}$$

Because the numerator cannot be factored further, we leave our answer in this form. We could also multiply out the denominator, but factored form is usually more useful.

TRY YOUR TURN 2

**YOUR TURN 2** Perform each of the following operations.

**(a)** $\dfrac{z^2 + 5z + 6}{2z^2 - 5z - 3} \cdot \dfrac{2z^2 - z - 1}{z^2 + 2z - 3}.$

**(b)** $\dfrac{a - 3}{a^2 + 3a + 2} + \dfrac{5a}{a^2 - 4}.$

# R.3   EXERCISES

**Write each rational expression in lowest terms.**

**1.** $\dfrac{5v^2}{35v}$

**2.** $\dfrac{25p^3}{10p^2}$

**3.** $\dfrac{8k + 16}{9k + 18}$

**4.** $\dfrac{2(t - 15)}{(t - 15)(t + 2)}$

**5.** $\dfrac{4x^3 - 8x^2}{4x^2}$

**6.** $\dfrac{36y^2 + 72y}{9y}$

**7.** $\dfrac{m^2 - 4m + 4}{m^2 + m - 6}$

**8.** $\dfrac{r^2 - r - 6}{r^2 + r - 12}$

**9.** $\dfrac{3x^2 + 3x - 6}{x^2 - 4}$

**10.** $\dfrac{z^2 - 5z + 6}{z^2 - 4}$

**11.** $\dfrac{m^4 - 16}{4m^2 - 16}$

**12.** $\dfrac{6y^2 + 11y + 4}{3y^2 + 7y + 4}$

**Perform the indicated operations.**

**13.** $\dfrac{9k^2}{25} \cdot \dfrac{5}{3k}$

**14.** $\dfrac{15p^3}{9p^2} \div \dfrac{6p}{10p^2}$

**15.** $\dfrac{3a + 3b}{4c} \cdot \dfrac{12}{5(a + b)}$

**16.** $\dfrac{a - 3}{16} \div \dfrac{a - 3}{32}$

**17.** $\dfrac{2k - 16}{6} \div \dfrac{4k - 32}{3}$

**18.** $\dfrac{9y - 18}{6y + 12} \cdot \dfrac{3y + 6}{15y - 30}$

**19.** $\dfrac{4a + 12}{2a - 10} \div \dfrac{a^2 - 9}{a^2 - a - 20}$

**20.** $\dfrac{6r - 18}{9r^2 + 6r - 24} \cdot \dfrac{12r - 16}{4r - 12}$

**21.** $\dfrac{k^2 + 4k - 12}{k^2 + 10k + 24} \cdot \dfrac{k^2 + k - 12}{k^2 - 9}$

**22.** $\dfrac{m^2 + 3m + 2}{m^2 + 5m + 4} \div \dfrac{m^2 + 5m + 6}{m^2 + 10m + 24}$

**23.** $\dfrac{2m^2 - 5m - 12}{m^2 - 10m + 24} \div \dfrac{4m^2 - 9}{m^2 - 9m + 18}$

**24.** $\dfrac{4n^2 + 4n - 3}{6n^2 - n - 15} \cdot \dfrac{8n^2 + 32n + 30}{4n^2 + 16n + 15}$

**25.** $\dfrac{a + 1}{2} - \dfrac{a - 1}{2}$

**26.** $\dfrac{3}{p} + \dfrac{1}{2}$

**27.** $\dfrac{6}{5y} - \dfrac{3}{2}$

**28.** $\dfrac{1}{6m} + \dfrac{2}{5m} + \dfrac{4}{m}$

**29.** $\dfrac{1}{m - 1} + \dfrac{2}{m}$

**30.** $\dfrac{5}{2r + 3} - \dfrac{2}{r}$

**31.** $\dfrac{8}{3(a - 1)} + \dfrac{2}{a - 1}$

**32.** $\dfrac{2}{5(k - 2)} + \dfrac{3}{4(k - 2)}$

**33.** $\dfrac{4}{x^2 + 4x + 3} + \dfrac{3}{x^2 - x - 2}$

**34.** $\dfrac{y}{y^2 + 2y - 3} - \dfrac{1}{y^2 + 4y + 3}$

**35.** $\dfrac{3k}{2k^2 + 3k - 2} - \dfrac{2k}{2k^2 - 7k + 3}$

**36.** $\dfrac{4m}{3m^2 + 7m - 6} - \dfrac{m}{3m^2 - 14m + 8}$

**37.** $\dfrac{2}{a + 2} + \dfrac{1}{a} + \dfrac{a - 1}{a^2 + 2a}$

**38.** $\dfrac{5x + 2}{x^2 - 1} + \dfrac{3}{x^2 + x} - \dfrac{1}{x^2 - x}$

▮▮▮ YOUR TURN ANSWERS

**1.** $(z + 2)/(2z + 1)$

**2a.** $(z + 2)/(z - 3)$

**2b.** $6(a^2 + 1)/[(a - 2)(a + 2)(a + 1)]$

# R.4 Equations

## Linear Equations

Equations that can be written in the form $ax + b = 0$, where $a$ and $b$ are real numbers, with $a \neq 0$, are **linear equations**. Examples of linear equations include $5y + 9 = 16$, $8x = 4$, and $-3p + 5 = -8$. Equations that are *not* linear include absolute value equations such as $|x| = 4$. The following properties are used to solve linear equations.

### Properties of Equality

For all real numbers $a$, $b$, and $c$:

1. If $a = b$, then $a + c = b + c$.     **Addition property of equality**
   (The same number may be added to both sides of an equation.)

2. If $a = b$, then $ac = bc$.     **Multiplication property of equality**
   (Both sides of an equation may be multiplied by the same number.)

▮▮▮ EXAMPLE 1  **Solving Linear Equations**

Solve the following equations.

**(a)** $x - 2 = 3$

**SOLUTION**  The goal is to isolate the variable. Using the addition property of equality yields

$$x - 2 + 2 = 3 + 2, \quad \text{or} \quad x = 5.$$

**(b)** $\frac{x}{2} = 3$

   **SOLUTION**   Using the multiplication property of equality yields

$$2 \cdot \frac{x}{2} = 2 \cdot 3, \quad \text{or} \quad x = 6.$$

The following example shows how these properties are used to solve linear equations. The goal is to isolate the variable. The solutions should always be checked by substitution into the original equation.

> ### EXAMPLE 2   Solving a Linear Equation
>
> Solve $2x - 5 + 8 = 3x + 2(2 - 3x)$.
> **SOLUTION**

$$\begin{aligned} 2x - 5 + 8 &= 3x + 4 - 6x && \text{Distributive property} \\ 2x + 3 &= -3x + 4 && \text{Combine like terms.} \\ 5x + 3 &= 4 && \text{Add } 3x \text{ to both sides.} \\ 5x &= 1 && \text{Add } -3 \text{ to both sides.} \\ x &= \frac{1}{5} && \text{Multiply both sides by } \tfrac{1}{5}. \end{aligned}$$

**YOUR TURN 1** Solve
$3x - 7 = 4(5x + 2) - 7x$.

Check by substituting in the original equation. The left side becomes $2(1/5) - 5 + 8$ and the right side becomes $3(1/5) + 2[2 - 3(1/5)]$. Verify that both of these expressions simplify to 17/5.            **TRY YOUR TURN 1**

## Quadratic Equations
An equation with 2 as the highest exponent of the variable is a *quadratic equation*. A **quadratic equation** has the form $ax^2 + bx + c = 0$, where $a$, $b$, and $c$ are real numbers and $a \neq 0$. A quadratic equation written in the form $ax^2 + bx + c = 0$ is said to be in **standard form**.

The simplest way to solve a quadratic equation, but one that is not always applicable, is by factoring. This method depends on the **zero-factor property**.

> ### Zero-Factor Property
> If $a$ and $b$ are real numbers, with $ab = 0$, then either
>
> $$a = 0 \quad \text{or} \quad b = 0 \quad \text{(or both)}.$$

> ### EXAMPLE 3   Solving a Quadratic Equation
>
> Solve $6r^2 + 7r = 3$.
> **SOLUTION**   First write the equation in standard form.

$$6r^2 + 7r - 3 = 0$$

Now factor $6r^2 + 7r - 3$ to get

$$(3r - 1)(2r + 3) = 0.$$

By the zero-factor property, the product $(3r - 1)(2r + 3)$ can equal 0 if and only if

$$3r - 1 = 0 \quad \text{or} \quad 2r + 3 = 0.$$

**YOUR TURN 2** Solve $2m^2 + 7m = 15$.

Solve each of these equations separately to find that the solutions are 1/3 and −3/2. Check these solutions by substituting them into the original equation.        TRY YOUR TURN 2

> **CAUTION**  Remember, the zero-factor property requires that the product of two (or more) factors be equal to *zero*, not some other quantity. It would be incorrect to use the zero-factor property with an equation in the form $(x + 3)(x - 1) = 4$, for example.

If a quadratic equation cannot be solved easily by factoring, use the *quadratic formula*. (The derivation of the quadratic formula is given in most algebra books.)

> **Quadratic Formula**
> The solutions of the quadratic equation $ax^2 + bx + c = 0$, where $a \neq 0$, are given by
> $$x = \frac{-b \pm \sqrt{b^2 - 4ac}}{2a}.$$

## EXAMPLE 4   Quadratic Formula

Solve $x^2 - 4x - 5 = 0$ by the quadratic formula.

**SOLUTION**  The equation is already in standard form (it has 0 alone on one side of the equal sign), so the values of $a$, $b$, and $c$ from the quadratic formula are easily identified. The coefficient of the squared term gives the value of $a$; here, $a = 1$. Also, $b = -4$ and $c = -5$, where $b$ is the coefficient of the linear term and $c$ is the constant coefficient. (Be careful to use the correct signs.) Substitute these values into the quadratic formula.

$$x = \frac{-(-4) \pm \sqrt{(-4)^2 - 4(1)(-5)}}{2(1)} \qquad a = 1, b = -4, c = -5$$

$$x = \frac{4 \pm \sqrt{16 + 20}}{2} \qquad (-4)^2 = (-4)(-4) = 16$$

$$x = \frac{4 \pm 6}{2} \qquad \sqrt{16 + 20} = \sqrt{36} = 6$$

The $\pm$ sign represents the two solutions of the equation. To find both of the solutions, first use + and then use −.

$$x = \frac{4 + 6}{2} = \frac{10}{2} = 5 \qquad \text{or} \qquad x = \frac{4 - 6}{2} = \frac{-2}{2} = -1$$

The two solutions are 5 and −1.

> **CAUTION**  Notice in the quadratic formula that the square root is added to or subtracted from the value of $-b$ *before* dividing by $2a$.

## EXAMPLE 5   Quadratic Formula

Solve $x^2 + 1 = 4x$.

**SOLUTION**  First, add $-4x$ on both sides of the equal sign in order to get the equation in standard form.

$$x^2 - 4x + 1 = 0$$

Now identify the values of $a$, $b$, and $c$. Here $a = 1$, $b = -4$, and $c = 1$. Substitute these numbers into the quadratic formula.

$$x = \frac{-(-4) \pm \sqrt{(-4)^2 - 4(1)(1)}}{2(1)}$$

$$= \frac{4 \pm \sqrt{16 - 4}}{2}$$

$$= \frac{4 \pm \sqrt{12}}{2}$$

Simplify the solutions by writing $\sqrt{12}$ as $\sqrt{4 \cdot 3} = \sqrt{4} \cdot \sqrt{3} = 2\sqrt{3}$. Substituting $2\sqrt{3}$ for $\sqrt{12}$ gives

$$x = \frac{4 \pm 2\sqrt{3}}{2}$$

$$= \frac{2(2 \pm \sqrt{3})}{2} \qquad \text{Factor } 4 \pm 2\sqrt{3}.$$

$$= 2 \pm \sqrt{3}. \qquad \text{Reduce to lowest terms.}$$

The two solutions are $2 + \sqrt{3}$ and $2 - \sqrt{3}$.

The exact values of the solutions are $2 + \sqrt{3}$ and $2 - \sqrt{3}$. The $\sqrt{\phantom{x}}$ key on a calculator gives decimal approximations of these solutions (to the nearest thousandth):

$$2 + \sqrt{3} \approx 2 + 1.732 = 3.732*$$
$$2 - \sqrt{3} \approx 2 - 1.732 = 0.268 \qquad \text{TRY YOUR TURN 3}$$

**YOUR TURN 3** Solve $z^2 + 6 = 8z$.

**NOTE** Sometimes the quadratic formula will give a result with a negative number under the radical sign, such as $3 \pm \sqrt{-5}$. A solution of this type is a complex number. Since this text deals only with real numbers, such solutions cannot be used.

## Equations with Fractions
When an equation includes fractions, first eliminate all denominators by multiplying both sides of the equation by a common denominator, a number that can be divided (with no remainder) by each denominator in the equation. When an equation involves fractions with variable denominators, it is *necessary* to check all solutions in the original equation to be sure that no solution will lead to a zero denominator.

### EXAMPLE 6 Solving Rational Equations
Solve each equation.

(a) $\dfrac{r}{10} - \dfrac{2}{15} = \dfrac{3r}{20} - \dfrac{1}{5}$

**SOLUTION** The denominators are 10, 15, 20, and 5. Each of these numbers can be divided into 60, so 60 is a common denominator. Multiply both sides of the equation by 60 and use the distributive property. (If a common denominator cannot be found easily, all the denominators in the problem can be multiplied together to produce one.)

$$\frac{r}{10} - \frac{2}{15} = \frac{3r}{20} - \frac{1}{5}$$

$$60\left(\frac{r}{10} - \frac{2}{15}\right) = 60\left(\frac{3r}{20} - \frac{1}{5}\right) \qquad \text{Multiply by the common denominator.}$$

$$60\left(\frac{r}{10}\right) - 60\left(\frac{2}{15}\right) = 60\left(\frac{3r}{20}\right) - 60\left(\frac{1}{5}\right) \qquad \text{Distributive property}$$

$$6r - 8 = 9r - 12$$

Add $-9r$ and 8 to both sides.

*The symbol $\approx$ means "is approximately equal to."

$$6r - 8 + (-9r) + 8 = 9r - 12 + (-9r) + 8$$
$$-3r = -4$$
$$r = \frac{4}{3} \qquad\qquad \text{Multiply each side by } -\tfrac{1}{3}.$$

Check by substituting into the original equation.

**(b)** $\dfrac{3}{x^2} - 12 = 0$

**SOLUTION** Begin by multiplying both sides of the equation by $x^2$ to get $3 - 12x^2 = 0$. This equation could be solved by using the quadratic formula with $a = -12$, $b = 0$, and $c = 3$. Another method that works well for the type of quadratic equation in which $b = 0$ is shown below.

$$3 - 12x^2 = 0$$
$$3 = 12x^2 \qquad \text{Add } 12x^2.$$
$$\frac{1}{4} = x^2 \qquad \text{Multiply by } \tfrac{1}{12}.$$
$$\pm\frac{1}{2} = x \qquad \text{Take square roots.}$$

Verify that there are two solutions, $-1/2$ and $1/2$.

**(c)** $\dfrac{2}{k} - \dfrac{3k}{k + 2} = \dfrac{k}{k^2 + 2k}$

**SOLUTION** Factor $k^2 + 2k$ as $k(k + 2)$. The least common denominator for all the fractions is $k(k + 2)$. Multiplying both sides by $k(k + 2)$ gives the following:

$$k(k + 2) \cdot \left(\frac{2}{k} - \frac{3k}{k + 2}\right) = k(k + 2) \cdot \frac{k}{k^2 + 2k}$$
$$2(k + 2) - 3k(k) = k$$
$$2k + 4 - 3k^2 = k \qquad\qquad \text{Distributive property}$$
$$-3k^2 + k + 4 = 0 \qquad\qquad \text{Add } -k; \text{ rearrange terms.}$$
$$3k^2 - k - 4 = 0 \qquad\qquad \text{Multiply by } -1.$$
$$(3k - 4)(k + 1) = 0 \qquad\qquad \text{Factor.}$$
$$3k - 4 = 0 \quad \text{or} \quad k + 1 = 0$$
$$k = \frac{4}{3} \quad k = -1$$

**YOUR TURN 4** Solve $\dfrac{1}{x^2 - 4} + \dfrac{2}{x - 2} = \dfrac{1}{x}$.

Verify that the solutions are $4/3$ and $-1$. **TRY YOUR TURN 4**

**CAUTION** It is possible to get, as a solution of a rational equation, a number that makes one or more of the denominators in the original equation equal to zero. That number is not a solution, so it is *necessary* to check all potential solutions of rational equations. These introduced solutions are called **extraneous solutions**.

**EXAMPLE 7** **Solving a Rational Equation**

Solve $\dfrac{2}{x - 3} + \dfrac{1}{x} = \dfrac{6}{x(x - 3)}$.

**SOLUTION**  The common denominator is $x(x - 3)$. Multiply both sides by $x(x - 3)$ and solve the resulting equation.

$$x(x - 3) \cdot \left( \frac{2}{x - 3} + \frac{1}{x} \right) = x(x - 3) \cdot \left[ \frac{6}{x(x - 3)} \right]$$

$$2x + x - 3 = 6$$

$$3x = 9$$

$$x = 3$$

Checking this potential solution by substitution into the original equation shows that 3 makes two denominators 0. Thus, 3 cannot be a solution, so there is no solution for this equation.

# R.4  EXERCISES

**Solve each equation.**

**1.** $2x + 8 = x - 4$

**2.** $5x + 2 = 8 - 3x$

**3.** $0.2m - 0.5 = 0.1m + 0.7$

**4.** $\frac{2}{3}k - k + \frac{3}{8} = \frac{1}{2}$

**5.** $3r + 2 - 5(r + 1) = 6r + 4$

**6.** $5(a + 3) + 4a - 5 = -(2a - 4)$

**7.** $2[3m - 2(3 - m) - 4] = 6m - 4$

**8.** $4[2p - (3 - p) + 5] = -7p - 2$

**Solve each equation by factoring or by using the quadratic formula. If the solutions involve square roots, give both the exact solutions and the approximate solutions to three decimal places.**

**9.** $x^2 + 5x + 6 = 0$

**10.** $x^2 = 3 + 2x$

**11.** $m^2 = 14m - 49$

**12.** $2k^2 - k = 10$

**13.** $12x^2 - 5x = 2$

**14.** $m(m - 7) = -10$

**15.** $4x^2 - 36 = 0$

**16.** $z(2z + 7) = 4$

**17.** $12y^2 - 48y = 0$

**18.** $3x^2 - 5x + 1 = 0$

**19.** $2m^2 - 4m = 3$

**20.** $p^2 + p - 1 = 0$

**21.** $k^2 - 10k = -20$

**22.** $5x^2 - 8x + 2 = 0$

**23.** $2r^2 - 7r + 5 = 0$

**24.** $2x^2 - 7x + 30 = 0$

**25.** $3k^2 + k = 6$

**26.** $5m^2 + 5m = 0$

**Solve each equation.**

**27.** $\frac{3x - 2}{7} = \frac{x + 2}{5}$

**28.** $\frac{x}{3} - 7 = 6 - \frac{3x}{4}$

**29.** $\frac{4}{x - 3} - \frac{8}{2x + 5} + \frac{3}{x - 3} = 0$

**30.** $\frac{5}{p - 2} - \frac{7}{p + 2} = \frac{12}{p^2 - 4}$

**31.** $\frac{2m}{m - 2} - \frac{6}{m} = \frac{12}{m^2 - 2m}$

**32.** $\frac{2y}{y - 1} = \frac{5}{y} + \frac{10 - 8y}{y^2 - y}$

**33.** $\frac{1}{x - 2} - \frac{3x}{x - 1} = \frac{2x + 1}{x^2 - 3x + 2}$

**34.** $\frac{5}{a} + \frac{-7}{a + 1} = \frac{a^2 - 2a + 4}{a^2 + a}$

**35.** $\frac{5}{b + 5} - \frac{4}{b^2 + 2b} = \frac{6}{b^2 + 7b + 10}$

**36.** $\frac{2}{x^2 - 2x - 3} + \frac{5}{x^2 - x - 6} = \frac{1}{x^2 + 3x + 2}$

**37.** $\frac{4}{2x^2 + 3x - 9} + \frac{2}{2x^2 - x - 3} = \frac{3}{x^2 + 4x + 3}$

**YOUR TURN ANSWERS**

**1.** $-3/2$

**2.** $3/2, -5$

**3.** $4 \pm \sqrt{10}$

**4.** $-1, -4$

# R.5  Inequalities

To write that one number is greater than or less than another number, we use the following symbols.

**Inequality Symbols**

$<$ means *is less than*

$\leq$ means *is less than or equal to*

$>$ means *is greater than*

$\geq$ means *is greater than or equal to*

## Linear Inequalities

An equation states that two expressions are equal; an **inequality** states that they are unequal. A **linear inequality** is an inequality that can be simplified to the form $ax < b$. (Properties introduced in this section are given only for $<$, but they are equally valid for $>$, $\leq$, or $\geq$.) Linear inequalities are solved with the following properties.

### Properties of Inequality

For all real numbers $a$, $b$, and $c$:

1. If $a < b$, then $a + c < b + c$.
2. If $a < b$ and if $c > 0$, then $ac < bc$.
3. If $a < b$ and if $c < 0$, then $ac > bc$.

Pay careful attention to property 3; it says that if both sides of an inequality are multiplied by a negative number, the direction of the inequality symbol must be reversed.

### EXAMPLE 1    Solving a Linear Inequality

Solve $4 - 3y \leq 7 + 2y$.

**SOLUTION**    Use the properties of inequality.

$$4 - 3y + (-4) \leq 7 + 2y + (-4) \qquad \text{Add } -4 \text{ to both sides.}$$
$$-3y \leq 3 + 2y$$

Remember that *adding* the same number to both sides never changes the direction of the inequality symbol.

$$-3y + (-2y) \leq 3 + 2y + (-2y) \qquad \text{Add } -2y \text{ to both sides.}$$
$$-5y \leq 3$$

Multiply both sides by $-1/5$. Since $-1/5$ is negative, change the direction of the inequality symbol.

**YOUR TURN 1** Solve $3z - 2 > 5z + 7$.

$$-\frac{1}{5}(-5y) \geq -\frac{1}{5}(3)$$
$$y \geq -\frac{3}{5}$$

TRY YOUR TURN 1

> **CAUTION**    It is a common error to forget to reverse the direction of the inequality sign when multiplying or dividing by a negative number. For example, to solve $-4x \leq 12$, we must multiply by $-1/4$ on both sides *and* reverse the inequality symbol to get $x \geq -3$.

The solution $y \geq -3/5$ in Example 1 represents an interval on the number line. **Interval notation** often is used for writing intervals. With interval notation, $y \geq -3/5$ is written as $[-3/5, \infty)$. This is an example of a **half-open interval**, since one endpoint, $-3/5$, is included. The **open interval** $(2, 5)$ corresponds to $2 < x < 5$, with neither endpoint included. The **closed interval** $[2, 5]$ includes both endpoints and corresponds to $2 \leq x \leq 5$.

The **graph** of an interval shows all points on a number line that correspond to the numbers in the interval. To graph the interval $[-3/5, \infty)$, for example, use a solid circle at $-3/5$, since $-3/5$ is part of the solution. To show that the solution includes all real numbers greater than or equal to $-3/5$, draw a heavy arrow pointing to the right (the positive direction). See Figure 1.

FIGURE 1

### EXAMPLE 2   Graphing a Linear Inequality

Solve $-2 < 5 + 3m < 20$. Graph the solution.

**SOLUTION**   The inequality $-2 < 5 + 3m < 20$ says that $5 + 3m$ is *between* $-2$ and $20$. Solve this inequality with an extension of the properties given above. Work as follows, first adding $-5$ to each part.

$$-2 + (-5) < 5 + 3m + (-5) < 20 + (-5)$$
$$-7 < 3m < 15$$

Now multiply each part by 1/3.

$$-\frac{7}{3} < m < 5$$

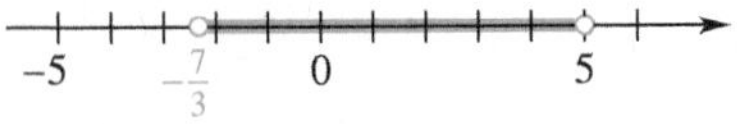

A graph of the solution is given in Figure 2; here open circles are used to show that $-7/3$ and 5 are *not* part of the graph.*

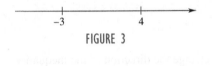

FIGURE 2

## Quadratic Inequalities   A **quadratic inequality** has the form $ax^2 + bx + c > 0$ (or $<$, or $\le$, or $\ge$). The highest exponent is 2. The next few examples show how to solve quadratic inequalities.

### EXAMPLE 3   Solving a Quadratic Inequality

Solve the quadratic inequality $x^2 - x < 12$.

**SOLUTION**   Write the inequality with 0 on one side, as $x^2 - x - 12 < 0$. This inequality is solved with values of $x$ that make $x^2 - x - 12$ negative $(< 0)$. The quantity $x^2 - x - 12$ changes from positive to negative or from negative to positive at the points where it equals 0. For this reason, first solve the *equation* $x^2 - x - 12 = 0$.

$$x^2 - x - 12 = 0$$
$$(x - 4)(x + 3) = 0$$
$$x = 4 \quad \text{or} \quad x = -3$$

Locating $-3$ and 4 on a number line, as shown in Figure 3, determines three intervals A, B, and C. Decide which intervals include numbers that make $x^2 - x - 12$ negative by substituting any number from each interval into the polynomial. For example,

choose $-4$ from interval A: $(-4)^2 - (-4) - 12 = 8 > 0$;

choose 0 from interval B: $0^2 - 0 - 12 = -12 < 0$;

choose 5 from interval C: $5^2 - 5 - 12 = 8 > 0$.

Only numbers in interval B satisfy the given inequality, so the solution is $(-3, 4)$. A graph of this solution is shown in Figure 4.   **TRY YOUR TURN 2**

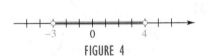

A       B       C

FIGURE 3

FIGURE 4

**YOUR TURN 2** Solve $3y^2 \le 16y + 12$.

### EXAMPLE 4   Solving a Polynomial Inequality

Solve the inequality $x^3 + 2x^2 - 3x \ge 0$.

**SOLUTION**   This is not a quadratic inequality because of the $x^3$ term, but we solve it in a similar way by first factoring the polynomial.

$$x^3 + 2x^2 - 3x = x(x^2 + 2x - 3) \qquad \text{Factor out the common factor.}$$
$$= x(x - 1)(x + 3) \qquad \text{Factor the quadratic.}$$

*Some textbooks use brackets in place of solid circles for the graph of a closed interval, and parentheses in place of open circles for the graph of an open interval.

Now solve the corresponding equation.

$$x(x - 1)(x + 3) = 0$$

$$x = 0 \quad \text{or} \quad x - 1 = 0 \quad \text{or} \quad x + 3 = 0$$
$$x = 1 \qquad\qquad x = -3$$

These three solutions determine four intervals on the number line: $(-\infty, -3)$, $(-3, 0)$, $(0, 1)$, and $(1, \infty)$. Substitute a number from each interval into the original inequality to determine that the solution consists of the numbers between $-3$ and $0$ (including the endpoints) and all numbers that are greater than or equal to 1. See Figure 5. In interval notation, the solution is

**FIGURE 5**

$$[-3, 0] \cup [1, \infty).*$$

## Inequalities with Fractions

Inequalities with fractions are solved in a similar manner as quadratic inequalities.

### EXAMPLE 5  Solving a Rational Inequality

Solve $\dfrac{2x - 3}{x} \geq 1$.

**SOLUTION**   First solve the corresponding equation.

$$\frac{2x - 3}{x} = 1$$

$$2x - 3 = x$$

$$x = 3$$

The solution, $x = 3$, determines the intervals on the number line where the fraction may change from greater than 1 to less than 1. This change also may occur on either side of a number that makes the denominator equal 0. Here, the $x$-value that makes the denominator 0 is $x = 0$. Test each of the three intervals determined by the numbers 0 and 3.

$$\text{For } (-\infty, 0), \text{ choose } -1: \frac{2(-1) - 3}{-1} = 5 \geq 1.$$

$$\text{For } (0, 3), \quad \text{choose } 1: \frac{2(1) - 3}{1} = -1 \ngeq 1.$$

$$\text{For } (3, \infty), \quad \text{choose } 4: \frac{2(4) - 3}{4} = \frac{5}{4} \geq 1.$$

**FIGURE 6**

The symbol $\ngeq$ means "is *not* greater than or equal to." Testing the endpoints 0 and 3 shows that the solution is $(-\infty, 0) \cup [3, \infty)$, as shown in Figure 6.

> **CAUTION**   A common error is to try to solve the inequality in Example 5 by multiplying both sides by $x$. The reason this is wrong is that we don't know in the beginning whether $x$ is positive, negative, or 0. If $x$ is negative, the $\geq$ would change to $\leq$ according to the third property of inequality listed at the beginning of this section.

*The symbol $\cup$ indicates the *union* of two sets, which includes all elements in either set.

### EXAMPLE 6 Solving a Rational Inequality

Solve $\dfrac{(x - 1)(x + 1)}{x} \leq 0$.

**SOLUTION** We first solve the corresponding equation.

$$\frac{(x - 1)(x + 1)}{x} = 0$$

$$(x - 1)(x + 1) = 0 \qquad \text{Multiply both sides by } x.$$

$$x = 1 \quad \text{or} \quad x = -1 \qquad \text{Use the zero-factor property.}$$

Setting the denominator equal to 0 gives $x = 0$, so the intervals of interest are $(-\infty, -1)$, $(-1, 0)$, $(0, 1)$, and $(1, \infty)$. Testing a number from each region in the original inequality and checking the endpoints, we find the solution is

$$(-\infty, -1] \cup (0, 1],$$

as shown in Figure 7.

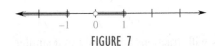

**FIGURE 7**

---

**CAUTION** Remember to solve the equation formed by setting the *denominator* equal to zero. Any number that makes the denominator zero always creates two intervals on the number line. For instance, in Example 6, substituting $x = 0$ makes the denominator of the rational inequality equal to 0, so we know that there may be a sign change from one side of 0 to the other (as was indeed the case).

---

### EXAMPLE 7 Solving a Rational Inequality

Solve $\dfrac{x^2 - 3x}{x^2 - 9} < 4$.

**SOLUTION** Solve the corresponding equation.

$$\frac{x^2 - 3x}{x^2 - 9} = 4$$

$$x^2 - 3x = 4x^2 - 36 \qquad \text{Multiply by } x^2 - 9.$$

$$0 = 3x^2 + 3x - 36 \qquad \text{Get 0 on one side.}$$

$$0 = x^2 + x - 12 \qquad \text{Multiply by } \tfrac{1}{3}.$$

$$0 = (x + 4)(x - 3) \qquad \text{Factor.}$$

$$x = -4 \quad \text{or} \quad x = 3$$

Now set the denominator equal to 0 and solve that equation.

$$x^2 - 9 = 0$$

$$(x - 3)(x + 3) = 0$$

$$x = 3 \quad \text{or} \quad x = -3$$

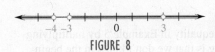

**FIGURE 8**

**YOUR TURN 3** Solve

$$\frac{k^2 - 35}{k} \geq 2.$$

The intervals determined by the three (different) solutions are $(-\infty, -4)$, $(-4, -3)$, $(-3, 3)$, and $(3, \infty)$. Testing a number from each interval in the given inequality shows that the solution is

$$(-\infty, -4) \cup (-3, 3) \cup (3, \infty),$$

as shown in Figure 8. For this example, none of the endpoints are part of the solution because $x = 3$ and $x = -3$ make the denominator zero and $x = -4$ produces an equality.

**TRY YOUR TURN 3**

# R.5 EXERCISES

**Write each expression in interval notation. Graph each interval.**

**1.** $x < 4$

**2.** $x \geq -3$

**3.** $1 \leq x < 2$

**4.** $-2 \leq x \leq 3$

**5.** $-9 > x$

**6.** $6 \leq x$

**Using the variable $x$, write each interval as an inequality.**

**7.** $[-7, -3]$

**8.** $[4, 10)$

**9.** $(-\infty, -1]$

**10.** $(3, \infty)$

**11.**

**12.**

**13.**

**14.**

**Solve each inequality and graph the solution.**

**15.** $6p + 7 \leq 19$

**16.** $6k - 4 < 3k - 1$

**17.** $m - (3m - 2) + 6 < 7m - 19$

**18.** $-2(3y - 8) \geq 5(4y - 2)$

**19.** $3p - 1 < 6p + 2(p - 1)$

**20.** $x + 5(x + 1) > 4(2 - x) + x$

**21.** $-11 < y - 7 < -1$

**22.** $8 \leq 3r + 1 \leq 13$

**23.** $-2 < \dfrac{1 - 3k}{4} \leq 4$

**24.** $-1 \leq \dfrac{5y + 2}{3} \leq 4$

**25.** $\dfrac{3}{5}(2p + 3) \geq \dfrac{1}{10}(5p + 1)$

**26.** $\dfrac{8}{3}(z - 4) \leq \dfrac{2}{9}(3z + 2)$

**Solve each inequality. Graph each solution.**

**27.** $(m - 3)(m + 5) < 0$

**28.** $(t + 6)(t - 1) \geq 0$

**29.** $y^2 - 3y + 2 < 0$

**30.** $2k^2 + 7k - 4 > 0$

**31.** $x^2 - 16 > 0$

**32.** $2k^2 - 7k - 15 \leq 0$

**33.** $x^2 - 4x \geq 5$

**34.** $10r^2 + r \leq 2$

**35.** $3x^2 + 2x > 1$

**36.** $3a^2 + a > 10$

**37.** $9 - x^2 \leq 0$

**38.** $p^2 - 16p > 0$

**39.** $x^3 - 4x \geq 0$

**40.** $x^3 + 7x^2 + 12x \leq 0$

**41.** $2x^3 - 14x^2 + 12x < 0$

**42.** $3x^3 - 9x^2 - 12x > 0$

**Solve each inequality.**

**43.** $\dfrac{m - 3}{m + 5} \leq 0$

**44.** $\dfrac{r + 1}{r - 1} > 0$

**45.** $\dfrac{k - 1}{k + 2} > 1$

**46.** $\dfrac{a - 5}{a + 2} < -1$

**47.** $\dfrac{2y + 3}{y - 5} \leq 1$

**48.** $\dfrac{a + 2}{3 + 2a} \leq 5$

**49.** $\dfrac{2k}{k - 3} \leq \dfrac{4}{k - 3}$

**50.** $\dfrac{5}{p + 1} > \dfrac{12}{p + 1}$

**51.** $\dfrac{2x}{x^2 - x - 6} \geq 0$

**52.** $\dfrac{8}{p^2 + 2p} > 1$

**53.** $\dfrac{z^2 + z}{z^2 - 1} \geq 3$

**54.** $\dfrac{a^2 + 2a}{a^2 - 4} \leq 2$

**YOUR TURN ANSWERS**

**1.** $z < -9/2$

**2.** $[-2/3, 6]$

**3.** $[-5, 0) \cup [7, \infty)$

# R.6 Exponents

## Integer Exponents

Recall that $a^2 = a \cdot a$, while $a^3 = a \cdot a \cdot a$, and so on. In this section, a more general meaning is given to the symbol $a^n$.

**Definition of Exponent**

If $n$ is a natural number, then

$$a^n = a \cdot a \cdot a \cdot \cdots \cdot a,$$

where $a$ appears as a factor $n$ times.

In the expression $a^n$, the power $n$ is the **exponent** and $a$ is the **base**. This definition can be extended by defining $a^n$ for zero and negative integer values of $n$.

### Zero and Negative Exponents

If $a$ is any nonzero real number, and if $n$ is a positive integer, then

$$a^0 = 1 \quad \text{and} \quad a^{-n} = \frac{1}{a^n}.$$

(The symbol $0^0$ is meaningless.)

### EXAMPLE 1  Exponents

(a) $6^0 = 1$

(b) $(-9)^0 = 1$

(c) $3^{-2} = \frac{1}{3^2} = \frac{1}{9}$

(d) $9^{-1} = \frac{1}{9^1} = \frac{1}{9}$

(e) $\left(\frac{3}{4}\right)^{-1} = \frac{1}{(3/4)^1} = \frac{1}{3/4} = \frac{4}{3}$

The following properties follow from the definitions of exponents given above.

### Properties of Exponents

For any integers $m$ and $n$, and any real numbers $a$ and $b$ for which the following exist:

1. $a^m \cdot a^n = a^{m+n}$

2. $\dfrac{a^m}{a^n} = a^{m-n}$

3. $(a^m)^n = a^{mn}$

4. $(ab)^m = a^m \cdot b^m$

5. $\left(\dfrac{a}{b}\right)^m = \dfrac{a^m}{b^m}$

Note that $(-a)^n = a^n$ if $n$ is an even integer, but $(-a)^n = -a^n$ if $n$ is an odd integer.

### EXAMPLE 2  Simplifying Exponential Expressions

Use the properties of exponents to simplify each expression. Leave answers with positive exponents. Assume that all variables represent positive real numbers.

(a) $7^4 \cdot 7^6 = 7^{4+6} = 7^{10}$ (or 282,475,249)    Property 1

(b) $\dfrac{9^{14}}{9^6} = 9^{14-6} = 9^8$ (or 43,046,721)    Property 2

(c) $\dfrac{r^9}{r^{17}} = r^{9-17} = r^{-8} = \dfrac{1}{r^8}$    Property 2

(d) $(2m^3)^4 = 2^4 \cdot (m^3)^4 = 16m^{12}$    Properties 3 and 4

(e) $(3x)^4 = 3^4 \cdot x^4 = 81x^4$    Property 4

(f) $\left(\dfrac{x^2}{y^3}\right)^6 = \dfrac{(x^2)^6}{(y^3)^6} = \dfrac{x^{2 \cdot 6}}{y^{3 \cdot 6}} = \dfrac{x^{12}}{y^{18}}$    Properties 3 and 5

(g) $\dfrac{a^{-3}b^5}{a^4b^{-7}} = \dfrac{b^{5-(-7)}}{a^{4-(-3)}} = \dfrac{b^{5+7}}{a^{4+3}} = \dfrac{b^{12}}{a^7}$    Property 2

**(h)** $p^{-1} + q^{-1} = \dfrac{1}{p} + \dfrac{1}{q} = \dfrac{1}{p} \cdot \dfrac{q}{q} + \dfrac{1}{q} \cdot \dfrac{p}{p} = \dfrac{q}{pq} + \dfrac{p}{pq} = \dfrac{p+q}{pq}$

**YOUR TURN 1** Simplify
$\left( \dfrac{y^2 z^{-4}}{y^{-3} z^4} \right)^{-2}.$

**(i)** $\dfrac{x^{-2} - y^{-2}}{x^{-1} - y^{-1}} = \dfrac{\dfrac{1}{x^2} - \dfrac{1}{y^2}}{\dfrac{1}{x} - \dfrac{1}{y}}$    Definition of $a^{-n}$

$= \dfrac{\dfrac{y^2 - x^2}{x^2 y^2}}{\dfrac{y - x}{xy}}$    Get common denominators and combine terms.

$= \dfrac{y^2 - x^2}{x^2 y^2} \cdot \dfrac{xy}{y - x}$    Invert and multiply.

$= \dfrac{(y - x)(y + x)}{x^2 y^2} \cdot \dfrac{xy}{y - x}$    Factor.

$= \dfrac{x + y}{xy}$    Simplify.      **TRY YOUR TURN 1**

**CAUTION**    If Example 2(e) were written $3x^4$, the properties of exponents would not apply. When no parentheses are used, the exponent refers only to the factor closest to it. Also notice in Examples 2(c), 2(g), 2(h), and 2(i) that a negative exponent does *not* indicate a negative number.

## Roots
For *even* values of $n$ and nonnegative values of $a$, the expression $a^{1/n}$ is defined to be the **positive $n$th root** of $a$ or the **principal $n$th root** of $a$. For example, $a^{1/2}$ denotes the positive second root, or **square root**, of $a$, while $a^{1/4}$ is the positive fourth root of $a$. When $n$ is *odd*, there is only one $n$th root, which has the same sign as $a$. For example, $a^{1/3}$, the **cube root** of $a$, has the same sign as $a$. By definition, if $b = a^{1/n}$, then $b^n = a$. On a calculator, a number is raised to a power using a key labeled $x^y$, $y^x$, or $\wedge$. For example, to take the fourth root of 6 on a TI-84 Plus calculator, enter $6 \wedge (1/4)$ to get the result 1.56508458.

**EXAMPLE 3**   **Calculations with Exponents**

**(a)** $121^{1/2} = 11$, since 11 is positive and $11^2 = 121$.

**(b)** $625^{1/4} = 5$, since $5^4 = 625$.

**(c)** $256^{1/4} = 4$

**(d)** $64^{1/6} = 2$

**(e)** $27^{1/3} = 3$

**(f)** $(-32)^{1/5} = -2$

**(g)** $128^{1/7} = 2$

**(h)** $(-49)^{1/2}$ is not a real number.

## Rational Exponents
In the following definition, the domain of an exponent is extended to include all rational numbers.

**Definition of $a^{m/n}$**

For all real numbers $a$ for which the indicated roots exist, and for any rational number $m/n$,

$$a^{m/n} = \left( a^{1/n} \right)^m.$$

## EXAMPLE 4   Calculations with Exponents

(a) $27^{2/3} = (27^{1/3})^2 = 3^2 = 9$

(b) $32^{2/5} = (32^{1/5})^2 = 2^2 = 4$

(c) $64^{4/3} = (64^{1/3})^4 = 4^4 = 256$

(d) $25^{3/2} = (25^{1/2})^3 = 5^3 = 125$

**NOTE**
$27^{2/3}$ could also be evaluated as $(27^2)^{1/3}$, but this is more difficult to perform without a calculator because it involves squaring 27 and then taking the cube root of this large number. On the other hand, when we evaluate it as $(27^{1/3})^2$, we know that the cube root of 27 is 3 without using a calculator, and squaring 3 is easy.

All the properties for integer exponents given in this section also apply to any rational exponent on a nonnegative real-number base.

## EXAMPLE 5   Simplifying Exponential Expressions

(a) $\dfrac{y^{1/3}y^{5/3}}{y^3} = \dfrac{y^{1/3+5/3}}{y^3} = \dfrac{y^2}{y^3} = y^{2-3} = y^{-1} = \dfrac{1}{y}$

(b) $m^{2/3}(m^{7/3} + 2m^{1/3}) = m^{2/3+7/3} + 2m^{2/3+1/3} = m^3 + 2m$

(c) $\left(\dfrac{m^7 n^{-2}}{m^{-5} n^2}\right)^{1/4} = \left(\dfrac{m^{7-(-5)}}{n^{2-(-2)}}\right)^{1/4} = \left(\dfrac{m^{12}}{n^4}\right)^{1/4} = \dfrac{(m^{12})^{1/4}}{(n^4)^{1/4}} = \dfrac{m^{12/4}}{n^{4/4}} = \dfrac{m^3}{n}$

In calculus, it is often necessary to factor expressions involving fractional exponents.

## EXAMPLE 6   Simplifying Exponential Expressions

Factor out the smallest power of the variable, assuming all variables represent positive real numbers.

(a) $4m^{1/2} + 3m^{3/2}$

**SOLUTION**   The smallest exponent is $1/2$. Factoring out $m^{1/2}$ yields

$$4m^{1/2} + 3m^{3/2} = m^{1/2}(4m^{1/2-1/2} + 3m^{3/2-1/2})$$
$$= m^{1/2}(4 + 3m).$$

Check this result by multiplying $m^{1/2}$ by $4 + 3m$.

(b) $9x^{-2} - 6x^{-3}$

**SOLUTION**   The smallest exponent here is $-3$. Since 3 is a common numerical factor, factor out $3x^{-3}$.

$$9x^{-2} - 6x^{-3} = 3x^{-3}(3x^{-2-(-3)} - 2x^{-3-(-3)}) = 3x^{-3}(3x - 2)$$

Check by multiplying. The factored form can be written without negative exponents as

$$\dfrac{3(3x - 2)}{x^3}.$$

(c) $(x^2 + 5)(3x - 1)^{-1/2}(2) + (3x - 1)^{1/2}(2x)$

**SOLUTION**   There is a common factor of 2. Also, $(3x - 1)^{-1/2}$ and $(3x - 1)^{1/2}$ have a common factor. Always factor out the quantity to the *smallest* exponent. Here $-1/2 < 1/2$, so the common factor is $2(3x - 1)^{-1/2}$ and the factored form is

**YOUR TURN 2** Factor $5z^{1/3} + 4z^{-2/3}$.

$$2(3x - 1)^{-1/2}[(x^2 + 5) + (3x - 1)x] = 2(3x - 1)^{-1/2}(4x^2 - x + 5).$$

TRY YOUR TURN 2

# R.6   EXERCISES

**Evaluate each expression. Write all answers without exponents.**

**1.** $8^{-2}$

**2.** $3^{-4}$

**3.** $5^0$

**4.** $\left(-\dfrac{3}{4}\right)^0$

**5.** $-(-3)^{-2}$

**6.** $-(-3^{-2})$

**7.** $\left(\dfrac{1}{6}\right)^{-2}$

**8.** $\left(\dfrac{4}{3}\right)^{-3}$

**Simplify each expression. Assume that all variables represent positive real numbers. Write answers with only positive exponents.**

**9.** $\dfrac{4^{-2}}{4}$

**10.** $\dfrac{8^9 \cdot 8^{-7}}{8^{-3}}$

**11.** $\dfrac{10^8 \cdot 10^{-10}}{10^4 \cdot 10^2}$

**12.** $\left(\dfrac{7^{-12} \cdot 7^3}{7^{-8}}\right)^{-1}$

**13.** $\dfrac{x^4 \cdot x^3}{x^5}$

**14.** $\dfrac{y^{10} \cdot y^{-4}}{y^6}$

**15.** $\dfrac{(4k^{-1})^2}{2k^{-5}}$

**16.** $\dfrac{(3z^2)^{-1}}{z^5}$

**17.** $\dfrac{3^{-1} \cdot x \cdot y^2}{x^{-4} \cdot y^5}$

**18.** $\dfrac{5^{-2}m^2y^{-2}}{5^2m^{-1}y^{-2}}$

**19.** $\left(\dfrac{a^{-1}}{b^2}\right)^{-3}$

**20.** $\left(\dfrac{c^3}{7d^{-2}}\right)^{-2}$

**Simplify each expression, writing the answer as a single term without negative exponents.**

**21.** $a^{-1} + b^{-1}$

**22.** $b^{-2} - a$

**23.** $\dfrac{2n^{-1} - 2m^{-1}}{m + n^2}$

**24.** $\left(\dfrac{m}{3}\right)^{-1} + \left(\dfrac{n}{2}\right)^{-2}$

**25.** $(x^{-1} - y^{-1})^{-1}$

**26.** $(x \cdot y^{-1} - y^{-2})^{-2}$

**Write each number without exponents.**

**27.** $121^{1/2}$

**28.** $27^{1/3}$

**29.** $32^{2/5}$

**30.** $-125^{2/3}$

**31.** $\left(\dfrac{36}{144}\right)^{1/2}$

**32.** $\left(\dfrac{64}{27}\right)^{1/3}$

**33.** $8^{-4/3}$

**34.** $625^{-1/4}$

**35.** $\left(\dfrac{27}{64}\right)^{-1/3}$

**36.** $\left(\dfrac{121}{100}\right)^{-3/2}$

**Simplify each expression. Write all answers with only positive exponents. Assume that all variables represent positive real numbers.**

**37.** $3^{2/3} \cdot 3^{4/3}$

**38.** $27^{2/3} \cdot 27^{-1/3}$

**39.** $\dfrac{4^{9/4} \cdot 4^{-7/4}}{4^{-10/4}}$

**40.** $\dfrac{3^{-5/2} \cdot 3^{3/2}}{3^{7/2} \cdot 3^{-9/2}}$

**41.** $\left(\dfrac{x^6y^{-3}}{x^{-2}y^5}\right)^{1/2}$

**42.** $\left(\dfrac{a^{-7}b^{-1}}{b^{-4}a^2}\right)^{1/3}$

**43.** $\dfrac{7^{-1/3} \cdot 7r^{-3}}{7^{2/3} \cdot (r^{-2})^2}$

**44.** $\dfrac{12^{3/4} \cdot 12^{5/4} \cdot y^{-2}}{12^{-1} \cdot (y^{-3})^{-2}}$

**45.** $\dfrac{3k^2 \cdot (4k^{-3})^{-1}}{4^{1/2} \cdot k^{7/2}}$

**46.** $\dfrac{8p^{-3} \cdot (4p^2)^{-2}}{p^{-5}}$

**47.** $\dfrac{a^{4/3} \cdot b^{1/2}}{a^{2/3} \cdot b^{-3/2}}$

**48.** $\dfrac{x^{3/2} \cdot y^{4/5} \cdot z^{-3/4}}{x^{5/3} \cdot y^{-6/5} \cdot z^{1/2}}$

**49.** $\dfrac{k^{-3/5} \cdot h^{-1/3} \cdot t^{2/5}}{k^{-1/5} \cdot h^{-2/3} \cdot t^{1/5}}$

**50.** $\dfrac{m^{7/3} \cdot n^{-2/5} \cdot p^{3/8}}{m^{-2/3} \cdot n^{3/5} \cdot p^{-5/8}}$

**Factor each expression.**

**51.** $3x^3(x^2 + 3x)^2 - 15x(x^2 + 3x)^2$

**52.** $6x(x^3 + 7)^2 - 6x^2(3x^2 + 5)(x^3 + 7)$

**53.** $10x^3(x^2 - 1)^{-1/2} - 5x(x^2 - 1)^{1/2}$

**54.** $9(6x + 2)^{1/2} + 3(9x - 1)(6x + 2)^{-1/2}$

**55.** $x(2x + 5)^2(x^2 - 4)^{-1/2} + 2(x^2 - 4)^{1/2}(2x + 5)$

**56.** $(4x^2 + 1)^2(2x - 1)^{-1/2} + 16x(4x^2 + 1)(2x - 1)^{1/2}$

**▬▬ YOUR TURN ANSWERS**

**1.** $z^{16}/y^{10}$

**2.** $z^{-2/3}(5z + 4)$

# R.7  Radicals

We have defined $a^{1/n}$ as the positive or principal $n$th root of $a$ for appropriate values of $a$ and $n$. An alternative notation for $a^{1/n}$ uses radicals.

> **Radicals**
> If $n$ is an even natural number and $a > 0$, or $n$ is an odd natural number, then
> $$a^{1/n} = \sqrt[n]{a}.$$

The symbol $\sqrt[n]{\ }$ is a **radical sign**, the number $a$ is the **radicand**, and $n$ is the **index** of the radical. The familiar symbol $\sqrt{a}$ is used instead of $\sqrt[2]{a}$.

### EXAMPLE 1  Radical Calculations

(a) $\sqrt[4]{16} = 16^{1/4} = 2$

(b) $\sqrt[5]{-32} = -2$

(c) $\sqrt[3]{1000} = 10$

(d) $\sqrt[6]{\dfrac{64}{729}} = \dfrac{2}{3}$

With $a^{1/n}$ written as $\sqrt[n]{a}$, the expression $a^{m/n}$ also can be written using radicals.

$$a^{m/n} = (\sqrt[n]{a})^m \quad \text{or} \quad a^{m/n} = \sqrt[n]{a^m}$$

The following properties of radicals depend on the definitions and properties of exponents.

### Properties of Radicals
For all real numbers $a$ and $b$ and natural numbers $m$ and $n$ such that $\sqrt[n]{a}$ and $\sqrt[n]{b}$ are real numbers:

1. $(\sqrt[n]{a})^n = a$

2. $\sqrt[n]{a^n} = \begin{cases} |a| & \text{if } n \text{ is even} \\ a & \text{if } n \text{ is odd} \end{cases}$

3. $\sqrt[n]{a} \cdot \sqrt[n]{b} = \sqrt[n]{ab}$

4. $\dfrac{\sqrt[n]{a}}{\sqrt[n]{b}} = \sqrt[n]{\dfrac{a}{b}} \quad (b \neq 0)$

5. $\sqrt[m]{\sqrt[n]{a}} = \sqrt[mn]{a}$

Property 3 can be used to simplify certain radicals. For example, since $48 = 16 \cdot 3$,
$$\sqrt{48} = \sqrt{16 \cdot 3} = \sqrt{16} \cdot \sqrt{3} = 4\sqrt{3}.$$

To some extent, simplification is in the eye of the beholder, and $\sqrt{48}$ might be considered as simple as $4\sqrt{3}$. In this textbook, we will consider an expression to be simpler when we have removed as many factors as possible from under the radical.

### EXAMPLE 2  Radical Calculations

(a) $\sqrt{1000} = \sqrt{100 \cdot 10} = \sqrt{100} \cdot \sqrt{10} = 10\sqrt{10}$

(b) $\sqrt{128} = \sqrt{64 \cdot 2} = 8\sqrt{2}$

(c) $\sqrt{2} \cdot \sqrt{18} = \sqrt{2 \cdot 18} = \sqrt{36} = 6$

(d) $\sqrt[3]{54} = \sqrt[3]{27 \cdot 2} = \sqrt[3]{27} \cdot \sqrt[3]{2} = 3\sqrt[3]{2}$

(e) $\sqrt{288m^5} = \sqrt{144 \cdot m^4 \cdot 2m} = 12m^2\sqrt{2m}$

(f) $2\sqrt{18} - 5\sqrt{32} = 2\sqrt{9 \cdot 2} - 5\sqrt{16 \cdot 2}$
$$= 2\sqrt{9} \cdot \sqrt{2} - 5\sqrt{16} \cdot \sqrt{2}$$
$$= 2(3)\sqrt{2} - 5(4)\sqrt{2} = -14\sqrt{2}$$

(g) $\sqrt{x^5} \cdot \sqrt[3]{x^5} = x^{5/2} \cdot x^{5/3} = x^{5/2+5/3} = x^{25/6} = \sqrt[6]{x^{25}} = x^4\sqrt[6]{x}$

**YOUR TURN 1**
Simplify $\sqrt{28x^9y^5}$.

TRY YOUR TURN 1

When simplifying a square root, keep in mind that $\sqrt{x}$ is nonnegative by definition. Also, $\sqrt{x^2}$ is not $x$, but $|x|$, the **absolute value of $x$**, defined as

$$|x| = \begin{cases} x & \text{if } x \geq 0 \\ -x & \text{if } x < 0. \end{cases}$$

For example, $\sqrt{(-5)^2} = |-5| = 5$. It is correct, however, to simplify $\sqrt{x^4} = x^2$. We need not write $|x^2|$ because $x^2$ is always nonnegative.

**EXAMPLE 3   Simplifying by Factoring**

Simplify $\sqrt{m^2 - 4m + 4}$.

**SOLUTION**   Factor the polynomial as $m^2 - 4m + 4 = (m - 2)^2$. Then by property 2 of radicals and the definition of absolute value,

$$\sqrt{(m-2)^2} = |m - 2| = \begin{cases} m - 2 & \text{if } m - 2 \geq 0 \\ -(m - 2) = 2 - m & \text{if } m - 2 < 0. \end{cases}$$

> **CAUTION**   Avoid the common error of writing $\sqrt{a^2 + b^2}$ as $\sqrt{a^2} + \sqrt{b^2}$. We must add $a^2$ and $b^2$ *before* taking the square root. For example, $\sqrt{16 + 9} = \sqrt{25} = 5$, *not* $\sqrt{16} + \sqrt{9} = 4 + 3 = 7$. This idea applies as well to higher roots. For example, in general,
>
> $$\sqrt[3]{a^3 + b^3} \neq \sqrt[3]{a^3} + \sqrt[3]{b^3},$$
>
> $$\sqrt[4]{a^4 + b^4} \neq \sqrt[4]{a^4} + \sqrt[4]{b^4}.$$
>
> Also,     $\sqrt{a + b} \neq \sqrt{a} + \sqrt{b}.$

# Rationalizing Denominators
The next example shows how to *rationalize* (remove all radicals from) the denominator in an expression containing radicals.

**EXAMPLE 4   Rationalizing Denominators**

Simplify each expression by rationalizing the denominator.

**(a)** $\dfrac{4}{\sqrt{3}}$

**SOLUTION**   To rationalize the denominator, multiply by $\sqrt{3}/\sqrt{3}$ (or 1) so the denominator of the product is a rational number.

$$\frac{4}{\sqrt{3}} \cdot \frac{\sqrt{3}}{\sqrt{3}} = \frac{4\sqrt{3}}{3} \qquad \sqrt{3} \cdot \sqrt{3} = \sqrt{9} = 3$$

**(b)** $\dfrac{2}{\sqrt[3]{x}}$

**SOLUTION**   Here, we need a perfect cube under the radical sign to rationalize the denominator. Multiplying by $\sqrt[3]{x^2}/\sqrt[3]{x^2}$ gives

$$\frac{2}{\sqrt[3]{x}} \cdot \frac{\sqrt[3]{x^2}}{\sqrt[3]{x^2}} = \frac{2\sqrt[3]{x^2}}{\sqrt[3]{x^3}} = \frac{2\sqrt[3]{x^2}}{x}.$$

**(c)** $\dfrac{1}{1 - \sqrt{2}}$

**YOUR TURN 2** Rationalize the denominator in

$\dfrac{5}{\sqrt{x} - \sqrt{y}}.$

**SOLUTION** The best approach here is to multiply both numerator and denominator by the number $1 + \sqrt{2}$. The expressions $1 + \sqrt{2}$ and $1 - \sqrt{2}$ are conjugates,* and their product is $1^2 - (\sqrt{2})^2 = 1 - 2 = -1$. Thus,

$$\frac{1}{1 - \sqrt{2}} = \frac{1(1 + \sqrt{2})}{(1 - \sqrt{2})(1 + \sqrt{2})} = \frac{1 + \sqrt{2}}{1 - 2} = -1 - \sqrt{2}.$$

**TRY YOUR TURN 2**

Sometimes it is advantageous to rationalize the *numerator* of a rational expression. The following example arises in calculus when evaluating a *limit*.

**EXAMPLE 5** Rationalizing Numerators

Rationalize each numerator.

**(a)** $\dfrac{\sqrt{x} - 3}{x - 9}.$

**SOLUTION** Multiply the numerator and denominator by the conjugate of the numerator, $\sqrt{x} + 3$.

$$\frac{\sqrt{x} - 3}{x - 9} \cdot \frac{\sqrt{x} + 3}{\sqrt{x} + 3} = \frac{(\sqrt{x})^2 - 3^2}{(x - 9)(\sqrt{x} + 3)} \qquad (a - b)(a + b) = a^2 - b^2$$

$$= \frac{x - 9}{(x - 9)(\sqrt{x} + 3)}$$

$$= \frac{1}{\sqrt{x} + 3}$$

**(b)** $\dfrac{\sqrt{3} + \sqrt{x + 3}}{\sqrt{3} - \sqrt{x + 3}}$

**SOLUTION** Multiply the numerator and denominator by the conjugate of the numerator, $\sqrt{3} - \sqrt{x + 3}$.

$$\frac{\sqrt{3} + \sqrt{x + 3}}{\sqrt{3} - \sqrt{x + 3}} \cdot \frac{\sqrt{3} - \sqrt{x + 3}}{\sqrt{3} - \sqrt{x + 3}} = \frac{3 - (x + 3)}{3 - 2\sqrt{3}\sqrt{x + 3} + (x + 3)}$$

$$= \frac{-x}{6 + x - 2\sqrt{3(x + 3)}}$$

# R.7 EXERCISES

**Simplify each expression by removing as many factors as possible from under the radical. Assume that all variables represent positive real numbers.**

**1.** $\sqrt[3]{125}$     **2.** $\sqrt[4]{1296}$

**3.** $\sqrt[5]{-3125}$     **4.** $\sqrt{50}$

**5.** $\sqrt{2000}$

**7.** $\sqrt{27} \cdot \sqrt{3}$

**9.** $7\sqrt{2} - 8\sqrt{18} + 4\sqrt{72}$

**10.** $4\sqrt{3} - 5\sqrt{12} + 3\sqrt{75}$

**11.** $4\sqrt{7} - \sqrt{28} + \sqrt{343}$

**6.** $\sqrt{32y^5}$

**8.** $\sqrt{2} \cdot \sqrt{32}$

*If $a$ and $b$ are real numbers, the *conjugate* of $a + b$ is $a - b$.

**12.** $3\sqrt{28} - 4\sqrt{63} + \sqrt{112}$

**13.** $\sqrt[3]{2} - \sqrt[3]{16} + 2\sqrt[3]{54}$

**14.** $2\sqrt[3]{5} - 4\sqrt[3]{40} + 3\sqrt[3]{135}$

**15.** $\sqrt{2x^3y^2z^4}$        **16.** $\sqrt{160r^7s^9t^{12}}$

**17.** $\sqrt[3]{128x^3y^8z^9}$        **18.** $\sqrt[4]{x^8y^7z^{11}}$

**19.** $\sqrt{a^3b^5} - 2\sqrt{a^7b^3} + \sqrt{a^3b^9}$

**20.** $\sqrt{p^7q^3} - \sqrt{p^5q^9} + \sqrt{p^9q}$

**21.** $\sqrt{a} \cdot \sqrt[3]{a}$

**22.** $\sqrt{b^3} \cdot \sqrt[4]{b^3}$

**Simplify each root, if possible.**

**23.** $\sqrt{16 - 8x + x^2}$

**24.** $\sqrt{9y^2 + 30y + 25}$

**25.** $\sqrt{4 - 25z^2}$

**26.** $\sqrt{9k^2 + h^2}$

**Rationalize each denominator. Assume that all radicands represent positive real numbers.**

**27.** $\dfrac{5}{\sqrt{7}}$        **28.** $\dfrac{5}{\sqrt{10}}$

**29.** $\dfrac{-3}{\sqrt{12}}$        **30.** $\dfrac{4}{\sqrt{8}}$

**31.** $\dfrac{3}{1 - \sqrt{2}}$        **32.** $\dfrac{5}{2 - \sqrt{6}}$

**33.** $\dfrac{6}{2 + \sqrt{2}}$        **34.** $\dfrac{\sqrt{5}}{\sqrt{5} + \sqrt{2}}$

**35.** $\dfrac{1}{\sqrt{r} - \sqrt{3}}$        **36.** $\dfrac{5}{\sqrt{m} - \sqrt{5}}$

**37.** $\dfrac{y - 5}{\sqrt{y} - \sqrt{5}}$        **38.** $\dfrac{\sqrt{z} - 1}{\sqrt{z} - \sqrt{5}}$

**39.** $\dfrac{\sqrt{x} + \sqrt{x + 1}}{\sqrt{x} - \sqrt{x + 1}}$        **40.** $\dfrac{\sqrt{p} + \sqrt{p^2 - 1}}{\sqrt{p} - \sqrt{p^2 - 1}}$

**Rationalize each numerator. Assume that all radicands represent positive real numbers.**

**41.** $\dfrac{1 + \sqrt{2}}{2}$        **42.** $\dfrac{3 - \sqrt{3}}{6}$

**43.** $\dfrac{\sqrt{x} + \sqrt{x + 1}}{\sqrt{x} - \sqrt{x + 1}}$        **44.** $\dfrac{\sqrt{p} - \sqrt{p - 2}}{\sqrt{p}}$

**YOUR TURN ANSWERS**

**1.** $2x^4y^2\sqrt{7xy}$        **2.** $5(\sqrt{x} + \sqrt{y})/(x - y)$

# Functions

1.1  Lines and Linear Functions

1.2  The Least Squares Line

1.3  Properties of Functions

1.4  Quadratic Functions; Translation and Reflection

1.5  Polynomial and Rational Functions

Chapter 1 Review

Extended Application: Using Extrapolation
to Predict Life Expectancy

Over short time intervals, many changes in biology are well modeled by linear functions. In an exercise in the second section of this chapter, we will examine a linear model that predicts the skeletal maturity of children given their age. Such functions are useful to pediatricians in determining the health of children.

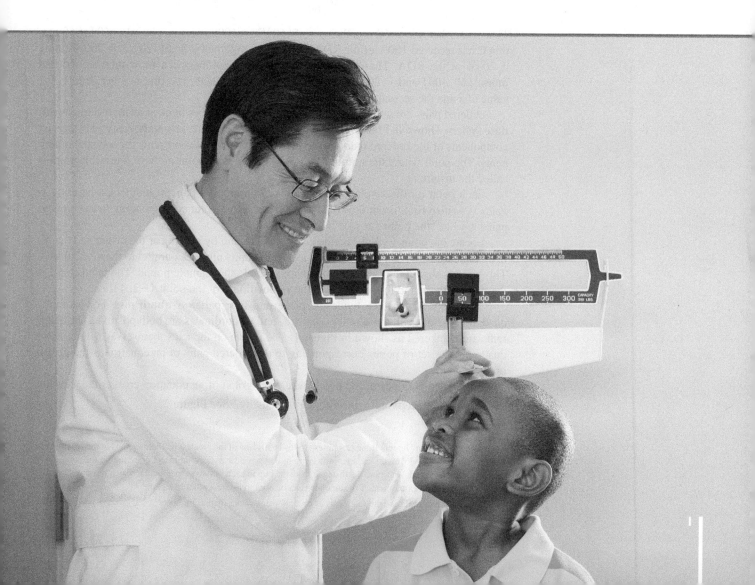

Before using mathematics to solve a real-world problem, we usually must set up a **mathematical model**, a mathematical description of the situation. Many mathematical models involve the concept of a *function*, which we will define and explore in this chapter. The simplest type of function is a *linear function*. Despite their simplicity, or perhaps because of it, linear functions are widely used in the life sciences, as we will see in the examples and exercises. Linear functions are also a special case of polynomial functions, which we will study later in this chapter.

# 1.1    Lines and Linear Functions

**APPLY IT**    How fast is the percentage of the U.S. adult population who smoke decreasing and can we predict the percentage of smokers in the future?
*In Example 11 of this section, we will answer these questions using the equation of a line.*

There are many situations in the life sciences in which two quantities are related. For example, the Recommended Daily Allowance (RDA) of vitamin C is 60 mg, so if an adult takes $x$ mg of vitamin C, the percent of the daily allowance that the person receives is given by

$$y = 100\left(\frac{x}{60}\right), \qquad \text{or} \qquad y = \frac{5x}{3}.$$

This means that if $x = 60$ mg, then $y = 5(60)/3 = 100$; a person who takes 60 mg of vitamin C has received 100% of the RDA. If $x = 120$ mg, then $y = 5(120)/3 = 200$; 120 mg is 200% of the RDA. These corresponding pairs of numbers can be written as **ordered pairs**, $(60, 100)$ and $(120, 200)$, whose order is important. The first number denotes the value of $x$ and the second number the value of $y$.

Ordered pairs are graphed with the perpendicular number lines of a **Cartesian coordinate system**, shown in Figure 1.* The horizontal number line, or $x$-**axis**, represents the first components of the ordered pairs, while the vertical or $y$-**axis** represents the second components. The point where the number lines cross is the zero point on both lines; this point is called the **origin**.

Each point on the $xy$-plane corresponds to an ordered pair of numbers, where the $x$-value is written first. From now on, we will refer to the point corresponding to the ordered pair $(x, y)$ as "the point $(x, y)$."

Locate the point $(-2, 4)$ on the coordinate system by starting at the origin and counting 2 units to the left on the horizontal axis and 4 units upward, parallel to the vertical axis. This point is shown in Figure 1, along with several other sample points. The number $-2$ is the $x$-**coordinate** and the number 4 is the $y$-**coordinate** of the point $(-2, 4)$.

The $x$-axis and $y$-axis divide the plane into four parts, or **quadrants**. For example, quadrant I includes all those points whose $x$- and $y$-coordinates are both positive. The quadrants are numbered as shown in Figure 1. The points on the axes themselves belong to no quadrant. The set of points corresponding to the ordered pairs of an equation is the **graph** of the equation.

The $x$- and $y$-values of the points where the graph of an equation crosses the axes are called the $x$-**intercept** and $y$-**intercept**, respectively.[†] See Figure 2.

---

*The name "Cartesian" honors René Descartes (1596–1650), one of the greatest mathematicians of the seventeenth century. According to legend, Descartes was lying in bed when he noticed an insect crawling on the ceiling and realized that if he could determine the distance from the bug to each of two perpendicular walls, he could describe its position at any given moment. The same idea can be used to locate a point in a plane.

[†]Some people prefer to define the intercepts as ordered pairs, rather than as numbers.

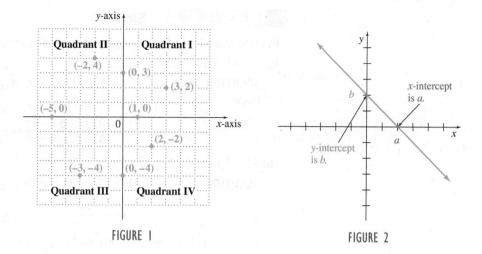

FIGURE I

FIGURE 2

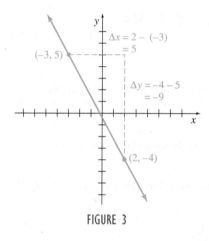

FIGURE 3

# Slope of a Line

An important characteristic of a straight line is its *slope*, a number that represents the "steepness" of the line. To see how slope is defined, look at the line in Figure 3. The line passes through the points $(x_1, y_1) = (-3, 5)$ and $(x_2, y_2) = (2, -4)$. The difference in the two *x*-values,

$$x_2 - x_1 = 2 - (-3) = 5$$

in this example, is called the **change in x**. The symbol $\Delta x$ (read "delta *x*") is used to represent the change in *x*. In the same way, $\Delta y$ represents the **change in y**. In our example,

$$\Delta y = y_2 - y_1$$
$$= -4 - 5$$
$$= -9.$$

These symbols, $\Delta x$ and $\Delta y$, are used in the following definition of slope.

## Slope of a Non-Vertical Line

The **slope** of a line is defined as the vertical change (the "rise") over the horizontal change (the "run") as one travels along the line. In symbols, taking two different points $(x_1, y_1)$ and $(x_2, y_2)$ on the line, the slope is

$$m = \frac{\textbf{Change in } y}{\textbf{Change in } x} = \frac{\Delta y}{\Delta x} = \frac{y_2 - y_1}{x_2 - x_1},$$

where $x_1 \neq x_2$.

By this definition, the slope of the line in Figure 3 is

$$m = \frac{\Delta y}{\Delta x} = \frac{-4 - 5}{2 - (-3)} = -\frac{9}{5}.$$

The slope of a line tells how fast *y* changes for each unit of change in *x*.

**NOTE**   Using similar triangles, it can be shown that the slope of a line is independent of the choice of points on the line. That is, the same slope will be obtained for *any* choice of two different points on the line.

### EXAMPLE 1    Slope

Find the slope of the line through each pair of points.

**(a)** $(7, 6)$ and $(-4, 5)$

**SOLUTION**  Let $(x_1, y_1) = (7, 6)$ and $(x_2, y_2) = (-4, 5)$. Use the definition of slope.

$$m = \frac{\Delta y}{\Delta x} = \frac{5 - 6}{-4 - 7} = \frac{-1}{-11} = \frac{1}{11}$$

**(b)** $(5, -3)$ and $(-2, -3)$

**SOLUTION**  Let $(x_1, y_1) = (5, -3)$ and $(x_2, y_2) = (-2, -3)$. Then

$$m = \frac{-3 - (-3)}{-2 - 5} = \frac{0}{-7} = 0.$$

Lines with zero slope are horizontal (parallel to the $x$-axis).

**(c)** $(2, -4)$ and $(2, 3)$

**SOLUTION**  Let $(x_1, y_1) = (2, -4)$ and $(x_2, y_2) = (2, 3)$. Then

$$m = \frac{3 - (-4)}{2 - 2} = \frac{7}{0},$$

which is undefined. This happens when the line is vertical (parallel to the $y$-axis).

**TRY YOUR TURN 1**

**YOUR TURN 1**  Find the slope of the line through (1, 5) and (4, 6).

| CAUTION | The phrase "no slope" should be avoided; specify instead whether the slope is zero or undefined. |

In finding the slope of the line in Example 1(a), we could have let $(x_1, y_1) = (-4, 5)$ and $(x_2, y_2) = (7, 6)$. In that case,

$$m = \frac{6 - 5}{7 - (-4)} = \frac{1}{11},$$

the same answer as before. The order in which coordinates are subtracted does not matter, as long as it is done consistently.

Figure 4 shows examples of lines with different slopes. Lines with positive slopes rise from left to right, while lines with negative slopes fall from left to right.

It might help you to compare slope with the percent grade of a hill. If a sign says a hill has a 10% grade uphill, this means the slope is 0.10, or 1/10, so the hill rises 1 foot for every 10 feet horizontally. A 15% grade downhill means the slope is $-0.15$.

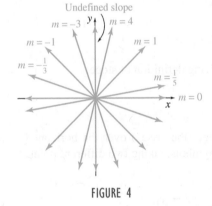

Undefined slope
$m = -3$   $m = 4$
$m = -1$   $m = 1$
$m = -\frac{1}{3}$
$m = \frac{1}{5}$
$m = 0$
$x$

FIGURE 4

## Equations of a Line

An equation in two first-degree variables, such as $4x + 7y = 20$, has a line as its graph, so it is called a **linear equation**. In the rest of this section, we consider various forms of the equation of a line.

Suppose a line has a slope $m$ and $y$-intercept $b$. This means that it passes through the point $(0, b)$. If $(x, y)$ is any other point on the line, then the definition of slope tells us that

$$m = \frac{y - b}{x - 0}.$$

**FOR REVIEW**
For review on solving a linear equation, see Section R.4.

We can simplify this equation by multiplying both sides by $x$ and adding $b$ to both sides. The result is

$$y = mx + b,$$

which we call the *slope-intercept* form of a line. This is the most common form for writing the equation of a line.

**Slope-Intercept Form**

If a line has slope $m$ and $y$-intercept $b$, then the equation of the line in **slope-intercept form** is

$$y = mx + b.$$

When $b = 0$, we say that $y$ is **proportional** to $x$.

**EXAMPLE 2**    **Equation of a Line**

Find an equation in slope-intercept form for each line.

**(a)** Through $(0, -3)$ with slope 3/4

**SOLUTION**    We recognize $(0, -3)$ as the $y$-intercept because it's the point with 0 as its $x$-coordinate, so $b = -3$. The slope is 3/4, so $m = 3/4$. Substituting these values into $y = mx + b$ gives

$$y = \frac{3}{4}x - 3.$$

**(b)** With $x$-intercept 7 and $y$-intercept 2

**SOLUTION**    Notice that $b = 2$. To find $m$, use the definition of slope after writing the $x$-intercept as $(7, 0)$ (because the $y$-coordinate is 0 where the line crosses the $x$-axis) and the $y$-intercept as $(0, 2)$.

$$m = \frac{0 - 2}{7 - 0} = -\frac{2}{7}$$

**YOUR TURN 2**  Find the equation of the line with $x$-intercept $-4$ and $y$-intercept 6.

Substituting these values into $y = mx + b$, we have

$$y = -\frac{2}{7}x + 2. \qquad \text{TRY YOUR TURN 2}$$

The slope-intercept form of the equation of a line involves the slope and the $y$-intercept. Sometimes, however, the slope of a line is known, together with one point (perhaps *not* the $y$-intercept) that the line passes through. The *point-slope form* of the equation of a line is used to find the equation in this case. Let $(x_1, y_1)$ be any fixed point on the line, and let $(x, y)$ represent any other point on the line. If $m$ is the slope of the line, then by the definition of slope,

$$\frac{y - y_1}{x - x_1} = m,$$

or

$$y - y_1 = m(x - x_1). \qquad \text{Multiply both sides by } x - x_1.$$

**Point-Slope Form**

If a line has slope $m$ and passes through the point $(x_1, y_1)$, then an equation of the line is given by

$$y - y_1 = m(x - x_1),$$

the **point-slope form** of the equation of a line.

The point-slope form also can be useful to find an equation of a line if we know two different points that the line passes through, as in the next example.

**EXAMPLE 3**   **Using Point-Slope Form to Find an Equation**

Find an equation of the line through $(5, 4)$ and $(-10, -2)$.

**SOLUTION**   Begin by using the definition of slope to find the slope of the line that passes through the given points.

$$\text{Slope} = m = \frac{-2 - 4}{-10 - 5} = \frac{-6}{-15} = \frac{2}{5}$$

Either $(5, 4)$ or $(-10, -2)$ can be used in the point-slope form with $m = 2/5$. If $(x_1, y_1) = (5, 4)$, then

$$y - y_1 = m(x - x_1)$$

$$y - 4 = \frac{2}{5}(x - 5) \qquad y_1 = 4,\ m = \tfrac{2}{5},\ x_1 = 5$$

$$y - 4 = \frac{2}{5}x - 2 \qquad \text{Distributive property}$$

$$y = \frac{2}{5}x + 2 \qquad \text{Add 4 to both sides.}$$

**YOUR TURN 3**   Find the equation of the line through $(2, 9)$ and $(5, 3)$. Put your answer in slope-intercept form.

Check that the same result is found if $(x_1, y_1) = (-10, -2)$.   **TRY YOUR TURN 3**

**EXAMPLE 4**   **Horizontal Line**

Find an equation of the line through $(8, -4)$ and $(-2, -4)$.

**SOLUTION**   Find the slope.

$$m = \frac{-4 - (-4)}{-2 - 8} = \frac{0}{-10} = 0$$

Choose, say, $(8, -4)$ as $(x_1, y_1)$.

$$y - y_1 = m(x - x_1)$$

$$y - (-4) = 0(x - 8) \qquad y_1 = -4,\ m = 0,\ x_1 = 8$$

$$y + 4 = 0 \qquad 0(x - 8) = 0$$

$$y = -4$$

Plotting the given ordered pairs and drawing a line through the points show that the equation $y = -4$ represents a horizontal line. See Figure 5(a). Every horizontal line has a slope of zero and an equation of the form $y = k$, where $k$ is the $y$-value of all ordered pairs on the line.

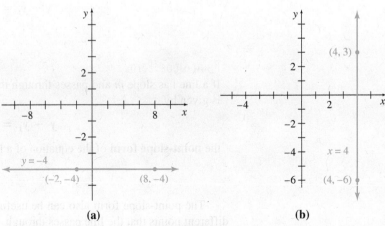

**(a)**                    **(b)**

FIGURE 5

**EXAMPLE 5**   **Vertical Line**

Find an equation of the line through $(4, 3)$ and $(4, -6)$.

**SOLUTION**   The slope of the line is

$$m = \frac{-6 - 3}{4 - 4} = \frac{-9}{0},$$

which is undefined. Since both ordered pairs have $x$-coordinate 4, the equation is $x = 4$. Because the slope is undefined, the equation of this line cannot be written in the slope-intercept form.

Again, plotting the given ordered pairs and drawing a line through them show that the graph of $x = 4$ is a vertical line. See Figure 5(b).

### Slope of Horizontal and Vertical Lines
The slope of a horizontal line is 0.

The slope of a vertical line is undefined.

The different forms of linear equations discussed in this section are summarized below. The slope-intercept and point-slope forms are equivalent ways to express the equation of a nonvertical line. The slope-intercept form is simpler for a final answer, but you may find the point-slope form easier to use when you know the slope of a line and a point through which the line passes. The slope-intercept form is often considered the standard form. Any line that is not vertical has a unique slope-intercept form but can have many point-slope forms for its equation.

### Equations of Lines

| Equation | Description |
|---|---|
| $y = mx + b$ | **Slope-intercept form:** slope $m$, $y$-intercept $b$ |
| $y - y_1 = m(x - x_1)$ | **Point-slope form:** slope $m$, line passes through $(x_1, y_1)$ |
| $x = k$ | **Vertical line:** $x$-intercept $k$, no $y$-intercept (except when $k = 0$), undefined slope |
| $y = k$ | **Horizontal line:** $y$-intercept $k$, no $x$-intercept (except when $k = 0$), slope 0 |

# Parallel and Perpendicular Lines   One application of slope involves deciding whether two lines are parallel, which means that they never intersect. Since two parallel lines are equally "steep," they should have the same slope. Also, two lines with the same "steepness" are parallel.

### Parallel Lines
Two lines are **parallel** if and only if they have the same slope, or if they are both vertical.

EXAMPLE 6   **Parallel Line**

Find the equation of the line that passes through the point $(3, 5)$ and is parallel to the line $2x + 5y = 4$.

**SOLUTION**   The slope of $2x + 5y = 4$ can be found by writing the equation in slope-intercept form. To put the equation in this form, solve for $y$.

$$2x + 5y = 4$$
$$y = -\frac{2}{5}x + \frac{4}{5} \quad \text{Subtract } 2x \text{ from both sides and divide both sides by 5.}$$

This result shows that the slope is $-2/5$, the coefficient of $x$. Since the lines are parallel, $-2/5$ is also the slope of the line whose equation we want. This line passes through $(3, 5)$. Substituting $m = -2/5$, $x_1 = 3$, and $y_1 = 5$ into the point-slope form gives

$$y - y_1 = m(x - x_1)$$
$$y - 5 = -\frac{2}{5}(x - 3) = -\frac{2}{5}x + \frac{6}{5}$$
$$y = -\frac{2}{5}x + \frac{6}{5} + 5$$
$$y = -\frac{2}{5}x + \frac{31}{5}. \quad \text{Multiply 5 by 5/5 to get a common denominator.}$$

**YOUR TURN 4**   Find the equation of the line that passes through the point $(4, 5)$ and is parallel to the line $3x - 6y = 7$. Put your answer in slope-intercept form.

TRY YOUR TURN 4

As already mentioned, two nonvertical lines are parallel if and only if they have the same slope. Two lines having slopes with a product of $-1$ are perpendicular. A proof of this fact, which depends on similar triangles from geometry, is given as Exercise 43 in this section.

**Perpendicular Lines**
Two lines are **perpendicular** if and only if the product of their slopes is $-1$, or if one is vertical and the other horizontal.

EXAMPLE 7   **Perpendicular Line**

Find the equation of the line $L$ passing through the point $(3, 7)$ and perpendicular to the line having the equation $5x - y = 4$.

**SOLUTION**   To find the slope, write $5x - y = 4$ in slope-intercept form:

$$y = 5x - 4.$$

The slope is 5. Since the lines are perpendicular, if line $L$ has slope $m$, then

$$5m = -1$$
$$m = -\frac{1}{5}.$$

Now substitute $m = -1/5$, $x_1 = 3$, and $y_1 = 7$ into the point-slope form.

$$y - 7 = -\frac{1}{5}(x - 3)$$
$$y - 7 = -\frac{1}{5}x + \frac{3}{5}$$
$$y = -\frac{1}{5}x + \frac{3}{5} + 7 \cdot \frac{5}{5} \quad \text{Add 7 to both sides and get a common denominator.}$$
$$y = -\frac{1}{5}x + \frac{38}{5}$$

**YOUR TURN 5**   Find the equation of the line passing through the point $(3, 2)$ and perpendicular to the line having the equation $2x + 3y = 4$.

TRY YOUR TURN 5

## Graph of a Line
It can be shown that every equation of the form $ax + by = c$ has a straight line as its graph, assuming $a$ and $b$ are not both 0. Although just two points are needed to determine a line, it is a good idea to plot a third point as a check.

### EXAMPLE 8    Graph of a Line

Graph $y = -\dfrac{3}{2}x + 6$.

**SOLUTION**    There are several ways to graph a line.

**Method 1**
**Plot Points**

Note that the $y$-intercept is 6, so the point $(0, 6)$ is on the line. Next, by substituting $x = 2$ and $x = 4$ into the equation, we find the points $(2, 3)$ and $(4, 0)$. (We could use any values for $x$, but we chose even numbers so the value of $y$ would be an integer.) These three points are plotted in Figure 6(a). A line is drawn through them in Figure 6(b). (We need to find only two points on the line. The third point gives a confirmation of our work.)

**Method 2**
**Find Intercepts**

We already found the $y$-intercept $(0, 6)$ by setting $x = 0$. To find the $x$-intercept, let $y = 0$.

$$0 = -\frac{3}{2}x + 6$$

$$\frac{3}{2}x = 6 \qquad \text{Add } \tfrac{3}{2}x \text{ to both sides.}$$

$$x = \frac{6}{3/2} = 6\left(\frac{2}{3}\right) = 4 \qquad \text{Divide both sides by } \tfrac{3}{2}.$$

This gives the point $(4, 0)$ that we found earlier. Once the $x$- and $y$-intercepts are found, we can draw a line through them.

**Method 3**
**Use the Slope and $y$-Intercept**

Observe that the slope of $-3/2$ means that every time $x$ increases by 2, $y$ decreases by 3. So start at $(0, 6)$ and go 2 across and 3 down to the point $(2, 3)$. Again, go 2 across and 3 down to the point $(4, 0)$. Draw the line through these points.

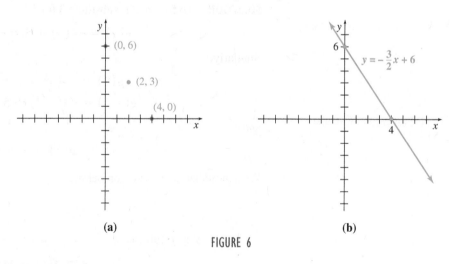

**(a)**                                     **(b)**

FIGURE 6

Not every line has two distinct intercepts; the graph in the next example does not cross the $x$-axis, and so it has no $x$-intercept.

### EXAMPLE 9    Graph of a Horizontal Line

Graph $y = -3$.

**SOLUTION**    The equation $y = -3$, or equivalently, $y = 0x - 3$, always gives the same $y$-value, $-3$, for any value of $x$. Therefore, no value of $x$ will make $y = 0$, so the graph has no $x$-intercept. As we saw in Example 4, the graph of such an equation is a horizontal line parallel to the $x$-axis. In this case the $y$-intercept is $-3$, as shown in Figure 7.

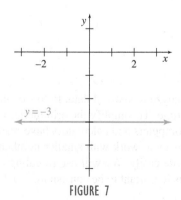

FIGURE 7

## Linear Functions

When $y$ can be expressed in terms of $x$ in the form $y = mx + b$, we say that $y$ is a **linear function** of $x$. This means that for any value of $x$ (the **independent variable**), we can use the equation to find the corresponding value of $y$ (the **dependent variable**). (In general, the independent variable in a linear function can be any real number, but in some applications, the independent variable may be restricted.) Examples of linear functions include $y = 2x + 3$, $y = -5$, and $2x - 3y = 7$, which can be written as $y = (2/3)x - (7/3)$. Equations in the form $x = k$, where $k$ is a constant, are not linear functions. All other linear equations define linear functions.

## $f(x)$ Notation

Letters such as $f$, $g$, or $h$ are often used to name functions. For example, $f$ might be used to name the function defined by

$$y = 5 - 3x.$$

To show that this function is named $f$, it is common to replace $y$ with $f(x)$ (read "$f$ of $x$") to get

$$f(x) = 5 - 3x.$$

By choosing 2 as a value of $x$, $f(x)$ becomes $5 - 3 \cdot 2 = 5 - 6 = -1$, written

$$f(2) = -1.$$

The corresponding ordered pair is $(2, -1)$. In a similar manner,

$$f(-4) = 5 - 3(-4) = 17, \quad f(0) = 5, \quad f(-6) = 23,$$

and so on.

### EXAMPLE 10    Function Notation

Let $g(x) = -4x + 5$. Find $g(3)$, $g(0)$, $g(-2)$, and $g(b)$.

**SOLUTION**    To find $g(3)$, substitute 3 for $x$.

$$g(3) = -4(3) + 5 = -12 + 5 = -7$$

Similarly,

$$g(0) = -4(0) + 5 = 0 + 5 = 5,$$
$$g(-2) = -4(-2) + 5 = 8 + 5 = 13,$$

and

$$g(b) = -4b + 5. \qquad \text{TRY YOUR TURN 6}$$

**YOUR TURN 6**  Calculate $g(-5)$.

We summarize the discussion below.

> ### Linear Function
> A relationship $f$ defined by
> $$y = f(x) = mx + b,$$
> for real numbers $m$ and $b$, is a **linear function**.

The next example uses the equation of a line to analyze real-world data. In this example, we are looking at how one variable changes over time. To simplify the arithmetic, we will *rescale* the variable representing time, although computers and calculators have made rescaling less important than in the past. Here it allows us to work with smaller numbers, and, as you will see, find the $y$-intercept of the line more easily. We will use rescaling on many examples throughout this book. When we do, it is important to be consistent.

**EXAMPLE 11**  **Prevalence of Cigarette Smoking**

APPLY IT

In recent years, the percentage of the U.S. population age 18 and older who smoke has decreased at a roughly constant rate, from 23.3% in 2000 to 19.3% in 2010. *Source: Centers for Disease Control and Prevention.*

**(a)** Find the function describing this linear relationship.

**SOLUTION**  Let $t$ represent time in years, with $t = 0$ representing 2000. With this rescaling, the year 2010 corresponds to $t = 10$. Let $y$ represent the percentage of the population who smoke. The two ordered pairs representing the given information are then $(0, 23.3)$ and $(10, 19.3)$. The slope of the line through these points is

$$m = \frac{19.3 - 23.3}{10 - 0} = \frac{-4}{10} = -0.4.$$

This means that, on average, the percentage of the adult population who smoke is decreasing by about 0.4% per year.

Notice that the point $(0, 23.3)$ is the y-intercept. Using $m = -0.4$ and $b = 23.3$ in the slope-intercept form gives the required function.

$$y = f(t) = -0.4t + 23.3$$

**(b)** One objective of Healthy People 2020 (a campaign of the U.S. Department of Health and Human Services) is to reduce the percentage of U.S. adults who smoke to 12% or less by the year 2020. If this decline in smoking continues at the same rate, will they meet this objective?

**SOLUTION**  Using the same rescaling, $t = 20$ corresponds to the year 2020. Substituting this value into the above function gives

$$y = f(20) = -0.4(20) + 23.3 = 15.3.$$

Continuing at this rate, an estimated 15.3% of the adult population will still smoke in 2020, and the objective of Healthy People 2020 will not be met.

Notice that if the formula from part (a) of Example 11 is valid for all nonnegative $t$, then eventually $y$ becomes 0:

$$-0.4t + 23.3 = 0$$
$$-0.4t = -23.3 \qquad \text{Subtract 23.3 from both sides.}$$
$$t = \frac{-23.3}{-0.4} = 58.25 \approx 58* \qquad \text{Divide both sides by } -0.4.$$

which indicates that 58 years from 2000 (in the year 2058), 0% of the U.S. adult population will smoke. Of course, it is still possible that in 2058 there will be adults who smoke; the trend of recent years may not continue. Most equations are valid for some specific set of numbers. It is highly speculative to extrapolate beyond those values.

On the other hand, people in business and government often need to make some prediction about what will happen in the future, so a tentative conclusion based on past trends may be better than no conclusion at all. There are also circumstances, particularly in the physical sciences, in which theoretical reasons imply that the trend will continue.

**EXAMPLE 12**  **Deer Ticks**

Deer ticks cause concern because they can carry the parasite that causes Lyme disease. One study found a relationship between the density of acorns produced in the fall and the density

*The symbol $\approx$ means "is approximately equal to."

of deer tick larvae the following spring. The relationship can be approximated by the linear function

$$f(x) = 34x + 230,$$

where $x$ is the number of acorns per square meter in the fall, and $f(x)$ is the number of deer tick larvae per 400 square meters the following spring. *Source: Science.*

**(a)** Approximately how many acorns per square meter would result in 1000 deer tick larvae per 400 square meters?

**SOLUTION**    Substitute 1000 for $f(x)$ in the equation and solve for $x$.

$$f(x) = 34x + 230$$
$$1000 = 34x + 230$$
$$770 = 34x \qquad \text{Subtract 230 from both sides.}$$
$$x \approx 22.6 \qquad \text{Divide by 34.}$$

This means that approximately 23 acorns per square meter in the fall would result in 1000 deer tick larvae per 400 square meters in the following spring.

**(b)** Find and interpret the slope of the line.

**SOLUTION**    The equation is given in slope-intercept form, so the slope is 34, the coefficient of $x$. The slope indicates that the number of deer tick larvae per 400 square meters in the spring will increase by 34 for each additional acorn per square meter in the fall.

## Cost Analysis    The cost of manufacturing an item commonly consists of two parts. The first is a **fixed cost** for designing the product, setting up a factory, training workers, and so on. Within broad limits, the fixed cost is constant for a particular product and does not change as more items are made. The second part is a *cost per item*, or **marginal cost**, for labor, materials, packing, shipping, and so on. The total value of this second cost *does* depend on the number of items made. Analysis of these costs is important in business. Medicine is a business, so the same ideas apply, as in the next example.

### Linear Cost Function
In a cost function of the form $C(x) = mx + b$, the $m$ represents the marginal cost and $b$ the fixed cost. Conversely, if the fixed cost of producing an item is $b$ and the marginal cost is $m$, then the **linear cost function** $C(x)$ for producing $x$ items is $C(x) = mx + b$.

### EXAMPLE 13    Cost Analysis

Suppose a hospital has to choose one of two X-ray machines to purchase. One costs $200,000 to purchase, and each image produced by the machine costs $10. The other sells for $250,000, but each image costs only $2.

**(a)** What is the fixed cost and marginal cost for the first machine?

**SOLUTION**    The fixed cost is $200,000 and the marginal cost is $10.

**(b)** How many X-ray images must be made before the cost of the two machines becomes equal?

**SOLUTION**    The cost function for the first machine can be written as

$$C_1(x) = 200,000 + 10x,$$

where $x$ is the number of X-ray images taken, while the cost function for the second machine can be written as

$$C_2(x) = 250,000 + 2x.$$

The break-even point (where the two functions are equal) is found by setting $C_1(x) = C_2(x)$.

$$200{,}000 + 10x = 250{,}000 + 2x$$

$$8x = 50{,}000$$

$$x = \frac{50{,}000}{8} = 6250$$

If the hospital purchasing agents expect the hospital to make fewer than 6250 X-ray images over the life of the machine, they should buy the first machine, assuming all else is equal. Otherwise, the second machine is less expensive to use in the long run.

## Temperature
One of the most common linear functions found in everyday situations deals with temperature. Recall that water freezes at 32° Fahrenheit and 0° Celsius, while it boils at 212° Fahrenheit and 100° Celsius.* The ordered pairs $(0, 32)$ and $(100, 212)$ are graphed in Figure 8 on axes showing Fahrenheit ($F$) as a function of Celsius ($C$). The line joining them is the graph of the function.

### EXAMPLE 14 Temperature

Derive an equation relating $F$ and $C$.

**SOLUTION** To derive the required linear equation, first find the slope using the given ordered pairs, $(0, 32)$ and $(100, 212)$.

$$m = \frac{212 - 32}{100 - 0} = \frac{180}{100} = \frac{9}{5}$$

The $F$-intercept of the graph is 32, so by the slope-intercept form, the equation of the line is

$$F = \frac{9}{5}C + 32.$$

With simple algebra this equation can be rewritten to give $C$ in terms of $F$:

$$C = \frac{5}{9}(F - 32).$$

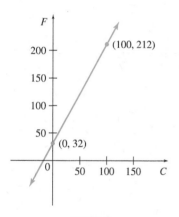

**FIGURE 8**

*Gabriel Fahrenheit (1686–1736), a German physicist, invented his scale with 0° representing the temperature of an equal mixture of ice and ammonium chloride (a type of salt), and 96° as the temperature of the human body. (It is often said, erroneously, that Fahrenheit set 100° as the temperature of the human body. Fahrenheit's own words are quoted in *A History of the Thermometer and Its Use in Meteorology* by W. E. Knowles, Middleton: The Johns Hopkins Press, 1966, p. 75.) The Swedish astronomer Anders Celsius (1701–1744) set 0° and 100° as the freezing and boiling points of water.

# 1.1 EXERCISES

**Find the slope of each line.**

1. Through $(4, 5)$ and $(-1, 2)$

2. Through $(5, -4)$ and $(1, 3)$

3. Through $(8, 4)$ and $(8, -7)$

4. Through $(1, 5)$ and $(-2, 5)$

5. $y = x$

6. $y = 3x - 2$

7. $5x - 9y = 11$

8. $4x + 7y = 1$

9. $x = 5$

10. The $x$-axis

11. $y = 8$

12. $y = -6$

13. A line parallel to $6x - 3y = 12$

14. A line perpendicular to $8x = 2y - 5$

**In Exercises 15–24, find an equation in slope-intercept form for each line.**

15. Through $(1, 3)$, $m = -2$

16. Through $(2, 4)$, $m = -1$

**17.** Through $(-5, -7)$, $m = 0$

**18.** Through $(-8, 1)$, with undefined slope

**19.** Through $(4, 2)$ and $(1, 3)$

**20.** Through $(8, -1)$ and $(4, 3)$

**21.** Through $(2/3, 1/2)$ and $(1/4, -2)$

**22.** Through $(-2, 3/4)$ and $(2/3, 5/2)$

**23.** Through $(-8, 4)$ and $(-8, 6)$

**24.** Through $(-1, 3)$ and $(0, 3)$

**In Exercises 25–34, find an equation for each line in the form $ax + by = c$, where $a$, $b$, and $c$ are integers with no factor common to all three and $a \geq 0$.**

**25.** $x$-intercept $-6$, $y$-intercept $-3$

**26.** $x$-intercept $-2$, $y$-intercept $4$

**27.** Vertical, through $(-6, 5)$

**28.** Horizontal, through $(8, 7)$

**29.** Through $(-4, 6)$, parallel to $3x + 2y = 13$

**30.** Through $(2, -5)$, parallel to $2x - y = -4$

**31.** Through $(3, -4)$, perpendicular to $x + y = 4$

**32.** Through $(-2, 6)$, perpendicular to $2x - 3y = 5$

**33.** The line with $y$-intercept 4 and perpendicular to $x + 5y = 7$

**34.** The line with $x$-intercept $-2/3$ and perpendicular to $2x - y = 4$

**35.** Do the points $(4, 3)$, $(2, 0)$, and $(-18, -12)$ lie on the same line? Explain why or why not. (*Hint:* Find the slopes between the points.)

**36.** Find $k$ so that the line through $(4, -1)$ and $(k, 2)$ is

  **a.** parallel to $2x + 3y = 6$,

  **b.** perpendicular to $5x - 2y = -1$.

**37.** Use slopes to show that the quadrilateral with vertices at $(1, 3)$, $(-5/2, 2)$, $(-7/2, 4)$, and $(2, 1)$ is a parallelogram.

**38.** Use slopes to show that the square with vertices at $(-2, 5)$, $(4, 5)$, $(4, -1)$, and $(-2, -1)$ has diagonals that are perpendicular.

**For the lines in Exercises 39 and 40, which of the following is closest to the slope of the line? (a) 1 (b) 2 (c) 3 (d) 21 (e) 22 (f) −3**

**39.**

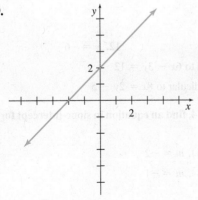

**40.**

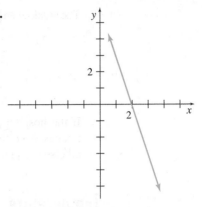

**In Exercises 41 and 42, estimate the slope of the lines.**

**41.**

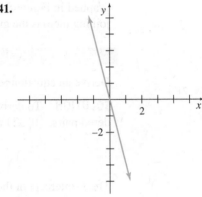

**42.**

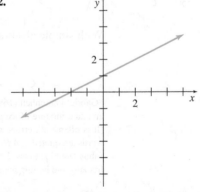

**43.** To show that two perpendicular lines, neither of which is vertical, have slopes with a product of $-1$, go through the following steps. Let line $L_1$ have equation $y = m_1 x + b_1$, and let $L_2$ have equation $y = m_2 x + b_2$, with $m_1 > 0$ and $m_2 < 0$. Assume that $L_1$ and $L_2$ are perpendicular, and use right triangle $MPN$ shown in the figure. Prove each of the following statements.

  **a.** $MQ$ has length $m_1$.

  **b.** $QN$ has length $-m_2$.

  **c.** Triangles $MPQ$ and $PNQ$ are similar.

**d.** $m_1/1 = 1/(-m_2)$ and $m_1m_2 = -1$

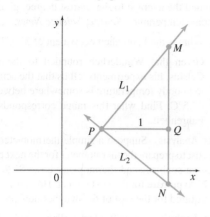

**44.** Consider the equation $\dfrac{x}{a} + \dfrac{y}{b} = 1$.

   **a.** Show that this equation represents a line by writing it in the form $y = mx + b$.

   **b.** Find the $x$- and $y$-intercepts of this line.

   **c.** Explain in your own words why the equation in this exercise is known as the intercept form of a line.

**Graph each equation.**

**45.** $y = x - 1$     **46.** $y = 4x + 5$

**47.** $y = -4x + 9$     **48.** $y = -6x + 12$

**49.** $2x - 3y = 12$     **50.** $3x - y = -9$

**51.** $3y - 7x = -21$     **52.** $5y + 6x = 11$

**53.** $y = -2$     **54.** $x = 4$

**55.** $x + 5 = 0$     **56.** $y + 8 = 0$

**57.** $y = 2x$     **58.** $y = -5x$

**59.** $x + 4y = 0$     **60.** $3x - 5y = 0$

**For Exercises 61–70, let $f(x) = 7 - 5x$ and $g(x) = 2x - 3$. Find the following.**

**61.** $f(2)$     **62.** $f(4)$

**63.** $f(-3)$     **64.** $f(-1)$

**65.** $g(1.5)$     **66.** $g(2.5)$

**67.** $g(-1/2)$     **68.** $g(-3/4)$

**69.** $f(t)$     **70.** $g(k^2)$

## LIFE SCIENCE APPLICATIONS

**71. Exercise Heart Rate**   To achieve the maximum benefit for the heart when exercising, your heart rate (in beats per minute) should be in the target heart rate zone. The lower limit of this zone is found by taking 70% of the difference between 220 and your age. The upper limit is found by using 85%. ***Source: Physical Fitness.***

   **a.** Find formulas for the upper and lower limits ($u$ and $l$) as linear equations involving the age $x$.

   **b.** What is the target heart rate zone for a 20-year-old?

   **c.** What is the target heart rate zone for a 40-year-old?

   **d.** Two women in an aerobics class stop to take their pulse and are surprised to find that they have the same pulse. One woman is 36 years older than the other and is working at the upper limit of her target heart rate zone. The younger woman is working at the lower limit of her target heart rate zone. What are the ages of the two women, and what is their pulse?

   **e.** Run for 10 minutes, take your pulse, and see if it is in your target heart rate zone. (After all, this is listed as an exercise!)

**72. HIV Infection**   The time interval between a person's initial infection with HIV and that person's eventual development of AIDS symptoms is an important issue. The method of infection with HIV affects the time interval before AIDS develops. One study of HIV patients who were infected by intravenous drug use found that 17% of the patients had AIDS after 4 years, and 33% had developed the disease after 7 years. The relationship between the time interval and the percentage of patients with AIDS can be modeled accurately with a linear equation. ***Source: Epidemiologic Review.***

   **a.** Write a linear equation $y = mt + b$ that models these data, using the ordered pairs $(4, 0.17)$ and $(7, 0.33)$.

   **b.** Use your equation from part a to predict the number of years before half of these patients will have AIDS.

**73. Life Expectancy**   Some scientists believe there is a limit to how long humans can live. One supporting argument is that during the past century, life expectancy from age 65 has increased more slowly than life expectancy from birth, so eventually these two will be equal, at which point, according to these scientists, life expectancy should increase no further. In 1900, life expectancy at birth was 46 yr, and life expectancy at age 65 was 76 yr. In 2008, these figures had risen to 78.1 and 83.8, respectively. In both cases, the increase in life expectancy has been linear. Using these assumptions and the data given, find the maximum life expectancy for humans. ***Source: Science.***

**74. Ponies Trotting**   A study found that the peak vertical force on a trotting Shetland pony increased linearly with the pony's speed, and that when the force reached a critical level, the pony switched from a trot to a gallop. For one pony, the critical force was 1.16 times its body weight. It experienced a force of 0.75 times its body weight at a speed of 2 meters per second and a force of 0.93 times its body weight at 3 meters per second. At what speed did the pony switch from a trot to a gallop? ***Source: Science.***

**75. Tobacco Deaths**   The U.S. Centers for Disease Control and Prevention projects that tobacco could soon be the leading cause of death in the world. In 1990, 35 million years of healthy life were lost globally due to tobacco. This quantity was going up linearly at a rate of about 28 million years each decade. In contrast, 100 million years of healthy life were lost due to diarrhea, with the rate going down linearly 22 million years each decade. ***Source: Science.***

   **a.** Write the years of healthy life ($y$) in millions lost globally to tobacco as a linear function of the years, $t$, since 1990.

   **b.** Write the years of healthy life ($y$) in millions lost to diarrhea as a linear function of the years, $t$, since 1990.

c. Using your answers to parts a and b, find in what year the amount of healthy life lost to tobacco was expected to first exceed that lost to diarrhea.

**76. Elk Energy** Researchers have found that the energy expenditure (in kilocalories per kg of mass per minute) of elk calves rises linearly as a function of their speed. A calf walking 63 meters per minute expends, on average, 0.109 kilocalorie per kg per minute, while a calf galloping 243 meters per minute expends, on average, 0.307 kilocalorie per kg per minute. *Source: Wildlife Feeding and Nutrition.*

a. Find an equation for $y$ as a linear function of $x$.

b. Find and interpret the $y$-intercept.

c. The maximum amount of energy that a calf can expend is about 0.36 kilocalorie per kg per minute. How fast is such a calf galloping?

**77. Ant Colonies** The number of male alates (winged, sexual ants) in an ant colony has been found to be approximately a linear function of the age of the colony, with a 5-year-old colony having 8.2 alates on average, and a 17-year-old colony having an average of 33.4 alates. *Source: Oecologia.*

a. Write the number of male alates $(y)$ as a linear function of the age of the colony in years $(t)$.

b. A one-year-old colony typically has no alates. What does the formula found in part a predict?

c. Assuming the linear function found in part a continues to be accurate, how old would we expect a colony to be before it has about 40 male alates?

**78. Global Warming** Due to global warming, the area covered by the ice cap on Mt. Kilimanjaro has decreased approximately linearly, from 12.1 square kilometers in 1912 to 1.9 square kilometers in 2007. *Source: Proceedings of the National Academy of Science of the United States.*

a. Write the area of the ice, y, as a linear function of the year $t$, where $t$ is the number of years since 1900.

b. In what year did the ice cover 6 square kilometers?

c. Assuming the linear function found in part a continues, in what year will the ice cap disappear?

**79. Global Warming** In 1990, the Intergovernmental Panel on Climate Change predicted that the average temperature on Earth would rise 0.3°C per decade in the absence of international controls on greenhouse emissions. Let $t$ measure the time in years since 1970, when the average global temperature was 15°C. *Source: Science News.*

a. Find a linear equation giving the average global temperature in degrees Celsius in terms of $t$, the number of years since 1970.

b. Scientists have estimated that the sea level will rise by 65 cm if the average global temperature rises to 19°C. According to your answer to part a, when would this occur?

**80. Body Temperature** You may have heard that the average temperature of the human body is 98.6°. Recent experiments show that the actual figure is closer to 98.2°. The figure of 98.6 comes from experiments done by Carl Wunderlich in 1868. But Wunderlich measured the temperatures in degrees Celsius and rounded the average to the nearest degree, giving 37°C as the average temperature. *Source: Science News.*

a. What is the Fahrenheit equivalent of 37°C?

b. Given that Wunderlich rounded to the nearest degree Celsius, his experiments tell us that the actual average human body temperature is somewhere between 36.5°C and 37.5°C. Find what this range corresponds to in degrees Fahrenheit.

**81. Cost Analysis** Suppose a simple thermometer costs $10, but the cost to prepare the thermometer for the next patient costs $2. A more sophisticated thermometer costs $120 but costs only $0.75 to prepare for the next patient. How many patients would be required for the cost of the two thermometers to be equal?

**82. Cost Analysis** Acme Products sells a machine for doing a certain type of blood test for $40,000, which costs $20 for each use. Amalgamated Medical Supplies sells a similar machine for only $32,000, but it costs $30 for each use. How many times must the machines be used for the total cost to be equal?

## OTHER APPLICATIONS

**83. Marriage** The following table lists the U.S. median age at first marriage for men and women. The age at which both groups marry for the first time seems to be increasing at a roughly linear rate in recent decades. Let $t$ correspond to the number of years since 1980. *Source: U.S. Census Bureau.*

| Age at First Marriage | | | | |
|---|---|---|---|---|
| **Year** | 1980 | 1990 | 2000 | 2010 |
| **Men** | 24.7 | 26.1 | 26.8 | 28.2 |
| **Women** | 22.0 | 23.9 | 25.1 | 26.1 |

a. Find a linear equation that approximates the data for men, using the data for the years 1980 and 2010.

b. Repeat part a using the data for women.

c. Which group seems to have the faster increase in median age at first marriage?

d. In what year will the men's median age at first marriage reach 30?

e. When the men's median age at first marriage is 30, what will the median age be for women?

**84. Immigration** In 1950, there were 249,187 immigrants admitted to the United States. In 2010, the number was 1,042,625. *Source: 2011 Yearbook of Immigration Statistics.*

a. Assuming that the change in immigration is linear, write an equation expressing the number of immigrants, $y$, in terms of $t$, the number of years after 1900.

b. Use your result in part a to predict the number of immigrants admitted to the United States in 2015.

c. Considering the value of the $y$-intercept in your answer to part a, discuss the validity of using this equation to model the number of immigrants throughout the entire 20th century.

85. **Child Mortality Rate**   The mortality rate for children under 5 years of age around the world has been declining in a roughly linear fashion in recent years. The rate per 1000 live births was 88 in 1990 and 57 in 2010. *Source: World Health Organization.*

    a. Determine a linear equation that approximates the mortality rate in terms of time $t$, where $t$ represents the number of years since 1900.

    b. If this trend continues, in what year will the mortality rate first drop to 40 or below per 1000 live births?

86. **Temperature**   Use the formula for conversion between Fahrenheit and Celsius derived in Example 14 to convert each of the following temperatures.

    a. 58°F to Celsius              b. −20°F to Celsius

    c. 50°C to Fahrenheit

87. **Temperature**   Find the temperature at which the Celsius and Fahrenheit temperatures are numerically equal.

**YOUR TURN ANSWERS**

1. 1/3            2. $y = (3/2)x + 6$       3. $y = -2x + 13$

4. $y = (1/2)x + 3$   5. $y = (3/2)x - 5/2$   6. 25

# 1.2 The Least Squares Line

**APPLY IT**   **How has the accidental death rate in the United States changed over time?**

*In Example 1 in this section, we show how to answer such questions using the method of least squares.*

| Accidental Death Rate | |
|---|---|
| Year | Death Rate |
| 1920 | 71.2 |
| 1930 | 80.5 |
| 1940 | 73.4 |
| 1950 | 60.3 |
| 1960 | 52.1 |
| 1970 | 56.2 |
| 1980 | 46.5 |
| 1990 | 36.9 |
| 2000 | 34.0 |
| 2010 | 39.1 |

We use past data to find trends and to make tentative predictions about the future. The only assumption we make is that the data are related linearly—that is, if we plot pairs of data, the resulting points will lie close to some line. This method cannot give exact answers. The best we can expect is that, if we are careful, we will get a reasonable approximation.

The table lists the number of accidental deaths per 100,000 people in the United States through the past century. *Source: National Center for Health Statistics.* If you were a manager at an insurance company, these data could be very important. You might need to make some predictions about how much you will pay out next year in accidental death benefits, and even a very tentative prediction based on past trends is better than no prediction at all.

The first step is to draw a **scatterplot**, as we have done in Figure 9. Notice that the points lie approximately along a line, which means that a linear function may give a good approximation of the data. If we select two points and find the line that passes through them, as we did in Section 1.1, we will get a different line for each pair of points, and in some cases the lines will be very different. We want to draw one line that is simultaneously close to all the points on the graph, but many such lines are possible, depending upon how we define the phrase "simultaneously close to all the points." How do we decide on the best possible line? Before going on, you might want to try drawing the line you think is best on Figure 9 on the next page.

The line used most often in applications is that in which the sum of the squares of the vertical distances from the data points to the line is as small as possible. Such a line is called the **least squares line**. The least squares line for the data in Figure 9 is drawn in Figure 10. How does the line compare with the one you drew on Figure 9? It may not be exactly the same, but should appear similar.

In Figure 10, the vertical distances from the points to the line are indicated by $d_1$, $d_2$, and so on, up through $d_{10}$ (read "d-sub-one, d-sub-two, d-sub-three," and so on). For $n$ points, corresponding to the $n$ pairs of data, the least squares line is found by minimizing the sum $(d_1)^2 + (d_2)^2 + (d_3)^2 + \cdots + (d_n)^2$.

We often use **summation notation** to write the sum of a list of numbers. The Greek letter sigma, $\Sigma$, is used to indicate "the sum of." For example, we write the sum $x_1 + x_2 + \cdots + x_n$, where $n$ is the number of data points, as

$$x_1 + x_2 + \cdots + x_n = \Sigma x.$$

Similarly, $\Sigma xy$ means $x_1y_1 + x_2y_2 + \cdots + x_ny_n$, and so on.

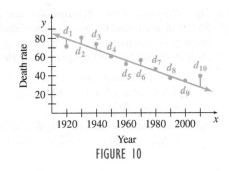

FIGURE 9   FIGURE 10

> **CAUTION** Note that $\sum x^2$ means $x_1^2 + x_2^2 + \cdots + x_n^2$, which is *not* the same as squaring $\sum x$. When we square $\sum x$, we write it as $(\sum x)^2$.

For the least squares line, the sum of the distances we are to minimize, $d_1^2 + d_2^2 + \cdots + d_n^2$, is written as

$$d_1^2 + d_2^2 + \cdots + d_n^2 = \sum d^2.$$

To calculate the distances, we let $(x_1, y_1), (x_2, y_2), \cdots, (x_n, y_n)$ be the actual data points and we let the least squares line be $Y = mx + b$. We use $Y$ in the equation instead of $y$ to distinguish the predicted values $(Y)$ from the $y$-value of the given data points. The predicted value of $Y$ at $x_1$ is $Y_1 = mx_1 + b$, and the distance, $d_1$, between the actual $y$-value $y_1$ and the predicted value $Y_1$ is

$$d_1 = |Y_1 - y_1| = |mx_1 + b - y_1|.$$

Likewise,

$$d_2 = |Y_2 - y_2| = |mx_2 + b - y_2|,$$

and

$$d_n = |Y_n - y_n| = |mx_n + b - y_n|.$$

The sum to be minimized becomes

$$\sum d^2 = (mx_1 + b - y_1)^2 + (mx_2 + b - y_2)^2 + \cdots + (mx_n + b - y_n)^2$$
$$= \sum (mx + b - y)^2,$$

where $(x_1, y_1), (x_2, y_2), \cdots, (x_n, y_n)$ are known and $m$ and $b$ are to be found.

The method of minimizing this sum requires advanced techniques and is not given here. To obtain the equation for the least squares line, a system of equations must be solved, producing the following formulas for determining the slope $m$ and $y$-intercept $b$.*

### Least Squares Line

The **least squares line** $Y = mx + b$ that gives the best fit to the data points $(x_1, y_1), (x_2, y_2), \ldots, (x_n, y_n)$ has slope $m$ and $y$-intercept $b$ given by

$$m = \frac{n(\sum xy) - (\sum x)(\sum y)}{n(\sum x^2) - (\sum x)^2} \quad \text{and} \quad b = \frac{\sum y - m(\sum x)}{n}.$$

*See Exercise 9 at the end of this section.

**EXAMPLE 1**   **Least Squares Line**

APPLY IT   Calculate the least squares line for the accidental death rate data.

**SOLUTION**

**Method 1**
**Calculating by Hand**

To find the least squares line for the given data, we first find the required sums. To reduce the size of the numbers, we rescale the year data. Let $x$ represent the years since 1900, so that, for example, $x = 20$ corresponds to the year 1920. Let $y$ represent the death rate. We then calculate the values in the $xy$, $x^2$, and $y^2$ columns and find their totals. (The column headed $y^2$ will be used later.) Note that the number of data points is $n = 10$.

| | | Least Squares Calculations | | |
|---|---|---|---|---|
| $x$ | $y$ | $xy$ | $x^2$ | $y^2$ |
| 20 | 71.2 | 1424 | 400 | 5069.44 |
| 30 | 80.5 | 2415 | 900 | 6480.25 |
| 40 | 73.4 | 2936 | 1600 | 5387.56 |
| 50 | 60.3 | 3015 | 2500 | 3636.09 |
| 60 | 52.1 | 3126 | 3600 | 2714.41 |
| 70 | 56.2 | 3934 | 4900 | 3158.44 |
| 80 | 46.5 | 3720 | 6400 | 2162.25 |
| 90 | 36.9 | 3321 | 8100 | 1361.61 |
| 100 | 34.0 | 3400 | 10,000 | 1156.00 |
| 110 | 39.1 | 4301 | 12,100 | 1528.81 |
| $\Sigma x = 650$ | $\Sigma y = 550.2$ | $\Sigma xy = 31{,}592$ | $\Sigma x^2 = 50{,}500$ | $\Sigma y^2 = 32{,}654.86$ |

Putting the column totals into the formula for the slope $m$, we get

$$m = \frac{n(\Sigma xy) - (\Sigma x)(\Sigma y)}{n(\Sigma x^2) - (\Sigma x)^2} \qquad \text{Formula for } m$$

$$= \frac{10(31{,}592) - (650)(550.2)}{10(50{,}500) - (650)^2} \qquad \text{Substitute from the table.}$$

$$= \frac{315{,}920 - 357{,}630}{505{,}000 - 422{,}500} \qquad \text{Multiply.}$$

$$= \frac{-41{,}710}{82{,}500} \qquad \text{Subtract.}$$

$$= -0.5055757576 \approx -0.506.$$

The significance of $m$ is that the death rate per 100,000 people is tending to drop (because of the negative) at a rate of 0.506 per year.

Now substitute the value of $m$ and the column totals in the formula for $b$:

$$b = \frac{\Sigma y - m(\Sigma x)}{n} \qquad \text{Formula for } b$$

$$= \frac{550.2 - (-0.5055757576)(650)}{10} \qquad \text{Substitute.}$$

$$= \frac{550.2 - (-328.6242424)}{10} \qquad \text{Multiply.}$$

$$= \frac{878.8242424}{10} = 87.88242424 \approx 87.9.$$

Substitute $m$ and $b$ into the least squares line, $Y = mx + b$; the least squares line that best fits the 10 data points has equation

$$Y = -0.506x + 87.9.$$

This gives a mathematical description of the relationship between the year and the number of accidental deaths per 100,000 people. The equation can be used to predict $y$ from a given value of $x$, as we will show in Example 2. As we mentioned before, however, caution must be exercised when using the least squares equation to predict data points that are far from the range of points on which the equation was modeled.

| CAUTION | In computing $m$ and $b$, we rounded the final answer to three digits because the original data were known only to three digits. It is important, however, *not* to round any of the intermediate results (such as $\sum x^2$) because round-off error may have a detrimental effect on the accuracy of the answer. Similarly, it is important not to use a rounded-off value of $m$ when computing $b$. |

**Method 2**
**Graphing Calculator**

The calculations for finding the least squares line are often tedious, even with the aid of a calculator. Fortunately, many calculators can calculate the least squares line with just a few keystrokes. For purposes of illustration, we will show how the least squares line in the previous example is found with a TI-84 Plus graphing calculator.

We begin by entering the data into the calculator. We will be using the first two lists, called $L_1$ and $L_2$. Choosing the STAT menu, then choosing the fourth entry ClrList, we enter $L_1$, $L_2$, to indicate the lists to be cleared. Now we press STAT again and choose the first entry EDIT, which brings up the blank lists. As before, we will only use the last two digits of the year, putting the numbers in $L_1$. We put the death rate in $L_2$, giving the two screens shown in Figure 11.

| L1 | **L2** | L3    2 |
|----|--------|---------|
| 20 | 71.2   | -------- |
| 30 | 80.5   |         |
| 40 | 73.4   |         |
| 50 | 60.3   |         |
| 60 | 52.1   |         |
| 70 | 56.2   |         |
| 80 | 46.5   |         |

L2 = {71.2, 80.5, 7...

| L1  | L2   | L3    1 |
|-----|------|---------|
| 60  | 52.1 |         |
| 70  | 56.2 |         |
| 80  | 46.5 |         |
| 90  | 36.9 |         |
| 100 | 34   |         |
| 110 | 39.1 |         |
| ----- | -------- |       |

L1(11) =

FIGURE 11

Quit the editor, press STAT again, and choose CALC instead of EDIT. Then choose item 4 LinReg(ax + b) to get the values of $a$ (the slope) and $b$ (the $y$-intercept) for the least squares line, as shown in Figure 12. With $a$ and $b$ rounded to three decimal places, the least squares line is $Y = -0.506x + 87.9$. A graph of the data points and the line is shown in Figure 13.

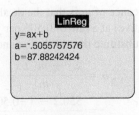

LinReg
y=ax+b
a=-.5055757576
b=87.88242424

FIGURE 12

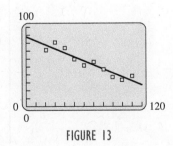

FIGURE 13

For more details on finding the least squares line with a graphing calculator, see the *Graphing Calculator and Excel Spreadsheet Manual* available with this book.

| | Method 3 |
|---|---|
| | Spreadsheet |

**YOUR TURN 1** Calculate the least squares line for the following data.

| x | 1 | 2 | 3 | 4 | 5 | 6 |
|---|---|---|---|---|---|---|
| y | 3 | 4 | 6 | 5 | 7 | 8 |

Many computer spreadsheet programs can also find the least squares line. Figure 14 shows the scatterplot and least squares line for the accidental death rate data using an Excel spreadsheet. The scatterplot was found using the Marked Scatter chart from the Gallery and the line was found using the Add Trendline command under the Chart menu. For details, see the *Graphing Calculator and Excel Spreadsheet Manual* available with this book.

TRY YOUR TURN 1

**Accidental Deaths**

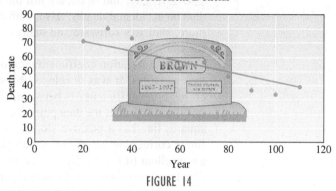

FIGURE 14

### EXAMPLE 2 Least Squares Line

What do we predict the accidental death rate to be in 2015?

**SOLUTION** Use the least squares line equation given above with $x = 115$.

$$Y = -0.506x + 87.9$$
$$= -0.506(115) + 87.9$$
$$= 29.7$$

The accidental death rate in 2015 is predicted to be about 29.7 per 100,000 population. In this case, we will have to wait until the 2015 data become available to see how accurate our prediction is. We have observed, however, that the accidental death rate began to go up after 2000 and was 39.1 per 100,000 population in 2010. This illustrates the danger of extrapolating beyond the data.

### EXAMPLE 3 Least Squares Line

In what year is the death rate predicted to drop below 26 per 100,000 population?

**SOLUTION** Let $Y = 26$ in the equation above and solve for $x$.

$$26 = -0.506x + 87.9$$
$$-61.9 = -0.506x \qquad \text{Subtract 87.9 from both sides.}$$
$$x \approx 122.3 \qquad \text{Divide both sides by } -0.506.$$

This implies that sometime in the year 2022 (122 years after 1900) the death rate drops below 26 per 100,000 population.

## Correlation

Although the least squares line can always be found, it may not be a good model. For example, if the data points are widely scattered, no straight line will model the data accurately. One measure of how well the original data fits a straight line is the **correlation coefficient**, denoted by $r$, which can be calculated by the following formula.

**Correlation Coefficient**

$$r = \frac{n(\Sigma xy) - (\Sigma x)(\Sigma y)}{\sqrt{n(\Sigma x^2) - (\Sigma x)^2} \cdot \sqrt{n(\Sigma y^2) - (\Sigma y)^2}}$$

Although the expression for $r$ looks daunting, remember that each of the summations, $\Sigma x$, $\Sigma y$, $\Sigma xy$, and so on, are just the totals from a table like the one we prepared for the data on accidental deaths. Also, with a calculator, the arithmetic is no problem! Furthermore, statistics software and many calculators can calculate the correlation coefficient for you.

The correlation coefficient measures the strength of the linear relationship between two variables. It was developed by statistics pioneer Karl Pearson (1857–1936). The correlation coefficient $r$ is between 1 and $-1$ or is equal to 1 or $-1$. Values of exactly 1 or $-1$ indicate that the data points lie *exactly* on the least squares line. If $r = 1$, the least squares line has a positive slope; $r = -1$ gives a negative slope. If $r = 0$, there is no linear correlation between the data points (but some *nonlinear* function might provide an excellent fit for the data). A correlation coefficient of zero may also indicate that the data fit a horizontal line. To investigate what is happening, it is always helpful to sketch a scatterplot of the data. Some scatterplots that correspond to these values of $r$ are shown in Figure 15.

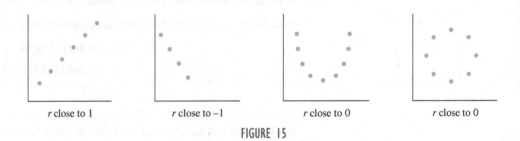

| $r$ close to 1 | $r$ close to $-1$ | $r$ close to 0 | $r$ close to 0 |

FIGURE 15

A value of $r$ close to 1 or $-1$ indicates the presence of a linear relationship. The exact value of $r$ necessary to conclude that there is a linear relationship depends upon $n$, the number of data points, as well as how confident we want to be of our conclusion. For details, consult a text on statistics.*

**EXAMPLE 4** **Correlation Coefficient**

Find $r$ for the data on accidental death rates in Example 1.

**SOLUTION**

**Method 1**
**Calculating by Hand**

From the table in Example 1,

$$\Sigma x = 650, \ \Sigma y = 550.2, \ \Sigma xy = 31,592, \ \Sigma x^2 = 50,500,$$
$$\Sigma y^2 = 32,654.86, \quad \text{and} \quad n = 10.$$

*For example, see *Introductory Statistics*, 9th edition, by Neil A. Weiss, Boston, Mass.: Pearson, 2012.

Substituting these values into the formula for $r$ gives

$$r = \frac{n(\Sigma xy) - (\Sigma x)(\Sigma y)}{\sqrt{n(\Sigma x^2) - (\Sigma x)^2} \cdot \sqrt{n(\Sigma y^2) - (\Sigma y)^2}} \qquad \text{Formula for } r$$

$$= \frac{10(31{,}592) - (650)(550.2)}{\sqrt{10(50{,}500) - (650)^2} \cdot \sqrt{10(32{,}654.86) - (550.2)^2}} \qquad \text{Substitute.}$$

$$= \frac{315{,}920 - 357{,}630}{\sqrt{505{,}000 - 422{,}500} \cdot \sqrt{326{,}548.6 - 302{,}720.04}} \qquad \text{Multiply.}$$

$$= \frac{-41{,}710}{\sqrt{82{,}500} \cdot \sqrt{23{,}828.56}} \qquad \text{Subtract.}$$

$$= \frac{-41{,}710}{44{,}337.97695} \qquad \text{Take square roots and multiply.}$$

$$= -0.9407285327 \approx -0.941.$$

This is a high correlation, which agrees with our observation that the data fit a line quite well.

**Method 2**
**Graphing Calculator**

Most calculators that give the least squares line will also give the correlation coefficient. To do this on the TI-84 Plus, press the second function CATALOG and go down the list to the entry DiagnosticOn. Press ENTER at that point, then press STAT, CALC, and choose item 4 to get the display in Figure 16. The result is the same as we got by hand. The command DiagnosticOn need only be entered once, and the correlation coefficient will always appear in the future.

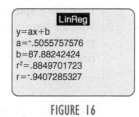

FIGURE 16

**Method 3**
**Spreadsheet**

Many computer spreadsheet programs have a built-in command to find the correlation coefficient. For example, in Excel, use the command "= CORREL(A1:A10,B1:B10)" to find the correlation of the 10 data points stored in columns A and B. For more details, see the *Graphing Calculator and Excel Spreadsheet Manual* available with this text.

**YOUR TURN 2** Find $r$ for the following data.

| $x$ | 1 | 2 | 3 | 4 | 5 | 6 |
|-----|---|---|---|---|---|---|
| $y$ | 3 | 4 | 6 | 5 | 7 | 8 |

TRY YOUR TURN 2

The square of the correlation coefficient gives the fraction of the variation in $y$ that is explained by the linear relationship between $x$ and $y$. Consider Example 4, where $r^2 = (-0.941)^2 = 0.885$. This means that 88.5% of the variation in $y$ is explained by the linear relationship found earlier in Example 1. The remaining 11.5% comes from the scattering of the points about the line.

TECHNOLOGY

**EXAMPLE 5** **Silicone Implants**

Silicones have long been used for fabricating medical devices on the presumption that they are biocompatible materials. This presumption is not entirely correct. Silicone prostheses, when implanted within the soft tissues of the breast, may evoke an inflammatory reaction.

In response to silicone exposure, production of protein factors that mediate the inflammatory response (known as inflammatory mediators) was observed in experimental studies. The levels of four inflammatory mediators from ten patients with silicone breast implants after in vitro culture for 24 hours were as shown below. ***Source: Journal of Investigative Surgery.***

| Patient | IL-2 | TNF-$\alpha$ | IL-6 | PGE$_2$ |
|---|---|---|---|---|
| 1 | 48 | ND | 231 | 68 |
| 2 | 219 | 78 | 308,287 | 2710 |
| 3 | 109 | 65 | 33,291 | 1804 |
| 4 | 2179 | 149 | 124,550 | 8053 |
| 5 | 219 | 451 | 17,075 | 7371 |
| 6 | 54 | 64 | 22,955 | 3418 |
| 7 | 6 | 79 | 95,102 | 9768 |
| 8 | 10 | 115 | 5649 | 441 |
| 9 | 42 | 618 | 840,585 | 9585 |
| 10 | 196 | 69 | 58,924 | 4536 |

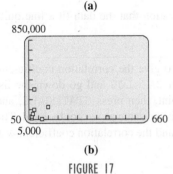

850,000

5,000 ⊢ 50 ⊣ 660

**(a)**

850,000

5,000 ⊢ 50 ⊣ 660

**(b)**

FIGURE 17

A graphing calculator was used to plot the last nine points in the table for TNF-$\alpha(x)$ and IL-6$(y)$. (There is no data point for patient 1, since there is no TNF-$\alpha$ level.) The graph is shown in Figure 17(a). The correlation for these two mediators is 0.6936. Notice that one data point is way off by itself, and is called an **outlier**. Sometimes, by removing such a point from the graph, we can achieve a higher correlation.[*] Figure 17(b) shows the scatterplot with the remaining eight points. However, the correlation is now $-0.2493$![†] With the ninth point, the points were closer to a line with a positive slope. Without the point, there is little linear correlation and it has become negative.

---

**CAUTION** A common error is to assume a cause-and-effect relationship from a high correlation. In Example 5, even if the high correlation were not caused by a single data point, it would be incorrect to conclude that increasing TNF-$\alpha$ causes IL-6 to increase. It is just as likely that increasing IL-6 causes TNF-$\alpha$ to increase, or that neither one affects the other, but both are affected by some third factor.

---

[*]Before discarding a point, we should investigate the reason it is an outlier.
[†]The observation that removing one point changes the correlation from positive to negative was made by Patrick Fleury, *Chance News* (an electronic newsletter), Vol. 4, No. 16, Dec. 1995.

# 1.2 EXERCISES

1. Suppose a positive linear correlation is found between two quantities. Does this mean that one of the quantities increasing causes the other to increase? If not, what does it mean?

2. Given a set of points, the least squares line formed by letting $x$ be the independent variable will not necessarily be the same as the least squares line formed by letting $y$ be the independent variable. Give an example to show why this is true.

3. For the following table of data,

| $x$ | 1 | 2 | 3 | 4 | 5 | 6 | 7 | 8 | 9 | 10 |
|---|---|---|---|---|---|---|---|---|---|---|
| $y$ | 0 | 0.5 | 1 | 2 | 2.5 | 3 | 3 | 4 | 4.5 | 5 |

   a. draw a scatterplot.

   b. calculate the correlation coefficient.

   c. calculate the least squares line and graph it on the scatterplot.

   d. predict the $y$-value when $x$ is 11.

**The following problem is reprinted from the November 1989**
*Actuarial Examination on Applied Statistical Methods. Source:*
*Society of Actuaries.*

**4.** You are given

| X | 6.8 | 7.0 | 7.1 | 7.2 | 7.4 |
|---|---|---|---|---|---|
| Y | 0.8 | 1.2 | 0.9 | 0.9 | 1.5 |

Determine $r^2$, the coefficient of determination for the regression of
$Y$ on $X$. Choose one of the following. (*Note:* The coefficient of determination is defined as the square of the correlation coefficient.)

    **a.** 0.3   **b.** 0.4   **c.** 0.5   **d.** 0.6   **e.** 0.7

**5.** Consider the following table of data.

| x | 1 | 1 | 2 | 2 | 9 |
|---|---|---|---|---|---|
| y | 1 | 2 | 1 | 2 | 9 |

  **a.** Calculate the least squares line and the correlation
coefficient.

  **b.** Repeat part a, but this time delete the last point.

  **c.** Draw a graph of the data, and use it to explain the dramatic
difference between the answers to parts a and b.

**6.** Consider the following table of data.

| x | 1 | 2 | 3 | 4 | 9 |
|---|---|---|---|---|---|
| y | 1 | 2 | 3 | 4 | −20 |

  **a.** Calculate the least squares line and the correlation
coefficient.

  **b.** Repeat part a, but this time delete the last point.

  **c.** Draw a graph of the data, and use it to explain the dramatic
difference between the answers to parts a and b.

**7.** Consider the following table of data.

| x | 1 | 2 | 3 | 4 |
|---|---|---|---|---|
| y | 1 | 1 | 1 | 1.1 |

  **a.** Calculate the correlation coefficient.

  **b.** Sketch a graph of the data.

  **c.** Based on how closely the data fits a straight line, is your
answer to part a surprising? Discuss the extent to which
the correlation coefficient describes how well the data fit a
horizontal line.

**8.** Consider the following table of data.

| x | 0 | 1 | 2 | 3 | 4 |
|---|---|---|---|---|---|
| y | 4 | 1 | 0 | 1 | 4 |

  **a.** Calculate the least squares line and the correlation coefficient.

  **b.** Sketch a graph of the data.

  **c.** Comparing your answers to parts a and b, does a correlation coefficient of 0 mean that there is no relationship
between the $x$ and $y$ values? Would some curve other than
a line fit the data better? Explain.

**9.** The formulas for the least squares line were found by solving
the system of equations

$$nb + (\Sigma x)m = \Sigma y$$
$$(\Sigma x)b + (\Sigma x^2)m = \Sigma xy.$$

Solve the above system for $b$ and $m$ to show that

$$m = \frac{n(\Sigma xy) - (\Sigma x)(\Sigma y)}{n(\Sigma x^2) - (\Sigma x)^2} \quad \text{and}$$

$$b = \frac{\Sigma y - m(\Sigma x)}{n}.$$

## LIFE SCIENCE APPLICATIONS

**10. Bird Eggs** The average length and width of various bird eggs
are given in the following table. *Source: National Council of
Teachers of Mathematics.*

| Bird Name | Width (cm) | Length (cm) |
|---|---|---|
| Canada goose | 5.8 | 8.6 |
| Robin | 1.5 | 1.9 |
| Turtledove | 2.3 | 3.1 |
| Hummingbird | 1.0 | 1.0 |
| Raven | 3.3 | 5.0 |

  **a.** Plot the points, putting the length on the $y$-axis and the
width on the $x$-axis. Do the data appear to be linear?

  **b.** Find the least squares line, and plot it on the same graph as
the data.

  **c.** Suppose there are birds with eggs even smaller than those
of hummingbirds. Would the equation found in part b continue to make sense for all positive widths, no matter how
small? Explain.

  **d.** Find the correlation coefficient.

**11. Size of Hunting Parties** In the 1960s, the famous researcher
Jane Goodall observed that chimpanzees hunt and eat meat as
part of their regular diet. Sometimes chimpanzees hunt alone,
while other times they form hunting parties. The table on the next
page summarizes research on chimpanzee hunting parties, giving
the size of the hunting party and the percentage of successful
hunts. *Source: American Scientist and Mathematics Teacher.*

  **a.** Plot the data. Do the data points lie in a linear pattern?

  **b.** Find the correlation coefficient. Combining this with your
answer to part a, does the percentage of successful hunts
tend to increase with the size of the hunting party?

  **c.** Find the equation of the least squares line, and graph it on
your scatterplot.

| Number of Chimps in Hunting Party | Percentage of Successful Hunts |
|:---:|:---:|
| 1 | 20 |
| 2 | 30 |
| 3 | 28 |
| 4 | 42 |
| 5 | 40 |
| 6 | 58 |
| 7 | 45 |
| 8 | 62 |
| 9 | 65 |
| 10 | 63 |
| 12 | 75 |
| 13 | 75 |
| 14 | 78 |
| 15 | 75 |
| 16 | 82 |

**12. Skeletal Maturity** Researchers measuring skeletal maturity used a maturity score of 1 to 45 points, with 45 representing complete maturity. The average maturity score for the hip joint and pelvis of children of various ages (in years) is given in the table. *Source: Standards in Pediatric Orthopedics.*

a. Find the coefficient of correlation. How closely do the data seem to fit a straight line?

b. Find the equation of the least squares line.

c. Plot the data and least squares line on the same graph. Discuss any ways in which the least squares line does not fit the data.

d. Assuming the linear trend continues, what would be the maturity score for an 18-year-old? Compare this with the actual score of 43.6.

| Age (x) | Maturity (y) |
|:---:|:---:|
| 1 | 8.4 |
| 2 | 10.7 |
| 3 | 12.9 |
| 4 | 15.2 |
| 5 | 17.5 |
| 6 | 19.1 |
| 7 | 20.4 |
| 8 | 21.9 |
| 9 | 23.4 |
| 10 | 25.0 |
| 11 | 27.2 |
| 12 | 29.9 |
| 13 | 32.6 |
| 14 | 36.7 |

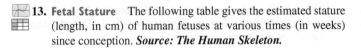

**13. Fetal Stature** The following table gives the estimated stature (length, in cm) of human fetuses at various times (in weeks) since conception. *Source: The Human Skeleton.*

| Age (x) | Stature (y) |
|:---:|:---:|
| 18 | 19.81 |
| 20 | 23.80 |
| 22 | 27.40 |
| 24 | 30.60 |
| 26 | 33.72 |
| 28 | 36.52 |
| 30 | 39.13 |
| 32 | 41.58 |
| 34 | 43.84 |
| 36 | 46.03 |
| 38 | 48.08 |
| 40 | 50.02 |

a. Find the coefficient of correlation. How closely do the data seem to fit a straight line?

b. Find the equation of the least squares line.

c. Plot the data and least squares line on the same graph. Does the line accurately fit the data?

d. According to the least squares line, how much is the fetal stature increasing per week, on average?

e. If this formula continues to be accurate, what would we expect for the stature of a 45-week-old fetus?

**14. Medical School Applications** The number of applications to medical schools in the United States for recent years is given in the following table. Years are represented by their last two digits and applications are given in thousands. *Source: Association of American Medical Colleges.*

| Year (x) | 01 | 02 | 03 | 04 | 05 | 06 | 07 | 08 | 09 | 10 | 11 | 12 |
|:---|:---:|:---:|:---:|:---:|:---:|:---:|:---:|:---:|:---:|:---:|:---:|:---:|
| Applications (y) | 35 | 34 | 35 | 36 | 37 | 39 | 42 | 42 | 42 | 43 | 44 | 45 |

a. Plot the data. Do the data points lie in a linear pattern?

b. Determine the least squares line for these data and graph it on the same coordinate axes. Does the line fit the data reasonably well?

c. Find the coefficient of correlation. Does it agree with your estimate of the fit in part b?

**15. Crickets Chirping**  Biologists have observed a linear relationship between the temperature and the frequency with which a cricket chirps. The following data were measured for the striped ground cricket. *Source: The Song of Insects.*

| Temperature °F ($x$) | Chirps per Second ($y$) |
|---|---|
| 88.6 | 20.0 |
| 71.6 | 16.0 |
| 93.3 | 19.8 |
| 84.3 | 18.4 |
| 80.6 | 17.1 |
| 75.2 | 15.5 |
| 69.7 | 14.7 |
| 82.0 | 17.1 |
| 69.4 | 15.4 |
| 83.3 | 16.2 |
| 79.6 | 15.0 |
| 82.6 | 17.2 |
| 80.6 | 16.0 |
| 83.5 | 17.0 |
| 76.3 | 14.4 |

a. Find the equation for the least squares line for the data.

b. Use the results of part a to determine how many chirps per second you would expect to hear from the striped ground cricket if the temperature were 73°F.

c. Use the results of part a to determine what the temperature is when the striped ground crickets are chirping at a rate of 18 times per sec.

d. Find the correlation coefficient.

**16. Athletic Records**  The table shows the men's and women's outdoor world records (in seconds) in the 800-m run. *Source: Nature, Track and Field Athletics, Statistics in Sports, and The World Almanac and Book of Facts.*

| Year | Men's Record | Women's Record |
|---|---|---|
| 1905 | 113.4 | — |
| 1915 | 111.9 | — |
| 1925 | 111.9 | 144 |
| 1935 | 109.7 | 135.6 |
| 1945 | 106.6 | 132 |
| 1955 | 105.7 | 125 |
| 1965 | 104.3 | 118 |
| 1975 | 103.7 | 117.48 |
| 1985 | 101.73 | 113.28 |
| 1995 | 101.11 | 113.28 |
| 2005 | 101.11 | 113.28 |

Let $x$ be the year, with $x = 0$ corresponding to 1900.

a. Find the equation for the least squares line for the men's record ($y$) in terms of the year ($x$).

b. Find the equation for the least squares line for the women's record.

c. Suppose the men's and women's records continue to improve as predicted by the equations found in parts a and b. In what year will the women's record catch up with the men's record? Do you believe that will happen? Why or why not?

d. There have been no improvements in the women's record since 1983. How does this affect your answer to part c?

e. Calculate the correlation coefficient for both the men's and the women's record. What do these numbers tell you?

f. Draw a plot of the data, and discuss to what extent a linear function describes the trend in the data.

**17. Running**  If you think a marathon is a long race, consider the Hardrock 100, a 100.5 mile running race held in southwestern Colorado. The chart below lists the times that the 2008 winner, Kyle Skaggs, arrived at various mileage points along the way. *Source: www.run100s.com.*

a. What was Skaggs's average speed?

b. Graph the data, plotting time on the $x$-axis and distance on the $y$-axis. You will need to convert the time from hours and minutes into hours. Do the data appear to lie approximately on a straight line?

c. Find the equation for the least squares line, fitting distance as a linear function of time.

d. Calculate the correlation coefficient. Does it indicate a good fit of the least squares line to the data?

e. Based on your answer to part d, what is a good value for Skaggs's average speed? Compare this with your answer to part a. Which answer do you think is better? Explain your reasoning.

| Time (hr:min) | Miles |
|---|---|
| 0 | 0 |
| 2:19 | 11.5 |
| 3:43 | 18.9 |
| 5:36 | 27.8 |
| 7:05 | 32.8 |
| 7:30 | 36.0 |
| 8:30 | 43.9 |
| 10:36 | 51.5 |
| 11:56 | 58.4 |
| 15:14 | 71.8 |
| 17:49 | 80.9 |
| 18:58 | 85.2 |
| 20:50 | 91.3 |
| 23:23 | 100.5 |

## OTHER APPLICATIONS

**18. Poverty Levels** The following table lists how poverty level income cutoffs (in dollars) for a family of four have changed over time. *Source: U.S. Census Bureau.*

| Year | Income |
|------|--------|
| 1980 | 8414 |
| 1985 | 10,989 |
| 1990 | 13,359 |
| 1995 | 15,569 |
| 2000 | 17,604 |
| 2005 | 19,961 |
| 2010 | 22,314 |

Let $x$ represent the year, with $x = 0$ corresponding to 1980 and $y$ represent the income in thousands of dollars. (*Note:* $\Sigma x = 105$, $\Sigma x^2 = 2275$, $\Sigma y = 108{,}210$, $\Sigma y^2 = 1{,}818{,}667{,}092$, and $\Sigma xy = 1{,}942{,}595$.)

  a. Plot the data. Do the data appear to lie along a straight line?

  b. Calculate the correlation coefficient. Does your result agree with your answer to part a?

  c. Find the equation of the least squares line.

  d. Use your answer from part c to predict the poverty level in the year 2018.

**19. Ideal Partner Height** In an introductory statistics course at Cornell University, 147 undergraduates were asked their own height and the ideal height for their ideal spouse or partner. For this exercise, we are including the data for only a representative sample of 10 of the students, as given in the following table. All heights are in inches. *Source: Chance.*

| Height | Ideal Partner's Height |
|--------|------------------------|
| 59 | 66 |
| 62 | 71 |
| 66 | 72 |
| 68 | 73 |
| 71 | 75 |
| 67 | 63 |
| 70 | 63 |
| 71 | 67 |
| 73 | 66 |
| 75 | 66 |

  a. Find the regression line and correlation coefficient for these data. What strange phenomenon is suggested by the slope?

  b. The first five data pairs are for female students and the second five for male students. Find the regression line and correlation coefficient for each set of data.

  c. Plot all the data on one graph, using different types of points to distinguish the data for the males and for the females. Using this plot and the results from part b, explain the strange phenomenon that you observed in part a.

**20. SAT Scores** At Hofstra University, all students take the math SAT before entrance, and most students take a mathematics placement test before registration. Recently, one professor collected the following data for 19 students in his Finite Mathematics class:

| Math SAT | Placement Test | Math SAT | Placement Test | Math SAT | Placement Test |
|----------|----------------|----------|----------------|----------|----------------|
| 540 | 20 | 580 | 8 | 440 | 10 |
| 510 | 16 | 680 | 15 | 520 | 11 |
| 490 | 10 | 560 | 8 | 620 | 11 |
| 560 | 8 | 560 | 13 | 680 | 8 |
| 470 | 12 | 500 | 14 | 550 | 8 |
| 600 | 11 | 470 | 10 | 620 | 7 |
| 540 | 10 | | | | |

  a. Find an equation for the least squares line. Let $x$ be the math SAT and $y$ be the placement test score.

  b. Use your answer from part a to predict the mathematics placement test score for a student with a math SAT score of 420.

  c. Use your answer from part a to predict the mathematics placement test score for a student with a math SAT score of 620.

  d. Calculate the correlation coefficient.

  e. Based on your answer to part d, what can you conclude about the relationship between a student's math SAT and mathematics placement test score?

**21. Length of a Pendulum** Grandfather clocks use pendulums to keep accurate time. The relationship between the length of a pendulum $L$ and the time $T$ for one complete oscillation can be determined from the data in the table. *Source: Gary Rockswold.*

| $L$ (ft) | $T$ (sec) |
|----------|-----------|
| 1.0 | 1.11 |
| 1.5 | 1.36 |
| 2.0 | 1.57 |
| 2.5 | 1.76 |
| 3.0 | 1.92 |
| 3.5 | 2.08 |
| 4.0 | 2.22 |

  a. Plot the data from the table with $L$ as the horizontal axis and $T$ as the vertical axis.

  b. Find the least squares line equation and graph it simultaneously, if possible, with the data points. Does it seem to fit the data?

  c. Find the correlation coefficient and interpret it. Does it confirm your answer to part b?

**22. Football**  The following data give the expected points for a football team with first down and 10 yards to go from various points on the field. **Source: Operations Research.** (*Note:* $\Sigma x = 500$, $\Sigma x^2 = 33{,}250$, $\Sigma y = 20.668$, $\Sigma y^2 = 91.927042$, and $\Sigma xy = 399.16$.)

| Yards from Goal (x) | Expected Points (y) |
| --- | --- |
| 5 | 6.041 |
| 15 | 4.572 |
| 25 | 3.681 |
| 35 | 3.167 |
| 45 | 2.392 |
| 55 | 1.538 |
| 65 | 0.923 |
| 75 | 0.236 |
| 85 | −0.637 |
| 95 | −1.245 |

**a.** Calculate the correlation coefficient. Does there appear to be a linear correlation?

**b.** Find the equation of the least squares line.

**c.** Use your answer from part a to predict the expected points when a team is at the 50-yd line.

**YOUR TURN ANSWERS**

**1.** $Y = 0.94x + 2.2$

**2.** $r = 0.94$

# 1.3  Properties of Functions

APPLY IT **How does a woman's basal body temperature vary with time?**
*In this section we will explore this question using the concept of a function.*

Figure 18 illustrates a function that, unlike those studied in Section 1.1, is *nonlinear.* Its graph is not a straight line. Linear functions are simple to study, and they can be used to approximate many functions over short intervals. But most functions exhibit behavior that, in the long run, does not follow a straight line. In the rest of this chapter and the next we will study some of the most common nonlinear functions.

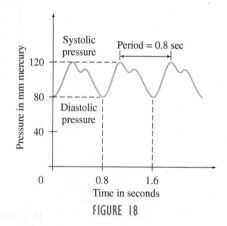

FIGURE 18

As saw in Section 1.1, the linear cost function $C(x) = 200{,}000 + 10x$ for an X-ray machine is related to the number of images produced. The number of images produced is the independent variable and the total cost is the dependent variable because it depends on the number produced. When a specific number of images (say, 1000) is substituted for $x$, the cost $C(x)$ has one specific value $(200{,}000 + 10 \cdot 1000)$. Because of this, the variable $C(x)$ is said to be a *function* of $x$.

### Function

A **function** is a rule that assigns to each element from one set exactly one element from another set.

In most cases in this book, the "rule" mentioned in the box is expressed as an equation, such as $C(x) = 200,000 + 10x$. When an equation is given for a function, we say that the equation *defines* the function. Whenever $x$ and $y$ are used in this book to define a function, $x$ represents the independent variable and $y$ the dependent variable. Of course, letters other than $x$ and $y$ could be used and are often more meaningful. For example, if the independent variable represents the number of minutes and the dependent variable represents the number of seconds, we might write $s = 60m$.

The independent variable in a function can take on any value within a specified set of values called the *domain*.

### Domain and Range

The set of all possible values of the independent variable in a function is called the **domain** of the function, and the resulting set of possible values of the dependent variable is called the **range**.

**APPLY IT**    A woman trying to get pregnant, or using natural methods to avoid getting pregnant, might be very interested in her basal body temperature (her temperature at rest). To predict when she is ovulating, a woman can take her temperature each day on waking.* Ovulation begins when her temperature begins to rise, usually about halfway through her cycle. A typical graph for a woman's temperature as a function of her cycle day is shown in Figure 19. Let us label this function $y = f(x)$, where $y$ is the temperature of the woman's body and $x$ is the cycle day. Notice that the function increases and decreases during the month, so it is not linear. Such a function, whose graph is not a straight line, is called a *nonlinear function*.

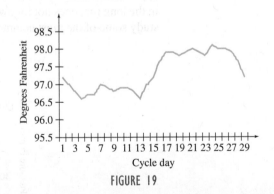

FIGURE 19

The concepts you learned in the section on linear functions apply to this and other nonlinear functions as well. The independent variable here is $x$, the time in days; the dependent variable is $y$, the temperature at any time. The domain is $\{x \mid 1 \le x \le 29\}$, or $[1, 29]$; $x = 1$ corresponds to the first day of this woman's cycle, and $x = 29$ corresponds to the last day. By looking for the lowest and highest values of the function, we estimate the range to be approximately $\{y \mid 96.6 \le y \le 98.1\}$, or $[96.6, 98.1]$. As with linear functions, the domain is mapped along the horizontal axis and the range along the vertical axis.

*For more information, see www.webmd.com/infertility-and-reproduction/fertility-tests-for-women.

We do not have a formula for $f(x)$. (If a woman had such a formula, she would no longer have to take her temperature.) Instead, we can use the graph to estimate values of the function. To estimate $f(21)$, for example, we draw a vertical line from day 21, as shown in Figure 20. The $y$-coordinate appears to be 98.0, so we estimate $f(21) = 98.0$. Similarly, if we wanted to solve the equation $f(x) = 97.0$, we would look for points on the graph that have a $y$-coordinate of 97.0. As Figure 21 shows, this occurs at three points, around days 2, 7, and 14. Thus $f(x) = 97.0$ when $x = 2$, 7, and 14.

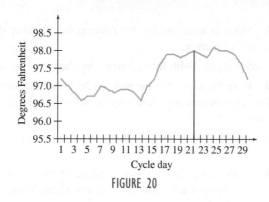

FIGURE 20

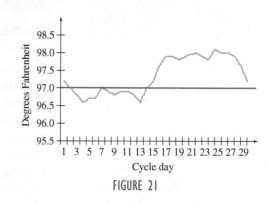

FIGURE 21

| Day ($x$) | Temperature ($y$) |
|-----------|-------------------|
| 1 | 97.2 |
| 2 | 97.0 |
| 3 | 96.8 |
| 4 | 96.6 |
| 5 | 96.7 |
| 6 | 96.7 |
| 7 | 97.0 |

This function can also be given as a table. The table in the margin shows the value of the function for several values of $x$.

Notice from the table that $f(5) = f(6) = 96.7$. This illustrates an important property of functions: Several different values of the independent variable can have the same value for the dependent variable. On the other hand, we cannot have several different $y$-values corresponding to the same value of $x$; if we did, this would not be a function.

What is $f(2.5)$? We do not know. Between days 2 and 3, the woman's temperature presumably went up and back down, as normally happens on a daily basis. Unless the woman monitors her temperature continually throughout the day, we do not have access to this information.

Functions arise in numerous applications, and an understanding of them is critical for understanding calculus. The following example shows some of the ways functions can be represented, and will help you in determining whether a relationship between two variables is a function or not.

### EXAMPLE 1    Functions

Which of the following are functions?

**(a)**

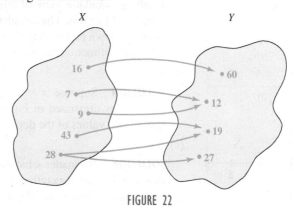

FIGURE 22

**SOLUTION**    Figure 22 shows that an $x$-value of 28 corresponds to *two* $y$-values, 19 and 27. In a function, each $x$ must correspond to exactly one $y$, so this correspondence is not a function.

**(b)** The $x^2$ key on a calculator

**SOLUTION**    This correspondence between input and output is a function because the calculator produces just one $x^2$ (one $y$-value) for each $x$-value entered. Notice also that two $x$-values, such as 3 and $-3$, produce the same $y$-value of 9, but this does not violate the definition of a function.

**(c)**

| $x$ | 1 | 1 | 2 | 2 | 3 | 3 |
|-----|---|---|---|---|---|---|
| $y$ | 3 | $-3$ | 5 | $-5$ | 8 | $-8$ |

**SOLUTION**    Since at least one $x$-value corresponds to more than one $y$-value, this table does not define a function.

**(d)** The set of ordered pairs with first elements mothers and second elements their children

**SOLUTION**    Here the mother is the independent variable and the child is the dependent variable. For a given mother, there may be several children, so this correspondence is not a function.

**(e)** The set of ordered pairs with first elements children and second elements their birth mothers

**SOLUTION**    In this case the child is the independent variable and the mother is the dependent variable. Since each child has only one birth mother, this is a function.

## EXAMPLE 2    Functions

Decide whether each equation or graph represents a function. (Assume that $x$ represents the independent variable here, an assumption we shall make throughout this book.) Give the domain and range of any functions.

**(a)** $y = 11 - 4x^2$

**SOLUTION**    For a given value of $x$, calculating $11 - 4x^2$ produces exactly one value of $y$. (For example, if $x = -7$, then $y = 11 - 4(-7)^2 = -185$, so $f(-7) = -185$.) Since one value of the independent variable leads to exactly one value of the dependent variable, $y = 11 - 4x^2$ meets the definition of a function.

Because $x$ can take on any real-number value, the domain of this function is the set of all real numbers. Finding the range is more difficult. One way to find it would be to ask what possible values of $y$ could come out of this function. Notice that the value of $y$ is 11 minus a quantity that is always 0 or positive, since $4x^2$ can never be negative. There is no limit to how large $4x^2$ can be, so the range is $(-\infty, 11]$.

Another way to find the range would be to examine the graph. Figure 23 shows a graphing calculator view of this function, and we can see that the function takes on $y$-values of 11 or less. The calculator cannot tell us, however, whether the function continues to go down past the viewing window, or turns back up. To find out, we need to study this type of function more carefully, as we will do in the next section.

**(b)** $y^2 = x$

**SOLUTION**    Suppose $x = 4$. Then $y^2 = x$ becomes $y^2 = 4$, from which $y = 2$ or $y = -2$, as illustrated in Figure 24. Since one value of the independent variable can lead to two values of the dependent variable, $y^2 = x$ does not represent a function.

**(c)** $y = 7$

**SOLUTION**    No matter what the value of $x$, the value of $y$ is always 7. This is indeed a function; it assigns exactly one element, 7, to each value of $x$. Such a function is known as a **constant function**. The domain is the set of all real numbers, and the range is the set $\{7\}$. Its graph is the horizontal line that intersects the $y$-axis at $y = 7$, as shown in Figure 25. Every constant function has a horizontal line for its graph.

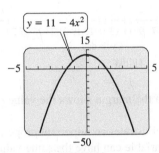

$y = 11 - 4x^2$

FIGURE 23

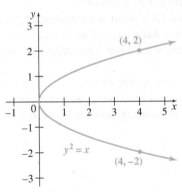

FIGURE 24

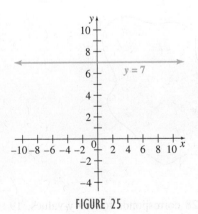

FIGURE 25

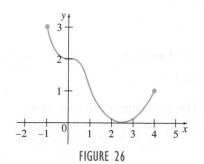

**FIGURE 26**

**(d)** The graph in Figure 26.

**SOLUTION**  For each value of $x$, there is only one value of $y$. For example, the point $(-1, 3)$ on the graph shows that $f(-1) = 3$. Therefore, the graph represents a function. From the graph, we see that the values of $x$ go from $-1$ to $4$, so the domain is $[-1, 4]$. By looking at the values of $y$, we see that the range is $[0, 3]$.

The following agreement on domains is customary.

## Agreement on Domains

Unless otherwise stated, assume that the domain of all functions defined by an equation is the greatest subset of real numbers that are meaningful replacements for the independent variable.

For example, suppose

$$y = \frac{-4x}{2x - 3}.$$

Any real number can be used for $x$ except $x = 3/2$, which makes the denominator equal 0. By the agreement on domains, the domain of this function is the set of all real numbers except $3/2$, which we denote $\{x \mid x \neq 3/2\}$, $\{x \neq 3/2\}$, or $(-\infty, 3/2) \cup (3/2, \infty)$.*

| CAUTION | When finding the domain of a function, there are two operations to avoid: (1) dividing by zero; and (2) taking the square root (or any even root) of a negative number. Later sections will present other functions, such as logarithms, which require further restrictions on the domain. For now, just remember these two restrictions on the domain. |

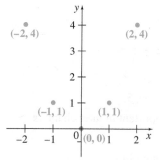

**FIGURE 27**

## EXAMPLE 3    Domain and Range

Find the domain and range for each function defined as follows.

**(a)** $f(x) = x^2$

**SOLUTION**  Any number may be squared, so the domain is the set of all real numbers, written $(-\infty, \infty)$. Since $x^2 \geq 0$ for every value of $x$, the range is $[0, \infty)$.

**(b)** $y = x^2$, with the domain specified as $\{-2, -1, 0, 1, 2\}$.

**SOLUTION**  With the domain specified, the range is the set of values found by applying the function to the domain. Since $f(0) = 0, f(-1) = f(1) = 1$, and $f(-2) = f(2) = 4$, the range is $\{0, 1, 4\}$. The graph of the set of ordered pairs is shown in Figure 27.

**(c)** $y = \sqrt{6 - x}$

**SOLUTION**  For $y$ to be a real number, $6 - x$ must be nonnegative. This happens only when $6 - x \geq 0$, or $6 \geq x$, making the domain $(-\infty, 6]$. The range is $[0, \infty)$ because $\sqrt{6 - x}$ is always nonnegative.

**(d)** $y = \sqrt{2x^2 + 5x - 12}$

**SOLUTION**  The domain includes only those values of $x$ satisfying $2x^2 + 5x - 12 \geq 0$. Using the methods for solving a quadratic inequality produces the domain

$$(-\infty, -4] \cup [3/2, \infty).$$

As in part (c), the range is $[0, \infty)$.

**FOR REVIEW**

Section R.5 demonstrates the method for solving a quadratic inequality. To solve $2x^2 + 5x - 12 \geq 0$, factor the quadratic to get $(2x - 3)(x + 4) \geq 0$. Setting each factor equal to 0 gives $x = 3/2$ or $x = -4$, leading to the intervals $(-\infty, -4]$, $[-4, 3/2]$, and $[3/2, \infty)$. Testing a number from each interval shows that the solution is $(-\infty, -4] \cup [3/2, \infty)$.

*The *union* of sets $A$ and $B$, written $A \cup B$, is defined as the set of all elements in $A$ or $B$ or both.

(e) $y = \dfrac{2}{x^2 - 9}$

**SOLUTION** Since the denominator cannot be zero, $x \neq 3$ and $x \neq -3$. The domain is

$$(-\infty, -3) \cup (-3, 3) \cup (3, \infty).$$

Because the numerator can never be zero, $y \neq 0$. The denominator can take on any real number except for 0, allowing $y$ to take on any value except for 0, so the range is $(-\infty, 0) \cup (0, \infty)$. **TRY YOUR TURN 1**

To understand how a function works, think of a function $f$ as a machine—for example, a calculator or computer—that takes an input $x$ from the domain and uses it to produce an output $f(x)$ (which represents the $y$-value), as shown in Figure 28. In the basal body temperature example, when we put 21 into the machine, we get an output of 98.0, since $f(21) = 98.0$.

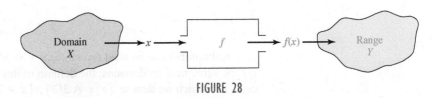

**FIGURE 28**

**EXAMPLE 4  Evaluating Functions**

Let $g(x) = -x^2 + 4x - 5$. Find the following.

**(a)** $g(3)$

**SOLUTION** Replace $x$ with 3.

$$g(3) = -3^2 + 4 \cdot 3 - 5 = -9 + 12 - 5 = -2$$

**(b)** $g(a)$

**SOLUTION** Replace $x$ with $a$ to get $g(a) = -a^2 + 4a - 5$.

This replacement of one variable with another is important in later chapters.

**(c)** $g(x + h)$

**SOLUTION** Replace $x$ with the expression $x + h$ and simplify.

$$\begin{aligned}
g(x + h) &= -(x + h)^2 + 4(x + h) - 5 \\
&= -(x^2 + 2xh + h^2) + 4(x + h) - 5 \\
&= -x^2 - 2xh - h^2 + 4x + 4h - 5
\end{aligned}$$

**(d)** $g\left(\dfrac{2}{r}\right)$

**SOLUTION** Replace $x$ with $2/r$ and simplify.

$$g\left(\frac{2}{r}\right) = -\left(\frac{2}{r}\right)^2 + 4\left(\frac{2}{r}\right) - 5 = -\frac{4}{r^2} + \frac{8}{r} - 5$$

**(e)** Find all values of $x$ such that $g(x) = -12$.

**SOLUTION** Set $g(x)$ equal to $-12$, and then add 12 to both sides to make one side equal to 0.

$$-x^2 + 4x - 5 = -12$$
$$-x^2 + 4x + 7 = 0$$

This equation does not factor, but can be solved with the quadratic formula, which says that if $ax^2 + bx + c = 0$, where $a \neq 0$, then

$$x = \frac{-b \pm \sqrt{b^2 - 4ac}}{2a}.$$

In this case, with $a = -1$, $b = 4$, and $c = 7$, we have

$$x = \frac{-4 \pm \sqrt{16 - 4(-1)7}}{2(-1)}$$

$$= \frac{-4 \pm \sqrt{44}}{-2}$$

$$= 2 \pm \sqrt{11}$$

$$\approx -1.317 \quad \text{or} \quad 5.317. \qquad \text{TRY YOUR TURN 2}$$

**YOUR TURN 2** Given the function $f(x) = 2x^2 - 3x - 4$, find each of the following.
**(a)** $f(x + h)$  **(b)** All values of $x$ such that $f(x) = -5$.

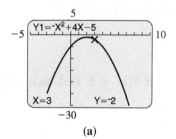

(a)

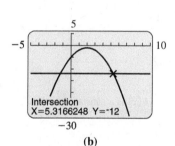

(b)

FIGURE 29

**TECHNOLOGY NOTE**

We can verify the results of parts (a) and (e) of the previous example using a graphing calculator. In Figure 29(a), after graphing $f(x) = -x^2 + 4x - 5$, we have used the "value" feature on the TI-84 Plus to support our answer from part (a). In Figure 29(b) we have used the "intersect" feature to find the intersection of $y = g(x)$ and $y = -12$. The result is $x = 5.3166248$, which is one of our two answers to part (e). The graph clearly shows that there is another answer on the opposite side of the $y$-axis.

**CAUTION**  Notice from Example 4(c) that $g(x + h)$ is *not* the same as $g(x) + h$, which equals $-x^2 + 4x - 5 + h$. There is a significant difference between applying a function to the quantity $x + h$ and applying a function to $x$ and adding $h$ afterward.

If you tend to get confused when replacing $x$ with $x + h$, as in Example 4(c), you might try replacing the $x$ in the original function with a box, like this:

$$g\left(\boxed{\phantom{x+h}}\right) = -\left(\boxed{\phantom{x+h}}\right)^2 + 4\left(\boxed{\phantom{x+h}}\right) - 5$$

Then, to compute $g(x + h)$, just enter $x + h$ into the box:

$$g\left(\boxed{x + h}\right) = -\left(\boxed{x + h}\right)^2 + 4\left(\boxed{x + h}\right) - 5$$

and proceed as in Example 4(c).

Notice in the basal body temperature example that to find the value of the function for a given value of $x$, we drew a vertical line from the value of $x$ and found where it intersected the graph. If a graph is to represent a function, each value of $x$ from the domain must lead to exactly one value of $y$. In the graph in Figure 30, the domain value $x_1$ leads to *two* $y$-values, $y_1$ and $y_2$. Since the given $x$-value corresponds to two different $y$-values, this is not the graph of a function. This example suggests the **vertical line test** for the graph of a function.

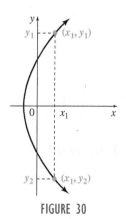

FIGURE 30

**Vertical Line Test**

If a vertical line intersects a graph in more than one point, the graph is not the graph of a function.

A graph represents a function if and only if every vertical line intersects the graph in no more than one point.

### EXAMPLE 5    Vertical Line Test

Use the vertical line test to determine which of the graphs in Example 2 represent functions.

**SOLUTION**    Every vertical line intersects the graphs in Figures 23, 25, and 26 in at most one point, so these are the graphs of functions. It is possible for a vertical line to intersect the graph in Figure 24 twice, so this is not a function.

## Composite Functions    Since we can substitute any expression, such as $x + h$, into a function, we can substitute one function into another. Such an operation is called *function composition*.

> **Function Composition**
> The **composition** of two functions $f$ and $g$ is the new function $h$, where
> $$h(x) = f(g(x)).$$
> The notation for function composition is $h = f \circ g$, or $h(x) = (f \circ g)(x)$.

### EXAMPLE 6    Function Composition

Given the functions $f(x) = x^2 + x$, $g(x) = 5x - 4$, and $h(x) = 3$, find each of the following.

**(a)** $(f \circ g)(2)$

**SOLUTION**
$$
\begin{aligned}
(f \circ g)(2) &= f(g(2)) \\
&= f(5 \cdot 2 - 4) \\
&= f(6) \\
&= 6^2 + 6 = 42
\end{aligned}
$$

**(b)** $h(f(5))$

**SOLUTION**    $h(f(5)) = h(5^2 + 5) = h(30) = 3$    $h(x)$ is a constant function.

**(c)** $(f \circ g)(x)$

**SOLUTION**
$$
\begin{aligned}
(f \circ g)(x) &= f(g(x)) \\
&= (g(x))^2 + g(x) \\
&= (5x - 4)^2 + (5x - 4) \\
&= 25x^2 - 35x + 12
\end{aligned}
$$

**(d)** $(g \circ f)(x)$

**SOLUTION**
$$
\begin{aligned}
(g \circ f)(x) &= g(f(x)) \\
&= 5f(x) - 4 \\
&= 5(x^2 + x) - 4 \\
&= 5x^2 + 5x - 4 \qquad \text{TRY YOUR TURN 3}
\end{aligned}
$$

**YOUR TURN 3**    Given the functions $f(x) = 3x + 6$ and $g(x) = x^2 - 1$, find **(a)** $f(g(1))$, **(b)** $(f \circ g)(x)$, and **(c)** $(g \circ f)(x)$.

In Section 1.1 you saw several examples of linear models. In Example 7, we use a quadratic equation to model the area of a lot.

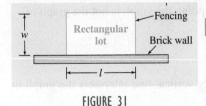

FIGURE 31

### EXAMPLE 7    Area

A fence is to be built against a brick wall to form a rectangular lot, as shown in Figure 31. Only three sides of the fence need to be built, because the wall forms the fourth side. The contractor will use 200 m of fencing. Let the length of the wall be $l$ and the width $w$, as shown in Figure 31.

**(a)** Find the area of the lot as a function of the length $l$.

**SOLUTION** The area formula for a rectangle is area $=$ length $\times$ width, or

$$A = lw.$$

We want the area as a function of the length only, so we must eliminate the width. We use the fact that the total amount of fencing is the sum of the three sections, one length and two widths, so $200 = l + 2w$. Solve this for $w$:

$$200 = l + 2w$$

$$200 - l = 2w \qquad \text{Subtract } l \text{ from both sides.}$$

$$100 - l/2 = w. \qquad \text{Divide both sides by 2.}$$

Substituting this into the formula for area gives

$$A = l(100 - l/2).$$

**(b)** Find the domain of the function in part (a).

**SOLUTION** The length cannot be negative, so $l \geq 0$. Similarly, the width cannot be negative, so $100 - l/2 \geq 0$, from which we find $l \leq 200$. Therefore, the domain is $[0, 200]$.

**(c)** Sketch a graph of the function in part (a).

**SOLUTION** The result from a graphing calculator is shown in Figure 32. Notice that at the endpoints of the domain, when $l = 0$ and $l = 200$, the area is 0. This makes sense: If the length or width is 0, the area will be 0 as well. In between, as the length increases from 0 to 100 m, the area increases, and seems to reach a peak of 5000 m$^2$ when $l = 100$ m. After that, the area decreases as the length continues to increase because the width is becoming smaller.

In the next section, we will study this type of function in more detail and determine exactly where the maximum occurs.

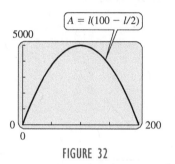

FIGURE 32

# 1.3 EXERCISES

**Which of the following rules define $y$ as a function of $x$?**

**1.**

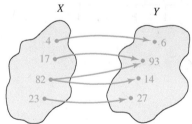

**2.**

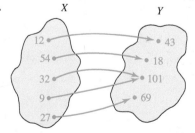

**3.**

| $x$ | $y$ |
|---|---|
| 3 | 9 |
| 2 | 4 |
| 1 | 1 |
| 0 | 0 |
| −1 | 1 |
| −2 | 4 |
| −3 | 9 |

**4.**

| $x$ | $y$ |
|---|---|
| 9 | 3 |
| 4 | 2 |
| 1 | 1 |
| 0 | 0 |
| 1 | −1 |
| 4 | −2 |
| 9 | −3 |

**5.** $y = x^3 + 2$

**6.** $y = \sqrt{x}$

**7.** $x = |y|$

**8.** $x = y^2 + 4$

**List the ordered pairs obtained from each equation, given $\{-2, -1, 0, 1, 2, 3\}$ as the domain. Graph each set of ordered pairs. Give the range.**

**9.** $y = 2x + 3$

**10.** $y = -3x + 9$

**11.** $2y - x = 5$

**12.** $6x - y = -1$

**13.** $y = x(x + 2)$

**14.** $y = (x - 2)(x + 2)$

**15.** $y = x^2$

**16.** $y = -4x^2$

**Give the domain of each function defined as follows.**

**17.** $f(x) = 2x$

**18.** $f(x) = 2x + 3$

**19.** $f(x) = x^4$

**20.** $f(x) = (x + 3)^2$

**21.** $f(x) = \sqrt{4 - x^2}$

**22.** $f(x) = |3x - 6|$

**23.** $f(x) = (x - 3)^{1/2}$

**24.** $f(x) = (3x + 5)^{1/2}$

**25.** $f(x) = \dfrac{2}{1 - x^2}$

**26.** $f(x) = \dfrac{-8}{x^2 - 36}$

**27.** $f(x) = -\sqrt{\dfrac{2}{x^2 - 16}}$

**28.** $f(x) = -\sqrt{\dfrac{5}{x^2 + 36}}$

**29.** $f(x) = \sqrt{x^2 - 4x - 5}$

**30.** $f(x) = \sqrt{15x^2 + x - 2}$

**31.** $f(x) = \dfrac{1}{\sqrt{3x^2 + 2x - 1}}$

**32.** $f(x) = \sqrt{\dfrac{x^2}{3 - x}}$

**Give the domain and the range of each function. Where arrows are drawn, assume the function continues in the indicated direction.**

**33.**

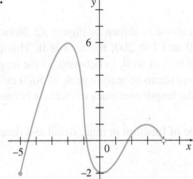

**34.**

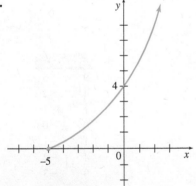

**35.**

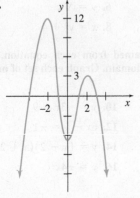

**36.**

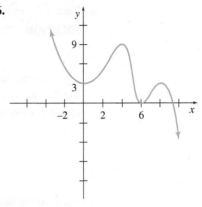

**In Exercises 37–40, give the domain and range. Then, use each graph to find (a) $f(-2)$, (b) $f(0)$, (c) $f(1/2)$, and (d) any values of $x$ such that $f(x) = 1$.**

**37.**

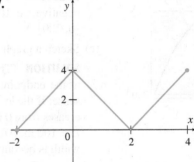

**38.**

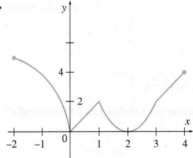

**39.**

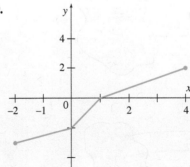

**40.**

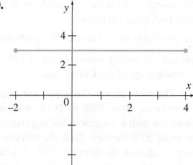

For each function, find (a) $f(4)$, (b) $f(-1/2)$, (c) $f(a)$, (d) $f(2/m)$, and (e) any values of $x$ such that $f(x) = 1$.

**41.** $f(x) = 3x^2 - 4x + 1$    **42.** $f(x) = (x + 3)(x - 4)$

**43.** $f(x) = \begin{cases} \dfrac{2x + 1}{x - 4} & \text{if } x \neq 4 \\ 7 & \text{if } x = 4 \end{cases}$

**44.** $f(x) = \begin{cases} \dfrac{x - 4}{2x + 1} & \text{if } x \neq -\dfrac{1}{2} \\ 10 & \text{if } x = -\dfrac{1}{2} \end{cases}$

Let $f(x) = 6x^2 - 2$ and $g(x) = x^2 - 2x + 5$ to find the following values.

**45.** $f(t + 1)$                **46.** $f(2 - r)$

**47.** $g(r + h)$               **48.** $g(z - p)$

**49.** $g\left(\dfrac{3}{q}\right)$              **50.** $g\left(-\dfrac{5}{z}\right)$

For each function defined as follows, find (a) $f(x + h)$, (b) $f(x + h) - f(x)$, and (c) $[f(x + h) - f(x)]/h$.

**51.** $f(x) = 2x + 1$          **52.** $f(x) = x^2 - 3$

**53.** $f(x) = 2x^2 - 4x - 5$   **54.** $f(x) = -4x^2 + 3x + 2$

**55.** $f(x) = \dfrac{1}{x}$               **56.** $f(x) = -\dfrac{1}{x^2}$

Decide whether each graph represents a function.

**57.**
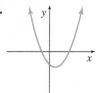

**58.**

**59.**

**60.**

**61.**

**62.**

For each pair of functions, find (a) $f(g(1))$, (b) $g(f(1))$, (c) $(f \circ g)(x)$, and (d) $(g \circ f)(x)$.

**63.** $f(x) = 3x - 7$ and $g(x) = 4 - x$

**64.** $f(x) = x^2 - 1$ and $g(x) = -2$

**65.** $f(x) = 2x^2 + 5x + 1$ and $g(x) = 3x - 1$

**66.** $f(x) = \dfrac{1}{3x + 2}$ and $g(x) = x^2 + 4x$

## LIFE SCIENCE APPLICATIONS

**67. Whales Diving** The figure shows the depth of a diving sperm whale as a function of time, as recorded by researchers at the Woods Hole Oceanographic Institution in Massachusetts. *Source: Peter Tyack, Woods Hole Oceanographic Institution.*

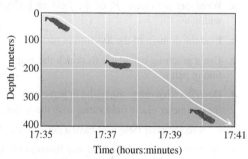

Find the depth of the whale at the following times.

**a.** 17 hours and 37 minutes

**b.** 17 hours and 39 minutes

**68. Metabolic Rate** The basal metabolic rate (in kcal/day) for large anteaters is given by

$$y = f(x) = 19.7x^{0.753},$$

where $x$ is the anteater's weight in kilograms.* *Source: Wildlife Feeding and Nutrition.*

**a.** Find the basal metabolic rate for anteaters with the following weights.

**i.** 5 kg               **ii.** 25 kg

**b.** Suppose the anteater's weight is given in pounds rather than kilograms. Given that 1 lb = 0.454 kg, find a function $x = g(z)$ giving the anteater's weight in kilograms if $z$ is the animal's weight in pounds.

**c.** Write the basal metabolic rate as a function of the weight in pounds in the form $y = az^b$ by calculating $f(g(z))$.

*Technically, kilograms are a measure of mass, not weight. Weight is a measure of the force of gravity, which varies with the distance from the center of Earth. For objects on the surface of Earth, weight and mass are often used interchangeably, and we will do so in this text.

**69. Swimming Energy** The energy expenditure (in kcal/km) for animals swimming at the surface of the water is given by

$$y = f(x) = 0.01x^{0.88},$$

where $x$ is the animal's weight in grams. *Source: Wildlife Feeding and Nutrition.*

**a.** Find the energy for the following animals swimming at the surface of the water.

   **i.** A muskrat weighing 800 g

   **ii.** A sea otter weighing 20,000 g

**b.** Suppose the animal's weight is given in kilograms rather than grams. Given that 1 kg = 1000 g, find a function $x = g(z)$ giving the animal's weight in grams if $z$ is the animal's weight in kilograms.

**c.** Write the energy expenditure as a function of the weight in kilograms in the form $y = az^b$ by calculating $f(g(z))$.

## OTHER APPLICATIONS

**70. Perimeter** A rectangular field is to have an area of 500 m².

**a.** Write the perimeter, $P$, of the field as a function of the width, $w$.

**b.** Find the domain of the function in part a.

**c.** Use a graphing calculator to sketch the graph of the function in part a.

**d.** Describe what the graph found in part c tells you about how the perimeter of the field varies with the width.

**71. Area** A rectangular field is to have a perimeter of 6000 ft.

**a.** Write the area, $A$, of the field as a function of the width, $w$.

**b.** Find the domain of the function in part a.

**c.** Use a graphing calculator to sketch the graph of the function in part a.

**d.** Describe what the graph found in part c tells you about how the area of the field varies with the width.

**72. Energy Consumption** The U.S. Energy Information Administration has used recent consumption data to project the energy consumption (in quadrillion Btu) for the United States, China, and India until 2035. The following graph illustrates these data. *Source: U.S. Energy Information Administration.*

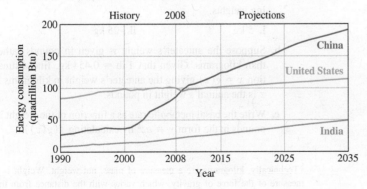

**a.** Estimate the energy consumption in 2015 for the United States. Repeat for China and India.

**b.** In what year will China first consume 150 quadrillion Btu?

**c.** In what year did the energy consumption of China equal the energy consumption of the United States?

**73. Internet Users** The table shows the estimated number of Internet users worldwide from 2008 to 2012. Although the table shows a function with a $y$-variable for just five $x$-values representing the years 2008 through 2012, the function can be defined for every $x$-value in the interval $2008 \le x \le 2012$. Let $y = f(x)$ represent the number of Internet users and $x$ represent the years. *Source: Internet World Stats.*

| Year | Millions of Users |
|------|-------------------|
| 2008 | 1574 |
| 2009 | 1802 |
| 2010 | 1971 |
| 2011 | 2267 |
| 2012 | 2405 |

**a.** What is the independent variable?

**b.** What is the dependent variable?

**c.** Find $f(2010)$.

**d.** Give the domain and range of the function.

**YOUR TURN ANSWERS**

**1.** $(-\infty, -2) \cup (2, \infty), (0, \infty)$

**2. (a)** $2x^2 + 4xh + 2h^2 - 3x - 3h - 4$

  **(b)** 1 and $\frac{1}{2}$

**3. (a)** 6

  **(b)** $3x^2 + 3$

  **(c)** $9x^2 + 36x + 35$

# 1.4 Quadratic Functions; Translation and Reflection

**APPLY IT** **At what point in the development of a fetus is the resistance in the splenic artery at a maximum?**

*In Exercise 51 in this section we will see how knowledge of* quadratic functions *will help provide an answer to the question above.*

**FOR REVIEW**
In this section you will need to know how to solve a quadratic equation by factoring and by the quadratic formula, which are covered in Sections R.2 and R.4. Factoring is usually easiest; when a polynomial is set equal to zero and factored, then a solution is found by setting any one factor equal to zero. But factoring is not always possible. The quadratic formula will provide the solution to *any* quadratic equation.

A linear function is defined by

$$f(x) = ax + b,$$

for real numbers $a$ and $b$. In a *quadratic function* the independent variable is squared. A quadratic function is an especially good model for many situations with a maximum or a minimum function value. Next to linear functions, they are the simplest type of function, and well worth studying thoroughly.

## Quadratic Function

A **quadratic function** is defined by

$$\mathbf{f(x) = ax^2 + bx + c,}$$

where $a$, $b$, and $c$ are real numbers, with $a \neq 0$.

FIGURE 33

The simplest quadratic function has $f(x) = x^2$, with $a = 1$, $b = 0$, and $c = 0$. This function describes situations where the dependent variable $y$ is proportional to the *square* of the independent variable $x$. The graph of this function is shown in Figure 33. This graph is called a **parabola**. Every quadratic function has a parabola as its graph. The lowest (or highest) point on a parabola is the **vertex** of the parabola. The vertex of the parabola in Figure 33 is $(0, 0)$.

If the graph in Figure 33 were folded in half along the $y$-axis, the two halves of the parabola would match exactly. This means that the graph of a quadratic function is *symmetric* with respect to a vertical line through the vertex; this line is the **axis** of the parabola.

There are many real-world instances of parabolas. For example, cross sections of spotlight reflectors or radar dishes form parabolas. Also, a projectile thrown in the air follows a parabolic path. For such applications, we need to study more complicated quadratic functions than $y = x^2$, as in the next several examples.

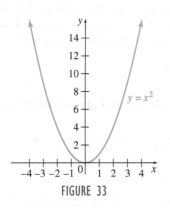

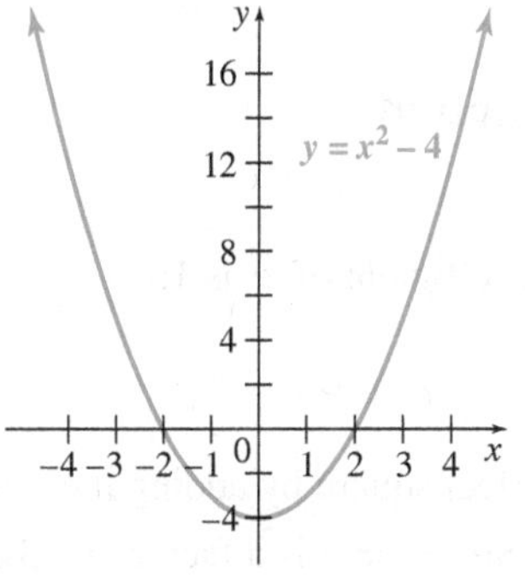

FIGURE 34

### EXAMPLE 1 Graphing a Quadratic Function

Graph $y = x^2 - 4$.

**SOLUTION** Each value of $y$ will be 4 less than the corresponding value of $y$ in $y = x^2$. The graph of $y = x^2 - 4$ has the same shape as that of $y = x^2$ but is 4 units lower. See Figure 34. The vertex of the parabola (on this parabola, the *lowest* point) is at $(0, -4)$. The $x$-intercepts can be found by letting $y = 0$ to get

$$0 = x^2 - 4,$$

from which $x = 2$ and $x = -2$ are the $x$-intercepts. The axis of the parabola is the vertical line $x = 0$.

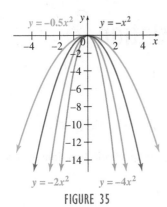

$y = -0.5x^2$    $y = -x^2$

$y = -2x^2$    $y = -4x^2$

**FIGURE 35**

Example 1 suggests that the effect of $c$ in $ax^2 + bx + c$ is to lower the graph if $c$ is negative and to raise the graph if $c$ is positive. This is true for any function; the movement up or down is referred to as a **vertical translation** of the function.

### EXAMPLE 2  Graphing Quadratic Functions

Graph $y = ax^2$ with $a = -0.5$, $a = -1$, $a = -2$, and $a = -4$.

**SOLUTION**  Figure 35 shows all four functions plotted on the same axes. We see that since $a$ is negative, the graph opens downward. When $a$ is between $-1$ and $1$ (that is, when $a = -0.5$), the graph is wider than the original graph, because the values of $y$ are smaller in magnitude. On the other hand, when $a$ is greater than 1 or less than $-1$, the graph is steeper.

Example 2 shows that the sign of $a$ in $ax^2 + bx + c$ determines whether the parabola opens upward or downward. Multiplying $f(x)$ by a negative number flips the graph of $f$ upside down. This is called a **vertical reflection** of the graph. The magnitude of $a$ determines how steeply the graph increases or decreases.

### EXAMPLE 3  Graphing Quadratic Functions

Graph $y = (x - h)^2$ for $h = 3$, 0, and $-4$.

**SOLUTION**  Figure 36 shows all three functions on the same axes. Notice that since the number is subtracted before the squaring occurs, the graph does not move up or down but instead moves left or right. Evaluating $f(x) = (x - 3)^2$ at $x = 3$ gives the same result as evaluating $f(x) = x^2$ at $x = 0$. Therefore, when we subtract the positive number 3 from $x$, the graph shifts 3 units to the right, so the vertex is at $(3, 0)$. Similarly, when we subtract the negative number $-4$ from $x$—in other words, when the function becomes $f(x) = (x + 4)^2$—the graph shifts to the left 4 units.

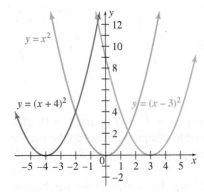

$y = x^2$

$y = (x + 4)^2$    $y = (x - 3)^2$

**FIGURE 36**

The left or right shift of the graph illustrated in Figure 36 is called a **horizontal translation** of the function.

If a quadratic equation is given in the form $ax^2 + bx + c$, we can identify the translations and any vertical reflection by rewriting it in the form

$$y = a(x - h)^2 + k.$$

In this form, we can identify the vertex as $(h, k)$. A quadratic equation not given in this form can be converted by a process called **completing the square**. The next example illustrates the process.

### EXAMPLE 4  Graphing a Quadratic Function

Graph $y = -3x^2 - 2x + 1$.

**Method I**
**Completing the Square**

**SOLUTION**  To begin, factor $-3$ from the $x$-terms so the coefficient of $x^2$ is 1:

$$y = -3\left(x^2 + \frac{2}{3}x\right) + 1.$$

Next, we make the expression inside the parentheses a perfect square by adding the square of one-half of the coefficient of $x$, which is $\left(\frac{1}{2} \cdot \frac{2}{3}\right)^2 = \frac{1}{9}$. Since there is a factor of $-3$ outside the parentheses, we are actually adding $-3 \cdot \left(\frac{1}{9}\right)$. To make sure that the value of the function is not changed, we must also add $3 \cdot \left(\frac{1}{9}\right)$ to the function. Actually, we are simply adding $-3 \cdot \left(\frac{1}{9}\right) + 3 \cdot \left(\frac{1}{9}\right) = 0$, and not changing the function. To summarize our steps,

$$y = -3\left(x^2 + \frac{2}{3}x\right) + 1 \qquad \text{Factor out } -3.$$

$$= -3\left(x^2 + \frac{2}{3}x + \frac{1}{9}\right) + 1 + 3\left(\frac{1}{9}\right) \qquad \begin{array}{l}\text{Add and subtract } -3 \text{ times} \\ \left(\frac{1}{2}\text{ the coefficient of } x\right)^2.\end{array}$$

$$= -3\left(x + \frac{1}{3}\right)^2 + \frac{4}{3}. \qquad \text{Factor and combine terms.}$$

The function is now in the form $y = a(x - h)^2 + k$. Since $h = -1/3$ and $k = 4/3$, the graph is the graph of the parabola $y = x^2$ translated 1/3 unit to the left and 4/3 units upward. This puts the vertex at $(-1/3, 4/3)$. Since $a = -3$ is negative, the graph will be flipped upside down. The 3 will cause the parabola to be stretched vertically by a factor of 3. These results are shown in Figure 37.

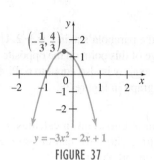

$$y = -3x^2 - 2x + 1$$

**FIGURE 37**

**Method 2**
**The Quadratic Formula**

Instead of completing the square to find the vertex of the graph of a quadratic function given in the form $y = ax^2 + bx + c$, we can develop a formula for the vertex. By the quadratic formula, if $ax^2 + bx + c = 0$, where $a \neq 0$, then

$$x = \frac{-b \pm \sqrt{b^2 - 4ac}}{2a}.$$

Notice that this is the same as

$$x = \frac{-b}{2a} \pm \frac{\sqrt{b^2 - 4ac}}{2a} = \frac{-b}{2a} \pm Q,$$

where $Q = \sqrt{b^2 - 4ac}/(2a)$. Since a parabola is symmetric with respect to its axis, the vertex is halfway between its two roots. Halfway between $x = -b/(2a) + Q$ and $x = -b/(2a) - Q$ is $x = -b/(2a)$. Once we have the $x$-coordinate of the vertex, we can easily find the $y$-coordinate by substituting the $x$-coordinate into the original equation. For the function in this example, use the quadratic formula to verify that the $x$-intercepts are at $x = -1$ and $x = 1/3$, and the vertex is at $(-1/3, 4/3)$. The $y$-intercept (where $x = 0$) is 1. The graph is in Figure 37.

**YOUR TURN 1** For the function $y = 2x^2 - 6x - 1$,
**(a)** complete the square, **(b)** find the $y$-intercept, **(c)** find the $x$-intercepts, **(d)** find the vertex, and **(e)** sketch the graph.

**TRY YOUR TURN 1**

**Graph of the Quadratic Function**

The graph of the quadratic function $f(x) = ax^2 + bx + c$ has its vertex at

$$\left(\frac{-b}{2a}, f\left(\frac{-b}{2a}\right)\right).$$

The graph opens upward if $a > 0$ and downward if $a < 0$.

Another situation that may arise is the absence of any $x$-intercepts, as in the next example.

**EXAMPLE 5** **Graphing a Quadratic Function**

Graph $y = x^2 + 4x + 6$.

**SOLUTION** This does not appear to factor, so we'll try the quadratic formula.

$$x = \frac{-b \pm \sqrt{b^2 - 4ac}}{2a} \qquad a = 1, b = 4, c = 6$$

$$= \frac{-4 \pm \sqrt{4^2 - 4(1)(6)}}{2(1)} = \frac{-4 \pm \sqrt{-8}}{2}$$

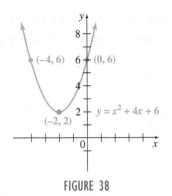

FIGURE 38

As soon as we see the negative value under the square root sign, we know the solutions are complex numbers. Therefore, there are no $x$-intercepts. Nevertheless, the vertex is still at

$$x = \frac{-b}{2a} = \frac{-4}{2} = -2.$$

Substituting this into the equation gives

$$y = (-2)^2 + 4(-2) + 6 = 2.$$

The $y$-intercept is at $(0, 6)$, which is 2 units to the right of the parabola's axis $x = -2$. Using the symmetry of the figure, we can also plot the mirror image of this point on the opposite side of the parabola's axis: At $x = -4$ (2 units to the left of the axis), $y$ is also equal to 6. Plotting the vertex, the $y$-intercept, and the point $(-4, 6)$ gives the graph in Figure 38.

The concept of maximizing or minimizing a function is important in calculus, as we will see in future chapters. The following example illustrates how we can determine the maximum or minimum value of a quadratic function without graphing.

**EXAMPLE 6** Carbon Dioxide Emissions

The amount of carbon dioxide emissions (in teragrams*) from human activities in the United States can be approximated with the following quadratic function

$$f(x) = -21.5t^2 + 242.8t + 5424.4,$$

where $t$ is the number of years since 2000. Using this model, estimate the maximum carbon dioxide emissions and determine in what year the maximum occurred. ***Source: U.S. Environmental Protection Agency.***

**SOLUTION** We see by the negative in the $t^2$-term that this function defines a parabola opening downward, so the maximum emissions is at the vertex. The $t$-coordinate of the vertex is

$$t = \frac{-b}{2a} = \frac{-242.8}{2(-21.5)} \approx 5.6.$$

The $y$-coordinate is then

$$y = f(5.6) = -21.5(5.6^2) + 242.8(5.6) + 5424.4 = 6109.84.$$

Therefore, the maximum carbon dioxide emissions was approximately 6110 teragrams, which occurred in about 2006.

Section 1.2 showed how the equation of a line that closely approximates a set of data points is found using linear regression. Some graphing calculators with statistics capability perform other kinds of regression. For example, **quadratic regression** gives the coefficients of a quadratic equation that models a given set of points.

**EXAMPLE 7** Vaccination Coverage

The following table gives the percent of children in North Dakota, 19 to 35 months of age, who had received the recommended combined $4:3:1:3:3:1^\dagger$ series of vaccinations. ***Source: U.S. Department of Health and Human Services.***

---

*Note that 1 teragram is equal to $10^{12}$ grams.

†The combined series $4:3:1:3:3:1$ of vaccinations consists of 4 or more doses of diphtheria and tetanus toxoids and pertussis vaccine (DTP), diphtheria and tetanus toxoids (DT), or diphtheria and tetanus toxoids and acellular pertussis vaccine (DTaP); 3 or more doses of any poliovirus vaccine; 1 or more doses of a measles-containing vaccine (MCV); 3 or more doses of *Haemophilus influenzae* type b vaccine (Hib); 3 or more doses of hepatitis B vaccine; and 1 or more doses of varicella vaccine.

| Year | Percent |
|------|---------|
| 2004 | 71 |
| 2005 | 79 |
| 2006 | 80 |
| 2007 | 77 |
| 2008 | 70 |
| 2009 | 56 |

**(a)** Plot the data by letting $x = 0$ correspond to the year 2000. Would a linear or quadratic function model these data best?

**SOLUTION**

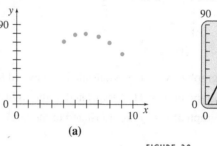
(a)

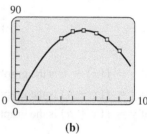
(b)

FIGURE 39

The scatterplot in Figure 39(a) suggests that a quadratic function with a negative value of $a$ (so the graph opens downward) would be a reasonable model for the data. The maximum value (vertex) appears to be $(6, 80)$.

**(b)** Find a quadratic function defined by $f(x) = a(x - h)^2 + k$ that models the data. Use $(6, 80)$ as the vertex, $(h, k)$. Then choose a second point, such as $(8, 70)$, to determine $a$.

**SOLUTION**   Substitute $h = 6, k = 80, x = 8$, and $y = f(x) = 70$ into the function and solve for $a$.

$$f(x) = a(x - h)^2 + k$$
$$70 = a(8 - 6)^2 + 80$$
$$70 = 4a + 80$$
$$4a = -10$$
$$a = -2.5$$

Substituting $a$, $h$, and $k$ into the model gives the quadratic function:

$$f(x) = -2.5(x - 6)^2 + 80.$$

Note that other choices for the second point will lead to slightly different equations.

**TECHNOLOGY NOTE**

Another way to find a quadratic function that fits a set of data is to use a graphing calculator. Begin by entering the data into the calculator. We will use the first two lists, $L_1$ and $L_2$. Press STAT and choose the first entry EDIT, which brings up the blank lists. (If the lists are not blank, be sure to clear them using the ClrList command.) Put the year in $L_1$, letting $x = 0$ correspond to 2000. Put the percent in $L_2$.

Quit the editor, press STAT again, and then select the CALC menu. QuadReg is item 5. The command QuadReg $L_1$, $L_2$, $Y_1$ finds the regression equation for the data in $L_1$ and $L_2$ and stores the function in $Y_1$. The quadratic function is

$$f(x) = -2.54x^2 + 29.96x - 8.07,$$

which models the data quite well, as shown in Figure 39(b).

Below, we provide guidelines for sketching graphs that involve translations and reflections.

## Translations and Reflections of Functions

Let $f$ be any function, and let $h$ and $k$ be positive constants (Figure 40).

The graph of $y = f(x) + k$ is the graph of $y = f(x)$ translated upward by $k$ units (Figure 41).

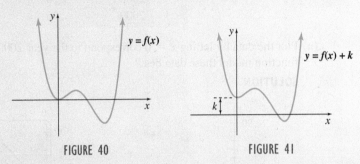

FIGURE 40                                     FIGURE 41

The graph of $y = f(x) - k$ is the graph of $y = f(x)$ translated downward by $k$ units (Figure 42).
The graph of $y = f(x - h)$ is the graph of $y = f(x)$ translated to the right by $h$ units (Figure 43).
The graph of $y = f(x + h)$ is the graph of $y = f(x)$ translated to the left by $h$ units (Figure 44).

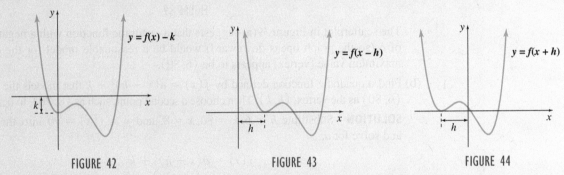

FIGURE 42                    FIGURE 43                    FIGURE 44

The graph of $y = -f(x)$ is the graph of $y = f(x)$ reflected vertically across the $x$-axis, that is, turned upside down (Figure 45).
The graph of $y = f(-x)$ is the graph of $y = f(x)$ reflected horizontally across the $y$-axis, that is, its mirror image (Figure 46).

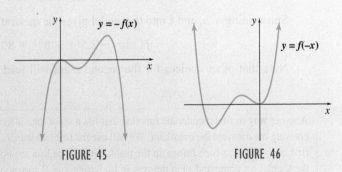

FIGURE 45                                     FIGURE 46

Notice in Figure 46 another type of reflection, known as a **horizontal reflection**. Multiplying $x$ or $f(x)$ by a constant $a$ to get $y = f(ax)$ or $y = a \cdot f(x)$ does not change the general appearance of the graph, except to compress or stretch it. When $a$ is negative, it also causes a reflection, as shown in the last two figures in the summary for $a = -1$. Also see Exercises 39–42 in this section.

**EXAMPLE 8** **Translations and Reflections of a Graph**

Graph $f(x) = -\sqrt{4-x} + 3$.

**SOLUTION** Begin with the simplest possible function, then add each variation in turn. Start with the graph of $f(x) = \sqrt{x}$. As Figure 47 reveals, this is just one-half of the graph of $f(x) = x^2$ lying on its side.

Now add another component of the original function, the negative in front of the $x$, giving $f(x) = \sqrt{-x}$. This is a horizontal reflection of the $f(x) = \sqrt{x}$ graph, as shown in Figure 48. Next, include the 4 under the square root sign. To get $4 - x$ into the form $f(x - h)$ or $f(x + h)$, we need to factor out the negative: $\sqrt{4-x} = \sqrt{-(x-4)}$. Now the 4 is subtracted, so this function is a translation to the right of the function $f(x) = \sqrt{-x}$ by 4 units, as Figure 49 indicates.

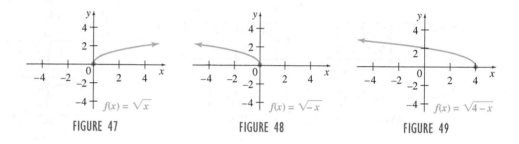

FIGURE 47          FIGURE 48          FIGURE 49

The effect of the negative in front of the radical is a vertical reflection, as in Figure 50, which shows the graph of $f(x) = -\sqrt{4-x}$. Finally, adding the constant 3 raises the entire graph by 3 units, giving the graph of $f(x) = -\sqrt{4-x} + 3$ in Figure 51(a).

**TRY YOUR TURN 2**

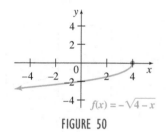

FIGURE 50

**YOUR TURN 2** Graph each of the following:
(a) $f(x) = \sqrt{x-2} + 1$
(b) $g(x) = -\sqrt{x} - 1$.

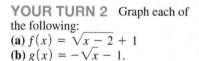

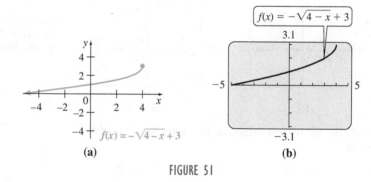

(a)          (b)

FIGURE 51

**TECHNOLOGY NOTE**

If you viewed a graphing calculator image such as Figure 51(b), you might think the function continues to go up and to the right. By realizing that $(4, 3)$ is the vertex of the sideways parabola, we see that this is the rightmost point on the graph. Another approach is to find the domain of $f$ by setting $4 - x \geq 0$, from which we conclude that $x \leq 4$. This demonstrates the importance of knowing the algebraic techniques in order to interpret a graphing calculator image correctly.

# 1.4   EXERCISES

1. How does the value of $a$ affect the graph of $y = ax^2$? Discuss the case for $a \geq 1$ and for $0 \leq a \leq 1$.

2. How does the value of $a$ affect the graph of $y = ax^2$ if $a \leq 0$?

**In Exercises 3–8, match the correct graph A–F to the function without using your calculator. Then, if you have a graphing calculator, use it to check your answers. Each graph in this group shows $x$ and $y$ in $[-10, 10]$.**

3. $y = x^2 - 3$

4. $y = (x - 3)^2$

5. $y = (x - 3)^2 + 2$

6. $y = (x + 3)^2 + 2$

7. $y = -(3 - x)^2 + 2$

8. $y = -(x + 3)^2 + 2$

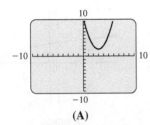
(A)

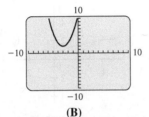
(B)

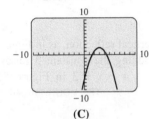
(C)

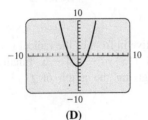
(D)

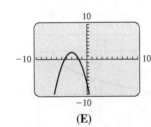
(E)

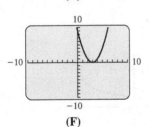
(F)

**Complete the square and determine the vertex for each of the following.**

9. $y = 3x^2 + 9x + 5$

10. $y = 4x^2 - 20x - 7$

11. $y = -2x^2 + 8x - 9$

12. $y = -5x^2 - 8x + 3$

**In Exercises 13–24, graph each parabola and give its vertex, axis, $x$-intercepts, and $y$-intercept.**

13. $y = x^2 + 5x + 6$

14. $y = x^2 + 4x - 5$

15. $y = -2x^2 - 12x - 16$

16. $y = -3x^2 - 6x + 4$

17. $f(x) = 2x^2 + 8x - 8$

18. $f(x) = -x^2 + 6x - 6$

19. $f(x) = 2x^2 - 4x + 5$

20. $f(x) = \frac{1}{2}x^2 + 6x + 24$

21. $f(x) = -2x^2 + 16x - 21$

22. $f(x) = \frac{3}{2}x^2 - x - 4$

23. $f(x) = \frac{1}{3}x^2 - \frac{8}{3}x + \frac{1}{3}$

24. $f(x) = -\frac{1}{2}x^2 - x - \frac{7}{2}$

**In Exercises 25–30, follow the directions for Exercises 3–8.**

25. $y = \sqrt{x + 2} - 4$

26. $y = \sqrt{x - 2} - 4$

27. $y = \sqrt{-x + 2} - 4$

28. $y = \sqrt{-x - 2} - 4$

29. $y = -\sqrt{x + 2} - 4$

30. $y = -\sqrt{x - 2} - 4$

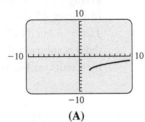
(A)

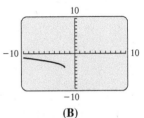
(B)

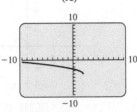
(C)

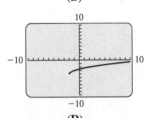
(D)

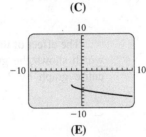
(E)

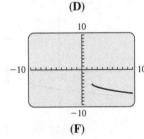
(F)

**Given the following graph, sketch by hand the graph of the function described, giving the new coordinates for the three points labeled on the original graph.**

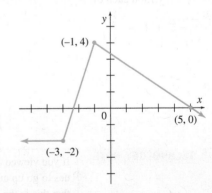

31. $y = -f(x)$

32. $y = f(x - 2) + 2$

33. $y = f(-x)$

34. $y = f(2 - x) + 2$

**Use the ideas in this section to graph each function without a calculator.**

35. $f(x) = \sqrt{x - 2} + 2$

36. $f(x) = \sqrt{x + 2} - 3$

37. $f(x) = -\sqrt{2 - x} - 2$

38. $f(x) = -\sqrt{2 - x} + 2$

Using the graph of $f(x)$ in Figure 40, show the graph of $f(ax)$ where $a$ satisfies the given condition.

**39.** $0 < a < 1$ **40.** $1 < a$

**41.** $-1 < a < 0$ **42.** $a < -1$

Using the graph of $f(x)$ in Figure 40, show the graph of $af(x)$ where $a$ satisfies the given condition.

**43.** $0 < a < 1$ **44.** $1 < a$

**45.** $-1 < a < 0$ **46.** $a < -1$

**47.** If $r$ is an $x$-intercept of the graph of $y = f(x)$, what is an $x$-intercept of the graph of each of the following?

  **a.** $y = -f(x)$ **b.** $y = f(-x)$

  **c.** $y = -f(-x)$

**48.** If $b$ is the $y$-intercept of the graph of $y = f(x)$, what is the $y$-intercept of the graph of each of the following?

  **a.** $y = -f(x)$ **b.** $y = f(-x)$

  **c.** $y = -f(-x)$

## LIFE SCIENCE APPLICATIONS

**49. Length of Life** According to recent data from the Teachers Insurance and Annuity Association (TIAA), the survival function for life after 65 is approximately given by

$$S(x) = 1 - 0.058x - 0.076x^2,$$

where $x$ is measured in decades. This function gives the probability that an individual who reaches the age of 65 will live at least $x$ decades ($10x$ years) longer. *Source: Ralph DeMarr.*

  **a.** Find the median length of life for people who reach 65, that is, the age for which the survival rate is 0.50.

  **b.** Find the age beyond which virtually nobody lives. (There are, of course, exceptions.)

**50. Tooth Length** The length (in mm) of the mesiodistal crown of the first molar for human fetuses can be approximated by

$$L(t) = -0.01t^2 + 0.788t - 7.048,$$

where $t$ is the number of weeks since conception. *Source: American Journal of Physical Anthropology.*

  **a.** What does this formula predict for the length at 14 weeks? 24 weeks?

  **b.** What does this formula predict for the maximum length, and when does that occur? Explain why the formula does not make sense past that time.

**51. APPLY IT Splenic Artery Resistance** Blood flow to the fetal spleen is of research interest because several diseases are associated with increased resistance in the splenic artery (the artery that goes to the spleen). Researchers have found that the index of splenic artery resistance in the fetus can be described by the function

$$y = 0.057x - 0.001x^2,$$

where $x$ is the number of weeks of gestation. *Source: American Journal of Obstetrics and Gynecology.*

  **a.** At how many weeks is the splenic artery resistance a maximum?

  **b.** What is the maximum splenic artery resistance?

  **c.** At how many weeks is the splenic artery resistance equal to 0, according to this formula? Is your answer reasonable for this function? Explain.

**52. Cancer** From 1975 to 2009, the age-adjusted incidence rate of invasive lung and bronchial cancer among women can be closely approximated by

$$f(t) = -0.038083t^2 + 2.0899t + 24.262,$$

where $t$ is the number of years since 1975. *Source: National Cancer Institute.* Based on this model, in what year did the incidence rate reach a maximum? On what years was the rate increasing? Decreasing?

**53. Gender Ratio** The number of males per 100 females, age 65 or over, in the United States for some recent years is shown in the following table. *Source: The U.S. Census Bureau.*

| Year | Males per 100 Females |
|------|-----------------------|
| 1960 | 82.8 |
| 1970 | 72.1 |
| 1980 | 67.6 |
| 1990 | 67.2 |
| 2000 | 70.0 |
| 2010 | 77.0 |

  **a.** Plot the data, letting $x$ be the years since 1900.

  **b.** Would a linear or quadratic function best model this data? Explain.

  **c.** If your graphing calculator has a quadratic regression feature, find the quadratic function that best fits the data. Graph this function on the same calculator window as the data. (See Example 7(c).)

  **d.** Choose the lowest point in the table above as the vertex and (110, 77.0) as a second point to find a quadratic function defined by $f(x) = a(x - h)^2 + k$ that models the data.

  **e.** Graph the function from part d on the same calculator window as the data and function from part c. Do the graphs of the two functions differ by much?

  **f.** Predict the number of males per 100 females in 2004 using the two functions from parts c and d, and compare with the actual figure of 71.7.

**54. Angioplasty** In an effort to improve angioplasty—in which a wire is threaded through a coronary artery to guide a balloon which then opens a blocked artery—researchers have modeled the guide wire using a quadratic equation

$$y = px^2 + qx + r.$$

Part of the process involves taking pairs of points $(x_0, y_0)$ and $(x_1, y_1)$ on an X-ray image and finding the values of $p, q,$ and $r$ as described below. *Source: IEEE Transactions on Biomedical Engineering.*

**a.** Substitute the two points into the quadratic equation, and call the resulting equations $E_0$ and $E_1$. Subtract $E_1 - E_0$ to eliminate $r$, and solve the resulting equation for $p$ to get

$$p = \frac{y_1 - y_0 - q(x_1 - x_0)}{x_1^2 - x_0^2}.$$

**b.** Substitute the result from part a into equation $E_0$ and solve for $r$ to obtain

$$r = y_0 - x_0^2\left(\frac{y_1 - y_0}{x_1^2 - x_0^2}\right) + qx_0^2\left(\frac{x_1 - x_0}{x_1^2 - x_0^2}\right) - qx_0.$$

**c.** Substitute the values for $p$ and $r$ from parts a and c into the original quadratic equation to get

$$y = \left(\frac{y_1 - y_0}{x_1^2 - x_0^2}\right)x^2$$
$$+ q\left[x - x_0 - \left(\frac{x_1 - x_0}{x_1^2 - x_0^2}\right)x^2 + \left(\frac{x_1 - x_0}{x_1^2 - x_0^2}\right)x_0^2\right]$$
$$+ y_0 - \left(\frac{y_1 - y_0}{x_1^2 - x_0^2}\right)x_0^2.$$

**d.** Substitute

$$K = \frac{y_1 - y_0}{x_1^2 - x_0^2},$$

$$L = Kx_0^2,$$

$$\text{and} \quad M = \frac{x_1 - x_0}{x_1^2 - x_0^2} = \frac{1}{x_1 + x_0}$$

into the result from part c to obtain

$$y = Kx^2 + q(x - x_0 - Mx^2 + Mx_0^2) + y_0 - L.$$

**e.** Finally, substitute

$$A = L - y_0,$$
$$P = Mx_0^2,$$
$$\text{and} \quad R = x_0 - P,$$

into the result from part d, and solve for $q$ to obtain

$$q = \frac{y - Kx^2 + A}{x - R - Mx^2},$$

the final equation used by the researchers.

**f.** Suppose $(x_0, y_0) = (1, 2)$ and $(x_1, y_1) = (3, 5)$. Find the equation for $q$ in terms of $x$ and $y$.

## OTHER APPLICATIONS

**55. Age of Marriage** The following table gives the median age at their first marriage of women in the United States for some selected years. *Source: U.S. Census Bureau.*

| Year | Age |
|------|------|
| 1940 | 21.5 |
| 1950 | 20.3 |
| 1960 | 20.3 |
| 1970 | 20.8 |
| 1980 | 22.0 |
| 1990 | 23.9 |
| 2000 | 25.1 |
| 2010 | 26.5 |

**a.** Plot the data using $x = 40$ for 1940, and so on.

**b.** Would a linear or quadratic function best model these data? Explain.

**c.** If your graphing calculator has a regression feature, find the quadratic function that best fits the data. Graph this function on the same calculator window as the data. (See Example 7(c).)

**d.** Find a quadratic function defined by $f(x) = a(x - h)^2 + k$ that models the data using $(60, 20.3)$ as the vertex and then choosing $(110, 26.5)$ as a second point to determine the value of $a$.

**e.** Graph the function from part d on the same calculator window as the data and function from part c. Do the graphs of the two functions differ by much?

**56. Accident Rate** According to data from the National Highway Traffic Safety Administration, the accident rate as a function of the age of the driver in years $x$ can be approximated by the function

$$f(x) = 60.0 - 2.28x + 0.0232x^2$$

for $16 \le x \le 85$. Find the age at which the accident rate is a minimum and the minimum rate. *Source: Ralph DeMarr.*

**57. Maximizing the Height of an Object** If an object is thrown upward with an initial velocity of 32 ft/second, then its height after $t$ seconds is given by

$$h = 32t - 16t^2.$$

**a.** Find the maximum height attained by the object.

**b.** Find the number of seconds it takes the object to hit the ground.

**58. Stopping Distance** According to data from the National Traffic Safety Institute, the stopping distance $y$ in feet of a car traveling $x$ mph can be described by the equation $y = 0.056057x^2 + 1.06657x$. *Source: National Traffic Safety Institute.*

**a.** Find the stopping distance for a car traveling 25 mph.

**b.** How fast can you drive if you need to be certain of stopping within 150 ft?

**59. Maximizing Area** Glenview Community College wants to construct a rectangular parking lot on land bordered on one side by a highway. It has 380 ft of fencing to use along the other three sides. What should be the dimensions of the lot if the enclosed area is to be a maximum? (*Hint:* Let $x$ represent the width of the lot, and let $380 - 2x$ represent the length.)

**60. Maximizing Area** What would be the maximum area that could be enclosed by the college's 380 ft of fencing if it decided to close the entrance by enclosing all four sides of the lot? (See Exercise 59.)

**In Exercises 61 and 62, draw a sketch of the arch or culvert on coordinate axes, with the horizontal and vertical axes through the vertex of the parabola. Use the given information to label points on the parabola. Then give the equation of the parabola and answer the question.**

**61. Parabolic Arch** An arch is shaped like a parabola. It is 30 m wide at the base and 15 m high. How wide is the arch 10 m from the ground?

**62. Parabolic Culvert** A culvert is shaped like a parabola, 18 ft across the top and 12 ft deep. How wide is the culvert 8 ft from the top?

**■ YOUR TURN ANSWERS**

**1. (a)** $y = 2(x - 3/2)^2 - 11/2$  **(b)** $-1$  **(c)** $\left(3 \pm \sqrt{11}\right)/2$  **(d)** $(3/2, -11/2)$  **(e)**

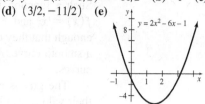

**2. (a)**      **(b)**

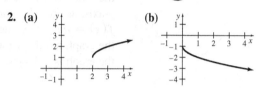

# 1.5  Polynomial and Rational Functions

**APPLY IT** How can we model the incidence of tuberculosis in the United States? How much does it cost to remove a pollutant from the environment?

*In Examples 5 and 8 of this section, we will explore these questions using* polynomial *and* rational *functions.*

## Polynomial Functions

Earlier, we discussed linear and quadratic functions and their graphs. Both of these functions are special types of *polynomial functions*.

> **Polynomial Function**
>
> A **polynomial function** of **degree** $n$, where $n$ is a nonnegative integer, is defined by
>
> $$f(x) = a_n x^n + a_{n-1} x^{n-1} + \cdots + a_1 x + a_0,$$
>
> where $a_n, a_{n-1}, \ldots, a_1$, and $a_0$ are real numbers, called **coefficients**, with $a_n \neq 0$. The number $a_n$ is called the **leading coefficient**.

For $n = 1$, a polynomial function takes the form

$$f(x) = a_1 x + a_0,$$

a linear function. A linear function, therefore, is a polynomial function of degree 1. (Note, however, that a linear function of the form $f(x) = a_0$ for a real number $a_0$ is a polynomial function of degree 0, the constant function.) A polynomial function of degree 2 is a quadratic function.

Accurate graphs of polynomial functions of degree 3 or higher require methods of calculus to be discussed later. Meanwhile, a graphing calculator is useful for obtaining such graphs, but care must be taken in choosing a viewing window that captures the significant behavior of the function.

The simplest polynomial functions of higher degree are those of the form $f(x) = x^n$. Such a function is known as a **power function**. Figure 52 below shows the graphs of $f(x) = x^3$ and $f(x) = x^5$, as well as tables of their values. These functions are simple enough that they can be drawn by hand by plotting a few points and connecting them with a smooth curve. An important property of all polynomials is that their graphs are smooth curves.

The graphs of $f(x) = x^4$ and $f(x) = x^6$, shown in Figure 53 along with tables of their values, can be sketched in a similar manner. These graphs have symmetry about the $y$-axis, as does the graph of $f(x) = ax^2$ for a nonzero real number $a$. As with the graph of $f(x) = ax^2$, the value of $a$ in $f(x) = ax^n$ affects the direction of the graph. When $a > 0$, the graph has the same general appearance as the graph of $f(x) = x^n$. However, if $a < 0$, the graph is reflected vertically.

| $f(x) = x^3$ | | $f(x) = x^5$ | |
|---|---|---|---|
| $x$ | $f(x)$ | $x$ | $f(x)$ |
| $-2$ | $-8$ | $-1.5$ | $-7.6$ |
| $-1$ | $-1$ | $-1$ | $-1$ |
| $0$ | $0$ | $0$ | $0$ |
| $1$ | $1$ | $1$ | $1$ |
| $2$ | $8$ | $1.5$ | $7.6$ |

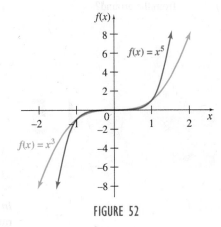

FIGURE 52

| $f(x) = x^4$ | | $f(x) = x^6$ | |
|---|---|---|---|
| $x$ | $f(x)$ | $x$ | $f(x)$ |
| $-2$ | $16$ | $-1.5$ | $11.4$ |
| $-1$ | $1$ | $-1$ | $1$ |
| $0$ | $0$ | $0$ | $0$ |
| $1$ | $1$ | $1$ | $1$ |
| $2$ | $16$ | $1.5$ | $11.4$ |

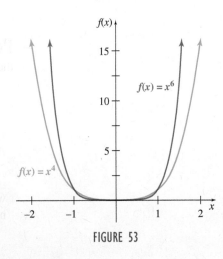

FIGURE 53

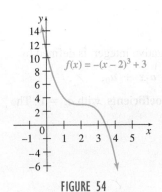

FIGURE 54

**EXAMPLE 1** **Translations and Reflections**

Graph $f(x) = -(x - 2)^3 + 3$.

**SOLUTION** Using the principles of translation and reflection from the previous section, we recognize that this is similar to the graph of $y = x^3$, but reflected vertically (because of the negative in front of $(x - 2)^3$), and with its center moved 2 units to the right and 3 units up. The result is shown in Figure 54.       **TRY YOUR TURN 1**

**YOUR TURN 1** Graph $f(x) = 64 - x^6$.

A polynomial of degree 3, such as that in the previous example and in the next, is known as a **cubic polynomial**. A polynomial of degree 4, such as that in Example 3, is known as a **quartic polynomial**.

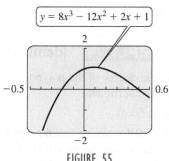

TECHNOLOGY

## EXAMPLE 2  Graphing a Polynomial

Graph $f(x) = 8x^3 - 12x^2 + 2x + 1$.

**SOLUTION**  Figure 55 shows the function graphed on the $x$- and $y$-intervals $[-0.5, 0.6]$ and $[-2, 2]$. In this view, it appears similar to a parabola opening downward. Zooming out to $[-1, 2]$ by $[-8, 8]$, we see in Figure 56 that the graph goes upward as $x$ gets large. There are also two **turning points** near $x = 0$ and $x = 1$. (In a later chapter, we will introduce another term for such turning points: *relative extrema*.) By zooming in with the graphing calculator, we can find these turning points to be at approximately $(0.09175, 1.08866)$ and $(0.90825, -1.08866)$.

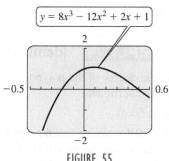

FIGURE 55

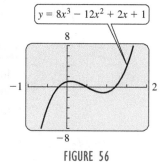

FIGURE 56

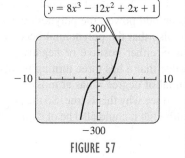

FIGURE 57

Zooming out still further, we see the function on $[-10, 10]$ by $[-300, 300]$ in Figure 57. From this viewpoint, we don't see the turning points at all, and the graph seems similar in shape to that of $y = x^3$. This is an important point: When $x$ is large in magnitude, either positive or negative, $8x^3 - 12x^2 + 2x + 1$ behaves a lot like $8x^3$, because the other terms are small in comparison with the cubic term. So this viewpoint tells us something useful about the function, but it is less useful than the previous graph for determining the turning points.

After the previous example, you may wonder how to be sure you have the viewing window that exhibits all the important properties of a function. We will find an answer to this question in later chapters using the techniques of calculus. Meanwhile, let us consider one more example to get a better idea how the graphs of polynomials look.

TECHNOLOGY

## EXAMPLE 3  Graphing a Polynomial

Graph $f(x) = -3x^4 + 14x^3 - 54x + 3$.

**SOLUTION**  Figure 58 shows a graphing calculator view on $[-3, 5]$ by $[-50, 50]$. If you have a graphing calculator, we recommend that you experiment with various viewpoints and verify for yourself that this viewpoint captures the important behavior of the function. Notice that it has three turning points. Notice also that as $|x|$ gets large, the graph turns downward. This is because as $|x|$ becomes large, the $x^4$-term dominates the other terms, which are small in comparison, and the $x^4$-term has a negative coefficient.

FIGURE 58

As suggested by the graphs above, the domain of a polynomial function is the set of all real numbers. The range of a polynomial function of odd degree is also the set of all real numbers. Some typical graphs of polynomial functions of odd and even degree are shown in

Figure 59. The first two graphs suggest that for every polynomial function $f$ of odd degree, there is at least one real value of $x$ for which $f(x) = 0$. Such a value of $x$ is called a **real zero** of $f$; these values are also the $x$-intercepts of the graph.

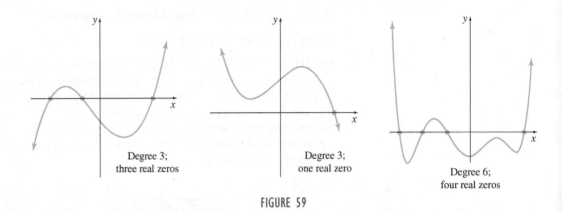

Degree 3;
three real zeros

Degree 3;
one real zero

Degree 6;
four real zeros

FIGURE 59

### EXAMPLE 4  Identifying the Degree of a Polynomial

Identify the degree of the polynomial in each of the figures, and give the sign (+ or −) for the leading coefficient.

**(a)** Figure 60(a)

**SOLUTION**  Notice that the polynomial has a range $[k, \infty)$. This must be a polynomial of even degree, because if the highest power of $x$ is an odd power, the polynomial can equal any real number, positive or negative. Notice also that the polynomial becomes a large positive number as $x$ gets large in magnitude, either positive or negative, so the leading coefficient must be positive. Finally, notice that it has three turning points. Observe from the previous examples that a polynomial of degree $n$ has at most $n - 1$ turning points. In a later chapter, we will use calculus to see why this is true. So the polynomial graphed in Figure 60(a) might be degree 4, although it could also be of degree 6, 8, etc. We can't be sure from the graph alone.

**(b)** Figure 60(b)

**SOLUTION**  Because the range is $(-\infty, \infty)$, this must be a polynomial of odd degree. Notice also that the polynomial becomes a large negative number as $x$ becomes a large positive number, so the leading coefficient must be negative. Finally, notice that it has four turning points, so it might be degree 5, although it could also be of degree 7, 9, etc.

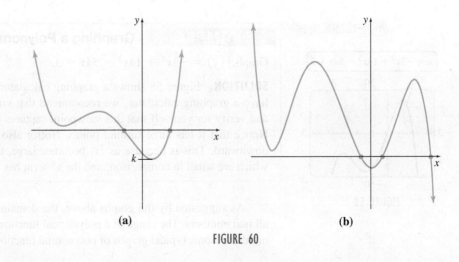

(a)

(b)

FIGURE 60

## Properties of Polynomial Functions

**1.** A polynomial function of degree $n$ can have at most $n - 1$ turning points. Conversely, if the graph of a polynomial function has $n$ turning points, it must have degree at least $n + 1$.

**2.** In the graph of a polynomial function of even degree, both ends go up or both ends go down. For a polynomial function of odd degree, one end goes up and one end goes down.

**3.** If the graph goes up as $x$ becomes a large positive number, the leading coefficient must be positive. If the graph goes down as $x$ becomes a large positive number, the leading coefficient is negative.

 **TECHNOLOGY**

**APPLY IT**

---

### EXAMPLE 5    Tuberculosis in the United States

The table at left gives the number of tuberculosis (TB) cases in the United States for some recent years. ***Source: The Centers for Disease Control and Prevention.***

**(a)** Draw a scatterplot, letting $x = 0$ represent 1985.

**SOLUTION**    The scatterplot is shown in Figure 61(a).

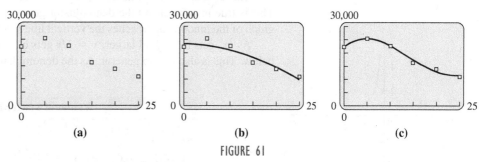

FIGURE 61

**(b)** Use the regression feature of a graphing calculator to get a quadratic function that approximates the data. Graph the function on the same window as the scatterplot.

**SOLUTION**    The quadratic equation is $y = -20.75x^2 - 32.44x + 23{,}857.3$ and is shown in Figure 61(b).

**(c)** Use the cubic regression to get a cubic function that approximates the data. Graph the cubic function on the same window as the scatterplot. Which of the two functions appears to be a better fit for the data?

**SOLUTION**    The cubic equation is $y = 3.848x^3 - 165.1x^2 + 1285.6x + 22{,}414.2$ and is shown in Figure 61(c). The cubic function appears to be a better fit for the data.

| Year | Number of Cases |
|------|-----------------|
| 1985 | 22,201 |
| 1990 | 25,701 |
| 1995 | 22,727 |
| 2000 | 16,309 |
| 2005 | 14,068 |
| 2010 | 11,171 |

## Rational Functions

Many situations require mathematical models that are quotients. A common model for such situations is a *rational function*.

### Rational Function

A **rational function** is defined by

$$f(x) = \frac{p(x)}{q(x)},$$

where $p(x)$ and $q(x)$ are polynomial functions and $q(x) \neq 0$.

Since any values of $x$ such that $q(x) = 0$ are excluded from the domain, a rational function often has a graph with one or more breaks.

EXAMPLE 6    **Graphing a Rational Function**

Graph $y = \dfrac{1}{x}$.

**SOLUTION**    This function is undefined for $x = 0$, since 0 is not allowed as the denominator of a fraction. For this reason, the graph of this function will not intersect the vertical line $x = 0$, which is the y-axis. Since $x$ can equal any value except 0, the values of $x$ can approach 0 as closely as desired from either side of 0.

| | Values of $1/x$ for Small $|x|$ | | | | | | | |
|---|---|---|---|---|---|---|---|---|
| | | | | *x* approaches 0. ↓ | | | | |
| $x$ | $-0.5$ | $-0.2$ | $-0.1$ | $-0.01$ | $0.01$ | $0.1$ | $0.2$ | $0.5$ |
| $y = \dfrac{1}{x}$ | $-2$ | $-5$ | $-10$ | $-100$ | $100$ | $10$ | $5$ | $2$ |
| | | | | ↑ $|y|$ gets larger and larger. | | | | |

The table above suggests that as $x$ gets closer and closer to 0, $|y|$ gets larger and larger. This is true in general: As the denominator gets smaller, the fraction gets larger. Thus, the graph of the function approaches the vertical line $x = 0$ (the y-axis) without ever touching it.

As $|x|$ gets larger and larger, $y = 1/x$ gets closer and closer to 0, as shown in the table below. This is also true in general: As the denominator gets larger, the fraction gets smaller.

| | Values of $1/x$ for Large $|x|$ | | | | | | | |
|---|---|---|---|---|---|---|---|---|
| $x$ | $-100$ | $-10$ | $-4$ | $-1$ | $1$ | $4$ | $10$ | $100$ |
| $y = \dfrac{1}{x}$ | $-0.01$ | $-0.1$ | $-0.25$ | $-1$ | $1$ | $0.25$ | $0.1$ | $0.01$ |

The graph of the function approaches the horizontal line $y = 0$ (the x-axis). The information from both tables supports the graph in Figure 62.

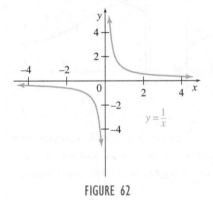

FIGURE 62

In Example 6, the vertical line $x = 0$ and the horizontal line $y = 0$ are *asymptotes*, defined as follows.

**Asymptotes**

If a function gets larger and larger in magnitude without bound as $x$ approaches the number $k$, then the line $x = k$ is a **vertical asymptote**.

If the values of $y$ approach a number $k$ as $|x|$ gets larger and larger, the line $y = k$ is a **horizontal asymptote**.

There is an easy way to find any vertical asymptotes of a rational function. First, find the roots of the denominator. If a number $k$ makes the denominator 0 but does not make the numerator 0, then the line $x = k$ is a vertical asymptote. If, however, a number $k$ makes both the denominator and the numerator 0, then further investigation will be necessary, as we will see in the next example. Later we will show another way to find asymptotes using the concept of a *limit*.

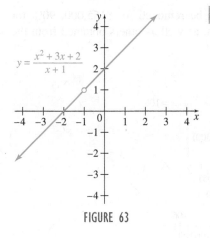

$y = \dfrac{x^2 + 3x + 2}{x + 1}$

**FIGURE 63**

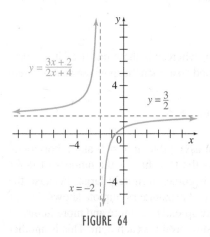

$y = \dfrac{3x + 2}{2x + 4}$

$y = \dfrac{3}{2}$

$x = -2$

**FIGURE 64**

**YOUR TURN 2**

Graph $y = \dfrac{4x - 6}{x - 3}$.

**EXAMPLE 7**    **Graphing a Rational Function**

Graph the following rational functions:

**(a)** $y = \dfrac{x^2 + 3x + 2}{x + 1}$.

**SOLUTION**    The value $x = -1$ makes the denominator 0, and so $-1$ is not in the domain of this function. Note that the value $x = -1$ also makes the numerator 0. In fact, if we factor the numerator and simplify the function, we get

$$y = \frac{x^2 + 3x + 2}{x + 1} = \frac{(x + 2)(x + 1)}{(x + 1)} = x + 2 \quad \text{for} \quad x \neq -1.$$

The graph of this function, therefore, is the graph of $y = x + 2$ with a hole at $x = -1$, as shown in Figure 63.

**(b)** $y = \dfrac{3x + 2}{2x + 4}$.

**SOLUTION**    The value $x = -2$ makes the denominator 0, but not the numerator, so the line $x = -2$ is a vertical asymptote. To find a horizontal asymptote, let $x$ get larger and larger, so that $3x + 2 \approx 3x$ because the 2 is very small compared with $3x$. Similarly, for $x$ very large, $2x + 4 \approx 2x$. Therefore, $y = (3x + 2)/(2x + 4) \approx (3x)/(2x) = 3/2$. This means that the line $y = 3/2$ is a horizontal asymptote. (A more precise way of approaching this idea will be seen in a later chapter when limits at infinity are discussed.)

The intercepts should also be noted. When $x = 0$, the y-intercept is $y = 2/4 = 1/2$. To make a fraction 0, the numerator must be 0; so to make $y = 0$, it is necessary that $3x + 2 = 0$. Solve this for $x$ to get $x = -2/3$ (the x-intercept). We can also use these values to determine where the function is positive and where it is negative. Using the techniques described in Chapter R, verify that the function is negative on $(-2, -2/3)$ and positive on $(-\infty, -2) \cup (-2/3, \infty)$. With this information, the two asymptotes to guide us, and the fact that there are only two intercepts, we suspect the graph is as shown in Figure 64. A graphing calculator can support this.    **TRY YOUR TURN 2**

Rational functions occur often in practical applications. In many situations involving environmental pollution, much of the pollutant can be removed from the air or water at a fairly reasonable cost, but the last small part of the pollutant can be very expensive to remove. Cost as a function of the percentage of pollutant removed from the environment can be calculated for various percentages of removal, with a curve fitted through the resulting data points. This curve then leads to a mathematical model of the situation. Rational functions are often a good choice for these **cost-benefit models** because they rise rapidly as they approach a vertical asymptote.

**EXAMPLE 8**    **Cost-Benefit Analysis**

Suppose a cost-benefit model is given by

$$y = \frac{18x}{106 - x},$$

where $y$ is the cost (in thousands of dollars) of removing $x$ percent of a certain pollutant. The domain of $x$ is the set of all numbers from 0 to 100 inclusive; any amount of pollutant from 0% to 100% can be removed. Find the cost to remove the following amounts of the pollutant: 100%, 95%, 90%, and 80%. Graph the function.

**APPLY IT**

**SOLUTION**    Removal of 100% of the pollutant would cost

$$y = \frac{18(100)}{106 - 100} = 300,$$

or $300,000. Check that 95% of the pollutant can be removed for $155,000, 90% for $101,000, and 80% for $55,000. Using these points, as well as others obtained from the function, gives the graph shown in Figure 65.

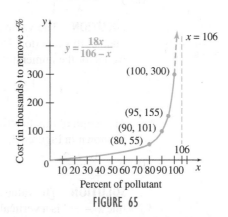

FIGURE 65

If a cost function has the form $C(x) = mx + b$, where $x$ is the number of items produced, $m$ is the marginal cost per item, and $b$ is the fixed cost, then the **average cost** per item is given by

$$\overline{C}(x) = \frac{C(x)}{x} = \frac{mx + b}{x}.$$

Notice that this is a rational function with a vertical asymptote at $x = 0$ and a horizontal asymptote at $y = m$. The vertical asymptote reflects the fact that, as the number of items produced approaches 0, the average cost per item becomes infinitely large, because the fixed costs are spread over fewer and fewer items. The horizontal asymptote shows that, as the number of items becomes large, the fixed costs are spread over more and more items, so most of the average cost per item is the marginal cost to produce each item. This is another example of how asymptotes give important information in real applications.

# 1.5    EXERCISES

1. Explain how translations and reflections can be used to graph $y = -(x - 1)^4 + 2$.

2. Describe an asymptote, and explain when a rational function will have (a) a vertical asymptote and (b) a horizontal asymptote.

**Use the principles of the previous section with the graphs of this section to sketch a graph of the given function.**

3. $f(x) = (x - 2)^3 + 3$
4. $f(x) = (x + 1)^3 - 2$
5. $f(x) = -(x + 3)^4 + 1$
6. $f(x) = -(x - 1)^4 + 2$

**In Exercises 7–15, match the correct graph A–I to the function without using your calculator. Then, after you have answered all of them, if you have a graphing calculator, use your calculator to check your answers. Each graph is plotted on $[-6, 6]$ by $[-50, 50]$.**

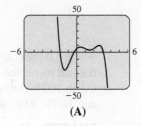

(A)

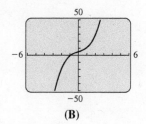

(B)

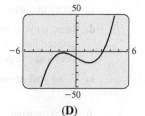

(C)          (D)

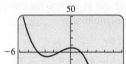
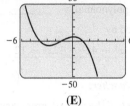

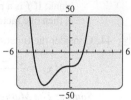
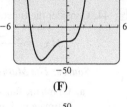

(E)          (F)

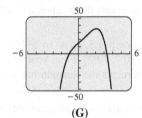
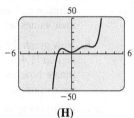

(G)          (H)

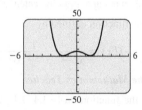

(I)

**7.** $y = x^3 - 7x - 9$        **8.** $y = -x^3 + 4x^2 + 3x - 8$

**9.** $y = -x^3 - 4x^2 + x + 6$    **10.** $y = 2x^3 + 4x + 5$

**11.** $y = x^4 - 5x^2 + 7$      **12.** $y = x^4 + 4x^3 - 20$

**13.** $y = -x^4 + 2x^3 + 10x + 15$

**14.** $y = 0.7x^5 - 2.5x^4 - x^3 + 8x^2 + x + 2$

**15.** $y = -x^5 + 4x^4 + x^3 - 16x^2 + 12x + 5$

In Exercises 16–20, match the correct graph A–E to the function without using your calculator. Then, after you have answered all of them, if you have a graphing calculator, use your calculator to check your answers. Each graph in this group is plotted on $[-6, 6]$ by $[-6, 6]$. *Hint:* Consider the asymptotes. (If you try graphing these graphs in Connected mode rather than Dot mode, you will see some lines that are not part of the graph, but the result of the calculator connecting disconnected parts of the graph.)

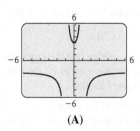
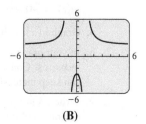

(A)          (B)

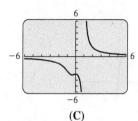
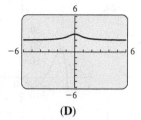

(C)          (D)

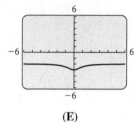

(E)

**16.** $y = \dfrac{2x^2 + 3}{x^2 - 1}$        **17.** $y = \dfrac{2x^2 + 3}{x^2 + 1}$

**18.** $y = \dfrac{-2x^2 - 3}{x^2 - 1}$       **19.** $y = \dfrac{-2x^2 - 3}{x^2 + 1}$

**20.** $y = \dfrac{2x^2 + 3}{x^3 - 1}$

Each of the following is the graph of a polynomial function. Give the possible values for the degree of the polynomial, and give the sign ($+$ or $-$) for the leading coefficient.

**21.**

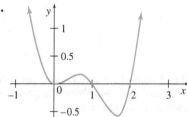

**22.**

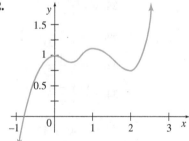

**23.**

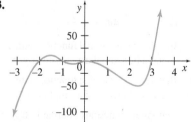

**24.**

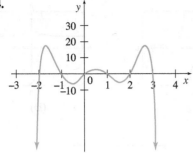

**25.**

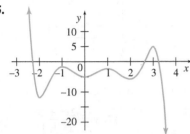

**26.**

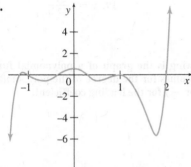

**Find any horizontal and vertical asymptotes and any holes that may exist for each rational function. Draw the graph of each function, including any *x*- and *y*-intercepts.**

**27.** $y = \dfrac{-4}{x + 2}$

**28.** $y = \dfrac{-1}{x + 3}$

**29.** $y = \dfrac{2}{3 + 2x}$

**30.** $y = \dfrac{8}{5 - 3x}$

**31.** $y = \dfrac{2x}{x - 3}$

**32.** $y = \dfrac{4x}{3 - 2x}$

**33.** $y = \dfrac{x + 1}{x - 4}$

**34.** $y = \dfrac{x - 4}{x + 1}$

**35.** $y = \dfrac{3 - 2x}{4x + 20}$

**36.** $y = \dfrac{6 - 3x}{4x + 12}$

**37.** $y = \dfrac{-x - 4}{3x + 6}$

**38.** $y = \dfrac{-2x + 5}{x + 3}$

**39.** $y = \dfrac{x^2 + 7x + 12}{x + 4}$

**40.** $y = \dfrac{9 - 6x + x^2}{3 - x}$

**41.** Write an equation that defines a rational function with a vertical asymptote at $x = 1$ and a horizontal asymptote at $y = 2$.

**42.** Write an equation that defines a rational function with a vertical asymptote at $x = -2$ and a horizontal asymptote at $y = 0$.

**43.** Consider the polynomial functions defined by $f(x) = (x - 1)(x - 2)(x + 3)$, $g(x) = x^3 + 2x^2 - x - 2$, and $h(x) = 3x^3 + 6x^2 - 3x - 6$.

  **a.** What is the value of $f(1)$?

  **b.** For what values, other than 1, is $f(x) = 0$?

**c.** Verify that $g(-1) = g(1) = g(-2) = 0$.

**d.** Based on your answer from part c, what do you think is the factored form of $g(x)$? Verify your answer by multiplying it out and comparing with $g(x)$.

**e.** Using your answer from part d, what is the factored form of $h(x)$?

**f.** Based on what you have learned in this exercise, fill in the blank: If $f$ is a polynomial and $f(a) = 0$ for some number $a$, then one factor of the polynomial is _____.

**44.** Consider the function defined by

$$f(x) = \frac{x^7 - 4x^5 - 3x^4 + 4x^3 + 12x^2 - 12}{x^7}.$$

***Source: The Mathematics Teacher.***

**a.** Graph the function on $[-6, 6]$ by $[-6, 6]$. From your graph, estimate how many *x*-intercepts the function has and what their values are.

**b.** Now graph the function on $[-1.5, -1.4]$ by $[-10^{-4}, 10^{-4}]$ and on $[1.4, 1.5]$ by $[-10^{-5}, 10^{-5}]$. From your graphs, estimate how many *x*-intercepts the function has and what their values are.

**c.** From your results in parts a and b, what advice would you give a friend on using a graphing calculator to find *x*-intercepts?

**45.** Consider the function defined by

$$f(x) = \frac{1}{x^5 - 2x^3 - 3x^2 + 6}.$$

***Source: The Mathematics Teacher.***

**a.** Graph the function on $[-3.4, 3.4]$ by $[-3, 3]$. From your graph, estimate how many vertical asymptotes the function has and where they are located.

**b.** Now graph the function on $[-1.5, -1.4]$ by $[-10, 10]$ and on $[1.4, 1.5]$ by $[-1000, 1000]$. From your graphs, estimate how many vertical asymptotes the function has and where they are located.

**c.** From your results in parts a and b, what advice would you give a friend on using a graphing calculator to find vertical asymptotes?

## LIFE SCIENCE APPLICATIONS

**46. Cardiac Output** A technique for measuring cardiac output depends on the concentration of a dye after a known amount is injected into a vein near the heart. In a normal heart, the concentration of the dye at time $x$ (in seconds) is given by the function

$$g(x) = -0.006x^4 + 0.140x^3 - 0.053x^2 + 1.79x.$$

**a.** Graph $g(x)$ on $[0, 6]$ by $[0, 20]$.

**b.** In your graph from part a, notice that the function initially increases. Considering the form of $g(x)$, do you think it can keep increasing forever? Explain.

**c.** Write a short paragraph about the extent to which the concentration of dye might be described by the function $g(x)$.

**47. Alcohol Concentration**  The polynomial function

$$A(t) = 0.003631t^3 - 0.03746t^2 + 0.1012t + 0.009$$

gives the approximate blood alcohol concentration in a 170-lb woman $t$ hours after drinking 2 oz of alcohol on an empty stomach, for $t$ in the interval $[0, 5]$. *Source: Medical Aspects of Alcohol Determination in Biological Specimens.*

  **a.** Graph $A(t)$ on $0 \le t \le 5$.

  **b.** Using the graph from part a, estimate the time of maximum alcohol concentration.

  **c.** In many states, a person is legally drunk if the blood alcohol concentration exceeds 0.08%. Use the graph from part a to estimate the period in which this 170-lb woman is legally drunk.

**48. Population Variation**  During the early part of the 20th century, the deer population of the Kaibab Plateau in Arizona experienced a rapid increase, because hunters had reduced the number of natural predators. The increase in population depleted the food resources and eventually caused the population to decline. For the period from 1905 to 1930, the deer population was approximated by

$$D(x) = -0.125x^5 + 3.125x^4 + 4000,$$

where $x$ is time in years from 1905.

  **a.** Graph $D(x)$ on $0 \le x \le 30$.

  **b.** From the graph, over what period of time (from 1905 to 1930) was the population increasing? Relatively stable? Decreasing?

**49. Brain Mass**  The mass (in grams) of the human brain during the last trimester of gestation and the first two years after birth can be approximated by the function

$$m(c) = \frac{c^3}{100} - \frac{1500}{c},$$

where $c$ is the circumference of the head in centimeters. *Source: Early Human Development.*

  **a.** Find the approximate mass of brains with a head circumference of 30, 40, or 50 cm.

  **b.** Clearly, the formula is invalid for any values of $c$ yielding negative values of $w$. For what values of $c$ is this true?

  **c.** Use a graphing calculator to sketch this graph on the interval $20 \le c \le 50$.

  **d.** Suppose an infant brain has mass of 700 g. Use features on a graphing calculator to find what the circumference of the head is expected to be.

**50. Contact Lenses**  The strength of a contact lens is given in units known as diopters, as well as in mm of arc. The following is taken from a chart used by optometrists to convert diopters to mm of arc. *Source: Bausch & Lomb.*

| Diopters | mm of Arc |
| --- | --- |
| 36.000 | 9.37 |
| 36.125 | 9.34 |
| 36.250 | 9.31 |
| 36.375 | 9.27 |
| 36.500 | 9.24 |
| 36.625 | 9.21 |
| 36.750 | 9.18 |
| 36.875 | 9.15 |
| 37.000 | 9.12 |

  **a.** Notice that as the diopters increase, the mm of arc decrease. Find a value of $k$ so the function $a = f(d) = k/d$ gives $a$, the mm of arc, as a function of $d$, the strength in diopters. (Round $k$ to the nearest integer. For a more accurate answer, average all the values of $k$ given by each pair of data.)

  **b.** An optometrist wants to order 40.50 diopter lenses for a patient. The manufacturer needs to know the strength in mm of arc. What is the strength in mm of arc?

**51. Population Biology**  The function

$$f(x) = \frac{\lambda x}{1 + (ax)^b}$$

is used in population models to give the size of the next generation $(f(x))$ in terms of the current generation $(x)$. *Source: Models in Ecology.*

  **a.** What is a reasonable domain for this function, considering what $x$ represents?

  **b.** Graph this function for $\lambda = a = b = 1$.

  **c.** Graph this function for $\lambda = a = 1$ and $b = 2$.

  **d.** What is the effect of making $b$ larger?

**52. Growth Model**  The function

$$f(x) = \frac{Kx}{A + x}$$

is used in biology to give the growth rate of a population in the presence of a quantity $x$ of food. This is called Michaelis-Menten kinetics. *Source: Mathematical Models in Biology.*

  **a.** What is a reasonable domain for this function, considering what $x$ represents?

  **b.** Graph this function for $K = 5$ and $A = 2$.

  **c.** Show that $y = K$ is a horizontal asymptote.

  **d.** What do you think $K$ represents?

  **e.** Show that $A$ represents the quantity of food for which the growth rate is half of its maximum.

**53. Soil Ingestion**  The amount of acid-insoluble ash in scat from white-footed mice can be approximated by

$$y = \frac{b(1-x) + cx}{1 - a(1-x)},$$

where $x$ is the fraction of soil in the diet, $a$ is the digestibility of food, $b$ is the concentration of acid-insoluble ash in the food, and $c$ is the concentration of acid-insoluble ash in the soil. *Source: Journal of Wildlife Management.*

 **a.** Explain in your own words what the numerator of this fraction represents.

**b.** Explain in your own words what the denominator of this fraction represents.

**c.** The researchers estimated the values of the parameters as $a = 0.76$, $b = 0.025$, and $c = 0.92$. Use these values to simplify the function above.

**d.** Use a graphing calculator to sketch the graph of the function in part c on the interval $0 \leq x \leq 0.15$.

**e.** Suppose the mouse scat contains 20% ash, that is, $y = 0.20$. Use algebra to find the concentration of soil in the diet, and then verify your answer with a graphing calculator.

## OTHER APPLICATIONS

 **54. Batting Power**  The rate at which energy is transferred from a batter to a baseball bat was found for one batter to be approximated by the function

$$y = 0.0078109 + 2.08079t - 393.385t^2 + 127,454t^3$$
$$- 3.64084 \times 10^6 t^4 + 4.07979 \times 10^7 t^5 - 1.99528$$
$$\times 10^8 t^6 + 3.54337 \times 10^8 t^7,$$

where $t$ is the time into the swing in seconds and $y$ is the power in horsepower. *Source: Rose-Hulman.*

**a.** Use a graphing calculator to sketch this function.

**b.** From your graph in part a, estimate over what values of $t$ this function is valid.

**c.** From your graph in part a, estimate at what time into the swing the power is a maximum. What is the maximum power?

**55. Cost–Benefit Model**  Suppose a cost–benefit model is given by

$$y = \frac{6.7x}{100 - x},$$

where $y$ is the cost in thousands of dollars of removing $x$ percent of a given pollutant.

**a.** Find the cost of removing each percent of pollutants: 50%; 70%; 80%; 90%; 95%; 98%; 99%.

**b.** Is it possible, according to this function, to remove *all* the pollutant?

**c.** Graph the function.

**56. Cost–Benefit Model**  Suppose a cost–benefit model is given by

$$y = \frac{6.5x}{102 - x},$$

where $y$ is the cost in thousands of dollars of removing $x$ percent of a certain pollutant.

**a.** Find the cost of removing each percent of pollutants: 0%; 50%; 80%; 90%; 95%; 99%; 100%.

**b.** Graph the function.

**57. Head Start**  The enrollment in Head Start for some recent years is included in the table. *Source: Head Start.*

| Year | Enrollment |
|------|-----------|
| 1966 | 733,000 |
| 1970 | 477,400 |
| 1980 | 376,300 |
| 1990 | 540,930 |
| 1995 | 750,696 |
| 2000 | 857,664 |
| 2005 | 906,993 |
| 2010 | 904,118 |

**a.** Plot the points from the table using 0 for 1960, and so on.

**b.** Use the quadratic regression feature of a graphing calculator to get a quadratic function that approximates the data. Graph the function on the same window as the scatterplot.

**c.** Use cubic regression to get a cubic function that approximates the data. Graph the function on the same window as the scatterplot.

**d.** Which of the two functions in parts b and c appears to be a better fit for the data? Explain your reasoning.

**58. Length of a Pendulum**  A simple pendulum swings back and forth in regular time intervals. Grandfather clocks use pendulums to keep accurate time. The relationship between the length of a pendulum $L$ and the period (time) $T$ for one complete oscillation can be expressed by the function $L = kT^n$, where $k$ is a constant and $n$ is a positive integer to be determined. The data below were taken for different lengths of pendulums.* *Source: Gary Rockswold.*

| T (sec) | L (ft) |
|---------|--------|
| 1.11 | 1.0 |
| 1.36 | 1.5 |
| 1.57 | 2.0 |
| 1.76 | 2.5 |
| 1.92 | 3.0 |
| 2.08 | 3.5 |
| 2.22 | 4.0 |

*See Exercise 21, Section 1.2.

**a.** Find the value of $k$ for $n = 1, 2,$ and 3, using the data for the 4-ft pendulum.

**b.** Use a graphing calculator to plot the data in the table and to graph the function $L = kT^n$ for the three values of $k$ (and their corresponding values of $n$) found in part a. Which function best fits the data?

**c.** Use the best-fitting function from part a to predict the period of a pendulum having a length of 5 ft.

**d.** If the length of pendulum doubles, what happens to the period?

**e.** If you have a graphing calculator or computer program with a quadratic regression feature, use it to find a quadratic function that approximately fits the data. How does this answer compare with the answer to part b?

**59. Coal Consumption**   The table gives U.S. coal consumption for selected years. *Source: U.S. Department of Energy.*

| Year | Millions of Short Tons |
|------|------------------------|
| 1950 | 494.1 |
| 1960 | 398.1 |
| 1970 | 523.2 |
| 1980 | 702.7 |
| 1985 | 818.0 |
| 1990 | 902.9 |
| 1995 | 962.1 |
| 2000 | 1084.1 |
| 2005 | 1128.3 |
| 2010 | 1051.3 |

**a.** Draw a scatterplot, letting $x = 0$ represent 1950.

**b.** Use the quadratic regression feature of a graphing calculator to get a quadratic function that approximates the data.

**c.** Graph the function from part b on the same window as the scatterplot.

**d.** Use cubic regression to get a cubic function that approximates the data.

**e.** Graph the cubic function from part d on the same window as the scatterplot.

**f.** Which of the two functions in parts b and d appears to be a better fit for the data? Explain your reasoning.

**YOUR TURN ANSWERS**

**1.**
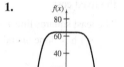

**2.**
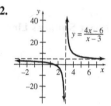

# CHAPTER REVIEW

## SUMMARY

In this chapter, we have studied various types of functions: linear, quadratic, polynomial, and rational. Linear functions are the simplest; we can find the equation of such a function when given a point and the slope or when given two points. The method of least squares allows us to derive linear functions that approximately describe sets of data. Quadratic functions are next in order of complication; they, too, can be analyzed completely without calculus. Linear and quadratic functions are special cases of polynomial functions. We have briefly introduced polynomial functions of higher degree and rational functions; a complete analysis requires techniques discussed in later chapters. The functions in this chapter have a broad range of applications, as demonstrated in the applications exercises. Other types of functions will be studied in the next chapter.

**Slope of a Line**   The slope of a line is defined as the vertical change (the "rise") over the horizontal change (the "run") as one travels along the line. In symbols, taking two different points $(x_1, y_1)$ and $(x_2, y_2)$ on the line, the slope is

$$m = \frac{\Delta y}{\Delta x} = \frac{y_2 - y_1}{x_2 - x_1},$$

where $x_1 \neq x_2$.

| Equations of Lines | Equation | Description |
|---|---|---|
| | $y = mx + b$ | Slope-intercept form: slope $m$ and $y$-intercept $b$. |
| | $y - y_1 = m(x - x_1)$ | Point-slope form: slope $m$ and line passes through $(x_1, y_1)$. |
| | $x = k$ | Vertical line: $x$-intercept $k$, no $y$-intercept (except when $k = 0$), undefined slope. |
| | $y = k$ | Horizontal line: $y$-intercept $k$, no $x$-intercept (except when $k = 0$), slope 0. |

**Parallel Lines**  Two lines are parallel if and only if they have the same slope, or if they are both vertical.

**Perpendicular Lines**  Two lines are perpendicular if and only if the product of their slopes is $-1$, or if one is vertical and the other horizontal.

**Linear Function**  A relationship $f$ defined by

$$y = f(x) = mx + b,$$

for real numbers $m$ and $b$, is a linear function.

**Linear Cost Function**  In a cost function of the form $C(x) = mx + b$, the $m$ represents the marginal cost and $b$ represents the fixed cost.

**Least Squares Line**  The least squares line $Y = mx + b$ that gives the best fit to the data points $(x_1, y_1)$, $(x_2, y_2)$, ..., $(x_n, y_n)$ has slope $m$ and $y$-intercept $b$ given by the equations

$$m = \frac{n(\Sigma xy) - (\Sigma x)(\Sigma y)}{n(\Sigma x^2) - (\Sigma x)^2}$$

$$b = \frac{\Sigma y - m(\Sigma x)}{n}$$

**Correlation Coefficient**
$$r = \frac{n(\Sigma xy) - (\Sigma x)(\Sigma y)}{\sqrt{n(\Sigma x^2) - (\Sigma x)^2}\sqrt{n(\Sigma y^2) - (\Sigma y)^2}}$$

**Function**  A function is a rule that assigns to each element from one set exactly one element from another set.

**Domain and Range**  The set of all possible values of the independent variable in a function is called the domain of the function, and the resulting set of possible values of the dependent variable is called the range.

**Vertical Line Test**  A graph represents a function if and only if every vertical line intersects the graph in no more than one point.

**Quadratic Function**  A quadratic function is defined by

$$f(x) = ax^2 + bx + c,$$

where $a$, $b$, and $c$ are real numbers, with $a \neq 0$.

**Graph of a Quadratic Function**  The graph of the quadratic function $f(x) = ax^2 + bx + c$ is a parabola with its vertex at

$$\left(\frac{-b}{2a}, f\left(\frac{-b}{2a}\right)\right).$$

The graph opens upward if $a > 0$ and downward if $a < 0$.

**Polynomial Function**  A polynomial function of degree $n$, where $n$ is a nonnegative integer, is defined by

$$f(x) = a_n x^n + a_{n-1} x^{n-1} + \cdots + a_1 x + a_0,$$

where $a_n$, $a_{n-1}$, ..., $a_1$, and $a_0$ are real numbers, called coefficients, with $a_n \neq 0$. The number $a_n$ is called the leading coefficient.

**Properties of Polynomial Functions**
1. A polynomial function of degree $n$ can have at most $n - 1$ turning points. Conversely, if the graph of a polynomial function has $n$ turning points, it must have degree at least $n + 1$.
2. In the graph of a polynomial function of even degree, both ends go up or both ends go down. For a polynomial function of odd degree, one end goes up and one end goes down.
3. If the graph goes up as $x$ becomes a large positive number, the leading coefficient must be positive. If the graph goes down as $x$ becomes large, the leading coefficient is negative.

**Rational Function**   A rational function is defined by

$$f(x) = \frac{p(x)}{q(x)},$$

where $p(x)$ and $q(x)$ are polynomial functions and $q(x) \neq 0$.

**Asymptotes**   If a function gets larger and larger in magnitude without bound as $x$ approaches the number $k$, then the line $x = k$ is a vertical asymptote.

If the values of $y$ approach a number $k$ as $|x|$ gets larger and larger, the line $y = k$ is a horizontal asymptote.

**Graphs of Basic Functions**

Quadratic

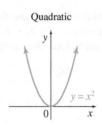

$y = x^2$

Absolute Value

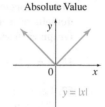

$y = |x|$

Square Root

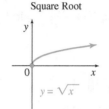

$y = \sqrt{x}$

Rational

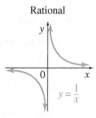

$y = \frac{1}{x}$

# KEY TERMS

*To understand the concepts presented in this chapter, you should know the meaning and use of the following terms. For easy reference, the section in the chapter where a word (or expression) was first used is provided.*

mathematical model

**1.1**
ordered pair
Cartesian coordinate system
axes
origin
coordinates
quadrants
graph
intercepts
slope
linear equation
slope-intercept form
proportional
point-slope form
parallel

perpendicular
linear function
independent variable
dependent variable
fixed cost
marginal cost
linear cost function

**1.2**
scatterplot
least squares line
summation notation
correlation coefficient
outlier

**1.3**
function
domain

range
constant function
vertical line test
composition

**1.4**
quadratic function
parabola
vertex
axis
vertical translation
vertical reflection
horizontal translation
completing the square
quadratic regression
horizontal reflection

**1.5**
polynomial function
degree
coefficient
leading coefficient
power function
cubic polynomial
quartic polynomial
turning point
real zero
rational function
vertical asymptote
horizontal asymptote
cost-benefit model
average cost

# REVIEW EXERCISES

## CONCEPT CHECK

**Determine whether each statement is true or false, and explain why.**

1. A given line can have more than one slope.

2. The equation $y = 3x + 4$ represents the equation of a line with slope 4.

3. The line $y = -2x + 5$ intersects the point $(3, -1)$.

4. The line that intersects the points $(2, 3)$ and $(2, 5)$ is a horizontal line.

5. The line that intersects the points $(4, 6)$ and $(5, 6)$ is a horizontal line.

6. The $x$-intercept of the line $y = 8x + 9$ is 9.

7. The function $f(x) = \pi x + 4$ represents a linear function.

8. The function $f(x) = 2x^2 + 3$ represents a linear function.

9. The lines $y = 3x + 17$ and $y = -3x + 8$ are perpendicular.

10. The lines $4x + 3y = 8$ and $4x + y = 5$ are parallel.

11. A correlation coefficient of zero indicates a perfect fit with the data.

12. It is not possible to get a correlation coefficient of $-1.5$ for a set of data.

13. A linear function is an example of a polynomial function.

14. A rational function is an example of a polynomial function.

15. The function $f(x) = 3x^2 - 75x + 2$ is a quadratic function.

16. The function $f(x) = x^2 - 6x + 4$ has a vertex at $x = 3$.

17. The function $f(x) = \dfrac{1}{x - 6}$ has a vertical asymptote at $y = 6$.

18. The domain of the function $f(x) = \dfrac{1}{x^2 - 4}$ includes all real numbers except $x = 2$.

## PRACTICE AND EXPLORATIONS

19. What is marginal cost? Fixed cost?

20. What six quantities are needed to compute a correlation coefficient?

21. What is a function? A linear function? A quadratic function? A rational function?

22. How do you find a vertical asymptote?

23. How do you find a horizontal asymptote?

24. What can you tell about the graph of a polynomial function of degree $n$ before you plot any points?

**Find the slope for each line that has a slope.**

25. Through $(-3, 7)$ and $(2, 12)$

26. Through $(4, -1)$ and $(3, -3)$

27. Through the origin and $(11, -2)$

28. Through the origin and $(0, 7)$

29. $4x + 3y = 6$  30. $4x - y = 7$

31. $y + 4 = 9$  32. $3y - 1 = 14$

33. $y = 5x + 4$  34. $x = 5y$

**Find an equation in the form $y = mx + b$ for each line.**

35. Through $(5, -1)$; slope $= 2/3$

36. Through $(8, 0)$; slope $= -1/4$

37. Through $(-6, 3)$ and $(2, -5)$

38. Through $(2, -3)$ and $(-3, 4)$

39. Through $(2, -10)$, perpendicular to a line with undefined slope

40. Through $(-2, 5)$; slope $= 0$

**Find an equation for each line in the form $ax + by = c$, where $a$, $b$, and $c$ are integers with no factor common to all three and $a \geq 0$.**

41. Through $(3, -4)$, parallel to $4x - 2y = 9$

42. Through $(0, 5)$, perpendicular to $8x + 5y = 3$

43. Through $(-1, 4)$; undefined slope

44. Through $(7, -6)$, parallel to a line with undefined slope

45. Through $(3, -5)$, parallel to $y = 4$

46. Through $(-3, 5)$, perpendicular to $y = -2$

**Graph each linear equation defined as follows.**

47. $y = 4x + 3$  48. $y = 6 - 2x$

49. $3x - 5y = 15$  50. $4x + 6y = 12$

51. $x - 3 = 0$  52. $y = 1$

53. $y = 2x$  54. $x + 3y = 0$

**List the ordered pairs obtained from the following if the domain of $x$ for each exercise is $\{-3, -2, -1, 0, 1, 2, 3\}$. Graph each set of ordered pairs. Give the range.**

55. $y = (2x - 1)(x + 1)$  56. $y = \dfrac{x}{x^2 + 1}$

57. Let $f(x) = 5x^2 - 3$ and $g(x) = -x^2 + 4x + 1$. Find the following.

 a. $f(-2)$  b. $g(3)$  c. $f(-k)$  d. $g(3m)$

 e. $f(x + h)$  f. $g(x + h)$  g. $\dfrac{f(x + h) - f(x)}{h}$

 h. $\dfrac{g(x + h) - g(x)}{h}$

58. Let $f(x) = 2x^2 + 5$ and $g(x) = 3x^2 + 4x - 1$. Find the following.

 a. $f(-3)$  b. $g(2)$  c. $f(3m)$  d. $g(-k)$

 e. $f(x + h)$  f. $g(x + h)$  g. $\dfrac{f(x + h) - f(x)}{h}$

 h. $\dfrac{g(x + h) - g(x)}{h}$

**Find the domain of each function defined as follows.**

59. $y = \dfrac{3x - 4}{x}$

60. $y = \dfrac{\sqrt{x - 2}}{2x + 3}$

**Graph the following by hand.**

61. $y = 2x^2 + 3x - 1$  62. $y = -\dfrac{1}{4}x^2 + x + 2$

63. $y = -x^2 + 4x + 2$  64. $y = 3x^2 - 9x + 2$

65. $f(x) = x^3 - 3$  66. $f(x) = 1 - x^4$

67. $y = -(x - 1)^4 + 4$  68. $y = -(x + 2)^3 - 2$

69. $f(x) = \dfrac{8}{x}$  70. $f(x) = \dfrac{2}{3x - 6}$

71. $f(x) = \dfrac{4x - 2}{3x + 1}$  72. $f(x) = \dfrac{6x}{x + 2}$

**For each pair of functions, find $(f \circ g)(x)$ and $(g \circ f)(x)$.**

73. $f(x) = 4x^2 + 3x$ and $g(x) = \dfrac{x}{x + 1}$

74. $f(x) = 3x + 4$ and $g(x) = x^2 - 6x - 7$

## LIFE SCIENCE APPLICATIONS

**75. Registered Nurses** According to the Bureau of Labor Statistics' Employment Projections, the number of employed registered nurses will grow from 2.74 million in 2010 to 3.45 million in 2020. Assume that the growth is linear. *Source: U.S. Bureau of Labor Statistics.*

**a.** Determine a linear equation that approximates the number of employed nurses in terms of time $t$, where $t$ represents the number of years since 2000.

**b.** Based on the answer to part a, what was the number of employed nurses in 2014?

**c.** Describe the rate at which the number of employed nurses is changing.

**76. Meat Consumption** The U.S. per capita consumption (in pounds) of beef, pork, and chicken for 1990 and 2010 are given in the following table. Assume that the changes in consumption are approximately linear. *Source: U.S. Department of Agriculture.*

|  | 1990 | 2010 |
|---|---|---|
| **Beef** | 64.5 | 56.7 |
| **Pork** | 47.8 | 44.3 |
| **Chicken** | 42.4 | 58.0 |

**a.** Determine three linear functions, $b(t)$, $p(t)$, and $c(t)$, that approximate the pounds of beef, pork, and chicken, respectively, consumed for time $t$, where $t$ is the number of years since 1990.

**b.** Describe the rates at which the consumption of each type of meat is changing.

**c.** In what year did the consumption of chicken surpass the consumption of pork?

**d.** If this trend were to continue, approximate the consumption of each type of meat in 2015.

**77. Ecosystem Diversity** A study at the Cedar Creek Natural History Area of Minnesota found that an average local diversity of $x$ plant species per 0.5 square meter required an average regional diversity given by

$$y = \frac{x - 1.1}{0.124}$$

in species per 0.5 hectare. *Source: Science.*

**a.** What is the slope of this line? Describe in words what your answer means.

**b.** What average regional diversity is required for an average local diversity of 6 species per 0.5 square meter?

**c.** What average local diversity will lead to an average regional diversity of 70 species per hectare?

**d.** The coefficient of correlation for this study is given as $r = 0.82$. Describe in your own words what this means.

**78. World Health** In general, people tend to live longer in countries that have a greater supply of food. Listed below is the 2009 daily calorie supply and 2009 life expectancy at birth for 10 randomly selected countries. *Source: Food and Agriculture Organization.*

| Country | Calories ($x$) | Life Expectancy ($y$) |
|---|---|---|
| Belize | 2680 | 75.6 |
| Cambodia | 2382 | 62.7 |
| France | 3531 | 81.2 |
| India | 2321 | 64.7 |
| Mexico | 3146 | 76.5 |
| New Zealand | 3172 | 80.4 |
| Peru | 2563 | 73.6 |
| Sweden | 3125 | 81.1 |
| Tanzania | 2137 | 56.6 |
| United States | 3668 | 78.2 |

**a.** Find the correlation coefficient. Do the data seem to fit a straight line?

**b.** Draw a scatterplot of the data. Combining this with your results from part a, do the data seem to fit a straight line?

**c.** Find the equation of the least squares line.

**d.** Use your answer from part c to predict the life expectancy in Canada, which has a daily calorie supply of 3399. Compare your answer with the actual value of 80.8 years.

**e.** Briefly explain why countries with a higher daily calorie supply might tend to have a longer life expectancy. Is this trend likely to continue to higher calorie levels? Do you think that an American who eats 5000 calories a day is likely to live longer than one who eats 3600 calories? Why or why not?

**f.** (For the ambitious!) Find the correlation coefficient and least squares line using the data for a larger sample of countries, as found in an almanac or other reference. Is the result in general agreement with the previous results?

**79. Blood Sugar and Cholesterol Levels** The following data show the connection between blood sugar levels and cholesterol levels for eight different patients.

| Patient | Blood Sugar Level ($x$) | Cholesterol Level ($y$) |
|---|---|---|
| 1 | 130 | 170 |
| 2 | 138 | 160 |
| 3 | 142 | 173 |
| 4 | 159 | 181 |
| 5 | 165 | 201 |
| 6 | 200 | 192 |
| 7 | 210 | 240 |
| 8 | 250 | 290 |

For the data given in the preceding table, $\Sigma x = 1394$, $\Sigma y = 1607$, $\Sigma xy = 291{,}990$, $\Sigma x^2 = 255{,}214$, and $\Sigma y^2 = 336{,}155$.

**a.** Find the equation of the least squares line.

**b.** Predict the cholesterol level for a person whose blood sugar level is 190.

**c.** Find the correlation coefficient.

**80. Eating Behavior**  In a study of eating behavior, the cumulative intake (in grams) of one individual was found to vary with time according to the formula

$$I = 27 + 72t - 1.5t^2,$$

where $t$ is the time in minutes. *Source: Appetite.*

**a.** At what time is the intake for this individual a maximum?

**b.** What is the maximum intake?

**c.** Over what domain is this function valid? Explain why the function makes no sense outside of this domain.

**81. Fever**  A certain viral infection causes a fever that typically lasts 6 days. A model of the fever (in °F) on day $x$, $1 \le x \le 6$, is

$$F(x) = -\frac{2}{3}x^2 + \frac{14}{3}x + 96.$$

According to the model, on what day should the maximum fever occur? What is the maximum fever?

**82. Blood Volume**  A formula proposed by Hurley for the red cell volume (RCV) in milliliters for males is

$$RCV = 1486S^2 - 4106S + 4514,$$

where $S$ is the surface area in square meters. A formula given by Pearson et al. is

$$RCV = 1486S - 825.$$

*Source: British Journal of Haematology.*

**a.** Verify that these two formulas never give the same answer

  **i.** using the quadratic formula;

  **ii.** using a graphing calculator.

**b.** The formula for plasma volume for females given by Hurley is

$$PV = 1278S^{1.289},$$

while the formula given by Pearson et al. is

$$PV = 1395S,$$

where $PV$ is measured in milliliters and $S$ in square meters. Find the values of $S$ for which these two formulas give the same answer. What is the predicted plasma volume at each of these values of $S$?

**83. Sunscreen**  An article in a medical journal says that a sunscreen with a sun protection factor (SPF) of 2 provides 50% protection against ultraviolet B (UVB) radiation, an SPF of 4 provides 75% protection, and an SPF of 8 provides 87.5% protection (which the article rounds to 87%). *Source: Family Practice.*

**a.** 87.5% protection means that 87.5% of the UVB radiation is screened out. Write as a fraction the amount of radiation that is let in, and then describe how this fraction, in general, relates to the SPF rating.

**b.** Plot UVB percent protection ($y$) against $x$, where $x = 1/SPF$.

**c.** Based on your graph from part b, give an equation relating UVB protection to SPF rating.

**d.** An SPF of 8 has double the chemical concentration of an SPF 4. Find the increase in the percent protection.

**e.** An SPF of 30 has double the chemical concentration of an SPF 15. Find the increase in the percent protection.

**f.** Based on your answers from parts d and e, what happens to the increase in the percent protection as the SPF continues to double?

**84. Population Growth**  In 1960 in an article in *Science* magazine, H. Van Forester, P. M. Mora, and W. Amiot predicted that world population would be infinite in the year 2026. Their projection was based on the rational function defined by

$$p(t) = \frac{1.79 \times 10^{11}}{(2026.87 - t)^{0.99}},$$

where $p(t)$ gives population in year $t$. This function has provided a relatively good fit to the population until very recently. *Source: Science.*

**a.** Estimate world population in 2010 using this function, and compare it with the estimate of 6.909 billion. *Source: United Nations.*

**b.** What does the function predict for world population in 2020? 2025?

**c.** Discuss why this function is not realistic, despite its good fit to past data.

**85. Pollution**  The cost to remove $x$ percent of a pollutant is

$$y = \frac{7x}{100 - x},$$

in thousands of dollars. Find the cost of removing each of the following percents of the pollutant.

**a.** 80%  **b.** 50%  **c.** 90%

**d.** Graph the function.

**e.** Can all of the pollutant be removed?

**86. Diabetes**  The following table lists the annual number (in thousands) of new cases of diagnosed diabetes among adults, aged 18–70 years, in the United States. *Source: Centers for Disease Control and Prevention.*

| Year | Number (in thousands) |
|---|---|
| 1980 | 493 |
| 1985 | 564 |
| 1990 | 603 |
| 1995 | 796 |
| 2000 | 1104 |
| 2005 | 1468 |
| 2010 | 1735 |

**a.** Plot the data on a graphing calculator, letting $t = 0$ correspond to the year 1980.

**b.** Using the regression feature on your calculator, find a linear, a quadratic, and a cubic function that models these data.

**c.** Plot the three functions with the data on the same coordinate axes. Which function, or functions, best capture the behavior of the data over the years plotted?

**d.** Find the number of cases predicted by all three functions for 2015.

**87. Organic Farmland**  As organic products grow in popularity, the amount of organic farmland has also increased. The number of U.S. certified acres (in thousands) for the production of grains from 2000–2008 can be modeled by the function

$$f(t) = 7.95t^2 - 8.85t + 447.9,$$

where $t = 0$ corresponds to 2000. *Source: U.S. Department of Agriculture.*

**a.** Discuss the general trend in organic farmland for grains.

**b.** Find a function $g(t)$ that models the same farmland acreage except that $t$ is the actual year between 2000 and 2008. For example, $g(2003) = f(3)$. (*Hint:* Use a horizontal translation.)

## OTHER APPLICATIONS

**88. Marital Status**  More people are staying single longer in the United States. In 1995, the number of never-married adults, age 15 and over, was 55.0 million. By 2012, it was 77.3 million. Assume the data increase linearly, and write an equation that defines a linear function for this data. Let $t$ represent the number of years since 1990. *Source: U.S. Census Bureau.*

**89. Poverty**  The following table gives the number of families under the poverty level in the United States in recent years. *Source: U.S. Census Bureau.*

| Year | Families Below Poverty Level (in thousands) |
|---|---|
| 2000 | 6400 |
| 2001 | 6813 |
| 2002 | 7229 |
| 2003 | 7607 |
| 2004 | 7623 |
| 2005 | 7657 |
| 2006 | 7668 |
| 2007 | 7835 |
| 2008 | 8147 |
| 2009 | 8792 |
| 2010 | 9400 |

**a.** Find a linear equation for the number of families below poverty level (in thousands) in terms of $t$, the number of years since 2000, using the data for 2000 and 2010.

**b.** Repeat part a, using the data for 2005 and 2010.

**c.** Using all the data, find the equation of the least squares line.

**d.** Calculate the correlation coefficient. Do the data seem to fit a straight line?

**e.** Plot the data and the three lines from parts a–c on a graphing calculator. Discuss which of the three lines found in parts a–c best describes the data, as well as to what extent a linear model accurately describes the data.

**f.** Use cubic regression to get a cubic function that approximates the data. Graph the function with the data points. Discuss to what extent a cubic model accurately describes the data.

**90. Governors' Salaries**  In general, the larger a state's population, the more the governor earns. Listed in the table below are the estimated 2012 populations (in millions) and the salary of the governor (in thousands of dollars) for eight randomly selected states. *Source: U.S. Census Bureau and Time Almanac 2013.*

| State | AZ | MD | MA | ND | NY | PA | TN | WY |
|---|---|---|---|---|---|---|---|---|
| Population (x) | 6.55 | 5.88 | 6.65 | 0.70 | 19.57 | 12.76 | 6.46 | 0.58 |
| Governor's Salary (y) | 95 | 150 | 140 | 110 | 179 | 183 | 170 | 105 |

**a.** Find the correlation coefficient. Do the data seem to fit a straight line?

**b.** Draw a scatterplot of the data. Compare this with your answer from part a.

**c.** Find the equation for the least squares line.

**d.** Based on your answer to part c, how much does a governor's salary increase, on average, for each additional million in population?

**e.** Use your answer from part c to predict the governor's salary in your state. Based on your answers from parts a and b, would this prediction be very accurate? Compare with the actual salary, as listed in an almanac or other reference.

**f.** (For the ambitious!) Find the correlation coefficient and least squares line using the data for all 50 states, as found in an almanac or other reference. Is the result in general agreement with the previous results?

**91. Average Speed**  Suppose a plane flies from one city to another that is a distance $d$ miles away. The plane flies at a constant speed $v$ relative to the wind, but there is a constant wind speed of $w$. Therefore, the speed in one direction is $v + w$, and the speed in the other direction is $v - w$. *Source: The Mathematics Teacher.*

**a.** Using the formula distance = rate × time, show that the time to make the round trip is

$$\frac{d}{v+w} + \frac{d}{v-w}.$$

**b.** Using the result from part a, show that the average speed for the round trip is

$$v_{aver} = \frac{2d}{\dfrac{d}{v+w} + \dfrac{d}{v-w}}.$$

**c.** Simplify your result from part b to get

$$v_{aver} = v - \frac{w^2}{v}.$$

**d.** Consider $v$ in the result in part c to be a constant. What wind speed results in the greatest average speed? Explain why your result makes sense.

**92. Average Speed** Suppose the plane in the previous exercise makes the trip one way at a speed of $v$, and the return trip at a speed $xv$. *Source: The AMATYC Review.*

**a.** Explain in words what $x = 0.9$ and $x = 1.1$ represent.

**b.** Show that the average velocity can be written as

$$v_{aver} = \left(\frac{2x}{x+1}\right)v.$$

(*Hint:* Use the steps in parts a–c of the previous exercise.)

**c.** With $v$ held constant, the equation in part b defines $v_{aver}$ as a function of $x$. Discuss the behavior of this function. In particular, consider the horizontal asymptote and what it says about the average velocity.

**93. Planets** The following table contains the average distance $D$ from the sun for the eight planets and their period $P$ of revolution around the sun in years. *Source: The Natural History of the Universe.*

| Planet | Distance ($D$) | Period ($P$) |
|---|---|---|
| Mercury | 0.39 | 0.24 |
| Venus | 0.72 | 0.62 |
| Earth | 1 | 1 |
| Mars | 1.52 | 1.89 |
| Jupiter | 5.20 | 11.9 |
| Saturn | 9.54 | 29.5 |
| Uranus | 19.2 | 84.0 |
| Neptune | 30.1 | 164.8 |

The distances are given in astronomical units (A.U.); 1 A.U. is the average distance from Earth to the sun. For example, since Jupiter's distance is 5.2 A.U., its distance from the sun is 5.2 times farther than Earth's.

**a.** Find functions of the form $P = kD^n$ for $n = 1, 1.5$, and $2$ that fit the data at Neptune.

**b.** Use a graphing calculator to plot the data in the table and to graph the three functions found in part a. Which function best fits the data?

**c.** Use the best-fitting function from part b to predict the period of Pluto (which was removed from the list of planets in 2006), which has a distance from the sun of 39.5 A.U. Compare your answer to the true value of 248.5 years.

**d.** If you have a graphing calculator or computer program with a power regression feature, use it to find a power function (a function of the form $P = kD^n$) that approximately fits the data. How does this answer compare with the answer to part b?

# EXTENDED APPLICATION

## USING EXTRAPOLATION TO PREDICT LIFE EXPECTANCY

One reason for developing a mathematical model is to make predictions. If your model is a least squares line, you can predict the $y$-value corresponding to some new $x$ by substituting this $x$ into an equation of the form $Y = mx + b$.

(We use a capital $Y$ to remind us that we're getting a predicted value rather than an actual data value.) Data analysts distinguish between two very different kinds of prediction, *interpolation* and *extrapolation*. An interpolation uses a new $x$ inside the $x$ range of your original data. For example, if you have inflation data at 5-year intervals from 1950 to 2015, estimating the rate of inflation in 1957 is an interpolation problem. But if you use the same data to estimate what the inflation rate was in 1920, or what it will be in 2025, you are extrapolating.

In general, interpolation is much safer than extrapolation, because data that are approximately linear over a short interval may be nonlinear over a larger interval. One way to detect nonlinearity is to look at *residuals*, which are the differences between the actual data values and the values predicted by the line of best fit. Here is a simple example:

The regression equation for the linear fit in Figure 66 is $Y = 3.431 + 1.334x$. Since the *r*-value for this regression line is 0.93, our linear model fits the data very well. But we might notice that the predictions are a bit low at the ends and high in the middle. We can get a better look at this pattern by plotting the residuals. To find them, we put each value of the independent variable into the regression equation, calculate the predicted value $Y$, and subtract it from the actual *y*-value. The residual plot is shown in Figure 67, with the vertical axis rescaled to exaggerate the pattern. The residuals indicate that our data have a nonlinear, U-shaped component that is not captured by the linear fit. Extrapolating from this data set is probably not a good idea; our linear prediction for the value of *y* when *x* is 10 may be much too low.

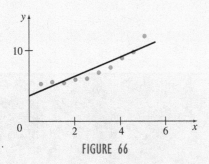

FIGURE 66

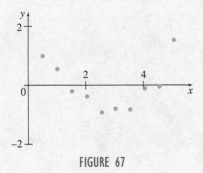

FIGURE 67

## EXERCISES

The following table gives the life expectancy at birth of females born in the United States in various years from 1970 to 2010. *Source: National Center for Health Statistics.*

| Year of Birth | Life Expectancy (years) |
|---|---|
| 1970 | 74.7 |
| 1975 | 76.6 |
| 1980 | 77.4 |
| 1985 | 78.2 |
| 1990 | 78.8 |
| 1995 | 78.9 |
| 2000 | 79.3 |
| 2005 | 79.9 |
| 2010 | 81.0 |

1. Find an equation for the least squares line for these data, using year of birth as the independent variable.

2. Use your regression equation to guess a value for the life expectancy of females born in 1900.

3. Compare your answer with the actual life expectancy for females born in 1900, which was 48.3 years. Are you surprised?

4. Find the life expectancy predicted by your regression equation for each year in the table, and subtract it from the actual value in the second column. This gives you a table of residuals. Plot your residuals as points on a graph.

5. What will happen if you try linear regression on the *residuals*? If you're not sure, use your calculator or software to find the regression equation for the residuals. Why does this result make sense?

6. Now look at the residuals as a fresh data set, and see if you can sketch the graph of a smooth function that fits the residuals well. How easy do you think it will be to predict the life expectancy at birth of females born in 2015?

7. Since most of the females born in 1995 are still alive, how did the Public Health Service come up with a life expectancy of 78.9 years for these women?

8. Go to the website WolframAlpha.com and enter: "linear fit { 1970, 74.7 }, { 1975, 76.6 }, etc.," putting in all the data from the table. Discuss how the solution compares with the solutions provided by a graphing calculator and by Microsoft Excel.

## DIRECTIONS FOR GROUP PROJECT

*Assume that you and your group (3–5 students) are preparing a report for a local health agency that is interested in using linear regression to predict life expectancy. Using the questions above as a guide, write a report that addresses the spirit of each question and any issues related to that question. The report should be mathematically sound, grammatically correct, and professionally crafted. Provide recommendations as to whether the health agency should proceed with the linear equation or whether it should seek other means of making such predictions.*

# 2

# Exponential, Logarithmic, and Trigonometric Functions

2.1 **Exponential Functions**

2.2 **Logarithmic Functions**

2.3 **Applications: Growth and Decay**

2.4 **Trigonometric Functions**

**Chapter 2 Review**

**Extended Application: Power Functions**

Ecologists want to know how long it takes for a population of animals to double. Microbiologists want to know the same thing about a population of bacteria. Oncologists want to know how long it will be until a tumor doubles in size. As we will see in Section 2 of this chapter, all of these questions can be answered with the same mathematical methods involving exponential functions and logarithms.

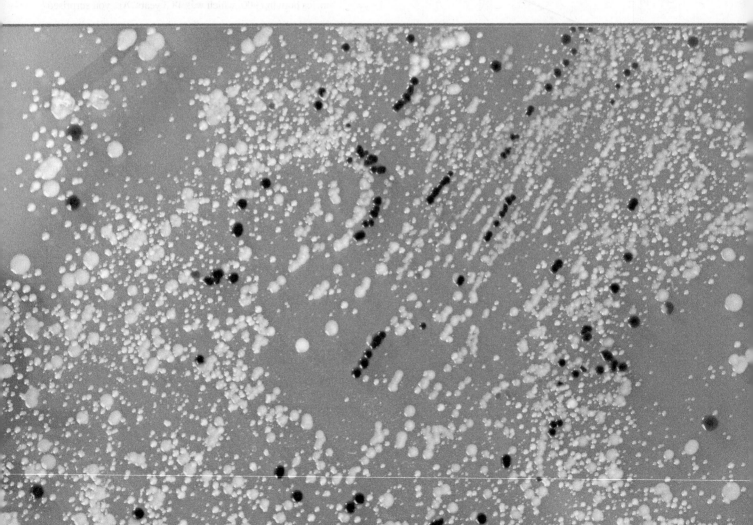

n the previous chapter, we studied *algebraic functions*: those that can be formed by addition, subtraction, multiplication, division, and raising to a numerical power. But some of the most important functions in the life sciences do not fall into this category; they are *transcendental*, because they transcend, or go beyond, the algebraic functions. In this chapter, we will study three types of transcendental functions. The first, exponential functions, arise whenever a population grows at a rate proportional to the size of the population. The second, logarithmic functions, are the inverse of exponential functions and are just as important. The third, trigonometric functions, are useful for describing periodic phenomena, such as a heartbeat or seasonal migration.

# 2.1  Exponential Functions

**APPLY IT**   **What is the oxygen consumption of yearling salmon?**
*In Example 6, we will see that the answer to this question depends on exponential functions.*

In the previous chapter we discussed functions involving expressions such as $x^2$, $(2x + 1)^3$, or $x^{-1}$, where the variable or variable expression is the base of an exponential expression, and the exponent is a constant. In an exponential function, the variable is in the exponent and the base is a constant.

---FOR REVIEW---

To review the properties of exponents used in this section, see Section R.6.

---

### Exponential Function

An **exponential function** with base $a$ is defined as

$$f(x) = a^x, \quad \text{where} \quad a > 0 \text{ and } a \neq 1.$$

(If $a = 1$, the function is the constant function $f(x) = 1$.)

Exponential functions may be the single most important type of functions used in practical applications. They are used to describe growth and decay, which are important ideas in management, social science, and biology.

Figure 1 shows a graph of the exponential function defined by $f(x) = 2^x$. You could plot such a curve by hand by noting that $2^{-2} = 1/4$, $2^{-1} = 1/2$, $2^0 = 1$, $2^1 = 2$, and $2^2 = 4$, and then drawing a smooth curve through the points $(-2, 1/4)$, $(-1, 1/2)$, $(0, 1)$, $(1, 2)$, and $(2, 4)$. This graph is typical of the graphs of exponential functions of the form $y = a^x$, where $a > 1$. The $y$-intercept is $(0, 1)$. Notice that as $x$ gets larger and larger, the function also gets larger. As $x$ gets more and more negative, the function becomes smaller and smaller, approaching but never reaching 0. Therefore, the $x$-axis is a horizontal asymptote, but the function only approaches the left side of the asymptote. In contrast, rational functions approach both the left and right sides of the asymptote. The graph suggests that the domain is the set of all real numbers and the range is the set of all positive numbers.

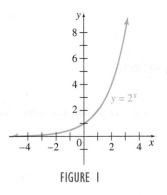

FIGURE 1

### EXAMPLE 1   Graphing an Exponential Function

Graph $f(x) = 2^{-x}$.

**SOLUTION**   The graph, shown in Figure 2, is the horizontal reflection of the graph of $f(x) = 2^x$ given in Figure 1. Since $2^{-x} = 1/2^x = (1/2)^x$, this graph is typical of the graphs of exponential functions of the form $y = a^x$ where $0 < a < 1$. The domain includes all real numbers and the range includes all positive numbers. The y-intercept is $(0, 1)$. Notice that this function, with $f(x) = 2^{-x} = (1/2)^x$, is decreasing over its domain.

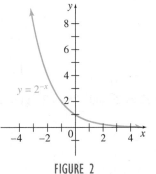

FIGURE 2

In the definition of an exponential function, notice that the base $a$ is restricted to positive values, with negative or zero bases not allowed. For example, the function $y = (-4)^x$ could not include such numbers as $x = 1/2$ or $x = 1/4$ in the domain because the $y$-values would not be real numbers. The resulting graph would be at best a series of separate points having little practical use.

**FOR REVIEW**

Recall from Section 1.4 that the graph of $f(-x)$ is the reflection of the graph of $f(x)$ about the $y$-axis.

## EXAMPLE 2   Graphing an Exponential Function

Graph $f(x) = -2^x + 3$.

**SOLUTION**    The graph of $y = -2^x$ is the vertical reflection of the graph of $y = 2^x$, so this is a decreasing function. (Notice that $-2^x$ is not the same as $(-2)^x$. In $-2^x$, we raise 2 to the $x$ power and then take the negative.) The 3 indicates that the graph should be translated vertically 3 units, as compared to the graph of $y = -2^x$. Since $y = -2^x$ would have $y$-intercept $(0, -1)$, this function has $y$-intercept $(0, 2)$, which is up 3 units. For negative values of $x$, the graph approaches the line $y = 3$, which is a horizontal asymptote. The graph is shown in Figure 3.

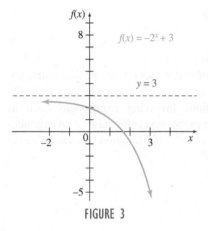

FIGURE 3

## Exponential Equations    In Figures 1 and 2, which are typical graphs of exponential functions, a given value of $x$ leads to exactly one value of $a^x$. Because of this, an equation with a variable in the exponent, called an **exponential equation**, often can be solved using the following property.

> If $a > 0$, $a \neq 1$, and $a^x = a^y$, then $x = y$.

The value $a = 1$ is excluded, since $1^2 = 1^3$, for example, even though $2 \neq 3$. To solve $2^{3x} = 2^7$ using this property, work as follows.

$$2^{3x} = 2^7$$
$$3x = 7$$
$$x = \frac{7}{3}.$$

## EXAMPLE 3   Solving Exponential Equations

**(a)** Solve $9^x = 27$.

**SOLUTION**    First rewrite both sides of the equation so the bases are the same. Since $9 = 3^2$ and $27 = 3^3$,

**FOR REVIEW**

Recall from Section R.6 that $(a^m)^n = a^{mn}$.

$$9^x = 27$$
$$(3^2)^x = 3^3$$
$$3^{2x} = 3^3 \qquad \text{Multiply exponents.}$$
$$2x = 3$$
$$x = \frac{3}{2}.$$

**(b)** Solve $32^{2x-1} = 128^{x+3}$.

**SOLUTION**    Since the bases must be the same, write 32 as $2^5$ and 128 as $2^7$, giving

$$32^{2x-1} = 128^{x+3}$$
$$(2^5)^{2x-1} = (2^7)^{x+3}$$
$$2^{10x-5} = 2^{7x+21}. \qquad \text{Multiply exponents.}$$

Now use the property from above to get

$$10x - 5 = 7x + 21$$
$$3x = 26$$
$$x = \frac{26}{3}.$$

**YOUR TURN 1** Solve
$25^{x/2} = 125^{x+3}$.

Verify this solution in the original equation.

TRY YOUR TURN 1

## Compound Interest

To understand the most fundamental type of growth function in biology, it will help to start with an example from the mathematics of finance. As we shall see in the next section, money and populations sometimes grow in a similar way.

The cost of borrowing money or the return on an investment is called **interest**. The amount borrowed or invested is the **principal**, *P*. The **rate of interest** *r* is given as a percent per year, and *t* is the **time**, measured in years.

### Simple Interest

The product of the principal *P*, rate *r*, and time *t* gives **simple interest**, *I*:

$$I = Prt.$$

With **compound interest**, interest is charged (or paid) on interest as well as on the principal. To find a formula for compound interest, first suppose that *P* dollars, the principal, is deposited at a rate of interest *r* per year. The interest earned during the first year is found using the formula for simple interest.

$$\text{First-year interest} = P \cdot r \cdot 1 = Pr.$$

At the end of one year, the amount on deposit will be the sum of the original principal and the interest earned, or

$$P + Pr = P(1 + r). \tag{1}$$

If the deposit earns compound interest, the interest earned during the second year is found from the total amount on deposit at the end of the first year. Thus, the interest earned during the second year (again found by the formula for simple interest), is

$$[P(1 + r)](r)(1) = P(1 + r)r, \tag{2}$$

so the total amount on deposit at the end of the second year is the sum of amounts from (1) and (2) above, or

$$P(1 + r) + P(1 + r)r = P(1 + r)(1 + r) = P(1 + r)^2.$$

In the same way, the total amount on deposit at the end of three years is

$$P(1 + r)^3.$$

After *t* years, the total amount on deposit, called the *compound amount*, is $P(1 + r)^t$.

When interest is compounded more than once a year, the compound interest formula is adjusted. For example, if interest is to be paid quarterly (four times a year), 1/4 of the interest rate is used each time interest is calculated, so the rate becomes *r*/4, and the number of compounding periods in *t* years becomes 4*t*. Generalizing from this idea gives the following formula.

### Compound Amount

If *P* dollars is invested at a yearly rate of interest *r* per year, compounded *m* times per year for *t* years, the **compound amount** is

$$A = P\left(1 + \frac{r}{m}\right)^{tm} \text{ dollars.}$$

---

EXAMPLE 4    **Compound Interest**

Inga Moffitt invests a bonus of $9000 at 6% annual interest compounded semiannually for 4 years. How much interest will she earn?

**SOLUTION**    Use the formula for compound interest with $P = 9000$, $r = 0.06$, $m = 2$, and $t = 4$.

$$A = P\left(1 + \frac{r}{m}\right)^{tm}$$

$$= 9000\left(1 + \frac{0.06}{2}\right)^{4(2)}$$

$$= 9000(1.03)^8$$

$$\approx 11{,}400.93 \qquad \text{Use a calculator.}$$

**YOUR TURN 2**   Find the interest earned on $4400 at 3.25% interest compounded quarterly for 5 years.

The investment plus the interest is $11,400.93. The interest is $11,400.93 − $9000 = $2400.93.                                                                          TRY YOUR TURN 2

**NOTE**   When using a calculator to compute the compound interest, store each partial result in the calculator and avoid rounding until the final answer.

## The Number $e$

Perhaps the single most useful base for an exponential function is the number $e$, an irrational number that occurs often in practical applications. The famous Swiss mathematician Leonhard Euler (pronounced "oiler") (1707–1783) was the first person known to have referred to this number as $e$, and the notation has continued to this day. To see how the number $e$ occurs in an application, begin with the formula for compound interest,

$$P\left(1 + \frac{r}{m}\right)^{tm}.$$

Suppose that a lucky investment produces annual interest of 100%, so that $r = 1.00 = 1$. Suppose also that you can deposit only $1 at this rate, and for only one year. Then $P = 1$ and $t = 1$. Substituting these values into the formula for compound interest gives

$$P\left(1 + \frac{r}{m}\right)^{t(m)} = 1\left(1 + \frac{1}{m}\right)^{1(m)} = \left(1 + \frac{1}{m}\right)^{m}.$$

As interest is compounded more and more often, $m$ gets larger and the value of this expression will increase. For example, if $m = 1$ (interest is compounded annually),

$$\left(1 + \frac{1}{m}\right)^{m} = \left(1 + \frac{1}{1}\right)^{1} = 2^1 = 2,$$

| X | Y1 |
|---|---|
| 1 | 2 |
| 8 | 2.5658 |
| 50 | 2.6916 |
| 100 | 2.7048 |
| 1000 | 2.7169 |
| 10000 | 2.7181 |
| 100000 | 2.7183 |

X=100000

**FIGURE 4**

so that your $1 becomes $2 in one year. Using a graphing calculator, we produced Figure 4 (where $m$ is represented by X and $(1 + 1/m)^m$ by $Y_1$) to see what happens as $m$ becomes larger and larger. A spreadsheet can also be used to produce this table.

The table suggests that as $m$ increases, the value of $(1 + 1/m)^m$ gets closer and closer to a fixed number, called $e$. As we shall see in the next chapter, this is an example of a limit.

### Definition of $e$

As $m$ becomes larger and larger, $\left(1 + \dfrac{1}{m}\right)^{m}$ becomes closer and closer to the number $e$, whose approximate value is 2.718281828.

The value of $e$ is approximated here to 9 decimal places. Euler approximated $e$ to 18 decimal places. Many calculators give values of $e^x$, usually with a key labeled $e^x$. Some

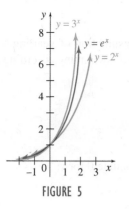

FIGURE 5

require two keys, either INV LN or 2nd LN. (We will define ln $x$ in the next section.) In Figure 5, the functions $y = 2^x$, $y = e^x$, and $y = 3^x$ are graphed for comparison. Notice that the graph of $e^x$ is between the graphs of $2^x$ and $3^x$, because $e$ is between 2 and 3. For $x > 0$, the graphs show that $3^x > e^x > 2^x$. All three functions have $y$-intercept $(0, 1)$. It is difficult to see from the graph, but $3^x < e^x < 2^x$ when $x < 0$.

The number $e$ is often used as the base in an exponential equation because it provides a good model for many natural, as well as economic, phenomena. In the exercises for this section, we will look at several examples of such applications.

## Continuous Compounding

In economics, the formula for **continuous compounding** is a good example of an exponential growth function. Recall the formula for compound amount

$$A = P\left(1 + \frac{r}{m}\right)^{tm},$$

where $m$ is the number of times annually that interest is compounded. As $m$ becomes larger and larger, the compound amount also becomes larger but not without bound. Recall that as $m$ becomes larger and larger, $(1 + 1/m)^m$ becomes closer and closer to $e$. Similarly,

$$\left(1 + \frac{1}{(m/r)}\right)^{m/r}$$

becomes closer and closer to $e$. Let us rearrange the formula for compound amount to take advantage of this fact.

$$A = P\left(1 + \frac{r}{m}\right)^{tm}$$

$$= P\left(1 + \frac{1}{(m/r)}\right)^{tm}$$

$$= P\left[\left(1 + \frac{1}{(m/r)}\right)^{m/r}\right]^{rt} \qquad \frac{m}{r} \cdot rt = tm$$

This last expression becomes closer and closer to $Pe^{rt}$ as $m$ becomes larger and larger, which describes what happens when interest is compounded continuously. Essentially, the number of times annually that interest is compounded becomes infinitely large. We thus have the following formula for the compound amount when interest is compounded continuously.

### Continuous Compounding

If a deposit of $P$ dollars is invested at a rate of interest $r$ compounded continuously for $t$ years, the compound amount is

$$A = Pe^{rt} \text{ dollars.}$$

### EXAMPLE 5    Continuous Compound Interest

If $600 is invested in an account earning 2.75% compounded continuously, how much would be in the account after 5 years?

**SOLUTION**  In the formula for continuous compounding, let $P = 600$, $t = 5$, and $r = 0.0275$ to get

$$A = 600e^{5(0.0275)} \approx 688.44,$$

or $688.44.

TRY YOUR TURN 3

**YOUR TURN 3**  Find the amount after 4 years if $800 is invested in an account earning 3.15% compounded continuously.

In situations that involve growth or decay of a population, the size of the population at a given time $t$ often is determined by an exponential function of $t$. The next example illustrates a typical application of this kind.

### EXAMPLE 6    Oxygen Consumption

Biologists studying salmon have found that the oxygen consumption of yearling salmon (in appropriate units) increases exponentially with the speed of swimming according to the function defined by

$$f(x) = 100e^{0.6x},$$

where $x$ is the speed in feet per second. Find the following.

**(a)** The oxygen consumption when the fish are still

**SOLUTION**    When the fish are still, their speed is 0. Substitute 0 for $x$:

$$f(0) = 100e^{(0.6)(0)} = 100e^0$$
$$= 100 \cdot 1 = 100. \qquad e^0 = 1$$

When the fish are still, their oxygen consumption is 100 units.

**(b)** The oxygen consumption at a speed of 2 ft per second

**SOLUTION**    Find $f(2)$ as follows.

$$f(2) = 100e^{(0.6)(2)} = 100e^{1.2} \approx 332$$

At a speed of 2 ft per second, oxygen consumption is about 332 units rounded to the nearest integer. Because the function is only an approximation of the real situation, further accuracy is not realistic.

---

### EXAMPLE 7    Food Surplus

A magazine article argued that the cause of the obesity epidemic in the United States is the decreasing cost of food (in real terms) due to the increasing surplus of food. ***Source: The New York Times Magazine.*** As one piece of evidence, the table at left was provided, which we have updated, showing U.S. corn production (in billions of bushels) for selected years.

**(a)** Plot the data. Does the production appear to grow linearly or exponentially?

**SOLUTION**    Figure 6 shows a graphing calculator plot of the data, which suggests that corn production is growing exponentially.

**(b)** Find an exponential function in the form of $p(t) = p_0 a^{t-1930}$ that models these data, where $t$ is the year and $p(t)$ is the production of corn. Use the data for 1930 and 2010.

**SOLUTION**    Since $p(1930) = p_0 a^0 = p_0$, we have $p_0 = 1.757$. Using $t = 2010$, we have

$$p(2010) = 1.757 a^{2010-1930} = 1.757 a^{80} = 12.447$$

$$a^{80} = \frac{12.447}{1.757} \qquad \text{Divide by 1.757.}$$

$$a = \left(\frac{12.447}{1.757}\right)^{1/80} \qquad \text{Take the 80th root.}$$

$$\approx 1.0248.$$

Thus, $p(t) = 1.757(1.0248)^{t-1930}$. Figure 7 shows that this function fits the data well.

**(c)** Determine the expected annual percentage increase in corn production during this time period.

**SOLUTION**    Since $a$ is approximately 1.0248, the production of corn each year is 1.0248 times its value the previous year, for a rate of increase of about $0.0248 = 2.48\%$ per year.

**FOR REVIEW**

Refer to the discussion on linear regression in Section 1.2. A similar process is used to fit data points to other types of functions. Many of the functions in this chapter's applications were determined in this way, including that given in Example 6.

| Year | Production (billions of bushels) |
|------|----------------------------------|
| 1930 | 1.757 |
| 1940 | 2.207 |
| 1950 | 2.764 |
| 1960 | 3.907 |
| 1970 | 4.152 |
| 1980 | 6.639 |
| 1990 | 7.934 |
| 2000 | 9.968 |
| 2010 | 12.447 |

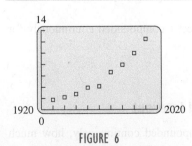

FIGURE 6

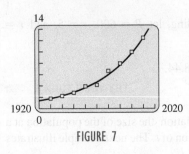

FIGURE 7

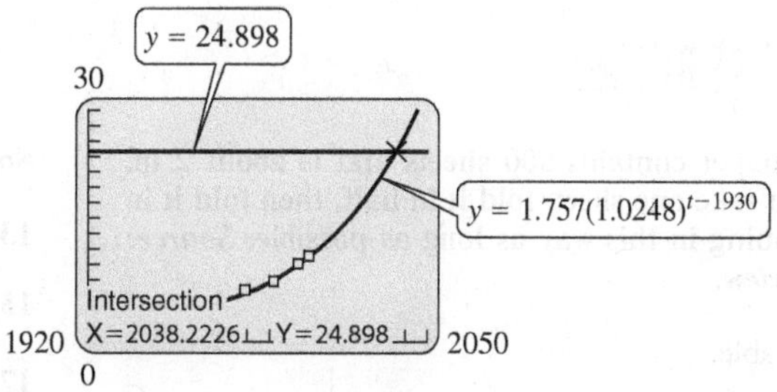

FIGURE 8

**(d)** Graph $p$ and estimate the year when corn production will be double what it was in 2010.

**SOLUTION**   Figure 8 shows the graphs of $p(t)$ and $y = 2 \cdot 12.447 = 24.898$ on the same coordinate axes. (Note that the scale in Figure 8 is different from the scale in Figures 6 and 7 so that larger values of $t$ and $p(t)$ are visible.) Their graphs intersect at approximately 2038, which is thus the year when corn production will be double its 2010 level. In the next section, we will see another way to solve such problems that does not require the use of a graphing calculator.

Another method of checking whether an exponential function fits the data is to see if points whose $x$-coordinates are equally spaced have $y$-coordinates with a constant ratio. This must be true for an exponential function because if $f(x) = a \cdot b^x$, then $f(x_1) = a \cdot b^{x_1}$ and $f(x_2) = a \cdot b^{x_2}$, so

$$\frac{f(x_2)}{f(x_1)} = \frac{a \cdot b^{x_2}}{a \cdot b^{x_1}} = b^{x_2 - x_1}.$$

This last expression is constant if $x_2 - x_1$ is constant, that is, if the $x$-coordinates are equally spaced.

In the previous example, all data points have $t$-coordinates 10 years apart, so we can compare the ratios of corn production for any of these pairs of years. Here are the ratios for 1930–1940 and for 1990–2000:

$$\frac{2.207}{1.757} = 1.256$$

$$\frac{9.968}{7.934} = 1.256.$$

These ratios are identical to 3 decimal places, so an exponential function fits the data very well. Not all ratios are this close; using the values at 1970 and 1980, we have $6.639/4.152 = 1.599$. From Figure 7, we can see that this is because the 1970 value is below the exponential curve and the 1980 value is above the curve.

**TECHNOLOGY NOTE**

> Another way to find an exponential function that fits a set of data is to use a graphing calculator or computer program with an exponential regression feature. This fits an exponential function through a set of points using the least squares method, introduced in Section 1.2 for fitting a line through a set of points. On a TI-84 Plus, for example, enter the year into the list $L_1$ and the corn production into $L_2$. For simplicity, subtract 1930 from each year, so that 1930 corresponds to $t = 0$. Selecting ExpReg from the STAT CALC menu yields $y = 1.733(1.0253)^t$, which is close to the function we found in Example 7(b).

# 2.1   EXERCISES

A ream of 20-lb paper contains 500 sheets and is about 2 in. high. Suppose you take one sheet, fold it in half, then fold it in half again, continuing in this way as long as possible. *Source: The AMATYC Review.*

1. Complete the table.

| Number of Folds | 1 | 2 | 3 | 4 | 5 | ... | 10 | ... | 50 |
|---|---|---|---|---|---|---|---|---|---|
| Layers of Paper | | | | | | | | | |

2. After folding 50 times (if this were possible), what would be the height (in miles) of the folded paper?

For Exercises 3–11, match the correct graph A–F to the function without using your calculator. Notice that there are more functions than graphs; some of the functions are equivalent. After you have answered all of them, use a graphing calculator to check your answers. Each graph in this group is plotted on the window $[-2, 2]$ by $[-4, 4]$.

3. $y = 3^x$

4. $y = 3^{-x}$

5. $y = \left(\dfrac{1}{3}\right)^{1-x}$

6. $y = 3^{x+1}$

7. $y = 3(3)^x$

8. $y = \left(\dfrac{1}{3}\right)^x$

9. $y = 2 - 3^{-x}$

10. $y = -2 + 3^{-x}$

11. $y = 3^{x-1}$

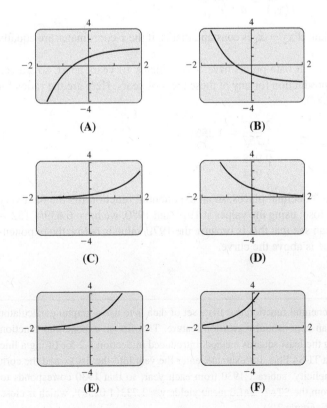

(A)

(B)

(C)

(D)

(E)

(F)

12. In Exercises 3–11, there were more formulas for functions than there were graphs. Explain how this is possible.

**Solve each equation.**

13. $2^x = 32$

14. $4^x = 64$

15. $3^x = \dfrac{1}{81}$

16. $e^x = \dfrac{1}{e^5}$

17. $4^x = 8^{x+1}$

18. $25^x = 125^{x+2}$

19. $16^{x+3} = 64^{2x-5}$

20. $(e^3)^{-2x} = e^{-x+5}$

21. $e^{-x} = (e^4)^{x+3}$

22. $2^{|x|} = 8$

23. $5^{-|x|} = \dfrac{1}{25}$

24. $2^{x^2-4x} = \left(\dfrac{1}{16}\right)^{x-4}$

25. $5^{x^2+x} = 1$

26. $8^{x^2} = 2^{x+4}$

27. $27^x = 9^{x^2+x}$

28. $e^{x^2+5x+6} = 1$

**Graph each of the following.**

29. $y = 5e^x + 2$

30. $y = -2e^x - 3$

31. $y = -3e^{-2x} + 2$

32. $y = 4e^{-x/2} - 1$

33. In our definition of an exponential function, we ruled out negative values of $a$. The author of a textbook on mathematical economics, however, obtained a "graph" of $y = (-2)^x$ by plotting the following points and drawing a smooth curve through them.

| $x$ | $-4$ | $-3$ | $-2$ | $-1$ | 0 | 1 | 2 | 3 |
|---|---|---|---|---|---|---|---|---|
| $y$ | 1/16 | $-1/8$ | 1/4 | $-1/2$ | 1 | $-2$ | 4 | $-8$ |

The graph oscillates very neatly from positive to negative values of $y$. Comment on this approach. (This exercise shows the dangers of point plotting when drawing graphs.)

34. Explain why the exponential equation $4^x = 6$ cannot be solved using the method described in Example 3.

35. Explain why $3^x > e^x > 2^x$ when $x > 0$, but $3^x < e^x < 2^x$ when $x < 0$.

36. A friend claims that as $x$ becomes large, the expression $1 + 1/x$ gets closer and closer to 1, and 1 raised to any power is still 1. Therefore, $f(x) = (1 + 1/x)^x$ gets closer and closer to 1 as $x$ gets larger. Use a graphing calculator to graph $f$ on $0.1 \le x \le 50$. How might you use this graph to explain to the friend why $f(x)$ does not approach 1 as $x$ becomes large? What does it approach?

## LIFE SCIENCE APPLICATIONS

37. **Population Growth**   Since 1960, the growth in world population (in millions) closely fits the exponential function defined by

$$A(t) = 3100e^{0.0166t},$$

where $t$ is the number of years since 1960. *Source: United Nations.*

**a.** World population was about 3686 million in 1970. How closely does the function approximate this value?

**b.** Use the function to approximate world population in 2000. (The actual 2000 population was about 6115 million.)

**c.** Estimate world population in the year 2015.

**38. Growth of Bacteria** Salmonella bacteria, found on almost all chicken and eggs, grow rapidly in a nice warm place. If just a few hundred bacteria are left on the cutting board when a chicken is cut up, and they get into the potato salad, the population begins compounding. Suppose the number present in the potato salad after $t$ hours is given by

$$f(t) = 500 \cdot 2^{3t}.$$

**a.** If the potato salad is left out on the table, how many bacteria are present 1 hour later?

**b.** How many were present initially?

**c.** How often do the bacteria double?

**d.** How quickly will the number of bacteria increase to 32,000?

**39. Minority Population** According to the U.S. Census Bureau, the United States is becoming more diverse. Based on U.S. Census population projections for 2000 to 2050, the projected Hispanic population (in millions) can be modeled by the exponential function

$$h(t) = 37.79(1.021)^t,$$

where $t = 0$ corresponds to 2000 and $0 \le t \le 50$. *Source: U.S. Census Bureau.*

**a.** Find the projected Hispanic population for 2005. Compare this to the actual value of 42.69 million.

**b.** The U.S. Asian population is also growing exponentially, and the projected Asian population (in millions) can be modeled by the exponential function

$$a(t) = 11.14(1.023)^t,$$

where $t = 0$ corresponds to 2000 and $0 \le t \le 50$. Find the projected Asian population for 2005, and compare this to the actual value of 12.69 million.

**c.** Determine the expected annual percentage increase for Hispanics and for Asians. Which minority population, Hispanic or Asian, is growing at a faster rate?

**d.** The U.S. black population is growing at a linear rate, and the projected black population (in millions) can be modeled by the linear function

$$b(t) = 0.5116t + 35.43,$$

where $t = 0$ corresponds to 2000 and $0 \le t \le 50$. Find the projected black population for 2005 and compare this projection to the actual value of 37.91 million.

 **e.** Graph the projected population function for Hispanics and estimate when the Hispanic population will be double its actual value for 2005. Then do the same for the Asian and black populations. Comment on the accuracy of these numbers.

**40. Physician Demand** The demand for physicians is expected to increase in the future, as shown in the accompanying table. *Source: Association of American Medical Colleges.*

| Year | Demand for Physicians (in thousands) |
|---|---|
| 2006 | 680.5 |
| 2015 | 758.6 |
| 2020 | 805.8 |
| 2025 | 859.3 |

**a.** Plot the data, letting $t = 0$ correspond to 2000. Does fitting an exponential curve to the data seem reasonable?

**b.** Use the data for 2006 and 2015 to find a function of the form $f(t) = Ce^{kt}$ that goes through these two points.

**c.** Use your function from part b to predict the demand for physicians in 2020 and 2025. How well do these predictions fit the data?

**d.** If you have a graphing calculator or computer program with an exponential regression feature, use it to find an exponential function that approximately fits the data. How does this answer compare with the answer to part b?

**41. Bald Eagle Population** In 1967, bald eagles south of the 40th parallel were listed under the Endangered Species Preservation Act. Conservation methods, including habitat protection, have helped bald eagles make a remarkable recovery. The bald eagle was removed from the list of threatened and endangered species in June 2007. The following table gives the number of bald eagle breeding pairs in the lower 48 states for selected years. *Source: U.S. Fish & Wildlife Service.*

| Year | Number of Pairs |
|---|---|
| 1963 | 487 |
| 1974 | 791 |
| 1981 | 1188 |
| 1990 | 3035 |
| 2000 | 6471 |
| 2006 | 9789 |

**a.** Plot the data, letting $t = 0$ correspond to 1963. Do the number of pairs appear to grow linearly or exponentially?

**b.** Find an exponential function in the form of $f(t) = f_0 a^t$ that fits the data at 1963 and 2006, where $t$ is the number of years since 1963 and $f(t)$ is the number of breeding pairs.

**c.** Approximate the average annual percentage increase in the number of breeding pairs of bald eagles during this time period.

**d.** If you have a graphing calculator or computer program with an exponential regression feature, use it to find an exponential function that approximately fits the data. How does this answer compare with the answer to part b?

**42. Bacteria in Sausage** The number of *Enterococcus faecium* bacteria contaminating bologna sausage at 32°C can be described by the equation

$$N = N_0 \exp\{9.8901 \exp[-\exp(2.5420 - 0.2167t)]\},$$

where $t$ is the time in hours and $N_0$ is the number at time $t = 0$. **Source: Applied and Environmental Microbiology.** (We have used the notation exp $x$ to represent $e^x$ because otherwise all the powers raised to powers would make the expression difficult to read.) Suppose $N_0 = 100$.

a. Find the number of bacteria after 10, 20, 30, 40, and 50 hours.

b. Use a graphing calculator to graph the function on the interval $[0, 50]$.

c. Using your answers to parts a and b, what seems to happen to the number of bacteria after a long period of time?

**43. Heart Attack Risk** The following data give the 10-year risk of a heart attack for men based on their score on a test of risk factors. (For example, a man increases his score by 2 points if his HDL cholesterol is less than 40 mg/dL.) **Source: The New York Times.**

| Score ($x$) | Percent Risk ($y$) |
|---|---|
| 0 | 1 |
| 5 | 2 |
| 10 | 6 |
| 15 | 20 |
| 16 | 25 |
| 17 | 30 |

a. Graph the function

$$f(x) = 0.8454e^{0.2081x}$$

on the same axes as the data. How well does this function describe the data?

b. Using the function in part a, find the risk of a man with a score of 3.

c. A man aged 20 to 34 increases his score by 8 points if he is a smoker. If his score was 3 without considering his smoking, by what percent does his risk go up if he smokes?

d. Redo part c under the assumption that the man's score, without considering his smoking, was 6.

e. The formula for women is

$$g(x) = 0.1210e^{0.2249x}.$$

Answer part b for a woman.

f. A woman aged 20 to 34 increases her score by 9 points if she is a smoker. If her score was 3 without considering her smoking, by what percent does her risk go up if she smokes?

g. Redo part f under the assumption that the woman's score, without considering her smoking, was 6.

**44. Plasma Volume** A formula for plasma volume for males proposed by Hurley is

$$PV = 995e^{0.6085S},$$

where $PV$ is measured in milliliters and $S$ in square meters. **Source: Journal of Nuclear Medicine.** A formula proposed by Pearson et al. is

$$PV = 1578S.$$

**Source: British Journal of Haematology.** Use a graphing calculator to verify that these two formulas never give the same answer.

**45. Cortisone Concentration** A mathematical model for the concentration of administered cortisone in humans over a 24-hour period uses the function

$$C = \frac{D \times a}{V(a - b)}(e^{-bt} - e^{-at}),$$

where $C$ is the concentration, $D$ is the dose given at time $t = 0$, $V$ is the volume of distribution (volume divided by bioavailability), $a$ is the absorption rate, $b$ is the elimination rate, and $t$ is the time in hours. **Source: Journal of Pharmacokinetics and Biopharmaceutics.**

a. What is the value of $C$ at $t = 0$? Explain why this makes sense.

b. What happens to the concentration as a large amount of time passes? Explain why this makes sense.

c. The researchers used the values $D = 500$ micrograms, $a = 8.5$, $b = 0.09$, and $V = 3700$ liters. Use these values and a graphing calculator to estimate when the concentration is greatest.

## OTHER APPLICATIONS

**46. Carbon Dioxide** The table gives the estimated global carbon dioxide ($CO_2$) emissions from fossil-fuel burning, cement production, and gas flaring over the last century. The $CO_2$ estimates are expressed in millions of metric tons. **Source: Carbon Dioxide Information Analysis Center.**

| Year | $CO_2$ Emissions (millions of metric tons) |
|---|---|
| 1900 | 534 |
| 1910 | 819 |
| 1920 | 932 |
| 1930 | 1053 |
| 1940 | 1299 |
| 1950 | 1630 |
| 1960 | 2569 |
| 1970 | 4053 |
| 1980 | 5316 |
| 1990 | 6151 |
| 2000 | 6750 |
| 2008 | 8749 |

a. Plot the data, letting $t = 0$ correspond to 1900. Do the emissions appear to grow linearly or exponentially?

b. Find an exponential function in the form of $f(t) = f_0 a^t$ that fits these data at 1900 and 2008, where $t$ is the number of years since 1900 and $f(t)$ is the $CO_2$ emissions.

c. Approximate the average annual percentage increase in $CO_2$ emissions during this time period.

d. Graph $f(t)$ and estimate the first year when emissions will be at least double what they were in 2008.

**47. Radioactive Decay** Suppose the quantity (in grams) of a radio-active substance present at time $t$ is

$$Q(t) = 1000(5^{-0.3t}),$$

where $t$ is measured in months.

a. How much will be present in 6 months?

b. How long will it take to reduce the substance to 8 g?

**48. Atmospheric Pressure** The atmospheric pressure (in milli-bars) at a given altitude (in meters) is listed in the table. *Source: Elements of Meteorology.*

| Altitude | Pressure |
|----------|----------|
| 0 | 1013 |
| 1000 | 899 |
| 2000 | 795 |
| 3000 | 701 |
| 4000 | 617 |
| 5000 | 541 |
| 6000 | 472 |
| 7000 | 411 |
| 8000 | 357 |
| 9000 | 308 |
| 10,000 | 265 |

a. Find functions of the form $P = ae^{kx}$, $P = mx + b$, and $P = 1/(ax + b)$ that fit the data at $x = 0$ and $x = 10{,}000$, where $P$ is the pressure and $x$ is the altitude.

b. Plot the data in the table and graph the three functions found in part a. Which function best fits the data?

c. Use the best-fitting function from part b to predict pressure at 1500 m and 11,000 m. Compare your answers to the true values of 846 millibars and 227 millibars, respectively.

d. If you have a graphing calculator or computer program with an exponential regression feature, use it to find an exponential function that approximately fits the data. How does this answer compare with the answer to part b?

**49. Computer Chips** The power of personal computers has increased dramatically as a result of the ability to place an in-creasing number of transistors on a single processor chip. The following table lists the number of transistors on some popular computer chips made by Intel. *Source: Intel.*

| Year | Chip | Transistors (in millions) |
|------|------|---------------------------|
| 1985 | 386 | 0.275 |
| 1989 | 486 | 1.2 |
| 1993 | Pentium | 3.1 |
| 1997 | Pentium II | 7.5 |
| 1999 | Pentium III | 9.5 |
| 2000 | Pentium 4 | 42 |
| 2005 | Pentium D | 291 |
| 2007 | Penryn | 820 |
| 2009 | Nehalem | 1900 |

a. Let $t$ be the year, where $t = 0$ corresponds to 1985, and $y$ be the number of transistors (in millions). Find functions of the form $y = mt + b$, $y = at^2 + b$, and $y = ab^t$ that fit the data at 1985 and 2009.

b. Use a graphing calculator to plot the data in the table and to graph the three functions found in part a. Which function best fits the data?

c. Use the best-fitting function from part b to predict the number of transistors on a chip in the year 2015.

d. If you have a graphing calculator or computer program with an exponential regression feature, use it to find an exponential function that approximately fits the data. How does this answer compare with the answer to part b?

e. In 1965 Gordon Moore wrote a paper predicting how the power of computer chips would grow in the future. Moore's law says that the number of transistors that can be put on a chip doubles roughly every 18 months. Discuss the extent to which the data in this exercise confirms or refutes Moore's law.

**50. Wind Energy** The following table gives the total world wind energy capacity (in megawatts) in recent years. *Source: World Wind Energy Association.*

| Year | Capacity (MW) |
|------|---------------|
| 2001 | 24,322 |
| 2002 | 31,181 |
| 2003 | 39,295 |
| 2004 | 47,681 |
| 2005 | 59,012 |
| 2006 | 74,112 |
| 2007 | 93,919 |
| 2008 | 120,913 |
| 2009 | 159,762 |
| 2010 | 196,986 |
| 2011 | 237,016 |

a. Let $t$ be the number of years since 2000, and $C$ the capac-ity (in MW). Find functions of the form $C = mt + b$, $C = at^2 + b$, and $C = ab^t$ that fit the data at 2001 and 2011.

b. Use a graphing calculator to plot the data in the table and to graph the three functions found in part a. Which function best fits the data?

c. If you have a graphing calculator or computer program with an exponential regression feature, use it to find an exponen-tial function that approximately fits the data in the table. How does this answer compare with the answer to part b?

d. Using the three functions from part b and the function from part c, predict the total world wind capacity in 2012. Com-pare these with the World Wind Energy Association's pre-diction of 273,000.

**51. Interest** Find the interest earned on $10,000 invested for 5 years at 4% interest compounded as follows.

a. Annually    b. Semiannually (twice a year)

c. Quarterly    d. Monthly    e. Continuously

**52. Interest** Suppose $26,000 is borrowed for 4 years at 6% interest. Find the interest paid over this period if the interest is compounded as follows.

  a. Annually

  b. Semiannually

  c. Quarterly

  d. Monthly

  e. Continuously

**53. Interest** Ron Hampton needs to choose between two investments: One pays 6% compounded annually, and the other pays 5.9% compounded monthly. If he plans to invest $18,000 for 2 years, which investment should he choose? How much extra interest will he earn by making the better choice?

**54. Interest** Find the interest rate required for an investment of $5000 to grow to $7500 in 5 years if interest is compounded as follows.

  a. Annually

  b. Quarterly

**55. Inflation** Assuming continuous compounding, what will it cost to buy a $10 item in 3 years at the following inflation rates?

  a. 3%      b. 4%      c. 5%

**56. Interest** Ali Williams invests a $25,000 inheritance in a fund paying 5.5% per year compounded continuously. What will be the amount on deposit after each time period?

  a. 1 year      b. 5 years      c. 10 years

**57. Interest** Leigh Jacks plans to invest $500 into a money market account. Find the interest rate that is needed for the money to grow to $1200 in 14 years if the interest is compounded quarterly.

**58. Interest** Kristi Perez puts $10,500 into an account to save money to buy a car in 12 years. She expects the car of her dreams to cost $30,000 by then. Find the interest rate that is necessary if the interest is computed using the following methods.

  a. Compounded quarterly

  b. Compounded daily (Use 365 days per year.)

**59. Inflation** If money loses value at the rate of 8% per year, the value of $1 in $t$ years is given by

$$y = (1 - 0.08)^t = (0.92)^t.$$

  a. Use a calculator to help complete the following table.

| $t$ | 0 | 1 | 2 | 3 | 4 | 5 | 6 | 7 | 8 | 9 | 10 |
|---|---|---|---|---|---|---|---|---|---|---|---|
| $y$ | 1 | | | | | 0.66 | | | | | 0.43 |

  b. Graph $y = (0.92)^t$.

  c. Suppose a house costs $165,000 today. Use the results of part a to estimate the cost of a similar house in 10 years.

  d. Find the cost of a $50 textbook in 8 years.

**60. Interest** On January 1, 2000, Jack deposited $1000 into Bank X to earn interest at the rate of $j$ per annum compounded semiannually. On January 1, 2005, he transferred his account to Bank Y to earn interest at the rate of $k$ per annum compounded quarterly. On January 1, 2008, the balance at Bank Y was $1990.76. If Jack could have earned interest at the rate of $k$ per annum compounded quarterly from January 1, 2000, through January 1, 2008, his balance would have been $2203.76. Which of the following represents the ratio $k/j$? *Source: Society of Actuaries.*

  a. 1.25      b. 1.30      c. 1.35

  d. 1.40      e. 1.45

**YOUR TURN ANSWERS**

1. $-9/2$      2. $772.97      3. $907.43

# 2.2 Logarithmic Functions

**APPLY IT** If a population of squirrels grows 5% per year, how long will it take for the population to double?

The amount of time it will take for a population to double under given conditions is called the **doubling time**. To analyze how 400 squirrels become 800 in $t$ years, assuming 5% annual growth, we can use the same equation that we used for money growing at 5% interest compounded annually. In mathematical symbols,

$$800 = 400(1.05)^t.$$

Dividing by 400 gives

$$2 = (1.05)^t.$$

This equation would be easier to solve if the variable were not in the exponent. **Logarithms** are defined for just this purpose. In Example 8, we will use logarithms to answer the question posed above.

Definition of Logarithm
For $a > 0$, $a \neq 1$, and $x > 0$,

$$y = \log_a x \qquad \text{means} \qquad a^y = x.$$

(Read $y = \log_a x$ as "$y$ is the logarithm of $x$ to the base a.") For example, the exponential statement $2^4 = 16$ can be translated into the logarithmic statement $4 = \log_2 16$. Also, in the problem discussed above, $(1.05)^t = 2$ can be rewritten with this definition as $t = \log_{1.05} 2$. A logarithm is an exponent: $\log_a x$ **is the exponent used with the base $a$ to get $x$.**

### EXAMPLE 1   Equivalent Expressions

This example shows the same statements written in both exponential and logarithmic forms.

| *Exponential Form* | *Logarithmic Form* |
|---|---|
| **(a)** $3^2 = 9$ | $\log_3 9 = 2$ |
| **(b)** $(1/5)^{-2} = 25$ | $\log_{1/5} 25 = -2$ |
| **(c)** $10^5 = 100{,}000$ | $\log_{10} 100{,}000 = 5$ |
| **(d)** $4^{-3} = 1/64$ | $\log_4(1/64) = -3$ |
| **(e)** $2^{-4} = 1/16$ | $\log_2(1/16) = -4$ |
| **(f)** $e^0 = 1$ | $\log_e 1 = 0$ |

**YOUR TURN 1** Write the equation $5^{-2} = 1/25$ in logarithmic form.

TRY YOUR TURN 1

### EXAMPLE 2   Evaluating Logarithms

Evaluate each of the following logarithms.

**(a)** $\log_4 64$

**SOLUTION**   We seek a number $x$ such that $4^x = 64$. Since $4^3 = 64$, we conclude that $\log_4 64 = 3$.

**(b)** $\log_2(-8)$

**SOLUTION**   We seek a number $x$ such that $2^x = -8$. Since $2^x$ is positive for all real numbers $x$, we conclude that $\log_2(-8)$ is undefined. (Actually, $\log_2(-8)$ can be defined if we use complex numbers, but in this textbook, we restrict ourselves to real numbers.)

**(c)** $\log_5 80$

**YOUR TURN 2** Evaluate $\log_3(1/81)$.

**SOLUTION**   We know that $5^2 = 25$ and $5^3 = 125$, so $\log_5 25 = 2$ and $\log_5 125 = 3$. Therefore, $\log_5 80$ must be somewhere between 2 and 3. We will find a more accurate answer in Example 4.

TRY YOUR TURN 2

## Logarithmic Functions
For a given positive value of $x$, the definition of logarithm leads to exactly one value of $y$, so $y = \log_a x$ defines the *logarithmic function* of base $a$ (the base $a$ must be positive, with $a \neq 1$).

Logarithmic Function
If $a > 0$ and $a \neq 1$, then the **logarithmic function** of base $a$ is defined by

$$f(x) = \log_a x$$

for $x > 0$.

The graphs of the exponential function with $f(x) = 2^x$ and the logarithmic function with $g(x) = \log_2 x$ are shown in Figure 9 on the next page. The graphs show that $f(3) = 2^3 = 8$, while $g(8) = \log_2 8 = 3$. Thus, $f(3) = 8$ and $g(8) = 3$. Also, $f(2) = 4$ and

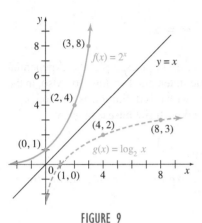

**FIGURE 9**

$g(4) = 2$. In fact, for any number $m$, if $f(m) = p$, then $g(p) = m$. Functions related in this way are called **inverse functions** of each other. The graphs also show that the domain of the exponential function (the set of real numbers) is the range of the logarithmic function. Also, the range of the exponential function (the set of positive real numbers) is the domain of the logarithmic function. Every logarithmic function is the inverse of some exponential function. This means that we can graph logarithmic functions by rewriting them as exponential functions using the definition of logarithm. The graphs in Figure 9 show a characteristic of a pair of inverse functions: their graphs are mirror images about the line $y = x$. Therefore, since exponential functions pass through the point $(0, 1)$, logarithmic functions pass through the point $(1, 0)$. Notice that because the exponential function has the $x$-axis as a horizontal asymptote, the logarithmic function has the $y$-axis as a vertical asymptote. A more complete discussion of inverse functions is given in most standard intermediate algebra and college algebra books.

The graph of $\log_2 x$ is typical of logarithms with bases $a > 1$. When $0 < a < 1$, the graph is the vertical reflection of the logarithm graph in Figure 9. Because logarithms with bases less than 1 are rarely used, we will not explore them here.

> **CAUTION** The domain of $\log_a x$ consists of all $x > 0$. In other words, you cannot take the logarithm of zero or a negative number. This also means that in a function such as $g(x) = \log_a(x - 2)$, the domain is given by $x - 2 > 0$, or $x > 2$.

## Properties of Logarithms

The usefulness of logarithmic functions depends in large part on the following **properties of logarithms**.

> **Properties of Logarithms**
>
> Let $x$ and $y$ be any positive real numbers and $r$ be any real number. Let $a$ be a positive real number, $a \neq 1$. Then
>
> **a.** $\log_a xy = \log_a x + \log_a y$
> **b.** $\log_a \dfrac{x}{y} = \log_a x - \log_a y$
> **c.** $\log_a x^r = r \log_a x$
> **d.** $\log_a a = 1$
> **e.** $\log_a 1 = 0$
> **f.** $\log_a a^r = r$.

To prove property (a), let $m = \log_a x$ and $n = \log_a y$. Then, by the definition of logarithm,

$$a^m = x \quad \text{and} \quad a^n = y.$$

Hence,

$$a^m a^n = xy.$$

By a property of exponents, $a^m a^n = a^{m+n}$, so

$$a^{m+n} = xy.$$

Now use the definition of logarithm to write

$$\log_a xy = m + n.$$

Since $m = \log_a x$ and $n = \log_a y$,

$$\log_a xy = \log_a x + \log_a y.$$

Proofs of properties (b) and (c) are left for the exercises. Properties (d) and (e) depend on the definition of a logarithm. Property (f) follows from properties (c) and (d).

**EXAMPLE 3**   **Properties of Logarithms**

If all the following variable expressions represent positive numbers, then for $a > 0$, $a \neq 1$, the statements in (a)–(c) are true.

**(a)** $\log_a x + \log_a(x - 1) = \log_a x(x - 1)$

**(b)** $\log_a \dfrac{x^2 - 4x}{x + 6} = \log_a(x^2 - 4x) - \log_a(x + 6)$

**(c)** $\log_a(9x^5) = \log_a 9 + \log_a(x^5) = \log_a 9 + 5 \cdot \log_a x$    TRY YOUR TURN 3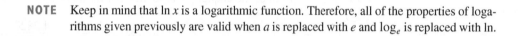

**YOUR TURN 3**  Write the expression $\log_a(x^2/y^3)$ as a sum, difference, or product of simpler logarithms.

# Evaluating Logarithms

The invention of logarithms is credited to John Napier (1550–1617), who first called logarithms "artificial numbers." Later he joined the Greek words *logos* (ratio) and *arithmos* (number) to form the word used today. The development of logarithms was motivated by a need for faster computation. Tables of logarithms and slide rule devices were developed by Napier, Henry Briggs (1561–1631), Edmund Gunter (1581–1626), and others.

For many years logarithms were used primarily to assist in involved calculations. Current technology has made this use of logarithms obsolete, but logarithmic functions play an important role in many applications of mathematics. Since our number system has base 10, logarithms to base 10 were most convenient for numerical calculations and so base 10 logarithms were called **common logarithms**. Common logarithms are still useful in other applications. For simplicity,

$$\log_{10} x \text{ is abbreviated } \log x.$$

Most practical applications of logarithms use the number $e$ as base. (Recall that to 7 decimal places, $e = 2.7182818$.) Logarithms to base e are called **natural logarithms**, and

$$\log_e x \text{ is abbreviated } \ln x$$

(read "el-en x"). A graph of $f(x) = \ln x$ is given in Figure 10.

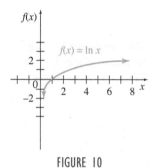

FIGURE 10

**NOTE**  Keep in mind that ln $x$ is a logarithmic function. Therefore, all of the properties of logarithms given previously are valid when $a$ is replaced with $e$ and $\log_e$ is replaced with ln.

Although common logarithms may seem more "natural" than logarithms to base $e$, there are several good reasons for using natural logarithms instead. The most important reason is discussed later, in Section 4.5 on Derivatives of Logarithmic Functions.

A calculator can be used to find both common and natural logarithms. For example, using a calculator and 4 decimal places, we get the following values.

$$\log 2.34 = 0.3692, \quad \log 594 = 2.7738, \quad \text{and} \quad \log 0.0028 = -2.5528.$$
$$\ln 2.34 = 0.8502, \quad \ln 594 = 6.3869, \quad \text{and} \quad \log 0.0028 = -5.8781.$$

Notice that logarithms of numbers less than 1 are negative when the base is greater than 1. A look at the graph of $y = \log_2 x$ or $y = \ln x$ will show why.

Sometimes it is convenient to use logarithms to bases other than 10 or $e$. For example, some computer science applications use base 2. In such cases, the following theorem is useful for converting from one base to another.

## Change-of-Base Theorem for Logarithms

If $x$ is any positive number and if $a$ and $b$ are positive real numbers, $a \neq 1$, $b \neq 1$, then

$$\log_a x = \frac{\log_b x}{\log_b a}.$$

To prove this result, use the definition of logarithm to write $y = \log_a x$ as $x = a^y$ or $x = a^{\log_a x}$ (for positive $x$ and positive $a$, $a \neq 1$). Now take base $b$ logarithms of both sides of this last equation.

$$\log_b x = \log_b a^{\log_a x}$$

$$\log_b x = (\log_a x)(\log_b a), \qquad \log_a x^r = r \log_a x$$

$$\log_a x = \frac{\log_b x}{\log_b a} \qquad \text{Solve for } \log_a x.$$

If the base $b$ is equal to $e$, then by the change-of-base theorem,

$$\log_a x = \frac{\log_e x}{\log_e a}.$$

Using $\ln x$ for $\log_e x$ gives the special case of the theorem using natural logarithms.

For any positive numbers $a$ and $x$, $a \neq 1$,

$$\log_a x = \frac{\ln x}{\ln a}.$$

The change-of-base theorem for logarithms is useful when graphing $y = \log_a x$ on a graphing calculator for a base $a$ other than $e$ or 10. For example, to graph $y = \log_2 x$, let $y = \ln x/\ln 2$. The change-of-base theorem is also needed when using a calculator to evaluate a logarithm with a base other than $e$ or 10.

### EXAMPLE 4  Evaluating Logarithms

Evaluate $\log_5 80$.

**SOLUTION**  As we saw in Example 2, this number is between 2 and 3. Using the second form of the change-of-base theorem for logarithms with $x = 80$ and $a = 5$ gives

$$\log_5 80 = \frac{\ln 80}{\ln 5} \approx \frac{4.3820}{1.6094} \approx 2.723.$$

**YOUR TURN 4** Evaluate $\log_3 50$.

To check, use a calculator to verify that $5^{2.723} \approx 80$.

TRY YOUR TURN 4

**CAUTION**  As mentioned earlier, when using a calculator, do not round off intermediate results. Keep all numbers in the calculator until you have the final answer. In Example 4, we showed the rounded intermediate values of $\ln 80$ and $\ln 5$, but we used the unrounded quantities when doing the division.

## Logarithmic Equations
Equations involving logarithms are often solved by using the fact that exponential functions and logarithmic functions are inverses, so a logarithmic equation can be rewritten (with the definition of logarithm) as an exponential equation. In other cases, the properties of logarithms may be useful in simplifying a **logarithmic equation**.

### EXAMPLE 5  Solving Logarithmic Equations

Solve each equation.

**(a)** $\log_x \frac{8}{27} = 3$

**SOLUTION** Using the definition of logarithm, write the expression in exponential form. To solve for $x$, take the cube root on both sides.

$$x^3 = \frac{8}{27}$$

$$x = \frac{2}{3}$$

**(b)** $\log_4 x = \dfrac{5}{2}$

**SOLUTION** In exponential form,

$$4^{5/2} = x$$
$$\left(4^{1/2}\right)^5 = x$$
$$2^5 = x$$
$$32 = x.$$

**(c)** $\log_2 x - \log_2(x - 1) = 1$

**SOLUTION** By a property of logarithms,

$$\log_2 x - \log_2(x - 1) = \log_2 \frac{x}{x - 1},$$

so the original equation becomes

$$\log_2 \frac{x}{x - 1} = 1.$$

Now write this equation in exponential form, and solve.

$$\frac{x}{x - 1} = 2^1 = 2$$

Solve this equation.

$$\frac{x}{x - 1}(x - 1) = 2(x - 1) \qquad \text{Multiply both sides by } x - 1.$$
$$x = 2(x - 1)$$
$$x = 2x - 2$$
$$-x = -2$$
$$x = 2$$

**(d)** $\log x + \log(x - 3) = 1$

**SOLUTION** Similar to part (c), we have

$$\log x + \log(x - 3) = \log[x(x - 3)] = 1.$$

Since the logarithm base is 10,

$$x(x - 3) = 10 \qquad 10^1 = 10$$
$$x^2 - 3x - 10 = 0 \qquad \text{Subtract 10 from both sides.}$$
$$(x - 5)(x + 2) = 0. \qquad \text{Factor.}$$

**YOUR TURN 5** Solve for $x$:
$\log_2 x + \log_2(x + 2) = 3$.

This leads to two solutions: $x = 5$ and $x = -2$. But notice that $-2$ is not a valid value for $x$ in the original equation, since the logarithm of a negative number is undefined. Therefore the only solution is $x = 5$.    **TRY YOUR TURN 5**

> **CAUTION** It is important to check solutions when solving equations involving logarithms because $\log_a u$, where $u$ is an expression in $x$, has domain given by $u > 0$.

# Exponential Equations

In the previous section exponential equations like $(1/3)^x = 81$ were solved by writing each side of the equation as a power of 3. That method cannot be used to solve an equation such as $3^x = 5$, however, since 5 cannot easily be written as a power of 3. Such equations can be solved approximately with a graphing calculator, but an algebraic method is also useful, particularly when the equation involves variables such as $a$ and $b$ rather than just numbers such as 3 and 5. A general method for solving these equations is shown in the following example.

### EXAMPLE 6  Solving Exponential Equations

Solve each equation.

**(a)** $3^x = 5$

**SOLUTION**  Taking natural logarithms (logarithms to any base could be used) on both sides gives

$$\ln 3^x = \ln 5$$
$$x \ln 3 = \ln 5 \qquad \qquad \text{\small{ln } } u^r = r \ln u$$
$$x = \frac{\ln 5}{\ln 3} \approx 1.465$$

**(b)** $3^{2x} = 4^{x+1}$

**SOLUTION**  Taking natural logarithms on both sides gives

$$\ln 3^{2x} = \ln 4^{x+1}$$
$$2x \ln 3 = (x + 1) \ln 4 \qquad \quad \text{\small{ln } } u^r = r \ln u$$
$$(2 \ln 3)x = (\ln 4)x + \ln 4$$
$$(2 \ln 3)x - (\ln 4)x = \ln 4 \qquad \qquad \text{\small{Subtract }} (\ln 4)x \text{ \small{from both sides.}}$$
$$(2 \ln 3 - \ln 4)x = \ln 4 \qquad \qquad \text{\small{Factor } } x.$$
$$x = \frac{\ln 4}{2 \ln 3 - \ln 4}. \qquad \quad \text{\small{Divide both sides by 2 ln 3 − ln 4.}}$$

Use a calculator to evaluate the logarithms, then divide, to get

$$x \approx \frac{1.3863}{2(1.0986) - 1.3863} \approx 1.710.$$

**(c)** $5e^{0.01x} = 9$

**SOLUTION**

$$e^{0.01x} = \frac{9}{5} = 1.8 \qquad \qquad \text{\small{Divide both sides by 5.}}$$
$$\ln e^{0.01x} = \ln 1.8 \qquad \qquad \text{\small{Take natural logarithms on both sides.}}$$
$$0.01x = \ln 1.8 \qquad \qquad \text{\small{ln } } e^u = u$$
$$x = \frac{\ln 1.8}{0.01} \approx 58.779 \qquad \qquad \textbf{\small{TRY YOUR TURN 6}}$$

**YOUR TURN 6**  Solve for x:
$2^{x+1} = 3^x$.

Just as $\log_a x$ can be written as a base $e$ logarithm, any exponential function $y = a^x$ can be written as an exponential function with base $e$. For example, there exists a real number $k$ such that

$$2 = e^k.$$

Raising both sides to the power $x$ gives

$$2^x = e^{kx},$$

so that powers of 2 can be found by evaluating appropriate powers of $e$. To find the necessary number $k$, solve the equation $2 = e^k$ for $k$ by first taking logarithms on both sides.

$$2 = e^k$$

$$\ln 2 = \ln e^k$$

$$\ln 2 = k \ln e$$

$$\ln 2 = k \qquad \text{ln } e = 1$$

Thus, $k = \ln 2$. In Section 4.4 on Derivatives of Exponential Functions, we will see why this change of base is useful. A general statement can be drawn from this example.

### Change-of-Base Theorem for Exponentials
For every positive real number $a$,

$$a^x = e^{(\ln a)x}.$$

Another way to see why the change-of-base theorem for exponentials is true is to first observe that $e^{\ln a} = a$. Combining this with the fact that $e^{ab} = (e^a)^b$, we have $e^{(\ln a)x} = (e^{\ln a})^x = a^x$.

### EXAMPLE 7    Change-of-Base Theorem

(a) Write $7^x$ using base $e$ rather than base 7.

**SOLUTION**    According to the change-of-base theorem,

$$7^x = e^{(\ln 7)x}.$$

Using a calculator to evaluate $\ln 7$, we could also approximate this as $e^{1.9459x}$.

(b) Approximate the function $f(x) = e^{2x}$ as $f(x) = a^x$ for some base $a$.

**SOLUTION**    We do not need the change-of-base theorem here. Just use the fact that

$$e^{2x} = (e^2)^x \approx 7.389^x,$$

**YOUR TURN 7** Approximate $e^{0.025x}$ in the form $a^x$.

where we have used a calculator to approximate $e^2$.    **TRY YOUR TURN 7**

The final examples in this section illustrate the use of logarithms in practical problems.

### EXAMPLE 8    Doubling Time

APPLY IT    Complete the solution of the problem posed at the beginning of this section.

**SOLUTION**    Recall that if a population will double after $t$ years at an annual growth rate of 5%, $t$ is given by the equation

$$2 = (1.05)^t.$$

We solve this equation by first taking natural logarithms on both sides.

$$\ln 2 = \ln(1.05)^t$$

$$\ln 2 = t \ln 1.05 \qquad \ln x^r = r \ln x$$

$$t = \frac{\ln 2}{\ln 1.05} \approx 14.2$$

It will take a little over 14 years for the population to double.

The problem solved in Example 8 can be generalized for the growth equation

$$y = y_0(1 + r)^t.$$

Solving for $t$ as in Example 8 (with $y = 2$ and $y_0 = 1$) gives the doubling time in years as $t = (\ln 2)/(\ln(1 + r))$.

**Doubling Time**

For a population growing at an annual rate of $100r\%$ per year, the years for the population to double is given by

$$t = \frac{\ln 2}{\ln(1 + r)}.$$

We will study growth equations more in the next section.

**EXAMPLE 9**  **Index of Diversity**

One measure of the diversity of the species in an ecological community is given by the **index of diversity** $H$, where

$$H = -[P_1 \ln P_1 + P_2 \ln P_2 + \cdots + P_n \ln P_n],$$

and $P_1, P_2, \ldots, P_n$ are the proportions of a sample belonging to each of $n$ species found in the sample. *Source: Statistical Ecology.* For example, in a community with two species, where there are 90 of one species and 10 of the other, $P_1 = 90/100 = 0.9$ and $P_2 = 10/100 = 0.1$, with

$$H = -[0.9 \ln 0.9 + 0.1 \ln 0.1] \approx 0.325.$$

Verify that $H \approx 0.673$ if there are 60 of one species and 40 of the other. As the proportions of $n$ species get closer to $1/n$ each, the index of diversity increases to a maximum of $\ln n$.

# 2.2  EXERCISES

**Write each exponential equation in logarithmic form.**

**1.** $5^3 = 125$

**2.** $7^2 = 49$

**3.** $3^4 = 81$

**4.** $2^7 = 128$

**5.** $3^{-2} = \frac{1}{9}$

**6.** $\left(\frac{5}{4}\right)^{-2} = \frac{16}{25}$

**Write each logarithmic equation in exponential form.**

**7.** $\log_2 32 = 5$

**8.** $\log_3 81 = 4$

**9.** $\ln \frac{1}{e} = -1$

**10.** $\log_2 \frac{1}{8} = -3$

**11.** $\log 100{,}000 = 5$

**12.** $\log 0.001 = -3$

**Evaluate each logarithm without using a calculator.**

**13.** $\log_8 64$

**14.** $\log_9 81$

**15.** $\log_4 64$

**16.** $\log_3 27$

**17.** $\log_2 \frac{1}{16}$

**18.** $\log_3 \frac{1}{81}$

**19.** $\log_2 \sqrt[3]{\frac{1}{4}}$

**20.** $\log_8 \sqrt[4]{\frac{1}{2}}$

**21.** $\ln e$

**22.** $\ln e^3$

**23.** $\ln e^{5/3}$

**24.** $\ln 1$

**25.** Is the "logarithm to the base 3 of 4" written as $\log_4 3$ or $\log_3 4$?

**26.** Write a few sentences describing the relationship between $e^x$ and $\ln x$.

**Use the properties of logarithms to write each expression as a sum, difference, or product of simpler logarithms. For example,**
$$\log_2(\sqrt{3}x) = \tfrac{1}{2}\log_2 3 + \log_2 x.$$

**27.** $\log_5(3k)$

**28.** $\log_9(4m)$

**29.** $\log_3 \frac{3p}{5k}$

**30.** $\log_7 \frac{15p}{7y}$

**31.** $\ln \frac{3\sqrt{5}}{\sqrt[3]{6}}$

**32.** $\ln \frac{9\sqrt[3]{5}}{\sqrt[4]{3}}$

**Suppose $\log_b 2 = a$ and $\log_b 3 = c$. Use the properties of logarithms to find the following.**

**33.** $\log_b 32$

**34.** $\log_b 18$

**35.** $\log_b(72b)$

**36.** $\log_b(9b^2)$

**Use natural logarithms to evaluate each logarithm to the nearest thousandth.**

**37.** $\log_5 30$

**38.** $\log_{12} 210$

**39.** $\log_{1.2} 0.95$

**40.** $\log_{2.8} 0.12$

**Solve each equation in Exercises 41–64. Round decimal answers to four decimal places.**

**41.** $\log_x 36 = -2$

**42.** $\log_9 27 = m$

**43.** $\log_8 16 = z$

**44.** $\log_y 8 = \frac{3}{4}$

**45.** $\log_r 5 = \dfrac{1}{2}$      **46.** $\log_4(5x + 1) = 2$

**47.** $\log_5(9x - 4) = 1$

**48.** $\log_4 x - \log_4(x + 3) = -1$

**49.** $\log_9 m - \log_9(m - 4) = -2$

**50.** $\log(x + 5) + \log(x + 2) = 1$

**51.** $\log_3(x - 2) + \log_3(x + 6) = 2$

**52.** $\log_3(x^2 + 17) - \log_3(x + 5) = 1$

**53.** $\log_2(x^2 - 1) - \log_2(x + 1) = 2$

**54.** $\ln(5x + 4) = 2$

**55.** $\ln x + \ln 3x = -1$

**56.** $\ln(x + 1) - \ln x = 1$      **57.** $2^x = 6$

**58.** $5^x = 12$      **59.** $e^{k-1} = 6$

**60.** $e^{2y} = 15$      **61.** $3^{x+1} = 5^x$

**62.** $2^{x+1} = 6^{x-1}$      **63.** $5(0.10)^x = 4(0.12)^x$

**64.** $1.5(1.05)^x = 2(1.01)^x$

**Write each expression using base $e$ rather than base 10.**

**65.** $10^{x+1}$      **66.** $10^{x^2}$

**Approximate each expression in the form $a^x$ without using $e$.**

**67.** $e^{3x}$      **68.** $e^{-4x}$

**Find the domain of each function.**

**69.** $f(x) = \log(5 - x)$      **70.** $f(x) = \ln(x^2 - 9)$

**71.** Lucky Larry was faced with solving

$$\log(2x + 1) - \log(3x - 1) = 0.$$

Larry just dropped the logs and proceeded:

$$(2x + 1) - (3x - 1) = 0$$
$$-x + 2 = 0$$
$$x = 2.$$

Although Lucky Larry is wrong in dropping the logs, his procedure will always give the correct answer to an equation of the form

$$\log A - \log B = 0,$$

where $A$ and $B$ are any two expressions in $x$. Prove that this last equation leads to the equation $A - B = 0$, which is what you get when you drop the logs. *Source: The AMATYC Review.*

**72.** Find all errors in the following calculation.

$$(\log(x + 2))^2 = 2 \log(x + 2)$$
$$= 2(\log x + \log 2)$$
$$= 2(\log x + 100)$$

**73.** Prove: $\log_a\left(\dfrac{x}{y}\right) = \log_a x - \log_a y$.

**74.** Prove: $\log_a x^r = r \log_a x$.

## LIFE SCIENCE APPLICATIONS

**75. Population Growth** Suppose a population grows at each of the following annual rates. Find the time it would take for the population to double.

    **a.** 3%      **b.** 6%      **c.** 8%

**76. Population Growth** Suppose each spring a population of house sparrows produces a new brood so that the population is 6% larger.

    **a.** How many years are required for the population to at least double? (Note that the population only increases in size each spring.)

    **b.** In how many years will the population at least triple?

**77. Minority Population** The U.S. Census Bureau has reported that the United States is becoming more diverse. In Exercise 39 of the previous section, the projected Hispanic population (in millions) was modeled by the exponential function

$$h(t) = 37.79(1.021)^t$$

where $t = 0$ corresponds to 2000 and $0 \le t \le 50$. *Source: U.S. Census Bureau.*

    **a.** Estimate in what year the Hispanic population will double the 2005 population of 42.69 million. Use the algebra of logarithms to solve this problem.

    **b.** The projected U.S. Asian population (in millions) was modeled by the exponential function

$$a(t) = 11.14(1.023)^t,$$

where $t = 0$ corresponds to 2000 and $0 \le t \le 50$. Estimate in what year the Asian population will double the 2005 population of 12.69 million.

**78. Bald Eagle Population** The U.S. Fish & Wildlife Service has monitored the number of bald eagle breeding pairs in the lower 48 states. In Exercise 41b of the previous section, the number of pairs was modeled by the exponential function

$$f(t) = 487(1.0723)^t,$$

where $t = 0$ corresponded to 1963. *Source: U.S. Fish & Wildlife Service.* If this trend continues, when will the number of bald eagle breeding pairs be double the 2006 level of 9789?

**79. Population Growth** Two species of squirrels inhabit a national park. There are currently 4500 gray squirrels, whose population increases at a rate of 4% a year. There are only 3000 black squirrels, but they increase at a rate of 6% a year. In how many years will the black squirrels outnumber the gray squirrels? Use the algebra of logarithms to solve this problem, and support your answer by using a graphing calculator to see where the two population functions intersect.

**80. Population Growth** In July 1994, the population of New York state was 18.2 million and increasing at a rate of 0.1% per year. The population of Florida was 14.0 million and increasing at a rate of 1.7% per year. *Source: U.S. Census Bureau.*

    **a.** Find exponential functions to model the population of New York and of Florida.

**b.** If this trend continued, estimate in what year the population of Florida would have exceeded the population of New York. Use the algebra of logarithms to solve this problem, and verify your answer by using a graphing calculator or spreadsheet to see where the two population functions intersect.

**c.** In July 2011, the population of New York was 19.5 million and increasing at a rate of 0.4%. The population of Florida was 19.1 million and increasing at a rate of 1.2%. If this trend continued, estimate in what year the population of Florida would have exceeded the population of New York.

**d.** Comment on the accuracy of predicting far beyond your data set.

**Index of Diversity** For Exercises 81–83, refer to Example 9.

**81.** Suppose a sample of a small community shows two species with 50 individuals each.

   **a.** Find the index of diversity $H$.

   **b.** What is the maximum value of the index of diversity for two species?

   **c.** Does your answer for part a equal ln 2? Explain why.

**82.** A virgin forest in northwestern Pennsylvania has 4 species of large trees with the following proportions of each: hemlock, 0.521; beech, 0.324; birch, 0.081; maple, 0.074. Find the index of diversity $H$.

**83.** Find the value of the index of diversity for populations with $n$ species and $1/n$ of each if

   **a.** $n = 3$;         **b.** $n = 4$.

   **c.** Verify that your answers for parts a and b equal ln 3 and ln 4, respectively.

**84. Allometric Growth** The allometric formula is used to describe a wide variety of growth patterns. It says that $y = nx^m$, where $x$ and $y$ are variables, and $n$ and $m$ are constants. For example, the famous biologist J. S. Huxley used this formula to relate the weight of the large claw of the fiddler crab to the weight of the body without the claw. *Source: Problems of Relative Growth.* Show that if $x$ and $y$ are given by the allometric formula, then $X = \log_b x$, $Y = \log_b y$, and $N = \log_b n$ are related by the linear equation

$$Y = mX + N.$$

**85. Body Surface Area** A formula used for the body surface area (in square cm) of infants is

$$A = 4.688w^{0.8168 - 0.0154 \log w},$$

where $w$ is the weight in grams.* Find the surface area of an infant with each of the following weights. *Source: British Journal of Cancer.*

   **a.** 4000 g         **b.** 8000 g

*Technically, grams are a measure of mass, not weight. Weight is a measure of the force of gravity, which varies with the distance from the center of the earth. For objects on the surface of the earth, weight and mass are often used interchangeably, and we will do so in this text.

**c.** Use a graphing calculator to find the weight of an infant with a surface area of 4000 square cm.

**86. Insect Species** An article in *Science* stated that the number of insect species of a given mass is proportional to $m^{-0.6}$, where $m$ is the mass in grams. *Source: Science.* A graph accompanying the article shows the common logarithm of the mass on the horizontal axis and the common logarithm of the number of species on the vertical axis. Explain why the graph is a straight line. What is the slope of the line?

**87. Drug Concentration** When a pharmaceutical drug is injected into the bloodstream, its concentration at time $t$ can be approximated by $C(t) = C_0 e^{-kt}$, where $C_0$ is the concentration at $t = 0$. Suppose the drug is ineffective below a concentration $C_1$ and harmful above a concentration $C_2$. Then it can be shown that the drug should be given at intervals of time $T$, where

$$T = \frac{1}{k} \ln \frac{C_2}{C_1}.$$

*Source: Applications of Calculus to Medicine.*

A certain drug is harmful at a concentration five times the concentration below which it is ineffective. At noon an injection of the drug results in a concentration of 2 mg per liter of blood. Three hours later the concentration is down to 1 mg per liter. How often should the drug be given?

**88. Tumor Growth** An article on growth rates of cancer gives a formula for the doubling time of a tumor, when the volume of the tumor at two different times is known. In this exercise we will derive the formula. *Source: Cancer.*

   **a.** Suppose at time $t_1$ the tumor has volume $V_1$ and at time $t_2$ it has volume $V_2$. Put this information into the equation

$$V = V_0(1 + r)^t$$

   to get two equations. Then divide the second equation by the first to get

$$\frac{V_2}{V_1} = (1 + r)^{t_2 - t_1}.$$

   **b.** By taking the natural logarithm of the answer to part a and simplifying, show that

$$\ln(1 + r) = \frac{\ln V_2 - \ln V_1}{t_2 - t_1}.$$

   **c.** Put the result from part b into the equation for doubling time given in the text after Example 8 to derive the formula for doubling time given in the paper:

$$t = \frac{(t_2 - t_1) \ln 2}{\ln V_2 - \ln V_1}.$$

   **d.** In the paper, the average follow-up time for a carotid body tumor was 4.5 years, after which the tumor had grown 55% on average. Letting $t_1 = 0$ and $V_2 = 1.55V_1$, find the doubling time for the tumor.

**89. Physician Demand** Suppose that a quantity can be described by the exponential growth equation

$$y(t) = y_0 e^{kt}.$$

**a.** Show that the quantity

$$\frac{1}{t} \ln \frac{y(t)}{y_0}$$

is a constant.

**b.** For the physician demand data in Section 2.1, Exercise 40, let $t = 0$ correspond to 2006. Calculate the expression in part a for the other times. To what extent is this expression constant?

The graph for Exercise 90 is plotted on a logarithmic scale where differences between successive measurements are not always the same. Data that do not plot in a linear pattern on the usual Cartesian axes often form a linear pattern when plotted on a logarithmic scale. Notice that on the horizontal scale, the distance from 5 to 10 is not the same as the distance from 10 to 15, and so on. This is characteristic of a graph drawn on logarithmic scales.

**90. Metabolism Rate** The accompanying graph shows the basal metabolism rate (in cm³ of oxygen per gram per hour) for marsupial carnivores, which include the Tasmanian devil. This rate is inversely proportional to body mass raised to the power 0.25. *Source: The Quarterly Review of Biology.*

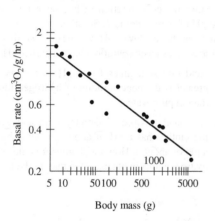

Body mass (g)

**a.** Estimate the metabolism rate for a marsupial carnivore with body mass of 10 g. Do the same for one with body mass of 1000 g.

**b.** Verify that if the relationship between $x$ and $y$ is of the form $y = ax^b$, then there will be a linear relationship between $\ln x$ and $\ln y$. (*Hint*: Apply ln to both sides of $y = ax^b$.)

**c.** If a function of the form $y = ax^b$ contains the points $(x_1, y_1)$ and $(x_2, y_2)$, then values for $a$ and $b$ can be found by dividing $y_1 = ax_1^b$ by $y_2 = ax_2^b$, solving the resulting equation for $b$, and putting the result back into either equation to solve for $a$. Use this procedure and the results from part a to find an equation of the form $y = ax^b$ that gives the basal metabolism rate as a function of body mass.

**d.** Use the result of part c to predict the basal metabolism rate of a marsupial carnivore whose body mass is 100 g.

**91. Toronto's Jewish Population** The table gives the population of Toronto's Jewish community at various times. *Source: The Mathematics Teacher.*

| Year | Population |
|------|------------|
| 1901 | 3103 |
| 1911 | 18,294 |
| 1921 | 34,770 |
| 1931 | 46,751 |
| 1941 | 52,798 |
| 1951 | 66,773 |
| 1961 | 85,000 |
| 1971 | 97,000 |
| 1981 | 128,650 |
| 1991 | 162,605 |
| 2001 | 179,100 |
| 2011 | 196,408 |

**a.** Plot the population on the $y$-axis against the year on the $x$-axis. Let $x$ represent the years since 1900. Do the data appear to lie along a straight line?

**b.** Plot the natural logarithm of the population against the year. Does the graph appear to be more linear than the graph in part a?

**c.** Find an equation for the least squares line for the data plotted in part b.

**d.** If your graphing calculator has an exponential regression feature, find the exponential function that best fits the given data according to the least squares method.

**e.** Take the natural logarithm of the equation found in part d, and verify that the result is the same as the equation found in part c. (In Section 11.1 on Solutions of Elementary and Separable Differential Equations, we will see another type of function that is a better fit to these data.)

## OTHER APPLICATIONS

**92. Music Theory** A music theorist associates the fundamental frequency of a pitch $f$ with a real number defined by

$$p = 69 + 12 \log_2(f/440).$$

*Source: Science.*

**a.** Standard concert pitch for an A is 440 cycles per second. Find the associated value of $p$.

**b.** An A one octave higher than standard concert pitch is 880 cycles per second. Find the associated value of $p$.

**93. Evolution of Languages** The number of years $N(r)$ since two independently evolving languages split off from a common ancestral language is approximated by

$$N(r) = -5000 \ln r,$$

where $r$ is the proportion of the words from the ancestral language that are common to both languages now. Find the following.

**a.** $N(0.9)$    **b.** $N(0.5)$    **c.** $N(0.3)$

**d.** How many years have elapsed since the split if 70% of the words of the ancestral language are common to both languages today?

**e.** If two languages split off from a common ancestral language about 1000 years ago, find $r$.

**For Exercises 94–97, recall that log $x$ represents the common (base 10) logarithm of $x$.**

**94. Intensity of Sound** The loudness of sounds is measured in a unit called a *decibel*. To do this, a very faint sound, called the *threshold sound*, is assigned an intensity $I_0$. If a particular sound has intensity $I$, then the decibel rating of this louder sound is

$$10 \log \frac{I}{I_0}.$$

Find the decibel ratings of the following sounds having intensities as given. Round answers to the nearest whole number.

**a.** Whisper, $115I_0$

**b.** Busy street, $9,500,000I_0$

**c.** Heavy truck, 20 m away, $1,200,000,000I_0$

**d.** Rock music concert, $895,000,000,000I_0$

**e.** Jetliner at takeoff, $109,000,000,000,000I_0$

**f.** In a noise ordinance instituted in Stamford, Connecticut, the threshold sound $I_0$ was defined as 0.0002 microbars. *Source: The New York Times.* Use this definition to express the sound levels in parts c and d in microbars.

**95. Intensity of Sound** A story on the National Public Radio program *All Things Considered* discussed a proposal to lower the noise limit in Austin, Texas, from 85 decibels to 75 decibels. A manager for a restaurant was quoted as saying, "If you cut from 85 to 75, . . . you're basically cutting the sound down in half." Is this correct? If not, to what fraction of its original level is the sound being cut? *Source: National Public Radio.*

**96. Earthquake Intensity** The magnitude of an earthquake, measured on the Richter scale, is given by

$$R(I) = \log \frac{I}{I_0},$$

where $I$ is the amplitude registered on a seismograph located 100 km from the epicenter of the earthquake, and $I_0$ is the amplitude of a certain small size earthquake. Find the Richter scale ratings of earthquakes with the following amplitudes.

**a.** $1,000,000I_0$      **b.** $100,000,000I_0$

**c.** On June 15, 1999, the city of Puebla in central Mexico was shaken by an earthquake that measured 6.7 on the Richter scale. Express this reading in terms of $I_0$. *Source: Exploring Colonial Mexico.*

**d.** On September 19, 1985, Mexico's largest recent earthquake, measuring 8.1 on the Richter scale, killed about 10,000 people. Express the magnitude of an 8.1 reading in terms of $I_0$. *Source: History.com.*

**e.** Compare your answers to parts c and d. How much greater was the force of the 1985 earthquake than the 1999 earthquake?

**f.** The relationship between the energy $E$ of an earthquake and the magnitude on the Richter scale is given by

$$R(E) = \frac{2}{3} \log \left( \frac{E}{E_0} \right),$$

where $E_0$ is the energy of a certain small earthquake. Compare the energies of the 1999 and 1985 earthquakes.

**g.** According to a newspaper article, "Scientists say such an earthquake of magnitude 7.5 could release 15 times as much energy as the magnitude 6.7 trembler that struck the Northridge section of Los Angeles" in 1994. *Source: The New York Times.* Using the formula from part f, verify this quote by computing the magnitude of an earthquake with 15 times the energy of a magnitude 6.7 earthquake.

**97. Acidity of a Solution** A common measure for the acidity of a Solution is its pH. It is defined by pH $= -\log[H^+]$, where $H^+$ measures the concentration of hydrogen ions in the solution. The pH of pure water is 7. Solutions that are more acidic than pure water have a lower pH, while solutions that are less acidic (referred to as basic solutions) have a higher pH.

**a.** Acid rain sometimes has a pH as low as 4. How much greater is the concentration of hydrogen ions in such rain than in pure water?

**b.** A typical mixture of laundry soap and water for washing clothes has a pH of about 11, while black coffee has a pH of about 5. How much greater is the concentration of hydrogen ions in black coffee than in the laundry mixture?

**YOUR TURN ANSWERS**

| | |
|---|---|
| 1. $\log_5(1/25) = -2$ | 2. $-4$ |
| 3. $2\log_a x - 3\log_a y$ | 4. 3.561 |
| 5. 2 | 6. $(\ln 2)/\ln(3/2) \approx 1.7095$ |
| 7. $1.0253^x$ | |

# 2.3 Applications: Growth and Decay

**APPLY IT** How can the exponential growth of the world population be modeled?

*This question, which will be answered in Exercise 6, is one of many situations that occur in biology, economics, and the social sciences, in which a quantity changes at a rate proportional to the amount of the quantity present.*

In cases such as population growth described above, the amount present at time $t$ is a function of $t$, called the **exponential growth and decay function**. (The derivation of this equation is presented later in Section 11.1 on Differential Equations.)

## Exponential Growth and Decay Function

Let $y_0$ be the amount or number of some quantity present at time $t = 0$. The quantity is said to grow or decay exponentially, if, for some constant $k$, the amount present at time $t$ is given by

$$y = y_0 e^{kt}.$$

If $k > 0$, then $k$ is called the **growth constant**; if $k < 0$, then $k$ is called the **decay constant**. A common example is the growth of bacteria in a culture. The more bacteria present, the faster the population increases.

### EXAMPLE 1 Yeast Production

Yeast in a sugar solution is growing at a rate such that 1 g becomes 1.5 g after 20 hours. Find the growth function, assuming exponential growth.

**SOLUTION** The values of $y_0$ and $k$ in the exponential growth function $y = y_0 e^{kt}$ must be found. Since $y_0$ is the amount present at time $t = 0$, $y_0 = 1$. To find $k$, substitute $y = 1.5$, $t = 20$, and $y_0 = 1$ into the equation.

$$y = y_0 e^{kt}$$
$$1.5 = 1 e^{k(20)}$$

Now take natural logarithms on both sides and use the power rule for logarithms and the fact that $\ln e = 1$.

$$1.5 = e^{20k}$$
$$\ln 1.5 = \ln e^{20k} \qquad \text{Take ln of both sides.}$$
$$\ln 1.5 = 20k \qquad \ln e^x = x$$
$$\frac{\ln 1.5}{20} = k \qquad \text{Divide both sides by 20.}$$
$$k \approx 0.02 \text{ (to the nearest hundredth)}$$

**YOUR TURN 1** Find the growth function if 5 g grows exponentially to 18 g after 16 hours.

The exponential growth function is $y = e^{0.02t}$, where $y$ is the number of grams of yeast present after $t$ hours.                    **TRY YOUR TURN 1**

The decline of a population or decay of a substance may also be described by the exponential growth function. In this case the decay constant $k$ is negative, since an increase in time leads to a decrease in the quantity present. Radioactive substances provide a good example of exponential decay. By definition, the **half-life** of a radioactive substance is the time it takes for exactly half of the initial quantity to decay.

### EXAMPLE 2 Carbon Dating

Carbon-14 is a radioactive form of carbon that is found in all living plants and animals. After a plant or animal dies, the carbon-14 disintegrates. Scientists determine the age of the remains by comparing its carbon-14 with the amount found in living plants and animals. The amount of carbon-14 present after $t$ years is given by the exponential equation

$$A(t) = A_0 e^{kt},$$

with $k = -[(\ln 2)/5600]$.

(a) Find the half-life of carbon-14.

**SOLUTION** Let $A(t) = (1/2)A_0$ and $k = -[(\ln 2)/5600]$.

$$\frac{1}{2}A_0 = A_0 e^{-[(\ln 2)/5600]t}$$

$$\frac{1}{2} = e^{-[(\ln 2)/5600]t} \qquad \text{Divide by } A_0.$$

$$\ln\frac{1}{2} = \ln e^{-[(\ln 2)/5600]t} \qquad \text{Take ln of both sides.}$$

$$\ln\frac{1}{2} = -\frac{\ln 2}{5600}t \qquad \ln e^x = x$$

$$-\frac{5600}{\ln 2}\ln\frac{1}{2} = t \qquad \text{Multiply by } -\frac{5600}{\ln 2}.$$

$$-\frac{5600}{\ln 2}(\ln 1 - \ln 2) = t \qquad \ln\frac{x}{y} = \ln x - \ln y$$

$$-\frac{5600}{\ln 2}(-\ln 2) = t \qquad \ln 1 = 0$$

$$5600 = t$$

The half-life is 5600 years.

(b) Charcoal from an ancient fire pit on Java had 1/4 the amount of carbon-14 found in a living sample of wood of the same size. Estimate the age of the charcoal.

**SOLUTION** Let $A(t) = (1/4)A_0$ and $k = -[(\ln 2)/5600]$.

$$\frac{1}{4}A_0 = A_0 e^{-[(\ln 2)/5600]t}$$

$$\frac{1}{4} = e^{-[(\ln 2)/5600]t}$$

$$\ln\frac{1}{4} = \ln e^{-[(\ln 2)/5600]t}$$

$$\ln\frac{1}{4} = -\frac{\ln 2}{5600}t$$

$$-\frac{5600}{\ln 2}\ln\frac{1}{4} = t$$

$$t = 11,200$$

**YOUR TURN 2** Estimate the age of a sample with 1/10 the amount of carbon-14 as a live sample.

The charcoal is about 11,200 years old. **TRY YOUR TURN 2**

By following the steps in Example 2, we get the general equation giving the half-life $T$ in terms of the decay constant $k$ as

$$T = -\frac{\ln 2}{k}.$$

For example, the decay constant for potassium-40, where $t$ is in billions of years, is approximately $-0.5545$ so its half-life is

$$T = -\frac{\ln 2}{(-0.5545)}$$

$$\approx 1.25 \text{ billion years.}$$

We can rewrite the growth and decay function as

$$y = y_0 e^{kt} = y_0(e^k)^t = y_0 a^t,$$

where $a = e^k$. This is sometimes a helpful way to look at an exponential growth or decay function.

### EXAMPLE 3  Radioactive Decay

Rewrite the function for radioactive decay of carbon-14 in the form $A(t) = A_0 a^{f(t)}$.

**SOLUTION**  From the previous example, we have

$$A(t) = A_0 e^{kt} = A_0 e^{-[(\ln 2)/5600]t}$$
$$= A_0 \left(e^{\ln 2}\right)^{-t/5600}$$
$$= A_0 2^{-t/5600} = A_0 (2^{-1})^{t/5600} = A_0 \left(\frac{1}{2}\right)^{t/5600}.$$

This last expression shows clearly that every time t increases by 5600 years, the amount of carbon-14 decreases by a factor of 1/2.

## Limited Growth Functions  The exponential growth functions discussed so far all continued to grow without bound. More realistically, many populations grow exponentially for a while, but then the growth is slowed by some external constraint that eventually limits the growth. For example, an animal population may grow to the point where its habitat can no longer support the population and the growth rate begins to dwindle until a stable population size is reached. Models that reflect this pattern are called **limited growth functions**. The next example discusses a function of this type.

### EXAMPLE 4  Sculpin Growth

A fish in the Bering Sea known as the threaded sculpin has been found to grow according to a limited growth function called a **von Bertalanffy growth curve**. The length of a male sculpin (in mm) is approximated by

$$L(t) = 165.5\left(1 - e^{-0.375299(t+0.0704311)}\right),$$

where $t$ is the age of the fish in years. *Source: Fishery Bulletin.*

**(a)** What happens to the length of the fish as $t$ gets larger and larger?

> **SOLUTION**  As $t$ gets larger, $e^{-0.375299(t+0.0704311)}$ becomes closer to 0, so $L(x)$ approaches 165.5. The limit on the length of a male sculpin is 165.5 mm. Note that this limit represents a horizontal asymptote on the graph of $L$ shown in Figure 11.

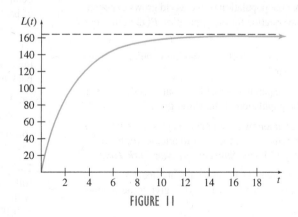

FIGURE 11

**(b)** How many years will it take for a male sculpin to be 20 times its length when it was hatched?

**SOLUTION** The length when the fish was hatched is $L(0)$. Using a calculator, we find that $L(0) = 4.3173$, and $20L(0) = 86.346$. Let $L(t) = 86.346$ and solve for $t$.

$$L(t) = 165.5\left(1 - e^{-0.375299(t+0.0704311)}\right)$$

$$86.346 = 165.5\left(1 - e^{-0.375299(t+0.0704311)}\right)$$

$$0.52173 = 1 - e^{-0.375299(t+0.0704311)} \qquad \text{Divide by 165.5.}$$

$$-0.47827 = -e^{-0.375299(t+0.0704311)} \qquad \text{Subtract 1.}$$

$$0.47827 = e^{-0.375299(t+0.0704311)}$$

$$\ln 0.47827 = \ln e^{-0.375299(t+0.0704311)} \qquad \text{Take natural logarithms.}$$

$$-0.73758 = -0.375299(t + 0.0704311) \qquad \ln e^u = u$$

$$1.9653 = t + 0.0704311 \qquad \text{Divide by } -0.375299.$$

$$t \approx 1.89$$

In about 1.89 years the fish will be 20 times as long as when it was hatched.

# 2.3 EXERCISES

1. In the exponential growth or decay function $y = y_0e^{kt}$, what does $y_0$ represent? What does $k$ represent?

2. In the exponential growth or decay function, explain the circumstances that cause $k$ to be positive or negative.

3. What is meant by the half-life of a quantity?

4. Show that if a radioactive substance has a half-life of $T$, then the corresponding constant $k$ in the exponential decay function is given by $k = -(\ln 2)/T$.

5. Show that if a radioactive substance has a half-life of $T$, then the corresponding exponential decay function can be written as $y = y_0(1/2)^{(t/T)}$.

## LIFE SCIENCE APPLICATIONS

6. **APPLY IT Population Growth** The population of the world in the year 1650 was about 500 million, and in the year 2010 was 6756 million. *Source: U.S. Census Bureau.*

   a. Assuming that the population of the world grows exponentially, find the equation for the population $P(t)$ in millions in the year $t$.

   b. Use your answer from part a to find the population of the world in the year 1.

   c. Is your answer to part b reasonable? What does this tell you about how the population of the world grows?

7. **Giardia** When a person swallows giardia cysts, stomach acids and pancreatic enzymes cause the cysts to release trophozoites, which divide every 12 hours. *Source: The New York Times.*

   a. Suppose the number of trophozoites at time $t = 0$ is $y_0$. Write a function in the form $y = y_0e^{kt}$ giving the number after $t$ hours.

   b. Write the function from part a in the form $y = y_02^{f(t)}$.

   c. The article cited above said that a single trophozoite can multiply to a million in just 10 days and a billion in 15 days. Verify this fact.

8. **Growth of Bacteria** A culture contains 25,000 bacteria, with the population increasing exponentially. The culture contains 40,000 bacteria after 10 hours.

   a. Write a function in the form $y = y_0e^{kt}$ giving the number of bacteria after $t$ hours.

   b. Write the function from part a in the form $y = y_0a^t$.

   c. How long will it be until there are 60,000 bacteria?

9. **Decrease in Bacteria** When an antibiotic is introduced into a culture of 50,000 bacteria, the number of bacteria decreases exponentially. After 9 hours, there are only 20,000 bacteria.

   a. Write an exponential equation to express the growth function $y$ in terms of time $t$ in hours.

   b. In how many hours will half the number of bacteria remain?

10. **Growth of Bacteria** The growth of bacteria in food products makes it necessary to time-date some products (such as milk) so that they will be sold and consumed before the bacterial count is too high. Suppose for a certain product that the number of bacteria present is given by

$$f(t) = 500e^{0.1t},$$

under certain storage conditions, where $t$ is time in days after packing of the product and the value of $f(t)$ is in millions.

   a. If the product cannot be safely eaten after the bacterial count reaches 3000 million, how long will this take?

   b. If $t = 0$ corresponds to January 1, what date should be placed on the product?

11. **Cancer Research** An article on cancer treatment contains the following statement: A 37% 5-year survival rate for women with ovarian cancer yields an estimated annual mortality rate of 0.1989. The authors of this article assume that the number of survivors is described by the exponential decay function given at the beginning of this section, where $y$ is the number of survivors and $k$ is the mortality rate. Verify that the given survival rate leads to the given mortality rate. *Source: American Journal of Obstetrics and Gynecology.*

**12. Chromosomal Abnormality**  The graph below shows how the risk of chromosomal abnormality in a child rises with the age of the mother. *Source: downsyndrome.about.com.*

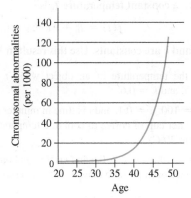

**a.** Read from the graph the risk of chromosomal abnormality (per 1000) at ages 20, 35, 42, and 49.

**b.** Assuming the graph to be of the form $y = Ce^{kt}$, find $k$ using $t = 20$ and $t = 35$.

**c.** Still assuming the graph to be of the form $y = Ce^{kt}$, find $k$ using $t = 42$ and $t = 49$.

**d.** Based on your results from parts a–c, is it reasonable to assume the graph is of the form $y = Ce^{kt}$? Explain.

**e.** In situations such as parts a–c, where an exponential function does not fit because different data points give different values for the growth constant $k$, it is often appropriate to describe the data using an equation of the form $y = Ce^{kt^n}$. Parts b and c show that $n = 1$ results in a smaller constant using the interval $[20, 35]$ than using the interval $[42, 49]$. Repeat parts b and c using $n = 2, 3$, etc., until the interval $[20, 35]$ yields a larger value of $k$ than the interval $[42, 49]$, and then estimate what $n$ should be.

**13. Botany**  A group of Tasmanian botanists have claimed that a King's holly shrub, the only one of its species in the world, is also the oldest living plant. Using carbon-14 dating of charcoal found along with fossilized leaf fragments, they arrived at an age of 43,000 years for the plant, whose exact location in southwest Tasmania is being kept a secret. What percent of the original carbon-14 in the charcoal was present? *Source: Science.*

**14. Decay of Radioactivity**  A large cloud of radioactive debris from a nuclear explosion has floated over the Pacific Northwest, contaminating much of the hay supply. Consequently, farmers in the area are concerned that the cows who eat this hay will give contaminated milk. (The tolerance level for radioactive iodine in milk is 0.) The percent of the initial amount of radioactive iodine still present in the hay after $t$ days is approximated by $P(t)$, which is given by the mathematical model

$$P(t) = 100e^{-0.1t}.$$

**a.** Find the percent remaining after 4 days.

**b.** Find the percent remaining after 10 days.

**c.** Some scientists feel that the hay is safe after the percent of radioactive iodine has declined to 10% of the original amount. Solve the equation $10 = 100e^{-0.1t}$ to find the number of days before the hay may be used.

**d.** Other scientists believe that the hay is not safe until the level of radioactive iodine has declined to only 1% of the original level. Find the number of days that this would take.

**15. Radioactive Decay**  Potassium-40, with a half-life of 1.25 billion years, has been used by geochronologists trying to sort out the mass extinction of 250 million years ago. What percent of the potassium-40 remains from a creature that died 250 million years ago?

**16. Melanoma**  An article on skin cancer said that the number of diagnosed melanoma cases is growing by 4% a year. *Source: Science News.* It also said that the chance of a U.S. resident getting cancer increased from 1 in 250 in 1980 to 1 in 84 in 1997. Is this a 4% increase annually? If not, what percent increase annually is this?

## OTHER APPLICATIONS

**17. Carbon Dating**  Refer to Example 2. A sample from a refuse deposit near the Strait of Magellan had 60% of the carbon-14 found in a contemporary living sample. How old was the sample?

**Half-Life**  Find the half-life of each radioactive substance. See Example 2.

**18.** Plutonium-241; $A(t) = A_0 e^{-0.053t}$

**19.** Radium-226; $A(t) = A_0 e^{-0.00043t}$

**20. Half-Life**  The half-life of plutonium-241 is approximately 13 years.

**a.** How much of a sample weighing 4 g will remain after 100 years?

**b.** How much time is necessary for a sample weighing 4 g to decay to 0.1 g?

**21. Half-Life**  The half-life of radium-226 is approximately 1620 years.

**a.** How much of a sample weighing 4 g will remain after 100 years?

**b.** How much time is necessary for a sample weighing 4 g to decay to 0.1 g?

**22. Radioactive Decay**  500 g of iodine-131 is decaying exponentially. After 3 days 386 g of iodine-131 is left.

**a.** Write a function in the form $y = y_0 e^{kt}$ giving the number of grams of iodine-131 after $t$ days.

**b.** Write the function from part a in the form $y = y_0 (386/500)^{f(t)}$.

**c.** Use your answer from part a to find the half-life of iodine-131.

**23. Radioactive Decay**  25 g of polonium-210 is decaying exponentially. After 50 days 19.5 g of polonium-210 is left.

**a.** Write a function in the form $y = y_0 e^{kt}$ giving the number of grams of polonium-210 after $t$ days.

**b.** Write the function from part a in the form $y = y_0 a^{t/50}$.

**c.** Use your answer from part a to find the half-life of polonium-210.

**24. Nuclear Energy** Nuclear energy derived from radioactive isotopes can be used to supply power to space vehicles. The output of the radioactive power supply for a certain satellite is given by the function $y = 40e^{-0.004t}$, where $y$ is in watts and $t$ is the time in days.

    **a.** How much power will be available at the end of 180 days?

    **b.** How long will it take for the amount of power to be half of its original strength?

    **c.** Will the power ever be completely gone? Explain.

**25. Chemical Dissolution** The amount of chemical that will dissolve in a solution increases exponentially as the temperature is increased. At 0°C, 10 g of the chemical dissolves, and at 10°C, 11 g dissolves.

    **a.** Write an equation to express the amount of chemical dissolved, $y$, in terms of temperature, $t$, in degrees Celsius.

    **b.** At what temperature will 15 g dissolve?

**Newton's Law of Cooling** Newton's law of cooling says that the rate at which a body cools is proportional to the difference in temperature between the body and an environment into which it is introduced. **This leads to an equation where the temperature $f(t)$ of the body at time $t$ after being introduced into an environment having constant temperature $T_0$ is**

$$f(t) = T_0 + Ce^{-kt},$$

**where $C$ and $k$ are constants. Use this result in Exercises 26–28.**

**26.** Find the temperature of an object when $t = 9$ if $T_0 = 18$, $C = 5$, and $k = 0.6$.

**27.** If $C = 100$, $k = 0.1$, and $t$ is time in minutes, how long will it take a hot cup of coffee to cool to a temperature of 25°C in a room at 20°C?

**28.** If $C = -14.6$ and $k = 0.6$ and $t$ is time in hours, how long will it take a frozen pizza to thaw to 10°C in a room at 18°C?

**YOUR TURN ANSWERS**

    **1.** $y = 5e^{0.08t}$                   **2.** 18,600 years old

# 2.4  Trigonometric Functions

**APPLY IT** *How does the pressure on the eardrum change when a pure musical tone is played?*

*In Example 9 of this section, we will answer this question using trigonometry.*

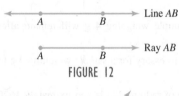

FIGURE 12

The angle is one of the basic concepts of trigonometry. The definition of an angle depends on that of a ray: A **ray** is the portion of a line that starts at a given point and continues indefinitely in one direction. Figure 12 shows a line through the two points $A$ and $B$. The portion of the line $AB$ that starts at $A$ and continues through and past $B$ is called ray $AB$. Point $A$ is the **endpoint** of the ray.

An **angle** is formed by rotating a ray about its endpoint. The initial position of the ray is called the **initial side** of the angle, and the endpoint of the ray is called the **vertex** of the angle. The location of the ray at the end of its rotation is called the **terminal side** of the angle. See Figure 13.

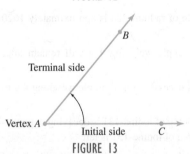

FIGURE 13

An angle can be named by its vertex. For example, the angle in Figure 13 can be called angle $A$. An angle also can be named by using three letters, with the vertex letter in the middle. For example, the angle in Figure 13 could be named angle $BAC$ or angle $CAB$.

An angle is in **standard position** if its vertex is at the origin of a coordinate system and if its initial side is along the positive $x$-axis. The angles in Figures 14 and 15 are in standard position. An angle in standard position is said to be in the quadrant of its terminal side. For example, the angle in Figure 14(a) is in quadrant I, while the angle in Figure 14(b) is in quadrant II.

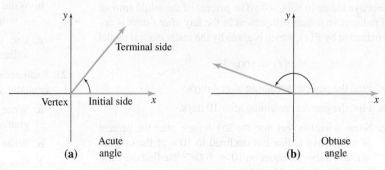

FIGURE 14

Notice that the angles in Figures 14 and 15 are formed with a counterclockwise rotation from the positive *x*-axis. This is true for any positive angle. A negative angle is measured clockwise from the positive *x*-axis, as we shall see in Example 5.

## Degree Measure

The sizes of angles are often indicated in *degrees*. **Degree measure** has remained unchanged since the Babylonians developed it over 4000 years ago. In degree measure, 360 degrees represents a complete rotation of a ray. *One degree*, written 1°, is 1/360 of a rotation. Also, 90° is 90/360 or 1/4 of a rotation, and 180° is 180/360 or 1/2 of a rotation. See Figure 15.

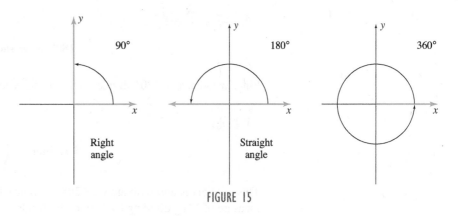

FIGURE 15

An angle having a degree measure between 0° and 90° is called an **acute angle**. An angle of 90° is a **right angle**. An angle having measure more than 90° but less than 180° is an **obtuse angle**, while an angle of 180° is a **straight angle**. See Figures 14 and 15.

A complete rotation of a ray results in an angle of measure 360°. But there is no reason why the rotation need stop at 360°. By continuing the rotation, angles of measure greater than 360° can be produced. The angles in Figure 16 have measures 60° and 420°. These two angles have the same initial side and the same terminal side, but different amounts of rotation.

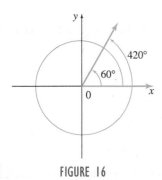

FIGURE 16

## Radian Measure

While degree measure works well for some applications, using degree measurement in calculus is complicated. Fortunately, there is an alternative system, called **radian measure**, that helps to keep the formulas for derivatives and antiderivatives as simple as possible. To see how this system for measuring angles is obtained, look at angle *θ* (the Greek letter *theta*) in Figure 17(a). The angle *θ* is in standard position; Figure 17(a) also shows a circle of radius 1, known as the **unit circle**, centered at the origin.

The vertex of *θ* is at the center of the circle in Figure 17(a). Angle *θ* cuts a piece of the circle called an **arc**. The length of this arc is the measure of the angle in radians. In other words, an angle in radians is the length of arc formed by the angle on a unit circle. The term *radian* comes from the phrase *radial angle*. Two 19th-century scientists, mathematician Thomas Muir and physicist James Thomson, are credited with the development of the radian as a unit of angular measure, although the concept originated over 100 years earlier.

On a circle, the length of an arc is proportional to the radius of the circle. Thus, for a specific angle *θ*, as shown in Figure 17(b), the ratio of the length of arc *s* to the length of the radius *r* of the circle is the same, regardless of the radius of the circle. This allows us to define radian measure on a circle of arbitrary radius as follows:

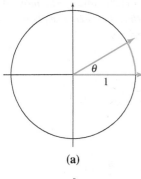

(a)

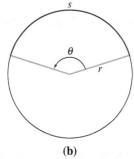

(b)

FIGURE 17

$$\text{Radian measure of } \theta = \frac{\text{Length of arc}}{\text{Length of Radius}} = \frac{s}{r}.$$

Note that the formula gives the same radian measure of an angle, regardless of the size of the circle, and that one radian results when the angle cuts an arc on the circle equal in length to the radius of the circle.

Since the circumference of a circle is $2\pi$ times the radius of the circle, the radius could be marked off $2\pi$ times around the circle. Therefore, an angle of 360°—that is, a complete circle—cuts off an arc equal in length to $2\pi$ times the radius of the circle, or

$$360° = 2\pi \text{ radians}.$$

This result gives a basis for comparing degree and radian measure.

Since an angle of 180° is half the size of an angle of 360°, an angle of 180° would have half the radian measure of an angle of 360°, or

$$180° = \frac{1}{2}(2\pi) \text{ radians} = \pi \text{ radians}.$$

## Degree and Radians

$$180° = \pi \text{ radians}$$

Since $\pi$ radians = 180°, divide both sides by $\pi$ to find the degree measure of 1 radian.

## 1 Radian

$$1 \text{ radian} = \left(\frac{180°}{\pi}\right)$$

This quotient is approximately 57.29578°. Since 180° = $\pi$ radians, we can find the radian measure of 1° by dividing by 180° on both sides.

## 1 Degree

$$1° = \frac{\pi}{180} \text{ radians}$$

One degree is approximately equal to 0.0174533 radian.

Graphing calculators and many scientific calculators have the capability of changing from degree to radian measure or from radian to degree measure. If your calculator has this capability, you can practice using it with the angle measures in Example 1. *The most important thing to remember when using a calculator to work with angle measures is to be sure the calculator mode is set for degrees or radians, as appropriate.*

### EXAMPLE 1    Equivalent Angles

Convert degree measure to radians and radian measure to degrees.

**(a)** 45°

**SOLUTION**    Since $1° = \pi/180$ radians,

$$45° = 45\left(\frac{\pi}{180}\right) \text{ radians} = \frac{45\pi}{180} \text{ radians} = \frac{\pi}{4} \text{ radians}.$$

The word *radian* is often omitted, so the answer could be written as just $45° = \pi/4$.

**(b)** $\dfrac{9\pi}{4}$

**SOLUTION**    Since 1 radian = 180°/$\pi$,

$$\frac{9\pi}{4} \text{ radians} = \frac{9\pi}{4}\left(\frac{180°}{\pi}\right) = 405°.$$

**YOUR TURN 1**    **(a)** Convert 210° to radians.    **(b)** Convert $3\pi/4$ radians to degrees.

TRY YOUR TURN 1

The following table shows the equivalent radian and degree measure for several angles that we will encounter frequently.

| Degrees and Radians of Common Angles | | | | | | | | |
|---|---|---|---|---|---|---|---|---|
| *Degrees* | 0° | 30° | 45° | 60° | 90° | 180° | 270° | 360° |
| *Radians* | 0 | $\pi/6$ | $\pi/4$ | $\pi/3$ | $\pi/2$ | $\pi$ | $3\pi/2$ | $2\pi$ |

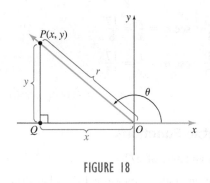

**FIGURE 18**

# The Trigonometric Functions

To define the six basic trigonometric functions, we start with an angle $\theta$ in standard position, as shown in Figure 18. Next, we choose an arbitrary point $P$ having coordinates $(x, y)$, located on the terminal side of angle $\theta$. (The point $P$ must not be the vertex of $\theta$.)

Drawing a line segment perpendicular to the $x$-axis from $P$ to point $Q$ forms a right triangle having vertices at $O$ (the origin), $P$, and $Q$. The distance from $P$ to $O$ is $r$. Since the distance from $P$ to $O$ can never be negative, $r > 0$. The six **trigonometric functions** of angle $\theta$ are defined as follows.

## Trigonometric Functions

Let $(x, y)$ be a point other than the origin on the terminal side of an angle $\theta$ in standard position. Let $r$ be the distance from the origin to $(x, y)$. Then

$$\textbf{sine } \theta = \sin\theta = \frac{y}{r} \qquad\qquad \textbf{cosecant } \theta = \csc\theta = \frac{r}{y} \;\; (y \neq 0)$$

$$\textbf{cosine } \theta = \cos\theta = \frac{x}{r} \qquad\qquad \textbf{secant } \theta = \sec\theta = \frac{r}{x} \;\; (x \neq 0)$$

$$\textbf{tangent } \theta = \tan\theta = \frac{y}{x} \;\; (x \neq 0) \qquad \textbf{cotangent } \theta = \cot\theta = \frac{x}{y} \;\; (y \neq 0).$$

From these definitions, it is easy to prove the following elementary trigonometric identities.

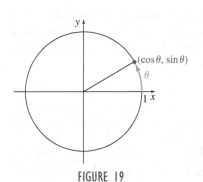

**FIGURE 19**

## Elementary Trigonometric Identities

$$\csc\theta = \frac{1}{\sin\theta} \qquad \sec\theta = \frac{1}{\cos\theta} \qquad \cot\theta = \frac{1}{\tan\theta}$$

$$\tan\theta = \frac{\sin\theta}{\cos\theta} \qquad \cot\theta = \frac{\cos\theta}{\sin\theta} \qquad \sin^2\theta + \cos^2\theta = 1$$

These identities are meaningless when the denominator is zero.

If we let $r = 1$ in the definitions of the trigonometric functions, then we can think of $\theta$ as the length of the arc, and $\cos\theta$ and $\sin\theta$ as the $x$- and $y$-coordinates, respectively, of a point on the unit circle, as shown in Figure 19.

**EXAMPLE 2** **Values of Trigonometric Functions**

The terminal side of an angle $\alpha$ (the Greek letter *alpha*) passes through the point $(8, 15)$. Find the values of the six trigonometric functions of angle $\alpha$.

**SOLUTION** Figure 20 shows angle $\alpha$ with terminal side through point $(8, 15)$ and the triangle formed by dropping a perpendicular from the point $(8, 15)$ to the $x$-axis. To find the distance $r$, use the Pythagorean theorem:* In a triangle with a right angle, if the longest side of the triangle (called the hypotenuse) is $r$ and the shorter sides (called the legs) are $x$ and $y$, then

$$r^2 = x^2 + y^2,$$

or

$$r = \sqrt{x^2 + y^2}.$$

**FIGURE 20**

*Although one of the most famous theorems in mathematics is named after the Greek mathematician Pythagoras, there is much evidence that the relationship between the sides of a right triangle was known long before his time. The Babylonian mathematical tablet identified as *Plimpton 322* has been interpreted by many to be essentially a list of *Pythagorean triples*—sets of three numbers $a$, $b$, and $c$ that satisfy the equation $a^2 + b^2 = c^2$.

(Recall that $\sqrt{b}$ represents the *positive* square root of $b$.)

Substituting the known values $x = 8$ and $y = 15$ in the equation gives

$$r = \sqrt{8^2 + 15^2} = \sqrt{64 + 225} = \sqrt{289} = 17.$$

We have $x = 8$, $y = 15$, and $r = 17$. The values of the six trigonometric functions of angle $\alpha$ are found by using the definitions.

$$\sin \alpha = \frac{y}{r} = \frac{15}{17} \quad \tan \alpha = \frac{y}{x} = \frac{15}{8} \quad \sec \alpha = \frac{r}{x} = \frac{17}{8}$$

$$\cos \alpha = \frac{x}{r} = \frac{8}{17} \quad \cot \alpha = \frac{x}{y} = \frac{8}{15} \quad \csc \alpha = \frac{r}{y} = \frac{17}{15}$$

**YOUR TURN 2** Find the values of the six trigonometric functions of an angle $\alpha$ whose terminal side goes through the point $(9, 40)$.

TRY YOUR TURN 2

**EXAMPLE 3**   **Values of Trigonometric Functions**

Find the values of the six trigonometric functions for an angle of $\pi/2$.

**SOLUTION**   Select any point on the terminal side of an angle of measure $\pi/2$ radians (or $90°$). See Figure 21. Selecting the point $(0, 1)$ gives $x = 0$ and $y = 1$. Check that $r = 1$ also. Then

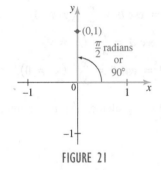

FIGURE 21

$$\sin \frac{\pi}{2} = \frac{1}{1} = 1 \qquad \cot \frac{\pi}{2} = \frac{0}{1} = 0$$

$$\cos \frac{\pi}{2} = \frac{0}{1} = 0 \qquad \csc \frac{\pi}{2} = \frac{1}{1} = 1.$$

The values of $\tan(\pi/2)$ and $\sec(\pi/2)$ are undefined because the denominator is 0 in each ratio.

Methods similar to the procedure in Example 3 can be used to find the values of the six trigonometric functions for the angles with measure $0$, $\pi$, and $3\pi/2$. These results are summarized in the following table. The table shows that the results for $2\pi$ are the same as those for 0.

| **Trigonometric Functions at Multiples of $\pi/2$** | | | | | | | |
|---|---|---|---|---|---|---|---|
| **$\theta$ (in radians)** | **$\theta$ (in degrees)** | **$\sin \theta$** | **$\cos \theta$** | **$\tan \theta$** | **$\cot \theta$** | **$\sec \theta$** | **$\csc \theta$** |
| 0 | 0° | 0 | 1 | 0 | Undefined | 1 | Undefined |
| $\pi/2$ | 90° | 1 | 0 | Undefined | 0 | Undefined | 1 |
| $\pi$ | 180° | 0 | −1 | 0 | Undefined | −1 | Undefined |
| $3\pi/2$ | 270° | −1 | 0 | Undefined | 0 | Undefined | −1 |
| $2\pi$ | 360° | 0 | 1 | 0 | Undefined | 1 | Undefined |

**NOTE** When considering the trigonometric functions, it is customary to use $x$ (rather than $\theta$) for the domain elements, as we did with earlier functions, and to write $y = \sin x$ instead of $y = \sin \theta$.

## Special Angles
The values of the trigonometric functions for most angles must be found by using a calculator with trigonometric keys. For a few angles called **special angles**, however, the function values can be found exactly. These values are found with the aid of two kinds of right triangles that will be described in this section.

### 30° − 60° − 90° Triangle

In a right triangle having angles of 30°, 60°, and 90°, the hypotenuse is always twice as long as the shortest side, and the middle side has a length that is $\sqrt{3}$ times as long as that of the shortest side. Also, the shortest side is opposite the 30° angle.

## EXAMPLE 4    Values of Trigonometric Functions

Find the values of the trigonometric functions for an angle of $\pi/6$ radians.

**SOLUTION**   Since $\pi/6$ radians $= 30°$, find the necessary values by placing a 30° angle in standard position, as in Figure 22. Choose a point $P$ on the terminal side of the angle so that $r = 2$. From the description of 30°–60°–90° triangles, $P$ will have coordinates $(\sqrt{3}, 1)$, with $x = \sqrt{3}$, $y = 1$, and $r = 2$. Using the definitions of the trigonometric functions gives the following results.

$$\sin\frac{\pi}{6} = \frac{1}{2} \qquad \tan\frac{\pi}{6} = \frac{1}{\sqrt{3}} = \frac{\sqrt{3}}{3} \qquad \sec\frac{\pi}{6} = \frac{2}{\sqrt{3}} = \frac{2\sqrt{3}}{3}$$

$$\cos\frac{\pi}{6} = \frac{\sqrt{3}}{2} \qquad \cot\frac{\pi}{6} = \sqrt{3} \qquad \csc\frac{\pi}{6} = 2$$

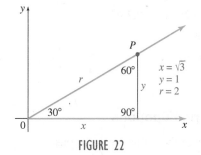

FIGURE 22

We can find the trigonometric function values for 45° angles by using the properties of a right triangle having two sides of equal length.

### 45° − 45° − 90° Triangle

In a right triangle having angles of 45°, 45°, and 90°, the hypotenuse has a length that is $\sqrt{2}$ times as long as the length of either of the shorter (equal) sides.

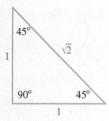

For a derivation of the properties of the 30°–60°–90° and 45°–45°–90° triangles, see Exercises 75 and 76.

## EXAMPLE 5    Values of Trigonometric Functions

Find the trigonometric function values for an angle of $-\pi/4$.

**SOLUTION**   Place an angle of $-\pi/4$ radians, or $-45°$, in standard position, as in Figure 23. Choose point $P$ on the terminal side so that $r = \sqrt{2}$. By the description of 45°–45°–90° triangles, $P$ has coordinates $(1, -1)$, with $x = 1$, $y = -1$, and $r = \sqrt{2}$.

$$\sin\left(-\frac{\pi}{4}\right) = -\frac{1}{\sqrt{2}} = -\frac{\sqrt{2}}{2} \quad \tan\left(-\frac{\pi}{4}\right) = -1 \quad \sec\left(-\frac{\pi}{4}\right) = \sqrt{2}$$

$$\cos\left(-\frac{\pi}{4}\right) = \frac{1}{\sqrt{2}} = \frac{\sqrt{2}}{2} \qquad \cot\left(-\frac{\pi}{4}\right) = -1 \quad \csc\left(-\frac{\pi}{4}\right) = -\sqrt{2}$$

FIGURE 23

**YOUR TURN 3**  Find the trigonometric function values for an angle of $7\pi/6$.

TRY YOUR TURN 3

For angles other than the special angles of 30°, 45°, 60°, and their multiples, a calculator should be used. Many calculators have keys labeled sin, cos, and tan. To get the other trigonometric functions, use the fact that sec $x$ = 1/cos $x$, csc $x$ = 1/sin $x$, and cot $x$ = 1/tan $x$. (The $x^{-1}$ key is also useful here.)

| CAUTION | Whenever you use a calculator to compute trigonometric functions, check whether the calculator is set on radians or degrees. If you want one and your calculator is set on the other, you will get erroneous answers. Most calculators have a way of switching back and forth; check the calculator manual for details. On the TI-84 Plus, press the MODE button and then select Radian or Degree. |

### EXAMPLE 6   Values of Trigonometric Functions

Use a calculator to verify the following results.

(a) sin 10° ≈ 0.1736

(b) cos 48° ≈ 0.6691

(c) tan 82° ≈ 7.1154

(d) sin 0.2618 ≈ 0.2588

(e) cot 1.2043 = 1/tan 1.2043 ≈ 1/2.6053 ≈ 0.3838

(f) sec 0.7679 = 1/cos 0.7679 ≈ 1/0.71937 ≈ 1.3901

**YOUR TURN 4** Use a calculator to find each of the following.
(a) cos 6°   (b) sec 4

TRY YOUR TURN 4

### EXAMPLE 7   Values of Trigonometric Functions

Find all values of $x$ between 0 and $2\pi$ that satisfy each of the following equations.

(a) sin $x$ = 1/2

**SOLUTION**   The sine function is positive in quadrants I and II. In quadrant I, we draw a triangle with an angle whose sine is 1/2, as shown in Figure 24(a). We recognize this triangle as the 30°–60°–90° triangle, with angle $x = \pi/6$. In Figure 24(b), we show the same triangle in quadrant II. The angle is now $x = \pi - \pi/6 = 5\pi/6$. There are two solutions between 0 and $2\pi$, namely, $x = \pi/6$ and $x = 5\pi/6$.

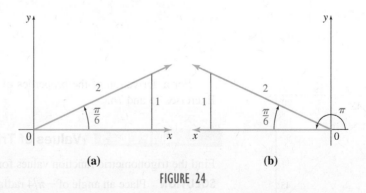

(a)                           (b)

FIGURE 24

(b) sec $x$ = −2

**YOUR TURN 5** Find all values of $x$ between 0 and $2\pi$ that satisfy the equation cos $x$ = −√2/2.

**SOLUTION**   The secant function is negative in quadrants II and III. In quadrant II, we draw a triangle with an angle whose secant is −2, as shown in Figure 25(a). We recognize the 30°–60°–90° triangle once again, with angle $x = \pi - \pi/3 = 2\pi/3$. In Figure 25(b), we show the same triangle in quadrant III. The angle is now $x = \pi + \pi/3 = 4\pi/3$. There are two solutions between 0 and $2\pi$, namely, $x = 2\pi/3$ and $x = 4\pi/3$.

TRY YOUR TURN 5

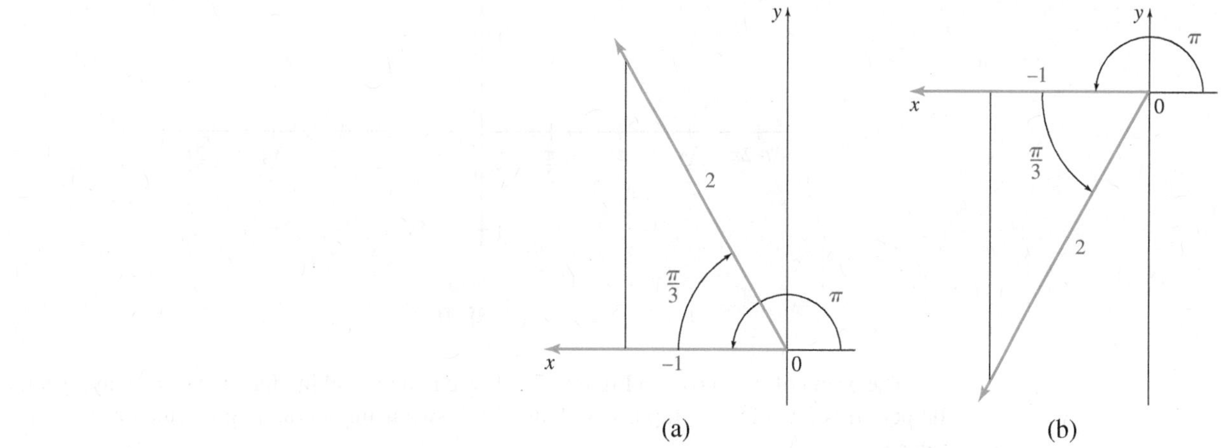

(a)          (b)

FIGURE 25

# Graphs of the Trigonometric Functions
Because of the way the trigonometric functions are defined (using a circle), the same function values will be obtained for any two angles that differ by $2\pi$ radians (or 360°). For example,

$$\sin(x + 2\pi) = \sin x \quad \text{and} \quad \cos(x + 2\pi) = \cos x$$

for any value of $x$. Because of this property, the trigonometric functions are *periodic functions*.

## Periodic Function
A function $y = f(x)$ is **periodic** if there exists a positive real number $a$ such that

$$f(x) = f(x + a)$$

for all values of $x$ in the domain of the function. The smallest positive value of $a$ is called the **period** of the function.*

Intuitively, a function with period $a$ repeats itself over intervals of length $a$. Once we know what the graph looks like over one period of length $a$, we know what the entire graph looks like by simply repeating. Because sine is periodic with period $2\pi$, the graph is found by first finding the graph on the interval between 0 and $2\pi$ and then repeating as many times as necessary.

To find values of $y = \sin x$ for values of $x$ between 0 and $2\pi$, think of a point moving counterclockwise around a circle, tracing out an arc for angle $x$. The value of $\sin x$ gradually increases from 0 to 1 as $x$ increases from 0 to $\pi/2$. The value of $\sin x$ then decreases back to 0 as $x$ goes from $\pi/2$ to $\pi$. For $\pi < x < 2\pi$, $\sin x$ is negative. A few typical values from these intervals are given in the following table, where decimals have been rounded to the nearest tenth.

| **Values of the Sine Function** | | | | | | | | |
|---|---|---|---|---|---|---|---|---|
| $x$ | 0 | $\pi/4$ | $\pi/2$ | $3\pi/4$ | $\pi$ | $5\pi/4$ | $3\pi/2$ | $7\pi/4$ | $2\pi$ |
| $\sin x$ | 0 | 0.7 | 1 | 0.7 | 0 | $-0.7$ | $-1$ | $-0.7$ | 0 |

Plotting the points from the table of values and connecting them with a smooth curve gives the solid portion of the graph in Figure 26 on the next page. Since $y = \sin x$ is periodic, the graph continues in both directions indefinitely, as suggested by the dashed lines. The solid portion of the graph in Figure 26 gives the graph over one period.

*Some authors define the period of the function as any value of $a$ that satisfies $f(x) = f(x + a)$.

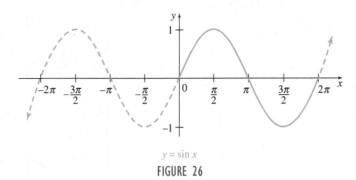

$y = \sin x$

FIGURE 26

The graph of $y = \cos x$ in Figure 27 below can be found in much the same way. Again, the period is $2\pi$. (These graphs could also be drawn using a graphing calculator or a computer.)

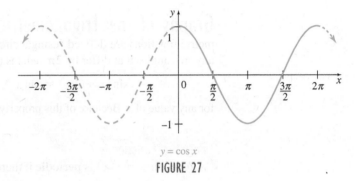

$y = \cos x$

FIGURE 27

Finally, Figure 28 shows the graph of $y = \tan x$. Since $\tan x$ is undefined (because of zero denominators) for $x = \pi/2, 3\pi/2, -\pi/2$, and so on, the graph has vertical asymptotes at these values. As the graph suggests, the tangent function is periodic, with a period of $\pi$.

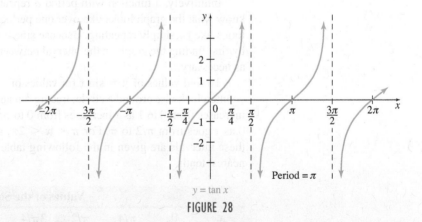

$y = \tan x$

FIGURE 28

The graphs of the secant, cosecant, and cotangent functions are not used as often as these three, so they are not given here.

## Translating Graphs of Sine and Cosine Functions   In Section 1.4
we saw that the graph of the function $y = f(x - c) + d$ was simply the graph of $y = f(x)$ translated $c$ units horizontally and $d$ units vertically. The same facts hold true

with trigonometric functions. The constants $a$, $b$, $c$, and $d$ affect the graphs of the functions $y = a \sin(bx - c) + d$ and $y = a \cos(bx - c) + d$ in a similar manner.

In addition, the constants $a$, $b$, and $c$ have particular properties. Since the sine and cosine functions range between $-1$ and $1$, the value of $a$, whose absolute value is called the **amplitude**, can be interpreted as half the difference between the maximum and minimum values of the function.

The period of the function is determined by the constant $b$, which we will assume to be greater than 0. Recall that the period of both $\sin x$ and $\cos x$ is $2\pi$. The value of $b > 0$ will increase or decrease the period, depending on its value. A similar phenomenon occurs when $b < 0$, but it is not covered in this textbook. Thus, the graph of $y = \sin(bx)$ will look like that of $y = \sin x$, but with a period of $T = 2\pi/b$. The results are similar for $y = \cos(bx)$.

The reciprocal of the period is the **frequency**. That is, the frequency is $b/(2\pi)$. The period tells how long one cycle is, while the frequency tells how many cycles occur per unit of $x$.

Because $y = a \sin(bx - c) = a \sin[b(x - c/b)]$, the quantity $c/b$ determines the number of units that the graph of $\sin bx$ (or $\cos bx$) is shifted horizontally. The quantity $c/b$ is known as the **phase shift**. The constant $d$ determines the vertical shift of $\sin x$ or $\cos x$. In the life sciences, $c/b$ and $d$ are often referred to as the **acrophase** and the **MESOR** (for Midline-Estimating Statistic of Rhythm).

## Transformation of Trigonometric Functions

For the function $y = a \sin(bx - c) + d$ or $y = a \cos(bx - c) + d$,

$$|a| = \text{the amplitude,}$$

$$\frac{2\pi}{b} = \text{the period,}$$

$$\frac{b}{2\pi} = \text{the frequency,}$$

$$\frac{c}{b} = \text{the phase shift, or acrophase,}$$

$$d = \text{the vertical shift, or MESOR.}$$

### EXAMPLE 8    Graphing Trigonometric Functions

Graph each function.

**(a)** $y = \sin 3x$

**SOLUTION**    The graph of this function has amplitude $a = 1$ and no vertical or horizontal shifts. The period of this function is $T = 2\pi/b = 2\pi/3$. Hence, the graph of $y = \sin 3x$ is the same as $y = \sin x$ except that the period is different. See Figure 29.

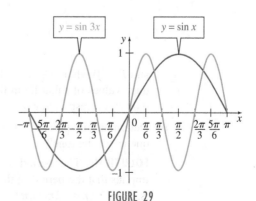

FIGURE 29

**(b)** $f(x) = 4 \cos\left(\frac{1}{2}x + \frac{\pi}{4}\right) - 1$

**SOLUTION** The amplitude is $a = 4$. The graph of $f(x)$ is shifted down 1 unit vertically. Since $\cos[(1/2)x + \pi/4] = \cos[(1/2)x - (-\pi/4)]$, the phase shift is $c/b = (-\pi/4)/(1/2) = -\pi/2$. This shifts the graph $\pi/2$ units to the left, relative to the graph $g(x) = \cos[(1/2)x]$. The period of $f(x)$ is $2\pi/(1/2) = 4\pi$. Making these translations on $y = \cos x$ leads to Figure 30.

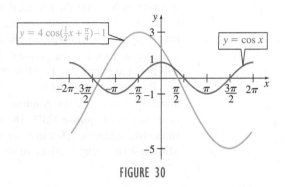

FIGURE 30

## EXAMPLE 9 Music

APPLY IT

A change in pressure on the eardrum occurs when a pure musical tone is played. For some tones, the pressure on the eardrum follows the sine curve

$$P(t) = 0.004 \sin\left(2\pi ft + \frac{\pi}{7}\right),$$

where $P$ is the pressure in pounds per square foot at time $t$ seconds and $f$ is the frequency of the sound wave in cycles per second. ***Source: The Physics and Psychophysics of Music: An Introduction.*** When $P(t)$ is positive there is an increase in pressure and the eardrum is pushed inward; when $P(t)$ is negative there is a decrease in pressure and the eardrum is pushed outward.

**(a)** Graph the pressure on the eardrum for Middle C, which has a frequency of $f = 261.63$ cycles per second, on $[0, 0.005]$.

**SOLUTION** Figure 31 gives a graphing calculator graph of the function $P(t) = 0.004 \sin(2\pi ft + \pi/7) = 0.004 \sin(523.26\pi t + \pi/7)$.

**(b)** Determine analytically, for Middle C, the values of $t$ for which $P = 0$ on $[0, 0.005]$.

**SOLUTION** Since the sine function is zero for multiples of $\pi$, we can determine the value(s) of $t$ where $P = 0$ by setting $523.26\pi t + \pi/7 = n\pi$, where $n$ is an integer, and solving for $t$. After some algebraic manipulations,

$$t = \frac{n - \frac{1}{7}}{523.26}$$

and $P = 0$ when $n = 0, \pm 1, \pm 2, \ldots$. However, only values of $n = 1$ or $n = 2$ produce values of $t$ that lie in the interval $[0, 0.005]$. Thus, $P = 0$ when $t \approx 0.0016$ and $0.0035$, corresponding to $n = 1$ and $n = 2$, respectively.

**(c)** Determine the period $T$ of $P(t)$. What is the relationship between the period and frequency of the tone?

**SOLUTION** The period is $T = 2\pi/b = 2\pi/(523.26\pi) = 1/261.63 \approx 0.004$. This implies that the period of the pressure equation is the reciprocal of the frequency. That is, $T = 2\pi/b = 2\pi/(2\pi f) = 1/f$. Note that this small period implies that the eardrum is actually vibrating very quickly, making nearly 262 cycles per second.

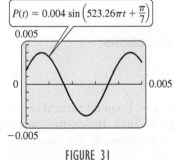

FIGURE 31

**TECHNOLOGY**

## EXAMPLE 10  Sunrise

The table lists the approximate number of minutes after midnight, eastern standard time, that the sun rises in Boston for specific days of the year. *Source: The Old Farmer's Almanac.*

| Time of Boston Sunrise | |
| --- | --- |
| Day of the Year | Sunrise (minutes after midnight) |
| 21 | 428 |
| 52 | 393 |
| 81 | 345 |
| 112 | 293 |
| 142 | 257 |
| 173 | 247 |
| 203 | 266 |
| 234 | 298 |
| 265 | 331 |
| 295 | 365 |
| 326 | 403 |
| 356 | 431 |

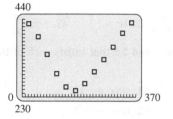

440

0        230        370

**FIGURE 32**

**(a)** Plot the data. Is it reasonable to assume that the times of sunrise are periodic?

**SOLUTION**   Figure 32 shows a graphing calculator plot of the data. Because of the cyclical nature of the days of the year, it is reasonable to assume that the data are periodic.

**(b)** Find a trigonometric function of the form $s(t) = a \sin(bt + c) + d$ that models these data when $t$ is the day of the year and $s(t)$ is the number of minutes past midnight, eastern standard time, that the sun rises. Use the data from the table.

**SOLUTION**   The function $s(t)$, derived by a TI-84 Plus using the sine regression function under the STAT-CALC menu, is given by

$$s(t) = 92.1414 \sin(0.016297t + 1.80979) + 342.934.$$

Figure 33 shows that this function fits the data well.

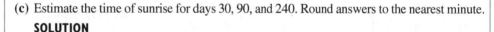

$s(t) = 92.1414 \sin(0.016297t + 1.80979) + 342.934$

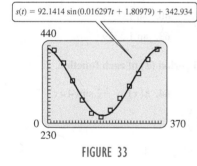

440

0        230        370

**FIGURE 33**

**(c)** Estimate the time of sunrise for days 30, 90, and 240. Round answers to the nearest minute.

**SOLUTION**

$$s(30) = 92.1414 \sin(0.016297(30) + 1.80979) + 342.934$$
$$\approx 412 \text{ minutes} = 412/60 \text{ hours} \approx 6.867 \text{ hours}$$
$$= 6 \text{ hours} + 0.867(60) \text{ minutes} = 6{:}52 \text{ A.M.}$$

Similarly,

$$s(90) \approx 331 \text{ minutes} = 5{:}31 \text{ A.M.}$$
$$s(240) \approx 294 \text{ minutes.}$$

Before finding the time of day, remember to add 60 minutes for daylight savings. Therefore, $s(240)$ is about 5:54 A.M.

**(d)** Estimate the days of the year that the sun rises at 5:45 A.M.

**SOLUTION**   Figure 34 shows the graphs of $s(t)$ and $y = 345$ (corresponding to a sunrise of 5:45 A.M.). These graphs first intersect on day 80. However, because of daylight savings time, to find the second value we find where the graphs of $s(t)$ and $y = 345 - 60 = 285$ intersect. These graphs intersect on day 233. Thus, the sun rises at approximately 5:45 A.M. on the 80th and 233rd days of the year.

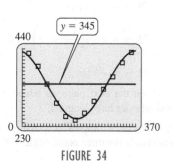

$y = 345$

440

0        230        370

**FIGURE 34**

**(e)** What is the period of the function found in part (b)?

**SOLUTION**    The period of the function given above is $T = 2\pi/0.016297 \approx 385.5$ days. This is close to the true period of about 365 days. The discrepancy could be due to many factors. For example, the underlying function may be more complex than a simple sine function.

# 2.4    EXERCISES

**Convert the following degree measures to radians. Leave answers as multiples of $\pi$.**

**1.** $60°$     **2.** $90°$     **3.** $150°$     **4.** $135°$

**5.** $270°$     **6.** $320°$     **7.** $495°$     **8.** $510°$

**Convert the following radian measures to degrees.**

**9.** $\dfrac{5\pi}{4}$     **10.** $\dfrac{2\pi}{3}$     **11.** $-\dfrac{13\pi}{6}$     **12.** $-\dfrac{\pi}{4}$

**13.** $\dfrac{8\pi}{5}$     **14.** $\dfrac{5\pi}{9}$     **15.** $\dfrac{7\pi}{12}$     **16.** $5\pi$

**Find the values of the six trigonometric functions for the angles in standard position having the points in Exercises 17–20 on their terminal sides.**

**17.** $(-3, 4)$     **18.** $(-12, -5)$

**19.** $(7, -24)$     **20.** $(20, 15)$

In quadrant I, $x$, $y$, and $r$ are all positive, so that all six trigonometric functions have positive values. In quadrant II, $x$ is negative and $y$ is positive ($r$ is always positive). Thus, in quadrant II, sine is positive, cosine is negative, and so on. For Exercises 21–24, complete the following table of values for the signs of the trigonometric functions.

| | *Quadrant of $\theta$* | $\sin\theta$ | $\cos\theta$ | $\tan\theta$ | $\cot\theta$ | $\sec\theta$ | $\csc\theta$ |
|---|---|---|---|---|---|---|---|
| **21.** | I | + | | | | | |
| **22.** | II | | | | | | |
| **23.** | III | | | | | | |
| **24.** | IV | | | | | | |

**For Exercises 25–32, complete the following table. Use the $30°-60°-90°$ and $45°-45°-90°$ triangles. Do not use a calculator.**

| $\theta$ | $\sin\theta$ | $\cos\theta$ | $\tan\theta$ | $\cot\theta$ | $\sec\theta$ | $\csc\theta$ |
|---|---|---|---|---|---|---|
| **25.** $30°$ | $1/2$ | $\sqrt{3}/2$ | | | $2\sqrt{3}/3$ | |
| **26.** $45°$ | | | $1$ | $1$ | | |
| **27.** $60°$ | | $1/2$ | $\sqrt{3}$ | | $2$ | |
| **28.** $120°$ | $\sqrt{3}/2$ | | $-\sqrt{3}$ | | | $2\sqrt{3}/3$ |
| **29.** $135°$ | $\sqrt{2}/2$ | $-\sqrt{2}/2$ | | | $-\sqrt{2}$ | $\sqrt{2}$ |
| **30.** $150°$ | | $-\sqrt{3}/2$ | $-\sqrt{3}/3$ | | | $2$ |
| **31.** $210°$ | $-1/2$ | | $\sqrt{3}/3$ | $\sqrt{3}$ | | $-2$ |
| **32.** $240°$ | $-\sqrt{3}/2$ | $-1/2$ | | | $-2$ | $-2\sqrt{3}/3$ |

**Find the following function values without using a calculator.**

**33.** $\sin\dfrac{\pi}{3}$     **34.** $\cos\dfrac{\pi}{6}$     **35.** $\tan\dfrac{\pi}{4}$     **36.** $\cot\dfrac{\pi}{3}$

**37.** $\csc\dfrac{\pi}{6}$     **38.** $\sin\dfrac{3\pi}{2}$     **39.** $\cos 3\pi$     **40.** $\sec\pi$

**41.** $\sin\dfrac{7\pi}{4}$     **42.** $\tan\dfrac{5\pi}{2}$     **43.** $\sec\dfrac{5\pi}{4}$     **44.** $\cos 5\pi$

**45.** $\cot -\dfrac{3\pi}{4}$     **46.** $\tan -\dfrac{5\pi}{6}$     **47.** $\sin -\dfrac{7\pi}{6}$     **48.** $\cos -\dfrac{\pi}{6}$

**Find all values of $x$ between 0 and $2\pi$ that satisfy each of the following equations.**

**49.** $\cos x = 1/2$          **50.** $\sin x = -1/2$

**51.** $\tan x = -1$          **52.** $\tan x = \sqrt{3}$

**53.** $\sec x = -2/\sqrt{3}$          **54.** $\sec x = \sqrt{2}$

**Use a calculator to find the following function values.**

**55.** $\sin 39°$          **56.** $\cos 67°$

**57.** $\tan 123°$          **58.** $\tan 54°$

**59.** $\sin 0.3638$          **60.** $\tan 1.0123$

**61.** $\cos 1.2353$          **62.** $\sin 1.5359$

**Find the amplitude ($a$) and period ($T$) of each function.**

**63.** $f(x) = \cos(3x)$          **64.** $h(x) = -\dfrac{1}{2}\sin(4\pi x)$

**65.** $g(t) = -2\sin\left(\dfrac{\pi}{4}t + 2\right)$

**66.** $s(t) = 3\sin(880\pi t - 7)$

**Graph each function defined as follows over a two-period interval.**

**67.** $y = 2\sin x$          **68.** $y = 2\cos x$

**69.** $y = -\sin x$          **70.** $y = -\dfrac{1}{2}\cos x$

**71.** $y = 2\cos\left(3x - \dfrac{\pi}{4}\right) + 1$     **72.** $y = 4\sin\left(\dfrac{1}{2}x + \pi\right) + 2$

**73.** $y = \dfrac{1}{2}\tan x$          **74.** $y = -3\tan x$

**75.** Consider the triangle shown on the next page, in which the three angles $\theta$ are equal and all sides have length 2.

   **a.** Using the fact that the sum of the angles in a triangle is $180°$, what are the measures of the three equal angles $\theta$?

**b.** Suppose the triangle is cut in half as shown by a vertical line. What are the measures of the angles in the blue triangle on the left?

**c.** What are the measures of the sides of the blue triangle on the left? (*Hint*: Once you've found the length of the base, use the Pythagorean theorem to find the height.)

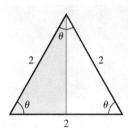

**76.** Consider the right triangle shown, in which the two sides have length 1.

**a.** Using the Pythagorean theorem, what is the length of the hypotenuse?

**b.** Using the fact that the sum of the angles in a triangle is 180°, what are the measures of the three angles?

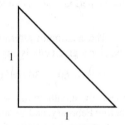

## LIFE SCIENCE APPLICATIONS

**77. Monkey Eyes** In a study of how monkeys' eyes pursue a moving object, an image was moved sinusoidally through a monkey's field of vision with an amplitude of 2° and a period of 0.350 second. *Source: Journal of Neurophysiology.*

**a.** Find an equation giving the position of the image in degrees as a function of time in seconds.

**b.** After how many seconds does the image reach its maximum amplitude?

**c.** What is the position of the object after 2 seconds?

**78. Transylvania Hypothesis** The "Transylvania hypothesis" claims that the full moon has an effect on health-related behavior. A study investigating this effect found a significant relationship between the phase of the moon and the number of general practice consultations nationwide, given by

$$y = 100 + 1.8 \cos\left[\frac{(t-6)\pi}{14.77}\right],$$

where $y$ is the number of consultations as a percentage of the daily mean and $t$ is the days since the last full moon. *Source: Family Practice.*

**a.** What is the period of this function? What is the significance of this period?

**b.** There was a full moon on January 16, 2014. On what day in January 2014 does this formula predict the maximum number of consultations? What percent increase would be predicted for that day?

**c.** What does the formula predict for January 31, 2014?

**79. Air Pollution** The amount of pollution in the air fluctuates with the seasons. It is lower after heavy spring rains and higher after periods of little rain. In addition to this seasonal fluctuation, the long-term trend in many areas is upward. An idealized graph of this situation is shown in the figure. Trigonometric functions can be used to describe the fluctuating part of the pollution levels. Powers of the number $e$ can be used to show the long-term growth. In fact, the pollution level in a certain area might be given by

$$P(t) = 7(1 - \cos 2\pi t)(t + 10) + 100e^{0.2t},$$

where $t$ is time in years, with $t = 0$ representing January 1 of the base year. Thus, July 1 of the same year would be represented by $t = 0.5$, while October 1 of the following year would be represented by $t = 1.75$. Find the pollution levels on the following dates.

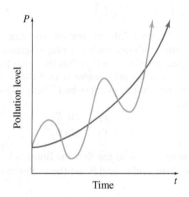

**a.** January 1, base year

**b.** July 1, base year

**c.** January 1, following year

**d.** July 1, following year

**80. Air Pollution** Using a computer or a graphing calculator, sketch the function for air pollution given in Exercise 79 over the interval $[0, 6]$.

**81. Alzheimer's Disease** A study on the circadian rhythms of patients with Alzheimer's disease found that the body temperature of patients could be described by a function of the form

$$T = T_0 + a \cos\left(\frac{2\pi(t - k)}{24}\right),$$

where $t$ is the time in hours since midnight. For the patients without Alzheimer's, the average values of $T_0$ (the MESOR), $a$ (the amplitude), and $k$ (the acrophase) were 36.91°C, 0.32°C, and 14.92 hours, while for the patients with the disease, the values were 37.29°C, 0.46°C, and 16.37 hours. *Source: American Journal of Psychiatry.*

a. Graph the functions giving the temperature for each of the two groups using a graphing calculator. Do these two functions ever cross?

b. At what time is the temperature highest for the patients without Alzheimer's?

c. At what time is the temperature highest for the patients with Alzheimer's?

**82. Femoral Angles** The true angle of torsion $\theta$ of the femur is given by

$$\tan \theta = (\tan \theta_2)(\cos a - \cot B_2 \sin a),$$

where $\theta_2$ is the measured angle of torsion, $a$ is the angle of abduction, and $B_2$ is the measured angle of inclination. Find the true angle of torsion for each of the following values of the other angles. (*Hint:* After finding the value of the right-hand side, use the $\tan^{-1}$ button on your calculator to find the value of $\theta$. Remember to set your calculator on degrees for this exercise.) *Source: Standards in Pediatric Orthopedics.*

a. $\theta_2 = 23°, a = 10°, B_2 = 170°$

b. $\theta_2 = 20°, a = 10°, B_2 = 160°$

## OTHER APPLICATIONS

**Light Rays**   When a light ray travels from one medium, such as air, to another medium, such as water or glass, the speed of the light changes, and the direction that the ray is traveling changes. (This is why a fish under water is in a different position from the place at which it appears to be.) These changes are given by Snell's law,

$$\frac{c_1}{c_2} = \frac{\sin \theta_1}{\sin \theta_2},$$

where $c_1$ is the speed in the first medium, $c_2$ is the speed in the second medium, and $\theta_1$ and $\theta_2$ are the angles shown in the figure.

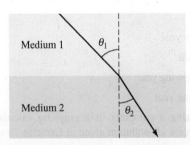

In Exercises 83 and 84, assume that $c_1 = 3 \times 10^8$ m per second, and find the speed of light in the second medium.

**83.** $\theta_1 = 39°, \theta_2 = 28°$

**84.** $\theta_1 = 46°, \theta_2 = 31°$

**Sound**   Pure sounds produce single sine waves on an oscilloscope. Find the period of each sine wave in the photographs in Exercises 85 and 86. On the vertical scale each square represents 0.5, and on the horizontal scale each square represents 30°.

**85.**

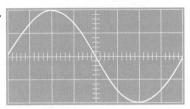

**86.**

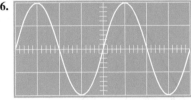

**87. Sound**   Suppose the A key above Middle C is played as a pure tone. For this tone,

$$P(t) = 0.002 \sin(880\pi t),$$

where $P(t)$ is the change of pressure (in pounds per square foot) on a person's eardrum at time $t$ (in seconds). *Source: The Physics and Psychophysics of Music: An Introduction.*

a. Graph this function on $[0, 0.003]$.

b. Determine analytically the values of $t$ for which $P = 0$ on $[0, 0.003]$ and check graphically.

c. Determine the period $T$ of $P(t)$ and the frequency of the A note.

**88. Temperature**   The maximum afternoon temperature (in degrees Fahrenheit) in a given city is approximated by

$$T(t) = 60 - 30 \cos(t/2),$$

where $t$ represents the month, with $t = 0$ representing January, $t = 1$ representing February, and so on. Use a calculator to find the maximum afternoon temperature for the following months.

a. February

b. April

c. September

d. July

e. December

**89. Temperature**   A mathematical model for the temperature in Fairbanks is

$$T(t) = 37 \sin\left[\frac{2\pi}{365}(t - 101)\right] + 25,$$

where $T(t)$ is the temperature (in degrees Fahrenheit) on day $t$, with $t = 0$ corresponding to January 1 and $t = 364$ corresponding to December 31. *Source: Mathematics Teacher.* Use a calculator to estimate the temperature for a–d.

a. March 16 (Day 74)

b. May 2 (Day 121)

c. Day 250

d. Day 325

e. Find maximum and minimum values of $T$.

f. Find the period, $T$.

**90. Sunset** The number of minutes after noon, eastern standard time, that the sun sets in Boston for specific days of the year is approximated in the following table. *Source: The Old Farmer's Almanac.*

| Day of the Year | Sunset (minutes after noon) |
|---|---|
| 21 | 283 |
| 52 | 323 |
| 81 | 358 |
| 112 | 393 |
| 142 | 425 |
| 173 | 445 |
| 203 | 434 |
| 234 | 396 |
| 265 | 343 |
| 295 | 292 |
| 326 | 257 |
| 356 | 255 |

a. Plot the data. Is it reasonable to assume that the times of sunset are periodic?

b. Use a calculator with trigonometric regression to find a trigonometric function of the form $s(t) = a \sin(bt + c) + d$ that models these data when $t$ is the day of the year and $s(t)$ is the number of minutes past noon, eastern standard time, that the sun sets.

c. Estimate the time of sunset for days 60, 120, and 240. Round answers to the nearest minute. (*Hint:* Don't forget about daylight savings time.)

d. Use part b to estimate the days of the year that the sun sets at 6:00 P.M. In reality, the days are close to 82 and 290.

**91. Measurement** A surveyor standing 65 m from the base of a building measures the angle to the top of the building and finds it to be 42.8°. (See the figure.) Use trigonometry to find the height of the building.

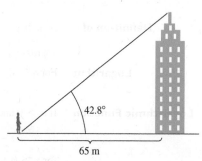

42.8°

65 m

**92. Measurement** Jenny Crum stands on a cliff at the edge of a canyon. On the opposite side of the canyon is another cliff equal in height to the one she is on. (See the figure.) By dropping a rock and timing its fall, she determines that it is 105 ft to the bottom of the canyon. She also determines that the angle to the base of the opposite cliff is 27°. How far is it to the opposite side of the canyon?

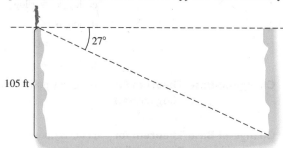

27°

105 ft

**YOUR TURN ANSWERS**

1. a. $7\pi/6$  b. 135°

2. $\sin \alpha = 40/41$, $\cos \alpha = 9/41$, $\tan \alpha = 40/9$, $\cot \alpha = 9/40$, $\sec \alpha = 41/9$, $\csc \alpha = 41/40$

3. $\sin(7\pi/6) = -1/2$, $\cos(7\pi/6) = -\sqrt{3}/2$, $\tan(7\pi/6) = 1/\sqrt{3} = \sqrt{3}/3$, $\cot(7\pi/6) = \sqrt{3}$, $\sec(7\pi/6) = -2/\sqrt{3} = -2\sqrt{3}/3$, $\csc(7\pi/6) = -2$

4. a. 0.9945  b. −1.5299

5. $3\pi/4, 5\pi/4$

# 2 CHAPTER REVIEW

## SUMMARY

In this chapter we have studied three more families of functions: exponential, logarithmic, and trigonometric. When added to the repertoire of functions in the last chapter, they give us a substantial number of mathematical tools for analyzing real-life applications. We now know when the use of one function is more appropriate than another, and we can predict how the dependent variable will react to changes in the independent variable. In particular, we have applied this knowledge to the study of growth and decay. In the next chapters, we will build on this knowledge.

**Exponential Function** An exponential function with base $a$ is defined as

$$f(x) = a^x, \text{ where } a > 0 \text{ and } a \neq 1.$$

**Definition of e**  As $m$ becomes larger and larger, $\left(1 + \frac{1}{m}\right)^m$ becomes closer and closer to the number $e$, whose approximate value is 2.718281828.

**Logarithm**  For $a > 0$, $a \neq 1$, and $x > 0$,

$$y = \log_a x \text{ means } a^y = x.$$

**Logarithmic Function**  If $a > 0$ and $a \neq 1$, then the logarithmic function of base $a$ is defined by

$$f(x) = \log_a x,$$

for $x > 0$.

**Properties of Logarithms**  Let $x$ and $y$ be any positive real numbers and $r$ be any real number. Let $a$ be a positive real number, $a \neq 1$. Then

a. $\log_a xy = \log_a x + \log_a y$

b. $\log_a \frac{x}{y} = \log_a x - \log_a y$

c. $\log_a x^r = r \log_a x$

d. $\log_a a = 1$

e. $\log_a 1 = 0$

f. $\log_a a^r = r$.

**Change-of-Base Theorem for Logarithms**  If $x$ is any positive number and if $a$ and $b$ are positive real numbers, $a \neq 1$, $b \neq 1$, then

$$\log_a x = \frac{\log_b x}{\log_b a} = \frac{\ln x}{\ln a}.$$

**Change-of-Base Theorem for Exponentials**  For every positive real number $a$,

$$a^x = e^{(\ln a)x}.$$

**Exponential Growth and Decay Function**  Let $y_0$ be the amount or number of some quantity present at time $t = 0$. The quantity is said to grow or decay exponentially if, for some constant $k$, the amount present at any time $t$ is given by

$$y = y_0 e^{kt}.$$

**1 Radian**  1 radian $= \left(\frac{180°}{\pi}\right)$

**1 Degree**  $1° = \frac{\pi}{180}$ radians

**Degrees/Radians**

| Degrees | 0° | 30° | 45° | 60° | 90° | 180° | 270° | 360° |
|---|---|---|---|---|---|---|---|---|
| Radians | 0 | $\pi/6$ | $\pi/4$ | $\pi/3$ | $\pi/2$ | $\pi$ | $3\pi/2$ | $2\pi$ |

**Trigonometric Functions**  Let $(x, y)$ be a point other than the origin on the terminal side of an angle $\theta$ in standard position. Let $r$ be the distance from the origin to $(x, y)$. Then

$$\sin \theta = \frac{y}{r} \qquad \csc \theta = \frac{r}{y} \quad (y \neq 0)$$

$$\cos \theta = \frac{x}{r} \qquad \sec \theta = \frac{r}{x} \quad (x \neq 0)$$

$$\tan \theta = \frac{y}{x} \quad (x \neq 0) \qquad \cot \theta = \frac{x}{y} \quad (y \neq 0).$$

**Elementary Trigonometric Identities**

$$\csc \theta = \frac{1}{\sin \theta} \qquad \sec \theta = \frac{1}{\cos \theta} \qquad \cot \theta = \frac{1}{\tan \theta}$$

$$\tan \theta = \frac{\sin \theta}{\cos \theta} \qquad \cot \theta = \frac{\cos \theta}{\sin \theta} \qquad \sin^2 \theta + \cos^2 \theta = 1$$

**Values of Trigonometric Functions for Common Angles**

| $\theta$ (in radians) | $\theta$ (in degrees) | $\sin \theta$ | $\cos \theta$ | $\tan \theta$ | $\cot \theta$ | $\sec \theta$ | $\csc \theta$ |
|---|---|---|---|---|---|---|---|
| 0 | 0° | 0 | 1 | 0 | Undefined | 1 | Undefined |
| $\pi/2$ | 90° | 1 | 0 | Undefined | 0 | Undefined | 1 |
| $\pi$ | 180° | 0 | -1 | 0 | Undefined | -1 | Undefined |
| $3\pi/2$ | 270° | -1 | 0 | Undefined | 0 | Undefined | -1 |
| $2\pi$ | 360° | 0 | 1 | 0 | Undefined | 1 | Undefined |

**Periodic Function**   A function $y = f(x)$ is periodic if there exists a positive real number $a$ such that

$$f(x) = f(x + a)$$

for all values of $x$ in the domain. The smallest positive value of $a$ is called the period of the function.

## Graphs of Basic Functions

Quadratic

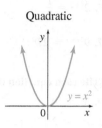

Absolute Value

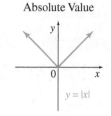

Square Root

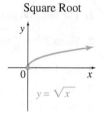

Rational

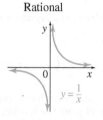

Exponential

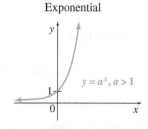

Logarithmic

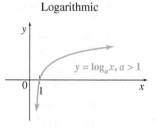

Trigonometric

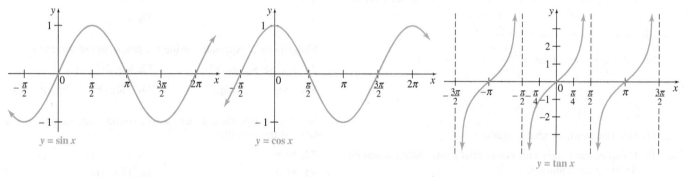

# KEY TERMS

*To understand the concepts presented in this chapter, you should know the meaning and use of the following terms. For easy reference, the section in the chapter where a word (or expression) was first used is provided.*

**2.1**
exponential function
exponential equation
interest
principal
rate of interest
time
simple interest
compound interest
compound amount
$e$
continuous compounding

**2.2**
doubling time
logarithm
logarithmic function
inverse function

properties of logarithms
common logarithms
natural logarithms
change-of-base theorem for
    logarithms
logarithmic equation
change-of-base theorem for
    exponentials
index of diversity

**2.3**
exponential growth and decay
    function
growth constant
decay constant
half-life
limited growth function
von Bertalanffy growth curve

**2.4**
ray
endpoint
angle
initial side
vertex
terminal side
standard position
degree measure
acute angle
right angle
obtuse angle
straight angle
radian measure
unit circle
arc
trigonometric functions

sine
cosine
tangent
cotangent
secant
cosecant
special angles
periodic functions
period
amplitude
frequency
phase shift
acrophase
MESOR

# REVIEW EXERCISES

## CONCEPT CHECK

**Determine whether each of the following statements is true or false, and explain why.**

1. The function $g(x) = x^\pi$ is an exponential function.

2. Since $3^{-2} = \dfrac{1}{9}$ we can conclude that $\log_3 \dfrac{1}{9} = -2$.

3. $\log_1 1 = 0$

4. $\ln(5 + 7) = \ln 5 + \ln 7$

5. $(\ln 3)^4 = 4 \ln 3$

6. $\log_{10} 0 = 1$

7. $e^{\ln 2} = 2$

8. $e^{\ln(-2)} = -2$

9. $\dfrac{\ln 4}{\ln 8} = \ln 4 - \ln 8$

10. The function $g(x) = e^x$ grows faster than the function $f(x) = \ln x$.

11. The half-life of a radioactive substance is the time required for half of the initial quantity to decay.

12. The function $f(x) = \cos x$ is periodic with a period of $\pi$.

13. All six of the basic trigonometric functions are periodic.

14. It is reasonable to expect that the Dow Jones Industrial Average is periodic and can be modeled using a sine function.

15. $\cos(a + b) = \cos a + \cos b$

16. $\sin^2\left(\dfrac{\pi}{7}\right) + \cos^2\left(2\pi + \dfrac{\pi}{7}\right) = 1$

## PRACTICE AND EXPLORATIONS

17. Describe in words what a logarithm is.

18. Compare and contrast the exponential growth function and the limited growth function.

19. What is the relationship between the degree measure and the radian measure of an angle?

20. Under what circumstances should radian measure be used instead of degree measure? Degree measure instead of radian measure?

21. Describe in words how each of the six trigonometric functions is defined.

22. At what angles (given as rational multiples of $\pi$) can you determine the exact values for the trigonometric functions?

**Find the domain of each function defined as follows.**

23. $y = \ln(x^2 - 9)$

24. $y = \dfrac{1}{e^x - 1}$

25. $y = \dfrac{1}{\sin x - 1}$

26. $y = \tan x$

**Graph the following by hand.**

27. $y = 4^x$

28. $y = 4^{-x} + 3$

29. $y = \left(\dfrac{1}{5}\right)^{2x-3}$

30. $y = \left(\dfrac{1}{2}\right)^{x-1}$

31. $y = \log_2(x - 1)$

32. $y = 1 + \log_3 x$

33. $y = -\ln(x + 3)$

34. $y = 2 - \ln x^2$

**Solve each equation.**

35. $2^{x+2} = \dfrac{1}{8}$

36. $\left(\dfrac{9}{16}\right)^x = \dfrac{3}{4}$

37. $9^{2y+3} = 27^y$

38. $\dfrac{1}{2} = \left(\dfrac{b}{4}\right)^{1/4}$

**Write each equation using logarithms.**

39. $3^5 = 243$

40. $5^{1/2} = \sqrt{5}$

41. $e^{0.8} = 2.22554$

42. $10^{1.07918} = 12$

**Write each equation using exponents.**

43. $\log_2 32 = 5$

44. $\log_9 3 = \dfrac{1}{2}$

45. $\ln 82.9 = 4.41763$

46. $\log 3.21 = 0.50651$

**Evaluate each expression without using a calculator. Then support your work using a calculator and the change-of-base theorem for logarithms.**

47. $\log_3 81$

48. $\log_{32} 16$

49. $\log_4 8$

50. $\log_{100} 1000$

**Simplify each expression using the properties of logarithms.**

51. $\log_5 3k + \log_5 7k^3$

52. $\log_3 2y^3 - \log_3 8y^2$

53. $4 \log_3 y - 2 \log_3 x$

54. $3 \log_4 r^2 - 2 \log_4 r$

**Solve each equation. If necessary, round each answer to the nearest thousandth.**

55. $6^p = 17$

56. $3^{z-2} = 11$

57. $2^{1-m} = 7$

58. $12^{-k} = 9$

59. $e^{-5-2x} = 5$

60. $e^{3x-1} = 14$

61. $\left(1 + \dfrac{m}{3}\right)^5 = 15$

62. $\left(1 + \dfrac{2p}{5}\right)^2 = 3$

63. $\log_k 64 = 6$

64. $\log_3(2x + 5) = 5$

65. $\log(4p + 1) + \log p = \log 3$

66. $\log_2(5m - 2) - \log_2(m + 3) = 2$

67. Give the following properties of the exponential function $f(x) = a^x; a > 0, a \neq 1$.

    a. Domain               b. Range

    c. $y$-intercept      d. Asymptote(s)

    e. Increasing if $a$ is _____     f. Decreasing if $a$ is _____

68. Give the following properties of the logarithmic function $f(x) = \log_a x; a > 0, a \neq 1$.

    a. Domain               b. Range

    c. $x$-intercept      d. Asymptote(s)

    e. Increasing if $a$ is _____     f. Decreasing if $a$ is _____

**69.** Compare your answers for Exercises 67 and 68. What similarities do you notice? What differences?

**Convert the following degree measures to radians. Leave answers as multiples of $\pi$.**

**70.** 90°  **71.** 160°

**72.** 225°  **73.** 270°

**74.** 360°  **75.** 405°

**Convert the following radian measures to degrees.**

**76.** $5\pi$  **77.** $\dfrac{3\pi}{4}$

**78.** $\dfrac{9\pi}{20}$  **79.** $\dfrac{3\pi}{10}$

**80.** $\dfrac{13\pi}{20}$  **81.** $\dfrac{13\pi}{15}$

**Find each function value *without* using a calculator.**

**82.** $\sin 60°$  **83.** $\tan 120°$

**84.** $\cos(-45°)$  **85.** $\sec 150°$

**86.** $\csc 120°$  **87.** $\cot 300°$

**88.** $\sin \dfrac{\pi}{6}$  **89.** $\cos \dfrac{7\pi}{3}$

**90.** $\sec \dfrac{5\pi}{3}$  **91.** $\csc \dfrac{7\pi}{3}$

**Find each function value.**

**92.** $\sin 47°$  **93.** $\cos 72°$

**94.** $\tan 115°$  **95.** $\sin(-123°)$

**96.** $\sin 2.3581$  **97.** $\cos 0.8215$

**98.** $\cos 0.5934$  **99.** $\tan 1.2915$

**Graph one period of each function.**

**100.** $y = 4\cos x$  **101.** $y = \dfrac{1}{2}\tan x$

**102.** $y = -\tan x$  **103.** $y = -\dfrac{2}{3}\sin x$

## LIFE SCIENCE APPLICATIONS

**104. Population Growth**  A population of 15,000 small deer in a specific region has grown exponentially to 17,000 in 4 years.

  **a.** Write an exponential equation to express the population growth $y$ in terms of time $t$ in years.

  **b.** At this rate, how long will it take for the population to reach 45,000?

**105. Intensity of Light**  The intensity of light (in appropriate units) passing through water decreases exponentially with the depth it penetrates beneath the surface according to the function

$$I(x) = 10e^{-0.3x},$$

where $x$ is the depth in meters. A certain water plant requires light of an intensity of 1 unit. What is the greatest depth of water in which it will grow?

**106. Drug Concentration**  The concentration of a certain drug in the bloodstream at time $t$ (in minutes) is given by

$$c(t) = e^{-t} - e^{-2t}.$$

Use a graphing calculator to find the maximum concentration and the time when it occurs.

**107. Glucose Concentration**  When glucose is infused into a person's bloodstream at a constant rate of $c$ grams per minute, the glucose is converted and removed from the bloodstream at a rate proportional to the amount present. The amount of glucose in grams in the bloodstream at time $t$ (in minutes) is given by

$$g(t) = \frac{c}{a} + \left(g_0 - \frac{c}{a}\right)e^{-at},$$

where $a$ is a positive constant. Assume $g_0 = 0.08$, $c = 0.1$, and $a = 1.3$.

  **a.** At what time is the amount of glucose a maximum? What is the maximum amount of glucose in the bloodstream?

  **b.** When is the amount of glucose in the bloodstream 0.1 g?

  **c.** What happens to the amount of glucose in the bloodstream after a very long time?

**108. Mutation**  Researchers studying the relationship between the generation time of a species and the mutation rate for genes that cause deleterious effects gathered the following data. *Source: Science.*

| Species | Generation Time ($x$) in Years | Genomic Mutation Rate ($y$) per Generation |
|---|---|---|
| D. melanogaster/ D. pseudoobscura | 0.1 | 0.070 |
| D. melanogaster/ D. simulans | 0.1 | 0.058 |
| D. picticornis/ D. silvestris | 0.2 | 0.071 |
| Mouse/rat | 0.5 | 0.50 |
| Chicken/ old world quail | 2 | 0.49 |
| Dog/cat | 4 | 1.6 |
| Sheep/cow | 6 | 0.90 |
| Macaque/New World monkey | 11 | 1.9 |
| Human/chimpanzee | 25 | 3.0 |

  **a.** The researchers plotted the common logarithm of $y$ against the common logarithm of $x$. Do so with a graphing calculator. Do the data follow a linear trend?

  **b.** Find the least squares line of the common logarithm of $y$ against the common logarithm of $x$. Plot this line on the same graph as in part a.

c. Use your answer from part b to find an equation describing $y$ in terms of $x$.

d. Find the coefficient of correlation.

**109. Cancer Growth**    Cancer cells often have a diameter of about 20 microns, or $2(10)^{-5}$ m.

a. Although cancer cells are not exactly spheres, they are close enough to spheres to make this simplification. What is the volume of a cancer cell?

b. Suppose a cancer cell doubles every day. Starting with one cancer cell, find a formula for the total volume of the cancer cells after $t$ days.

c. Assume that a tumor of cancer cells is roughly a sphere. Also assume that about 74% of the tumor is made up of cancerous cells, while the rest consists of blood vessels and space between the cells. How many days must pass before the single cancer cell grows to a tumor with a diameter of 1 cm $(10^{-2}\,\text{m})$ ?

**110. Polar Bear Mass**    One formula for estimating the mass (in kg) of a polar bear is given by

$$m(g) = e^{0.02+0.062g-0.000165g^2},$$

where $g$ is the axillary girth in cm. It seems reasonable that as girth increases, so does the mass. What is the largest girth for which this formula gives a reasonable answer? What is the predicted mass of a polar bear with this girth? *Source: Journal of Wildlife Management.*

**111. Respiratory Rate**    Researchers have found that the 95th percentile (the value at which 95% of the data are at or below) for respiratory rates (in breaths per minute) during the first 3 years of infancy are given by

$$y = 10^{1.82411-0.0125995x+0.00013401x^2}$$

for awake infants and

$$y = 10^{1.72858-0.0139928x+0.00017646x^2}$$

for sleeping infants, where $x$ is the age in months. *Source: Pediatrics.*

a. What is the domain for each function?

b. For each respiratory rate, is the rate decreasing or increasing over the first 3 years of life? (*Hint:* Is the graph of the quadratic in the exponent opening upward or downward? Where is the vertex?)

c. Verify your answer to part b using a graphing calculator.

d. For a 1-year-old infant in the 95th percentile, how much higher is the waking respiratory rate than the sleeping respiratory rate?

**112. Species**    Biologists have long noticed a relationship between the area of a piece of land and the number of species found there. The following data show a sample of the British Isles and how many vascular plants are found on each. *Source: Journal of Biogeography.*

| Isle | Area (km²) | Species |
|---|---|---|
| Ailsa | 0.8 | 75 |
| Fair | 5.2 | 174 |
| Iona | 9.1 | 388 |
| Man | 571.6 | 765 |
| N. Ronaldsay | 7.3 | 131 |
| Skye | 1735.3 | 594 |
| Stronsay | 35.2 | 62 |
| Wight | 380.7 | 1008 |

a. One common model for this relationship is logarithmic. Using the logarithmic regression feature on a graphing calculator, find a logarithmic function that best fits the data.

b. An alternative to the logarithmic model is a power function of the form $S = b(A^c)$. Using the power regression feature on a graphing calculator, find a power function that best fits the data.

c. Graph both functions from parts a and b along with the data. Give advantages and drawbacks of both models.

d. Use both functions to predict the number of species found on the isle of Shetland, with an area of 984.2 km². Compare with the actual number of 421.

e. Describe one or more situations in which being able to predict the number of species could be useful.

**113. Blood Pressure**    A person's blood pressure at time $t$ (in seconds) is given by

$$P(t) = 90 + 15 \sin 144\pi t.$$

Find the maximum and minimum values of $P$ on the interval $[0, 1/72]$. Graph one period of $y = P(t)$.

**114. Circadian Testosterone**    The following model has been used to describe the circadian rhythms in testosterone concentration:

$$C(t) = a + (C_1 - a)e^{3[\cos(b(t-t_1))-1]} + (C_0 - a)e^{3[\cos(b(t-t_0))-1]}$$

where $C$ is the concentration in nanograms per deciliter, $t$ is the time in hours, and $a$, $b$, $C_0$, $C_1$, $t_0$, and $t_1$ are constants. *Source: Journal of Clinical Pharmacology.*

a. The value of $b$ is $2\pi/24$. What is the period of this function?

b. The values of $a$, $C_0$, $C_1$, $t_0$, and $t_1$ for healthy young men are estimated to be 546, 511, 634, 20.27, and 6.05, respectively. Graph this function using a graphing calculator.

c. What is the value of $C(t_0)$? What is noteworthy about this answer?

d. What is the value of $C(t_1)$? What is noteworthy about this answer?

e. Is $C(t_0)$ exactly equal to $C_0$ or $C(t_1)$ exactly equal to $C_1$? Explain.

**115. World Grain Harvest** World grain harvests have been increasing linearly over recent years. The harvest was about 0.8 billion tons in 1960 and about 2.2 billion tons in 2010. *Source: U.S. Department of Agriculture.*

**a.** Write the world grain harvest $(g)$ in billions of tons as a function of $t$, the number of years since 1900.

 **b.** World population in billions can be described by the function

$$p(t) = \frac{9.803}{1 + 27.28\, e^{-0.03391t}} + 0.99,$$

where $t$ is the number of years since 1900. *Source: Science.* Per capita grain harvests can be described as world grain harvests divided by the population. Use a graphing calculator to plot world per capita grain harvests from 1950 to 2020 $(50 \leq t \leq 120)$. Describe any general trends.

**116. Cortisone Concentration** Researchers have found that the average baseline cortisone concentration in volunteers over a 24-hour period could be modeled by a sum of trigonometric functions. *Source: Journal of Pharmacokinetics and Biopharmaceutics.* They used the functions

$$R_1(t) = 67.75 + 47.72 \cos\left(\frac{\pi t}{12}\right) + 11.79 \sin\left(\frac{\pi t}{12}\right)$$

$$R_2(t) = R_1(t) + 14.29 \cos\left(\frac{\pi t}{6}\right) - 21.09 \sin\left(\frac{\pi t}{6}\right)$$

$$R_3(t) = R_2(t) - 2.86 \cos\left(\frac{\pi t}{4}\right) - 14.31 \sin\left(\frac{\pi t}{4}\right),$$

where each function approximates the cortisol concentration in nanograms per milliliter, and $t$ is the time in hours. Notice that all three functions are periodic over a 24-hour period.

**a.** Plot each of these functions on the window $0 \leq t \leq 24$, $0 \leq y \leq 160$. What happens as the function contains more terms?

**b.** The actual concentration at 12 hours is about 42 nanograms per milliliter. What value does each of the functions give? Which is most accurate?

## OTHER APPLICATIONS

**Interest** **Find the amount of interest earned by each deposit.**

**117.** $6902 if interest is 6% compounded semiannually for 8 years

**118.** $2781.36 if interest is 4.8% compounded quarterly for 6 years

**Interest** **Find the compound amount if $12,104 is invested at 6.2% compounded continuously for each period.**

**119.** 2 years          **120.** 4 years

**Interest** **Find the compound amounts for the following deposits if interest is compounded continuously.**

**121.** $1500 at 6% for 9 years

**122.** $12,000 at 5% for 8 years

**123.** How long will it take for $1000 deposited at 6% compounded semiannually to double? To triple?

**124.** How long will it take for $2100 deposited at 4% compounded quarterly to double? To triple?

**125. Oil Production** The production of an oil well has decreased exponentially from 128,000 barrels per year 5 years ago to 100,000 barrels per year at present.

**a.** Letting $t = 0$ represent the present time, write an exponential equation for production $y$ in terms of time $t$ in years.

**b.** Find the time it will take for production to fall to 70,000 barrels per year.

**126. Dating Rocks** Geologists sometimes measure the age of rocks by using "atomic clocks." By measuring the amounts of potassium-40 and argon-40 in a rock, the age $t$ of the specimen (in years) is found with the formula

$$t = (1.26 \times 10^9) \frac{\ln[1 + 8.33(A/K)]}{\ln 2},$$

where $A$ and $K$, respectively, are the numbers of atoms of argon-40 and potassium-40 in the specimen.

**a.** How old is a rock in which $A = 0$ and $K > 0$?

**b.** The ratio $A/K$ for a sample of granite from New Hampshire is 0.212. How old is the sample?

**c.** Let $A/K = r$. What happens to $t$ as $r$ gets larger? Smaller?

**127. Temperature** The table lists the average monthly temperatures in Vancouver, Canada. *Source: Weather.com.*

| Month | Jan | Feb | Mar | Apr | May | June |
|---|---|---|---|---|---|---|
| Temperature | 37 | 41 | 43 | 48 | 54 | 59 |

| Month | July | Aug | Sep | Oct | Nov | Dec |
|---|---|---|---|---|---|---|
| Temperature | 63 | 63 | 58 | 50 | 43 | 38 |

These average temperatures cycle yearly and change only slightly over many years. Because of the repetitive nature of temperatures from year to year, they can be modeled with a sine function. Some graphing calculators have a sine regression feature. If the table is entered into a calculator, the points can be plotted automatically, as shown in the early chapters of this book with other types of functions.

**a.** Use a graphing calculator to plot the ordered pairs (month, temperature) in the interval $[0, 12]$ by $[30, 70]$.

**b.** Use a graphing calculator with a sine regression feature to find an equation of the sine function that models these data.

**c.** Graph the equation from part b.

**d.** Calculate the period for the function found in part b. Is this period reasonable?

# POWER FUNCTIONS

We have seen several applications of power functions, which have the general form $y = ax^b$. Power functions are so named because the independent variable is raised to a power. (These should not be confused with exponential functions, in which the independent variable appears in the power.) We explored some special cases of power functions, such as $b = 2$ (a simple quadratic function) and $b = 1/2$ (a square root function). But applications of power functions vary greatly and are not limited to these special cases.

For example, in Exercise 90 in Section 2.2, we saw that the basal metabolism rate of marsupial carnivores is a power function of the body mass. In that exercise we also saw a way to verify that empirical data can be modeled with a power function. By taking the natural logarithm of both sides of the equation for a power function,

$$y = ax^b, \tag{1}$$

we obtain the equation

$$\ln y = \ln a + b \ln x. \tag{2}$$

Letting $Y = \ln y$, $X = \ln x$, and $A = \ln a$ results in the linear equation

$$Y = A + bX. \tag{3}$$

Plotting the logarithm of the original data reveals whether a straight line approximates the data well. If it does, then a power function is a good fit to the original data.

Here is another example. In an attempt to measure how the pace of city life is related to the population of the city, two researchers estimated the average speed of pedestrians in 15 cities by measuring the mean time it took them to walk 50 feet. Their results are shown in the table in the next column. *Source: Nature.*

Figure 35(a) shows the speed (stored in the list $L_2$ on a TI-84 Plus) plotted against the population (stored in $L_1$). The natural logarithm of the data was then calculated and stored using the commands $\ln(L_1) \rightarrow L_3$ and $\ln(L_2) \rightarrow L_4$. A plot of the data in $L_3$ and $L_4$ is shown in Figure 35(b). Notice that the data lie fairly closely along a straight line, confirming that a power function is an appropriate model for the original data. (These calculations and plots could also be carried out on a spreadsheet.)

A power function that best fits the data according to the least squares principle of Section 1.2 is found with the TI-84 Plus command PwrReg $L_1$, $L_2$, $Y_1$. The result is

$$y = 1.363x^{0.09799}, \tag{4}$$

| City | Population (x) | Speed (ft/sec) (y) |
|------|---------------|--------------------|
| Brno, Czechoslovakia | 341,948 | 4.81 |
| Prague, Czechoslovakia | 1,092,759 | 5.88 |
| Corte, Corsica | 5491 | 3.31 |
| Bastia, France | 49,375 | 4.90 |
| Munich, Germany | 1,340,000 | 5.62 |
| Psychro, Crete | 365 | 2.67 |
| Itea, Greece | 2500 | 2.27 |
| Iráklion, Greece | 78,200 | 3.85 |
| Athens, Greece | 867,023 | 5.21 |
| Safed, Israel | 14,000 | 3.70 |
| Dimona, Israel | 23,700 | 3.27 |
| Netanya, Israel | 70,700 | 4.31 |
| Jerusalem, Israel | 304,500 | 4.42 |
| New Haven, U.S.A. | 138,000 | 4.39 |
| Brooklyn, U.S.A. | 2,602,000 | 5.05 |

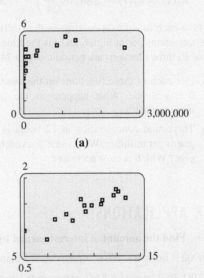

(a)

(b)

**FIGURE 35**

with a correlation coefficient of $r = 0.9081$. (For more on the correlation coefficient, see Section 1.2.) Because this value is close to 1, it indicates a good fit. Similarly, we can find a line that fits the data in Figure 35(b) with the command LinReg (ax + b) $L_3$, $L_4$, $Y_2$. The result is

$$Y = 0.30985 + 0.09799X, \tag{5}$$

again with $r = 0.9081$. The identical correlation coefficient is not a surprise, since the two commands accomplish essentially the same thing. Comparing Equations (4) and (5) with Equations (1), (2), and (3), notice that $b = 0.09799$ in both Equations (4) and (5), and that $A = 0.30985 \approx \ln a = \ln 1.363$. (The slight difference is due to rounding.) Equations (4) and (5) are plotted on the same window as the data in Figure 36(a) and (b), respectively.

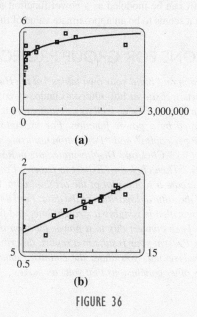

(a)

(b)

FIGURE 36

These results raise numerous questions worth exploring. What does this analysis tell you about the connection between the pace of city life and the population of a city? What might be some reasons for this connection?

A third example was considered in Review Exercise 112, where we explored the relationship between the area of each of the British isles and the number of species of vascular plants on the isles. In Figure 37(a) we have plotted the complete set of data from the original article (except for the Isle of Britain, whose area is so large that it doesn't fit on a graph that shows the other data in detail). *Source: Journal of Biogeography.* In Figure 37(b) we have plotted the natural logarithm of the data (again leaving out Britain). Notice that despite the large amount of scatter in the data, there is a linear trend. Figure 37(a) includes the best-fitting power function

$$y = 125.9x^{0.2088}, \tag{6}$$

while Figure 37(b) includes the best-fitting line

$$Y = 4.836 + 0.2088X. \tag{7}$$

(We have included the data for Britain in both of these calculations.) Notice as before that the exponent of the power function equals the slope of the linear function, and that $A = 4.836 \approx \ln a = \ln 125.9$. The correlation is 0.6917 for both, indicating that there is a trend, but that the data is somewhat scattered about the best-fitting curve or line.

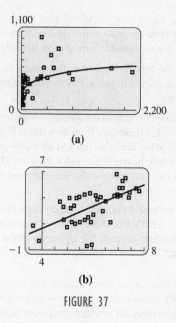

(a)

(b)

FIGURE 37

There are many more examples of data that fit a power function. Bohorquez et al. have found that the frequency of attacks in a war is a power function of the number of people killed in the attacks. *Source: Nature.* Amazingly, this holds true for a wide variety of different wars, with the average value of the exponent as $b \approx 2.5$. For a fragmented, fluid enemy with more groups, the value of $b$ tends to be larger, and for a robust, stronger enemy with fewer groups, the value of $b$ tends to be smaller.* You will explore other applications of power functions in the exercises.

## EXERCISES

1. Gwartney et al. listed the following data relating the price of a monthly cellular bill (in dollars) and the demand (in millions of subscribers).[†] *Source: Economics: Private and Public Choice.*

| Quantity (in millions) | Price (in dollars) |
|---|---|
| 2.1 | 123 |
| 3.5 | 107 |
| 5.3 | 92 |
| 7.6 | 79 |
| 11.0 | 73 |
| 16.0 | 63 |
| 24.1 | 56 |

*For a TED video on this phenomenon, see www.ted.com/talks/sean_gourley_on_the_mathematics_of_war.html.

[†]*The authors point out that these are actual prices and quantities annually for 1988 to 1994. If they could assume that other demand determinants, such as income, had remained constant during that period, this would give an accurate measurement of the demand function.*

a. Using a graphing calculator, plot the natural logarithm of the price against the natural logarithm of the quantity. Does the relationship appear to be linear?

b. Find the best-fitting line to the natural logarithm of the data, as plotted in part a. Plot this line on the same axes as the data.

c. Plot the price against the quantity. What is different about the trend in these data from the trend in Figures 35(a) and 37(a)? What does this tell you about the exponent of the best-fitting power function for these data? What conclusions can you make about how demand varies with the price?

d. Find the best-fitting power function for the data plotted in part c. Verify that this function is equivalent to the least squares line through the logarithm of the data found in part b.

2. For many years researchers thought that the basal metabolic rate (BMR) in mammals was a power function of the mass, with disagreement on whether the power was 0.67 or 0.75. More recently, White et al. proposed that the power may vary for mammals with different evolutionary lineages. *Source: Evolution.* The following table shows a portion of their data containing the natural logarithm of the mass (in grams) and the natural logarithm of the BMR (in mL of oxygen per hour) for 12 species from the genus Peromyscus, or deer mouse.

| Species | ln(mass) | ln(BMR) |
|---|---|---|
| *Peromyscus boylii* | 3.1442 | 3.9943 |
| *Peromyscus californicus* | 3.8618 | 3.9506 |
| *Peromyscus crinitus* | 2.7663 | 3.2237 |
| *Peromyscus eremicus* | 3.0681 | 3.4998 |
| *Peromyscus gossypinus* | 3.0681 | 3.6104 |
| *Peromyscus leucopus* | 3.1355 | 3.8111 |
| *Peromyscus maniculatus* | 3.0282 | 3.6835 |
| *Peromyscus megalops* | 4.1927 | 4.5075 |
| *Peromyscus oreas* | 3.2019 | 3.7729 |
| *Peromyscus polionotus* | 2.4849 | 3.0671 |
| *Peromyscus sitkensis* | 3.3439 | 3.8447 |
| *Peromyscus truei* | 3.5041 | 4.0378 |

a. Plot ln(BMR) against ln(mass) using a graphing calculator. Does the relationship appear to be linear?

b. Find the least squares line for the data plotted in part a. Plot the line on the same axes as the data.

c. Calculate the mass and BMR for each species, and then find the best-fitting power function for these data. Plot this function on the same axes as the mass and BMR data.

d. What would you conclude about whether the deer mouse BMR can be modeled as a power function of the mass? What seems to be an approximate value of the power?

## DIRECTIONS FOR GROUP PROJECT

*Go to the section on "build your own tables" of the Human Development Reports website at http://hdrstats.undp.org/en/buildtables. Select a group of countries, as well as two indicators that you think might be related by a power function. For example, you might choose "GDP per capita" and "Population not using an improved water source (%)." Click on "Display indicators in Row" and then "Show results." Then click on "Export to Excel." From the Excel spreadsheet, create a scatterplot of the original data, as well as a scatterplot of the natural logarithm of the data. Find data for which the natural logarithm is roughly a straight line, and find the least squares line. Then convert this to a power function modeling the original data. Present your results in a report, describing in detail what your analysis tells you about the countries under consideration and any other conclusions that you can make.*

# 3

# The Derivative

3.1  Limits

3.2  Continuity

3.3  Rates of Change

3.4  Definition of the Derivative

3.5  Graphical Differentiation

Chapter 3 Review
Extended Application: A Model for Drugs
Administered Intravenously

The populations of plant and animal species are dynamic, increasing and decreasing in size. Using examples in the third section of this chapter, we explore two rates of change related to populations. In the first examples, we calculate an average rate of change; in the last examples, we calculate the rate of change at a particular time. This latter rate is an example of a derivative, the subject of this chapter.

The algebraic problems considered in earlier chapters dealt with *static* situations:

- In how many years will a particular population reach 10,000?
- What is the oxygen consumption of yearling salmon?
- What is the target heart rate zone for a 20-year-old?

Calculus, on the other hand, deals with *dynamic* situations:

- How fast is the flu spreading throughout an area?
- At what rate is the population growing?
- At what time after a meal does the maximum thermal effect of food occur?

The techniques of calculus allow us to answer these questions, which deal with rates of change.

The key idea underlying calculus is the concept of limit, so we will begin by studying limits.

# 3.1    Limits

APPLY IT    **What happens to the oxygen concentration in a pond over the long run?**
*We will find an answer to this question in Example 11 using the concept of limit.*

The limit is one of the tools that we use to describe the behavior of a function as the values of $x$ approach, or become closer and closer to, some particular number.

### EXAMPLE 1    Finding a Limit

What happens to $f(x) = x^2$ when $x$ is a number *very close* to (but not equal to) 2?

**SOLUTION**    We can construct a table with $x$-values getting closer and closer to 2 and find the corresponding values of $f(x)$.

| | $x$ approaches 2 from left | | | | ↓ | $x$ approaches 2 from right | | | |
|---|---|---|---|---|---|---|---|---|---|
| $x$ | 1.9 | 1.99 | 1.999 | 1.9999 | 2 | 2.0001 | 2.001 | 2.01 | 2.1 |
| $f(x)$ | 3.61 | 3.9601 | 3.996001 | 3.99960001 | 4 | 4.00040001 | 4.004001 | 4.0401 | 4.41 |
| | | | $f(x)$ approaches 4 | | ↑ | $f(x)$ approaches 4 | | | |

The table suggests that, as $x$ gets closer and closer to 2 from either side, $f(x)$ gets closer and closer to 4. In fact, you can use a calculator to show the values of $f(x)$ can be made as close as you want to 4 by taking values of $x$ close enough to 2. This is not surprising since the value of the function at $x = 2$ is $f(x) = 4$. We can observe this fact by looking at the graph $y = x^2$, as shown in Figure 1. In such a case, we say "the limit of $f(x)$ as $x$ approaches 2 is 4," which is written as

$$\lim_{x \to 2} f(x) = 4.$$

**YOUR TURN 1**
Find $\lim_{x \to 1} (x^2 + 2)$.

TRY YOUR TURN 1

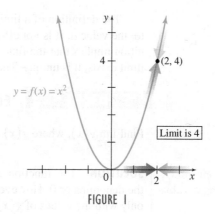

FIGURE 1

The phrase "$x$ approaches 2 from the left" is written $x \rightarrow 2^-$. Similarly, "$x$ approaches 2 from the right" is written $x \rightarrow 2^+$. These expressions are used to write **one-sided limits**. The **limit from the left** (as $x$ approaches 2 from the negative direction) is written

$$\lim_{x \to 2^-} f(x) = 4,$$

and shown in red in Figure 1. The **limit from the right** (as $x$ approaches 2 from the positive direction) is written

$$\lim_{x \to 2^+} f(x) = 4,$$

and shown in blue in Figure 1. A **two-sided limit**, such as

$$\lim_{x \to 2} f(x) = 4,$$

exists only if both one-sided limits exist and are the same; that is, if $f(x)$ approaches the same number as $x$ approaches a given number from *either* side.

**CAUTION** | Notice that $\lim_{x \to a^-} f(x)$ does not mean to take negative values of $x$, nor does it mean to choose values of $x$ to the right of $a$ and then move in the negative direction. It means to use values less than $a$ $(x < a)$ that get closer and closer to $a$.

The previous example suggests the following informal definition.

### Limit of a Function

Let $f$ be a function and let $a$ and $L$ be real numbers. If

1. as $x$ takes values closer and closer (but not equal) to $a$ on both sides of $a$, the corresponding values of $f(x)$ get closer and closer (and perhaps equal) to $L$; and

2. the value of $f(x)$ can be made as close to $L$ as desired by taking values of $x$ close enough to $a$;

then $L$ is the **limit** of $f(x)$ as $x$ approaches $a$, written

$$\lim_{x \to a} f(x) = L.$$

This definition is informal because the expressions "closer and closer to" and "as close as desired" have not been defined. A more formal definition would be needed to prove the rules for limits given later in this section.*

*The limit is the key concept from which all the ideas of calculus flow. Calculus was independently discovered by the English mathematician Isaac Newton (1642–1727) and the German mathematician Gottfried Wilhelm Leibniz (1646–1716). For the next century, supporters of each accused the other of plagiarism, resulting in a lack of communication between mathematicians in England and on the European continent. Neither Newton nor Leibniz developed a mathematically rigorous definition of the limit (and we have no intention of doing so here). More than 100 years passed before the French mathematician Augustin-Louis Cauchy (1789–1857) accomplished this feat.

The definition of a limit describes what happens to $f(x)$ when $x$ is near, but not equal to, the value $a$. It is not affected by how (or even whether) $f(a)$ is defined. Also the definition implies that the function values cannot approach two different numbers, so that if a limit exists, it is unique. These ideas are illustrated in the following examples.

### EXAMPLE 2    Finding a Limit

Find $\lim\limits_{x \to 2} g(x)$, where $g(x) = \dfrac{x^3 - 2x^2}{x - 2}$.

**Method 1**
**Using a Table**

**SOLUTION**    The function $g(x)$ is undefined when $x = 2$, since the value $x = 2$ makes the denominator 0. However, in determining the limit as $x$ approaches 2 we are concerned only with the values of $g(x)$ when $x$ is close to but *not equal to* 2. To determine if the limit exists, consider the value of $g$ at some numbers close to but not equal to 2, as shown in the following table.

| | x approaches 2 from left | | | | | x approaches 2 from right | | | |
|---|---|---|---|---|---|---|---|---|---|
| $x$ | 1.9 | 1.99 | 1.999 | 1.9999 | 2 | 2.0001 | 2.001 | 2.01 | 2.1 |
| $g(x)$ | 3.61 | 3.9601 | 3.996001 | 3.99960001 | | 4.00040001 | 4.004001 | 4.0401 | 4.41 |

$f(x)$ approaches 4      $f(x)$ approaches 4

Undefined

Notice that this table is almost identical to the previous table, except that $g$ is undefined at $x = 2$. This suggests that $\lim\limits_{x \to 2} g(x) = 4$, in spite of the fact that the function $g$ does not exist at $x = 2$.

**Method 2**
**Using Algebra**

A second approach to this limit is to analyze the function. By factoring the numerator,

$$x^3 - 2x^2 = x^2(x - 2),$$

$g(x)$ simplifies to

$$g(x) = \frac{x^2(x - 2)}{x - 2} = x^2, \qquad \text{provided } x \neq 2.$$

The graph of $g(x)$, as shown in Figure 2, is almost the same as the graph of $y = x^2$, except that it is undefined at $x = 2$ (illustrated by the "hole" in the graph).

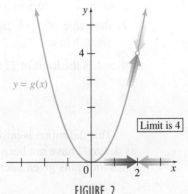

$y = g(x)$

Limit is 4

**FIGURE 2**

**YOUR TURN 2**
Find $\lim\limits_{x \to 2} \dfrac{x^2 - 4}{x - 2}$.

Since we are looking at the limit as $x$ approaches 2, we look at values of the function for $x$ close to but not equal to 2. Thus, the limit is

$$\lim_{x \to 2} g(x) = \lim_{x \to 2} x^2 = 4.$$

**TRY YOUR TURN 2**

TECHNOLOGY NOTE

We can use the TRACE feature on a graphing calculator to determine the limit. Figure 3 shows the graph of the function in Example 2 drawn with a graphing calculator. Notice that the function has a small gap at the point $(2, 4)$, which agrees with our previous observation that the function is undefined at $x = 2$, where the limit is 4. (Due to the limitations of the graphing calculator, this gap may vanish when the viewing window is changed very slightly.)

The result after pressing the TRACE key is shown in Figure 4. The cursor is already located at $x = 2$; if it were not, we could use the right or left arrow key to move the cursor there. The calculator does not give a $y$-value because the function is undefined at $x = 2$. Moving the cursor back a step gives $x = 1.98$, $y = 3.92$. Moving the cursor forward two steps gives $x = 2.02$, $y = 4.09$. It seems that as $x$ approaches 2, $y$ approaches 4, or at least something close to 4. Zooming in on the point $(2, 4)$ (such as using the window $[1.9, 2.1)$ by $[3.9, 4.1]$) allows the limit to be estimated more accurately and helps ensure that the graph has no unexpected behavior very close to $x = 2$.

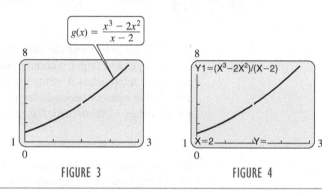

FIGURE 3                    FIGURE 4

## EXAMPLE 3  Finding a Limit

Determine $\lim\limits_{x \to 2} h(x)$ for the function $h$ defined by

$$h(x) = \begin{cases} x^2, & \text{if } x \ne 2, \\ 1, & \text{if } x = 2. \end{cases}$$

**YOUR TURN 3**
Find $\lim\limits_{x \to 3} f(x)$ if

$$f(x) = \begin{cases} 2x - 1 & \text{if } x \ne 3 \\ 1 & \text{if } x = 3. \end{cases}$$

**SOLUTION**  A function defined by two or more cases is called a **piecewise function**. The domain of $h$ is all real numbers, and its graph is shown in Figure 5. Notice that $h(2) = 1$, but $h(x) = x^2$ when $x \ne 2$. To determine the limit as $x$ approaches 2, we are concerned only with the values of $h(x)$ when $x$ is close but not equal to 2. Once again,

$$\lim_{x \to 2} h(x) = \lim_{x \to 2} x^2 = 4.$$   **TRY YOUR TURN 3**

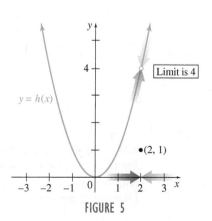

FIGURE 5

EXAMPLE 4  Finding a Limit

Find $\lim\limits_{x \to -2} f(x)$, where

$$f(x) = \frac{3x + 2}{2x + 4}.$$

**SOLUTION**  The graph of the function is shown in Figure 6. A table with the values of $f(x)$ as $x$ gets closer and closer to $-2$ is given below.

| | x approaches −2 from left | | | | x approaches −2 from right | | | |
|---|---|---|---|---|---|---|---|---|
| **x** | −2.1 | −2.01 | −2.001 | −2.0001 | −1.9999 | −1.999 | −1.99 | −1.9 |
| **f(x)** | 21.5 | 201.5 | 2001.5 | 20,001.5 | −19,998.5 | −1998.5 | −198.5 | −18.5 |

Both the graph and the table suggest that as $x$ approaches $-2$ from the left, $f(x)$ becomes larger and larger without bound. This happens because as $x$ approaches $-2$, the denominator approaches 0, while the numerator approaches $-4$, and $-4$ divided by a smaller and smaller number becomes larger and larger. When this occurs, we say that "the limit as $x$ approaches $-2$ from the left is infinity," and we write

$$\lim_{x \to -2^-} f(x) = \infty.$$

Because $\infty$ is not a real number, the limit in this case does not exist.

In the same way, the behavior of the function as $x$ approaches $-2$ from the right is indicated by writing

$$\lim_{x \to -2^+} f(x) = -\infty,$$

**YOUR TURN 4**

Find $\lim\limits_{x \to 0} \dfrac{2x - 1}{x}$.

since $f(x)$ becomes more and more negative without bound. Since there is no real number that $f(x)$ approaches as $x$ approaches $-2$ (from either side), nor does $f(x)$ approach either $\infty$ or $-\infty$, we simply say

$$\lim_{x \to -2} \frac{3x + 2}{2x + 4} \text{ does not exist.}$$   TRY YOUR TURN 4

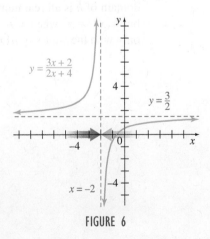

$y = \dfrac{3x + 2}{2x + 4}$

$y = \dfrac{3}{2}$

$x = -2$

FIGURE 6

**NOTE**  In general, if both the limit from the left and from the right approach $\infty$, so that $\lim\limits_{x \to a} f(x) = \infty$, the limit would not exist because $\infty$ is not a real number. It is customary, however, to give $\infty$ as the answer, since it describes how the function is behaving near $x = a$. Likewise, if $\lim\limits_{x \to a} f(x) = -\infty$, we give $-\infty$ as the answer.

We have shown three methods for determining limits: (1) using a table of numbers, (2) using algebraic simplification, and (3) tracing the graph on a graphing calculator. Which method you choose depends on the complexity of the function and the accuracy required by the application. Algebraic simplification gives the exact answer, but it can be difficult or even impossible to use in some situations. Calculating a table of numbers or tracing the graph may be easier when the function is complicated, but be careful, because the results could be inaccurate, inconclusive, or misleading. A graphing calculator does not tell us what happens between or beyond the points that are plotted.

### EXAMPLE 5  Finding a Limit

Find $\lim_{x \to 0} \dfrac{|x|}{x}$.

**SOLUTION**

**Method 1**
**Algebraic Approach**

The function $f(x) = |x|/x$ is not defined when $x = 0$. When $x > 0$, the definition of absolute value says that $|x| = x$, so $f(x) = |x|/x = x/x = 1$. When $x < 0$, then $|x| = -x$ and $f(x) = |x|/x = -x/x = -1$. Therefore,

$$\lim_{x \to 0^+} f(x) = 1 \qquad \text{and} \qquad \lim_{x \to 0^-} f(x) = -1.$$

Since the limits from the left and from the right are different, the limit does not exist.

**Method 2**
**Graphing Calculator Approach**

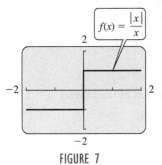

$f(x) = \dfrac{|x|}{x}$

FIGURE 7

A calculator graph of $f$ is shown in Figure 7.

As $x$ approaches 0 from the right, $x$ is always positive and the corresponding value of $f(x)$ is 1, so

$$\lim_{x \to 0^+} f(x) = 1.$$

But as $x$ approaches 0 from the left, $x$ is always negative and the corresponding value of $f(x)$ is $-1$, so

$$\lim_{x \to 0^-} f(x) = -1.$$

As in the algebraic approach, the limits from the left and from the right are different, so the limit does not exist.

The discussion up to this point can be summarized as follows.

**Existence of Limits**

The limit of $f$ as $x$ approaches $a$ may not exist.

1. If $f(x)$ becomes infinitely large in magnitude (positive or negative) as $x$ approaches the number $a$ from either side, we write $\lim_{x \to a} f(x) = \infty$ or $\lim_{x \to a} f(x) = -\infty$. In either case, the limit does not exist.

2. If $f(x)$ becomes infinitely large in magnitude (positive) as $x$ approaches $a$ from one side and infinitely large in magnitude (negative) as $x$ approaches $a$ from the other side, then $\lim_{x \to a} f(x)$ does not exist.

3. If $\lim_{x \to a^-} f(x) = L$ and $\lim_{x \to a^+} f(x) = M$, and $L \neq M$, then $\lim_{x \to a} f(x)$ does not exist.

Figure 8 illustrates these three facts.

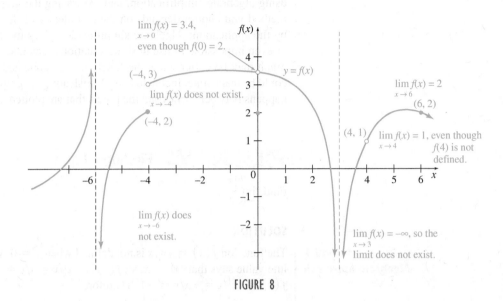

**FIGURE 8**

# Rules for Limits

As shown by the preceding examples, tables and graphs can be used to find limits. However, it is usually more efficient to use the rules for limits given below. (Proofs of these rules require a formal definition of limit, which we have not given.)

---

### Rules for Limits

Let $a$, $A$, and $B$ be real numbers, and let $f$ and $g$ be functions such that

$$\lim_{x \to a} f(x) = A \qquad \text{and} \qquad \lim_{x \to a} g(x) = B.$$

1. If $k$ is a constant, then $\lim_{x \to a} k = k$ and $\lim_{x \to a} [k \cdot f(x)] = k \cdot \lim_{x \to a} f(x) = k \cdot A$.

2. $\lim_{x \to a} [f(x) \pm g(x)] = \lim_{x \to a} f(x) \pm \lim_{x \to a} g(x) = A \pm B$

   (The limit of a sum or difference is the sum or difference of the limits.)

3. $\lim_{x \to a} [f(x) \cdot g(x)] = \left[\lim_{x \to a} f(x)\right] \cdot \left[\lim_{x \to a} g(x)\right] = A \cdot B$

   (The limit of a product is the product of the limits.)

4. $\lim_{x \to a} \dfrac{f(x)}{g(x)} = \dfrac{\lim_{x \to a} f(x)}{\lim_{x \to a} g(x)} = \dfrac{A}{B}$  if $B \neq 0$

   (The limit of a quotient is the quotient of the limits, provided the limit of the denominator is not zero.)

5. If $p(x)$ is a polynomial, then $\lim_{x \to a} p(x) = p(a)$.

6. For any real number $k$, $\lim_{x \to a} [f(x)]^k = \left[\lim_{x \to a} f(x)\right]^k = A^k$, provided this limit exists.*

7. $\lim_{x \to a} f(x) = \lim_{x \to a} g(x)$ if $f(x) = g(x)$ for all $x \neq a$.

8. For any real number $b > 0$, $\lim_{x \to a} b^{f(x)} = b^{[\lim_{x \to a} f(x)]} = b^A$.

9. For any real number $b$ such that $0 < b < 1$ or $1 < b$,

   $\lim_{x \to a} [\log_b f(x)] = \log_b[\lim_{x \to a} f(x)] = \log_b A$ if $A > 0$.

---

*This limit does not exist, for example, when $A < 0$ and $k = 1/2$, or when $A = 0$ and $k \leq 0$.

This list may seem imposing, but these limit rules, once understood, agree with common sense. For example, Rule 3 says that if $f(x)$ becomes close to $A$ as $x$ approaches $a$, and if $g(x)$ becomes close to $B$, then $f(x) \cdot g(x)$ should become close to $A \cdot B$, which seems plausible.

**EXAMPLE 6** **Rules for Limits**

Suppose $\lim_{x \to 2} f(x) = 3$ and $\lim_{x \to 2} g(x) = 4$. Use the limit rules to find the following limits.

**(a)** $\lim_{x \to 2} [f(x) + 5g(x)]$

**SOLUTION**

$$\lim_{x \to 2} [f(x) + 5g(x)] = \lim_{x \to 2} f(x) + \lim_{x \to 2} 5g(x) \quad \text{Rule 2}$$

$$= \lim_{x \to 2} f(x) + 5 \lim_{x \to 2} g(x) \quad \text{Rule 1}$$

$$= 3 + 5(4)$$

$$= 23$$

**(b)** $\lim_{x \to 2} \dfrac{[f(x)]^2}{\ln g(x)}$

**SOLUTION**

$$\lim_{x \to 2} \frac{[f(x)]^2}{\ln g(x)} = \frac{\lim_{x \to 2} [f(x)]^2}{\lim_{x \to 2} \ln g(x)} \quad \text{Rule 4}$$

$$= \frac{[\lim_{x \to 2} f(x)]^2}{\ln[\lim_{x \to 2} g(x)]} \quad \text{Rule 6 and Rule 9}$$

$$= \frac{3^2}{\ln 4}$$

$$\approx \frac{9}{1.38629} \approx 6.492 \quad \text{TRY YOUR TURN 5}$$

**YOUR TURN 5** Using the limits defined in Example 6, find $\lim_{x \to 2} [f(x) + g(x)]^2$.

**EXAMPLE 7** **Finding a Limit**

Find $\lim_{x \to 3} \dfrac{x^2 - x - 1}{\sqrt{x + 1}}$.

**SOLUTION**

$$\lim_{x \to 3} \frac{x^2 - x - 1}{\sqrt{x + 1}} = \frac{\lim_{x \to 3} (x^2 - x - 1)}{\lim_{x \to 3} \sqrt{x + 1}} \quad \text{Rule 4}$$

$$= \frac{\lim_{x \to 3} (x^2 - x - 1)}{\sqrt{\lim_{x \to 3} (x + 1)}} \quad \text{Rule 6} \ (\sqrt{a} = a^{1/2})$$

$$= \frac{3^2 - 3 - 1}{\sqrt{3 + 1}} \quad \text{Rule 5}$$

$$= \frac{5}{\sqrt{4}}$$

$$= \frac{5}{2}$$

As Examples 6 and 7 suggest, the rules for limits actually mean that many limits can be found simply by evaluation. This process is valid for polynomials, rational functions, exponential functions, logarithmic functions, and roots and powers, as long as this does not involve an illegal operation, such as division by 0 or taking the logarithm of a negative number. Division by 0 presents particular problems that can often be solved by algebraic simplification, as the following examples show.

**EXAMPLE 8   Finding a Limit**

Find $\lim\limits_{x \to 2} \dfrac{x^2 + x - 6}{x - 2}$.

**SOLUTION**   Rule 4 cannot be used here, since

$$\lim_{x \to 2} (x - 2) = 0.$$

The numerator also approaches 0 as $x$ approaches 2, and 0/0 is meaningless. For $x \neq 2$, we can, however, simplify the function by rewriting the fraction as

$$\frac{x^2 + x - 6}{x - 2} = \frac{(x + 3)(x - 2)}{x - 2} = x + 3.$$

Now Rule 7 can be used.

**YOUR TURN 6**   Find $\lim\limits_{x \to -3} \dfrac{x^2 - x - 12}{x + 3}$.

$$\lim_{x \to 2} \frac{x^2 + x - 6}{x - 2} = \lim_{x \to 2} (x + 3) = 2 + 3 = 5$$

**TRY YOUR TURN 6**

**NOTE**   Mathematicians often refer to a limit that gives 0/0, as in Example 8, as an *indeterminate form*. This means that when the numerator and denominator are polynomials, they must have a common factor, which is why we factored the numerator in Example 8.

**EXAMPLE 9   Finding a Limit**

Find $\lim\limits_{x \to 4} \dfrac{\sqrt{x} - 2}{x - 4}$.

**SOLUTION**   As $x \to 4$, the numerator approaches 0 and the denominator also approaches 0, giving the meaningless expression 0/0. In an expression such as this involving square roots, rather than trying to factor, you may find it simpler to use algebra to rationalize the numerator by multiplying both the numerator and the denominator by $\sqrt{x} + 2$. This gives

$$\frac{\sqrt{x} - 2}{x - 4} \cdot \frac{\sqrt{x} + 2}{\sqrt{x} + 2} = \frac{(\sqrt{x})^2 - 2^2}{(x - 4)(\sqrt{x} + 2)} \qquad (a - b)(a + b) = a^2 - b^2$$

$$= \frac{x - 4}{(x - 4)(\sqrt{x} + 2)} = \frac{1}{\sqrt{x} + 2}$$

if $x \neq 4$. Now use the rules for limits.

**YOUR TURN 7**   Find $\lim\limits_{x \to 1} \dfrac{\sqrt{x} - 1}{x - 1}$.

$$\lim_{x \to 4} \frac{\sqrt{x} - 2}{x - 4} = \lim_{x \to 4} \frac{1}{\sqrt{x} + 2} = \frac{1}{\sqrt{4} + 2} = \frac{1}{2 + 2} = \frac{1}{4}$$

**TRY YOUR TURN 7**

CAUTION | Simply because the expression in a limit is approaching 0/0, as in Examples 8 and 9, does *not* mean that the limit is 0 or that the limit does not exist. For such a limit, try to simplify the expression using the following principle: **To calculate the limit of $f(x)/g(x)$ as $x$ approaches $a$, where $f(a) = g(a) = 0$, you should attempt to factor $x - a$ from both the numerator and the denominator.**

### EXAMPLE 10  Finding a Limit

Find $\displaystyle\lim_{x\to 1} \frac{x^2 - 2x + 1}{(x - 1)^3}$.

**Method 1**
**Algebraic Approach**

**SOLUTION**

Again, Rule 4 cannot be used, since $\displaystyle\lim_{x\to 1} (x - 1)^3 = 0$. If $x \neq 1$, the function can be rewritten as

$$\frac{x^2 - 2x + 1}{(x - 1)^3} = \frac{(x - 1)^2}{(x - 1)^3} = \frac{1}{x - 1}.$$

Then

$$\lim_{x\to 1} \frac{x^2 - 2x + 1}{(x - 1)^3} = \lim_{x\to 1} \frac{1}{x - 1}$$

by Rule 7. None of the rules can be used to find

$$\lim_{x\to 1} \frac{1}{x - 1},$$

but as $x$ approaches 1, the denominator approaches 0 while the numerator stays at 1, making the result larger and larger in magnitude. If $x > 1$, both the numerator and denominator are positive, so $\displaystyle\lim_{x\to 1^+} 1/(x - 1) = \infty$. If $x < 1$, the denominator is negative, so $\displaystyle\lim_{x\to 1^-} 1/(x - 1) = -\infty$. Therefore,

$$\lim_{x\to 1} \frac{x^2 - 2x + 1}{(x - 1)^3} = \lim_{x\to 1} \frac{1}{x - 1} \text{ does not exist.}$$

**Method 2**
**Graphing Calculator Approach**

Using the TABLE feature on a TI-84 Plus, we can produce the table of numbers shown in Figure 9, where $Y_1$ represents the function $y = 1/(x - 1)$. Figure 10 shows a graphing calculator view of the function on $[0, 2]$ by $[-10, 10]$. The behavior of the function indicates a vertical asymptote at $x = 1$, with the limit approaching $-\infty$ from the left and $\infty$ from the right, so

$$\lim_{x\to 1} \frac{x^2 - 2x + 1}{(x - 1)^3} = \lim_{x\to 1} \frac{1}{x - 1} \text{ does not exist.}$$

Both the table and the graph can be easily generated using a spreadsheet. Consult the *Graphing Calculator and Excel Spreadsheet Manual*, available with this text, for details.

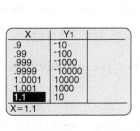

FIGURE 9

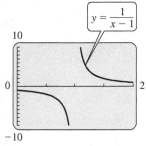

FIGURE 10

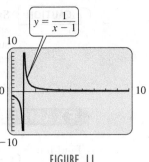

FIGURE 11

CAUTION | A graphing calculator can give a deceptive view of a function. Figure 11 shows the result if we graph the previous function on $[0, 10]$ by $[-10, 10]$. Near $x = 1$, the graph appears to be a steep line connecting the two pieces. The graph in Figure 10 is more representative of the function near $x = 1$. When using a graphing calculator, you may need to experiment with the viewing window, guided by what you have learned about functions and limits, to get a good picture of a function. On many calculators, extraneous lines connecting parts of the graph can be avoided by using DOT mode rather than CONNECTED mode.

**NOTE** Another way to understand the behavior of the function in the previous example near $x = 1$ is to recall from Section 1.5 on Polynomial and Rational Functions that a rational function often has a vertical asymptote at a value of $x$ where the denominator is 0, although it may not if the numerator there is also 0. In this example, we see after simplifying that the function has a vertical asymptote at $x = 1$ because that would make the denominator of $1/(x - 1)$ equal to 0, while the numerator is 1.

## Limits at Infinity

Sometimes it is useful to examine the behavior of the values of $f(x)$ as $x$ gets larger and larger (or more and more negative). The phrase "$x$ approaches infinity," written $x \to \infty$, expresses the fact that $x$ becomes larger without bound. Similarly, the phrase "$x$ approaches negative infinity" (symbolically, $x \to -\infty$) means that $x$ becomes more and more negative without bound (such as $-10, -1000, -10,000$, etc.). The next example illustrates a **limit at infinity**.

### EXAMPLE 11   Oxygen Concentration

Suppose a small pond normally contains 12 units of dissolved oxygen in a fixed volume of water. Suppose also that at time $t = 0$ a quantity of organic waste is introduced into the pond, with the oxygen concentration $t$ weeks later given by

$$f(t) = \frac{12t^2 - 15t + 12}{t^2 + 1}.$$

As time goes on, what will be the ultimate concentration of oxygen? Will it return to 12 units?

APPLY IT    **SOLUTION**   After 2 weeks, the pond contains

$$f(2) = \frac{12 \cdot 2^2 - 15 \cdot 2 + 12}{2^2 + 1} = \frac{30}{5} = 6$$

units of oxygen, and after 4 weeks, it contains

$$f(4) = \frac{12 \cdot 4^2 - 15 \cdot 4 + 12}{4^2 + 1} \approx 8.5$$

units. Choosing several values of $t$ and finding the corresponding values of $f(t)$, or using a graphing calculator or computer, leads to the table and graph in Figure 12.

The graph suggests that, as time goes on, the oxygen level gets closer and closer to the original 12 units. If so, the line $y = 12$ is a horizontal asymptote. The table suggests that

$$\lim_{t \to \infty} f(t) = 12.$$

Thus, the oxygen concentration will approach 12, but it will never be *exactly* 12.

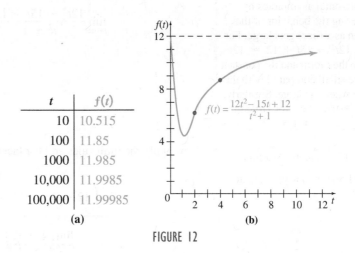

| $t$ | $f(t)$ |
|---|---|
| 10 | 10.515 |
| 100 | 11.85 |
| 1000 | 11.985 |
| 10,000 | 11.9985 |
| 100,000 | 11.99985 |

(a)          (b)

FIGURE 12

As we saw in the previous example, *limits at infinity or negative infinity, if they exist, correspond to horizontal asymptotes of the graph of the function.* In the previous chapters, we saw one way to find horizontal asymptotes. We will now show a more precise way, based upon some simple limits at infinity. The graphs of $f(x) = 1/x$ (in red) and $g(x) = 1/x^2$ (in blue) shown in Figure 13, as well as the table there, indicate that $\lim_{x \to \infty} 1/x = 0$, $\lim_{x \to -\infty} 1/x = 0$, $\lim_{x \to \infty} 1/x^2 = 0$, and $\lim_{x \to -\infty} 1/x^2 = 0$, suggesting the following rule.

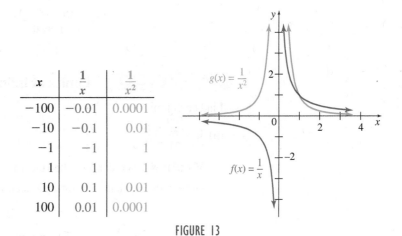

| $x$ | $\dfrac{1}{x}$ | $\dfrac{1}{x^2}$ |
|---|---|---|
| $-100$ | $-0.01$ | $0.0001$ |
| $-10$ | $-0.1$ | $0.01$ |
| $-1$ | $-1$ | $1$ |
| $1$ | $1$ | $1$ |
| $10$ | $0.1$ | $0.01$ |
| $100$ | $0.01$ | $0.0001$ |

FIGURE 13

**Limits at Infinity**

For any positive real number $n$,

$$\lim_{x \to \infty} \frac{1}{x^n} = 0 \quad \text{and} \quad \lim_{x \to -\infty} \frac{1}{x^n} = 0.^*$$

*If $x$ is negative, $x^n$ does not exist for certain values of $n$, so the second limit is undefined for those values of $n$.

The rules for limits given earlier remain unchanged when $a$ is replaced with $\infty$ or $-\infty$.

To evaluate the limit at infinity of a rational function, divide the numerator and denominator by the largest power of the variable that appears in the denominator, $t^2$ here, and then use these results. In the previous example, we find that

$$\lim_{t\to\infty}\frac{12t^2-15t+12}{t^2+1}=\lim_{t\to\infty}\frac{\dfrac{12t^2}{t^2}-\dfrac{15t}{t^2}+\dfrac{12}{t^2}}{\dfrac{t^2}{t^2}+\dfrac{1}{t^2}}$$

$$=\lim_{t\to\infty}\frac{12-15\cdot\dfrac{1}{t}+12\cdot\dfrac{1}{t^2}}{1+\dfrac{1}{t^2}}.$$

**FOR REVIEW**

In Section 1.5, we saw a way to find horizontal asymptotes by considering the behavior of the function as $x$ (or $t$) gets large. For large $t$, $12t^2-15t+12\approx 12t^2$, because the $t$-term and the constant term are small compared with the $t^2$-term when $t$ is large. Similarly, $t^2+1\approx t^2$. Thus, for large $t$,

$f(t)=\dfrac{12t^2-15t+12}{t^2+1}\approx$

$\dfrac{12t^2}{t^2}=12$. Thus the function $f$ has a horizontal asymptote at $y=12$.

Now apply the limit rules and the fact that $\lim\limits_{t\to\infty}1/t^n=0$.

$$\frac{\lim\limits_{t\to\infty}\left(12-15\cdot\dfrac{1}{t}+12\cdot\dfrac{1}{t^2}\right)}{\lim\limits_{t\to\infty}\left(1+\dfrac{1}{t^2}\right)}$$

$$=\frac{\lim\limits_{t\to\infty}12-\lim\limits_{t\to\infty}15\cdot\dfrac{1}{t}+\lim\limits_{t\to\infty}12\cdot\dfrac{1}{t^2}}{\lim\limits_{t\to\infty}1+\lim\limits_{t\to\infty}\dfrac{1}{t^2}}\qquad \text{Rules 4 and 2}$$

$$=\frac{12-15\left(\lim\limits_{t\to\infty}\dfrac{1}{t}\right)+12\left(\lim\limits_{t\to\infty}\dfrac{1}{t^2}\right)}{1+\lim\limits_{t\to\infty}\dfrac{1}{t^2}}\qquad \text{Rule 1}$$

$$=\frac{12-15\cdot 0+12\cdot 0}{1+0}=12.\qquad \text{Limits at infinity}$$

**EXAMPLE 12** **Limits at Infinity**

Find each limit.

**(a)** $\lim\limits_{x\to\infty}\dfrac{8x+6}{3x-1}$

**SOLUTION** We can use the rule $\lim\limits_{x\to\infty}1/x^n=0$ to find this limit by first dividing the numerator and denominator by $x$, as follows.

$$\lim_{x\to\infty}\frac{8x+6}{3x-1}=\lim_{x\to\infty}\frac{\dfrac{8x}{x}+\dfrac{6}{x}}{\dfrac{3x}{x}-\dfrac{1}{x}}=\lim_{x\to\infty}\frac{8+6\cdot\dfrac{1}{x}}{3-\dfrac{1}{x}}=\frac{8+0}{3-0}=\frac{8}{3}$$

**(b)** $\lim\limits_{x\to\infty}\dfrac{3x+2}{4x^3-1}=\lim\limits_{x\to\infty}\dfrac{3\cdot\dfrac{1}{x^2}+2\cdot\dfrac{1}{x^3}}{4-\dfrac{1}{x^3}}=\dfrac{0+0}{4-0}=\dfrac{0}{4}=0$

Here, the highest power of $x$ in the denominator is $x^3$, which is used to divide each term in the numerator and denominator.

**(c)** $\displaystyle\lim_{x\to\infty}\frac{3x^2+2}{4x-3}=\lim_{x\to\infty}\frac{3x+\frac{2}{x}}{4-\frac{3}{x}}$

The highest power of $x$ in the denominator is $x$ (to the first power). There is a higher power of $x$ in the numerator, but we don't divide by this. Notice that the denominator approaches 4, while the numerator becomes infinitely large, so

$$\lim_{x\to\infty}\frac{3x^2+2}{4x-3}=\infty.$$

**(d)** $\displaystyle\lim_{x\to\infty}\frac{5x^2-4x^3}{3x^2+2x-1}=\lim_{x\to\infty}\frac{5-4x}{3+\frac{2}{x}-\frac{1}{x^2}}$

The highest power of $x$ in the denominator is $x^2$. The denominator approaches 3, while the numerator becomes a negative number that is larger and larger in magnitude, so

$$\lim_{x\to\infty}\frac{5x^2-4x^3}{3x^2+2x-1}=-\infty.$$

**TRY YOUR TURN 8**

> **YOUR TURN 8**
>
> Find $\displaystyle\lim_{x\to\infty}\frac{2x^2+3x-4}{6x^2-5x+7}$.

The method used in Example 12 is a useful way to rewrite expressions with fractions so that the rules for limits at infinity can be used.

> **Finding Limits at Infinity**
>
> If $f(x)=p(x)/q(x)$, for polynomials $p(x)$ and $q(x)$, $q(x)\neq 0$, $\displaystyle\lim_{x\to-\infty}f(x)$ and $\displaystyle\lim_{x\to\infty}f(x)$ can be found as follows.
>
> **1.** Divide $p(x)$ and $q(x)$ by the highest power of $x$ in $q(x)$.
> **2.** Use the rules for limits, including the rules for limits at infinity,
>
> $$\lim_{x\to\infty}\frac{1}{x^n}=0 \qquad\text{and}\qquad \lim_{x\to-\infty}\frac{1}{x^n}=0,$$
>
> to find the limit of the result from Step 1.

For an alternate approach to finding limits at infinity, see Exercise 87.

**EXAMPLE 13**  **Arctic Foxes**

The age–weight relationship of female Arctic foxes caught in Svalbard, Norway, can be estimated by the function

$$M(t)=3102e^{-e^{-0.022(t-56)}},$$

where $t$ is the age of the fox in days and $M(t)$ is the weight of the fox in grams. Use $M(t)$ to estimate the largest size that a female fox can attain. ***Source: Journal of Mammalogy.***

**SOLUTION**  In Section 2.1, we saw that an exponential function with base $a$, such that $0<a<1$, is a decreasing function with the horizontal asymptote $y=0$. Therefore, as time $t$ increases

$$\lim_{t\to\infty}\left(-e^{-0.022(t-56)}\right)=-\lim_{t\to\infty}\left(\frac{1}{e}\right)^{0.022(t-56)}=0,$$

so

$$\lim_{t\to\infty}M(t)=3102e^{\lim_{t\to\infty}\left(-e^{-0.022(t-56)}\right)}=3102e^0=3102.$$

Thus the maximum weight of a female Arctic fox predicted by this function is 3102 g.

# 3.1 EXERCISES

**In Exercises 1–4, choose the best answer for each limit.**

1. If $\lim\limits_{x\to 2^-} f(x) = 5$ and $\lim\limits_{x\to 2^+} f(x) = 6$, then $\lim\limits_{x\to 2} f(x)$
   a. is 5.
   b. is 6.
   c. does not exist.
   d. is infinite.

2. If $\lim\limits_{x\to 2^-} f(x) = \lim\limits_{x\to 2^+} f(x) = -1$, but $f(2) = 1$, then $\lim\limits_{x\to 2} f(x)$
   a. is −1.
   b. does not exist.
   c. is infinite.
   d. is 1.

3. If $\lim\limits_{x\to 4^-} f(x) = \lim\limits_{x\to 4^+} f(x) = 6$, but $f(4)$ does not exist, then $\lim\limits_{x\to 4} f(x)$
   a. does not exist.
   b. is 6.
   c. is −∞.
   d. is ∞.

4. If $\lim\limits_{x\to 1^-} f(x) = -\infty$ and $\lim\limits_{x\to 1^+} f(x) = -\infty$, then $\lim\limits_{x\to 1} f(x)$
   a. is ∞.
   b. is −∞.
   c. does not exist.
   d. is 1.

**Decide whether each limit exists. If a limit exists, estimate its value.**

5. a. $\lim\limits_{x\to 3} f(x)$
   b. $\lim\limits_{x\to 0} f(x)$

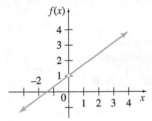

6. a. $\lim\limits_{x\to 2} F(x)$
   b. $\lim\limits_{x\to -1} F(x)$

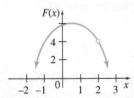

7. a. $\lim\limits_{x\to 0} f(x)$
   b. $\lim\limits_{x\to 2} f(x)$

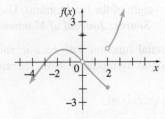

8. a. $\lim\limits_{x\to 3} g(x)$
   b. $\lim\limits_{x\to 5} g(x)$

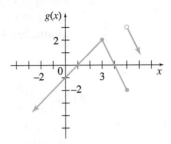

**In Exercises 9 and 10, use the graph to find (i) $\lim\limits_{x\to a^-} f(x)$, (ii) $\lim\limits_{x\to a^+} f(x)$, (iii) $\lim\limits_{x\to a} f(x)$, and (iv) $f(a)$ if it exists.**

9. a. $a = -2$
   b. $a = -1$

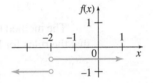

10. a. $a = 1$
    b. $a = 2$

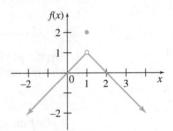

**Decide whether each limit exists. If a limit exists, find its value.**

11. $\lim\limits_{x\to\infty} f(x)$

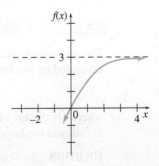

**12.** $\lim\limits_{x\to-\infty} g(x)$

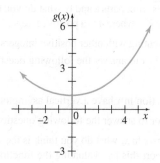

**13.** Explain why $\lim\limits_{x\to2} F(x)$ in Exercise 6 exists, but $\lim\limits_{x\to-2} f(x)$ in Exercise 9 does not.

**14.** In Exercise 10, why does $\lim\limits_{x\to1} f(x) = 1$, even though $f(1) = 2$?

**15.** Use the table of values to estimate $\lim\limits_{x\to1} f(x)$.

| $x$ | 0.9 | 0.99 | 0.999 | 0.9999 | 1.0001 | 1.001 | 1.01 | 1.1 |
|---|---|---|---|---|---|---|---|---|
| $f(x)$ | 3.9 | 3.99 | 3.999 | 3.9999 | 4.0001 | 4.001 | 4.01 | 4.1 |

**Complete the tables and use the results to find the indicated limits.**

**16.** If $f(x) = 2x^2 - 4x + 7$, find $\lim\limits_{x\to1} f(x)$.

| $x$ | 0.9 | 0.99 | 0.999 | | 1.001 | | 1.01 | 1.1 |
|---|---|---|---|---|---|---|---|---|
| $f(x)$ | | | 5.000002 | | 5.000002 | | | |

**17.** If $k(x) = \dfrac{x^3 - 2x - 4}{x - 2}$, find $\lim\limits_{x\to2} k(x)$.

| $x$ | 1.9 | 1.99 | 1.999 | 2.001 | 2.01 | 2.1 |
|---|---|---|---|---|---|---|
| $k(x)$ | | | | | | |

**18.** If $f(x) = \dfrac{2x^3 + 3x^2 - 4x - 5}{x + 1}$, find $\lim\limits_{x\to-1} f(x)$.

| $x$ | -1.1 | -1.01 | -1.001 | -0.999 | -0.99 | -0.9 |
|---|---|---|---|---|---|---|
| $f(x)$ | | | | | | |

**19.** If $h(x) = \dfrac{\sqrt{x} - 2}{x - 1}$, find $\lim\limits_{x\to1} h(x)$.

| $x$ | 0.9 | 0.99 | 0.999 | 1.001 | 1.01 | 1.1 |
|---|---|---|---|---|---|---|
| $h(x)$ | | | | | | |

**20.** If $f(x) = \dfrac{\sqrt{x} - 3}{x - 3}$, find $\lim\limits_{x\to3} f(x)$.

| $x$ | 2.9 | 2.99 | 2.999 | 3.001 | 3.01 | 3.1 |
|---|---|---|---|---|---|---|
| $f(x)$ | | | | | | |

**21.** If $f(x) = \dfrac{\sin x}{x}$, find $\lim\limits_{x\to0} f(x)$.

| $x$ | -0.1 | -0.01 | -0.001 | 0.001 | 0.01 | 0.1 |
|---|---|---|---|---|---|---|
| $f(x)$ | | | | | | |

**22.** If $g(x) = \dfrac{\cos x - 1}{x}$, find $\lim\limits_{x\to0} g(x)$.

| $x$ | -0.1 | -0.01 | -0.001 | 0.001 | 0.01 | 0.1 |
|---|---|---|---|---|---|---|
| $g(x)$ | | | | | | |

**Let $\lim\limits_{x\to4} f(x) = 9$ and $\lim\limits_{x\to4} g(x) = 27$. Use the limit rules to find each limit.**

**23.** $\lim\limits_{x\to4}[f(x) - g(x)]$

**24.** $\lim\limits_{x\to4}[g(x) \cdot f(x)]$

**25.** $\lim\limits_{x\to4} \dfrac{f(x)}{g(x)}$

**26.** $\lim\limits_{x\to4} \log_3 f(x)$

**27.** $\lim\limits_{x\to4} \sqrt{f(x)}$

**28.** $\lim\limits_{x\to4} \sqrt[3]{g(x)}$

**29.** $\lim\limits_{x\to4} 2^{f(x)}$

**30.** $\lim\limits_{x\to4}[1 + f(x)]^2$

**31.** $\lim\limits_{x\to4} \dfrac{f(x) + g(x)}{2g(x)}$

**32.** $\lim\limits_{x\to4} \dfrac{5g(x) + 2}{1 - f(x)}$

**33.** $\lim\limits_{x\to4}\left[\sin\left(\dfrac{\pi}{18} \cdot f(x)\right)\right]$

**34.** $\lim\limits_{x\to4}\left[\cos\left(\pi \cdot \dfrac{f(x)}{g(x)}\right)\right]$

**Use the properties of limits to help decide whether each limit exists. If a limit exists, find its value.**

**35.** $\lim\limits_{x\to3} \dfrac{x^2 - 9}{x - 3}$

**36.** $\lim\limits_{x\to-2} \dfrac{x^2 - 4}{x + 2}$

**37.** $\lim\limits_{x\to1} \dfrac{5x^2 - 7x + 2}{x^2 - 1}$

**38.** $\lim\limits_{x\to-3} \dfrac{x^2 - 9}{x^2 + x - 6}$

**39.** $\lim\limits_{x\to-2} \dfrac{x^2 - x - 6}{x + 2}$

**40.** $\lim\limits_{x\to5} \dfrac{x^2 - 3x - 10}{x - 5}$

**41.** $\lim\limits_{x\to0} \dfrac{1/(x + 3) - 1/3}{x}$

**42.** $\lim\limits_{x\to0} \dfrac{-1/(x + 2) + 1/2}{x}$

**43.** $\lim\limits_{x\to25} \dfrac{\sqrt{x} - 5}{x - 25}$

**44.** $\lim\limits_{x\to36} \dfrac{\sqrt{x} - 6}{x - 36}$

**45.** $\lim\limits_{h\to0} \dfrac{(x + h)^2 - x^2}{h}$

**46.** $\lim\limits_{h\to0} \dfrac{(x + h)^3 - x^3}{h}$

**47.** $\lim\limits_{x\to0} \dfrac{1 - \cos^2 x}{\sin^2 x}$

**48.** $\lim\limits_{x\to\pi/2} \dfrac{\tan^2 x + 1}{\sec^2 x}$

**49.** $\lim\limits_{x\to\infty} \dfrac{3x}{7x - 1}$

**50.** $\lim\limits_{x\to-\infty} \dfrac{8x + 2}{4x - 5}$

**51.** $\lim\limits_{x\to-\infty} \dfrac{3x^2 + 2x}{2x^2 - 2x + 1}$

**52.** $\lim\limits_{x\to\infty} \dfrac{x^2 + 2x - 5}{3x^2 + 2}$

**53.** $\lim\limits_{x\to\infty} \dfrac{3x^3 + 2x - 1}{2x^4 - 3x^3 - 2}$

**54.** $\lim\limits_{x\to\infty} \dfrac{2x^2 - 1}{3x^4 + 2}$

**55.** $\lim\limits_{x\to\infty} \dfrac{2x^3 - x - 3}{6x^2 - x - 1}$

**56.** $\lim\limits_{x\to\infty} \dfrac{x^4 - x^3 - 3x}{7x^2 + 9}$

**57.** $\lim\limits_{x\to\infty} \dfrac{2x^2 - 7x^4}{9x^2 + 5x - 6}$

**58.** $\lim\limits_{x\to\infty} \dfrac{-5x^3 - 4x^2 + 8}{6x^2 + 3x + 2}$

**59.** Let $f(x) = \begin{cases} x^3 + 2 & \text{if } x \neq -1 \\ 5 & \text{if } x = -1. \end{cases}$  Find $\lim\limits_{x\to -1} f(x)$.

**60.** Let $g(x) = \begin{cases} 0 & \text{if } x = -2 \\ \frac{1}{2}x^2 - 3 & \text{if } x \neq -2. \end{cases}$  Find $\lim\limits_{x\to -2} g(x)$.

**61.** Let $f(x) = \begin{cases} x - 1 & \text{if } x < 3 \\ 2 & \text{if } 3 \leq x \leq 5 \\ x + 3 & \text{if } x > 5. \end{cases}$

  **a.** Find $\lim\limits_{x\to 3} f(x)$.     **b.** Find $\lim\limits_{x\to 5} f(x)$.

**62.** Let $g(x) = \begin{cases} 5 & \text{if } x < 0 \\ x^2 - 2 & \text{if } 0 \leq x \leq 3 \\ 7 & \text{if } x > 3. \end{cases}$

  **a.** Find $\lim\limits_{x\to 0} g(x)$.     **b.** Find $\lim\limits_{x\to 3} g(x)$.

**In Exercises 63–66, calculate the limit in the specified exercise, using a table such as in Exercises 15–22. Verify your answer by using a graphing calculator to zoom in on the point on the graph.**

**63.** Exercise 35          **64.** Exercise 36

**65.** Exercise 37          **66.** Exercise 38

**67.** Let $F(x) = \dfrac{3x}{(x + 2)^3}$.

  **a.** Find $\lim\limits_{x\to -2} F(x)$.
  **b.** Find the vertical asymptote of the graph of $F(x)$.
  **c.** Compare your answers for parts a and b. What can you conclude?

**68.** Let $G(x) = \dfrac{-6}{(x - 4)^2}$.

  **a.** Find $\lim\limits_{x\to 4} G(x)$.
  **b.** Find the vertical asymptote of the graph of $G(x)$.
  **c.** Compare your answers for parts a and b. Are they related? How?

**69.** How can you tell that the graph in Figure 10 is more representative of the function $f(x) = 1/(x - 1)$ than the graph in Figure 11?

**70.** A friend who is confused about limits wonders why you investigate the value of a function closer and closer to a point, instead of just finding the value of a function at the point. How would you respond?

**71.** Use a graph of $f(x) = e^x$ to answer the following questions.

  **a.** Find $\lim\limits_{x\to -\infty} e^x$. (See Example 13.)
  **b.** Where does the function $e^x$ have a horizontal asymptote?

**72.** Use a graphing calculator to answer the following questions.

  **a.** From a graph of $y = xe^{-x}$, what do you think is the value of $\lim\limits_{x\to\infty} xe^{-x}$? Support this by evaluating the function for several large values of $x$.

  **b.** Repeat part a, this time using the graph of $y = x^2 e^{-x}$.

  **c.** Based on your results from parts a and b, what do you think is the value of $\lim\limits_{x\to\infty} x^n e^{-x}$, where $n$ is a positive integer? Support this by experimenting with other positive integers $n$.

**73.** Use a graph of $f(x) = \ln x$ to answer the following questions.

  **a.** Find $\lim\limits_{x\to 0^+} \ln x$.

  **b.** Where does the function $\ln x$ have a vertical asymptote?

**74.** Use a graphing calculator to answer the following questions.

  **a.** From a graph of $y = x \ln x$, what do you think is the value of $\lim\limits_{x\to 0^+} x \ln x$? Support this by evaluating the function for several small values of $x$.

  **b.** Repeat part a, this time using the graph of $y = x(\ln x)^2$.

  **c.** Based on your results from parts a and b, what do you think is the value of $\lim\limits_{x\to 0^+} x(\ln x)^n$, where $n$ is a positive integer? Support this by experimenting with other positive integers $n$.

**75.** Explain in your own words why the rules for limits at infinity should be true.

**76.** Explain in your own words what Rule 4 for limits means.

**Find each of the following limits (a) by investigating values of the function near the $x$-value where the limit is taken, and (b) using a graphing calculator to view the function near that value of $x$.**

**77.** $\lim\limits_{x\to 1} \dfrac{x^4 + 4x^3 - 9x^2 + 7x - 3}{x - 1}$

**78.** $\lim\limits_{x\to 2} \dfrac{x^4 + x - 18}{x^2 - 4}$

**79.** $\lim\limits_{x\to -1} \dfrac{x^{1/3} + 1}{x + 1}$

**80.** $\lim\limits_{x\to 4} \dfrac{x^{3/2} - 8}{x + x^{1/2} - 6}$

**Use a graphing calculator to graph the function. (a) Determine the limit from the graph. (b) Explain how your answer could be determined from the expression for $f(x)$.**

**81.** $\lim\limits_{x\to\infty} \dfrac{\sqrt{9x^2 + 5}}{2x}$

**82.** $\lim\limits_{x\to -\infty} \dfrac{\sqrt{9x^2 + 5}}{2x}$

**83.** $\lim\limits_{x\to -\infty} \dfrac{\sqrt{36x^2 + 2x + 7}}{3x}$

**84.** $\lim\limits_{x\to\infty} \dfrac{\sqrt{36x^2 + 2x + 7}}{3x}$

**85.** $\lim\limits_{x\to\infty} \dfrac{(1 + 5x^{1/3} + 2x^{5/3})^3}{x^5}$

**86.** $\lim\limits_{x\to -\infty} \dfrac{(1 + 5x^{1/3} + 2x^{5/3})^3}{x^5}$

**87.** Explain why the following rules can be used to find $\lim\limits_{x\to\infty} [p(x)/q(x)]$:

  **a.** If the degree of $p(x)$ is less than the degree of $q(x)$, the limit is 0.

  **b.** If the degree of $p(x)$ is equal to the degree of $q(x)$, the limit is $A/B$, where $A$ and $B$ are the leading coefficients of $p(x)$ and $q(x)$, respectively.

  **c.** If the degree of $p(x)$ is greater than the degree of $q(x)$, the limit is $\infty$ or $-\infty$.

**88.** Does a value of $k$ exist such that the following limit exists?

$$\lim_{x\to2}\frac{3x^2+kx-2}{x^2-3x+2}$$

If so, find the value of $k$ and the corresponding limit. If not, explain why not.

## LIFE SCIENCE APPLICATIONS

**89. Consumer Demand** When the price of an essential commodity (such as healthcare) decreases rapidly, consumption increases slowly at first. If the price continues to drop, however, a "tipping" point may be reached, at which consumption takes a sudden substantial increase. Suppose that the accompanying graph shows the consumption of a certain antibiotic, $C(t)$, in millions of units. We assume that the price is decreasing rapidly. Here $t$ is time in months after the price began decreasing. Use the graph to find the following.

**a.** $\lim_{t\to12}C(t)$    **b.** $\lim_{t\to16}C(t)$

**c.** $C(16)$    **d.** the tipping point (in months)

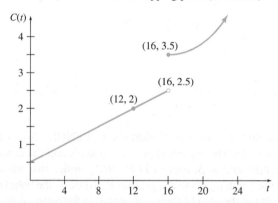

**90. Dialysis Patients** Because of new patients, transplantations, and deaths, the number of patients receiving dialysis during some part of each month in a small clinic varies over a one-year period as indicated by the following graph. Let $N(t)$ represent the number of patients in month $t$. Find the following.

**a.** $\lim_{t\to4}N(t)$    **b.** $\lim_{t\to6^+}N(t)$

**c.** $\lim_{t\to6^-}N(t)$    **d.** $\lim_{t\to6}N(t)$

**e.** $N(6)$

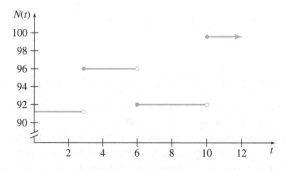

**91. Average Cost** The cost (in dollars) to a certain hospital for a particular medical test is $C(n)=15{,}000+60n$, where $n$ is the number of tests given. The average cost per test, denoted by $\overline{C}(n)$, is found by dividing $C(n)$ by $n$. Find and interpret $\lim_{n\to\infty}\overline{C}(n)$.

**92. Drug Concentration** The concentration of a drug in a patient's bloodstream $h$ hours after it was injected is given by

$$A(h)=\frac{0.17h}{h^2+2}.$$

Find and interpret $\lim_{h\to\infty}A(h)$.

**93. Alligator Teeth** Researchers have developed a mathematical model that can be used to estimate the number of teeth $N(t)$ at time $t$ (days of incubation) for *Alligator mississippiensis*, where

$$N(t)=71.8e^{-8.96e^{-0.0685t}}.$$

*Source: Journal of Theoretical Biology.*

**a.** Find $N(65)$, the number of teeth of an alligator that hatched after 65 days.

**b.** Find $\lim_{t\to\infty}N(t)$ and use this value as an estimate of the number of teeth of a newborn alligator. (*Hint:* See Exercise 71.) Does this estimate differ significantly from the estimate of part a?

**94. Sediment** To develop strategies to manage water quality in polluted lakes, biologists must determine the depths of sediments and the rate of sedimentation. It has been determined that the depth of sediment $D(t)$ (in centimeters) with respect to time (in years before 1990) for Lake Coeur d'Alene, Idaho, can be estimated by the equation

$$D(t)=155(1-e^{-0.0133t}).$$

*Source: Mathematics Teacher.*

**a.** Find $D(20)$ and interpret.

**b.** Find $\lim_{t\to\infty}D(t)$ and interpret.

**95. Cell Surface Receptors** In an article on the local clustering of cell surface receptors, the researcher analyzed limits of the function

$$C=F(S)=\frac{1-S}{S(1+kS)^{f-1}},$$

where $C$ is the concentration of free ligand in the medium, $S$ is the concentration of free receptors, $f$ is the number of functional groups of cells, and $k$ is a positive constant. *Source: Mathematical Biosciences.* Find each of the following limits.

**a.** $\lim_{S\to1}F(S)$    **b.** $\lim_{S\to0}F(S)$

**96. Nervous System** In a model of the nervous system, the intensity of excitation of a nerve pathway is given by

$$I=E[1-e^{-a(S-h)/E}],$$

where $E$ is the maximum possible excitation, $S$ is the intensity of a stimulus, $h$ is a threshold stimulus, and $a$ is a constant. Find $\lim_{S\to\infty}I$. *Source: Mathematical Biology of Social Behavior.*

## OTHER APPLICATIONS

**97. Legislative Voting**   Members of a legislature often must vote repeatedly on the same bill. As time goes on, members may change their votes. Suppose that $p_0$ is the probability that an individual legislator favors an issue before the first roll call vote, and suppose that $p$ is the probability of a change in position from one vote to the next. Then the probability that the legislator will vote "yes" on the $n$th roll call is given by

$$p_n = \frac{1}{2} + \left(p_0 - \frac{1}{2}\right)(1 - 2p)^n.$$

For example, the chance of a "yes" on the third roll call vote is

$$p_3 = \frac{1}{2} + \left(p_0 - \frac{1}{2}\right)(1 - 2p)^3.$$

*Source: Mathematics in the Behavioral and Social Sciences.*

Suppose that there is a chance of $p_0 = 0.7$ that Congressman Stephens will favor the budget appropriation bill before the first roll call, but only a probability of $p = 0.2$ that he will change his mind on the subsequent vote. Find and interpret the following.

**a.** $p_2$     **b.** $p_4$

**c.** $p_8$     **d.** $\lim\limits_{n \to \infty} p_n$

**98. Employee Productivity**   A company training program has determined that, on the average, a new employee produces $P(t)$ items per day after $t$ days of on-the-job training, where

$$P(t) = \frac{63t}{t + 8}.$$

Find and interpret $\lim\limits_{t \to \infty} P(t)$.

**YOUR TURN ANSWERS**

**1.** 3   **2.** 4   **3.** 5   **4.** Does not exist.   **5.** 49   **6.** −7
**7.** 1/2   **8.** 1/3

# 3.2   Continuity

**APPLY IT**   How does the recommended dosage of medication change with respect to weight?

**APPLY IT**   Figure 14 shows how the recommended dosage of Children's TYLENOL® Soft Chews changes with respect to a child's weight. The weight is given in pounds and the dosage is given as the number of tablets recommended. *Source: TYLENOL®*. Notice that whenever the dosage changes, the height of the graph jumps to a higher level. To study the behavior of this function at these points, we will use the idea of limits, presented in the previous section. Let us denote this function $g(w)$, where $w$ is the weight, in pounds, of the child.

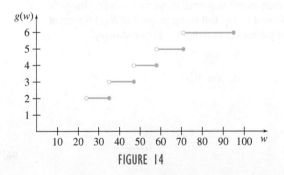

FIGURE 14

Notice from the graph that $\lim\limits_{w \to 59^-} g(w) = 4$ and that $\lim\limits_{w \to 59^+} g(w) = 5$, so $\lim\limits_{w \to 59} g(w)$ does not exist. Notice also that $g(59) = 4$. A point such as this, where a function has a sudden sharp break, is a point where the function is *discontinuous*.

Intuitively speaking, a function is *continuous* at a point if you can draw the graph of the function in the vicinity of that point without lifting your pencil from the paper. As we already mentioned, this would not be possible in Figure 14. Conversely, a function is discontinuous at any $x$-value where the pencil *must* be lifted from the paper in order to draw the graph on both sides of the point. A more precise definition is as follows.

## Continuity at $x = c$

A function $f$ is **continuous** at $x = c$ if the following three conditions are satisfied:

1. $f(c)$ is defined,
2. $\lim\limits_{x \to c} f(x)$ exists, and
3. $\lim\limits_{x \to c} f(x) = f(c)$.

If $f$ is not continuous at $c$, it is **discontinuous** there.

The following example shows how to check a function for continuity at a specific point. We use a three-step test, and if any step of the test fails, the function is not continuous at that point.

### EXAMPLE 1  Continuity

Determine if each function is continuous at the indicated $x$-value.

**(a)** $f(x)$ in Figure 15 at $x = 3$

**SOLUTION**

***Step 1*** Does the function exist at $x = 3$?

The open circle on the graph of Figure 15 at the point where $x = 3$ means that $f(x)$ does not exist at $x = 3$. Since the function does not pass the first test, it is discontinuous at $x = 3$, and there is no need to proceed to Step 2.

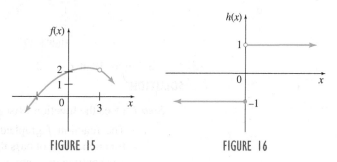

FIGURE 15　　　　　　FIGURE 16

**(b)** $h(x)$ in Figure 16 at $x = 0$

**SOLUTION**

***Step 1*** Does the function exist at $x = 0$?

According to the graph in Figure 16, $h(0)$ exists and is equal to $-1$.

***Step 2*** Does the limit exist at $x = 0$?

As $x$ approaches 0 from the left, $h(x)$ is $-1$. As $x$ approaches 0 from the right, however, $h(x)$ is 1. In other words,

$$\lim_{x \to 0^-} h(x) = -1,$$

while

$$\lim_{x \to 0^+} h(x) = 1.$$

Since no single number is approached by the values of $h(x)$ as $x$ approaches 0, the limit $\lim\limits_{x \to 0} h(x)$ does not exist. Since the function does not pass the second test, it is discontinuous at $x = 0$, and there is no need to proceed to Step 3.

**(c)** $g(x)$ in Figure 17 at $x = 4$

**SOLUTION**

**Step 1** Is the function defined at $x = 4$?

In Figure 17, the heavy dot above 4 shows that $g(4)$ is defined. In fact, $g(4) = 1$.

**Step 2** Does the limit exist at $x = 4$?

The graph shows that

$$\lim_{x \to 4^-} g(x) = -2, \text{ and } \lim_{x \to 4^+} g(x) = -2.$$

Therefore, the limit exists at $x = 4$ and

$$\lim_{x \to 4} g(x) = -2.$$

**Step 3** Does $g(4) = \lim_{x \to 4} g(x)$?

Using the results of Step 1 and Step 2, we see that $g(4) \neq \lim_{x \to 4} g(x)$.

Since the function does not pass the third test, it is discontinuous at $x = 4$.

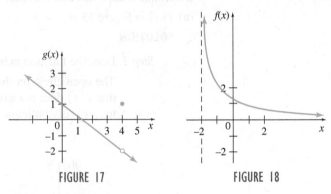

FIGURE 17                    FIGURE 18

**(d)** $f(x)$ in Figure 18 at $x = -2$.

**SOLUTION**

**Step 1** Does the function exist at $x = -2$?

The function $f$ graphed in Figure 18 is not defined at $x = -2$. Since the function does not pass the first test, it is discontinuous at $x = -2$. (Function $f$ is continuous at any value of $x$ greater than $-2$, however.)

Notice that the function in part (a) of Example 1 could be made continuous simply by defining $f(3) = 2$. Similarly, the function in part (c) could be made continuous by redefining $g(4) = -2$. In such cases, when the function can be made continuous at a specific point simply by defining or redefining it at that point, the function is said to have a **removable discontinuity**.

A function is said to be **continuous on an open interval** if it is continuous at every $x$-value in the interval. Continuity on a closed interval is slightly more complicated because we must decide what to do with the endpoints. We will say that a function $f$ is **continuous from the right** at $x = c$ if $\lim_{x \to c^+} f(x) = f(c)$. A function $f$ is **continuous from the left** at $x = c$ if $\lim_{x \to c^-} f(x) = f(c)$. With these ideas, we can now define continuity on a closed interval.

## Continuity on a Closed Interval

A function is **continuous on a closed interval** $[a, b]$ if

1. it is continuous on the open interval $(a, b)$,
2. it is continuous from the right at $x = a$, and
3. it is continuous from the left at $x = b$.

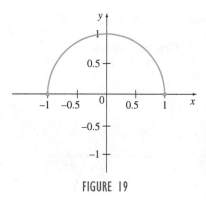

FIGURE 19

For example, the function $f(x) = \sqrt{1 - x^2}$, shown in Figure 19, is continuous on the closed interval $[-1, 1]$. By defining continuity on a closed interval in this way, we need not worry about the fact that $\sqrt{1 - x^2}$ does not exist to the left of $x = -1$ or to the right of $x = 1$.

The table below lists some key functions and tells where each is continuous.

| | **Continuous Functions** | |
|---|---|---|
| **Type of Function** | **Where It Is Continuous** | **Graphic Example** |
| *Polynomial Function* $y = a_n x^n + a_{n-1} x^{n-1} + \cdots + a_1 x + a_0$, where $a_n$ $a_{n-1}, \cdots, a_1, a_0$ are real numbers, not all 0 | For all $x$ | |
| *Rational Function* $y = \dfrac{p(x)}{q(x)}$, where $p(x)$ and $q(x)$ are polynomials, with $q(x) \neq 0$ | For all $x$ where $q(x) \neq 0$ | |
| *Root Function* $y = \sqrt{ax + b}$, where $a$ and $b$ are real numbers, with $a \neq 0$ and $ax + b \geq 0$ | For all $x$ where $ax + b \geq 0$ | |
| *Exponential Function* $y = a^x$ where $a > 0$ | For all $x$ | |
| *Logarithmic Function* $y = \log_a x$ where $a > 0$, $a \neq 1$ | For all $x > 0$ | |

*(continued)*

| | Continuous Functions (cont.) | |
|---|---|---|
| Type of Function | Where It Is Continuous | Graphic Example |

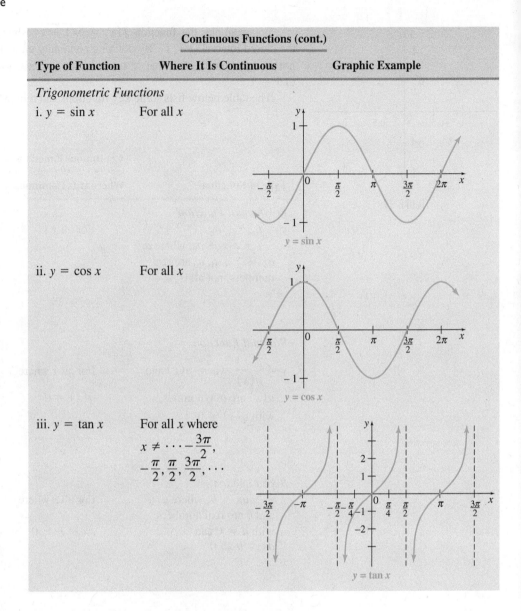

*Trigonometric Functions*

i. $y = \sin x$     For all $x$

ii. $y = \cos x$     For all $x$

iii. $y = \tan x$     For all $x$ where
$$x \neq \cdots -\frac{3\pi}{2},$$
$$-\frac{\pi}{2}, \frac{\pi}{2}, \frac{3\pi}{2}, \cdots$$

Continuous functions are nice to work with because finding $\lim_{x \to c} f(x)$ is simple if $f$ is continuous: just evaluate $f(c)$.

When a function is given by a graph, any discontinuities are clearly visible. When a function is given by a formula, it is usually continuous at all $x$-values except those where the function is undefined or possibly where there is a change in the defining formula for the function, as shown in the following examples.

### EXAMPLE 2    Continuity

Find all values of $x$ where the function is discontinuous.

**(a)** $f(x) = \dfrac{4x - 3}{2x - 7}$

**SOLUTION**    This rational function is discontinuous wherever the denominator is zero. There is a discontinuity when $x = 7/2$.

**(b)** $g(x) = e^{2x-3}$

**SOLUTION**    This exponential function is continuous for all $x$.    **TRY YOUR TURN 1**

**YOUR TURN 1**    Find all values of $x$ where the function is discontinuous.
$$f(x) = \sqrt{5x + 3}$$

**EXAMPLE 3** **Continuity**

Find all values of $x$ where the following piecewise function is discontinuous.

$$f(x) = \begin{cases} x + 1 & \text{if } x < 1 \\ x^2 - 3x + 4 & \text{if } 1 \le x \le 3. \\ 5 - x & \text{if } x > 3 \end{cases}$$

**SOLUTION** Since each piece of this function is a polynomial, the only $x$-values where $f$ might be discontinuous here are 1 and 3. We investigate at $x = 1$ first. From the left, where $x$-values are less than 1,

$$\lim_{x \to 1^-} f(x) = \lim_{x \to 1^-} (x + 1) = 1 + 1 = 2.$$

From the right, where $x$-values are greater than 1,

$$\lim_{x \to 1^+} f(x) = \lim_{x \to 1^+} (x^2 - 3x + 4) = 1^2 - 3 + 4 = 2.$$

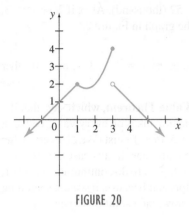

FIGURE 20

Furthermore, $f(1) = 1^2 - 3 + 4 = 2$, so $\lim_{x \to 1} f(x) = f(1) = 2$. Thus $f$ is continuous at $x = 1$, since $f(1) = \lim_{x \to 1} f(x)$.

Now let us investigate $x = 3$. From the left,

$$\lim_{x \to 3^-} f(x) = \lim_{x \to 3^-} (x^2 - 3x + 4) = 3^2 - 3(3) + 4 = 4.$$

From the right,

$$\lim_{x \to 3^+} f(x) = \lim_{x \to 3^+} (5 - x) = 5 - 3 = 2.$$

**YOUR TURN 2** Find all values of $x$ where the piecewise function is discontinuous.

$$f(x) = \begin{cases} 5x - 4 & \text{if } x < 0 \\ x^2 & \text{if } 0 \le x \le 3 \\ x + 6 & \text{if } x > 3 \end{cases}$$

Because $\lim_{x \to 3^-} f(x) \ne \lim_{x \to 3^+} f(x)$, the limit $\lim_{x \to 3} f(x)$ does not exist, so $f$ is discontinuous at $x = 3$, regardless of the value of $f(3)$.

The graph of $f(x)$ can be drawn by considering each of the three parts separately. In the first part, the line $y = x + 1$ is drawn including only the section of the line to the left of $x = 1$. The other two parts are drawn similarly, as illustrated in Figure 20. We can see by the graph that the function is continuous at $x = 1$ and discontinuous at $x = 3$, which confirms our solution above.          **TRY YOUR TURN 2**

▦ **TECHNOLOGY NOTE**

Some graphing calculators have the ability to draw piecewise functions. On the TI-84 Plus, letting

$$Y_1 = (X + 1)(X < 1) + (X^2 - 3X + 4)(1 \le X)(X \le 3) + (5 - X)(X > 3)$$

produces the graph shown in Figure 21(a).

**CAUTION** It is important here that the graphing mode be set on DOT rather than CONNECTED. Otherwise, the calculator will show a line segment at $x = 3$ connecting the parabola to the line, as in Figure 21(b), although such a segment does not really exist.

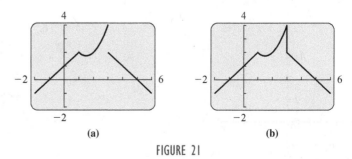

(a)                    (b)

FIGURE 21

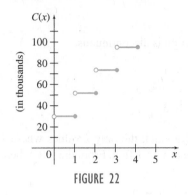

FIGURE 22

### EXAMPLE 4  Cost Analysis

A medical device can be leased for a flat fee of $8000 plus $22,000 per year or fraction of a year. Let $C(x)$, in thousands of dollars, represent the cost of leasing the device for $x$ years.

**(a)** Graph $C$.

**SOLUTION**  The charge for one year is $8000 plus $22,000, or $30,000. In fact, if $0 < x \le 1$, then $C(x) = 30$ (thousand). To rent the device for more than one year, but not more than two years, the charge is $8000 + 2 \cdot 22,000 = 52,000$ dollars. For any value of $x$ satisfying $1 < x \le 2$, the cost is $C(x) = 52$ (thousand). Also, if $2 < x \le 3$, then $C(x) = 74$ (thousand). These results lead to the graph in Figure 22.

**(b)** Find any values of $x$ where $C$ is discontinuous.

**SOLUTION**  As the graph suggests, $C$ is discontinuous at $x = 1, 2, 3, 4$, and all other positive integers.

One application of continuity is the **Intermediate Value Theorem**, which says that if a function is continuous on a closed interval $[a, b]$, the function takes on every value between $f(a)$ and $f(b)$. For example, if $f(1) = -3$ and $f(2) = 5$, then $f$ must take on every value between $-3$ and $5$ as $x$ varies over the interval $[1, 2]$. In particular (in this case), there must be a value of $x$ in the interval $(1, 2)$ such that $f(x) = 0$. If $f$ were discontinuous, however, this conclusion would not necessarily be true. This is important because, if we are searching for a solution to $f(x) = 0$ in $[1, 2]$, we would like to know that a solution exists.

## 3.2 EXERCISES

In Exercises 1–6, find all values $x = a$ where the function is discontinuous. For each point of discontinuity, give (a) $f(a)$ if it exists, (b) $\lim\limits_{x \to a^-} f(x)$, (c) $\lim\limits_{x \to a^+} f(x)$, (d) $\lim\limits_{x \to a} f(x)$, and (e) identify which conditions for continuity are not met. Be sure to note when the limit doesn't exist.

**1.**

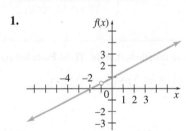

**2.**

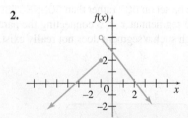

**3.**

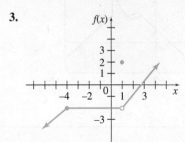

**4.**

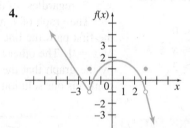

**5.**

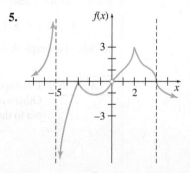

**6.**

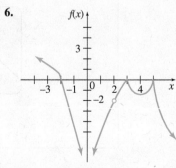

**Find all values of $x$ where the function is discontinuous. For each value of $x$, give the limit of the function at that value of $x$. Be sure to note when the limit doesn't exist.**

**7.** $f(x) = \dfrac{5 + x}{x(x - 2)}$

**8.** $f(x) = \dfrac{-2x}{(2x + 1)(3x + 6)}$

**9.** $f(x) = \dfrac{x^2 - 4}{x - 2}$

**10.** $f(x) = \dfrac{x^2 - 25}{x + 5}$

**11.** $p(x) = x^2 - 4x + 11$

**12.** $q(x) = -3x^3 + 2x^2 - 4x + 1$

**13.** $p(x) = \dfrac{|x + 2|}{x + 2}$

**14.** $r(x) = \dfrac{|5 - x|}{x - 5}$

**15.** $k(x) = e^{\sqrt{x-1}}$

**16.** $j(x) = e^{1/x}$

**17.** $r(x) = \ln\left|\dfrac{x}{x - 1}\right|$

**18.** $j(x) = \ln\left|\dfrac{x + 2}{x - 3}\right|$

**19.** $f(x) = \sin\left(\dfrac{x}{x + 2}\right)$

**20.** $g(x) = \tan(\pi x)$

**In Exercises 21–26, (a) graph the given function, (b) find all values of $x$ where the function is discontinuous, and (c) find the limit from the left and from the right at any values of $x$ found in part b.**

**21.** $f(x) = \begin{cases} 1 & \text{if } x < 2 \\ x + 3 & \text{if } 2 \le x \le 4 \\ 7 & \text{if } x > 4 \end{cases}$

**22.** $f(x) = \begin{cases} x - 1 & \text{if } x < 1 \\ 0 & \text{if } 1 \le x \le 4 \\ x - 2 & \text{if } x > 4 \end{cases}$

**23.** $g(x) = \begin{cases} 11 & \text{if } x < -1 \\ x^2 + 2 & \text{if } -1 \le x \le 3 \\ 11 & \text{if } x > 3 \end{cases}$

**24.** $g(x) = \begin{cases} 0 & \text{if } x < 0 \\ x^2 - 5x & \text{if } 0 \le x \le 5 \\ 5 & \text{if } x > 5 \end{cases}$

**25.** $h(x) = \begin{cases} 4x + 4 & \text{if } x \le 0 \\ x^2 - 4x + 4 & \text{if } x > 0 \end{cases}$

**26.** $h(x) = \begin{cases} x^2 + x - 12 & \text{if } x \le 1 \\ 3 - x & \text{if } x > 1 \end{cases}$

**In Exercises 27–30, find the value of the constant $k$ that makes the function continuous.**

**27.** $f(x) = \begin{cases} kx^2 & \text{if } x \le 2 \\ x + k & \text{if } x > 2 \end{cases}$

**28.** $g(x) = \begin{cases} x^3 + k & \text{if } x \le 3 \\ kx - 5 & \text{if } x > 3 \end{cases}$

**29.** $g(x) = \begin{cases} \dfrac{2x^2 - x - 15}{x - 3} & \text{if } x \ne 3 \\ kx - 1 & \text{if } x = 3 \end{cases}$

**30.** $h(x) = \begin{cases} \dfrac{3x^2 + 2x - 8}{x + 2} & \text{if } x \ne -2 \\ 3x + k & \text{if } x = -2 \end{cases}$

**31.** Explain in your own words what the Intermediate Value Theorem says and why it seems plausible.

**32.** Explain why $\lim\limits_{x \to 2}(3x^2 + 8x)$ can be evaluated by substituting $x = 2$.

**In Exercises 33–34, (a) use a graphing calculator to tell where the rational function $P(x)/Q(x)$ is discontinuous, and (b) verify your answer from part (a) by using the graphing calculator to plot $Q(x)$ and determine where $Q(x) = 0$. You will need to choose the viewing window carefully.**

**33.** $f(x) = \dfrac{x^2 + x + 2}{x^3 - 0.9x^2 + 4.14x - 5.4}$

**34.** $f(x) = \dfrac{x^2 + 3x - 2}{x^3 - 0.9x^2 + 4.14x + 5.4}$

**35.** Let $g(x) = \dfrac{x + 4}{x^2 + 2x - 8}$. Determine all values of $x$ at which $g$ is discontinuous, and for each of these values of $x$, define $g$ in such a manner so as to remove the discontinuity, if possible. Choose one of the following. *Source: Society of Actuaries.*

    **a.** $g$ is discontinuous only at $-4$ and $2$.
        Define $g(-4) = -\frac{1}{6}$ to make $g$ continuous at $-4$.
        $g(2)$ cannot be defined to make $g$ continuous at $2$.

    **b.** $g$ is discontinuous only at $-4$ and $2$.
        Define $g(-4) = -\frac{1}{6}$ to make $g$ continuous at $-4$.
        Define $g(2) = 6$ to make $g$ continuous at $2$.

    **c.** $g$ is discontinuous only at $-4$ and $2$.
        $g(-4)$ cannot be defined to make $g$ continuous at $-4$.
        $g(2)$ cannot be defined to make $g$ continuous at $2$.

    **d.** $g$ is discontinuous only at $2$.
        Define $g(2) = 6$ to make $g$ continuous at $2$.

    **e.** $g$ is discontinuous only at $2$.
        $g(2)$ cannot be defined to make $g$ continuous at $2$.

**36.** Tell at what values of $x$ the function $f(x)$ in Figure 8 from the previous section is discontinuous. Explain why it is discontinuous at each of these values.

CHAPTER 3 The Derivative

## LIFE SCIENCE APPLICATIONS

**37. Tumor Growth** A very aggressive tumor is growing according to the function $N(t) = 2^t$, where $N(t)$ represents the number of cells at time $t$ (in months). The tumor continues to grow for 40 months, at which time the tumor is diagnosed.

**a.** How many tumor cells are there at diagnosis?

**b.** Suppose the patient receives chemotherapy immediately after the tumor has grown for 40 months, and 99.9% of the cells are instantaneously killed. Graph the function both before and after the chemotherapy. (*Hint:* To draw a manageable graph, use $\log_2 N(t)$ for the vertical axis.)

**c.** Is the function described in the graph in part b continuous? If not, then find the value(s) of $t$ where the function is discontinuous.

**38. Pregnancy** A woman's weight naturally increases during the course of a pregnancy. When she delivers, her weight immediately decreases by the approximate weight of the child. Suppose that a 120-lb woman gains 27 lb during pregnancy, delivers a 7-lb baby, and then, through diet and exercise, loses the remaining weight during the next 20 weeks.

**a.** Graph the weight gain and loss during the pregnancy and the 20 weeks following the birth of the baby. Assume that the pregnancy lasts 40 weeks, that delivery occurs immediately after this time interval, and that the weight gain/loss before and after birth is linear.

**b.** Is this a continuous function? If not, then find the value(s) of $t$ where the function is discontinuous.

**39. Pharmacology** When a patient receives an injection, the amount of the injected substance in the body immediately goes up. Comment on whether the function that describes the amount of the drug with respect to time is a continuous function.

**40. Pharmacology** When a person takes a pill, the amount of the drug contained in the pill is immediately inside the body. Comment on whether the function that describes the amount of the drug inside the body with respect to time is a continuous function.

**41. Poultry Farming** Researchers at Iowa State University and the University of Arkansas have developed a piecewise function that can be used to estimate the body weight (in grams) of a male broiler during the first 56 days of life according to

$$W(t) = \begin{cases} 48 + 3.64t + 0.6363t^2 + 0.00963t^3 & \text{if } 1 \le t \le 28, \\ -1004 + 65.8t & \text{if } 28 < t \le 56, \end{cases}$$

where $t$ is the age of the chicken (in days). *Source: Poultry Science.*

**a.** Determine the weight of a male broiler that is 25 days old.

**b.** Is $W(t)$ a continuous function?

**c.** Use a graphing calculator to graph $W(t)$ on $[1, 56]$ by $[0, 3000]$. Comment on the accuracy of the graph.

**d.** Comment on why researchers would use two different types of functions to estimate the weight of a chicken at various ages.

**42. Electrocardiogram** Electrocardiograms (EKGs or ECGs) are used by physicians to study the relative health of a patient's heart. The figures below indicate part of a complete EKG for a particular healthy patient. *Source: RnCeus.com.* The $x$-axis is time and the $y$-axis is voltage.

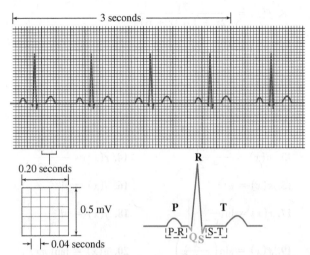

**a.** A patient's heart rate can be determined by the formula $R = 300/N$, where $N$ is the number of large squares between QRS complexes, or cycles of the heart. Find the heart rate for this particular patient.

**b.** Comment on whether the graph above actually represents a continuous function.

**43. Production** The graph shows the profit from the daily production of $x$ thousand kilograms of an industrial disinfectant. Use the graph to find the following limits.

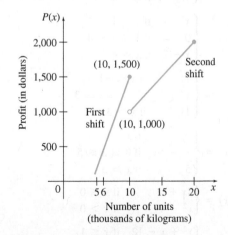

**a.** $\lim_{x \to 6} P(x)$     **b.** $\lim_{x \to 10^-} P(x)$

**c.** $\lim_{x \to 10^+} P(x)$     **d.** $\lim_{x \to 10} P(x)$

**e.** Where is the function discontinuous? What might account for such a discontinuity?

**f.** Use the graph to estimate the number of units of the chemical that must be produced before the second shift is as profitable as the first.

**44. Cost Analysis** The cost of ambulance transport depends on the distance, $x$, in miles that the patient is moved. Let $C(x)$ represent the cost to transport a patient $x$ miles. One firm charges a flat fee of $250, plus the charge as follows.

| Cost per Mile | Distance in Miles |
|---|---|
| $10.00 | $0 < x \le 150$ |
| $ 9.00 | $150 < x \le 400$ |
| $ 8.50 | $400 < x$ |

(Assume $x$ can take on any positive real-number value.) Find the cost to transport a patient the following distances.

a. 130 mi      b. 150 mi

c. 210 mi      d. 400 mi

e. 500 mi

f. Where is $C$ discontinuous?

## OTHER APPLICATIONS

**45. Cost Analysis** A company charges $1.25 per lb for a certain fertilizer on all orders 100 lb or less, and $1 per lb for orders over 100 lb. Let $F(x)$ represent the cost for buying $x$ lb of the

fertilizer. (Assume $x$ can take on any positive real-number value.) Find the cost of buying the following.

a. 80 lb      b. 150 lb      c. 100 lb

d. Where is $F$ discontinuous?

**46. Postage** To send international first class mail (large envelopes) from the United States to Australia in 2013, it cost $2.05 for the first ounce, $0.85 for each additional ounce up to a total of 8 oz, and $1.70 for each additional four ounces after that up to a total of 64 oz. Let $C(x)$ be the cost to mail $x$ ounces. Find the following. *Source: U.S. Postal Service.*

a. $\lim_{x\to3^-} C(x)$    b. $\lim_{x\to3^+} C(x)$    c. $\lim_{x\to3} C(x)$

d. $C(3)$    e. $\lim_{x\to14^+} C(x)$    f. $\lim_{x\to14^-} C(x)$

g. $\lim_{x\to14} C(x)$    h. $C(14)$

i. Find all values on the interval $(0, 64)$ where the function $C$ is discontinuous.

# 3.3 Rates of Change

What is the relationship between the age of a moose and its mass?
*This question will be answered in Example 1 of this section as we develop a method for finding the rate of change of one variable with respect to a unit change in another variable.*

**Average Rate of Change** One of the main applications of calculus is determining how one variable changes in relation to another. A marketing manager wants to know how profit changes with respect to the amount spent on advertising, while a physician wants to know how a patient's reaction to a drug changes with respect to the dose.

For example, suppose we take a trip from San Francisco driving south. Every half-hour we note how far we have traveled, with the following results for the first three hours.

| Distance Traveled | | | | | | | |
|---|---|---|---|---|---|---|---|
| Time in Hours | 0 | 0.5 | 1 | 1.5 | 2 | 2.5 | 3 |
| Distance in Miles | 0 | 30 | 55 | 80 | 104 | 124 | 138 |

If $s$ is the function whose rule is

$$s(t) = \text{Distance from San Francisco at time } t,$$

then the table shows, for example, that $s(0) = 0$, $s(1) = 55$, $s(2.5) = 124$, and so on. The distance traveled during, say, the second hour can be calculated by $s(2) - s(1) = 104 - 55 = 49$ miles.

Distance equals time multiplied by rate (or speed); so the distance formula is $d = rt$. Solving for rate gives $r = d/t$, or

$$\text{Average speed} = \frac{\text{Distance}}{\text{Time}}.$$

For example, the average speed over the time interval from $t = 0$ to $t = 3$ is

$$\text{Average speed} = \frac{s(3) - s(0)}{3 - 0} = \frac{138 - 0}{3} = 46,$$

or 46 mph. We can use this formula to find the average speed for any interval of time during the trip, as shown below.

| Average Speed | |
|---|---|
| **Time Interval** | **Average Speed** $= \dfrac{\text{Distance}}{\text{Time}}$ |
| $t = 0.5$ to $t = 1$ | $\dfrac{s(1) - s(0.5)}{1 - 0.5} = \dfrac{25}{0.5} = 50$ |
| $t = 0.5$ to $t = 1.5$ | $\dfrac{s(1.5) - s(0.5)}{1.5 - 0.5} = \dfrac{50}{1} = 50$ |
| $t = 1$ to $t = 2$ | $\dfrac{s(2) - s(1)}{2 - 1} = \dfrac{49}{1} = 49$ |
| $t = 1$ to $t = 3$ | $\dfrac{s(3) - s(1)}{3 - 1} = \dfrac{83}{2} = 41.5$ |
| $t = a$ to $t = b$ | $\dfrac{s(b) - s(a)}{b - a}$ |

---

**FOR REVIEW**

Recall from Section 1.1 the formula for the slope of a line through two points $(x_1, y_1)$ and $(x_2, y_2)$:

$$\frac{y_2 - y_1}{x_2 - x_1}.$$

Find the slopes of the lines through the following points.

(0.5, 30)   and   (1, 55)
(0.5, 30)   and   (1.5, 80)
(1, 55)   and   (2, 104)

Compare your answers to the average speeds shown in the table.

---

The analysis of the average speed or *average rate of change* of distance $s$ with respect to $t$ can be extended to include any function defined by $f(x)$ to get a formula for the average rate of change of $f$ with respect to $x$.

---

**Average Rate of Change**

The **average rate of change** of $f(x)$ with respect to $x$ as $x$ changes from $a$ to $b$ is

$$\frac{f(b) - f(a)}{b - a}.$$

---

**NOTE**   The formula for the average rate of change is the same as the formula for the slope of the line through $(a, f(a))$ and $(b, f(b))$. This connection between slope and rate of change will be examined more closely in the next section.

In Figure 23 we have plotted the distance vs. time for our trip from San Francisco, connecting the points with straight line segments. Because the change in $y$ gives the change in distance, and the change in $x$ gives the change in time, the slope of each line segment gives the average speed over that time interval:

$$\text{Slope} = \frac{\text{Change in } y}{\text{Change in } x} = \frac{\text{Change in distance}}{\text{Change in time}} = \text{Average speed}.$$

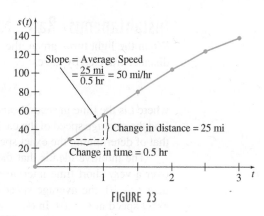

**FIGURE 23**

### EXAMPLE 1   Alaskan Moose

Researchers from the Moose Research Center in Soldotna, Alaska, have developed a mathematical relationship between the age of a captive female moose and its mass. The function is

$$M(t) = 369(0.93)^t t^{0.36},$$

where $M$ is the mass of the moose (in kg) and $t$ is the age (in years) of the moose. Find the average rate of change in the mass of a female moose between the ages of two and three years. *Source: Journal of Wildlife Management.*

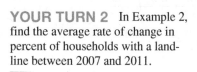

**YOUR TURN 1**   Find the average rate of change in the mass of a captive female moose between the ages of two and four years.

**SOLUTION**   On the interval from $t = 2$ to $t = 3$, the average rate of change is

$$\frac{M(3) - M(2)}{3 - 2} = \frac{369(0.93)^3(3)^{0.36} - 369(0.93)^2(2)^{0.36}}{3 - 2} \approx 31.194.$$

Therefore, the average amount of mass gained by a captive female moose between the ages of two and three is about 31.194 kg per year.    **TRY YOUR TURN 1**

### EXAMPLE 2   Household Telephones

Some U.S. households are substituting wireless telephones for traditional landline telephones. The graph in Figure 24 shows the percent of households in the United States with landline telephones for the years 2005 to 2011. Find the average rate of change in the percent of households with a landline between 2005 and 2011. *Source: Centers for Disease Control and Prevention.*

**SOLUTION**   Let $L(t)$ be the percent of U.S. households with landlines in the year $t$. Then the average rate of change between 2005 and 2011 was

**YOUR TURN 2**   In Example 2, find the average rate of change in percent of households with a landline between 2007 and 2011.

$$\frac{L(2011) - L(2005)}{2011 - 2005} = \frac{63.8 - 89.7}{6} = \frac{-25.9}{6} \approx -4.32,$$

or $-4.32\%$. On average, the percent of U.S. households with landline telephones decreased by about 4.32% per year during this time period.    **TRY YOUR TURN 2**

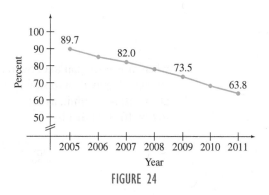

**FIGURE 24**

## Instantaneous Rate of Change

Suppose a car is stopped at a traffic light. When the light turns green, the car begins to move along a straight road. Assume that the distance traveled by the car is given by the function

$$s(t) = 3t^2, \quad \text{for} \quad 0 \le t \le 15,$$

where $t$ is the time in seconds and $s(t)$ is the distance in feet. We have already seen how to find the *average* speed of the car over any time interval. We now turn to a different problem, that of determining the exact speed of the car at a particular instant, say, $t = 10$.

The intuitive idea is that the exact speed at $t = 10$ is very close to the average speed over a very short time interval near $t = 10$. If we take shorter and shorter time intervals near $t = 10$, the average speeds over these intervals should get closer and closer to the exact speed at $t = 10$. In other words, the exact speed at $t = 10$ is the limit of the average speeds over shorter and shorter time intervals near $t = 10$. The following chart illustrates this idea. The values in the chart are found using $s(t) = 3t^2$, so that, for example, $s(10) = 3(10)^2 = 300$ and $s(10.1) = 3(10.1)^2 = 306.03$.

| Approximation of Speed at 10 Seconds | |
|---|---|
| **Interval** | **Average Speed** |
| $t = 10$ to $t = 10.1$ | $\dfrac{s(10.1) - s(10)}{10.1 - 10} = \dfrac{306.03 - 300}{0.1} = 60.3$ |
| $t = 10$ to $t = 10.01$ | $\dfrac{s(10.01) - s(10)}{10.01 - 10} = \dfrac{300.6003 - 300}{0.01} = 60.03$ |
| $t = 10$ to $t = 10.001$ | $\dfrac{s(10.001) - s(10)}{10.001 - 10} = \dfrac{300.060003 - 300}{0.001} = 60.003$ |

The results in the chart suggest that the exact speed at $t = 10$ is 60 ft/sec. We can confirm this by computing the average speed from $t = 10$ to $t = 10 + h$, where $h$ is a small, but nonzero, number that represents a small change in time. (The chart does this for $h = 0.1$, $h = 0.01$, and $h = 0.001$.) The average speed from $t = 10$ to $t = 10 + h$ is then

$$\frac{s(10 + h) - s(10)}{(10 + h) - 10} = \frac{3(10 + h)^2 - 3(10)^2}{h}$$

$$= \frac{3(100 + 20h + h^2) - 300}{h}$$

$$= \frac{300 + 60h + 3h^2 - 300}{h}$$

$$= \frac{60h + 3h^2}{h}$$

$$= \frac{h(60 + 3h)}{h}$$

$$= 60 + 3h,$$

where $h$ is not equal to 0. Saying that the time interval from 10 to $10 + h$ gets shorter and shorter is equivalent to saying that $h$ gets closer and closer to 0. Therefore, the exact speed at $t = 10$ is the limit, as $h$ approaches 0, of the average speed over the interval from $t = 10$ to $t = 10 + h$; that is,

$$\lim_{h \to 0} \frac{s(10 + h) - s(10)}{h} = \lim_{h \to 0} (60 + 3h)$$

$$= 60 \text{ ft/sec.}$$

This example can be easily generalized to any function $f$. Let $a$ be a specific $x$-value, such as 10 in the example. Let $h$ be a (small) number, which represents the distance between the two values of $x$, namely, $a$ and $a + h$. The average rate of change of $f$ as $x$ changes from $a$ to $a + h$ is

$$\frac{f(a + h) - f(a)}{(a + h) - a} = \frac{f(a + h) - f(a)}{h},$$

which is often called the **difference quotient**. Observe that the difference quotient is equivalent to the average rate of change formula, which can be verified by letting $b = a + h$ in the average rate of change formula. Furthermore, the exact rate of change of $f$ at $x = a$, called the *instantaneous rate of change of f at x = a*, is the limit of this difference quotient.

## Instantaneous Rate of Change

The **instantaneous rate of change** for a function $f$ when $x = a$ is

$$\lim_{h \to 0} \frac{f(a + h) - f(a)}{h},$$

provided this limit exists.

CAUTION   Remember that $f(x + h) \neq f(x) + f(h)$. To find $f(x + h)$, replace $x$ with $(x + h)$ in the expression for $f(x)$. For example, if $f(x) = x^2$,

$$f(x + h) = (x + h)^2 = x^2 + 2xh + h^2,$$

but

$$f(x) + f(h) = x^2 + h^2.$$

In the example just discussed, with the car starting from the traffic light, we saw that the instantaneous rate of change gave the speed of the car. But speed is always positive, while instantaneous rate of change can be positive or negative. Therefore, we will refer to **velocity** when we want to consider not only how fast something is moving but also in what direction it is moving. In any motion along a straight line, one direction is arbitrarily labeled as positive, so when an object moves in the opposite direction, its velocity is negative. In general, velocity is the same as the instantaneous rate of change of a function that gives position in terms of time.

In Figure 25, we have plotted the function $s(t) = 3t^2$, giving distance as a function of time. We have also plotted in green a line through the points $(10, s(10))$ and $(15, s(15))$.

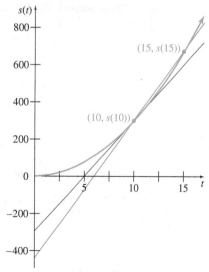

FIGURE 25

As we observed earlier, the slope of this line is the same as the average speed between $t = 10$ and $t = 15$. Finally, in red, we have plotted the line that results when the second point, $(15, s(15))$, moves closer and closer to the first point until the two coincide. The slope of this line corresponds to the instantaneous velocity at $t = 10$. We will explore these ideas further in the next section. Meanwhile, you might think about how to calculate the equations of these lines.

An alternate, but equivalent, approach is to let $a + h = b$ in the definition for instantaneous rate of change, so that $h = b - a$. This makes the instantaneous rate of change formula look more like the average rate of change formula.

---

**Instantaneous Rate of Change (Alternate Form)**

The **instantaneous rate of change** for a function $f$ when $x = a$ can be written as

$$\lim_{b \to a} \frac{f(b) - f(a)}{b - a},$$

provided this limit exists.

---

### EXAMPLE 3    Velocity

The distance in feet of an object from a starting point is given by $s(t) = 2t^2 - 5t + 40$, where $t$ is time in seconds.

**(a)** Find the average velocity of the object from 2 seconds to 4 seconds.

**SOLUTION**    The average velocity is

$$\frac{s(4) - s(2)}{4 - 2} = \frac{52 - 38}{2} = \frac{14}{2} = 7,$$

or 7 ft per second.

**(b)** Find the instantaneous velocity at 4 seconds.

**Method 1**
**Standard Form**

**SOLUTION**    For $t = 4$, the instantaneous velocity is

$$\lim_{h \to 0} \frac{s(4 + h) - s(4)}{h}$$

ft per second. We first calculate $s(4 + h)$ and $s(4)$, that is,

$$\begin{aligned} s(4 + h) &= 2(4 + h)^2 - 5(4 + h) + 40 \\ &= 2(16 + 8h + h^2) - 20 - 5h + 40 \\ &= 32 + 16h + 2h^2 - 20 - 5h + 40 \\ &= 2h^2 + 11h + 52, \end{aligned}$$

and

$$s(4) = 2(4)^2 - 5(4) + 40 = 52.$$

Therefore, the instantaneous velocity at $t = 4$ is

$$\lim_{h \to 0} \frac{(2h^2 + 11h + 52) - 52}{h} = \lim_{h \to 0} \frac{2h^2 + 11h}{h} = \lim_{h \to 0} \frac{h(2h + 11)}{h}$$
$$= \lim_{h \to 0} (2h + 11) = 11,$$

or 11 ft per second.

**Method 2**
**Alternate Form**

**SOLUTION** For $t = 4$, the instantaneous velocity is

$$\lim_{b \to 4} \frac{s(b) - s(4)}{b - 4}$$

ft per second. We first calculate $s(b)$ and $s(4)$, that is,

$$s(b) = 2b^2 - 5b + 40$$

and

$$s(4) = 2(4)^2 - 5(4) + 40 = 52.$$

The instantaneous rate of change is then

$$\lim_{b \to 4} \frac{2b^2 - 5b + 40 - 52}{b - 4} = \lim_{b \to 4} \frac{2b^2 - 5b - 12}{b - 4} \qquad \text{Simplify the numerator.}$$

$$= \lim_{b \to 4} \frac{(2b + 3)(b - 4)}{b - 4} \qquad \text{Factor the numerator.}$$

$$= \lim_{b \to 4} (2b + 3) \qquad \text{Cancel the } b - 4.$$

$$= 11, \qquad \text{Calculate the limit.}$$

**YOUR TURN 3** For the function in Example 3, find the instantaneous velocity at 2 seconds.

or 11 ft per second. **TRY YOUR TURN 3**

---

**EXAMPLE 4** **Mass of Bighorn Yearlings**

Researchers have determined that the body mass of yearling bighorn sheep on Ram Mountain in Alberta, Canada, can be estimated by

$$M(t) = 27.5 + 0.3t - 0.001t^2$$

where $M(t)$ is the mass of the sheep (in kg) and $t$ is the number of days since May 25. *Source: Canadian Journal of Zoology.*

**(a)** Find the average rate of change of the mass of a bighorn yearling between 70 and 75 days past May 25.

**SOLUTION** Use the formula for average rate of change. The mass of the yearling 70 days after May 25 is

$$M(70) = 27.5 + 0.3(70) - 0.001(70)^2 = 43.6 \text{ kg.}$$

Similarly, the mass of the yearling 75 days after May 25 is

$$M(75) = 27.5 + 0.3(75) - 0.001(75)^2 = 44.375 \text{ kg.}$$

The average rate of change of mass in this time interval is

$$\frac{M(75) - M(70)}{75 - 70} = \frac{44.375 - 43.6}{75 - 70} = 0.155.$$

Thus, on the average, the mass increases at the rate of 0.155 kg/day as the age of the sheep increases from 70 to 75 days past May 25.

**(b)** Find the instantaneous rate of change of mass for a yearling sheep whose age is 70 days past May 25.

**SOLUTION** The instantaneous rate of change for $t = 70$ is given by

$$\lim_{h \to 0} \frac{M(70 + h) - M(70)}{h}$$

$$= \lim_{h \to 0} \frac{[27.5 + 0.3(70 + h) - 0.001(70 + h)^2] - [27.5 + 0.3(70) - 0.001(70)^2]}{h}$$

$$= \lim_{h \to 0} \frac{27.5 + 21 + 0.3h - 4.9 - 0.14h - 0.001h^2 - 43.6}{h}$$

$$= \lim_{h \to 0} \frac{0.16h - 0.001h^2}{h} \qquad \text{Combine terms.}$$

$$= \lim_{h \to 0} \frac{h(0.16 - 0.001h)}{h} \qquad \text{Factor.}$$

$$= \lim_{h \to 0} (0.16 - 0.001h) \qquad \text{Divide by } h.$$

$$= 0.16 \qquad \text{Calculate the limit.}$$

When 70 days have passed since May 25, the mass of the yearling bighorn sheep is increasing at the rate of 0.16 kg/day. We can use this notion to predict that the sheep will have a mass of $43.6 + 0.16 = 43.76$ kg on day 71.

Another name for the instantaneous rate found in Example 4(b) is the *marginal value*. The concept of marginal analysis is quite important in the field of economics, where people are interested in calculating marginal cost, marginal demand, marginal profit, and marginal revenue. These values are obtained in the same manner as described above.

**EXAMPLE 5** Alaskan Moose

Estimate the instantaneous rate of change of the mass of a captive female moose at the beginning of year 2.

**SOLUTION** We saw in Example 1 that the estimated mass of a captive female moose (in kg) is given by $M(t) = 369(0.93)^t t^{0.36}$, where $t$ is the age (in years) of the moose. Unlike the previous example, in which the function was a polynomial, the function in this example is exponential, making it harder to compute the limit directly using the formula for instantaneous rate of change at $t = 2$:

$$\lim_{h \to 0} \frac{369(0.93)^{2+h}(2 + h)^{0.36} - 369(0.93)^2 2^{0.36}}{h}.$$

**YOUR TURN 4** Estimate the instantaneous rate of change of the mass of a captive female moose at the beginning of year 3.

Instead, we will approximate the instantaneous rate of change by using smaller and smaller values of $h$. See the following table. The limit seems to be approaching 44.003. Thus, the instantaneous rate of change in the mass of a captive two-year-old female moose is an increase of about 44 kg per year. **TRY YOUR TURN 4**

| | **Limit Calculations** |
|---|---|
| $h$ | $\dfrac{369(0.93)^{2+h}(2 + h)^{0.36} - 369(0.93)^2 2^{0.36}}{h}$ |
| 0.1 | 42.438 |
| 0.01 | 43.843 |
| 0.001 | 43.987 |
| 0.0001 | 44.002 |
| 0.00001 | 44.003 |
| 0.000001 | 44.003 |

The table could be created using the TABLE Feature on a TI-84 Plus calculator by entering $Y_1$ as the function from Example 5 and $Y_2 = (Y_1(2 + X) - Y_1(2))/X$. (The calculator requires us to use X in place of $h$ in the formula for instantaneous rate of change.) The result is shown in Figure 26. This table can also be generated using a spreadsheet.

| X | $Y_2$ |
|---|---|
| .1 | 42.438 |
| .01 | 43.843 |
| .001 | 43.987 |
| 1E⁻4 | 44.002 |
| 1E⁻5 | 44.003 |
| 1E⁻6 | 44.003 |
| X = | |

**FIGURE 26**

When we have data, rather than a formula, we can find the average velocity or rate of change, but we can only approximate the instantaneous velocity or rate of change.

### EXAMPLE 6   Velocity

One day Musk, the friendly pit bull, escaped from the yard and ran across the street to see a neighbor, who was 50 ft away. An estimate of the distance Musk ran as a function of time is given by the following table.

| Distance Traveled | | | | | |
|---|---|---|---|---|---|
| $t$ (sec) | 0 | 1 | 2 | 3 | 4 |
| $s$ (ft) | 0 | 10 | 25 | 42 | 50 |

**(a)** Find Musk's average velocity during her 4-second trip.

**SOLUTION**   The total distance she traveled is 50 ft, and the total time is 4 seconds, so her average velocity is 50/4 = 12.5 feet per second.

**(b)** Estimate Musk's velocity at 2 seconds.

**SOLUTION**   We could estimate her velocity by taking the short time interval from 2 to 3 seconds, for which the velocity is

$$\frac{42 - 25}{1} = 17 \text{ ft per second.}$$

Alternatively, we could estimate her velocity by taking the short time interval from 1 to 2 seconds, for which the velocity is

$$\frac{25 - 10}{1} = 15 \text{ ft per second.}$$

A better estimate is found by averaging these two values, to get

$$\frac{17 + 15}{2} = 16 \text{ ft per second.}$$

Another way to get this same answer is to take the time interval from 1 to 3 seconds, for which the velocity is

$$\frac{42 - 10}{2} = 16 \text{ ft per second.}$$

This answer is reasonable if we assume Musk's velocity changes at a fairly steady rate, and does not increase or decrease drastically from one second to the next. It is impossible to calculate Musk's exact velocity without knowing her position at times arbitrarily close to 2 seconds, or without a formula for her position as a function of time, or without a radar gun or speedometer on her. (In any case, she was very happy when she reached the neighbor.)

# 3.3    EXERCISES

**Find the average rate of change for each function over the given interval.**

1. $y = x^2 + 2x$   between $x = 1$ and $x = 3$

2. $y = -4x^2 - 6$   between $x = 2$ and $x = 6$

3. $y = -3x^3 + 2x^2 - 4x + 1$   between $x = -2$ and $x = 1$

4. $y = 2x^3 - 4x^2 + 6x$   between $x = -1$ and $x = 4$

5. $y = \sqrt{x}$   between $x = 1$ and $x = 4$

6. $y = \sqrt{3x - 2}$   between $x = 1$ and $x = 2$

7. $y = \dfrac{1}{x - 1}$   between   $x = -2$ and $x = 0$

8. $y = \dfrac{-5}{2x - 3}$   between   $x = 2$ and $x = 4$

9. $y = e^x$   between $x = -2$ and $x = 0$

10. $y = \ln x$   between $x = 2$ and $x = 4$

11. $y = \sin x$   between   $x = 0$ and $x = \pi/4$

12. $y = \cos x$   between   $x = \pi/2$ and $x = \pi$

**Suppose the position of an object moving in a straight line is given by $s(t) = t^2 + 5t + 2$. Find the instantaneous velocity at each time.**

13. $t = 6$

14. $t = 1$

**Suppose the position of an object moving in a straight line is given by $s(t) = 5t^2 - 2t - 7$. Find the instantaneous velocity at each time.**

15. $t = 2$

16. $t = 3$

**Suppose the position of an object moving in a straight line is given by $s(t) = t^3 + 2t + 9$. Find the instantaneous velocity at each time.**

17. $t = 1$

18. $t = 4$

**Find the instantaneous rate of change for each function at the given value.**

19. $f(x) = x^2 + 2x$   at $x = 0$

20. $s(t) = -4t^2 - 6$   at $t = 2$

21. $g(t) = 1 - t^2$   at $t = -1$

22. $F(x) = x^2 + 2$   at $x = 0$

**Use the formula for instantaneous rate of change, approximating the limit by using smaller and smaller values of $h$, to find the instantaneous rate of change for each function at the given value.**

23. $f(x) = \sin x$   at $x = 0$

24. $f(x) = \cos x$   at $x = \pi/2$

25. $f(x) = x^x$   at $x = 2$

26. $f(x) = x^x$   at $x = 3$

27. $f(x) = x^{\ln x}$   at $x = 2$

28. $f(x) = x^{\ln x}$   at $x = 3$

29. Explain the difference between the average rate of change of $y$ as $x$ changes from $a$ to $b$, and the instantaneous rate of change of $y$ at $x = a$.

30. If the instantaneous rate of change of $f(x)$ with respect to $x$ is positive when $x = 1$, is $f$ increasing or decreasing there?

## LIFE SCIENCE APPLICATIONS

31. **Flu Epidemic**   Epidemiologists in College Station, Texas, estimate that $t$ days after the flu begins to spread in town, the percent of the population infected by the flu is approximated by

$$p(t) = t^2 + t$$

for $0 \le t \le 5$.

   **a.** Find the average rate of change of $p$ with respect to $t$ over the interval from 1 to 4 days.

   **b.** Find and interpret the instantaneous rate of change of $p$ with respect to $t$ at $t = 3$.

32. **World Population Growth**   The future size of the world population depends on how soon it reaches replacement-level fertility, the point at which each woman bears on average about 2.1 children. The graph shows projections for reaching that point in different years. *Source: Population Reference Bureau.*

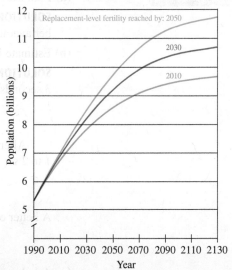

**Ultimate World Population Size Under Different Assumptions**

   **a.** Estimate the average rate of change in population for each projection from 1990 to 2050. Which projection shows the smallest rate of change in world population?

   **b.** Estimate the average rate of change in population from 2090 to 2130 for each projection. Interpret your answer.

33. **Bacterial Population**   The graph shows the population in millions of bacteria $t$ minutes after an antibiotic is introduced into a

culture. Find and interpret the average rate of change of population with respect to time for the following time intervals.

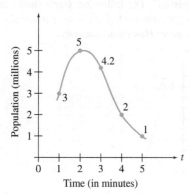

Time (in minutes)

**a.** 1 to 2    **b.** 2 to 3    **c.** 3 to 4    **d.** 4 to 5

**e.** How long after the antibiotic was introduced did the population begin to decrease?

**f.** At what time did the rate of decrease of the population slow down?

**34. Thermic Effect of Food**   The metabolic rate of a person who has just eaten a meal tends to go up and then, after some time has passed, returns to a resting metabolic rate. This phenomenon is known as the thermic effect of food. Researchers have indicated that the thermic effect of food (in kJ/hr) for a particular person is

$$F(t) = -10.28 + 175.9te^{-t/1.3},$$

where $t$ is the number of hours that have elapsed since eating a meal. *Source: American Journal of Clinical Nutrition.*

**a.** Graph the function on $[0, 6]$ by $[-20, 100]$.

**b.** Find the average rate of change of the thermic effect of food during the first hour after eating.

**c.** Use a graphing calculator to find the instantaneous rate of change of the thermic effect of food exactly 1 hour after eating.

**d.** Use a graphing calculator to estimate when the function stops increasing and begins to decrease.

**35. Molars**   The crown length (as shown below) of first molars in fetuses is related to the postconception age of the tooth as

$$L(t) = -0.01t^2 + 0.788t - 7.048,$$

where $L(t)$ is the crown length, in millimeters, of the molar $t$ weeks after conception. *Source: American Journal of Physical Anthropology.*

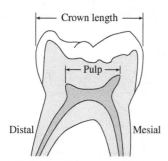

**a.** Find the average rate of growth in crown length during weeks 22 through 28.

**b.** Find the instantaneous rate of growth in crown length when the tooth is exactly 22 weeks of age.

**c.** Graph the function on $[0, 50]$ by $[0, 9]$. Does a function that increases and then begins to decrease make sense for this particular application? What do you suppose is happening during the first 11 weeks? Does this function accurately model crown length during those weeks?

**36. Mass of Bighorn Yearlings**   The body mass of yearling bighorn sheep on Ram Mountain in Alberta, Canada, can be estimated by

$$M(t) = 27.5 + 0.3t - 0.001t^2$$

where $M(t)$ is measured in kilograms and $t$ is days since May 25. *Source: Canadian Journal of Zoology.*

**a.** Find the average rate of change of the mass of a bighorn yearling between 105 and 115 days past May 25.

**b.** Find the instantaneous rate of change of the mass for a bighorn yearling sheep whose age is 105 days past May 25.

**c.** Graph the function $M(t)$ on $[5, 125]$ by $[25, 65]$.

**d.** Does the behavior of the function past 125 days accurately model the mass of the sheep? Why or why not?

**37. Minority Population**   The U.S. population is becoming more diverse. Based on the U.S. Census population projections for 2000 to 2050, the projected Hispanic population (in millions) can be modeled by the exponential function

$$H(t) = 37.791(1.021)^t,$$

where $t = 0$ corresponds to 2000 and $0 \le t \le 50$. *Source: U.S. Census Bureau.*

**a.** Use $H$ to estimate the average rate of change in the Hispanic population from 2000 to 2010.

**b.** Estimate the instantaneous rate of change in the Hispanic population in 2010.

**38. Minority Population**   Based on the U.S. Census population projections for 2000 to 2050, the projected Asian population (in millions) can be modeled by the exponential function

$$A(t) = 11.14(1.023)^t,$$

where $t = 0$ corresponds to 2000 and $0 \le t \le 50$. *Source: U.S. Census Bureau.*

**a.** Use $A$ to estimate the average rate of change in the Asian population from 2000 to 2010.

**b.** Estimate the instantaneous rate of change in the Asian population in 2010.

**39. Drug Use**   The chart on the next page shows how the percentage of eighth graders, tenth graders, and twelfth graders who have used marijuana in their lifetime has varied in recent years. *Source: The Monitoring the Future Study.*

**a.** Find the average annual rate of change in the percent of eighth graders who have used marijuana in their lifetime over the four-year period 2004–2008 and the four-year period 2008–2012. Then calculate the annual rate of change for 2004–2012.

**b.** Repeat part a for tenth graders.

c. Repeat part a for twelfth graders.

d. Discuss any similarities and differences between your answers to parts a through c, as well as possible reasons for these differences and similarities.

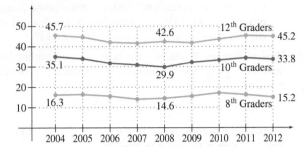

**40. Human Strength** According to research done by T. Curetun and Barry Ak at the University of Illinois Sports Fitness School, the distance (in inches) that typical boys of various ages can throw an 8-lb shotput can be estimated by the following graph. Use the graph to estimate and interpret the average rate of increase in shotput distance for boys from 9 to 11 years old. *Source: Standards in Pediatric Orthopedics: Tables, Charts, and Graphs Illustrating Growth.*

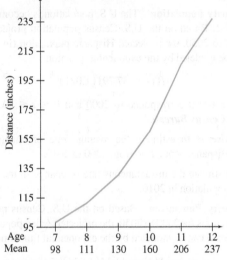

Source: Hensinger, Robert, *Standards in Pediatric Orthopedics: Tables, Charts, and Graphs Illustrating Grwoth*, New York, Raven Press, 1986, p.378.

**41. Medicare Trust Fund** The graph shows the money remaining in the Medicare Trust Fund at the end of the fiscal year. *Source: Social Security Administration.*

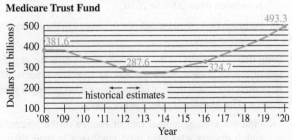

Using the Consumer Price Index for Urban Wage Earners and Clerical Workers

Find the approximate average rate of change in the trust fund for each time period.

a. From 2008 to 2012    b. From 2012 to 2020

## OTHER APPLICATIONS

**42. Immigration** The following graph shows how immigration (in thousands) to the United States has varied over the past century. *Source: Homeland Security.*

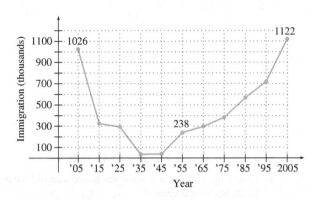

a. Find the average annual rate of change in immigration for the first half of the century (from 1905 to 1955).

b. Find the average annual rate of change in immigration for the second half of the century (from 1955 to 2005).

c. Find the average annual rate of change in immigration for the entire century (from 1905 to 2005).

d. Average your answers to parts a and b, and compare the result with your answer from part c. Will these always be equal for any two time periods?

e. If the annual average rate of change for the entire century continues, predict the number of immigrants in 2009. Compare your answer to the actual number of 1,130,818 immigrants.

**43. Temperature** The graph shows the temperature $T$ in degrees Celsius as a function of the altitude $h$ in feet when an inversion layer is over Southern California. (An inversion layer is formed when air at a higher altitude, say 3000 ft, is warmer than air at sea level, even though air normally is cooler with increasing altitude.) Estimate and interpret the average rate of change in temperature for the following changes in altitude.

a. 1000 to 3000 ft          b. 1000 to 5000 ft

c. 3000 to 9000 ft          d. 1000 to 9000 ft

e. At what altitude at or below 7000 ft is the temperature highest? Lowest? How would your answer change if 7000 ft is changed to 10,000 ft?

f. At what altitude is the temperature the same as it is at 1000 ft?

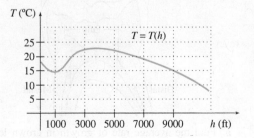

**44. Velocity** A car is moving along a straight test track. The position in feet of the car, $s(t)$, at various times $t$ is measured, with the following results.

| $t$ (sec) | 0 | 2 | 4 | 6 | 8 | 10 |
|---|---|---|---|---|---|---|
| $s(t)$ (ft) | 0 | 10 | 14 | 20 | 30 | 36 |

Find and interpret the average velocities for the following changes in $t$.

**a.** 0 to 2 seconds      **b.** 2 to 4 seconds

**c.** 4 to 6 seconds      **d.** 6 to 8 seconds

**e.** Estimate the instantaneous velocity at 4 seconds

     **i.** by finding the average velocity between 2 and 6 seconds, and

     **ii.** by averaging the answers for the average velocity in the two seconds before and the two seconds after (that is, the answers to parts b and c).

**f.** Estimate the instantaneous velocity at 6 seconds using the two methods in part e.

**g.** Notice in parts e and f that your two answers are the same. Discuss whether this will always be the case, and why or why not.

**45. Velocity** Consider the example at the beginning of this section regarding the car traveling from San Francisco.

**a.** Estimate the instantaneous velocity at 1 hour. Assume that the velocity changes at a steady rate from one half-hour to the next.

**b.** Estimate the instantaneous velocity at 2 hours.

**46. Velocity** The distance of a particle from some fixed point is given by

$$s(t) = t^2 + 5t + 2,$$

where $t$ is time measured in seconds. Find the average velocity of the particle over the following intervals.

**a.** 4 to 6 seconds

**b.** 4 to 5 seconds

**c.** Find the instantaneous velocity of the particle when $t = 4$.

**47. Interest** If $1000 is invested in an account that pays 5% compounded annually, the total amount, $A(t)$, in the account after $t$ years is

$$A(t) = 1000(1.05)^t.$$

**a.** Find the average rate of change per year of the total amount in the account for the first five years of the investment (from $t = 0$ to $t = 5$).

**b.** Find the average rate of change per year of the total amount in the account for the second five years of the investment (from $t = 5$ to $t = 10$).

**c.** Estimate the instantaneous rate of change for $t = 5$.

**48. Interest** If $1000 is invested in an account that pays 5% compounded continuously, the total amount, $A(t)$, in the account after $t$ years is

$$A(t) = 1000e^{0.05t}.$$

**a.** Find the average rate of change per year of the total amount in the account for the first five years of the investment (from $t = 0$ to $t = 5$).

**b.** Find the average rate of change per year of the total amount in the account for the second five years of the investment (from $t = 5$ to $t = 10$).

**c.** Estimate the instantaneous rate of change for $t = 5$.

**YOUR TURN ANSWERS**

1. Increase, on average, by 22.5 kg per year
2. Decrease, on average, of 4.55% per year
3. 3 ft per second
4. Increase is about 21 kg per year.

# 3.4 Definition of the Derivative

**APPLY IT** How does the risk of chromosomal abnormality in a child change with the mother's age?

*We will answer this question in Example 3, using the concept of the derivative.*

In the previous section, the formula

$$\lim_{h \to 0} \frac{f(a + h) - f(a)}{h}$$

was used to calculate the instantaneous rate of change of a function $f$ at the point where $x = a$. Now we will give a geometric interpretation of this limit.

## The Tangent Line

In geometry, a *tangent line* to a circle is defined as a line that touches the circle at only one point, as at the point $P$ in Figure 27 (which shows the top half of a circle). If you think of this half-circle as part of a curving road on which you are driving at night, then the tangent line indicates the direction of the light beam from your headlights as you pass through the point $P$. (We are not considering the new type of headlights on some cars that follow the direction of the curve.) Intuitively, the tangent line to an arbitrary curve at a point $P$ on the curve should touch the curve at $P$, but not at any points nearby, and should indicate the direction of the curve. In Figure 28, for example, the lines through $P_1$ and $P_3$ are tangent lines, while the lines through $P_2$ and $P_5$ are not. The tangent lines just touch the curve and indicate the direction of the curve, while the other lines pass through the curve heading in some other direction. To decide about the line at $P_4$, we need to define the idea of a tangent line to the graph of a function more carefully.

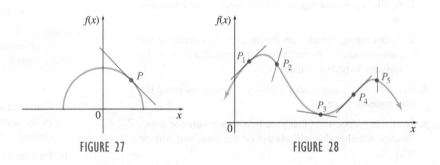

FIGURE 27          FIGURE 28

To see how we might define the slope of a line tangent to the graph of a function $f$ at a given point, let $R$ be a fixed point with coordinates $(a, f(a))$ on the graph of a function $y = f(x)$, as in Figure 29 below. Choose a different point $S$ on the graph and draw the line through $R$ and $S$; this line is called a **secant line**. If $S$ has coordinates $(a + h, f(a + h))$, then by the definition of slope, the slope of the secant line $RS$ is given by

$$\text{Slope of secant} = \frac{\Delta y}{\Delta x} = \frac{f(a + h) - f(a)}{a + h - a} = \frac{f(a + h) - f(a)}{h}.$$

This slope corresponds to the average rate of change of $y$ with respect to $x$ over the interval from $a$ to $a + h$. As $h$ approaches 0, point $S$ will slide along the curve, getting closer and closer to the fixed point $R$. See Figure 30, which shows successive positions $S_1$, $S_2$, $S_3$, and $S_4$ of the point $S$. If the slopes of the corresponding secant lines approach a limit as $h$ approaches 0, then this limit is defined to be the slope of the tangent line at point $R$.

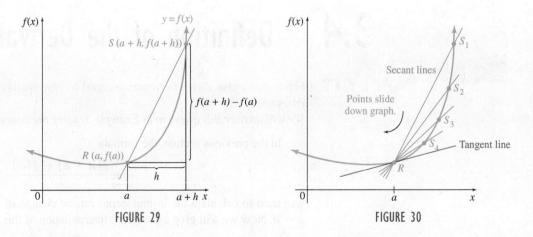

FIGURE 29          FIGURE 30

## Slope of the Tangent Line

The **tangent line** of the graph of $y = f(x)$ at the point $(a, f(a))$ is the line through this point having slope

$$\lim_{h \to 0} \frac{f(a + h) - f(a)}{h},$$

provided this limit exists. If this limit does not exist, then there is no tangent at the point.

Notice that the definition of the slope of the tangent line is identical to that of the instantaneous rate of change discussed in the previous section and is calculated by the same procedure.

The slope of the tangent line at a point is also called the **slope of the curve** at the point and corresponds to the instantaneous rate of change of $y$ with respect to $x$ at the point. It indicates the direction of the curve at that point.

---

**FOR REVIEW**

In Section 1.1, we saw that the equation of a line can be found with the point-slope form $y - y_1 = m(x - x_1)$, if the slope $m$ and the coordinates $(x_1, y_1)$ of a point on the line are known. Use the point-slope form to find the equation of the line with slope 3 that goes through the point $(-1, 4)$.

Let $m = 3$, $x_1 = -1$, $y_1 = 4$. Then

$$y - y_1 = m(x - x_1)$$
$$y - 4 = 3(x - (-1))$$
$$y - 4 = 3x + 3$$
$$y = 3x + 7.$$

---

## EXAMPLE 1  Tangent Line

Consider the graph of $f(x) = x^2 + 2$.

**(a)** Find the slope and equation of the secant line through the points where $x = -1$ and $x = 2$.

**SOLUTION**   Use the formula for slope as the change in $y$ over the change in $x$, where $y$ is given by $f(x)$. Since $f(-1) = (-1)^2 + 2 = 3$ and $f(2) = 2^2 + 2 = 6$, we have

$$\text{Slope of secant line} = \frac{f(2) - f(-1)}{2 - (-1)} = \frac{6 - 3}{3} = 1.$$

The slope of the secant line through $(-1, f(-1)) = (-1, 3)$ and $(2, f(2)) = (2, 6)$ is 1.

The equation of the secant line can be found with the point-slope form of the equation of a line from Chapter 1. We'll use the point $(-1, 3)$, although we could have just as well used the point $(2, 6)$.

$$y - y_1 = m(x - x_1)$$
$$y - 3 = 1[x - (-1)]$$
$$y - 3 = x + 1$$
$$y = x + 4$$

Figure 31 shows a graph of $f(x) = x^2 + 2$, along with a graph of the secant line (in green) through the points where $x = -1$ and $x = 2$.

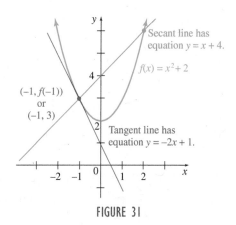

FIGURE 31

(b) Find the slope and equation of the tangent line at $x = -1$.

**SOLUTION**   Use the definition given previously, with $f(x) = x^2 + 2$ and $a = -1$. The slope of the tangent line is given by

$$\text{Slope of tangent} = \lim_{h \to 0} \frac{f(a + h) - f(a)}{h}$$

$$= \lim_{h \to 0} \frac{[(-1 + h)^2 + 2] - [(-1)^2 + 2]}{h}$$

$$= \lim_{h \to 0} \frac{[1 - 2h + h^2 + 2] - [1 + 2]}{h}$$

$$= \lim_{h \to 0} \frac{-2h + h^2}{h}$$

$$= \lim_{h \to 0} \frac{h(-2 + h)}{h}$$

$$= \lim_{h \to 0} (-2 + h) = -2.$$

The slope of the tangent line at $(-1, f(-1)) = (-1, 3)$ is $-2$.

The equation of the tangent line can be found with the point-slope form of the equation of a line from Chapter 1.

$$y - y_1 = m(x - x_1)$$
$$y - 3 = -2[x - (-1)]$$
$$y - 3 = -2(x + 1)$$
$$y - 3 = -2x - 2$$
$$y = -2x + 1$$

**YOUR TURN 1**   For the graph of $f(x) = x^2 - x$, (a) find the equation of the secant line through the points where $x = -2$ and $x = 1$, and (b) find the equation of the tangent line at $x = -2$.

The tangent line at $x = -1$ is shown in red in Figure 31 on the previous page.

TRY YOUR TURN 1

Figure 32 shows the result of zooming in on the point $(-1, 3)$ in Figure 31. Notice that in this closeup view, the graph and its tangent line appear virtually identical. This gives us another interpretation of the tangent line. Suppose, as we zoom in on a function, the graph appears to become a straight line. Then this line is the tangent line to the graph at that point. In other words, the tangent line captures the behavior of the function very close to the point under consideration. (This assumes, of course, that the function when viewed close up is approximately a straight line. As we will see later in this section, this may not occur.)

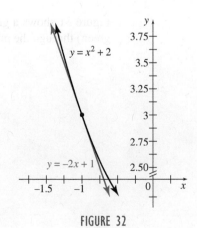

FIGURE 32

If it exists, the tangent line at $x = a$ is a good approximation of the graph of a function near $x = a$.

Consequently, another way to approximate the slope of the curve is to zoom in on the function using a graphing calculator until it appears to be a straight line (the tangent line). Then find the slope using any two points on that line.

### EXAMPLE 2  Slope (Using a Graphing Calculator)

Use a graphing calculator to find the slope of the graph of $f(x) = x^x$ at $x = 1$.

**SOLUTION**  The slope would be challenging to evaluate algebraically using the limit definition. Instead, using a graphing calculator on the window $[0, 2]$ by $[0, 2]$, we see the graph in Figure 33.

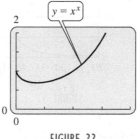

FIGURE 33

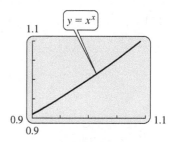

FIGURE 34

Zooming in gives the view in Figure 34. Using the TRACE key, we find two points on the line to be $(1, 1)$ and $(1.0021277, 1.0021322)$. Therefore, the slope is approximately

$$\frac{1.0021322 - 1}{1.0021277 - 1} \approx 1.$$

In addition to the method used in Example 2, there are other ways to use a graphing calculator to determine slopes of tangent lines and estimate instantaneous rates. Some alternate methods are listed below.

1. The Tangent command (under the DRAW menu) on a TI-84 Plus allows the tangent line to be drawn on a curve, giving an easy way to generate a graph with its tangent line similar to Figure 31.

2. Rather than using a graph, we could use a TI-84 Plus to create a table, as we did in the previous section, to estimate the instantaneous rate of change. Letting $Y_1 = X^{\wedge}X$ and $Y_2 = (Y_1(1 + X) - Y_1(1))/X$, along with specific table settings, results in the table shown in Figure 35. Based on this table, we estimate that the slope of the graph of $f(x) = x^x$ at $x = 1$ is 1.

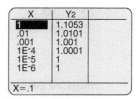

FIGURE 35

3. An even simpler method on a TI-84 Plus is to use the dy/dx command (under the CALC menu) or the nDeriv command (under the MATH menu). We will use this method in Example 4(b). But be careful, because sometimes these commands give erroneous results. For an example, see the Caution at the end of this section. For more details on the dy/dx command or the nDeriv command, see the *Graphing Calculator and Excel Spreadsheet Manual* available with this book.

### EXAMPLE 3   Genetics

Figure 36 shows how the risk of chromosomal abnormality in a child increases with the age of the mother. Find the rate that the risk is rising when the mother is 40 years old. *Source: downsyndrome.about.com.*

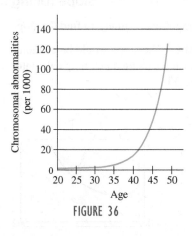

FIGURE 36

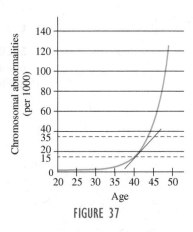

FIGURE 37

**APPLY IT**

**SOLUTION**   In Figure 37, we have added the tangent line to the graph at the point where the age of the mother is 40. At that point, the risk is approximately 15 per 1000. Extending the line, we estimate that when the age is 45, the $y$-coordinate of the line is roughly 35. Thus, the slope of the line is

$$\frac{35 - 15}{45 - 40} = \frac{20}{5} = 4.$$

Therefore, at the age of 40, the risk of chromosomal abnormality in the child is increasing at the rate of about 4 per 1000 for each additional year of the mother's age.

## The Derivative

If $y = f(x)$ is a function and $a$ is a number in its domain, then we shall use the symbol $f'(a)$ to denote the special limit

$$\lim_{h \to 0} \frac{f(a + h) - f(a)}{h},$$

provided that it exists. This means that for each number $a$ we can assign the number $f'(a)$ found by calculating this limit. This assignment defines an important new function.

### Derivative

The **derivative** of the function $f$ at $x$ is defined as

$$f'(x) = \lim_{h \to 0} \frac{f(x + h) - f(x)}{h},$$

provided this limit exists.

The notation $f'(x)$ is read "$f$-prime of $x$." The function $f'(x)$ is called the derivative of $f$ with respect to $x$. If $x$ is a value in the domain of $f$ and if $f'(x)$ exists, then $f$ is **differentiable** at $x$. The process that produces $f'$ is called **differentiation**.

The derivative function has several interpretations, two of which we have discussed.

**1.** The function $f'(x)$ represents the *instantaneous rate of change* of $y = f(x)$ with respect to $x$. This instantaneous rate of change could be interpreted as marginal cost,

**NOTE**
The derivative is a *function of x*, since $f'(x)$ varies as $x$ varies. This differs from both the slope of the tangent line and the instantaneous rate of change, either of which is represented by the number $f'(a)$ that corresponds to a number $a$. Otherwise, the formula for the derivative is identical to the formula for the slope of the tangent line given earlier in this section and to the formula for instantaneous rate of change given in the previous section.

revenue, or profit (if the original function represented cost, revenue, or profit) or velocity (if the original function described displacement along a line). From now on we will use *rate of change* to mean *instantaneous* rate of change.

**2.** The function $f'(x)$ represents the *slope* of the graph of $f(x)$ at any point $x$. If the derivative is evaluated at the point $x = a$, then it represents the slope of the curve, or the slope of the tangent line, at that point.

The following table compares the different interpretations of the difference quotient and the derivative.

| The Difference Quotient and the Derivative | |
|---|---|
| **Difference Quotient** | **Derivative** |
| $$\dfrac{f(x + h) - f(x)}{h}$$ | $$\lim_{h \to 0} \dfrac{f(x + h) - f(x)}{h}$$ |
| ■ Slope of the secant line | ■ Slope of the tangent line |
| ■ Average rate of change | ■ Instantaneous rate of change |
| ■ Average velocity | ■ Instantaneous velocity |
| ■ Average rate of change in cost, revenue, or profit | ■ Marginal cost, revenue, or profit |

Just as we had an alternate definition in the previous section by using $b$ instead of $a + h$, we now have an alternate definition by using $b$ in place of $x + h$.

**Derivative (Alternate Form)**
The **derivative** of function $f$ at $x$ can be written as

$$f'(x) = \lim_{b \to x} \frac{f(b) - f(x)}{b - x},$$

provided this limit exists.

The next few examples show how to use the definition to find the derivative of a function by means of a four-step procedure.

**EXAMPLE 4**  **Derivative**

Let $f(x) = x^2$.

**(a)** Find the derivative.

**Method I**
**Original Definition**

**SOLUTION**  By definition, for all values of $x$ where the following limit exists, the derivative is given by

$$f'(x) = \lim_{h \to 0} \frac{f(x + h) - f(x)}{h}.$$

Use the following sequence of steps to evaluate this limit.

***Step 1***  Find $f(x + h)$.

Replace $x$ with $x + h$ in the equation for $f(x)$. Simplify the result.

$$f(x) = x^2$$
$$f(x + h) = (x + h)^2$$
$$= x^2 + 2xh + h^2$$

(Note that $f(x + h) \neq f(x) + h$, since $f(x) + h = x^2 + h$.)

**Step 2** Find $f(x + h) - f(x)$.
Since $f(x) = x^2$,

$$f(x + h) - f(x) = (x^2 + 2xh + h^2) - x^2 = 2xh + h^2.$$

**Step 3** Find and simplify the quotient $\dfrac{f(x + h) - f(x)}{h}$. We find that

$$\frac{f(x + h) - f(x)}{h} = \frac{2xh + h^2}{h} = \frac{h(2x + h)}{h} = 2x + h,$$

except that $2x + h$ is defined for all real numbers $h$, while $[f(x + h) - f(x)]/h$ is not defined at $h = 0$. But this makes no difference in the limit, which ignores the value of the expression at $h = 0$.

**Step 4** Finally, find the limit as $h$ approaches 0. In this step, $h$ is the variable and $x$ is fixed.

$$f'(x) = \lim_{h \to 0} \frac{f(x + h) - f(x)}{h}.$$
$$= \lim_{h \to 0} (2x + h)$$
$$= 2x + 0 = 2x$$

**Method 2**
**Alternate Form**

**SOLUTION** Use

$$f(b) = b^2$$

and

$$f(x) = x^2.$$

We apply the alternate definition of the derivative as follows.

$$\lim_{b \to x} \frac{f(b) - f(x)}{b - x} = \lim_{b \to x} \frac{b^2 - x^2}{b - x}$$
$$= \lim_{b \to x} \frac{(b + x)(b - x)}{b - x} \qquad \text{Factor the numerator.}$$
$$= \lim_{b \to x} (b + x) \qquad \text{Divide by } b - x.$$
$$= x + x \qquad \text{Calculate the limit.}$$
$$= 2x$$

The alternate method appears shorter here because factoring $b^2 - x^2$ may seem simpler than calculating $f(x + h) - f(x)$. In other problems, however, factoring may be harder, in which case the first method may be preferable. Thus, from now on, we will use only the first method.

**(b)** Calculate and interpret $f'(3)$.

**Method 1**
**Algebraic Method**

**SOLUTION** Since $f'(x) = 2x$, we have

$$f'(3) = 2 \cdot 3 = 6.$$

The number 6 is the slope of the tangent line to the graph of $f(x) = x^2$ at the point where $x = 3$, that is, at $(3, f(3)) = (3, 9)$. See Figure 38(a).

**Method 2**
**Graphing Calculator**

As we mentioned earlier, some graphing calculators can calculate the value of the derivative at a given $x$-value. For example, the TI-84 Plus uses the `nDeriv` command as shown in Figure 38(b), with the expression for $f(x)$, the variable, and the value of $a$ entered to find $f'(3)$ for $f(x) = x^2$.

**YOUR TURN 2** Let $f(x) = x^2 - x$. Find the derivative, and then find $f'(-2)$.

TRY YOUR TURN 2

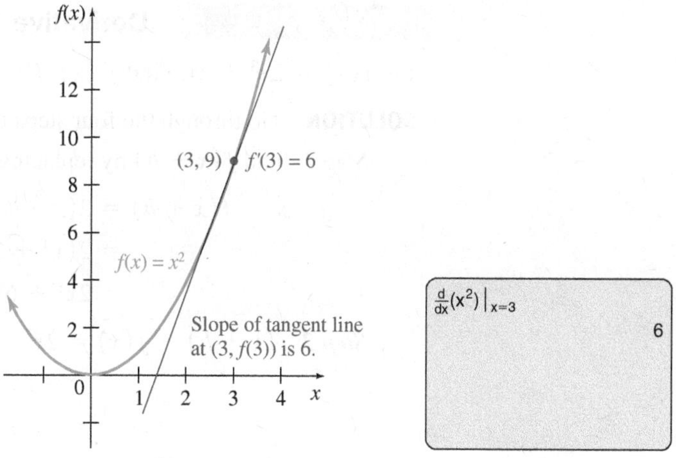

$$\frac{d}{dx}(x^2)\Big|_{x=3}$$

6

**(a)**                        **(b)**

**FIGURE 38**

**CAUTION**    **1.** In Example 4(a) notice that $f(x + h)$ is *not* equal to $f(x) + h$. In fact,

$$f(x + h) = (x + h)^2 = x^2 + 2xh + h^2,$$

but

$$f(x) + h = x^2 + h.$$

**2.** In Example 4(b), do not confuse $f(3)$ and $f'(3)$. The value $f(3)$ is the $y$-value that corresponds to $x = 3$. It is found by substituting 3 for $x$ in $f(x)$; $f(3) = 3^2 = 9$. On the other hand, $f'(3)$ is the slope of the tangent line to the curve at $x = 3$; as Example 4(b) shows, $f'(3) = 2 \cdot 3 = 6$.

## Finding $f'(x)$ from the Definition of Derivative

The four steps used to find the derivative $f'(x)$ for a function $y = f(x)$ are summarized here.

**1.** Find $f(x + h)$.

**2.** Find and simplify $f(x + h) - f(x)$.

**3.** Divide by $h$ to get $\dfrac{f(x + h) - f(x)}{h}$.

**4.** Let $h \to 0$; $f'(x) = \lim\limits_{h \to 0} \dfrac{f(x + h) - f(x)}{h}$, if this limit exists.

We now have four equivalent expressions for the change in $x$, but each has its uses, as the following box shows. We emphasize that these expressions all represent the same concept.

## Equivalent Expressions for the Change in $x$

| | |
|---|---|
| $x_2 - x_1$ | Useful for describing the equation of a line through two points |
| $b - a$ | A way to write $x_2 - x_1$ without the subscripts |
| $\Delta x$ | Useful for describing slope without referring to the individual points |
| $h$ | A way to write $\Delta x$ with just one symbol |

### EXAMPLE 5   Derivative

Let $f(x) = 2x^3 + 4x$. Find $f'(x)$, $f'(2)$, and $f'(-3)$.

**SOLUTION**   Go through the four steps to find $f'(x)$.

*Step 1* Find $f(x + h)$ by replacing $x$ with $x + h$.

$$f(x + h) = 2(x + h)^3 + 4(x + h)$$
$$= 2(x^3 + 3x^2h + 3xh^2 + h^3) + 4(x + h)$$
$$= 2x^3 + 6x^2h + 6xh^2 + 2h^3 + 4x + 4h$$

*Step 2* $f(x + h) - f(x) = 2x^3 + 6x^2h + 6xh^2 + 2h^3 + 4x + 4h$
$$- 2x^3 - 4x$$
$$= 6x^2h + 6xh^2 + 2h^3 + 4h$$

*Step 3* $\dfrac{f(x + h) - f(x)}{h} = \dfrac{6x^2h + 6xh^2 + 2h^3 + 4h}{h}$
$$= \dfrac{h(6x^2 + 6xh + 2h^2 + 4)}{h}$$
$$= 6x^2 + 6xh + 2h^2 + 4$$

*Step 4* Now use the rules for limits to get

$$f'(x) = \lim_{h \to 0} \frac{f(x + h) - f(x)}{h}$$
$$= \lim_{h \to 0} (6x^2 + 6xh + 2h^2 + 4)$$
$$= 6x^2 + 6x(0) + 2(0)^2 + 4$$
$$f'(x) = 6x^2 + 4.$$

**YOUR TURN 3** Let $f(x) = x^3 - 1$. Find $f'(x)$ and $f'(-1)$.

Use this result to find $f'(2)$ and $f'(-3)$.

$$f'(2) = 6 \cdot 2^2 + 4 = 28$$
$$f'(-3) = 6 \cdot (-3)^2 + 4 = 58 \qquad \text{TRY YOUR TURN 3}$$

---

**TECHNOLOGY NOTE**

One way to support the result in Example 5 is to plot $[f(x + h) - f(x)]/h$ on a graphing calculator with a small value of $h$. Figure 39 shows a graphing calculator screen of $y = [f(x + 0.1) - f(x)]/0.1$, where $f$ is the function $f(x) = 2x^3 + 4x$, and $y = 6x^2 + 4$, which was just found to be the derivative of $f$. The two functions, plotted on the window $[-2, 2]$ by $[0, 30]$, appear virtually identical. If $h = 0.01$ had been used, the two functions would be indistinguishable.

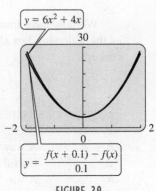

$y = 6x^2 + 4x$

$y = \dfrac{f(x + 0.1) - f(x)}{0.1}$

**FIGURE 39**

### EXAMPLE 6   Derivative

Let $f(x) = \dfrac{4}{x}$. Find $f'(x)$.

**SOLUTION**

*Step 1*  $f(x + h) = \dfrac{4}{x + h}$

*Step 2*  $f(x + h) - f(x) = \dfrac{4}{x + h} - \dfrac{4}{x}$

$\qquad\qquad\qquad\quad = \dfrac{4x - 4(x + h)}{x(x + h)}$    Find a common denominator.

$\qquad\qquad\qquad\quad = \dfrac{4x - 4x - 4h}{x(x + h)}$    Simplify the numerator.

$\qquad\qquad\qquad\quad = \dfrac{-4h}{x(x + h)}$

*Step 3*  $\dfrac{f(x + h) - f(x)}{h} = \dfrac{\dfrac{-4h}{x(x + h)}}{h}$

$\qquad\qquad\qquad\qquad\quad = \dfrac{-4h}{x(x + h)} \cdot \dfrac{1}{h}$    Invert and multiply.

$\qquad\qquad\qquad\qquad\quad = \dfrac{-4}{x(x + h)}$

*Step 4*  $f'(x) = \lim\limits_{h \to 0} \dfrac{f(x + h) - f(x)}{h}$

$\qquad\qquad\quad = \lim\limits_{h \to 0} \dfrac{-4}{x(x + h)}$

$\qquad\qquad\quad = \dfrac{-4}{x(x + 0)}$

$\qquad f'(x) = \dfrac{-4}{x(x)} = \dfrac{-4}{x^2}$    **TRY YOUR TURN 4**

**YOUR TURN 4**  Let
$f(x) = -\dfrac{2}{x}$. Find $f'(x)$.

Notice that in Example 6 neither $f(x)$ nor $f'(x)$ is defined when $x = 0$. Look at a graph of $f(x) = 4/x$ to see why this is true.

### EXAMPLE 7   Weight Gain

A mathematics professor found that, after introducing his dog Django to a new brand of food, Django's weight began to increase. After $x$ weeks on the new food, Django's weight (in pounds) was approximately given by $w(x) = \sqrt{x} + 40$ for $0 \le x \le 6$. Find the rate of change of Django's weight after $x$ weeks.

**SOLUTION**

*Step 1*  $w(x + h) = \sqrt{x + h} + 40$

*Step 2*  $w(x + h) - w(x) = \sqrt{x + h} + 40 - \left(\sqrt{x} + 40\right)$

$\qquad\qquad\qquad\qquad\quad = \sqrt{x + h} - \sqrt{x}$

*Step 3*  $\dfrac{w(x + h) - w(x)}{h} = \dfrac{\sqrt{x + h} - \sqrt{x}}{h}$

In order to be able to divide by $h$, multiply both numerator and denominator by $\sqrt{x + h} + \sqrt{x}$; that is, rationalize the *numerator*.

$$\frac{w(x + h) - w(x)}{h} = \frac{\sqrt{x + h} - \sqrt{x}}{h} \cdot \frac{\sqrt{x + h} + \sqrt{x}}{\sqrt{x + h} + \sqrt{x}}$$   Rationalize the numerator.

$$= \frac{(\sqrt{x + h})^2 - (\sqrt{x})^2}{h(\sqrt{x + h} + \sqrt{x})}$$   $(a - b)(a + b) = a^2 - b^2.$

$$= \frac{x + h - x}{h(\sqrt{x + h} + \sqrt{x})}$$

$$= \frac{h}{h(\sqrt{x + h} + \sqrt{x})}$$   Simplify.

$$= \frac{1}{\sqrt{x + h} + \sqrt{x}}$$   Divide by $h$.

**Step 4** $w'(x) = \lim\limits_{h \to 0} \dfrac{1}{\sqrt{x + h} + \sqrt{x}} = \dfrac{1}{\sqrt{x} + \sqrt{x}} = \dfrac{1}{2\sqrt{x}}$

**YOUR TURN 5** Let $f(x) = 2\sqrt{x}$. Find $f'(x)$.

This tells us, for example, that after 4 weeks, when Django's weight is $w(4) = \sqrt{4} + 40 = 42$ lb, her weight is increasing at a rate of $w'(4) = 1/(2\sqrt{4}) = 1/4$ lb per week.   **TRY YOUR TURN 5**

### EXAMPLE 8   Eating Behavior

As we saw in the review exercises for Chapter 1, studies on the eating behavior of humans revealed that for one subject, the cumulative intake of food during a meal can be described by

$$I(t) = 27 + 72t - 1.5t^2,$$

where $t$ is the number of minutes since the meal began, and $I(t)$ represents the amount, in grams, that the person has eaten at time $t$. Find the rate of change of the intake of food for this particular person 1 minute into a meal and 10 minutes into a meal. *Source: Appetite.*

**SOLUTION**   The rate of change of intake is given by the derivative of the intake function,

$$I'(t) = \lim_{h \to 0} \frac{I(t + h) - I(t)}{h}.$$

Going through the steps for finding $I'(t)$ gives

$$I'(t) = 72 - 3t.$$

When $t = 1$,

$$I'(1) = 72 - 3(1) = 69.$$

This rate of change of intake per minute gives the *marginal value* of the function at $t = 1$, which means that the person will eat approximately 69 grams of food during the second minute of the meal.

When 10 minutes have passed, the marginal value of the function is

$$I'(10) = 72 - 3(10) = 42,$$

or 42 g/min. This indicates that, as the meal progresses, the rate of intake of food decreases. In fact, the rate of intake is zero 24 minutes after the meal began.

We can use the notation for the derivative to write the equation of the tangent line. Using the point-slope form, $y - y_1 = m(x - x_1)$, and letting $y_1 = f(x_1)$ and $m = f'(x_1)$, we have the following formula.

Equation of the Tangent Line
The tangent line to the graph of $y = f(x)$ at the point $(x_1, f(x_1))$ is given by the equation

$$y - f(x_1) = f'(x_1)(x - x_1),$$

provided $f'(x)$ exists.

### EXAMPLE 9  Tangent Line

Find the equation of the tangent line to the graph of $f(x) = 4/x$ at $x = 2$.

**SOLUTION**   From the answer to Example 6, we have $f'(x) = -4/x^2$, so $f'(x_1) = f'(2) = -4/2^2 = -1$. Also $f(x_1) = f(2) = 4/2 = 2$. Therefore the equation of the tangent line is

$$y - 2 = (-1)(x - 2),$$

or

$$y = -x + 4$$

after simplifying.

TRY YOUR TURN 6

**YOUR TURN 6**   Find the equation of the tangent line to the graph $f(x) = 2\sqrt{x}$ at $x = 4$.

## Existence of the Derivative

The definition of the derivative included the phrase "provided this limit exists." If the limit used to define the derivative does not exist, then of course the derivative does not exist. For example, a derivative cannot exist at a point where the function itself is not defined. If there is no function value for a particular value of $x$, there can be no tangent line for that value. This was the case in Example 6—there was no tangent line (and no derivative) when $x = 0$.

Derivatives also do not exist at "corners" or "sharp points" on a graph. For example, the function graphed in Figure 40 is the *absolute value function*, defined previously as

$$f(x) = \begin{cases} x & \text{if } x \geq 0 \\ -x & \text{if } x < 0, \end{cases}$$

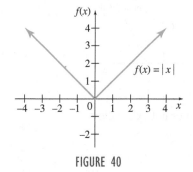

FIGURE 40

and written $f(x) = |x|$. By the definition of derivative, the derivative at any value of $x$ is given by

$$f'(x) = \lim_{h \to 0} \frac{f(x + h) - f(x)}{h},$$

provided this limit exists. To find the derivative at 0 for $f(x) = |x|$, replace $x$ with 0 and $f(x)$ with $|0|$ to get

$$f'(0) = \lim_{h \to 0} \frac{|0 + h| - |0|}{h} = \lim_{h \to 0} \frac{|h|}{h}.$$

In Example 5 in the first section of this chapter, we showed that

$$\lim_{h \to 0} \frac{|h|}{h} \text{ does not exist;}$$

therefore, the derivative does not exist at 0. However, the derivative does exist for all values of $x$ other than 0.

**CAUTION**   The command `nDeriv(abs(X),X,0)` on a TI-84 Plus calculator gives the answer 0, which is wrong. It does this by investigating a point slightly to the left of 0 and slightly to the right of 0. Since the function has the same value at these two points, it assumes that the function must be flat around 0, which is false in this case because of the sharp corner at 0. Be careful about naively trusting your calculator; think about whether the answer is reasonable.

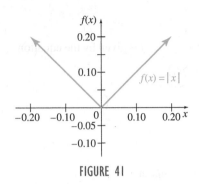

**FIGURE 41**

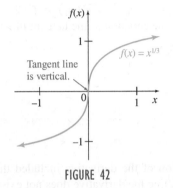

Tangent line is vertical.

**FIGURE 42**

In Figure 41, we have zoomed in on the origin in Figure 40. Notice that the graph looks essentially the same. The corner is still sharp, and the graph does not resemble a straight line any more than it originally did. As we observed earlier, the derivative only exists at a point when the function more and more resembles a straight line as we zoom in on the point.

A graph of the function $f(x) = x^{1/3}$ is shown in Figure 42. As the graph suggests, the tangent line is vertical when $x = 0$. Since a vertical line has an undefined slope, the derivative of $f(x) = x^{1/3}$ cannot exist when $x = 0$. Use the fact that $\lim_{h \to 0} h^{1/3}/h = \lim_{h \to 0} 1/h^{2/3}$ does not exist and the definition of the derivative to verify that $f'(0)$ does not exist for $f(x) = x^{1/3}$.

Figure 43 summarizes the various ways that a derivative can fail to exist. Notice in Figure 43 that at a point where the function is discontinuous, such as $x_3$, $x_4$, and $x_6$, the derivative does not exist. A function must be continuous at a point for the derivative to exist there. But just because a function is continuous at a point does not mean the derivative necessarily exists. For example, observe that the function in Figure 43 is continuous at $x_1$ and $x_2$, but the derivative does not exist at those values of $x$ because of the sharp corners, making a tangent line impossible. This is exactly what happens with the function $f(x) = |x|$ at $x = 0$, as we saw in Figure 40. Also, the function is continuous at $x_5$, but the derivative doesn't exist there because the tangent line is vertical, and the slope of a vertical line is undefined.

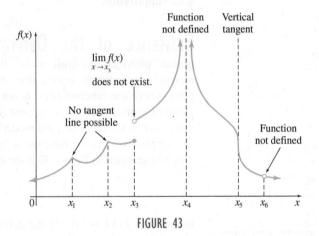

**FIGURE 43**

We summarize conditions for the derivative to exist or not exist below.

### Existence of the Derivative

The derivative exists when a function $f$ satisfies *all* of the following conditions at a point.

1. $f$ is continuous,
2. $f$ is smooth, and
3. $f$ does not have a vertical tangent line.

The derivative does *not* exist when *any* of the following conditions are true for a function at a point.

1. $f$ is discontinuous,
2. $f$ has a sharp corner, or
3. $f$ has a vertical tangent line.

**EXAMPLE 10** **Astronomy**

In August 2013, Japanese amateur astronomer Koichi Itagaki discovered Nova Delphinus 2013. A nova is an exploding star whose brightness suddenly increases and then gradually fades. Astronomers use magnitudes of brightness to compare night sky objects. A lower star number indicates brighter objects. Prior to the explosion, the star was undetectable to

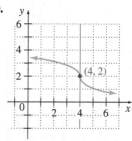

FIGURE 44

the naked eye and had a brightness measure of $+17$, which is very dim. During a 24 hour period the brightness of the star increased to about $+4.5$, at which time the explosion occurred. Gradually the brightness of the star has decreased. Suppose that the graph of the intensity of light emitted by Nova Delphinus 2013 is given as the function of time shown in Figure 44. Find where the function is not differentiable. *Source: Space.com.*

**SOLUTION** Notice that although the graph is a continuous curve, it is not differentiable at the point of explosion.

# 3.4 EXERCISES

1. By considering, but not calculating, the slope of the tangent line, give the derivative of the following.

   **a.** $f(x) = 5$   **b.** $f(x) = x$   **c.** $f(x) = -x$

   **d.** The line $x = 3$   **e.** The line $y = mx + b$

2. **a.** Suppose $g(x) = \sqrt[3]{x}$. Use the graph of $g(x)$ to find $g'(0)$.

    **b.** Explain why the derivative of a function does not exist at a point where the tangent line is vertical.

3. If $f(x) = \dfrac{x^2 - 1}{x + 2}$, where is $f$ not differentiable?

4. If the rate of change of $f(x)$ is zero when $x = a$, what can be said about the tangent line to the graph of $f(x)$ at $x = a$?

**Estimate the slope of the tangent line to each curve at the given point $(x, y)$.**

5.
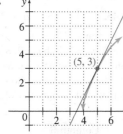

6.
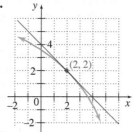

7.
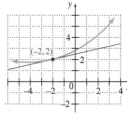

8.
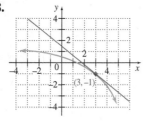

9.
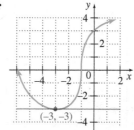

10.
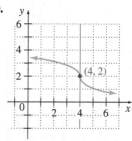

**Using the definition of the derivative, find $f'(x)$. Then find $f'(-2)$, $f'(0)$, and $f'(3)$ when the derivative exists. (*Hint for Exercises 17 and 18:* In Step 3, multiply numerator and denominator by $\sqrt{x + h} + \sqrt{x}$.)**

11. $f(x) = 3x - 7$      12. $f(x) = -2x + 5$

13. $f(x) = -4x^2 + 9x + 2$      14. $f(x) = 6x^2 - 5x - 1$

15. $f(x) = 12/x$      16. $f(x) = 3/x$

17. $f(x) = \sqrt{x}$      18. $f(x) = -3\sqrt{x}$

19. $f(x) = 2x^3 + 5$      20. $f(x) = 4x^3 - 3$

**For each function, find (a) the equation of the secant line through the points where $x$ has the given values, and (b) the equation of the tangent line when $x$ has the first value.**

21. $f(x) = x^2 + 2x;\quad x = 3,\quad x = 5$

22. $f(x) = 6 - x^2;\quad x = -1,\quad x = 3$

23. $f(x) = 5/x;\quad x = 2,\quad x = 5$

24. $f(x) = -3/(x + 1);\quad x = 1,\quad x = 5$

25. $f(x) = 4\sqrt{x};\quad x = 9,\quad x = 16$

26. $f(x) = \sqrt{x};\quad x = 25,\quad x = 36$

Use a graphing calculator to find $f'(2)$, $f'(16)$, and $f'(-3)$ for the following when the derivative exists.

**27.** $f(x) = -4x^2 + 11x$     **28.** $f(x) = 6x^2 - 4x$

**29.** $f(x) = e^x$     **30.** $f(x) = \ln|x|$

**31.** $f(x) = -\dfrac{2}{x}$     **32.** $f(x) = \dfrac{6}{x}$

**33.** $f(x) = \sqrt{x}$     **34.** $f(x) = -3\sqrt{x}$

Find the $x$-values where the following do not have derivatives.

**35.** 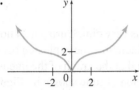     **36.**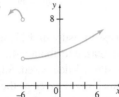

**37.**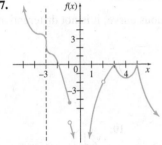

**38.**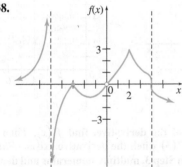

**39.** For the function shown in the sketch, give the intervals or points on the $x$-axis where the rate of change of $f(x)$ with respect to $x$ is

   **a.** positive;  **b.** negative;  **c.** zero.

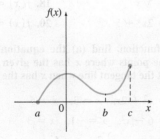

In Exercises 40 and 41, tell which graph, a or b, represents velocity and which represents distance from a starting point. (*Hint:* Consider where the derivative is zero, positive, or negative.)

**40. a.**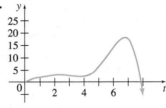

   **b.**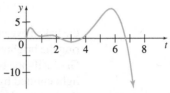

**41. a.**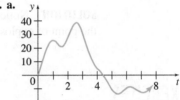

   **b.**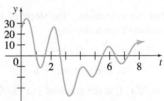

In Exercises 42–45, find the derivative of the function at the given point.

   **a.** Approximate the definition of the derivative with small values of $h$.

   **b.** Use a graphing calculator to zoom in on the function until it appears to be a straight line, and then find the slope of that line.

**42.** $f(x) = x^x$;  $a = 2$     **43.** $f(x) = x^x$;  $a = 3$

**44.** $f(x) = x^{1/x}$;  $a = 2$     **45.** $f(x) = x^{1/x}$;  $a = 3$

**46.** For each function in Column A, graph $[f(x + h) - f(x)]/h$ for a small value of $h$ on the window $[-2, 2]$ by $[-2, 8]$. Then graph each function in Column B on the same window. Compare the first set of graphs with the second set to choose from Column B the derivative of each of the functions in Column A.

| Column A | Column B |
|---|---|
| $\ln|x|$ | $e^x$ |
| $e^x$ | $3x^2$ |
| $x^3$ | $\dfrac{1}{x}$ |

**47.** Explain why

$$\frac{f(x + h) - f(x - h)}{2h}$$

should give a reasonable approximation of $f'(x)$ when $f'(x)$ exists and $h$ is small.

**48. a.** For the function $f(x) = -4x^2 + 11x$, find the value of $f'(3)$, as well as the approximation using

$$\frac{f(x + h) - f(x)}{h}$$

and using the formula in Exercise 47 with $h = 0.1$.

**b.** Repeat part a using $h = 0.01$.

**c.** Repeat part a using the function $f(x) = -2/x$ and $h = 0.1$.

**d.** Repeat part c using $h = 0.01$.

**e.** Repeat part a using the function $f(x) = \sqrt{x}$ and $h = 0.1$.

**f.** Repeat part e using $h = 0.01$.

**g.** Using the results of parts a through f, discuss which approximation formula seems to give better accuracy.

## LIFE SCIENCE APPLICATIONS

**49. Water Snakes** Scientists in Arkansas have collected data revealing that there is a linear relationship between the length (snout to vent) of female water snakes and the length of their catfish prey such that

$$l(x) = -3.6 + 0.17x,$$

where $x$ is the snout-vent length (cm) of the snake and $l(x)$ is the predicted length (cm) of the catfish. *Source: Copeia.*

**a.** Use the function above to estimate the size of catfish that a snake which has a snout-vent length of 70 cm will choose for prey.

**b.** Calculate the derivative of this function and interpret the results.

**c.** Estimate the snout-vent length of the smallest snake for which the function above makes sense.

**50. Beaked Sea Snake** Scientists in Malaysia have collected data revealing that there is a linear relationship between the length (snout to vent) of the beaked sea snake and the length of its catfish prey such that

$$l(x) = 2.41 + 0.12x,$$

where $x$ is the snout-vent length (cm) of the snake and $l(x)$ is the predicted length (cm) of the catfish. *Source: Biotropica.*

**a.** Use the function above to estimate the size of catfish that a snake which has a snout-vent length of 70 cm will choose for prey.

**b.** Calculate the derivative of this function and interpret the results.

**51. Eating Behavior** As indicated in Example 8, the eating behavior of a typical human during a meal can be described by

$$I(t) = 27 + 72t - 1.5t^2,$$

where $t$ is the number of minutes since the meal began, and $I(t)$ represents the amount (in grams) that the person has eaten at time $t$. *Source: Appetite.*

**a.** Find the rate of change of the intake of food for this particular person 5 minutes into a meal and interpret.

**b.** Verify that the rate in which food is consumed is zero 24 minutes after the meal starts.

**c.** Comment on the assumptions and usefulness of this function after 24 minutes. Given this fact, determine a logical domain for this function.

**52. Quality Control of Cheese** It is often difficult to evaluate the quality of products that undergo a ripening or maturation process. Researchers have successfully used ultrasonic velocity to determine the maturation time of Mahon cheese. The age can be determined by

$$M(v) = 0.0312443v^2 - 101.39v + 82,264, \quad v \geq 1620,$$

where $M(v)$ is the estimated age of the cheese (in days) for a velocity $v$ (m per second). *Source: Journal of Food Science.*

**a.** If Mahon cheese ripens in 150 days, determine the velocity of the ultrasound that one would expect to measure. (*Hint:* Set $M(v) = 150$ and solve for $v$.)

**b.** Determine the derivative of this function when $v = 1700$ m per second and interpret.

**53. Shellfish Population** In one research study, the population of a certain shellfish in an area at time $t$ was closely approximated by the following graph. Estimate and interpret the derivative at each of the marked points.

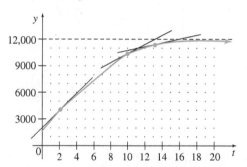

**54. Flight Speed** Many birds, such as cockatiels or the Arctic terns shown below, have flight muscles whose expenditure of power varies with the flight speed in a manner similar to the graph shown on the next page. The horizontal axis of the graph shows flight speed in meters per second, and the vertical axis shows power in watts per kilogram. *Source: Biolog-e: The Undergraduate Bioscience Research Journal.*

**a.** The speed $V_{mp}$ minimizes energy costs per unit of time. What is the slope of the line tangent to the curve at the point corresponding to $V_{mp}$? What is the physical significance of the slope at that point?

**b.** The speed $V_{mr}$ minimizes the energy costs per unit of distance covered. Estimate the slope of the curve at the point corresponding to $V_{mr}$. Give the significance of the slope at that point.

**c.** By looking at the shape of the curve, describe how the power level decreases and increases for various speeds.

**d.** Notice that the slope of the lines found in parts a and b represents the power divided by speed. Power is measured in energy per unit time per unit weight of the bird, and speed is distance per unit time, so the slope represents energy per unit distance per unit weight of the bird. If a line is drawn from the origin to a point on the graph, at which point is the slope of the line (representing energy per unit distance per unit weight of the bird) smallest? How does this compare with your answers to parts a and b?

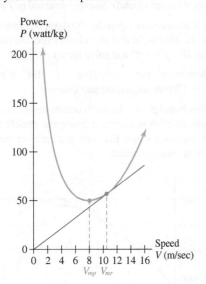

**OTHER APPLICATIONS**

**55. Temperature** The graph shows the temperature in degrees Celsius as a function of the altitude $h$ in feet when an inversion layer is over Southern California. (See Exercise 43 in the previous section.) Estimate and interpret the derivatives of $T(h)$ at the marked points.

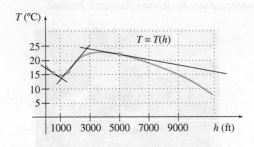

**56. Oven Temperature** The graph shows the temperature one Christmas day of a Heat-Kit Bakeoven, a wood-burning oven for baking. *Source: Heatkit.com.* The oven was lit at 8:30 A.M. Let $T(x)$ be the temperature $x$ hours after 8:30 A.M.

**a.** Find and interpret $T'(0.5)$.

**b.** Find and interpret $T'(3)$.

**c.** Find and interpret $T'(4)$.

**d.** At a certain time a Christmas turkey was put into the oven, causing the oven temperature to drop. Estimate when the turkey was put into the oven.

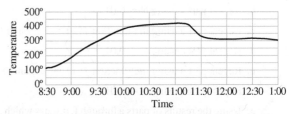

**57. Baseball** The graph shows how the velocity of a baseball that was traveling at 40 miles per hour when it was hit by a Little League baseball player varies with respect to the weight of the bat. *Source: Biological Cybernetics.*

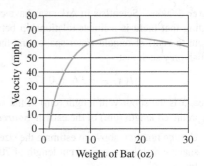

**a.** Estimate and interpret the derivative for a 16-oz and 25-oz bat.

**b.** What is the optimal bat weight for this player?

**58. Baseball** The graph shows how the velocity of a baseball that was traveling at 90 miles per hour when it was hit by a Major League baseball player varies with respect to the weight of the bat. *Source: Biological Cybernetics.*

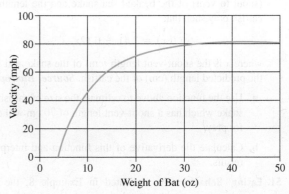

**a.** Estimate and interpret the derivative for a 40-oz and 30-oz bat.

**b.** What is the optimal bat weight for this player?

**59. Social Security Assets** The table gives actual and projected year-end assets in Social Security trust funds, in trillions of current dollars, where Year represents the number of years since 2000. *Source: Social Security Administration.* The polynomial function defined by

$$f(t) = 0.0000329t^3 - 0.00450t^2 + 0.0613t + 2.34$$

models the data quite well.

**a.** To verify the fit of the model, find $f(10)$, $f(20)$, and $f(30)$.

**b.** Use a graphing calculator with a command such as nDeriv to find the slope of the tangent line to the graph of $f$ at the following $t$-values: 0, 10, 20, 30, and 35.

**c.** Use your results in part b to describe the graph of $f$ and interpret the corresponding changes in Social Security assets.

| Year | Trillions of Dollars |
|------|---------------------|
| 10 | 2.45 |
| 20 | 2.34 |
| 30 | 1.03 |
| 40 | −0.57 |
| 50 | −1.86 |

**YOUR TURN ANSWERS**

**1. a.** $y = -2x + 2$  **b.** $y = -5x - 4$
**2.** $f'(x) = 2x - 1$ and $f'(-2) = -5$
**3.** $f'(x) = 3x^2$ and $f'(-1) = 3$
**4.** $f'(x) = 2/x^2$  **5.** $f'(x) = 1/\sqrt{x}$
**6.** $y = (1/2)x + 2$

# 3.5 Graphical Differentiation

APPLY IT **Given a graph of the weight of a young female as a function of age, how can we find the graph of the rate of change in weight of the young female each year?**

*We will answer this question in Example 1 using graphical differentiation.*

In the previous section, we estimated the derivative at various points of a graph by estimating the slope of the tangent line at those points. We will now extend this process to show how to sketch the graph of the derivative given the graph of the original function. This is important because, in many applications, a graph is all we have, and it is easier to find the derivative graphically than to find a formula that fits the graph and take the derivative of that formula.

**EXAMPLE 1** **Weight of a Young Female**

In Figure 45(a), from a book on human growth, the graph shows the typical weight (in kg) of an English girl for the first 18 years of life. The graph in Figure 45(b) shows the rate of change in weight of the girl each year. Verify that the graph of the rate of change is the graph of the derivative of the weight function. *Source: Human Growth After Birth.*

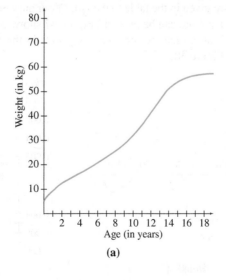

(a)

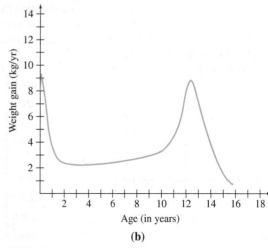

(b)

FIGURE 45

**APPLY IT**

**SOLUTION** Let $t$ refer to the age (in years) of the girl. Observe that the slope of the weight function is always positive. However, notice that as age increases, the weight gain is not constant. For example, right after birth, the slope of the function is steeper than it is from about one year to ten years. But, from about age 11 to age 13, the function increases rapidly, illustrating a second growth spurt. Notice that these growth spurts can also be seen by looking at the graph in Figure 45(b).

In Example 1, we saw how the general shape of the graph of the derivative could be found from the graph of the original function. To get a more accurate graph of the derivative, we need to estimate the slope of the tangent line at various points, as we did in the previous section.

### EXAMPLE 2 Temperature

Figure 46 gives the temperature in degrees Celsius as a function, $T(h)$, of the altitude $h$ in feet when an inversion layer is over Southern California. Sketch the graph of the derivative of the function. (This graph appeared in Exercise 55 in the previous section.)

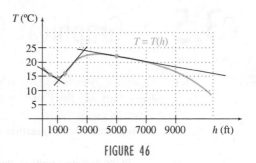

FIGURE 46

**SOLUTION** First, observe that when $h = 1000$ and $h = 3500$, $T(h)$ has horizontal tangent lines, so $T'(1000) = 0$ and $T'(3500) = 0$.

Notice that the tangent lines would have a negative slope for $0 < h < 1000$. Thus, the graph of the derivative should be negative (below the $x$-axis) there. Then, for $1000 < h < 3500$, the tangent lines have positive slope, so the graph of the derivative should be positive (above the $x$-axis) there. Notice from the graph of $T(h)$ that the slope is largest when $h = 1500$. Finally, for $h > 3500$, the tangent lines would be negative again, forcing the graph of the derivative back down below the $x$-axis to take on negative values.

Now that we have a general shape of the graph, we can estimate the value of the derivative at several points to improve the accuracy of the graph. To estimate the derivative, find two points on each tangent line and compute the slope between the two points. The estimates at selected points are given in the table to the left. (Your answers may be slightly different, since estimation from a picture can be inexact.) Figure 47 shows a graph of these values of $T'(h)$.

Using all of these facts, we connect the points in the graph $T'(h)$ smoothly, with the result shown in Figure 48.                                        **TRY YOUR TURN 1**

**Estimates of $T'(h)$**

| $h$ | $T'(h)$ |
|------|---------|
| 500  | −0.005  |
| 1000 | 0       |
| 1500 | 0.008   |
| 3500 | 0       |
| 5000 | −0.00125 |

**YOUR TURN 1** Sketch the graph of the derivative of the function $f(x)$.

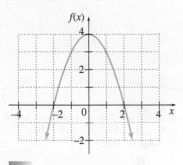

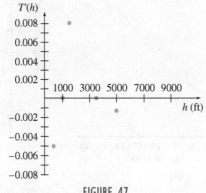

FIGURE 47

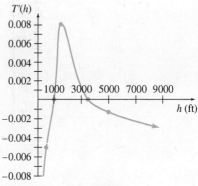

FIGURE 48

**CAUTION**   Remember that when you graph the derivative, you are graphing the *slope* of the original function. Do not confuse the slope of the original function with the *y*-value of the original function. In fact, the slope of the original function is equal to the *y*-value of its derivative.

Sometimes the original function is not smooth or even continuous, so the graph of the derivative may also be discontinuous.

### EXAMPLE 3   Graphing a Derivative

Sketch the graph of the derivative of the function shown in Figure 49.

**SOLUTION**   Notice that when $x < -2$, the slope is 1, and when $-2 < x < 0$, the slope is $-1$. At $x = -2$, the derivative does not exist due to the sharp corner in the graph. The derivative also does not exist at $x = 0$ because the function is discontinuous there. Using this information, the graph of $f'(x)$ on $x < 0$ is shown in Figure 50.

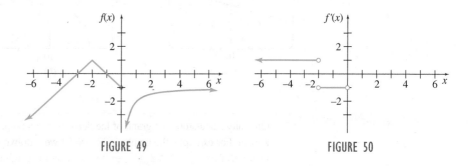

FIGURE 49                    FIGURE 50

FIGURE 51

**YOUR TURN 2**   Sketch the graph of the derivative of the function $g(x)$.

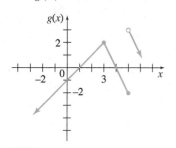

For $x > 0$, the derivative is positive. If you draw a tangent line at $x = 1$, you should find that the slope of this line is roughly 1. As $x$ approaches 0 from the right, the derivative becomes larger and larger. As $x$ approaches infinity, the tangent lines become more and more horizontal, so the derivative approaches 0. The resulting sketch of the graph of $y = f'(x)$ is shown in Figure 51.   **TRY YOUR TURN 2**

Finding the derivative graphically may seem difficult at first, but with practice you should be able to quickly sketch the derivative of any function graphed. Your answers to the exercises may not look exactly like those in the back of the book, because estimating the slope accurately can be difficult, but your answers should have the same general shape.

Figures 52(a), (b), and (c) on the next page show the graphs of $y = x^2$, $y = x^4$, and $y = x^{4/3}$ on a graphing calculator. When finding the derivative graphically, all three seem to have the same behavior: negative derivative for $x < 0$, 0 derivative at $x = 0$, and positive derivative for $x > 0$. Beyond these general features, however, the derivatives look quite different, as you can see from Figures 53(a), (b), and (c), which show the graphs of the derivatives. When finding derivatives graphically, detailed information can only be found by very carefully measuring the slope of the tangent line at a large number of points.

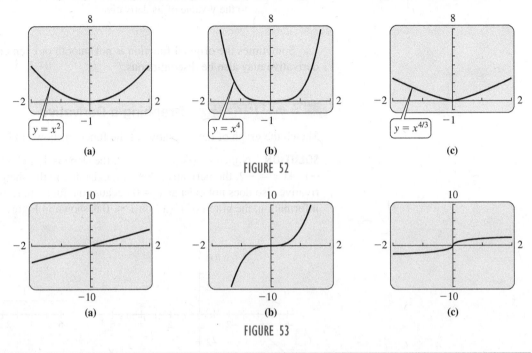

FIGURE 52

FIGURE 53

TECHNOLOGY NOTE    On many calculators, the graph of the derivative can be plotted if a formula for the original function is known. For example, the graphs in Figure 53 were drawn on a TI-84 Plus by using the nDeriv command. Define $Y_2 = \frac{d}{dx}(Y_1)\big|_{x=x}$ after entering the original function into $Y_1$. You can use this feature to practice finding the derivative graphically. Enter a function into $Y_1$, sketch the graph on the graphing calculator, and use it to draw by hand the graph of the derivative. Then use nDeriv to draw the graph of the derivative, and compare it with your sketch.

## EXAMPLE 4    Graphical Differentiation

Figure 54 shows the graph of a function $f$ and its derivative function $f'$. Use slopes to decide which graph is that of $f$ and which is the graph of $f'$.

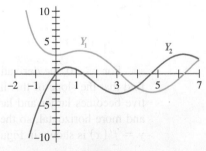

FIGURE 54

**SOLUTION**    Look at the places where each graph crosses the $x$-axis; that is, the $x$-intercepts, since $x$-intercepts occur on the graph of $f'$ whenever the graph of $f$ has a horizontal tangent line or slope of zero. Also, a decreasing graph corresponds to negative slope or a negative derivative, while an increasing graph corresponds to positive slope or a positive derivative. $Y_1$ has zero slope near $x = 0$, $x = 1$, and $x = 5$; $Y_2$ has $x$-intercepts near these values of $x$. $Y_1$ decreases on $(-2, 0)$ and $(1, 5)$; $Y_2$ is negative on those intervals. $Y_1$ increases on $(0, 1)$ and $(5, 7)$; $Y_2$ is positive there. Thus, $Y_1$ is the graph of $f$ and $Y_2$ is the graph of $f'$.

# 3.5   EXERCISES

1. Explain how to graph the derivative of a function given the graph of the function.

2. Explain how to graph a function given the graph of the derivative function. Note that the answer is not unique.

**Each graphing calculator window shows the graph of a function $f(x)$ and its derivative function $f'(x)$. Decide which is the graph of the function and which is the graph of the derivative.**

3.
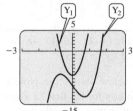

4.
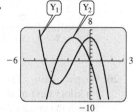

5.
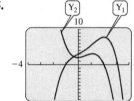

6.
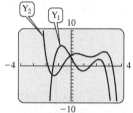

**Sketch the graph of the derivative for each function shown.**

7.
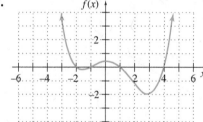

8.
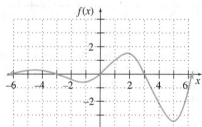

9.
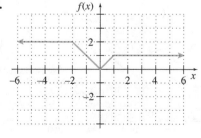

10.
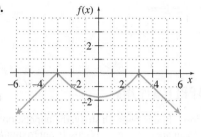

11.
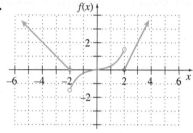

12.
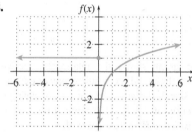

13.
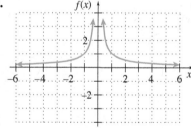

14.
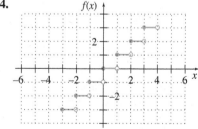

15.
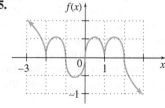

**16.**

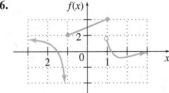

## LIFE SCIENCE APPLICATIONS

**17. Shellfish Population** In one research study, the population of a certain shellfish in an area at time $t$ was closely approximated by the following graph. Sketch a graph of the growth rate of the population.

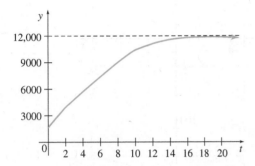

**18. Flight Speed** The graph shows the relationship between the speed of a bird in flight and the required power expended by flight muscles. Sketch the graph of the rate of change of the power as a function of the speed. *Source: Biolog-e: The Undergraduate Bioscience Research Journal.*

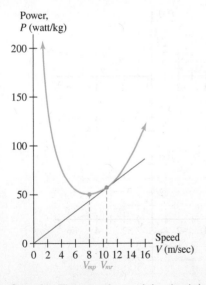

**19. Human Growth** The growth remaining in sitting height at consecutive skeletal age levels for boys is indicated in the next column. Sketch a graph showing the rate of change of growth remaining for the indicated years. Use the graph and your sketch to estimate the remaining growth and the rate of change of remaining growth for a 14-year-old boy. *Source: Standards in Pediatric Orthopedics: Tables, Charts, and Graphs Illustrating Growth.*

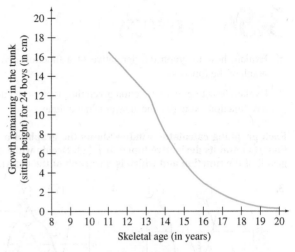

**20. Weight Gain** The graph below shows the typical weight (in kilograms) of an English boy for his first 18 years of life. Sketch the graph of the rate of change of weight with respect to time. *Source: Human Growth After Birth.*

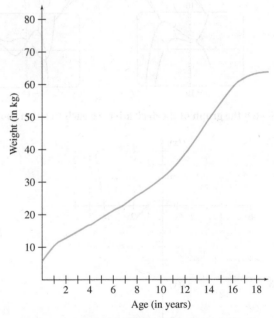

**21. Insecticide** The graph shows how the number of arthropod species resistant to insecticides has varied with time. Sketch a graph of the rate of change of the insecticide-resistant species as a function of time. *Source: Science.*

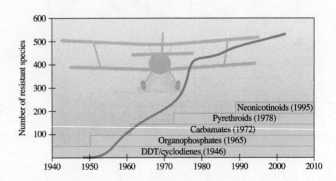

**22. Body Mass Index** The following graph shows how the body mass index-for-age percentile for boys varies from the age of 2 to 20 years. *Source: Centers for Disease Control.*

**a.** Sketch a graph of the rate of change of the 95th percentile as a function of age.

**b.** Sketch a graph of the rate of change of the 50th percentile as a function of age.

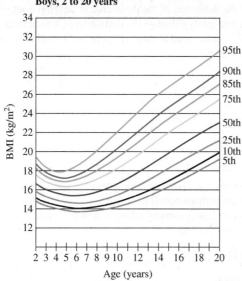

Body Mass Index-for-Age Percentiles:
Boys, 2 to 20 years

**23. DNA** The curves in the figure show the thermal denaturation profile of pea (upper curve) and *Escherichia coli* (lower curve) DNA, giving the relative absorbance as a function of the temperature in degrees Celsius. *Source: Proceedings of the National Academy of Sciences of the United States.*

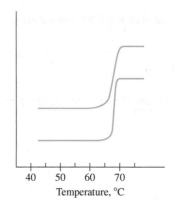

The researchers found that examining the graph of the derivative was useful in characterizing the profiles. Sketch a graph of the derivative of each profile.

## OTHER APPLICATIONS

**24. Consumer Demand** When the price of an essential commodity rises rapidly, consumption drops slowly at first. If the price continues to rise, however, a "tipping" point may be reached,

at which consumption takes a sudden substantial drop. Suppose the accompanying graph shows the consumption of gasoline, $G(t)$, in millions of gallons, in a certain area. We assume that the price is rising rapidly. Here $t$ is the time in months after the price began rising. Sketch a graph of the rate of change in consumption as a function of time.

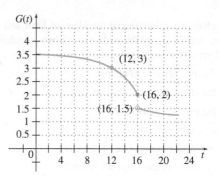

**25. Sales** The graph shows annual sales (in thousands of dollars) of an Xbox game at a particular store. Sketch a graph of the rate of change of sales as a function of time.

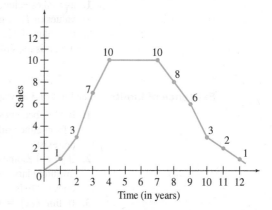

**1.**

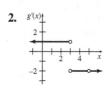

**2.**

# 3

# CHAPTER REVIEW

## SUMMARY

In this chapter we introduced the ideas of limit and continuity of functions and then used these ideas to explore calculus. We saw that the difference quotient can represent

- the average rate of change,
- the slope of the secant line, and
- the average velocity.

We saw that the derivative can represent

- the instantaneous rate of change,
- the slope of the tangent line, and
- the instantaneous velocity.

We also learned how to estimate the value of the derivative using graphical differentiation. In the next chapter, we will take a closer look at the definition of the derivative to develop a set of rules to quickly and easily calculate the derivative of a wide range of functions without the need to directly apply the definition of the derivative each time.

**Limit of a Function**   Let $f$ be a function and let $a$ and $L$ be real numbers. If

1. as $x$ takes values closer and closer (but not equal) to $a$ on both sides of $a$, the corresponding values of $f(x)$ get closer and closer (and perhaps equal) to $L$; and
2. the value of $f(x)$ can be made as close to $L$ as desired by taking values of $x$ close enough to $a$; then $L$ is the limit of $f(x)$ as $x$ approaches $a$, written

$$\lim_{x \to a} f(x) = L.$$

**Existence of Limits**   The limit of $f$ as $x$ approaches $a$ may not exist.

1. If $f(x)$ becomes infinitely large in magnitude (positive or negative) as $x$ approaches the number $a$ from either side, we write $\lim_{x \to a} f(x) = \infty$ or $\lim_{x \to a} f(x) = -\infty$. In either case, the limit does not exist.
2. If $f(x)$ becomes infinitely large in magnitude (positive) as $x$ approaches $a$ from one side and infinitely large in magnitude (negative) as $x$ approaches $a$ from the other side, then $\lim_{x \to a} f(x)$ does not exist.
3. If $\lim_{x \to a^-} f(x) = L$ and $\lim_{x \to a^+} f(x) = M$, and $L \neq M$, then $\lim_{x \to a} f(x)$ does not exist.

**Limits at Infinity**   For any positive real number $n$,

$$\lim_{x \to \infty} \frac{1}{x^n} = \lim_{x \to -\infty} \frac{1}{x^n} = 0.$$

**Finding Limits at Infinity**   If $f(x) = p(x)/q(x)$ for polynomials $p(x)$ and $q(x)$, $\lim_{x \to \infty} f(x)$ and $\lim_{x \to -\infty} f(x)$ can be found by dividing $p(x)$ and $q(x)$ by the highest power of $x$ in $q(x)$.

**Continuity**   A function $f$ is continuous at $c$ if

1. $f(c)$ is defined,
2. $\lim_{x \to c} f(x)$ exists, and
3. $\lim_{x \to c} f(x) = f(c)$.

**Average Rate of Change**   The average rate of change of $f(x)$ with respect to $x$ as $x$ changes from $a$ to $b$ is

$$\frac{f(b) - f(a)}{b - a}.$$

**Difference Quotient**   The average rate of change can also be written as

$$\frac{f(x + h) - f(x)}{h}.$$

**Derivative**   The derivative of $f(x)$ with respect to $x$ is

$$f'(x) = \lim_{h \to 0} \frac{f(x + h) - f(x)}{h}.$$

# KEY TERMS

*To understand the concepts presented in this chapter, you should know the meaning and use of the following terms.*
*For easy reference, the section in the chapter where a word (or expression) was first used is provided.*

**3.1**
limit
limit from the left/right
one-/two-sided limit
piecewise function
limit at infinity

**3.2**
continuous
discontinuous
removable discontinuity
continuous on an open/closed
   interval
continuous from the right/left
Intermediate Value Theorem

**3.3**
average rate of change
difference quotient
instantaneous rate of change
velocity

**3.4**
secant line
tangent line
slope of the curve
derivative
differentiable
differentiation

# REVIEW EXERCISES

## CONCEPT CHECK

**Determine whether each of the following statements is true or false, and explain why.**

**1.** The limit of a product is the product of the limits when each of the limits exists.

**2.** The limit of a function may not exist at a point even though the function is defined there.

**3.** If a rational function has a polynomial in the denominator of higher degree than the polynomial in the numerator, then the limit at infinity must equal zero.

**4.** If the limit of a function exists at a point, then the function is continuous there.

**5.** Trigonometric functions are continuous everywhere.

**6.** A rational function is continuous everywhere.

**7.** The derivative gives the average rate of change of a function.

**8.** The derivative gives the instantaneous rate of change of a function.

**9.** The instantaneous rate of change is a limit.

**10.** The derivative is a function.

**11.** The slope of the tangent line gives the average rate of change.

**12.** The derivative of a function exists wherever the function is continuous.

## PRACTICE AND EXPLORATIONS

**13.** Is a derivative always a limit? Is a limit always a derivative? Explain.

**14.** Is every continuous function differentiable? Is every differentiable function continuous? Explain.

**15.** Describe how to tell when a function is discontinuous at the real number $x = a$.

**16.** Give two applications of the derivative
$$f'(x) = \lim_{h \to 0} \frac{f(x + h) - f(x)}{h}.$$

**Decide whether the limits in Exercises 17–36 exist. If a limit exists, find its value.**

**17. a.** $\lim\limits_{x \to -3^-} f(x)$  **b.** $\lim\limits_{x \to -3^+} f(x)$  **c.** $\lim\limits_{x \to -3} f(x)$  **d.** $f(-3)$

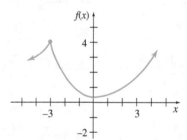

**18. a.** $\lim\limits_{x \to -1^-} g(x)$  **b.** $\lim\limits_{x \to -1^+} g(x)$  **c.** $\lim\limits_{x \to -1} g(x)$  **d.** $g(-1)$

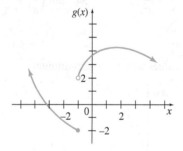

**19. a.** $\lim\limits_{x \to 4^-} f(x)$  **b.** $\lim\limits_{x \to 4^+} f(x)$  **c.** $\lim\limits_{x \to 4} f(x)$  **d.** $f(4)$

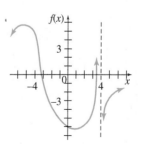

**20. a.** $\lim\limits_{x\to 2^-} h(x)$ **b.** $\lim\limits_{x\to 2^+} h(x)$ **c.** $\lim\limits_{x\to 2} h(x)$ **d.** $h(2)$

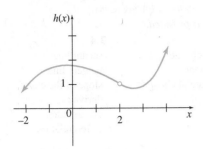

**21.** $\lim\limits_{x\to -\infty} g(x)$

**22.** $\lim\limits_{x\to\infty} f(x)$

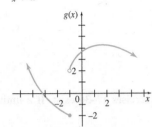 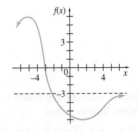

**23.** $\lim\limits_{x\to 6} \dfrac{2x+7}{x+3}$

**24.** $\lim\limits_{x\to -3} \dfrac{2x+5}{x+3}$

**25.** $\lim\limits_{x\to 4} \dfrac{x^2-16}{x-4}$

**26.** $\lim\limits_{x\to 2} \dfrac{x^2+3x-10}{x-2}$

**27.** $\lim\limits_{x\to -4} \dfrac{2x^2+3x-20}{x+4}$

**28.** $\lim\limits_{x\to 3} \dfrac{3x^2-2x-21}{x-3}$

**29.** $\lim\limits_{x\to 9} \dfrac{\sqrt{x}-3}{x-9}$

**30.** $\lim\limits_{x\to 16} \dfrac{\sqrt{x}-4}{x-16}$

**31.** $\lim\limits_{x\to\infty} \dfrac{2x^2+5}{5x^2-1}$

**32.** $\lim\limits_{x\to\infty} \dfrac{x^2+6x+8}{x^3+2x+1}$

**33.** $\lim\limits_{x\to -\infty} \left(\dfrac{3}{8}+\dfrac{3}{x}-\dfrac{6}{x^2}\right)$

**34.** $\lim\limits_{x\to -\infty} \left(\dfrac{9}{x^4}+\dfrac{10}{x^2}-6\right)$

**35.** $\lim\limits_{x\to\pi/2} \dfrac{1-\sin^2 x}{\cos^2 x}$

**36.** $\lim\limits_{x\to\pi} \dfrac{1+\cos x}{1-\cos^2 x}$

Identify the $x$-values where $f$ is discontinuous.

**37.**

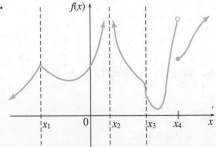

**38.**

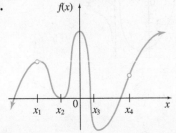

Find all values $x = a$ where the function is discontinuous. For each such value, give $f(a)$ and $\lim\limits_{x\to a} f(x)$ or state that it does not exist.

**39.** $f(x) = \dfrac{-5+x}{3x(3x+1)}$

**40.** $f(x) = \dfrac{7-3x}{(1-x)(3+x)}$

**41.** $f(x) = \dfrac{x-6}{x+5}$

**42.** $f(x) = \dfrac{x^2-9}{x+3}$

**43.** $f(x) = x^2+3x-4$

**44.** $f(x) = 2x^2-5x-3$

**45.** $f(x) = \cos\left(\dfrac{x}{x-1}\right)$

**46.** $f(x) = e^{\sin x}$

In Exercises 47 and 48, (a) graph the given function, (b) find all values of $x$ where the function is discontinuous, and (c) find the limit from the left and from the right at any values of $x$ found in part b.

**47.** $f(x) = \begin{cases} 1-x & \text{if } x < 1 \\ 2 & \text{if } 1 \le x \le 2 \\ 4-x & \text{if } x > 2 \end{cases}$

**48.** $f(x) = \begin{cases} 2 & \text{if } x < 0 \\ -x^2+x+2 & \text{if } 0 \le x \le 2 \\ 1 & \text{if } x > 2 \end{cases}$

Find each limit (a) by investigating values of the function near the point where the limit is taken and (b) by using a graphing calculator to view the function near the point.

**49.** $\lim\limits_{x\to 1} \dfrac{x^4+2x^3+2x^2-10x+5}{x^2-1}$

**50.** $\lim\limits_{x\to -2} \dfrac{x^4+3x^3+7x^2+11x+2}{x^3+2x^2-3x-6}$

Find the average rate of change for the following on the given interval. Then find the instantaneous rate of change at the first $x$-value.

**51.** $y = 6x^3+2$ from $x = 1$ to $x = 4$

**52.** $y = -2x^3-3x^2+8$ from $x = -2$ to $x = 6$

**53.** $y = \dfrac{-6}{3x-5}$ from $x = 4$ to $x = 9$

**54.** $y = \dfrac{x+4}{x-1}$ from $x = 2$ to $x = 5$

For each function, find (a) the equation of the secant line through the points where $x$ has the given values, and (b) the equation of the tangent line when $x$ has the first value.

**55.** $f(x) = 3x^2-5x+7; x = 2, x = 4$

**56.** $f(x) = \dfrac{1}{x}; x = 1/2, x = 3$

**57.** $f(x) = \dfrac{12}{x-1}; x = 3, x = 7$

**58.** $f(x) = 2\sqrt{x-1}; x = 5, x = 10$

Use the definition of the derivative to find the derivative of the following.

**59.** $y = 4x^2+3x-2$

**60.** $y = 5x^2-6x+7$

In Exercises 61 and 62, find the derivative of the function at the given point (a) by approximating the definition of the derivative with small values of $h$ and (b) by using a graphing calculator to zoom in on the function until it appears to be a straight line, and then finding the slope of that line.

**61.** $f(x) = (\ln x)^x$; $x_0 = 3$     **62.** $f(x) = x^{\ln x}$; $x_0 = 2$

**Sketch the graph of the derivative for each function shown.**

**63.**

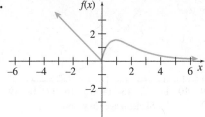

**64.**

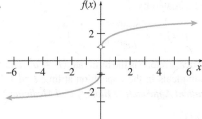

**65.** Let $f$ and $g$ be differentiable functions such that

$$\lim_{x \to \infty} f(x) = c$$

$$\lim_{x \to \infty} g(x) = d$$

where $c \neq d$. Determine

$$\lim_{x \to \infty} \frac{cf(x) - dg(x)}{f(x) - g(x)}.$$

(Choose one of the following.) *Source: Society of Actuaries.*

**a.** 0     **b.** $\dfrac{cf'(0) - dg'(0)}{f'(0) - g'(0)}$

**c.** $f'(0) - g'(0)$   **d.** $c - d$   **e.** $c + d$

## LIFE SCIENCE APPLICATIONS

**66. TYLENOL®** The table shows the recommended dosage of Children's TYLENOL® Suspension Liquid with respect to weight. Let $D(x)$ represent the dosage (in tsp) for a child weighing $x$ pounds.

| $x$, Weight in pounds | $D(x)$, Dosage in teaspoons |
|---|---|
| $24 \leq x < 36$ | 1 |
| $36 \leq x < 48$ | 1.5 |
| $48 \leq x < 60$ | 2 |
| $60 \leq x < 72$ | 2.5 |
| $72 \leq x < 95$ | 3 |

**a.** Draw a graph of this function.

**b.** Find $\lim\limits_{x \to 48^-} D(x)$.

**c.** Find $\lim\limits_{x \to 48^+} D(x)$.

**d.** Find $\lim\limits_{x \to 48} D(x)$.

**e.** Find the dosage for a child who weighs 48 pounds, $D(48)$.

**f.** Is this function continuous for children who weigh between 24 and 95 pounds? If not, what are the weights at which the function is discontinuous?

**67. Alzheimer's Disease** The graph below shows the projected number of people aged 65 and over in the United States with Alzheimer's disease. *Source: Alzheimer's Disease Facts and Figures.* Estimate and interpret the derivative in each of the following years.

**a.** 2000     **b.** 2040

**c.** Find the average rate of change between 2000 and 2040 in the number of people aged 65 and over in the United States with Alzheimer's disease.

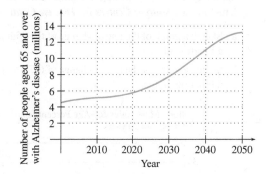

**68. Spread of a Virus** The spread of a virus is modeled by

$$V(t) = -t^2 + 6t - 4,$$

where $V(t)$ is the number of people (in hundreds) with the virus and $t$ is the number of weeks since the first case was observed.

**a.** Graph $V(t)$.

**b.** What is a reasonable domain of $t$ for this problem?

**c.** When does the number of cases reach a maximum? What is the maximum number of cases?

**d.** Find the rate of change function.

**e.** What is the rate of change in the number of cases at the maximum?

**f.** Give the sign (+ or −) of the rate of change up to the maximum and after the maximum.

**69. Whales Diving** The figure on the next page, already shown in Section 1.3 on Properties of Functions, shows the depth of a sperm whale as a function of time, recorded by researchers at the Woods Hole Oceanographic Institution in Massachusetts. *Source: Peter Tyack, Woods Hole Oceanographic Institution.*

a. Find the rate that the whale was descending at the following times.

   **i.** 17 hours and 37 minutes

   **ii.** 17 hours and 39 minutes

b. Sketch a graph of the rate the whale was descending as a function of time.

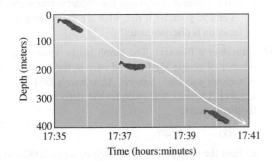

**70. Body Mass Index** The following graph shows how the body mass index-for-age percentile for girls varies from the age of 2 to 20 years. *Source: Centers for Disease Control.*

a. Sketch a graph of the rate of change of the 95th percentile as a function of age.

b. Sketch a graph of the rate of change of the 50th percentile as a function of age.

**Body Mass Index-for-Age Percentiles:
Girls, 2 to 20 years**

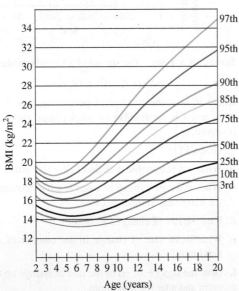

**71. Human Growth** The growth remaining in sitting height at consecutive skeletal age levels for girls is indicated in the next column. Sketch a graph showing the rate of change of growth remaining for the indicated years. Use the graph and your sketch to estimate the remaining growth and the rate of change of remaining growth for a 10-year-old girl. *Source: Standards in Pediatric Orthopedics: Tables, Charts, and Graphs Illustrating Growth.*

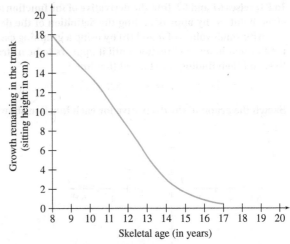

**72. Extinction** The probability of a population going extinct by time $t$ can be estimated by

$$p(t) = \left( \frac{a[e^{(b-a)t} - 1]}{be^{(b-a)t} - a} \right)^N,$$

where $a$ is the death rate, $b$ is the birth rate $(b \neq a)$, and $N$ is the number of individuals in the population at time $t = 0$. *Source: The Mathematical Approach to Biology and Medicine.*

a. Find $\lim_{t \to \infty} p(t)$

   **i.** if $b > a$;

   **ii.** if $b < a$.

b. If $a = b$, the probability of a population going extinct by time $t$ can be estimated by

$$p(t) = \left( \frac{at}{at + 1} \right)^N.$$

Find $\lim_{t \to \infty} p(t)$.

## OTHER APPLICATIONS

**73. Temperature** Suppose a gram of ice is at a temperature of $-100°C$. The graph shows the temperature of the ice as increasing numbers of calories of heat are applied. It takes 80 calories to melt one gram of ice at $0°C$ into water, and 540 calories to boil one gram of water at $100°C$ into steam.

a. Where is this graph discontinuous?

b. Where is this graph not differentiable?

c. Sketch the graph of the derivative.

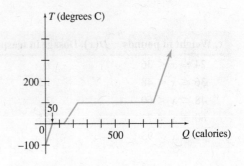

**74. Average Cost**   The graph shows the total cost $C(x)$ to produce $x$ tons of lumber. (Recall that average cost is given by total cost divided by the number produced, or $\overline{C}(x) = C(x)/x$.)

 **a.** Draw a line through $(0, 0)$ and $(5, C(5))$. Explain why the slope of this line represents the average cost per ton when 5 tons of lumber are produced.

 **b.** Find the value of $x$ for which the average cost is smallest.

 **c.** What can you say about the marginal cost at the point where the average cost is smallest?

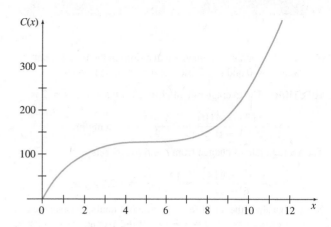

# EXTENDED APPLICATION

# A MODEL FOR DRUGS ADMINISTERED INTRAVENOUSLY

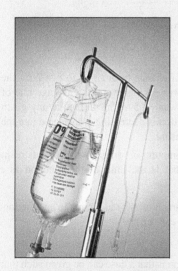

When a drug is administered intravenously, it enters the bloodstream immediately, producing an immediate effect for the patient. The drug can be either given as a single rapid injection or given at a constant drip rate. The latter is commonly referred to as an intravenous (IV) infusion. Common drugs administered intravenously include morphine for pain, diazepam (or Valium) to control a seizure, and digoxin for heart failure.

## SINGLE RAPID INJECTION

With a single rapid injection, the amount of drug in the bloodstream reaches its peak immediately and then the body eliminates the drug exponentially. The larger the amount of drug there is in the body, the faster the body eliminates it. If a lesser amount of drug is in the body, it is eliminated more slowly.

The amount of drug in the bloodstream $t$ hours after a single rapid injection can be modeled using an exponential decay function, like those found in Section 2.3 on Applications: Growth and Decay, as follows:

$$A(t) = De^{kt},$$

where $D$ is the size of the dose administered and $k$ is the exponential decay constant for the drug.

### EXAMPLE 1   Rapid Injection

The drug labetalol is used for the control of blood pressure in patients with severe hypertension. The half-life of labetalol is 4 hours. Suppose a 35-mg dose of the drug is administered to a patient by rapid injection.

**(a)** Find a model for the amount of drug in the bloodstream $t$ hours after the drug is administered.

**SOLUTION**   Since $D = 35$ mg, the function has the form

$$A(t) = 35e^{kt}.$$

Recall from Section 2.3 that the general equation giving the half-life $T$ in terms of the decay constant $k$ was

$$T = -\frac{\ln 2}{k}.$$

Solving this equation for $k$, we get

$$k = -\frac{\ln 2}{T}.$$

Since the half-life of this drug is 4 hours,

$$k = -\frac{\ln 2}{4} \approx -0.17.$$

Therefore, the model is

$$A(t) = 35e^{-0.17t}.$$

The graph of $A(t)$ is given in Figure 55.

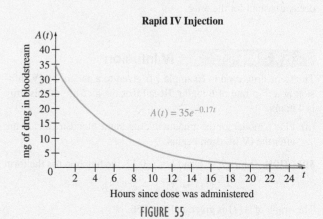

**Rapid IV Injection**

$A(t) = 35e^{-0.17t}$

Hours since dose was administered

FIGURE 55

**(b)** Find the average rate of change of drug in the bloodstream between $t = 0$ and $t = 2$. Repeat for $t = 4$ and $t = 6$.

**SOLUTION** The average rate of change from $t = 0$ to $t = 2$ is

$$\frac{A(2) - A(0)}{2 - 0} \approx \frac{25 - 35}{2} = -5 \text{ mg/hr.}$$

The average rate of change from $t = 4$ to $t = 6$ is

$$\frac{A(6) - A(4)}{6 - 4} \approx \frac{13 - 18}{2} = -2.5 \text{ mg/hr.}$$

Notice that since the half-life of the drug is 4 hours, the average rate of change from $t = 4$ to $t = 6$ is half of the average rate of change from $t = 0$ to $t = 2$. What would the average rate of change be from $t = 8$ to $t = 10$?

**(c)** What happens to the amount of drug in the bloodstream as $t$ increases? (i.e., What is the limit of the function as $t$ approaches $\infty$?)

**SOLUTION** Looking at the graph of $A(t)$, we can see that

$$\lim_{t \to \infty} A(t) = 0.$$

An advantage of an intravenous rapid injection is that the amount of drug in the body reaches a high level immediately. Suppose, however, that the effective level of this drug is between 30 mg and 40 mg. From the graph, we can see that it takes only an hour after the dose is given for the amount of drug in the body to fall below the effective level.

## INTRAVENOUS INFUSION

With an IV infusion, the amount of drug in the bloodstream starts at zero and increases until the rate the drug is entering the body equals the rate the drug is being eliminated from the body. At this point, the amount of drug in the bloodstream levels off. This model is a limited growth function, like those from Chapter 2.

The amount of drug in the bloodstream $t$ hours after an IV infusion begins can be modeled using a limited growth function, as follows.

$$A(t) = \frac{r}{-k}(1 - e^{kt}),$$

where $r$ is the rate of infusion per hour and $k$ is the exponential decay constant for the drug.

### EXAMPLE 2 IV Infusion

The same drug used in Example 1 is given to a patient by IV infusion at a drip rate of 6 mg/hr. Recall that the half-life of this drug is 4 hours.

**(a)** Find a model for the amount of drug in the bloodstream $t$ hours after the IV infusion begins.

**SOLUTION** Since $r = 6$ and $k = -0.17$, the function has the form

$$A(t) = 35(1 - e^{-0.17t}).$$

The graph of $A(t)$ is given in Figure 56.

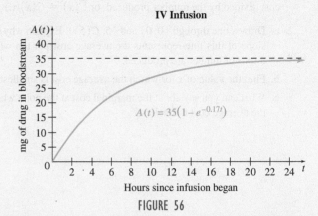

**IV Infusion**

$A(t) = 35(1 - e^{-0.17t})$

Hours since infusion began

**FIGURE 56**

**(b)** Find the average rate of change of drug in the bloodstream between $t = 0$ and $t = 2$. Repeat for $t = 4$ and $t = 6$.

**SOLUTION** The average rate of change from $t = 0$ to $t = 2$ is

$$\frac{A(2) - A(0)}{2 - 0} \approx \frac{10 - 0}{2} = 5 \text{ mg/hr.}$$

The average rate of change from $t = 4$ to $t = 6$ is

$$\frac{A(6) - A(4)}{6 - 4} \approx \frac{22 - 17}{2} = 2.5 \text{ mg/hr.}$$

Recall that the average rate of change from $t = 0$ to $t = 2$ for the rapid injection of this drug was $-5$ mg/hr and the average rate of change from $t = 4$ to $t = 6$ was $-2.5$ mg/hr. In fact, at any given time, the rapid injection function is decreasing at the same rate the IV infusion function is increasing.

**(c)** What happens to the amount of drug in the bloodstream as $t$ increases? (i.e., What is the limit of the function as $t$ approaches $\infty$?)

**SOLUTION** Looking at the graph of $A(t)$ in Figure 56 and the formula for $A(t)$ in part (a), we can see that

$$\lim_{t \to \infty} A(t) = 35.$$

An advantage of an IV infusion is that a dose can be given such that the limit of $A(t)$ as $t$ approaches $\infty$ is an effective level. Once the amount of drug has reached this effective level, it will remain there as long as the infusion continues. However, using this method of administration, it may take a while for the amount of drug in the body to reach an effective level. For our example, the effective level is between 30 mg and 40 mg. Looking at the graph, you can see that it takes about 11 hours to reach an effective level. If this patient were experiencing dangerously high blood pressure, you wouldn't want to wait 11 hours for the drug to reach an effective level.

## SINGLE RAPID INJECTION FOLLOWED BY AN INTRAVENOUS INFUSION

Giving a patient a single rapid injection immediately followed by an intravenous infusion allows a patient to experience the advantages of both methods. The single rapid injection immediately produces an effective drug level in the patient's bloodstream. While the amount of drug in the bloodstream from the rapid infusion is decreasing, the amount of drug in the system from the IV infusion is increasing.

The amount of drug in the bloodstream $t$ hours after the injection is given and infusion has started can be calculated by finding the sum of the two models.

$$A(t) = De^{kt} + \frac{r}{-k}(1 - e^{kt})$$

## EXAMPLE 3  Combination Model

A 35-mg dose of labetalol is administered to a patient by rapid injection. Immediately thereafter, the patient is given an IV infusion at a drip rate of 6 mg/hr. Find a model for the amount of drug in the bloodstream $t$ hours after the drug is administered.

**SOLUTION**  Recall from Example 1, the amount of drug in the bloodstream $t$ hours after the rapid injection was found to be

$$A(t) = 35e^{-0.17t}.$$

From Example 2, the amount of drug in the bloodstream $t$ hours after the IV infusion began was found to be

$$A(t) = 35(1 - e^{-0.17t}).$$

Therefore, $t$ hours after administering both the rapid injection and the IV infusion, the amount of drug in the bloodstream is

$$A(t) = 35e^{-0.17t} + 35(1 - e^{-0.17t})$$
$$= 35 \text{ mg}.$$

The graph of $A(t)$ is given in Figure 57.

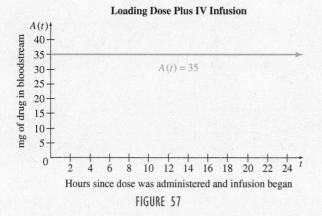

**Loading Dose Plus IV Infusion**

Hours since dose was administered and infusion began

FIGURE 57

Notice that the constant multiple of the rapid injection function, 35, is equal to the constant multiple of the IV infusion function. When this is the case, the sum of the two functions will be that constant.

## EXAMPLE 4  Combination Model

A drug with a half-life of 3 hours is found to be effective when the amount of drug in the bloodstream is 58 mg. A 58-mg loading dose is given by rapid injection followed by an IV infusion. What

should the rate of infusion be to maintain this level of drug in the bloodstream?

**SOLUTION**  Recall that the amount of drug in the bloodstream $t$ hours after both a rapid injection and IV infusion are administered is given by

$$A(t) = De^{kt} + \frac{r}{-k}(1 - e^{kt}).$$

The rapid injection dose, $D$, is 58 mg. The half-life of the drug is three hours; therefore,

$$k = -\frac{\ln 2}{3} \approx -0.23.$$

It follows that

$$A(t) = 58e^{-0.23t} + \frac{r}{0.23}(1 - e^{-0.23t}).$$

Since we want the sum of the rapid injection function and the IV infusion function to be 58 mg, it follows that

$$\frac{r}{0.23} = 58.$$

Solving for $r$, we get

$$r = 13.34 \text{ mg/hr.}$$

## EXERCISES

1. A 500-mg dose of a drug is administered by rapid injection to a patient. The half-life of the drug is 9 hours.

   a. Find a model for the amount of drug in the bloodstream $t$ hours after the drug is administered.

   b. Find the average rate of change of drug in the bloodstream between $t = 0$ and $t = 2$. Repeat for $t = 9$ and $t = 11$.

2. A drug is given to a patient by IV infusion at a drip rate of 350 mg/hr. The half-life of this drug is 3 hours.

   a. Find a model of the amount of drug in the bloodstream $t$ hours after the IV infusion begins.

   b. Find the average rate of change of drug in the bloodstream between $t = 0$ and $t = 3$. Repeat for $t = 3$ and $t = 6$.

3. A drug with a half-life of 9 hours is found to be effective when the amount of drug in the bloodstream is 250 mg. A 250-mg loading dose is given by rapid injection followed by an IV infusion. What should the rate of infusion be to maintain this level of drug in the bloodstream?

4. Use the table feature on a graphing calculator or a spreadsheet to develop a table that shows how much of the drug is present in a patient's system at the end of each 1/2 hour time interval for 24 hours for the model found in Exercise 1a. A chart such as this provides the health care worker with immediate information about patient drug levels.

**5.** Use the table feature on a graphing calculator or a spreadsheet to develop a table that shows how much of the drug is present in a patient's system at the end of each 1/2 hour time interval for 10 hours for the model found in Exercise 2a. A chart such as this provides the healthcare worker with immediate information about patient drug levels.

**6.** Use the table feature on a graphing calculator or a spreadsheet to develop a table that shows how much of the drug is present in a patient's system at the end of each 1/2 hour time interval for 10 hours for the model found in Exercise 3. Are your results surprising?

## DIRECTIONS FOR GROUP PROJECT

*Choose a drug that is commonly prescribed by physicians for a common ailment. Develop an analysis for this drug that is similar to the analysis for labetalol in Examples 1 through 3. You can obtain information on the drug from the Internet or from advertisements found in various media. Once you complete the analysis, prepare a professional presentation that can be delivered at a public forum. The presentation should summarize the facts presented in this extended application but at a level that is understandable to a typical layperson.*

# 4 Calculating the Derivative

4.1    Techniques for Finding Derivatives

4.2    Derivatives of Products and Quotients

4.3    The Chain Rule

4.4    Derivatives of Exponential Functions

4.5    Derivatives of Logarithmic Functions

4.6    Derivatives of Trigonometric Functions

Chapter 4 Review

Extended Application: Managing Renewable Resources

By differentiating the function defining a mathematical model we can see how the model's output changes with the input. In an exercise in Section 2 we explore a rational-function model for the length of the rest period needed to recover from vigorous exercise such as riding a bike. The derivative indicates how the rest required changes with the work expended in kilocalories per minute.

In the previous chapter, we found the derivative to be a useful tool for describing the rate of change, velocity, and the slope of a curve. Taking the derivative by using the definition, however, can be difficult. To take full advantage of the power of the derivative, we need faster ways of calculating the derivative. That is the goal of this chapter.

# 4.1  Techniques for Finding Derivatives

**APPLY IT**  How fast is the number of Americans who are expected to be over 100 years old growing? What is the rate of change of aortic pressure after a surgical procedure?

*These questions can be answered by finding the derivative of an appropriate function. We shall return to them at the end of this section in Examples 6 and 7.*

Using the definition to calculate the derivative of a function is a very involved process even for simple functions. In this section we develop rules that make the calculation of derivatives much easier. Keep in mind that even though the process of finding a derivative will be greatly simplified with these rules, *the interpretation of the derivative will not change.* But first, a few words about notation are in order.

In addition to $f'(x)$, there are several other commonly used notations for the derivative.

**Notations for the Derivative**

The derivative of $y = f(x)$ may be written in any of the following ways:

$$f'(x), \qquad \frac{dy}{dx}, \qquad \frac{d}{dx}[f(x)], \qquad \text{or} \qquad D_x[f(x)].$$

The $dy/dx$ notation for the derivative (read "the derivative of $y$ with respect to $x$") is sometimes referred to as *Leibniz notation*, named after one of the co-inventors of calculus, Gottfried Wilhelm von Leibniz (1646–1716). (The other was Sir Isaac Newton, 1642–1727.)

With the above notations, the derivative of $y = f(x) = 2x^3 + 4x$, for example, which was found in Example 5 of Section 3.4 to be $f'(x) = 6x^2 + 4$, would be written

$$\frac{dy}{dx} = 6x^2 + 4$$

$$\frac{d}{dx}(2x^3 + 4x) = 6x^2 + 4$$

$$D_x(2x^3 + 4x) = 6x^2 + 4.$$

A variable other than $x$ is often used as the independent variable. For example, if $y = f(t)$ gives population growth as a function of time, then the derivative of $y$ with respect to $t$ could be written

$$f'(t), \qquad \frac{dy}{dt}, \qquad \frac{d}{dt}[f(t)], \qquad \text{or} \qquad D_t[f(t)].$$

Other variables also may be used to name the function, as in $g(x)$ or $h(t)$.

Now we will use the definition

$$f'(x) = \lim_{h \to 0} \frac{f(x+h) - f(x)}{h}$$

to develop some rules for finding derivatives more easily than by the four-step process given in the previous chapter.

The first rule tells how to find the derivative of a constant function defined by $f(x) = k$, where $k$ is a constant real number. Since $f(x + h)$ is also $k$, by definition $f'(x)$ is

$$f'(x) = \lim_{h \to 0} \frac{f(x+h) - f(x)}{h}$$

$$= \lim_{h \to 0} \frac{k - k}{h} = \lim_{h \to 0} \frac{0}{h} = \lim_{h \to 0} 0 = 0,$$

establishing the following rule.

## Constant Rule

If $f(x) = k$, where $k$ is any real number, then

$$f'(x) = 0.$$

(The derivative of a constant is 0.)

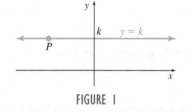

FIGURE 1

This rule is logical because the derivative represents rate of change, and a constant function, by definition, does not change. Figure 1 illustrates this constant rule geometrically; it shows a graph of the horizontal line $y = k$. At any point $P$ on this line, the tangent line at $P$ is the line itself. Since a horizontal line has a slope of 0, the slope of the tangent line is 0. This agrees with the result above: The derivative of a constant is 0.

**EXAMPLE 1** **Derivative of a Constant**

(a) If $f(x) = 9$, then $f'(x) = 0$.

(b) If $h(t) = \pi$, then $D_t[h(t)] = 0$.

(c) If $y = 2^3$, then $dy/dx = 0$.

Functions of the form $y = x^n$, where $n$ is a fixed real number, are very common in applications. To obtain a rule for finding the derivative of such a function, we can use the definition to work out the derivatives for various special values of $n$. This was done in Section 3.4 in Example 4 to show that for $f(x) = x^2$, $f'(x) = 2x$.

For $f(x) = x^3$, the derivative is found as follows.

$$f'(x) = \lim_{h \to 0} \frac{f(x+h) - f(x)}{h}$$

$$= \lim_{h \to 0} \frac{(x+h)^3 - x^3}{h}$$

$$= \lim_{h \to 0} \frac{(x^3 + 3x^2h + 3xh^2 + h^3) - x^3}{h}$$

The binomial theorem (discussed in most intermediate and college algebra texts) was used to expand $(x + h)^3$ in the last step. Now, the limit can be determined.

$$f'(x) = \lim_{h \to 0} \frac{3x^2h + 3xh^2 + h^3}{h}$$

$$= \lim_{h \to 0} (3x^2 + 3xh + h^2)$$

$$= 3x^2$$

The results in the following table were found in a similar way, using the definition of the derivative. (These results are modifications of some of the examples and exercises from the previous chapter.)

| Derivative of $f(x) = x^n$ | | |
|---|---|---|
| **Function** | $n$ | **Derivative** |
| $f(x) = x$ | 1 | $f'(x) = 1 = 1x^0$ |
| $f(x) = x^2$ | 2 | $f'(x) = 2x = 2x^1$ |
| $f(x) = x^3$ | 3 | $f'(x) = 3x^2$ |
| $f(x) = x^4$ | 4 | $f'(x) = 4x^3$ |
| $f(x) = x^{-1}$ | $-1$ | $f'(x) = -1 \cdot x^{-2} = \dfrac{-1}{x^2}$ |
| $f(x) = x^{1/2}$ | 1/2 | $f'(x) = \dfrac{1}{2}x^{-1/2} = \dfrac{1}{2x^{1/2}}$ |

These results suggest the following rule.

## Power Rule

If $f(x) = x^n$ for any real number $n$, then

$$f'(x) = nx^{n-1}.$$

(The derivative of $f(x) = x^n$ is found by multiplying by the exponent $n$ and decreasing the exponent on $x$ by 1.)

While the power rule is true for every real-number value of $n$, a proof is given here only for positive integer values of $n$. This proof follows the steps used above in finding the derivative of $f(x) = x^3$.

For any real numbers $p$ and $q$, by the binomial theorem,

$$(p + q)^n = p^n + np^{n-1}q + \frac{n(n-1)}{2}p^{n-2}q^2 + \cdots + npq^{n-1} + q^n.$$

Replacing $p$ with $x$ and $q$ with $h$ gives

$$(x + h)^n = x^n + nx^{n-1}h + \frac{n(n-1)}{2}x^{n-2}h^2 + \cdots + nxh^{n-1} + h^n,$$

from which

$$(x + h)^n - x^n = nx^{n-1}h + \frac{n(n-1)}{2}x^{n-2}h^2 + \cdots + nxh^{n-1} + h^n.$$

Dividing each term by $h$ yields

$$\frac{(x + h)^n - x^n}{h} = nx^{n-1} + \frac{n(n-1)}{2}x^{n-2}h + \cdots + nxh^{n-2} + h^{n-1}.$$

Use the definition of derivative, and the fact that each term except the first contains $h$ as a factor and thus approaches 0 as $h$ approaches 0, to get

$$f'(x) = \lim_{h \to 0} \frac{(x + h)^n - x^n}{h}$$

$$= nx^{n-1} + \frac{n(n-1)}{2}x^{n-2} \cdot 0 + \cdots + nx \cdot 0^{n-2} + 0^{n-1}$$

$$= nx^{n-1}.$$

This shows that the derivative of $f(x) = x^n$ is $f'(x) = nx^{n-1}$, proving the power rule for positive integer values of $n$.

### EXAMPLE 2  Power Rule

**(a)** If $f(x) = x^6$, find $f'(x)$.

**SOLUTION**   $f'(x) = 6x^{6-1} = 6x^5$

**(b)** If $y = t = t^1$, find $\dfrac{dy}{dt}$.

**SOLUTION**   $\dfrac{dy}{dt} = 1t^{1-1} = t^0 = 1$

**(c)** If $y = 1/x^3$, find $dy/dx$.

**SOLUTION**   Use a negative exponent to rewrite this equation as $y = x^{-3}$; then

$$\frac{dy}{dx} = -3x^{-3-1} = -3x^{-4} \qquad \text{or} \qquad \frac{-3}{x^4}.$$

**(d)** Find $D_x(x^{4/3})$.

**SOLUTION**   $D_x(x^{4/3}) = \dfrac{4}{3}x^{4/3-1} = \dfrac{4}{3}x^{1/3}$

**(e)** If $y = \sqrt{z}$, find $dy/dz$.

**SOLUTION**   Rewrite this as $y = z^{1/2}$; then

$$\frac{dy}{dz} = \frac{1}{2}z^{1/2-1} = \frac{1}{2}z^{-1/2} \qquad \text{or} \qquad \frac{1}{2z^{1/2}} \qquad \text{or} \qquad \frac{1}{2\sqrt{z}}.$$

**TRY YOUR TURN 1**

**FOR REVIEW**

At this point you may wish to turn back to Sections R.6 and R.7 for a review of negative exponents and rational exponents. The relationship between powers, roots, and rational exponents is explained there.

**YOUR TURN 1**   If $f(t) = \dfrac{1}{\sqrt{t}}$, find $f'(t)$.

The next rule shows how to find the derivative of the product of a constant and a function.

### Constant Times a Function

Let $k$ be a real number. If $g'(x)$ exists, then the derivative of $f(x) = k \cdot g(x)$ is

$$f'(x) = k \cdot g'(x).$$

(The derivative of a constant times a function is the constant times the derivative of the function.)

This rule is proved with the definition of the derivative and rules for limits.

$$
\begin{aligned}
f'(x) &= \lim_{h \to 0} \frac{kg(x+h) - kg(x)}{h} \\
&= \lim_{h \to 0} k\frac{[g(x+h) - g(x)]}{h} \qquad \text{Factor out } k. \\
&= k \lim_{h \to 0} \frac{g(x+h) - g(x)}{h} \qquad \text{Limit rule 1} \\
&= k \cdot g'(x) \qquad \text{Definition of derivative}
\end{aligned}
$$

### EXAMPLE 3  Derivative of a Constant Times a Function

**(a)** If $y = 8x^4$, find $\dfrac{dy}{dx}$.

**SOLUTION**   $\dfrac{dy}{dx} = 8(4x^3) = 32x^3$

(b) If $y = -\dfrac{3}{4}x^{12}$, find $dy/dx$.

**SOLUTION** $\dfrac{dy}{dx} = -\dfrac{3}{4}(12x^{11}) = -9x^{11}$

(c) Find $D_t(-8t)$.

**SOLUTION** $D_t(-8t) = -8(1) = -8$

(d) Find $D_p(10p^{3/2})$.

**SOLUTION** $D_p(10p^{3/2}) = 10\left(\dfrac{3}{2}p^{1/2}\right) = 15p^{1/2}$

(e) If $y = \dfrac{6}{x}$, find $\dfrac{dy}{dx}$.

**SOLUTION** Rewrite this as $y = 6x^{-1}$; then

$$\dfrac{dy}{dx} = 6(-1x^{-2}) = -6x^{-2} \qquad \text{or} \qquad \dfrac{-6}{x^2}.$$

**YOUR TURN 2** If $y = 3\sqrt{x}$, find $dy/dx$.

TRY YOUR TURN 2

**EXAMPLE 4** Beagles

Researchers have determined that the daily energy requirements of female beagles who are at least 1 year old change with respect to age according to the function

$$E(t) = 753t^{-0.1321},$$

where $E(t)$ is the daily energy requirements (in kJ/W$^{0.67}$) for a dog that is $t$ years old. *Source: Journal of Nutrition.*

(a) Graph this function. As a female beagle gets older, how do the daily energy requirements change? What happens to the function for very young animals?

**SOLUTION** As shown in Figure 2, the daily energy requirements are decreasing with respect to time. This indicates that older dogs do not have the same energy requirements as younger ones. The function approaches infinity for newborn beagles. This indicates that the function should not be used to predict energy requirements of very young beagles.

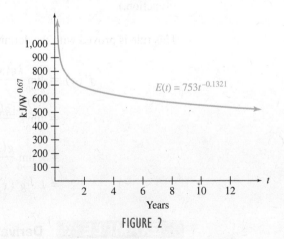

FIGURE 2

(b) Find $E'(t)$.

**SOLUTION** Using the rules of differentiation we find that

$$E'(t) = 753(-0.1321)t^{-0.1321-1} = -99.4713t^{-1.1321}.$$

**(c)** Determine the rate of change of the daily energy requirements of a 2-year-old female beagle.

**SOLUTION** $E'(2) = -99.4713(2)^{-0.1.1321} \approx -45.4$

Thus, the daily energy requirements of a 2-year-old female beagle are decreasing at the rate of 45.4 kJ/$W^{0.67}$ per year. ▄

The final rule in this section is for the derivative of a function that is a sum or difference of terms.

## Sum or Difference Rule

If $f(x) = u(x) \pm v(x)$, and if $u'(x)$ and $v'(x)$ exist, then

$$f'(x) = u'(x) \pm v'(x).$$

(The derivative of a sum or difference of functions is the sum or difference of the derivatives.)

The proof of the sum part of this rule is as follows: If $f(x) = u(x) + v(x)$, then

$$f'(x) = \lim_{h \to 0} \frac{[u(x + h) + v(x + h)] - [u(x) + v(x)]}{h}$$

$$= \lim_{h \to 0} \frac{[u(x + h) - u(x)] + [v(x + h) - v(x)]}{h}$$

$$= \lim_{h \to 0} \left[ \frac{u(x + h) - u(x)}{h} + \frac{v(x + h) - v(x)}{h} \right]$$

$$= \lim_{h \to 0} \frac{u(x + h) - u(x)}{h} + \lim_{h \to 0} \frac{v(x + h) - v(x)}{h}$$

$$= u'(x) + v'(x).$$

A similar proof can be given for the difference of two functions.

### EXAMPLE 5   Derivative of a Sum

Find the derivative of each function.

**(a)** $y = 6x^3 + 15x^2$

**SOLUTION** Let $u(x) = 6x^3$ and $v(x) = 15x^2$; then $y = u(x) + v(x)$. Since $u'(x) = 18x^2$ and $v'(x) = 30x$,

$$\frac{dy}{dx} = 18x^2 + 30x.$$

**(b)** $p(t) = 12t^4 - 6\sqrt{t} + \dfrac{5}{t}$

**SOLUTION** Rewrite $p(t)$ as $p(t) = 12t^4 - 6t^{1/2} + 5t^{-1}$; then

$$p'(t) = 48t^3 - 3t^{-1/2} - 5t^{-2}.$$

Also, $p'(t)$ may be written as $p'(t) = 48t^3 - \dfrac{3}{\sqrt{t}} - \dfrac{5}{t^2}$.

**(c)** $f(x) = \dfrac{x^3 + 3\sqrt{x}}{x}$

**SOLUTION** Rewrite $f(x)$ as $f(x) = \dfrac{x^3}{x} + \dfrac{3x^{1/2}}{x} = x^2 + 3x^{-1/2}$. Then

$$D_x[f(x)] = 2x - \frac{3}{2}x^{-3/2},$$

or

$$D_x[f(x)] = 2x - \frac{3}{2\sqrt{x^3}}.$$

**YOUR TURN 3** If
$$h(t) = -3t^2 + 2\sqrt{t} + \frac{5}{t^4} - 7,$$
find $h'(t)$.

**(d)** $f(x) = (4x^2 - 3x)^2$

**SOLUTION** Rewrite $f(x)$ as $f(x) = 16x^4 - 24x^3 + 9x^2$ using the fact that $(a - b)^2 = a^2 - 2ab + b^2$; then

$$f'(x) = 64x^3 - 72x^2 + 18x.$$   **TRY YOUR TURN 3**

---

**TECHNOLOGY NOTE**

Some computer programs and calculators have built-in methods for taking derivatives symbolically, which is what we have been doing in this section, as opposed to approximating the derivative numerically by using a small number for $h$ in the definition of the derivative. In the computer program Maple, we would do part (a) of Example 5 by entering

> diff(6*x^3+15*x^2,x);

where the $x$ after the comma indicates the derivative is with respect to the variable $x$. Maple would respond with

18*x^2+30*x.

Similarly, on the TI-89, we would enter d(6x^3+15x^2,x) and the calculator would give "18·x^2+30·x."

Other graphing calculators, such as the TI-84 Plus, do not have built-in methods for taking derivatives symbolically. As we saw in the last chapter, however, they do have the ability to calculate the derivative of a function at a particular point and to simultaneously graph a function and its derivative.

Recall that, on the TI-84 Plus, we could use the nDeriv Command, as shown in Figure 3, to approximate the value of the derivative when $x = 1$. Figure 4(a) and Figure 4(b) indicate how to input the functions into the calculator and the corresponding graphs of both the function and its derivative. Consult the *Graphing Calculator and Excel Spreadsheet Manual*, available with this book, for assistance.

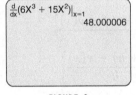

FIGURE 3

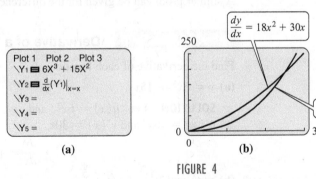

**(a)**   **(b)**

FIGURE 4

---

The rules developed in this section make it possible to find the derivative of a function more directly, so that applications of the derivative can be dealt with more effectively.

**EXAMPLE 6**   **Centenarians**

The number of Americans (in thousands) who are expected to be over 100 years old can be approximated by the function

$$f(t) = 0.00943t^3 - 0.470t^2 + 11.085t + 23.441,$$

where $t$ is the year, with $t = 0$ corresponding to 2000, and $0 \le t \le 50$. *Source: U.S. Census Bureau.*

APPLY IT

**(a)** Find a formula giving the rate of change of the number of Americans over 100 years old.

**SOLUTION** Using the techniques for finding the derivative, we have

$$f'(t) = 0.02829t^2 - 0.940t + 11.085.$$

This tells us the rate of change in the number of Americans over 100 years old.

**(b)** Find the rate of change in the number of Americans who are expected to be over 100 years old in the year 2020.

**SOLUTION** The year 2020 corresponds to $t = 20$.

$$f'(20) = 0.02829(20)^2 - 0.940(20) + 11.085 = 3.601$$

The number of Americans over 100 years old is expected to grow at a rate of about 3.6 thousand, or about 3600, per year in the year 2020.

---

**EXAMPLE 7** **Cardiology**

The aortic pressure–diameter relation in a single patient who underwent cardiac catheterization can be represented by

$$D(p) = 0.000002p^3 - 0.0008p^2 + 0.1141p + 16.683, \quad 55 \leq p \leq 130,$$

where $D(p)$ is the aortic diameter (in mm) and $p$ is the aortic pressure (in mm Hg). *Source: Circulation.*

APPLY IT

**(a)** Find a formula that gives the rate of change of aortic diameter with respect to aortic pressure.

**SOLUTION** Using the techniques for finding the derivative, we have

$$D'(p) = 0.000006p^2 - 0.0016p + 0.1141.$$

This tells us that the rate of change in aortic diameter with respect to pressure is expected to follow a quadratic function.

**(b)** Find the rate of change in the aortic diameter when the pressure is 100 mm Hg.

**SOLUTION** A pressure of 100 mm Hg corresponds to $p = 100$.

$$D'(100) = 0.000006(100)^2 - 0.0016(100) + 0.1141 = 0.0141$$

Thus the aortic diameter is expected to increase at a rate of 0.0141 mm per mm Hg when the pressure is 100 mm Hg.

---

# 4.1 EXERCISES

**Find the derivative of each function defined as follows.**

**1.** $y = 12x^3 - 8x^2 + 7x + 5$

**2.** $y = 8x^3 - 5x^2 - \dfrac{x}{12}$

**3.** $y = 3x^4 - 6x^3 + \dfrac{x^2}{8} + 5$

**4.** $y = 5x^4 + 9x^3 + 12x^2 - 7x$

**5.** $f(x) = 6x^{3.5} - 10x^{0.5}$

**6.** $f(x) = -2x^{1.5} + 12x^{0.5}$

**7.** $y = 8\sqrt{x} + 6x^{3/4}$

**8.** $y = -100\sqrt{x} - 11x^{2/3}$

**9.** $y = 10x^{-3} + 5x^{-4} - 8x$

**10.** $y = 5x^{-5} - 6x^{-2} + 13x^{-1}$

**11.** $f(t) = \dfrac{7}{t} - \dfrac{5}{t^3}$

**12.** $f(t) = \dfrac{14}{t} + \dfrac{12}{t^4} + \sqrt{2}$

**13.** $y = \dfrac{6}{x^4} - \dfrac{7}{x^3} + \dfrac{3}{x} + \sqrt{5}$

**14.** $y = \dfrac{3}{x^6} + \dfrac{1}{x^5} - \dfrac{7}{x^2}$

**15.** $p(x) = -10x^{-1/2} + 8x^{-3/2}$

**16.** $h(x) = x^{-1/2} - 14x^{-3/2}$

**17.** $y = \dfrac{6}{\sqrt[4]{x}}$

**18.** $y = \dfrac{-2}{\sqrt[3]{x}}$

**19.** $f(x) = \dfrac{x^3 + 5}{x}$

**20.** $g(x) = \dfrac{x^3 - 4x}{\sqrt{x}}$

**21.** $g(x) = (8x^2 - 4x)^2$

**22.** $h(x) = (x^2 - 1)^3$

**23.** Which of the following describes the derivative function $f'(x)$ of a quadratic function $f(x)$?

    **a.** Quadratic          **b.** Linear

    **c.** Constant          **d.** Cubic (third degree)

**24.** Which of the following describes the derivative function $f'(x)$ of a cubic (third-degree) function $f(x)$?

    **a.** Quadratic          **b.** Linear

    **c.** Constant          **d.** Cubic

**25.** Explain the relationship between the slope and the derivative of $f(x)$ at $x = a$.

**26.** Which of the following do *not* equal $\dfrac{d}{dx}(4x^3 - 6x^{-2})$?

    **a.** $\dfrac{12x^2 + 12}{x^3}$        **b.** $\dfrac{12x^5 + 12}{x^3}$

    **c.** $12x^2 + \dfrac{12}{x^3}$        **d.** $12x^3 + 12x^{-3}$

**Find each derivative.**

**27.** $D_x\left[9x^{-1/2} + \dfrac{2}{x^{3/2}}\right]$      **28.** $D_x\left[\dfrac{8}{\sqrt[4]{x}} - \dfrac{3}{\sqrt{x^3}}\right]$

**29.** $f'(-2)$   if   $f(x) = \dfrac{x^4}{6} - 3x$

**30.** $f'(3)$   if   $f(x) = \dfrac{x^3}{9} - 7x^2$

**In Exercises 31–34, find the slope of the tangent line to the graph of the given function at the given value of $x$. Find the equation of the tangent line in Exercises 31 and 32.**

**31.** $y = x^4 - 5x^3 + 2$;   $x = 2$

**32.** $y = -3x^5 - 8x^3 + 4x^2$;   $x = 1$

**33.** $y = -2x^{1/2} + x^{3/2}$;   $x = 9$

**34.** $y = -x^{-3} + x^{-2}$;   $x = 2$

**35.** Find all points on the graph of $f(x) = 9x^2 - 8x + 4$ where the slope of the tangent line is 0.

**36.** Find all points on the graph of $f(x) = x^3 + 9x^2 + 19x - 10$ where the slope of the tangent line is $-5$.

**In Exercises 37–40, for each function find all values of $x$ where the tangent line is horizontal.**

**37.** $f(x) = 2x^3 + 9x^2 - 60x + 4$

**38.** $f(x) = x^3 + 15x^2 + 63x - 10$

**39.** $f(x) = x^3 - 4x^2 - 7x + 8$

**40.** $f(x) = x^3 - 5x^2 + 6x + 3$

**41.** At what points on the graph of $f(x) = 6x^2 + 4x - 9$ is the slope of the tangent line $-2$?

**42.** At what point on the graph of $f(x) = -5x^2 + 4x - 2$ is the slope of the tangent line 6?

**43.** At what points on the graph of $f(x) = 2x^3 - 9x^2 - 12x + 5$ is the slope of the tangent line 12?

**44.** At what points on the graph of $f(x) = x^3 + 6x^2 + 21x + 2$ is the slope of the tangent line 9?

**45.** If $g'(5) = 12$ and $h'(5) = -3$, find $f'(5)$ for $f(x) = 3g(x) - 2h(x) + 3$.

**46.** If $g'(2) = 7$ and $h'(2) = 14$, find $f'(2)$ for $f(x) = \dfrac{1}{2}g(x) + \dfrac{1}{4}h(x)$.

**47.** Use the information given in the figure to find the following values.

    **a.** $f(1)$     **b.** $f'(1)$     **c.** The domain of $f$     **d.** The range of $f$

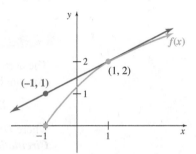

**48.** In Exercises 43–46 of Section 1.4, the effect of $a$ when graphing $y = af(x)$ was discussed. Now describe how this relates to the fact that $D_x[af(x)] = af'(x)$.

**49.** Show that, for any constant $k$,

$$\frac{d}{dx}\left[\frac{f(x)}{k}\right] = \frac{f'(x)}{k}.$$

**50.** Use the differentiation feature on your graphing calculator to solve the problems (to 2 decimal places) below, where $f(x)$ is defined as follows:

$$f(x) = 1.25x^3 + 0.01x^2 - 2.9x + 1.$$

    **a.** Find $f'(4)$.

    **b.** Find all values of $x$ where $f'(x) = 0$.

## LIFE SCIENCE APPLICATIONS

**51. Cancer** Insulation workers who were exposed to asbestos and employed before 1960 experienced an increased likelihood of lung cancer. If a group of insulation workers has a cumulative total of 100,000 years of work experience with their first date of employment $t$ years ago, then the number of lung cancer cases occurring within the group can be modeled using the function

$$N(t) = 0.00437t^{3.2}.$$

Find the rate of growth of the number of workers with lung cancer in a group as described by the following first dates of employment. *Source: Observation and Inference: An Introduction to the Methods of Epidemiology.*

    **a.** 5 years ago             **b.** 10 years ago

**52. Blood Sugar Level**  Insulin affects the glucose, or blood sugar, level of some diabetics according to the function

$$G(x) = -0.2x^2 + 450,$$

where $G(x)$ is the blood sugar level 1 hour after $x$ units of insulin are injected. (This mathematical model is only approximate, and it is valid only for values of $x$ less than about 40.) Find the blood sugar level after the following numbers of units of insulin are injected.

**a.** 0                     **b.** 25

Find the rate of change of blood sugar level after injection of the following numbers of units of insulin.

**c.** 10                **d.** 25

**53. Insect Mating Patterns**  In an experiment testing methods of sexually attracting male insects to sterile females, equal numbers of males and females of a certain species are permitted to intermingle. (Such an experiment tests ways to control insect populations.) Assume that

$$M(t) = 4t^{3/2} + 2t^{1/2}$$

approximates the number of matings observed among the insects in an hour, where $t$ is the temperature in degrees Celsius. (This formula is valid only for certain temperature ranges.) Find the number of matings at the following temperatures.

**a.** 16°C              **b.** 25°C

**c.** Find the rate of change in the number of matings when the temperature is 16°C.

**54. Track and Field**  In 1906 Kennelly developed a simple formula for predicting an upper limit on the fastest time that humans could ever run distances from 100 yards to 10 miles. His formula is given by

$$t = 0.0588s^{1.125},$$

where $s$ is the distance in meters and $t$ is the time to run that distance in seconds. *Source: Proceedings of the American Academy of Arts and Sciences.*

**a.** Find Kennelly's estimate for the fastest mile. (*Hint:* 1 mile ≈ 1609 meters.)

**b.** Find $dt/ds$ when $s = 100$ and interpret your answer.

**c.** Compare this and other estimates to the current world records. Have these estimates been surpassed?

**55. Bighorn Sheep**  The cumulative horn volume for certain types of bighorn rams, found in the Rocky Mountains, can be described by the quadratic function

$$V(t) = -2159 + 1313t - 60.82t^2,$$

where $V(t)$ is the horn volume (in cm³) and $t$ is the year of growth, $2 \leq t \leq 9$. *Source: Conservation Biology.*

**a.** Find the horn volume for a 3-year-old ram.

**b.** Find the rate at which the horn volume of a 3-year-old ram is changing.

**56. Brain Mass**  The brain mass of a human fetus during the last trimester can be accurately estimated from the circumference of the head by

$$m(c) = \frac{c^3}{100} - \frac{1500}{c},$$

where $m(c)$ is the mass of the brain (in grams) and $c$ is the circumference (in centimeters) of the head. *Source: Early Human Development.*

**a.** Estimate the brain mass of a fetus that has a head circumference of 30 cm.

**b.** Find the rate of change of the brain mass for a fetus that has a head circumference of 30 cm and interpret your results.

**57. Human Cough**  To increase the velocity of the air flowing through the trachea when a human coughs, the body contracts the windpipe, producing a more effective cough. Tuchinsky formulated that the velocity of air that is flowing through the trachea during a cough is

$$V = C(R_0 - R)R^2,$$

where $C$ is a constant based on individual body characteristics, $R_0$ is the radius of the windpipe before the cough, and $R$ is the radius of the windpipe during the cough. It can be shown that the maximum velocity of the cough occurs when $dV/dR = 0$. Find the value of $R$ that maximizes the velocity.* *Source: COMAP, Inc.*

**58. Body Mass Index**  The body mass index (BMI) is a number that can be calculated for any individual as follows: Multiply weight by 703 and divide by the person's height squared. That is,

$$BMI = \frac{703w}{h^2},$$

where $w$ is in pounds and $h$ is in inches. The National Heart, Lung, and Blood Institute uses the BMI to determine whether a person is "overweight" ($25 \leq BMI < 30$) or "obese" ($BMI \geq 30$). *Source: The National Institutes of Health.*

**a.** Calculate the BMI for Lebron James, basketball player for the Miami Heat, who is 250 lb and 6'8'' tall.

**b.** How much weight would Lebron James have to lose until he reaches a BMI of 24.9 and is no longer "overweight"? Comment on whether BMI cutoffs are appropriate for athletes with considerable muscle mass.

**c.** For a 125-lb female, what is the rate of change of BMI with respect to height? (*Hint:* Take the derivative of the function: $f(h) = 703(125)/h^2$.)

**d.** Calculate and interpret the meaning of $f'(65)$.

**e.** Use the TABLE feature on your graphing calculator to construct a table for BMI for various weights and heights.

---

*Interestingly, Tuchinsky also states that X-rays indicate that the body naturally contracts the windpipe to this radius during a cough.

**59. Heart** The left ventricular length (viewed from the front of the heart) of a human fetus that is at least 18 weeks old can be estimated by

$$l(t) = -2.318 + 0.2356t - 0.002674t^2,$$

where $l(t)$ is the ventricular length (in centimeters) and $t$ is the age (in weeks) of the fetus. **Source: American Journal of Cardiology.**

  **a.** Determine a meaningful domain for this function.

  **b.** Find $l'(t)$.

  **c.** Find $l'(25)$.

**60. Velocity of Marine Organism** The typical velocity (in centimeters per second) of a marine organism of length $l$ (in centimeters) is given by $v = 2.69l^{1.86}$. Find the rate of change of the velocity with respect to the length of the organism. **Source: Mathematical Topics in Population Biology Morphogenesis and Neurosciences.**

**61. Insect Species** The number of species of a particular type of insect depends on the size of the insect according to the formula $f(m) = 40m^{-0.6}$, where $m$ is the mass in grams. Find and interpret $f'(0.01)$. **Source: Science.**

**62. Bird Eggs** Using results from Exercise 10 in Section 1.2 and Exercise 40 in Section 8.3, we can approximate the volume (in cubic cm) of a bird egg of width $w$ (in cm) according to the formula

$$V = \frac{\pi(1.585w^3 - 0.487w^2)}{6}.$$

Find the rate of change of the volume with respect to the width for each of the following.

  **a.** A robin, whose egg has an average width of 1.5 cm

  **b.** A Canada goose, whose egg has an average width of 5.8 cm

**63. Popcorn** Researchers have determined that the amount of moisture present in a kernel of popcorn affects the volume of the popped corn and can be described for certain sizes of kernels by the function

$$v(x) = -35.98 + 12.09x - 0.4450x^2,$$

where $x$ is moisture content (percent, wet basis) and $v(x)$ is the expansion volume (in cm³/g). **Source: Cereal Chemistry.**

  **a.** Graph this function on [8, 20] by [40, 50].

  **b.** Determine the moisture content where the maximum volume occurs.

  **c.** What is the value of the derivative of the above function at this value?

**64. Dog's Human Age** From the data printed in the following table from the *Minneapolis Star Tribune* on September 20, 1998, a dog's age when compared to a human's age can be modeled using either a linear formula or a quadratic formula as follows:

$$y_1 = 4.13x + 14.63$$

$$y_2 = -0.033x^2 + 4.647x + 13.347,$$

where $y_1$ and $y_2$ represent a dog's human age for each formula and $x$ represents a dog's actual age. **Source: Mathematics Teacher.**

| Dog Age | Human Age |
| --- | --- |
| 1 | 16 |
| 2 | 24 |
| 3 | 28 |
| 5 | 36 |
| 7 | 44 |
| 9 | 52 |
| 11 | 60 |
| 13 | 68 |
| 15 | 76 |

  **a.** Find $y_1$ and $y_2$ when $x = 5$.

  **b.** Find $dy_1/dx$ and $dy_2/dx$ when $x = 5$ and interpret your answers.

  **c.** If the first two points are eliminated from the table, find the equation of a line that perfectly fits the reduced set of data. Interpret your findings.

  **d.** Of the three formulas, which do you prefer?

**65. Tree Growth** Researchers have found that the height (in cm) of a tree as a function of its diameter can be given by

$$H = 137 + aD - bD^2,$$

where $a$ and $b$ are constants that depend on the type of tree. **Source: Journal of Ecology.** When the diameter reaches its maximum value $D_{max}$, the height reaches its maximum value $H_{max}$, and $dH/dD = 0$. Derive the following equations given by the researchers:

$$a = \frac{2(H_{max} - 137)}{D_{max}}; \quad b = \frac{H_{max} - 137}{D_{max}^2}.$$

# OTHER APPLICATIONS

**Velocity** We saw in the previous chapter that if a function $s(t)$ gives the position of an object at time $t$, the derivative gives the velocity; that is, $v(t) = s'(t)$. For each position function in Exercises 66–69, find (a) $v(t)$ and (b) the velocity when $t = 0, t = 5$, and $t = 10$.

**66.** $s(t) = 11t^2 + 4t + 2$

**67.** $s(t) = 18t^2 - 13t + 8$

**68.** $s(t) = 4t^3 + 8t^2 + t$

**69.** $s(t) = -3t^3 + 4t^2 - 10t + 5$

**70. Velocity** If a rock is dropped from a 144-ft building, its position (in feet above the ground) is given by $s(t) = -16t^2 + 144$, where $t$ is the time in seconds since it was dropped.

  **a.** What is its velocity 1 second after being dropped? 2 seconds after being dropped?

  **b.** When will it hit the ground?

  **c.** What is its velocity upon impact?

**71. Velocity** A ball is thrown vertically upward from the ground at a velocity of 64 ft per second. Its distance from the ground at $t$ seconds is given by $s(t) = -16t^2 + 64t$.

   **a.** How fast is the ball moving 2 seconds after being thrown? 3 seconds after being thrown?

   **b.** How long after the ball is thrown does it reach its maximum height?

   **c.** How high will it go?

**72. Dead Sea** Researchers who have been studying the alarming rate at which the level of the Dead Sea has been dropping have shown that the density $d(x)$ (in g per cm$^3$) of the Dead Sea brine during evaporation can be estimated by the function

$$d(x) = 1.66 - 0.90x + 0.47x^2,$$

where $x$ is the fraction of the remaining brine, $0 \le x \le 1$. *Source: Geology.*

   **a.** Estimate the density of the brine when 50% of the brine remains.

   **b.** Find and interpret the instantaneous rate of change of the density when 50% of the brine remains.

**73. Postal Rates** U.S. postal rates have steadily increased since 1932. Using data depicted in the table for the years 1932–2013, the cost in cents to mail a single letter can be modeled using a quadratic formula as follows:

$$C(t) = 0.007756t^2 - 0.04258t + 0.7585$$

where $t$ is the number of years since 1932. *Source: U.S. Postal Service.*

| Year | Cost | Year | Cost |
|------|------|------|------|
| 1932 | 3  | 1988 | 25 |
| 1958 | 4  | 1991 | 29 |
| 1963 | 5  | 1995 | 32 |
| 1968 | 6  | 1999 | 33 |
| 1971 | 8  | 2001 | 34 |
| 1974 | 10 | 2002 | 37 |
| 1975 | 13 | 2006 | 39 |
| 1978 | 15 | 2007 | 41 |
| 1981 | 18 | 2008 | 42 |
| 1981 | 20 | 2009 | 44 |
| 1985 | 22 | 2012 | 45 |
|      |    | 2013 | 46 |

   **a.** Find the predicted cost of mailing a letter in 1982 and 2002 and compare these estimates with the actual rates.

   **b.** Find the rate of change of the postage cost for the years 1982 and 2002 and interpret your results.

   **c.** Using the regression feature on a graphing calculator, find a cubic function that models these data, letting $t = 0$ correspond to the year 1932. Then use your answer to find the rate of change of the postage cost for the years 1982 and 2002.

   **d.** Discuss whether the quadratic or cubic function best describes the data. Do the answers from part b or from part c best describe the rate that postage was going up in the years 1982 and 2002?

   **e.** Explore other functions that could be used to model the data, using the various regression features on a graphing calculator, and discuss to what extent any of them are useful descriptions of the data.

**74. Money** The total amount of money in circulation for the years 1990–2012 can be closely approximated by

$$M(t) = 0.05022t^3 - 1.127t^2 + 38.63t + 247.2$$

where $t$ represents the number of years since 1990 and $M(t)$ is in billions of dollars. Find the derivative of $M(t)$ and use it to find the rate of change of money in circulation in the following years. *Source: Federal Reserve Board.*

   **a.** 1990           **b.** 2000

   **c.** 2010          **d.** 2012

   **e.** What do your answers to parts a–d tell you about the amount of money in circulation in those years?

**75. AP Examinations** The probability (as a percent) of scoring 3 or more on the Calculus AB Advanced Placement Examination can be very closely predicted as a function of a student's PSAT/NMSQT Score $x$ by the function

$$P(x) = -0.00209x^3 + 0.3387x^2 - 15.15x + 208.6.$$

*Source: The College Board.* Find the probability of scoring 3 or more for each of the following PSAT/NMSQT scores, as well as the rate that this probability is increasing with respect to the PSAT/NMSQT score.

   **a.** 45           **b.** 75

   **c.** Based on your results from parts a and b, what recommendations would you make about who should take the Calculus AB Examination?

**YOUR TURN ANSWERS**

**1.** $f'(t) = -\dfrac{1}{2}t^{-3/2}$ or $f'(t) = -\dfrac{1}{2t^{3/2}}$

**2.** $\dfrac{dy}{dx} = \dfrac{3}{2}x^{-1/2}$ or $\dfrac{dy}{dx} = \dfrac{3}{2\sqrt{x}}$

**3.** $h'(t) = -6t + t^{-1/2} - 20t^{-5}$ or $h'(t) = -6t + \dfrac{1}{\sqrt{t}} - \dfrac{20}{t^5}$

# 4.2 Derivatives of Products and Quotients

APPLY IT How can we describe the behavior of the immune system when it is being attacked by a parasite?

*We show how the derivative is used to solve a problem like this in Example 5, later in this section.*

In the previous section we saw that the derivative of a sum of two functions is found from the sum of the derivatives. What about products? Is the derivative of a product equal to the product of the derivatives? For example, if

$$u(x) = 2x + 3 \quad \text{and} \quad v(x) = 3x^2,$$

then

$$u'(x) = 2 \quad \text{and} \quad v'(x) = 6x.$$

Let $f(x)$ be the product of $u$ and $v$; that is, $f(x) = (2x + 3)(3x^2) = 6x^3 + 9x^2$. By the rules of the preceding section, $f'(x) = 18x^2 + 18x = 18x(x + 1)$. On the other hand, $u'(x) \cdot v'(x) = 2(6x) = 12x \neq f'(x)$. In this example, the derivative of a product is *not* equal to the product of the derivatives, nor is this usually the case.

The rule for finding derivatives of products is as follows.

**Product Rule**

If $f(x) = u(x) \cdot v(x)$, and if $u'(x)$ and $v'(x)$ both exist, then

$$f'(x) = u(x) \cdot v'(x) + v(x) \cdot u'(x).$$

(The derivative of a product of two functions is the first function times the derivative of the second plus the second function times the derivative of the first.)

To sketch the method used to prove the product rule, let

$$f(x) = u(x) \cdot v(x).$$

Then $f(x + h) = u(x + h) \cdot v(x + h)$, and, by definition, $f'(x)$ is given by

$$f'(x) = \lim_{h \to 0} \frac{f(x + h) - f(x)}{h}$$

$$= \lim_{h \to 0} \frac{u(x + h) \cdot v(x + h) - u(x) \cdot v(x)}{h}.$$

**FOR REVIEW**
This proof uses several of the rules for limits given in the first section of the previous chapter. You may want to review them at this time.

Now subtract and add $u(x + h) \cdot v(x)$ in the numerator, giving

$$f'(x) = \lim_{h \to 0} \frac{u(x + h) \cdot v(x + h) - u(x + h) \cdot v(x) + u(x + h) \cdot v(x) - u(x) \cdot v(x)}{h}$$

$$= \lim_{h \to 0} \frac{u(x + h)[v(x + h) - v(x)] + v(x)[u(x + h) - u(x)]}{h}$$

$$= \lim_{h \to 0} u(x + h)\left[\frac{v(x + h) - v(x)}{h}\right] + \lim_{h \to 0} v(x)\left[\frac{u(x + h) - u(x)}{h}\right]$$

$$= \lim_{h \to 0} u(x + h) \cdot \lim_{h \to 0} \frac{v(x + h) - v(x)}{h} + \lim_{h \to 0} v(x) \cdot \lim_{h \to 0} \frac{u(x + h) - u(x)}{h}. \quad (1)$$

If $u'(x)$ and $v'(x)$ both exist, then

$$\lim_{h \to 0} \frac{u(x + h) - u(x)}{h} = u'(x) \quad \text{and} \quad \lim_{h \to 0} \frac{v(x + h) - v(x)}{h} = v'(x).$$

The fact that $u'(x)$ exists can be used to prove

$$\lim_{h \to 0} u(x + h) = u(x),$$

and since no $h$ is involved in $v(x)$,

$$\lim_{h \to 0} v(x) = v(x).$$

Substituting these results into Equation (1) gives

$$f'(x) = u(x) \cdot v'(x) + v(x) \cdot u'(x),$$

the desired result.

To help see why the product rule is true, consider the special case in which $u$ and $v$ are positive functions. Then $u(x) \cdot v(x)$ represents the area of a rectangle, as shown in Figure 5. If we assume that $u$ and $v$ are increasing, then $u(x + h) \cdot v(x + h)$ represents the area of a slightly larger rectangle when $h$ is a small positive number, as shown in the figure. The change in the area of the rectangle is given by the pink rectangle, with an area of $u(x)$ times the amount $v$ has changed, plus the blue rectangle, with an area of $v(x)$ times the amount $u$ has changed, plus the small green rectangle. As $h$ becomes smaller and smaller, the green rectangle becomes negligibly small, and the change in the area is essentially $u(x)$ times the change in $v$ plus $v(x)$ times the change in $u$.

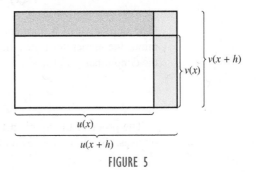

FIGURE 5

### EXAMPLE 1 Product Rule

Let $f(x) = (2x + 3)(3x^2)$. Use the product rule to find $f'(x)$.

**SOLUTION** Here $f$ is given as the product of $u(x) = 2x + 3$ and $v(x) = 3x^2$. By the product rule and the fact that $u'(x) = 2$ and $v'(x) = 6x$,

$$f'(x) = u(x) \cdot v'(x) + v(x) \cdot u'(x)$$
$$= (2x + 3)(6x) + (3x^2)(2)$$
$$= 12x^2 + 18x + 6x^2 = 18x^2 + 18x = 18x(x + 1).$$

This result is the same as that found at the beginning of the section.

### EXAMPLE 2 Product Rule

Find the derivative of $y = (\sqrt{x} + 3)(x^2 - 5x)$.

**SOLUTION** Let $u(x) = \sqrt{x} + 3 = x^{1/2} + 3$, and $v(x) = x^2 - 5x$. Then

$$\frac{dy}{dx} = u(x) \cdot v'(x) + v(x) \cdot u'(x)$$

$$= (x^{1/2} + 3)(2x - 5) + (x^2 - 5x)\left(\frac{1}{2} x^{-1/2}\right).$$

Simplify by multiplying and combining terms.

$$\frac{dy}{dx} = (2x)(x^{1/2}) + 6x - 5x^{1/2} - 15 + (x^2)\left(\frac{1}{2}x^{-1/2}\right) - (5x)\left(\frac{1}{2}x^{-1/2}\right)$$

$$= 2x^{3/2} + 6x - 5x^{1/2} - 15 + \frac{1}{2}x^{3/2} - \frac{5}{2}x^{1/2}$$

$$= \frac{5}{2}x^{3/2} + 6x - \frac{15}{2}x^{1/2} - 15$$

**YOUR TURN 1**   Find the derivative of $y = (x^3 + 7)(4 - x^2)$.

TRY YOUR TURN 1

We could have found the derivatives above by multiplying the original functions, then finding the derivative. In this case, the product rule would not have been needed. In the next section, however, we shall see products of functions where the product rule is essential.

What about *quotients* of functions? To find the derivative of the quotient of two functions, use the next rule.

**Quotient Rule**

If $f(x) = u(x)/v(x)$, if all indicated derivatives exist, and if $v(x) \neq 0$, then

$$f'(x) = \frac{v(x) \cdot u'(x) - u(x) \cdot v'(x)}{[v(x)]^2}.$$

(The derivative of a quotient is the denominator times the derivative of the numerator minus the numerator times the derivative of the denominator, all divided by the square of the denominator.)

The proof of the quotient rule is similar to that of the product rule and is left for the exercises. (See Exercises 37 and 38.)

**FOR REVIEW**

You may want to consult the Rational Expressions section of the Algebra Reference chapter (Section R.3) to help you work with the fractions in the section.

**CAUTION**   Just as the derivative of a product is *not* the product of the derivatives, the derivative of a quotient is *not* the quotient of the derivatives. If you are asked to take the derivative of a product or a quotient, it is essential that you recognize that the function contains a product or quotient and then use the appropriate rule.

**EXAMPLE 3   Quotient Rule**

Find $f'(x)$ if $f(x) = \frac{2x - 1}{4x + 3}$.

**SOLUTION**   Let $u(x) = 2x - 1$, with $u'(x) = 2$. Also, let $v(x) = 4x + 3$, with $v'(x) = 4$. Then, by the quotient rule,

$$f'(x) = \frac{v(x) \cdot u'(x) - u(x) \cdot v'(x)}{[v(x)]^2}$$

$$= \frac{(4x + 3)(2) - (2x - 1)(4)}{(4x + 3)^2}$$

$$= \frac{8x + 6 - 8x + 4}{(4x + 3)^2}$$

$$= \frac{10}{(4x + 3)^2}.$$

**YOUR TURN 2**   Find $f'(x)$ if $f(x) = \frac{3x + 2}{5 - 2x}$.

TRY YOUR TURN 2

CAUTION   In the second step of Example 3, we had the expression

$$\frac{(4x + 3)(2) - (2x - 1)(4)}{(4x + 3)^2}.$$

Students often incorrectly "cancel" the $4x + 3$ in the numerator with one factor of the denominator. Because the numerator is a *difference* of two products, however, you must multiply and combine terms *before* looking for common factors in the numerator and denominator.

### EXAMPLE 4   Product and Quotient Rules

Find $D_x\left[\dfrac{(3 - 4x)(5x + 1)}{7x - 9}\right].$

**SOLUTION**   This function has a product within a quotient. Instead of multiplying the factors in the numerator first (which is an option), we can use the quotient rule together with the product rule, as follows. Use the quotient rule first to get

$$D_x\left[\frac{(3 - 4x)(5x + 1)}{7x - 9}\right] = \frac{(7x - 9)D_x[(3 - 4x)(5x + 1)] - [(3 - 4x)(5x + 1)D_x(7x - 9)]}{(7x - 9)^2}.$$

Now use the product rule to find $D_x[(3 - 4x)(5x + 1)]$ in the numerator.

$$= \frac{(7x - 9)[(3 - 4x)5 + (5x + 1)(-4)] - (3 + 11x - 20x^2)(7)}{(7x - 9)^2}$$

$$= \frac{(7x - 9)(15 - 20x - 20x - 4) - (21 + 77x - 140x^2)}{(7x - 9)^2}$$

$$= \frac{(7x - 9)(11 - 40x) - 21 - 77x + 140x^2}{(7x - 9)^2}$$

$$= \frac{-280x^2 + 437x - 99 - 21 - 77x + 140x^2}{(7x - 9)^2}$$

$$= \frac{-140x^2 + 360x - 120}{(7x - 9)^2}$$

**YOUR TURN 3**   Find $D_x\left[\dfrac{(5x - 3)(2x + 7)}{3x + 7}\right].$

TRY YOUR TURN 3

### EXAMPLE 5   Immune System Activity

The activity of an immune system invaded by a parasite can be described by

$$I(n) = \frac{an^2}{b + n^2},$$

where $n$ measures the number of larvae in the host, $a$ is the maximum functional activity of the host's immune response, and $b$ is a measure of the sensitivity of the immune system. Find and interpret $I'(n)$. *Source: Mathematical Biology.*

APPLY IT   **SOLUTION**   Although it appears that there are three independent variables in the function $I(n)$, both $a$ and $b$ are constants which are specific to the particular host. When taking the derivative of a function with unspecified constants we simply treat those constants in the same manner as we treat any other number. Therefore, the derivative of $I(n)$ is given by

$$I'(n) = \frac{(b + n^2)(2an) - an^2(2n)}{(b + n^2)^2}$$

$$= \frac{2abn + 2an^3 - 2an^3}{(b + n^2)^2}$$

$$= \frac{2abn}{(b + n^2)^2}.$$

In this case, $I'(n)$ is the rate of change of the activity of the immune system with respect to the number of larvae.

# 4.2 EXERCISES

**Use the product rule to find the derivative of the following.**
**(*Hint for Exercises 3–6:* Write the quantity as a product.)**

**1.** $y = (3x^2 + 2)(2x - 1)$

**2.** $y = (5x^2 - 1)(4x + 3)$

**3.** $y = (2x - 5)^2$

**4.** $y = (7x - 6)^2$

**5.** $k(t) = (t^2 - 1)^2$

**6.** $g(t) = (3t^2 + 2)^2$

**7.** $y = (x + 1)(\sqrt{x} + 2)$

**8.** $y = (2x - 3)(\sqrt{x} - 1)$

**9.** $p(y) = (y^{-1} + y^{-2})(2y^{-3} - 5y^{-4})$

**10.** $q(x) = (x^{-2} - x^{-3})(3x^{-1} + 4x^{-4})$

**Use the quotient rule to find the derivative of the following.**

**11.** $f(x) = \dfrac{6x + 1}{3x + 10}$

**12.** $f(x) = \dfrac{8x - 11}{7x + 3}$

**13.** $y = \dfrac{5 - 3t}{4 + t}$

**14.** $y = \dfrac{9 - 7t}{1 - t}$

**15.** $y = \dfrac{x^2 + x}{x - 1}$

**16.** $y = \dfrac{x^2 - 4x}{x + 3}$

**17.** $f(t) = \dfrac{4t^2 + 11}{t^2 + 3}$

**18.** $y = \dfrac{-x^2 + 8x}{4x^2 - 5}$

**19.** $g(x) = \dfrac{x^2 - 4x + 2}{x^2 + 3}$

**20.** $k(x) = \dfrac{x^2 + 7x - 2}{x^2 - 2}$

**21.** $p(t) = \dfrac{\sqrt{t}}{t - 1}$

**22.** $r(t) = \dfrac{\sqrt{t}}{2t + 3}$

**23.** $y = \dfrac{5x + 6}{\sqrt{x}}$

**24.** $y = \dfrac{4x - 3}{\sqrt{x}}$

**25.** $h(z) = \dfrac{z^{2.2}}{z^{3.2} + 5}$

**26.** $g(y) = \dfrac{y^{1.4} + 1}{y^{2.5} + 2}$

**27.** $f(x) = \dfrac{(3x^2 + 1)(2x - 1)}{5x + 4}$

**28.** $g(x) = \dfrac{(2x^2 + 3)(5x + 2)}{6x - 7}$

**29.** If $g(3) = 4$, $g'(3) = 5$, $f(3) = 9$, and $f'(3) = 8$, find $h'(3)$ when $h(x) = f(x)g(x)$.

**30.** If $g(3) = 4$, $g'(3) = 5$, $f(3) = 9$, and $f'(3) = 8$, find $h'(3)$ when $h(x) = f(x)/g(x)$.

**31.** Find the error in the following work.

$$D_x\left(\frac{2x + 5}{x^2 - 1}\right) = \frac{(2x + 5)(2x) - (x^2 - 1)2}{(x^2 - 1)^2}$$

$$= \frac{4x^2 + 10x - 2x^2 + 2}{(x^2 - 1)^2}$$

$$= \frac{2x^2 + 10x + 2}{(x^2 - 1)^2}$$

**32.** Find the error in the following work.

$$D_x\left(\frac{x^2 - 4}{x^3}\right) = x^3(2x) - (x^2 - 4)(3x^2) = 2x^4 - 3x^4 + 12x^2$$

$$= -x^4 + 12x^2$$

**33.** Find an equation of the line tangent to the graph of $f(x) = x/(x - 2)$ at $(3, 3)$.

**34.** Find an equation of the line tangent to the graph of $f(x) = (2x - 1)(x + 4)$ at $(1, 5)$.

**35.** Consider the function

$$f(x) = \frac{3x^3 + 6}{x^{2/3}}.$$

   **a.** Find the derivative using the quotient rule.

   **b.** Find the derivative by first simplifying the function to

$$f(x) = \frac{3x^3}{x^{2/3}} + \frac{6}{x^{2/3}} = 3x^{7/3} + 6x^{-2/3}$$

   and using the rules from the previous section.

   **c.** Compare your answers from parts a and b and explain any discrepancies.

**36.** What is the result of applying the product rule to the function

$$f(x) = kg(x),$$

where $k$ is a constant? Compare with the rule for differentiating a constant times a function from the previous section.

**37.** Following the steps used to prove the product rule for derivatives, prove the quotient rule for derivatives.

**38.** Use the fact that $f(x) = u(x)/v(x)$ can be rewritten as $f(x)v(x) = u(x)$ and the product rule for derivatives to verify the quotient rule for derivatives. (*Hint:* After applying the product rule, substitute $u(x)/v(x)$ for $f(x)$ and simplify.)

**For each function, find the value(s) of $x$ in which $f'(x) = 0$, to 3 decimal places.**

**39.** $f(x) = (x^2 - 2)(x^2 - \sqrt{2})$

**40.** $f(x) = \dfrac{x - 2}{x^2 + 4}$

## LIFE SCIENCE APPLICATIONS

**41. Muscle Reaction** When a certain drug is injected into a muscle, the muscle responds by contracting. The amount of contraction, *s* (in millimeters), is related to the concentration of the drug, *x* (in milliliters), by

$$s(x) = \frac{x}{m + nx},$$

where *m* and *n* are constants.

**a.** Find $s'(x)$.

**b.** Find the rate of contraction when the concentration of the drug is 50 ml, $m = 10$, and $n = 3$.

**42. Growth Models** In Exercise 52 of Section 1.5, the formula for the growth rate of a population in the presence of a quantity *x* of food was given as

$$f(x) = \frac{Kx}{A + x}.$$

This was referred to as Michaelis-Menten kinetics.

**a.** Find the rate of change of the growth rate with respect to the amount of food.

**b.** The quantity *A* in the formula for $f(x)$ represents the quantity of food for which the growth rate is half of its maximum. Using your answer from part a, find the rate of change of the growth rate when $x = A$.

**43. Bacterial Population** Assume that the total number (in millions) of bacteria present in a culture at a certain time *t* (in hours) is given by

$$N(t) = 3t(t - 10)^2 + 40.$$

**a.** Find $N'(t)$.

Find the rate at which the population of bacteria is changing at the following times.

**b.** 8 hours

**c.** 11 hours

**d.** The answer in part b is negative, and the answer in part c is positive. What does this mean in terms of the population of bacteria?

**44. Work/Rest Cycles** Murrell's formula for calculating the total amount of rest, in minutes, required after performing a particular type of work activity for 30 minutes is given by the formula

$$R(w) = \frac{30(w - 4)}{w - 1.5},$$

where *w* is the work expended in kilocalories per minute, kcal/min. *Source: Human Factors in Engineering and Design.*

**a.** A value of 5 for *w* indicates light work, such as riding a bicycle on a flat surface at 10 mph. Find $R(5)$.

**b.** A value of 7 for *w* indicates moderate work, such as mowing grass with a pushmower on level ground. Find $R(7)$.

**c.** Find $R'(5)$ and $R'(7)$ and compare your answers. Explain whether these answers make sense.

**45. Optimal Foraging** Using data collected by zoologist Reto Zach, the work done by a crow to break open a whelk (large marine snail) can be estimated by the function

$$W = \left(1 + \frac{20}{H - 0.93}\right)H,$$

where *H* is the height (in meters) of the whelk when it is dropped. *Source: Mathematics Teacher.*

**a.** Find $dW/dH$.

**b.** One can show that the amount of work is minimized when $dW/dH = 0$. Find the value of *H* that minimizes *W*.

**c.** Interestingly, Zach observed the crows dropping the whelks from an average height of 5.23 m. What does this imply?

**46. Cell Traction Force** In a matrix of cells, the cell traction force per unit mass is given by

$$T(n) = \frac{an}{1 + bn^2},$$

where *n* is the number of cells, *a* is the measure of the traction force generated by a cell, and *b* measures how force is reduced due to neighboring cells. Find and interpret $T'(n)$. *Source: Mathematical Biology.*

**47. Calcium Kinetics** A function used to describe the kinetics of calcium in the cytogel (the gel of cell cytoplasm in the epithelium, an external tissue of epidermal cells) is given by

$$R(c) = \frac{ac^2}{1 + bc^2} - kc,$$

where $R(c)$ measures the release of calcium; *c* is the amount of free calcium outside the vesicles in which it is stored; and *a*, *b*, and *k* are positive constants. Find and interpret $R'(c)$. *Source: Mathematical Biology.*

**48. Cellular Chemistry** The release of a chemical neurotransmitter as a function of the amount *C* of intracellular calcium is given by

$$L(C) = \frac{aC^n}{k + C^n},$$

where *n* measures the degree of cooperativity, *k* measures saturation, and *a* is the maximum of the process. Find and interpret $L'(C)$. *Source: Mathematical Topics in Population Biology, Morphogenesis and Neurosciences.*

**49. Genetics** In the haploid model of natural selection, there is one copy of every gene within the genome. Consider two groups of individuals, one carrying the allele $A$ and the other the allele $a$. Then the proportion that carry the allele $A$ can be written as

$$p = \frac{n_A}{n_A + n_a}.$$

*Source: A Biologist's Guide to Mathematical Modeling in Ecology and Evolution.* Show that if both $n_A$ and $n_a$ are changing with time, then the rate of change of the proportion that carry the allele $A$ is

$$\frac{dp}{dt} = \frac{n_a \dfrac{dn_A}{dt} - n_A \dfrac{dn_a}{dt}}{(n_A + n_a)^2}.$$

**50. Memory Retention** Some psychologists contend that the number of facts of a certain type that are remembered after $t$ hours is given by

$$f(t) = \frac{90t}{99t - 90}.$$

Find the rate at which the number of facts remembered is changing after the following numbers of hours.

   **a.** 1            **b.** 10

## OTHER APPLICATIONS

**51. Vehicle Waiting Time** The average number of vehicles waiting in a line to enter a parking ramp can be modeled by the function

$$f(x) = \frac{x^2}{2(1 - x)},$$

where $x$ is a quantity between 0 and 1 known as the traffic intensity. Find the rate of change of the number of vehicles in line with respect to the traffic intensity for the following values of the intensity. *Source: Principles of Highway Engineering and Traffic Control.*

   **a.** $x = 0.1$           **b.** $x = 0.6$

**52. Employee Training** A company that manufactures bicycles has determined that a new employee can assemble $M(d)$ bicycles per day after $d$ days of on-the-job training, where

$$M(d) = \frac{100d^2}{3d^2 + 10}.$$

   **a.** Find the rate of change function for the number of bicycles assembled with respect to time.

   **b.** Find and interpret $M'(2)$ and $M'(5)$.

### YOUR TURN ANSWERS

1. $\dfrac{dy}{dx} = -5x^4 + 12x^2 - 14x$

2. $f'(x) = \dfrac{19}{(5 - 2x)^2}$

3. $\dfrac{30x^2 + 140x + 266}{(3x + 7)^2}$

# 4.3 The Chain Rule

**APPLY IT** Suppose we know how fast the radius of a circular oil slick is growing, and we know how much the area of the oil slick is growing per unit of change in the radius. How fast is the area growing?

*We will answer this question in Example 4 using the chain rule for derivatives.*

Before discussing the chain rule, we consider the composition of functions. Many of the most useful functions for modeling are created by combining simpler functions. Viewing complex functions as combinations of simpler functions often makes them easier to understand and use.

## Composition of Functions
Suppose a function $f$ assigns to each element $x$ in set $X$ some element $y = f(x)$ in set $Y$. Suppose also that a function $g$ takes each element in set $Y$ and assigns to it a value $z = g[f(x)]$ in set $Z$. By using both $f$ and $g$, an element $x$ in $X$ is assigned to an element $z$ in $Z$, as illustrated in Figure 6. The result of this process is a new function called the *composition* of functions $g$ and $f$ and defined as follows.

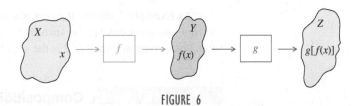

FIGURE 6

## Composite Function

Let $f$ and $g$ be functions. The **composite function**, or **composition**, of $g$ and $f$ is the function whose values are given by $g[f(x)]$ for all $x$ in the domain of $f$ such that $f(x)$ is in the domain of $g$. (Read $g[f(x)]$ as "$g$ of $f$ of $x$".)

### EXAMPLE 1 Composite Functions

Let $f(x) = 2x - 1$ and $g(x) = \sqrt{3x + 5}$. Find the following.

**(a)** $g[f(4)]$

**SOLUTION** Find $f(4)$ first.

$$f(4) = 2 \cdot 4 - 1 = 8 - 1 = 7$$

Then

$$g[f(4)] = g[7] = \sqrt{3 \cdot 7 + 5} = \sqrt{26}.$$

**(b)** $f[g(4)]$

**SOLUTION** Since $g(4) = \sqrt{3 \cdot 4 + 5} = \sqrt{17}$,

$$f[g(4)] = 2 \cdot \sqrt{17} - 1 = 2\sqrt{17} - 1.$$

**(c)** $f[g(-2)]$

**SOLUTION** $f[g(-2)]$ does not exist, since $-2$ is not in the domain of $g$.

**YOUR TURN 1** For the functions in Example 1, find $f[g(0)]$ and $g[f(0)]$.

TRY YOUR TURN 1

### EXAMPLE 2 Composition of Functions

Let $f(x) = 2x^2 + 5x$ and $g(x) = 4x + 1$. Find the following.

**(a)** $f[g(x)]$

**SOLUTION** Using the given functions, we have

$$
\begin{aligned}
f[g(x)] &= f[4x + 1] \\
&= 2(4x + 1)^2 + 5(4x + 1) \\
&= 2(16x^2 + 8x + 1) + 20x + 5 \\
&= 32x^2 + 16x + 2 + 20x + 5 \\
&= 32x^2 + 36x + 7.
\end{aligned}
$$

**(b)** $g[f(x)]$

**SOLUTION** By the definition above, with $f$ and $g$ interchanged,

$$
\begin{aligned}
g[f(x)] &= g[2x^2 + 5x] \\
&= 4(2x^2 + 5x) + 1 \\
&= 8x^2 + 20x + 1.
\end{aligned}
$$

**YOUR TURN 2** Let $f(x) = 2x - 3$ and $g(x) = x^2 + 1$. Find $g[f(x)]$.

TRY YOUR TURN 2

As Example 2 shows, it is not always true that $f[g(x)] = g[f(x)]$. In fact, it is rare to find two functions $f$ and $g$ such that $f[g(x)] = g[f(x)]$. The domain of both composite functions given in Example 2 is the set of all real numbers.

### EXAMPLE 3  Composition of Functions

Write each function as the composition of two functions $f$ and $g$ so that $h(x) = f[g(x)]$.

**(a)** $h(x) = 2(4x + 1)^2 + 5(4x + 1)$

**SOLUTION**  Let $f(x) = 2x^2 + 5x$ and $g(x) = 4x + 1$. Then $f[g(x)] = f(4x + 1) = 2(4x + 1)^2 + 5(4x + 1)$. Notice that $h(x)$ here is the same as $f[g(x)]$ in Example 2(a).

**(b)** $h(x) = \sqrt{1 - x^2}$

**SOLUTION**  One way to do this is to let $f(x) = \sqrt{x}$ and $g(x) = 1 - x^2$. Another choice is to let $f(x) = \sqrt{1 - x}$ and $g(x) = x^2$. Verify that with either choice, $f[g(x)] = \sqrt{1 - x^2}$. For the purposes of this section, the first choice is better; it is useful to think of $f$ as being the function on the outer layer and $g$ as the function on the inner layer. With this function $h$, we see a square root on the outer layer, and when we peel that away we see $1 - x^2$ on the inside.

**TRY YOUR TURN 3**

**YOUR TURN 3**  Write $h(x) = (2x - 3)^3$ as a composition of two functions $f$ and $g$ so that $h(x) = f[g(x)]$.

## The Chain Rule
Suppose $f(x) = x^2$ and $g(x) = 5x^3 + 2$. What is the derivative of $h(x) = f[g(x)] = (5x^3 + 2)^2$? At first you might think the answer is just $h'(x) = 2(5x^3 + 2) = 10x^3 + 4$ by using the power rule. You can check this answer by multiplying out $h(x) = (5x^3 + 2)^2 = 25x^6 + 20x^3 + 4$. Now calculate $h'(x) = 150x^5 + 60x^2$. The guess using the power rule was clearly wrong! The error is that the power rule applies to $x$ raised to a power, not to some other function of $x$ raised to a power.

How, then, could we take the derivative of $p(x) = (5x^3 + 2)^{20}$? This seems far too difficult to multiply out. Fortunately, there is a way. Notice from the previous paragraph that $h'(x) = 150x^5 + 60x^2 = 2(5x^3 + 2)15x^2$. So the original guess was almost correct, except it was missing the factor of $15x^2$, which just happens to be $g'(x)$. This is not a coincidence. To see why the derivative of $f[g(x)]$ involves taking the derivative of $f$ and then multiplying by the derivative of $g$, let us consider a realistic example, the question from the beginning of this section.

### EXAMPLE 4  Area of an Oil Slick

A leaking oil well off the Gulf Coast is spreading a circular film of oil over the water surface. At any time $t$ (in minutes) after the beginning of the leak, the radius of the circular oil slick (in feet) is given by

$$r(t) = 4t.$$

Find the rate of change of the area of the oil slick with respect to time.

**APPLY IT**  **SOLUTION**  We first find the rate of change in the radius over time by finding $dr/dt$:

$$\frac{dr}{dt} = 4.$$

This value indicates that the radius is increasing by 4 ft each minute.

The area of the oil slick is given by

$$A(r) = \pi r^2, \quad \text{with} \quad \frac{dA}{dr} = 2\pi r.$$

The derivative, $dA/dr$, gives the rate of change in area per unit increase in the radius.

As these derivatives show, the radius is increasing at a rate of 4 ft/min, and for each foot that the radius increases, the area increases by $2\pi r$ ft². It seems reasonable, then, that the area is increasing at a rate of

$$2\pi r \text{ ft}^2/\text{ft} \times 4 \text{ ft/min} = 8\pi r \text{ ft}^2/\text{min}.$$

That is,

$$\frac{dA}{dt} = \frac{dA}{dr} \cdot \frac{dr}{dt} = 2\pi r \cdot 4 = 8\pi r.$$

Notice that because area $(A)$ is a function of radius $(r)$, which is a function of time $(t)$, area as a function of time is a composition of two functions, written $A(r(t))$. The last step, then, can also be written as

$$\frac{dA}{dt} = \frac{d}{dt} A[r(t)] = A'[r(t)] \cdot r'(t) = 2\pi r \cdot 4 = 8\pi r.$$

Finally, we can substitute $r(t) = 4t$ to get the derivative in terms of $t$:

$$\frac{dA}{dt} = 8\pi r = 8\pi(4t) = 32\pi t.$$

The rate of change of the area of the oil slick with respect to time is $32\pi t$ ft²/min.

To check the result of Example 4, use the fact that $r = 4t$ and $A = \pi r^2$ to get the same result:

$$A = \pi(4t)^2 = 16\pi t^2, \quad \text{with} \quad \frac{dA}{dt} = 32\pi t.$$

The product used in Example 4,

$$\frac{dA}{dt} = \frac{dA}{dr} \cdot \frac{dr}{dt},$$

is an example of the **chain rule**, which is used to find the derivative of a composite function.

## Chain Rule

If $y$ is a function of $u$, say $y = f(u)$, and if $u$ is a function of $x$, say $u = g(x)$, then $y = f(u) = f[g(x)]$, and

$$\frac{dy}{dx} = \frac{dy}{du} \cdot \frac{du}{dx}.$$

One way to remember the chain rule is to pretend that $dy/du$ and $du/dx$ are fractions, with $du$ "canceling out." The proof of the chain rule requires advanced concepts and, therefore, is not given here.

### EXAMPLE 5  Chain Rule

Find $dy/dx$ if $y = (3x^2 - 5x)^{1/2}$.

**SOLUTION**  Let $y = u^{1/2}$, and $u = 3x^2 - 5x$. Then

$$\frac{dy}{dx} = \frac{dy}{du} \cdot \frac{du}{dx}$$

$$= \frac{1}{2}u^{-1/2} \cdot (6x - 5).$$

**YOUR TURN 4** Find $dy/dx$ if $y = (5x^2 - 6x)^{-2}$.

Replacing $u$ with $3x^2 - 5x$ gives

$$\frac{dy}{dx} = \frac{1}{2}(3x^2 - 5x)^{-1/2}(6x - 5) = \frac{6x - 5}{2(3x^2 - 5x)^{1/2}}.$$

**TRY YOUR TURN 4**

The following alternative version of the chain rule is stated in terms of composite functions.

### Chain Rule (Alternative Form)

If $y = f[g(x)]$, then

$$\frac{dy}{dx} = f'[g(x)] \cdot g'(x).$$

(To find the derivative of $f[g(x)]$, find the derivative of $f(x)$, replace each $x$ with $g(x)$, and then multiply the result by the derivative of $g(x)$.)

In words, the chain rule tells us to first take the derivative of the outer function, then multiply it by the derivative of the inner function.

### EXAMPLE 6  Chain Rule

Use the chain rule to find $D_x(x^2 + 5x)^8$.

**SOLUTION** As in Example 3(b), think of this as a function with layers. The outer layer is something being raised to the 8th power, so let $f(x) = x^8$. Once this layer is peeled away, we see that the inner layer is $x^2 + 5x$, so $g(x) = x^2 + 5x$. Then $(x^2 + 5x)^8 = f[g(x)]$ and

$$D_x(x^2 + 5x)^8 = f'[g(x)]g'(x).$$

Here $f'(x) = 8x^7$, with $f'[g(x)] = 8[g(x)]^7 = 8(x^2 + 5x)^7$ and $g'(x) = 2x + 5$.

$$D_x(x^2 + 5x)^8 = f'[g(x)]g'(x)$$

$$= 8[g(x)]^7 g'(x)$$

$$= 8(x^2 + 5x)^7(2x + 5)$$

**TRY YOUR TURN 5**

**YOUR TURN 5** Find $D_x(x^2 - 7)^{10}$.

CAUTION (a) A common error is to forget to multiply by $g'(x)$ when using the chain rule. Remember, the derivative must involve a "chain," or product, of derivatives.

(b) Another common mistake is to write the derivative as $f'[g'(x)]$. Remember to leave $g(x)$ unchanged in $f'[g(x)]$ and then to multiply by $g'(x)$.

One way to avoid both of the errors described above is to remember that the chain rule is a two-step process. In Example 6, the first step was taking the derivative of the power, and the second step was multiplying by $g'(x)$. Forgetting to multiply by $g'(x)$ would be an erroneous one-step process. The other erroneous one-step process is to take the derivative inside the power, getting $f'[g'(x)]$, or $8(2x + 5)^7$ in Example 6.

Sometimes both the chain rule and either the product or quotient rule are needed to find a derivative, as the next examples show.

**EXAMPLE 7** **Derivative Rules**

Find the derivative of $y = 4x(3x + 5)^5$.

**SOLUTION** Write $4x(3x + 5)^5$ as the product

$$(4x) \cdot (3x + 5)^5.$$

To find the derivative of $(3x + 5)^5$, let $g(x) = 3x + 5$, with $g'(x) = 3$. Now use the product rule and the chain rule.

$$\overset{\text{Derivative of } (3x + 5)^5 \qquad\qquad \text{Derivative of } 4x}{}$$

$$\frac{dy}{dx} = 4x\overbrace{[5(3x + 5)^4 \cdot 3]} + (3x + 5)^5\overbrace{(4)}$$

**YOUR TURN 6** Find the derivative of $y = x^2(5x - 1)^3$.

$$= 60x(3x + 5)^4 + 4(3x + 5)^5$$

$$= 4(3x + 5)^4[15x + (3x + 5)^1] \qquad \text{Factor out the greatest common factor, } 4(3x + 5)^4.$$

$$= 4(3x + 5)^4(18x + 5) \qquad\qquad \text{Simplify inside brackets.}$$

**TRY YOUR TURN 6**

**EXAMPLE 8** **Derivative Rules**

Find $D_x\left[\dfrac{(3x + 2)^7}{x - 1}\right]$.

**SOLUTION** Use the quotient rule and the chain rule.

$$D_x\left[\frac{(3x + 2)^7}{x - 1}\right] = \frac{(x - 1)[7(3x + 2)^6 \cdot 3] - (3x + 2)^7(1)}{(x - 1)^2}$$

$$= \frac{21(x - 1)(3x + 2)^6 - (3x + 2)^7}{(x - 1)^2}$$

$$= \frac{(3x + 2)^6[21(x - 1) - (3x + 2)]}{(x - 1)^2} \qquad \text{Factor out the greatest common factor, } (3x + 2)^6.$$

**YOUR TURN 7**

Find $D_x\left[\dfrac{(4x - 1)^3}{x + 3}\right]$.

$$= \frac{(3x + 2)^6[21x - 21 - 3x - 2]}{(x - 1)^2} \qquad \text{Distribute.}$$

$$= \frac{(3x + 2)^6(18x - 23)}{(x - 1)^2} \qquad\qquad \text{Simplify inside brackets.}$$

**TRY YOUR TURN 7**

Some applications requiring the use of the chain rule are illustrated in the next two examples.

**EXAMPLE 9** **Tasmanian Devil**

Named for the only place where it is currently known to live, the extremely voracious Tasmanian Devil often preys on animals larger than itself. Researchers have identified the mathematical relationship

$$L(w) = 2.265w^{2.543},$$

where $w$ (in kg) is the weight and $L(w)$ (in mm) is the length of the Tasmanian Devil. *Source: Journal of Mammalogy.* Suppose that the weight of a particular Tasmanian Devil can be estimated by the function

$$w(t) = 0.125 + 0.18t,$$

where $w(t)$ is the weight (in kg) and $t$ is the age, in weeks, of a Tasmanian Devil that is less than one year old. How fast is the length of a 30-week-old Tasmanian Devil changing?

**SOLUTION**   We want to find $dL/dt$, the rate of change of the length of a Tasmanian Devil. By the chain rule,

$$\frac{dL}{dt} = \frac{dL}{dw} \cdot \frac{dw}{dt}.$$

First find $dL/dw$, as follows.

$$\frac{dL}{dw} = (2.265)(2.543)w^{2.543-1} \approx 5.76w^{1.543}$$

Also,

$$\frac{dw}{dt} = 0.18.$$

Therefore,

$$\frac{dL}{dt} = 5.76w^{1.543} \cdot 0.18.$$

Since the Tasmanian Devil is 30 weeks old, $t = 30$ and

$$w(30) = 0.125 + 0.18(30) = 5.525 \text{ kg.}$$

Putting this all together, we have

$$\frac{dL}{dt} \approx 5.76(5.525)^{1.543}(0.18) \approx 14.5.$$

The length of a 30-week-old Tasmanian Devil is increasing at the rate of about 14.5 mm per week.

Alternatively, we can find $dL/dt$ by first substituting the formula for $w(t)$ into $L(w)$. Then directly take the derivative of $L(t)$ with respect to $t$.

## EXAMPLE 10   Compound Interest

Suppose a sum of $500 is deposited in an account with an interest rate of $r$ percent per year compounded monthly. At the end of 10 years, the balance in the account (as illustrated in Figure 7) is given by

$$A = 500\left(1 + \frac{r}{1200}\right)^{120}.$$

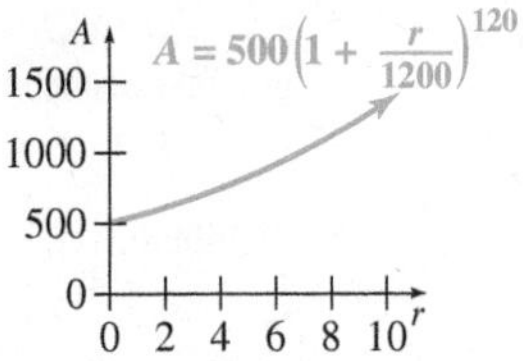

Find the rate of change of $A$ with respect to $r$ if $r = 5$ or 7.*

FIGURE 7

**SOLUTION**   First find $dA/dr$ using the chain rule.

$$\frac{dA}{dr} = (120)(500)\left(1 + \frac{r}{1200}\right)^{119}\left(\frac{1}{1200}\right)$$

$$= 50\left(1 + \frac{r}{1200}\right)^{119}$$

If $r = 5$,

$$\frac{dA}{dr} = 50\left(1 + \frac{5}{1200}\right)^{119}$$

$$\approx 82.01,$$

or $82.01 per percentage point. If $r = 7$,

$$\frac{dA}{dr} = 50\left(1 + \frac{7}{1200}\right)^{119}$$

$$\approx 99.90,$$

or $99.90 per percentage point.

*Notice that $r$ is given here as an integer percent, rather than as a decimal, which is why the formula for compound interest has 1200 where you would expect to see 12. This leads to a simpler interpretation of the derivative.

**NOTE** One lesson to learn from this section is that a derivative is always with respect to some variable. In the oil slick example, notice that the derivative of the area with respect to the radius is $2\pi r$, while the derivative of the area with respect to time is $8\pi r$. As another example, consider the velocity of a conductor walking at 2 mph on a train car. Her velocity with respect to the ground may be 50 mph, but the earth on which the train is running is moving about the sun at 1.6 million mph. The derivative of her position function might be 2, 50, or 1.6 million mph, depending on what variable it is with respect to.

# 4.3 EXERCISES

Let $f(x) = 5x^2 - 2x$ and $g(x) = 8x + 3$. Find the following.

**1.** $f[g(2)]$       **2.** $f[g(-5)]$      **3.** $g[f(2)]$

**4.** $g[f(-5)]$      **5.** $f[g(k)]$      **6.** $g[f(5z)]$

In Exercises 7–14, find $f[g(x)]$ and $g[f(x)]$.

**7.** $f(x) = \frac{x}{8} + 7;\quad g(x) = 6x - 1$

**8.** $f(x) = -8x + 9;\quad g(x) = \frac{x}{5} + 4$

**9.** $f(x) = \frac{1}{x};\quad g(x) = x^2$

**10.** $f(x) = \frac{2}{x^4};\quad g(x) = 2 - x$

**11.** $f(x) = \sqrt{x + 2};\quad g(x) = 8x^2 - 6$

**12.** $f(x) = 9x^2 - 11x;\quad g(x) = 2\sqrt{x + 2}$

**13.** $f(x) = \sqrt{x + 1};\quad g(x) = \frac{-1}{x}$

**14.** $f(x) = \frac{8}{x};\quad g(x) = \sqrt{3 - x}$

Write each function as the composition of two functions. (There may be more than one way to do this.)

**15.** $y = (5 - x^2)^{3/5}$      **16.** $y = (3x^2 - 7)^{2/3}$

**17.** $y = -\sqrt{13 + 7x}$      **18.** $y = \sqrt{9 - 4x}$

**19.** $y = (x^2 + 5x)^{1/3} - 2(x^2 + 5x)^{2/3} + 7$

**20.** $y = (x^{1/2} - 3)^2 + (x^{1/2} - 3) + 5$

Find the derivative of each function defined as follows.

**21.** $y = (8x^4 - 5x^2 + 1)^4$      **22.** $y = (2x^3 + 9x)^5$

**23.** $k(x) = -2(12x^2 + 5)^{-6}$      **24.** $f(x) = -7(3x^4 + 2)^{-4}$

**25.** $s(t) = 45(3t^3 - 8)^{3/2}$      **26.** $s(t) = 12(2t^4 + 5)^{3/2}$

**27.** $g(t) = -3\sqrt{7t^3 - 1}$      **28.** $f(t) = 8\sqrt{4t^2 + 7}$

**29.** $m(t) = -6t(5t^4 - 1)^4$      **30.** $r(t) = 4t(2t^5 + 3)^4$

**31.** $y = (3x^4 + 1)^4(x^3 + 4)$      **32.** $y = (x^3 + 2)(x^2 - 1)^4$

**33.** $q(y) = 4y^2(y^2 + 1)^{5/4}$      **34.** $p(z) = z(6z + 1)^{4/3}$

**35.** $y = \frac{-5}{(2x^3 + 1)^2}$      **36.** $y = \frac{1}{(3x^2 - 4)^5}$

**37.** $r(t) = \frac{(5t - 6)^4}{3t^2 + 4}$      **38.** $p(t) = \frac{(2t + 3)^3}{4t^2 - 1}$

**39.** $y = \frac{3x^2 - x}{(2x - 1)^5}$      **40.** $y = \frac{x^2 + 4x}{(3x^3 + 2)^4}$

**41.** In your own words explain how to form the composition of two functions.

**42.** The generalized power rule says that if $g(x)$ is a function of $x$ and $y = [g(x)]^n$ for any real number $n$, then

$$\frac{dy}{dx} = n \cdot [g(x)]^{n-1} \cdot g'(x).$$

Explain why the generalized power rule is a consequence of the chain rule and the power rule.

Consider the following table of values of the functions $f$ and $g$ and their derivatives at various points.

| $x$ | 1 | 2 | 3 | 4 |
|---|---|---|---|---|
| $f(x)$ | 2 | 4 | 1 | 3 |
| $f'(x)$ | $-6$ | $-7$ | $-8$ | $-9$ |
| $g(x)$ | 2 | 3 | 4 | 1 |
| $g'(x)$ | 2/7 | 3/7 | 4/7 | 5/7 |

Find the following using the table above.

**43. a.** $D_x(f[g(x)])$ at $x = 1$    **b.** $D_x(f[g(x)])$ at $x = 2$

**44. a.** $D_x(g[f(x)])$ at $x = 1$    **b.** $D_x(g[f(x)])$ at $x = 2$

In Exercises 45–48, find the equation of the tangent line to the graph of the given function at the given value of $x$.

**45.** $f(x) = \sqrt{x^2 + 16};\quad x = 3$

**46.** $f(x) = (x^3 + 7)^{2/3};\quad x = 1$

**47.** $f(x) = x(x^2 - 4x + 5)^4;\quad x = 2$

**48.** $f(x) = x^2\sqrt{x^4 - 12};\quad x = 2$

In Exercises 49 and 50, find all values of $x$ for the given function where the tangent line is horizontal.

**49.** $f(x) = \sqrt{x^3 - 6x^2 + 9x + 1}$

**50.** $f(x) = \frac{x}{(x^2 + 4)^4}$

**51.** Katie and Sarah are working on taking the derivative of
$$f(x) = \frac{2x}{3x + 4}.$$
Katie uses the quotient rule to get
$$f'(x) = \frac{(3x + 4)2 - 2x(3)}{(3x + 4)^2} = \frac{8}{(3x + 4)^2}.$$

Sarah converts it into a product and uses the product rule and the chain rule:
$$f(x) = 2x(3x + 4)^{-1}$$
$$f'(x) = 2x(-1)(3x + 4)^{-2}(3) + 2(3x + 4)^{-1}$$
$$= 2(3x + 4)^{-1} - 6x(3x + 4)^{-2}.$$

Explain the discrepancies between the two answers. Which procedure do you think is preferable?

**52.** Margy and Nate are working on taking the derivative of
$$f(x) = \frac{2}{(3x + 1)^4}.$$
Margy uses the quotient rule and chain rule as follows:
$$f'(x) = \frac{(3x + 1)^4 \cdot 0 - 2 \cdot 4(3x + 1)^3 \cdot 3}{(3x + 1)^8}$$
$$= \frac{-24(3x + 1)^3}{(3x + 1)^8} = \frac{-24}{(3x + 1)^5}.$$

Nate rewrites the function and uses the power rule and chain rule as follows:
$$f(x) = 2(3x + 1)^{-4}$$
$$f'(x) = (-4)2(3x + 1)^{-5} \cdot 3 = \frac{-24}{(3x + 1)^5}.$$

Compare the two procedures. Which procedure do you think is preferable?

## LIFE SCIENCE APPLICATIONS

**53. Fish Population** Suppose the population $P$ of a certain species of fish depends on the number $x$ (in hundreds) of a smaller fish that serves as its food supply, so that

$$P(x) = 2x^2 + 1.$$

Suppose, also, that the number of the smaller species of fish depends on the amount $a$ (in appropriate units) of its food supply, a kind of plankton. Specifically,

$$x = f(a) = 3a + 2.$$

A biologist wants to find the relationship between the population $P$ of the large fish and the amount $a$ of plankton available, that is, $P[f(a)]$. What is the relationship?

**54. Oil Pollution** An oil well off the Gulf Coast is leaking, with the leak spreading oil over the surface as a circle. At any time $t$ (in minutes) after the beginning of the leak, the radius of the circular oil slick on the surface is $r(t) = t^2$ feet. Let $A(r) = \pi r^2$ represent the area of a circle of radius $r$.

**a.** Find and interpret $A[r(t)]$.

**b.** Find and interpret $D_t A[r(t)]$ when $t = 100$.

**55. African Wild Dog** The African wild dog is currently one of the most endangered carnivores in the world. After years of attempts to rescue the dog from extinction, scientists have collected a lot of data on the habits of this nomadic creature. The collected data show the following mathematical relationship between the weight and the length of the dog:

$$L(w) = 2.472w^{2.571},$$

where $w$ (in kg) is the weight and $L(w)$ (in mm) is the length of the mammal. *Source: Journal of Mammalogy.* Suppose that the weight of a particular wild dog can be estimated by

$$w(t) = 0.265 + 0.21t,$$

where $w(t)$ is the weight (in kg) and $t$ is the age, in weeks, of an African wild dog that is less than one year old. How fast is the length of a 25-week-old African wild dog changing?

**56. Thermal Inversion** When there is a thermal inversion layer over a city (as happens often in Los Angeles), pollutants cannot rise vertically but are trapped below the layer and must disperse horizontally. Assume that a factory smokestack begins emitting a pollutant at 8 A.M. Assume that the pollutant disperses horizontally, forming a circle. If $t$ represents the time (in hours) since the factory began emitting pollutants ($t = 0$ represents 8 A.M.), assume that the radius of the circle of pollution is $r(t) = 2t$ miles. Let $A(r) = \pi r^2$ represent the area of a circle of radius $r$.

**a.** Find and interpret $A[r(t)]$.

**b.** Find and interpret $D_t A[r(t)]$ when $t = 4$.

**57. Bacterial Population** The total number of bacteria (in millions) present in a culture is given by

$$N(t) = 2t(5t + 9)^{1/2} + 12,$$

where $t$ represents time (in hours) after the beginning of an experiment. Find the rate of change of the population of bacteria with respect to time for the following numbers of hours.

**a.** 0        **b.** 7/5        **c.** 8

**58. Calcium Usage** To test an individual's use of calcium, a researcher injects a small amount of radioactive calcium into the person's bloodstream. The calcium remaining in the bloodstream is measured each day for several days. Suppose the amount of the calcium remaining in the bloodstream (in milligrams per cubic centimeter) $t$ days after the initial injection is approximated by

$$C(t) = \frac{1}{2}(2t + 1)^{-1/2}.$$

Find the rate of change of the calcium level with respect to time for the following numbers of days.

**a.** 0        **b.** 4        **c.** 7.5

**d.** Is $C$ always increasing or always decreasing? How can you tell?

**59. Drug Reaction** The strength of a person's reaction to a certain drug is given by

$$R(Q) = Q\left(C - \frac{Q}{3}\right)^{1/2},$$

where $Q$ represents the quantity of the drug given to the patient and $C$ is a constant.

**a.** The derivative $R'(Q)$ is called the *sensitivity* to the drug. Find $R'(Q)$.

**b.** Find the sensitivity to the drug if $C = 59$ and a patient is given 87 units of the drug.

**c.** Is the patient's sensitivity to the drug increasing or decreasing when $Q = 87$?

**60. Insect Competition** In the Hassell model of interspecies insect population,

$$f(x) = \frac{Rx}{(1 + x)^b}$$

gives the size of the next generation's population $f(x)$ as a function of this year's population $x$, where $R > 1$ is the number of adults in one generation in the absence of competition, and $b \geq 0$ represents the amount that increased mortality is compensated for by an increase in the population. ***Source: Essential Mathematical Biology.***

**a.** Find the equilibrium value $x^*$, that is, the value that makes $f(x^*) = x^*$ so the population is stable. Ignore the trivial value $x = 0$.

**b.** Find the value of $f'(x^*)$.

**c.** The equilibrium value $x^*$ is stable if $|f'(x^*)| < 1$. Find the condition on $R$ so the equilibrium value is stable when (i) $0 < b \leq 2$; (ii) $b > 2$.

**61. Extinction** The probability of a population going extinct by time $t$ when the birth and death rates are the same can be estimated by

$$p(t) = \left(\frac{at}{at + 1}\right)^N,$$

where $a$ is the birth and death rate, and $N$ is the number of individuals in the population at time $t = 0$. Find and interpret $p'(t)$. ***Source: The Mathematical Approach to Biology and Medicine.***

## OTHER APPLICATIONS

**62. Candy** The volume and surface area of a "jawbreaker" for any radius is given by the formulas

$$V(r) = \frac{4}{3}\pi r^3 \quad \text{and} \quad S(r) = 4\pi r^2,$$

respectively. Roger Guffey estimates the radius of a jawbreaker while in a person's mouth to be

$$r(t) = 6 - \frac{3}{17}t,$$

where $r(t)$ is in millimeters and $t$ is in minutes. ***Source: Mathematics Teacher.***

**a.** What is the life expectancy of a jawbreaker?

**b.** Find $dV/dt$ and $dS/dt$ when $t = 17$ and interpret your answer.

**c.** Construct an analogous experiment using some other type of food or verify the results of this experiment.

**63. Interest** A sum of \$1500 is deposited in an account with an interest rate of $r$ percent per year, compounded daily. At the end of 5 years, the balance in the account is given by

$$A = 1500\left(1 + \frac{r}{36,500}\right)^{1825}.$$

Find the rate of change of $A$ with respect to $r$ for the following interest rates.

**a.** 6%

**b.** 8%

**c.** 9%

**64. Depreciation** A certain truck depreciates according to the formula

$$V = \frac{60,000}{1 + 0.3t + 0.1t^2},$$

where $V$ is the value of the truck (in dollars), $t$ is time measured in years, and $t = 0$ represents the time of purchase (in years). Find the rate at which the value of the truck is changing at the following times.

**a.** 2 years          **b.** 4 years

**65. Zenzizenzizenzic** Zenzizenzizenzic is an obsolete word with the distinction of containing the most $z$'s of any word found in the Oxford English Dictionary. It was used in mathematics, before powers were written as superscript numbers, to represent the square of the square of the square of a number. In symbols, zenzizenzizenzic is written as $((x^2)^2)^2$. ***Source: The Phrontistery.***

**a.** Use the chain rule twice to find the derivative.

**b.** Use the properties of exponents to first simplify the expression, and then find the derivative.

**66. Zenzizenzicube** Zenzizenzicube is another obsolete word (see Exercise 65) that represents the square of the square of a cube. In symbols, zenzizenzicube is written as $((x^3)^2)^2$. ***Source: The Phrontistery.***

**a.** Use the chain rule twice to find the derivative.

**b.** Use the properties of exponents to first simplify the expression, and then find the derivative.

**YOUR TURN ANSWERS**

1. $2\sqrt{5} - 1; \sqrt{2}$

2. $4x^2 - 12x + 10$

3. One possible answer is $g(x) = 2x - 3$ and $f(x) = x^3$.

4. $\dfrac{dy}{dx} = \dfrac{-2(10x - 6)}{(5x^2 - 6x)^3}$

5. $20x(x^2 - 7)^9$

6. $\dfrac{dy}{dx} = x(5x - 1)^2(25x - 2)$

7. $\dfrac{(4x - 1)^2(8x + 37)}{(x + 3)^2}$

# 4.4 Derivatives of Exponential Functions

**What is the relationship between the age and weight of a cactus wren?**
*We will use a derivative to answer this question in Example 5 at the end of this section.*

---

**FOR REVIEW**

Recall from Section 2.1 that $e$ is a special irrational number whose value is approximately 2.718281828. It arises in many applications, such as continuously compounded interest, and it can be defined as

$$\lim_{m \to \infty} \left(1 + \frac{1}{m}\right)^m.$$

---

| Approximation of $\lim\limits_{h \to 0} \dfrac{e^h - 1}{h}$ | |
|:---:|:---:|
| $h$ | $\dfrac{e^h - 1}{h}$ |
| $-0.1$ | $0.9516$ |
| $-0.01$ | $0.9950$ |
| $-0.001$ | $0.9995$ |
| $-0.0001$ | $1.0000$ |
| $0.00001$ | $1.0000$ |
| $0.0001$ | $1.0001$ |
| $0.001$ | $1.0005$ |
| $0.01$ | $1.0050$ |
| $0.1$ | $1.0517$ |

We can find the derivative of the exponential function by using the definition of the derivative. Thus

$$\frac{d(e^x)}{dx} = \lim_{h \to 0} \frac{e^{x+h} - e^x}{h}$$

$$= \lim_{h \to 0} \frac{e^x e^h - e^x}{h} \quad \text{Property 1 of exponents}$$

$$= e^x \lim_{h \to 0} \frac{e^h - 1}{h}. \quad \text{Property 1 of limits}$$

In the last step, since $e^x$ does not involve $h$, we were able to bring $e^x$ in front of the limit. The result says that the derivative of $e^x$ is $e^x$ times a constant, namely, $\lim\limits_{h \to 0} (e^h - 1)/h$. To investigate this limit, we evaluate the expression for smaller and smaller values of $h$, as shown in the table in the margin. Based on the table, it appears that $\lim\limits_{h \to 0} (e^h - 1)/h = 1$. This is proved in more advanced courses. We, therefore, have the following formula.

**Derivative of $e^x$**

$$\frac{d}{dx}(e^x) = e^x$$

To find the derivative of the exponential function with a base other than $e$, use the change-of-base theorem for exponentials to rewrite $a^x$ as $e^{(\ln a)x}$. Thus, for any positive constant $a \neq 1$,

$$\frac{d(a^x)}{dx} = \frac{d[e^{(\ln a)x}]}{dx} \quad \text{Change-of-base theorem for exponentials}$$

$$= e^{(\ln a)x} \ln a \quad \text{Chain rule}$$

$$= (\ln a)a^x. \quad \text{Change-of-base theorem again}$$

**Derivative of $a^x$**

For any positive constant $a \neq 1$,

$$\frac{d}{dx}(a^x) = (\ln a)a^x.$$

(The derivative of an exponential function is the original function times the natural logarithm of the base.)

We now see why $e$ is the best base with which to work: It has the simplest derivative of all the exponential functions. Even if we choose a different base, $e$ appears in the derivative anyway through the $\ln a$ term. (Recall that $\ln a$ is the logarithm of $a$ to the base $e$.) In fact, of all the functions we have studied, $e^x$ is the simplest to differentiate, because its derivative is just itself.*

*There is a joke about a deranged mathematician who frightened other inmates at an insane asylum by screaming at them, "I'm going to differentiate you!" But one inmate remained calm and simply responded, "I don't care; I'm $e^x$."

The chain rule can be used to find the derivative of the more general exponential function $y = a^{g(x)}$. Let $y = f(u) = a^u$ and $u = g(x)$, so that $f[g(x)] = a^{g(x)}$. Then

$$f'[g(x)] = f'(u) = (\ln a)a^u = (\ln a)a^{g(x)},$$

and by the chain rule,

$$\frac{dy}{dx} = f'[g(x)] \cdot g'(x)$$

$$= (\ln a)a^{g(x)} \cdot g'(x).$$

As before, this formula becomes simpler when we use natural logarithms because $\ln e = 1$. We summarize these results next.

---

**Derivative of $a^{g(x)}$ and $e^{g(x)}$**

$$\frac{d}{dx}\left(a^{g(x)}\right) = (\ln a)\,a^{g(x)}g'(x)$$

and

$$\frac{d}{dx}\left(e^{g(x)}\right) = e^{g(x)}g'(x)$$

---

You need not memorize the previous two formulas. They are simply the result of applying the chain rule to the formula for the derivative of $a^x$.

---

**CAUTION**   Notice the difference between the derivative of a variable to a constant power, such as $D_x x^3 = 3x^2$, and a constant to a variable power, like $D_x 3^x = (\ln 3)3^x$. Remember, $D_x 3^x \neq x3^{x-1}$.

---

**EXAMPLE 1**   **Derivatives of Exponential Functions**

Find the derivative of each function.

**(a)** $y = e^{5x}$

**SOLUTION**   Let $g(x) = 5x$, so $g'(x) = 5$. Then

$$\frac{dy}{dx} = 5e^{5x}.$$

**(b)** $s = 3^t$

**SOLUTION**

$$\frac{ds}{dt} = (\ln 3)3^t$$

**(c)** $y = 10e^{3x^2}$

**SOLUTION**

$$\frac{dy}{dx} = 10(e^{3x^2})(6x) = 60xe^{3x^2}$$

**(d)** $s = 8 \cdot 10^{1/t}$

**SOLUTION**

$$\frac{ds}{dt} = 8(\ln 10)10^{1/t}\left(\frac{-1}{t^2}\right)$$

$$= \frac{-8(\ln 10)10^{1/t}}{t^2}$$

TRY YOUR TURN 1

**YOUR TURN 1**   Find $dy/dx$ for
**a.** $y = 4^{3x}$,   **b.** $y = e^{7x^3+5}$.

The rules for the derivatives of exponential functions may be used in conjunction with the rules we previously learned, as shown in the following examples.

**EXAMPLE 2** Derivatives of Exponential Functions

Find the derivative of each function.

**(a)** $f(x) = x^3 e^{x^2}$

**SOLUTION** Use the product rule, with $u = x^3$ and $v = e^{x^2}$.

$$f'(x) = x^3 \cdot e^{x^2} \cdot 2x + e^{x^2} \cdot 3x^2 \qquad \text{2x comes from the chain rule for the derivative of } e^{x^2}.$$
$$= 2x^4 e^{x^2} + 3x^2 e^{x^2}$$
$$= x^2 e^{x^2}(2x^2 + 3) \qquad \text{Factor out the greatest common factor.}$$

**(b)** $f(t) = \dfrac{31.4}{1 + 12.5e^{-0.393t}}$

**SOLUTION** Use the quotient rule.

$$f'(t) = \frac{(1 + 12.5e^{-0.393t})(0) - 31.4(-0.393 \cdot 12.5e^{-0.393t})}{(1 + 12.5e^{-0.393t})^2}$$

$$= \frac{154.2525e^{-0.393t}}{(1 + 12.5e^{-0.393t})^2} \qquad \text{TRY YOUR TURN 2}$$

**YOUR TURN 2** Find $f'(x)$ for
**a.** $f(x) = (x^2 + 3)e^{5x}$,
**b.** $f(x) = \dfrac{100}{5 + 2e^{-0.01x}}$.

In Example 2(b), we could have taken the derivative by writing $f(t) = 31.4(1 + 12.5e^{-0.393t})^{-1}$, from which we have

$$f'(t) = -31.4(1 + 12.5e^{-0.393t})^{-2} \cdot 12.5e^{-0.393t} \cdot (-0.393).$$

This simplifies to the same expression as in Example 2(b).

**EXAMPLE 3** Derivative of an Exponential Function

Let $y = e^{x^2+1}\sqrt{5x + 2}$. Find $\dfrac{dy}{dx}$.

**SOLUTION** Rewrite $y$ as $y = e^{x^2+1}(5x + 2)^{1/2}$, and then use the product rule and the chain rule.

$$\frac{dy}{dx} = e^{x^2+1} \cdot \frac{1}{2}(5x + 2)^{-1/2} \cdot 5 + (5x + 2)^{1/2}e^{x^2+1} \cdot 2x$$

$$= e^{x^2+1}(5x + 2)^{-1/2}\left[\frac{5}{2} + (5x + 2) \cdot 2x\right] \qquad \text{Factor out the greatest common factor, } e^{x^2+1}(5x + 2)^{-1/2}.$$

$$= e^{x^2+1}(5x + 2)^{-1/2}\left[\frac{5 + 4x(5x + 2)}{2}\right] \qquad \text{Least common denominator}$$

$$= e^{x^2+1}(5x + 2)^{-1/2}\left[\frac{5 + 20x^2 + 8x}{2}\right]$$

$$= \frac{e^{x^2+1}(20x^2 + 8x + 5)}{2\sqrt{5x + 2}} \qquad \text{Simplify.}$$

**YOUR TURN 3** Let $y = (x^2 + 1)^2 e^{2x}$. Find $dy/dx$.

TRY YOUR TURN 3

**EXAMPLE 4** Radioactivity

The amount in grams in a sample of uranium-239 after $t$ years is given by

$$A(t) = 100e^{-0.362t}.$$

Find the rate of change of the amount present after 3 years.

**SOLUTION** The rate of change is given by the derivative $dA/dt$.

$$\frac{dA}{dt} = 100(e^{-0.362t})(-0.362) = -36.2e^{-0.362t}$$

**YOUR TURN 4** The quantity (in grams) of a radioactive substance present after $t$ years is $Q(t) = 100e^{-0.421t}$. Find the rate of change of the quantity present after 2 years.

After 3 years ($t = 3$), the rate of change is

$$\frac{dA}{dt} = -36.2e^{-0.362(3)} = -36.2e^{-1.086} \approx -12.2$$

grams per year.                                                                 TRY YOUR TURN 4

Frequently a population will start growing slowly, then grow more rapidly, and then gradually level off. Such growth can often be approximated by a mathematical model known as the **logistic function**:

$$G(t) = \frac{mG_0}{G_0 + (m - G_0)e^{-kmt}},$$

where $t$ represents time in appropriate units, $G_0$ is the initial number present, $m$ is the maximum possible size of the population, $k$ is a positive constant, and $G(t)$ is the population at time $t$. It is sometimes simpler to divide the numerator and denominator of the logistic function by $G_0$, writing the result as

$$G(t) = \frac{m}{1 + \left(\dfrac{m}{G_0} - 1\right)e^{-kmt}}.$$

Notice that

$$\lim_{t \to \infty} G(t) = \frac{m}{1 + 0} = m$$

because $\lim_{t \to \infty} e^{-kmt} = 0$.

### EXAMPLE 5   Cactus Wrens (Arizona State Bird)

Researchers have found that the weight of a cactus wren (in grams) can be modeled with a logistic function. Let $t$ be the age of a wren (in days). Suppose that the initial weight of a wren is 2.33 g and, on day 2, its weight is 4.69 g. The researchers also estimate that the maximum weight of a wren is 31.4 g. *Source: Ecology.*

**(a)** Find the growth function $G(t)$ for the cactus wren.

APPLY IT       **SOLUTION**   We already know that $S_0 = 2.33$ and $m = 31.4$, so

$$G(t) = \frac{31.4}{1 + \left(\dfrac{31.4}{2.33} - 1\right)e^{-k \cdot 31.4t}}$$

$$= \frac{31.4}{1 + 12.5e^{-k \cdot 31.4t}}$$

To find $k$, use the fact that $G(2) = 4.69$.

$$4.69 = \frac{31.4}{1 + 12.5e^{-k \cdot 31.4 \cdot 2}}$$

$$4.69 = \frac{31.4}{1 + 12.5e^{-k \cdot 62.8}}$$

| | |
|---|---|
| $4.69(1 + 12.5e^{-k \cdot 62.8}) = 31.4$ | Cross multiply. |
| $4.69 + 58.625e^{-k \cdot 62.8} = 31.4$ | Distribute. |
| $58.625e^{-k \cdot 62.8} = 26.71$ | Subtract 4.69 from both sides. |
| $e^{-k \cdot 62.8} = 0.4556$ | Divide both sides by 58.625. |
| $-k \cdot 62.8 = \ln 0.4556$ | Take the natural logarithm of both sides. |
| $k = (-\ln 0.4556)/62.8$ | Divide both sides by $-62.8$. |
| $k \approx 0.0125$ | |

Substituting $k = 0.0125$, the growth function $G(t)$ for the cactus wren is

$$G(t) = \frac{31.4}{1 + 12.5e^{-0.393t}}$$

**(b)** Find the rate of change of the weight of the wren at day 1. Repeat for day 5 and day 10.

**SOLUTION**   The derivative of this growth function, which gives the rate of change of the weight of the wren, was found in Example 2(b). Using that derivative,

$$G'(1) = \frac{154.2525e^{-0.393 \cdot 1}}{(1 + 12.5e^{-0.393 \cdot 1})^2} \approx 1.17.$$

The rate of change of weight at 1 day is about 1.17 g per day. The positive number indicates that the weight is increasing at this time.

At day 5, $G'(5) \approx 2.85$, which indicates the weight is increasing by about 2.85 g per day. At day 10, $G'(10) \approx 1.95$, and the weight is increasing by about 1.95 g per day.

**(c)** What happens to the weight of the wren over time?

**SOLUTION**   As time increases, $t \to \infty$, and

$$e^{-0.393t} = \frac{1}{e^{0.393t}} \to 0.$$

Thus,

$$\lim_{t \to \infty} G(t) = \frac{31.4}{1 + 12.5 \cdot 0} = 31.4.$$

Over time, the weight of a cactus wren approaches 31.4 g, the maximum weight. The graph of the function $G(t)$ is illustrated in Figure 8. Notice the rate of growth increases for a while and then gradually decreases to 0, as $G(t)$ approaches a horizontal asymptote of 31.4 g.

| $t$ | $G(t)$ |
|-----|--------|
| 0   | 2.33   |
| 5   | 11.41  |
| 10  | 25.21  |
| 15  | 30.36  |
| 20  | 31.25  |
| 25  | 31.38  |

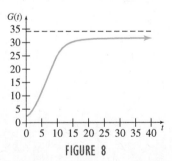

FIGURE 8

It is sometimes convenient to work with the derivative of the logistic function. One form of the derivative is

$$\frac{dG}{dt} = r\left(1 - \frac{G}{m}\right)G, \qquad \text{with } G(0) = G_0,$$

where $r$ is a constant, $G$ is the population at time $t$, and $m$ is the maximum size of the population. (See Exercise 39.) This concept will be explored in this chapter's extended application.

# 4.4 EXERCISES

**Find derivatives of the functions defined as follows.**

**1.** $y = e^{4x}$

**2.** $y = e^{-2x}$

**3.** $y = -8e^{3x}$

**4.** $y = 1.2e^{5x}$

**5.** $y = -16e^{2x+1}$

**6.** $y = -4e^{-0.3x}$

**7.** $y = e^{x^2}$

**8.** $y = e^{-x^2}$

**9.** $y = 3e^{2x^2}$

**10.** $y = -5e^{4x^3}$

**11.** $y = 4e^{2x^2-4}$

**12.** $y = -3e^{3x^2+5}$

**13.** $y = xe^x$

**14.** $y = x^2e^{-2x}$

**15.** $y = (x + 3)^2 e^{4x}$

**16.** $y = (3x^3 - 4x)e^{-5x}$

**17.** $y = \dfrac{x^2}{e^x}$

**18.** $y = \dfrac{e^x}{2x + 1}$

**19.** $y = \dfrac{e^x + e^{-x}}{x}$

**20.** $y = \dfrac{e^x - e^{-x}}{x}$

**21.** $p = \dfrac{10{,}000}{9 + 4e^{-0.2t}}$

**22.** $p = \dfrac{500}{12 + 5e^{-0.5t}}$

**23.** $f(z) = (2z + e^{-z^2})^2$

**24.** $f(t) = (e^{t^2} + 5t)^3$

**25.** $y = 7^{3x+1}$

**26.** $y = 4^{-5x+2}$

**27.** $y = 3 \cdot 4^{x^2+2}$

**28.** $y = -10^{3x^2-4}$

**29.** $s = 2 \cdot 3^{\sqrt{t}}$

**30.** $s = 5 \cdot 2^{\sqrt{t-2}}$

**31.** $y = \dfrac{te^t + 2}{e^{2t} + 1}$

**32.** $y = \dfrac{t^2e^{2t}}{t + e^{3t}}$

**33.** $f(x) = e^{x\sqrt{3x+2}}$

**34.** $f(x) = e^{x^2/(x^3+2)}$

**35.** Prove that if $y = y_0e^{kt}$, where $y_0$ and $k$ are constants, then $dy/dt = ky$. (This says that for exponential growth and decay, the rate of change of the population is proportional to the size of the population, and the constant of proportionality is the growth or decay constant.)

**36.** Use a graphing calculator to sketch the graph of $y = [f(x + h) - f(x)]/h$ using $f(x) = e^x$ and $h = 0.0001$. Compare it with the graph of $y = e^x$ and discuss what you observe.

**37.** Use graphical differentiation to verify that $\dfrac{d}{dx}(e^x) = e^x$.

**38. a.** Using a graphing calculator, sketch the graph of $y = 2^x$ on the interval $-1 \le x \le 1$, and have the calculator compute the derivative at $x = 0$. Compare the result with what you would predict based on the formula for the derivative of $a^x$.

**b.** Repeat part a, this time using the graph $y = 3^x$.

**c.** You should have found that the answer to part a is less than 1 and the answer to part b is greater than 1. Using trial and error, find a function $y = a^x$ that gives a derivative of 1 at $x = 0$. What do you find about the value of $a$?

**39.** Show that the logistic function,

$$G(t) = \dfrac{m}{1 + \left(\dfrac{m}{G_0} + 1\right)e^{-kmt}},$$

presented in Example 5 satisfies the equation

$$\dfrac{dG}{dt} = r\left(1 - \dfrac{G}{m}\right)G, \quad \text{with } G(0) = G_0,$$

where $r$ is a constant, $G$ is the population at time $t$, and $m$ is the maximum size of the population.

## LIFE SCIENCE APPLICATIONS

**40. Population Growth** In Section 2.1, Exercise 37, the growth in world population (in millions) was approximated by the exponential function

$$A(t) = 3100e^{0.0166t},$$

where $t$ is the number of years since 1960. Find the instantaneous rate of change in the world population at the following times. *Source: United Nations.*

  **a.** 2010              **b.** 2015

**41. Minority Population** In Section 2.1, Exercise 39, we saw that the projected Hispanic population in the United States (in millions) can be approximated by the function

$$h(t) = 37.79(1.021)^t$$

where $t = 0$ corresponds to 2000 and $0 \le t \le 50$. *Source: U.S. Census Bureau.*

  **a.** Estimate the Hispanic population in the United States for the year 2015.

  **b.** What is the instantaneous rate of change of the Hispanic population in the United States when $t = 15$? Interpret your answer.

**42. Insect Growth** The growth of a population of rare South American beetles is given by the logistic function with $k = 0.00001$ and $t$ in months. Assume that there are 200 beetles initially and that the maximum population size is 10,000.

  **a.** Find the growth function $G(t)$ for these beetles.

Find the population and rate of growth of the population after the following times.

  **b.** 6 months     **c.** 3 years     **d.** 7 years

  **e.** What happens to the rate of growth over time?

**43. Clam Population** The population of a bed of clams in the Great South Bay off Long Island is described by a logistic function with $t$ in years. Assume that there are 400 clams initially and that the maximum population size is 5200. After two years, the population is about 1000 clams.

  **a.** Find the growth function $G(t)$ for the clams.

Find the population and rate of growth of the population after the following times.

  **b.** 1 year     **c.** 4 years     **d.** 10 years

  **e.** What happens to the rate of growth over time?

**44. Whooping Cranes** Based on data from the U.S. Fish and Wildlife Service, the population of whooping cranes in the Aransas–Wood Buffalo National Park can be approximated by a logistic function with $k = 6.01 \cdot 10^{-5}$, with a population in

1958 of 32 and a maximum population of 787. **Source: U.S. Fish and Wildlife Service.**

**a.** Find the growth function $G(t)$ for the whooping crane population, where $t$ is the time since 1938, when the park first started counting the cranes.

**b.** Find the initial population $G_0$.

Find the population and rate of growth in the following years.

**c.** 1945        **d.** 1985        **e.** 2005

**f.** What happens to the rate of growth over time?

**45. Yeast Growth**    A population of the yeast *Saccharomyces cerevisiae* was put in a flask with nutrients and found to grow logistically with a limiting population of $m = 3.7 \times 10^8$ and a growth rate of $km = 0.55$ (where time was in hours). **Source: Genetical Research.** One population was found to be $2.1 \times 10^8$, while, at the same time, another was found to be $2.6 \times 10^8$. How much time was required for the first population to reach the initial size of the second? How fast was it growing at that time?

**46. Pollution Concentration**    The concentration of pollutants (in grams per liter) in the east fork of the Big Weasel River is approximated by

$$P(x) = 0.04e^{-4x},$$

where $x$ is the number of miles downstream from a paper mill that the measurement is taken. Find the following values.

**a.** The concentration of pollutants 0.5 mile downstream

**b.** The concentration of pollutants 1 mile downstream

**c.** The concentration of pollutants 2 miles downstream

Find the rate of change of concentration with respect to distance for the following distances.

**d.** 0.5 mile        **e.** 1 mile        **f.** 2 miles

**47. Mortality**    The percentage of people of any particular age group that will die in a given year may be approximated by the formula

$$P(t) = 0.00239e^{0.0957t},$$

where $t$ is the age of the person in years. **Source: U.S. Vital Statistics.**

**a.** Find $P(25)$, $P(50)$, and $P(75)$.

**b.** Find $P'(25)$, $P'(50)$, and $P'(75)$.

**c.** Interpret your answers for parts a and b. Are there any limitations of this formula?

**48. Breast Cancer**    It has been observed that the following formula accurately models the relationship between the size of a breast tumor and the amount of time that it has been growing.

$$V(t) = 1100[1023e^{-0.02415t} + 1]^{-4},$$

where $t$ is in months and $V(t)$ is measured in cubic centimeters. **Source: Cancer.**

**a.** Find the tumor volume at 240 months.

**b.** Assuming that the shape of a tumor is spherical, find the radius of the tumor from part a. (*Hint:* The volume of a sphere is given by the formula $V = (4/3)\pi r^3$.)

**c.** If a tumor of size 0.5 cm$^3$ is detected, according to the formula, how long has it been growing? What does this imply?

**d.** Find $\lim_{t \to \infty} V(t)$ and interpret this value. Explain whether this makes sense.

**e.** Calculate the rate of change of tumor volume at 240 months and interpret.

**49. Arctic Foxes**    The age–weight relationship of female Arctic foxes caught in Svalbard, Norway, can be estimated by the function

$$M(t) = 3102e^{-e^{-0.022(t-56)}},$$

where $t$ is the age of the fox in days and $M(t)$ is the weight of the fox in grams. **Source: Journal of Mammalogy.**

**a.** Estimate the weight of a female fox that is 200 days old.

**b.** Use $M(t)$ to estimate the largest size that a female fox can attain. (*Hint:* Find $\lim_{t \to \infty} M(t)$.)

**c.** Estimate the age of a female fox when it has reached 80% of its maximum weight.

**d.** Estimate the rate of change in weight of an Arctic fox that is 200 days old. (*Hint:* Recall that $D_t[e^{f(t)}] = f'(t)e^{f(t)}$.)

**e.** Use a graphing calculator to graph $M(t)$ and then describe the growth pattern.

**f.** Use the table function on a graphing calculator or a spreadsheet to develop a chart that shows the estimated weight and growth rate of female foxes for days 50, 100, 150, 200, 250, and 300.

**50. Holstein Dairy Cattle**    Researchers have used the Richardson model to develop the following function that can be used to accurately predict the weight of Holstein cows (females) of various ages:

$$W(t) = 619(1 - 0.905e^{-0.002t})^{1.2386},$$

where $W(t)$ is the weight of a Holstein cow (in kg) that is $t$ days old. **Source: Canadian Journal of Animal Science.**

**a.** What is the maximum weight of the average Holstein cow?

**b.** At what age does a cow reach 90% of this maximum weight?

**c.** Find $W'(1000)$ and interpret.

**51. Ayrshire Dairy Cattle**    The researchers from the previous exercise also developed a function that can be used to accurately predict the weight of Ayrshire cows (females). In this case,

$$W(t) = 532(1 - 0.911e^{-0.0021t})^{1.2466}.$$

**Source: Canadian Journal of Animal Science.**

**a.** What is the maximum weight of the average Ayrshire cow?

**b.** At what age does a cow reach 90% of this maximum weight?

**c.** Find $W'(1000)$ and interpret.

**d.** Graph the function in this exercise along with the function from the previous exercise on $[0, 2500]$ by $[0, 700]$. Describe the growth pattern of each cow.

**52. Cutlassfish**    The cutlassfish is one of the most important resources of the commercial marine fishing industry in China. Researchers have developed a von Bertalanffy growth model that uses the age of a certain species of cutlassfish to estimate length such that

$$L(t) = 589\left[1 - e^{(-0.168(t+2.682))}\right],$$

where $L(t)$ is the length of the fish (in mm) at time $t$ (yr). *Source: Fish Bulletin.*

a. What happens to the length of the average cutlassfish of this species over time?

b. Determine the age of a fish that has grown to 95% of its maximum length.

c. Find $L'(4)$ and interpret the result.

d. Graph the function on $[0, 20]$ by $[0, 600]$.

**53. Beef Cattle** Researchers have compared two models that are used to predict the weight of beef cattle of various ages,

$$W_1(t) = 509.7(1 - 0.941e^{-0.00181t})$$

and

$$W_2(t) = 498.4(1 - 0.889e^{-0.00219t})^{1.25},$$

where $W_1(t)$ and $W_2(t)$ represent the weight (in kilograms) of a $t$-day-old beef cow. *Source: Journal of Animal Science.*

a. What is the maximum weight predicted by each function for the average beef cow? Is this difference significant?

b. According to each function, find the age that the average beef cow reaches 90% of its maximum weight.

c. Find $W_1'(750)$ and $W_2'(750)$. Compare your results.

d. Graph the two functions on $[0, 2500]$ by $[0, 525]$ and comment on the differences in growth patterns for each of these functions.

e. Graph the derivative of these two functions on $[0, 2500]$ by $[0, 1]$ and comment on any differences you notice between these functions.

**54. Dialysis** One measure of whether a dialysis patient has been adequately dialyzed is by the urea reduction ratio (URR). It is generally agreed that a patient has been adequately dialyzed when URR exceeds a value of 0.65. The value of URR can be calculated for a particular patient using the following formula by Gotch:

$$URR = 1 - \left\{(0.96)^{0.14t-1} + \frac{8t}{126t + 900}[1 - (0.96)^{0.14t-1}]\right\},$$

where $t$ is measured in minutes. *Source: American Journal of Nephrology.*

a. Find the value of URR after a patient receives dialysis for 180 minutes. Has the patient received adequate dialysis?

b. Find the value of URR after a patient receives dialysis for 240 minutes. Has the patient received adequate dialysis?

c. Calculate the instantaneous rate of change of URR when time on dialysis is 240 minutes, and interpret.

**55. Medical Literature** It has been observed that there has been an increase in the proportion of medical research papers that use the word "novel" in the title or abstract, and that this proportion can be accurately modeled by the function

$$p(t) = 0.001131e^{0.1268t},$$

where $t$ is the number of years since 1970. *Source: Nature.*

a. Find $p(40)$.

b. If this phenomenon continues, estimate the year in which every medical article will contain the word "novel" in its title or abstract.

c. Estimate the rate of increase in the proportion of medical papers using this word in the year 2010.

d. Explain some factors that may be contributing to researchers using this word.

**56. Cholesterol** Researchers have found that the risk of coronary heart disease rises as blood cholesterol increases. This risk may be approximated by the function

$$R(c) = 3.19(1.006)^c, \quad 100 \le c \le 300,$$

where $R$ is the risk in terms of coronary heart disease incidence per 1000 per year, and $c$ is the cholesterol in mg/dL. Suppose a person's cholesterol is 180 mg/dL and going up at a rate of 15 mg/dL per year. At what rate is the person's risk of coronary heart disease going up? *Source: Circulation.*

**57. Nervous System** In a model of the nervous system, the intensity of excitation, $I$, of a nerve pathway is given by

$$I = E[1 - e^{-a(S-h)/E}],$$

where $E$ is the maximum possible excitation, $S$ is the intensity of a stimulus, $h$ is a threshold stimulus, and $a$ is a constant. Find the rate of change of the intensity of the excitation with respect to the intensity of the stimulus. *Source: Mathematical Biology of Social Behavior.*

**58. Extinction** The probability of a population going extinct by time $t$ can be estimated by

$$p(t) = \left(\frac{a[e^{(b-a)t} - 1]}{be^{(b-a)t} - a}\right)^N,$$

where $a$ is the death rate, $b$ is the birth rate $(b \ne a)$, and $N$ is the number of individuals in the population at time $t = 0$. Find and interpret $p'(t)$. *Source: The Mathematical Approach to Biology and Medicine.*

**59. Track and Field** In 1958, L. Lucy developed a method for predicting the world record for any given year that a human could run a distance of 1 mile. His formula is given as follows:

$$t(n) = 218 + 31(0.933)^n,$$

where $t(n)$ is the world record (in seconds) for the mile run in year $1950 + n$. Thus, $n = 5$ corresponds to the year 1955. *Source: Statistics in Sports.*

a. Find the estimate for the world record in the year 2010.

b. Calculate the instantaneous rate of change for the world record at the end of year 2010 and interpret.

c. Find $\lim\limits_{n \to \infty} t(n)$ and interpret. How does this compare with the current world record?

**60. Involutional Psychosis** Researchers have found that the incidence of involutional psychosis can be described by the equation

$$I(t) = A(1 - e^{-k(t-b)^r})^n,$$

where $t$ is age in years. *Source: An Introduction to the Mathematics of Medicine and Biology.* The values of $r$ and $n$ were found to be 4 and 3, respectively. For men, the values of $A$, $k$, and $b$ were found to be 0.0124, $3.2 \times 10^{-7}$, and 8. For women, the values are 0.0076, $1.6 \times 10^{-7}$, and 4.

    **a.** Find the incidence and rate of change of the incidence for men at (i) 20 years and (ii) 60 years.

    **b.** Repeat part a for women.

**61. Radioactive Iron** Researchers found that the fraction of radioactive iron in one patient's blood plasma could be described by

$$f(t) = 0.981e^{-6.74t} + 0.0163e^{-1.019t} + 0.0031e^{-0.210t},$$

where $t$ is the time in days after the iron was injected. *Source: The Journal of Clinical Investigation.*

    **a.** Find the fraction of radioactive iron remaining and the rate that it was declining at (i) 6 hours, (ii) 1 day, and (iii) 5 days.

    **b.** What is happening to the fraction of radioactive iron in the plasma as time goes on?

    **c.** What is happening to the rate of change of the fraction of radioactive iron in the plasma as time goes on?

    **d.** Calculate the value of each of the terms of $f(t)$ at (i) 6 hours, (ii) 1 day, and (iii) 5 days.

    **e.** From your answers to part d, which term is most important at (i) 6 hours, (ii) 1 day, and (iii) 5 days? Explain why this is happening, based on the values of the coefficients of each term and the coefficient of $t$ in each term.

**62. Radioactive Albumin** Researchers found that the fraction of radioactive albumin in one patient's blood plasma could be described by

$$f(t) = 0.42e^{-0.0364t} + 0.255e^{-0.642t} + 0.325e^{-2.98t},$$

where $t$ is the time in days after the iron was injected. *Source: The Journal of Laboratory and Clinical Medicine.*

    **a.** Find the fraction of radioactive albumin remaining and the rate that it was declining at (i) 12 hours, (ii) 1 day, and (iii) 10 days.

    **b.** What is happening to the fraction of radioactive albumin in the plasma as time goes on?

    **c.** What is happening to the rate of change of the fraction of radioactive albumin in the plasma as time goes on?

    **d.** Calculate the value of the first term of $f(t)$, as well as the portion of $f(t)$ that it represents, at (i) 12 hours, (ii) 1 day, and (iii) 10 days. Explain what is happening.

## OTHER APPLICATIONS

**63. Survival of Manuscripts** Paleontologist John Cisne has demonstrated that the survival of ancient manuscripts can be modeled by the logistic equation. For example, the number of copies of the Venerable Bede's *De Temporum Ratione* was found to approach a limiting value over the five centuries after its publication in the year 725. Let $G(t)$ represent the proportion of manuscripts known to exist after $t$ centuries out of the limiting value, so that $m = 1$. Cisne found that for Venerable Bede's

*De Temporum Ratione*, $k = 3.5$ and $G_0 = 0.00369$. *Source: Science.*

    **a.** Find the growth function $G(t)$ for the proportion of copies of *De Temporum Ratione* found.

Find the proportion of manuscripts and their rate of growth after the following number of centuries.

    **b.** 1      **c.** 2      **d.** 3

    **e.** What happens to the rate of growth over time?

**64. Online Learning** The growth of the number of students taking at least one online course can be approximated by a logistic function with $k = 0.0440$, where $t$ is the number of years since 2002. In 2002 (when $t = 0$), the number of students enrolled in at least one online course was 1.603 million. Assume that the number will level out at around 6.8 million students. *Source: The Sloan Consortium.*

    **a.** Find the growth function $G(t)$ for students enrolled in at least one online course.

Find the number of students enrolled in at least one online course and the rate of growth in the number in the following years.

    **b.** 2004      **c.** 2006      **d.** 2010

    **e.** What happens to the rate of growth over time?

**65. Habit Strength** According to work by the psychologist C. L. Hull, the strength of a habit is a function of the number of times the habit is repeated. If $N$ is the number of repetitions and $H(N)$ is the strength of the habit, then

$$H(N) = 1000(1 - e^{-kN}),$$

where $k$ is a constant. Find $H'(N)$ if $k = 0.1$ and the number of times the habit is repeated is as follows.

    **a.** 10      **b.** 100      **c.** 1000

    **d.** Show that $H'(N)$ is always positive. What does this mean?

**66. Radioactive Decay** The amount (in grams) of a sample of lead-214 present after $t$ years is given by

$$A(t) = 500e^{-0.31t}.$$

Find the rate of change of the quantity present after each of the following years.

    **a.** 4      **b.** 6      **c.** 10

    **d.** What is happening to the rate of change of the amount present as the number of years increases?

    **e.** According to the model, will the substance ever be gone completely?

**67. Electricity** In a series resistance-capacitance DC circuit, the instantaneous charge $Q$ on the capacitor as a function of time (where $t = 0$ is the moment the circuit is energized by closing a switch) is given by the equation

$$Q(t) = CV(1 - e^{-t/RC}),$$

where $C$, $V$, and $R$ are constants. Further, the instantaneous charging current $I_C$ is the rate of change of charge on the capacitor, or $I_C = dQ/dt$. *Source: Kevin Friedrich.*

**a.** Find the expression for $I_C$ as a function of time.

**b.** If $C = 10^{-5}$ farads, $R = 10^7$ ohms, and $V = 10$ volts, what is the charging current after 200 seconds? (*Hint:* When placed into the function in part a, the units can be combined into amps.)

**68. Heat Index** The heat index is a measure of how hot it really feels under different combinations of temperature and humidity. The heat index, in degrees Fahrenheit, can be approximated by

$$H(T) = T - 0.9971e^{0.02086T}[1 - e^{0.0445(D-57.2)}],$$

where the temperature $T$ and dewpoint $D$ are both expressed in degrees Fahrenheit. *Source: American Meteorological Society.*

**a.** Assume the dewpoint is $D = 85°F$. Find the function $H(T)$.

**b.** Using the function you found in part a, find the heat index when the temperature $T$ is 80°F.

**c.** Find the rate of change of the heat index when $T = 80° F$.

**69. Dead Grandmother Syndromes** Biology professor Mike Adams has found an increase over the years in the number of students claiming just before an exam that a relative (usually a grandmother) has died. Based on his data, he found that the average number of family deaths per 100 students could be approximated by

$$f(t) = 5.4572 \times 10^{-59}10^{2.923 \times 10^{-2}t},$$

where $t$ is the year. Find the average number of deaths per 100 students in 1990 and the rate that the function was increasing in that year. *Source: Annals of Improbable Research.*

**YOUR TURN ANSWERS**

**1. a.** $\dfrac{dy}{dx} = 3(\ln 4)4^{3x}$   **b.** $\dfrac{dy}{dx} = 21x^2e^{7x^3+5}$

**2. a.** $f'(x) = (5x^2 + 2x + 15)e^{5x}$

   **b.** $f'(x) = \dfrac{2e^{-0.01x}}{(5 + 2e^{-0.01x})^2}$

**3.** $\dfrac{dy}{dx} = 2e^{2x}(x^2 + 1)(x + 1)^2$

**4.** $-18.1$ grams per year

# 4.5 Derivatives of Logarithmic Functions

**APPLY IT** How does the average velocity of pedestrians in a city vary with the population size?

*We will use the derivative to answer this question in Example 4.*

Recall that in Section 2.2 on Logarithmic Functions, we showed that the logarithmic function and the exponential function are inverses of each other. In the last section we showed that the derivative of $a^x$ is $(\ln a)a^x$. We can use this information and the chain rule to find the derivative of $\log_a x$. We begin by solving the general logarithmic function for $x$.

$$f(x) = \log_a x$$
$$a^{f(x)} = x \qquad \text{Definition of the logarithm}$$

Now consider the left and right sides of the last equation as functions of $x$ that are equal, so their derivatives with respect to $x$ will also be equal. Notice in the first step that we need to use the chain rule when differentiating $a^{f(x)}$.

$$(\ln a)a^{f(x)}f'(x) = 1 \qquad \text{Derivative of the exponential function}$$
$$(\ln a)xf'(x) = 1 \qquad \text{Substitute } a^{f(x)} = x.$$

Finally, divide both sides of this equation by $(\ln a)x$ to get

$$f'(x) = \frac{1}{(\ln a)x}.$$

**Derivative of $\log_a x$**

$$\frac{d}{dx}[\log_a x] = \frac{1}{(\ln a)x}$$

(The derivative of a logarithmic function is the reciprocal of the product of the variable and the natural logarithm of the base.)

As with the exponential function, this formula becomes particularly simple when we let $a = e$, because of the fact that $\ln e = 1$.

---

**Derivative of $\ln x$**

$$\frac{d}{dx}[\ln x] = \frac{1}{x}$$

---

This fact can be further justified geometrically. Notice what happens to the slope of the line $y = 2x + 4$ if the $x$-axis and $y$-axis are switched. That is, if we replace $x$ with $y$ and $y$ with $x$, then the resulting line $x = 2y + 4$ or $y = x/2 - 2$ is a reflection of the line $y = 2x + 4$ across the line $y = x$, as seen in Figure 9. Furthermore, the slope of the new line is the reciprocal of the slope of the original line. In fact, the reciprocal property holds for all lines.

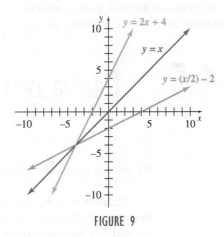

**FIGURE 9**

In Section 2.2 on Logarithmic Functions, we showed that switching the $x$ and $y$ variables changes the exponential graph into a logarithmic graph, a defining property of functions that are inverses of each other. We also showed in the previous section that the slope of the tangent line of $e^x$ at any point is $e^x$—that is, the $y$-coordinate itself. So, if we switch the $x$ and $y$ variables, the new slope of the tangent line will be $1/y$, except that it is no longer $y$; it is $x$. Thus, the slope of the tangent line of $y = \ln x$ must be $1/x$ and hence $D_x \ln x = 1/x$.

---

**EXAMPLE 1** Derivatives of Logarithmic Functions

Find the derivative of each function.

**(a)** $f(x) = \ln 6x$

**SOLUTION** Use the properties of logarithms and the rules for derivatives.

$$f'(x) = \frac{d}{dx}(\ln 6x)$$

$$= \frac{d}{dx}(\ln 6 + \ln x)$$

$$= \frac{d}{dx}(\ln 6) + \frac{d}{dx}(\ln x) = 0 + \frac{1}{x} = \frac{1}{x}$$

**YOUR TURN 1** Find the derivative of $f(x) = \log_3 x$.

**(b)** $y = \log x$

**SOLUTION** Recall that when the base is not specified, we assume that the logarithm is a common logarithm, which has a base of 10.

$$\frac{dy}{dx} = \frac{1}{(\ln 10)x}$$

**TRY YOUR TURN 1**

Applying the chain rule to the formulas for the derivative of logarithmic functions gives us

$$\frac{d}{dx}\log_a g(x) = \frac{1}{\ln a} \cdot \frac{g'(x)}{g(x)}$$

and

$$\frac{d}{dx}\ln g(x) = \frac{g'(x)}{g(x)}.$$

### EXAMPLE 2 Derivatives of Logarithmic Functions

Find the derivative of each function.

**(a)** $f(x) = \ln(x^2 + 1)$

**SOLUTION** Here $g(x) = x^2 + 1$ and $g'(x) = 2x$. Thus,

$$f'(x) = \frac{d}{dx}\ln g(x) = \frac{g'(x)}{g(x)} = \frac{2x}{x^2 + 1}.$$

**(b)** $y = \log_2(3x^2 - 4x)$

**YOUR TURN 2** Find the derivative of
**a.** $y = \ln(2x^3 - 3)$,
**b.** $f(x) = \log_4(5x + 3x^3)$.

**SOLUTION**

$$\frac{dy}{dx} = \frac{1}{\ln 2} \cdot \frac{6x - 4}{3x^2 - 4x}$$

$$= \frac{6x - 4}{(\ln 2)(3x^2 - 4x)}$$

**TRY YOUR TURN 2**

If $y = \ln(-x)$, where $x < 0$, the chain rule with $g(x) = -x$ and $g'(x) = -1$ gives

$$\frac{dy}{dx} = \frac{g'(x)}{g(x)} = \frac{-1}{-x} = \frac{1}{x}.$$

The derivative of $y = \ln(-x)$ is the same as the derivative of $y = \ln x$. For this reason, these two results can be combined into one rule using the absolute value of $x$. A similar situation holds true for $y = \ln[g(x)]$ and $y = \ln[-g(x)]$, as well as for $y = \log_a[g(x)]$ and $y = \log_a[-g(x)]$. These results are summarized as follows.

**Derivative of $\log_a|x|$, $\log_a|g(x)|$, $\ln|x|$, and $\ln|g(x)|$**

$$\frac{d}{dx}[\log_a|x|] = \frac{1}{(\ln a)x} \qquad \frac{d}{dx}[\log_a|g(x)|] = \frac{1}{\ln a} \cdot \frac{g'(x)}{g(x)}$$

$$\frac{d}{dx}[\ln|x|] = \frac{1}{x} \qquad \frac{d}{dx}[\ln|g(x)|] = \frac{g'(x)}{g(x)}$$

**NOTE** You need not memorize the previous four formulas. They are simply the result of the chain rule applied to the formula for the derivative of $y = \log_a x$, as well as the fact that when $\log_a x = \ln x$, so that $a = e$, then $\ln a = \ln e = 1$. An absolute value inside of a logarithm has no effect on the derivative, other than making the result valid for more values of $x$.

### EXAMPLE 3    Derivatives of Logarithmic Functions

Find the derivative of each function.

**(a)** $y = \ln |5x|$

**SOLUTION**    Let $g(x) = 5x$, so that $g'(x) = 5$. From the previous formula,

$$\frac{dy}{dx} = \frac{g'(x)}{g(x)} = \frac{5}{5x} = \frac{1}{x}.$$

Notice that the derivative of $\ln |5x|$ is the same as the derivative of $\ln |x|$. Also notice that we would have found the exact same answer for the derivative of $y = \ln 5x$ (without the absolute value), but the result would not apply to negative values of $x$. Also, in Example 1, the derivative of $\ln 6x$ was the same as that for $\ln x$. This suggests that for any constant $a$,

$$\frac{d}{dx} \ln |ax| = \frac{d}{dx} \ln |x|$$

$$= \frac{1}{x}.$$

Exercise 46 asks for a proof of this result.

**(b)** $f(x) = 3x \ln x^2$

**SOLUTION**    This function is the product of the two functions $3x$ and $\ln x^2$, so use the product rule.

$$f'(x) = (3x)\left[\frac{d}{dx} \ln x^2\right] + (\ln x^2)\left[\frac{d}{dx} 3x\right]$$

$$= 3x\left(\frac{2x}{x^2}\right) + (\ln x^2)(3)$$

$$= 6 + 3 \ln x^2$$

By the power rule for logarithms,

$$f'(x) = 6 + \ln(x^2)^3$$

$$= 6 + \ln x^6.$$

Alternatively, write the answer as $f'(x) = 6 + 6 \ln x$, except that this last form requires $x > 0$, while negative values of $x$ are acceptable in $6 + \ln x^6$.

Another method would be to use a rule of logarithms to simplify the function to $f(x) = 3x \cdot 2 \ln x = 6x \ln x$ and then to take the derivative.

**(c)** $s(t) = \dfrac{\log_8(t^{3/2} + 1)}{t}$

**SOLUTION**    Use the quotient rule and the chain rule.

$$s'(t) = \frac{t \cdot \dfrac{1}{(t^{3/2} + 1)\ln 8} \cdot \dfrac{3}{2}t^{1/2} - \log_8(t^{3/2} + 1) \cdot 1}{t^2}$$

This expression can be simplified slightly by multiplying the numerator and the denominator by $2(t^{3/2} + 1) \ln 8$.

$$s'(t) = \frac{t \cdot \dfrac{1}{(t^{3/2} + 1)\ln 8} \cdot \dfrac{3}{2}t^{1/2} - \log_8(t^{3/2} + 1)}{t^2} \cdot \frac{2(t^{3/2} + 1)\ln 8}{2(t^{3/2} + 1)\ln 8}$$

$$= \frac{3t^{3/2} - 2(t^{3/2} + 1)(\ln 8)\log_8(t^{3/2} + 1)}{2t^2(t^{3/2} + 1)\ln 8}$$

**YOUR TURN 3**    Find the derivative of each function.

**a.** $y = \ln|2x + 6|$

**b.** $f(x) = x^2 \ln 3x$

**c.** $s(t) = \dfrac{\ln(t^2 - 1)}{t + 1}$

TRY YOUR TURN 3

### EXAMPLE 4  Pedestrian Speed

In the Extended Application for Chapter 2, we found that the average speed (in feet per second) of pedestrians depends on the population $(x)$ of the city in which they are walking. These data could also be modeled by the function

$$f(x) = 0.873 \log x - 0.0255.$$

Find and interpret $f(1,000,000)$ and $f'(1,000,000)$.

**APPLY IT**  **SOLUTION**  Recognizing this function as a common (base 10) logarithm, we have $f(1,000,000) = 0.873 \log 1,000,000 - 0.0255 = 5.2125$. In a city of 1,000,000 people, pedestrians walk, on average, about 5.2 feet per second.

$$f'(x) = \frac{0.873}{(\ln 10)x},$$

so $f'(1,000,000) \approx 3.791 \cdot 10^{-7}$. This means that for each additional person in a city of 1,000,000, the average pedestrian speed increases by about $3.8 \cdot 10^{-7}$ feet per second. Perhaps a simpler interpretation is that for a city of 1,000,000, every additional 10,000 people causes an increase in the average pedestrian velocity of about $3.8 \cdot 10^{-7} \cdot 10,000 = 0.0038$ feet per second, a very small quantity. The increase in speed with larger populations is only noticeable when comparing cities that vary greatly in size.

### EXAMPLE 5  Black Crappie

Scientists in Florida have collected data which reveal that there is a logarithmic relationship between the length of a black crappie at age three and the total number of crappie in a lake. This relationship can be represented by

$$l(x) = 171.5 + 943.2 \log(x) - 458.15 \log(x^2),$$

where $x$ is the number of black crappie per hectare (ha—2.47 acres) and $l(x)$ (in mm) is the predicted length at age three. ***Source: North American Journal of Fisheries Management.***

**(a)** Use the properties of logarithms to simplify this function and predict the length of three-year-old black crappie when the lake density is 100 crappie/ha.

**SOLUTION**  Since $\log(x^2) = 2 \log(x)$, we can rewrite $l(x)$ as

$$l(x) = 171.5 + 943.2 \log(x) - 458.15 \log(x^2)$$
$$= 171.5 + 943.2 \log(x) - 2(458.15) \log(x)$$
$$= 171.5 + 26.9 \log(x).$$

We can see that the properties of logarithms can be useful to simplify a function. Also,

$$l(100) = 171.5 + 26.9 \log(100) = 225.3.$$

Thus, the predicted length of black crappie at age three when the density of the lake is 100 crappie/ha is 225.3 mm.

**(b)** Using the simplified function, calculate $l'(100)$ and comment on the results.

**SOLUTION**  The derivative of $l(x)$ is

$$l'(x) = \frac{26.9}{\ln(10)x},$$

so $l'(100) = 26.9/[\ln(10) \cdot 100] \approx 0.1168$. This means that for each additional black crappie per ha in a lake with 100 crappie per ha, the average length at age three will increase by approximately 0.1168 mm. As one can see in Figure 10 on the next page, the graph of $l(x)$ is always rising. We would certainly not expect this trend to continue for high densities of black crappie, since crowding tends to decrease the size of a population of any species of animal. Thus, we should be very cautious in using this function to predict black crappie lengths for high densities of crappie.

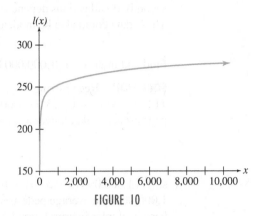

FIGURE 10

# 4.5   EXERCISES

**Find the derivative of each function.**

**1.** $y = \ln(8x)$

**2.** $y = \ln(-4x)$

**3.** $y = \ln(8 - 3x)$

**4.** $y = \ln(1 + x^3)$

**5.** $y = \ln|4x^2 - 9x|$

**6.** $y = \ln|-8x^3 + 2x|$

**7.** $y = \ln\sqrt{x + 5}$

**8.** $y = \ln\sqrt{2x + 1}$

**9.** $y = \ln(x^4 + 5x^2)^{3/2}$

**10.** $y = \ln(5x^3 - 2x)^{3/2}$

**11.** $y = -5x \ln(3x + 2)$

**12.** $y = (3x + 7) \ln(2x - 1)$

**13.** $s = t^2 \ln|t|$

**14.** $y = x \ln|2 - x^2|$

**15.** $y = \dfrac{2 \ln(x + 3)}{x^2}$

**16.** $v = \dfrac{\ln u}{u^3}$

**17.** $y = \dfrac{\ln x}{4x + 7}$

**18.** $y = \dfrac{-2 \ln x}{3x - 1}$

**19.** $y = \dfrac{3x^2}{\ln x}$

**20.** $y = \dfrac{x^3 - 1}{2 \ln x}$

**21.** $y = (\ln|x + 1|)^4$

**22.** $y = \sqrt{\ln|x - 3|}$

**23.** $y = \ln|\ln x|$

**24.** $y = (\ln 4)(\ln|3x|)$

**25.** $y = e^{x^2} \ln x$

**26.** $y = e^{2x-1} \ln(2x - 1)$

**27.** $y = \dfrac{e^x}{\ln x}$

**28.** $p(y) = \dfrac{\ln y}{e^y}$

**29.** $g(z) = (e^{2z} + \ln z)^3$

**30.** $s(t) = \sqrt{e^{-t} + \ln 2t}$

**31.** $y = \log(6x)$

**32.** $y = \log(4x - 3)$

**33.** $y = \log|1 - x|$

**34.** $y = \log|3x|$

**35.** $y = \log_5 \sqrt{5x + 2}$

**36.** $y = \log_7 \sqrt{4x - 3}$

**37.** $y = \log_3(x^2 + 2x)^{3/2}$

**38.** $y = \log_2(2x^2 - x)^{5/2}$

**39.** $w = \log_8(2^p - 1)$

**40.** $z = 10^y \log y$

**41.** $f(x) = e^{\sqrt{x}} \ln(\sqrt{x} + 5)$

**42.** $f(x) = \ln(xe^{\sqrt{x}} + 2)$

**43.** $f(t) = \dfrac{\ln(t^2 + 1) + t}{\ln(t^2 + 1) + 1}$

**44.** $f(t) = \dfrac{2t^{3/2}}{\ln(2t^{3/2} + 1)}$

**45.** Why do we use the absolute value of $x$ or of $g(x)$ in the derivative formulas for the natural logarithm?

**46.** Prove $\dfrac{d}{dx} \ln|ax| = \dfrac{d}{dx} \ln|x|$ for any constant $a$.

**47.** A friend concludes that because $y = \ln 6x$ and $y = \ln x$ have the same derivative—namely, $dy/dx = 1/x$—these two functions must be the same. Explain why this is incorrect.

**48.** Use a graphing calculator to sketch the graph of $y = [f(x + h) - f(x)]/h$ using $f(x) = \ln|x|$ and $h = 0.0001$. Compare it with the graph of $y = 1/x$ and discuss what you observe.

**49.** Using the fact that

$$\ln[u(x)v(x)] = \ln u(x) + \ln v(x),$$

use the chain rule and the formula for the derivative of $\ln x$ to derive the product rule. In other words, find $[u(x)v(x)]'$ without assuming the product rule.

**50.** Using the fact that

$$\ln \dfrac{u(x)}{v(x)} = \ln u(x) - \ln v(x),$$

use the chain rule and the formula for the derivative of $\ln x$ to derive the quotient rule. In other words, find $[u(x)/v(x)]'$ without assuming the quotient rule.

**51.** Use graphical differentiation to verify that $\dfrac{d}{dx}(\ln x) = \dfrac{1}{x}$.

**52.** Use the fact that $d \ln x/dx = 1/x$, as well as the change-of-base theorem for logarithms, to prove that

$$\dfrac{d \log_a x}{dx} = \dfrac{1}{x \ln a}.$$

**53.** Let

$$h(x) = u(x)^{v(x)}.$$

**a.** Using the fact that

$$\ln\left[u(x)^{v(x)}\right] = v(x)\ln u(x),$$

use the chain rule, the product rule, and the formula for the derivative of $\ln x$ to show that

$$\frac{d}{dx}\ln h(x) = \frac{v(x)u'(x)}{u(x)} + \left(\ln u(x)\right)v'(x).$$

**b.** Use the result from part a and the fact that

$$\frac{d}{dx}\ln h(x) = \frac{h'(x)}{h(x)}$$

to show that

$$\frac{d}{dx}h(x) = u(x)^{v(x)}\left[\frac{v(x)u'(x)}{u(x)} + \left(\ln u(x)\right)v'(x)\right].$$

The idea of taking the logarithm of a function before differentiating is known as logarithmic differentiation.

**Use the ideas from Exercise 53 to find the derivative of each of the following functions.**

**54.** $h(x) = x^x$

**55.** $h(x) = (x^2 + 1)^{5x}$

**56.** This exercise shows another way to derive the formula for the derivative of the natural logarithm function using the definition of the derivative.

**a.** Using the definition of the derivative, show that

$$\frac{d(\ln x)}{dx} = \lim_{h \to 0} \ln\left(1 + \frac{h}{x}\right)^{1/h}.$$

**b.** Eliminate $h$ from the result in part a using the substitution $h = x/m$ to show that

$$\frac{d(\ln x)}{dx} = \lim_{m \to \infty} \ln\left[\left(1 + \frac{1}{m}\right)^m\right]^{1/x}.$$

**c.** What property should the function $g$ have that would yield

$$\lim_{m \to \infty} g(h(m)) = g(\lim_{m \to \infty} h(m))?$$

Assuming that the natural logarithm has this property, and using a result about $(1 + 1/m)^m$ from Section 2.1, show that

$$\frac{d(\ln x)}{dx} = \ln e^{1/x} = \frac{1}{x}.$$

## LIFE SCIENCE APPLICATIONS

**57. Insect Mating** Consider an experiment in which equal numbers of male and female insects of a certain species are permitted to intermingle. Assume that

$$M(t) = (0.1t + 1)\ln\sqrt{t}$$

represents the number of matings observed among the insects in an hour, where $t$ is the temperature in degrees Celsius. (*Note: The formula is an approximation at best and holds only for specific temperature intervals.*)

**a.** Find the number of matings when the temperature is 15°C.

**b.** Find the number of matings when the temperature is 25°C.

**c.** Find the rate of change of the number of matings when the temperature is 15°C.

**58. Population Growth** Suppose that the population of a certain collection of rare Brazilian ants is given by

$$P(t) = (t + 100)\ln(t + 2),$$

where $t$ represents the time in days. Find the rates of change of the population on the second day and on the eighth day.

**59. Body Surface Area** There is a mathematical relationship between an infant's weight and total body surface area (BSA), given by

$$A(w) = 4.688w^{0.8168 - 0.0154 \log_{10} w},$$

where $w$ is the weight (in grams) and $A(w)$ is the BSA in square centimeters. *Source: British Journal of Cancer.*

**a.** Find the BSA for an infant who weighs 4000 g.

**b.** Find $A'(4000)$ and interpret your answer.

**c.** Use a graphing calculator to graph $A(w)$ on $[2000, 10{,}000]$ by $[0, 6000]$.

**60. Bologna Sausage** Scientists have developed a model to predict the growth of bacteria in bologna sausage at 32°C. The number of bacteria is given by

$$\ln\left(\frac{N(t)}{N_0}\right) = 9.8901e^{-e^{2.54197 - 0.2167t}},$$

where $N_0$ is the number of bacteria present at the beginning of the experiment and $N(t)$ is the number of bacteria present at time $t$ (in hours). *Source: Applied and Environmental Microbiology.*

**a.** Use the properties of logarithms to find an expression for $N(t)$. Assume that $N_0 = 1000$.

**b.** Use a graphing calculator to estimate the derivative of $N(t)$ when $t = 20$ and interpret.

**c.** Let $S(t) = \ln(N(t)/N_0)$. Graph $S(t)$ on $[0, 35]$ by $[0, 12]$.

**d.** Graph $N(t)$ on $[0, 35]$ by $[0, 20{,}000{,}000]$ and compare the graphs from parts c and d.

**e.** Find $\lim_{t \to \infty} S(t)$ and then use this limit to find $\lim_{t \to \infty} N(t)$.

**61. Pronghorn Fawns** The field metabolic rate (FMR), or the total energy expenditure per day in excess of growth, can be calculated for pronghorn fawns using Nagy's formula,

$$F(x) = 0.774 + 0.727 \log x,$$

where $x$ is the mass (in grams) of the fawn and $F(x)$ is the energy expenditure (in kJ/day). *Source: Animal Behavior.*

**a.** Determine the total energy expenditure per day in excess of growth for a pronghorn fawn that weighs 25,000 g.

**b.** Find $F'(25{,}000)$ and interpret the result.

**c.** Graph the function on $[5000, 30{,}000]$ by $[3, 5]$.

**62. Fruit Flies** A study of the relation between the rate of reproduction in *Drosophila* (fruit flies) bred in bottles and the density of the mated population found that the number of imagoes (sexually mature adults) per mated female per day ($y$) can be approximated by

$$\log y = 1.54 - 0.008x - 0.658 \log x,$$

where $x$ is the mean density of the mated population (measured as flies per bottle) over a 16-day period. *Source: Elements of Mathematical Biology.*

a. Show that the above equation is equivalent to

$$y = 34.7(1.0186)^{-x}x^{-0.658}.$$

b. Using your answer from part a, find the number of imagoes per mated female per day when the density is

i. 20 flies per bottle;

ii. 40 flies per bottle.

c. Using your answer from part a, find the rate of change in the number of imagoes per mated female per day with respect to the density when the density is

i. 20 flies per bottle;

ii. 40 flies per bottle.

**63. Jukes-Cantor Distance** The Jukes-Cantor distance between two DNA sequences is defined as

$$J(p) = -\frac{3}{4}\ln\left(1 - \frac{4}{3}p\right),$$

where $p$ is the fraction of sites that disagree when comparing the two sequences. *Source: Mathematical Models in Biology.* Find the Jukes-Cantor distance and the rate that this distance is changing with respect to $p$ for each of the following values of $p$.

a. 0.03

b. 0.10

## OTHER APPLICATIONS

**64. Poverty** The passage of the Social Security Amendments of 1965 resulted in the creation of the Medicare and Medicaid programs. Since then, the percent of persons 65 years and over with family income below the poverty level has declined. The percent can be approximated by the following function:

$$P(t) = 30.60 - 5.79 \ln t,$$

where $t$ is the number of years since 1965. Find the percent of persons 65 years and over with family income below the poverty level and the rate of change in the following years. *Source: U.S. Census.*

a. 1970

b. 1990

c. 2010

d. What happens to the rate of change over time?

**65. Richter Scale** The Richter scale provides a measure of the magnitude of an earthquake. In fact, the largest Richter number $M$ ever recorded for an earthquake was 8.9 from the 1933 earthquake in Japan. The following formula shows a relationship between the amount of energy released and the Richter number.

$$M = \frac{2}{3}\log\frac{E}{0.007},$$

where $E$ is measured in kilowatt-hours. *Source: Mathematics Teacher.*

a. For the 1933 earthquake in Japan, what value of $E$ gives a Richter number $M = 8.9$?

b. If the average household uses 247 kWh per month, how many months would the energy released by an earthquake of this magnitude power 10 million households?

c. Find the rate of change of the Richter number $M$ with respect to energy when $E = 70{,}000$ kWh.

d. What happens to $dM/dE$ as $E$ increases?

**66. Street Crossing** Consider a child waiting at a street corner for a gap in traffic that is large enough so that he can safely cross the street. A mathematical model for traffic shows that if the expected waiting time for the child is to be at most 1 minute, then the maximum traffic flow, in cars per hour, is given by

$$f(x) = \frac{29{,}000(2.322 - \log x)}{x},$$

where $x$ is the width of the street in feet. Find the maximum traffic flow and the rate of change of the maximum traffic flow with respect to street width for the following values of the street width. *Source: An Introduction to Mathematical Modeling.*

a. 30 ft

b. 40 ft

■ **YOUR TURN ANSWERS**

1. $f'(x) = \dfrac{1}{(\ln 3)x}$

2. a. $\dfrac{dy}{dx} = \dfrac{6x^2}{2x^3 - 3}$

 b. $f'(x) = \dfrac{5 + 9x^2}{(\ln 4)(5x + 3x^3)}$

3. a. $\dfrac{dy}{dx} = \dfrac{1}{x + 3}$

 b. $f'(x) = x + 2x \ln 3x$

 c. $s'(t) = \dfrac{2t - (t - 1)\ln(t^2 - 1)}{(t - 1)(t + 1)^2}$

# 4.6 Derivatives of Trigonometric Functions

APPLY IT **How does the pressure on the eardrum change with time?**
*In Example 8 in this section, we will use trigonometry and derivatives to answer this question.*

In this section, we derive formulas for the derivatives of some of the trigonometric functions. All these derivatives can be found from the formula for the derivative of $y = \sin x$.

We will need to use the following identities, which are listed without proof, to find the derivatives of the trigonometric functions.

## Basic Identities

$$\sin^2 x + \cos^2 x = 1$$

$$\tan x = \frac{\sin x}{\cos x}$$

$$\sin(x + y) = \sin x \cos y + \cos x \sin y$$

$$\sin(x - y) = \sin x \cos y - \cos x \sin y$$

$$\cos(x + y) = \cos x \cos y - \sin x \sin y$$

$$\cos(x - y) = \cos x \cos y + \sin x \sin y$$

The derivative of $y = \sin x$ also depends on the value of

$$\lim_{x \to 0} \frac{\sin x}{x}.$$

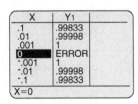

| X | Y₁ |
|---|---|
| .1 | .99833 |
| .01 | .99998 |
| .001 | 1 |
| 0 | ERROR |
| -.001 | 1 |
| -.01 | .99998 |
| -.1 | .99833 |

X=0

**FIGURE 11**

To estimate this limit, find the quotient $(\sin x)/x$ for various values of $x$ close to 0. (Be sure that your calculator is set for radian measure.) For example, we used the TABLE feature of the TI-84 Plus calculator to get the values of this quotient shown in Figure 11 as $x$ approaches 0 from either side. Note that, although the calculator shows the quotient equal to 1 for $x = \pm 0.001$, it is an approximation—the value is not exactly 1. Why does the calculator show ERROR when $x = 0$?

These results suggest, and it can be proved, that

$$\lim_{x \to 0} \frac{\sin x}{x} = 1.$$

In Example 1, this limit is used to obtain another limit. Then the derivative of $y = \sin x$ can be found.

**EXAMPLE 1** **Trigonometric Limit**

Find $\lim\limits_{h \to 0} \dfrac{\cos h - 1}{h}$.

**SOLUTION** Use the previous limit and some trigonometric identities.

$$\lim_{h \to 0} \frac{\cos h - 1}{h} = \lim_{h \to 0} \frac{(\cos h - 1)}{h} \cdot \frac{(\cos h + 1)}{(\cos h + 1)} \qquad \text{Multiply by } 1 = \frac{\cos h + 1}{\cos h + 1}.$$

$$= \lim_{h \to 0} \frac{\cos^2 h - 1}{h(\cos h + 1)}$$

$$= \lim_{h \to 0} \frac{-\sin^2 h}{h(\cos h + 1)} \qquad \cos^2 h - 1 = -\sin^2 h$$

$$= \lim_{h \to 0} (-\sin h)\left(\frac{\sin h}{h}\right)\left(\frac{1}{\cos h + 1}\right)$$

$$= (0)(1)\left(\frac{1}{1 + 1}\right) = 0$$

Therefore,

$$\lim_{h \to 0} \frac{\cos h - 1}{h} = 0.$$

We can now find the derivative of $y = \sin x$ by using the general definition for the derivative of a function $f$ given in Chapter 3:

$$f'(x) = \lim_{h \to 0} \frac{f(x + h) - f(x)}{h},$$

provided this limit exists. By this definition, the derivative of $f(x) = \sin x$ is

$$f'(x) = \lim_{h \to 0} \frac{\sin(x + h) - \sin x}{h}$$

$$= \lim_{h \to 0} \frac{\sin x \cdot \cos h + \cos x \cdot \sin h - \sin x}{h} \qquad \text{Identity for } \sin(x + h)$$

$$= \lim_{h \to 0} \frac{(\sin x \cdot \cos h - \sin x) + \cos x \cdot \sin h}{h} \qquad \text{Rearrange terms.}$$

$$= \lim_{h \to 0} \frac{\sin x(\cos h - 1) + \cos x \cdot \sin h}{h} \qquad \text{Factor.}$$

$$f'(x) = \lim_{h \to 0}\left(\sin x \frac{\cos h - 1}{h}\right) + \lim_{h \to 0}\left(\cos x \frac{\sin h}{h}\right) \qquad \text{Limit rule for sums}$$

$$= (\sin x)(0) + (\cos x)(1)$$

$$= \cos x.$$

This result is summarized below.

**Derivative of sin x**

$$D_x(\sin x) = \cos x$$

We can use the chain rule to find derivatives of other sine functions, as shown in the following examples.

**EXAMPLE 2** Derivatives of sin x

Find the derivative of each function.

**(a)** $y = \sin 6x$

**SOLUTION**  By the chain rule,

$$\frac{dy}{dx} = (\cos 6x) \cdot D_x(6x)$$

$$= (\cos 6x) \cdot 6$$

$$\frac{dy}{dx} = 6 \cos 6x.$$

**(b)** $y = 5 \sin(9x^2 + 2) + \cos\left(\dfrac{\pi}{7}\right)$

**SOLUTION**  By the chain rule,

$$\frac{dy}{dx} = [5 \cos(9x^2 + 2)] \cdot D_x(9x^2 + 2) + 0 \qquad \cos\left(\frac{\pi}{7}\right) \text{ is a constant.}$$

$$= [5 \cos(9x^2 + 2)]18x$$

**YOUR TURN 1**  Find the derivative of $y = 5 \sin(3x^4)$.

$$\frac{dy}{dx} = 90x \cos(9x^2 + 2).$$

TRY YOUR TURN 1

### EXAMPLE 3  Chain Rule

Find $D_x(\sin^4 x)$.

**YOUR TURN 2**  Find the derivative of $y = 2 \sin^3(\sqrt{x})$.

**SOLUTION**  The expression $\sin^4 x$ means $(\sin x)^4$. By the chain rule,

$$D_x(\sin^4 x) = 4 \cdot \sin^3 x \cdot D_x(\sin x)$$

$$= 4 \sin^3 x \cos x. \qquad \text{TRY YOUR TURN 2}$$

The derivative of $y = \cos x$ is found from trigonometric identities and from the fact that $D_x(\sin x) = \cos x$. First, use the identity for $\sin(x - y)$ to get

$$\sin\left(\frac{\pi}{2} - x\right) = \sin\frac{\pi}{2} \cdot \cos x - \cos\frac{\pi}{2} \cdot \sin x$$

$$= 1 \cdot \cos x - 0 \cdot \sin x$$

$$= \cos x.$$

In the same way, $\cos\left(\dfrac{\pi}{2} - x\right) = \sin x$. Therefore,

$$D_x(\cos x) = D_x\left[\sin\left(\frac{\pi}{2} - x\right)\right].$$

By the chain rule,

$$D_x\left[\sin\left(\frac{\pi}{2} - x\right)\right] = \cos\left(\frac{\pi}{2} - x\right) \cdot D_x\left(\frac{\pi}{2} - x\right)$$

$$= \cos\left(\frac{\pi}{2} - x\right) \cdot (-1) \qquad \pi/2 \text{ is constant.}$$

$$= -\cos\left(\frac{\pi}{2} - x\right)$$

$$= -\sin x.$$

Derivative of $\cos x$

$$D_x(\cos x) = -\sin x$$

### EXAMPLE 4    Derivatives of $\cos x$

Find each derivative.

(a) $D_x[\cos(3x)] = -\sin(3x) \cdot D_x(3x) = -3 \sin 3x$

(b) $D_x(\cos^4 x) = 4 \cos^3 x \cdot D_x(\cos x) = 4 \cos^3 x(-\sin x)$
$$= -4 \sin x \cos^3 x$$

(c) $D_x(3x \cos x)$

**SOLUTION**    Use the product rule.

$$D_x(3x \cos x) = 3x(-\sin x) + (\cos x)(3)$$
$$= -3x \sin x + 3 \cos x$$

**YOUR TURN 3**    Find the derivative of $y = x \cos(x^2)$.

TRY YOUR TURN 3

As mentioned in the list of basic identities at the beginning of this section, $\tan x = (\sin x)/\cos x$. The derivative of $y = \tan x$ can be found by using the quotient rule to find the derivative of $y = (\sin x)/\cos x$.

$$D_x(\tan x) = D_x\left(\frac{\sin x}{\cos x}\right) = \frac{\cos x \cdot D_x(\sin x) - \sin x \cdot D_x(\cos x)}{\cos^2 x}$$

$$= \frac{\cos x(\cos x) - \sin x(-\sin x)}{\cos^2 x}$$

$$= \frac{\cos^2 x + \sin^2 x}{\cos^2 x}$$

$$= \frac{1}{\cos^2 x} = \sec^2 x \qquad \cos^2 x + \sin^2 x = 1$$

The last step follows from the definitions of the trigonometric functions, which could be used to show that $1/\cos x = \sec x$. A similar calculation leads to the derivative of $\cot x$.

Derivatives of $\tan x$ and $\cot x$

$$D_x(\tan x) = \sec^2 x$$
$$D_x(\cot x) = -\csc^2 x$$

### EXAMPLE 5    Derivatives of $\tan x$ and $\cot x$

Find each derivative.

(a) $D_x(\tan 9x) = \sec^2 9x \cdot D_x(9x) = 9 \sec^2 9x$

(b) $D_x(\cot^6 x) = 6 \cot^5 x \cdot D_x(\cot x) = -6 \cot^5 x \csc^2 x$

**YOUR TURN 4**    Find the derivative of $y = x \tan^2 x$.

(c) $D_x(\ln|6 \tan x|) = \dfrac{D_x(6 \tan x)}{6 \tan x} = \dfrac{6 \sec^2 x}{6 \tan x} = \dfrac{\sec^2 x}{\tan x}$

TRY YOUR TURN 4

Using the facts that $\sec x = 1/\cos x$ and $\csc x = 1/\sin x$, it is possible to use the quotient rule to find the derivative of each of these functions. In Exercises 34 and 35 at the end of this section, you will be asked to verify the following.

Derivatives of $\sec x$ and $\csc x$

$$D_x (\sec x) = \sec x \tan x$$
$$D_x (\csc x) = -\csc x \cot x$$

### EXAMPLE 6 Derivatives of $\sec x$ and $\csc x$

Find each derivative.

(a) $D_x(x^2 \sec x) = x^2 \sec x \tan x + 2x \sec x$

**YOUR TURN 5** Find the derivative of $y = \sec^2(\sqrt{x})$.

(b) $D_x(\csc e^{2x}) = -\csc e^{2x} \cot e^{2x} \cdot D_x(e^{2x})$

$\qquad = -\csc e^{2x} \cot e^{2x} \cdot (2e^{2x})$

$\qquad = -2e^{2x} \csc e^{2x} \cot e^{2x}$

TRY YOUR TURN 5

### EXAMPLE 7 Derivatives of Trigonometric Functions

Find the derivative of each function at the specified value of $x$.

(a) $f(x) = \sin(\pi e^x)$, when $x = 0$

**SOLUTION** Using the chain rule, the derivative of $f(x)$ is

$$f'(x) = \cos(\pi e^x) \cdot \pi e^x.$$

Thus,

$$f'(0) = \cos(\pi e^0) \cdot \pi e^0 = (-1)\pi(1) = -\pi.$$

(b) $g(x) = e^x \sin(\pi x)$, when $x = 0$

**SOLUTION** Using the product rule, the derivative of $g(x)$ is

$$g'(x) = e^x \cos(\pi x)\pi + \sin(\pi x)e^x.$$

Thus,

$$g'(0) = e^0\cos(\pi 0)\pi + \sin(\pi 0)e^0$$
$$= 1 \cdot 1 \cdot \pi + 0 \cdot 1 = \pi.$$

(c) $h(x) = \tan(\cot x)$, when $x = \dfrac{\pi}{4}$

**SOLUTION** Using the chain rule, the derivative of $h(x)$ is

$$h'(x) = \sec^2[\cot(x)] \cdot [-\csc^2(x)].$$

Thus,

$$h'\left(\frac{\pi}{4}\right) = \sec^2\left[\cot\left(\frac{\pi}{4}\right)\right] \cdot \left[-\csc^2\left(\frac{\pi}{4}\right)\right] = \sec^2(1) \cdot (-2) \approx -6.851.$$

(d) $k(x) = \tan x \cot x$, when $x = \dfrac{\pi}{4}$

**YOUR TURN 6** Find the derivative of $f(x) = \sin(\cos x)$ when $x = \pi/2$.

**SOLUTION** Since $k(x) = \tan x \cot x = \dfrac{\sin x}{\cos x} \cdot \dfrac{\cos x}{\sin x} = 1$, $k'(x) = 0$. In particular,

$$k'\left(\frac{\pi}{4}\right) = 0.$$

TRY YOUR TURN 6

EXAMPLE 8 **Pressure on the Eardrum**

In Section 2.4, Example 9, we saw that a change in pressure on the eardrum occurs when a pure musical tone is played. For Middle C, the pressure on the eardrum follows the sine curve

$$P(t) = 0.004 \sin\left(523.26\pi t + \frac{\pi}{7}\right),$$

where $P$ is the pressure in pounds per square foot at time $t$ seconds. When $P(t)$ is positive, the eardrum is pushed inward; when $P(t)$ is negative, the eardrum is pushed outward. The period of this function was found to be $T \approx 0.004$ second, which implies the eardrum is vibrating nearly 262 times per second. *Source: The Physics and Psychophysics of Music: An Introduction.*

**(a)** For Middle C, find the pressure after 0.001 second.

**SOLUTION** Let $t = 0.001$, and

$$P(0.001) = 0.004 \sin\left(523.26\pi \cdot 0.001 + \frac{\pi}{7}\right)$$
$$\approx 0.0035.$$

Thus, 0.001 second after the note is played, the pressure is pushing the eardrum inward at 0.0035 pound per square foot.

**(b)** Find the pressure after 0.003 second.

**SOLUTION** When $t = 0.003$,

$$P(0.003) \approx -0.0031,$$

and the pressure is pushing the eardrum outward 0.0031 pound per square foot.

**(c)** For Middle C, find the instantaneous rate of change after 0.001 second.

APPLY IT    **SOLUTION** The derivative of $P(t)$ is

$$P'(t) = 0.004 \cos\left(523.26\pi t + \frac{\pi}{7}\right) \cdot 523.26\pi$$
$$\approx 2.093\pi \cos\left(523.26\pi t + \frac{\pi}{7}\right).$$

After 0.001 second, $t = 0.001$,

$$P'(0.001) = 2.093\pi \cos\left(523.26\pi \cdot 0.001 + \frac{\pi}{7}\right) \approx -3.28.$$

After 0.001 second, the pressure is positive, but it is decreasing at the rate of 3.28 pounds per square foot per second.

**(d)** Find the instantaneous rate of change after 0.003 second.

**SOLUTION** After 0.003 second, $P'(0.003) \approx 4.07$. The pressure is negative, but it is increasing at the rate of 4.07 pounds per square foot per second.

# 4.6    EXERCISES

**Find the derivatives of the functions defined as follows.**

**1.** $y = \dfrac{1}{2} \sin 8x$

**2.** $y = -\cos 2x + \cos \dfrac{\pi}{6}$

**3.** $y = 12 \tan(9x + 1)$

**4.** $y = -4 \cos(7x^2 - 4)$

**5.** $y = \cos^4 x$

**6.** $y = -9 \sin^5 x$

**7.** $y = \tan^8 x$

**8.** $y = 3 \cot^5 x$

**9.** $y = -6x \sin 2x$

**10.** $y = 2x \sec 4x$

**11.** $y = \dfrac{\csc x}{x}$

**12.** $y = \dfrac{\tan x}{x - 1}$

**13.** $y = \sin e^{4x}$

**14.** $y = \cos(4e^{2x})$

**15.** $y = e^{\cos x}$

**16.** $y = -2e^{\cot x}$

**17.** $y = \sin(\ln 3x^4)$

**18.** $y = \cos(\ln |2x^3|)$

**19.** $y = \ln |\sin x^2|$

**20.** $y = \ln |\tan^2 x|$

**21.** $y = \dfrac{2 \sin x}{3 - 2 \sin x}$

**22.** $y = \dfrac{3 \cos x}{5 - \cos x}$

**23.** $y = \sqrt{\dfrac{\sin x}{\sin 3x}}$

**24.** $y = \sqrt{\dfrac{\cos 4x}{\cos x}}$

**25.** $y = 3 \tan\left(\dfrac{1}{4}x\right) + 4 \cot 2x - 5 \csc x + e^{-2x}$

**26.** $y = [\sin 3x + \cot(x^3)]^8$

**In Exercises 27–32, recall that the slope of the tangent line to a graph is given by the derivative of the function. Find the slope of the tangent line to the graph of each equation at the given point. You may wish to use a graphing calculator to support your answers.**

**27.** $y = \sin x$;   $x = 0$

**28.** $y = \sin x$;   $x = \pi/4$

**29.** $y = \cos x$;   $x = -5\pi/6$

**30.** $y = \cos x$;   $x = -\pi/4$

**31.** $y = \tan x$;   $x = 0$

**32.** $y = \cot x$;   $x = \pi/4$

**33.** Find the derivative of $\cot x$ by using the quotient rule and the fact that $\cot x = \cos x / \sin x$.

**34.** Verify that the derivative of $\sec x$ is $\sec x \tan x$. (*Hint:* Use the fact that $\sec x = 1/\cos x$.)

**35.** Verify that the derivative of $\csc x$ is $-\csc x \cot x$. (*Hint:* Use the fact that $\csc x = 1/\sin x$.)

**36.** In the discussion of the limit of the quotient $(\sin x)/x$, explain why the calculator gave ERROR for the value of $(\sin x)/x$ when $x = 0$.

## LIFE SCIENCE APPLICATIONS

**37. Carbon Dioxide Levels**   At Mauna Loa, Hawaii, atmospheric carbon dioxide levels in parts per million (ppm) have been measured regularly since 1958. The function defined by

$$L(t) = 0.022t^2 + 0.55t + 316 + 3.5 \sin(2\pi t)$$

can be used to model these levels, where $t$ is in years and $t = 0$ corresponds to 1960. *Source: Greenhouse Earth.*

**a.** Graph $L(t)$ on $[0, 30]$.

**b.** Find $L(25)$, $L(35.5)$, and $L(50.2)$.

**c.** Find $L'(50.2)$ and interpret.

**38. Carbon Dioxide Levels**   At Barrow, Alaska, atmospheric carbon dioxide levels (in parts per million) can be modeled using the function defined by

$$C(t) = 0.04t^2 + 0.6t + 330 + 7.5 \sin(2\pi t),$$

where $t$ is in years and $t = 0$ corresponds to 1960. *Source: Introductory Astronomy and Astrophysics.*

**a.** Graph $C(t)$ on $[0, 25]$.

**b.** Find $C(25)$, $C(35.5)$, and $C(50.2)$.

**c.** Find $C'(50.2)$ and interpret.

**d.** $C$ is the sum of a quadratic function and a sine function. What is the significance of each of these functions? Discuss what physical phenomena may be responsible for each function.

**39. Population Growth**   Many biological populations, both plant and animal, experience seasonal growth. For example, an animal population might flourish during the spring and summer and die back in the fall. The population, $f(t)$, at time $t$, is often modeled by

$$f(t) = f(0)e^{c \sin(t)},$$

where $f(0)$ is the size of the population when $t = 0$. Suppose that $f(0) = 1000$ and $c = 2$. Find the functional values in parts a–d.

**a.** $f(0.2)$

**b.** $f(1)$

**c.** $f'(0)$

**d.** $f'(0.2)$

**e.** Use a graphing calculator to graph $f(t)$ on $[0, 11]$.

**40. Sound**   If a string with a fundamental frequency of 110 hertz is plucked in the middle, it will vibrate at the odd harmonics of $110, 330, 550, \ldots$ hertz but not at the even harmonics of $220, 440, 660, \ldots$ hertz. The resulting pressure $P$ on the eardrum caused by the string can be approximated using the equation

$$P(t) = 0.003 \sin(220\pi t) + \frac{0.003}{3} \sin(660\pi t)$$

$$+ \frac{0.003}{5} \sin(1100\pi t) + \frac{0.003}{7} \sin(1540\pi t),$$

where $P$ is in pounds per square foot at a time of $t$ seconds after the string is plucked. *Source: Fundamentals of Musical Acoustics and The Physics and Psychophysics of Music: An Introduction.*

**a.** Graph $P(t)$ on $[0, 0.01]$.

**b.** Find $P'(0.002)$ and interpret.

## OTHER APPLICATIONS

**41. Motion of a Particle**   A particle moves along a straight line. The distance of the particle from the origin at time $t$ is given by

$$s(t) = \sin t + 2 \cos t.$$

Find the velocity at the following times. (Recall that $v(t) = s'(t)$.)

**a.** $t = 0$     **b.** $t = \pi/4$     **c.** $t = 3\pi/2$

**42. Revenue from Seasonal Merchandise**   The revenue received from the sale of electric fans is seasonal, with maximum revenue in the summer. Let the revenue (in dollars) received from the sale of fans be approximated by

$$R(t) = 120 \cos 2\pi t + 150,$$

where $t$ is time in years, measured from July 1.

  **a.** Find $R'(t)$.

  **b.** Find $R'(t)$ for August 1. (*Hint:* August 1 is 1/12 of a year from July 1.)

  **c.** Find $R'(t)$ for January 1.

  **d.** Find $R'(t)$ for June 1.

  **e.** Discuss whether the answers in parts b–d are reasonable for this model.

**YOUR TURN ANSWERS.**

1. $\dfrac{dy}{dx} = 60x^3 \cos(3x^4)$

2. $\dfrac{dy}{dx} = 3 \sin^2(\sqrt{x}) \cos(\sqrt{x})/\sqrt{x}$

3. $\dfrac{dy}{dx} = -2x^2 \sin(x^2) + \cos(x^2)$

4. $\dfrac{dy}{dx} = 2x \tan x \sec^2 x + \tan^2 x$

5. $\dfrac{dy}{dx} = \sec^2(\sqrt{x}) \tan(\sqrt{x})/\sqrt{x}$

6. $-1$

# 4  CHAPTER REVIEW

## SUMMARY

In this chapter we used the definition of the derivative to develop techniques for finding derivatives of several types of functions. With the help of the rules that were developed, such as the power rule, product rule, quotient rule, and chain rule, we can now directly compute the derivative of a large variety of functions. In particular, we developed rules for finding derivatives of exponential, logarithmic, and trigonometric functions. We also began to see the wide range of applications that these functions have in the life sciences. In the next chapter we will apply these techniques to study the behavior of certain functions, and we will learn that differentiation can be used to find maximum and minimum values of continuous functions.

Assume all indicated derivatives exist.

**Constant Function**   If $f(x) = k$, where $k$ is any real number, then $f'(x) = 0$.

**Power Rule**   If $f(x) = x^n$, for any real number $n$, then $f'(x) = n \cdot x^{n-1}$.

**Constant Times a Function**   Let $k$ be a real number. Then the derivative of $y = k \cdot f(x)$ is $dy/dx = k \cdot f'(x)$.

**Sum or Difference Rule**   If $y = u(x) \pm v(x)$, then $\dfrac{dy}{dx} = u'(x) \pm v'(x)$.

**Product Rule**   If $f(x) = u(x) \cdot v(x)$, then

$$f'(x) = u(x) \cdot v'(x) + v(x) \cdot u'(x).$$

**Quotient Rule**   If $f(x) = \dfrac{u(x)}{v(x)}$, then

$$f'(x) = \frac{v(x) \cdot u'(x) - u(x) \cdot v'(x)}{[v(x)]^2}.$$

**Chain Rule**   If $y$ is a function of $u$, say $y = f(u)$, and if $u$ is a function of $x$, say $u = g(x)$, then $y = f(u) = f[g(x)]$, and

$$\frac{dy}{dx} = \frac{dy}{du} \cdot \frac{du}{dx}.$$

**Chain Rule (Alternative Form)**   Let $y = f[g(x)]$. Then $\dfrac{dy}{dx} = f'[g(x)] \cdot g'(x)$.

**Exponential Functions** $\dfrac{d}{dx}(e^x) = e^x$  $\dfrac{d}{dx}(a^x) = (\ln a)a^x$

$\dfrac{d}{dx}(e^{g(x)}) = e^{g(x)}g'(x)$  $\dfrac{d}{dx}(a^{g(x)}) = (\ln a)a^{g(x)}g'(x)$

**Logarithmic Functions** $\dfrac{d}{dx}(\ln|x|) = \dfrac{1}{x}$  $\dfrac{d}{dx}(\log_a|x|) = \dfrac{1}{(\ln a)x}$

$\dfrac{d}{dx}(\ln|g(x)|) = \dfrac{g'(x)}{g(x)}$  $\dfrac{d}{dx}(\log_a|g(x)|) = \dfrac{1}{\ln a}\cdot\dfrac{g'(x)}{g(x)}$

**Basic Trigonometric Derivatives**
$D_x(\sin x) = \cos x$  $D_x(\cos x) = -\sin x$
$D_x(\tan x) = \sec^2 x$  $D_x(\cot x) = -\csc^2 x$
$D_x(\sec x) = \sec x \tan x$  $D_x(\csc x) = -\csc x \cot x$

# KEY TERMS

**4.3**
composite function
composition
chain rule

**4.4**
logistic function

# REVIEW EXERCISES

## CONCEPT CHECK

**Determine whether each of the following statements is true or false, and explain why.**

1. The derivative of $\pi^3$ is $3\pi^2$.
2. The derivative of a sum is the sum of the derivatives.
3. The derivative of a product is the product of the derivatives.
4. The derivative of a quotient is the quotient of the derivatives.
5. The chain rule is used to take the derivative of a product of functions.
6. The only function that is its own derivative is $e^x$.
7. The derivative of $10^x$ is $x10^{x-1}$.
8. The derivative of $\ln|x|$ is the same as the derivative of $\ln x$.
9. The derivative of $\ln kx$ is the same as the derivative of $\ln x$.
10. The derivative of $\log x$ is the same as the derivative of $\ln x$.
11. $D_x \tan(x^2) = \sec^2(x^2)$
12. $D_x \tan^2 x = 2 \tan x \sec^2 x$

## PRACTICE AND EXPLORATIONS

**Use the rules for derivatives to find the derivative of each function defined as follows.**

13. $y = 5x^3 - 7x^2 - 9x + \sqrt{5}$
14. $y = 7x^3 - 4x^2 - 5x + \sqrt{2}$
15. $y = 9x^{8/3}$
16. $y = -4x^{-3}$
17. $f(x) = 3x^{-4} + 6\sqrt{x}$
18. $f(x) = 19x^{-1} - 8\sqrt{x}$
19. $k(x) = \dfrac{3x}{4x+7}$
20. $r(x) = \dfrac{-8x}{2x+1}$

21. $y = \dfrac{x^2 - x + 1}{x - 1}$
22. $y = \dfrac{2x^3 - 5x^2}{x + 2}$
23. $f(x) = (3x^2 - 2)^4$
24. $k(x) = (5x^3 - 1)^6$
25. $y = \sqrt{2t^7 - 5}$
26. $y = -3\sqrt{8t^4 - 1}$
27. $y = 3x(2x + 1)^3$
28. $y = 4x^2(3x - 2)^5$
29. $r(t) = \dfrac{5t^2 - 7t}{(3t + 1)^3}$
30. $s(t) = \dfrac{t^3 - 2t}{(4t - 3)^4}$
31. $p(t) = t^2(t^2 + 1)^{5/2}$
32. $g(t) = t^3(t^4 + 5)^{7/2}$
33. $y = -6e^{2x}$
34. $y = 8e^{0.5x}$
35. $y = e^{-2x^3}$
36. $y = -4e^{x^2}$
37. $y = 5xe^{2x}$
38. $y = -7x^2e^{-3x}$
39. $y = \ln(2 + x^2)$
40. $y = \ln(5x + 3)$
41. $y = \dfrac{\ln|3x|}{x - 3}$
42. $y = \dfrac{\ln|2x - 1|}{x + 3}$
43. $y = \dfrac{xe^x}{\ln(x^2 - 1)}$
44. $y = \dfrac{(x^2 + 1)e^{2x}}{\ln x}$
45. $s = (t^2 + e^t)^2$
46. $q = (e^{2p+1} - 2)^4$
47. $y = 3 \cdot 10^{-x^2}$
48. $y = 10 \cdot 2^{\sqrt{x}}$
49. $g(z) = \log_2(z^3 + z + 1)$
50. $h(z) = \log(1 + e^z)$
51. $f(x) = e^{2x}\ln(xe^x + 1)$
52. $f(x) = \dfrac{e^{\sqrt{x}}}{\ln(\sqrt{x} + 1)}$
53. $y = 2\tan 5x$
54. $y = -4\sin 7x$
55. $y = \cot(6 - 3x^2)$
56. $y = \tan(4x^2 + 3)$
57. $y = 2\sin^4(4x^2)$
58. $y = 2\cos^5 x$
59. $y = \cos(1 + x^2)$
60. $y = \cot\left(\dfrac{1}{2}x^4\right)$

**61.** $y = e^{-2x} \sin x$

**62.** $y = x^2 \csc x$

**63.** $y = \dfrac{\cos^2 x}{1 - \cos x}$

**64.** $y = \dfrac{\sin x - 1}{\sin x + 1}$

**65.** $y = \dfrac{\tan x}{1 + x}$

**66.** $y = \dfrac{6 - x}{\sec x}$

**67.** $y = \ln |5 \sin x|$

**68.** $y = \ln |\cos x|$

**Consider the following table of values of the functions $f$ and $g$ and their derivatives at various points.**

| $x$ | 1 | 2 | 3 | 4 |
|---|---|---|---|---|
| $f(x)$ | 3 | 4 | 2 | 1 |
| $f'(x)$ | −5 | −6 | −7 | −11 |
| $g(x)$ | 4 | 1 | 2 | 3 |
| $g'(x)$ | 2/9 | 3/10 | 4/11 | 6/13 |

**Find the following using the table.**

**69. a.** $D_x(f[g(x)])$ at $x = 2$   **b.** $D_x(f[g(x)])$ at $x = 3$

**70. a.** $D_x(g[f(x)])$ at $x = 2$   **b.** $D_x(g[f(x)])$ at $x = 3$

**Find the slope of the tangent line to the given curve at the given value of $x$. Find the equation of each tangent line.**

**71.** $y = x^2 - 6x;\quad x = 2$

**72.** $y = 8 - x^2;\quad x = 1$

**73.** $y = \dfrac{3}{x - 1};\quad x = -1$

**74.** $y = \dfrac{x}{x^2 - 1};\quad x = 2$

**75.** $y = \sqrt{6x - 2};\quad x = 3$

**76.** $y = -\sqrt{8x + 1};\quad x = 3$

**77.** $y = e^x;\quad x = 0$

**78.** $y = xe^x;\quad x = 1$

**79.** $y = \ln x;\quad x = 1$

**80.** $y = x \ln x;\quad x = e$

**81.** $y = x \cos x;\quad x = 0$

**82.** $y = \tan(\pi x);\quad x = 1$

**83.** Consider the graphs of the function $y = \sqrt{2x - 1}$ and the straight line $y = x + k$. Discuss the number of points of intersection versus the change in the value of $k$. **Source: Japanese University Entrance Examination Problems in Mathematics.**

**84. a.** Verify that

$$\frac{d \ln f(x)}{dx} = \frac{f'(x)}{f(x)}.$$

This expression is called the *relative rate of change*. It expresses the rate of change of $f$ relative to the size of $f$. Stephen B. Maurer denotes this expression by $\hat{f}$ and notes that economists commonly work with relative rates of change. **Source: The College Mathematics Journal.**

 **b.** Verify that

$$\widehat{fg} = \hat{f} + \hat{g}.$$

Interpret this equation in terms of relative rates of change.

**c.** In his article, Maurer uses the result of part b to solve the following problem:

"Last year, the population grew by 1% and the average income per person grew by 2%. By what approximate percent did the national income grow?"

Explain why the result from part b implies that the answer to this question is approximately 3%.

**85.** Suppose that the student body in your college grows by 2% and the tuition goes up by 3%. Use the result from the previous exercise to calculate the approximate amount that the total tuition collected goes up, and compare this with the actual amount.

**86.** Why is $e$ a convenient base for exponential and logarithmic functions?

## LIFE SCIENCE APPLICATIONS

**87. Walleye** Scientists in Wisconsin have collected data which reveal that there is a linear relationship between the total number of walleye per acre and an angler's catch rate per hour such that

$$c(x) = 0.19x,$$

where $x$ is the number of walleye per acre and $c(x)$ is the predicted number of fish that an angler will catch per hour. **Source: North American Journal of Fisheries Management.**

**a.** Use the function above to estimate the number of fish that an angler will catch in two hours if the density of walleye is 20 walleye per acre.

**b.** Calculate the derivative of this function and interpret the results.

**88. Holstein Cows** A common practice among farmers is to use heart-girth (distance around the animal just behind the front legs) measurements to estimate the weight of a cow. Scientists have found that there is a very high linear correlation between heart-girth and weight for Holstein cows such that

$$W(x) = 570 + 5.6(x - 190),\quad x \ge 170,$$

where $x$ is the heart-girth measurement (in cm) and $W(x)$ is the body weight (in kg). **Source: Breeding and Improvement of Farm Animals.**

**a.** Estimate the body weight of a cow that has a heart-girth measurement of 200 cm.

**b.** Develop a measuring tape that estimates the weight of a Holstein cow.

**c.** Discuss the relationship between the slope of this line and the derivative of the function $W(x)$.

**89. Shad** Gill-nets were used by scientists to estimate the number of various species of shad in Lake Texoma, located on the border between Texas and Oklahoma. During this investigation a quadratic relationship was found between the total number of fish and the time to lift vertical gill nets, remove the fish, and record data, such that

$$T(n) = 0.25 + (1.93 \times 10^{-2})n - (4.78 \times 10^{-5})n^2,$$

where $T(n)$ is the processing time (in hr) and $n$ is the total number of fish caught in the net. **Source: North American Journal of Fisheries Management.**

**a.** Graph this function on $[0, 100]$ by $[0, 2]$.

**b.** Find an $n$-value beyond which this function would no longer make sense. Why?

**c.** Find the derivative of $T(n)$ when $n = 40$ and interpret.

**90. Exponential Growth**   Suppose a population is growing exponentially with an annual growth constant $k = 0.05$. How fast is the population growing when it is 1,000,000? Use the derivative to calculate your answer, and then explain how the answer can be obtained without using the derivative.

**91. Logistic Growth**   Suppose a population can be modeled with the logistic function, $G(t) = m/[1 + (m/G_0 - 1)e^{-kmt}]$, with $k = 5 \times 10^{-6}$, $m = 30{,}000$, and $G_0 = 2000$. Assume time is measured in years.

  **a.** Find the growth function $G(t)$ for this population.

  **b.** Find the population and rate of growth of the population after 6 years.

**92. Fish**   The length of the monkeyface prickleback, a West Coast game fish, can be approximated by

$$L = 71.5(1 - e^{-0.1t})$$

and the weight by

$$W = 0.01289 \cdot L^{2.9},$$

where $L$ is the length in centimeters, $t$ is the age in years, and $W$ is the weight in grams. **Source: California Fish and Game.**

  **a.** Find the approximate length of a 5-year-old monkeyface.

  **b.** Find how fast the length of a 5-year-old monkeyface is growing.

  **c.** Find the approximate weight of a 5-year-old monkeyface. (*Hint:* Use your answer from part a.)

  **d.** Find the rate of change of the weight with respect to length for a 5-year-old monkeyface.

  **e.** Using the chain rule and your answers to parts b and d, find how fast the weight of a 5-year-old monkeyface is growing.

**93. Arctic Foxes**   The age–weight relationship of male Arctic foxes caught in Svalbard, Norway, can be estimated by the function

$$M(t) = 3583e^{-e^{-0.020(t-66)}},$$

where $t$ is the age of the fox in days and $M(t)$ is the weight of the fox in grams. **Source: Journal of Mammalogy.**

  **a.** Estimate the weight of a male fox that is 250 days old.

  **b.** Use $M(t)$ to estimate the largest size that a male fox can attain. (*Hint:* Find $\lim_{t \to \infty} M(t)$.)

  **c.** Estimate the age of a male fox when it has reached 50% of its maximum weight.

  **d.** Estimate the rate of change in weight of a male Arctic fox that is 250 days old. (*Hint:* Recall that $D_t e^{f(t)} = f'(t)e^{f(t)}$.)

  **e.** Use a graphing calculator to graph $M(t)$ and then describe the growth pattern.

  **f.** Use the table function on a graphing calculator or a spreadsheet to develop a chart that shows the estimated weight and growth rate of male foxes for days 50, 100, 150, 200, 250, and 300.

**94. Asian Population**   In Section 2.1, Exercise 39, we found that the projected Asian population in the United States, in millions, can be approximated by

$$a(t) = 11.14(1.023)^t,$$

where $t$ is the years since 2000. Find the instantaneous rate of change in the projected Asian population in the United States in each of the following years. **Source: U.S. Census.**

  **a.** 2005            **b.** 2025

## OTHER APPLICATIONS

**95. Compound Interest**   If a sum of $1000 is deposited into an account that pays $r\%$ interest compounded quarterly, the balance after 12 years is given by

$$A = 1000\left(1 + \frac{r}{400}\right)^{48}.$$

Find and interpret $\dfrac{dA}{dr}$ when $r = 5$.

**96. Continuous Compounding**   If a sum of $1000 is deposited into an account that pays $r\%$ interest compounded continuously, the balance after 12 years is given by

$$A = 1000e^{12r/100}.$$

Find and interpret $\dfrac{dA}{dr}$ when $r = 5$.

**97. Wind Energy**   In Section 2.1, Exercise 50, we found that the total world wind energy capacity (in megawatts) in recent years could be approximated by the function

$$C(t) = 19{,}370(1.2557)^t,$$

where $t$ is the number of years since 2000. Find the rate of change in the energy capacity for the following years. **Source: World Wind Energy Association.**

  **a.** 2005       **b.** 2010       **c.** 2015

**98. Cats**   The distance from Lisa Wunderle's cat, Belmar, to a piece of string he is stalking is given in feet by

$$f(t) = \frac{8}{t + 1} + \frac{20}{t^2 + 1},$$

where $t$ is the time in seconds since he begins.

  **a.** Find Belmar's average velocity between 1 second and 3 seconds.

  **b.** Find Belmar's instantaneous velocity at 3 seconds.

**99. Food Surplus**   In Section 2.1, Example 7, we found that the production of corn (in billions of bushels) in the United States since 1930 could be approximated by

$$p(t) = 1.757(1.0248)^{t-1930}$$

where $t$ is the year. Find and interpret $p'(2010)$.

**100. Dating a Language**   Over time, the number of original basic words in a language tends to decrease as words become obsolete or are replaced with new words. Linguists have used calculus to study this phenomenon and have developed a methodology for dating a language, called *glottochronology*. Experiments have indicated that a good estimate of the number of words that remain in use at a given time is given by

$$N(t) = N_0 e^{-0.217t},$$

where $N(t)$ is the number of words in a particular language, $t$ is measured in the number of millennium, and $N_0$ is the original number of words in the language. *Source: The UMAP Journal.*

**a.** In 1950, C. Feng and M. Swadesh established that of the original 210 basic ancient Chinese words from 950 A.D., 167 were still being used. Letting $t = 0$ correspond to 950, with $N_0 = 210$, find the number of words predicted to have been in use in 1950 A.D., and compare it with the actual number in use.

**b.** Estimate the number of words that will remain in the year 2050 ($t = 1.1$).

**c.** Find $N'(1.1)$ and interpret your answer.

**101. Driving Fatalities** A study by the National Highway Traffic Safety Administration found that driver fatality rates were highest for the youngest and oldest drivers. The rates per 1000 licensed drivers for every 100 million miles may be approximated by the function

$$f(x) = k(x - 49)^6 + 0.8,$$

where $x$ is the driver's age in years and $k$ is the constant $3.8 \times 10^{-9}$. Find and interpret the rate of change of the fatality rate when the driver is

**a.** 20 years old; **b.** 60 years old.

*Source: National Highway Traffic Safety Administration.*

**102. Tennis** It is possible to model the flight of a tennis ball that has just been served down the center of the court by the equation

$$y = x \tan \alpha - \frac{16x^2}{V^2} \sec^2 \alpha + h,$$

where $y$ is the height in feet of a tennis ball that is being served at an angle $\alpha$ relative to the horizontal axis, $x$ is the horizontal distance that the ball has traveled in feet, $h$ is the height of the ball when it leaves the server's racket, and $V$ is the velocity of the tennis ball when it leaves the server's racket. *Source: UMAP Journal.*

**a.** If a tennis ball is served from a height of 9 feet and the net is 3 feet high and 39 feet away from the server, does the tennis ball that is hit with a velocity of 50 miles per hour (approximately 73 ft/sec) make it over the net if it is served at an angle of $\pi/24$?

**b.** When $y = 0$, the corresponding value of $x$ gives the total distance that the tennis ball has traveled while in flight (provided that it cleared the net). For a serving height of 9 feet, the equation for calculating the distance traveled is given by

$$x = \frac{V^2 \sin \alpha \cos \alpha \pm V^2 \cos^2 \alpha \sqrt{\tan^2 \alpha + \frac{576}{V^2} \sec^2 \alpha}}{32}.$$

Use the table function on a graphing calculator or a spreadsheet to determine a range of angles for which the tennis ball will clear the net and travel between 39 and 60 feet when it is hit with an initial velocity of 44 ft/sec.

**c.** Because calculating $dx/d\alpha$ is so complicated analytically, use a graphing calculator to estimate this derivative when the initial velocity is 44 ft/sec and $\alpha = \pi/8$. Interpret your answer.

# EXTENDED APPLICATION
## MANAGING RENEWABLE RESOURCES

As the population of the earth continues to grow, it has become more and more important to manage all of the resources that sustain life. For example, efforts to manage the world's fisheries must provide an adequate supply of fish, while ensuring that each species remains viable for future generations. Of course, the dire alternative to mismanagement is widespread starvation, species extinction, and considerable human instability. In this Extended Application, we introduce some of the mathematics that is used to understand the natural resources that civilization must properly manage.

The food that we eat is clearly vital to sustaining life. The good news is that much of our food supply is renewable and endless, provided that civilization understands the dynamics of each food source and takes steps to ensure that overharvesting, pollution, and urban sprawl do not drive these resources into collapse. Given a particular renewable resource, it is vital that managers determine harvest levels that are sustainable over the long run. Through the years, researchers have attempted to determine the maximum level of harvesting that can be carried out over a long period of time without driving a particular food source to extinction. This level of harvesting is called the *maximum sustainable harvest, MSH*. *Source: An Invitation to Biomathematics.*

When a species is introduced and no harvesting occurs, it will grow to a size that is nearly constant and regulated by the carrying capacity $K$ of the environment. At this point, the rate of change of the population is near zero, as the birth and death rates are equal, predictable, and stable. When harvesting occurs, population levels can change drastically, depending on the particular level of harvest. On the other hand, it is possible in some populations to harvest a certain amount of that resource with the knowledge that new births will exceed deaths and push the population back to its carrying capacity. It turns out that we can use the concept of differentiation to estimate maximum sustainable harvest.

To see how this works, let's assume that a population grows according to the logistic model, written as a derivative. In this case, the mathematical model can be written as

$$\frac{dP}{dt} = r\left(1 - \frac{P}{K}\right)P, \text{ with } P(0) = P_0 \quad \textbf{(1)}$$

where $r$ is a constant, $P$ is the population at time $t$, and $K$ is the carrying capacity of the environment.

## PROPORTIONAL HARVESTING

When harvesting occurs and some proportion of the population is removed, our logistic model changes to

$$\frac{dP}{dt} = r\left(1 - \frac{P}{K}\right)P - sP, \qquad (2)$$

where $r$ is a population growth rate constant, $s$ is the fraction of the population that is harvested, $P$ is the population at time $t$, and $K$ is the carrying capacity of the environment. A population is referred to as being in *equilibrium* when its rate of change is 0. Notice that in Equation (2), the population has reached a state of equilibrium when

$$\frac{dP}{dt} = r\left(1 - \frac{P}{K}\right)P - sP = 0,$$

or when

$$sP = r\left(1 - \frac{P}{K}\right)P.$$

Solving for $P$ gives the equilibrium population. That is,

$$P = K\left(1 - \frac{s}{r}\right). \qquad (3)$$

This implies that at equilibrium, the proportion of the population that is harvested can be determined by multiplying both sides of Equation (3) by $s$. That is,

$$sP = sK\left(1 - \frac{s}{r}\right). \qquad (4)$$

Note that $sP$ will be nonnegative when $\frac{s}{r} < 1$, or when $s < r$ — that is, when the proportion of the population harvested is less than the rate of growth of the population. Observe that if $r > s$, the population will eventually disappear.

We can use Equation (4) to determine maximum sustainable harvest by finding when $sP$ reaches its maximum level. Note that the right-hand side of Equation (4) is a quadratic function of $s$, which attains a maximum value when $s = r/2$.

Thus, the maximum sustainable harvest for proportional harvesting is found by substituting $s = r/2$ into Equation (4). That is,

$$MSH = \frac{r}{2}P = \frac{r}{2}K\left(1 - \frac{r/2}{r}\right) = \frac{rK}{4}. \qquad (5)$$

## CONSTANT HARVESTING

When a constant number of individuals are harvested each year, our logistic model becomes

$$\frac{dP}{dt} = r\left(1 - \frac{P}{K}\right)P - B, \qquad (6)$$

where $r$ is a population growth rate constant, $B$ is the number of individuals from the population that are harvested, $P$ is the population at time $t$, and $K$ is the carrying capacity of the environment. Notice that in Equation (6), equilibrium is reached when the rate of change of the population is zero; that is when,

$$\frac{dP}{dt} = r\left(1 - \frac{P}{K}\right)P - B = 0. \qquad (7)$$

Observe that the middle piece of this equation is quadratic with respect to the population size $P$. This implies that it is equal to zero for two specific populations. At first glance, these two values present an opportunity for two choices for harvest. We must, however, be sure that the sign of the derivative near these two values points us in the right direction. For example, in Equation (6), suppose that two populations $p_1$ and $p_2$ produce two states of equilibrium, as shown by Figure 12.

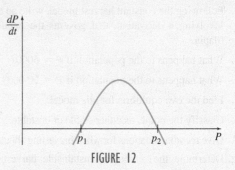

FIGURE 12

In Figure 12, the horizontal axis is population $P$, and the vertical axis is $\frac{dP}{dt}$. For values of $P$ near, but below, $p_2$, notice that the derivative is positive, indicating that the population will increase to the equilibrium point $p_2$. For values of $P$ near, but above, $p_2$ the derivative is negative, indicating that the population will decrease to the point $p_2$. This type of point is referred to as a *stable equilibrium*. The same analysis on $p_1$ produces a different result. For example, the derivative is negative for values of $P$ that are near, but below, $p_1$, indicating that the population will decrease and move away from the equilibrium point. For values of $P$ that are near, but above, $p_1$, the derivative is positive, indicating that the population will increase beyond $p_1$. This type of equilibrium point is referred to as an *unstable equilibrium*. It is very important to manage resources so that the population stays relatively close to a stable equilibrium. In this situation, waiting until the population gets near a stable equilibrium such as $p_2$ is essential. To determine the maximum sustainable harvest, simply solve Equation (7) for $B$ and then optimize the right-hand side of the equation. (Recall that the extremum of a quadratic occurs at the vertex.)

## EXERCISES

1. Suppose that a particular plot of land can sustain 500 deer and that the population of this particular species of deer can be modeled according to the logistic model as

$$\frac{dP}{dt} = 0.2\left(1 - \frac{P}{500}\right)P.$$

Each year, a proportion of the herd deer is sold to petting zoos.

   **a.** Find a function that gives the equilibrium population for various harvesting proportions.

   **b.** Determine the maximum number of deer that should be sold to petting zoos each year. (*Hint:* Find the maximum sustainable harvest.)

**2.** Suppose that a particular pond can sustain up to 30,000 tilapia and that the undisturbed population of tilapia in this environment can be modeled according to the logistic model as

$$\frac{dP}{dt} = 4\left(1 - \frac{P}{30,000}\right)P.$$

Suppose that the farmer wishes to sell 15,000 fish per year. In this case, the harvest model will be constant.

**a.** Following the constant harvest model, write an equation, involving a derivative, that governs the population of tilapia.

**b.** What happens to the population if $P = 5000$?

**c.** What happens to the population if $P = 20,000$?

**d.** Find the two equilibria for this model.

**e.** Classify the points as either stable or unstable.

**f.** Give recommendations for when harvesting should begin.

**g.** Determine the maximum sustainable harvest for this model.

# DIRECTIONS FOR GROUP PROJECT

*Use the logistic function to develop a mathematical model for a particular renewable resource. Use data collected from the Internet or other sources to provide realistic values of the carrying capacity of a particular plot of land, pond, and so on. Analyze the derivative of the model to produce an estimate of the maximum sustainable harvest for different harvest models.*

*Once your model is developed, create a class presentation that describes the model, your assumptions, results, and future research. Include graphical displays of the results.*

# 5 Graphs and the Derivative

5.1 Increasing and Decreasing Functions

5.2 Relative Extrema

5.3 Higher Derivatives, Concavity, and the Second Derivative Test

5.4 Curve Sketching

Chapter 5 Review

Extended Application: A Drug Concentration Model for Orally Administered Medications

Derivatives provide useful information about the behavior of functions and the shapes of their graphs. The first derivative describes the rate of increase or decrease, while the second derivative indicates how the rate of increase or decrease is changing. In an exercise at the end of this chapter, we will see what changes in the sign of the second derivative tell us about the shape of the graph that shows a weightlifter's performance as a function of age.

The graph in Figure 1 shows the relationship between the number of sleep-related accidents and traffic density during a 24-hour period. The blue line indicates the hourly distribution of sleep-related accidents. The green line indicates the hourly distribution of traffic density. The red line indicates the relative risk of sleep-related accidents. For example, the relative risk graph shows us that a person is nearly seven times as likely to have an accident at 4:00 A.M. than at 10:00 P.M. *Source: Sleep.*

Given a graph like the one in Figure 1, we can often locate maximum and minimum values simply by looking at the graph. It is difficult to get *exact* values or *exact* locations of maxima and minima from a graph, however, and many functions are difficult to graph without the aid of technology. In Chapter 1, we saw how to find exact maximum and minimum values for quadratic functions by identifying the vertex. A more general approach is to use the derivative of a function to determine precise maximum and minimum values of the function. The procedure for doing this is described in this chapter, which begins with a discussion of increasing and decreasing functions.

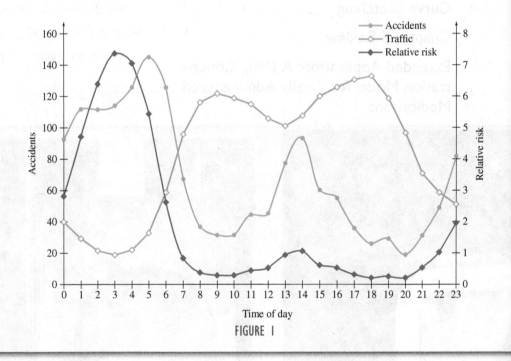

FIGURE 1

# 5.1    Increasing and Decreasing Functions

**APPLY IT**    **What is the yearly horn growth rate of bighorn rams?**
*We will answer this question in Example 5 after further investigating increasing and decreasing functions.*

A function is *increasing* if the graph goes *up* from left to right and *decreasing* if its graph goes *down* from left to right. Examples of increasing functions are shown in Figures 2(a)–(c), and examples of decreasing functions in Figures 2(d)–(f).

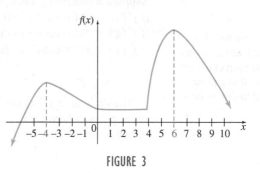

FIGURE 2

## Increasing and Decreasing Functions

Let $f$ be a function defined on some interval. Then for any two numbers $x_1$ and $x_2$ in the interval, $f$ is **increasing** on the interval if

$$f(x_1) < f(x_2) \quad \text{whenever} \quad x_1 < x_2,$$

and $f$ is **decreasing** on the interval if

$$f(x_1) > f(x_2) \quad \text{whenever} \quad x_1 < x_2.$$

**EXAMPLE 1**   **Increasing and Decreasing**

Where is the function graphed in Figure 3 increasing? Where is it decreasing?

**YOUR TURN 1**   Find where the function is increasing and decreasing.

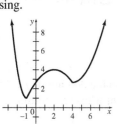

FIGURE 3

**SOLUTION**   Moving from left to right, the function is increasing for $x$-values up to $-4$, then decreasing for $x$-values from $-4$ to $0$, constant (neither increasing nor decreasing) for $x$-values from $0$ to $4$, increasing for $x$-values from $4$ to $6$, and decreasing for all $x$-values greater than $6$. In interval notation, the function is increasing on $(-\infty, -4)$ and $(4, 6)$, decreasing on $(-4, 0)$ and $(6, \infty)$, and constant on $(0, 4)$.   **TRY YOUR TURN 1**

How can we tell from the equation that defines a function where the graph increases and where it decreases? The derivative can be used to answer this question. Remember that the derivative of a function at a point gives the slope of the line tangent to the function at that point. Recall also that a line with a positive slope rises from left to right and a line with a negative slope falls from left to right.

The graph of a typical function, $f$, is shown in Figure 4. Think of the graph of $f$ as a roller coaster track moving from left to right along the graph. Now, picture one of the cars on the roller coaster. As shown in Figure 5, when the car is on level ground or parallel to level ground, its floor is horizontal, but as the car moves up the slope, its floor tilts upward. When the car reaches a peak, its floor is again horizontal, but it then begins to tilt downward (very steeply) as the car rolls downhill. The floor of the car as it moves from left to right along the track represents the tangent line at each point. Using this analogy, we can see that the slope of the tangent line will be *positive* when the car travels uphill and $f$ is *increasing*, and the slope of the tangent line will be *negative* when the car travels downhill and $f$ is *decreasing*. (In this case it is also true that the slope of the tangent line will be zero at "peaks" and "valleys.")

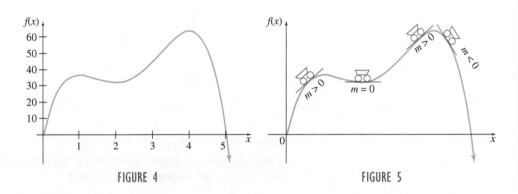

FIGURE 4                    FIGURE 5

Thus, on intervals where $f'(x) > 0$, the function $f(x)$ will increase, and on intervals where $f'(x) < 0$, the function $f(x)$ will decrease. We can determine where $f(x)$ peaks by finding the intervals on which it increases and decreases.

Our discussion suggests the following test.

## Test for Intervals Where $f(x)$ Is Increasing and Decreasing

Suppose a function $f$ has a derivative at each point in an open interval; then

if $f'(x) > 0$ for each $x$ in the interval, $f$ is *increasing* on the interval; ↗

if $f'(x) < 0$ for each $x$ in the interval, $f$ is *decreasing* on the interval; ↘

if $f'(x) = 0$ for each $x$ in the interval, $f$ is *constant* on the interval. →

**NOTE**
The third condition must hold for an entire open interval, not a single point. It would not be correct to say that because $f'(x) = 0$ at a point, then $f(x)$ is constant at that point.

The derivative $f'(x)$ can change signs from positive to negative (or negative to positive) at points where $f'(x) = 0$ and at points where $f'(x)$ does not exist. The values of $x$ where this occurs are called *critical numbers*.

## Critical Numbers

The **critical numbers** for a function $f$ are those numbers $c$ in the domain of $f$ for which $f'(c) = 0$ or $f'(c)$ does not exist. A **critical point** is a point whose $x$-coordinate is the critical number $c$ and whose $y$-coordinate is $f(c)$.

It is shown in more advanced classes that if the critical numbers of a function are used to determine open intervals on a number line, then the sign of the derivative at any point in an interval will be the same as the sign of the derivative at any other point in the interval. This suggests that the test for increasing and decreasing functions be applied as follows (assuming that no open intervals exist where the function is constant).

## Applying the Test

1. Locate the critical numbers for $f$ on a number line, as well as any points where $f$ is undefined. These points determine several open intervals.
2. Choose a value of $x$ in each of the intervals determined in Step 1. Use these values to determine whether $f'(x) > 0$ or $f'(x) < 0$ in that interval.
3. Use the test on the previous page to determine whether $f$ is increasing or decreasing on the interval.

### EXAMPLE 2    Increasing and Decreasing

Find the intervals in which the following function is increasing or decreasing. Locate all points where the tangent line is horizontal. Graph the function.

$$f(x) = x^3 + 3x^2 - 9x + 4$$

**SOLUTION**    The derivative is $f'(x) = 3x^2 + 6x - 9$. To find the critical numbers, set this derivative equal to 0 and solve the resulting equation by factoring.

$$3x^2 + 6x - 9 = 0$$
$$3(x^2 + 2x - 3) = 0$$
$$3(x + 3)(x - 1) = 0$$
$$x = -3 \quad \text{or} \quad x = 1$$

The tangent line is horizontal at $x = -3$ or $x = 1$. Since there are no values of $x$ where $f'(x)$ fails to exist, the only critical numbers are $-3$ and $1$. To determine where the function is increasing or decreasing, locate $-3$ and $1$ on a number line, as in Figure 6. (Be sure to place the values on the number line in numerical order.) These points determine three intervals: $(-\infty, -3)$, $(-3, 1)$, and $(1, \infty)$.

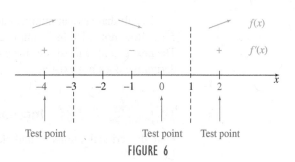

FIGURE 6

Now choose any value of $x$ in the interval $(-\infty, -3)$. Choosing $x = -4$ and evaluating $f'(-4)$ using the factored form of $f'(x)$ gives

$$f'(-4) = 3(-4 + 3)(-4 - 1) = 3(-1)(-5) = 15,$$

which is positive. You could also substitute $x = -4$ in the unfactored form of $f'(x)$, but using the factored form makes it easier to see whether the result is positive or negative, depending upon whether you have an even or an odd number of negative factors. Since one value of $x$ in this interval makes $f'(x) > 0$, all values will do so, and therefore, $f$ is increasing on $(-\infty, -3)$. Selecting 0 from the middle interval gives $f'(0) = -9$, so $f$ is decreasing on $(-3, 1)$. Finally, choosing 2 in the right-hand region gives $f'(2) = 15$, with $f$ increasing on $(1, \infty)$. The arrows in each interval in Figure 6 indicate where $f$ is increasing or decreasing.

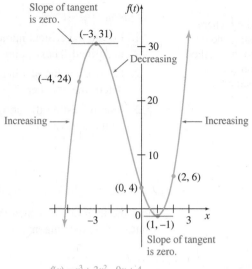

Slope of tangent
is zero.

$(-3, 31)$

Decreasing

$(-4, 24)$

Increasing

Increasing

$(2, 6)$

$(0, 4)$

$-3$    $0$    $3$    $x$

$(1, -1)$

Slope of tangent
is zero.

$f(x) = x^3 + 3x^2 - 9x + 4$

**FIGURE 7**

**YOUR TURN 2**

Find the intervals in which $f(x) = -x^3 - 2x^2 + 15x + 10$ is increasing or decreasing. Graph the function.

We now have an additional tool for graphing functions: the test for determining where a function is increasing or decreasing. (Other tools are discussed in the next few sections.) To graph the function, plot a point at each of the critical numbers by finding $f(-3) = 31$ and $f(1) = -1$. Also plot points for $x = -4, 0$, and $2$, the test values of each interval. Use these points along with the information about where the function is increasing and decreasing to get the graph in Figure 7.    **TRY YOUR TURN 2**

| **CAUTION** | Be careful to use $f(x)$, not $f'(x)$, to find the $y$-value of the points to plot.

Recall that critical numbers are numbers $c$ in the domain of $f$ for which $f'(c) = 0$ or $f'(x)$ does not exist. In Example 2, there are no critical values, $c$, where $f'(c)$ fails to exist. The next example illustrates the case where a function has a critical number at $c$ because the derivative does not exist at $c$.

**EXAMPLE 3**    **Increasing and Decreasing**

Find the critical numbers and decide where $f$ is increasing and decreasing if $f(x) = (x - 1)^{2/3}$.

**SOLUTION**    We first find $f'(x)$ using the power rule and the chain rule.

$$f'(x) = \frac{2}{3}(x - 1)^{-1/3}(1) = \frac{2}{3(x - 1)^{1/3}}$$

To find the critical numbers, we first find any values of $x$ that make $f'(x) = 0$, but here $f'(x)$ is never 0. Next, we find any values of $x$ where $f'(x)$ fails to exist. This occurs whenever the denominator of $f'(x)$ is 0, so set the denominator equal to 0 and solve.

$$3(x - 1)^{1/3} = 0 \quad \text{Divide by 3.}$$
$$[(x - 1)^{1/3}]^3 = 0^3 \quad \text{Raise both sides to the 3rd power.}$$
$$x - 1 = 0$$
$$x = 1$$

Since $f'(1)$ does not exist but $f(1)$ is defined, $x = 1$ is a critical number, the only critical number. This point divides the number line into two intervals: $(-\infty, 1)$ and $(1, \infty)$. Draw a number line for $f'$, and use a test point in each of the intervals to find where $f$ is increasing and decreasing.

$$f'(0) = \frac{2}{3(0-1)^{1/3}} = \frac{2}{-3} = -\frac{2}{3}$$

$$f'(2) = \frac{2}{3(2-1)^{1/3}} = \frac{2}{3}$$

**YOUR TURN 3** Find where $f$ is increasing and decreasing if $f(x) = (2x + 4)^{2/5}$. Graph the function.

Since $f$ is defined for all $x$, these results show that $f$ is decreasing on $(-\infty, 1)$ and increasing on $(1, \infty)$. The graph of $f$ is shown in Figure 8. **TRY YOUR TURN 3**

In Example 3, we found a critical number where $f'(x)$ failed to exist. This occurred when the denominator of $f'(x)$ was zero. Be on the alert for such values of $x$. Also be alert for values of $x$ that would make the expression under a square root, or other even root, negative. For example, if $f(x) = \sqrt{x}$, then $f'(x) = 1/(2\sqrt{x})$. Notice that $f'(x)$ does not exist for $x \le 0$, but the values of $x < 0$ are not critical numbers because those values of $x$ are not in the domain of $f$. The function $f(x) = \sqrt{x}$ does have a critical point at $x = 0$, because 0 is in the domain of $f$.

Sometimes a function may not have any critical numbers, but we are still able to determine where the function is increasing and decreasing, as shown in the next example.

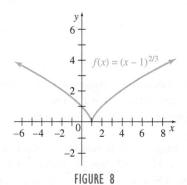

**FIGURE 8**

## EXAMPLE 4   Increasing and Decreasing (No Critical Numbers)

Find the intervals for which the following function increases and decreases. Graph the function.

$$f(x) = \frac{x - 1}{x + 1}$$

**SOLUTION**   Notice that the function $f$ is undefined when $x = -1$, so $-1$ is not in the domain of $f$. To determine any critical numbers, first use the quotient rule to find $f'(x)$.

$$f'(x) = \frac{(x+1)(1) - (x-1)(1)}{(x+1)^2}$$

$$= \frac{x + 1 - x + 1}{(x+1)^2} = \frac{2}{(x+1)^2}$$

This derivative is never 0, but it fails to exist at $x = -1$, where the function is undefined. Since $-1$ is not in the domain of $f$, there are no critical numbers for $f$.

We can still apply the first derivative test, however, to find where $f$ is increasing and decreasing. The number $-1$ (where $f$ is undefined) divides the number line into two intervals: $(-\infty, -1)$ and $(-1, \infty)$. Draw a number line for $f'$, and use a test point in each of these intervals to find that $f'(x) > 0$ for all $x$ except $-1$. (This can also be determined by observing that $f'(x)$ is the quotient of 2, which is positive, and $(x + 1)^2$, which is always positive when $x \ne -1$.) This means that the function $f$ is increasing on both $(-\infty, -1)$ and $(-1, \infty)$.

To graph the function, we find any asymptotes. Since the value $x = -1$ makes the denominator 0 but not the numerator, the line $x = -1$ is a vertical asymptote. To find the horizontal asymptote, we find

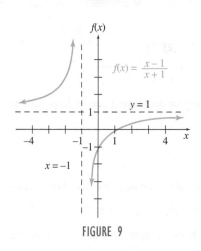

**FIGURE 9**

**YOUR TURN 4** Find where $f$ is increasing and decreasing if $f(x) = \dfrac{-2x}{x+2}$. Graph the function.

$$\lim_{x \to \infty} \frac{x - 1}{x + 1} = \lim_{x \to \infty} \frac{1 - 1/x}{1 + 1/x} \qquad \text{Divide numerator and denominator by } x.$$

$$= 1.$$

We get the same limit as $x$ approaches $-\infty$, so the graph has the line $y = 1$ as a horizontal asymptote. Using this information, as well as the $x$-intercept $(1, 0)$ and the $y$-intercept $(0, -1)$, gives the graph in Figure 9. **TRY YOUR TURN 4**

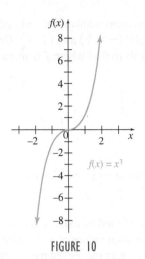

FIGURE 10

**CAUTION** It is important to note that the reverse of the test for increasing and decreasing functions is not true—it is possible for a function to be increasing on an interval even though the derivative is not positive at every point in the interval. A good example is given by $f(x) = x^3$, which is increasing on every interval, even though $f'(x) = 0$ when $x = 0$. See Figure 10.

Similarly, it is incorrect to assume that the sign of the derivative in regions separated by critical numbers must alternate between $+$ and $-$. If this were always so, it would lead to a simple rule for finding the sign of the derivative: just check one test point, and then make the other regions alternate in sign. But this is not true if one of the factors in the derivative is raised to an even power. In the function $f(x) = x^3$ just considered, $f'(x) = 3x^2$ is positive on both sides of the critical number $x = 0$.

**TECHNOLOGY NOTE**

A graphing calculator can be used to find the derivative of a function at a particular x-value. The screen in Figure 11 supports our results in Example 2 for the test values, $-4$ and $2$. The results are not exact because the calculator uses a numerical method to approximate the derivative at the given x-value.

Some graphing calculators can find where a function changes from increasing to decreasing by finding a maximum or minimum. The calculator windows in Figure 12 show this feature for the function in Example 2. Note that these, too, are approximations. This concept will be explored further in the next section.

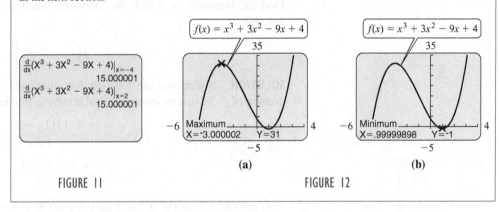

FIGURE 11                              FIGURE 12

Knowing the intervals where a function is increasing or decreasing can be important in applications, as shown by the next examples.

**EXAMPLE 5  Bighorn Sheep**

The growth rate of horn volume for rams of a certain group can be described by the function

$$v(t) = -1717 + 1499t - 263.7t^2 + 13.53t^3, \qquad 2 \le t \le 9,$$

where $t$ is the age (in yr) of the ram and $v(t)$ is the horn volume growth rate in $cm^3/yr$. *Source: Conservation Biology.*

**(a)** Determine the interval(s) on which the growth rate of horn volume is increasing. Where is it decreasing?

**APPLY IT**

**SOLUTION**  To find any intervals where this function is increasing, first set $v'(t) = 0$.

$$v'(t) = 1499 - 527.4t + 40.59t^2 = 0$$

Solving this with the quadratic formula gives the approximate solutions of $t = 4.20$ and $t = 8.79$. Use these two numbers to determine the three intervals on a number line, as shown in Figure 13. Choose $t = 3$, $t = 5$, and $t = 8.9$ as test points.

$$v'(3) = 1499 - 527.4(3) + 40.59(3^2) = 282.11$$
$$v'(5) = 1499 - 527.4(5) + 40.59(5^2) = -123.25$$
$$v'(8.9) = 1499 - 527.4(8.9) + 40.59(8.9^2) = 20.2739$$

This means that the growth rate of horn volume is increasing at a rate of $282.11 \text{ cm}^3/\text{yr}$ per year for rams of age three. When the ram is five years old, the growth rate of horn volume is decreasing at a rate of $123.25 \text{ cm}^3/\text{yr}$ per year. When the ram is almost nine years old the derivative implies that the growth rate begins to increase again. Thus, the analysis of this function indicates that the rate of horn volume is increasing in rams that are less than 4.20 years old, the rate is decreasing for rams that are between 4.20 and 8.79 years old, and it increases for rams that are older than this, as indicated in Figure 14. *Note:* Although this function indicates that the rate begins to increase for rams that are older than 8.79 years, one must be very skeptical as to the accuracy of the function at this point. It is more likely the case that the function, although accurate for most of the interval $[2, 9]$, begins to lose its accuracy near the right endpoint.

**(b)** How do these intervals relate to the cumulative horn volume over time?

**SOLUTION** The function $v(t)$ gives the growth rate of horn volume, which is itself the derivative of horn volume. It may appear surprising that the rate of growth is negative for years 4.20 through 8.79. This, however, does not imply that the total volume of the horns of a ram is decreasing during these years. It simply implies that the rate at which the horns are growing is decreasing. Similarly, in years 2 through 4.20, the yearly horn growth is increasing. This means that the horns are growing faster each successive year in this interval. Thus, a bighorn ram experiences increased horn growth over the first four years of life. After this, the function suggests that the horns continue to grow but the amount of growth decreases with age.

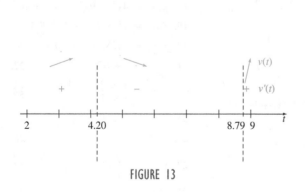

FIGURE 13

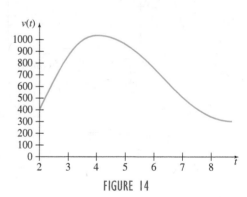

FIGURE 14

## EXAMPLE 6  Recollection of Facts

In the exercises in the previous chapter, the function

$$f(t) = \frac{90t}{99t - 90}$$

gave the number of facts recalled after $t$ hours for $t > 10/11$. Find the intervals in which $f(t)$ is increasing or decreasing.

**SOLUTION** First find the derivative, $f'(t)$. Using the quotient rule,

$$f'(t) = \frac{(99t - 90)(90) - 90t(99)}{(99t - 90)^2}$$

$$= \frac{8910t - 8100 - 8910t}{(99t - 90)^2} = \frac{-8100}{(99t - 90)^2}.$$

Since $(99t - 90)^2$ is positive everywhere in the domain of the function and since the numerator is a negative constant, $f'(t) < 0$ for all $t$ in the domain of $f(t)$. Thus $f(t)$ always decreases and, as expected, the number of words recalled decreases steadily over time.

# 5.1  EXERCISES

**Find the open intervals where the functions graphed as follows are (a) increasing, or (b) decreasing.**

**1.**

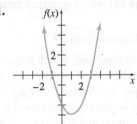

**2.**

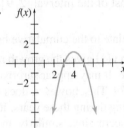

**3.**

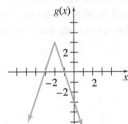

**4.**

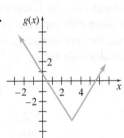

**5.**

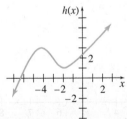

**6.**

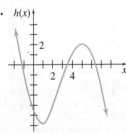

**7.**

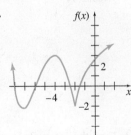

**8.**

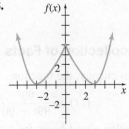

**For each of the exercises listed below, suppose that the function that is graphed is not $f(x)$, but $f'(x)$. Find the open intervals where $f(x)$ is (a) increasing or (b) decreasing.**

**9.** Exercise 1        **10.** Exercise 2

**11.** Exercise 7       **12.** Exercise 8

**For each function, find (a) the critical numbers; (b) the open intervals where the function is increasing; and (c) the open intervals where it is decreasing.**

**13.** $y = 2.3 + 3.4x - 1.2x^2$

**14.** $y = 1.1 - 0.3x - 0.3x^2$

**15.** $f(x) = \frac{2}{3}x^3 - x^2 - 24x - 4$

**16.** $f(x) = \frac{2}{3}x^3 - x^2 - 4x + 2$

**17.** $f(x) = 4x^3 - 15x^2 - 72x + 5$

**18.** $f(x) = 4x^3 - 9x^2 - 30x + 6$

**19.** $f(x) = x^4 + 4x^3 + 4x^2 + 1$

**20.** $f(x) = 3x^4 + 8x^3 - 18x^2 + 5$

**21.** $y = -3x + 6$          **22.** $y = 6x - 9$

**23.** $f(x) = \dfrac{x + 2}{x + 1}$     **24.** $f(x) = \dfrac{x + 3}{x - 4}$

**25.** $y = \sqrt{x^2 + 1}$         **26.** $y = x\sqrt{9 - x^2}$

**27.** $f(x) = x^{2/3}$           **28.** $f(x) = (x + 1)^{4/5}$

**29.** $y = x - 4\ln(3x - 9)$   **30.** $f(x) = \ln\dfrac{5x^2 + 4}{x^2 + 1}$

**31.** $f(x) = xe^{-3x}$         **32.** $f(x) = xe^{x^2 - 3x}$

**33.** $f(x) = x^2 2^{-x}$        **34.** $f(x) = x2^{-x^2}$

**35.** $y = x^{2/3} - x^{5/3}$     **36.** $y = x^{1/3} + x^{4/3}$

**37.** $y = \sin x$            **38.** $y = 5\tan x$

**39.** $y = 3\sec x$          **40.** $y = 4\cos 3x$

**41.** A friend looks at the graph of $y = x^2$ and observes that if you start at the origin, the graph increases whether you go to the right or the left, so the graph is increasing everywhere. Explain why this reasoning is incorrect.

**42.** Use the techniques of this chapter to find the vertex and intervals where $f$ is increasing and decreasing, given

$$f(x) = ax^2 + bx + c,$$

where we assume $a > 0$. Verify that this agrees with what we found in Chapter 1.

**43.** Repeat Exercise 42 under the assumption $a < 0$.

**44.** Where is the function defined by $f(x) = e^x$ increasing? Decreasing? Where is the tangent line horizontal?

**45.** Repeat Exercise 44 with the function defined by $f(x) = \ln x$.

**46.** **a.** For the function in Exercise 15, find the average of the critical numbers.

   **b.** For the function in Exercise 15, use a graphing calculator to find the roots of the function, and then find the average of those roots.

   **c.** Compare your answers to parts a and b. What do you notice?

   **d.** Repeat part a for the function in Exercise 17.

   **e.** Repeat part b for the function in Exercise 17.

   **f.** Compare your answers to parts d and e. What do you notice?

   It can be shown that the average of the roots of a polynomial (including the complex roots, if there are any) and the critical numbers of a polynomial (including complex roots of $f'(x) = 0$, if there are any) are always equal. *Source: The Mathematics Teacher.*

**For each of the following functions, use a graphing calculator to find the open intervals where $f(x)$ is (a) increasing, or (b) decreasing.**

**47.** $f(x) = e^{0.001x} - \ln x$

**48.** $f(x) = \ln(x^2 + 1) - x^{0.3}$

**49.** $f(x) = x^2 + 3 \cos x$

**50.** $f(x) = x^4 - \sin x$

## LIFE SCIENCE APPLICATIONS

**51. Air Pollution**  The graph shows the amount of air pollution removed by trees in the Chicago urban region for each month of the year. From the graph we see, for example, that the ozone level starting in May increases up to June, and then abruptly decreases. *Source: USDA Forest Service.*

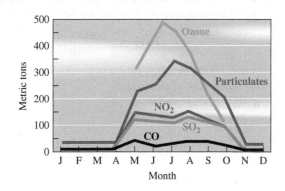

**a.** Are these curves the graphs of functions?

**b.** Look at the graph for particulates. Where is the function increasing? Decreasing? Constant?

**c.** On what intervals do all four lower graphs indicate that the corresponding functions are constant? Why do you think the functions are constant on those intervals?

**52. Spread of Infection**  The number of people $P(t)$ (in hundreds) infected $t$ days after an epidemic begins is approximated by

$$P(t) = \frac{10 \ln(0.19t + 1)}{0.19t + 1}.$$

When will the number of people infected start to decline?

**53. Alcohol Concentration**  In Exercise 47 in Section 1.5 on Polynomial and Rational Functions, we gave the function defined by

$$A(t) = 0.003631t^3 - 0.03746t^2 + 0.1012t + 0.009$$

as the approximate blood alcohol concentration in a 170-lb woman $t$ hours after drinking 2 oz of alcohol on an empty stomach, for $t$ in the interval $[0, 5]$. *Source: Medicolegal Aspects of Alcohol Determination in Biological Specimens.*

**a.** On what time intervals is the alcohol concentration increasing?

**b.** On what intervals is it decreasing?

**54. Drug Concentration**  The percent of concentration of a drug in the bloodstream $t$ hours after the drug is administered is given by

$$K(t) = \frac{4t}{3t^2 + 27}.$$

**a.** On what time intervals is the concentration of the drug increasing?

**b.** On what intervals is it decreasing?

**55. Drug Concentration**  Suppose a certain drug is administered to a patient, with the percent of concentration of the drug in the bloodstream $t$ hours later given by

$$K(t) = \frac{5t}{t^2 + 1}.$$

**a.** On what time intervals is the concentration of the drug increasing?

**b.** On what intervals is it decreasing?

**56. Cardiology**  The aortic pressure–diameter relation in a particular patient who underwent cardiac catheterization can be modeled by the polynomial

$$D(p) = 0.000002p^3 - 0.0008p^2 + 0.1141p + 16.683,$$

$$55 \le p \le 130,$$

where $D(p)$ is the aortic diameter (in millimeters) and $p$ is the aortic pressure (in mm Hg). Determine where this function is increasing and where it is decreasing within the interval given above. *Source: Circulation.*

**57. Thermic Effect of Food** The metabolic rate of a person who has just eaten a meal tends to go up and then, after some time has passed, returns to a resting metabolic rate. This phenomenon is known as the thermic effect of food. Researchers have indicated that the thermic effect of food for one particular person is

$$F(t) = -10.28 + 175.9te^{-t/1.3},$$

where $F(t)$ is the thermic effect of food (in kJ/hr) and $t$ is the number of hours that have elapsed since eating a meal. *Source: American Journal of Clinical Nutrition.*

a. Find $F'(t)$.

b. Determine where this function is increasing and where it is decreasing. Interpret your answers.

**58. Holstein Dairy Cattle** As we saw in Chapter 4, researchers have developed the following function that can be used to accurately predict the weight of Holstein cows (females) of various ages:

$$W_1(t) = 619(1 - 0.905e^{-0.002t})^{1.2386},$$

where $W_1(t)$ is the weight of the Holstein cow (in kilograms) that is $t$ days old. Where is this function increasing? *Source: Canadian Journal of Animal Science.*

**59. Ayrshire Dairy Cattle** As we saw in Chapter 4, researchers have developed a function that can be used to accurately predict the weight of Ayrshire cows (females). In this case,

$$W_2(t) = 532(1 - 0.911e^{-0.0021t})^{1.2466}.$$

*Source: Canadian Journal of Animal Science.*

a. Where is this function increasing?

b. Use a graphing calculator to graph this function with the function from the previous exercise on $[0, 2000]$ by $[0, 620]$.

c. Compare the two functions, indicating which type of cow grows the fastest and which type reaches the greatest weight.

**60. Vectorial Capacity** An organism (such as a mosquito) that carries pathogens from one host to another is called a vector. The vectorial capacity, defined as the capacity of the vector population to transmit the disease in terms of the potential number of secondary inoculations originating per unit time from an infective person, is given by

$$C = \frac{ma^2p^n}{-\ln p},$$

where $m$ is the number of vectors per host, $a$ is the number of blood meals taken on a host per vector per day (biting rate), $p$ is the proportion of vectors surviving per day, and $n$ is the length (in days) of the parasite incubation period in the vectors. *Source: Frontiers in Mathematical Biology.*

a. Calculate $dC/dp$. (*Hint:* Assume that $m$, $a$, and $n$ are constants.) Is vectorial capacity increasing or decreasing with respect to $p$?

b. Let $u$ be the death rate for the vectors, defined as $u = -\ln p$. Show that

$$C = \frac{ma^2e^{-un}}{u}.$$

c. Using your answer from part b, calculate $dC/du$. Is vectorial capacity increasing or decreasing with respect to $u$?

d. Explain how you could conclude that vectorial capacity is decreasing with respect to $u$ by the definition of $u$ and the fact that vectorial capacity is increasing with respect to $p$. Explain why both of these answers make sense biologically.

**61. Alzheimer's Disease** A study on the circadian rhythms of patients with Alzheimer's disease found that the body temperature of patients can be modeled by the function

$$C(t) = 37.29 + 0.46 \cos\left(\frac{2\pi(t - 16.37)}{24}\right),$$

where $t$ is the time in hours since midnight. *Source: American Journal of Psychiatry.* During a 24-hour period, where is this function increasing and where is it decreasing?

**62. Plant Growth** Researchers have found that the radius $R$ to which a plant will grow is affected by the number of plants $N$ in the area, as described by the equation

$$R = \sqrt{\frac{N_0 - N}{2\pi DN}},$$

where $D$ is the density of the plants in an area and $N_0$ is the number of trees in the area initially. *Source: Ecology.*

a. Find $\frac{dR}{dN}$.

b. Based on the sign of your answer to part a, what can you say about how the maximum radius of a plant is affected as the number of plants in the area increases?

c. What happens to $\frac{dR}{dN}$ as $N$ approaches 0? As $N$ approaches $N_0$?

## OTHER APPLICATIONS

**63. Population** The standard normal probability function is used to describe many different populations. Its graph is the well-known normal curve. This function is defined by

$$f(x) = \frac{1}{\sqrt{2\pi}}e^{-x^2/2}.$$

Give the intervals where the function is increasing and decreasing. Is your answer geometrically obvious?

**64. Housing Starts** A county realty group estimates that the number of housing starts per year over the next three years will be

$$H(r) = \frac{300}{1 + 0.03r^2},$$

where $r$ is the mortgage rate (in percent).

a. Where is $H(r)$ increasing?

b. Where is $H(r)$ decreasing?

1. Increasing on $(-1, 2)$ and $(4, \infty)$. Decreasing on $(-\infty, -1)$ and $(2, 4)$.

2. Increasing on $(-3, 5/3)$. Decreasing on $(-\infty, -3)$ and $(5/3, \infty)$.

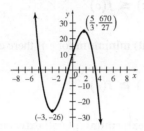

3. Increasing on $(-2, \infty)$ and decreasing on $(-\infty, -2)$.

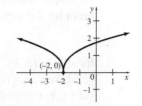

4. Never increasing. Decreasing on $(-\infty, -2)$ and $(-2, \infty)$.

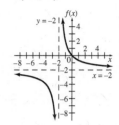

# 5.2 Relative Extrema

**APPLY IT** How much time does it take for a flu epidemic to peak?

*We will answer this question in Example 1 by investigating the idea of a relative maximum.*

As we have seen throughout this text, the graph of a function may have peaks and valleys. It is important in many applications to determine where these points occur. For example, if the function represents the profit of a company, these peaks and valleys indicate maximum profits and losses. When the function is given as an equation, we can use the derivative to determine these points, as shown in the first example.

**EXAMPLE 1** **Flu Epidemic**

Epidemiologists have discovered that a flu epidemic is overtaking Youngstown, Ohio. Based on preliminary data, the number of people (in thousands) who have the flu can be described by the equation

$$f(t) = -\frac{3}{20}t^2 + 6t + 20, \qquad 0 \le t \le 30,$$

where $t$ is the number of days since the epidemic was first discovered. Assuming that this trend continues, on what day is the epidemic expected to peak? How many people are expected to have the flu on that day?

**APPLY IT** **SOLUTION** The epidemic will peak when the number of people who have the flu is at a maximum. To find this time, find $f'(t)$.

$$f'(t) = -\frac{3}{10}t + 6 = -0.3t + 6$$

The derivative $f'(t)$ is greater than 0 when $-0.3t + 6 > 0, -3t > -60,$ or $t < 20$. Similarly, $f'(t) < 0$ when $-0.3t + 6 < 0, -3t < -60,$ or $t > 20$. Thus, the epidemic increases for the first 20 days and decreases for the last 10 days. The epidemic peaks on day 20. At that time there will be $f(20) = 80$ or 80,000 people who have the flu.

The maximum level of the epidemic (80,000) in the example above is a *relative maximum*, defined as follows.

### Relative Maximum or Minimum

Let $c$ be a number in the domain of a function $f$. Then $f(c)$ is a **relative** (or **local**) **maximum** for $f$ if there exists an open interval $(a, b)$ containing $c$ such that

$$f(x) \leq f(c)$$

for all $x$ in $(a, b)$.

Likewise, $f(c)$ is a **relative** (or **local**) **minimum** for $f$ if there exists an open interval $(a, b)$ containing $c$ such that

$$f(x) \geq f(c)$$

for all $x$ in $(a, b)$.

A function has a **relative** (or **local**) **extremum** (plural: **extrema**) at $c$ if it has either a relative maximum or a relative minimum there.

If $c$ is an endpoint of the domain of $f$, we only consider $x$ in the half-open interval that is in the domain.*

The intuitive idea is that a relative maximum is the greatest value of the function in some region right around the point, although there may be greater values elsewhere. For example, the highest value of your systolic blood pressure this week is a relative maximum, although it may have reached a higher value earlier this year. Similarly, a relative minimum is the least value of a function in some region around the point.

**NOTE** Recall from Section 1.5 on Polynomials and Rational Functions that a relative extremum that is not an endpoint is also referred to as a turning point.

A simple way to view these concepts is that a relative maximum is a peak, and a relative minimum is the bottom of a valley, although either a relative minimum or maximum can also occur at the endpoint of the domain.

### EXAMPLE 2   Relative Extrema

Identify the $x$-values of all points where the graph in Figure 15 has relative extrema.

**YOUR TURN 1**   Identify the $x$-values of all points where the graph has relative extrema.

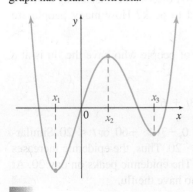

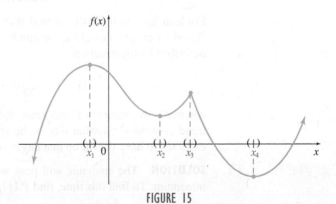

**FIGURE 15**

**SOLUTION**   The parentheses around $x_1$ show an open interval containing $x_1$ such that $f(x) \leq f(x_1)$, so there is a relative maximum of $f(x_1)$ at $x = x_1$. Notice that many other open intervals would work just as well. Similar intervals around $x_2$, $x_3$, and $x_4$ can be used to find a relative maximum of $f(x_3)$ at $x = x_3$ and relative minima of $f(x_2)$ at $x = x_2$ and $f(x_4)$ at $x = x_4$.                    **TRY YOUR TURN 1**

*There is disagreement on calling an endpoint a maximum or minimum. We define it this way because this is an applied calculus book, and in an application it would be considered a maximum or minimum value of the function.

The function graphed in Figure 16 has relative maxima when $x = x_1$ or $x = x_3$ and relative minima when $x = x_2$ or $x = x_4$. The tangent lines at the points having $x$-values $x_1$ and $x_2$ are shown in the figure. Both tangent lines are horizontal and have slope 0. There is no single tangent line at the point where $x = x_3$.

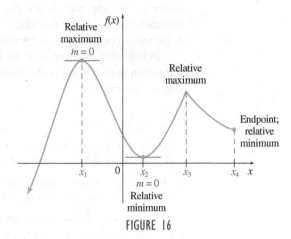

FIGURE 16

Since the derivative of a function gives the slope of a line tangent to the graph of the function, to find relative extrema we first identify all critical numbers and endpoints. A relative extremum *may* exist at a critical number. (A rough sketch of the graph of the function near a critical number often is enough to tell whether an extremum has been found.) These facts about extrema are summarized below.

If a function $f$ has a relative extremum at $c$, then $c$ is a critical number or $c$ is an endpoint of the domain.

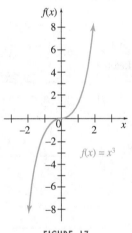

FIGURE 17

**CAUTION** Be very careful not to get this result backward. It does *not* say that a function has relative extrema at all critical numbers of the function. For example, Figure 17 shows the graph of $f(x) = x^3$. The derivative, $f'(x) = 3x^2$, is 0 when $x = 0$, so that 0 is a critical number for that function. However, as suggested by the graph of Figure 17, $f(x) = x^3$ has neither a relative maximum nor a relative minimum at $x = 0$ (or anywhere else, for that matter). A critical number is a candidate for the location of a relative extremum, but only a candidate.

## First Derivative Test

Suppose all critical numbers have been found for some function $f$. How is it possible to tell from the equation of the function whether these critical numbers produce relative maxima, relative minima, or neither? One way is suggested by the graph in Figure 18.

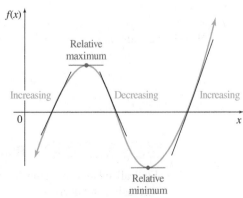

FIGURE 18

As shown in Figure 18, on the left of a relative maximum the tangent lines to the graph of a function have positive slopes, indicating that the function is increasing. At the relative maximum, the tangent line is horizontal. On the right of the relative maximum the tangent lines have negative slopes, indicating that the function is decreasing. Around a relative minimum the opposite occurs. As shown by the tangent lines in Figure 18, the function is decreasing on the left of the relative minimum, has a horizontal tangent at the minimum, and is increasing on the right of the minimum.

Putting this together with the methods from Section 1 for identifying intervals where a function is increasing or decreasing gives the following **first derivative test** for locating relative extrema.

## First Derivative Test

Let $c$ be a critical number for a function $f$. Suppose that $f$ is continuous on $(a, b)$ and differentiable on $(a, b)$ except possibly at $c$, and that $c$ is the only critical number for $f$ in $(a, b)$.

1. $f(c)$ is a relative maximum of $f$ if the derivative $f'(x)$ is positive in the interval $(a, c)$ and negative in the interval $(c, b)$.

2. $f(c)$ is a relative minimum of $f$ if the derivative $f'(x)$ is negative in the interval $(a, c)$ and positive in the interval $(c, b)$.

The sketches in the following table show how the first derivative test works. Assume the same conditions on $a$, $b$, and $c$ for the table as those given for the first derivative test.

| f(x) has: | Sign of f' in (a, c) | Sign of f' in (c, b) | Sketches |
|---|---|---|---|
| Relative maximum | + | − | |
| Relative minimum | − | + | |
| No relative extrema | + | + | |
| No relative extrema | − | − | |

**EXAMPLE 3**  **Relative Extrema**

Find all relative extrema for the following functions, as well as where each function is increasing and decreasing.

(a) $f(x) = 2x^3 - 3x^2 - 72x + 15$

**Method 1**
**First Derivative Test**

**SOLUTION**
The derivative is $f'(x) = 6x^2 - 6x - 72$. There are no points where $f'(x)$ fails to exist, so the only critical numbers will be found where the derivative equals 0. Setting the derivative equal to 0 gives

$$6x^2 - 6x - 72 = 0$$
$$6(x^2 - x - 12) = 0$$
$$6(x - 4)(x + 3) = 0$$
$$x - 4 = 0 \quad \text{or} \quad x + 3 = 0$$
$$x = 4 \quad \text{or} \quad x = -3.$$

As in the previous section, the critical numbers 4 and $-3$ are used to determine the three intervals $(-\infty, -3)$, $(-3, 4)$, and $(4, \infty)$ shown on the number line in Figure 19.

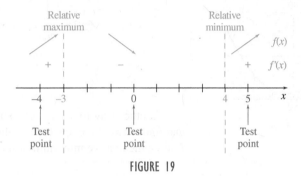

**FIGURE 19**

Any number from each of the three intervals can be used as a test point to find the sign of $f'$ in each interval. Using $-4$, 0, and 5 gives the following information.

$$f'(-4) = 6(-8)(-1) > 0$$
$$f'(0) = 6(-4)(3) < 0$$
$$f'(5) = 6(1)(8) > 0$$

Thus, the derivative is positive on $(-\infty, -3)$, negative on $(-3, 4)$, and positive on $(4, \infty)$. By Part 1 of the first derivative test, this means that the function has a relative maximum of $f(-3) = 150$ when $x = -3$; by Part 2, $f$ has a relative minimum of $f(4) = -193$ when $x = 4$. The function is increasing on $(-\infty, -3)$ and $(4, \infty)$ and decreasing on $(-3, 4)$. The graph is shown in Figure 20. **TRY YOUR TURN 2**

**YOUR TURN 2** Find all relative extrema of $f(x) = -x^3 - 2x^2 + 15x + 10$.

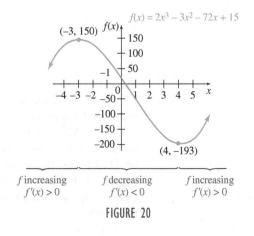

**FIGURE 20**

Method 2
Graphing Calculator

Many graphing calculators can locate a relative extremum when supplied with an interval containing the extremum. For example, after graphing the function $f(x) = 2x^3 - 3x^2 - 72x + 15$ on a TI-84 Plus, we selected "maximum" from the CALC menu and entered a left bound of $-4$ and a right bound of $0$. The calculator asks for an initial guess, but in this example it doesn't matter what we enter. The result of this process, as well as a similar process for finding the relative minimum, is shown in Figure 21. (Some calculators give slightly different values.)

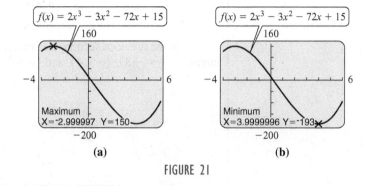

**(a)**          **(b)**

FIGURE 21

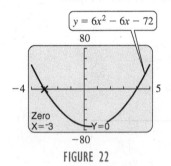

FIGURE 22

Another way to verify the extrema with a graphing calculator is to graph $y = f'(x)$ and find where the graph crosses the $x$-axis. Figure 22 shows the result of this approach for finding the relative minimum of the previous function.

**(b)** $f(x) = 6x^{2/3} - 4x$

**SOLUTION** Find $f'(x)$.

$$f'(x) = 4x^{-1/3} - 4 = \frac{4}{x^{1/3}} - 4$$

The derivative fails to exist when $x = 0$, but the function itself is defined when $x = 0$, making 0 a critical number for $f$. To find other critical numbers, set $f'(x) = 0$.

$$f'(x) = 0$$

$$\frac{4}{x^{1/3}} - 4 = 0$$

$$\frac{4}{x^{1/3}} = 4$$

$4 = 4x^{1/3}$    Multiply both sides by $x^{1/3}$.

$1 = x^{1/3}$    Divide both sides by 4.

$1 = x$    Cube both sides.

The critical numbers 0 and 1 are used to locate the intervals $(-\infty, 0)$, $(0, 1)$, and $(1, \infty)$ on a number line as in Figure 23. Evaluating $f'(x)$ at the test points $-1$, $1/2$, and 2 and using the first derivative test shows that $f$ has a relative maximum at $x = 1$; the value of this relative maximum is $f(1) = 2$. Also, $f$ has a relative minimum at $x = 0$; this relative minimum is $f(0) = 0$. The function is increasing on $(0, 1)$ and decreasing on $(-\infty, 0)$ and $(1, \infty)$. Notice that the graph, shown in Figure 24, has a sharp point at the critical number where the derivative does not exist. In the last section of this chapter we will show how to verify other features of the graph.

**YOUR TURN 3** Find all relative extrema of $f(x) = x^{2/3} - x^{5/3}$.

TRY YOUR TURN 3

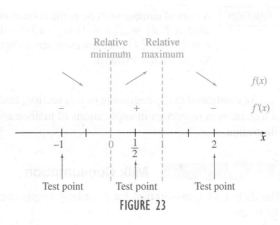

FIGURE 23

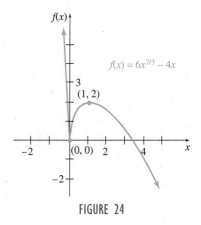

FIGURE 24

**(c)** $f(x) = xe^{2-x^2}$

**SOLUTION** The derivative, found by using the product rule and the chain rule, is

$$f'(x) = x(-2x)e^{2-x^2} + e^{2-x^2}$$
$$= e^{2-x^2}(-2x^2 + 1).$$

This expression exists for all $x$ in the domain of $f$. Since $e^{2-x^2}$ is always positive, the derivative is 0 when

$$-2x^2 + 1 = 0$$
$$1 = 2x^2$$
$$\frac{1}{2} = x^2$$
$$x = \pm\sqrt{1/2}$$
$$x = \pm\frac{1}{\sqrt{2}} \approx \pm 0.707.$$

There are two critical points, $-1/\sqrt{2}$ and $1/\sqrt{2}$. Using test points of $-1$, 0, and 1 gives the results shown in Figure 25.

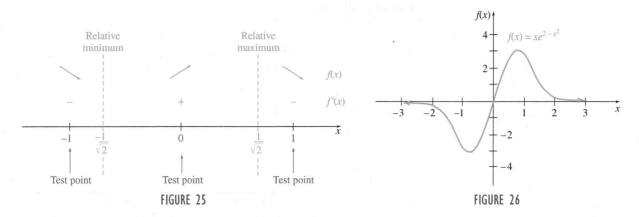

FIGURE 25

FIGURE 26

**YOUR TURN 4** Find all relative extrema of $f(x) = x^2 e^x$.

The function has a relative minimum at $-1/\sqrt{2}$ of $f(-1/\sqrt{2}) \approx -3.17$ and a relative maximum at $1/\sqrt{2}$ of $f(1/\sqrt{2}) \approx 3.17$. It is decreasing on the interval $(-\infty, -1/\sqrt{2})$, increasing on the interval $(-1/\sqrt{2}, 1/\sqrt{2})$, and decreasing on the interval $(1/\sqrt{2}, \infty)$. The graph is shown in Figure 26.　　**TRY YOUR TURN 4**

CAUTION | A critical number must be in the domain of the function. For example, the derivative of $f(x) = x/(x - 4)$ is $f'(x) = -4/(x - 4)^2$, which fails to exist when $x = 4$. But $f(4)$ does not exist, so 4 is not a critical number, and the function has no relative extrema.

As mentioned at the beginning of this section, finding the maximum or minimum value of a quantity is important in applications of mathematics. The final example gives a further illustration.

## EXAMPLE 4   Milk Consumption

The daily milk consumption for Aberdeen Angus and Hereford calves can be modeled by the function

$$M(t) = 5.955t^{0.247}e^{-0.027t}, \qquad 1 \le t \le 26,$$

where $M(t)$ is the milk consumption (in kg) and $t$ is the age of the calf (in weeks). *Source: Animal Production.* Find the time in which the maximum daily consumption occurs.

**SOLUTION**   To maximize $M(t)$, find $M'(t)$. Then find the critical numbers.

$$M'(t) = 5.955[t^{0.247}(-0.027e^{-0.027t}) + 0.247t^{-0.753}e^{-0.247}]$$
$$= 5.955e^{-0.027t}(-0.027t^{0.247} + 0.247t^{-0.753}) \qquad \text{Factor out } e^{-0.027t}.$$
$$= 5.955t^{0.247}e^{-0.027t}(-0.027 + 0.247t^{-1}) = 0$$

Since the expression in the parentheses above is the only term that can force the derivative to be zero, we set it equal to zero. That is,

$$-0.027 + 0.247t^{-1} = 0.$$

Solving for $t$, we find that $M'(t) = 0$ when $t \approx 9.15$. Since $M'(t)$ exists for all $t$ in this interval, 9.15 is the only critical number. To verify that $t = 9.15$ is a maximum, evaluate the derivative on either side of $t = 9.15$.

$$M'(5) \approx 0.1734 \quad \text{and} \quad M'(25) \approx -0.1150$$

As illustrated by these calculations and Figure 27, $M(t)$ is increasing up to $t = 9.15$, then decreasing, so there is a maximum value at $t \approx 9.15$ of $M(9.15) \approx 8.04$. The maximum daily milk consumption will be approximately 8.04 kg and will occur during the ninth week of a calf's life.

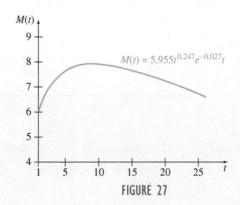

FIGURE 27

CAUTION | Be careful to give the $y$-value of the point where an extremum occurs. Although we solve the equation $f'(x) = 0$ for $x$ to find the extremum, the maximum or minimum value of the function is the corresponding $y$-value. Thus, in Example 4, we found that at $t = 9.15$, the maximum daily milk consumption is 8.04 (not 9.15).

Examples that involve the maximization of a quadratic function, such as the opening example, could be solved by the methods described in Chapter 1. But those involving more complicated functions, such as the milk consumption example, are difficult to analyze without the tools of calculus.

Finding extrema for realistic problems requires an accurate mathematical model of the problem. In particular, it is important to be aware of restrictions on the values of the variables. For example, if $N(t)$ closely approximates the number of cells in a tumor at time $t$, $N(t)$ must certainly be restricted to the positive integers, or perhaps to a few common fractional values.

On the other hand, to apply the tools of calculus to obtain an extremum for some function, the function must be defined and be meaningful at every real number in some interval. Because of this, the answer obtained from a mathematical model might be a number that is not feasible in the actual problem.

Usually, the requirement that a continuous function be used, rather than one that can take on only certain selected values, is of theoretical interest only. In most cases, the methods of calculus give acceptable results as long as the assumptions of continuity and differentiability are not totally unreasonable. If they lead to the conclusion, say, that a tumor has $80\sqrt{2}$ cells, it is usually only necessary to investigate acceptable values close to $80\sqrt{2}$.

# 5.2 EXERCISES

**Find the locations and values of all relative extrema for the functions with graphs as follows. Compare with Exercises 1–8 in the preceding section.**

**1.**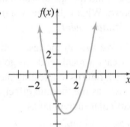

**2.**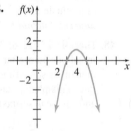

**3.**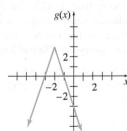

**4.**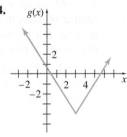

**5.**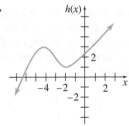

**6.**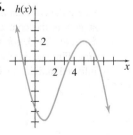

**7.**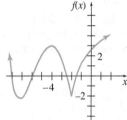

**8.**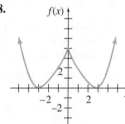

**For each of the exercises listed below, suppose that the function that is graphed is not $f(x)$ but $f'(x)$. Find the locations of all relative extrema, and tell whether each extremum is a relative maximum or minimum.**

**9.** Exercise 1      **10.** Exercise 2

**11.** Exercise 7      **12.** Exercise 8

**Find the $x$-value of all points where the functions defined as follows have any relative extrema. Find the value(s) of any relative extrema.**

**13.** $f(x) = x^2 - 10x + 33$

**14.** $f(x) = x^2 + 8x + 5$

**15.** $f(x) = x^3 + 6x^2 + 9x - 8$

**16.** $f(x) = x^3 + 3x^2 - 24x + 2$

**17.** $f(x) = -\frac{4}{3}x^3 - \frac{21}{2}x^2 - 5x + 8$

**18.** $f(x) = -\frac{2}{3}x^3 - \frac{1}{2}x^2 + 3x - 4$

**19.** $f(x) = x^4 - 18x^2 - 4$

**20.** $f(x) = x^4 - 8x^2 + 9$

**21.** $f(x) = 3 - (8 + 3x)^{2/3}$

**22.** $f(x) = \dfrac{(5 - 9x)^{2/3}}{7} + 1$

**23.** $f(x) = 2x + 3x^{2/3}$

**24.** $f(x) = 3x^{5/3} - 15x^{2/3}$

**25.** $f(x) = x - \dfrac{1}{x}$

**26.** $f(x) = x^2 + \dfrac{1}{x}$

**27.** $f(x) = \dfrac{x^2 - 2x + 1}{x - 3}$

**28.** $f(x) = \dfrac{x^2 - 6x + 9}{x + 2}$

**29.** $f(x) = x^2 e^x - 3$

**30.** $f(x) = 3xe^x + 2$

**31.** $f(x) = 2x + \ln x$

**32.** $f(x) = \dfrac{x^2}{\ln x}$

**33.** $f(x) = \dfrac{2^x}{x}$

**34.** $f(x) = x + 8^{-x}$

**35.** $f(x) = \sin(\pi x)$

**36.** $f(x) = \cos(2x)$

**Use the derivative to find the vertex of each parabola.**

**37.** $y = -2x^2 + 12x - 5$

**38.** $y = ax^2 + bx + c$

 **Graph each function on a graphing calculator, and then use the graph to find all relative extrema (to three decimal places). Then confirm your answer by finding the derivative and using the calculator to solve the equation $f'(x) = 0$.**

**39.** $f(x) = x^5 - x^4 + 4x^3 - 30x^2 + 5x + 6$

**40.** $f(x) = -x^5 - x^4 + 2x^3 - 25x^2 + 9x + 12$

**41.** Graph $f(x) = 2|x + 1| + 4|x - 5| - 20$ with a graphing calculator in the window $[-10, 10]$ by $[-15, 30]$. Use the graph and the function to determine the $x$-values of all extrema.

**42.** Consider the function

$$g(x) = \dfrac{1}{x^{12}} - 2\left(\dfrac{1000}{x}\right)^6.$$

*Source: Mathematics Teacher.*

**a.** Using a graphing calculator, try to find any local minima, or tell why finding a local minimum is difficult for this function.

**b.** Find any local minima using the techniques of calculus.

**c.** Based on your results in parts a and b, describe circumstances under which relative extrema are easier to find using the techniques of calculus than using a graphing calculator.

**In each figure, the graph of the derivative of $y = f(x)$ is shown. Find the locations of all relative extrema of $f(x)$.**

**43.**

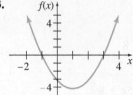

**44.**

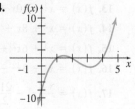

## LIFE SCIENCE APPLICATIONS

**45. Activity Level** In the summer the activity level of a certain type of lizard varies according to the time of day. A biologist has determined that the activity level is given by the function

$$a(t) = 0.008t^3 - 0.288t^2 + 2.304t + 7,$$

where $t$ is the number of hours after 12 noon. When is the activity level highest? When is it lowest?

**46. Milk Consumption** The average individual daily milk consumption for herds of Charolais, Angus, and Hereford calves can be described by the function

$$M(t) = 6.281t^{0.242}e^{-0.025t}, \quad 1 \le t \le 26,$$

where $M(t)$ is the milk consumption (in kilograms) and $t$ is the age of the calf (in weeks). *Source: Animal Production.*

**a.** Find the time in which the maximum daily consumption occurs and the maximum daily consumption.

**b.** If the general formula for this model is given by

$$M(t) = at^b e^{-ct},$$

find the time where the maximum consumption occurs and the maximum consumption. (*Hint:* Express your answer in terms of $a$, $b$, and $c$.)

**47. Alaskan Moose** The mathematical relationship between the age of a captive female moose and its mass can be described by the function

$$M(t) = 369(0.93)^t t^{0.36}, \quad t \le 12,$$

where $M(t)$ is the mass of the moose (in kilograms) and $t$ is the age (in years) of the moose. Find the age at which the mass of a female moose is maximized. What is the maximum mass? *Source: Journal of Wildlife Management.*

**48. Thermic Effect of Food** As we saw in the last section, the metabolic rate after a person eats a meal tends to go up and then, after some time has passed, returns to a resting metabolic rate. This phenomenon is known as the thermic effect of food and can be described for a particular individual as

$$F(t) = -10.28 + 175.9te^{-t/1.3},$$

where $F(t)$ is the thermic effect of food (in kJ/hr), and $t$ is the number of hours that have elapsed since eating a meal. Find the time after the meal when the thermic effect of the food is maximized. *Source: American Journal of Clinical Nutrition.*

**49. Epidemic** The rate of change in the number of individuals infected in an epidemic as a function of time is given by the epidemic curve, which has the form

$$f(t) = \dfrac{k(n - 1)n^2 e^{nt}}{[n - 1 + e^{nt}]^2},$$

where the constant $n$ is the size of the population and the constant $k$ is the contact rate between individuals of the group. *Source: The Mathematical Approach to Biology and Medicine.* Find the time at which the rate of change in the number of infected individuals is a maximum.

**50. Population Growth** As we saw in Exercise 39 in Section 4.6 on Derivatives of Trigonometric Functions, many biological populations can be modeled by

$$f(t) = f(0)e^{c \sin t},$$

where $f(0)$ is the size of the population when $t = 0$. Suppose that $f(0) = 1000$ and $c = 2$. Find the maximum and minimum values of $f(t)$ and the values of $t$ where they occur.

**51. Cohesiveness** To understand the spacing between fish in a school or birds in a flock, scientists have defined cohesiveness as

$$C(x) = \frac{R}{x^n} - \frac{A}{x^m},$$

where $A$ and $R$ are magnitudes of attraction and repulsion, $x$ is the distance between individuals, and $m$ and $n$ determine how fast the interactions fall off with distance. *Source: Journal of Mathematical Biology.* Suppose that $R > A$ and $n > m$.

**a.** Show that $C(x) = 0$ when

$$x = \left(\frac{R}{A}\right)^{1/(n-m)}.$$

**b.** Show that $C(x)$ has a relative minimum when

$$x = \left(\frac{nR}{mA}\right)^{1/(n-m)}.$$

**c.** For $R = 5, A = 3, n = 1/2$, and $m = 1/3$, find the value of $x$ for which $C(x) = 0$, and the value for which $C(x)$ has a relative minimum.

**52. Cohesiveness** An alternative form of cohesiveness (see the previous exercise) is given by

$$C(x) = Re^{-x/r} - Ae^{-x/a}.$$

*Source: Journal of Mathematical Biology.* Suppose that $R > A$ and $r < a$.

**a.** Show that $C(x) = 0$ when

$$x = \frac{\ln (A/R)}{1/a - 1/r}.$$

**b.** Show that $C(x)$ has a relative minimum when

$$x = \frac{\ln [(rA)/(aR)]}{1/a - 1/r}.$$

**c.** For $R = 5, A = 3, r = 1/5$, and $a = 1/2$, find the value of $x$ for which $C(x) = 0$, and the value for which $C(x)$ has a relative minimum.

**53. Blood Vessel System** The body's system of blood vessels is made up of arteries, arterioles, capillaries, and veins. The transport of blood from the heart through all organs of the body and back to the heart should be as efficient as possible. One way this can be done is by having large enough blood vessels to avoid turbulence, with blood cells small enough to minimize viscosity.

We will find the value of angle $\theta$ (see the figure) such that total resistance to the flow of blood is minimized. Assume that a main vessel of radius $r_1$ runs along the horizontal line from $A$ to $B$. A side artery, of radius $r_2$, heads for a point $C$. Choose point $B$ so that $CB$ is perpendicular to $AB$. Let $CB = s$ and let $D$ be the point where the axis of the branching vessel cuts the axis of the main vessel.

According to Poiseuille's law, the resistance $R$ in the system is proportional to the length $L$ of the vessel and inversely proportional to the fourth power of the radius $r$. That is,

$$R = k \cdot \frac{L}{r^4}, \tag{1}$$

where $k$ is a constant determined by the viscosity of the blood. Let $AB = L_0, AD = L_1$, and $DC = L_2$. *Source: Introduction to Mathematics for Life Scientists.*

**a.** Use right triangle $BDC$ to find $\sin \theta$.

**b.** Solve the result of part a for $L_2$.

**c.** Find $\cot \theta$ in terms of $s$ and $L_0 - L_1$.

**d.** Solve the result of part c for $L_1$.

**e.** Write an expression similar to Equation (1) for the resistance $R_1$ along $AD$.

**f.** Write a formula for the resistance along $DC$.

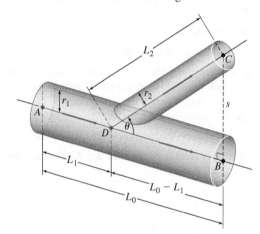

**g.** The total resistance $R$ is given by the sum of the resistances along $AD$ and $DC$. Use your answers to parts e and f to write an expression for $R$.

**h.** In your formula for $R$, replace $L_1$ with the result of part d and $L_2$ with the result of part b. Simplify your answer.

**i.** Find $dR/d\theta$. Simplify your answer. (Remember that $k$, $L_1$, $L_0$, $s$, $r_1$, and $r_2$ are constants.)

**j.** Set $dR/d\theta$ equal to 0.

**k.** Multiply through by $(\sin^2 \theta)/s$.

**l.** Solve for $\cos \theta$.

**m.** Verify by the first derivative test that the solution from the previous part gives a minimum.

**n.** Suppose $r_1 = 1$ cm and $r_2 = 1/4$ cm. Find $\cos \theta$ and then find $\theta$.

**o.** Find $\theta$ if $r_1 = 1.4$ cm and $r_2 = 0.8$ cm.

## OTHER APPLICATIONS

**54. Attitude Change** Social psychologists have found that as the discrepancy between the views of a speaker and those of an audience increases, the attitude change in the audience also increases to a point but decreases when the discrepancy becomes too large, particularly if the communicator is viewed by the audience as having low credibility. Suppose that the degree of change can be approximated by the function

$$D(x) = -x^4 + 8x^3 + 80x^2,$$

where $x$ is the discrepancy between the views of the speaker and those of the audience, as measured by scores on a questionnaire. Find the amount of discrepancy the speaker should aim for to maximize the attitude change in the audience. ***Source: Journal of Personality and Social Psychology.***

**55. Height** After a great deal of experimentation, two Atlantic Institute of Technology senior physics majors determined that when a bottle of French champagne is shaken several times, held upright, and uncorked, its cork travels according to

$$s(t) = -16t^2 + 40t + 3,$$

where $s$ is its height (in feet) above the ground $t$ seconds after being released.

  **a.** How high will it go?

  **b.** How long is it in the air?

**56. Piston Velocity** The distance $s$ of a piston from the center of the crankshaft as it rotates in a 1937 John Deere B engine with respect to the angle $\theta$ of the connecting rod, as indicated by the figure, is given by the formula

$$s(\theta) = 2.625 \cos \theta + 2.625(15 + \cos^2 \theta)^{1/2},$$

where $s$ is measured in inches and $\theta$ in radians. ***Source: The AMATYC Review.***

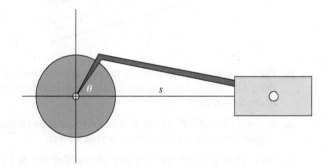

  **a.** Find $s\left(\dfrac{\pi}{2}\right)$.

  **b.** Find $\dfrac{ds}{d\theta}$.

  **c.** Find the value(s) of $\theta$ where the maximum velocity of the piston occurs.

**57. Engine Velocity** As shown in Exercise 56, a formula that can be used to determine the distance of a piston with respect to the crankshaft for a 1937 John Deere B engine is

$$s(\theta) = 2.625 \cos(\theta) + 2.625(15 + \cos^2 \theta)^{1/2},$$

where $s$ is measured in inches and $\theta$ in radians. ***Source: The AMATYC Review.***

  **a.** Given that the angle $\theta$ is changing with respect to time, that is, it is a function of $t$, use the chain rule to find the derivative of $s$ with respect to $t$, $ds/dt$.

  **b.** Use the answer for part a, with $\theta = 4.944$ and $d\theta/dt = 1340$ rev per minute, to find the maximum velocity of the engine. Express your answer in miles per hour. (*Hint:* 1340 rev per minute = 505,168.1 rad per hour. Use this value and then convert your answer from inches to miles, where 1 mile = 5280 ft.)

**▬ YOUR TURN ANSWERS**

1. Relative maximum of $f(x_2)$ at $x = x_2$; relative minima of $f(x_1)$ at $x = x_1$ and $f(x_3)$ at $x = x_3$.

2. Relative maximum of $f(5/3) = 670/27 \approx 24.8$ at $x = 5/3$ and relative minimum of $f(-3) = -26$ at $x = -3$.

3. Relative maximum of $f\left(\dfrac{2}{5}\right) = \dfrac{3}{5}\left(\dfrac{2}{5}\right)^{2/3} \approx 0.3257$ at $x = 2/5$ and relative minimum of $f(0) = 0$ at $x = 0$.

4. Relative maximum of $f(-2) = 4e^{-2} \approx 0.5413$ at $x = -2$ and relative minimum of $f(0) = 0$ at $x = 0$.

# 5.3 Higher Derivatives, Concavity, and the Second Derivative Test

APPLY IT If a diet plan guarantees weight loss, is it always possible to reach your weight loss goal?
*We will address this question in Example 1.*

In the first section of this chapter, we used the derivative to determine intervals where a function is increasing or decreasing. For example, if the function represents a person's weight, we can use the derivative to determine when the person's weight is increasing. In addition, it would be important for us to know how the *rate of increase* is changing. We can determine how the *rate of increase* (or the *rate of decrease*) is changing by determining the rate of change of the derivative of the function. In other words, we can find the derivative of the derivative, called the **second derivative**, as shown in the following example.

### EXAMPLE 1   Weight

Suppose that your friend, who is concerned that you are 100 pounds overweight, has studied the diet plan you started two months ago and is trying to get you to continue with the diet. He shows you the following function, which represents your weight $w(t)$ (in pounds):

$$w(t) = 300 - 15 \ln(t + 1),$$

where $t$ is the number of months since starting the diet. He points out that the derivative of the function is always negative, so your weight is always decreasing. He claims that you cannot help but lose all the weight you desire. Should you take his advice and use this diet plan?

**APPLY IT**

**SOLUTION**   It is true that the weight function decreases for all $t$. The derivative is

$$w'(t) = \frac{-15}{t + 1},$$

which is always negative for $t > 0$. The catch lies in *how fast* the function is decreasing. The derivative $w'(t) = -15/(t + 1)$ tells how fast your weight is decreasing at any number of months, $t$, since you began the diet. For example, when $t = 2$ months, $w'(2) = -5$, and the weight is decreasing at the rate of 5 lb per month. When $t = 4$ months, $w'(4) = -3$; the weight is decreasing at 3 lb per month. At $t = 14$ months, $w'(14) = -1$, or 1 lb per month. By the time you've been on the diet for $t = 29$ months, the weight decreases at 0.5 lb per month, and the *rate of decrease* looks as though it will continue to increase.

The rate of decrease in $w'(t)$ is given by the derivative of $w'(t)$, called the second derivative of $w(t)$ and denoted by $w''(t)$. Since $w'(t) = -15/(t + 1)$,

$$w''(t) = \frac{15}{(t + 1)^2}.$$

The function $w''(t)$ is positive for all positive values of $t$ and therefore confirms the suspicion that the rate of decrease in weight does indeed increase for all $t \geq 0$. Your weight will not increase, but the amount of weight you lose will certainly not be what your friend predicts. For example, at 30 months, your weight will be about 249 lb. A year later, it would be about 244 lb. It would take you approximately 786 months to lose 100 lb. The only people who could benefit from this diet are those who don't have a lot of weight to lose.

As mentioned earlier, the second derivative of a function $f$, written $f''$, gives the rate of change of the *derivative* of $f$. Before continuing to discuss applications of the second derivative, we need to introduce some additional terminology and notation.

## Higher Derivatives   If a function $f$ has a derivative $f'$, then the derivative of $f'$, if it exists, is the second derivative of $f$, written $f''$. The derivative of $f''$, if it exists, is called the **third derivative** of $f$, and so on. By continuing this process, we can find **fourth derivatives** and other higher derivatives. For example, if $f(x) = x^4 + 2x^3 + 3x^2 - 5x + 7$, then

$$f'(x) = 4x^3 + 6x^2 + 6x - 5, \qquad \text{First derivative of } f$$
$$f''(x) = 12x^2 + 12x + 6, \qquad \text{Second derivative of } f$$
$$f'''(x) = 24x + 12, \qquad \text{Third derivative of } f$$

and

$$f^{(4)}(x) = 24. \qquad \text{Fourth derivative of } f$$

## Notation for Higher Derivatives

The second derivative of $y = f(x)$ can be written using any of the following notations:

$$f''(x), \quad \frac{d^2y}{dx^2}, \quad \text{or} \quad D_x^2[f(x)].$$

The third derivative can be written in a similar way. For $n \geq 4$, the $n$th derivative is written $f^{(n)}(x)$.

---

**CAUTION** Notice the difference in notation between $f^{(4)}(x)$, which indicates the fourth derivative of $f(x)$, and $f^4(x)$, which indicates $f(x)$ raised to the fourth power.

---

### EXAMPLE 2 Second Derivative

Let $f(x) = x^3 + 6x^2 - 9x + 8$.

**(a)** Find $f''(x)$.

**SOLUTION** To find the second derivative of $f(x)$, find the first derivative, and then take its derivative.

$$f'(x) = 3x^2 + 12x - 9$$
$$f''(x) = 6x + 12$$

**(b)** Find $f''(0)$.

**YOUR TURN 1** Find $f''(1)$ if $f(x) = 5x^4 - 4x^3 + 3x$.

**SOLUTION** Since $f''(x) = 6x + 12$,

$$f''(0) = 6(0) + 12 = 12. \qquad \text{TRY YOUR TURN 1}$$

### EXAMPLE 3 Second Derivative

Find the second derivative for the functions defined as follows.

**(a)** $f(x) = (x^2 - 1)^2$

**SOLUTION** Here, using the chain rule,

$$f'(x) = 2(x^2 - 1)(2x) = 4x(x^2 - 1).$$

Use the product rule to find $f''(x)$.

$$f''(x) = 4x(2x) + (x^2 - 1)(4)$$
$$= 8x^2 + 4x^2 - 4$$
$$= 12x^2 - 4$$

**(b)** $g(x) = 4x(\ln x)$

**SOLUTION** Use the product rule.

$$g'(x) = 4x \cdot \frac{1}{x} + (\ln x) \cdot 4 = 4 + 4(\ln x)$$

$$g''(x) = 0 + 4 \cdot \frac{1}{x} = \frac{4}{x}$$

**(c)** $h(x) = \dfrac{x}{e^x}$

**SOLUTION** Here, we need the quotient rule.

$$h'(x) = \frac{e^x - xe^x}{(e^x)^2} = \frac{e^x(1 - x)}{(e^x)^2} = \frac{1 - x}{e^x}$$

$$h''(x) = \frac{e^x(-1) - (1 - x)e^x}{(e^x)^2} = \frac{e^x(-1 - 1 + x)}{(e^x)^2} = \frac{-2 + x}{e^x}$$

**YOUR TURN 2** Find the
second derivative for
**(a)** $f(x) = (x^3 + 1)^2$
**(b)** $g(x) = xe^x$
**(c)** $h(x) = \dfrac{\ln x}{x}$
**(d)** $p(x) = \sin^2 x$.

**(d)** $p(x) = \sin(x^2)$

**SOLUTION** We need the chain rule to take the first derivative. In addition, we need the product rule to take the second derivative.

$$p'(x) = \cos(x^2) \cdot 2x = 2x\cos(x^2)$$
$$p''(x) = 2x(-\sin(x^2) \cdot 2x) + \cos(x^2) \cdot 2$$
$$= -4x^2\sin(x^2) + 2\cos(x^2)$$

**TRY YOUR TURN 2**

Earlier, we saw that the first derivative of a function represents the rate of change of the function. The second derivative, then, represents the rate of change of the first derivative. If a function describes the position of a vehicle (along a straight line) at time $t$, then the first derivative gives the velocity of the vehicle. That is, if $y = s(t)$ describes the position (along a straight line) of the vehicle at time $t$, then $v(t) = s'(t)$ gives the velocity at time $t$.

We also saw that *velocity* is the rate of change of distance with respect to time. Recall, the difference between velocity and speed is that velocity may be positive or negative, whereas speed is always positive. A negative velocity indicates travel in a negative direction (backing up) with regard to the starting point; positive velocity indicates travel in the positive direction (going forward) from the starting point.

The instantaneous rate of change of velocity is called **acceleration**. Since instantaneous rate of change is the same as the derivative, acceleration is the derivative of velocity. Thus if $a(t)$ represents the acceleration at time $t$, then

$$a(t) = \frac{d}{dt}v(t) = s''(t).$$

If the velocity is positive and the acceleration is positive, the velocity is increasing, so the vehicle is speeding up. If the velocity is positive and the acceleration is negative, the vehicle is slowing down. A negative velocity and a positive acceleration mean the vehicle is backing up and slowing down. If both the velocity and acceleration are negative, the vehicle is speeding up in the negative direction.

**EXAMPLE 4** **Velocity and Acceleration**

Suppose a car is moving in a straight line, with its position from a starting point (in feet) at time $t$ (in seconds) given by

$$s(t) = t^3 - 2t^2 - 7t + 9.$$

Find the following.

**(a)** The velocity at any time $t$

**SOLUTION** The velocity is given by

$$v(t) = s'(t) = 3t^2 - 4t - 7$$

feet per second.

**(b)** The acceleration at any time $t$

**SOLUTION** Acceleration is given by

$$a(t) = v'(t) = s''(t) = 6t - 4$$

feet per second per second.

**(c)** The time intervals (for $t \geq 0$) when the car is going forward or backing up

**SOLUTION** We first find when the velocity is 0, that is, when the car is stopped.

$$v(t) = 3t^2 - 4t - 7 = 0$$
$$(3t - 7)(t + 1) = 0$$
$$t = 7/3 \quad \text{or} \quad t = -1$$

We are interested in $t \geq 0$. Choose a value of $t$ in each of the intervals $(0, 7/3)$ and $(7/3, \infty)$ to see that the velocity is negative in $(0, 7/3)$ and positive in $(7/3, \infty)$. The car is backing up for the first 7/3 seconds, then going forward.

(d) The time intervals (for $t \geq 0$) when the car is speeding up or slowing down

**SOLUTION** The car will speed up when the velocity and acceleration are the same sign and slow down when they have opposite signs. Here, the acceleration is positive when $6t - 4 > 0$, that is, $t > 2/3$ seconds, and negative for $t < 2/3$ seconds. Since the velocity is negative in $(0, 7/3)$ and positive in $(7/3, \infty)$, the car is speeding up for $0 < t < 2/3$ seconds, slowing down for $2/3 < t < 7/3$ seconds, and speeding up again for $t > 7/3$ seconds. See the sign graphs.

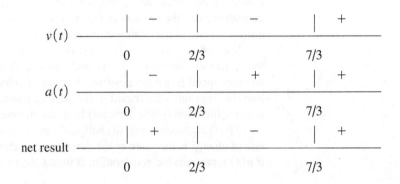

**YOUR TURN 3** Find the velocity and acceleration of the car if the distance (in feet) is given by $s(t) = t^3 - 3t^2 - 24t + 10$, at time $t$ (in seconds). When is the car going forward or backing up? When is the car speeding up or slowing down?

TRY YOUR TURN 3

## Concavity of a Graph
The first derivative has been used to show where a function is increasing or decreasing and where the extrema occur. The second derivative gives the rate of change of the first derivative; it indicates *how fast* the function is increasing or decreasing. The rate of change of the derivative (the second derivative) affects the *shape* of the graph. Intuitively, we say that a graph is *concave upward* on an interval if it "holds water" and *concave downward* if it "spills water." See Figure 28.

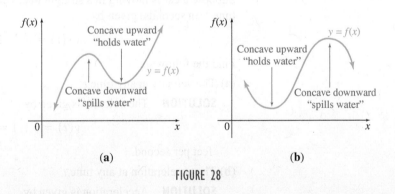

FIGURE 28

More precisely, a function is **concave upward** on an interval $(a, b)$ if the graph of the function lies above its tangent line at each point of $(a, b)$. A function is **concave downward** on $(a, b)$ if the graph of the function lies below its tangent line at each point of $(a, b)$. A point where a graph changes **concavity** is called an **inflection point**. See Figure 29.

Users of soft contact lenses recognize concavity as the way to tell if a lens is inside out. As Figure 30 shows, a correct contact lens has a profile that is entirely concave upward. The profile of an inside-out lens has inflection points near the edges, where the profile begins to turn concave downward very slightly.

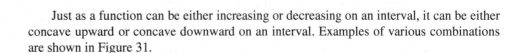

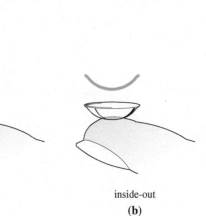

FIGURE 29

FIGURE 30

Just as a function can be either increasing or decreasing on an interval, it can be either concave upward or concave downward on an interval. Examples of various combinations are shown in Figure 31.

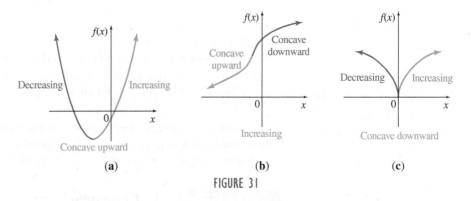

FIGURE 31

Figure 32 shows two functions that are concave upward on an interval $(a, b)$. Several tangent lines are also shown. In Figure 32(a), the slopes of the tangent lines (moving from left to right) are first negative, then 0, and then positive. In Figure 32(b), the slopes are all positive, but they get larger.

In both cases, the slopes are *increasing*. The slope at a point on a curve is given by the derivative. Since a function is increasing if its derivative is positive, its slope is increasing if the derivative of the slope function is positive. Since the derivative of a derivative is the second derivative, a function is concave upward on an interval if its second derivative is positive at each point of the interval.

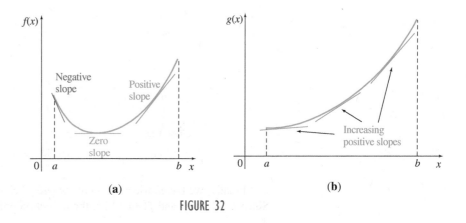

FIGURE 32

A similar result is suggested by Figure 33 for functions whose graphs are concave downward. In both graphs, the slopes of the tangent lines are *decreasing* as we move from left to right. Since a function is decreasing if its derivative is negative, a function is concave downward on an interval if its second derivative is negative at each point of the interval. These observations suggest the following test.

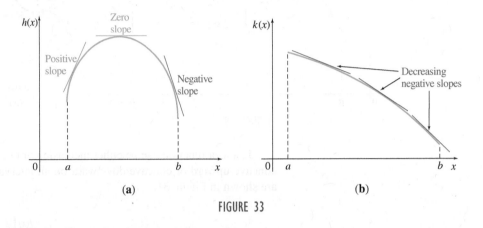

**FIGURE 33**

## Test for Concavity

Let $f$ be a function with derivatives $f'$ and $f''$ existing at all points in an interval $(a, b)$. Then $f$ is concave upward on $(a, b)$ if $f''(x) > 0$ for all $x$ in $(a, b)$ and concave downward on $(a, b)$ if $f''(x) < 0$ for all $x$ in $(a, b)$.

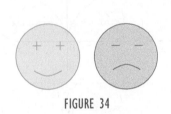

**FIGURE 34**

An easy way to remember this test is by the faces shown in Figure 34. When the second derivative is positive at a point $(+ +)$, the graph is concave upward $(\smile)$. When the second derivative is negative at a point $(- -)$, the graph is concave downward $(\frown)$.

### EXAMPLE 5  Concavity

Find all intervals where $f(x) = x^4 - 8x^3 + 18x^2$ is concave upward or downward, and find all inflection points.

**SOLUTION**   The first derivative is $f'(x) = 4x^3 - 24x^2 + 36x$, and the second derivative is $f''(x) = 12x^2 - 48x + 36$. We factor $f''(x)$ as $12(x - 1)(x - 3)$, and then create a number line for $f''(x)$ as we did in the previous two sections for $f'(x)$.

We see from Figure 35 that $f''(x) > 0$ on the intervals $(-\infty, 1)$ and $(3, \infty)$, so $f$ is concave upward on these intervals. Also, $f''(x) < 0$ on the interval $(1, 3)$, so $f$ is concave downward on this interval.

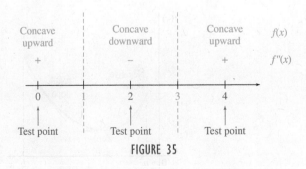

**FIGURE 35**

Finally, we have inflection points where $f''$ changes sign, namely at $x = 1$ and $x = 3$. Since $f(1) = 11$ and $f(3) = 27$, the inflection points are $(1, 11)$ and $(3, 27)$.

**YOUR TURN 4** Find all intervals where $f(x) = x^5 - 30x^3$ is concave upward or downward, and find all inflection points.

Although we were seeking information only about concavity and inflection points in this example, it is also worth noting that $f'(x) = 4x^3 - 24x^2 + 36x = 4x(x - 3)^2$, which has roots at $x = 0$ and $x = 3$. Verify that there is a relative minimum at $(0, 0)$, but that $(3, 27)$ is neither a relative minimum nor a relative maximum. The function is graphed in Figure 36.          **TRY YOUR TURN 4**

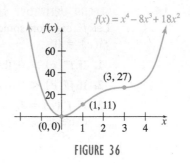

FIGURE 36

Example 5 suggests the following result.

At an inflection point for a function $f$, the second derivative is 0 or does not exist.

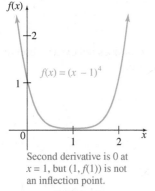

Second derivative is 0 at $x = 1$, but $(1, f(1))$ is not an inflection point.

**FIGURE 37**

**CAUTION**
1. Be careful with the previous statement. Finding a value of $x$ where $f''(x) = 0$ does not mean that an inflection point has been located. For example, if $f(x) = (x - 1)^4$, then $f''(x) = 12(x - 1)^2$, which is 0 at $x = 1$. The graph of $f(x) = (x - 1)^4$ is always concave upward, however, so it has no inflection point. See Figure 37.
2. Note that the concavity of a function might change not only at a point where $f''(x) = 0$ but also where $f''(x)$ does not exist. For example, this happens at $x = 0$ for $f(x) = x^{1/3}$.

**TECHNOLOGY NOTE**

Most graphing calculators do not have a feature for finding inflection points. Nevertheless, a graphing calculator sketch can be useful for verifying that your calculations for finding inflection points and intervals where the function is concave up or down are correct.

## Second Derivative Test

The idea of concavity can often be used to decide whether a given critical number produces a relative maximum or a relative minimum. This test, an alternative to the first derivative test, is based on the fact that a curve with a horizontal tangent at a point $c$ and concave downward on an open interval containing $c$ also has a relative maximum at $c$. A relative minimum occurs when a graph has a horizontal tangent at a point $d$ and is concave upward on an open interval containing $d$. See Figure 38.

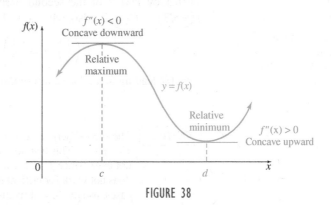

FIGURE 38

A function $f$ is concave upward on an interval if $f''(x) > 0$ for all $x$ in the interval, while $f$ is concave downward on an interval if $f''(x) < 0$ for all $x$ in the interval. These ideas lead to the **second derivative test** for relative extrema.

### Second Derivative Test

Let $f''$ exist on some open interval containing $c$ (except possibly at $c$ itself), and let $f'(c) = 0$.

1. If $f''(c) > 0$, then $f(c)$ is a relative minimum.
2. If $f''(c) < 0$, then $f(c)$ is a relative maximum.
3. If $f''(c) = 0$ or $f''(c)$ does not exist, then the test gives no information about extrema, so use the first derivative test.

**NOTE** In Case 3 of the second derivative test (when $f''(c) = 0$ or does not exist), observe that if $f''(x)$ changes sign at $c$, there is an inflection point at $x = c$.

### EXAMPLE 6    Second Derivative Test

Find all relative extrema for

$$f(x) = 4x^3 + 7x^2 - 10x + 8.$$

**SOLUTION**    First, find the points where the derivative is 0. Here $f'(x) = 12x^2 + 14x - 10$. Solve the equation $f'(x) = 0$ to get

$$12x^2 + 14x - 10 = 0$$
$$2(6x^2 + 7x - 5) = 0$$
$$2(3x + 5)(2x - 1) = 0$$

$$3x + 5 = 0 \quad \text{or} \quad 2x - 1 = 0$$
$$3x = -5 \qquad\qquad 2x = 1$$
$$x = -\frac{5}{3} \qquad\qquad x = \frac{1}{2}.$$

Now use the second derivative test. The second derivative is $f''(x) = 24x + 14$. Evaluate $f''(x)$ first at $-5/3$, getting

$$f''\left(-\frac{5}{3}\right) = 24\left(-\frac{5}{3}\right) + 14 = -40 + 14 = -26 < 0,$$

so that by Part 2 of the second derivative test, $-5/3$ leads to a relative maximum of $f(-5/3) = 691/27$. Also, when $x = 1/2$,

**YOUR TURN 5** Find all relative extrema of $f(x) = -2x^3 + 3x^2 + 72x$.

$$f''\left(\frac{1}{2}\right) = 24\left(\frac{1}{2}\right) + 14 = 12 + 14 = 26 > 0,$$

with 1/2 leading to a relative minimum of $f(1/2) = 21/4$.    **TRY YOUR TURN 5**

**CAUTION**    The second derivative test works only for those critical numbers $c$ that make $f'(c) = 0$. This test does not work for critical numbers $c$ for which $f'(c)$ does not exist (since $f''(c)$ would not exist either). Also, the second derivative test does not work for critical numbers $c$ that make $f''(c) = 0$. In both of these cases, use the first derivative test.

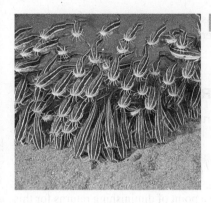

**EXAMPLE 7**   **Catfish Farming**

The graph in Figure 39 shows the population of catfish in a commercial catfish farm as a function of time. As the graph shows, the population increases rapidly up to a point and then increases at a slower rate. We saw several types of functions that produce such a graph in earlier sections. The horizontal dashed line shows that the population will approach some upper limit determined by the capacity of the farm. The point at which the rate of population growth starts to slow is the point of inflection for the graph.

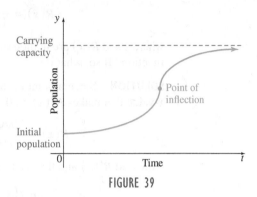

**FIGURE 39**

To produce the maximum yield of catfish, harvesting should take place at the point of fastest possible growth of the population; here, this is at the point of inflection. The rate of change of the population, given by the first derivative, is increasing up to the inflection point (on the interval where the second derivative is positive) and decreasing past the inflection point (on the interval where the second derivative is negative).

The *law of diminishing returns* in economics is related to the idea of concavity. The function graphed in Figure 40 gives the output $y$ from a given input $x$. If the input were advertising costs for some product, for example, the output might be the corresponding revenue from sales.

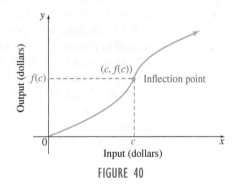

**FIGURE 40**

The graph in Figure 40 shows an inflection point at $(c, f(c))$. For $x < c$, the graph is concave upward, so the rate of change of the slope is increasing. This indicates that the output $y$ is increasing at a faster rate with each additional dollar spent. When $x > c$, however, the graph is concave downward, the rate of change of the slope is decreasing, and the increase in $y$ is smaller with each additional dollar spent. Thus, further input beyond $c$ dollars produces diminishing returns. The point of inflection at $(c, f(c))$ is called the **point of diminishing returns**. Beyond this point there is a smaller and smaller return for each dollar invested.

As another example of diminishing returns, consider a farm with a fixed amount of land, machinery, fertilizer, and so on. Adding workers increases production a lot at first, then less and less with each additional worker.

### EXAMPLE 8    Point of Diminishing Returns

The revenue $R(x)$ generated from sales of a certain drug is related to the amount $x$ spent on advertising by

$$R(x) = \frac{1}{15{,}000}(600x^2 - x^3), \quad 0 \le x \le 600,$$

where $x$ and $R(x)$ are in thousands of dollars. Is there a point of diminishing returns for this function? If so, what is it?

**SOLUTION**   Since a point of diminishing returns occurs at an inflection point, look for an $x$-value that makes $R''(x) = 0$. Write the function as

$$R(x) = \frac{600}{15{,}000}x^2 - \frac{1}{15{,}000}x^3 = \frac{1}{25}x^2 - \frac{1}{15{,}000}x^3.$$

Now find $R'(x)$ and then $R''(x)$.

$$R'(x) = \frac{2x}{25} - \frac{3x^2}{15{,}000} = \frac{2}{25}x - \frac{1}{5000}x^2$$

$$R''(x) = \frac{2}{25} - \frac{1}{2500}x$$

Set $R''(x)$ equal to 0 and solve for $x$.

$$\frac{2}{25} - \frac{1}{2500}x = 0$$

$$-\frac{1}{2500}x = -\frac{2}{25}$$

$$x = \frac{5000}{25} = 200$$

Test a number in the interval $(0, 200)$ to see that $R''(x)$ is positive there. Then test a number in the interval $(200, 600)$ to find $R''(x)$ negative in that interval. Since the sign of $R''(x)$ changes from positive to negative at $x = 200$, the graph changes from concave upward to concave downward at that point, and there is a point of diminishing returns at the inflection point $\left(200, 1066\frac{2}{3}\right)$. Investments in advertising beyond \$200,000 return less and less for each dollar invested. Verify that $R'(200) = 8$. This means that when \$200,000 is invested, another \$1000 invested returns approximately \$8000 in additional revenue. Thus it may still be economically sound to invest in advertising beyond the point of diminishing returns.

# 5.3    EXERCISES

**Find $f''(x)$ for each function. Then find $f''(0)$ and $f''(2)$.**

**1.** $f(x) = 5x^3 - 7x^2 + 4x + 3$

**2.** $f(x) = 4x^3 + 5x^2 + 6x - 7$

**3.** $f(x) = 4x^4 - 3x^3 - 2x^2 + 6$

**4.** $f(x) = -x^4 + 7x^3 - \dfrac{x^2}{2}$

**5.** $f(x) = 3x^2 - 4x + 8$

**6.** $f(x) = 8x^2 + 6x + 5$

**7.** $f(x) = \dfrac{x^2}{1 + x}$

**8.** $f(x) = \dfrac{-x}{1 - x^2}$

**9.** $f(x) = \sqrt{x^2 + 4}$

**10.** $f(x) = \sqrt{2x^2 + 9}$

**11.** $f(x) = 32x^{3/4}$

**12.** $f(x) = -6x^{1/3}$

**13.** $f(x) = 5e^{-x^2}$

**14.** $f(x) = 0.5e^{x^2}$

**15.** $f(x) = \dfrac{\ln x}{4x}$

**16.** $f(x) = \ln x + \dfrac{1}{x}$

**17.** $f(x) = \cos(x^3)$

**18.** $f(x) = \cos^3 x$

**Find $f'''(x)$, the third derivative of $f$, and $f^{(4)}(x)$, the fourth derivative of $f$, for each function.**

**19.** $f(x) = 7x^4 + 6x^3 + 5x^2 + 4x + 3$

**20.** $f(x) = -2x^4 + 7x^3 + 4x^2 + x$

**21.** $f(x) = 5x^5 - 3x^4 + 2x^3 + 7x^2 + 4$

**22.** $f(x) = 2x^5 + 3x^4 - 5x^3 + 9x - 2$

**23.** $f(x) = \dfrac{x - 1}{x + 2}$

**24.** $f(x) = \dfrac{x + 1}{x}$

**25.** $f(x) = \dfrac{3x}{x - 2}$

**26.** $f(x) = \dfrac{x}{2x + 1}$

**27.** Let $f$ be an $n$th-degree polynomial of the form
$f(x) = x^n + a_{n-1}x^{n-1} + \cdots + a_1 x + a_0$.

  **a.** Find $f^{(n)}(x)$.    **b.** Find $f^{(k)}(x)$ for $k > n$.

**28.** Let $f(x) = \ln x$.

  **a.** Compute $f'(x)$, $f''(x)$, $f'''(x)$, $f^{(4)}(x)$, and $f^{(5)}(x)$.

  **b.** Guess a formula for $f^{(n)}(x)$, where $n$ is any positive integer.

**29.** For $f(x) = e^x$, find $f''(x)$ and $f'''(x)$. What is the $n$th derivative of $f$ with respect to $x$?

**30.** For $f(x) = a^x$, find $f''(x)$ and $f'''(x)$. What is the $n$th derivative of $f$ with respect to $x$?

**31.** For $f(x) = \sin x$, find $f'(x)$, $f''(x)$, $f'''(x)$, and $f^{(4)}(x)$. What is $f^{(4n)}(x)$, where $n$ is a nonnegative integer?

**32.** For $f(x) = \cos x$, find $f'(x)$, $f''(x)$, $f'''(x)$, and $f^{(4)}(x)$. What is $f^{(4n)}(x)$, where $n$ is a nonnegative integer?

**In Exercises 33–56, find the open intervals where the functions are concave upward or concave downward. Find any inflection points.**

**33.**

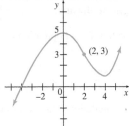

**34.**

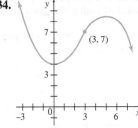

**35.**

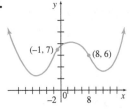

**36.**

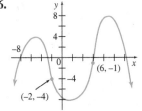

**37.**

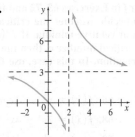

**38.**

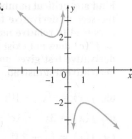

**39.** $f(x) = x^2 + 10x - 9$

**40.** $f(x) = 8 - 6x - x^2$

**41.** $f(x) = -2x^3 + 9x^2 + 168x - 3$

**42.** $f(x) = -x^3 - 12x^2 - 45x + 2$

**43.** $f(x) = \dfrac{3}{x - 5}$

**44.** $f(x) = \dfrac{-2}{x + 1}$

**45.** $f(x) = x(x + 5)^2$

**46.** $f(x) = -x(x - 3)^2$

**47.** $f(x) = 18x - 18e^{-x}$

**48.** $f(x) = 2e^{-x^2}$

**49.** $f(x) = x^{8/3} - 4x^{5/3}$

**50.** $f(x) = x^{7/3} + 56x^{4/3}$

**51.** $f(x) = \ln(x^2 + 1)$

**52.** $f(x) = x^2 + 8\ln|x + 1|$

**53.** $f(x) = x^2 \log|x|$

**54.** $f(x) = 5^{-x^2}$

**55.** $f(x) = \sin(2x)$

**56.** $f(x) = \tan(\pi x)$

**For each of the exercises listed below, suppose that the function that is graphed is not $f(x)$, but $f'(x)$. Find the open intervals where the original function is concave upward or concave downward, and find the location of any inflection points.**

**57.** Exercise 33    **58.** Exercise 34

**59.** Exercise 35    **60.** Exercise 36

**61.** Give an example of a function $f(x)$ such that $f'(0) = 0$ but $f''(0)$ does not exist. Is there a relative minimum or maximum or an inflection point at $x = 0$?

**62. a.** Graph the two functions $f(x) = x^{7/3}$ and $g(x) = x^{5/3}$ on the window $[-2, 2]$ by $[-2, 2]$.

  **b.** Verify that both $f$ and $g$ have an inflection point at $(0, 0)$.

  **c.** How is the value of $f''(0)$ different from $g''(0)$?

  **d.** Based on what you have seen so far in this exercise, is it always possible to tell the difference between a point where the second derivative is 0 or undefined based on the graph? Explain.

**63.** Describe the slope of the tangent line to the graph of $f(x) = e^x$ for the following.

  **a.** $x \to -\infty$    **b.** $x \to 0$

**64.** What is true about the slope of the tangent line to the graph of $f(x) = \ln x$ as $x \to \infty$? As $x \to 0$?

Find any critical numbers for $f$ in Exercises 65–72 and then use the second derivative test to decide whether the critical numbers lead to relative maxima or relative minima. If $f''(c) = 0$ or $f''(c)$ does not exist for a critical number $c$, then the second derivative test gives no information. In this case, use the first derivative test instead.

**65.** $f(x) = -x^2 - 10x - 25$     **66.** $f(x) = x^2 - 12x + 36$

**67.** $f(x) = 3x^3 - 3x^2 + 1$     **68.** $f(x) = 2x^3 - 4x^2 + 2$

**69.** $f(x) = (x + 3)^4$     **70.** $f(x) = x^3$

**71.** $f(x) = x^{7/3} + x^{4/3}$     **72.** $f(x) = x^{8/3} + x^{5/3}$

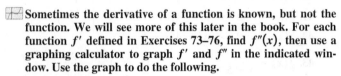 Sometimes the derivative of a function is known, but not the function. We will see more of this later in the book. For each function $f'$ defined in Exercises 73–76, find $f''(x)$, then use a graphing calculator to graph $f'$ and $f''$ in the indicated window. Use the graph to do the following.

**a.** Give the (approximate) $x$-values where $f$ has a maximum or minimum.

**b.** By considering the sign of $f'(x)$, give the (approximate) intervals where $f(x)$ is increasing and decreasing.

**c.** Give the (approximate) $x$-values of any inflection points.

**d.** By considering the sign of $f''(x)$, give the intervals where $f$ is concave upward or concave downward.

**73.** $f'(x) = x^3 - 6x^2 + 7x + 4$;   $[-5, 5]$ by $[-5, 15]$

**74.** $f'(x) = 10x^2(x - 1)(5x - 3)$;   $[-1, 1.5]$ by $[-20, 20]$

**75.** $f'(x) = \dfrac{1 - x^2}{(x^2 + 1)^2}$;   $[-3, 3]$ by $[-1.5, 1.5]$

**76.** $f'(x) = x^2 + x \ln x$;   $[0, 1]$ by $[-2, 2]$

**77.** Suppose a friend makes the following argument. A function $f$ is increasing and concave downward. Therefore, $f'$ is positive and decreasing, so it eventually becomes 0 and then negative, at which point $f$ decreases. Show that your friend is wrong by giving an example of a function that is always increasing and concave downward.

## LIFE SCIENCE APPLICATIONS

**78. Population Growth**   When a hardy new species is introduced into an area, the population often increases as shown. Explain the significance of the following function values on the graph.

**a.** $f_0$     **b.** $f(a)$     **c.** $f_M$

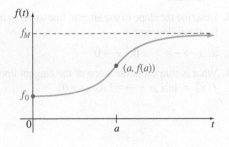

**79. Bacteria Population**   Assume that the number of bacteria $R(t)$ (in millions) present in a certain culture at time $t$ (in hours) is given by

$$R(t) = t^2(t - 18) + 96t + 1000.$$

**a.** At what time before 8 hours will the population be maximized?

**b.** Find the maximum population.

**80. Ozone Depletion**   According to an article in the *New York Times,* "Government scientists reported last week that they had detected a slowdown in the rate at which chemicals that deplete the earth's protective ozone layer are accumulating in the atmosphere." Letting $c(t)$ be the amount of ozone-depleting chemicals at time $t$, what does this statement tell you about $c(t)$, $c'(t)$, and $c''(t)$? *Source: The New York Times.*

**81. Drug Concentration**   The percent of concentration of a certain drug in the bloodstream $t$ hours after the drug is administered is given by

$$K(t) = \frac{3t}{t^2 + 4}.$$

For example, after 1 hour the concentration is given by

$$K(1) = \frac{3(1)}{1^2 + 4} = \frac{3}{5}\% = 0.6\% = 0.006.$$

**a.** Find the time at which the concentration is a maximum.

**b.** Find the maximum concentration.

**82. Drug Concentration**   The percent of concentration of a drug in the bloodstream $t$ hours after the drug is administered is given by

$$K(t) = \frac{4t}{3t^2 + 27}.$$

**a.** Find the time at which the concentration is a maximum.

**b.** Find the maximum concentration.

The next two exercises are a continuation of exercises first given in Section 4.4 on Derivatives of Exponential Functions. Find the inflection point of the graph of each logistic function. This is the point at which the growth rate begins to decline.

**83. Insect Growth**   The growth function for a population of beetles is given by

$$G(t) = \frac{10{,}000}{1 + 49e^{-0.1t}}.$$

**84. Clam Population Growth**   The population of a bed of clams is described by

$$G(t) = \frac{5200}{1 + 12e^{-0.52t}}.$$

*Hints for Exercises 85 and 86:* Leave $B$, $c$, and $k$ as constants until you are ready to calculate your final answer.

**85. Clam Growth** Researchers used a version of the Gompertz curve to model the growth of razor clams during the first seven years of the clams' lives with the equation

$$L(t) = Be^{-ce^{-kt}},$$

where $L(t)$ gives the length (in centimeters) after $t$ years, $B = 14.3032$, $c = 7.267963$, and $k = 0.670840$. Find the inflection point and describe what it signifies. *Source: Journal of Experimental Biology.*

**86. Breast Cancer Growth** Researchers used a version of the Gompertz curve to model the growth of breast cancer tumors with the equation

$$N(t) = e^{c(1-e^{-kt})},$$

where $N(t)$ is the number of cancer cells after $t$ days, $c = 27.3$, and $k = 0.011$. Find the inflection point and describe what it signifies. *Source: Cancer Research.*

**87. Popcorn** Researchers have determined that the amount of moisture present in a kernel of popcorn affects the volume of the popped corn and can be modeled for certain sizes of kernels by the function

$$v(x) = -35.98 + 12.09x - 0.4450x^2,$$

where $x$ is moisture content (%, wet basis) and $v(x)$ is the expansion volume (in cm³/gram). Describe the concavity of this function. *Source: Cereal Chemistry.*

**88. Alligator Teeth** Researchers have developed a mathematical model that can be used to estimate the number of teeth $N(t)$ at time $t$ (days of incubation) for *Alligator mississippiensis*, where

$$N(t) = 71.8e^{-8.96e^{-0.0685t}}.$$

Find the inflection point and describe its importance to this research. *Source: Journal of Theoretical Biology.*

**89. Thoroughbred Horses** The association between velocity during exercise and blood lactate concentration (a measure of anaerobic exercise) after submaximal 800-m exercise of thoroughbred racehorses on sand and grass tracks has been studied. Researchers in Australia have determined that the lactate–velocity relationship can be modeled by the functions

$$l_1(v) = 0.08e^{0.33v} \quad \text{and}$$

$$l_2(v) = -0.87v^2 + 28.17v - 211.41,$$

where $l_1(v)$ and $l_2(v)$ are the lactate concentrations (in mmol/L) and $v$ is the velocity (in m/s) of the horse during workout on sand and grass tracks, respectively. *Source: The Veterinary Journal.*

**a.** What is the concavity of each function on this interval?

**b.** Comment on why the concavity might be different for different types of tracks.

**90. Plant Growth** Researchers have found that the probability $P$ that a plant will grow to radius $R$ can be described by the equation

$$P(R) = \frac{1}{1 + 2\pi DR^2},$$

where $D$ is the density of the plants in an area. *Source: Ecology.* A graph in their publication shows an inflection point around $R = 0.022$. Find an expression for the value of $R$ in terms of $D$

at the inflection point, and find the value of $D$ corresponding to an inflection point at $R = 0.022$.

**91. Phytoplankton Growth** Researchers have described the maximum growth rate of phytoplankton as a function of the cell diameter $x$ by the equation

$$g(x) = \frac{x}{ax^2 + bx + c},$$

where $a$, $b$, and $c$ are positive constants and $x > 0$. *Source: The American Naturalist.* For simplicity, let $a = b = c = 1$.

**a.** The researchers state that when $x$ is small, $g'(x) > 0$, but when $x$ is large, $g'(x) < 0$. Find any intervals for which $g'(x) > 0$, and any for which $g'(x) < 0$.

**b.** Calculate $g''(x)$.

**c.** The researchers state that when $x$ is small, $g''(x) < 0$, but when $x$ is large, $g''(x) > 0$. Using a graphing calculator, find approximate intervals for which $g''(x) < 0$, and any for which $g''(x) > 0$.

**92. Foraging** The population growth rate of a foraging animal can be described by the equation

$$G(T) = bC_0RNT - C_1PNT^2 - DN,$$

where $T$ is the fraction of time that the animal spends foraging $(0 \le T \le 1)$; $N$ is the population size of the foraging animal; $R$ is the population size of the resource being foraged; $P$ is the population size of a predator; and $b$, $C_0$, $C_1$, and $D$ are constants. *Source: The American Naturalist.*

**a.** Find the value of $T$ for which $G'(T) = 0$.

**b.** Use the second derivative test to verify that there is a relative maximum at the value of $T$ found in part a.

**c.** Suppose the value of $T$ found in part a is outside the interval $[0, 1]$. What would then be the appropriate value of $T$?

**93. Foraging** In studying how a predator forages a prey, an important concept is the functional response, defined as

$$f(R) = \frac{TCR}{1 + hCR},$$

where $R$ is the prey population density, $C$ is the attack rate constant, $h$ is the handling time for a single prey, and $T$ is the optimal foraging time, which is itself a function of $R$. *Source: Ecology.* A type 2 response, which is the type most commonly observed in the laboratory, is when the curve is concave down. Show that this occurs when

$$\frac{d^2T}{dR^2}R(1 + hCR)^2 + \frac{dT}{dR}2(1 + hCR) - 2ThC < 0.$$

**94. Foraging** The functional response, defined in the previous exercise, can also take the form

$$f(R) = \frac{0.5\sqrt{R}}{1 + \sqrt{R}},$$

where $R$ is the prey population density $(R > 0)$. *Source: Ecology.*

**a.** Find $f'(R)$.     **b.** Find $f''(R)$.

**c.** From your answers to parts a and b, what can you say about whether $f$ is increasing or decreasing, and whether it is concave up or concave down?

## OTHER APPLICATIONS

**95. Crime** In 1995, the rate of violent crimes in New York City continued to decrease, but at a slower rate than in previous years. Letting $f(t)$ be the rate of violent crime as a function of time, what does this tell you about $f(t)$, $f'(t)$, and $f''(t)$? *Source: The New York Times.*

**96. Velocity and Acceleration** When an object is dropped straight down, the distance (in feet) that it travels in $t$ seconds is given by

$$s(t) = -16t^2.$$

Find the velocity at each of the following times.

  **a.** After 3 seconds

  **b.** After 5 seconds

  **c.** After 8 seconds

  **d.** Find the acceleration. (The answer here is a constant—the acceleration due to the influence of gravity alone near the surface of Earth.)

**97. Height of a Ball** If a cannonball is shot directly upward with a velocity of 256 ft per second, its height above the ground after $t$ seconds is given by $s(t) = 256t - 16t^2$. Find the velocity and the acceleration after $t$ seconds. What is the maximum height the cannonball reaches? When does it hit the ground?

**98. Velocity and Acceleration of a Car** A car rolls down a hill. Its distance (in feet) from its starting point is given by $s(t) = 1.5t^2 + 4t$, where $t$ is in seconds.

  **a.** How far will the car move in 10 seconds?

  **b.** What is the velocity at 5 seconds? At 10 seconds?

  **c.** How can you tell from $v(t)$ that the car will not stop?

  **d.** What is the acceleration at 5 seconds? At 10 seconds?

  **e.** What is happening to the velocity and the acceleration as $t$ increases?

**99. Velocity and Acceleration** A car is moving along a straight stretch of road. The acceleration of the car is given by the graph shown. Assume that the velocity of the car is always positive.

At what time was the car moving most rapidly? Explain. *Source: Larry Taylor.*

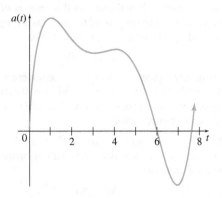

**Point of Diminishing Returns** In Exercises 100–103, find the point of diminishing returns $(x, y)$ for the given functions, where $R(x)$ represents revenue (in thousands of dollars) and $x$ represents the amount spent on advertising (in thousands of dollars).

**100.** $R(x) = 10{,}000 - x^3 + 42x^2 + 800x, \quad 0 \le x \le 20$

**101.** $R(x) = \dfrac{4}{27}(-x^3 + 66x^2 + 1050x - 400), \quad 0 \le x \le 25$

**102.** $R(x) = -0.3x^3 + x^2 + 11.4x, \quad 0 \le x \le 6$

**103.** $R(x) = -0.6x^3 + 3.7x^2 + 5x, \quad 0 \le x \le 6$

**YOUR TURN ANSWERS**

**1.** 36  **2. a.** $f''(x) = 30x^4 + 12x$  **b.** $g''(x) = 2e^x + xe^x$

  **c.** $h''(x) = \dfrac{-3 + 2\ln x}{x^3}$  **d.** $p''(x) = 2\cos^2 x - 2\sin^2 x$

**3.** $v(t) = 3t^2 - 6t - 24$ and $a(t) = 6t - 6$. Car backs up for the first 4 seconds and then goes forward. It speeds up for $0 < t < 1$, slows down for $1 < t < 4$, and then speeds up for $t > 4$.

**4.** Concave up on $(-3, 0)$ and $(3, \infty)$; concave down on $(-\infty, -3)$ and $(0, 3)$; inflection points are $(-3, 567)$, $(0, 0)$, and $(3, -567)$.

**5.** Relative maximum of $f(4) = 208$ at $x = 4$ and relative minimum of $f(-3) = -135$ at $x = -3$.

# 5.4 Curve Sketching

**APPLY IT** How can we use differentiation to help us sketch the graph of a function, and describe its behavior?

**APPLY IT** In the following examples, the test for concavity, the test for increasing and decreasing functions, and the concept of limits at infinity will help us sketch the graphs and describe the behavior of a variety of functions. This process, called **curve sketching**, has decreased somewhat in importance in recent years due to the widespread use of graphing calculators. We believe, however, that this topic is worth studying for the following reasons.

For one thing, a graphing calculator picture can be misleading, particularly if important points lie outside the viewing window. Even if all important features are within the viewing window, there is still the problem that the calculator plots and connects points and misses what goes on between those points. As an example of the difficulty in choosing an appropriate window without a knowledge of calculus, see Exercise 42 in the second section of this chapter.

Furthermore, curve sketching may be the best way to learn the material in the previous three sections. You may feel confident that you understand what increasing and concave upward mean, but using those concepts in a graph will put your understanding to the test.

Curve sketching may be done with the following steps.

## Curve Sketching

To sketch the graph of a function $f$:

1. Consider the domain of the function, and note any restrictions. (That is, avoid dividing by 0, taking a square root, or any even root, of a negative number, or taking the logarithm of 0 or a negative number.)

2. Find the $y$-intercept (if it exists) by substituting $x = 0$ into $f(x)$. Find any $x$-intercepts by solving $f(x) = 0$ if this is not too difficult.

3. a. If $f$ is a rational function, find any vertical asymptotes by investigating where the denominator is 0, and find any horizontal asymptotes by finding the limits as $x \rightarrow \infty$ and $x \rightarrow -\infty$.

   b. If $f$ is an exponential function, find any horizontal asymptotes; if $f$ is a logarithmic function, find any vertical asymptotes.

4. Investigate symmetry. If $f(-x) = f(x)$, the graph is symmetric about the $y$-axis. This means that the left side of the graph is the mirror image of the right side. If $f(-x) = -f(x)$, the graph is symmetric about the origin. This means that the left side of the graph can be found by rotating the right side 180° about the origin.

5. Find $f'(x)$. Locate any critical points by solving the equation $f'(x) = 0$ and determining where $f'(x)$ does not exist, but $f(x)$ does. Find any relative extrema and determine where $f$ is increasing or decreasing.

6. Find $f''(x)$. Locate potential inflection points by solving the equation $f''(x) = 0$ and determining where $f''(x)$ does not exist. Determine where $f$ is concave upward or concave downward.

7. Plot the intercepts, the critical points, the inflection points, the asymptotes, and other points as needed. Take advantage of any symmetry found in Step 4.

8. Connect the points with a smooth curve using the correct concavity, being careful not to connect points where the function is not defined.

9. Check your graph using a graphing calculator. If the picture looks very different from what you've drawn, see in what ways the picture differs and use that information to help find your mistake.

There are four possible combinations for a function to be increasing or decreasing and concave up or concave down, as shown in the following table.

| | Concavity Summary | |
|---|---|---|
| $f''(x)$     $f'(x)$ | **+** (Function Is Increasing) | **−** (Function Is Decreasing) |
| **+** (function is concave up) | | |
| **−** (function is concave down) | | |

EXAMPLE 1    **Polynomial Function Graph**

Graph $f(x) = 2x^3 - 3x^2 - 12x + 1$.

**SOLUTION**    The domain is $(-\infty, \infty)$. The $y$-intercept is located at $y = f(0) = 1$. Finding the $x$-intercepts requires solving the equation $f(x) = 0$. But this is a third-degree equation; since we have not covered a procedure for solving such equations, we will skip this step. This is neither a rational nor an exponential function, so we also skip Step 3. Observe that $f(-x) = 2(-x)^3 - 3(-x)^2 - 12(-x) + 1 = -2x^3 - 3x^2 + 12x + 1$, which is neither $f(x)$ nor $-f(x)$, so there is no symmetry about the $y$-axis or origin.

To find the intervals where the function is increasing or decreasing, find the first derivative.

$$f'(x) = 6x^2 - 6x - 12$$

This derivative is 0 when

$$6(x^2 - x - 2) = 0$$
$$6(x - 2)(x + 1) = 0$$
$$x = 2 \quad \text{or} \quad x = -1.$$

These critical numbers divide the number line in Figure 41 into three regions. Testing a number from each region in $f'(x)$ shows that $f$ is increasing on $(-\infty, -1)$ and $(2, \infty)$ and decreasing on $(-1, 2)$. This is shown with the arrows in Figure 41. By the first derivative test, $f$ has a relative maximum when $x = -1$ and a relative minimum when $x = 2$. The relative maximum is $f(-1) = 8$, while the relative minimum is $f(2) = -19$.

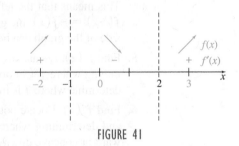

FIGURE 41

Now use the second derivative to find the intervals where the function is concave upward or downward. Here

$$f''(x) = 12x - 6,$$

which is 0 when $x = 1/2$. Testing a point with $x$ less than $1/2$, and one with $x$ greater than $1/2$, shows that $f$ is concave downward on $(-\infty, 1/2)$ and concave upward on $(1/2, \infty)$. The graph has an inflection point at $(1/2, f(1/2))$, or $(1/2, -11/2)$. This information is summarized in the following table.

| | **Graph Summary** | | | |
|---|---|---|---|---|
| **Interval** | $(-\infty, -1)$ | $(-1, 1/2)$ | $(1/2, 2)$ | $(2, \infty)$ |
| *Sign of f'* | + | − | − | + |
| *Sign of f''* | − | − | + | + |
| *f Increasing or Decreasing* | Increasing | Decreasing | Decreasing | Increasing |
| *Concavity of f* | Downward | Downward | Upward | Upward |
| *Shape of Graph* | ⌢ | ⌢ | ⌣ | ⌣ |

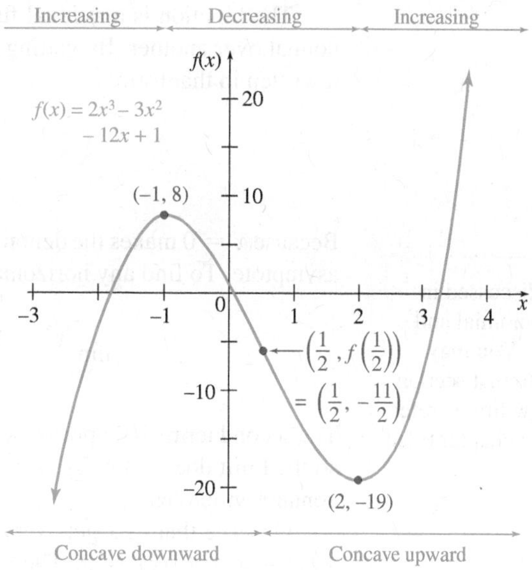

Increasing
Decreasing
Increasing
f(x)
20
10
0
-10
-20
$f(x) = 2x^3 - 3x^2 - 12x + 1$
(−1, 8)
-3
-1
1
2
3
4
x
$\left(\frac{1}{2}, f\left(\frac{1}{2}\right)\right) = \left(\frac{1}{2}, -\frac{11}{2}\right)$
(2, −19)
Concave downward
Concave upward

FIGURE 42

Use this information and the critical points to get the graph shown in Figure 42. Notice that the graph appears to be symmetric about its inflection point. It can be shown that is always true for third-degree polynomials. In other words, if you put your pencil point at the inflection point and then spin the book 180° about the pencil point, the graph will appear to be unchanged.

**TRY YOUR TURN 1**

**YOUR TURN 1** Graph $f(x) = -x^3 + 3x^2 + 9x - 10$.

**TECHNOLOGY NOTE**

A graphing calculator picture of the function in Figure 42 on the arbitrarily chosen window $[-3, 3]$ by $[-7, 7]$ gives a misleading picture, as Figure 43(a) shows. Knowing where the turning points lie tells us that a better window would be $[-3, 4]$ by $[-20, 20]$, with the results shown in Figure 43(b).

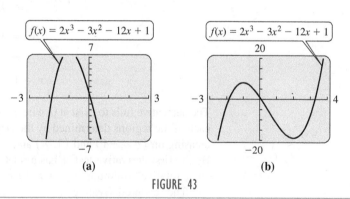

$f(x) = 2x^3 - 3x^2 - 12x + 1$
7
-3
3
-7
(a)
$f(x) = 2x^3 - 3x^2 - 12x + 1$
20
-3
4
-20
(b)

FIGURE 43

## EXAMPLE 2 Rational Function Graph

Graph $f(x) = x + \frac{1}{x}$.

**SOLUTION** Notice that $x = 0$ is not in the domain of the function, so there is no $y$-intercept. To find the $x$-intercept, solve $f(x) = 0$.

$$x + \frac{1}{x} = 0$$

$$x = -\frac{1}{x}$$

$$x^2 = -1$$

Since $x^2$ is always positive, there is also no $x$-intercept.

The function is a rational function, but it is not written in the usual form of one polynomial over another. By getting a common denominator and adding the fractions, it can be rewritten in that form:

$$f(x) = x + \frac{1}{x} = \frac{x^2 + 1}{x}.$$

**FOR REVIEW**

Asymptotes were discussed in Section 1.5 on Polynomial and Rational Functions. You may wish to refer back to that section to review. To review limits, refer to Section 3.1 in the chapter titled The Derivative.

Because $x = 0$ makes the denominator (but not the numerator) 0, the line $x = 0$ is a vertical asymptote. To find any horizontal asymptotes, we investigate

$$\lim_{x \to \infty} \frac{x^2 + 1}{x} = \lim_{x \to \infty}\left(\frac{x^2}{x} + \frac{1}{x}\right) = \lim_{x \to \infty}\left(x + \frac{1}{x}\right).$$

The second term, $1/x$, approaches 0 as $x \to \infty$, but the first term, $x$, becomes infinitely large, so the limit does not exist. Verify that $\lim_{x \to -\infty} f(x)$ also does not exist, so there are no horizontal asymptotes.

Observe that as $x$ gets very large, the second term $(1/x)$ in $f(x)$ gets very small, so $f(x) = x + (1/x) \approx x$. The graph gets closer and closer to the straight line $y = x$ as $x$ becomes larger and larger. This is what is known as an **oblique asymptote**.

Observe that

$$f(-x) = (-x) + \frac{1}{-x} = -\left(x + \frac{1}{x}\right) = -f(x),$$

so the graph is symmetric about the origin. This means that the left side of the graph can be found by rotating the right side 180° about the origin.

Here $f'(x) = 1 - (1/x^2)$, which is 0 when

$$\frac{1}{x^2} = 1$$

$$x^2 = 1$$

$$x^2 - 1 = (x - 1)(x + 1) = 0$$

$$x = 1 \quad \text{or} \quad x = -1.$$

The derivative fails to exist at 0, where the vertical asymptote is located. Evaluating $f'(x)$ in each of the regions determined by the critical numbers and the asymptote shows that $f$ is increasing on $(-\infty, -1)$ and $(1, \infty)$ and decreasing on $(-1, 0)$ and $(0, 1)$. See Figure 44(a). By the first derivative test, $f$ has a relative maximum of $y = f(-1) = -2$ when $x = -1$, and a relative minimum of $y = f(1) = 2$ when $x = 1$.

The second derivative is

$$f''(x) = \frac{2}{x^3},$$

which is never equal to 0 and does not exist when $x = 0$. (The function itself also does not exist at 0.) Because of this, there may be a change of concavity, but not an inflection point, when $x = 0$. The second derivative is negative when $x$ is negative, making $f$ concave downward on $(-\infty, 0)$. Also, $f''(x) > 0$ when $x > 0$, making $f$ concave upward on $(0, \infty)$. See Figure 44(b).

Use this information, the asymptotes, and the critical points to get the graph shown in Figure 45.          TRY YOUR TURN 2

**YOUR TURN 2**   Graph

$f(x) = 4x + \dfrac{1}{x}.$

| | | | Graph Summary | |
| --- | --- | --- | --- | --- |
| Interval | $(-\infty, -1)$ | $(-1, 0)$ | $(0, 1)$ | $(1, \infty)$ |
| Sign of $f'$ | $+$ | $-$ | $-$ | $+$ |
| Sign of $f''$ | $-$ | $-$ | $+$ | $+$ |
| $f$ Increasing or Decreasing | Increasing | Decreasing | Decreasing | Increasing |
| Concavity of $f$ | Downward | Downward | Upward | Upward |
| Shape of Graph | ⌒ | ⌒ | ⌣ | ⌣ |

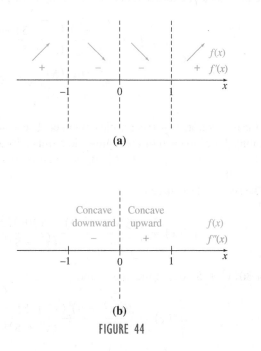

**(a)**

**(b)**

FIGURE 44

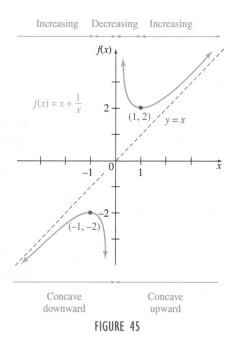

FIGURE 45

## EXAMPLE 3  Rational Function Graph

Graph $f(x) = \dfrac{3x^2}{x^2 + 5}$.

**SOLUTION**  The $y$-intercept is located at $y = f(0) = 0$. Verify that this is also the only $x$-intercept. There is no vertical asymptote, because $x^2 + 5 \neq 0$ for any value of $x$. Find any horizontal asymptote by calculating $\lim\limits_{x \to \infty} f(x)$ and $\lim\limits_{x \to -\infty} f(x)$. First, divide both the numerator and the denominator of $f(x)$ by $x^2$.

$$\lim_{x \to \infty} \frac{3x^2}{x^2 + 5} = \lim_{x \to \infty} \frac{\dfrac{3x^2}{x^2}}{\dfrac{x^2}{x^2} + \dfrac{5}{x^2}} = \frac{3}{1 + 0} = 3$$

Verify that the limit of $f(x)$ as $x \to -\infty$ is also 3. Thus, the horizontal asymptote is $y = 3$.

Observe that

$$f(-x) = \frac{3(-x)^2}{(-x)^2 + 5} = \frac{3x^2}{x^2 + 5} = f(x),$$

so the graph is symmetric about the $y$-axis. This means that the left side of the graph is the mirror image of the right side.

We now compute $f'(x)$:

$$f'(x) = \frac{(x^2 + 5)(6x) - (3x^2)(2x)}{(x^2 + 5)^2}.$$

Notice that $6x$ can be factored out of each term in the numerator:

$$f'(x) = \frac{(6x)[(x^2 + 5) - x^2]}{(x^2 + 5)^2}$$

$$= \frac{(6x)(5)}{(x^2 + 5)^2} = \frac{30x}{(x^2 + 5)^2}.$$

From the numerator, $x = 0$ is a critical number. The denominator is always positive. (Why?) Evaluating $f'(x)$ in each of the regions determined by $x = 0$ shows that $f$ is decreasing on $(-\infty, 0)$ and increasing on $(0, \infty)$. By the first derivative test, $f$ has a relative minimum when $x = 0$.

The second derivative is

$$f''(x) = \frac{(x^2 + 5)^2(30) - (30x)(2)(x^2 + 5)(2x)}{(x^2 + 5)^4}.$$

Factor $30(x^2 + 5)$ out of the numerator:

$$f''(x) = \frac{30(x^2 + 5)[(x^2 + 5) - (x)(2)(2x)]}{(x^2 + 5)^4}.$$

Divide a factor of $(x^2 + 5)$ out of the numerator and denominator, and simplify the numerator:

$$f''(x) = \frac{30[(x^2 + 5) - (x)(2)(2x)]}{(x^2 + 5)^3}$$

$$= \frac{30[(x^2 + 5) - (4x^2)]}{(x^2 + 5)^3}$$

$$= \frac{30(5 - 3x^2)}{(x^2 + 5)^3}.$$

**YOUR TURN 3**  Graph
$$f(x) = \frac{4x^2}{x^2 + 4}.$$

The numerator of $f''(x)$ is 0 when $x = \pm\sqrt{5/3} \approx \pm 1.29$. Testing a point in each of the three intervals defined by these points shows that $f$ is concave downward on $(-\infty, -1.29)$ and $(1.29, \infty)$, and concave upward on $(-1.29, 1.29)$. The graph has inflection points at $(\pm\sqrt{5/3}, f(\pm\sqrt{5/3})) \approx (\pm 1.29, 0.75)$.

Use this information, the asymptote, the critical point, and the inflection points to get the graph shown in Figure 46.    **TRY YOUR TURN 3**

| Graph Summary | | | | |
|---|---|---|---|---|
| Interval | $(-\infty, -1.29)$ | $(-1.29, 0)$ | $(0, 1.29)$ | $(1.29, \infty)$ |
| Sign of $f'$ | − | − | + | + |
| Sign of $f''$ | − | + | + | − |
| $f$ Increasing or Decreasing | Decreasing | Decreasing | Increasing | Increasing |
| Concavity of $f$ | Downward | Upward | Upward | Downward |
| Shape of Graph | ⌐ | ⌣ | ⌣ | ⌐ |

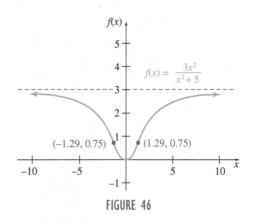

FIGURE 46

<hr>

**EXAMPLE 4** **Graph with Logarithm**

Graph $f(x) = \dfrac{\ln x}{x^2}$.

**SOLUTION** The domain is $x > 0$, so there is no $y$-intercept. The $x$-intercept is 1, because $\ln 1 = 0$. We know that $y = \ln x$ has a vertical asymptote at $x = 0$, because $\lim\limits_{x \to 0^+} \ln x = -\infty$. Dividing by $x^2$ when $x$ is small makes $(\ln x)/x^2$ even more negative than $\ln x$. Therefore, $(\ln x)/x^2$ has a vertical asymptote at $x = 0$ as well. The first derivative is

$$f'(x) = \frac{x^2 \cdot \frac{1}{x} - 2x \ln x}{(x^2)^2} = \frac{x(1 - 2\ln x)}{x^4} = \frac{1 - 2\ln x}{x^3}$$

by the quotient rule. Setting the numerator equal to 0 and solving for $x$ gives

$$1 - 2\ln x = 0$$
$$1 = 2\ln x$$
$$\ln x = 0.5$$
$$x = e^{0.5} \approx 1.65.$$

Since $f'(1)$ is positive and $f'(2)$ is negative, $f$ increases on $(0, 1.65)$ then decreases on $(1.65, \infty)$, with a maximum value of $f(1.65) \approx 0.18$.

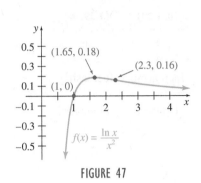

FIGURE 47

To find any inflection points, we set $f''(x) = 0$.

$$f''(x) = \frac{x^3\left(-2 \cdot \dfrac{1}{x}\right) - (1 - 2\ln x) \cdot 3x^2}{(x^3)^2}$$

$$= \frac{-2x^2 - 3x^2 + 6x^2\ln x}{x^6} = \frac{-5 + 6\ln x}{x^4}$$

$$\frac{-5 + 6\ln x}{x^4} = 0$$

$$-5 + 6\ln x = 0 \qquad \text{Set the numerator equal to 0.}$$

$$6\ln x = 5 \qquad \text{Add 5 to both sides.}$$

$$\ln x = 5/6 \qquad \text{Divide both sides by 6.}$$

$$x = e^{5/6} \approx 2.3 \qquad e^{\ln x} = x.$$

There is an inflection point at $(2.3, f(2.3)) \approx (2.3, 0.16)$. Verify that $f''(1)$ is negative and $f''(3)$ is positive, so the graph is concave downward on $(1, 2.3)$ and upward on $(2.3, \infty)$. This information is summarized in the following table and could be used to sketch the graph. A graph of the function is shown in Figure 47.     **TRY YOUR TURN 4**

**YOUR TURN 4**   Graph $f(x) = (x + 2)e^{-x}$. (Recall that $\lim\limits_{x \to \infty} x^n e^{-x} = 0$.)

|  | **Graph Summary** | | |
|---|---|---|---|
| **Interval** | **(0, 1.65)** | **(1.65, 2.3)** | **(2.3, ∞)** |
| *Sign of f'* | + | − | − |
| *Sign of f"* | − | − | + |
| *f Increasing or Decreasing* | Increasing | Decreasing | Decreasing |
| *Concavity of f* | Downward | Downward | Upward |
| *Shape of Graph* | ⌒ | ⌒ | ⌣ |

As we saw earlier, a graphing calculator, when used with care, can be helpful in studying the behavior of functions. This section has illustrated that calculus is also a great help. The techniques of calculus show where the important points of a function, such as the relative extrema and the inflection points, are located. Furthermore, they tell how the function behaves between and beyond the points that are graphed, something a graphing calculator cannot always do.

# 5.4   EXERCISES

1. By sketching a graph of the function or by investigating values of the function near 0, find $\lim\limits_{x \to 0} x \ln |x|$. (This result will be useful in Exercise 21.)

2. Describe how you would find the equation of the horizontal asymptote for the graph of
$$f(x) = \frac{3x^2 - 2x}{2x^2 + 5}.$$

**Graph each function, considering the domain, critical points, symmetry, regions where the function is increasing or decreasing,** inflection points, regions where the function is concave upward or concave downward, intercepts where possible, and asymptotes where applicable. (*Hint:* In Exercise 21, use the result of Exercise 1 in this section. In Exercises 25–27, recall from Exercise 72 in Section 3.1 on Limits that $\lim\limits_{x \to \infty} x^n e^{-x} = 0$.)

3. $f(x) = -2x^3 - 9x^2 + 108x - 10$

4. $f(x) = x^3 - \dfrac{15}{2}x^2 - 18x - 1$

5. $f(x) = -3x^3 + 6x^2 - 4x - 1$

6. $f(x) = x^3 - 6x^2 + 12x - 11$

7. $f(x) = x^4 - 24x^2 + 80$

8. $f(x) = -x^4 + 6x^2$

9. $f(x) = x^4 - 4x^3$

10. $f(x) = x^5 - 15x^3$

11. $f(x) = 2x + \dfrac{10}{x}$

12. $f(x) = 16x + \dfrac{1}{x^2}$

13. $f(x) = \dfrac{-x + 4}{x + 2}$

14. $f(x) = \dfrac{3x}{x - 2}$

15. $f(x) = \dfrac{1}{x^2 + 4x + 3}$

16. $f(x) = \dfrac{-8}{x^2 - 6x - 7}$

17. $f(x) = \dfrac{x}{x^2 + 1}$

18. $f(x) = \dfrac{1}{x^2 + 4}$

19. $f(x) = \dfrac{1}{x^2 - 9}$

20. $f(x) = \dfrac{-2x}{x^2 - 4}$

21. $f(x) = x \ln |x|$

22. $f(x) = x - \ln |x|$

23. $f(x) = \dfrac{\ln x}{x}$

24. $f(x) = \dfrac{\ln x^2}{x^2}$

25. $f(x) = xe^{-x}$

26. $f(x) = x^2 e^{-x}$

27. $f(x) = (x - 1)e^{-x}$

28. $f(x) = e^x + e^{-x}$

29. $f(x) = x^{2/3} - x^{5/3}$

30. $f(x) = x^{1/3} + x^{4/3}$

31. $f(x) = x + \cos x$

32. $f(x) = x + \sin x$

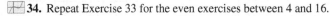

33. The default window on many calculators is $[-10, 10]$ by $[-10, 10]$. For the odd exercises between 3 and 15, tell which would give a poor representation in this window. (*Note:* Your answers may differ from ours, depending on what you consider "poor.")

34. Repeat Exercise 33 for the even exercises between 4 and 16.

35. Repeat Exercise 33 for the odd exercises between 17 and 29.

36. Repeat Exercise 33 for the even exercises between 18 and 30.

**In Exercises 37–41, sketch the graph of a single function that has all of the properties listed.**

37. **a.** Continuous and differentiable everywhere except at $x = 1$, where it has a vertical asymptote

   **b.** $f'(x) < 0$ everywhere it is defined

   **c.** A horizontal asymptote at $y = 2$

   **d.** $f''(x) < 0$ on $(-\infty, 1)$ and $(2, 4)$

   **e.** $f''(x) > 0$ on $(1, 2)$ and $(4, \infty)$

38. **a.** Continuous for all real numbers

   **b.** $f'(x) < 0$ on $(-\infty, -6)$ and $(1, 3)$

   **c.** $f'(x) > 0$ on $(-6, 1)$ and $(3, \infty)$

   **d.** $f''(x) > 0$ on $(-\infty, -6)$ and $(3, \infty)$

   **e.** $f''(x) < 0$ on $(-6, 3)$

   **f.** A $y$-intercept at $(0, 2)$

39. **a.** Continuous and differentiable for all real numbers

   **b.** $f'(x) > 0$ on $(-\infty, -3)$ and $(1, 4)$

   **c.** $f'(x) < 0$ on $(-3, 1)$ and $(4, \infty)$

   **d.** $f''(x) < 0$ on $(-\infty, -1)$ and $(2, \infty)$

   **e.** $f''(x) > 0$ on $(-1, 2)$

   **f.** $f'(-3) = f'(4) = 0$

   **g.** $f''(x) = 0$ at $(-1, 3)$ and $(2, 4)$

40. **a.** Continuous for all real numbers

   **b.** $f'(x) > 0$ on $(-\infty, -2)$ and $(0, 3)$

   **c.** $f'(x) < 0$ on $(-2, 0)$ and $(3, \infty)$

   **d.** $f''(x) < 0$ on $(-\infty, 0)$ and $(0, 5)$

   **e.** $f''(x) > 0$ on $(5, \infty)$

   **f.** $f'(-2) = f'(3) = 0$

   **g.** $f'(0)$ doesn't exist

   **h.** Differentiable everywhere except at $x = 0$

   **i.** An inflection point at $(5, 1)$

41. **a.** Continuous for all real numbers

   **b.** Differentiable everywhere except at $x = 4$

   **c.** $f(1) = 5$

   **d.** $f'(1) = 0$ and $f'(3) = 0$

   **e.** $f'(x) > 0$ on $(-\infty, 1)$ and $(4, \infty)$

   **f.** $f'(x) < 0$ on $(1, 3)$ and $(3, 4)$

   **g.** $\displaystyle\lim_{x \to 4^-} f'(x) = -\infty$ and $\displaystyle\lim_{x \to 4^+} f'(x) = \infty$

   **h.** $f''(x) > 0$ on $(2, 3)$

   **i.** $f''(x) < 0$ on $(-\infty, 2)$, $(3, 4)$, and $(4, \infty)$

42. On many calculators, graphs of rational functions produce lines at vertical asymptotes. For example, graphing $y = (x - 1)/(x + 1)$ on the window $[-4.9, 4.9]$ by $[-4.9, 4.9]$ produces such a line at $x = -1$ on the TI-84 Plus and TI-89. But with the window $[-4.7, 4.7]$ by $[-4.7, 4.7]$ on a TI-84 Plus, or $[-7.9, 7.9]$ by $[-7.9, 7.9]$ on a TI-89, the spurious line does not appear. Experiment with this function on your calculator, trying different windows, and try to figure out an explanation for this phenomenon. (*Hint:* Consider the number of pixels on the calculator screen.)

## LIFE SCIENCE APPLICATIONS

43. **Thoroughbred Horses** As we saw in the previous section, the association between velocity during exercise and blood lactate concentration after submaximal 800-m exercise of thoroughbred racehorses on sand and grass tracks has been studied. The lactate–velocity relationship can be described by the functions

$$l_1(v) = 0.08e^{0.33v} \quad \text{and}$$
$$l_2(v) = -0.87v^2 + 28.17v - 211.41,$$

where $l_1(v)$ and $l_2(v)$ are the lactate concentrations (in mmol/L) and $v$ is the velocity (in m/sec) of the horse during workout on sand and grass tracks, respectively. Sketch the graph of both functions for $13 \le v \le 17$. ***Source: The Veterinary Journal.***

**44. Neuron Communications** In the FitzHugh-Nagumo model of how neurons communicate, the rate of change of the electric potential $v$ with respect to time is given as a function of $v$ by $f(v) = v(a - v)(v - 1)$, where $a$ is a positive constant. Sketch a graph of this function when $a = 0.25$ and $0 \le v \le 1$. *Source: Mathematical Biology.*

**45. Immune System Activity** As we saw in Chapter 4, the activity of an immune system invaded by a parasite can be described by

$$I(n) = \frac{an^2}{b + n^2}$$

where $n$ measures the number of larvae in the host, $a$ is the maximum functional activity of the host's immune response, and $b$ is a measure of the sensitivity of the immune system. *Source: Mathematical Biology.* Sketch a graph of this function when $a = 0.5$ and $b = 10$.

**46. Neuron Membrane Potential** A model of neuron firing involves the function

$$J(z) = \frac{1}{1 + e^{-1.3(z-4)}} - \frac{1}{1 + e^{5.2}},$$

where $z$ is the neuron membrane potential, and the net threshold for a neuron to fire is at $z = 4$. *Source: Mathematical Topics in Population Biology, Morphogenesis and Neurosciences.* Sketch a graph of this function.

**47. Cellular Chemistry** As we saw in Chapter 4, the release of a chemical neurotransmitter as a function of the amount $C$ of intracellular calcium is given by

$$L(C) = \frac{aC^n}{k + C^n},$$

where $n$ measures the degree of cooperativity, $k$ measures saturation, and $a$ is the maximum of the process. *Source: Mathematical Topics in Population Biology, Morphogenesis and Neurosciences.* Sketch a graph of this function when $n = 4$, $k = 3$, and $a = 100$.

**48. Fruit Flies** In Chapter 4, we showed that the number of imagoes (sexually mature adult fruit flies) per mated female per day ($y$) can be approximated by

$$y = 34.7(1.0186)^{-x}x^{-0.658},$$

where $x$ is the mean density of the mated population (measured as flies per bottle) over a sixteen-day period. *Source: Elements of Mathematical Biology.* Sketch the graph of the function.

---

**YOUR TURN ANSWERS**

**1.**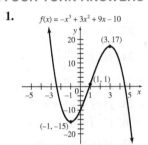

**2.**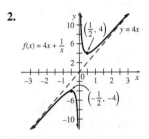

**3.**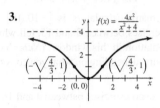

**4.**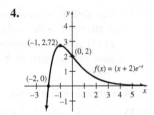

---

# 5 CHAPTER REVIEW

## SUMMARY

In this chapter we have explored various concepts related to the graph of a function:
- increasing and decreasing,
- critical numbers (numbers $c$ in the domain of $f$ for which $f'(x) = 0$ or $f'(x)$ does not exist),
- critical points (whose $x$-coordinate is a critical number $c$ and whose $y$-coordinate is $f(c)$),
- relative maxima and minima (together known as relative extrema),
- concavity, and
- inflection points (where the concavity changes).

The first and second derivative tests provide ways to locate relative extrema. The last section brings all these concepts together. Also, we investigated two applications of the second derivative:
- acceleration (the second derivative of the position function), and
- the point of diminishing returns (an inflection point on an input/output graph).

**Test for Increasing/Decreasing**   On any open interval,

if $f'(x) > 0$, then $f$ is increasing;
if $f'(x) < 0$, then $f$ is decreasing;
if $f'(x) = 0$, then $f$ is constant.

**First Derivative Test**   If $c$ is a critical number for $f$ on the open interval $(a, b)$, $f$ is continuous on $(a, b)$, and $f$ is differentiable on $(a, b)$ (except possibly at $c$), then

1. $f(c)$ is a relative maximum if $f'(x) > 0$ on $(a, c)$ and $f'(x) < 0$ on $(c, b)$;
2. $f(c)$ is a relative minimum if $f'(x) < 0$ on $(a, c)$ and $f'(x) > 0$ on $(c, b)$.

**Test for Concavity**   On any open interval,

if $f''(x) > 0$, then $f$ is concave upward;
if $f''(x) < 0$, then $f$ is concave downward.

**Second Derivative Test**   Suppose $f''$ exists on an open interval containing $c$ and $f'(c) = 0$.

1. If $f''(c) > 0$, then $f(c)$ is a relative minimum.
2. If $f''(c) < 0$, then $f(c)$ is a relative maximum.
3. If $f''(c) = 0$ or $f''(c)$ does not exist, then the test gives no information about extrema, so use the first derivative test.

**Curve Sketching**   To sketch the graph of a function $f$:

1. Consider the domain of the function, and note any restrictions. (That is, avoid dividing by 0, taking a square root, or any even root, of a negative number, or taking the logarithm of 0 or a negative number.)
2. Find the $y$-intercept (if it exists) by substituting $x = 0$ into $f(x)$. Find any $x$-intercepts by solving $f(x) = 0$ if this is not too difficult.
3. **a.** If $f$ is a rational function, find any vertical asymptotes by investigating where the denominator is 0, and find any horizontal asymptotes by finding the limits as $x \to \infty$ and $x \to -\infty$.
   **b.** If $f$ is an exponential function, find any horizontal asymptotes; if $f$ is a logarithmic function, find any vertical asymptotes.
4. Investigate symmetry. If $f(-x) = f(x)$, the graph is symmetric about the $y$-axis. If $f(-x) = -f(x)$, the graph is symmetric about the origin.
5. Find $f'(x)$. Locate any critical points by solving the equation $f'(x) = 0$ and determining where $f'(x)$ does not exist, but $f(x)$ does. Find any relative extrema and determine where $f$ is increasing or decreasing.
6. Find $f''(x)$. Locate potential inflection points by solving the equation $f''(x) = 0$ and determining where $f''(x)$ does not exist. Determine where $f$ is concave upward or concave downward.
7. Plot the intercepts, the critical points, the inflection points, the asymptotes, and other points as needed. Take advantage of any symmetry found in Step 4.
8. Connect the points with a smooth curve using the correct concavity, being careful not to connect points where the function is not defined.
9. Check your graph using a graphing calculator. If the picture looks very different from what you've drawn, see in what ways the picture differs and use that information to help find your mistake.

# KEY TERMS

**5.1**
increasing function
decreasing function
critical number
critical point

**5.2**
relative (or local) maximum

relative (or local) minimum
relative (or local) extremum
first derivative test

**5.3**
second derivative
third derivative

fourth derivative
acceleration
concave upward and downward
concavity
inflection point

second derivative test
point of diminishing returns

**5.4**
curve sketching
oblique asymptote

# REVIEW EXERCISES

## CONCEPT CHECK

**For Exercises 1–12 determine whether each of the following statements is true or false, and explain why.**

1. A critical number $c$ is a number in the domain of a function $f$ for which $f'(c) = 0$ or $f'(c)$ does not exist.

2. If $f'(x) > 0$ on an interval, the function is positive on that interval.

3. If $c$ is a critical number, then the function must have a relative maximum or minimum at $c$.

4. If $f$ is continuous on $(a, b), f'(x) < 0$ on $(a, c)$, and $f'(x) > 0$ on $(c, b)$, then $f$ has a relative minimum at $c$.

5. If $f'(c)$ exists, $f''(c)$ also exists.

6. The acceleration is the second derivative of the position function.

7. If $f''(x) > 0$ on an interval, the function is increasing on that interval.

8. If $f''(c) = 0$, the function has an inflection point at $c$.

9. If $f''(c) = 0$, the function does not have a relative maximum or minimum at $c$.

10. Every rational function has either a vertical or a horizontal asymptote.

11. If a function that is symmetric about the origin has a $y$-intercept, it must pass through the origin.

12. If $f'(c) = 0$, where $c$ is a value in interval $(a, b)$, then $f$ is a constant on the interval $(a, b)$.

## PRACTICE AND EXPLORATIONS

13. When given the equation for a function, how can you determine where it is increasing and where it is decreasing?

14. When given the equation for a function, how can you determine where the relative extrema are located? Give two ways to test whether a relative extremum is a minimum or a maximum.

15. Does a relative maximum of a function always have the largest $y$-value in the domain of the function? Explain your answer.

16. What information about a graph can be found from the second derivative?

**Find the open intervals where $f$ is increasing or decreasing.**

17. $f(x) = x^2 + 9x + 8$

18. $f(x) = -2x^2 + 7x + 14$

19. $f(x) = -x^3 + 2x^2 + 15x + 16$

20. $f(x) = 4x^3 + 8x^2 - 16x + 11$

21. $f(x) = \dfrac{16}{9 - 3x}$

22. $f(x) = \dfrac{15}{2x + 7}$

23. $f(x) = \ln|x^2 - 1|$

24. $f(x) = 8xe^{-4x}$

25. $f(x) = -\tan 2x$

26. $f(x) = -3 \sin 4x$

**Find the locations and values of all relative maxima and minima.**

27. $f(x) = -x^2 + 4x - 8$

28. $f(x) = x^2 - 6x + 4$

29. $f(x) = 2x^2 - 8x + 1$

30. $f(x) = -3x^2 + 2x - 5$

31. $f(x) = 2x^3 + 3x^2 - 36x + 20$

32. $f(x) = 2x^3 + 3x^2 - 12x + 5$

33. $f(x) = \dfrac{xe^x}{x - 1}$

34. $f(x) = \dfrac{\ln(3x)}{2x^2}$

35. $f(x) = 2 \cos \pi x$

36. $f(x) = -5 \sin 3x$

**Find the second derivative of each function, and then find $f''(1)$ and $f''(-3)$.**

37. $f(x) = 3x^4 - 5x^2 - 11x$

38. $f(x) = 9x^3 + \dfrac{1}{x}$

39. $f(x) = \dfrac{4x + 2}{3x - 6}$

40. $f(x) = \dfrac{1 - 2x}{4x + 5}$

41. $f(t) = \sqrt{t^2 + 1}$

42. $f(t) = -\sqrt{5 - t^2}$

43. $f(x) = \tan 7x$

44. $f(x) = x \cos 3x$

**Graph each function, considering the domain, critical points, symmetry, regions where the function is increasing or decreasing, inflection points, regions where the function is concave up or concave down, intercepts where possible, and asymptotes where applicable.**

45. $f(x) = -2x^3 - \dfrac{1}{2}x^2 + x - 3$

46. $f(x) = -\dfrac{4}{3}x^3 + x^2 + 30x - 7$

47. $f(x) = x^4 - \dfrac{4}{3}x^3 - 4x^2 + 1$

48. $f(x) = -\dfrac{2}{3}x^3 + \dfrac{9}{2}x^2 + 5x + 1$

49. $f(x) = \dfrac{x - 1}{2x + 1}$

50. $f(x) = \dfrac{2x - 5}{x + 3}$

51. $f(x) = -4x^3 - x^2 + 4x + 5$

52. $f(x) = x^3 + \dfrac{5}{2}x^2 - 2x - 3$

53. $f(x) = x^4 + 2x^2$

54. $f(x) = 6x^3 - x^4$

55. $f(x) = \dfrac{x^2 + 4}{x}$

56. $f(x) = x + \dfrac{8}{x}$

57. $f(x) = \dfrac{2x}{3 - x}$

58. $f(x) = \dfrac{-4x}{1 + 2x}$

59. $f(x) = xe^{2x}$

60. $f(x) = x^2 e^{2x}$

61. $f(x) = \ln(x^2 + 4)$

62. $f(x) = x^2 \ln x$

**63.** $f(x) = 4x^{1/3} + x^{4/3}$     **64.** $f(x) = 5x^{2/3} + x^{5/3}$

**65.** $f(x) = x - \sin x$     **66.** $f(x) = x - \cos x$

**In Exercises 67 and 68, sketch the graph of a single function that has all of the properties listed.**

**67. a.** Continuous everywhere except at $x = -4$, where there is a vertical asymptote

**b.** A $y$-intercept at $y = -2$

**c.** $x$-intercepts at $x = -3$, 1, and 4

**d.** $f'(x) < 0$ on $(-\infty, -5)$, $(-4, -1)$, and $(2, \infty)$

**e.** $f'(x) > 0$ on $(-5, -4)$ and $(-1, 2)$

**f.** $f''(x) > 0$ on $(-\infty, -4)$ and $(-4, -3)$

**g.** $f''(x) < 0$ on $(-3, -1)$ and $(-1, \infty)$

**h.** Differentiable everywhere except at $x = -4$ and $x = -1$

**68. a.** Continuous and differentiable everywhere except at $x = -3$, where it has a vertical asymptote

**b.** A horizontal asymptote at $y = 1$

**c.** An $x$-intercept at $x = -2$

**d.** A $y$-intercept at $y = 4$

**e.** $f'(x) > 0$ on the intervals $(-\infty, -3)$ and $(-3, 2)$

**f.** $f'(x) < 0$ on the interval $(2, \infty)$

**g.** $f''(x) > 0$ on the intervals $(-\infty, -3)$ and $(4, \infty)$

**h.** $f''(x) < 0$ on the interval $(-3, 4)$

**i.** $f'(2) = 0$

**j.** An inflection point at $(4, 3)$

## LIFE SCIENCE APPLICATIONS

**69. Weightlifting** An abstract for an article states, "We tentatively conclude that Olympic weightlifting ability in trained subjects undergoes a nonlinear decline with age, in which the second derivative of the performance versus age curve repeatedly changes sign." *Source: Medicine and Science in Sports and Exercise.*

**a.** What does this quote tell you about the first derivative of the performance versus age curve?

**b.** Describe what you know about the performance versus age curve based on the information in the quote.

**70. Scaling Laws** Many biological variables depend on body mass, with a functional relationship of the form

$$Y = Y_0 M^b,$$

where $M$ represents body mass, $b$ is a multiple of 1/4, and $Y_0$ is a constant. For example, when $Y$ represents metabolic rate, $b = 3/4$. When $Y$ represents heartbeat, $b = -1/4$. When $Y$ represents life span, $b = 1/4$. *Source: Science.*

**a.** Determine which of metabolic rate, heartbeat, and life span are increasing or decreasing functions of mass. Also determine which have graphs that are concave upward and which have graphs that are concave downward.

**b.** Verify that all functions of the form given above satisfy the equation

$$\frac{dY}{dM} = \frac{b}{M}Y.$$

This means that the rate of change of $Y$ is proportional to $Y$ and inversely proportional to body mass.

**71. Blood Volume** A formula proposed by Hurley for the red cell volume (RCV) in milliliters for males is

$$RCV = 1486S^2 - 4106S + 4514,$$

where $S$ is the surface area (in square meters). A formula given by Pearson et al. is

$$RCV = 1486S - 825.$$

*Source: Journal of Nuclear Medicine and British Journal of Haematology.*

**a.** For the value of $S$ for which the $RCV$ values given by the two formulas are closest, find the rate of change of $RCV$ with respect to $S$ for both formulas. What does this number represent?

**b.** The formula for plasma volume for males given by Hurley is

$$PV = 995e^{0.6085S},$$

while the formula given by Pearson et al. is

$$PV = 1578S,$$

where $PV$ is measured in milliliters and $S$ in square meters. Find the value of $S$ for which the $PV$ values given by the two formulas are the closest. Then find the value of $PV$ that each formula gives for this value of $S$.

**c.** For the value of $S$ found in part b, find the rate of change of $PV$ with respect to $S$ for both formulas. What does this number represent?

**d.** Notice in parts a and c that both formulas give the same instantaneous rate of change at the value of $S$ for which the function values are closest. Prove that if two functions $f$ and $g$ are differentiable and never cross but are closest together when $x = x_0$, then $f'(x_0) = g'(x_0)$.

**72. Cell Surface Receptors** In an article on the local clustering of cell surface receptors, the researcher sought to verify that the function

$$y(S) = S[1 + C(1 + kS)^{f-1}]$$

is equal to 1 for a unique value of $S$ (the concentration of free receptors), where $C$ is the concentration of free ligand in the medium, $f$ is the number of functional groups of cells, and $k$ is a positive constant. *Source: Mathematical Biosciences.*

**a.** Verify that $y(0) = 0$ and $y(1) > 1$.

**b.** Show that

$$y'(S) = 1 + C(1 + fkS)(1 + kS)^{f-2}.$$

Notice from this that $y'(S) > 1$.

**c.** Explain why the results of parts a and b imply that there is a unique value of $S$ where $y(S) = 1$.

**d.** If $k = 0$, what is the value of $S$ such that $y(S) = 1$?

**e.** Explain why, if $k > 0$, the value of $S$ such that $y(S) = 1$ must be between 0 and $1/(1 + C)$.

**f.** The researcher also calculated $f''(S)$. Show that

$$y''(S) = \begin{cases} 2Ck & \text{if } f = 2 \\ (f - 1)kC(2 + fkS)(1 + kS)^{f-3} & \text{if } f \geq 3. \end{cases}$$

**g.** Show that solving $y(S) = 1$ for $C$ yields the equation

$$C = \frac{1 - S}{S(1 + kS)^{f-1}},$$

which the researcher denotes $C = F(S)$.

**h.** Show that

$$F'(S) = -\frac{1 + kS[S + f(1 - S)]}{S^2(1 + kS)^f}.$$

**i.** For $0 < S < 1$, is $F$ increasing or decreasing? Why?

**73. Cell Surface Receptors** In the same article on the local clustering of cell surface receptors (see previous exercise), the researcher sought to maximize the function

$$z = (1 - S)[1 - (1 + kS)^{-(f-1)}],$$

where $z$ is the fraction of receptor sites bound to multiply bound ligand, $S$ is the concentration of free receptors, $f$ is the number of functional groups of cells, and $k$ is a positive constant. ***Source: Mathematical Biosciences.***

**a.** To maximize $z$, the researcher calculated $dz/dS$. Show that

$$\frac{dz}{dS} = -1 + (1 + kS)^{-f}[1 + kS + k(f - 1)(1 - S)].$$

**b.** Show that when $f = 2$, $dz/dS = 0$ when

$$S = (-1 + \sqrt{1 + k})/k.$$

**74. Cell Traction Force** In Chapter 4 we saw that, in a matrix of cells, the cell traction force per unit of mass is given by

$$T(n) = \frac{an}{1 + bn^2},$$

where $n$ is the number of cells, $a$ is the measure of the traction force generated by a cell, and $b$ measures how force is reduced due to neighboring cells. ***Source: Mathematical Biology.*** Sketch a graph of this function when $a = 20$ and $b = 1$.

**75. Calcium Kinetics** In Chapter 4 we saw that a function used to describe the kinetics of calcium in the cytogel (the gel of cell cytoplasm in the epithelium, an external tissue of epidermal cells) is given by

$$R(c) = \frac{ac^2}{1 + bc^2} - kc,$$

where $R(c)$ measures the release of calcium; $c$ is the amount of free calcium outside the vesicles in which it is stored; and $a$, $b$, and $k$ are positive constants. ***Source: Mathematical Biology.***

**a.** Sketch a graph of this function when $a = 10$, $b = 0.08$, and $k = 7$.

**b.** Since $R$ cannot be negative, for what values of $c$ is this function valid?

**76. Swing of a Runner's Arm** A runner's arm swings rhythmically according to the equation

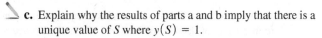

$$y = \frac{\pi}{8} \cos\left[3\pi\left(t - \frac{1}{3}\right)\right],$$

where $y$ denotes the angle between the actual position of the upper arm and the downward vertical position (as shown in the figure) and where $t$ denotes time (in seconds). ***Source: Calculus for the Life Sciences.***

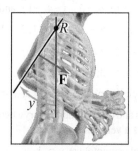

**a.** Graph $y$ as a function of $t$.

**b.** Calculate the velocity and the acceleration of the arm.

**c.** Verify that the angle $y$ and the acceleration $d^2y/dt^2$ are related by the equation

$$\frac{d^2y}{dt^2} + 9\pi^2y = 0.$$

**d.** Apply the fact that the force exerted by the muscle as the arm swings is proportional to the acceleration of $y$, with a positive constant of proportionality, to find the direction of the force (counterclockwise or clockwise) at $t = 1$ second, $t = 4/3$ seconds, and $t = 5/3$ seconds. What is the position of the arm at each of these times?

**77. Swing of a Jogger's Arm** A jogger's arm swings according to the equation

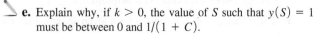

$$y = \frac{1}{5}\sin[\pi(t - 1)].$$

***Source: Calculus for the Life Sciences.*** Proceed as directed in parts a–d of the preceding exercise, with the following exceptions: in part c, replace the equation with

$$\frac{d^2y}{dt^2} + \pi^2y = 0,$$

and in part d, consider the times $t = 1.5$ seconds, $t = 2.5$ seconds, and $t = 3.5$ seconds.

## OTHER APPLICATIONS

**78. Learning** Researchers used a version of the Gompertz curve to model the rate that children learn with the equation

$$y(t) = A^{c^t},$$

where $y(t)$ is the portion of children of age $t$ years passing a certain mental test, $A = 0.3982 \times 10^{-291}$, and $c = 0.4252$.

Find the inflection point and describe what it signifies. (*Hint:* Leave $A$ and $c$ as constants until you are ready to calculate your final answer. If $A$ is too small for your calculator to handle, use common logarithms and properties of logarithms to calculate $(\log A)/(\log e)$.) *Source: School and Society.*

79. **Population** Under the scenario that the fertility rate in the European Union (EU) remains at 1.8 until 2020, when it rises to replacement level, the predicted population (in millions) of the 15 member countries of the EU can be approximated over the next century by

$$P(t) = 325 + 7.475(t + 10)e^{-(t+10)/20},$$

where $t$ is the number of years since 2000. *Source: Science.*

**a.** In what year is the population predicted to be largest? What is the population predicted to be in that year?

**b.** In what year is the population declining most rapidly?

**c.** What is the population approaching as time goes on?

80. **Nuclear Weapons** The graph shows the total inventory of nuclear weapons held by the United States and by the Soviet

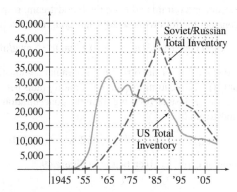

Union and its successor states from 1945 to 2010. *Source: Federation of American Scientists.*

**a.** In what years was the U.S. total inventory of weapons at a relative maximum?

**b.** When the U.S. total inventory of weapons was at the largest relative maximum, is the graph for the Soviet stockpile concave up or concave down? What does this mean?

81. **Velocity and Acceleration** A projectile is shot straight up with an initial velocity of 512 ft per second. Its height above the ground after $t$ seconds is given by $s(t) = 512t - 16t^2$.

**a.** Find the velocity and acceleration after $t$ seconds.

**b.** What is the maximum height attained?

**c.** When does the projectile hit the ground and with what velocity?

82. **Flying Gravel** The grooves or tread in a tire occasionally pick up small pieces of gravel, which then are often thrown into the air as they work loose from the tire. When following behind a vehicle on a highway with loose gravel, it is possible to determine a safe distance to travel behind the vehicle so that your automobile is not hit with flying debris by analyzing the function

$$y = x \tan \alpha - \frac{16x^2}{V^2} \sec^2 \alpha,$$

where $y$ is the height (in feet) of a piece of gravel that leaves the bottom of a tire at an angle $\alpha$ relative to the roadway, $x$ is the horizontal distance (in feet) of the gravel, and $V$ is the velocity of the automobile (in feet per second). *Source: UMAP Journal.*

**a.** If a car is traveling 30 mph (44 ft per second), find the height of a piece of gravel thrown from a car tire at an angle of $\pi/4$, when the stone is 40 ft from the car.

**b.** Setting $y = 0$ and solving for $x$ gives the distance that the gravel will fly. Show that the function that gives a relationship between $x$ and the angle $\alpha$ for $y = 0$ is given by

$$x = \frac{V^2}{32} \sin(2\alpha).$$

(*Hint:* $2 \sin \alpha \cos \alpha = \sin(2\alpha)$.)

**c.** Using part b, if the gravel is thrown from the car at an angle of $\pi/3$ and initial velocity of 44 ft per second, determine how far the gravel will travel.

**d.** Find $dx/d\alpha$ and use it to determine the value of $\alpha$ that gives the maximum distance that a stone could fly.

**e.** Find the maximum distance that a stone can fly from a car that is traveling 60 mph.

# EXTENDED APPLICATION

# A DRUG CONCENTRATION MODEL FOR ORALLY ADMINISTERED MEDICATIONS

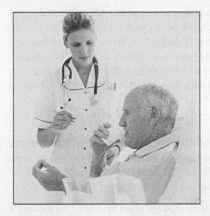

Finding a range for the concentration of a drug in the bloodstream that is both safe and effective is one of the primary goals in pharmaceutical research and development. This range is called the *therapeutic window*. When determining the proper dosage (both the size of the dose and the frequency of administration), it is important to understand the behavior of the drug once it enters the body. Using data gathered during research we can create a mathematical model that predicts the concentration of the drug in the bloodstream at any given time.

We will look at two examples that explore a mathematical model for the concentration of a particular drug in the bloodstream. We will find the maximum and minimum concentrations of the drug given the size of the dose and the frequency of administration. We will then determine what dose should be administered to maintain concentrations within a given therapeutic window.

The drug tolbutamide is used for the management of mild to moderately severe type 2 diabetes. Suppose a 1000-mg dose of this drug is taken every 12 hours for three days. The concentration of the drug in the bloodstream, $t$ hours after the initial dose is taken, is shown in Figure 48.

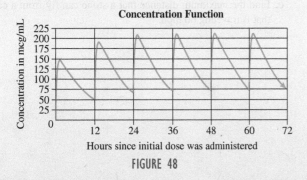

**Concentration Function**

FIGURE 48

Looking at the graph, you can see that after a few doses have been administered, the maximum values of the concentration function begin to level off. The function also becomes periodic, repeating itself between every dose. At this point, the concentration is said to be at steady-state. The time it takes to reach steady-state depends on the elimination half-life of the drug (the time it takes for half the dose to be eliminated from the body). The elimination half-life of

the drug used for this function is about 7 hours. Generally speaking, we say that steady-state is reached after about 5 half-lives.

We will define the *steady-state concentration function*, $C_{ss}(t)$, to be the concentration of drug in the bloodstream $t$ hours after a dose has been administered once steady-state has been reached.

The steady-state concentration function can be written as the difference of two exponential decay functions, or

$$C_{ss}(t) = c_1 e^{kt} - c_2 e^{k_a t}.$$

The constants $c_1$ and $c_2$ are influenced by several factors, including the size of the dose and how widely the particular drug disperses through the body. The constants $k_a$ and $k$ are decay constants reflecting the rate at which the drug is being absorbed into the bloodstream and eliminated from the bloodstream, respectively.

Consider the following steady-state concentration function:

$$C_{ss}(t) = 0.2473 D e^{-0.1t} - 0.1728 D e^{-2.8t} \text{ mcg/mL}$$

where $D$ is the size of the dose (in milligrams) administered every 12 hours. The concentration is given in micrograms per milliliter.

If a single dose is 1000 mg, then the concentration of drug in the bloodstream is

$$C_{ss}(t) = 247.3 e^{-0.1t} - 172.8 e^{-2.8t} \text{ mcg/mL}.$$

The graph of $C_{ss}(t)$ is given below in Figure 49.

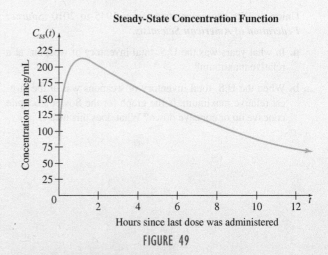

**Steady-State Concentration Function**

FIGURE 49

## EXAMPLE 1   Drug Concentration

Find the maximum and minimum concentrations for the steady-state concentration function

$$C_{ss}(t) = 247.3 e^{-0.1t} - 172.8 e^{-2.8t} \text{ mcg/mL}.$$

**SOLUTION** The maximum concentration occurs when $C'_{ss}(t) = 0$. Calculating the derivative, we get:

$$C'_{ss}(t) = 247.3(-0.1)e^{-0.1t} - 172.8(-2.8)e^{-2.8t}$$
$$= -24.73e^{-0.1t} + 483.84e^{-2.8t}.$$

If we factor out $e^{-0.1t}$, we can find where the derivative is equal to zero.

$$C'_{ss}(t) = e^{-0.1t}(-24.73 + 483.84e^{-2.7t}) = 0$$

$C'_{ss}(t) = 0$ when

$$-24.73 + 483.84e^{-2.7t} = 0.$$

Solving this equation for $t$, we get

$$t = \frac{\ln\left(\dfrac{24.73}{483.84}\right)}{-2.7} \approx 1.1 \text{ hours.}$$

Therefore, the maximum concentration is

$$C_{ss}(1.1) = 247.3e^{-0.1(1.1)} - 172.8e^{-2.8(1.1)} \approx 214 \text{ mcg/mL.}$$

Looking at the graph of $C_{ss}(t)$ in Figure 49, you can see that the minimum concentration occurs at the endpoints (when $t = 0$ and $t = 12$; immediately after a dose is administered and immediately before a next dose is to be administered, respectively).

Therefore, the minimum concentration is

$$C_{ss}(0) = 247.3e^{-0.1(0)} - 172.8e^{-2.8(0)} = 247.3 - 172.8$$
$$= 74.5 \text{ mcg/mL.}$$

Verify that $C_{ss}(12)$ gives the same value.

If the therapeutic window for this drug is 70–240 mcg/mL, then, once steady-state has been reached, the concentration remains safe and effective as long as treatment continues.

Suppose, however, that a new study found that this drug is effective only if the concentration remains between 100 and 400 mcg/mL. How could you adjust the dose so that the maximum and minimum steady-state concentrations fall within this range?

### EXAMPLE 2   Therapeutic Window

Find a range for the size of doses such that the steady-state concentration remains within the therapeutic window of 100 to 400 mcg/mL.

**SOLUTION** Recall that the steady-state concentration function is

$$C_{ss}(t) = 0.2473De^{-0.1t} - 0.1728De^{-2.8t} \text{ mcg/mL,}$$

where $D$ is the size of the dose given (in milligrams) every 12 hours.

From Example 1, we found that the minimum concentration occurs when $t = 0$. Therefore, we want the minimum concentration, $C_{ss}(0)$, to be greater than or equal to 100 mcg/mL.

$$C_{ss}(0) = 0.2473De^{-0.1(0)} - 0.1728De^{-2.8(0)} \geq 100$$

or

$$0.2473D - 0.1728D \geq 100$$

Solving for $D$, we get

$$0.0745D \geq 100$$
$$D \geq 1342 \text{ mg.}$$

In Example 1, we also found that the maximum concentration occurs when $t = 1.1$ hours. If we change the size of the dose, the maximum concentration will change; however, the time it takes to reach the maximum concentration does not change. Can you see why this is true?

Since the maximum concentration occurs when $t = 1.1$, we want $C_{ss}(1.1)$, the maximum concentration, to be less than or equal to 400 mcg/mL.

$$C_{ss}(1.1) = 0.2473De^{-0.1(1.1)} - 0.1728De^{-2.8(1.1)} \leq 400$$

or

$$0.2215D - 0.0079D \leq 400$$

Solving for $D$, we get

$$0.2136D \leq 400$$
$$D \leq 1873 \text{ mg.}$$

Therefore, if the dose is between 1342 mg and 1873 mg, the steady-state concentration remains within the new therapeutic window.

## EXERCISES

Use the following information to answer Exercises 1–3.

A certain drug is given to a patient every 12 hours. The steady-state concentration function is given by

$$C_{ss}(t) = 1.99De^{-0.14t} - 1.62De^{-2.08t} \text{ mcg/mL,}$$

where $D$ is the size of the dose in milligrams.

1. If a 500-mg dose is given every 12 hours, find the maximum and minimum steady-state concentrations.

2. If the dose is increased to 1500 mg every 12 hours, find the maximum and minimum steady-state concentrations.

3. What dose should be given every 12 hours to maintain a steady-state concentration between 80 and 400 mcg/mL?

## DIRECTIONS FOR GROUP PROJECT

*Because of declining health, many elderly people rely on prescription medications to stabilize or improve their medical condition. Your group has been assigned the task of developing a brochure to be made available at senior citizens' centers and physicians' offices that describes drug concentrations in the body for orally administered medications. The brochure should summarize the facts presented in this extended application but at a level that is understandable to a typical layperson. The brochure should be designed to look professional with a marketing flair.*

# 6

# Applications of the Derivative

6.1   Absolute Extrema

6.2   Applications of Extrema

6.3   Implicit Differentiation

6.4   Related Rates

6.5   Differentials: Linear Approximation

Chapter 6 Review

Extended Application: A Total Cost Model for a Training Program

When two or more variables are related by a single equation, their rates are also related. For example, the metabolic rates for animals from weasels to elk are directly related to the body mass of the animal. In an exercise in Section 4 we will use the mathematics of related rates to determine how a change in an elk's metabolic rate is related to its change in mass.

The previous chapter included examples in which we used the derivative to find the maximum or minimum value of a function. This problem is ubiquitous; consider the efforts people expend trying to maximize their income, or to minimize their costs or the time required to complete a task. In this chapter we will treat the topic of optimization in greater depth.

The derivative is applicable in far wider circumstances, however. In roughly 500 B.C., Heraclitus said, "Nothing endures but change," and his observation has relevance here. If change is continuous, rather than in sudden jumps, the derivative can be used to describe the rate of change. This explains why calculus has been applied to so many fields.

# 6.1 Absolute Extrema

**APPLY IT** During a 10-year period, when did the percent of college students using illicit drugs reach its maximum, and what was the percent in that year? *We will answer this question in Example 3.*

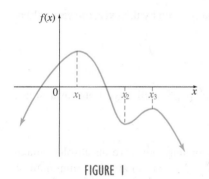

FIGURE 1

A function may have more than one relative maximum. It may be important, however, in some cases to determine if one function value is greater than any other. In other cases, we may want to know whether one function value is less than any other. For example, in Figure 1, $f(x_1) \geq f(x)$ for all $x$ in the domain. There is no function value that is less than all others, however, because $f(x) \rightarrow -\infty$ as $x \rightarrow \infty$ or as $x \rightarrow -\infty$.

The greatest possible value of a function is called the *absolute maximum* and the least possible value of a function is called the *absolute minimum*. As Figure 1 shows, one or both of these may not exist on the domain of the function, $(-\infty, \infty)$ here. Absolute extrema often coincide with relative extrema, as with $f(x_1)$ in Figure 1. Although a function may have several relative maxima or relative minima, it never has more than one *absolute maximum* or *absolute minimum*, although the absolute maximum or minimum might occur at more than one value of $x$.

## Absolute Maximum or Minimum

Let $f$ be a function defined on some interval. Let $c$ be a number in the interval. Then $f(c)$ is the **absolute maximum** of $f$ on the interval if

$$f(x) \leq f(c)$$

for every $x$ in the interval, and $f(c)$ is the **absolute minimum** of $f$ on the interval if

$$f(x) \geq f(c)$$

for every $x$ in the interval.

A function has an **absolute extremum** (plural: **extrema**) at $c$ if it has either an absolute maximum or an absolute minimum there.

| CAUTION | Notice that, just like a relative extremum, an absolute extremum is a $y$-value, not an $x$-value. |

Now look at Figure 2, which shows three functions defined on closed intervals. In each case there is an absolute maximum value and an absolute minimum value. These absolute extrema may occur at the endpoints or at relative extrema. As the graphs in Figure 2 show, an absolute extremum is either the largest or the smallest function value occurring on a closed interval, while a relative extremum is the largest or smallest function value in some (perhaps small) open interval.

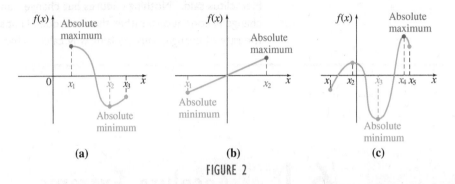

FIGURE 2

Although a function can have only one absolute minimum value and only one absolute maximum value, it can have many points where these values occur. (Note that the absolute maximum value and absolute minimum value are numbers, not points.) As an extreme example, consider the function $f(x) = 2$. The absolute minimum value of this function is clearly 2, as is the absolute maximum value. Both the absolute minimum and the absolute maximum occur at every real number $x$.

One reason for the importance of absolute extrema is given by the **extreme value theorem** (which is proved in more advanced courses).

## Extreme Value Theorem

A function $f$ that is continuous on a closed interval $[a, b]$ will have both an absolute maximum and an absolute minimum on the interval.

A continuous function on an open interval may or may not have an absolute maximum or minimum. For example, the function in Figure 3(a) has an absolute minimum on the interval $(a, b)$ at $x_1$, but it does not have an absolute maximum. Instead, it becomes arbitrarily large as $x$ approaches $a$ or $b$. Also, a discontinuous function on a closed interval may or may not have an absolute minimum or maximum. The function in Figure 3(b) has an absolute minimum at $x = a$, yet it has no absolute maximum. It may appear at first to have an absolute maximum at $x_1$, but notice that $f(x_1)$ has a smaller value than $f$ at values of $x$ less than $x_1$.

The extreme value theorem guarantees the existence of absolute extrema for a continuous function on a closed interval. To find these extrema, use the following steps.

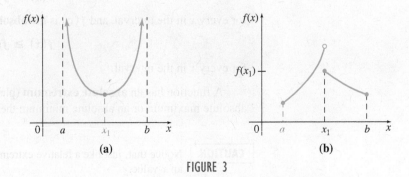

FIGURE 3

## Finding Absolute Extrema

To find absolute extrema for a function $f$ continuous on a closed interval $[a, b]$:

1. Find all critical numbers for $f$ in $(a, b)$.
2. Evaluate $f$ for all critical numbers in $(a, b)$.
3. Evaluate $f$ for the endpoints $a$ and $b$ of the interval $[a, b]$.
4. The largest value found in Step 2 or 3 is the absolute maximum for $f$ on $[a, b]$, and the smallest value found is the absolute minimum for $f$ on $[a, b]$.

### EXAMPLE 1 Absolute Extrema

Find the absolute extrema of the function

$$f(x) = x^{8/3} - 16x^{2/3}$$

on the interval $[-1, 8]$.

**SOLUTION** First look for critical numbers in the interval $(-1, 8)$.

$$f'(x) = \frac{8}{3}x^{5/3} - \frac{32}{3}x^{-1/3}$$

$$= \frac{8}{3}x^{-1/3}(x^2 - 4) \qquad \text{Factor.}$$

$$= \frac{8}{3}\left(\frac{x^2 - 4}{x^{1/3}}\right)$$

Set $f'(x) = 0$ and solve for $x$. Notice that $f'(x) = 0$ at $x = 2$ and $x = -2$, but $-2$ is not in the interval $(-1, 8)$, so we ignore it. The derivative is undefined at $x = 0$, but the function is defined there, so 0 is also a critical number.

Evaluate the function at the critical numbers and the endpoints.

| $x$-Value | Extrema Candidates<br>Value<br>of Function | |
| :---: | :---: | :--- |
| $-1$ | $-15$ | |
| 0 | 0 | |
| 2 | $-19.05$ | ← Absolute minimum |
| 8 | 192 | ← Absolute maximum |

The absolute maximum, 192, occurs when $x = 8$, and the absolute minimum, approximately $-19.05$, occurs when $x = 2$. A graph of $f$ is shown in Figure 4 on the next page.

**TRY YOUR TURN 1**

---

**FOR REVIEW**

Recall from Section R.6 that an exponential expression can be simplified by factoring out the smallest power of the variable. The expression $\frac{8}{3}x^{5/3} - \frac{32}{3}x^{-1/3}$ has a common factor of $\frac{8}{3}x^{-1/3}$.

---

**YOUR TURN 1** Find the absolute extrema of the function $f(x) = 3x^{2/3} - 3x^{5/3}$ on the interval $[0, 8]$.

---

**TECHNOLOGY NOTE**

In Example 1, a graphing calculator that gives the maximum and minimum values of a function on an interval, such as the fMax or fMin feature of the TI-84 Plus, could replace the table. Alternatively, we could first graph the function on the given interval and then select the feature that gives the maximum or minimum value of the graph of the function instead of completing the table.

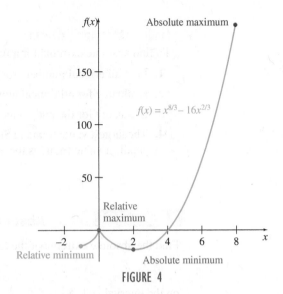

FIGURE 4

### EXAMPLE 2    Absolute Extrema

Find the locations and values of the absolute extrema, if they exist, for the function

$$f(x) = 3x^4 - 4x^3 - 12x^2 + 2.$$

**SOLUTION**    In this example, the extreme value theorem does not apply, since the domain is an open interval, $(-\infty, \infty)$, which has no endpoints. Begin as before by finding any critical numbers.

$$f'(x) = 12x^3 - 12x^2 - 24x = 0$$
$$12x(x^2 - x - 2) = 0$$
$$12x(x + 1)(x - 2) = 0$$
$$x = 0 \quad \text{or} \quad x = -1 \quad \text{or} \quad x = 2$$

There are no values of $x$ where $f'(x)$ does not exist. Evaluate the function at the critical numbers.

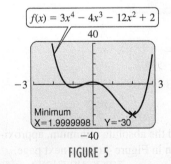

FIGURE 5

| Extrema Candidates | |
|---|---|
| $x$-Value | Value of Function |
| $-1$ | $-3$ |
| $0$ | $2$ |
| $2$ | $-30$ ← Absolute minimum |

For an open interval, rather than evaluating the function at the endpoints, we evaluate the limit of the function when the endpoints are approached. Because the positive $x^4$-term dominates the other terms as $x$ becomes large,

$$\lim_{x \to \infty} (3x^4 - 4x^3 - 12x^2 + 2) = \infty.$$

**YOUR TURN 2**  Find the locations and values of the absolute extrema, if they exist, for the function $f(x) = -x^4 - 4x^3 + 8x^2 + 20$.

The limit is also $\infty$ as $x$ approaches $-\infty$. Since the function can be made arbitrarily large, it has no absolute maximum. The absolute minimum, $-30$, occurs at $x = 2$. This result can be confirmed with a graphing calculator, as shown in Figure 5.    **TRY YOUR TURN 2**

In many of the applied extrema problems in the next section, a continuous function on an open interval has just one critical number. In that case, we can use the following theorem, which also applies to closed intervals.

## Critical Point Theorem

Suppose a function $f$ is continuous on an interval $I$ and that $f$ has exactly one critical number in the interval $I$, located at $x = c$.

If $f$ has a relative maximum at $x = c$, then this relative maximum is the absolute maximum of $f$ on the interval $I$.

If $f$ has a relative minimum at $x = c$, then this relative minimum is the absolute minimum of $f$ on the interval $I$.

The critical point theorem is of no help in the previous two examples because they each had more than one critical point on the interval under consideration. But the theorem could be useful for some of Exercises 33–40 at the end of this section, and we will make good use of it in the next section.

### EXAMPLE 3   Illicit Drug Use

Based on data from the National Institute on Drug Abuse, the percent $y$ of college students who used any illicit drug between 1991 and 2011 can be approximated by the function

$$f(t) = 0.00331t^3 - 0.146t^2 + 1.96t + 28.2,$$

where $t$ is the number of years since 1991. **Source: National Institute on Drug Abuse.** In what year during this period did illicit drug use reach its absolute maximum? Based on this model, what percent of college students used illicit drugs in that year?

**APPLY IT**   **SOLUTION**   The function is defined on the interval $[0, 20]$. We first look for critical numbers in this interval. Here $f'(t) = 0.00993t^2 - 0.292t + 1.96$. We set this derivative equal to zero and use the quadratic formula to solve for $t$.

$$0.00993t^2 - 0.292t + 1.96 = 0$$

$$t = \frac{0.292 \pm \sqrt{(-0.292)^2 - 4(0.00993)(1.96)}}{2(0.00993)}$$

$$t = 19.0 \quad \text{or} \quad t = 10.4$$

Both values are in the interval $[0, 20]$. Now evaluate the function at the critical numbers and the endpoints 0 and 20.

| Extrema Candidates | |
|---|---|
| $t$-Value | Value of Function |
| 0 | 28.2 |
| 10.4 | 36.5 ← Absolute maximum |
| 19.0 | 35.4 |
| 20 | 35.5 |

About 10.4 years after 1991, that is, some time during 2001, illicit drug use by college students reached its absolute maximum of about 36.5% in the given period. It is also worth noting that the minimum use of 28.2% occurred at the beginning of this period, in 1991.

## Graphical Optimization

Figure 6 shows the production output for a family-owned business that produces landscape mulch. As the number of workers (measured in hours of labor) varies, the total production of mulch also varies. Notice that the maximum output of 7500 occurs when 320 hours of labor are used. A manager, however, may want to know how many hours of labor to use in order to maximize the output per hour of labor. For any point on the curve, the y-coordinate measures the output and the x-coordinate measures the hours of labor, so the y-coordinate divided by the x-coordinate gives the output per hour of labor. This quotient is also the slope of the line through the origin and the point on the curve. Therefore, to maximize the output per hour of labor, we need to find where this slope is greatest. As shown in Figure 6, this occurs when approximately 270 hours of labor are used. Notice that this is also where the line from the origin to the curve is tangent to the curve. Another way of looking at this is to say that the point on the curve where the tangent line passes through the origin is the point that maximizes the output per hour of labor.

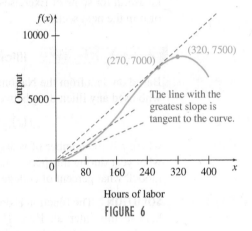

FIGURE 6

We can show that, in general, when $y = f(x)$ represents the output as a function of input, the maximum output per unit input occurs when the line from the origin to a point on the graph of the function is tangent to the function. Our goal is to maximize

$$g(x) = \frac{\text{output}}{\text{input}} = \frac{f(x)}{x}.$$

Taking the derivative and setting it equal to 0 gives

$$g'(x) = \frac{xf'(x) - f(x)}{x^2} = 0$$

$$xf'(x) = f(x)$$

$$f'(x) = \frac{f(x)}{x}.$$

Notice that $f'(x)$ gives the slope of the tangent line at the point, and $f(x)/x$ gives the slope of the line from the origin to the point. When these are equal, as in Figure 6, the output per input is maximized. In other examples, the point on the curve where the tangent line passes through the origin gives a minimum. For a life science example of this, see Exercise 54 in Section 3.4 on the Definition of the Derivative.

# 6.1 EXERCISES

**Find the locations of any absolute extrema for the functions with graphs as follows.**

**1.**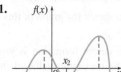

**2.**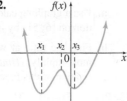

**3.**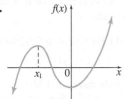

**4.**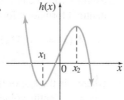

**5.**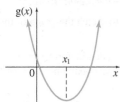

**6.**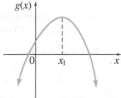

**7.**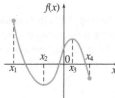

**8.**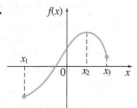

**9.** What is the difference between a relative extremum and an absolute extremum?

**10.** Can a relative extremum be an absolute extremum? Is a relative extremum necessarily an absolute extremum?

**Find the absolute extrema if they exist, as well as all values of $x$ where they occur, for each function, and specified domain. If you have one, use a graphing calculator to verify your answers.**

**11.** $f(x) = x^3 - 6x^2 + 9x - 8;\ \ [0, 5]$

**12.** $f(x) = x^3 - 3x^2 - 24x + 5;\ \ [-3, 6]$

**13.** $f(x) = \frac{1}{3}x^3 + \frac{3}{2}x^2 - 4x + 1;\ \ [-5, 2]$

**14.** $f(x) = \frac{1}{3}x^3 - \frac{1}{2}x^2 - 6x + 3;\ \ [-4, 4]$

**15.** $f(x) = x^4 - 18x^2 + 1;\ \ [-4, 4]$

**16.** $f(x) = x^4 - 32x^2 - 7;\ \ [-5, 6]$

**17.** $f(x) = \frac{1 - x}{3 + x};\ \ [0, 3]$

**18.** $f(x) = \frac{8 + x}{8 - x};\ \ [4, 6]$

**19.** $f(x) = \frac{x - 1}{x^2 + 1};\ \ [1, 5]$

**20.** $f(x) = \frac{x}{x^2 + 2};\ \ [0, 4]$

**21.** $f(x) = (x^2 - 4)^{1/3};\ \ [-2, 3]$

**22.** $f(x) = (x^2 - 16)^{2/3};\ \ [-5, 8]$

**23.** $f(x) = 5x^{2/3} + 2x^{5/3};\ \ [-2, 1]$

**24.** $f(x) = x + 3x^{2/3};\ \ [-10, 1]$

**25.** $f(x) = x^2 - 8\ln x;\ \ [1, 4]$

**26.** $f(x) = \frac{\ln x}{x^2};\ \ [1, 4]$

**27.** $f(x) = x + e^{-3x};\ \ [-1, 3]$

**28.** $f(x) = x^2 e^{-0.5x};\ \ [2, 5]$

**29.** $f(x) = \frac{x}{2} - \sin x;\ [0, \pi]$

**30.** $f(x) = \tan x - 2x;\ [0, \pi/3]$

**Graph each function on the indicated domain, and use the capabilities of your calculator to find the location and value of the absolute extrema.**

**31.** $f(x) = \frac{-5x^4 + 2x^3 + 3x^2 + 9}{x^4 - x^3 + x^2 + 7};\ \ [-1, 1]$

**32.** $f(x) = \frac{x^3 + 2x + 5}{x^4 + 3x^3 + 10};\ \ [-3, 0]$

**Find the absolute extrema if they exist, as well as all values of $x$ where they occur.**

**33.** $f(x) = 2x + \frac{8}{x^2} + 1,\ \ x > 0$

**34.** $f(x) = 12 - x - \frac{9}{x},\ \ x > 0$

**35.** $f(x) = -3x^4 + 8x^3 + 18x^2 + 2$

**36.** $f(x) = x^4 - 4x^3 + 4x^2 + 1$

**37.** $f(x) = \frac{x - 1}{x^2 + 2x + 6}$

**38.** $f(x) = \frac{x}{x^2 + 1}$

**39.** $f(x) = \frac{\ln x}{x^3}$

**40.** $f(x) = x \ln x$

**41.** Find the absolute maximum and minimum of $f(x) = 2x - 3x^{2/3}$ **(a)** on the interval $[-1, 0.5]$; **(b)** on the interval $[0.5, 2]$.

**42.** Let $f(x) = e^{-2x}$. For $x > 0$, let $P(x)$ be the perimeter of the rectangle with vertices $(0, 0)$, $(x, 0)$, $(x, f(x))$, and $(0, f(x))$. Which of the following statements is true? *Source: Society of Actuaries.*

**a.** The function $P$ has an absolute minimum but not an absolute maximum on the interval $(0, \infty)$.

**b.** The function $P$ has an absolute maximum but not an absolute minimum on the interval $(0, \infty)$.

**c.** The function $P$ has both an absolute minimum and an absolute maximum on the interval $(0, \infty)$.

**d.** The function $P$ has neither an absolute maximum nor an absolute minimum on the interval $(0, \infty)$, but the graph of the function $P$ does have an inflection point with positive $x$-coordinate.

e. The function $P$ has neither an absolute maximum nor an absolute minimum on the interval $(0, \infty)$, and the graph of the function $P$ does not have an inflection point with positive $x$-coordinate.

## LIFE SCIENCE APPLICATIONS

43. **Pollution** A marshy region used for agricultural drainage has become contaminated with selenium. It has been determined that flushing the area with clean water will reduce the selenium for a while, but it will then begin to build up again. A biologist has found that the percent of selenium in the soil $x$ months after the flushing begins is given by

$$f(x) = \frac{x^2 + 36}{2x}, \quad 1 \le x \le 12.$$

When will the selenium be reduced to a minimum? What is the minimum percent?

44. **Salmon Spawning** The number of salmon swimming upstream to spawn is approximated by

$$S(x) = -x^3 + 3x^2 + 360x + 5000, \quad 6 \le x \le 20,$$

where $x$ represents the temperature of the water in degrees Celsius. Find the water temperature that produces the maximum number of salmon swimming upstream.

45. **Molars** Researchers have determined that the crown length of first molars in fetuses is related to the postconception age of the tooth as

$$L(t) = -0.01t^2 + 0.788t - 7.048,$$

where $L(t)$ is the crown length (in millimeters) of the molar $t$ weeks after conception. Find the maximum length of the crown of first molars during weeks 22 through 28. ***Source: American Journal of Physical Anthropology.***

46. **Fungal Growth** Because of the time that many people spend indoors, there is a concern about the health risk of being exposed to harmful fungi that thrive in buildings. The risk appears to increase in damp environments. Researchers have discovered that by controlling both the temperature and the relative humidity in a building, the growth of the fungus A. *versicolor* can be limited. The relationship between temperature and relative humidity, which limits growth, can be described by

$$R(T) = -0.00007T^3 + 0.0401T^2 - 1.6572T + 97.086,$$
$$15 \le T \le 46,$$

where $R(T)$ is the relative humidity (in percent) and $T$ is the temperature (in degrees Celsius). Find the temperature at which the relative humidity is minimized. ***Source: Applied and Environmental Microbiology.***

47. **Dentin Growth** The growth of dentin in the molars of mice has been studied by researchers in Copenhagen. They determined that the growth curve that best fits dentinal formation for the second molar is

$$M(t) = -0.5035295 + 0.0229883t + 0.0108021t^2$$
$$- 0.0003139t^3 + 0.0000025t^4, \quad 6 \le t \le 48,$$

where $t$ is the age of the mouse (in days), and $M(t)$ is the cumulative dentin volume (in $10^{-1}$ mm$^3$). ***Source: Journal of Craniofacial Genetics and Developmental Biology.***

a. Use a graphing calculator to sketch the graph of this function on $[6, 51]$ by $[0, 4.5]$.

b. Find the time at which the dentin formation is growing most rapidly. (*Hint:* Find the maximum value of the derivative of this function.)

48. **Dentin Growth** The researchers mentioned in the previous exercise also determined that the growth curve that best fits dentinal formation for the third molar is

$$M(t) = -0.3347806 + 0.0236529t + 0.0012409t^2$$
$$- 0.0000207t^3, \quad 9 \le t \le 51,$$

where $t$ is the age of the mouse (in days), and $M(t)$ is the cumulative dentin volume (in $10^{-1}$ mm$^3$). ***Source: Journal of Craniofacial Genetics and Developmental Biology.***

a. Use a graphing calculator to sketch the graph of this function on $[9, 51]$ by $[0, 1.5]$.

b. Find the time at which the dentin formation is growing most rapidly. (*Hint:* Find the maximum value of the derivative of this function.)

c. Discuss whether this function should be used for mice that are older than 48 days.

49. **Human Skin Surface** The surface of the skin is made up of a network of intersecting lines that form polygons. Researchers have discovered a functional relationship between the age of a male and the number of polygons per area of skin according to

$$P(t) = 241.75 - 10.372t + 0.31247t^2 - 0.0044474t^3$$
$$+ 0.0000221195t^4, 0 \le t \le 95,$$

where $t$ is the age of the person (in years) and $P(t)$ is the number of polygons for a particular surface area of skin. ***Source: Gerontology.***

a. Use a graphing calculator to sketch the graph of $P(t)$ on $[0, 95]$ by $[0, 300]$.

b. Find the maximum and minimum number of polygons per area predicted by the model.

c. Discuss the accuracy of this model for older people.

50. **Satisfaction** Suppose some substance (such as a preferred food) gives satisfaction to an individual, but the substance requires effort to obtain, so that after a while the individual is no longer interested in expending more effort to obtain the substance. A mathematical model of this situation is given by

$$S = a \ln kx - bx,$$

where $S$ is the amount of satisfaction, $x$ is the amount of the substance, and $a$, $b$, and $k$ are constants. ***Source: Mathematical Biology of Social Behavior.*** Find the amount of the substance that gives the maximum amount of satisfaction.

# OTHER APPLICATIONS

**Area**  A piece of wire 12 ft long is cut into two pieces. (See the figure.) One piece is made into a circle and the other piece is made into a square. Let the piece of length $x$ be formed into a circle. We allow $x$ to equal 0 or 12, so all the wire may be used for the square or for the circle.

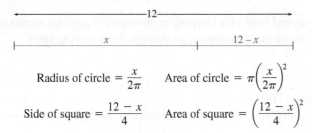

$$\text{Radius of circle} = \frac{x}{2\pi} \qquad \text{Area of circle} = \pi\left(\frac{x}{2\pi}\right)^2$$

$$\text{Side of square} = \frac{12 - x}{4} \qquad \text{Area of square} = \left(\frac{12 - x}{4}\right)^2$$

**51.** Where should the cut be made in order to minimize the sum of the areas enclosed by both figures?

**52.** Where should the cut be made in order to make the sum of the areas maximum? (*Hint:* Remember to use the endpoints of a domain when looking for absolute maxima and minima.)

**53.** For the solution to Exercise 51, show that the side of the square equals the diameter of the circle, that is, that the circle can be inscribed in the square.*

**54. Information Content**  Suppose dots and dashes are transmitted over a telegraph line so that dots occur a fraction $p$ of the time (where $0 < p < 1$) and dashes occur a fraction $1 - p$ of the time. The *information content* of the telegraph line is given by $I(p)$, where

$$I(p) = -p \ln p - (1 - p) \ln(1 - p).$$

  **a.** Show that $I'(p) = -\ln p + \ln(1 - p)$.

  **b.** Set $I'(p) = 0$ and find the value of $p$ that maximizes the information content.

  **c.** How might the result in part b be used?

**Cost**  Each graph gives the cost as a function of production level. Use the method of graphical optimization to estimate the production level that results in the minimum cost per item produced.

**55.**

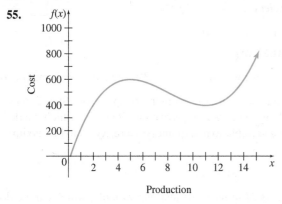

*For a generalization of this phenomenon, see Cade, Pat and Russell A. Gordon, "An Apothem Apparently Appears," *College Mathematics Journal,* Vol. 36, No. 1, Jan. 2005, pp. 52–55.

**56.**

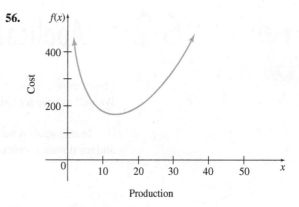

**Profit**  Each graph gives the profit as a function of production level. Use graphical optimization to estimate the production level that gives the maximum profit per item produced.

**57.**

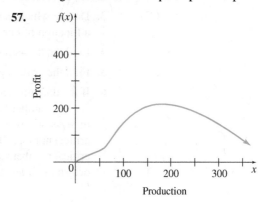

**58.**

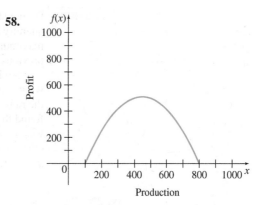

**▬ YOUR TURN ANSWERS**

1. Absolute maximum of about 0.977 occurs when $x = 2/5$ and absolute minimum of $-84$ when $x = 8$.

2. Absolute maximum, 148, occurs at $x = -4$. No absolute minimum.

# 6.2 Applications of Extrema

APPLY IT **How can a thirsty animal reach a spring in the shortest possible time?**
*We will use the techniques of calculus to answer this question in Example 2.*

In this section we give several examples showing applications of calculus to maximum and minimum problems. To solve these examples, go through the following steps.

## Solving an Applied Extrema Problem

1. Read the problem carefully. Make sure you understand what is given and what is unknown.
2. If possible, sketch a diagram. Label the various parts.
3. Decide which variable must be maximized or minimized. Express that variable as a function of *one* other variable.
4. Find the domain of the function.
5. Find the critical points for the function from Step 3.
6. If the domain is a closed interval, evaluate the function at the endpoints and at each critical number to see which yields the absolute maximum or minimum. If the domain is an open interval, apply the critical point theorem when there is only one critical number. If there is more than one critical number, evaluate the function at the critical numbers and find the limit as the endpoints of the interval are approached to determine if an absolute maximum or minimum exists at one of the critical points.

**CAUTION** Do not skip Step 6 in the preceding box. If a problem asks you to maximize a quantity and you find a critical point at Step 5, do not automatically assume the maximum occurs there, for it may occur at an endpoint, as in Exercise 52 of the previous section, or it may not exist at all.

An infamous case of such an error occurred in a 1945 study of "flying wing" aircraft designs similar to the Stealth bomber. In seeking to maximize the range of the aircraft (how far it can fly on a tank of fuel), the study's authors found that a critical point occurred when almost all of the volume of the plane was in the wing. They claimed that this critical point was a maximum. But another engineer later found that this critical point, in fact, *minimized* the range of the aircraft! *Source: Science.*

## EXAMPLE 1 Maximization

Find two nonnegative numbers $x$ and $y$ for which $2x + y = 30$, such that $xy^2$ is maximized.

**SOLUTION** Step 1, reading and understanding the problem, is up to you. Step 2 does not apply in this example; there is nothing to draw. We proceed to Step 3, in which we decide what is to be maximized and assign a variable to that quantity. Here, $xy^2$ is to be maximized, so let

$$M = xy^2.$$

According to Step 3, we must express $M$ in terms of just *one* variable, which can be done using the equation $2x + y = 30$ by solving for either $x$ or $y$. Solving for $y$ gives

$$2x + y = 30$$
$$y = 30 - 2x.$$

Substitute $30 - 2x$ for $y$ in the expression for $M$ to get

$$M = x(30 - 2x)^2$$
$$= x(900 - 120x + 4x^2)$$
$$= 900x - 120x^2 + 4x^3.$$

We are now ready for Step 4, when we find the domain of the function. Because of the nonnegativity requirement, $x$ must be at least 0. Since $y$ must also be at least 0, we require $30 - 2x \geq 0$, so $x \leq 15$. Thus $x$ is confined to the interval $[0, 15]$.

Moving on to Step 5, we find the critical points for $M$ by finding $dM/dx$, and then solving the equation $dM/dx = 0$ for $x$.

$$\frac{dM}{dx} = 900 - 240x + 12x^2 = 0$$

$$12(75 - 20x + x^2) = 0 \qquad \text{Factor out the 12.}$$

$$12(5 - x)(15 - x) = 0 \qquad \text{Factor the quadratic.}$$

$$x = 5 \quad \text{or} \quad x = 15$$

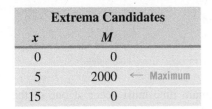

| Extrema Candidates | |
|---|---|
| $x$ | $M$ |
| 0 | 0 |
| 5 | 2000 ← Maximum |
| 15 | 0 |

Finally, at Step 6, we find $M$ for the critical numbers $x = 5$ and $x = 15$, as well as for $x = 0$, an endpoint of the domain. The other endpoint, $x = 15$, has already been included as a critical number. We see in the table that the maximum value of the function occurs when $x = 5$. Since $y = 30 - 2x = 30 - 2(5) = 20$, the values that maximize $xy^2$ are $x = 5$ and $y = 20$.  **TRY YOUR TURN 1**

**YOUR TURN 1** Find two nonnegative numbers $x$ and $y$ for which $x + 3y = 30$, such that $x^2y$ is maximized.

**NOTE** A critical point is only a candidate for an absolute maximum or minimum. The absolute maximum or minimum might occur at a different critical point or at an endpoint.

### EXAMPLE 2   Minimizing Time

A thirsty animal wants to get to a spring as soon as possible. It can get to the spring by traveling east along a clear path for 300 meters, and then north through the woods for 800 meters. The animal can run 160 meters per minute along the clear path, but only 70 meters per minute through the woods. Running directly through the woods toward the spring minimizes the distance, but the animal will be going slowly the whole time. It could instead run 300 meters along the clear path before entering the woods, maximizing the total distance but minimizing the time in the woods. Perhaps the fastest route is a combination, as shown in Figure 7. Find the path that will get the animal to the spring in the minimum time. (Although it may seem odd for an animal to follow the path given by calculus, a mathematician has shown that a dog chasing a ball does exactly that. *Source: The College Math Journal.*)

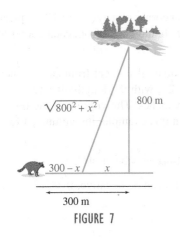

$\sqrt{800^2 + x^2}$  800 m

$300 - x$  $x$

300 m

**FIGURE 7**

**APPLY IT**

**SOLUTION**   As in Example 1, the first step is to read and understand the problem. If the statement of the problem is not clear to you, go back and reread it until you understand it before moving on.

We have already started Step 2 by providing Figure 7. Let $x$ be the distance shown in Figure 7, so the distance the animal runs on the clear path is $300 - x$. By the Pythagorean theorem, the distance it runs through the woods is $\sqrt{800^2 + x^2}$.

The first part of Step 3 is noting that we are trying to minimize the total amount of time, which is the sum of the time on the clear path and the time through the woods. We must express this time as a function of $x$. Since time = distance/speed, the total time is

$$T(x) = \frac{300 - x}{160} + \frac{\sqrt{800^2 + x^2}}{70}.$$

To complete Step 4, notice in this equation that $0 \leq x \leq 300$.

We now move to Step 5, in which we find the critical points by calculating the derivative and setting it equal to 0. Since $\sqrt{800^2 + x^2} = (800^2 + x^2)^{1/2}$,

$$T'(x) = -\frac{1}{160} + \frac{1}{70}\left(\frac{1}{2}\right)(800^2 + x^2)^{-1/2}(2x) = 0.$$

| Extrema Candidates | | |
|:---:|:---:|:---:|
| $x$ | $T(x)$ | |
| 0 | 13.30 | |
| 300 | 12.21 | ← Minimum |

$$\frac{x}{70\sqrt{800^2 + x^2}} = \frac{1}{160}$$  Cross multiply and divide by 10.

$$16x = 7\sqrt{800^2 + x^2}$$  Square both sides.

$$256x^2 = 49(800^2 + x^2) = 49 \cdot 800^2 + 49x^2$$

$$207x^2 = 49 \cdot 800^2$$  Subtract $49x^2$ from both sides.

$$x^2 = \frac{49 \cdot 800^2}{207}$$

$$x = \frac{7 \cdot 800}{\sqrt{207}} \approx 389$$  Take the square root of both sides.

**YOUR TURN 2** Suppose the animal in Example 2 can run only 40 m per minute through the woods. Find the path that will get it to the spring in the minimum time.

Since 389 is not in the interval $[0, 300]$, the minimum time must occur at one of the endpoints.

We now complete Step 6 by creating a table with $T(x)$ evaluated at the endpoints. We see from the table that the time is minimized when $x = 300$, that is, when the animal heads straight for the spring. **TRY YOUR TURN 2**

---

**EXAMPLE 3**  **Maximizing Volume**

An open box is to be made by cutting a square from each corner of a 12-in. by 12-in. piece of metal and then folding up the sides. What size square should be cut from each corner to produce a box of maximum volume?

**SOLUTION**  Let $x$ represent the length of a side of the square that is cut from each corner, as shown in Figure 8(a). The width of the box is $12 - 2x$, with the length also $12 - 2x$. As shown in Figure 8(b), the depth of the box will be $x$ inches. The volume of the box is given by the product of the length, width, and height. In this example, the volume, $V(x)$, depends on $x$:

$$V(x) = x(12 - 2x)(12 - 2x) = 144x - 48x^2 + 4x^3.$$

Clearly, $0 \le x$, and since neither the length nor the width can be negative, $0 \le 12 - 2x$, so $x \le 6$. Thus, the domain of $V$ is the interval $[0, 6]$.

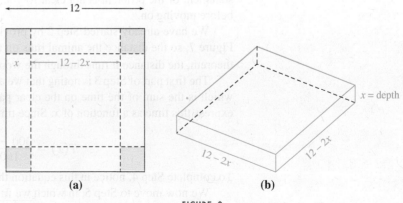

(a)                    (b)

**FIGURE 8**

| Extrema Candidates | | |
|---|---|---|
| $x$ | $V(x)$ | |
| 0 | 0 | |
| 2 | 128 | ← Maximum |
| 6 | 0 | |

**YOUR TURN 3** Repeat Example 3 using an 8-m by 8-m piece of metal.

The derivative is $V'(x) = 144 - 96x + 12x^2$. Set this derivative equal to 0.

$$12x^2 - 96x + 144 = 0$$
$$12(x^2 - 8x + 12) = 0$$
$$12(x - 2)(x - 6) = 0$$
$$x - 2 = 0 \quad \text{or} \quad x - 6 = 0$$
$$x = 2 \qquad\qquad x = 6$$

Find $V(x)$ for $x$ equal to 0, 2, and 6 to find the depth that will maximize the volume. The table indicates that the box will have maximum volume when $x = 2$ and that the maximum volume will be 128 in$^3$.   **TRY YOUR TURN 3**

---

**EXAMPLE 4**   Minimizing Area

A pharmaceutical company wants to manufacture cylindrical aluminum cans with a volume of 1000 cm$^3$ (1 liter). What should the radius and height of the can be to minimize the amount of aluminum used?

**SOLUTION**   The two variables in this problem are the radius and the height of the can, which we shall label $r$ and $h$, as in Figure 9. Minimizing the amount of aluminum used requires minimizing the surface area of the can, which we will designate $S$. The surface area consists of a top and a bottom, each of which is a circle with an area $\pi r^2$, plus the side. If the side were sliced vertically and unrolled, it would form a rectangle with height $h$ and width equal to the circumference of the can, which is $2\pi r$. Thus the surface area is given by

$$S = 2\pi r^2 + 2\pi rh.$$

The right side of the equation involves two variables. We need to get a function of a single variable. We can do this by using the information about the volume of the can:

$$V = \pi r^2 h = 1000.$$

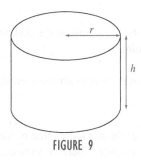

**FIGURE 9**

(Here we have used the formula for the volume of a cylinder.) Solve this for $h$:

$$h = \frac{1000}{\pi r^2}.$$

(Solving for $r$ would have involved a square root and a more complicated function.)
We now substitute this expression for $h$ into the equation for $S$ to get

$$S = 2\pi r^2 + 2\pi r\frac{1000}{\pi r^2} = 2\pi r^2 + \frac{2000}{r}.$$

There are no restrictions on $r$ other than that it be a positive number, so the domain of $S$ is $(0, \infty)$.

Find the critical points for $S$ by finding $dS/dr$, then solving the equation $dS/dr = 0$ for $r$.

$$\frac{dS}{dr} = 4\pi r - \frac{2000}{r^2} = 0$$
$$4\pi r^3 = 2000$$
$$r^3 = \frac{500}{\pi}$$

Take the cube root of both sides to get

$$r = \left(\frac{500}{\pi}\right)^{1/3} \approx 5.419$$

centimeters. Substitute this expression into the equation for $h$ to get

$$h = \frac{1000}{\pi(5.419)^2} \approx 10.84$$

centimeters. Notice that the height of the can is twice its radius.

There are several ways to carry out Step 6 to verify that we have found the minimum. Because there is only one critical number, the critical point theorem applies.

**Method 1**
**Critical Point Theorem**
**with First Derivative Test**

Verify that when $r < 5.419$, then $dS/dr < 0$, and when $r > 5.419$, then $dS/dr > 0$. Since the function is decreasing before 5.419 and increasing after 5.419, there must be a relative minimum at $r = 5.419$ cm. By the critical point theorem, there is an absolute minimum there.

**Method 2**
**Critical Point Theorem**
**with Second Derivative Test**

We could also use the critical point theorem with the second derivative test.

$$\frac{d^2S}{dr^2} = 4\pi + \frac{4000}{r^3}$$

Notice that for positive $r$, the second derivative is always positive, so there is a relative minimum at $r = 5.419$ cm. By the critical point theorem, there is an absolute minimum there.

**Method 3**
**Limits at Endpoints**

We could also find the limit as the endpoints are approached.

$$\lim_{r \to 0} S = \lim_{r \to \infty} S = \infty$$

The surface area becomes arbitrarily large as $r$ approaches the endpoints of the domain, so the absolute minimum surface area must be at the critical point.

**YOUR TURN 4** Repeat Example 4 if the volume is to be 500 cm³.

The graphing calculator screen in Figure 10 confirms that there is an absolute minimum at $r = 5.419$ cm.    **TRY YOUR TURN 4**

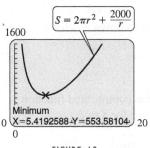

FIGURE 10

Notice that if the previous example had asked for the height and radius that maximize the amount of aluminum used, the problem would have no answer. There is no maximum for a function that can be made arbitrarily large.

**EXAMPLE 5   Volume**

The owners of a boarding stable wish to construct a watering trough from which the horses can drink. They have 9-ft by 9-ft pieces of metal, which they can bend into three parts to make the bottom and sides of the trough, as shown in Figure 11(a). They can then weld pieces of scrap metal to the ends to form a trough. At what angle $\theta$ should they bend the metal to create the largest possible volume? What is the largest possible volume?

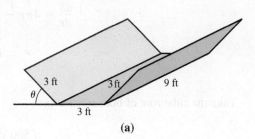
(a)

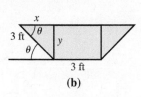
(b)

FIGURE 11

**SOLUTION**  The volume of the trough is its length, 9 ft, times the cross-sectional area. (This is true for any shape with parallel ends and straight sides. For example, the volume of a cylinder is the height times the area of the circular ends, or $h\pi r^2$.) Notice from Figure 11(b) that the cross-sectional area can be broken up into a rectangle with base 3 and height $y$, and two triangles, each with base $x$, height $y$, and angle $\theta$. Since $x/3 = \cos\theta$, we have $x = 3\cos\theta$. Similarly, since $y/3 = \sin\theta$, we have $y = 3\sin\theta$. Therefore,

$$A = 3y + 2 \cdot \frac{1}{2}xy$$
$$= 3(3\sin\theta) + (3\cos\theta)(3\sin\theta)$$
$$= 9(\sin\theta + \cos\theta\sin\theta).$$

Because the volume is the length times the area,

$$V = 9A = 9 \cdot 9(\sin\theta + \cos\theta\sin\theta)$$
$$= 81(\sin\theta + \cos\theta\sin\theta),$$

where $0 \le \theta \le \pi/2$. To find the maximum volume, set the derivative equal to 0.

$$\frac{dV}{d\theta} = 81[\cos\theta + \cos\theta\cos\theta + \sin\theta(-\sin\theta)] \qquad \text{Product rule}$$
$$= 81(\cos\theta + \cos^2\theta - \sin^2\theta)$$
$$= 81[\cos\theta + \cos^2\theta - (1 - \cos^2\theta)] \qquad \text{Use } \sin^2 x + \cos^2 x = 1.$$
$$= 81(2\cos^2\theta + \cos\theta - 1) \qquad \text{Rearrange terms.}$$
$$= 81(2\cos\theta - 1)(\cos\theta + 1) \qquad \text{Factor.}$$

Notice in the third line of the above derivation that we used a trigonometric identity to put the expression entirely in terms of $\cos\theta$. To make $dV/d\theta = 0$, set either factor equal to 0.

$$2\cos\theta - 1 = 0 \qquad \cos\theta + 1 = 0$$
$$\cos\theta = \frac{1}{2} \qquad \cos\theta = -1$$
$$\theta = \frac{\pi}{3} \qquad \text{No solution on } [0, \pi/2]$$

The only value of 0 for which $dV/d\theta = 0$ is $\theta = \pi/3$, where $\cos\theta = 1/2$, $\sin\theta = \sqrt{3}/2$, and

$$V = 81\left(\frac{\sqrt{3}}{2} + \frac{1}{2} \cdot \frac{\sqrt{3}}{2}\right)$$
$$= \frac{243\sqrt{3}}{4} \approx 105.2 \text{ ft}^3$$

We must also check the endpoints. At $\theta = 0$, we have $V = 0$, while at $\theta = \pi/2$, we have $V = 81$. Thus the maximum volume of about about 105.2 ft$^3$ is achieved with $\theta = \pi/3$.

## Maximum Sustainable Harvest

For most living things, reproduction is *seasonal* —it can take place only at selected times of the year. Large whales, for example, reproduce every two years during a relatively short time span of about two months. Shown on the time axis in Figure 12 are the reproductive periods. Let $S$ = number of adults present during the reproductive period and let $R$ = number of adults that return the next season to reproduce. ***Source: Mathematics for the Biosciences.***

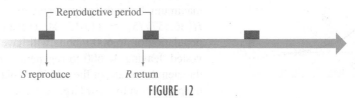

FIGURE 12

If we find a relationship between $R$ and $S$, $R = f(S)$, then we have formed a **spawner-recruit** function or **parent-progeny** function. These functions are notoriously hard to develop because of the difficulty of obtaining accurate counts and because of the many hypotheses that can be made about the life stages. We will simply suppose that the function $f$ takes various forms.

If $R > S$, we can presumably harvest

$$H = R - S = f(S) - S$$

individuals, leaving $S$ to reproduce. Next season, $R = f(S)$ will return and the harvesting process can be repeated, as shown in Figure 13.

Let $S_0$ be the number of spawners that will allow as large a harvest as possible without threatening the population with extinction. Then $H(S_0)$ is called the **maximum sustainable harvest**. (See the Extended Application in Chapter 4 for more about the maximum sustainable harvest.)

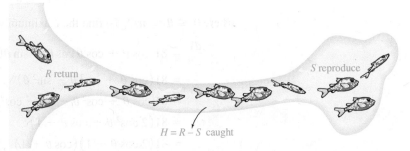

R return    S reproduce

$H = R - S$ caught

FIGURE 13

### EXAMPLE 6 Maximum Sustainable Harvest

Suppose the spawner-recruit function for Idaho rabbits is $f(S) = 2.17\sqrt{S}\ln(S + 1)$, where $S$ is measured in thousands of rabbits. Find $S_0$ and the maximum sustainable harvest, $H(S_0)$.

**SOLUTION**  $S_0$ is the value of $S$ that maximizes $H$. Since

$$H(S) = f(S) - S$$
$$= 2.17\sqrt{S}\ln(S + 1) - S,$$
$$H'(S) = 2.17\left(\frac{\ln(S + 1)}{2\sqrt{S}} + \frac{\sqrt{S}}{S + 1}\right) - 1.$$

Now we want to set this derivative equal to 0 and solve for $S$.

$$0 = 2.17\left(\frac{\ln(S + 1)}{2\sqrt{S}} + \frac{\sqrt{S}}{S + 1}\right) - 1.$$

This equation cannot be solved analytically, so we will graph $H'(S)$ with a graphing calculator and find any $S$-values where $H'(S)$ is 0. (An alternative approach is to use the equation solver some graphing calculators have.) The graph with the value where $H'(S)$ is 0 is shown in Figure 14.

From the graph we see that $H'(S) = 0$ when $S = 36.557775$, so the number of rabbits needed to sustain the population is about 36,600. A graph of $H$ will show that this is a maximum. From the graph, using the capability of the calculator, we find that the harvest is $H(36.557775) \approx 11.015504$. These results indicate that after one reproductive season, a population of 36,600 rabbits will have increased to 47,600. Of these, 11,000 may be harvested, leaving 36,600 to regenerate the population. Any harvest larger than 11,000 will threaten the future of the rabbit population, while a harvest smaller than 11,000 will allow the population to grow larger each season. Thus 11,000 is the maximum sustainable harvest for this population.

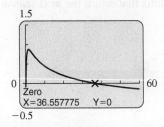

1.5

0    60

Zero
X=36.557775    Y=0

−0.5

FIGURE 14

# 6.2 EXERCISES

**In Exercises 1–4, use the steps shown in Exercise 1 to find non-negative numbers $x$ and $y$ that satisfy the given requirements. Give the optimum value of the indicated expression.**

1. $x + y = 180$ and the product $P = xy$ is as large as possible.

   a. Solve $x + y = 180$ for $y$.

   b. Substitute the result from part a into $P = xy$, the equation for the variable that is to be maximized.

   c. Find the domain of the function $P$ found in part b.

   d. Find $dP/dx$. Solve the equation $dP/dx = 0$.

   e. Evaluate $P$ at any solutions found in part d, as well as at the endpoints of the domain found in part c.

   f. Give the maximum value of $P$, as well as the two numbers $x$ and $y$ whose product is that value.

2. The sum of $x$ and $y$ is 140 and the sum of the squares of $x$ and $y$ is minimized.

3. $x + y = 90$ and $x^2y$ is maximized.

4. $x + y = 105$ and $xy^2$ is maximized.

## LIFE SCIENCE APPLICATIONS

5. **Disease** Epidemiologists have found a new communicable disease running rampant in College Station, Texas. They estimate that $t$ days after the disease is first observed in the community, the percent of the population infected by the disease is approximated by

$$p(t) = \frac{20t^3 - t^4}{1000}$$

for $0 \le t \le 20$.

   a. After how many days is the percent of the population infected a maximum?

   b. What is the maximum percent of the population infected?

6. **Disease** Another disease hits the chronically ill town of College Station, Texas. This time the percent of the population infected by the disease $t$ days after it hits town is approximated by $p(t) = 10te^{-t/8}$ for $0 \le t \le 40$.

   a. After how many days is the percent of the population infected a maximum?

   b. What is the maximum percent of the population infected?

**Maximum Sustainable Harvest** **Find the maximum sustainable harvest in Exercises 7 and 8. See Example 6.**

7. $f(S) = 12S^{0.25}$

8. $f(S) = \dfrac{25S}{S + 2}$

9. **Pollution** A lake polluted by bacteria is treated with an antibacterial chemical. After $t$ days, the number $N$ of bacteria per milliliter of water is approximated by

$$N(t) = 20\left(\frac{t}{12} - \ln\left(\frac{t}{12}\right)\right) + 30$$

for $1 \le t \le 15$.

   a. When during this time will the number of bacteria be a minimum?

   b. What is the minimum number of bacteria during this time?

   c. When during this time will the number of bacteria be a maximum?

   d. What is the maximum number of bacteria during this time?

10. **Maximum Sustainable Harvest** The population of salmon next year is given by $f(S) = Se^{r(1 - S/P)}$, where $S$ is this year's salmon population, $P$ is the equilibrium population, and $r$ is a constant that depends upon how fast the population grows. The number of salmon that can be fished next year while keeping the population the same is $H(S) = f(S) - S$. The maximum value of $H(S)$ is the maximum sustainable harvest. *Source: Journal of the Fisheries Research Board of Canada.*

   a. Show that the maximum sustainable harvest occurs when $f'(S) = 1$. (*Hint:* To maximize, set $H'(S) = 0$.)

   b. Let the value of $S$ found in part a be denoted by $S_0$. Show that the maximum sustainable harvest is given by

$$S_0\left(\frac{1}{1 - rS_0/P} - 1\right).$$

   (*Hint:* Set $f'(S_0) = 1$ and solve for $e^{r(1 - S_0/P)}$. Then find $H(S_0)$ and substitute the expression for $e^{r(1 - S_0/P)}$.)

**Maximum Sustainable Harvest** **In Exercises 11 and 12, refer to Exercise 10. Find $f'(S_0)$ and solve the equation $f'(S_0) = 1$, using a calculator to find the intersection of the graphs of $f'(S_0)$ and $y = 1$.**

11. Find the maximum sustainable harvest if $r = 0.1$ and $P = 100$.

12. Find the maximum sustainable harvest if $r = 0.4$ and $P = 500$.

13. **Pigeon Flight** Homing pigeons avoid flying over large bodies of water, preferring to fly around them instead. (One possible explanation is the fact that extra energy is required to fly over water because air pressure drops over water in the daytime.) Assume that a pigeon released from a boat 1 mile from the shore of a lake (point $B$ in the figure) flies first to point $P$ on the shore and then along the straight edge of the lake to reach its home at $L$. If $L$ is 2 miles from point $A$, the point on the shore closest to the boat, and if a pigeon needs 4/3 as much energy per mile to fly over water as over land, find the location of point $P$, which minimizes energy used.

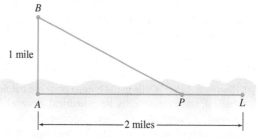

14. **Pigeon Flight** Repeat Exercise 13, but assume a pigeon needs 10/9 as much energy to fly over water as over land.

15. **Harvesting Cod** A recent article described the population $f(S)$ of cod in the North Sea next year as a function of this year's population $S$ (in thousands of tons) by various mathematical models.

$$\text{Shepherd:} \quad f(S) = \frac{aS}{1 + (S/b)^c};$$

$$\text{Ricker:} \quad f(S) = aSe^{-bS};$$

$$\text{Beverton-Holt:} \quad f(S) = \frac{aS}{1 + (S/b)},$$

where $a$, $b$, and $c$ are constants. *Source: Nature.*

   a. Find a replacement of variables in the Ricker model above that will make it the same as another form of the Ricker model described in Exercise 10 of this section, $f(S) = Se^{r(1-S/P)}$.

   b. Find $f'(S)$ for all three models.

   c. Find $f'(0)$ for all three models. From your answer, describe in words the geometric meaning of the constant $a$.

   d. The values of $a$, $b$, and $c$ reported in the article for the Shepherd model are 3.026, 248.72, and 3.24, respectively. Find the value of this year's population that maximizes next year's population using the Shepherd model.

   e. The values of $a$ and $b$ reported in the article for the Ricker model are 4.151 and 0.0039, respectively. Find the value of this year's population that maximizes next year's population using the Ricker model.

   f. Explain why, for the Beverton–Holt model, there is no value of this year's population that maximizes next year's population.

16. **Bird Migration** Suppose a migrating bird flies at a velocity $v$, and suppose the amount of time the bird can fly depends on its velocity according to the function $T(v)$. *Source: A Concrete Approach to Mathematical Modelling.*

   a. If $E$ is the bird's initial energy, then the bird's effective power is given by $kE/T$, where $k$ is the fraction of the power that can be converted into mechanical energy. According to principles of aerodynamics,

$$\frac{kE}{T} = aSv^3 + I,$$

   where $a$ is a constant, $S$ is the wind speed, and $I$ is the induced power, or rate of working against gravity. Using this result and the fact that distance is velocity multiplied by time, show that the distance that the bird can fly is given by

$$D(v) = \frac{kEv}{aSv^3 + I}.$$

   b. Show that the migrating bird can fly a maximum distance by flying at a velocity

$$v = \left(\frac{I}{2aS}\right)^{1/3}.$$

17. **Cylindrical Cells** Cells in the human body are not necessarily spherical; some cancer cells are cylindrical. *Source: American Journal of Surgical Pathology.* Suppose that the volume of such a cell is $65 \ \mu m^3$. Find the radius and height of the cell with the minimum surface area.

18. **Honeycomb** Honey bees form combs with a hexagonal structure, in which the ends of each cell form a three-sided pyramid. For many years it was wondered why the sides come together at an angle of about $109°$, until it was discovered that this minimized the amount of wax being used. *Source: On Growth and Form.* The surface area of each cell is

$$A(\theta) = H + 3s\left(2h - \frac{1}{2}s \cot\theta\right) + \frac{3s^2\sqrt{3}}{2\sin\theta},$$

where $\theta$ is the angle between the central axis of the pyramid and one of its sides, $H$ is the area of the hexagonal base (a constant as far as $\theta$ is concerned), $s$ is the side length of the pyramid, and $h$ is the height of the hexagonal structure.

   a. Show that the angle $\theta$ that minimizes the area on the interval $[0, \pi/2]$ occurs when $\cos\theta = 1/\sqrt{3}$ by finding where $A'(\theta) = 0$.

   b. Use the second derivative to verify that your answer from part a is a minimum.

   c. The angle between two sides of the pyramid is $2\theta$. Verify that this is approximately $109°$.

19. **Biomechanics** Researchers have determined that the distance that a shot-putter can throw a shot depends on the height, angle, initial velocity, and release angle of the shot. For some shot-putters, the researchers determined that the distance a shot will travel can be estimated by

$$d(\theta) = 5.1\sin(2\theta)\left[1 + \sqrt{1 + \frac{0.41}{\sin^2\theta}}\right],$$

where $\theta$ is the release angle (in degrees) and $d(\theta)$ is the distance the shot travels (in m). *Source: Journal of Sports Sciences.* Determine the release angle that maximizes the distance traveled by the shot and the corresponding maximum distance.

## OTHER APPLICATIONS

20. **Area** A campground owner has 1400 m of fencing. He wants to enclose a rectangular field bordering a river, with no fencing needed along the river. (See the sketch.) Let $x$ represent the width of the field.

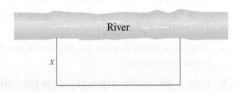

   a. Write an expression for the length of the field.

   b. Find the area of the field (area = length × width).

   c. Find the value of $x$ leading to the maximum area.

   d. Find the maximum area.

21. **Area** Find the dimensions of the rectangular field of maximum area that can be made from 300 m of fencing material. (This fence has four sides.)

**22. Area** An ecologist is conducting a research project on breeding pheasants in captivity. She first must construct suitable pens. She wants a rectangular area with two additional fences across its width, as shown in the sketch. Find the maximum area she can enclose with 3600 m of fencing.

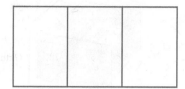

**23. Area** A farmer is constructing a rectangular pen with one additional fence across its width. Find the maximum area that can be enclosed with 2400 m of fencing.

**24. Cost with Fixed Area** A fence must be built in a large field to enclose a rectangular area of 25,600 m². One side of the area is bounded by an existing fence; no fence is needed there. Material for the fence costs $3 per meter for the two ends and $1.50 per meter for the side opposite the existing fence. Find the cost of the least expensive fence.

**25. Cost with Fixed Area** A fence must be built to enclose a rectangular area of 20,000 ft². Fencing material costs $2.50 per foot for the two sides facing north and south and $3.20 per foot for the other two sides. Find the cost of the least expensive fence.

**26. Packaging Design** An exercise equipment manufacturing firm needs to design an open-topped box with a square base. The box must hold 32 in³. Find the dimensions of the box that can be built with the minimum amount of materials. (See the figure.)

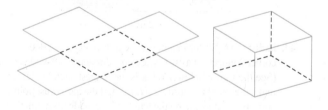

**27. Packaging Design** A company wishes to manufacture a box with a volume of 36 ft³ that is open on top and is twice as long as it is wide. Find the dimensions of the box produced from the minimum amount of material.

**28. Container Design** An open box will be made by cutting a square from each corner of a 3-ft by 8-ft piece of cardboard and then folding up the sides. What size square should be cut from each corner in order to produce a box of maximum volume?

**29. Container Design** Consider the problem of cutting corners out of a rectangle and folding up the sides to make a box. Specific examples of this problem are discussed in Example 3 and Exercise 28.

  **a.** In the solution to Example 3, compare the area of the base of the box with the area of the walls.

  **b.** Repeat part a for the solution to Exercise 28.

  **c.** Make a conjecture about the area of the base compared with the area of the walls for the box with the maximum volume.

**30. Packaging Cost** A closed box with a square base is to have a volume of 16,000 cm³. The material for the top and bottom of the box costs $3 per square centimeter, while the material for the sides costs $1.50 per square centimeter. Find the dimensions of the box that will lead to the minimum total cost. What is the minimum total cost?

**31. Use of Materials** A mathematics book is to contain 36 in² of printed matter per page, with margins of 1 in. along the sides and $1\frac{1}{2}$ in. along the top and bottom. Find the dimensions of the page that will require the minimum amount of paper. (See the figure.)

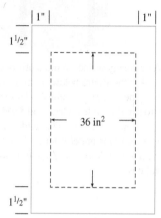

**32. Can Design**

  **a.** For the can problem in Example 4, the minimum surface area required that the height be twice the radius. Show that this is true for a can of arbitrary volume *V*.

  **b.** Do many cans in grocery stores have a height that is twice the radius? If not, discuss why this may be so.

**33. Container Design** Your company needs to design cylindrical metal containers with a volume of 16 cubic feet. The top and bottom will be made of a sturdy material that costs $2 per square foot, while the material for the sides costs $1 per square foot. Find the radius, height, and cost of the least expensive container.

**In Exercises 34–36, use a graphing calculator to determine where the derivative is equal to zero.**

**34. Can Design** Modify the can problem in Example 4 so the cost must be minimized. Assume that aluminum costs 3¢ per square centimeter, and that there is an additional cost of 2¢ per cm times the perimeter of the top, and a similar cost for the bottom, to seal the top and bottom of the can to the side.

**35. Can Design** In this modification of the can problem in Example 4, the cost must be minimized. Assume that aluminum costs 3¢ per square centimeter, and that there is an additional cost of 1¢ per cm times the height of the can to make a vertical seam on the side.

**36. Can Design** This problem is a combination of Exercises 34 and 35. We will again minimize the cost of the can, assuming that aluminum costs 3¢ per square centimeter. In addition, there is a cost of 2¢ per cm to seal the top and bottom of the can to the side, plus 1¢ per cm to make a vertical seam.

**37. Packaging Design** A cylindrical box will be tied up with ribbon as shown in the figure. The longest piece of ribbon available is 130 cm long, and 10 cm of that are required for the bow. Find the radius and height of the box with the largest possible volume.

**38. Cost** A company wishes to run a utility cable from point $A$ on the shore (see the figure below) to an installation at point $B$ on the island. The island is 6 miles from the shore. It costs $400 per mile to run the cable on land and $500 per mile underwater. Assume that the cable starts at $A$ and runs along the shoreline, then angles and runs underwater to the island. Find the point at which the line should begin to angle in order to yield the minimum total cost.

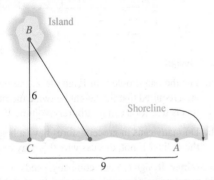

**39. Cost** Repeat Exercise 38, but make point $A$ 7 miles from point $C$.

**40. Travel Time** A hunter is at a point along a river bank. He wants to get to his cabin, located 3 miles north and 8 miles west. (See the figure.) He can travel 5 mph along the river but only 2 mph on this very rocky land. How far upriver should he go in order to reach the cabin in minimum time?

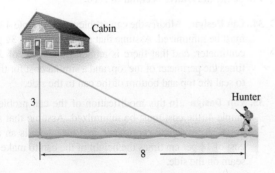

**41. Travel Time** Repeat Exercise 40, but assume the cabin is 19 miles north and 8 miles west.

**42. Postal Regulations** The U.S. Postal Service stipulates that any boxes sent through the mail must have a length plus girth totaling no more than 108 in. (See the figure.) Find the dimensions of the box with maximum volume that can be sent through the U.S. mail, assuming that the width and the height of the box are equal. *Source: U.S. Postal Service.*

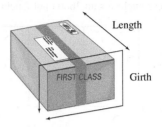

**43. Ladder** A thief tries to enter a building by placing a ladder over a 9-ft-high fence so it rests against the building, which is 2 ft back from the fence. (See the figure.) What is the length of the shortest ladder that can be used? (*Hint:* Let $\theta$ be the angle between the ladder and the ground. Express the length of the ladder in terms of $\theta$, and then find the value of $\theta$ that minimizes the length of the ladder.)

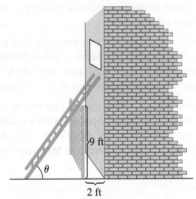

**44. Ladder** A janitor in a hospital needs to carry a ladder around a corner connecting a 10-ft-wide corridor and a 5-ft-wide corridor. (See the figure below.) What is the longest such ladder that can make it around the corner? (*Hint:* Find the narrowest point in the corridor by minimizing the length of the ladder as a function of $\theta$, the angle the ladder makes with the 5-ft-wide corridor.)

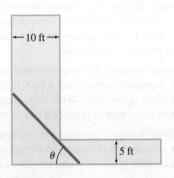

**YOUR TURN ANSWERS**

1. $x = 20$ and $y = 10/3$
2. Go 93 m along the trail and then head into the woods.
3. Box will have maximum volume when $x = 4/3$ m and the maximum volume is 1024/27 m³.
4. Radius is 4.3 cm and height is 8.6 cm.

# 6.3 Implicit Differentiation

**APPLY IT** How fast are bacteria growing in a sausage?
*We will answer this question in Example 4.*

In almost all of the examples and applications so far, all functions have been defined in the form

$$y = f(x),$$

with $y$ given **explicitly** in terms of $x$, or as an **explicit function** of $x$. For example,

$$y = 3x - 2, \qquad y = x^2 + x + 6, \qquad \text{and} \qquad y = -x^3 + 2$$

are all explicit functions of $x$. The equation $4xy - 3x = 6$ can be expressed as an explicit function of $x$ by solving for $y$. This gives

$$4xy - 3x = 6$$
$$4xy = 3x + 6$$
$$y = \frac{3x + 6}{4x}.$$

On the other hand, some equations in $x$ and $y$ cannot be readily solved for $y$, and some equations cannot be solved for $y$ at all. For example, while it would be possible (but tedious) to use the quadratic formula to solve for $y$ in the equation $y^2 + 2yx + 4x^2 = 0$, it is not possible to solve for $y$ in the equation $y^5 + 8y^3 + 6y^2x^2 + 2yx^3 + 6 = 0$. In equations such as these last two, $y$ is said to be given **implicitly** in terms of $x$.

In such cases, it may still be possible to find the derivative $dy/dx$ by a process called **implicit differentiation**. In doing so, we assume that there exists some function or functions $f$, which we may or may not be able to find, such that $y = f(x)$ and $dy/dx$ exists. It is useful to use $dy/dx$ here rather than $f'(x)$ to make it clear which variable is independent and which is dependent.

---

**FOR REVIEW**

In Chapter 1, we pointed out that when $y$ is given as a function of $x$, $x$ is the independent variable and $y$ is the dependent variable. We later defined the derivative $dy/dx$ when $y$ is a function of $x$. In an equation such as $3xy + 4y^2 = 10$, either variable can be considered the independent variable. If a problem asks for $dy/dx$, consider $x$ the independent variable; if it asks for $dx/dy$, consider $y$ the independent variable. A similar rule holds when other variables are used.

---

**EXAMPLE 1**  Implicit Differentiation

Find $dy/dx$ if $3xy + 4y^2 = 10$.

**SOLUTION**  Differentiate with respect to $x$ on both sides of the equation.

$$3xy + 4y^2 = 10$$
$$\frac{d}{dx}(3xy + 4y^2) = \frac{d}{dx}(10) \qquad \textbf{(1)}$$

Now differentiate each term on the left side of the equation. Think of $3xy$ as the product $(3x)(y)$ and use the product rule and the chain rule. Since

$$\frac{d}{dx}(3x) = 3 \quad \text{and} \quad \frac{d}{dx}(y) = \frac{dy}{dx},$$

the derivative of $(3x)(y)$ is

$$(3x)\frac{dy}{dx} + (y)3 = 3x\frac{dy}{dx} + 3y.$$

To differentiate the second term, $4y^2$, use the chain rule, since $y$ is assumed to be some function of $x$.

$$\frac{d}{dx}(4y^2) = 4\overbrace{(2y^1)\frac{dy}{dx}}^{\text{Derivative of }y^2} = 8y\frac{dy}{dx}$$

On the right side of Equation (1), the derivative of 10 is 0. Taking the indicated derivatives in Equation (1) term by term gives

$$3x\frac{dy}{dx} + 3y + 8y\frac{dy}{dx} = 0.$$

Now solve this result for $dy/dx$.

$$(3x + 8y)\frac{dy}{dx} = -3y$$

$$\frac{dy}{dx} = \frac{-3y}{3x + 8y}$$

**YOUR TURN 1** Find $dy/dx$ if $x^2 + y^2 = xy$.

TRY YOUR TURN 1

**NOTE** Because we are treating $y$ as a function of $x$, notice that each time an expression has $y$ in it, we use the chain rule.

### EXAMPLE 2 Implicit Differentiation

Find $dy/dx$ for $x + \sqrt{x}\sqrt{y} = y^2$.

**SOLUTION** Take the derivative on both sides with respect to $x$.

$$\frac{d}{dx}(x + \sqrt{x}\sqrt{y}) = \frac{d}{dx}(y^2)$$

Since $\sqrt{x} \cdot \sqrt{y} = x^{1/2} \cdot y^{1/2}$, use the product rule and the chain rule as follows.

$$\overbrace{1}^{\substack{\text{Derivative}\\\text{of }x}} + \overbrace{x^{1/2}\left(\frac{1}{2}y^{-1/2} \cdot \frac{dy}{dx}\right) + y^{1/2}\left(\frac{1}{2}x^{-1/2}\right)}^{\substack{\text{Derivative}\\\text{of }x^{1/2}y^{1/2}}} = \overbrace{2y\frac{dy}{dx}}^{\substack{\text{Derivative}\\\text{of }y^2}}$$

$$1 + \frac{x^{1/2}}{2y^{1/2}} \cdot \frac{dy}{dx} + \frac{y^{1/2}}{2x^{1/2}} = 2y\frac{dy}{dx}$$

Multiply both sides by $2x^{1/2} \cdot y^{1/2}$.

$$2x^{1/2} \cdot y^{1/2} + x\frac{dy}{dx} + y = 4x^{1/2} \cdot y^{3/2} \cdot \frac{dy}{dx}$$

Combine terms and solve for $dy/dx$.

$$2x^{1/2} \cdot y^{1/2} + y = (4x^{1/2} \cdot y^{3/2} - x)\frac{dy}{dx}$$

$$\frac{dy}{dx} = \frac{2x^{1/2} \cdot y^{1/2} + y}{4x^{1/2} \cdot y^{3/2} - x}$$

$$= \frac{2\sqrt{x}\sqrt{y} + y}{4y\sqrt{x}\sqrt{y} - x}$$

**YOUR TURN 2** Find $dy/dx$ for $xe^y + x^2 = \ln y$.

TRY YOUR TURN 2

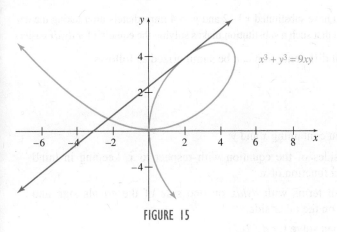

FIGURE 15

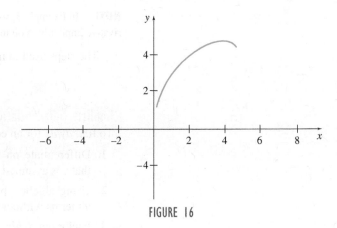

FIGURE 16

## EXAMPLE 3 Tangent Line

The graph of $x^3 + y^3 = 9xy$, shown in Figure 15, is a *folium of Descartes.** Find the equation of the tangent line at the point $(2, 4)$, shown in Figure 15.

**SOLUTION** Since this is not the graph of a function, $y$ is not a function of $x$, and $dy/dx$ is not defined. But if we restrict the curve to the vicinity of $(2, 4)$, as shown in Figure 16, the curve does represent the graph of a function, and we can calculate $dy/dx$ by implicit differentiation.

$$3x^2 + 3y^2 \cdot \frac{dy}{dx} = 9x\frac{dy}{dx} + 9y \qquad \text{Chain rule and product rule}$$

$$3y^2 \cdot \frac{dy}{dx} - 9x\frac{dy}{dx} = 9y - 3x^2 \qquad \text{Move all } dy/dx \text{ terms to the same side of the equation.}$$

$$\frac{dy}{dx}(3y^2 - 9x) = 9y - 3x^2 \qquad \text{Factor.}$$

$$\frac{dy}{dx} = \frac{9y - 3x^2}{3y^2 - 9x}$$

$$= \frac{3(3y - x^2)}{3(y^2 - 3x)} = \frac{3y - x^2}{y^2 - 3x}$$

To find the slope of the tangent line at the point $(2, 4)$, let $x = 2$ and $y = 4$. The slope is

$$m = \frac{3y - x^2}{y^2 - 3x} = \frac{3(4) - 2^2}{4^2 - 3(2)} = \frac{8}{10} = \frac{4}{5}.$$

The equation of the tangent line is then found by using the point-slope form of the equation of a line.

$$y - y_1 = m(x - x_1)$$

$$y - 4 = \frac{4}{5}(x - 2)$$

$$y - 4 = \frac{4}{5}x - \frac{8}{5}$$

$$y = \frac{4}{5}x + \frac{12}{5}$$

**YOUR TURN 3** The graph of $y^4 - x^4 - y^2 + x^2 = 0$ is called the *devil's curve*. Find the equation of the tangent line at the point $(1, 1)$.

The tangent line is graphed in Figure 15.     **TRY YOUR TURN 3**

*Information on this curve and others is available on the Famous Curves section of the MacTutor History of Mathematics Archive website at www-history.mcs.st-and.ac.uk/~history. See Exercises 37–40 for more curves.

**NOTE**   In Example 3, we could have substituted $x = 2$ and $y = 4$ immediately after taking the derivative implicitly. You may find that such a substitution makes solving the equation for $dy/dx$ easier.

The steps used in implicit differentiation can be summarized as follows.

---

**Implicit Differentiation**

To find $dy/dx$ for an equation containing $x$ and $y$:

1. Differentiate on both sides of the equation with respect to $x$, keeping in mind that $y$ is assumed to be a function of $x$.
2. Using algebra, place all terms with $dy/dx$ on one side of the equals sign and all terms without $dy/dx$ on the other side.
3. Factor out $dy/dx$, and then solve for $dy/dx$.

---

When an applied problem involves an equation that is not given in explicit form, implicit differentiation can be used to locate maxima and minima or to find rates of change.

**EXAMPLE 4**   **Bacteria in Sausage**

As we saw in Chapter 4, researchers in Italy have developed a modified Gompertz model to predict that the growth of *Enterococcus faecium* bacteria in bologna sausage at 32°C can be described by the equation

$$\ln\left(\frac{N(t)}{N_0}\right) = 9.8901 e^{-e^{2.54197 - 0.2167t}},$$

where $N_0$ is the number of bacteria present at the beginning of the experiment and $N(t)$ is the number of bacteria present at time $t$ (in hr). *Source: Applied and Environmental Microbiology.* Use implicit differentiation to determine $N'(3)$. Assume that $N_0 = 1000$.

APPLY IT   **SOLUTION**   We could first solve the above equation for $N(t)$ and then find $N'(t)$. It is also possible and convenient to use implicit differentiation to determine $N'(t)$. Using implicit differentiation, we have

$$\frac{\dfrac{N'(t)}{N_0}}{\dfrac{N(t)}{N_0}} = 9.8901\,(0.2167)\left(e^{2.54197 - 0.2167t}\right)\left(e^{-e^{2.54197 - 0.2167t}}\right).$$

Next, we simplify and solve for $N'(t)$.

$$N'(t) = N(t)(2.14318)\left(e^{2.54197 - 0.2167t}\right)\left(e^{-e^{2.54197 - 0.2167t}}\right)$$

To calculate $N'(3)$ we first need to find $N(3)$.

$$\ln\left(\frac{N(3)}{1000}\right) = 9.8901 e^{-e^{2.54197 - 0.2167(3)}} \approx 0.013034$$

Thus,

$$N(3) = 1000 e^{0.013034} \approx 1013.1189$$

and

$$N'(3) = 1013.1189(2.14318)\left(e^{2.54197 - 0.2167(3)}\right)\left(e^{-e^{2.54197 - 0.2167(3)}}\right) \approx 19.$$

Thus, we would expect that the number of bacteria will increase by approximately 19 during the third hour.

# 6.3 EXERCISES

**Find *dy/dx* by implicit differentiation for the following.**

**1.** $6x^2 + 5y^2 = 36$

**2.** $7x^2 - 4y^2 = 24$

**3.** $8x^2 - 10xy + 3y^2 = 26$

**4.** $7x^2 = 5y^2 + 4xy + 1$

**5.** $5x^3 = 3y^2 + 4y$

**6.** $3x^3 - 8y^2 = 10y$

**7.** $3x^2 = \dfrac{2 - y}{2 + y}$

**8.** $2y^2 = \dfrac{5 + x}{5 - x}$

**9.** $2\sqrt{x} + 4\sqrt{y} = 5y$

**10.** $4\sqrt{x} - 8\sqrt{y} = 6y^{3/2}$

**11.** $x^4y^3 + 4x^{3/2} = 6y^{3/2} + 5$

**12.** $(xy)^{4/3} + x^{1/3} = y^6 + 1$

**13.** $e^{x^2y} = 5x + 4y + 2$

**14.** $x^2e^y + y = x^3$

**15.** $x + \ln y = x^2y^3$

**16.** $y \ln x + 2 = x^{3/2}y^{5/2}$

**17.** $\sin(xy) = x$

**18.** $\tan y + x = 4$

**Find the equation of the tangent line at the given point on each curve.**

**19.** $x^2 + y^2 = 25$;  $(-3, 4)$

**20.** $x^2 + y^2 = 100$;  $(8, -6)$

**21.** $x^2y^2 = 1$;  $(-1, 1)$

**22.** $x^2y^3 = 8$;  $(-1, 2)$

**23.** $2y^2 - \sqrt{x} = 4$;  $(16, 2)$

**24.** $y + \dfrac{\sqrt{x}}{y} = 3$;  $(4, 2)$

**25.** $e^{x^2+y^2} = xe^{5y} - y^2e^{5x/2}$;  $(2, 1)$

**26.** $2xe^{xy} = e^{x^3} + ye^{x^2}$;  $(1, 1)$

**27.** $\ln(x + y) = x^3y^2 + \ln(x^2 + 2) - 4$;  $(1, 2)$

**28.** $\ln(x^2 + y^2) = \ln 5x + \dfrac{y}{x} - 2$;  $(1, 2)$

**29.** $x - \sin(\pi y) = 1$;  $(1, 0)$

**30.** $x^2 + \tan\left(\dfrac{\pi}{4}xy\right) = 2$;  $(1, 1)$

**In Exercises 31–36, find the equation of the tangent line at the given value of *x* on each curve.**

**31.** $y^3 + xy - y = 8x^4$;  $x = 1$

**32.** $y^3 + 2x^2y - 8y = x^3 + 19$;  $x = 2$

**33.** $y^3 + xy^2 + 1 = x + 2y^2$;  $x = 2$

**34.** $y^4(1 - x) + xy = 2$;  $x = 1$

**35.** $2y^3(x - 3) + x\sqrt{y} = 3$;  $x = 3$

**36.** $\dfrac{y}{18}(x^2 - 64) + x^{2/3}y^{1/3} = 12$;  $x = 8$

Information on curves in Exercises 37–40, as well as many other curves, is available on the Famous Curves section of the MacTutor History of Mathematics Archive website at www-history.mcs.st-and.ac.uk/~history.

**37.** The graph of $x^{2/3} + y^{2/3} = 2$, shown in the figure, is an *astroid*. Find the equation of the tangent line at the point $(1, 1)$.

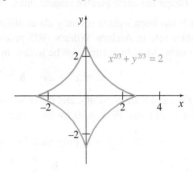

**38.** The graph of $3(x^2 + y^2)^2 = 25(x^2 - y^2)$, shown in the figure, is a *lemniscate of Bernoulli*. Find the equation of the tangent line at the point $(2, 1)$.

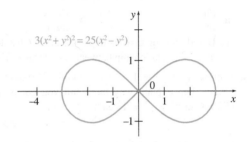

**39.** The graph of $y^2(x^2 + y^2) = 20x^2$, shown in the figure, is a *kappa curve*. Find the equation of the tangent line at the point $(1, 2)$.

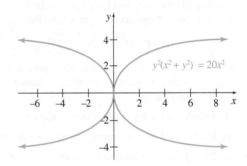

**40.** The graph of $2(x^2 + y^2)^2 = 25xy^2$, shown in the figure, is a *double folium*. Find the equation of the tangent line at the point $(2, 1)$.

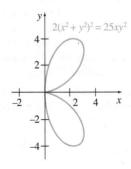

**41.** The graph of $x^2 + y^2 = 100$ is a circle having center at the origin and radius 10.

  **a.** Write the equations of the tangent lines at the points where $x = 6$.

  **b.** Graph the circle and the tangent lines.

**42.** Much has been written recently about elliptic curves because of their role in Andrew Wiles's 1995 proof of Fermat's Last Theorem. An elliptic curve can be written in the form

$$y^2 = x^3 + ax + b,$$

where $a$ and $b$ are constants, and the cubic function on the right has distinct roots. Find $dy/dx$ for this curve.

**43.** Let $\sqrt{u} + \sqrt{2v + 1} = 5$. Find each derivative.

  **a.** $\dfrac{du}{dv}$   **b.** $\dfrac{dv}{du}$

  **c.** Based on your answers to parts a and b, what do you notice about the relationship between $du/dv$ and $dv/du$?

**44.** Let $e^{u^2-v} - v = 1$. Find each derivative.

  **a.** $\dfrac{du}{dv}$   **b.** $\dfrac{dv}{du}$

  **c.** Based on your answers to parts a and b, what do you notice about the relationship between $du/dv$ and $dv/du$?

**45.** Suppose $x^2 + y^2 + 1 = 0$. Use implicit differentiation to find $dy/dx$. Then explain why the result you got is meaningless. (*Hint:* Can $x^2 + y^2 + 1$ equal 0?)

## LIFE SCIENCE APPLICATIONS

**46. Respiratory Rate** Researchers have found a correlation between respiratory rate and body mass in the first three years of life. This correlation can be expressed by the function

$$\log R(w) = 1.83 - 0.43 \log(w),$$

where $w$ is the body weight (in kilograms) and $R(w)$ is the respiratory rate (in breaths per minute). *Source: Archives of Disease in Children.*

  **a.** Find $R'(w)$ using implicit differentiation.

  **b.** Find $R'(w)$ by first solving the equation for $R(w)$.

  **c.** Discuss the two procedures. Is there a situation when you would want to use one method over another?

**47. Biochemical Reaction** A simple biochemical reaction with three molecules has solutions that oscillate toward a steady state when positive constants $a$ and $b$ are below the curve $b - a = (b + a)^3$. Find the largest possible value of $a$ for which the reaction has solutions that oscillate toward a steady state. (*Hint:* Find where $da/db = 0$. Derive values for $a + b$ and $a - b$, and then solve the equations in two unknowns.) *Source: Mathematical Biology.*

**48. Species** The relationship between the number of species in a genus ($x$) and the number of genera ($y$) comprising $x$ species is given by

$$xy^a = k,$$

where $a$ and $k$ are constants. Find $dy/dx$. *Source: Elements of Mathematical Biology.*

**49. Genotype Fitness** The fitness $\lambda$ of a genotype has been related to the fecundity $B$ of sexually mature individuals by the equation

$$\lambda^a - \lambda^{a-1}\phi(B) - pB = 0,$$

where $\phi(B)$ is the annual survival rate of adults (as a function of fecundity) and $p$ is the prereproductive survival (a constant). *Source: The American Naturalist.*

  **a.** Using implicit differentiation, derive the equation

$$\frac{d\lambda}{dB} = \frac{\lambda^{a-1}\phi'(B) + p}{a\lambda^{a-1} - (a-1)\lambda^{a-2}\phi(B)}.$$

  **b.** Find an expression for $\phi'(B)$ that makes $d\lambda/dB = 0$. The researchers proved that this expression yields a maximum for the function $\lambda$, that is, maximizes fitness.

**50. Phytoplankton Growth** The equation

$$f(x)g(N) - m - s(x) = 0$$

describes the growth rate of phytoplankton at equilibrium, where $x$ is the phytoplankton cell size, $f$ is the maximum growth rate, $N$ is the nutrient concentration, $g$ represents the nutrient limitation experienced by phytoplankton, $m$ is the mortality rate (a constant), and $s$ is the loss due to sinking. *Source: The American Naturalist.* Addressing the question of how phytoplankton evolution affects nutrient concentration requires finding the rate of change of $N$ with respect to $x$. Using implicit differentiation, show that

$$\frac{dN}{dx} = \frac{s'(x) - f'(x)g(N)}{f(x)g'(N)}.$$

**51. Foraging** In studying how a predator forages a prey, an important concept is the functional response, defined as

$$f(R) = \frac{tCR}{1 + hCR},$$

Where $R$ is the prey population density, $C$ is the attack rate constant, $h$ is the handling time for a single prey, and $t$ is the optimal foraging time, which is itself a function of $R$. *Source: Ecology.* A type 2 response, which is the type most commonly observed in the laboratory, is when the curve is concave down. Show that this occurs when

$$\frac{d^2t}{dR^2}R(1 + hCR)^2 + \frac{dt}{dR}2(1 + hCR) - 2thC < 0.$$

## OTHER APPLICATIONS

**Velocity** The position of a particle at time $t$ is given by $s$. Find the velocity $ds/dt$.

**52.** $s^3 - 4st + 2t^3 - 5t = 0$

**53.** $2s^2 + \sqrt{st} - 4 = 3t$

**YOUR TURN ANSWERS**

**1.** $dy/dx = (y - 2x)/(2y - x)$

**2.** $\dfrac{dy}{dx} = \dfrac{ye^y + 2xy}{1 - xye^y}$

**3.** $y = x$

# 6.4  Related Rates

**APPLY IT**  When a skier's blood vessels contract because of the cold, how fast is the velocity of blood changing?
*We use related rates to answer this question in Example 6 of this section.*

It is common for variables to be functions of time; for example, sales of an item may depend on the season of the year, or a population of animals may be increasing at a certain rate several months after being introduced into an area. Time is often present implicitly in a mathematical model, meaning that derivatives with respect to time must be found by the method of implicit differentiation discussed in the previous section.

We start with a simple algebraic example.

## EXAMPLE 1  Related Rates

Suppose that $x$ and $y$ are both functions of $t$, which can be considered to represent time, and that $x$ and $y$ are related by the equation

$$xy^2 + y = x^2 + 17.$$

Suppose further that when $x = 2$ and $y = 3$, then $dx/dt = 13$. Find the value of $dy/dt$ at that moment.

**SOLUTION**  We start by taking the derivative of the relationship, using the product and chain rules. Keep in mind that both $x$ and $y$ are functions of $t$. The result is

$$x\left(2y\,\frac{dy}{dt}\right) + y^2\,\frac{dx}{dt} + \frac{dy}{dt} = 2x\,\frac{dx}{dt}.$$

Now substitute $x = 2$, $y = 3$, and $dx/dt = 13$ to get

$$2\left(6\,\frac{dy}{dt}\right) + 9(13) + \frac{dy}{dt} = 4(13),$$

$$12\frac{dy}{dt} + 117 + \frac{dy}{dt} = 52.$$

**YOUR TURN 1**  Suppose $x$ and $y$ are both functions of $t$ and $x^3 + 2xy + y^2 = 1$. If $x = 1$, $y = -2$, and $dx/dt = 6$, then find $dy/dt$.

Solve this last equation for $dy/dt$ to get

$$13\frac{dy}{dt} = -65,$$

$$\frac{dy}{dt} = -5. \qquad \text{TRY YOUR TURN 1}$$

Our next example is typical of word problems involving related rates.

## EXAMPLE 2  Bacterial Growth

Bacteria are reproducing in concentric rings in a petri dish. The circular area occupied by the bacteria is increasing at a rate of 3 cm²/day. Find the rate of change of the radius of the circular area when that radius is 4 cm.

**SOLUTION**  As shown in Figure 17, the area $A$ and the radius $r$ are related by

$$A = \pi r^2.$$

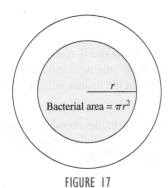

Bacterial area $= \pi r^2$

FIGURE 17

Take the derivative of each side with respect to time.

$$\frac{d}{dt}(A) = \frac{d}{dt}(\pi r^2)$$

$$\frac{dA}{dt} = 2\pi r\,\frac{dr}{dt} \qquad (1)$$

Since the area is increasing at the rate of 3 cm²/day,

$$\frac{dA}{dt} = 3.$$

The rate of change of the radius at the instant when that radius is 4 cm is given by $dr/dt$ evaluated at $r = 4$. Substituting into Equation (1) gives

$$3 = 2\pi \cdot 4 \cdot \frac{dr}{dt}$$

$$\frac{dr}{dt} = \frac{3}{2\pi \cdot 4} = \frac{3}{8\pi} \approx 0.119 \text{ cm/day}.$$

In Example 2, the derivatives (or rates of change) $dA/dt$ and $dr/dt$ are related by Equation (1); for this reason they are called **related rates**. As suggested by Example 2, four basic steps are involved in solving problems about related rates.

### Solving a Related Rates Problem

1. Identify all given quantities, as well as the quantities to be found. Draw a sketch when possible.

2. Write an equation relating the variables of the problem.

3. Use implicit differentiation to find the derivative of both sides of the equation in Step 2 with respect to time.

4. Solve for the derivative, giving the unknown rate of change, and substitute the given values.

**CAUTION**
1. Differentiate *first*, and *then* substitute values for the variables. If the substitutions were performed first, differentiating would not lead to useful results.

2. Some students confuse related rates problems with applied extrema problems, perhaps because they are both word problems. There is an easy way to tell the difference. In applied extrema problems, you are always trying to maximize or minimize something, that is, make it as large or as small as possible. In related rates problems, you are trying to find how fast something is changing; time is always the independent variable.

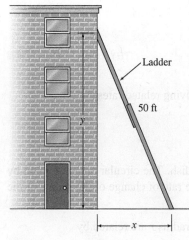
Ladder
50 ft
y
x
**FIGURE 18**

**EXAMPLE 3** Sliding Ladder

A 50-ft ladder is placed against a large building. The base of the ladder is resting on an oil spill, and it slips (to the right in Figure 18) at the rate of 3 ft per minute. Find the rate of change of the height of the top of the ladder above the ground at the instant when the base of the ladder is 30 ft from the base of the building.

**SOLUTION** Starting with Step 1, let $y$ be the height of the top of the ladder above the ground, and let $x$ be the distance of the base of the ladder from the base of the building. We are trying to find $dy/dt$ when $x = 30$. To perform Step 2, use the Pythagorean theorem to write

$$x^2 + y^2 = 50^2. \qquad (2)$$

Both $x$ and $y$ are functions of time $t$ (in minutes) after the moment that the ladder starts slipping. According to Step 3, take the derivative of both sides of Equation (2) with respect to time, getting

$$\frac{d}{dt}(x^2 + y^2) = \frac{d}{dt}(50^2)$$

$$2x\frac{dx}{dt} + 2y\frac{dy}{dt} = 0. \qquad (3)$$

To complete Step 4, we need to find the values of $x$, $y$, and $dx/dt$. Once we find these, we can substitute them into Equation (3) to find $dy/dt$.

Since the base is sliding at the rate of 3 ft per minute,

$$\frac{dx}{dt} = 3.$$

Also, the base of the ladder is 30 ft from the base of the building, so $x = 30$. Use this to find $y$.

$$50^2 = 30^2 + y^2$$
$$2500 = 900 + y^2$$
$$1600 = y^2$$
$$y = 40$$

In summary, $y = 40$ when $x = 30$. Also, the rate of change of $x$ over time $t$ is $dx/dt = 3$. Substituting these values into Equation (3) to find the rate of change of $y$ over time gives

$$2(30)(3) + 2(40)\frac{dy}{dt} = 0$$

$$180 + 80\frac{dy}{dt} = 0$$

$$80\frac{dy}{dt} = -180$$

$$\frac{dy}{dt} = \frac{-180}{80} = \frac{-9}{4} = -2.25.$$

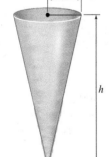

**YOUR TURN 2** A 25-ft ladder is placed against a building. The base of the ladder is slipping away from the building at a rate of 3 ft per minute. Find the rate at which the top of the ladder is sliding down the building when the bottom of the ladder is 7 ft from the base of the building.

At the instant when the base of the ladder is 30 ft from the base of the building, the top of the ladder is sliding down the building at the rate of 2.25 ft per minute. (The minus sign shows that the ladder is sliding *down*, so the distance $y$ is *decreasing*.)\*          TRY YOUR TURN 2

---

**EXAMPLE 4** **Icicle**

A cone-shaped icicle is dripping from the roof. The radius of the icicle is decreasing at a rate of 0.2 cm per hour, while the length is increasing at a rate of 0.8 cm per hour. If the icicle is currently 4 cm in radius and 20 cm long, is the volume of the icicle increasing or decreasing, and at what rate?

**SOLUTION** For this problem we need the formula for the volume of a cone:

$$V = \frac{1}{3}\pi r^2 h, \qquad (4)$$

where $r$ is the radius of the cone and $h$ is the height of the cone, which in this case is the length of the icicle, as in Figure 19.

FIGURE 19

---

\*The model in Example 3 breaks down as the top of the ladder nears the ground. As $y$ approaches 0, $dy/dt$ becomes infinitely large. In reality, the ladder loses contact with the wall before $y$ reaches 0.

In this problem, $V$, $r$, and $h$ are functions of the time $t$ in hours. Taking the derivative of both sides of Equation (4) with respect to time yields

$$\frac{dV}{dt} = \frac{1}{3}\pi\left[r^2\frac{dh}{dt} + (h)(2r)\frac{dr}{dt}\right]. \qquad \textbf{(5)}$$

Since the radius is decreasing at a rate of 0.2 cm per hour and the length is increasing at a rate of 0.8 cm per hour,

$$\frac{dr}{dt} = -0.2 \quad \text{and} \quad \frac{dh}{dt} = 0.8.$$

**YOUR TURN 3** Suppose that in Example 4 the volume of the icicle is decreasing at a rate of 10 cm$^3$ per hour and the radius is decreasing at a rate of 0.4 cm per hour. Find the rate of change of the length of the icicle when the radius is 4 cm and the length is 20 cm.

Substituting these, as well as $r = 4$ and $h = 20$, into Equation (5) yields

$$\frac{dV}{dt} = \frac{1}{3}\pi[4^2(0.8) + (20)(8)(-0.2)]$$

$$= \frac{1}{3}\pi(-19.2) \approx -20.$$

Because the sign of $dV/dt$ is negative, the volume of the icicle is decreasing at a rate of 20 cm$^3$ per hour. **TRY YOUR TURN 3**

### EXAMPLE 5 Revenue

A company is increasing production of peanuts at the rate of 50 cases per day. All cases produced can be sold. The daily demand function is given by

$$p = 50 - \frac{q}{200},$$

where $q$ is the number of units produced (and sold) and $p$ is price in dollars. (Even though the quantity that consumers demand depends on the selling price, economists write the price as a function of the quantity.) Find the rate of change of revenue with respect to time (in days) when the daily production is 200 units.

**SOLUTION** The revenue function,

$$R = qp = q\left(50 - \frac{q}{200}\right) = 50q - \frac{q^2}{200},$$

relates $R$ and $q$. The rate of change of $q$ over time (in days) is $dq/dt = 50$. The rate of change of revenue over time, $dR/dt$, is to be found when $q = 200$. Differentiate both sides of the equation

$$R = 50q - \frac{q^2}{200}$$

with respect to $t$.

$$\frac{dR}{dt} = 50\frac{dq}{dt} - \frac{1}{100}q\frac{dq}{dt} = \left(50 - \frac{1}{100}q\right)\frac{dq}{dt}$$

**YOUR TURN 4** Repeat Example 5 using the daily demand function given by $p = 2000 - \frac{q^2}{100}$.

Now substitute the known values for $q$ and $dq/dt$.

$$\frac{dR}{dt} = \left[50 - \frac{1}{100}(200)\right](50) = 2400$$

Thus revenue is increasing at the rate of $2400 per day. **TRY YOUR TURN 4**

### EXAMPLE 6 Blood Flow

Blood flows faster the closer it is to the center of a blood vessel. According to Poiseuille's laws, the velocity $V$ of blood is given by

$$V = k(R^2 - r^2),$$

where $R$ is the radius of the blood vessel, $r$ is the distance of a layer of blood flow from the center of the vessel, and $k$ is a constant, assumed here to equal 375. See Figure 20. Suppose a skier's blood vessel has radius $R = 0.08$ mm and that cold weather is causing the vessel to contract at a rate of $dR/dt = -0.01$ mm per minute. How fast is the velocity of blood changing?

**APPLY IT**   **SOLUTION**   Find $dV/dt$. Treat $r$ as a constant. Assume the given units are compatible.

$$V = 375(R^2 - r^2)$$

$$\frac{dV}{dt} = 375\left(2R\frac{dR}{dt} - 0\right) \quad \text{$r$ is a constant.}$$

$$\frac{dV}{dt} = 750R\frac{dR}{dt}$$

Here $R = 0.08$ and $dR/dt = -0.01$, so

$$\frac{dV}{dt} = 750(0.08)(-0.01) = -0.6.$$

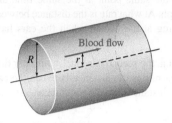

Blood flow

FIGURE 20

That is, the velocity of the blood is decreasing at a rate of $-0.6$ mm per minute each minute. The minus sign indicates that this is a deceleration (negative acceleration), since it represents a negative rate of change of velocity.

# 6.4   EXERCISES

**Assume $x$ and $y$ are functions of $t$. Evaluate $dy/dt$ for each of the following.**

**1.** $y^2 - 8x^3 = -55;$   $\dfrac{dx}{dt} = -4, x = 2, y = 3$

**2.** $8y^3 + x^2 = 1;$   $\dfrac{dx}{dt} = 2, x = 3, y = -1$

**3.** $2xy - 5x + 3y^3 = -51;$   $\dfrac{dx}{dt} = -6, x = 3, y = -2$

**4.** $4x^3 - 6xy^2 + 3y^2 = 228;$   $\dfrac{dx}{dt} = 3, x = -3, y = 4$

**5.** $\dfrac{x^2 + y}{x - y} = 9;$   $\dfrac{dx}{dt} = 2, x = 4, y = 2$

**6.** $\dfrac{y^3 - 4x^2}{x^3 + 2y} = \dfrac{44}{31};$   $\dfrac{dx}{dt} = 5, x = -3, y = -2$

**7.** $xe^y = 2 - \ln 2 + \ln x;$   $\dfrac{dx}{dt} = 6, x = 2, y = 0$

**8.** $y \ln x + xe^y = 1;$   $\dfrac{dx}{dt} = 5, x = 1, y = 0$

**9.** $\cos(\pi xy) + 2x + y^2 = 2;$   $\dfrac{dx}{dt} = -\dfrac{2}{\pi}, x = \dfrac{1}{2}, y = 1$

**10.** $\sin(x + y) + (1 + x)^2 + (2 + y)^2 = 5;$
$\dfrac{dx}{dt} = -10, x = 0, y = 0$

## LIFE SCIENCE APPLICATIONS

**11. Blood Velocity**   A cross-country skier has a history of heart problems. She takes nitroglycerin to dilate blood vessels, thus avoiding angina (chest pain) due to blood vessel contraction.

Use Poiseuille's law with $k = 555.6$ to find the rate of change of the blood velocity when $R = 0.02$ mm and $R$ is changing at 0.003 mm per minute. Assume $r$ is constant. (See Example 6.)

**12. Allometric Growth**   Suppose $x$ and $y$ are two quantities that vary with time according to the allometric formula $y = nx^m$. (See Exercise 84 in Section 2.2 on Exponential, Logarithmic, and Trigonometric Functions.) Show that the derivatives of $x$ and $y$ are related by the formula

$$\frac{1}{y}\frac{dy}{dt} = \frac{m}{x}\frac{dx}{dt}.$$

(*Hint:* Take natural logarithms of both sides before taking the derivatives.)

**13. Brain Mass**   The brain mass of a fetus can be estimated using the total mass of the fetus by the function

$$b = 0.22m^{0.87},$$

where $m$ is the mass of the fetus (in grams) and $b$ is the brain mass (in grams). Suppose the brain mass of a 25-g fetus is changing at a rate of 0.25 g per day. Use this to estimate the rate of change of the total mass of the fetus, $dm/dt$. *Source: Archives d'Anatomie, d'Histologie et d'Embryologie.*

**14. Birds**   The energy cost of bird flight as a function of body mass is given by

$$E = 429m^{-0.35},$$

where $m$ is the mass of the bird (in grams) and $E$ is the energy expenditure (in calories per gram per hour). Suppose that the mass of a 10-g bird is increasing at a rate of 0.001 g per hour. Find the rate at which the energy expenditure is changing with respect to time. *Source: Wildlife Feeding and Nutrition.*

**15. Metabolic Rate** The average daily metabolic rate for rodents can be expressed as a function of weight by

$$m = 85.65w^{0.54},$$

where $w$ is the weight of the rodent (in kg) and $m$ is the metabolic rate (in kcal/day). *Source: Wildlife Feeding and Nutrition.*

**a.** Suppose that the weight of the rodent is changing with respect to time at a rate $dw/dt$. Find $dm/dt$.

**b.** Determine $dm/dt$ for a 0.25-kg rodent that is gaining weight at a rate of 0.01 kg/day.

**16. Metabolic Rate** The average daily metabolic rate for captive animals from weasels to elk can be expressed as a function of mass by

$$r = 140.2m^{0.75},$$

where $m$ is the mass of the animal (in kilograms) and $r$ is the metabolic rate (in kcal per day). *Source: Wildlife Feeding and Nutrition.*

**a.** Suppose that the mass of a weasel is changing with respect to time at a rate $dm/dt$. Find $dr/dt$.

**b.** Determine $dr/dt$ for a 250-kg elk that is gaining mass at a rate of 2 kg per day.

**17. Lizards** The energy cost of horizontal locomotion as a function of the body mass of a lizard is given by

$$E = 26.5m^{-0.34},$$

where $m$ is the mass of the lizard (in kilograms) and $E$ is the energy expenditure (in kcal/kg/km). *Source: Wildlife Feeding and Nutrition.* Suppose that the mass of a 5-kg lizard is increasing at a rate of 0.05 kg per day. Find the rate at which the energy expenditure is changing with respect to time.

**18. Marsupials** The energy cost of horizontal locomotion as a function of the body weight of a marsupial is given by

$$E = 22.8w^{-0.34},$$

where $w$ is the weight of the animal (in kg) and $E$ is the energy expenditure (in kcal/kg/km). *Source: Wildlife Feeding and Nutrition.* Suppose that the weight of a 10-kg marsupial is increasing at a rate of 0.1 kg/day. Find the rate at which the energy expenditure is changing with respect to time.

## OTHER APPLICATIONS

**19. Crime Rate** Sociologists have found that crime rates are influenced by temperature. In a midwestern town of 100,000 people, the crime rate has been approximated as

$$C = \frac{1}{10}(T - 60)^2 + 100,$$

where $C$ is the number of crimes per month and $T$ is the average monthly temperature in degrees Fahrenheit. The average temperature for May was 76°, and by the end of May the temperature was rising at the rate of 8° per month. How fast is the crime rate rising at the end of May?

**20. Memorization Skills** Under certain conditions, a person can memorize $W$ words in $t$ minutes, where

$$W(t) = \frac{-0.02t^2 + t}{t + 1}.$$

Find $dW/dt$ when $t = 5$.

**21. Sliding Ladder** A 17-ft ladder is placed against a building. The base of the ladder is slipping away from the building at a rate of 9 ft per minute. Find the rate at which the top of the ladder is sliding down the building at the instant when the bottom of the ladder is 8 ft from the base of the building.

**22. Distance**

**a.** One car leaves a given point and travels north at 30 mph. Another car leaves the same point at the same time and travels west at 40 mph. At what rate is the distance between the two cars changing at the instant when the cars have traveled 2 hours?

**b.** Suppose that, in part a, the second car left 1 hour later than the first car. At what rate is the distance between the two cars changing at the instant when the second car has traveled 1 hour?

**23. Area** A rock is thrown into a still pond. The circular ripples move outward from the point of impact of the rock so that the radius of the circle formed by a ripple increases at the rate of 2 ft per minute. Find the rate at which the area is changing at the instant the radius is 4 ft.

**24. Volume** A spherical snowball is placed in the sun. The sun melts the snowball so that its radius decreases 1/4 in. per hour. Find the rate of change of the volume with respect to time at the instant the radius is 4 in.

**25. Ice Cube** An ice cube that is 3 cm on each side is melting at a rate of 2 cm³ per min. How fast is the length of the side decreasing?

**26. Volume** A sand storage tank used by the highway department for winter storms is leaking. As the sand leaks out, it forms a conical pile. The radius of the base of the pile increases at the rate of 0.75 in. per minute. The height of the pile is always twice the radius of the base. Find the rate at which the volume of the pile is increasing at the instant the radius of the base is 6 in.

**27. Shadow Length** A man 6 ft tall is walking away from a lamp post at the rate of 50 ft per minute. When the man is 8 ft from the lamp post, his shadow is 10 ft long. Find the rate at which the length of the shadow is increasing when he is 25 ft from the lamp post. (See the figure.)

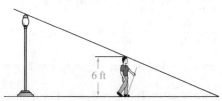

**28. Water Level** A trough has a triangular cross section. The trough is 6 ft across the top, 6 ft deep, and 16 ft long. Water is being pumped into the trough at the rate of 4 ft³ per minute. Find the rate at which the height of the water is increasing at the instant that the height is 4 ft.

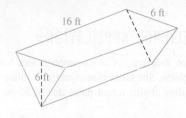

**29. Velocity** A pulley is on the edge of a dock, 8 ft above the water level. A rope is being used to pull in a boat. The rope is attached to the boat at water level. The rope is being pulled in at the rate of 1 ft per second. Find the rate at which the boat is approaching the dock at the instant the boat is 8 ft from the dock.

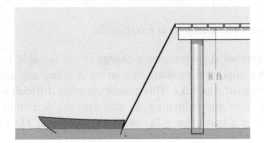

**30. Kite Flying** Christine O'Brien is flying her kite in a wind that is blowing it east at a rate of 50 ft per minute. She has already let out 200 ft of string, and the kite is flying 100 ft above her hand. How fast must she let out string at this moment to keep the kite flying with the same speed and altitude?

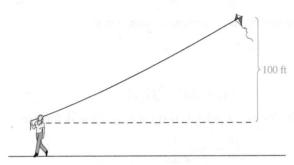

**31. Cost** A manufacturer of handcrafted wine racks has determined that the cost to produce $x$ units per month is given by $C = 0.2x^2 + 10,000$. How fast is the cost per month changing when production is changing at the rate of 12 units per month and the production level is 80 units?

**32. Cost/Revenue** The manufacturer in Exercise 31 has found that the cost $C$ and revenue $R$ (in dollars) in one month are related by the equation

$$C = \frac{R^2}{450,000} + 12,000.$$

Find the rate of change of revenue with respect to time when the cost is changing by $15 per month and the monthly revenue is $25,000.

**33. Revenue/Cost/Profit** Given the revenue and cost functions $R = 50x - 0.4x^2$ and $C = 5x + 15$ (in dollars), where $x$ is the daily production (and sales), find the following when 40 units are produced daily and the rate of change of production is 10 units per day.

  **a.** The rate of change of revenue with respect to time

  **b.** The rate of change of cost with respect to time

  **c.** The rate of change of profit with respect to time (Note that Profit = Revenue − Cost.)

**34. Revenue/Cost/Profit** Repeat Exercise 33, given that 80 units are produced daily and the rate of change of production is 12 units per day.

**35. Rotating Lighthouse** The beacon on a lighthouse 50 m from a straight shoreline rotates twice per minute. (See the figure.)

  **a.** How fast is the beam moving along the shoreline at the moment when the light beam and the shoreline are at right angles? Find an equation relating $\theta$, the angle between the beam of light and the line from the lighthouse to the shoreline, and $x$, the distance along the shoreline from the point on the shoreline closest to the lighthouse and the point where the beam hits the shoreline. You need to express $d\theta/dt$ in radians per minute.)

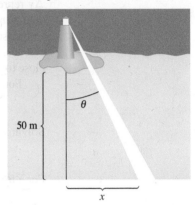

  **b.** In part a, how fast is the beam moving along the shoreline when the beam hits the shoreline 50 m from the point on the shoreline closest to the lighthouse?

**36. Rotating Camera** A television camera on a tripod 60 ft from a road is filming a car carrying the president of the United States. (See the figure.) The car is moving along the road at 600 ft per minute.

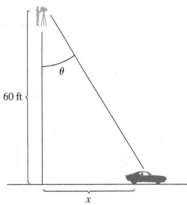

  **a.** How fast is the camera rotating (in revolutions per minute) when the car is at the point on the road closest to the camera? (See the hint for Exercise 35.)

  **b.** How fast is the camera rotating 6 seconds after the moment in part a?

**1.** $dy/dt = -3$            **2.** $-7/8$ ft/min
**3.** Increases 3.4 cm per hour
**4.** Revenue is increasing at the rate of $40,000 per day.

# 6.5 Differentials: Linear Approximation

**APPLY IT** How can the pulp cavity of the canine tooth of a gray wolf be used to estimate the age of the wolf?

*Using differentials, we will answer this question in Exercise 26.*

As mentioned earlier, the symbol $\Delta x$ represents a change in the variable $x$. Similarly, $\Delta y$ represents a change in $y$. An important problem that arises in many applications is to determine $\Delta y$ given specific values of $x$ and $\Delta x$. This quantity is often difficult to evaluate. In this section we show a method of approximating $\Delta y$ that uses the derivative $dy/dx$. In essence, we use the tangent line at a particular value of $x$ to approximate $f(x)$ for values close to $x$.

For values $x_1$ and $x_2$,

$$\Delta x = x_2 - x_1.$$

Solving for $x_2$ gives

$$x_2 = x_1 + \Delta x.$$

For a function $y = f(x)$, the symbol $\Delta y$ represents a change in $y$:

$$\Delta y = f(x_2) - f(x_1).$$

Replacing $x_2$ with $x_1 + \Delta x$ gives

$$\Delta y = f(x_1 + \Delta x) - f(x_1).$$

If $\Delta x$ is used instead of $h$, the derivative of a function $f$ at $x_1$ could be defined as

$$\frac{dy}{dx} = \lim_{\Delta x \to 0} \frac{\Delta y}{\Delta x}.$$

If the derivative exists, then

$$\frac{dy}{dx} \approx \frac{\Delta y}{\Delta x}$$

as long as $\Delta x$ is close to 0. Multiplying both sides by $\Delta x$ (assume $\Delta x \neq 0$) gives

$$\Delta y \approx \frac{dy}{dx} \cdot \Delta x.$$

Until now, $dy/dx$ has been used as a single symbol representing the derivative of $y$ with respect to $x$. In this section, separate meanings for $dy$ and $dx$ are introduced in such a way that their quotient, when $dx \neq 0$, is the derivative of $y$ with respect to $x$. These meanings of $dy$ and $dx$ are then used to find an approximate value of $\Delta y$.

To define $dy$ and $dx$, look at Figure 21 on the next page, which shows the graph of a function $y = f(x)$. The tangent line to the graph has been drawn at the point $P$. Let $\Delta x$ be any nonzero real number (in practical problems, $\Delta x$ is a small number) and locate the point $x + \Delta x$ on the $x$-axis. Draw a vertical line through $x + \Delta x$. Let this vertical line cut the tangent line at $M$ and the graph of the function at $Q$.

Define the new symbol $dx$ to be the same as $\Delta x$. Define the new symbol $dy$ to equal the length $MR$. The slope of $PM$ is $f'(x)$. By the definition of slope, the slope of $PM$ is also $dy/dx$, so that

$$f'(x) = \frac{dy}{dx},$$

or

$$dy = f'(x)dx.$$

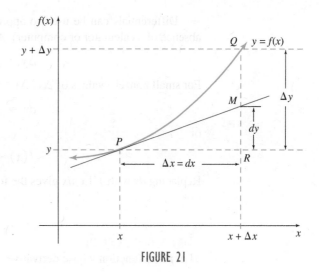

FIGURE 21

In summary, the definitions of the symbols $dy$ and $dx$ are as follows.

**Differentials**

For a function $y = f(x)$ whose derivative exists, the **differential** of $x$, written $dx$, is an arbitrary real number (usually small compared with $x$); the **differential** of $y$, written $dy$, is the product of $f'(x)$ and $dx$, or

$$dy = f'(x)dx.$$

The usefulness of the differential is suggested by Figure 21. As $dx$ approaches 0, the value of $dy$ gets closer and closer to that of $\Delta y$, so that for small nonzero values of $dx$,

$$dy \approx \Delta y,$$

or

$$\Delta y \approx f'(x)\, dx.$$

### EXAMPLE 1 Differential

Find $dy$ for the following functions.

**(a)** $y = 6x^2$

**SOLUTION** The derivative is $dy/dx = 12x$, so

$$dy = 12x\, dx.$$

**(b)** $y = 800x^{-3/4}$, $x = 16$, $dx = 0.01$

**SOLUTION**

$$dy = -600x^{-7/4}dx$$

$$= -600(16)^{-7/4}(0.01)$$

$$= -600\left(\frac{1}{2^7}\right)(0.01) = -0.046875$$

**YOUR TURN 1** Find $dy$ if $y = 300x^{-2/3}$, $x = 8$, and $dx = 0.05$.

TRY YOUR TURN 1

Differentials can be used to approximate function values for a given $x$-value (in the absence of a calculator or computer). As discussed above,

$$\Delta y = f(x + \Delta x) - f(x).$$

For small nonzero values of $\Delta x$, $\Delta y \approx dy$, so that

$$dy \approx f(x + \Delta x) - f(x),$$

or

$$f(x) + dy \approx f(x + \Delta x).$$

Replacing $dy$ with $f'(x)dx$ gives the following result.

> **Linear Approximation**
>
> Let $f$ be a function whose derivative exists. For small nonzero values of $\Delta x$,
>
> $$dy \approx \Delta y,$$
>
> and
>
> $$f(x + \Delta x) \approx f(x) + dy = f(x) + f'(x)\, dx.$$

**EXAMPLE 2** Approximation

Approximate $\sqrt{50}$.

**SOLUTION** We will use the linear approximation

$$f(x + \Delta x) \approx f(x) + f'(x)\, dx$$

for a small value of $\Delta x$ to form this estimate. We first choose a number $x$ that is close to 50 for which we know its square root. Since $\sqrt{49} = 7$, we let $f(x) = \sqrt{x}$, $x = 49$, and $\Delta x = dx = 1$. Using this information, with the fact that when $f(x) = \sqrt{x} = x^{1/2}$,

$$f'(x) = \frac{1}{2}x^{-1/2} = \frac{1}{2x^{1/2}},$$

we have

$$f(x + \Delta x) \approx f(x) + f'(x)\, dx = \sqrt{x} + \frac{1}{2\sqrt{x}}\, dx.$$

Substituting $x = 49$ and $dx = 1$ into the preceding formula gives

$$f(50) = f(49 + 1) \approx \sqrt{49} + \frac{1}{2\sqrt{49}}(1)$$

$$= 7 + \frac{1}{14}$$

$$= 7\frac{1}{14}.$$

**YOUR TURN 2** Approximate $\sqrt{99}$.

A calculator gives $7\frac{1}{14} \approx 7.07143$ and $\sqrt{50} \approx 7.07107$. Our approximation of $7\frac{1}{14}$ is close to the true answer and does not require a calculator. **TRY YOUR TURN 2**

While calculators have made differentials less important, the approximation of functions, including linear approximation, is still important in the branch of mathematics known as numerical analysis.

# Marginal Analysis

Differentials are used to find an approximation of the change in a value of the dependent variable corresponding to a given change in the independent variable. Marginal analysis is often used in economics to estimate the change in cost (or profit or revenue) for nonlinear functions. These ideas can also be used in the life sciences.

For example, suppose that the weight of a particular animal can be expressed as a function of time, $W(t)$. Then

$$dW = W'(t)dt = W'(t)\Delta t.$$

Since $\Delta W \approx dW$,

$$\Delta W \approx W'(t)\Delta t.$$

If the change in time, $\Delta t$, is equal to 1, then

$$W(t + 1) - W(t) = \Delta W$$
$$\approx W'(t)\Delta t$$
$$= W'(t),$$

which shows that the marginal weight $W'(t)$ approximates the weight gain during the next unit of time.

## EXAMPLE 3  Pigs

Researchers have observed that the weight of a male (boar) pig can be estimated by the function

$$W(t) = 10.1 + 160.1e^{-e^{-0.0212(t - 103.63)}},$$

where $t$ is the age of the boar (in days) and $W(t)$ is the weight (in kg). *Source: Animal Science.* If a particular boar is 70 days old, estimate how much it will weigh when it is 73 days old.

**SOLUTION**  Differentials can be used to find the approximate change in the pig's weight resulting from the extra days. This change can be approximated by $dW = W'(t)dt$ where $t = 70$ and $dt = 3$.

$$W'(t) = 160.1\left(e^{-e^{-0.0212(t - 103.63)}}\right)\left(-e^{-0.0212(t - 103.63)}\right)(-0.0212)$$
$$= 3.39412e^{-0.0212(t - 103.63)}e^{-e^{-0.0212(t - 103.63)}}$$

$$\Delta W \approx dW = 3.39412e^{-0.0212(t - 103.63)}e^{-e^{-0.0212(t - 103.63)}}dt$$
$$= 3.39412e^{-0.0212(70 - 103.63)}e^{-e^{-0.0212(70 - 103.63)}}(3)$$
$$\approx 2.70$$

Thus a 70-day-old boar is expected to grow approximately 2.70 kg during the next 3 days, which is fairly close to the actual value of about 2.79 kg.

# Error Estimation

The final example in this section shows how differentials are used to estimate errors that might enter into measurements of a physical quantity.

## EXAMPLE 4  Error Estimation

In a precision manufacturing process, ball bearings must be made with a radius of 0.6 mm, with a maximum error in the radius of $\pm 0.015$ mm. Estimate the maximum error in the volume of the ball bearing.

**SOLUTION**  The formula for the volume of a sphere is

$$V = \frac{4}{3}\pi r^3.$$

If an error of $\Delta r$ is made in measuring the radius of the sphere, the maximum error in the volume is

$$\Delta V = \frac{4}{3}\pi(r + \Delta r)^3 - \frac{4}{3}\pi r^3.$$

Rather than calculating $\Delta V$, approximate $\Delta V$ with $dV$, where

$$dV = 4\pi r^2 dr.$$

Replacing $r$ with 0.6 and $dr = \Delta r$ with $\pm 0.015$ gives

**YOUR TURN 3**  Repeat
Example 4 for $r = 1.25$ mm with
a maximum error in the radius of
$\pm 0.025$ mm.

$$dV = 4\pi(0.6)^2(\pm 0.015)$$
$$\approx \pm 0.0679.$$

The maximum error in the volume is about 0.07 mm³.

**TRY YOUR TURN 3**

# 6.5  EXERCISES

**For Exercises 1–8, find $dy$ for the given values of $x$ and $\Delta x$.**

**1.** $y = 2x^3 - 5x$;  $x = -2, \Delta x = 0.1$

**2.** $y = 4x^3 - 3x$;  $x = 3, \Delta x = 0.2$

**3.** $y = x^3 - 2x^2 + 3$;  $x = 1, \Delta x = -0.1$

**4.** $y = 2x^3 + x^2 - 4x$;  $x = 2, \Delta x = -0.2$

**5.** $y = \sqrt{3x + 2}$;  $x = 4, \Delta x = 0.15$

**6.** $y = \sqrt{4x - 1}$;  $x = 5, \Delta x = 0.08$

**7.** $y = \dfrac{2x - 5}{x + 1}$;  $x = 2, \Delta x = -0.03$

**8.** $y = \dfrac{6x - 3}{2x + 1}$;  $x = 3, \Delta x = -0.04$

**Use the differential to approximate each quantity. Then use a
calculator to approximate the quantity, and give the absolute
value of the difference in the two results to 4 decimal places.**

**9.** $\sqrt{145}$   **10.** $\sqrt{23}$

**11.** $\sqrt{0.99}$   **12.** $\sqrt{17.02}$

**13.** $e^{0.01}$   **14.** $e^{-0.002}$

**15.** $\ln 1.05$   **16.** $\ln 0.98$

**17.** $\sin(0.03)$   **18.** $\cos(-0.002)$

## LIFE SCIENCE APPLICATIONS

**19. Alcohol Concentration**  In Exercise 47 in Section 1.5 on
Polynomial and Rational Functions, we gave the function defined by

$$A(t) = 0.003631t^3 - 0.03746t^2 + 0.1012t + 0.009$$

as the approximate blood alcohol concentration in a 170-lb
woman $t$ hours after drinking 2 oz of alcohol on an empty stomach, for $t$ in the interval $[0, 5]$. *Source: Medicolegal Aspects of
Alcohol Determination in Biological Specimens.*

  **a.** Approximate the change in alcohol level from 1 to 1.2
  hours.

  **b.** Approximate the change in alcohol level from 3 to 3.2
  hours.

**20. Drug Concentration**  The concentration of a certain drug in the
bloodstream $t$ hours after being administered is approximately

$$C(t) = \frac{5t}{9 + t^2}.$$

Use the differential to approximate the changes in concentration for the following changes in $t$.

  **a.** 1 to 1.5   **b.** 2 to 2.25

**21. Bacteria Population**  The population of bacteria (in millions)
in a certain culture $t$ hours after an experimental nutrient is introduced into the culture is

$$P(t) = \frac{25t}{8 + t^2}.$$

Use the differential to approximate the changes in population
for the following changes in $t$.

  **a.** 2 to 2.5   **b.** 3 to 3.25

**22. Area of a Blood Vessel**  The radius of a blood vessel is
1.7 mm. A drug causes the radius to change to 1.6 mm. Find the
approximate change in the area of a cross section of the vessel.

**23. Volume of a Tumor**  A tumor is approximately spherical
in shape. If the radius of the tumor changes from 14 mm to
16 mm, find the approximate change in volume.

**24. Area of an Oil Slick** An oil slick is in the shape of a circle. Find the approximate increase in the area of the slick if its radius increases from 1.2 miles to 1.4 miles.

**25. Area of a Bacteria Colony** The shape of a colony of bacteria on a Petri dish is circular. Find the approximate increase in its area if the radius increases from 20 mm to 22 mm.

**26. APPLY IT Gray Wolves** Accurate methods of estimating the age of gray wolves are important to scientists who study wolf population dynamics. One method of estimating the age of a gray wolf is to measure the percent closure of the pulp cavity of a canine tooth and then estimate age by

$$A(p) = \frac{1.181p}{94.359 - p},$$

where $p$ is the percent closure and $A(p)$ is the age of the wolf (in years). *Source: Journal of Wildlife Management.*

**a.** What is a sensible domain for this function?

**b.** Use differentials to estimate how long it will take for a gray wolf that first measures a 60% closure to obtain a 65% closure. Compare this with the actual value of about 0.55 years.

**27. Pigs** Researchers have observed that the mass of a female (gilt) pig can be estimated by the function

$$M(t) = -3.5 + 197.5e^{-e^{-0.01394(t - 108.4)}},$$

where $t$ is the age of the pig (in days) and $M(t)$ is the mass of the pig (in kilograms). *Source: Animal Science.*

**a.** If a particular gilt is 80 days old, use differentials to estimate how much it will gain before it is 90 days old.

**b.** What is the actual gain in mass?

## OTHER APPLICATIONS

**28. Volume** A spherical balloon is being inflated. Find the approximate change in volume if the radius increases from 4 cm to 4.2 cm.

**29. Volume** A spherical snowball is melting. Find the approximate change in volume if the radius decreases from 3 cm to 2.8 cm.

**30. Volume** A cubical crystal is growing in size. Find the approximate change in the length of a side when the volume increases from 27 cubic mm to 27.1 cubic mm.

**31. Volume** An icicle is gradually increasing in length, while maintaining a cone shape with a length 15 times the radius. Find the approximate amount that the volume of the icicle increases when the length increases from 13 cm to 13.2 cm.

**32. Measurement Error** The edge of a square is measured as 3.45 in., with a possible error of $\pm 0.002$ in. Estimate the maximum error in the area of the square.

**33. Tolerance** A worker is cutting a square from a piece of sheet metal. The specifications call for an area that is 16 cm² with an error of no more than 0.01 cm². How much error could be tolerated in the length of each side to ensure that the area is within the tolerance?

**34. Measurement Error** The radius of a circle is measured as 4.87 in., with a possible error of $\pm 0.040$ in. Estimate the maximum error in the area of the circle.

**35. Measurement Error** A sphere has a radius of 5.81 in., with a possible error of $\pm 0.003$ in. Estimate the maximum error in the volume of the sphere.

**36. Tolerance** A worker is constructing a cubical box that must contain 125 ft³, with an error of no more than 0.3 ft³. How much error could be tolerated in the length of each side to ensure that the volume is within the tolerance?

**37. Measurement Error** A cone has a known height of 7.284 in. The radius of the base is measured as 1.09 in., with a possible error of $\pm 0.007$ in. Estimate the maximum error in the volume of the cone.

**38. Material Requirement** A cube 4 in. on an edge is given a protective coating 0.1 in. thick. About how much coating should a production manager order for 1000 such cubes?

**39. Material Requirement** Beach balls 1 ft in diameter have a thickness of 0.03 in. How much material would be needed to make 5000 beach balls?

**YOUR TURN ANSWERS**

**1.** $-5/16$
**2.** 9.95
**3.** About 0.5 mm³

# 6 CHAPTER REVIEW

## SUMMARY

In this chapter, we began by discussing how to find an absolute maximum or minimum. In contrast to a relative extremum, which is the largest or smallest value of a function on some open interval about the point, an absolute extremum is the largest or smallest value of the function on the entire interval under consideration. We then studied various applications with maximizing or minimizing as the goal. Implicit differentiation is more of a technique than an application, but it underlies related rate problems, in which one or more rates are given and another is to be found. Finally, we studied the differential as a way to find linear approximations of functions.

**Finding Absolute Extrema**    To find absolute extrema for a function $f$ continuous on a closed interval $[a, b]$:

1. Find all critical numbers for $f$ in $(a, b)$.
2. Evaluate $f$ for all critical numbers in $(a, b)$.
3. Evaluate $f$ for the endpoints $a$ and $b$ of the interval.
4. The largest value found in Step 2 or 3 is the maximum, and the smallest value is the minimum.

**Solving an Applied**
**Extrema Problem**

1. Read the problem carefully. Make sure you understand what is given and what is unknown.
2. If possible, sketch a diagram. Label the various parts.
3. Decide which variable must be maximized or minimized. Express that variable as a function of *one* other variable.
4. Find the domain of the function.
5. Find the critical points for the function from Step 3.
6. If the domain is a closed interval, evaluate the function at the endpoints and at each critical number to see which yields the absolute maximum or minimum. If the domain is an open interval, apply the critical point theorem when there is only one critical number. If there is more than one critical number, evaluate the function at the critical numbers and find the limit as the endpoints of the interval are approached to determine if an absolute maximum or minimum exists at one of the critical points.

**Implicit Differentiation**    To find $dy/dx$ for an equation containing $x$ and $y$:

1. Differentiate on both sides of the equation with respect to $x$, keeping in mind that $y$ is assumed to be a function of $x$.
2. Place all terms with $dy/dx$ on one side of the equals sign and all terms without $dy/dx$ on the other side.
3. Factor out $dy/dx$, and then solve for $dy/dx$.

**Solving a Related Rates Problem**

1. Identify all given quantities, as well as the quantities to be found. Draw a sketch when possible.
2. Write an equation relating the variables of the problem.
3. Use implicit differentiation to find the derivative of both sides of the equation in Step 2 with respect to time.
4. Solve for the derivative, giving the unknown rate of change, and substitute the given values.

**Differentials**    $dy = f'(x)dx$

**Linear Approximation**    $f(x + \Delta x) \approx f(x) + dy = f(x) + f'(x)dx$

## KEY TERMS

**6.1**
absolute maximum
absolute minimum
absolute extremum (or extrema)
extreme value theorem

critical point theorem
graphical optimization

**6.2**
spawner-recruit function
parent-progeny function

maximum sustainable
harvest

**6.3**
explicit function
implicit differentiation

**6.4**
related rates

**6.5**
differential

# REVIEW EXERCISES

## CONCEPT CHECK

**Determine whether each of the following statements is true or false, and explain why.**

**1.** The absolute maximum of a function always occurs where the derivative has a critical number.

**2.** A continuous function on a closed interval has an absolute maximum and minimum.

**3.** A continuous function on an open interval does not have an absolute maximum or minimum.

**4.** Implicit differentiation can be used to find $dy/dx$ when $x$ is defined in terms of $y$.

**5.** In a related rates problem, all derivatives are with respect to time.

**6.** In a related rates problem, there can be more than two quantities that vary with time.

**7.** A differential is a real number.

**8.** When the change in $x$ is small, the differential of $y$ is approximately the change in $y$.

## PRACTICE AND EXPLORATIONS

**Find the absolute extrema if they exist, and all values of $x$ where they occur on the given intervals.**

**9.** $f(x) = -x^3 + 6x^2 + 1$;  $[-1, 6]$

**10.** $f(x) = 4x^3 - 9x^2 - 3$;  $[-1, 2]$

**11.** $f(x) = x^3 + 2x^2 - 15x + 3$;  $[-4, 2]$

**12.** $f(x) = -2x^3 - 2x^2 + 2x - 1$;  $[-3, 1]$

**13.** $f(x) = \cos x + x$;  $[0, \pi]$

**14.** $f(x) = \sin x - \cos x + 1$;  $[0, \pi]$

**15.** When solving applied extrema problems, why is it necessary to check the endpoints of the domain?

**16.** Does a continuous function on an open interval always have an absolute maximum or an absolute minimum? Explain why, or give a counterexample.

**17.** Find the absolute maximum and minimum of $f(x) = \dfrac{2 \ln x}{x^2}$ on each interval.

    **a.** $[1, 4]$           **b.** $[2, 5]$

**18.** Find the absolute maximum and minimum of $f(x) = \dfrac{e^{2x}}{x^2}$ on each interval.

    **a.** $[1/2, 2]$       **b.** $[1, 3]$

**19.** When is it necessary to use implicit differentiation?

**20.** When a term involving $y$ is differentiated in implicit differentiation, it is multiplied by $dy/dx$. Why? Why aren't terms involving $x$ multiplied by $dx/dx$?

**Find $dy/dx$.**

**21.** $x^2 - 4y^2 = 3x^3y^4$

**22.** $x^2y^3 + 4xy = 2$

**23.** $2\sqrt{y - 1} = 9x^{2/3} + y$

**24.** $9\sqrt{x} + 4y^3 = 2\sqrt{y}$

**25.** $\dfrac{6 + 5x}{2 - 3y} = \dfrac{1}{5x}$

**26.** $\dfrac{x + 2y}{x - 3y} = y^{1/2}$

**27.** $\ln(xy + 1) = 2xy^3 + 4$

**28.** $\ln(x + y) = 1 + x^2 + y^3$

**29.** $x + \cos(x + y) = y^2$

**30.** $\tan(xy) + x^3 = y$

**31.** Find the equation of the line tangent to the graph of $\sqrt{2y} - 4xy = -22$ at the point $(3, 2)$.

**32.** Find an equation of the line tangent to the graph of $8y^3 - 4xy^2 = 20$ at the point $(-3, 1)$.

**33.** What is the difference between a related rates problem and an applied extremum problem?

**34.** Why is implicit differentiation used in related rates problems?

**Find $dy/dt$.**

**35.** $y = 8x^3 - 7x^2$;  $\dfrac{dx}{dt} = 4, x = 2$

**36.** $y = \dfrac{9 - 4x}{3 + 2x}$;  $\dfrac{dx}{dt} = -1, x = -3$

**37.** $y = \dfrac{1 + \sqrt{x}}{1 - \sqrt{x}}$;  $\dfrac{dx}{dt} = -4, x = 4$

**38.** $\dfrac{x^2 + 5y}{x - 2y} = 2$;  $\dfrac{dx}{dt} = 1, x = 2, y = 0$

**39.** $y = xe^{3x}$;  $\dfrac{dx}{dt} = -2, x = 1$

**40.** $y = \dfrac{1}{e^{x^2} + 1}$;  $\dfrac{dx}{dt} = 3, x = 1$

**41.** What is a differential? What is it used for?

**42.** Describe when linear approximations are most accurate.

**Evaluate $dy$.**

**43.** $y = \dfrac{3x - 7}{2x + 1}$;  $x = 2, \Delta x = 0.003$

**44.** $y = 8 - x^2 + x^3$;  $x = -1, \Delta x = 0.02$

**45.** $y = \sin x$;  $x = \dfrac{\pi}{3}, \Delta x = -0.01$

**46.** $y = \tan x$;  $x = \dfrac{\pi}{4}, \Delta x = -0.02$

**47.** Suppose $x$ and $y$ are related by the equation

$$-12x + x^3 + y + y^2 = 4.$$

**a.** Find all critical points on the curve.

**b.** Determine whether the critical points found in part a are relative maxima or relative minima by taking values of $x$ nearby and solving for the corresponding values of $y$.

**c.** Is there an absolute maximum or minimum for $x$ and $y$ in the relationship given in part a? Why or why not?

**48.** In Exercise 47, implicit differentiation was used to find the relative extrema. The exercise was contrived to avoid various difficulties that could have arisen. Discuss some of the difficulties that might be encountered in such problems, and how these difficulties might be resolved.

## LIFE SCIENCE APPLICATIONS

**49. Pollution** A circle of pollution is spreading from a broken underwater waste disposal pipe, with the radius increasing at the rate of 4 ft per minute. Find the rate of change of the area of the circle when the radius is 7 ft.

**50. Logistic Growth** Many populations grow according to the logistic equation

$$\frac{dx}{dt} = rx(N - x),$$

where $r$ is a constant involving the rate of growth and $N$ is the carrying capacity of the environment, beyond which the population decreases. Show that the graph of $x$ has an inflection point where $x = N/2$. (*Hint:* Use implicit differentiation. Then set $d^2x/dt^2 = 0$, and factor.)

**51. Dentin Growth** The dentinal formation of molars in mice has been studied by researchers in Copenhagen. They determined that the growth curve that best fits dentinal formation for the first molar is

$$M(t) = 1.3386309 - 0.4321173t + 0.0564512t^2$$
$$-0.0020506t^3 + 0.0000315t^4 - 0.0000001785t^5,$$
$$5 \leq t \leq 51,$$

where $t$ is the age of the mouse (in days), and $M(t)$ is the cumulative dentin volume (in $10^{-1}$ mm$^3$). *Source: Journal of Craniofacial Genetics and Developmental Biology.*

**a.** Use a graphing calculator to sketch the graph of this function on $[5, 51]$ by $[0, 7.5]$.

**b.** Find the time in which the dentin formation is growing most rapidly. (*Hint:* Find the maximum value of the derivative of this function.)

**52. Human Skin Surface** The surface of the skin is made up of a network of intersecting lines that form polygons. Researchers have discovered a functional relationship between the age of a female and the number of polygons per area of skin according to

$$P(t) = 237.09 - 8.0398t + 0.20813t^2 - 0.0027563t^3$$
$$+ 0.000013016t^4, \quad 0 \leq t \leq 95,$$

where $t$ is the age of the person (in years) and $P(t)$ is the number of polygons for a particular surface area of skin. *Source: Gerontology.*

**a.** Use a graphing calculator to sketch a graph of $P(t)$ on $[0, 95]$ by $[0, 300]$.

**b.** Find the maximum and minimum number of polygons per area predicted by the model.

**c.** Discuss the accuracy of this model for older people.

**53. Fungal Growth** As we saw earlier in this chapter, because of the time that many people spend indoors, there is a concern about the health risk of being exposed to harmful fungi that thrive in buildings. The risk appears to increase in damp environments. Researchers have discovered that by controlling both the temperature and the relative humidity in a building, the growth of the fungus *E. herbariorum* can be limited. The relationship between the temperature and relative humidity which limits growth can be described by

$$R(T) = -0.0011T^3 + 0.1005T^2 - 2.6876T + 98.171,$$
$$15 \leq T \leq 46,$$

where $R(T)$ is the relative humidity (in %) and $T$ is the temperature (in °C). *Source: Applied and Environmental Microbiology.* Find the temperature at which the relative humidity is minimized.

**54. Pigs** Researchers have observed that the total nitrogen content of a female (gilt) pig can be estimated by the function

$$W(t) = 0.109 + 4.47e^{-e^{-0.01708(t-107.4)}},$$

where $t$ is the age of the pig (in days) and $W(t)$ is the weight of the nitrogen in the pig (in kg). *Source: Animal Science.*

**a.** If a particular gilt is 70 days old, estimate how much nitrogen the pig will gain during the next two days.

**b.** What is the actual nitrogen gain?

## OTHER APPLICATIONS

**55. Sliding Ladder** A 50-ft ladder is placed against a building. The top of the ladder is sliding down the building at the rate of 2 ft per minute. Find the rate at which the base of the ladder is slipping away from the building at the instant that the base is 30 ft from the building.

**56. Spherical Radius** A large weather balloon is being inflated with air at the rate of 0.9 ft$^3$ per minute. Find the rate of change of the radius when the radius is 1.7 ft.

**57. Water Level** A water trough 2 ft across, 4 ft long, and 1 ft deep has ends in the shape of isosceles triangles. (See the figure.) It is being filled with 3.5 ft$^3$ of water per minute. Find the rate at which the depth of water in the tank is changing when the water is 1/3 ft deep.

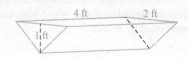

**58. Volume**   Approximate the volume of coating on a sphere of radius 4 in. if the coating is 0.02 in. thick.

**59. Area**   A square has an edge of 9.2 in., with a possible error in the measurement of $\pm 0.04$ in. Estimate the possible error in the area of the square.

**60. Package Dimensions**   UPS has the following rule regarding package dimensions. The length can be no more than 108 in., and the length plus the girth (twice the sum of the width and the height) can be no more than 130 in. If the width of a package is 4 in. more than its height and it has the maximum length plus girth allowed, find the length that produces maximum volume.

**61. Pursuit**   A boat moves north at a constant speed. A second boat, moving at the same speed, pursues the first boat in such a way that it always points directly at the first boat. When the first boat is at the point $(0, 1)$, the second boat is at the point $(6, 2.5)$, with the positive $y$-axis pointing north. It can then be shown that the curve traced by the second boat, known as a pursuit curve, is given by

$$y = \frac{x^2}{16} - 2 \ln x + \frac{1}{4} + 2 \ln 6.$$

Find the $y$-coordinate of the southernmost point of the second boat's path. *Source: Differential Equations: Theory and Applications.*

**62. Playground Area**   The city park department is planning an enclosed play area in a new park. One side of the area will be against an existing building, with no fence needed there. Find the dimensions of the maximum rectangular area that can be made with 900 m of fence.

**63. Surfing**   A mathematician is surfing in Long Beach, New York. He is standing on the shore and wants to paddle out to a spot 40 ft from shore; the closest point on the shore to that spot is 40 ft from where he is now standing. (See the figure.) If he can walk 5 ft per second along the shore and paddle 3 ft per second once he's in the water, how far along the shore should he walk before paddling toward the desired destination if he wants to complete the trip in the shortest possible time? What is the shortest possible time?

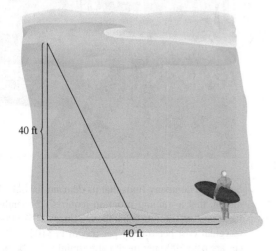

**64.** Repeat Exercise 63, but the closest point on the shore to the desired destination is now 25 ft from where he is standing.

**65. Packaging Design**   The packaging department of a corporation is designing a box with a square base and no top. The volume is to be 32 m³. To reduce cost, the box is to have minimum surface area. What dimensions (height, length, and width) should the box have?

**66. Packaging Design**   A company plans to package its product in a cylinder that is open at one end. The cylinder is to have a volume of $27\pi$ in³. What radius should the circular bottom of the cylinder have to minimize the cost of the material?

**67. Packaging Design**   Fruit juice will be packaged in cylindrical cans with a volume of 40 in³ each. The top and bottom of the can cost 4¢ per in², while the sides cost 3¢ per in². Find the radius and height of the can of minimum cost.

**68. Area**   A 6-ft board is placed against a wall as shown in the figure below, forming a triangular-shaped area beneath it. At what angle $\theta$ should the board be placed to make the triangular area as large as possible?

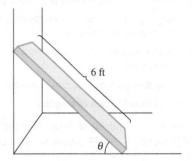

# EXTENDED APPLICATION

## A TOTAL COST MODEL FOR A TRAINING PROGRAM

I n this application, we set up a mathematical model for determining the total costs in setting up a training program, such as a hospital might use. Then we use calculus to find the time interval between training programs that produces the minimum total cost. The model assumes that the demand for trainees is constant and that the fixed cost of training a batch of trainees is known. Also, it is assumed that people who are trained, but for whom no job is readily available, will be paid a fixed amount per month while waiting for a job to open up.

The model uses the following variables.

$D$ = demand for trainees per month

$N$ = number of trainees per batch

$C_1$ = Fixed cost of training a batch of trainees

$C_2$ = marginal cost of training per trainee per month

$C_3$ = salary paid monthly to a trainee who has not yet been given a job after training

$m$ = time interval in months between successive batches of trainees

$t$ = length of training program in months

$Z(m)$ = total monthly cost of program

The total cost of training a batch of trainees is given by $C_1 + NtC_2$. However, $N = mD$, so that the total cost per batch is $C_1 + mDtC_2$.

After training, personnel are given jobs at the rate of $D$ per month. Thus, $N - D$ of the trainees will not get a job the first month, $N - 2D$ will not get a job the second month, and so on. The $N - D$ trainees who do not get a job the first month produce total costs of $(N - D)C_3$, those not getting jobs during the second month produce costs of $(N - 2D)C_3$, and so on. Since $N = mD$, the costs during the first month can be written as

$$(N - D)C_3 = (mD - D)C_3 = (m - 1)DC_3,$$

while the costs during the second month are $(m - 2)DC_3$, and so on. The total cost for keeping the trainees without a job is thus

$$(m - 1)DC_3 + (m - 2)DC_3$$
$$+ (m - 3)DC_3 + \cdots + 2DC_3 + DC_3,$$

which can be factored to give

$$DC_3[(m - 1) + (m - 2) + (m - 3) + \cdots + 2 + 1].$$

The expression in brackets is the sum of the terms of an arithmetic sequence, discussed in most algebra texts. Using formulas for arithmetic sequences, the expression in brackets can be shown to equal $m(m - 1)/2$, so that we have

$$DC_3\left[\frac{m(m - 1)}{2}\right] \qquad (1)$$

as the total cost for keeping jobless trainees.

The total cost per batch is the sum of the training cost per batch, $C_1 + mDtC_2$, and the cost of keeping trainees without a proper job, given by Equation (1). Since we assume that a batch of trainees is trained every $m$ months, the total cost per month, $Z(m)$, is given by

$$Z(m) = \frac{C_1 + mDtC_2}{m} + \frac{DC_3\left[\dfrac{m(m - 1)}{2}\right]}{m}$$

$$= \frac{C_1}{m} + DtC_2 + DC_3\left(\frac{m - 1}{2}\right).$$

*Source: P. L. Goyal and S. K. Goyal.*

## EXERCISES

1. Find $Z'(m)$.

2. Solve the equation $Z'(m) = 0$.

   As a practical matter, it is usually required that $m$ be a whole number. If $m$ does not come out to be a whole number, then $m^+$ and $m^-$, the two whole numbers closest to $m$, must be chosen. Calculate both $Z(m^+)$ and $Z(m^-)$; the smaller of the two provides the optimum value of $Z$.

3. Suppose a company finds that its demand for trainees is 3 per month, that a training program requires 12 months, that the fixed cost of training a batch of trainees is $15,000, that the marginal cost per trainee per month is $100, and that trainees are paid $900 per month after training but before going to work. Use your result from Exercise 2 and find $m$.

4. Since $m$ is not a whole number, find $m^+$ and $m^-$.

5. Calculate $Z(m^+)$ and $Z(m^-)$.

6. What is the optimum time interval between successive batches of trainees? How many trainees should be in a batch?

7. The parameters of this model are likely to change over time; it is essential that such changes be incorporated into the model as they change. One way to anticipate this is to create a spreadsheet that gives the manager the total cost of training a batch of trainees for various scenarios. Using the data from Exercise 3 as a starting point, create a spreadsheet that varies these numbers and calculates the total cost of training a group of employees for each scenario. Graph the total cost of training with respect to changes in the various costs associated with training.

## DIRECTIONS FOR GROUP PROJECT

*Suppose you have read an article in the paper announcing that a new high-tech company is locating in your town. Given that the company is manufacturing very specialized equipment, you realize that it must develop a program to train all new employees. Because you would like to get an internship at this new company, use the information above to develop a hypothetical training program that optimizes the time interval between successive batches of trainees and the number of trainees that should be in each session. Assume that you know the new CEO because you and three of your friends have served her pizza at various times at the local pizza shop (your current jobs) and that she is willing to listen to a proposal that describes your training program. Prepare a presentation for your interview that will describe your training program. Use presentation software such as Microsoft PowerPoint.*

# 7

# Integration

7.1 Antiderivatives

7.2 Substitution

7.3 Area and the Definite Integral

7.4 The Fundamental Theorem of Calculus

7.5 The Area Between Two Curves

Chapter 7 Review

Extended Application: Estimating
Depletion Dates for Minerals

If we know the rate at which a quantity is changing, we can find the total change over a period of time by integrating. An exercise in Section 4 illustrates this process with a model for the growth of the horns in bighorn rams. The integral of the rate function tells us how much the ram's horn increases in size over a given time. This same concept allows us to determine how far a car has gone, given its speed as a function of time, or how much a culture of bacteria will grow.

U p to this point in calculus you have solved problems such as

$$f(x) = x^5; \text{ find } f'(x).$$

In this chapter you will be asked to solve problems that are the reverse of these, that is, problems of the form

$$f'(x) = 5x^4; \text{ find } f(x).$$

The derivative and its applications, which you studied in previous chapters, are part of what is called *differential calculus*. The next two chapters are devoted to the other main branch of calculus, *integral calculus*. Integrals have many applications: finding areas, determining the lengths of curved paths, solving complicated probability problems, and calculating the location of an object (such as the distance of a space shuttle from Earth) when its velocity and initial position are known. The Fundamental Theorem of Calculus, presented later in this chapter, will reveal a surprisingly close connection between differential and integral calculus.

# 7.1 Antiderivatives

**APPLY IT** How long will it take a population of bacteria to reach a certain size, given that we know its growth rote?
*Using antiderivatives, we will answer this question in Example 10.*

Functions used in applications in previous chapters have provided information about a *total amount* of a quantity, such as cost, revenue, profit, temperature, gallons of oil, or distance. Derivatives of these functions provided information about the rate of change of these quantities and allowed us to answer important questions about the extrema of the functions. It is not always possible to find ready-made functions that provide information about the total amount of a quantity, but it is often possible to collect enough data to come up with a function that gives the *rate of change* of a quantity. We know that derivatives give the rate of change when the total amount is known. The reverse of finding a derivative is known as **antidifferentiation**. The goal is to find an *antiderivative*, defined as follows.

**Antiderivative**

If $F'(x) = f(x)$, then $F(x)$ is an **antiderivative** of $f(x)$.

**EXAMPLE 1** **Antiderivative**

**(a)** If $F(x) = 10x$, then $F'(x) = 10$, so $F(x) = 10x$ is an antiderivative of $f(x) = 10$.

**(b)** For $F(x) = x^2$, $F'(x) = 2x$, making $F(x) = x^2$ an antiderivative of $f(x) = 2x$.

## EXAMPLE 2   Antiderivative

Find an antiderivative of $f(x) = 5x^4$.

**SOLUTION**   To find a function $F(x)$ whose derivative is $5x^4$, work backwards. Recall that the derivative of $x^n$ is $nx^{n-1}$. If

$$nx^{n-1} \text{ is } 5x^4,$$

then $n - 1 = 4$ and $n = 5$, so $x^5$ is an antiderivative of $5x^4$.   **TRY YOUR TURN 1**

**YOUR TURN 1**   Find an antiderivative of $f(x) = 8x^7$.

## EXAMPLE 3   Population

Suppose a population is growing at a rate given by $f(t) = e^t$, where $t$ is time in years from some initial date. Find a function giving the population at time $t$.

**SOLUTION**   Let the population function be $F(t)$. Then

$$f(t) = F'(t) = e^t.$$

The derivative of the function defined by $F(t) = e^t$ is $F'(t) = e^t$, so one possible population function with the given growth rate is $F(t) = e^t$.

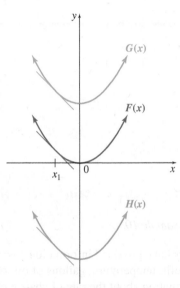

Slopes of the tangent lines at $x = x_1$ are the same.

FIGURE 1

The function from Example 1(b), defined by $F(x) = x^2$, is not the only function whose derivative is $f(x) = 2x$. For example,

$$F(x) = x^2, \qquad G(x) = x^2 + 2, \qquad \text{and} \qquad H(x) = x^2 - 4$$

are all antiderivatives of $f(x) = 2x$, and any two of them differ only by a constant. These three functions, shown in Figure 1, have the same derivative, $f(x) = 2x$, and the slopes of their tangent lines at any particular value of $x$ are the same. In fact, for any real number $C$, the function $F(x) = x^2 + C$ has $f(x) = 2x$ as its derivative. This means that there is a *family* or *class* of functions having $2x$ as a derivative. As the next theorem states, if two functions $F(x)$ and $G(x)$ are antiderivatives of $f(x)$, then $F(x)$ and $G(x)$ can differ only by a constant.

If $F(x)$ and $G(x)$ are both antiderivatives of a function $f(x)$ on an interval, then there is a constant $C$ such that

$$F(x) - G(x) = C.$$

(Two antiderivatives of a function can differ only by a constant.) The arbitrary real number $C$ is called an integration constant.

The family of all antiderivatives of the function $f$ is indicated by

$$\int f(x)\, dx = F(x) + C.$$

The symbol $\int$ is the **integral sign**, $f(x)$ is the **integrand**, and $\int f(x)\, dx$ is called an **indefinite integral**, the most general antiderivative of $f$.

### Indefinite Integral
If $F'(x) = f(x)$, then

$$\int f(x)\, dx = F(x) + C,$$

for any real number $C$.

For example, using this notation,

$$\int 2x\, dx = x^2 + C.$$

The $dx$ in the indefinite integral indicates that $\int f(x)\, dx$ is the "integral of $f(x)$ *with respect to* $x$" just as the symbol $dy/dx$ denotes the "derivative of $y$ with respect to $x$." For example, in the indefinite integral $\int 2ax\, dx$, $dx$ indicates that $a$ is to be treated as a constant and $x$ as the variable, so that

$$\int 2ax\, dx = \int a(2x)dx = ax^2 + C.$$

On the other hand,

$$\int 2ax\, da = a^2x + C = xa^2 + C.$$

A more complete interpretation of $dx$ will be discussed later.

The symbol $\int f(x)\, dx$ was created by G. W. Leibniz (1646–1716) in the latter part of the seventeenth century. The $\int$ is an elongated S from *summa*, the Latin word for *sum*. The word *integral* as a term in the calculus was coined by Jakob Bernoulli (1654–1705), a Swiss mathematician who corresponded frequently with Leibniz. The relationship between sums and integrals will be clarified in a later section.

Finding an antiderivative is the reverse of finding a derivative. Therefore, each rule for derivatives leads to a rule for antiderivatives. For example, the power rule for derivatives tells us that

$$\frac{d}{dx}x^5 = 5x^4.$$

Consequently,

$$\int 5x^4\, dx = x^5 + C,$$

the result found in Example 2. Note that the derivative of $x^n$ is found by multiplying $x$ by $n$ and reducing the exponent on $x$ by 1. To find an indefinite integral—that is, to undo what was done—*increase* the exponent by 1 and *divide* by the new exponent, $n + 1$.

**FOR REVIEW**

Recall that $\dfrac{d}{dx}x^n = nx^{n-1}$.

### Power Rule
For any real number $n \neq -1$,

$$\int x^n\, dx = \frac{x^{n+1}}{n+1} + C.$$

(The antiderivative of $f(x) = x^n$ for $n \neq -1$ is found by increasing the exponent $n$ by 1 and dividing $x$ raised to the new power by the new value of the exponent.)

This rule can be verified by differentiating the expression on the right side of the power rule:

$$\frac{d}{dx}\left(\frac{x^{n+1}}{n+1} + C\right) = \frac{n+1}{n+1}x^{(n+1)-1} + 0 = x^n.$$

(If $n = -1$, the expression in the denominator is 0, and the above rule cannot be used. Finding an antiderivative for this case is discussed later.)

### EXAMPLE 4 Power Rule

Use the power rule to find each indefinite integral.

(a) $\int t^3 \, dt$

**SOLUTION** Use the power rule with $n = 3$.

$$\int t^3 \, dt = \frac{t^{3+1}}{3+1} + C = \frac{t^4}{4} + C$$

To check the solution, find the derivative of $t^4/4 + C$. The derivative is $t^3$, the original function.

(b) $\int \frac{1}{t^2} \, dt$

**SOLUTION** First, write $1/t^2$ as $t^{-2}$. Then

$$\int \frac{1}{t^2} \, dt = \int t^{-2} \, dt = \frac{t^{-2+1}}{-2+1} + C = \frac{t^{-1}}{-1} + C = -\frac{1}{t} + C.$$

Verify the solution by differentiating $-(1/t) + C$ to get $1/t^2$.

(c) $\int \sqrt{u} \, du$

**SOLUTION** Since $\sqrt{u} = u^{1/2}$,

$$\int \sqrt{u} \, du = \int u^{1/2} \, du = \frac{u^{3/2}}{3/2} + C = \frac{2}{3} u^{3/2} + C.$$

To check this, differentiate $(2/3)u^{3/2} + C$; the derivative is $u^{1/2}$, the original function.

(d) $\int dx$

**SOLUTION** Write $dx$ as $1 \cdot dx$, and use the fact that $x^0 = 1$ for any nonzero number $x$ to get

$$\int dx = \int 1 \, dx = \int x^0 \, dx = \frac{x^1}{1} + C = x + C.$$

**YOUR TURN 2** Find $\int \frac{1}{t^4} \, dt$.

TRY YOUR TURN 2

As shown earlier, the derivative of the product of a constant and a function is the product of the constant and the derivative of the function. A similar rule applies to indefinite integrals. Also, since derivatives of sums or differences are found term by term, indefinite integrals also can be found term by term.

---

**FOR REVIEW**

Recall that $\frac{d}{dx}[f(x) \pm g(x)] = [f'(x) \pm g'(x)]$ and $\frac{d}{dx}[kf(x)] = kf'(x)$.

---

### Constant Multiple Rule and Sum or Difference Rule

If all indicated integrals exist,

$$\int k \cdot f(x) \, dx = k \int f(x) \, dx, \qquad \text{for any real number } k,$$

and

$$\int [f(x) \pm g(x)] \, dx = \int f(x) \, dx \pm \int g(x) \, dx.$$

(The antiderivative of a constant times a function is the constant times the antiderivative of the function. The antiderivative of a sum or difference of functions is the sum or difference of the antiderivatives.)

**CAUTION** The constant multiple rule requires $k$ to be a *number*. The rule does not apply to a *variable*. For example,

$$\int x\sqrt{x-1}\, dx \neq x \int \sqrt{x-1}\, dx.$$

**EXAMPLE 5**  **Rules of Integration**

Use the rules to find each integral.

**(a)** $\int 2v^3\, dv$

**SOLUTION**  By the constant multiple rule and the power rule,

$$\int 2v^3\, dv = 2\int v^3\, dv = 2\left(\frac{v^4}{4}\right) + C = \frac{v^4}{2} + C.$$

Because $C$ represents any real number, it is not necessary to multiply it by 2 in the next-to-last step.

**(b)** $\int \frac{12}{z^5}\, dz$

**SOLUTION**  Rewrite $12/z^5$ as $12z^{-5}$, then find the integral.

$$\int \frac{12}{z^5}\, dz = \int 12z^{-5}\, dz = 12\int z^{-5}\, dz = 12\left(\frac{z^{-4}}{-4}\right) + C$$

$$= -3z^{-4} + C = \frac{-3}{z^4} + C$$

**(c)** $\int (3z^2 - 4z + 5)\, dz$

**SOLUTION**  By extending the sum and difference rules to more than two terms, we get

$$\int (3z^2 - 4z + 5)\, dz = 3\int z^2\, dz - 4\int z\, dz + 5\int dz$$

$$= 3\left(\frac{z^3}{3}\right) - 4\left(\frac{z^2}{2}\right) + 5z + C$$

$$= z^3 - 2z^2 + 5z + C.$$

**YOUR TURN 3**

Find $\int (6x^2 + 8x - 9)\, dx$.

Only one constant $C$ is needed in the answer; the three constants from integrating term by term are combined.  **TRY YOUR TURN 3**

Remember to check your work by taking the derivative of the result. For instance, in Example 5(c) check that $z^3 - 2z^2 + 5z + C$ is the required indefinite integral by taking the derivative

$$\frac{d}{dz}(z^3 - 2z^2 + 5z + C) = 3z^2 - 4z + 5,$$

which agrees with the original information.

**EXAMPLE 6** **Rules of Integration**

Use the rules to find each integral.

**(a)** $\displaystyle\int \frac{x^2 + 1}{\sqrt{x}}\, dx$

**SOLUTION** First rewrite the integrand as follows.

$$\int \frac{x^2 + 1}{\sqrt{x}}\, dx = \int \left( \frac{x^2}{\sqrt{x}} + \frac{1}{\sqrt{x}} \right) dx \qquad \text{Rewrite as a sum of fractions.}$$

$$= \int \left( \frac{x^2}{x^{1/2}} + \frac{1}{x^{1/2}} \right) dx \qquad \sqrt{a} = a^{1/2}$$

$$= \int \left( x^{3/2} + x^{-1/2} \right) dx \qquad \text{Use } \frac{a^m}{a^n} = a^{m-n}.$$

Now find the antiderivative.

$$\int \left( x^{3/2} + x^{-1/2} \right) dx = \frac{x^{5/2}}{5/2} + \frac{x^{1/2}}{1/2} + C$$

$$= \frac{2}{5} x^{5/2} + 2x^{1/2} + C$$

**(b)** $\displaystyle\int (x^2 - 1)^2\, dx$

**SOLUTION** Square the binomial first, and then find the antiderivative.

$$\int (x^2 - 1)^2\, dx = \int (x^4 - 2x^2 + 1)\, dx$$

$$= \frac{x^5}{5} - \frac{2x^3}{3} + x + C$$

**YOUR TURN 4**

Find $\displaystyle\int \frac{x^3 - 2}{\sqrt{x}}\, dx$.

TRY YOUR TURN 4

It was shown earlier that the derivative of $f(x) = e^x$ is $f'(x) = e^x$, and the derivative of $f(x) = a^x$ is $f'(x) = (\ln a)a^x$. Also, the derivative of $f(x) = e^{kx}$ is $f'(x) = k \cdot e^{kx}$, and the derivative of $f(x) = a^{kx}$ is $f'(x) = k(\ln a)a^{kx}$. These results lead to the following formulas for indefinite integrals of exponential functions.

**Indefinite Integrals of Exponential Functions**

$$\int e^x\, dx = e^x + C$$

$$\int e^{kx}\, dx = \frac{e^{kx}}{k} + C, \quad k \neq 0$$

For $a > 0, a \neq 1$:

$$\int a^x\, dx = \frac{a^x}{\ln a} + C$$

$$\int a^{kx}\, dx = \frac{a^{kx}}{k(\ln a)} + C, \quad k \neq 0$$

(The antiderivative of the exponential function $e^x$ is itself. If $x$ has a coefficient of $k$, we must divide by $k$ in the antiderivative. If the base is not $e$, we must divide by the natural logarithm of the base.)

### EXAMPLE 7 Exponential Functions

**(a)** $\displaystyle\int 9e^t\,dt = 9\int e^t\,dt = 9e^t + C$

**(b)** $\displaystyle\int e^{9t}\,dt = \frac{e^{9t}}{9} + C$

**(c)** $\displaystyle\int 3e^{(5/4)u}\,du = 3\left(\frac{e^{(5/4)u}}{5/4}\right) + C$

$$= 3\left(\frac{4}{5}\right)e^{(5/4)u} + C$$

$$= \frac{12}{5}\,e^{(5/4)u} + C$$

**(d)** $\displaystyle\int 2^{-5x}\,dx = \frac{2^{-5x}}{-5(\ln 2)} + C = -\frac{2^{-5x}}{5(\ln 2)} + C$

The restriction $n \ne -1$ was necessary in the formula for $\int x^n\,dx$, since $n = -1$ made the denominator of $1/(n + 1)$ equal to 0. To find $\int x^n\,dx$ when $n = -1$, that is, to find $\int x^{-1}\,dx$, recall the differentiation formula for the logarithmic function: The derivative of $f(x) = \ln|x|$, where $x \ne 0$, is $f'(x) = 1/x = x^{-1}$. This formula for the derivative of $f(x) = \ln|x|$ gives a formula for $\int x^{-1}\,dx$.

#### Indefinite Integral of $x^{-1}$

$$\int x^{-1}\,dx = \int \frac{1}{x}\,dx = \ln|x| + C$$

(The antiderivative of $f(x) = x^n$ for $n = -1$ is the natural logarithm of the absolute value of $x$.)

**CAUTION** Don't neglect the absolute value sign in the natural logarithm when integrating $x^{-1}$. If $x$ can take on a negative value, $\ln x$ will be undefined there. Note, however, that the absolute value is redundant (but harmless) in an expression such as $\ln|x^2 + 1|$, since $x^2 + 1$ can never be negative.

### EXAMPLE 8 Integrals

**YOUR TURN 5**

Find $\displaystyle\int\left(\frac{3}{x} + e^{-3x}\right)dx.$

**(a)** $\displaystyle\int \frac{4}{x}\,dx = 4\int\frac{1}{x}\,dx = 4\ln|x| + C$

**(b)** $\displaystyle\int\left(-\frac{5}{x} + e^{-2x}\right)dx = -5\ln|x| - \frac{1}{2}e^{-2x} + C$

TRY YOUR TURN 5

The formulas for derivatives of the trigonometric functions lead to the following antiderivative formulas.

**Basic Trigonometric Integrals**

$$\int \sin x \, dx = -\cos x + C \qquad\qquad \int \cos x \, dx = \sin x + C$$

$$\int \sec^2 x \, dx = \tan x + C \qquad\qquad \int \csc^2 x \, dx = -\cot x + C$$

$$\int \sec x \tan x \, dx = \sec x + C \qquad\qquad \int \csc x \cot x \, dx = -\csc x + C$$

### EXAMPLE 9  Trigonometric Integrals

**(a)** $\displaystyle\int (2 \cos x + 3 \sin x) \, dx = 2 \sin x - 3 \cos x + C$

**(b)** $\displaystyle\int \frac{1}{\cos^2 x} \, dx = \int \sec^2 x \, dx = \tan x + C$

In all these examples, the antiderivative family of functions was found. In many applications, however, the given information allows us to determine the value of the integration constant $C$. The next examples illustrate this idea.

### EXAMPLE 10  Population

Suppose that a population of bacteria grows at a rate that slows down with time, so that the growth rate after $t$ days is given by

$$P'(t) = \frac{5000}{\sqrt{t}}$$

for $t \geq 1$. Suppose further that when $t = 1$, the population is 25,000.

**(a)** Find the function $P(t)$ giving the population at time $t$.

**SOLUTION**   Write $5000/\sqrt{t}$ as $5000/t^{1/2}$ or $5000t^{-1/2}$, and then use the definite integral rules to integrate the function.

$$\int \frac{5000}{\sqrt{t}} \, dt = \int 5000t^{-1/2} \, dt$$

$$= 5000(2t^{1/2}) + C$$

$$= 10{,}000t^{1/2} + C$$

To find the value of $C$, use the fact that $P(1) = 25{,}000$.

$$P(t) = 10{,}000t^{1/2} + C$$

$$25{,}000 = 10{,}000 \cdot 1 + C$$

$$C = 15{,}000$$

With this result, the population function is $P(t) = 10{,}000t^{1/2} + 15{,}000$.

**(b)** How many days will it take for the population to reach 115,000?

APPLY IT

**SOLUTION** Set $P(t)$ equal to 115,000.

$$10,000t^{1/2} + 15,000 = 115,000$$
$$10,000t^{1/2} = 100,000$$
$$t^{1/2} = 10$$
$$t = 100$$

The population will reach 115,000 after 100 days.

In the next example, integrals are used to find the position of a particle when the acceleration of the particle is given.

## EXAMPLE 11  Velocity and Acceleration

Recall that if the function $s(t)$ gives the position of a particle at time $t$, then its velocity $v(t)$ and its acceleration $a(t)$ are given by

$$v(t) = s'(t) \quad \text{and} \quad a(t) = v'(t) = s''(t).$$

**(a)** Suppose the velocity of an object is $v(t) = 6t^2 - 8t$ and that the object is at 5 when time is 0. Find $s(t)$.

**SOLUTION** Since $v(t) = s'(t)$, the function $s(t)$ is an antiderivative of $v(t)$.

$$s(t) = \int v(t)\, dt = \int (6t^2 - 8t)\, dt$$
$$= 2t^3 - 4t^2 + C$$

for some constant $C$. Find $C$ from the given information that $s = 5$ when $t = 0$.

$$s(t) = 2t^3 - 4t^2 + C$$
$$5 = 2(0)^3 - 4(0)^2 + C$$
$$5 = C$$
$$s(t) = 2t^3 - 4t^2 + 5$$

**(b)** Many experiments have shown that when an object is dropped, its acceleration (ignoring air resistance) is constant. This constant has been found to be approximately 32 ft per second every second; that is,

$$a(t) = -32.$$

The negative sign is used because the object is falling. Suppose an object is thrown down from the top of the 1100-ft-tall Willis Tower (formerly known as the Sears Tower) in Chicago. If the initial velocity of the object is $-20$ ft per second, find $s(t)$, the distance of the object from the ground at time $t$.

**SOLUTION** First find $v(t)$ by integrating $a(t)$:

$$v(t) = \int (-32)\, dt = -32t + k.$$

When $t = 0$, $v(t) = -20$:

$$-20 = -32(0) + k$$
$$-20 = k$$

and

$$v(t) = -32t - 20.$$

Be sure to evaluate the constant of integration $k$ before integrating again to get $s(t)$. Now integrate $v(t)$ to find $s(t)$.

$$s(t) = \int (-32t - 20)\, dt = -16t^2 - 20t + C$$

Since $s(t) = 1100$ when $t = 0$, we can substitute these values into the equation for $s(t)$ to get $C = 1100$, and

$$s(t) = -16t^2 - 20t + 1100$$

is the distance of the object from the ground after $t$ seconds.

(c) Use the equations derived in (b) to find the velocity of the object when it hit the ground and how long it took to strike the ground.

**SOLUTION** When the object strikes the ground, $s = 0$, so

$$0 = -16t^2 - 20t + 1100.$$

To solve this equation for $t$, factor out the common factor of $-4$ and then use the quadratic formula.

$$0 = -4(4t^2 + 5t - 275)$$
$$t = \frac{-5 \pm \sqrt{25 + 4400}}{8} \approx \frac{-5 \pm 66.5}{8}$$

Only the positive value of $t$ is meaningful here: $t \approx 7.69$. It took the object about 7.69 seconds to strike the ground. From the velocity equation, with $t = 7.69$, we find

$$v(t) = -32t - 20$$
$$v(7.69) = -32(7.69) - 20 \approx -266,$$

so the object was falling (as indicated by the negative sign) at about 266 ft per second when it hit the ground.

**YOUR TURN 6** Repeat Example 11(b) and 11(c) for the Burj Khalifa in Dubai, which is the tallest building in the world, standing 2717 ft. The initial velocity is $-20$ ft per second.

TRY YOUR TURN 6

### EXAMPLE 12 Slope

Find a function $f$ whose graph has slope $f'(x) = 6x^2 + 4$ and passes through the point $(1, 1)$.

**SOLUTION** Since $f'(x) = 6x^2 + 4$,

$$f(x) = \int (6x^2 + 4)\, dx = 2x^3 + 4x + C.$$

The graph of $f$ passes through $(1, 1)$, so $C$ can be found by substituting 1 for $x$ and 1 for $f(x)$.

$$1 = 2(1)^3 + 4(1) + C$$
$$1 = 6 + C$$
$$C = -5$$

**YOUR TURN 7** Find an equation of the curve whose tangent line has slope $f'(x) = 3x^{1/2} + 4$ and passes through the point $(1, -2)$.

Therefore, $f(x) = 2x^3 + 4x - 5$.

TRY YOUR TURN 7

# 7.1 EXERCISES

1. What must be true of $F(x)$ and $G(x)$ if both are antiderivatives of $f(x)$?

2. How is the antiderivative of a function related to the function?

3. In your own words, describe what is meant by an integrand.

4. Explain why the restriction $n \neq -1$ is necessary in the rule
$$\int x^n \, dx = \frac{x^{n+1}}{n+1} + C.$$

**Find the following.**

5. $\displaystyle\int 6 \, dk$

6. $\displaystyle\int 9 \, dy$

7. $\displaystyle\int (2z + 3) \, dz$

8. $\displaystyle\int (3x - 5) \, dx$

9. $\displaystyle\int (6t^2 - 8t + 7) \, dt$

10. $\displaystyle\int (5x^2 - 6x + 3) \, dx$

11. $\displaystyle\int (4z^3 + 3z^2 + 2z - 6) \, dz$

12. $\displaystyle\int (16y^3 + 9y^2 - 6y + 3) \, dy$

13. $\displaystyle\int (5\sqrt{z} + \sqrt{2}) \, dz$

14. $\displaystyle\int (t^{1/4} + \pi^{1/4}) \, dt$

15. $\displaystyle\int 5x(x^2 - 8) \, dx$

16. $\displaystyle\int x^2(x^4 + 4x + 3) \, dx$

17. $\displaystyle\int (4\sqrt{v} - 3v^{3/2}) \, dv$

18. $\displaystyle\int (15x\sqrt{x} + 2\sqrt{x}) \, dx$

19. $\displaystyle\int (10u^{3/2} - 14u^{5/2}) \, du$

20. $\displaystyle\int (56t^{5/2} + 18t^{7/2}) \, dt$

21. $\displaystyle\int \left(\frac{7}{z^2}\right) dz$

22. $\displaystyle\int \left(\frac{4}{x^3}\right) dx$

23. $\displaystyle\int \left(\frac{\pi^3}{y^3} - \frac{\sqrt{\pi}}{\sqrt{y}}\right) dy$

24. $\displaystyle\int \left(\sqrt{u} + \frac{1}{u^2}\right) du$

25. $\displaystyle\int (-9t^{-2.5} - 2t^{-1}) \, dt$

26. $\displaystyle\int (10x^{-3.5} + 4x^{-1}) \, dx$

27. $\displaystyle\int \frac{1}{3x^2} \, dx$

28. $\displaystyle\int \frac{2}{3x^4} \, dx$

29. $\displaystyle\int 3e^{-0.2x} \, dx$

30. $\displaystyle\int -4e^{0.2v} \, dv$

31. $\displaystyle\int \left(\frac{-3}{x} + 4e^{-0.4x} + e^{0.1}\right) dx$

32. $\displaystyle\int \left(\frac{9}{x} - 3e^{-0.4x}\right) dx$

33. $\displaystyle\int \frac{1 + 2t^3}{4t} \, dt$

34. $\displaystyle\int \frac{2y^{1/2} - 3y^2}{6y} \, dy$

35. $\displaystyle\int (e^{2u} + 4u) \, du$

36. $\displaystyle\int (v^2 - e^{3v}) \, dv$

37. $\displaystyle\int (x + 1)^2 \, dx$

38. $\displaystyle\int (2y - 1)^2 \, dy$

39. $\displaystyle\int \frac{\sqrt{x} + 1}{\sqrt[3]{x}} \, dx$

40. $\displaystyle\int \frac{1 - 2\sqrt[3]{z}}{\sqrt[3]{z}} \, dz$

41. $\displaystyle\int 10^x \, dx$

42. $\displaystyle\int 3^{2x} \, dx$

43. $\displaystyle\int (3 \cos x - 4 \sin x) \, dx$

44. $\displaystyle\int (9 \sin x + 8 \cos x) \, dx$

45. Find an equation of the curve whose tangent line has a slope of
$$f'(x) = x^{2/3},$$
given that the point $(1, 3/5)$ is on the curve.

46. The slope of the tangent line to a curve is given by
$$f'(x) = 6x^2 - 4x + 3.$$
If the point $(0, 1)$ is on the curve, find an equation of the curve.

## LIFE SCIENCE APPLICATIONS

47. **Biochemical Excretion** If the rate of excretion of a biochemical compound is given by
$$f'(t) = 0.01e^{-0.01t},$$
the total amount excreted by time $t$ (in minutes) is $f(t)$.

   a. Find an expression for $f(t)$.

   b. If 0 units are excreted at time $t = 0$, how many units are excreted in 10 minutes?

48. **Flour Beetles** A model for describing the population of adult flour beetles involves evaluating the integral
$$\int \frac{g(x)}{x} \, dx,$$
where $g(x)$ is the per-unit-abundance growth rate for a population of size $x$. The researchers consider the simple case in which $g(x) = a - bx$ for positive constants $a$ and $b$. Find the integral in this case. *Source: Ecology.*

49. **Concentration of a Solute** According to Fick's law, the diffusion of a solute across a cell membrane is given by
$$c'(t) = \frac{kA}{V}[C - c(t)], \qquad \text{(1)}$$
where $A$ is the area of the cell membrane, $V$ is the volume of the cell, $c(t)$ is the concentration inside the cell at time $t$, $C$ is the concentration outside the cell, and $k$ is a constant. If $c_0$ represents the concentration of the solute inside the cell when $t = 0$, then it can be shown that
$$c(t) = (c_0 - C)e^{-kAt/V} + C. \qquad \text{(2)}$$

   a. Use the last result to find $c'(t)$.

   b. Substitute back into Equation (1) to show that (2) is indeed the correct antiderivative of (1).

**50. Cell Growth** Under certain conditions, the number of cancer cells $N(t)$ at time $t$ increases at a rate

$$N'(t) = Ae^{kt},$$

where $A$ is the rate of increase at time 0 (in cells per day) and $k$ is a constant.

  **a.** Suppose $A = 50$, and at 5 days, the cells are growing at a rate of 250 per day. Find a formula for the number of cells after $t$ days, given that 300 cells are present at $t = 0$.

  **b.** Use your answer from part a to find the number of cells present after 12 days.

**51. Blood Pressure** The rate of change of the volume $V(t)$ of blood in the aorta at time $t$ is given by

$$V'(t) = -kP(t),$$

where $P(t)$ is the pressure in the aorta at time $t$ and $k$ is a constant that depends upon properties of the aorta. The pressure in the aorta is given by

$$P(t) = P_0 e^{-mt},$$

where $P_0$ is the pressure at time $t = 0$ and $m$ is another constant. Letting $V_0$ be the volume at time $t = 0$, find a formula for $V(t)$.

## OTHER APPLICATIONS

**52. Bachelor's Degrees** The number of bachelor's degrees conferred in the United States has been increasing steadily in recent decades. Based on data from the National Center for Education Statistics, the rate of change of the number of bachelor's degrees (in thousands) can be approximated by the function

$$B'(t) = 0.06048t^2 - 1.292t + 15.86,$$

where $t$ is the number of years since 1970. *Source: National Center for Education Statistics.*

  **a.** Find $B(t)$, given that about 839,700 degrees were conferred in 1970 ($t = 0$).

  **b.** Use the formula from part a to project the number of bachelor's degrees that will be conferred in 2015 ($t = 45$).

**53. Degrees in Dentistry** The number of degrees in dentistry (D.D.S. or D.M.D.) conferred to females in the United States has been increasing steadily in recent decades. Based on data from the National Center for Education Statistics, the rate of change of the number of bachelor's degrees can be approximated by the function

$$D'(t) = 29.25e^{0.03572t},$$

where $t$ is the number of years since 1980. *Source: National Center for Education Statistics.*

  **a.** Find $D(t)$, given that about 700 degrees in dentistry were conferred to females in 1980 ($t = 0$).

  **b.** Use the formula from part a to project the number of degrees in dentistry that will be conferred to females in 2015 ($t = 35$).

**54. Text Messaging** The approximate rate of change in the number (in billions) of monthly text messages is given by

$$f'(t) = 7.50t - 16.8,$$

where $t$ represents the number of years since 2000. In 2005 ($t = 5$) there were approximately 9.8 billion monthly text messages. *Source: Cellular Telecommunication & Internet Association.*

  **a.** Find the function that gives the total number (in billions) of monthly text messages in year $t$.

  **b.** According to this function, how many monthly text messages were there in 2009? Compare this with the actual number of 152.7 billion.

*Exercises 55–59 refer to Example 11 in this section.*

**55. Velocity** For a particular object, $a(t) = 5t^2 + 4$ and $v(0) = 6$. Find $v(t)$.

**56. Distance** Suppose $v(t) = 9t^2 - 3\sqrt{t}$ and $s(1) = 8$. Find $s(t)$.

**57. Time** An object is dropped from a small plane flying at 6400 ft. Assume that $a(t) = -32$ ft per second and $v(0) = 0$. Find $s(t)$. How long will it take the object to hit the ground?

**58. Distance** Suppose $a(t) = 18t + 8$, $v(1) = 15$, and $s(1) = 19$. Find $s(t)$.

**59. Distance** Suppose $a(t) = (15/2)\sqrt{t} + 3e^{-t}$, $v(0) = -3$, and $s(0) = 4$. Find $s(t)$.

**60. Motion Under Gravity** Show that an object thrown from an initial height $h_0$ with an initial velocity $v_0$ has a height at time $t$ given by the function

$$h(t) = \tfrac{1}{2}gt^2 + v_0 t + h_0,$$

where $g$ is the acceleration due to gravity, a constant with value $-32$ ft/sec$^2$.

**61. Rocket** A small rocket was launched straight up from a platform. After 5 seconds, the rocket reached a maximum height of 412 ft. Find the initial velocity and height of the rocket. (*Hint:* See the previous exercise.)

**62. Rocket Science** In the 1999 movie *October Sky*, Homer Hickum was accused of launching a rocket that started a forest fire. Homer proved his innocence by showing that his rocket could not have flown far enough to reach where the fire started. He used the following reasoning.

  **a.** Using the fact that $a(t) = -32$ (see Example 11(b)), find $v(t)$ and $s(t)$, given $v(0) = v_0$ and $s(0) = 0$. (The initial velocity was unknown, and the initial height was 0 ft.)

  **b.** Homer estimated that the rocket was in the air for 14 seconds. Use $s(14) = 0$ to find $v_0$.

  **c.** If the rocket left the ground at a 45° angle, the velocity in the horizontal direction would be equal to $v_0$, the velocity in the vertical direction, so the distance traveled horizontally would be $v_0 t$. (The rocket left the ground at a steeper angle, so this would overestimate the distance from starting to landing point.) Find the distance the rocket would travel horizontally during its 14-second flight.

1. $x^8$ or $x^8 + C$

2. $-\dfrac{1}{3t^3} + C$

3. $2x^3 + 4x^2 - 9x + C$

4. $\dfrac{2}{7}x^{7/2} - 4x^{1/2} + C$

5. $3 \ln|x| - \dfrac{1}{3} e^{-3x} + C$

6. $s(t) = -16\,t^2 - 20t + 2717$; 12.42 sec; 417 ft/sec

7. $f(x) = 2x^{3/2} + 4x - 8$

# 7.2  Substitution

APPLY IT    **If a pharmaceutical company knows a formula for the marginal revenue for a decongestant, how can it determine a formula for the demand?**
*Using the method of substitution, this question will be answered in Example 9.*

In earlier chapters you learned all the rules for finding derivatives of elementary functions. By correctly applying those rules, you can take the derivative of any function involving powers of $x$, and exponential, logarithmic, and trigonometric functions, combined in any way using the operations of arithmetic (addition, subtraction, multiplication, division, and exponentiation). By contrast, finding the antiderivative is much more complicated. There are a large number of techniques—more than we can cover in this book. Furthermore, for some functions all possible techniques fail. In the last section we saw how to integrate a few simple functions. In this section we introduce a technique known as *substitution* that will greatly expand the set of functions you can integrate.

The substitution technique depends on the idea of a differential, discussed in Chapter 6 on Applications of the Derivative. If $u = f(x)$, the *differential* of $u$, written $du$, is defined as

$$du = f'(x)\, dx.$$

> **FOR REVIEW**
>
> The chain rule, discussed in detail in Chapter 4 on Calculating the Derivative, states that
>
> $$\dfrac{d}{dx}[f(g(x))] = f'(g(x)) \cdot g'(x).$$

For example, if $u = 2x^3 + 1$, then $du = 6x^2\, dx$. In this chapter we will only use differentials as a convenient notational device when finding an antiderivative such as

$$\int (2x^3 + 1)^4 6x^2\, dx.$$

The function $(2x^3 + 1)^4 6x^2$ might remind you of the result when using the chain rule to take the derivative. We will now use differentials and the chain rule in reverse to find the antiderivative. Let $u = 2x^3 + 1$; then $du = 6x^2\, dx$. Now substitute $u$ for $2x^3 + 1$ and $du$ for $6x^2\, dx$ in the indefinite integral.

$$\int (2x^3 + 1)^4 6x^2\, dx = \int \overbrace{(2x^3 + 1)^4}^{u} \overbrace{(6x^2\, dx)}^{du}$$

$$= \int u^4\, du$$

With substitution we have changed a complicated integral into a simple one. This last integral can now be found by the power rule.

$$\int u^4\, du = \frac{u^5}{5} + C$$

Finally, substitute $2x^3 + 1$ for $u$ in the antiderivative to get

$$\int (2x^3 + 1)^4 6x^2\, dx = \frac{(2x^3 + 1)^5}{5} + C.$$

We can check the accuracy of this result by using the chain rule to take the derivative. We get

$$\frac{d}{dx}\left[\frac{(2x^3 + 1)^5}{5} + C\right] = \frac{1}{5} \cdot 5(2x^3 + 1)^4(6x^2) + 0$$

$$= (2x^3 + 1)^4 6x^2.$$

This method of integration is called **integration by substitution**. As shown above, it is simply the chain rule for derivatives in reverse. The results can always be verified by differentiation.

### EXAMPLE 1  Substitution

Find $\int 6x(3x^2 + 4)^7 \, dx$.

**SOLUTION**  If we choose $u = 3x^2 + 4$, then $du = 6x \, dx$ and the integrand can be written as the product of $(3x^2 + 4)^7$ and $6x \, dx$. Now substitute.

$$\int 6x(3x^2 + 4)^7 \, dx = \int (3x^2 + 4)^7(6x \, dx) = \int u^7 \, du$$

Find this last indefinite integral.

$$\int u^7 \, du = \frac{u^8}{8} + C$$

Now replace $u$ with $3x^2 + 4$.

$$\int 6x(3x^2 + 4)^7 \, dx = \frac{u^8}{8} + C = \frac{(3x^2 + 4)^8}{8} + C$$

To verify this result, find the derivative.

**YOUR TURN 1**

Find $\int 8x(4x^2 + 8)^6 dx$.

$$\frac{d}{dx}\left[\frac{(3x^2 + 4)^8}{8} + C\right] = \frac{8}{8}(3x^2 + 4)^7(6x) + 0 = (3x^2 + 4)^7 (6x)$$

The derivative is the original function, as required.

TRY YOUR TURN 1

### EXAMPLE 2  Substitution

Find $\int x^2\sqrt{x^3 + 1} \, dx$.

**SOLUTION**

**Method 1**
**Modifying the Integral**

An expression raised to a power is usually a good choice for $u$, so because of the square root or 1/2 power, let $u = x^3 + 1$; then $du = 3x^2 \, dx$. The integrand does not contain the constant 3, which is needed for $du$. To take care of this, multiply by 3/3, placing 3 inside the integral sign and 1/3 outside.

$$\int x^2\sqrt{x^3 + 1} \, dx = \frac{1}{3}\int 3x^2\sqrt{x^3 + 1} \, dx = \frac{1}{3}\int \sqrt{x^3 + 1} \, (3x^2 \, dx)$$

Now substitute $u$ for $x^3 + 1$ and $du$ for $3x^2 \, dx$, and then integrate.

$$\frac{1}{3}\int \sqrt{x^3 + 1} \, (3x^2 dx) = \frac{1}{3}\int \sqrt{u} \, du = \frac{1}{3}\int u^{1/2} \, du$$

$$= \frac{1}{3} \cdot \frac{u^{3/2}}{3/2} + C = \frac{2}{9}u^{3/2} + C$$

Since $u = x^3 + 1$,

$$\int x^2\sqrt{x^3 + 1} \, dx = \frac{2}{9}(x^3 + 1)^{3/2} + C.$$

**Method 2**
**Eliminating the Constant**

As in Method 1, we let $u = x^3 + 1$, so that $du = 3x^2\,dx$. Since there is no 3 in the integral, we divide the equation for $du$ by 3 to get

$$\frac{1}{3}\,du = x^2\,dx.$$

We then substitute $u$ for $x^3 + 1$ and $du/3$ for $x^2\,dx$ to get

$$\int \sqrt{u}\,\frac{1}{3}\,du = \frac{1}{3}\int u^{1/2}\,du$$

and proceed as we did in Method 1. The two methods are just slightly different ways of doing the same thing, but some people prefer one method over the other.

**YOUR TURN 2**

Find $\int x^3 \sqrt{3x^4 + 10}\,dx.$

TRY YOUR TURN 2

The substitution method given in the examples above *will not always work*. For example, you might try to find

$$\int x^3 \sqrt{x^3 + 1}\,dx$$

by substituting $u = x^3 + 1$, so that $du = 3x^2\,dx$. However, there is no *constant* that can be inserted inside the integral sign to give $3x^2$ alone. This integral, and a great many others, cannot be evaluated by substitution.

With practice, choosing $u$ will become easy if you keep two principles in mind.

**1.** $u$ should equal some expression in the integral that, when replaced with $u$, tends to make the integral simpler.

**2.** $u$ must be an expression whose derivative—disregarding any constant multiplier, such as the 3 in $3x^2$—is also present in the integral.

The substitution should include as much of the integral as possible, as long as its derivative is still present. In Example 1, we could have chosen $u = 3x^2$, but $u = 3x^2 + 4$ is better, because it has the same derivative as $3x^2$ and captures more of the original integral. If we carry this reasoning further, we might try $u = (3x^2 + 4)^4$, but this is a poor choice, for $du = 4(3x^2 + 4)^3(6x)\,dx$, an expression not present in the original integral.

**EXAMPLE 3** **Substitution**

Find $\int \dfrac{x + 3}{(x^2 + 6x)^2}\,dx.$

**SOLUTION** Let $u = x^2 + 6x$, so that $du = (2x + 6)\,dx = 2(x + 3)\,dx$. The integral is missing the 2, so multiply by $2 \cdot (1/2)$, putting 2 inside the integral sign and 1/2 outside.

$$\int \frac{x + 3}{(x^2 + 6x)^2}\,dx = \frac{1}{2}\int \frac{2(x + 3)}{(x^2 + 6x)^2}\,dx$$

$$= \frac{1}{2}\int \frac{du}{u^2} = \frac{1}{2}\int u^{-2}\,du = \frac{1}{2}\cdot\frac{u^{-1}}{-1} + C = \frac{-1}{2u} + C$$

Substituting $x^2 + 6x$ for $u$ gives

$$\int \frac{x + 3}{(x^2 + 6x)^2}\,dx = \frac{-1}{2(x^2 + 6x)} + C.$$

**YOUR TURN 3**

Find $\int \dfrac{x + 1}{(4x^2 + 8x)^3}\,dx.$

TRY YOUR TURN 3

In Example 3, the quantity $x^2 + 6x$ was raised to a power in the denominator. When such an expression is not raised to a power, the function can often be integrated using the fact that

$$\frac{d}{dx} \ln |f(x)| = \frac{1}{f(x)} \cdot f'(x).$$

This suggests that such integrals can be solved by letting $u$ equal the expression in the denominator, as long as the derivative of the denominator is present in the numerator (disregarding any constant multiplier as usual). The next example illustrates this idea.

### EXAMPLE 4 Substitution

Find $\int \dfrac{(2x - 3)\, dx}{x^2 - 3x}$.

**YOUR TURN 4**

Find $\int \dfrac{x + 3}{x^2 + 6x}\, dx$.

**SOLUTION** Let $u = x^2 - 3x$, so that $du = (2x - 3)\, dx$. Then

$$\int \frac{(2x - 3)\, dx}{x^2 - 3x} = \int \frac{du}{u} = \ln |u| + C = \ln |x^2 - 3x| + C.$$

TRY YOUR TURN 4

Recall that if $f(x)$ is a function, then by the chain rule, the derivative of the exponential function $y = e^{f(x)}$ is

$$\frac{d}{dx} e^{f(x)} = e^{f(x)} \cdot f'(x).$$

This suggests that the antiderivative of a function of the form $e^{f(x)}$ can be found by letting $u$ be the exponent, as long as $f'(x)$ is also present in the integral (disregarding any constant multiplier as usual).

### EXAMPLE 5 Substitution

Find $\int x^2 e^{x^3}\, dx$.

**SOLUTION** Let $u = x^3$, the exponent on $e$. Then $du = 3x^2\, dx$. Multiplying by 3/3 gives

$$\int x^2 e^{x^3}\, dx = \frac{1}{3} \int e^{x^3}(3x^2\, dx)$$

**YOUR TURN 5**

Find $\int x^3 e^{x^4}\, dx$.

$$= \frac{1}{3} \int e^u\, du = \frac{1}{3} e^u + C = \frac{1}{3} e^{x^3} + C.$$

TRY YOUR TURN 5

The techniques in the preceding examples can be summarized as follows.

**Substitution**

Each of the following forms can be integrated using the substitution $u = f(x)$.

| Form of the Integral | Result |
|---|---|
| 1. $\int [f(x)]^n f'(x)\, dx, \quad n \neq -1$ | $\int u^n\, du = \dfrac{u^{n+1}}{n+1} + C = \dfrac{[f(x)]^{n+1}}{n+1} + C$ |
| 2. $\int \dfrac{f'(x)}{f(x)}\, dx$ | $\int \dfrac{1}{u}\, du = \ln |u| + C = \ln |f(x)| + C$ |
| 3. $\int e^{f(x)} f'(x)\, dx$ | $\int e^u\, du = e^u + C = e^{f(x)} + C$ |

The next example shows a more complicated integral in which none of the previous forms apply, but for which substitution still works.

### EXAMPLE 6 Substitution

Find $\int x\sqrt{1-x}\,dx$.

**SOLUTION** Let $u = 1 - x$. To get the $x$ outside the radical in terms of $u$, solve $u = 1 - x$ for $x$ to get $x = 1 - u$. Then $dx = -du$ and we can substitute as follows.

$$\int x\sqrt{1-x}\,dx = \int (1-u)\sqrt{u}(-du) = \int (u-1)u^{1/2}\,du$$

$$= \int (u^{3/2} - u^{1/2})\,du = \frac{2}{5}u^{5/2} - \frac{2}{3}u^{3/2} + C$$

$$= \frac{2}{5}(1-x)^{5/2} - \frac{2}{3}(1-x)^{3/2} + C$$

**YOUR TURN 6**
Find $\int x\sqrt{3+x}\,dx$

TRY YOUR TURN 6

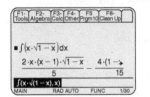

TECHNOLOGY NOTE

**FIGURE 2**

Some calculators, such as the TI-89 and TI-Nspire CAS, can find indefinite integrals automatically. Many computer algebra systems, such as Maple, Matlab, and Mathematica, also do this. The website www.wolframalpha.com can also be used to symbolically determine indefinite integrals and derivatives of functions. Figure 2 shows the integral in Example 6 performed on a TI-89. The answer looks different but is algebraically equivalent to the answer found in Example 6.

Substitution can also be used in trigonometric integrals.

### EXAMPLE 7 Integrals of Trigonometric Functions

Find each integral.

**(a)** $\int \sin 7x\,dx$

**SOLUTION** Use substitution. Let $u = 7x$, so that $du = 7\,dx$. Then

$$\int \sin 7x\,dx = \frac{1}{7}\int \sin 7x\,(7\,dx)$$

$$= \frac{1}{7}\int \sin u\,du$$

$$= -\frac{1}{7}\cos u + C$$

$$= -\frac{1}{7}\cos 7x + C.$$

**(b)** $\int \sin^2 x \cos x\,dx$

**SOLUTION** Let $u = \sin x$, with $du = \cos x\,dx$. This gives

$$\int \sin^2 x \cos x\,dx = \int u^2\,du = \frac{1}{3}u^3 + C.$$

Replacing $u$ with $\sin x$ gives

$$\int \sin^2 x \cos x\,dx = \frac{1}{3}\sin^3 x + C.$$

**(c)** $\displaystyle\int \frac{\sin x}{\sqrt{\cos x}}\, dx$

**SOLUTION** Rewrite the integrand as

$$\int (\cos x)^{-1/2} \sin x \, dx.$$

If $u = \cos x$, then $du = -\sin x \, dx$, with

$$\int (\cos x)^{-1/2} \sin x \, dx = -\int (\cos x)^{-1/2}(-\sin x \, dx)$$

$$= -\int u^{-1/2}\, du$$

$$= -2u^{1/2} + C$$

$$= -2\cos^{1/2} x + C.$$

**(d)** $\displaystyle\int e^{3x} \sec e^{3x} \tan e^{3x}\, dx$

**SOLUTION** Use substitution. Let $u = e^{3x}$, so that $du = 3e^{3x}\, dx$. Then,

$$\int e^{3x} \sec e^{3x} \tan e^{3x}\, dx = \frac{1}{3}\int \sec e^{3x} \tan e^{3x} \,(3e^{3x}\, dx)$$

$$= \frac{1}{3}\int \sec u \tan u \, du$$

$$= \frac{1}{3}\sec u + C$$

$$= \frac{1}{3}\sec e^{3x} + C.$$

> **YOUR TURN 7** Find each
> integral. **(a)** $\displaystyle\int \sin(x/2)\, dx$
> **(b)** $\displaystyle\int 6 \sec^2 x\sqrt{\tan x}\,dx$.

**TRY YOUR TURN 7**

As mentioned earlier, $\tan x = (\sin x)/\cos x$, so that

$$\int \tan x \, dx = \int \frac{\sin x}{\cos x}\, dx.$$

To find $\int \tan x \, dx$, let $u = \cos x$, with $du = -\sin x \, dx$. Then

$$\int \tan x \, dx = \int \frac{\sin x}{\cos x}\, dx = -\int \frac{du}{u} = -\ln |u| + C.$$

Replacing $u$ with $\cos x$ gives the formula for integrating $\tan x$. The integral for $\cot x$ is found in a similar way.

> **Integrals of $\tan x$ and $\cot x$**
>
> $$\int \tan x \, dx = -\ln |\cos x| + C$$
>
> $$\int \cot x \, dx = \ln |\sin x| + C$$

**EXAMPLE 8** Integrals of $\tan x$ and $\cot x$

Find each integral.

**(a)** $\displaystyle\int \tan 6x \, dx$

**SOLUTION** Let $u = 6x$, so that $du = 6\,dx$. Then

$$\int \tan 6x \, dx = \frac{1}{6} \int \tan 6x \, (6 \, dx)$$

$$= \frac{1}{6} \int \tan u \, du$$

$$= -\frac{1}{6} \ln |\cos u| + C$$

$$= -\frac{1}{6} \ln |\cos 6x| + C.$$

**(b)** $\displaystyle\int x \cot x^2 \, dx$

**SOLUTION** Let $u = x^2$, so that $du = 2x \, dx$. Then

$$\int x \cot x^2 \, dx = \frac{1}{2} \int (\cot x^2)(2x \, dx)$$

$$= \frac{1}{2} \int \cot u \, du$$

$$= \frac{1}{2} \ln |\sin u| + C$$

$$= \frac{1}{2} \ln |\sin x^2| + C.$$

**YOUR TURN 8**

Find $\displaystyle\int \tan(\sqrt{x})/\sqrt{x} \, dx$.

TRY YOUR TURN 8

The substitution method is useful if the integral can be written in one of the following forms, where $u$ is some function of $x$.

---

**Substitution Method**

In general, for the types of problems in this book, there are four cases. We choose $u$ to be one of the following:

1. the quantity under a root or raised to a power;
2. the quantity in the denominator;
3. the exponent on $e$;
4. the quantity inside a trigonometric function.

Remember that some integrands may need to be rearranged to fit one of these cases.

---

### EXAMPLE 9   Demand

The marketing department for a pharmaceutical company has studied the number of cases of a decongestant they can sell each week at various prices. When the price is given as a function of the number of cases sold, the function is known as the **demand function**. The research department determined the rate that the price changes with respect to the number of cases sold. Such a function is known as the **marginal demand**,* and in this instance was given by

$$p'(x) = \frac{-4{,}000{,}000}{(2x + 15)^3}.$$

*For more details on demand functions and marginal analysis, see *Calculus with Applications*, 10th ed., Lial, Margaret L., Raymond N. Greenwell, and Nathan P. Ritchey, Pearson, 2012.

Find the demand equation if the weekly demand for this type of decongestant is 10 cases when the price of a case of decongestant is $4000.

APPLY IT **SOLUTION** To find the demand function, first integrate $p'(x)$ as follows.

$$p(x) = \int p'(x)\,dx = \int \frac{-4{,}000{,}000}{(2x + 15)^3}\,dx$$

Let $u = 2x + 15$. Then $du = 2\,dx$, and

$$p(x) = -\frac{4{,}000{,}000}{2}\int (2x + 15)^{-3}\,2\,dx \qquad \text{Multiply by 2/2.}$$

$$= -2{,}000{,}000 \int u^{-3}\,du \qquad \text{Substitute.}$$

$$= (-2{,}000{,}000)\frac{u^{-2}}{-2} + C$$

$$= \frac{1{,}000{,}000}{u^2} + C$$

$$p(x) = \frac{1{,}000{,}000}{(2x + 15)^2} + C. \tag{1}$$

Find the value of $C$ by using the given information that $p = 4000$ when $x = 10$.

$$4000 = \frac{1{,}000{,}000}{(2 \cdot 10 + 15)^2} + C$$

$$4000 = \frac{1{,}000{,}000}{35^2} + C$$

$$4000 \approx 816 + C$$

$$3184 \approx C$$

Replacing C with 3184 in equation (1) gives the demand function,

$$p(x) = \frac{1{,}000{,}000}{(2x + 15)^2} + 3184.$$

With a little practice, you will find you can skip the substitution step for integrals such as that shown in Example 9, in which the derivative of $u$ is a constant. Recall from the chain rule that when you differentiate a function, such as $p'(x) = -4{,}000{,}000/(2x + 15)^3$ in the previous example, you multiply by 2, the derivative of $(2x + 15)$. So when taking the antiderivative, simply divide by 2:

$$\int -4{,}000{,}000(2x + 15)^{-3}\,dx = \frac{-4{,}000{,}000}{2} \cdot \frac{(2x + 15)^{-2}}{-2} + C$$

$$= \frac{1{,}000{,}000}{(2x + 15)^2} + C.$$

CAUTION This procedure is valid because of the constant multiple rule presented in the previous section, which says that constant multiples can be brought into or out of integrals, just as they can with derivatives. This procedure is *not* valid with any expression other than a constant.

EXAMPLE 10 **Popularity Index**

To determine the top 100 popular songs of each year since 1956, Jim Quirin and Barry Cohen developed a function that represents the rate of change on the charts of *Billboard* magazine required for a song to earn a "star" on the *Billboard* "Hot 100" survey. They developed the function

$$f(x) = \frac{A}{B + x},$$

where $f(x)$ represents the rate of change in position on the charts, $x$ is the position on the "Hot 100" survey, and $A$ and $B$ are positive constants. The function

$$F(x) = \int f(x)\, dx$$

is defined as the "Popularity Index." Find $F(x)$. *Source: Chartmasters' Rock 100.*

**SOLUTION**  Integrating $f(x)$ gives

$$F(x) = \int f(x)\, dx$$

$$= \int \frac{A}{B + x}\, dx$$

$$= A \int \frac{1}{B + x}\, dx.$$

Let $u = B + x$, so that $du = dx$. Then

$$F(x) = A \int \frac{1}{u}\, du = A \ln u + C$$

$$= A \ln(B + x) + C.$$

(The absolute value bars are not necessary, since $B + x$ is always positive here.)

# 7.2 EXERCISES

1. Integration by substitution is related to what differentiation method? What type of integrand suggests using integration by substitution?

2. The following integrals may be solved using substitution. Choose a function $u$ that may be used to solve each problem. Then find $du$.

   a. $\int (3x^2 - 5)^4\, 2x\, dx$    b. $\int \sqrt{1 - x}\, dx$

   c. $\int \frac{x^2}{2x^3 + 1}\, dx$    d. $\int 4x^3 e^{x^4}\, dx$

**Use substitution to find each indefinite integral.**

3. $\int 4(2x + 3)^4\, dx$    4. $\int (-4t + 1)^3\, dt$

5. $\int \frac{2\, dm}{(2m + 1)^3}$    6. $\int \frac{3\, du}{\sqrt{3u - 5}}$

7. $\int \frac{2x + 2}{(x^2 + 2x - 4)^4}\, dx$    8. $\int \frac{6x^2\, dx}{(2x^3 + 7)^{3/2}}$

9. $\int z\sqrt{4z^2 - 5}\, dz$    10. $\int r\sqrt{5r^2 + 2}\, dr$

11. $\int 3x^2 e^{2x^3}\, dx$    12. $\int re^{-r^2}\, dr$

13. $\int (1 - t)e^{2t - t^2}\, dt$    14. $\int (x^2 - 1)e^{x^3 - 3x}\, dx$

15. $\int \frac{e^{1/z}}{z^2}\, dz$    16. $\int \frac{e^{\sqrt{y}}}{2\sqrt{y}}\, dy$

17. $\int \frac{t}{t^2 + 2}\, dt$    18. $\int \frac{-4x}{x^2 + 3}\, dx$

19. $\int \frac{x^3 + 2x}{x^4 + 4x^2 + 7}\, dx$    20. $\int \frac{t^2 + 2}{t^3 + 6t + 3}\, dt$

21. $\int \frac{2x + 1}{(x^2 + x)^3}\, dx$    22. $\int \frac{y^2 + y}{(2y^3 + 3y^2 + 1)^{2/3}}\, dy$

23. $\int p(p + 1)^5\, dp$    24. $\int 4r\sqrt{8 - r}\, dr$

**25.** $\int \dfrac{u}{\sqrt{u-1}}\,du$

**26.** $\int \dfrac{2x}{(x+5)^6}\,dx$

**27.** $\int (\sqrt{x^2+12x})(x+6)\,dx$

**28.** $\int (\sqrt{x^2-6x})(x-3)\,dx$

**29.** $\int \dfrac{(1+3\ln x)^2}{x}\,dx$

**30.** $\int \dfrac{\sqrt{2+\ln x}}{x}\,dx$

**31.** $\int \dfrac{e^{2x}}{e^{2x}+5}\,dx$

**32.** $\int \dfrac{1}{x(\ln x)}\,dx$

**33.** $\int \dfrac{\log x}{x}\,dx$

**34.** $\int \dfrac{(\log_2(5x+1))^2}{5x+1}\,dx$

**35.** $\int x8^{3x^2+1}\,dx$

**36.** $\int \dfrac{10^{5\sqrt{x}+2}}{\sqrt{x}}\,dx$

**37.** $\int \cos 3x\,dx$

**38.** $\int \sin 5x\,dx$

**39.** $\int x\sin x^2\,dx$

**40.** $\int 2x\cos x^2\,dx$

**41.** $-\int 3\sec^2 3x\,dx$

**42.** $-\int 2\csc^2 8x\,dx$

**43.** $\int \sin^7 x\cos x\,dx$

**44.** $\int \sin^4 x\cos x\,dx$

**45.** $\int 3\sqrt{\cos x}\,(\sin x)\,dx$

**46.** $\int \dfrac{\cos x}{\sqrt{\sin x}}\,dx$

**47.** $\int \dfrac{\sin x}{1+\cos x}\,dx$

**48.** $\int \dfrac{\cos x}{1-\sin x}\,dx$

**49.** $\int 2x^7\cos x^8\,dx$

**50.** $\int (x+2)^4\sin(x+2)^5\,dx$

**51.** $\int \tan \dfrac{1}{3}x\,dx$

**52.** $\int \cot\left(-\dfrac{3}{8}x\right)dx$

**53.** $\int x^5\cot x^6\,dx$

**54.** $\int \dfrac{x}{4}\tan\left(\dfrac{x}{4}\right)^2 dx$

**55.** $\int e^x\sin e^x\,dx$

**56.** $\int e^{-x}\tan e^{-x}\,dx$

**57.** $\int e^x\csc e^x\cot e^x\,dx$

**58.** $\int x^4\sec x^5\tan x^5\,dx$

**59.** Stan and Ollie work on the integral

$$\int 3x^2 e^{x^3}\,dx.$$

Stan lets $u = x^3$ and proceeds to get

$$\int e^u\,du = e^u + C = e^{x^3} + C.$$

Ollie tries $u = e^{x^3}$ and proceeds to get

$$\int du = u + C = e^{x^3} + C.$$

Discuss which procedure you prefer, and why.

**60.** Stan and Ollie work on the integral

$$\int 2x(x^2+2)\,dx.$$

Stan lets $u = x^2 + 2$ and proceeds to get

$$\int u\,du = \dfrac{u^2}{2} + C = \dfrac{(x^2+2)^2}{2} + C.$$

Ollie multiplies out the function under the integral and gets

$$\int (2x^3+4x)\,dx = \dfrac{x^4}{2} + 2x^2 + C.$$

How can they both be right?

## LIFE SCIENCE APPLICATIONS

**61. Outpatient Visits** According to data from the American Hospital Association, the rate of change in the number of hospital outpatient visits, in millions, in the United States each year from 1980 to the present can be approximated by

$$f'(t) = 0.001483t(t-1980)^{0.75},$$

where $t$ is the year. *Source: Hospital Statistics.*

a. Using the fact that in 1980 there were 262,951,000 outpatient visits, find a formula giving the approximate number of outpatient visits as a function of time.

b. Use the answer to part a to forecast the number of outpatient visits in the year 2015.

**62. Population Growth** A population grows at a rate

$$P'(t) = 500te^{-t^2/5},$$

where $P(t)$ is the population after $t$ months.

a. Find a formula for the population size after $t$ months, given that the population is 2000 at $t = 0$.

b. Use your answer from part a to find the size of the population after 3 months.

**63. Epidemic** An epidemic is growing in a region according to the rate

$$N'(t) = \dfrac{100t}{t^2+2},$$

where $N(t)$ is the number of people infected after $t$ days.

a. Find a formula for the number of people infected after $t$ days, given that 37 people were infected at $t = 0$.

b. Use your answer from part a to find the number of people infected after 21 days.

## OTHER APPLICATIONS

**64. Transportation** According to data from the Bureau of Transportation Statistics, the rate of change in the number of local transit vehicles (buses, light rail, etc.), in thousands, in the United States from 1970 to the present can be approximated by

$$f'(t) = 4.0674 \times 10^{-4}t(t - 1970)^{0.4},$$

where $t$ is the year. *Source: National Transportation Statistics 2006.*

a. Using the fact that in 1970 there were 61,298 such vehicles, find a formula giving the approximate number of local transit vehicles as a function of time.

b. Use the answer to part a to forecast the number of local transit vehicles in the year 2015.

**Self-Answering Problems** The problems in Exercises 65–67 are called self-answering problems because the answers are embedded in the question. For example, how many ways can you arrange the letters in the word "six"? The answer is six. *Source: Math Horizons.*

**65.** At time $t = 0$, water begins pouring into an empty sink so that the volume of water is changing at a rate $V'(t) = \cos t$. For time $t = k$, where $0 \le k \le \pi/2$, determine the amount of water in the sink.

**66.** At time $t = 0$, water begins pouring into an empty tank so that the volume of water is changing at a rate $V'(t) = \sec^2 t$. For time $t = k$, where $0 \le k \le \pi/2$, determine the amount of water in the tank.

**67.** The cost of a widget varies according to the formula $C'(t) = -\sin t$. At time $t = 0$, the cost is $1. For arbitrary time $t$, determine a formula for the cost.

**■■■ YOUR TURN ANSWERS**

1. $\dfrac{(4x^2 + 8)^7}{7} + C$

2. $\dfrac{(3x^4 + 10)^{3/2}}{18} + C$

3. $-\dfrac{1}{16(4x^2 + 8x)^2} + C$

4. $\frac{1}{2}\ln|x^2 + 6x| + C$

5. $\frac{1}{4}e^{x^4} + C$

6. $\frac{2}{5}(3 + x)^{5/2} - 2(3 + x)^{3/2} + C$

7. **(a)** $-2\cos(x/2) + C$

   **(b)** $4(\tan x)^{3/2} + C$

8. $-2\ln|\cos(\sqrt{x})| + C$

# 7.3 Area and the Definite Integral

**APPLY IT** How many cattle and pigs were culled in the United Kingdom during an epidemic of foot-and-mouth disease?
*We will answer this question in Exercise 27 using a method introduced in this section.*

To calculate the areas of geometric figures such as rectangles, squares, triangles, and circles, we use specific formulas. In this section we consider the problem of finding the area of a figure or region that is bounded by curves, such as the shaded region in Figure 3.

The brilliant Greek mathematician Archimedes (about 287 B.C.–212 B.C.) is considered one of the greatest mathematicians of all time. His development of a rigorous method known as *exhaustion* to derive results was a forerunner of the ideas of integral calculus. Archimedes used a method that would later be verified by the theory of integration. His method involved viewing a geometric figure as a sum of other figures. For example, he thought of a plane surface area as a figure consisting of infinitely many parallel line segments. Among the results established by Archimedes' method was the fact that the area of a segment of a parabola (shown in color in Figure 3) is equal to 4/3 the area of a triangle with the same base and the same height.

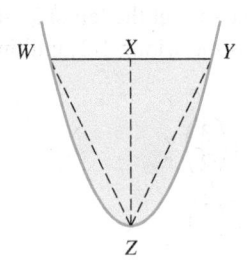

Area of parabolic segment

$= \frac{4}{3}$ (area of triangle $WYZ$)

**FIGURE 3**

## EXAMPLE 1 Approximation of Area

Consider the region bounded by the $y$-axis, the $x$-axis, and the graph of $f(x) = \sqrt{4 - x^2}$, shown in Figure 4.

**(a)** Approximate the area of the region using two rectangles. Determine the height of the rectangle by the value of the function at the *left* endpoint.

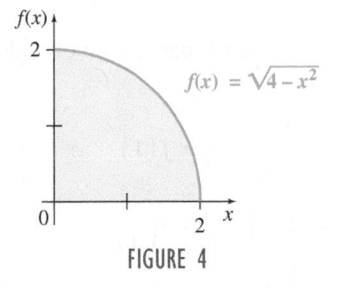

**FIGURE 4**

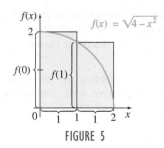

FIGURE 5

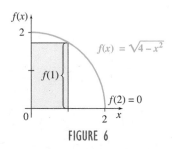

FIGURE 6

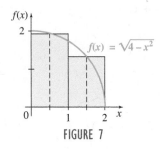

FIGURE 7

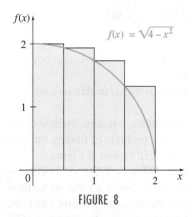

FIGURE 8

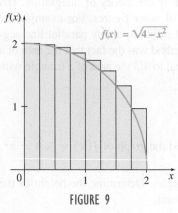

FIGURE 9

**SOLUTION** A very rough approximation of the area of this region can be found by using two rectangles whose heights are determined by the value of the function at the left endpoints, as in Figure 5. The height of the rectangle on the left is $f(0) = 2$ and the height of the rectangle on the right is $f(1) = \sqrt{3}$. The width of each rectangle is 1, making the total area of the two rectangles

$$1 \cdot f(0) + 1 \cdot f(1) = 2 + \sqrt{3} \approx 3.7321 \text{ square units.}$$

Note that $f(x)$ is a decreasing function, and that we will overestimate the area when we evaluate the function at the left endpoint to determine the height of the rectangle in that interval.

**(b)** Repeat part (a) using the value of the function at the *right* endpoint to determine the height of the rectangle.

**SOLUTION** Using the right endpoints, as in Figure 6, the area of the two rectangles is

$$1 \cdot f(1) + 1 \cdot f(2) = \sqrt{3} + 0 \approx 1.7321 \text{ square units.}$$

Note that we underestimate the area of this particular region when we use the right endpoints.

If the left endpoint gives an answer too big and the right endpoint an answer too small, it seems reasonable to average the two answers. This produces the method called the *trapezoidal rule*, discussed in more detail in the next chapter. In this example, we get

$$\frac{3.7321 + 1.7321}{2} = 2.7321 \text{ square units.}$$

**(c)** Repeat part (a) using the value of the function at the *midpoint* of each interval to determine the height of the rectangle.

**SOLUTION** In Figure 7, the rectangles are drawn with height determined by the midpoint of each interval. This method is called the **midpoint rule**, and gives

$$1 \cdot f\left(\frac{1}{2}\right) + 1 \cdot f\left(\frac{3}{2}\right) = \frac{\sqrt{15}}{2} + \frac{\sqrt{7}}{2} \approx 3.2594 \text{ square units.}$$

**(d)** We can improve the accuracy of the previous approximations by increasing the number of rectangles. Repeat part (a) using four rectangles.

**SOLUTION** Divide the interval from $x = 0$ to $x = 2$ into four equal parts, each of width $1/2$. The height of each rectangle is given by the value of $f$ at the left side of the rectangle, as shown in Figure 8. The area of each rectangle is the width, $1/2$, multiplied by the height. The total area of the four rectangles is

$$\frac{1}{2} \cdot f(0) + \frac{1}{2} \cdot f\left(\frac{1}{2}\right) + \frac{1}{2} \cdot f(1) + \frac{1}{2} \cdot f\left(\frac{3}{2}\right)$$

$$= \frac{1}{2}(2) + \frac{1}{2}\left(\frac{\sqrt{15}}{2}\right) + \frac{1}{2}(\sqrt{3}) + \frac{1}{2}\left(\frac{\sqrt{7}}{2}\right)$$

$$= 1 + \frac{\sqrt{15}}{4} + \frac{\sqrt{3}}{2} + \frac{\sqrt{7}}{4} \approx 3.4957 \text{ square units.}$$

This approximation looks better, but it is still greater than the actual area.

**(e)** Repeat part (a) using eight rectangles.

**SOLUTION** Divide the interval from $x = 0$ to $x = 2$ into 8 equal parts, each of width $1/4$ (see Figure 9). The total area of all of these rectangles is

$$\frac{1}{4} \cdot f(0) + \frac{1}{4} \cdot f\left(\frac{1}{4}\right) + \frac{1}{4} \cdot f\left(\frac{1}{2}\right) + \frac{1}{4} \cdot f\left(\frac{3}{4}\right) + \frac{1}{4} \cdot f(1)$$

$$+ \frac{1}{4} \cdot f\left(\frac{5}{4}\right) + \frac{1}{4} \cdot f\left(\frac{3}{2}\right) + \frac{1}{4} \cdot f\left(\frac{7}{4}\right)$$

$$\approx 3.3398 \text{ square units.}$$

The process used in Example 1 of approximating the area under a curve by using more and more rectangles to get a better and better approximation can be generalized. To do this, divide the interval from $x = 0$ to $x = 2$ into $n$ equal parts. Each of these $n$ intervals has width

$$\frac{2 - 0}{n} = \frac{2}{n},$$

so each rectangle has width $2/n$ and height determined by the function value at the left side of the rectangle, or the right side, or the midpoint. We could also average the left and right side values as before. Using a computer or graphing calculator to find approximations to the area for several values of $n$ gives the results in the following table.

| | Approximations to the Area | | | |
|---|---|---|---|---|
| $n$ | Left Sum | Right Sum | Trapezoidal | Midpoint |
| 2 | 3.7321 | 1.7321 | 2.7321 | 3.2594 |
| 4 | 3.4957 | 2.4957 | 2.9957 | 3.1839 |
| 8 | 3.3398 | 2.8398 | 3.0898 | 3.1567 |
| 10 | 3.3045 | 2.9045 | 3.1045 | 3.1524 |
| 20 | 3.2285 | 3.0285 | 3.1285 | 3.1454 |
| 50 | 3.1783 | 3.0983 | 3.1383 | 3.1426 |
| 100 | 3.1604 | 3.1204 | 3.1404 | 3.1419 |
| 500 | 3.1455 | 3.1375 | 3.1415 | 3.1416 |

The numbers in the last four columns of this table represent approximations to the area under the curve, above the $x$-axis, and between the lines $x = 0$ and $x = 2$. As $n$ becomes larger and larger, all four approximations become better and better, getting closer to the actual area. In this example, the exact area can be found by a formula from plane geometry. Write the given function as

$$y = \sqrt{4 - x^2},$$

then square both sides to get

$$y^2 = 4 - x^2$$
$$x^2 + y^2 = 4,$$

the equation of a circle centered at the origin with radius 2. The region in Figure 4 is the quarter of this circle that lies in the first quadrant. The actual area of this region is one-quarter of the area of the entire circle, or

$$\frac{1}{4}\pi(2)^2 = \pi \approx 3.1416.$$

As the number of rectangles increases without bound, the sum of the areas of these rectangles gets closer and closer to the actual area of the region, $\pi$. This can be written as

$$\lim_{n \to \infty} (\text{sum of areas of } n \text{ rectangles}) = \pi.$$

(The value of $\pi$ was originally found by a process similar to this.)*

---

*The number $\pi$ is the ratio of the circumference of a circle to its diameter. It is an example of an *irrational number*, and as such it cannot be expressed as a terminating or repeating decimal. Many approximations have been used for $\pi$ over the years. A passage in the Bible (1 Kings 7:23) indicates a value of 3. The Egyptians used the value 3.16, and Archimedes showed that its value must be between 22/7 and 223/71. A Hindu writer, Brahmagupta, used $\sqrt{10}$ as its value in the seventh century. The search for the digits of $\pi$ has continued into modern times. Fabrice Bellard, using a desktop computer, recently computed the value to nearly 2.7 trillion digits.

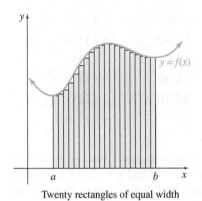

Ten rectangles of equal width

**(a)**

Twenty rectangles of equal width

**(b)**

FIGURE 10

Notice in the previous table that for a particular value of $n$, the midpoint rule gave the best answer (the one closest to the true value of $\pi \approx 3.1416$), followed by the trapezoidal rule, followed by the left and right sums. In fact, the midpoint rule with $n = 20$ gives a value $(3.1454)$ that is slightly more accurate than the left sum with $n = 500$ $(3.1455)$. It is usually the case that the midpoint rule gives a more accurate answer than either the left or the right sum.

Now we can generalize to get a method of finding the area bounded by the curve $y = f(x)$, the $x$-axis, and the vertical lines $x = a$ and $x = b$, as shown in Figure 10. To approximate this area, we could divide the region under the curve first into 10 rectangles (Figure 10(a)) and then into 20 rectangles (Figure 10(b)). The sum of the areas of the rectangles gives an approximation to the area under the curve when $f(x) \geq 0$. In the next section we will consider the case in which $f(x)$ might be negative.

To develop a process that would yield the *exact* area, begin by dividing the interval from $a$ to $b$ into $n$ pieces of equal width, using each of these $n$ pieces as the base of a rectangle (see Figure 11). Let $x_1$ be an arbitrary point in the first interval, $x_2$ be an arbitrary point in the second interval, and so on, up to the $n$th interval. In the graph of Figure 11, the symbol $\Delta x$ is used to represent the width of each of the intervals. Since the length of the entire interval is $b - a$, each of the $n$ pieces has length

$$\Delta x = \frac{b - a}{n}.$$

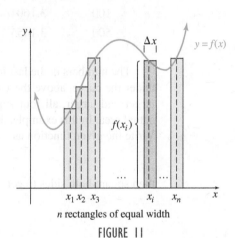

$n$ rectangles of equal width

FIGURE 11

The pink rectangle is an arbitrary rectangle called the $i$th rectangle. Its area is the product of its length and width. Since the width of the $i$th rectangle is $\Delta x$ and the length of the $i$th rectangle is given by the height $f(x_i)$,

$$\text{Area of the } i\text{th rectangle} = f(x_i) \cdot \Delta x.$$

**FOR REVIEW**

Recall from Chapter 1 that the symbol $\Sigma$ (sigma) indicates "the sum of." Here, we use $\sum_{i=1}^{n} f(x_i)\Delta x$ to indicate the sum

$f(x_1)\Delta x + f(x_2)\Delta x +$
$f(x_3)\Delta x + \cdots + f(x_n)\Delta x,$

where we replace $i$ with 1 in the first term, 2 in the second term, and so on, ending with $n$ replacing $i$ in the last term.

The total area under the curve is approximated by the sum of the areas of all $n$ of the rectangles. With sigma notation, the approximation of the total area becomes

$$\text{Area of all } n \text{ rectangles} = \sum_{i=1}^{n} f(x_i) \cdot \Delta x.$$

The exact area is defined to be the limit of this sum (if the limit exists) as the number of rectangles increases without bound:

$$\text{Exact area} = \lim_{n \to \infty} \sum_{i=1}^{n} f(x_i)\Delta x.$$

Whenever this limit exists, regardless of whether $f(x)$ is positive or negative, we will call it the *definite integral* of $f(x)$ from $a$ to $b$. It is written as follows.

## The Definite Integral

If $f$ is defined on the interval $[a, b]$, the **definite integral** of $f$ from $a$ to $b$ is given by

$$\int_a^b f(x)\,dx = \lim_{n \to \infty} \sum_{i=1}^{n} f(x_i)\Delta x,$$

provided the limit exists, where $\Delta x = (b - a)/n$ and $x_i$ is *any* value of $x$ in the $i$th interval.*

**NOTE** In the life sciences literature, the definite integral is often referred to as the area under the curve, abbreviated as AUC.

The definite integral can be approximated by

$$\sum_{i=1}^{n} f(x_i)\Delta x.$$

If $f(x) \geq 0$ on the interval $[a, b]$, the definite integral gives the area under the curve between $x = a$ and $x = b$. In the midpoint rule, $x_i$ is the midpoint of the $i$th interval. We may also let $x_i$ be the left endpoint, the right endpoint, or any other point in the $i$th interval.

In Example 1, the area bounded by the $x$-axis, the curve $y = \sqrt{4 - x^2}$, and the lines $x = 0$ and $x = 2$ could be written as the definite integral

$$\int_0^2 \sqrt{4 - x^2}\,dx = \pi.$$

**NOTE** Notice that unlike the indefinite integral, which is a set of *functions*, the definite integral represents a *number*. The next section will show how antiderivatives are used in finding the definite integral and, thus, the area under a curve.

Keep in mind that finding the definite integral of a function can be thought of as a mathematical process that gives the sum of an infinite number of individual parts (within certain limits). The definite integral represents area only if the function involved is *nonnegative* ($f(x) \geq 0$) at every $x$-value in the interval $[a, b]$. There are many other interpretations of the definite integral, and all of them involve this idea of approximation by appropriate sums. In the next section we will consider the definite integral when $f(x)$ might be negative.

As indicated in this definition, although the left endpoint of the $i$th interval has been used to find the height of the $i$th rectangle, any number in the $i$th interval can be used. (A more general definition is possible in which the rectangles do not necessarily all have the same width.) The $b$ above the integral sign is called the **upper limit** of integration, and the $a$ is the **lower limit** of integration. This use of the word *limit* has nothing to do with the limit of the sum; it refers to the limits, or boundaries, on $x$.

**TECHNOLOGY NOTE** Some calculators have a built-in function for evaluating the definite integral. For example, the TI-84 Plus uses the `fnInt` command, found in the `MATH` menu, as shown in Figure 12(a). Figure 12(b) shows the command used for Example 1 and gives the answer 3.141593074, with an error of approximately 0.0000004.

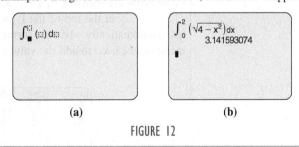

**(a)**　　　　　　　**(b)**

FIGURE 12

*The sum in the definition of the definite integral is an example of a Riemann sum, named for the German mathematician Georg Riemann (1826–1866), who at the age of 20 changed his field of study from theology and the classics to mathematics. Twenty years later he died of tuberculosis while traveling in Italy in search of a cure. The concepts of *Riemann sum* and *Riemann integral* are still studied in rigorous calculus textbooks.

EXAMPLE 2 Approximation of Area

Approximate $\int_0^4 2x\,dx$, the area of the region under the graph of $f(x) = 2x$, above the $x$-axis, and between $x = 0$ and $x = 4$, by using four rectangles of equal width whose heights are the values of the function at the midpoint of each subinterval.

**Method 1**
**Calculating by Hand**

**SOLUTION**

We want to find the area of the shaded region in Figure 13. The heights of the four rectangles given by $f(x_i)$ for $i = 1, 2, 3$, and 4 are as follows.

| | Rectangle Heights | |
|---|---|---|
| $i$ | $x_i$ | $f(x_i)$ |
| 1 | $x_1 = 0.5$ | $f(0.5) = 1.0$ |
| 2 | $x_2 = 1.5$ | $f(1.5) = 3.0$ |
| 3 | $x_3 = 2.5$ | $f(2.5) = 5.0$ |
| 4 | $x_4 = 3.5$ | $f(3.5) = 7.0$ |

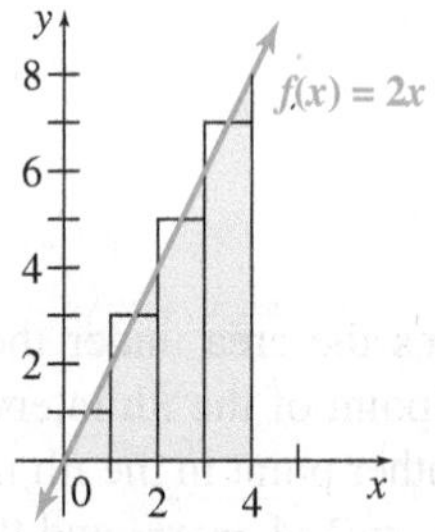

The width of each rectangle is $\Delta x = (4 - 0)/4 = 1$. The sum of the areas of the four rectangles is

$$\sum_{i=1}^{4} f(x_i)\Delta x = f(x_1)\Delta x + f(x_2)\Delta x + f(x_3)\Delta x + f(x_4)\Delta x$$
$$= f(0.5)\Delta x + f(1.5)\Delta x + f(2.5)\Delta x + f(3.5)\Delta x$$
$$= 1(1) + 3(1) + 5(1) + 7(1)$$
$$= 16.$$

Using the formula for the area of a triangle, $A = (1/2)bh$, with $b$, the length of the base, equal to 4 and $h$, the height, equal to 8, gives

$$A = \frac{1}{2}bh = \frac{1}{2}(4)(8) = 16,$$

the exact value of the area. The approximation equals the exact area in this case because our use of the midpoints of each subinterval distributed the error evenly above and below the graph.

**FIGURE 13**

**Method 2**
**Graphing Calculator**

A graphing calculator can be used to organize the information in this example. For example, the seq feature in the LIST OPS menu of the TI-84 Plus calculator can be used to store the values of $i$ in the list $L_1$. Using the STAT EDIT menu, the entries for $x_i$ can be generated by entering the formula -.5+L₁ as the heading of $L_2$. Similarly, entering the formula for $f(x_i)$, 2*L₂, at the top of list $L_3$ will generate the values of $f(x_i)$ in $L_3$. (The entries are listed automatically when the formula is entered.) Then the sum feature in the LIST MATH menu can be used to add the values in $L_3$. The resulting screens are shown in Figure 14.

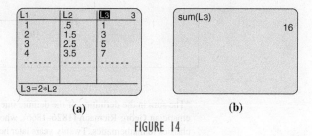

(a)     (b)

**FIGURE 14**

■■ Method 3
   Spreadsheet

**YOUR TURN 1**   Repeat

Example 1 to approximate $\int_1^5 4x\,dx$.

The calculations in this example can also be done on a spreadsheet. In Microsoft Excel, for example, store the values of $i$ in column A. Put the command "=A1-.5" into B1; copying this formula into the rest of column B gives the values of $x_i$. Similarly, use the formula for $f(x_i)$ to fill column C. Column D is the product of Column C and $\Delta x$. Sum column D to get the answer. For more details, see the *Graphing Calculator and Excel Spreadsheet Manual* available with this book.

**TRY YOUR TURN 1** ■■

**Total Change**   Suppose the function $f(x) = x^2 + 20$ gives the rate (in ml/hr) that an intravenous liquid flows into a patient at time $x$. Then $f(2) = 24$ means that the fluid is flowing at a rate of 24 ml/hr at 2 hours. If the fluid continued to flow at this rate, the patient would receive 24 ml over the next hour. But the rate is changing; because $f(3) = 29$, the fluid is flowing at a rate of 29 ml/hr at 3 hours.

To find the *total* amount of fluid the patient receives as $x$ (the time) changes from 2 to 3 (that is, the total change in the amount of fluid the patient has received), we could divide the interval from 2 to 3 into $n$ equal parts, using each part as the base of a rectangle as we did above. The area of each rectangle would approximate the amount of fluid the patient receives over the time interval indicated by the base of the rectangle. Then the sum of the areas of these rectangles would approximate the net total fluid flow from $x = 2$ to $x = 3$. The limit of this sum as $n \to \infty$ would give the exact total amount of fluid, or the total change in the amount of fluid the patient has received.

This result demonstrates another application of the definite integral. The area of the region that is under the graph of the rate of fluid flow function $f(x)$, and that is above the $x$-axis and between $x = a$ and $x = b$, gives the net total change in the fluid as $x$ goes from $a$ to $b$.

**Total Change in $F(x)$**
If $f(x)$ gives the rate of change of $F(x)$ for x in $[a,b]$, then the total change in $F(x)$ as $x$ goes from $a$ to $b$ is given by

$$\lim_{n\to\infty} \sum_{i=1}^{n} f(x_i)\Delta x = \int_a^b f(x)dx.$$

In other words, the total change in a quantity can be found from the function that gives the rate of change of the quantity, using the same methods used to approximate the area under a curve.

**EXAMPLE 3**   **Total Maintenance Charges**

Figure 15 shows the rate of change of the annual maintenance charges for a certain magnetic resonance imaging (MRI) scanner. To approximate the total maintenance charges over the 10-year life of the scanner, use approximating rectangles, dividing the interval from 0 to 10 into ten equal subdivisions. Each rectangle has width 1; using the left endpoint of each rectangle to determine the height of the rectangle, the approximation becomes

$$1 \cdot 0 + 1 \cdot 500 + 1 \cdot 750 + 1 \cdot 1800 + 1 \cdot 1900 + 1 \cdot 3000 + 1 \cdot 3100$$
$$+ 1 \cdot 3400 + 1 \cdot 4200 + 1 \cdot 5200 = 23{,}850.$$

About \$23,850 will be spent on maintenance over the 10-year life of the scanner.   ■

Recall, velocity is the rate of change in distance from time $a$ to time $b$. Thus the area under the velocity function defined by $v(t)$ from $t = a$ to $t = b$ gives the distance traveled in that time period.

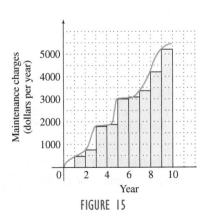

FIGURE 15

### EXAMPLE 4  Total Distance

A driver traveling on a business trip checks the speedometer each hour. The table shows the driver's velocity at several times.

Approximate the total distance traveled during the 3-hour period using the left endpoint of each interval, then the right endpoint.

| Velocity | | | | |
|---|---|---|---|---|
| Time (hr) | 0 | 1 | 2 | 3 |
| Velocity (mph) | 0 | 52 | 58 | 60 |

**YOUR TURN 2**  Repeat Example 4 for a driver traveling at the following velocities at various times.

| Time (hr) | 0 | 0.5 | 1 | 1.5 | 2 |
|---|---|---|---|---|---|
| Velocity (mph) | 0 | 50 | 56 | 40 | 48 |

**SOLUTION**  Using left endpoints, the total distance is

$$0 \cdot 1 + 52 \cdot 1 + 58 \cdot 1 = 110.$$

With right endpoints, we get

$$52 \cdot 1 + 58 \cdot 1 + 60 \cdot 1 = 170.$$

Again, left endpoints give a total that is too small, while right endpoints give a total that is too large. The average, 140 miles, is a better estimate of the total distance traveled.

**TRY YOUR TURN 2**

Before discussing further applications of the definite integral, we need a more efficient method for evaluating it. This method will be developed in the next section.

# 7.3  EXERCISES

1. Explain the difference between an indefinite integral and a definite integral.

2. Complete the following statement.

$$\int_0^4 (x^2 + 3)\, dx = \lim_{n \to \infty} \underline{\quad\quad}, \text{ where } \Delta x = \underline{\quad\quad}, \text{ and } x_i$$

is _____ .

3. Let $f(x) = 2x + 5$, $x_1 = 0$, $x_2 = 2$, $x_3 = 4$, $x_4 = 6$, and $\Delta x = 2$.

   a. Find $\sum_{i=1}^{4} f(x_i)\Delta x$.

   b. The sum in part a approximates a definite integral using rectangles. The height of each rectangle is given by the value of the function at the left endpoint. Write the definite integral that the sum approximates.

4. Let $f(x) = 1/x$, $x_1 = 1/2$, $x_2 = 1$, $x_3 = 3/2$, $x_4 = 2$, and $\Delta x = 1/2$.

   a. Find $\sum_{i=1}^{4} f(x_i)\Delta x$.

   b. The sum in part a approximates a definite integral using rectangles. The height of each rectangle is given by the value of the function at the left endpoint. Write the definite integral that the sum approximates.

In Exercises 5–14, approximate the area under the graph of $f(x)$ and above the $x$-axis using the following methods with $n = 4$. (a) Use left endpoints. (b) Use right endpoints. (c) Average the answers in parts a and b. (d) Use midpoints.

5. $f(x) = 2x + 5$  from $x = 2$ to $x = 4$

6. $f(x) = 3x + 2$  from $x = 1$ to $x = 3$

7. $f(x) = -x^2 + 4$  from $x = -2$ to $x = 2$

8. $f(x) = x^2$  from $x = 1$ to $x = 5$

9. $f(x) = e^x + 1$  from $x = -2$ to $x = 2$

10. $f(x) = e^x - 1$  from $x = 0$ to $x = 4$

11. $f(x) = \dfrac{2}{x}$  from $x = 1$ to $x = 9$

12. $f(x) = \dfrac{1}{x}$  from $x = 1$ to $x = 3$

13. $f(x) = \sin x$  from $x = 0$ to $x = \pi$

14. $f(x) = \ln x$  from $x = 1$ to $x = 5$

15. Consider the region below $f(x) = x/2$, above the $x$-axis, and between $x = 0$ and $x = 4$. Let $x_i$ be the midpoint of the $i$th subinterval.

   a. Approximate the area of the region using four rectangles.

   b. Find $\int_0^4 f(x)\, dx$ by using the formula for the area of a triangle.

**16.** Consider the region below $f(x) = 5 - x$, above the $x$-axis, and between $x = 0$ and $x = 5$. Let $x_i$ be the midpoint of the $i$th subinterval.

    **a.** Approximate the area of the region using five rectangles.

    **b.** Find $\int_0^5 (5 - x)\, dx$ by using the formula for the area of a triangle.

**17.** Find $\int_0^4 f(x)\, dx$ for each graph of $y = f(x)$.

    **a.** 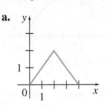     **b.**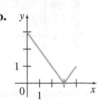

**18.** Find $\int_0^6 f(x)\, dx$ for each graph of $y = f(x)$, where $f(x)$ consists of line segments and circular arcs.

    **a.** 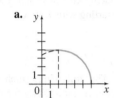     **b.**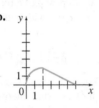

**Find the exact value of each integral using formulas from geometry.**

**19.** $\displaystyle\int_{-4}^{0} \sqrt{16 - x^2}\, dx$

**20.** $\displaystyle\int_{-3}^{3} \sqrt{9 - x^2}\, dx$

**21.** $\displaystyle\int_{2}^{5} (1 + 2x)\, dx$

**22.** $\displaystyle\int_{1}^{3} (5 + x)\, dx$

**23.** In this exercise, we investigate the value of $\int_0^1 x^2\, dx$ using larger and larger values of $n$ in the definition of the definite integral.

    **a.** First let $n = 10$, so $\Delta x = 0.1$. Fill a list on your calculator with values of $x^2$ as $x$ goes from 0.1 to 1. (On a TI-84 Plus, use the command `seq(X^2, X, .1,1, .1)` $\rightarrow$ L1.)

    **b.** Sum the values in the list formed in part a, and multiply by 0.1, to estimate $\int_0^1 x^2\, dx$ with $n = 10$. (On a TI-84 Plus, use the command `.1*sum(L1)`.)

    **c.** Repeat parts a and b with $n = 100$.

    **d.** Repeat parts a and b with $n = 500$.

    **e.** Based on your answers to parts b through d, what do you estimate the value of $\int_0^1 x^2\, dx$ to be?

**24.** Repeat Exercise 23 for $\int_0^1 x^3\, dx$.

**25.** The booklet *All About Lawns* published by Ortho Books gives instructions for approximating the area of an irregularly shaped region. (See figure.) *Source: All About Lawns.* The booklet suggests measuring a long axis of the area, such as the vertical line in the figure. Then, every 10 feet along this axis, measure the width at right angles to the axis line. Sum these widths and multiply by 10. In the figure, for example, suppose the lengths of lines $L_1$, $L_2$, and $L_3$ are 34 ft, 48 ft, and 30 ft, respectively. Then

$$\text{Area} = (34 + 48 + 30)10 = 1120 \text{ square ft.}$$

How does this method relate to the discussion in this section?

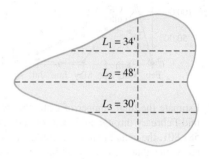

## LIFE SCIENCE APPLICATIONS

**In Exercises 26–31, estimate the area under each curve by summing the area of rectangles. Use the left endpoints, then the right endpoints, then give the average of those answers.**

**26. Oxygen Inhalation** The graph below shows the rate of inhalation of oxygen (in liters per minute) by a person riding a bicycle very rapidly for 10 minutes. Estimate the total volume of oxygen inhaled in the first 20 minutes after the beginning of the ride. Use rectangles with widths of 1 minute.

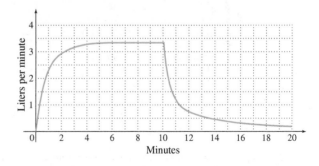

**27.** **APPLY IT** **Foot-and-Mouth Epidemic** In 2001, the United Kingdom suffered an epidemic of foot-and-mouth disease. The graph below shows the reported number of cattle (red) and pigs (blue) that were culled each month from mid-February through mid-October in an effort to stop the spread of the disease. *Source: Department of Environment, Food and Rural Affairs, United Kingdom.*

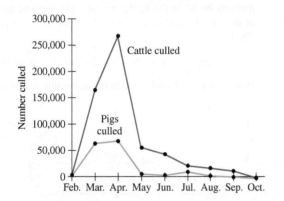

**a.** Estimate the total number of cattle that were culled from mid-February through mid-October and compare this with 581,801, the actual number of cattle that were culled. Use rectangles that are one month in width, starting with mid-February.

**b.** Estimate the total number of pigs that were culled from mid-February through mid-October and compare this with 146,145, the actual number of pigs that were culled. Use rectangles that are one month in width starting with mid-February.

**28.** **Alcohol Concentration** The following graph shows the approximate concentration of alcohol in a person's bloodstream $t$ hr after drinking 2 oz of alcohol. Estimate the total amount of alcohol in the bloodstream by estimating the area under the curve. Use rectangles with widths of 1 hr.

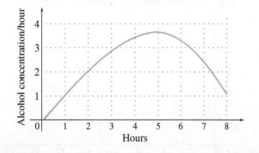

**29.** **Cliff Ecology** Cliffs form sheltered environments for a large number of plants and animals. An article on the ecology of cliffs included the diagram in the next column which gives the daily insolation (calories per square centimeter per hour) on two cliffs, one facing north and one facing south. *Source: American Scientist.*

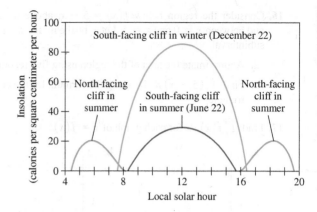

**a.** Estimate the total calories per square centimeter each day for the south-facing cliff in summer. Use rectangles with widths of 1 hr, starting with 8:15 A.M., and with the last rectangle (ending at 3:45) with a width of 1/2 hr.

**b.** Repeat part a for the south-facing cliff in winter. Use rectangles with widths of 1 hr, starting with 7:30 A.M.

## OTHER APPLICATIONS

**30.** **Wind Energy Consumption** The following graph shows the U.S. wind energy consumption (trillion BTU) for various years. Estimate the total consumption for the 12-year period from 2000 to 2012 using rectangles of width 3 years. *Source: Annual Energy Review.*

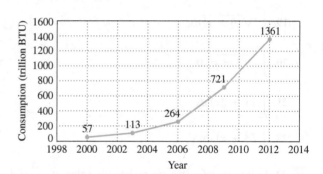

**31.** **Automobile Accidents** The graph shows the number of fatal automobile accidents in California for various years. Estimate the total number of accidents in the 8-year period from 2002 to 2010 using rectangles of width 2 years. *Source: California Highway Patrol.*

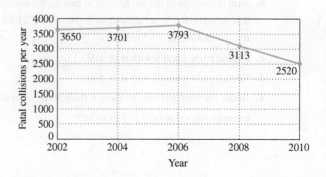

**Distance** The next two graphs are from the Road & Track website. The curves show the velocity at *t* seconds after the car accelerates from a dead stop. To find the total distance traveled by the car in reaching 130 mph, we must estimate the definite integral

$$\int_0^T v(t)\, dt,$$

where *T* represents the number of seconds it takes for the car to reach 130 mph.

Use the graphs to estimate this distance by adding the areas of rectangles and using the midpoint rule. To adjust your answer to miles per hour, divide by 3600 (the number of seconds in an hour). You then have the number of miles that the car traveled in reaching 130 mph. Finally, multiply by 5280 ft per mile to convert the answer to feet. *Source: Road & Track.*

**32.** Estimate the distance traveled by the Lamborghini Gallardo LP560-4 using the graph below. Use rectangles with widths of 3 seconds, except for the last rectangle, which should have a width of 2 seconds. The circle marks the point where the car has gone a quarter mile. Does this seem correct?

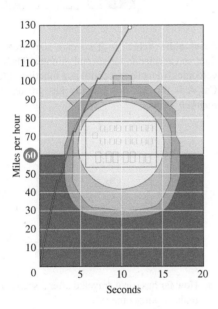

**33.** Estimate the distance traveled by the Alfa Romeo 8C Competizione using the graph in the next column. Use rectangles with widths of 4 seconds, except for the last rectangle, which should have a width of 3.5 seconds. The circle marks the point where the car has gone a quarter mile. Does this seem correct?

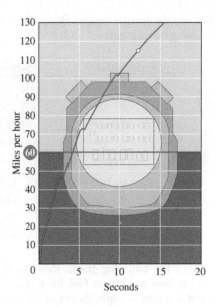

**Distance** When data are given in tabular form, you may need to vary the size of the interval to calculate the area under the curve. The next two exercises include data from *Car and Driver* magazine. To estimate the total distance traveled by the car (in feet) during the time it took to reach its maximum velocity, estimate the area under the velocity versus time graph, as in the previous two exercises. Use the left endpoint for each time interval (the velocity at the beginning of that interval) and then the right endpoint (the velocity at the end of the interval). Finally, average the two answers together. Calculating and adding up the areas of the rectangles is most easily done on a spreadsheet or graphing calculator. As in the previous two exercises, you will need to multiply by a conversion factor of 5280/3600 = 22/15, since the velocities are given in miles per hour, but the time is in seconds, and we want the answer in feet. *Source: Car and Driver.*

**34.** Estimate the distance traveled by the Mercedes-Benz S550, using the table below.

| Acceleration | Seconds |
|---|---|
| Zero to 30 mph | 2.0 |
| 40 mph | 2.9 |
| 50 mph | 4.1 |
| 60 mph | 5.3 |
| 70 mph | 6.9 |
| 80 mph | 8.7 |
| 90 mph | 10.7 |
| 100 mph | 13.2 |
| 110 mph | 16.1 |
| 120 mph | 19.3 |
| 130 mph | 23.4 |

**35.** Estimate the distance traveled by the Chevrolet Malibu Maxx SS, using the table below.

| Acceleration | Seconds |
|---|---|
| Zero to 30 mph | 2.4 |
| 40 mph | 3.5 |
| 50 mph | 5.1 |
| 60 mph | 6.9 |
| 70 mph | 8.9 |
| 80 mph | 11.2 |
| 90 mph | 14.9 |
| 100 mph | 19.2 |
| 110 mph | 24.4 |

**Heat Gain** The following graphs show the typical heat gain, in BTU per hour per square foot, for windows (one with plain glass and one that is triple glazed) in Pittsburgh in June, one facing east and one facing south. The horizontal axis gives the time of the day. Estimate the total heat gain per square foot by summing the areas of rectangles. Use rectangles with widths of 2 hours, and let the function value at the midpoint of the subinterval give the height of the rectangle. *Source: Sustainable by Design.*

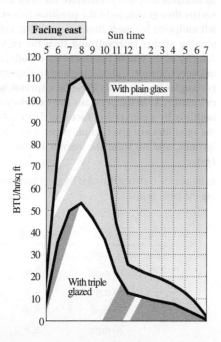

**36. a.** Estimate the total heat gain per square foot for a plain glass window facing east.

   **b.** Estimate the total heat gain per square foot for a triple glazed window facing east.

**37. a.** Estimate the total heat gain per square foot for a plain glass window facing south.

**b.** Estimate the total heat gain per square foot for a triple glazed window facing south.

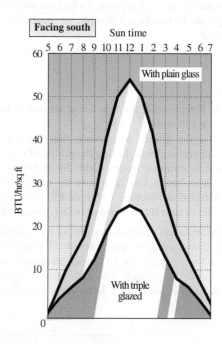

**38. Automobile Velocity** Two cars start from rest at a traffic light and accelerate for several minutes. The graph shows their velocities (in feet per second) as a function of time (in seconds). Car A is the one that initially has greater velocity. *Source: Stephen Monk.*

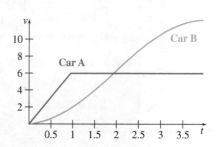

**a.** How far has car A traveled after 2 seconds? (*Hint:* Use formulas from geometry.)

**b.** When is car A farthest ahead of car B?

**c.** Estimate the farthest that car A gets ahead of car B. For car A, use formulas from geometry. For car B, use $n = 4$ and the value of the function at the midpoint of each interval.

**d.** Give a rough estimate of when car B catches up with car A.

**39. Distance** Musk the friendly pit bull has escaped again! Here is her velocity during the first 4 seconds of her romp.

| t (sec) | 0 | 1 | 2 | 3 | 4 |
|---|---|---|---|---|---|
| v (ft/sec) | 0 | 8 | 13 | 17 | 18 |

Give two estimates for the total distance Musk traveled during her 4-second trip, one using the left endpoint of each interval and one using the right endpoint.

**40. Distance** The speed of a particle in a test laboratory was noted every second for 3 seconds. The results are shown in the following table. Use the left endpoints and then the right endpoints to estimate the total distance the particle moved in the first three seconds.

| $t$ (sec)     | 0   | 1   | 2   | 3   |
|---------------|-----|-----|-----|-----|
| $v$ (ft/sec)  | 10  | 6.5 | 6   | 5.5 |

**41. Running** In 1987, Canadian Ben Johnson set a world record in the 100-m sprint. (The record was later taken away when he was found to have used an anabolic steroid to enhance his performance.) His speed at various times in the race is given in the following table*. ***Source: Information Graphics.***

| Time (sec) | Speed (mph) |
|------------|-------------|
| 0          | 0           |
| 1.84       | 12.9        |
| 3.80       | 23.8        |
| 6.38       | 26.3        |
| 7.23       | 26.3        |
| 8.96       | 26.0        |
| 9.83       | 25.7        |

**a.** Use the information in the table and left endpoints to estimate the distance that Johnson ran in miles. You will first need to calculate $\Delta t$ for each interval. At the end, you will need to divide by 3600 (the number of seconds in an hour), since the speed is in miles per hour.

**b.** Repeat part a, using right endpoints.

**c.** Wait a minute; we know that the distance Johnson ran is 100 m. Divide this by 1609, the number of meters in a mile, to find how far Johnson ran in miles. Is your answer from part a or part b closer to the true answer? Briefly explain why you think this answer should be more accurate.

**42. Traffic** The following graph shows the number of vehicles per hour crossing the Tappan Zee Bridge, which spans the Hudson River north of New York City. The graph shows the number of vehicles traveling eastbound (into the city) and westbound (out of the city) as a function of time. ***Source: The New York Times.***

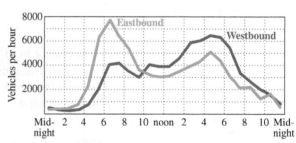

Source: New York Metropolitan Transportation Council

**a.** Using midpoints on intervals of one hour, estimate the total number of vehicles that cross the bridge going eastbound each day.

**b.** Repeat the instructions for part a for vehicles going westbound.

**c.** Discuss whether the answers to parts a and b should be equal, and try to explain any discrepancies.

**YOUR TURN ANSWERS**

**1.** 48

**2.** Left endpoint estimate is 73 miles, right endpoint estimate is 97 miles, and average is 85 miles.

*The world record of 9.58 seconds is currently held by Usain Bolt of Jamaica.

# 7.4 The Fundamental Theorem of Calculus

APPLY IT **What is the total energy expenditure of a pronghorn fawn between 1 and 12 weeks of age?**

*We will answer this question in Example 8.*

In the first section of this chapter, you learned about antiderivatives. In the previous section, you learned about the definite integral. In this section, we connect these two separate topics and present one of the most powerful theorems of calculus.

We have seen that, if $f(x) \geq 0$,

$$\int_a^b f(x)\, dx$$

gives the area between the graph of $f(x)$ and the $x$-axis, from $x = a$ to $x = b$. The definite integral was defined and evaluated in the previous section using the limit of a sum. In that section, we also saw that if $f(x)$ gives the rate of change of $F(x)$, the definite integral $\int_a^b f(x)\, dx$ gives the total change of $F(x)$ as $x$ changes from $a$ to $b$. If $f(x)$ gives the rate of change of $F(x)$, then $F(x)$ is an antiderivative of $f(x)$. Writing the total change in $F(x)$ from $x = a$ to $x = b$ as $F(b) - F(a)$ shows the connection between antiderivatives and definite integrals. This relationship is called the **Fundamental Theorem of Calculus**.

### Fundamental Theorem of Calculus

Let $f$ be continuous on the interval $[a, b]$, and let $F$ be *any* antiderivative of $f$. Then

$$\int_a^b f(x)\, dx = F(b) - F(a) = F(x)\Big|_a^b.$$

The symbol $F(x)\big|_a^b$ is used to represent $F(b) - F(a)$. It is important to note that the Fundamental Theorem does not require $f(x) > 0$. The condition $f(x) > 0$ is necessary only when using the Fundamental Theorem to find area. Also, note that the Fundamental Theorem does not *define* the definite integral; it just provides a method for evaluating it.

### EXAMPLE 1   Fundamental Theorem of Calculus

First find $\int 4t^3\, dt$ and then find $\int_1^2 4t^3\, dt$.

**SOLUTION**   By the power rule given earlier, the indefinite integral is

$$\int 4t^3\, dt = t^4 + C.$$

By the Fundamental Theorem, the value of the definite integral $\int_1^2 4t^3\, dt$ is found by evaluating $t^4\big|_1^2$, with no constant $C$ required.

**YOUR TURN 1**   Find $\displaystyle\int_1^3 3x^2\, dx$.

$$\int_1^2 4t^3\, dt = t^4\Big|_1^2 = 2^4 - 1^4 = 15 \qquad \text{TRY YOUR TURN 1}$$

Example 1 illustrates the difference between the definite integral and the indefinite integral. A definite integral is a real number; an indefinite integral is a family of functions in which all the functions are antiderivatives of a function $f$.

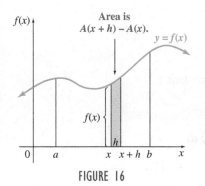

FIGURE 16

**NOTE** No constant $C$ is needed, as it is for the indefinite integral, because even if $C$ were added to an antiderivative $F$, it would be eliminated in the final answer:

$$\int_a^b f(x)\, dx = (F(x) + C)\Big|_a^b$$

$$= (F(b) + C) - (F(a) + C)$$

$$= F(b) - F(a).$$

In other words, any antiderivative will give the same answer, so for simplicity, we choose the one with $C = 0$.

To see why the Fundamental Theorem of Calculus is true for $f(x) > 0$ when $f$ is continuous, look at Figure 16. Define the function $A(x)$ as the area between the $x$-axis and the graph of $y = f(x)$ from $a$ to $x$. We first show that $A$ is an antiderivative of $f$; that is, $A'(x) = f(x)$.

To do this, let $h$ be a small positive number. Then $A(x + h) - A(x)$ is the shaded area in Figure 16. This area can be approximated with a rectangle having width $h$ and height $f(x)$. The area of the rectangle is $h \cdot f(x)$, and

$$A(x + h) - A(x) \approx h \cdot f(x).$$

Dividing both sides by $h$ gives

$$\frac{A(x + h) - A(x)}{h} \approx f(x).$$

This approximation improves as $h$ gets smaller and smaller. Taking the limit on the left as $h$ approaches 0 gives an exact result.

$$\lim_{h \to 0} \frac{A(x + h) - A(x)}{h} = f(x)$$

This limit is simply $A'(x)$, so

$$A'(x) = f(x).$$

This result means that $A$ is an antiderivative of $f$, as we set out to show.

$A(b)$ is the area under the curve from $a$ to $b$, and $A(a) = 0$, so the area under the curve can be written as $A(b) - A(a)$. From the previous section, we know that the area under the curve is also given by $\int_a^b f(x)\, dx$. Putting these two results together gives

$$\int_a^b f(x)\, dx = A(b) - A(a)$$

$$= A(x)\Big|_a^b$$

where $A$ is an antiderivative of $f$. From the note after Example 1, we know that any antiderivative will give the same answer, which proves the Fundamental Theorem of Calculus.

The Fundamental Theorem of Calculus certainly deserves its name, which sets it apart as the most important theorem of calculus. It is the key connection between differential calculus and integral calculus, which originally were developed separately without knowledge of this connection between them.

The variable used in the integrand does not matter; each of the following definite integrals represents the number $F(b) - F(a)$.

$$\int_a^b f(x)\, dx = \int_a^b f(t)\, dt = \int_a^b f(u)\, du$$

Key properties of definite integrals are listed on the next page. Some of them are just restatements of properties from Section 1.

## Properties of Definite Integrals

If all indicated definite integrals exist,

1. $\displaystyle\int_a^a f(x)\,dx = 0;$

2. $\displaystyle\int_a^b k \cdot f(x)\,dx = k \cdot \int_a^b f(x)\,dx$ for any real constant $k$
(constant multiple of a function);

3. $\displaystyle\int_a^b [f(x) \pm g(x)]\,dx = \int_a^b f(x)\,dx \pm \int_a^b g(x)\,dx$
(sum or difference of functions);

4. $\displaystyle\int_a^b f(x)\,dx = \int_a^c f(x)\,dx + \int_c^b f(x)\,dx$ for any real number $c$;

5. $\displaystyle\int_a^b f(x)\,dx = -\int_b^a f(x)\,dx.$

For $f(x) \geq 0$, since the distance from $a$ to $a$ is 0, the first property says that the "area" under the graph of $f$ bounded by $x = a$ and $x = a$ is 0. Also, since $\int_a^c f(x)\,dx$ represents the blue region in Figure 17 and $\int_c^b f(x)\,dx$ represents the pink region,

$$\int_a^b f(x)\,dx = \int_a^c f(x)\,dx + \int_c^b f(x)\,dx,$$

as stated in the fourth property. While the figure shows $a < c < b$, the property is true for any value of $c$ where both $f(x)$ and $F(x)$ are defined.

An algebraic proof is given here for the third property; proofs of the other properties are left for the exercises. If $F(x)$ and $G(x)$ are antiderivatives of $f(x)$ and $g(x)$, respectively,

$$\int_a^b [f(x) + g(x)]\,dx = [F(x) + G(x)]\Big|_a^b$$
$$= [F(b) + G(b)] - [F(a) + G(a)]$$
$$= [F(b) - F(a)] + [G(b) - G(a)]$$
$$= \int_a^b f(x)\,dx + \int_a^b g(x)\,dx.$$

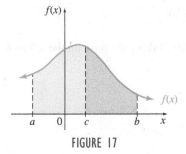

**FIGURE 17**

### EXAMPLE 2 Fundamental Theorem of Calculus

Find $\int_2^5 (6x^2 - 3x + 5)\,dx$.

**SOLUTION** Use the properties above and the Fundamental Theorem, along with properties from Section 1.

$$\int_2^5 (6x^2 - 3x + 5)\,dx = 6\int_2^5 x^2\,dx - 3\int_2^5 x\,dx + 5\int_2^5 dx$$
$$= 2x^3\Big|_2^5 - \frac{3}{2}x^2\Big|_2^5 + 5x\Big|_2^5$$
$$= 2(5^3 - 2^3) - \frac{3}{2}(5^2 - 2^2) + 5(5 - 2)$$
$$= 2(125 - 8) - \frac{3}{2}(25 - 4) + 5(3)$$
$$= 234 - \frac{63}{2} + 15 = \frac{435}{2} \qquad \textbf{TRY YOUR TURN 2}$$

**YOUR TURN 2**

Find $\displaystyle\int_3^5 (2x^3 - 3x + 4)\,dx$.

> **EXAMPLE 3** **Fundamental Theorem of Calculus**

$$\int_1^2 \frac{dy}{y} = \ln |y|\Big|_1^2 = \ln |2| - \ln |1|$$

**YOUR TURN 3** Find $\int_1^3 \frac{2}{y} dy$.

$$= \ln 2 - \ln 1 \approx 0.6931 - 0 = 0.6931$$

**TRY YOUR TURN 3**

> **EXAMPLE 4** **Substitution**

Evaluate $\int_0^5 x\sqrt{25 - x^2}\, dx$.

**SOLUTION**

**Method 1**
**Changing the Limits**

Use substitution. Let $u = 25 - x^2$, so that $du = -2x\, dx$. With a definite integral, the limits should be changed, too. The new limits on $u$ are found as follows.

$$\text{If } x = 5, \text{ then } u = 25 - 5^2 = 0.$$
$$\text{If } x = 0, \text{ then } u = 25 - 0^2 = 25.$$

Then

$$\int_0^5 x\sqrt{25 - x^2}\, dx = -\frac{1}{2}\int_0^5 \sqrt{25 - x^2}(-2x\, dx) \qquad \text{Multiply by } -2/-2.$$

$$= -\frac{1}{2}\int_{25}^0 \sqrt{u}\, du \qquad \text{Substitute and change limits.}$$

$$= -\frac{1}{2}\int_{25}^0 u^{1/2}\, du \qquad \sqrt{u} = u^{1/2}$$

$$= -\frac{1}{2}\cdot\frac{u^{3/2}}{3/2}\Big|_{25}^0 \qquad \text{Use the power rule.}$$

$$= -\frac{1}{2}\cdot\frac{2}{3}[0^{3/2} - 25^{3/2}]$$

$$= -\frac{1}{3}(-125) = \frac{125}{3}.$$

**Method 2**
**Evaluating the Antiderivative**

An alternative method that some people prefer is to evaluate the antiderivative first and then calculate the definite integral. To evaluate the antiderivative in this example, ignore the limits on the original integral and use the substitution $u = 25 - x^2$, so that $du = -2x\, dx$. Then

$$\int x\sqrt{25 - x^2}\, dx = -\frac{1}{2}\int \sqrt{25 - x^2}(-2x\, dx)$$

$$= -\frac{1}{2}\int \sqrt{u}\, du$$

$$= -\frac{1}{2}\int u^{1/2}\, du$$

$$= -\frac{1}{2}\frac{u^{3/2}}{3/2} + C$$

$$= -\frac{u^{3/2}}{3} + C$$

$$= -\frac{(25 - x^2)^{3/2}}{3} + C.$$

We will ignore the constant $C$ because it doesn't affect the answer, as we mentioned in the Note following Example 1.

Then, using the Fundamental Theorem of Calculus, we have

$$\int_0^5 x\sqrt{25 - x^2}\, dx = -\left.\frac{(25 - x^2)^{3/2}}{3}\right|_0^5$$

$$= 0 - \left[-\frac{(25)^{3/2}}{3}\right]$$

$$= \frac{125}{3}.$$

TRY YOUR TURN 4

**YOUR TURN 4**

Evaluate $\displaystyle\int_0^4 2x\sqrt{16 - x^2}\, dx$.

---

**CAUTION** Don't confuse these two methods. In Method 1, we never return to the original variable or the original limits of integration. In Method 2, it is essential to return to the original variable and to not change the limits. When using Method 1, we recommend labeling the limits with the appropriate variable to avoid confusion, so the substitution in Example 4 becomes

$$\int_{x=0}^{x=5} x\sqrt{25 - x^2}\, dx = -\frac{1}{2}\int_{u=25}^{u=0} \sqrt{u}\, du.$$

The Fundamental Theorem of Calculus is a powerful tool, but it has a limitation. The problem is that not every function has an antiderivative in terms of the functions and operations you have seen so far. One example of an integral that cannot be evaluated by the Fundamental Theorem of Calculus for this reason is

$$\int_a^b e^{-x^2/2}\, dx,$$

yet this integral is crucial in probability and statistics. Such integrals may be evaluated by numerical integration, which is covered in the next chapter. Fortunately for you, all the integrals in this section can be antidifferentiated using the techniques presented in the first two sections of this chapter.

## Area

In the previous section we saw that, if $f(x) \geq 0$ in $[a, b]$, the definite integral $\int_a^b f(x)\, dx$ gives the area below the graph of the function $y = f(x)$, above the $x$-axis, and between the lines $x = a$ and $x = b$.

To see how to work around the requirement that $f(x) \geq 0$, look at the graph of $f(x) = x^2 - 4$ in Figure 18. The area bounded by the graph of $f$, the $x$-axis, and the vertical lines $x = 0$ and $x = 2$ lies below the $x$-axis. Using the Fundamental Theorem gives

$$\int_0^2 (x^2 - 4)\, dx = \left.\left(\frac{x^3}{3} - 4x\right)\right|_0^2$$

$$= \left(\frac{8}{3} - 8\right) - (0 - 0) = -\frac{16}{3}.$$

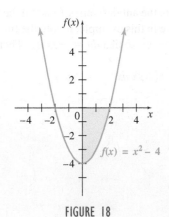

FIGURE 18

The result is a negative number because $f(x)$ is negative for values of $x$ in the interval $[0, 2]$. Since $\Delta x$ is always positive, if $f(x) < 0$ the product $f(x) \cdot \Delta x$ is negative, so $\int_0^2 f(x)\, dx$ is negative. Since area is nonnegative, the required area is given by $|-16/3|$ or $16/3$. Using a definite integral, the area could be written as

$$\left|\int_0^2 (x^2 - 4)\, dx\right| = \left|-\frac{16}{3}\right| = \frac{16}{3}.$$

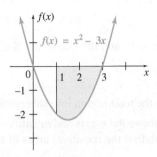

**FIGURE 19**

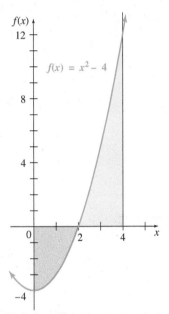

**FIGURE 20**

**YOUR TURN 5**   Repeat
Example 6 for the function
$f(x) = x^2 - 9$ from $x = 0$
to $x = 6$.

## EXAMPLE 5   Area

Find the area of the region between the $x$-axis and the graph of $f(x) = x^2 - 3x$ from $x = 1$ to $x = 3$.

**SOLUTION**   The region is shown in Figure 19. Since the region lies below the $x$-axis, the area is given by

$$\left| \int_1^3 (x^2 - 3x)\, dx \right|.$$

By the Fundamental Theorem,

$$\int_1^3 (x^2 - 3x)\, dx = \left( \frac{x^3}{3} - \frac{3x^2}{2} \right) \Big|_1^3 = \left( \frac{27}{3} - \frac{27}{2} \right) - \left( \frac{1}{3} - \frac{3}{2} \right) = -\frac{10}{3}.$$

The required area is $|-10/3| = 10/3$.

## EXAMPLE 6   Area

Find the area between the $x$-axis and the graph of $f(x) = x^2 - 4$ from $x = 0$ to $x = 4$.

**SOLUTION**   Figure 20 shows the required region. Part of the region is below the $x$-axis. The definite integral over that interval will have a negative value. To find the area, integrate the negative and positive portions separately and take the absolute value of the first result before combining the two results to get the total area. Start by finding the point where the graph crosses the $x$-axis. This is done by solving the equation

$$x^2 - 4 = 0.$$

The solutions of this equation are 2 and $-2$. The only solution in the interval $[0, 4]$ is 2. The total area of the region in Figure 20 is

$$\left| \int_0^2 (x^2 - 4)\, dx \right| + \int_2^4 (x^2 - 4)\, dx = \left| \left( \frac{1}{3}x^3 - 4x \right) \Big|_0^2 \right| + \left( \frac{1}{3}x^3 - 4x \right) \Big|_2^4$$

$$= \left| \frac{8}{3} - 8 \right| + \left( \frac{64}{3} - 16 \right) - \left( \frac{8}{3} - 8 \right)$$

$$= 16.$$

TRY YOUR TURN 5

Incorrectly using one integral over the entire interval to find the area in Example 6 would have given

$$\int_0^4 (x^2 - 4)\, dx = \left( \frac{x^3}{3} - 4x \right) \Big|_0^4 = \left( \frac{64}{3} - 16 \right) - 0 = \frac{16}{3},$$

which is not the correct area. This definite integral does not represent any area but is just a real number.

For instance, if $f(x)$ in Example 6 represents the annual rate of profit of a company, then 16/3 represents the total profit for the company over a 4-year period. The integral between 0 and 2 is $-16/3$; the negative sign indicates a loss for the first two years. The integral between 2 and 4 is 32/3, indicating a profit. The overall profit is $32/3 - 16/3 = 16/3$, although the total shaded area is $32/3 + |-16/3| = 16$.

### Finding Area

In summary, to find the area bounded by $f(x)$, $x = a$, $x = b$, and the $x$-axis, use the following steps.

1. Sketch a graph.
2. Find any $x$-intercepts of $f(x)$ in $[a, b]$. These divide the total region into subregions.
3. The definite integral will be *positive* for subregions above the $x$-axis and *negative* for subregions below the $x$-axis. Use separate integrals to find the (positive) areas of the subregions.
4. The total area is the sum of the areas of all of the subregions.

---

**EXAMPLE 7**   **Area Under the Curve**

Find the shaded area in Figure 21.

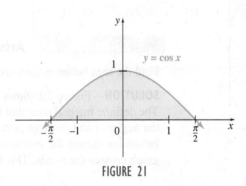

**FIGURE 21**

**SOLUTION**   The shaded area in Figure 21 is bounded by $y = \cos x$, $x = -\pi/2$, $x = \pi/2$, and the $x$-axis. By the Fundamental Theorem of Calculus, this area is given by

$$\int_{-\pi/2}^{\pi/2} \cos x \, dx = \sin x \Big|_{-\pi/2}^{\pi/2}$$

$$= \sin \frac{\pi}{2} - \sin\left(-\frac{\pi}{2}\right)$$

$$= 1 - (-1)$$

$$= 2.$$

**YOUR TURN 6**   Find the area under the curve $y = \sec^2(x/3)$ between $x = -\pi$ and $x = \pi$.

By symmetry, the same area could be found by evaluating

$$2\int_{0}^{\pi/2} \cos x \, dx.$$

TRY YOUR TURN 6

---

📐 **TECHNOLOGY NOTE**

The area in Example 7 could also be found using the definite integral feature of a graphing calculator, entering the expression $\cos x$, the variable $x$, and the limits of integration.

---

In the last section, we saw that the area under a rate of change function $f'(x)$ from $x = a$ to $x = b$ gives the total value of $f(x)$ on $[a, b]$. Now we can use the definite integral to solve these problems.

EXAMPLE 8   **Energy in Pronghorn Fawns**

The total energy expenditure of pronghorn fawns increases at a roughly linear rate with age, from an expenditure of about 2000 kJ (kilojoules) a day at 1 week to about 9000 kJ a day at 12 weeks. *Source: Animal Behavior.**\* Find the total energy expended by a typical pronghorn fawn over this 11-week (77-day) period.

APPLY IT   **SOLUTION**   At $t = 7$ days, $E = 2000$, while at $t = 12 \cdot 7 = 84$ days, $E = 9000$. To find $E$ as a linear function of $t$, first find the slope.

$$m = \frac{9000 - 2000}{84 - 7} = \frac{7000}{77} \approx 90.909$$

Using the point-slope form of the line,

$$y - 2000 = 90.909(x - 7)$$
$$y - 2000 = 90.909x - 636$$
$$y = 90.909x + 1364.$$

Now integrate this function to get the total amount of energy.

$$\int_{7}^{84} (90.909x + 1364)\, dx = \left[ 90.909\left(\frac{x^2}{2}\right) + 1364x \right]\Bigg|_{7}^{84}$$
$$= 435{,}303 - 11{,}775 = 423{,}528$$

A pronghorn fawn uses about 423,500 kJ of energy between 1 and 12 weeks.

TECHNOLOGY

EXAMPLE 9   **Precipitation in Vancouver, Canada**

The average monthly precipitation (in inches) for Vancouver, Canada, is found in the following table. *Source: Weather.com.*

| Month | Precipitation (inches) |
|---|---|
| January | 5.9 |
| February | 4.9 |
| March | 4.3 |
| April | 3.0 |
| May | 2.4 |
| June | 1.8 |
| July | 1.4 |
| August | 1.5 |
| September | 2.5 |
| October | 4.5 |
| November | 6.7 |
| December | 7.0 |

\*In Exercise 61 of Section 4.5, we saw how the energy depends on the mass of the fawn.

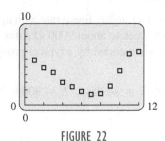

FIGURE 22

10

0

0                    12

FIGURE 23

**(a)** Plot the data, letting $t = 1$ correspond to January, $t = 2$ to February, and so on. Is it reasonable to assume that average monthly precipitation is periodic?

**SOLUTION** Figure 22 shows a graphing calculator plot of the data. Because of the cyclical nature of the four seasons, it is reasonable to assume that the data are periodic.

**(b)** Find a trigonometric function of the form $C(t) = a \sin(bt + c) + d$ that models this data when $t$ is the month and $C(t)$ is the amount of precipitation. Use the table.

**SOLUTION** The function $C(t)$, derived by the sine regression feature on a TI-84 Plus calculator, is given by

$$C(t) = 2.56349 \sin(0.528143t + 1.30663) + 3.79128.$$

Figure 23 shows that this function fits the data fairly well.

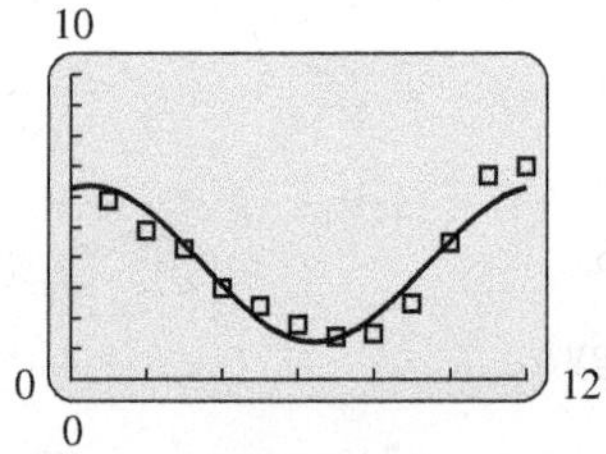

**(c)** Estimate the amount of precipitation for the month of October, and compare it to the actual value.

**SOLUTION** $C(10) \approx 4.6$ in. The actual value is 4.5 inches.

**(d)** Estimate the rate at which the amount of precipitation is changing in October.

**SOLUTION** The derivative of $C(t)$ is

$$C'(t) = (0.528143)2.56349 \cos(0.528143t + 1.30663)$$
$$= 1.35389 \cos(0.528143t + 1.30663).$$

In October, $t = 10$ and

$$C'(10) \approx 1.29 \text{ inches per month.}$$

**(e)** Estimate the total precipitation for the year and compare it to the actual value.

**SOLUTION** To estimate the total precipitation, we use integration as follows.

$$\int_0^{12} C(t)\, dt = \int_0^{12} (2.56349 \sin(0.528143t + 1.30663) + 3.79128)\, dt$$

$$= \left[ -\frac{2.56349}{0.528143} \cos(0.528143t + 1.30663) + 3.79128t \right]\Bigg|_0^{12}$$

$$\approx 45.75 \text{ inches}$$

The actual value is 45.9 inches.

**(f)** What would you expect the period of a function that models annual precipitation to be? What is the period of the function found in part (b)?

**SOLUTION** If we assume that the annual rainfall in Vancouver is periodic, we would expect the period to be 12 so that it repeats itself every 12 months. The period for the function given above is

$$T = 2\pi/b = 2\pi/0.528143 \approx 11.90,$$

or about 12 months.

# 7.4 EXERCISES

**Evaluate each definite integral.**

**1.** $\int_{-2}^{4} (-3)\, dp$

**2.** $\int_{-4}^{1} \sqrt{2}\, dx$

**3.** $\int_{-1}^{2} (5t - 3)\, dt$

**4.** $\int_{-2}^{2} (4z + 3)\, dz$

**5.** $\int_{0}^{2} (5x^2 - 4x + 2)\, dx$

**6.** $\int_{-2}^{3} (-x^2 - 3x + 5)\, dx$

**7.** $\int_{0}^{2} 3\sqrt{4u + 1}\, du$

**8.** $\int_{3}^{9} \sqrt{2r - 2}\, dr$

**9.** $\int_{0}^{4} 2(t^{1/2} - t)\, dt$

**10.** $\int_{0}^{4} -(3x^{3/2} + x^{1/2})\, dx$

**11.** $\int_{1}^{4} (5y\sqrt{y} + 3\sqrt{y})\, dy$

**12.** $\int_{4}^{9} (4\sqrt{r} - 3r\sqrt{r})\, dr$

**13.** $\int_{4}^{6} \frac{2}{(2x - 7)^2}\, dx$

**14.** $\int_{1}^{4} \frac{-3}{(2p + 1)^2}\, dp$

**15.** $\int_{1}^{5} (6n^{-2} - n^{-3})\, dn$

**16.** $\int_{2}^{3} (3x^{-3} - 5x^{-4})\, dx$

**17.** $\int_{-3}^{-2} \left(2e^{-0.1y} + \frac{3}{y}\right) dy$

**18.** $\int_{-2}^{-1} \left(\frac{-2}{t} + 3e^{0.3t}\right) dt$

**19.** $\int_{1}^{2} \left(e^{4u} - \frac{1}{(u + 1)^2}\right) du$

**20.** $\int_{0.5}^{1} (p^3 - e^{4p})\, dp$

**21.** $\int_{-1}^{0} y(2y^2 - 3)^5\, dy$

**22.** $\int_{0}^{3} m^2(4m^3 + 2)^3\, dm$

**23.** $\int_{1}^{64} \frac{\sqrt{z} - 2}{\sqrt[3]{z}}\, dz$

**24.** $\int_{1}^{8} \frac{3 - y^{1/3}}{y^{2/3}}\, dy$

**25.** $\int_{1}^{2} \frac{\ln x}{x}\, dx$

**26.** $\int_{1}^{3} \frac{\sqrt{\ln x}}{x}\, dx$

**27.** $\int_{0}^{8} x^{1/3} \sqrt{x^{4/3} + 9}\, dx$

**28.** $\int_{1}^{2} \frac{3}{x(1 + \ln x)}\, dx$

**29.** $\int_{0}^{1} \frac{e^{2t}}{(3 + e^{2t})^2}\, dt$

**30.** $\int_{0}^{1} \frac{e^{2z}}{\sqrt{1 + e^{2z}}}\, dz$

**31.** $\int_{0}^{\pi/4} \sin x\, dx$

**32.** $\int_{-\pi/2}^{0} \cos x\, dx$

**33.** $\int_{0}^{\pi/6} \tan x\, dx$

**34.** $\int_{\pi/4}^{\pi/2} \cot x\, dx$

**35.** $\int_{\pi/2}^{2\pi/3} \cos x\, dx$

**36.** $\int_{\pi/6}^{\pi/4} \sin x\, dx$

**In Exercises 37–48, use the definite integral to find the area between the x-axis and $f(x)$ over the indicated interval. Check first to see if the graph crosses the x-axis in the given interval.**

**37.** $f(x) = 2x - 14;\quad [6, 10]$

**38.** $f(x) = 4x - 32;\quad [5, 10]$

**39.** $f(x) = 2 - 2x^2;\quad [0, 5]$

**40.** $f(x) = 9 - x^2;\quad [0, 6]$

**41.** $f(x) = x^3;\quad [-1, 3]$

**42.** $f(x) = x^3 - 2x;\quad [-2, 4]$

**43.** $f(x) = e^x - 1;\quad [-1, 2]$

**44.** $f(x) = 1 - e^{-x};\quad [-1, 2]$

**45.** $f(x) = \frac{1}{x} - \frac{1}{e};\quad [1, e^2]$

**46.** $f(x) = 1 - \frac{1}{x};\quad [e^{-1}, e]$

**47.** $f(x) = \sin x;\quad [0, 3\pi/2]$

**48.** $f(x) = \tan x;\quad [-\pi/4, \pi/3]$

**Find the area of each shaded region.**

**49.**

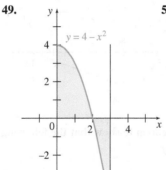

**50.**

**51.**

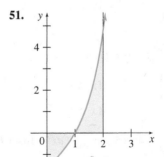

**52.**

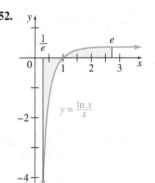

**53.** Assume $f(x)$ is continuous for $g \le x \le c$, as shown in the figure. Write an equation relating the three quantities

$$\int_a^b f(x)\,dx, \qquad \int_a^c f(x)\,dx, \qquad \int_b^c f(x)\,dx.$$

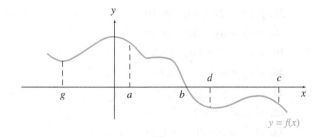

$y = f(x)$

**54.** Is the equation you wrote for Exercise 53 still true

   **a.** if $b$ is replaced by $d$?

   **b.** if $b$ is replaced by $g$?

**55.** The graph of $f(x)$, shown here, consists of two straight line segments and two quarter circles. Find the value of $\int_0^{16} f(x)\,dx$.

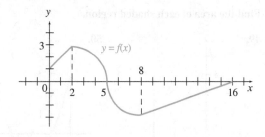

$y = f(x)$

**Use the Fundamental Theorem to show that the following are true.**

**56.** $\displaystyle\int_a^b kf(x)\,dx = k\int_a^b f(x)\,dx$

**57.** $\displaystyle\int_a^b f(x)\,dx = \int_a^c f(x)\,dx + \int_c^b f(x)\,dx$

**58.** $\displaystyle\int_a^b f(x)\,dx = -\int_b^a f(x)\,dx$

**59.** Use Exercise 57 to find $\int_{-1}^4 f(x)\,dx$, given

$$f(x) = \begin{cases} 2x + 3 & \text{if } x \le 0 \\ -\dfrac{x}{4} - 3 & \text{if } x > 0. \end{cases}$$

**60.** You are given $\int_0^1 e^{x^2}\,dx = 1.46265$ and $\int_0^2 e^{x^2}\,dx = 16.45263$. Use this information to find

   **a.** $\displaystyle\int_{-1}^1 e^{x^2}\,dx$;

   **b.** $\displaystyle\int_1^2 e^{x^2}\,dx$.

**61.** Let $g(t) = t^4$ and define $f(x) = \int_c^x g(t)\,dt$ with $c = 1$.

   **a.** Find a formula for $f(x)$.

   **b.** Verify that $f'(x) = g(x)$. The fact that

$$\frac{d}{dx}\int_c^x g(t)\,dt = g(x)$$

   is true for all continuous functions $g$ is an alternative version of the Fundamental Theorem of Calculus.

   **c.** Let us verify the result in part b for a function whose antiderivative cannot be found. Let $g(t) = e^{t^2}$ and let $c = 0$. Use the integration feature on a graphing calculator to find $f(x)$ for $x = 1$ and $x = 1.01$. Then use the definition of the derivative with $h = 0.01$ to approximate $f'(1)$, and compare it with $g(1)$.

**62.** Consider the function $f(x) = x(x^2 + 3)^7$.

   **a.** Use the Fundamental Theorem of Calculus to evaluate

$$\int_{-5}^5 f(x)\,dx.$$

   **b.** Use symmetry to describe how the integral from part a could be evaluated without using substitution or finding an antiderivative.

## LIFE SCIENCE APPLICATIONS

**63. Pollution** Pollution from a factory is entering a lake. The rate of concentration of the pollutant at time $t$ is given by

$$P'(t) = 140t^{5/2},$$

where $t$ is the number of years since the factory started introducing pollutants into the lake. Ecologists estimate that the lake can accept a total level of pollution of 4850 units before all the fish life in the lake ends. Can the factory operate for 4 years without killing all the fish in the lake?

**64. Spread of an Oil Leak** An oil tanker is leaking oil at the rate given (in barrels per hour) by

$$L'(t) = \frac{80\ln(t + 1)}{t + 1},$$

where $t$ is the time (in hours) after the tanker hits a hidden rock (when $t = 0$).

   **a.** Find the total number of barrels that the ship will leak on the first day.

   **b.** Find the total number of barrels that the ship will leak on the second day.

   **c.** What is happening over the long run to the amount of oil leaked per day?

**65. Tree Growth** After long study, tree scientists conclude that a eucalyptus tree will grow at the rate of $0.6 + 4/(t + 1)^3$ ft per year, where $t$ is time (in years).

   **a.** Find the number of feet that the tree will grow in the second year.

   **b.** Find the number of feet the tree will grow in the third year.

**66. Growth of a Substance** The rate at which a substance grows is given by

$$R'(x) = 150e^{0.2x},$$

where $x$ is the time (in days). What is the total accumulated growth during the first 3.5 days?

**67. Drug Reaction** For a certain drug, the rate of reaction in appropriate units is given by

$$R'(t) = \frac{5}{t+1} + \frac{2}{\sqrt{t+1}},$$

where $t$ is time (in hours) after the drug is administered. Find the total reaction to the drug over the following time periods.

**a.** From $t = 1$ to $t = 12$

**b.** From $t = 12$ to $t = 24$

**68. Human Mortality** If $f(x)$ is the instantaneous death rate for members of a population at time $x$, then the number of individuals who survive to age $T$ is given by

$$F(T) = \int_0^T f(x)\, dx.$$

In 1825 the biologist Benjamin Gompertz proposed that $f(x) = kb^x$. Find a formula for $F(T)$. *Source: Philosophical Transactions of the Royal Society of London.*

**69. Cell Division** Let the expected number of cells in a culture that have an $x$ percent probability of undergoing cell division during the next hour be denoted by $n(x)$.

**a.** Explain why $\int_{20}^{30} n(x)\, dx$ approximates the total number of cells with a 20% to 30% chance of dividing during the next hour.

**b.** Give an integral representing the number of cells that have less than a 60% chance of dividing during the next hour.

**c.** Let $n(x) = \sqrt{5x+1}$ give the expected number of cells (in millions) with $x$ percent probability of dividing during the next hour. Find the number of cells with a 5% to 10% chance of dividing.

**70. Bacterial Growth** A population of *E. coli* bacteria will grow at a rate given by

$$w'(t) = (3t+2)^{1/3},$$

where $w$ is the weight (in milligrams) after $t$ hours. Find the change in weight of the population from $t = 0$ to $t = 3$.

**71. Blood Flow** In an example from an earlier chapter, the velocity $v$ of the blood in a blood vessel was given as

$$v = k(R^2 - r^2),$$

where $R$ is the (constant) radius of the blood vessel, $r$ is the distance of the flowing blood from the center of the blood vessel, and $k$ is a constant. Total blood flow (in millimeters per minute) is given by

$$Q(R) = \int_0^R 2\pi vr\, dr.$$

**a.** Find the general formula for $Q$ in terms of $R$ by evaluating the definite integral given above.

**b.** Evaluate $Q(0.4)$.

**72. Rams' Horns** The average annual increment in the horn length (in centimeters) of bighorn rams born since 1986 can be approximated by

$$y = 0.1762t^2 - 3.986t + 22.68,$$

where $t$ is the ram's age (in years) for $t$ between 3 and 9. Integrate to find the total increase in the length of a ram's horn during this time. *Source: Journal of Wildlife Management.*

**73. Beagles** The daily energy requirements of female beagles who are at least 1 year old change with respect to time according to the function

$$E(t) = 753t^{-0.1321},$$

where $E(t)$ is the daily energy requirement (in kJ/$W^{0.67}$), where $W$ is the dog's weight (in kilograms) for a beagle that is $t$ years old. *Source: Journal of Nutrition.*

**a.** Assuming 365 days in a year, show that the energy requirement for a female beagle that is $t$ days old is given by

$$E(t) = 1642t^{-0.1321}.$$

**b.** Using the formula from part a, determine the total energy requirements (in kJ/$W^{0.67}$) for a female beagle between her first and third birthday.

**74. Sediment** The density of sediment (in grams per cubic centimeter) at the bottom of Lake Coeur d'Alene, Idaho, is given by

$$p(x) = p_0 e^{0.0133x},$$

where $x$ is the depth (in centimeters) and $p_0$ is the density at the surface. The total mass of a square-centimeter column of sediment above a depth of $h$ cm is given by

$$\int_0^h p(x)\, dx.$$

If $p_0 = 0.85$ g per cm$^3$, find the total mass above a depth of 100 cm. *Source: Mathematics Teacher.*

**75. Age Distribution** The U.S. Census Bureau gives an age distribution that is approximately modeled (in millions) by the function

$$f(x) = 40.2 + 3.50x - 0.897x^2,$$

where $x$ varies from 0 to 9 decades. The population of a given age group can be found by integrating this function over the interval for that age group. *Source: Ralph DeMarr and the U.S. Census Bureau.*

**a.** Find the integral of $f(x)$ over the interval $[0, 9]$. What does this integral represent?

**b.** Baby boomers are those born between 1945 and 1965, that is, those in the range of 4.5 to 6.5 decades in 2010. Estimate the number of baby boomers.

**76. Nervous System** In a model of the nervous system, the time required for $n$ circuits of nerves to be excited is given by

$$t = \int_0^n \frac{du}{K - au},$$

where $K$ and $a$ are constants. **Source: Mathematical Biology of Social Behavior.**

**a.** Find a formula for the number of neural circuits that are excited as a function of time.

**b.** What happens to the number of neural circuits that are excited as time progresses?

**77. Biochemical Reaction** In an example of the Field-Noyes model of a biochemical reaction, where $x$ denotes the concentration of $HBrO_2$ (bromous acid), and $T$ is the period of an oscillation,

$$\int_2^1 \left(x - \frac{1}{x}\right) dx = -\int_0^{T/2} dt.$$

**Source: Mathematical Biology.** Find the value of $T$.

**78. Drug Concentration** In a model of drug use, the concentration of a drug in a user's bloodstream is given by

$$C(t) = e^{-kt} \int_0^t e^{ky} D(y) dy,$$

where $k$ is the proportional rate that the drug is removed from the bloodstream, $D$ is the rate that the drug is introduced into the bloodstream, and $t$ is the time. **Source: Mathematical Biosciences.**

**a.** Under the simplifying assumption that $D(y) = D$ (a constant), show that

$$C(t) = \frac{D}{k}(1 - e^{-kt}).$$

**b.** Letting $A(t)$ represent the number of free (active) sites where the drug can cause a response in the individual and $N$ represent the total number of sites, free (active) and bound (inactive),

$$A(t) = Ne^{-\int_0^t mC(z)dz},$$

where $m$ is a constant. Using the expression for $C(t)$ from part a, show that

$$A(t) = N \exp\left[\frac{-mD}{k}\left(t + \frac{-1 + e^{-kt}}{k}\right)\right].$$

**79. Migratory Animals** The number of migratory animals (in hundreds) counted at a certain checkpoint is given by

$$T(t) = 50 + 50 \cos\left(\frac{\pi}{6}t\right),$$

where $t$ is time in months, with $t = 0$ corresponding to July. The figure in the next column shows a graph of $T$. Use a definite integral to find the number of animals passing the checkpoint in a year.

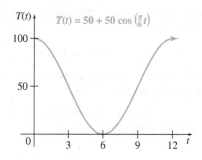

**80. Oscillating System** In a model of an oscillating system in biology,

$$\int_\theta^\phi (1 + I \cos 2x)^{-1} dx = \int_0^T dt,$$

where $I$ is a constant, $\theta$ is the old phase of the oscillator, $\phi$ is the new phase, and $T$ is the time that a stimulus is applied to the oscillator. **Source: Mathematical Biology.** Solve for $T$ when $I = 1$. (*Hint:* Use the trigonometric identity $\cos 2x = 2 \cos^2 x - 1$.)

## OTHER APPLICATIONS

**81. Income Distribution** Based on data from the U.S. Census Bureau, an approximate family income distribution for the United States is given by the function

$$f(x) = 0.0353x^3 - 0.541x^2 + 3.78x + 4.29,$$

where $x$ is the annual income in units of \$10,000, with $0 \le x \le 10$. For example, $x = 0.5$ represents an annual family income of \$5000. (*Note:* This function does not give a good representation for family incomes over \$100,000.) The percent of the families with an income in a given range can be found by integrating this function over that range. Find the percentage of families with an income between \$25,000 and \$50,000. **Source: Ralph DeMarr and the U.S. Census Bureau.**

**82. Oil Consumption** Suppose that the rate of consumption of a natural resource is $c'(t)$, where

$$c'(t) = ke^{rt}.$$

Here $t$ is time in years, $r$ is a constant, and $k$ is the consumption in the year when $t = 0$. In 2010, an oil company sold 1.2 billion barrels of oil. Assume that $r = 0.04$.

**a.** Write $c'(t)$ for the oil company, letting $t = 0$ represent 2010.

**b.** Set up a definite integral for the amount of oil that the company will sell in the next 10 years.

**c.** Evaluate the definite integral of part b.

**d.** The company has about 20 billion barrels of oil in reserve. To find the number of years that this amount will last, solve the equation

$$\int_0^T 1.2e^{0.04t}dt = 20.$$

**e.** Rework part d, assuming that $r = 0.02$.

**83. Oil Consumption** The rate of consumption of oil (in billions of barrels) by the company in Exercise 82 was given as

$$1.2e^{0.04t},$$

where $t = 0$ corresponds to 2010. Find the total amount of oil used by the company from 2010 to year $T$. At this rate, how much will be used in 5 years?

**84. Voltage** The electrical voltage from a standard wall outlet is given as a function of time $t$ by

$$V(t) = 170 \sin(120 \pi t).$$

This is an example of alternating current, which is electricity that reverses direction at regular intervals. The common method for measuring the level of voltage from an alternating current is the *root mean square*, which is given by

$$\text{Root mean square} = \sqrt{\frac{\int_0^T V^2(t) \, dt}{T}},$$

where $T$ is one period of the current.

   **a.** Verify that $T = 1/60$ second for $V(t)$ given above.

   **b.** You may have seen that the voltage from a standard wall outlet is 120 volts. Verify that this is the root mean square value for $V(t)$ given above. (*Hint:* Use the trigonometric identity $\sin^2 x = (1 - \cos 2x)/2$. This identity can be derived by letting $y = x$ in the basic identity for $\cos(x + y)$, and then eliminating $\cos^2 x$ by using the identity $\cos^2 x = 1 - \sin^2 x$.)

**85. Length of Day** The following function can be used to estimate the number of minutes of daylight in Boston for any given day of the year.

$$N(t) = 183.549 \sin(0.0172t - 1.329) + 728.124,$$

where $t$ is the day of the year. *Source: The Old Farmer's Almanac.* Use this function to estimate the total amount of daylight in a year and compare it to the total amount of daylight reported to be 4467.57 hours.

**86. Petroleum Consumption** The monthly residential consumption of natural gas (in trillions of BTU) in the United States for 2012 is found in the table in the next column. *Source: Energy Information Administration.*

   **a.** Plot the data, letting $t = 1$ correspond to January, $t = 2$ correspond to February, and so on. Is it reasonable to assume that natural gas consumption is periodic?

| Month | Natural Gas (trillion BTU) |
|---|---|
| January | 817.038 |
| February | 679.743 |
| March | 414.488 |
| April | 286.132 |
| May | 166.070 |
| June | 125.925 |
| July | 110.607 |
| August | 108.238 |
| September | 121.369 |
| October | 246.604 |
| November | 494.875 |
| December | 689.759 |

   **b.** Use a calculator with trigonometric regression to find a trigonometric function of the form

$$C(t) = a \sin(bt + c) + d$$

that models these data when $t$ is the month and $C(t)$ is the amount of natural gas consumed (in trillions of BTU). Graph this function on the same calculator window as the data.

   **c.** Estimate the consumption for the month of September and compare it to the actual value.

   **d.** Estimate the rate at which the consumption is changing in September.

   **e.** Estimate the total natural gas consumption for the year for residential customers and compare it to the actual value.

 **f.** What would you expect the period of a function that models annual natural gas consumption to be? What is the period of the function found in part b? Discuss possible reasons for the discrepancy in the values.

**YOUR TURN ANSWERS**

1. 26
2. 256
3. 2 ln 3 or ln 9
4. 128/3
5. 54
6. $6\sqrt{3}$

# 7.5 The Area Between Two Curves

APPLY IT If a hospital purchasing agent knows how the savings from a new imaging technology will decline over time and how the costs of that technology will increase, how can she determine when the net savings will cease, and what the total savings will be?
*We will answer this question in Example 4.*

Many important applications of integrals require finding the area between two graphs. The method used in previous sections to find the area between the graph of a function and the x-axis from $x = a$ to $x = b$ can be generalized to find such an area. For example, the area between the graphs of $f(x)$ and $g(x)$ from $x = a$ to $x = b$ in Figure 24(a) is the same as the area under the graph of $f(x)$, shown in Figure 24(b), minus the area under the graph of $g(x)$ (see Figure 24(c)). That is, the area between the graphs is given by

$$\int_a^b f(x)\, dx - \int_a^b g(x)\, dx,$$

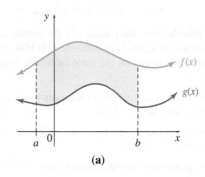
(a)

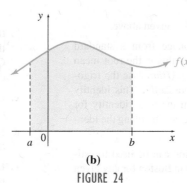
(b)

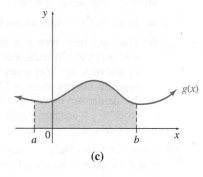
(c)

FIGURE 24

which can be written as

$$\int_a^b [f(x) - g(x)]\, dx.$$

## Area Between Two Curves

If $f$ and $g$ are continuous functions and $f(x) \geq g(x)$ on $[a, b]$, then the area between the curves $f(x)$ and $g(x)$ from $x = a$ to $x = b$ is given by

$$\int_a^b [f(x) - g(x)]\, dx.$$

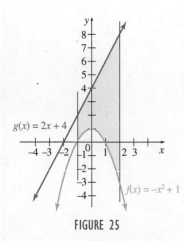
FIGURE 25

### EXAMPLE 1 Area

Find the area bounded by $f(x) = -x^2 + 1$, $g(x) = 2x + 4$, $x = -1$, and $x = 2$.

**SOLUTION** A sketch of the four equations is shown in Figure 25. In general, it is not necessary to spend time drawing a detailed sketch, but only to know whether the two functions intersect, and which function is greater between the intersections. To find out, set the two functions equal.

$$-x^2 + 1 = 2x + 4$$
$$0 = x^2 + 2x + 3$$

Verify by the quadratic formula that this equation has no real roots. Since the graph of $f$ is a parabola opening downward that does not cross the graph of $g$ (a line), the parabola must be entirely under the line, as shown in Figure 25. Therefore $g(x) \geq f(x)$ for $x$ in the interval $[-1, 2]$, and the area is given by

$$\int_{-1}^{2} [g(x) - f(x)] \, dx = \int_{-1}^{2} [(2x + 4) - (-x^2 + 1)] \, dx$$

$$= \int_{-1}^{2} (2x + 4 + x^2 - 1) \, dx$$

$$= \int_{-1}^{2} (x^2 + 2x + 3) \, dx$$

$$= \frac{x^3}{3} + x^2 + 3x \Big|_{-1}^{2}$$

$$= \left(\frac{8}{3} + 4 + 6\right) - \left(\frac{-1}{3} + 1 - 3\right)$$

$$= \frac{8}{3} + 10 + \frac{1}{3} + 2$$

$$= 15. \qquad \textbf{TRY YOUR TURN 1}$$

**YOUR TURN 1** Repeat Example 1 for $f(x) = 4 - x^2$, $g(x) = x + 2$, $x = -2$, and $x = 1$.

**NOTE** It is not necessary to draw the graphs to determine which function is greater. Since the functions in the previous example do not intersect, we can evaluate them at *any* point to make this determination. For example, $f(0) = 1$ and $g(0) = 4$. Because $g(x) > f(x)$ at $x = 4$, and the two functions are continuous and never intersect, $g(x) > f(x)$ for all $x$.

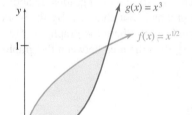

FIGURE 26

**EXAMPLE 2** Area

Find the area between the curves $y = x^{1/2}$ and $y = x^3$.

**SOLUTION** Let $f(x) = x^{1/2}$ and $g(x) = x^3$. As before, set the two equal to find where they intersect.

$$x^{1/2} = x^3$$
$$0 = x^3 - x^{1/2}$$
$$0 = x^{1/2}(x^{5/2} - 1)$$

The only solutions are $x = 0$ and $x = 1$. Verify that the graph of $f$ is concave downward, while the graph of $g$ is concave upward, so the graph of $f$ must be greater between 0 and 1. (This may also be verified by taking a point between 0 and 1, such as 0.5, and verifying that $0.5^{1/2} > 0.5^3$.) The graph is shown in Figure 26.

The area between the two curves is given by

$$\int_{a}^{b} [f(x) - g(x)] \, dx = \int_{0}^{1} (x^{1/2} - x^3) \, dx.$$

Using the Fundamental Theorem,

$$\int_{0}^{1} (x^{1/2} - x^3) \, dx = \left(\frac{x^{3/2}}{3/2} - \frac{x^4}{4}\right) \Big|_{0}^{1}$$

$$= \left(\frac{2}{3} x^{3/2} - \frac{x^4}{4}\right) \Big|_{0}^{1}$$

$$= \frac{2}{3}(1) - \frac{1}{4}$$

$$= \frac{5}{12}. \qquad \textbf{TRY YOUR TURN 2}$$

**YOUR TURN 2** Repeat Example 2 for $y = x^{1/4}$ and $y = x^2$.

TECHNOLOGY NOTE

A graphing calculator is very useful in approximating solutions of problems involving the area between two curves. First, it can be used to graph the functions and identify any intersection points. Then it can be used to approximate the definite integral that represents the area. (A function that gives a numerical approximation to the integral is located in the MATH menu of a TI-84 Plus calculator.) Figure 27 shows the results of using these steps for Example 2. The second window shows that the area closely approximates 5/12.

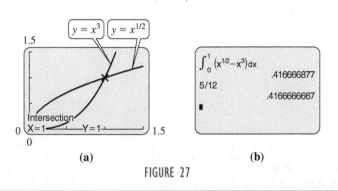

(a)                            (b)

FIGURE 27

The difference between two integrals can be used to find the area between the graphs of two functions even if one graph lies below the x-axis. In fact, if $f(x) \geq g(x)$ for all values of $x$ in the interval $[a, b]$, then the area between the two graphs is always given by

$$\int_a^b [f(x) - g(x)] \, dx.$$

To see this, look at the graphs in Figure 28(a), where $f(x) \geq g(x)$ for $x$ in $[a, b]$. Suppose a constant $C$ is added to both functions, with $C$ large enough so that both graphs lie above the x-axis, as in Figure 28(b). The region between the graphs is not changed. By the work above, this area is given by $\int_a^b [f(x) - g(x)] dx$ regardless of where the graphs of $f(x)$ and $g(x)$ are located. As long as $f(x) \geq g(x)$ on $[a, b]$, then the area between the graphs from $x = a$ to $x = b$ will equal $\int_a^b [f(x) - g(x)] dx$.

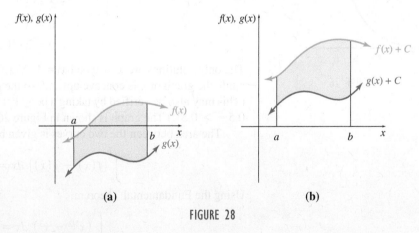

(a)                            (b)

FIGURE 28

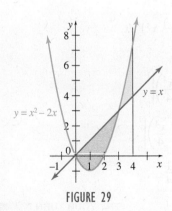

FIGURE 29

## EXAMPLE 3    Area

Find the area of the region enclosed by $y = x^2 - 2x$ and $y = x$ on $[0, 4]$.

**SOLUTION**    Verify that the two graphs cross at $x = 0$ and $x = 3$. Because the first graph is a parabola opening upward, the parabola must be below the line between 0 and 3 and above the line between 3 and 4. See Figure 29. (The greater function could also be identified by checking a point between 0 and 3, such as 1, and a point between 3 and 4, such as 3.5. For each of these values of $x$, we could calculate the corresponding value of $y$ for the two functions and see which is greater.) Because the graphs cross at $x = 3$, the area is found by taking the sum of two integrals as follows.

$$\text{Area} = \int_0^3 [x - (x^2 - 2x)]\, dx + \int_3^4 [(x^2 - 2x) - x]\, dx$$

$$= \int_0^3 (-x^2 + 3x)\, dx + \int_3^4 (x^2 - 3x)\, dx$$

$$= \left( \frac{-x^3}{3} + \frac{3x^2}{2} \right)\Big|_0^3 + \left( \frac{x^3}{3} - \frac{3x^2}{2} \right)\Big|_3^4$$

$$= \left( -9 + \frac{27}{2} - 0 \right) + \left( \frac{64}{3} - 24 - 9 + \frac{27}{2} \right)$$

$$= \frac{19}{3}$$

**YOUR TURN 3** Repeat Example 3 for $y = x^2 - 3x$ and $y = 2x$ on $[0, 6]$.

**TRY YOUR TURN 3**

In the remainder of this section we will consider some typical applications that require finding the area between two curves.

### EXAMPLE 4 Savings

A hospital is considering a new magnetic resonance imaging (MRI) machine. The new machine provides substantial initial savings, with the savings declining with time $t$ (in years) according to the rate-of-savings function

$$S'(t) = 100 - t^2,$$

where $S'(t)$ is in thousands of dollars per year. At the same time, the cost of operating the new machine increases with time $t$ (in years), according to the rate-of-cost function (in thousands of dollars per year)

$$C'(t) = t^2 + \frac{14}{3}t.$$

**(a)** For how many years will the hospital realize savings?

**APPLY IT**

**SOLUTION** Figure 30 shows the graphs of the rate-of-savings and rate-of-cost functions. The rate of cost (marginal cost) is increasing, while the rate of savings (marginal savings) is decreasing. The hospital should use this new process until the difference between these quantities is zero—that is, until the time at which these graphs intersect. The graphs intersect when

$$C'(t) = S'(t),$$

or

$$t^2 + \frac{14}{3}t = 100 - t^2.$$

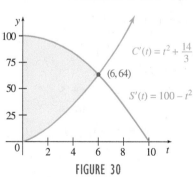

FIGURE 30

Solve this equation as follows.

$$0 = 2t^2 + \frac{14}{3}t - 100$$

$$0 = 3t^2 + 7t - 150 \qquad \text{Multiply by } \tfrac{3}{2}.$$

$$= (t - 6)(3t + 25) \qquad \text{Factor.}$$

Set each factor equal to 0 to get

$$t = 6 \quad \text{or} \quad t = -25/3.$$

Only 6 is a meaningful solution here. The hospital should use the new process for 6 years.

**(b)** What will be the net total savings during this period?

**SOLUTION** Since the total savings over the 6-year period is given by the area under the rate-of-savings curve and the total additional cost is given by the area under the rate-of-cost curve, the net total savings over the 6-year period is given by the area between the rate-of-cost and the rate-of-savings curves and the lines $t = 0$ and $t = 6$. This area can be evaluated with a definite integral as follows.

$$
\begin{aligned}
\text{Total savings} &= \int_0^6 \left[ (100 - t^2) - \left( t^2 + \frac{14}{3}t \right) \right] dt \\
&= \int_0^6 \left( 100 - \frac{14}{3}t - 2t^2 \right) dt \\
&= \left( 100t - \frac{7}{3}t^2 - \frac{2}{3}t^3 \right) \Big|_0^6 \\
&= 100(6) - \frac{7}{3}(36) - \frac{2}{3}(216) = 372
\end{aligned}
$$

The hospital will save a total of $372,000 over the 6-year period.

The answer to a problem will not always be an integer. Suppose in solving the quadratic equation in Example 4 we found the solutions to be $t = 6.7$ and $t = -7.3$. It may not be realistic to use a new machine for 6.7 years; it may be necessary to choose between 6 years and 7 years. Since the mathematical model produces a result that is not in the domain of the function in this case, it is necessary to find the total savings after 6 years and after 7 years and then select the better result.

# 7.5 EXERCISES

**Find the area between the curves in Exercises 1–28.**

1. $x = -2$, $x = 1$, $y = 2x^2 + 5$, $y = 0$

2. $x = 1$, $x = 2$, $y - 3x^3 + 2$, $y = 0$

3. $x = -3$, $x = 1$, $y = x^3 + 1$, $y = 0$

4. $x = -3$, $x = 0$, $y = 1 - x^2$, $y = 0$

5. $x = -2$, $x = 1$, $y = 2x$, $y = x^2 - 3$

6. $x = 0$, $x = 6$, $y = 5x$, $y = 3x + 10$

7. $y = x^2 - 30$, $y = 10 - 3x$

8. $y = x^2 - 18$, $y = x - 6$

9. $y = x^2$, $y = 2x$

10. $y = x^2$, $y = x^3$

11. $x = 1$, $x = 6$, $y = \frac{1}{x}$, $y = \frac{1}{2}$

12. $x = 0$, $x = 4$, $y = \frac{1}{x+1}$, $y = \frac{x-1}{2}$

13. $x = -1$, $x = 1$, $y = e^x$, $y = 3 - e^x$

14. $x = -1$, $x = 2$, $y = e^{-x}$, $y = e^x$

15. $x = -1$, $x = 2$, $y = 2e^{2x}$, $y = e^{2x} + 1$

16. $x = 2$, $x = 4$, $y = \frac{x-1}{4}$, $y = \frac{1}{x-1}$

17. $y = x^3 - x^2 + x + 1$, $y = 2x^2 - x + 1$

18. $y = 2x^3 + x^2 + x + 5$, $y = x^3 + x^2 + 2x + 5$

19. $y = x^4 + \ln(x + 10)$, $y = x^3 + \ln(x + 10)$

20. $y = x^5 - 2\ln(x + 5)$, $y = x^3 - 2\ln(x + 5)$

21. $y = x^{4/3}$, $y = 2x^{1/3}$

22. $y = \sqrt{x}$, $y = x\sqrt{x}$

23. $x = 0$, $x = 3$, $y = 2e^{3x}$, $y = e^{3x} + e^6$

24. $x = 0$, $x = 3$, $y = e^x$, $y = e^{4-x}$

25. $x = 0$, $x = \frac{\pi}{4}$, $y = \cos x$, $y = \sin x$

**26.** $x = 0, \quad x = \pi/4, \quad y = \sec^2 x, \quad y = \sin 2x$

**27.** $x = \pi/4, \quad y = \tan x, \quad y = \sin x$

**28.** $x = 0, \quad x = \pi, \quad y = \sin x, \quad y = 1 - \sin x$

**In Exercises 29 and 30, use a graphing calculator to find the values of $x$ where the curves intersect and then to find the area between the two curves.**

**29.** $y = e^x, \quad y = -x^2 - 2x$

**30.** $y = \ln x, \quad y = x^3 - 5x^2 + 6x - 1$

## LIFE SCIENCE APPLICATIONS

**31. Pollution** Pollution begins to enter a lake at time $t = 0$ at a rate (in gallons per hour) given by the formula

$$f(t) = 10(1 - e^{-0.5t}),$$

where $t$ is the time (in hours). At the same time, a pollution filter begins to remove the pollution at a rate

$$g(t) = 0.4t$$

as long as pollution remains in the lake.

  **a.** How much pollution is in the lake after 12 hours?

  **b.** Use a graphing calculator to find the time when the rate that pollution enters the lake equals the rate the pollution is removed.

  **c.** Find the amount of pollution in the lake at the time found in part b.

  **d.** Use a graphing calculator to find the time when all the pollution has been removed from the lake.

**32. Pollution** Repeat the steps of Exercise 31, using the functions

$$f(t) = 15(1 - e^{-0.05t})$$

and

$$g(t) = 0.3t.$$

## OTHER APPLICATIONS

**33. Distribution of Income** Suppose that all the people in a country are ranked according to their incomes, starting at the bottom. Let $x$ represent the fraction of the community making the lowest income $(0 \le x \le 1)$; $x = 0.4$, therefore, represents the lower 40% of all income producers. Let $I(x)$ represent the proportion of the total income earned by the lowest $x$ of all people. Thus, $I(0.4)$ represents the fraction of total income earned by the lowest 40% of the population. The curve described by this function is known as a *Lorenz curve*. Suppose

$$I(x) = 0.9x^2 + 0.1x.$$

Find and interpret the following.

  **a.** $I(0.1)$

  **b.** $I(0.4)$

If income were distributed uniformly, we would have $I(x) = x$. The area under this line of complete equality is 1/2. As $I(x)$ dips further below $y = x$, there is less equality of income distribution. This inequality can be quantified by the ratio of the area between $I(x)$ and $y = x$ to 1/2. This ratio is called the *Gini index of income inequality* and equals $2 \int_0^1 [x - I(x)] \, dx$.

  **c.** Graph $I(x) = x$ and $I(x) = 0.9x^2 + 0.1x$, for $0 \le x \le 1$, on the same axes.

  **d.** Find the area between the curves.

  **e.** For U.S. families, the Gini index was 0.386 in 1968 and 0.466 in 2008. Describe how the distribution of family incomes has changed over this time. *Source: U.S. Census.*

**34. Fuel Economy** In an article in the December 1994 *Scientific American* magazine, the authors estimated future gas use. Without a change in U.S. policy, auto fuel use is forecasted to rise along the projection shown at the right in the figure below. The shaded band predicts gas use if the technologies for increased fuel economy are phased in by the year 2010. The moderate estimate (center curve) corresponds to an average of 46 mpg for all cars on the road. *Source: Scientific American.*

  **a.** Discuss the interpretation of the shaded area and other regions of the graph that pertain to the topic in this section.

  **b.** According to the Energy Information Administration, the U.S. gasoline consumption in 2010 was 9,030,000 barrels per day. Discuss how this affects the areas considered in part a. *Source: U.S. Energy Information Administration.*

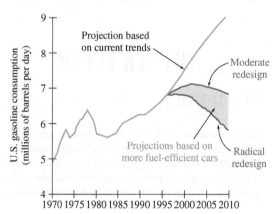

**35. Net Savings** Suppose a company wants to introduce a new machine that will produce a rate of annual savings (in dollars) given by

$$S'(t) = 150 - t^2,$$

where $t$ is the number of years of operation of the machine, while producing a rate of annual costs (in dollars) of

$$C'(t) = t^2 + \frac{11}{4}t.$$

  **a.** For how many years will it be profitable to use this new machine?

  **b.** What are the net total savings during the first year of use of the machine?

  **c.** What are the net total savings over the entire period of use of the machine?

**36. Net Savings** A new smog-control device will reduce the output of sulfur oxides from automobile exhausts. It is estimated that the rate of savings (in millions of dollars per year) to the community from the use of this device will be approximated by

$$S'(t) = -t^2 + 4t + 8,$$

after $t$ years of use of the device. The new device cuts down on the production of sulfur oxides, but it causes an increase in the production of nitrous oxides. The rate of additional costs (in millions of dollars per year) to the community after $t$ years is approximated by

$$C'(t) = \frac{3}{25}t^2.$$

**a.** For how many years will it pay to use the new device?

**b.** What will be the net savings over this period of time?

**37. Profit** Canham Enterprises had an expenditure rate of $E'(t) = e^{0.1t}$ dollars per day and an income rate of $I'(t) = 98.8 - e^{0.1t}$ dollars per day on a particular job, where $t$ was the number of days from the start of the job. The company's profit on that job will equal total income less total expenditures. Profit will be maximized if the job ends at the optimum time, which is the point where the two curves meet. Find the following.

**a.** The optimum number of days for the job to last

**b.** The total income for the optimum number of days

**c.** The total expenditures for the optimum number of days

**d.** The maximum profit for the job

**38. Net Savings** A factory of Hollis Sherman Industries has installed a new process that will produce an increased rate of revenue (in thousands of dollars per year) of

$$R'(t) = 104 - 0.4e^{t/2},$$

where $t$ is time measured in years. The new process produces additional costs (in thousands of dollars per year) at the rate of

$$C'(t) = 0.3e^{t/2}.$$

**a.** When will it no longer be profitable to use this new process?

**b.** Find the net total savings.

**1.** 9/2    **2.** 7/15    **3.** 71/3

# 7 CHAPTER REVIEW

## SUMMARY

Earlier chapters dealt with the derivative, one of the two main ideas of calculus. This chapter deals with integration, the second main idea. There are two aspects of integration. The first is indefinite integration, or finding an antiderivative; the second is definite integration, which can be used to find the area under a curve. The

Fundamental Theorem of Calculus unites these two ideas by showing that the way to find the area under a curve is to use the antiderivative. Substitution is a technique for finding antiderivatives. The idea of the definite integral can also be applied to finding the area between two curves.

**Antidifferentiation Formulas**

Power Rule $\quad \displaystyle\int x^n \, dx = \frac{x^{n+1}}{n+1} + C, \; n \neq -1$

Constant Multiple Rule $\quad \displaystyle\int k \cdot f(x) \, dx = k \int f(x) \, dx, \text{ for any real number } k$

Sum or Difference Rule $\quad \displaystyle\int [f(x) \pm g(x)] \, dx = \int f(x) \, dx \pm \int g(x) \, dx$

Integration of $x^{-1}$ $\quad \displaystyle\int x^{-1} \, dx = \ln |x| + C$

Integration of Exponential Functions $\quad \displaystyle\int e^{kx} \, dx = \frac{e^{kx}}{k} + C, \; k \neq 0$

$$\int a^{kx} \, dx = \frac{a^{kx}}{k(\ln a)} + C, \; k \neq 0$$

| Basic Trigonometric Integrals | $\int \sin x \, dx = -\cos x + C$ | $\int \cos x \, dx = \sin x + C$ |
|---|---|---|
| | $\int \sec^2 x \, dx = \tan x + C$ | $\int \csc^2 x \, dx = -\cot x + C$ |
| | $\int \sec x \tan x \, dx = \sec x + C$ | $\int \csc x \cot x \, dx = -\csc x + C$ |
| | $\int \tan x \, dx = -\ln|\cos x| + C$ | $\int \cot x \, dx = \ln|\sin x| + C$ |

**Substitution Method**   Choose $u$ to be one of the following:

1. the quantity under a root or raised to a power;
2. the quantity in the denominator;
3. the exponent on $e$;
4. the quantity inside a trigonometric function.

## Definite Integrals

**Definition of the Definite Integral**   $\int_a^b f(x)\,dx = \lim\limits_{n\to\infty} \sum\limits_{i=1}^{n} f(x_i)\Delta x$, where $\Delta x = (b-a)/n$ and $x_i$ is any value of $x$ in the $i$th interval.

If $f(x)$ gives the rate of change of $F(x)$ for $x$ in $[a, b]$, then this represents the total change in $F(x)$ as $x$ goes from $a$ to $b$.

**Properties of Definite Integrals**

1. $\displaystyle\int_a^a f(x)\,dx = 0$

2. $\displaystyle\int_a^b k \cdot f(x)\,dx = k\int_a^b f(x)\,dx$,   for any real number $k$

3. $\displaystyle\int_a^b [f(x) \pm g(x)]\,dx = \int_a^b f(x)\,dx \pm \int_a^b g(x)\,dx$

4. $\displaystyle\int_a^b f(x)\,dx = \int_a^c f(x)\,dx + \int_c^b f(x)\,dx$,   for any real number $c$

5. $\displaystyle\int_a^b f(x)\,dx = -\int_b^a f(x)\,dx$

**Fundamental Theorem of Calculus**   $\int_a^b f(x)\,dx = F(x)\big|_a^b = F(b) - F(a)$, where $f$ is continuous on $[a, b]$ and $F$ is any antiderivative of $f$

**Area Between Two Curves**   $\int_a^b [f(x) - g(x)]\,dx$, where $f$ and $g$ are continuous functions and $f(x) \geq g(x)$ on $[a, b]$

# KEY TERMS

| **7.1** | **7.2** | **7.3** | **7.4** |
|---|---|---|---|
| antidifferentiation | integration by substitution | midpoint rule | Fundamental Theorem of |
| antiderivative | demand function | definite integral | Calculus |
| integral sign | marginal demand | limits of integration | |
| integrand | | total change | |
| indefinite integral | | | |

# REVIEW EXERCISES

## CONCEPT CHECK

**Determine whether each of the following statements is true or false, and explain why.**

1. The indefinite integral is another term for the family of all antiderivatives of a function.

2. The indefinite integral of $x^n$ is $x^{n+1}/(n+1) + C$ for all real numbers $n$.

3. The indefinite integral $\int xf(x)\,dx$ is equal to $x\int f(x)\,dx$.

4. The velocity function is an antiderivative of the acceleration function.

5. Substitution can often be used to turn a complicated integral into a simpler one.

6. The definite integral gives the instantaneous rate of change of a function.

7. The definite integral gives an approximation to the area under a curve.

8. The definite integral of a positive function is the limit of the sum of the areas of rectangles.

9. The Fundamental Theorem of Calculus gives a relationship between the definite integral and an antiderivative of a function.

10. The definite integral of a function is always a positive quantity.

11. The area between two distinct curves is always a positive quantity.

## PRACTICE AND EXPLORATION

12. Explain the differences between an indefinite integral and a definite integral.

13. Explain under what circumstances substitution is useful in integration.

14. Explain why the limits of integration are changed when $u$ is substituted for an expression in $x$ in a definite integral.

**In Exercises 15–51, find each indefinite integral.**

15. $\int (2x + 3) \, dx$

16. $\int (5x - 1) \, dx$

17. $\int (x^2 - 3x + 2) \, dx$

18. $\int (6 - x^2) \, dx$

19. $\int 3\sqrt{x} \, dx$

20. $\int \dfrac{\sqrt{x}}{2} \, dx$

21. $\int (x^{1/2} + 3x^{-2/3}) \, dx$

22. $\int (2x^{4/3} + x^{-1/2}) \, dx$

23. $\int \dfrac{-4}{x^3} \, dx$

24. $\int \dfrac{5}{x^4} \, dx$

25. $\int -3e^{2x} \, dx$

26. $\int 5e^{-x} \, dx$

27. $\int xe^{3x^2} \, dx$

28. $\int 2xe^{x^2} \, dx$

29. $\int \dfrac{3x}{x^2 - 1} \, dx$

30. $\int \dfrac{-x}{2 - x^2} \, dx$

31. $\int \dfrac{x^2 \, dx}{(x^3 + 5)^4}$

32. $\int (x^2 - 5x)^4 (2x - 5) \, dx$

33. $\int \dfrac{x^3}{e^{3x^4}} \, dx$

34. $\int e^{3x^2 + 4} x \, dx$

35. $\int \dfrac{(3 \ln x + 2)^4}{x} \, dx$

36. $\int \dfrac{\sqrt{5 \ln x + 3}}{x} \, dx$

37. $\int \sin 2x \, dx$

38. $\int \cos 5x \, dx$

39. $\int \tan 7x \, dx$

40. $\int \sec^2 5x \, dx$

41. $\int 8 \sec^2 x \, dx$

42. $\int 4 \csc^2 x \, dx$

43. $\int x^2 \sin 4x^3 \, dx$

44. $\int 5x \sec 2x^2 \tan 2x^2 \, dx$

45. $\int \sqrt{\cos x} \sin x \, dx$

46. $\int \cos^8 x \sin x \, dx$

47. $\int x \tan 11x^2 \, dx$

48. $\int x^2 \cot 8x^3 \, dx$

49. $\int (\sin x)^{3/2} \cos x \, dx$

50. $\int (\cos x)^{-4/3} \sin x \, dx$

51. $\int \sec^2 5x \tan 5x \, dx$

52. Let $f(x) = 3x + 1$, $x_1 = -1$, $x_2 = 0$, $x_3 = 1$, $x_4 = 2$, and $x_5 = 3$. Find $\sum\limits_{i=1}^{5} f(x_i)$.

53. Find $\int_0^4 f(x) \, dx$ for each graph of $y = f(x)$.

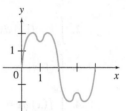

a.

b.

54. Approximate the area under the graph of $f(x) = 2x + 3$ and above the $x$-axis from $x = 0$ to $x = 4$ using four rectangles. Let the height of each rectangle be the function value on the left side.

55. Find $\int_0^4 (2x + 3) \, dx$ by using the formula for the area of a trapezoid: $A = (1/2)(B + b)h$, where $B$ and $b$ are the lengths of the parallel sides and $h$ is the distance between them. Compare with Exercise 54.

56. In Exercises 32 and 33 of Section 7.3 on Area and the Definite Integral, you calculated the distance that a car traveled by estimating the integral $\int_0^T v(t) \, dt$.

a. Let $s(t)$ represent the mileage reading on the odometer. Express the distance traveled between $t = 0$ and $t = T$ using the function $s(t)$.

b. Since your answer to part a and the original integral both represent the distance traveled by the car, the two can be set equal. Explain why the resulting equation is a statement of the Fundamental Theorem of Calculus.

57. What does the Fundamental Theorem of Calculus state?

**Find each definite integral.**

58. $\int_1^2 (3x^2 + 5) \, dx$

59. $\int_1^6 (2x^2 + x) \, dx$

60. $\int_1^5 (3x^{-1} + x^{-3}) \, dx$

61. $\int_1^3 (2x^{-1} + x^{-2}) \, dx$

**62.** $\displaystyle\int_0^1 x\sqrt{5x^2 + 4}\; dx$

**63.** $\displaystyle\int_0^2 x^2(3x^3 + 1)^{1/3}\; dx$

**64.** $\displaystyle\int_0^2 3e^{-2x}\; dx$

**65.** $\displaystyle\int_1^5 \frac{5}{2}e^{0.4x}\; dx$

**66.** $\displaystyle\int_0^{\pi/2} \cos x\; dx$

**67.** $\displaystyle\int_{-\pi}^{2\pi/3} -\sin x\; dx$

**68.** $\displaystyle\int_0^{2\pi} (10 + 10\cos x)\; dx$

**69.** $\displaystyle\int_0^{\pi/3} (3 - 3\sin x)\; dx$

**70.** Use the substitution $u = 4x^2$ and the equation of a semicircle to evaluate

$$\int_0^{1/2} x\sqrt{1 - 16x^4}\; dx.$$

**71.** Use the substitution $u = x^2$ and the equation of a semicircle to evaluate

$$\int_0^{\sqrt{2}} 4x\sqrt{4 - x^4}\,dx.$$

In Exercises 72 and 73, use substitution to change the integral into one that can be evaluated by a formula from geometry, and then find the value of the integral.

**72.** $\displaystyle\int_1^{e^5} \frac{\sqrt{25 - (\ln x)^2}}{x}\; dx$

**73.** $\displaystyle\int_1^{\sqrt{7}} 2x\sqrt{36 - (x^2 - 1)^2}\; dx$

In Exercises 74–77, find the area between the *x*-axis and $f(x)$ over each of the given intervals.

**74.** $f(x) = \sqrt{4x - 3};\quad [1, 3]$

**75.** $f(x) = (3x + 2)^6;\quad [-2, 0]$

**76.** $f(x) = xe^{x^2};\quad [0, 2]$

**77.** $f(x) = 1 + e^{-x};\quad [0, 4]$

Find the area of the region enclosed by each group of curves.

**78.** $f(x) = 5 - x^2,\quad g(x) = x^2 - 3$

**79.** $f(x) = x^2 - 4x,\quad g(x) = x - 6$

**80.** $f(x) = x^2 - 4x,\quad g(x) = x + 6,\quad x = -2,\quad x = 4$

**81.** $f(x) = 5 - x^2,\quad g(x) = x^2 - 3,\quad x = 0,\quad x = 4$

## LIFE SCIENCE APPLICATIONS

**82. Population Growth** The rate of change of the population of a rare species of Australian spider for one year is given by

$$f'(t) = 100 - t\sqrt{0.4t^2 + 1},$$

where $f(t)$ is the number of spiders present at time $t$ (in months). Find the total number of additional spiders in the first 10 months.

**83. Infection Rate** The rate of infection of a disease (in people per month) is given by the function

$$I'(t) = \frac{100t}{t^2 + 1},$$

where $t$ is the time (in months) since the disease broke out. Find the total number of infected people over the first four months of the disease.

**84. Insect Cannibalism** In certain species of flour beetles, the larvae cannibalize the unhatched eggs. In calculating the population cannibalism rate per egg, researchers needed to evaluate the integral

$$\int_0^A c(t)\; dt,$$

where $A$ is the length of the larval stage and $c(t)$ is the cannibalism rate per egg per larva of age $t$. The minimum value of $A$ for the flour beetle *Tribolium castaneum* is 17.6 days, which is the value we will use. The function $c(t)$ starts at day 0 with a value of 0, increases linearly to the value 0.024 at day 12, and then stays constant. *Source: Journal of Animal Ecology.* Find the values of the integral using

**a.** formulas from geometry;

**b.** the Fundamental Theorem of Calculus.

**85. Insulin in Sheep** A research group studied the effect of a large injection of glucose in sheep fed a normal diet compared with sheep that were fasting. A graph of the plasma insulin levels (in pM—picomolar, or $10^{-12}$ molar—concentration) for both groups is shown below. The red graph designates the fasting sheep and the green graph the sheep fed a normal diet. The researchers compared the area under the curves for the two groups. *Source: Endocrinology.*

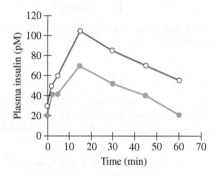

**a.** For the fasting sheep, estimate the area under the curve using rectangles, first by using the left endpoints, then the right endpoints, and then averaging the two. Note that the width of the rectangles will vary.

**b.** Repeat part a for the sheep fed a normal diet.

**c.** How much higher is the area under the curve for the fasting sheep compared with the normal sheep?

**86. Risk in Fisheries** We saw in the previous chapter that the maximum sustainable yield is an important quantity in managing a fishery. Suppose that $\theta$ is the maximum sustainable yield, but that we do not know it exactly, and must instead use an estimator, $T$, plus a correction factor, $x$. There is a risk that our estimator is too high or too low. The expected risk of the estimator, $T + x$, is

$$E = \int_{-\infty}^{\infty} \left( \frac{|t + x - \theta|}{\theta} + \frac{a(t + x)^2}{\theta^2} \right) h(t; \theta) \, dt,$$

where $h(t; \theta)$ is a probability density function of $t$ that depends on the parameter $\theta$ and $a$ is a constant. *Source: Mathematical Biosciences.* (For more information on estimation, probability density functions, and expected value, see Chapters 12 and 13.) Our goal is to find the value of $x$ that minimizes the risk.

a. For simplicity, let $\theta = 1$. The simplest function $h$ is a constant on an interval, set so that the integral of $h$ over that interval is 1. Let

$$h(t; \theta) = \begin{cases} 1 & \text{if } 0.5 \le t \le 1.5 \\ 0 & \text{otherwise.} \end{cases}$$

Show that

$$E = x^2 + \frac{1}{4} + \frac{a}{3}[(1.5 + x)^3 - (0.5 + x)^3].$$

(*Hint:* To integrate, note that $t + x - 1 < 0$ if $0.5 \le t \le 1 - x$, in which case $|t + x - 1| = -(t + x - 1)$, and $t + x - 1 \ge 0$ if $1 - x \le t \le 1.5$.)

b. We can minimize $E$ by setting $dE/dx = 0$. (How do we know this is a minimum and not a maximum?) Show that

$$x = \frac{-a}{1 + a},$$

that is, the expected risk is minimized using the estimator $T - a/(1 + a)$.

c. Suppose we leave $\theta$ as an unknown parameter in part a, and let

$$h(t; \theta) = \begin{cases} 1/\theta & \text{if } 0.5\theta \le t \le 1.5\theta \\ 0 & \text{otherwise.} \end{cases}$$

Show that in this case, the expected risk is minimized using the estimator $T - a\theta/(1 + a)$.

## OTHER APPLICATIONS

**87. Oil Production** The following table shows the amount of crude oil (in billions of barrels) produced in the United States in recent years. *Source: U.S. Energy Information Administration.*

| Year | Crude Oil Produced |
|------|-----|
| 2002 | 2.097 |
| 2003 | 2.060 |
| 2004 | 1.989 |
| 2005 | 1.893 |
| 2006 | 1.857 |
| 2007 | 1.853 |
| 2008 | 1.830 |
| 2009 | 1.954 |
| 2010 | 2.000 |
| 2011 | 2.063 |
| 2012 | 2.377 |

In this exercise we are interested in the total amount of crude oil produced over the 10-year period from mid-2002 to mid-2012, using the data for the 11 years above.

a. One approach is to sum up the numbers in the second column, but only count half of the first and last numbers. Give the answer to this calculation.

b. Approximate the amount of crude oil produced over the 10-year period 2002–2012 by taking the average of the left endpoint sum and the right endpoint sum. Explain why this is equivalent to the calculation done in part a. This is also equivalent to a formula known as the trapezoidal rule, discussed in the next chapter.

c. If your calculator has a cubic regression feature, find the best-fitting cubic function for these data, letting $t = 0$ correspond to 2000. Then integrate this equation over the interval $[2, 12]$ to estimate the amount of crude oil produced over this time period. Compare with your answer to part a.

**88. Automotive Accidents** The following table shows the amount of property damage (in dollars) due to automobile accidents in California in recent years. In this exercise we are interested in the total amount of property damage due to automobile accidents over the 8-year period from mid-2002 to mid-2010, using the data for the 9 years. *Source: The California Highway Patrol.*

| Year | Property Damage ($) |
|------|-----|
| 2002 | 335,869 |
| 2003 | 331,055 |
| 2004 | 331,208 |
| 2005 | 330,195 |
| 2006 | 325,453 |
| 2007 | 313,357 |
| 2008 | 278,986 |
| 2009 | 259,899 |
| 2010 | 252,876 |

a. One approach is to sum up the numbers in the second column, but only count half of the first and last numbers. Give the answer to this calculation.

b. Approximate the amount of property damage over the 8-year period 2002–2010 by taking the average of the left endpoint sum and the right endpoint sum. Explain why this is equivalent to the calculation done in part a. This is also equivalent to a formula known as the trapezoidal rule, discussed in the next chapter.

c. Find the equation of the least squares line for this data, letting $t = 0$ correspond to 2000. Then integrate this equation over the interval $[2, 10]$ to estimate the amount of property damage over this time period. Compare with your answer to part a.

**89. Linear Motion** A particle is moving along a straight line with velocity $v(t) = t^2 - 2t$. Its distance from the starting point after 3 seconds is 8 cm. Find $s(t)$, the distance of the particle from the starting point after $t$ seconds.

**90. Energy Consumption** The monthly residential consumption of energy in the U.S. for 2012 is found in the following table. *Source: Energy Information Administration.*

| Month | Energy (trillion BTU) |
|---|---|
| January | 990.527 |
| February | 833.163 |
| March | 560.826 |
| April | 412.426 |
| May | 297.500 |
| June | 253.015 |
| July | 240.486 |
| August | 248.035 |
| September | 249.354 |
| October | 378.094 |
| November | 631.203 |
| December | 838.265 |

a. Plot the data, letting $t = 1$ correspond to January, $t = 2$ to February, and so on. Is it reasonable to assume that the monthly consumption of energy is periodic?

b. Find the trigonometric function of the form

$$C(t) = a \sin (bt + c) + d$$

that models these data when $t$ is the month of the year and $C(t)$ is the energy consumption. Graph the function on the same calculator window as the data.

c. Estimate the total energy consumption for the year for residential customers in the United States and compare it to the actual value.

d. Calculate the period of the function found in part b. Is this period reasonable?

**91. Temperature** The table lists the average monthly temperatures in Vancouver, Canada. *Source: Weather.com.*

| Month | Jan | Feb | Mar | Apr | May | June |
|---|---|---|---|---|---|---|
| Temperature | 37 | 41 | 43 | 48 | 54 | 59 |

| Month | July | Aug | Sep | Oct | Nov | Dec |
|---|---|---|---|---|---|---|
| Temperature | 63 | 63 | 58 | 50 | 43 | 38 |

These average temperatures cycle yearly and change only slightly over many years. Because of the repetitive nature of temperatures from year to year, they can be modeled with a sine function. Some graphing calculators have a sine regression feature. If the table is entered into a calculator, the points can be plotted automatically, as shown in the early chapters of this book with other types of functions.

a. Use a graphing calculator to plot the ordered pairs (month, temperature) in the interval $[0, 12]$ by $[30, 70]$.

b. Use a graphing calculator with a sine regression feature to find an equation of the sine function that models these data.

c. Graph the equation from part b.

d. Calculate the period for the function found in part b. Is this period reasonable?

**92. Mercator's World Map** Before Gerardus Mercator designed his map of the world in 1569, sailors who traveled in a fixed compass direction could follow a straight line on a map only over short distances. Over long distances, such a course would be a curve on existing maps, which tried to make area on the map proportional to the actual area. Mercator's map greatly simplified navigation: Even over long distances, straight lines on the map corresponded to fixed compass bearings. This was accomplished by distorting distances. On Mercator's map, the distance of an object from the equator to a parallel at latitude $\theta$ is given by

$$D(\theta) = k \int_0^\theta \sec x \, dx,$$

where $k$ is a constant of proportionality. Calculus had not yet been discovered when Mercator designed his map; he approximated the distance between parallels of latitude by hand. *Source: Mathematics Magazine.*

a. Verify that

$$\frac{d}{dx} \ln |\sec x + \tan x| = \sec x.$$

b. Verify that

$$\frac{d}{dx} (-\ln |\sec x - \tan x|) = \sec x.$$

c. Using parts a and b, give two different formulas for $\int \sec x \, dx$. Explain how they can both be correct.

d. Los Angeles has a latitude of 34°03′N. (The 03′ represents 3 minutes of latitude. Each minute of latitude is 1/60 of a degree.) If Los Angeles is to be 7 in. from the equator on a Mercator map, how far from the equator should we place New York City, which has a latitude of 40°45′N?

e. Repeat part d for Miami, which has a latitude of 25°46′N.

f. If you do not live in Los Angeles, New York City, or Miami, repeat part d for your town or city.

**93. Net Savings** A company has installed new machinery that will produce a savings rate (in thousands of dollars per year) of

$$S'(t) = 225 - t^2,$$

where $t$ is the number of years the machinery is to be used. The rate of additional costs (in thousands of dollars per year) to the company due to the new machinery is expected to be

$$C'(t) = t^2 + 25t + 150.$$

For how many years should the company use the new machinery? Find the net savings (in thousands of dollars) over this period.

# EXTENDED APPLICATION

## ESTIMATING DEPLETION DATES FOR MINERALS

It is becoming more and more obvious that the earth contains only a finite quantity of minerals. The "easy and cheap" sources of minerals are being used up, forcing an ever more expensive search for new sources. For example, oil from the North Slope of Alaska would never have been used in the United States during the 1930s because a great deal of Texas and California oil was readily available.

We said in an earlier chapter that population tends to follow an exponential growth curve. Mineral usage also follows such a curve. Thus, if $q$ represents the rate of consumption of a certain mineral at time $t$, while $q_0$ represents consumption when $t = 0$, then

$$q = q_0 e^{kt},$$

where $k$ is the growth constant. For example, the world consumption of petroleum in 1970 was 16,900 million barrels. During this period energy use was growing rapidly, and by 1975 annual world consumption had risen to 21,300 million barrels. We can use these two values to make a rough estimate of the constant $k$, and we find that over this 5-year span the average value of $k$ was about 0.047, representing 4.7% annual growth. If we let $t = 0$ correspond to the base year 1970, then

$$q = 16,900 e^{0.047t}$$

is the rate of consumption at time $t$, assuming that all the trends of the early 1970s have continued. In 1970 a reasonable guess would have put the total amount of oil in provable reserves or likely to be discovered in the future at 1,500,000 million barrels. At the 1970–1975 rate of consumption, in how many years after 1970 would you expect the world's reserves to be depleted? We can use the integral calculus of this chapter to find out. *Source: Energy Information Administration.*

To begin, we need to know the total quantity of petroleum that would be used between time $t = 0$ and some future time $t = T$. Figure 31 shows a typical graph of the function $q = q_0 e^{kt}$.

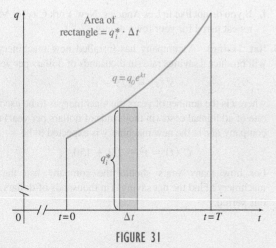

FIGURE 31

Following the work we did in Section 3, divide the time interval from $t = 0$ to $t = T$ into $n$ subintervals. Let each subinterval have width $\Delta t$. Let the rate of consumption for the $i$th subinterval be approximated by $q_i^*$. Thus, the approximate total consumption for the subinterval is given by

$$q_i^* \cdot \Delta t,$$

and the total consumption over the interval from time $t = 0$ to $t = T$ is approximated by

$$\sum_{i=1}^{n} q_i^* \cdot \Delta t.$$

The limit of this sum as $\Delta t$ approaches 0 gives the total consumption from time $t = 0$ to $t = T$. That is,

$$\text{Total consumption} = \lim_{\Delta t \to 0} \sum q_i^* \cdot \Delta t.$$

We have seen, however, that this limit is the definite integral of the function $q = q_0 e^{kt}$ from $t = 0$ to $t = T$, or

$$\text{Total consumption} = \int_0^T q_0 e^{kt} \, dt.$$

We can now evaluate this definite integral.

$$\int_0^T q_0 e^{kt} \, dt = q_0 \int_0^T e^{kt} \, dt = q_0 \left( \frac{e^{kt}}{k} \right) \Big|_0^T$$

$$= \frac{q_0}{k} e^{kt} \Big|_0^T = \frac{q_0}{k} e^{kT} - \frac{q_0}{k} e^0$$

$$= \frac{q_0}{k} e^{kT} - \frac{q_0}{k} (1)$$

$$= \frac{q_0}{k} (e^{kT} - 1) \qquad \textbf{(1)}$$

Now let us return to the numbers we gave for petroleum. We said that $q_0 = 16,900$ million barrels, where $q_0$ represents consumption in the base year of 1970. We have $k = 0.047$ with total petroleum reserves estimated at 1,500,000 million barrels. Thus, using Equation (1) we have

$$1,500,000 = \frac{16,900}{0.047} \left( e^{0.047T} - 1 \right).$$

Multiply both sides of the equation by 0.047.

$$70,500 = 16,900 \left( e^{0.047T} - 1 \right)$$

Divide both sides of the equation by 16,900.

$$4.2 = e^{0.047T} - 1$$

Add 1 to both sides.

$$5.2 = e^{0.047T}$$

Take natural logarithms of both sides.

$$\ln 5.2 = \ln e^{0.047T}$$

$$= 0.047T$$

Finally,

$$T = \frac{\ln 5.2}{0.047} \approx 35.$$

By this result, petroleum reserves would last only 35 years after 1970, that is, until about 2005.

In fact, in the early 1970s some analysts were predicting that reserves would be exhausted before the end of the century, and this was a reasonable guess. But since 1970, more reserves have been discovered. One way to refine our model is to look at the historical data over a longer time span. The following table gives average world annual petroleum consumption in millions of barrels at 5-year intervals from 1970 to 2010. **Source: Energy Information Administration.**

| Year | World Consumption (in millions of barrels) |
|------|--------------------------------------------|
| 1970 | 16,900 |
| 1975 | 21,300 |
| 1980 | 23,000 |
| 1985 | 21,900 |
| 1990 | 24,300 |
| 1995 | 25,600 |
| 2000 | 28,000 |
| 2005 | 30,700 |
| 2010 | 31,900 |

The first step in comparing these data with our exponential model is to estimate a value for the growth constant $k$. One simple way of doing this is to solve the equation

$$31,900 = 16,900 \cdot e^{k \cdot 40}.$$

Using natural logarithms just as we did in estimating the time to depletion for $k = 0.036$, we find that

$$k = \frac{\ln\left(\frac{31,900}{16,900}\right)}{40} \approx 0.016.$$

So the data from the Bureau of Transportation Statistics suggest a growth constant of about 1.6%. We can check the fit by plotting the function $16,900 \cdot e^{0.016t}$ along with a bar graph of the consumption data, shown in Figure 32. The fit looks reasonably good, but over this short range of 40 years, the exponential model is close to a linear model, and the growth in consumption is certainly not smooth.

The exponential model rests on the assumption of a constant growth rate. As already noted, we might expect instead that the growth rate would change as the world comes closer to exhausting its reserves. In particular, scarcity might drive up the price of oil and thus reduce consumption. We can use integration to explore an alternative model in which the factor $k$ changes over time, so that $k$ becomes $k(t)$, a function of time.

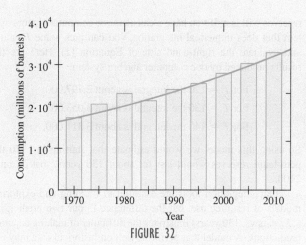

FIGURE 32

As an illustration, we explore a model in which the growth constant $k$ declines toward 0 over time. We'll use 1970 as our base year, so the variable $t$ will count years since 1970. We need a simple positive function $k(t)$ that tends toward 0 as $t$ gets large. To get some numbers to work with, assume that the growth rate was 2% in 1970 and declined to 1% by 1995. There are many possible choices for the function $k(t)$, but a convenient one is

$$k(t) = \frac{0.5}{t + 25}.$$

Using integration to turn the instantaneous rate of consumption into the total consumption up to time $T$, we can write

$$\text{Total consumption} = 16,900 \int_0^T e^{k(t) \cdot t} \, dt$$

$$= 16,900 \int_0^T e^{0.5t/(t+25)} \, dt.$$

We'd like to find out when the world will use up its estimated reserves, but as just noted, the estimates have increased since the 1970s. It is estimated that the current global petroleum reserves are 3,000,000 million barrels. **Source: Geotimes.** So we need to solve

$$3,000,000 = 16,900 \int_0^T e^{0.5t/(t+25)} \, dt. \qquad (2)$$

But this problem is much harder to solve than the corresponding problem for constant growth, because *there is no formula for evaluating this definite integral!* The function

$$g(t) = e^{0.5t/(t+25)}$$

doesn't have an antiderivative that we can write down in terms of functions that we know how to compute.

Here the numerical integration techniques discussed in the next chapter come to the rescue. We can use one of the integration rules to *approximate* the integral numerically for various values of $T$, and with some trial and error we can estimate how long the reserves will last. If you have a calculator or computer algebra system that does numerical integration, you can pick some $T$ values and evaluate the right-hand side of Equation (2). Here are the results produced by one computer algebra system:

For $T = 120$ the integral is about 2,797,000.

For $T = 130$ the integral is about 3,053,000.

For $T = 140$ the integral is about 3,311,000.

So using this model we would estimate that starting in 1970 the petroleum reserves would last for about 130 years, that is, until 2100.

Our integration tools are essential in building and exploring models of resource use, but the difference in our two predictions (35 years vs. 130 years) illustrates the difficulty of making accurate predictions. A model that performs well on historical data may not take the changing dynamics of resource use into account, leading to forecasts that are either unduly gloomy or too optimistic.

## EXERCISES

1. Find the number of years that the estimated petroleum reserves would last if used at the same rate as in the base year.

2. How long would the estimated petroleum reserves last if the growth constant was only 2% instead of 4.7%?

**Estimate the length of time until depletion for each mineral.**

3. Bauxite (the ore from which aluminum is obtained): estimated reserves in base year 15,000,000 thousand tons; rate of consumption in base year 63,000 thousand tons; growth constant 6%

4. Bituminous coal: estimated world reserves 2,000,000 million tons; rate of consumption in base year 2200 million tons; growth constant 4%

5. **a.** Verify that the function $k(t)$ defined on the previous page has the right values at $t = 0$ and $t = 25$.

   **b.** Find a similar function that has $k(0) = 0.03$ and $k(25) = 0.02$.

6. **a.** Use the function you defined in Exercise 5b to write an integral for world petroleum consumption from 1970 until $T$ years after 1970.

   **b.** If you have access to a numerical integrator, compute some values of your integral and estimate the time required to exhaust the reserve of 3,000,000 million barrels.

7. A reasonable assumption is that over time scarcity might drive up the price of oil and thus reduce consumption. Comment on the fact that the rate of oil consumption actually increased in 2002, connecting current events and economic forecasts to the short-term possibility of a reduction in consumption.

8. Develop a spreadsheet that shows the time to exhaustion for various values of $k$.

9. Go to the website WolframAlpha.com and enter "integrate." Follow the instructions to find the time to exhaustion for various values of $k$. Discuss how the solution compares with the solutions provided by a graphing calculator and by Microsoft Excel.

## DIRECTIONS FOR GROUP PROJECT

*Suppose that you and three other students are spending a summer as interns for a local congresswoman. During your internship you realize that the information contained in your calculus class could be used to help with a new bill under consideration. The primary purpose of the bill is to require, by law, that all cars manufactured after a certain date get at least 60 miles per gallon of gasoline. Prepare a report that uses the information above to make a case for or against a bill of this nature.*

# 8

# Further Techniques and Applications of Integration

8.1    Numerical Integration

8.2    Integration by Parts

8.3    Volume and Average Value

8.4    Improper Integrals

       Chapter 8 Review

       Extended Application: Flow Systems

It might seem that definite integrals with infinite limits have only theoretical interest, but in fact these *improper* integrals provide answers to many practical questions. An example in Section 4 models an environmental cleanup process in which the amount of pollution entering a stream decreases by a constant fraction each year. An improper integral gives the total amount of pollutant that will ever enter the river.

n the previous chapter we discussed indefinite and definite integrals, and presented rules for finding the antiderivatives of several types of functions. In this chapter we will show how numerical methods can be used for functions that cannot be integrated by the techniques previously presented. We will then develop a method of integrating functions known as integration by parts. We also show how to evaluate an integral that has one or both limits at infinity. These new techniques allow us to consider additional applications of integration, such as volumes of solids of revolution and the average value of a function.

# 8.1 Numerical Integration

**APPLY IT**   If we know the daily milk consumption by calves as a function of time, how can we calculate the total milk consumed?
*Using numerical integration, we will answer this question in Exercise 29 of this section.*

Some integrals cannot be evaluated by any technique. One solution to this problem was presented in Section 3 of the previous chapter, in which the area under a curve was approximated by summing the areas of rectangles. This method is seldom used in practice because better methods exist that are more accurate for the same amount of work. These methods are referred to as **numerical integration** methods. We discuss two such methods here: the trapezoidal rule and Simpson's rule.

## Trapezoidal Rule
Recall, the trapezoidal rule was mentioned briefly in Section 3 of the previous chapter, where we found approximations with it by averaging the sums of rectangles found by using left endpoints and then using right endpoints. In this section we derive an explicit formula for the trapezoidal rule in terms of function values.* To illustrate the derivation of the trapezoidal rule, consider the integral

$$\int_1^5 \frac{1}{x}\,dx.$$

The shaded region in Figure 1 shows the area representing that integral, the area under the graph $f(x) = 1/x$, above the x-axis, and between the lines $x = 1$ and $x = 5$.

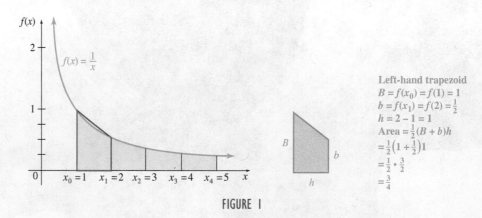

Left-hand trapezoid
$B = f(x_0) = f(1) = 1$
$b = f(x_1) = f(2) = \frac{1}{2}$
$h = 2 - 1 = 1$
$\text{Area} = \frac{1}{2}(B + b)h$
$= \frac{1}{2}\left(1 + \frac{1}{2}\right)1$
$= \frac{1}{2} \cdot \frac{3}{2}$
$= \frac{3}{4}$

FIGURE 1

*In American English a trapezoid is a four-sided figure with two parallel sides, contrasted with a trapezium, which has no parallel sides. In British English, however, it is just the opposite. What Americans call a trapezoid is called a trapezium in Great Britain.

Note that this function can be integrated using the Fundamental Theorem of Calculus. Since $\int (1/x)\,dx = \ln |x| + C$,

$$\int_1^5 \frac{1}{x}\,dx = \ln |x| \Big|_1^5 = \ln 5 - \ln 1 = \ln 5 - 0 = \ln 5 \approx 1.609438.$$

We can also approximate the integral using numerical integration. As shown in the figure, if the area under the curve is approximated with trapezoids rather than rectangles, the approximation should be improved.

As in earlier work, to approximate this area we divide the interval $[1, 5]$ into subintervals of equal widths. To get a first approximation to $\ln 5$ by the trapezoidal rule, find the sum of the areas of the four trapezoids shown in Figure 1. From geometry, the area of a trapezoid is half the product of the sum of the bases and the altitude. Each of the trapezoids in Figure 1 has altitude 1. (In this case, the bases of the trapezoid are vertical and the altitudes are horizontal.) Adding the areas gives

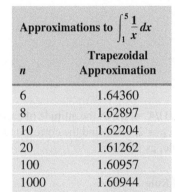

| Approximations to $\int_1^5 \frac{1}{x}\,dx$ | |
|---|---|
| $n$ | **Trapezoidal Approximation** |
| 6 | 1.64360 |
| 8 | 1.62897 |
| 10 | 1.62204 |
| 20 | 1.61262 |
| 100 | 1.60957 |
| 1000 | 1.60944 |

$$\ln 5 = \int_1^5 \frac{1}{x}\,dx \approx \frac{1}{2}\left(\frac{1}{1} + \frac{1}{2}\right)(1) + \frac{1}{2}\left(\frac{1}{2} + \frac{1}{3}\right)(1) + \frac{1}{2}\left(\frac{1}{3} + \frac{1}{4}\right)(1) + \frac{1}{2}\left(\frac{1}{4} + \frac{1}{5}\right)(1)$$

$$= \frac{1}{2}\left(\frac{3}{2} + \frac{5}{6} + \frac{7}{12} + \frac{9}{20}\right) \approx 1.68333.$$

To get a better approximation, divide the interval $[1, 5]$ into more subintervals. Generally speaking, the larger the number of subintervals, the better the approximation. The results for selected values of $n$ are shown to 5 decimal places. When $n = 1000$, the approximation agrees with the true value of $\ln 5 \approx 1.609438$ to 5 decimal places.

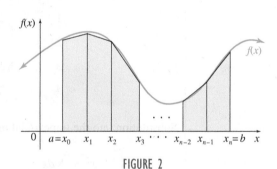

**FIGURE 2**

Generalizing from this example, let $f$ be a continuous function on an interval $[a, b]$. Divide the interval from $a$ to $b$ into $n$ equal subintervals by the points $a = x_0, x_1, x_2, \ldots, x_n = b$, as shown in Figure 2. Use the subintervals to make trapezoids that approximately fill in the region under the curve. The approximate value of the definite integral $\int_a^b f(x)\,dx$ is given by the sum of the areas of the trapezoids, or

$$\int_a^b f(x)\,dx \approx \frac{1}{2}[f(x_0) + f(x_1)]\left(\frac{b-a}{n}\right) + \frac{1}{2}[f(x_1) + f(x_2)]\left(\frac{b-a}{n}\right)$$

$$+ \cdots + \frac{1}{2}[f(x_{n-1}) + f(x_n)]\left(\frac{b-a}{n}\right)$$

$$= \left(\frac{b-a}{n}\right)\left[\frac{1}{2}f(x_0) + \frac{1}{2}f(x_1) + \frac{1}{2}f(x_1) + \frac{1}{2}f(x_2) + \frac{1}{2}f(x_2) + \cdots + \frac{1}{2}f(x_{n-1}) + \frac{1}{2}f(x_n)\right]$$

$$= \left(\frac{b-a}{n}\right)\left[\frac{1}{2}f(x_0) + f(x_1) + f(x_2) + \cdots + f(x_{n-1}) + \frac{1}{2}f(x_n)\right].$$

This result gives the following rule.

### Trapezoidal Rule

Let $f$ be a continuous function on $[a, b]$ and let $[a, b]$ be divided into $n$ equal subintervals by the points $a = x_0, x_1, x_2, \ldots, x_n = b$. Then, by the **trapezoidal rule**,

$$\int_a^b f(x)\, dx \approx \left(\frac{b - a}{n}\right)\left[\frac{1}{2}f(x_0) + f(x_1) + \cdots + f(x_{n-1}) + \frac{1}{2}f(x_n)\right].$$

### EXAMPLE 1   Trapezoidal Rule

Use the trapezoidal rule with $n = 4$ to approximate

$$\int_0^2 \sqrt{x^2 + 1}\, dx.$$

**Method 1**
**Calculating by Hand**

**SOLUTION**

Here $a = 0$, $b = 2$, and $n = 4$, with $(b - a)/n = (2 - 0)/4 = 1/2$ as the altitude of each trapezoid. Then $x_0 = 0$, $x_1 = 1/2$, $x_2 = 1$, $x_3 = 3/2$, and $x_4 = 2$. Now find the corresponding function values. The work can be organized into a table, as follows.

| | Calculations for Trapezoidal Rule | |
|---|---|---|
| $i$ | $x_i$ | $f(x_i)$ |
| 0 | 0 | $\sqrt{0^2 + 1} = 1$ |
| 1 | 1/2 | $\sqrt{(1/2)^2 + 1} \approx 1.11803$ |
| 2 | 1 | $\sqrt{1^2 + 1} \approx 1.41421$ |
| 3 | 3/2 | $\sqrt{(3/2)^2 + 1} \approx 1.80278$ |
| 4 | 2 | $\sqrt{2^2 + 1} \approx 2.23607$ |

Substitution into the trapezoidal rule gives

$$\int_0^2 \sqrt{x^2 + 1}\, dx$$

$$\approx \frac{2 - 0}{4}\left[\frac{1}{2}(1) + 1.11803 + 1.41421 + 1.80278 + \frac{1}{2}(2.23607)\right]$$

$$\approx 2.97653.$$

The approximation 2.97653 found above using the trapezoidal rule with $n = 4$ differs from the true value of 2.95789 by 0.01864. As mentioned earlier, this error would be reduced if larger values were used for $n$. For example, if $n = 8$, the trapezoidal rule gives an answer of 2.96254, which differs from the true value by 0.00465. Techniques for estimating such errors are considered in more advanced courses.

**Method 2**
**Graphing Calculator**

Just as we used a graphing calculator to approximate area using rectangles, we can also use it for the trapezoidal rule. As before, put the values of $i$ in $L_1$ and the values of $x_i$ in $L_2$. In the heading for $L_3$, put $\sqrt{L_2{}^2 + 1}$. Using the fact that $(b - a)/n = (2 - 0)/4 = 0.5$, the command .5*(.5*L₃(1) + sum(L₃,2,4) + .5*L₃(5)) gives the result 2.976528589. For more details, see the *Graphing Calculator and Excel Spreadsheet Manual* available with this book.

▦ **Method 3**
**Spreadsheet**

The trapezoidal rule can also be done on a spreadsheet. In Microsoft Excel, for example, store the values of 0 through $n$ in column A. After putting the left endpoint in E1 and $\Delta x$ in E2, put the command "=$E$1+A1*$E$2" into B1; copying this formula into the rest of column B gives the values of $x_i$. Similarly, use the formula for $f(x_i)$ to fill column C. Using the fact that $n = 5$ in this example, the command "$E$2*(.5*C1+sum(C2:C4)+.5*C5)" gives the result 2.976529. For more details, see the *Graphing Calculator and Excel Spreadsheet Manual* available with this book.

**YOUR TURN 1** Use the trapezoidal rule with $n = 4$ to approximate $\displaystyle\int_1^3 \sqrt{x^2 + 3}\,dx$.

**TRY YOUR TURN 1** ▰

The trapezoidal rule is not widely used because its results are not very accurate. In fact, the midpoint rule discussed in the previous chapter is usually more accurate than the trapezoidal rule. We will now consider a method that usually gives more accurate results than either the trapezoidal or midpoint rule.

## Simpson's Rule

Another numerical method, *Simpson's rule*, approximates consecutive portions of the curve with portions of parabolas rather than the line segments of the trapezoidal rule. Simpson's rule usually gives a better approximation than the trapezoidal rule for the same number of subintervals. As shown in Figure 3, one parabola is fitted through points $A$, $B$, and $C$, another through $C$, $D$, and $E$, and so on. Then the sum of the areas under these parabolas will approximate the area under the graph of the function. Because of the way the parabolas overlap, it is necessary to have an even number of intervals, and therefore an odd number of points, to apply Simpson's rule.

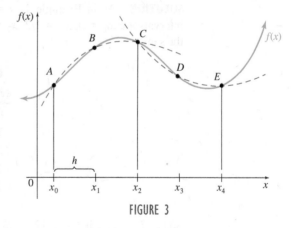

**FIGURE 3**

If $h$, the length of each subinterval, is $(b - a)/n$, the area under the parabola through points $A$, $B$, and $C$ can be found by a definite integral. The details are omitted; the result is

$$\frac{h}{3}[f(x_0) + 4f(x_1) + f(x_2)].$$

Similarly, the area under the parabola through points $C$, $D$, and $E$ is

$$\frac{h}{3}[f(x_2) + 4f(x_3) + f(x_4)].$$

When these expressions are added, the last term of one expression equals the first term of the next. For example, the sum of the two areas given above is

$$\frac{h}{3}[f(x_0) + 4f(x_1) + 2f(x_2) + 4f(x_3) + f(x_4)].$$

This illustrates the origin of the pattern of the terms in the following rule.

## Simpson's Rule

Let $f$ be a continuous function on $[a, b]$ and let $[a, b]$ be divided into an even number $n$ of equal subintervals by the points $a = x_0, x_1, x_2, \ldots, x_n = b$. Then by **Simpson's rule**,

$$\int_a^b f(x)\, dx \approx \left(\frac{b - a}{3n}\right)[f(x_0) + 4f(x_1) + 2f(x_2) + 4f(x_3) + \cdots$$

$$+ 2f(x_{n-2}) + 4f(x_{n-1}) + f(x_n)].$$

Thomas Simpson (1710–1761), a British mathematician, wrote texts on many branches of mathematics. Some of these texts went through as many as ten editions. His name became attached to this numerical method of approximating definite integrals even though the method preceded his work.

**CAUTION** In Simpson's rule, $n$ (the number of subintervals) must be even.

### EXAMPLE 2  Simpson's Rule

Use Simpson's rule with $n = 4$ to approximate

$$\int_0^2 \sqrt{x^2 + 1}\, dx,$$

which was approximated by the trapezoidal rule in Example 1.

**SOLUTION** As in Example 1, $a = 0$, $b = 2$, and $n = 4$, and the endpoints of the four intervals are $x_0 = 0$, $x_1 = 1/2$, $x_2 = 1$, $x_3 = 3/2$, and $x_4 = 2$. The table of values is also the same.

| Calculations for Simpson's Rule | | |
|---|---|---|
| $i$ | $x_i$ | $f(x_i)$ |
| 0 | 0 | 1 |
| 1 | 1/2 | 1.11803 |
| 2 | 1 | 1.41421 |
| 3 | 3/2 | 1.80278 |
| 4 | 2 | 2.23607 |

Since $(b - a)/(3n) = 2/12 = 1/6$, substituting into Simpson's rule gives

$$\int_0^2 \sqrt{x^2 + 1}\, dx \approx \frac{1}{6}[1 + 4(1.11803) + 2(1.41421) + 4(1.80278) + 2.23607] \approx 2.95796.$$

**YOUR TURN 2** Use Simpson's rule with $n = 4$ to approximate $\int_1^3 \sqrt{x^2 + 3}\, dx$.

This differs from the true value by 0.00007, which is less than the trapezoidal rule with $n = 8$. If $n = 8$ for Simpson's rule, the approximation is 2.95788, which differs from the true value by only 0.00001. **TRY YOUR TURN 2**

**NOTE**

1. Just as we can use a graphing calculator or a spreadsheet for the trapezoidal rule, we can also use such technology for Simpson's rule. For more details, see the *Graphing Calculator and Excel Spreadsheet Manual* available with this book.

2. Let $M$ represent the midpoint rule approximation and $T$ the trapezoidal rule approximation, using $n$ subintervals in each. Then the formula $S = (2M + T)/3$ gives the Simpson's rule approximation with $2n$ subintervals.

Numerical methods make it possible to approximate

$$\int_a^b f(x)\, dx$$

even when $f(x)$ is not known. The next example shows how this is done.

**EXAMPLE 3** **Total Distance**

As mentioned earlier, the velocity $v(t)$ gives the rate of change of distance $s(t)$ with respect to time $t$. Suppose a vehicle travels an unknown distance. The passengers keep track of the velocity at 10-minute intervals (every 1/6 of an hour) with the following results.

| Velocity of a Vehicle | | | | | | | |
|---|---|---|---|---|---|---|---|
| Time in Hours, $t$ | 1/6 | 2/6 | 3/6 | 4/6 | 5/6 | 1 | 7/6 |
| Velocity in Miles per Hour, $v(t)$ | 45 | 55 | 52 | 60 | 64 | 58 | 47 |

What is the total distance traveled in the 60-minute period from $t = 1/6$ to $t = 7/6$?

**SOLUTION**   The distance traveled in $t$ hours is $s(t)$, with $s'(t) = v(t)$. The total distance traveled between $t = 1/6$ and $t = 7/6$ is given by

$$\int_{1/6}^{7/6} v(t)\, dt.$$

Even though this integral cannot be evaluated, since we do not have an expression for $v(t)$, either the trapezoidal rule or Simpson's rule can be used to approximate its value and give the total distance traveled. In either case, let $n = 6$, $a = t_0 = 1/6$, and $b = t_6 = 7/6$. By the trapezoidal rule,

$$\int_{1/6}^{7/6} v(t)\, dt \approx \frac{7/6 - 1/6}{6}\left[\frac{1}{2}(45) + 55 + 52 + 60 + 64 + 58 + \frac{1}{2}(47)\right]$$

$$\approx 55.83.$$

By Simpson's rule,

$$\int_{1/6}^{7/6} v(t)\, dt \approx \frac{7/6 - 1/6}{3(6)}[45 + 4(55) + 2(52) + 4(60) + 2(64) + 4(58) + 47]$$

$$= \frac{1}{18}(45 + 220 + 104 + 240 + 128 + 232 + 47) \approx 56.44.$$

The distance traveled in the 1-hour period was about 56 miles.

As already mentioned, Simpson's rule generally gives a better approximation than the trapezoidal rule. As $n$ increases, the two approximations get closer and closer. For the same accuracy, however, a smaller value of $n$ generally can be used with Simpson's rule so that less computation is necessary. Simpson's rule is the method used by many calculators that have a built-in integration feature.

The branch of mathematics that studies methods of approximating definite integrals (as well as many other topics) is called *numerical analysis*. Numerical integration is useful even with functions whose antiderivatives can be determined if the antidifferentiation is complicated and a computer or calculator programmed with Simpson's rule is handy. You may want to program your calculator for both the trapezoidal rule and Simpson's rule. For some calculators, these programs are in the *Graphing Calculator and Excel Spreadsheet Manual* available with this book.

# 8.1 EXERCISES

In Exercises 1–12, use $n = 4$ to approximate the value of the given integrals by the following methods: (a) the trapezoidal rule, and (b) Simpson's rule. (c) Find the exact value by integration.

**1.** $\int_0^2 (3x^2 + 2)\, dx$

**2.** $\int_0^2 (2x^2 + 1)\, dx$

**3.** $\int_{-1}^3 \frac{3}{5 - x}\, dx$

**4.** $\int_1^5 \frac{6}{2x + 1}\, dx$

**5.** $\int_{-1}^2 (2x^3 + 1)\, dx$

**6.** $\int_0^3 (2x^3 + 1)\, dx$

**7.** $\int_1^5 \frac{1}{x^2}\, dx$

**8.** $\int_2^4 \frac{1}{x^3}\, dx$

**9.** $\int_0^1 4xe^{-x^2}\, dx$

**10.** $\int_0^4 x\sqrt{2x^2 + 1}\, dx$

**11.** $\int_0^\pi \sin x\, dx$

**12.** $\int_0^{\pi/4} \sec^2 x\, dx$

**13.** Find the area under the semicircle $y = \sqrt{4 - x^2}$ and above the x-axis by using $n = 8$ with the following methods.

  **a.** The trapezoidal rule       **b.** Simpson's rule

  **c.** Compare the results with the area found by the formula for the area of a circle. Which of the two approximation techniques was more accurate?

**14.** Find the area between the x-axis and the upper half of the ellipse $4x^2 + 9y^2 = 36$ by using $n = 12$ with the following methods.

  **a.** The trapezoidal rule       **b.** Simpson's rule

  (*Hint:* Solve the equation for y and find the area of the semiellipse.)

  **c.** Compare the results with the actual area, $3\pi \approx 9.4248$ (which can be found by methods not considered in this text). Which approximation technique was more accurate?

**15.** Suppose that $f(x) > 0$ and $f''(x) > 0$ for all x between a and b, where $a < b$. Which of the following cases is true of a trapezoidal approximation T for the integral $\int_a^b f(x)\, dx$? Explain.

  **a.** $T < \int_a^b f(x)\, dx$       **b.** $T > \int_a^b f(x)\, dx$

  **c.** Can't say which is larger

**16.** Refer to Exercise 15. Which of the three cases applies to these functions?

  **a.** $f(x) = x^2$;  $[0, 3]$       **b.** $f(x) = \sqrt{x}$;  $[0, 9]$

  **c.**

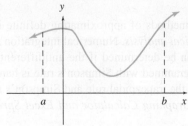

Exercises 17–20 require both the trapezoidal rule and Simpson's rule. They can be worked without calculator programs if such programs are not available, although they require more calculation than the other problems in this exercise set.

**Error Analysis** The difference between the true value of an integral and the value given by the trapezoidal rule or Simpson's rule is known as the error. In numerical analysis, the error is studied to determine how large n must be for the error to be smaller than some specified amount. For both rules, the error is inversely proportional to a power of n, the number of subdivisions. In other words, the error is roughly $k/n^p$, where k is a constant that depends on the function and the interval, and p is a power that depends only on the method used. With a little experimentation, you can find out what the power p is for the trapezoidal rule and for Simpson's rule.

**17. a.** Find the exact value of $\int_0^1 x^4\, dx$.

  **b.** Approximate the integral in part a using the trapezoidal rule with $n = 4, 8, 16$, and $32$. For each of these answers, find the absolute value of the error by subtracting the trapezoidal rule answer from the exact answer found in part a.

  **c.** If the error is $k/n^p$, then the error times $n^p$ should be approximately a constant. Multiply the errors in part b times $n^p$ for $p = 1, 2$, etc., until you find a power p yielding the same answer for all four values of n.

**18.** Based on the results of Exercise 17, what happens to the error in the trapezoidal rule when the number of intervals is doubled?

**19.** Repeat Exercise 17 using Simpson's rule.

**20.** Based on the results of Exercise 19, what happens to the error in Simpson's rule when the number of intervals is doubled?

**21.** For the integral in Exercise 7, apply the midpoint rule with $n = 4$ and Simpson's rule with $n = 8$ to verify the formula $S = (2M + T)/3$.

**22.** Repeat the instructions of Exercise 21 using the integral in Exercise 8.

## LIFE SCIENCE APPLICATIONS

**23. Drug Reaction Rate**  The reaction rate to a new drug is given by

$$y = e^{-t^2} + \frac{1}{t + 1},$$

where t is time (in hours) after the drug is administered. Find the total reaction to the drug from $t = 1$ to $t = 9$ by letting $n = 8$ and using the following methods.

  **a.** The trapezoidal rule

  **b.** Simpson's rule

**24. Growth Rate** The growth rate of a certain tree (in feet) is given by

$$y = \frac{2}{t + 2} + e^{-t^2/2},$$

where $t$ is time (in years). Find the total growth from $t = 1$ to $t = 7$ by using $n = 12$ with the following methods.

**a.** The trapezoidal rule  **b.** Simpson's rule

**Blood Level Curves** **In the study of bioavailability in pharmacy, a drug is given to a patient. The level of concentration of the drug is then measured periodically, producing blood level curves such as the ones shown in the figure.**

**The areas under the curves give the total amount of the drug available to the patient for each milliliter of blood. Use the trapezoidal rule with $n = 10$ to find the following areas.** *Source: Basics of Bioavailability.*

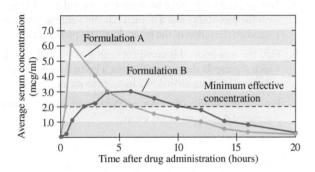

**25.** Find the total area under the curve for Formulation A. What does this area represent?

**26.** Find the total area under the curve for Formulation B. What does this area represent?

**27.** Find the area between the curve for Formulation A and the minimum effective concentration line. What does your answer represent?

**28.** Find the area between the curve for Formulation B and the minimum effective concentration line. What does this area represent?

**29.** APPLY IT **Calves** The daily milk consumption (in kilograms) for calves can be approximated by the function

$$y = b_0 w^{b_1} e^{-b_2 w},$$

where $w$ is the age of the calf (in weeks) and $b_0$, $b_1$, and $b_2$ are constants. *Source: Animal Production.*

**a.** The age in days is given by $t = 7w$. Use this fact to convert the function above to a function in terms of $t$.

**b.** For a group of Angus calves, $b_0 = 5.955$, $b_1 = 0.233$, and $b_2 = 0.027$. Use the trapezoidal rule with $n = 10$, and then Simpson's rule with $n = 10$, to find the total amount of milk consumed by one of these calves over the first 25 weeks of life.

**c.** For a group of Nelore calves, $b_0 = 8.409$, $b_1 = 0.143$, and $b_2 = 0.037$. Use the trapezoidal rule with $n = 10$, and then Simpson's rule with $n = 10$, to find the total amount of milk consumed by one of these calves over the first 25 weeks of life.

**30. Foot-and-Mouth Epidemic** In 2001, the United Kingdom suffered an epidemic of foot-and-mouth disease. The graph below shows the reported number of cattle (red) and pigs (blue) that were culled each month from mid-February through mid-October in an effort to stop the spread of the disease. In Section 7.3 on Area and the Definite Integral we estimated the number of cattle and pigs that were culled using rectangles. *Source: Department of Environment, Food and Rural Affairs, United Kingdom.*

**a.** Estimate the total number of cattle that were culled from mid-February through mid-October and compare this with 581,801, the actual number of cattle that were culled. Use Simpson's rule with interval widths of one month, starting with mid-February.

**b.** Estimate the total number of pigs that were culled from mid-February through mid-October and compare this with 146,145, the actual number of pigs that were culled. Use Simpson's rule with interval widths of one month, starting with mid-February.

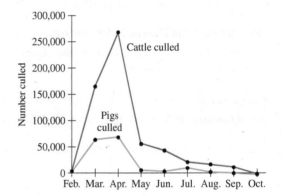

If you have a program for Simpson's rule in your graphing calculator, use it with $n = 20$ for Exercises 31–32.

**31. Milk Consumption** As we saw in an earlier chapter, the average individual daily milk consumption for herds of Charolais, Angus, and Hereford calves can be described by a mathematical function. Here we write the consumption in kg/day as a function of the age of the calf in days $(t)$ as

$$M(t) = 3.922t^{0.242}e^{-0.00357t}, \quad 7 \le t \le 182.$$

Find the total amount of milk consumed from 7 to 182 days for a calf. *Source: Animal Production.*

**32. Blood Pressure** Blood pressure in an artery changes rapidly over a very short time for a healthy young adult, from a high of about 120 to a low of about 80. Suppose the blood pressure function over an interval of 1.5 sec is given by

$$f(t) = 0.2t^5 - 0.68t^4 + 0.8t^3 - 0.39t^2 + 0.055t + 100,$$

where $t$ is the time in seconds after a peak reading. The area under the curve for one cycle is important in some blood pressure studies. Find the area under $f(t)$ from 0.1 sec to 1.1 sec.

## OTHER APPLICATIONS

**33. Total Sales** A sales manager presented the following results at a sales meeting.

| Year, t | 1 | 2 | 3 | 4 | 5 | 6 | 7 |
|---|---|---|---|---|---|---|---|
| Rate of Sales, $f(t)$ | 0.4 | 0.6 | 0.9 | 1.1 | 1.3 | 1.4 | 1.6 |

Find the total sales over the given period as follows.

**a.** Plot these points. Connect the points with line segments.

**b.** Use the trapezoidal rule to find the area bounded by the broken line of part a, the $t$-axis, the line $t = 1$, and the line $t = 7$.

**c.** Approximate the same area using Simpson's rule.

**34. Educational Psychology** The results from a research study in psychology were as follows.

| Number of Hours of Study, t | 1 | 2 | 3 | 4 | 5 | 6 | 7 |
|---|---|---|---|---|---|---|---|
| Rate of Extra Points Earned on a Test, $f(t)$ | 4 | 7 | 11 | 9 | 15 | 16 | 23 |

Repeat parts a–c of Exercise 33 for these data.

**35. Chemical Formation** The following table shows the results from a chemical experiment.

| Concentration of Chemical A, x | 1 | 2 | 3 | 4 | 5 | 6 | 7 |
|---|---|---|---|---|---|---|---|
| Rate of Formation of Chemical B, $f(x)$ | 12 | 16 | 18 | 21 | 24 | 27 | 32 |

Repeat parts a–c of Exercise 33 for these data.

**If you have a program for Simpson's rule in your graphing calculator, use it with $n = 20$ for Exercises 36–37.**

**36. Total Revenue** An electronics company analyst has determined that the rate per month at which revenue comes in from the calculator division is given by

$$R(t) = 105e^{0.01t} + 32,$$

where $t$ is the number of months the division has been in operation. Find the total revenue between the 12th and 36th months.

**37. Probability** The most important function in probability and statistics is the density function for the standard normal distribution, which is the familiar bell-shaped curve. The function is

$$f(x) = \frac{1}{\sqrt{2\pi}}e^{-x^2/2}.$$

**a.** The area under this curve between $x = -1$ and $x = 1$ represents the probability that a normal random variable is within 1 standard deviation of the mean. Find this probability.

**b.** Find the area under this curve between $x = -2$ and $x = 2$, which represents the probability that a normal random variable is within 2 standard deviations of the mean.

**c.** Find the probability that a normal random variable is within 3 standard deviations of the mean.

**YOUR TURN ANSWERS**

**1.** 5.3552   **2.** 5.3477

# 8.2 Integration by Parts

**APPLY IT** If we know the rate of growth of a patch of moss, how can we calculate the area the moss covers?

*We will use integration by parts to answer this question in Exercise 53.*

The technique of *integration by parts* often makes it possible to reduce a complicated integral to a simpler integral. We know that if $u$ and $v$ are both differentiable functions, then $uv$ is also differentiable and, by the product rule for derivatives,

$$\frac{d(uv)}{dx} = u\frac{dv}{dx} + v\frac{du}{dx}.$$

This expression can be rewritten, using differentials, as

$$d(uv) = u\,dv + v\,du.$$

Integrating both sides of this last equation gives

$$\int d(uv) = \int u\,dv + \int v\,du,$$

or

$$uv = \int u\,dv + \int v\,du.$$

Rearranging terms gives the following formula.

## Integration by Parts
If $u$ and $v$ are differentiable functions, then

$$\int u \, dv = uv - \int v \, du.$$

The process of finding integrals by this formula is called **integration by parts**. There are two ways to do integration by parts: the standard method and column integration. Both methods are illustrated in the following example.

### EXAMPLE 1   Integration by Parts

Find $\int xe^{5x} \, dx$.

**Method 1**
**Standard Method**

**SOLUTION**

Although this integral cannot be found by using any method studied so far, it can be found with integration by parts. First write the expression $xe^{5x} \, dx$ as a product of two functions $u$ and $dv$ in such a way that $\int dv$ can be found. One way to do this is to choose the two functions $x$ and $e^{5x}$. Both $x$ and $e^{5x}$ can be integrated, but $\int x \, dx$, which is $x^2/2$, is more complicated than $x$ itself, while the derivative of $x$ is 1, which is simpler than $x$. Since $e^{5x}$ remains the same (except for the coefficient) whether it is integrated or differentiated, it is best here to choose

$$dv = e^{5x} \, dx \qquad \text{and} \qquad u = x.$$

Then

$$du = dx,$$

and $v$ is found by integrating $dv$:

$$v = \int dv = \int e^{5x} \, dx = \frac{e^{5x}}{5}.$$

> **FOR REVIEW**
>
> In Section 7.2 on Substitution, we pointed out that when the chain rule is used to find the derivative of the function $e^{kx}$, we multiply by $k$, so when finding the antiderivative of $e^{kx}$, we divide by $k$. Thus $\int e^{5x} \, dx = e^{5x}/5 + C$. Keeping this technique in mind makes integration by parts simpler.

We need not introduce the constant of integration until the last step, because only one constant is needed. Now substitute into the formula for integration by parts and complete the integration.

$$\int u \, dv = uv - \int v \, du$$

$$\underbrace{\int xe^{5x} \, dx}_{u \; dv} = \underbrace{x\left(\frac{e^{5x}}{5}\right)}_{u \; v} - \underbrace{\int \frac{e^{5x}}{5} \, dx}_{v \; du}$$

$$= \frac{xe^{5x}}{5} - \frac{e^{5x}}{25} + C$$

$$= \frac{e^{5x}}{25}(5x - 1) + C \qquad \text{Factor out } e^{5x}/25.$$

The constant $C$ was added in the last step. As before, check the answer by taking its derivative.

**Method 2**
**Column Integration**

A technique called **column integration**, or *tabular integration*, is equivalent to integration by parts but helps in organizing the details.* We begin by creating two columns. The first column, labeled $D$, contains $u$, the part to be differentiated in the original integral. The second column, labeled $I$, contains the rest of the integral: that is, the part to be integrated, but without the $dx$. To create the remainder of the first column, write the derivative of the function in the first row underneath it in the second row. Now write the derivative of the function in the second row underneath it in the third row. Proceed in this manner down the first column, taking derivatives until you get a 0. Form the second column in a similar manner, except take an antiderivative at each row, until the second column has the same number of rows as the first.

To illustrate this process, consider our goal of finding $\int xe^{5x}\, dx$. Here $u = x$, so $e^{5x}$ is left for the second column. Taking derivatives down the first column and antiderivatives down the second column results in the following table.

| $D$ | $I$ |
|---|---|
| $x$ | $e^{5x}$ |
| $1$ | $e^{5x}/5$ |
| $0$ | $e^{5x}/25$ |

Next, draw a diagonal line from each term (except the last) in the left column to the term in the row below it in the right column. Label the first such line with "+", the next with "−", and continue alternating the signs as shown.

| $D$ | | $I$ |
|---|---|---|
| $x$ | + | $e^{5x}$ |
| $1$ | − | $e^{5x}/5$ |
| $0$ | | $e^{5x}/25$ |

Then multiply the terms on opposite ends of each diagonal line. Finally, sum up the products just formed, adding the "+" terms and subtracting the "−" terms.

$$\int xe^{5x}\, dx = x(e^{5x}/5) - 1(e^{5x}/25) + C$$

$$= \frac{xe^{5x}}{5} - \frac{e^{5x}}{25} + C$$

$$= \frac{e^{5x}}{25}(5x - 1) + C \qquad \text{Factor out } e^{5x}/25.$$

**YOUR TURN 1**
Find $\int xe^{-2x}\, dx$.

Compare these steps with those of Method 1 and convince yourself that the process is the same.

TRY YOUR TURN 1

## Conditions for Integration by Parts
Integration by parts can be used only if the integrand satisfies the following conditions.

1. The integrand can be written as the product of two factors, $u$ and $dv$.
2. It is possible to integrate $dv$ to get $v$ and to differentiate $u$ to get $du$.
3. The integral $\int v\, du$ can be found.

*This technique appeared in the 1988 movie *Stand and Deliver*.

## EXAMPLE 2    Integration by Parts

Find $\int \ln x \, dx$ for $x > 0$.

**Method 1**
**Standard Method**

**SOLUTION**    No rule has been given for integrating $\ln x$, so choose

$$dv = dx \qquad \text{and} \qquad u = \ln x.$$

Then

$$v = x \qquad \text{and} \qquad du = \frac{1}{x} dx,$$

and, since $uv = vu$, we have

$$\int \underbrace{\ln x}_{u} \, \underbrace{dx}_{dv} = \underbrace{x \ln x}_{v \cdot u} - \int \underbrace{x \cdot \frac{1}{x} dx}_{v \cdot du}$$

$$= x \ln x - \int dx$$

$$= x \ln x - x + C.$$

**Method 2**
**Column Integration**

Column integration works a little differently here. As in Method 1, choose $\ln x$ as the part to differentiate. The part to be integrated must be 1. (Think of $\ln x$ as $1 \cdot \ln x$.) No matter how many times $\ln x$ is differentiated, the result is never 0. In this case, stop as soon as the natural logarithm is gone.

| $D$ | $I$ |
|-----|-----|
| $\ln x$ | $1$ |
| $1/x$ | $x$ |

Draw diagonal lines with alternating $+$ and $-$ as before. On the last line, because the left column does not contain a 0, draw a horizontal line.

| $D$ | | $I$ |
|-----|-----|-----|
| $\ln x$ | $+$ | $1$ |
| $1/x$ | $-$ | $x$ |

The presence of a horizontal line indicates that the product is to be integrated, just as the original integral was represented by the first row of the two columns.

$$\int \ln x \, dx = (\ln x)x - \int \frac{1}{x} \cdot x \, dx$$

$$= x \ln x - \int dx$$

$$= x \ln x - x + C.$$

**YOUR TURN 2**
Find $\int \ln 2x \, dx$.

Note that when setting up the columns, a horizontal line is drawn only when a 0 does not eventually appear in the left column.

**TRY YOUR TURN 2**

Sometimes integration by parts must be applied more than once, as in the next example.

EXAMPLE 3  **Integration by Parts**

Find $\int (2x^2 + 5)e^{-3x}\, dx$.

**Method 1**
**Standard Method**

**SOLUTION**

Choose

$$dv = e^{-3x}\, dx \qquad \text{and} \qquad u = 2x^2 + 5.$$

Then

$$v = \frac{-e^{-3x}}{3} \qquad \text{and} \qquad du = 4x\, dx.$$

Substitute these values into the formula for integration by parts.

$$\int u\, dv = uv - \int v\, du$$

$$\int (2x^2 + 5)e^{-3x}\, dx = (2x^2 + 5)\left(\frac{-e^{-3x}}{3}\right) - \int \left(\frac{-e^{-3x}}{3}\right)4x\, dx$$

$$= -(2x^2 + 5)\left(\frac{e^{-3x}}{3}\right) + \frac{4}{3}\int xe^{-3x}\, dx$$

Now apply integration by parts to the last integral, letting

$$dv = e^{-3x}\, dx \qquad \text{and} \qquad u = x,$$

so

$$v = \frac{-e^{-3x}}{3} \qquad \text{and} \qquad du = dx.$$

$$\int (2x^2 + 5)e^{-3x}\, dx = -(2x^2 + 5)\left(\frac{e^{-3x}}{3}\right) + \frac{4}{3}\int xe^{-3x}\, dx$$

$$= -(2x^2 + 5)\left(\frac{e^{-3x}}{3}\right) + \frac{4}{3}\left[x\left(\frac{-e^{-3x}}{3}\right) - \int\left(\frac{-e^{-3x}}{3}\right)dx\right]$$

$$= -(2x^2 + 5)\left(\frac{e^{-3x}}{3}\right) + \frac{4}{3}\left[-\frac{x}{3}e^{-3x} - \left(\frac{e^{-3x}}{9}\right)\right] + C$$

$$= -(2x^2 + 5)\left(\frac{e^{-3x}}{3}\right) - \frac{4}{9}xe^{-3x} - \frac{4}{27}e^{-3x} + C$$

$$= [-(2x^2 + 5)(9) - 4x(3) - 4]\frac{e^{-3x}}{27} + C \qquad \text{Factor out } e^{-3x}/27.$$

$$= (-18x^2 - 12x - 49)\frac{e^{-3x}}{27} + C \qquad \text{Simplify.}$$

**Method 2**
**Column Integration**

Choose $2x^2 + 5$ as the part to be differentiated, and put $e^{-3x}$ in the integration column.

| D | I |
|---|---|
| $2x^2 + 5$ + | $e^{-3x}$ |
| $4x$ − | $-e^{-3x}/3$ |
| $4$ + | $e^{-3x}/9$ |
| $0$ | $-e^{-3x}/27$ |

Multiplying and adding as before yields

$$\int (2x^2 + 5)e^{-3x}\, dx = (2x^2 + 5)(-e^{-3x}/3) - 4x(e^{-3x}/9) + 4(-e^{-3x}/27) + C$$

$$= (-18x^2 - 12x - 49)\frac{e^{-3x}}{27} + C.$$

**YOUR TURN 3**

Find $\int (3x^2 + 4)e^{2x}\, dx$.

TRY YOUR TURN 3

With the functions discussed so far in this book, choosing $u$ and $dv$ (or the parts to be differentiated and integrated) is relatively simple. In general, the following strategy should be used.

First see if the integration can be performed using substitution. If substitution does not work:

■ See if $\ln x$ is in the integral. If it is, set $u = \ln x$ and $dv$ equal to the rest of the integral. (Equivalently, put $\ln x$ in the $D$ column and the rest of the function in the $I$ column.)

■ If $\ln x$ is not present, see if the integral contains $x^k$, where $k$ is any positive integer, or any other polynomial. If it does, set $u = x^k$ (or the polynomial) and $dv$ equal to the rest of the integral. (Equivalently, put $x^k$ in the $D$ column and the rest of the function in the $I$ column.)

---

**EXAMPLE 4** **Trigonometric Integral**

Find $\int 2x \sin x \, dx$.

Method 1
Standard Method

**SOLUTION**

Let $u = 2x$ and $dv = \sin x \, dx$. Then $du = 2dx$ and $v = -\cos x$. Use the formula for integration by parts,

$$\int u \, dv = uv - \int v \, du,$$

to get

$$\int 2x \sin x \, dx = -2x \cos x - \int (-\cos x)(2dx)$$

$$= -2x \cos x + 2 \int \cos x \, dx$$

$$= -2x \cos x + 2 \sin x + C.$$

Check the result by differentiating.

Method 2
Column Integration

Choose $2x$ as the part to be differentiated, and put $\sin x$ in the integration column.

| $D$ | | $I$ |
|---|---|---|
| $2x$ | $+$ | $\sin x$ |
| $2$ | $-$ | $-\cos x$ |
| $0$ | | $-\sin x$ |

Multiplying and adding as in the previous examples yields

$$\int 2x \sin x \, dx = -2x \cos x + 2 \sin x + C$$

TRY YOUR TURN 4

**YOUR TURN 4**

Find $\int -7x \cos x \, dx$.

---

**EXAMPLE 5** **Definite Integral**

Find $\displaystyle\int_1^e \frac{\ln x}{x^2} \, dx$.

**SOLUTION** First find the indefinite integral using integration by parts by the standard method. (You may wish to verify this using column integration.) Whenever $\ln x$ is present, it is selected as $u$, so let

$$u = \ln x \quad \text{and} \quad dv = \frac{1}{x^2} \, dx.$$

Then

$$du = \frac{1}{x} \, dx \quad \text{and} \quad v = -\frac{1}{x}.$$

Substitute these values into the formula for integration by parts, and integrate the second term on the right.

$$\int u \, dv = uv - \int v \, du$$

$$\int \frac{\ln x}{x^2} \, dx = (\ln x)\frac{-1}{x} - \int \left(-\frac{1}{x} \cdot \frac{1}{x}\right) dx$$

$$= -\frac{\ln x}{x} + \int \frac{1}{x^2} \, dx$$

$$= -\frac{\ln x}{x} - \frac{1}{x} + C$$

$$= \frac{-\ln x - 1}{x} + C$$

Now find the definite integral.

$$\int_1^e \frac{\ln x}{x^2} \, dx = \frac{-\ln x - 1}{x} \Big|_1^e$$

$$= \left(\frac{-1-1}{e}\right) - \left(\frac{0-1}{1}\right)$$

$$= \frac{-2}{e} + 1 \approx 0.2642411177$$

**YOUR TURN 5**

Find $\displaystyle\int_1^e x^2 \ln x \, dx$.

TRY YOUR TURN 5

**TECHNOLOGY NOTE**

Definite integrals can be found with a graphing calculator using the function integral feature or by finding the area under the graph of the function between the limits. For example, using the `fnInt` feature of the TI-84 Plus calculator to find the integral in Example 5 gives 0.2642411177. Using the area under the graph approach gives 0.26424112, the same result rounded.

Many integrals cannot be found by the methods presented so far. For example, consider the integral

$$\int \frac{1}{4 - x^2} \, dx.$$

Substitution of $u = 4 - x^2$ will not help, because $du = -2x \, dx$, and there is no $x$ in the numerator of the integral. We could try integration by parts, using $dv = dx$ and $u = (4 - x^2)^{-1}$. Integration gives $v = x$ and differentiation gives $du = 2x \, dx/(4 - x^2)^2$, with

$$\int \frac{1}{4 - x^2} \, dx = \frac{x}{4 - x^2} - \int \frac{2x^2}{(4 - x^2)^2} \, dx.$$

The integral on the right is more complicated than the original integral, however. A second use of integration by parts on the new integral would only make matters worse. Since we cannot choose $dv = (4 - x^2)^{-1} \, dx$ because it cannot be integrated by the methods studied so far, integration by parts is not possible for this problem.

This integration can be performed using one of the many techniques of integration beyond the scope of this text.* Tables of integrals can also be used, but technology is rapidly

---

*For example, see Thomas, George B., Maurice D. Weir, and Joel Hass, *Thomas' Calculus*, 12th ed., Pearson, 2010.

making such tables obsolete and even reducing the importance of techniques of integration. The following example shows how the table of integrals given in the appendix of this book may be used.

### EXAMPLE 6   Tables of Integrals

Find the following.

(a) $\int \dfrac{1}{4 - x^2}\, dx$

**SOLUTION**   Using formula 7 in the table of integrals in the appendix, with $a = 2$, gives

$$\int \frac{1}{4 - x^2}\, dx = \frac{1}{4} \cdot \ln \left| \frac{2 + x}{2 - x} \right| + C.$$

(b) $\int \sin(5x)\sin(3x)\, dx$

**SOLUTION**   Using formula 32 in the table of integrals, with $a = 5$ and $b = 3$, yields

$$\int \sin(5x)\sin(3x)\, dx = \frac{\sin(2x)}{4} - \frac{\sin(8x)}{16} + C.$$

**TRY YOUR TURN 6**

**YOUR TURN 6**

Find (a) $\displaystyle\int \dfrac{1}{x\sqrt{4 + x^2}}\, dx$, and

(b) $\displaystyle\int \sin(4x)\cos(2x)\, dx$.

---

**TECHNOLOGY NOTE**

We mentioned in the previous chapter how computer algebra systems and some calculators can perform integration. Using a TI-89, the answer to the integral in Example 6(a) is

$$\frac{\ln\left(\dfrac{|x + 2|}{|x - 2|}\right)}{4}.$$

(The $C$ is not included.) Verify that this is equivalent to the answer given in Example 6(a).

If you don't have a calculator or computer program that integrates symbolically, there is a Web site (http://integrals.wolfram.com), as of this writing, that finds indefinite integrals using the computer algebra system Mathematica. It includes instructions on how to enter your function. When the previous integral was entered, it returned the answer

$$\frac{1}{4}\big(\log(-x - 2) - \log(x - 2)\big).$$

Note that Mathematica does not include the $C$ or the absolute value, and that natural logarithms are written as log. Verify that this answer is equivalent to the answer given by the TI-89 and the answer given in Example 6(a).

Unfortunately, there are integrals that cannot be antidifferentiated by any technique, in which case numerical integration must be used. (See the previous section.) In this book, for simplicity, all integrals to be antidifferentiated can be done with substitution or by parts, except for Exercises 31–40 in this section.

# 8.2   EXERCISES

**Use integration by parts to find the integrals in Exercises 1–16.**

1. $\displaystyle\int xe^x\, dx$

2. $\displaystyle\int (x + 1)e^x\, dx$

3. $\displaystyle\int (5x - 9)e^{-3x}\, dx$

4. $\displaystyle\int (6x + 3)e^{-2x}\, dx$

5. $\displaystyle\int_0^1 \frac{2x + 1}{e^x}\, dx$

6. $\displaystyle\int_0^1 \frac{1 - x}{3e^x}\, dx$

7. $\displaystyle\int_1^4 \ln 2x\, dx$

8. $\displaystyle\int_1^2 \ln 5x\, dx$

**9.** $\int x \ln x \, dx$

**10.** $\int x^2 \ln x \, dx$

**11.** $\int -6x \cos 5x \, dx$

**12.** $\int 9x \sin 2x \, dx$

**13.** $\int 8x \sin x \, dx$

**14.** $\int -11x \cos x \, dx$

**15.** $\int -6x^2 \cos 8x \, dx$

**16.** $\int 10x^2 \sin \frac{x}{2} \, dx$

**17.** Find the area between $y = (x - 2)e^x$ and the $x$-axis from $x = 2$ to $x = 4$.

**18.** Find the area between $y = (x + 1) \ln x$ and the $x$-axis from $x = 1$ to $x = e$.

**Exercises 19–30 are mixed—some require integration by parts, while others can be integrated by using techniques discussed in the chapter on Integration.**

**19.** $\int x^2 e^{2x} \, dx$

**20.** $\int \frac{x^2 \, dx}{2x^3 + 1}$

**21.** $\int x^2 \sqrt{x + 4} \, dx$

**22.** $\int (2x - 1) \ln(3x) \, dx$

**23.** $\int (8x + 10) \ln(5x) \, dx$

**24.** $\int x^3 e^{x^4} \, dx$

**25.** $\int_1^2 (1 - x^2) e^{2x} \, dx$

**26.** $\int_0^1 \frac{x^2 \, dx}{2x^3 + 1}$

**27.** $\int_0^1 \frac{x^3 \, dx}{\sqrt{3 + x^2}}$

**28.** $\int_0^5 x \sqrt[3]{x^2 + 2} \, dx$

**29.** $\int e^x \cos e^x \, dx$

**30.** $\int x^2 \sin 4x \, dx$

**Use the table of integrals, or a computer or calculator with symbolic integration capabilities, to find each indefinite integral.**

**31.** $\int \frac{16}{\sqrt{x^2 + 16}} \, dx$

**32.** $\int \frac{10}{x^2 - 25} \, dx$

**33.** $\int \frac{3}{x \sqrt{121 - x^2}} \, dx$

**34.** $\int \frac{2}{3x(3x - 5)} \, dx$

**35.** $\int \frac{-6}{x(4x + 6)^2} \, dx$

**36.** $\int \sqrt{x^2 + 15} \, dx$

**37.** $\int e^{4x} \sin 3x \, dx$

**38.** $\int e^{5x} \cos 2x \, dx$

**39.** $\int \sec x \, dx$

**40.** $\int \csc x \, dx$

**41.** What rule of differentiation is related to integration by parts?

**42.** Explain why the two methods of solving Example 2 are equivalent.

**43.** Suppose that $u$ and $v$ are differentiable functions of $x$ with $\int_0^1 v \, du = 4$ and the following functional values.

| $x$ | $u(x)$ | $v(x)$ |
|-----|--------|--------|
| 0 | 2 | 1 |
| 1 | 3 | $-4$ |

Use this information to determine $\int_0^1 u \, dv$.

**44.** Suppose that $u$ and $v$ are differentiable functions of $x$ with $\int_1^{20} v \, du = -1$ and the following functional values.

| $x$ | $u(x)$ | $v(x)$ |
|-----|--------|--------|
| 1 | 5 | $-2$ |
| 20 | 15 | 6 |

Use this information to determine $\int_1^{20} u \, dv$.

**45.** Suppose we know that the functions $r$ and $s$ are everywhere differentiable and that $r(0) = 0$. Suppose we also know that for $0 \le x \le 2$, the area between the $x$-axis and the nonnegative function $h(x) = s(x) \frac{dr}{dx}$ is 5, and that on the same interval, the area between the $x$-axis and the nonnegative function $k(x) = r(x) \frac{ds}{dx}$ is 10. Determine $r(2)s(2)$.

**46.** Suppose we know that the functions $u$ and $v$ are everywhere differentiable and that $u(3) = 0$. Suppose we also know that for $1 \le x \le 3$, the area between the $x$-axis and the nonnegative function $h(x) = u(x) \frac{dv}{dx}$ is 15, and that on the same interval, the area between the $x$-axis and the nonnegative function $k(x) = v(x) \frac{du}{dx}$ is 20. Determine $u(1)v(1)$.

**47.** Use integration by parts to derive the following formula from the table of integrals.

$$\int x^n \cdot \ln |x| \, dx = x^{n+1} \left[ \frac{\ln |x|}{n + 1} - \frac{1}{(n + 1)^2} \right] + C, \quad n \ne -1$$

**48.** Use integration by parts to derive the following formula from the table of integrals.

$$\int x^n e^{ax}\,dx = \frac{x^n e^{ax}}{a} - \frac{n}{a}\int x^{n-1}e^{ax}\,dx + C, \quad a \neq 0$$

**49. a.** One way to integrate $\int x\sqrt{x+1}\,dx$ is to use integration by parts. Do so to find the antiderivative.

  **b.** Another way to evaluate the integral in part a is by using the substitution $u = x + 1$. Do so to find the antiderivative.

  **c.** Compare the results from the two methods. If they do not look the same, explain how this can happen. Discuss the advantages and disadvantages of each method.

**50.** Using integration by parts,

$$\int \frac{1}{x}\,dx = \int \frac{1}{x} \cdot 1\,dx$$

$$= \frac{1}{x} \cdot x - \int \left(-\frac{1}{x^2}\right)x\,dx$$

$$= 1 + \int \frac{1}{x}\,dx.$$

Subtracting $\int \frac{1}{x}\,dx$ from both sides we conclude that $0 = 1$. What is wrong with this logic? ***Source: Sam Northshield.***

## LIFE SCIENCE APPLICATIONS

**51. Reaction to a Drug** The rate of reaction to a drug is given by

$$r'(t) = 2t^2 e^{-t},$$

where $t$ is the number of hours since the drug was administered. Find the total reaction to the drug from $t = 1$ to $t = 6$.

**52. Growth of a Population** The rate of growth of a microbe population is given by

$$m'(t) = 27te^{3t},$$

where $t$ is time in days. What is the total accumulated growth during the first 2 days?

**53. APPLY IT Rate of Growth** The area covered by a patch of moss is growing at a rate of

$$A'(t) = \sqrt{t}\ln t$$

cm² per day, for $t \geq 1$. Find the additional amount of area covered by the moss between 4 and 9 days.

**54. Thermic Effect of Food** As we saw in an earlier chapter, a person's metabolic rate tends to go up after eating a meal and then, after some time has passed, it returns to a resting metabolic rate. This phenomenon is known as the thermic effect of food, and the effect (in kJ per hour) for one individual is

$$F(t) = -10.28 + 175.9te^{-t/1.3},$$

where $t$ is the number of hours that have elapsed since eating a meal. ***Source: American Journal of Clinical Nutrition.*** Find

the total thermic energy of a meal for the next six hours after a meal by integrating the thermic effect function between $t = 0$ and $t = 6$.

**55. Rumen Fermentation** The rumen is the first division of the stomach of a ruminant, or cud-chewing animal. An article on the rumen microbial system reports that the fraction of the soluble material passing from the rumen without being fermented during the first hour after its ingestion could be calculated by the integral

$$\int_0^1 ke^{-kt}(1-t)\,dt,$$

where $k$ measures the rate that the material is fermented. ***Source: Annual Review of Ecology and Systematics.***

  **a.** Determine the above integral, and evaluate it for the following values of $k$ used in the article: 1/12, 1/24, and 1/48 hour.

  **b.** The fraction of intermediate material left in the rumen at 1 hour that escapes digestion by passage between 1 and 6 hours is given by

$$\int_1^6 ke^{-kt}(6-t)/5\,dt.$$

  Determine this integral, and evaluate it for the values of $k$ given in part a.

## OTHER APPLICATION

**56. Rate of Change of Revenue** The rate of change of revenue (in dollars per calculator) from the sale of $x$ calculators is

$$R'(x) = (x+1)\ln(x+1).$$

Find the total revenue from the sale of the first 12 calculators. (*Hint:* In this exercise, it simplifies matters to write an antiderivative of $x + 1$ as $(x+1)^2/2$ rather than $x^2/2 + x$.)

**YOUR TURN ANSWERS**

1. $-e^{-2x}(2x+1)/4 + C$
2. $x\ln 2x - x + C$
3. $(6x^2 - 6x + 11)e^{2x}/4 + C$
4. $-7x\sin x - 7\cos x + C$
5. $(2e^3 + 1)/9$
6. (a) $(-1/2)\ln\left|(2 + \sqrt{4+x^2})/x\right| + C$
   (b) $-\cos(2x)/4 - \cos(6x)/12 + C$

# 8.3  Volume and Average Value

If we have a formula giving the cholesterol level of a patient as a function of time, how can we find the average cholesterol level over a certain period of time?

We will answer this question in Example 4 using concepts developed in this section, in which we will discover how to find the average value of a function, as well as how to compute the volume of a solid.

**Volume**  Figure 4 shows the regions below the graph of some function $y = f(x)$, above the x-axis, and between $x = a$ and $x = b$. We have seen how to use integrals to find the area of such a region. Suppose this region is revolved about the x-axis as shown in Figure 5. The resulting figure is called a **solid of revolution**. In many cases, the volume of a solid of revolution can be found by integration.

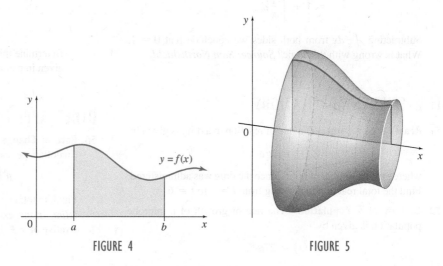

FIGURE 4                          FIGURE 5

To begin, divide the interval $[a, b]$ into $n$ subintervals of equal width $\Delta x$ by the points $a = x_0, x_1, x_2, \ldots, x_i, \ldots, x_n = b$. Then think of slicing the solid into $n$ slices of equal thickness $\Delta x$, as shown in Figure 6(a). If the slices are thin enough, each slice is very close to being a right circular cylinder, as shown in Figure 6(b). The formula for the volume of a right circular cylinder is $\pi r^2 h$, where $r$ is the radius of the circular base and $h$ is the height of the cylinder. As shown in Figure 7, the height of each slice is $\Delta x$. (The height is horizontal here, since the cylinder is on its side.) The radius of the circular base of each slice is $f(x_i)$. Thus, the volume of the slice is closely approximated by $\pi[f(x_i)]^2\Delta x$. The volume of the solid of revolution will be approximated by the sum of the volumes of the slices:

$$V \approx \sum_{i=1}^{n} \pi[f(x_i)]^2\Delta x.$$

By definition, the volume of the solid of revolution is the limit of this sum as the thickness of the slices approaches 0, or

$$V = \lim_{\Delta x \to 0} \sum_{i=1}^{n} \pi[f(x_i)]^2\Delta x.$$

This limit, like the one discussed earlier for area, is a definite integral.

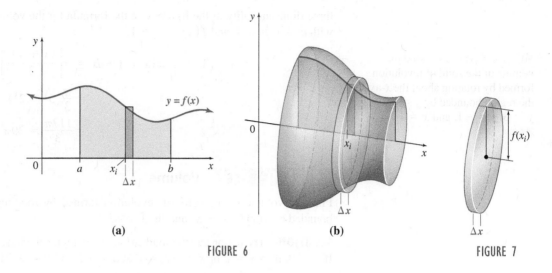

(a)

(b)

FIGURE 6

FIGURE 7

## Volume of a Solid of Revolution

If $f(x)$ is nonnegative and $R$ is the region between $f(x)$ and the $x$-axis from $x = a$ to $x = b$, the volume of the solid formed by rotating $R$ about the $x$-axis is given by

$$V = \lim_{\Delta x \to 0} \sum_{i=1}^{n} \pi [f(x_i)]^2 \Delta x = \int_a^b \pi [f(x)]^2 \, dx.$$

The technique of summing disks to approximate volumes was originated by Johannes Kepler (1571–1630), a famous German astronomer who discovered three laws of planetary motion. He estimated volumes of wine casks used at his wedding by means of solids of revolution.

### EXAMPLE 1   Volume

Find the volume of the solid of revolution formed by rotating about the $x$-axis the region bounded by $y = x + 1$, $y = 0$, $x = 1$, and $x = 4$.

**SOLUTION**   The region and the solid are shown in Figure 8. Notice that the orientation of the $x$-axis is slightly different in Figure 8(b) than in Figure 8(a) to emphasize the

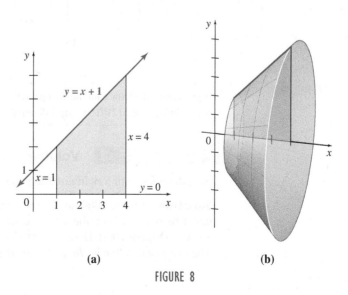

(a)

(b)

FIGURE 8

three-dimensionality of the figure. Use the formula for the volume of a solid of revolution, with $a = 1$, $b = 4$, and $f(x) = x + 1$.

**YOUR TURN 1** Find the volume of the solid of revolution formed by rotating about the $x$-axis the region bounded by $y = x^2 + 1$, $y = 0$, $x = -1$, and $x = 1$.

$$V = \int_1^4 \pi(x + 1)^2\, dx = \pi\left[\frac{(x + 1)^3}{3}\right]\Big|_1^4$$

$$= \frac{\pi}{3}(5^3 - 2^3)$$

$$= \frac{117\pi}{3} = 39\pi \qquad \text{TRY YOUR TURN 1}$$

### EXAMPLE 2   Volume

Find the volume of the solid of revolution formed by rotating about the $x$-axis the area bounded by $f(x) = 4 - x^2$ and the $x$-axis.

**SOLUTION**   The region and the solid are shown in Figure 9. Find $a$ and $b$ from the $x$-intercepts. If $y = 0$, then $x = 2$ or $x = -2$, so that $a = -2$ and $b = 2$. The volume is

$$V = \int_{-2}^2 \pi(4 - x^2)^2\, dx$$

$$= \int_{-2}^2 \pi(16 - 8x^2 + x^4)\, dx$$

$$= \pi\left(16x - \frac{8x^3}{3} + \frac{x^5}{5}\right)\Big|_{-2}^2$$

$$= \frac{512\pi}{15}.$$

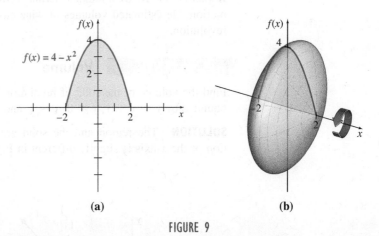

(a)                                    (b)

FIGURE 9

**TECHNOLOGY NOTE**   A graphing calculator with the `fnInt` feature gives the value as 107.2330292, which agrees with the approximation of $512\pi/15$ to the 7 decimal places shown.

### EXAMPLE 3   Volume

Find the volume of a right circular cone with height $h$ and base radius $r$.

**SOLUTION**   Figure 10(a) shows the required cone, while Figure 10(b) shows an area that could be rotated about the $x$-axis to get such a cone. The cone formed by the rotation is shown in Figure 10(c). Here $y = f(x)$ is the equation of the line through $(0, 0)$ and $(h, r)$. The slope of this line is $r/h$, and since the $y$-intercept is 0, the equation of the line is

$$y = \frac{r}{h}x.$$

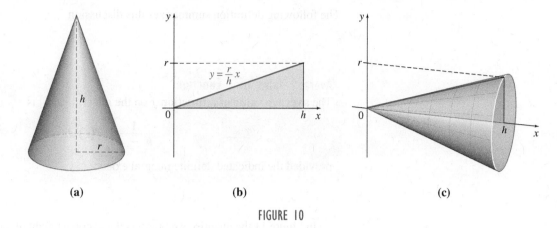

(a)                                      (b)                                      (c)

FIGURE 10

Then the volume is

$$V = \int_0^h \pi \left(\frac{r}{h}x\right)^2 dx = \pi \int_0^h \frac{r^2 x^2}{h^2} \, dx$$

$$= \pi \frac{r^2 x^3}{3h^2} \Big|_0^h \qquad \text{Since } r \text{ and } h \text{ are constants}$$

$$= \frac{\pi r^2 h}{3}.$$

This is the familiar formula for the volume of a right circular cone.

## Average Value of a Function

The average of the $n$ numbers $v_1, v_2, v_3, \ldots,$ $v_i, \ldots, v_n$ is given by

$$\frac{v_1 + v_2 + v_3 + \cdots + v_n}{n} = \frac{\displaystyle\sum_{i=1}^{n} v_i}{n}.$$

For example, to compute an average temperature, we could take readings at equally spaced intervals and average the readings.

The average value of a function $f$ on $[a, b]$ can be defined in a similar manner; divide the interval $[a, b]$ into $n$ subintervals, each of width $\Delta x$. Then choose an $x$-value, $x_i$, in each subinterval, and find $f(x_i)$. The average function value for the $n$ subintervals and the given choices of $x_i$ is

$$\frac{f(x_1) + f(x_2) + \cdots + f(x_n)}{n} = \frac{\displaystyle\sum_{i=1}^{n} f(x_i)}{n}.$$

Since $(b - a)/n = \Delta x$, multiply the expression on the right side of the equation by $(b - a)/(b - a)$ and rearrange the expression to get

$$\frac{b - a}{b - a} \cdot \frac{\displaystyle\sum_{i=1}^{n} f(x_i)}{n} = \frac{1}{b - a} \sum_{i=1}^{n} f(x_i) \left(\frac{b - a}{n}\right) = \frac{1}{b - a} \sum_{i=1}^{n} f(x_i) \Delta x.$$

Now, take the limit as $n \to \infty$. If the limit exists, then

$$\lim_{n \to \infty} \frac{1}{b - a} \sum_{i=1}^{n} f(x_i) \Delta x = \frac{1}{b - a} \lim_{n \to \infty} \sum_{i=1}^{n} f(x_i) \Delta x = \frac{1}{b - a} \int_a^b f(x) \, dx.$$

The following definition summarizes this discussion.

### Average Value of a Function

The **average value of a function** $f$ on the interval $[a, b]$ is

$$\frac{1}{b - a}\int_a^b f(x)\, dx,$$

provided the indicated definite integral exists.

In Figure 11 the quantity $\bar{y}$ represents the average height of the irregular region. The average height can be thought of as the height of a rectangle with base $b - a$. For $f(x) \geq 0$, this rectangle has area $\bar{y}(b - a)$, which equals the area under the graph of $f(x)$ from $x = a$ to $x = b$, so that

$$\bar{y}(b - a) = \int_a^b f(x)\, dx.$$

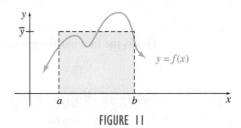

**FIGURE 11**

### EXAMPLE 4    Cholesterol Level

A physician finds that the cholesterol level of a patient on a diet can be approximated by

$$C(t) = 200 + 40e^{-0.1t},$$

where $t$ is the time (in months) since the patient began the diet. Find the average cholesterol level for the patient over the first six months.

**APPLY IT**

**SOLUTION**    Use the formula for average value with $a = 0$ and $b = 6$. The average cholesterol level is

$$\frac{1}{6 - 0}\int_0^6 (200 + 40e^{-0.1t})dt = \frac{1}{6}\left(200t + \frac{40}{-0.1}e^{-0.1t}\right)\Bigg|_0^6$$

$$= \frac{1}{6}(200t - 400e^{-0.1t})\Bigg|_0^6$$

**YOUR TURN 2**    Find the average value of the function $f(x) = x + \sqrt{x}$ on the interval $[1, 4]$.

$$= \frac{1}{6}(1200 - 400e^{-0.6} + 400)$$

$$\approx 230.$$

**TRY YOUR TURN 2**

# 8.3 EXERCISES

**Find the volume of the solid of revolution formed by rotating about the $x$-axis each region bounded by the given curves.**

1. $f(x) = x$, $y = 0$, $x = 0$, $x = 3$

2. $f(x) = 3x$, $y = 0$, $x = 0$, $x = 2$

3. $f(x) = 2x + 1$, $y = 0$, $x = 0$, $x = 4$

4. $f(x) = x - 4$, $y = 0$, $x = 4$, $x = 10$

5. $f(x) = \frac{1}{3}x + 2$, $y = 0$, $x = 1$, $x = 3$

6. $f(x) = \frac{1}{2}x + 4$, $y = 0$, $x = 0$, $x = 5$

7. $f(x) = \sqrt{x}$, $y = 0$, $x = 1$, $x = 4$

8. $f(x) = \sqrt{x + 5}$, $y = 0$, $x = 1$, $x = 3$

9. $f(x) = \sqrt{2x + 1}$, $y = 0$, $x = 1$, $x = 4$

10. $f(x) = \sqrt{4x + 2}$, $y = 0$, $x = 0$, $x = 2$

11. $f(x) = e^x$, $y = 0$, $x = 0$, $x = 2$

12. $f(x) = 2e^x$, $y = 0$, $x = -2$, $x = 1$

13. $f(x) = \dfrac{2}{\sqrt{x}}$, $y = 0$, $x = 1$, $x = 3$

14. $f(x) = \dfrac{2}{\sqrt{x + 2}}$, $y = 0$, $x = -1$, $x = 2$

15. $f(x) = x^2$, $y = 0$, $x = 1$, $x = 5$

16. $f(x) = \dfrac{x^2}{2}$, $y = 0$, $x = 0$, $x = 4$

17. $f(x) = 1 - x^2$, $y = 0$

18. $f(x) = 2 - x^2$, $y = 0$

19. $f(x) = \sec x$, $y = 0$, $x = 0$, $x = \pi/4$

20. $f(x) = \csc x$, $y = 0$, $x = \pi/4$, $x = \pi/2$

The function defined by $y = \sqrt{r^2 - x^2}$ has as its graph a semicircle of radius $r$ with center at $(0, 0)$ (see the figure). In Exercises 21–23, find the volume that results when each semicircle is rotated about the $x$-axis. (The result of Exercise 23 gives a formula for the volume of a sphere with radius $r$.)

21. $f(x) = \sqrt{1 - x^2}$

22. $f(x) = \sqrt{36 - x^2}$

23. $f(x) = \sqrt{r^2 - x^2}$

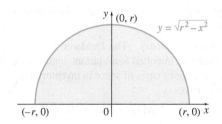

24. Find a formula for the volume of an ellipsoid. See Exercises 21–23 and the figures in the next column.

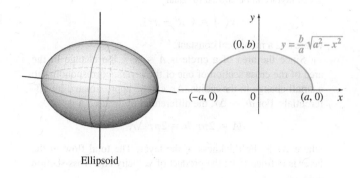

Ellipsoid

25. Use the methods of this section to find the volume of a cylinder with height $h$ and radius $r$.

**Find the average value of each function on the given interval.**

26. $f(x) = 2 - 3x^2$; $[1, 3]$

27. $f(x) = x^2 - 4$; $[0, 5]$

28. $f(x) = (2x - 1)^{1/2}$; $[1, 13]$

29. $f(x) = \sqrt{x + 1}$; $[3, 8]$

30. $f(x) = e^{0.1x}$; $[0, 10]$

31. $f(x) = e^{x/7}$; $[0, 7]$

32. $f(x) = x \ln x$; $[1, e]$

33. $f(x) = x^2 e^{2x}$; $[0, 2]$

34. $f(x) = \sin x$; $[0, \pi]$

35. $f(x) = \sec^2 x$; $[0, \pi/4]$

**In Exercises 36 and 37, use the integration feature on a graphing calculator to find the volume of the solid of revolution by rotating about the $x$-axis each region bounded by the given curves.**

36. $f(x) = \dfrac{1}{4 + x^2}$, $y = 0$, $x = -2$, $x = 2$

37. $f(x) = e^{-x^2}$, $y = 0$, $x = -1$, $x = 1$

## LIFE SCIENCE APPLICATIONS

38. **Blood Flow** The figure shows the blood flow in a small artery of the body. The flow of blood is *laminar* (in layers), with the velocity very low near the artery walls and highest in the center of the artery. In this model of blood flow, we calculate the total flow in the artery by thinking of the flow as being made up of many layers of concentric tubes sliding one on the other.

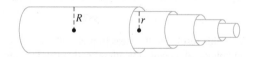

Suppose $R$ is the radius of an artery and $r$ is the distance from a given layer to the center. Then the velocity of blood in a given layer can be shown to equal

$$v(r) = k(R^2 - r^2),$$

where $k$ is a numerical constant.

Since the area of a circle is $A = \pi r^2$, the change in the area of the cross section of one of the layers, corresponding to a small change in the radius, $\Delta r$, can be approximated by differentials. For $dr = \Delta r$, the differential of the area $A$ is

$$dA = 2\pi r\, dr = 2\pi r\, \Delta r,$$

where $\Delta r$ is the thickness of the layer. The total flow in the layer is defined to be the product of velocity and cross-section area, or

$$F(r) = 2\pi rk(R^2 - r^2)\Delta r.$$

a. Set up a definite integral to find the total flow in the artery.

b. Evaluate this definite integral.

39. **Drug Reaction**  The intensity of the reaction to a certain drug, in appropriate units, is given by

$$R(t) = te^{-0.1t},$$

where $t$ is time (in hours) after the drug is administered. Find the average intensity during the following hours.

a. Second hour      b. Twelfth hour

c. Twenty-fourth hour

40. **Bird Eggs**  The average length and width of various bird eggs are given in the following table. *Source: NCTM.*

| Bird Name | Length (cm) | Width (cm) |
| --- | --- | --- |
| Canada goose | 8.6 | 5.8 |
| Robin | 1.9 | 1.5 |
| Turtledove | 3.1 | 2.3 |
| Hummingbird | 1.0 | 1.0 |
| Raven | 5.0 | 3.3 |

a. Assume for simplicity that a bird's egg is roughly the shape of an ellipsoid. Use the result of Exercise 24 to estimate the volume of an egg of each bird.

   i. Canada goose

  ii. Robin

 iii. Turtledove

 iv. Hummingbird

  v. Raven

b. In Exercise 10 of Section 1.2, we showed that the average length (in centimeters) of an egg of width $w$ cm is given by

$$l = 1.585w - 0.487.$$

Using this result and the ideas in part a, show that the average volume of an egg of width $w$ centimeters is given by

$$V = \pi(1.585w^3 - 0.487w^2)/6.$$

Use this formula to calculate the average volume for the bird eggs in part a, and compare with your results from part a.

## OTHER APPLICATIONS

41. **Production Rate**  Suppose the number of items a new worker on an assembly line produces daily after $t$ days on the job is given by

$$I(t) = 45\ln(t + 1).$$

Find the average number of items produced daily by this employee after the following numbers of days.

a. 5      b. 9      c. 30

42. **Typing Speed**  The function $W(t) = -3.75t^2 + 30t + 40$ describes a typist's speed (in words per minute) over a time interval $[0, 5]$.

a. Find $W(0)$.

b. Find the maximum $W$ value and the time $t$ when it occurs.

c. Find the average speed over $[0, 5]$.

43. **Earth's Volume**  Most people assume that the Earth has a spherical shape. It is actually more of an ellipsoid shape, but not an exact ellipsoid, since there are numerous mountains and valleys. Researchers have found that a *datum*, or a reference ellipsoid, that is offset from the center of the Earth can be used to accurately map different regions. According to one datum, called the Geodetic Reference System 1980, this reference ellipsoid assumes an equatorial radius of 6,378,137 m and a polar radius of 6,356,752.3141 m. *Source: Geodesy Information System.* Use the result of Exercise 24 to estimate the volume of the Earth.

44. **Average Price**  Otis Taylor plots the price per share of a stock that he owns as a function of time and finds that it can be approximated by the function

$$S(t) = t(25 - 5t) + 18,$$

where $t$ is the time (in years) since the stock was purchased. Find the average price of the stock over the first five years.

45. **Average Price**  A stock analyst plots the price per share of a certain common stock as a function of time and finds that it can be approximated by the function

$$S(t) = 37 + 6e^{-0.03t},$$

where $t$ is the time (in years) since the stock was purchased. Find the average price of the stock over the first six years.

46. **Average Inventory**  The Yasuko Okada Fragrance Company (YOFC) receives a shipment of 400 cases of specialty perfume early Monday morning of every week. YOFC sells the perfume to retail outlets in California at a rate of about 80 cases per day during each business day (Monday through Friday). What is the average daily inventory for YOFC? (*Hint:* Find a function that represents the inventory for any given business day and then integrate.)

47. **Average Inventory**  The DeMarco Pasta Company receives 600 cases of imported San Marzano tomato sauce every 30 days. The number of cases of sauce in inventory $t$ days after the shipment arrives is

$$N(t) = 600 - 20\sqrt{30t}.$$

Find the average daily inventory.

**YOUR TURN ANSWERS**

1. $56\pi/15$      2. $73/18$

# 8.4 Improper Integrals

*If we know the rate at which a pollutant is dumped into a stream, how can we compute the total amount released given that the rate of dumping is decreasing over time?*

In this section we will learn how to answer questions such as this one, which is answered in Example 3.

Sometimes it is useful to be able to integrate a function over an infinite period of time. For example, we might want to find the total amount of income generated by an apartment building into the indefinite future or the total amount of pollution into a bay from a source that is continuing indefinitely. In this section we define integrals with one or more infinite limits of integration that can be used to solve such problems.

The graph in Figure 12(a) shows the area bounded by the curve $f(x) = x^{-3/2}$, the $x$-axis, and the vertical line $x = 1$. Think of the shaded region below the curve as extending indefinitely to the right. Does this shaded region have an area?

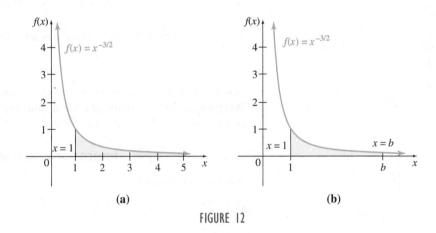

**FIGURE 12**

To see if the area of this region can be defined, introduce a vertical line at $x = b$, as shown in Figure 12(b). This vertical line gives a region with both upper and lower limits of integration. The area of this new region is given by the definite integral

$$\int_1^b x^{-3/2}\, dx.$$

By the Fundamental Theorem of Calculus,

$$\int_1^b x^{-3/2}\, dx = \left(-2x^{-1/2}\right)\Big|_1^b$$

$$= -2b^{-1/2} - \left(-2 \cdot 1^{-1/2}\right)$$

$$= -2b^{-1/2} + 2 = 2 - \frac{2}{b^{1/2}}.$$

Suppose we now let the vertical line $x = b$ in Figure 12(b) move farther to the right. That is, suppose $b \to \infty$. The expression $-2/b^{1/2}$ would then approach 0, and

$$\lim_{b \to \infty}\left(2 - \frac{2}{b^{1/2}}\right) = 2 - 0 = 2.$$

This limit is defined to be the *area* of the region shown in Figure 12(a), so that

$$\int_1^\infty x^{-3/2}\,dx = 2.$$

An integral of the form

$$\int_a^\infty f(x)\,dx, \qquad \int_{-\infty}^b f(x)\,dx, \qquad \text{or} \qquad \int_{-\infty}^\infty f(x)\,dx$$

is called an *improper integral*. These **improper integrals** are defined as follows.

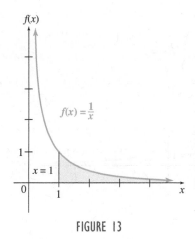

$f(x)$

$f(x) = \dfrac{1}{x}$

$x = 1$

**FIGURE 13**

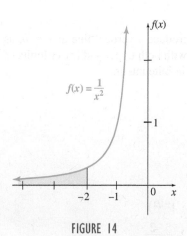

$f(x)$

$f(x) = \dfrac{1}{x^2}$

**FIGURE 14**

**YOUR TURN 1**
Find each integral.
(a) $\int_8^\infty \dfrac{1}{x^{1/3}}\,dx$ (b) $\int_8^\infty \dfrac{1}{x^{4/3}}\,dx$

**Improper Integrals**
If $f$ is continuous on the indicated interval and if the indicated limits exist, then

$$\int_a^\infty f(x)\,dx = \lim_{b\to\infty}\int_a^b f(x)\,dx,$$

$$\int_{-\infty}^b f(x)\,dx = \lim_{a\to-\infty}\int_a^b f(x)\,dx,$$

$$\int_{-\infty}^\infty f(x)\,dx = \int_{-\infty}^c f(x)\,dx + \int_c^\infty f(x)\,dx,$$

for real numbers $a$, $b$, and $c$, where c is arbitrarily chosen.

If the expressions on the right side exist, the integrals are **convergent**; otherwise, they are **divergent**. A convergent integral has a value that is a real number. A divergent integral does not, often because the area under the curve is infinitely large.

**EXAMPLE 1** Improper Integrals
Evaluate each integral.
(a) $\displaystyle\int_1^\infty \frac{dx}{x}$

**SOLUTION** A graph of this region is shown in Figure 13. By the definition of an improper integral,

$$\int_1^\infty \frac{dx}{x} = \lim_{b\to\infty}\int_1^b \frac{dx}{x}.$$

Find $\displaystyle\int_1^b \frac{dx}{x}$ by the Fundamental Theorem of Calculus.

$$\int_1^b \frac{dx}{x} = \ln|x|\Big|_1^b = \ln|b| - \ln|1| = \ln|b| - 0 = \ln|b|$$

As $b\to\infty$, $\ln|b|\to\infty$, so $\lim_{b\to\infty}\ln|b|$ does not exist. Since the limit does not exist, $\displaystyle\int_1^\infty \frac{dx}{x}$ is divergent.

(b) $\displaystyle\int_{-\infty}^{-2}\frac{1}{x^2}\,dx = \lim_{a\to-\infty}\int_a^{-2}\frac{1}{x^2}\,dx = \lim_{a\to-\infty}\left(\frac{-1}{x}\right)\Big|_a^{-2}$

$$= \lim_{a\to-\infty}\left(\frac{1}{2}+\frac{1}{a}\right) = \frac{1}{2}$$

A graph of this region is shown in Figure 14. Since the limit exists, this integral converges.

TRY YOUR TURN 1

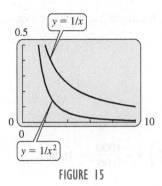

$y = 1/x$

0.5

0

0

$y = 1/x^2$

10

**FIGURE 15**

It may seem puzzling that the areas under the curves $f(x) = 1/x^{3/2}$ and $f(x) = 1/x^2$ are finite, while $f(x) = 1/x$ has an infinite amount of area. At first glance the graphs of these functions appear similar. The difference is that although all three functions get small as $x$ becomes infinitely large, $f(x) = 1/x$ does not become small enough fast enough. In the graphing calculator screen in Figure 15, notice how much faster $1/x^2$ becomes small compared with $1/x$.

| CAUTION | Since graphing calculators provide only approximations, using them to find improper integrals is tricky and requires skill and care. Although their approximations may be good in some cases, they are wrong in others, and they cannot tell us for certain that an improper integral does not exist. See Exercises 39–41. |

**EXAMPLE 2** Improper Integral

Find $\int_{-\infty}^{\infty} 4e^{-3x}\, dx$.

**SOLUTION** In the definition of an improper integral with limits of $-\infty$ and $\infty$, the value of $c$ is arbitrary, so we'll choose the simple value $c = 0$. We can then write the integral as

$$\int_{-\infty}^{\infty} 4e^{-3x}\, dx = \int_{-\infty}^{0} 4e^{-3x}\, dx + \int_{0}^{\infty} 4e^{-3x}\, dx$$

and evaluate each of the two improper integrals on the right. If they both converge, the original integral will equal their sum. To show you all the details while maintaining the suspense, we will evaluate the second integral first.

By definition,

$$\int_{0}^{\infty} 4e^{-3x}\, dx = \lim_{b\to\infty} \int_{0}^{b} 4e^{-3x}\, dx = \lim_{b\to\infty} \left(\frac{-4}{3}e^{-3x}\right)\Big|_{0}^{b}$$

$$= \lim_{b\to\infty} \left(\frac{-4}{3e^{3b}} + \frac{4}{3}\right) = 0 + \frac{4}{3} = \frac{4}{3}.$$

**FOR REVIEW**
Recall that
$$\lim_{p\to\infty} \frac{1}{e^p} = 0$$
and
$$\lim_{p\to-\infty} \frac{1}{e^p} = \infty.$$

Similarly, the first integral is evaluated as

$$\int_{-\infty}^{0} 4e^{-3x}\, dx = \lim_{b\to-\infty} \int_{b}^{0} 4e^{-3x}\, dx = \lim_{b\to-\infty} \left(\frac{-4}{3}e^{-3x}\right)\Big|_{b}^{0}$$

$$= \lim_{b\to-\infty} \left(-\frac{4}{3} + \frac{4}{3e^{3b}}\right) = \infty.$$

**YOUR TURN 2**
Find $\int_{0}^{\infty} 5e^{-2x}\, dx$.

Since one of the two improper integrals diverges, the original improper integral diverges. **TRY YOUR TURN 2**

The following examples describe applications of improper integrals.

**EXAMPLE 3** Pollution

The rate at which a pollutant is being dumped into a stream at time $t$ is given by $P_0 e^{-kt}$, where $P_0$ is the rate that the pollutant is initially released into the stream. Suppose $P_0 = 1000$ and $k = 0.06$. Find the total amount of the pollutant that will be released into the stream into the indefinite future.

**APPLY IT**

**SOLUTION** Find

$$\int_{0}^{\infty} P_0 e^{-kt}\, dt = \int_{0}^{\infty} 1000 e^{-0.06t}\, dt.$$

This integral is similar to one of the integrals used to solve Example 2 and may be evaluated by the same method.

$$\int_0^\infty 1000e^{-0.06t}\,dt = \lim_{b\to\infty}\int_0^b 1000e^{-0.06t}\,dt$$

$$= \lim_{b\to\infty}\left(\frac{1000}{-0.06}e^{-0.06t}\right)\Bigg|_0^b$$

$$= \lim_{b\to\infty}\left(\frac{1000}{-0.06e^{0.06b}} - \frac{1000}{-0.06}e^0\right) = \frac{-1000}{-0.06} \approx 16{,}667$$

A total of approximately 16,667 units of the pollutant will be released over time.

# 8.4 EXERCISES

**Determine whether each improper integral converges or diverges, and find the value of each that converges.**

**1.** $\displaystyle\int_3^\infty \frac{1}{x^2}\,dx$

**2.** $\displaystyle\int_3^\infty \frac{1}{(x+1)^3}\,dx$

**3.** $\displaystyle\int_4^\infty \frac{2}{\sqrt{x}}\,dx$

**4.** $\displaystyle\int_{27}^\infty \frac{2}{\sqrt[3]{x}}\,dx$

**5.** $\displaystyle\int_{-\infty}^{-1} \frac{2}{x^3}\,dx$

**6.** $\displaystyle\int_{-\infty}^{-4} \frac{3}{x^4}\,dx$

**7.** $\displaystyle\int_1^\infty \frac{1}{x^{1.0001}}\,dx$

**8.** $\displaystyle\int_1^\infty \frac{1}{x^{0.999}}\,dx$

**9.** $\displaystyle\int_{-\infty}^{-10} x^{-2}\,dx$

**10.** $\displaystyle\int_{-\infty}^{-1} (x-2)^{-3}\,dx$

**11.** $\displaystyle\int_{-\infty}^{-1} x^{-8/3}\,dx$

**12.** $\displaystyle\int_{-\infty}^{-27} x^{-5/3}\,dx$

**13.** $\displaystyle\int_0^\infty 8e^{-8x}\,dx$

**14.** $\displaystyle\int_0^\infty 50e^{-50x}\,dx$

**15.** $\displaystyle\int_{-\infty}^0 1000e^x\,dx$

**16.** $\displaystyle\int_{-\infty}^0 5e^{60x}\,dx$

**17.** $\displaystyle\int_{-\infty}^{-1} \ln|x|\,dx$

**18.** $\displaystyle\int_1^\infty \ln|x|\,dx$

**19.** $\displaystyle\int_0^\infty \frac{dx}{(x+1)^2}$

**20.** $\displaystyle\int_0^\infty \frac{dx}{(4x+1)^3}$

**21.** $\displaystyle\int_{-\infty}^{-1} \frac{2x-1}{x^2-x}\,dx$

**22.** $\displaystyle\int_1^\infty \frac{4x+6}{x^2+3x}\,dx$

**23.** $\displaystyle\int_2^\infty \frac{1}{x\ln x}\,dx$

**24.** $\displaystyle\int_2^\infty \frac{1}{x(\ln x)^2}\,dx$

**25.** $\displaystyle\int_0^\infty xe^{4x}\,dx$

**26.** $\displaystyle\int_{-\infty}^0 xe^{0.2x}\,dx$ (*Hint:* Recall that $\lim\limits_{x\to-\infty} xe^x = 0$.)

**27.** $\displaystyle\int_{-\infty}^\infty x^3 e^{-x^4}\,dx$ (*Hint:* Recall from Exercise 72 in Section 3.1 on Limits that $\lim\limits_{x\to\infty} x^n e^{-x} = 0$.)

**28.** $\displaystyle\int_{-\infty}^\infty e^{-|x|}\,dx$ (*Hint:* Recall that when $x < 0$, $|x| = -x$.)

**29.** $\displaystyle\int_{-\infty}^\infty \frac{x}{x^2+1}\,dx$

**30.** $\displaystyle\int_{-\infty}^\infty \frac{2x+4}{x^2+4x+5}\,dx$

**Find the area between the graph of the given function and the x-axis over the given interval, if possible.**

**31.** $f(x) = \dfrac{1}{x-1}$, for $(-\infty, 0]$

**32.** $f(x) = e^{-x}$, for $(-\infty, e]$

**33.** $f(x) = \dfrac{1}{(x-1)^2}$, for $(-\infty, 0]$

**34.** $f(x) = \dfrac{3}{(x-1)^3}$, for $(-\infty, 0]$

**35.** Find $\displaystyle\int_{-\infty}^\infty xe^{-x^2}\,dx$.

**36.** Find $\displaystyle\int_{-\infty}^\infty \frac{x}{(1+x^2)^2}\,dx$.

**37.** Show that $\displaystyle\int_1^\infty \frac{1}{x^p}\,dx$ converges if $p > 1$ and diverges if $p \le 1$.

**38.** Example 1(b) leads to a paradox. On the one hand, the unbounded region in that example has an area of 1/2, so theoretically it could be colored with ink. On the other hand, the boundary of that region is infinite, so it cannot be drawn with a finite amount of ink. This seems impossible, because coloring the region automatically colors the boundary. Explain why it is possible to color the region.

**39.** Consider the functions $f(x) = 1/\sqrt{1 + x^2}$ and $g(x) = 1/\sqrt{1 + x^4}$.

    **a.** Use your calculator to approximate $\int_1^b f(x)\, dx$ for $b = 20, 50, 100, 1000,$ and $10,000$.

    **b.** Based on your answers from part a, would you guess that $\int_1^\infty f(x)\, dx$ is convergent or divergent?

    **c.** Use your calculator to approximate $\int_1^b g(x)\, dx$ for $b = 20, 50, 100, 1000,$ and $10,000$.

    **d.** Based on your answers from part c, would you guess that $\int_1^\infty g(x)\, dx$ is convergent or divergent?

    **e.** Show how the answer to parts b and d might be guessed by comparing the integrals with others whose convergence or divergence is known. (*Hint:* For large $x$, the difference between $1 + x^2$ and $x^2$ is relatively small.)

    *Note:* The first integral is indeed divergent, and the second convergent, with an approximate value of 0.9270.

**40. a.** Use your calculator to approximate $\int_0^b e^{-x^2}\, dx$ for $b = 1, 5, 10,$ and $20$.

    **b.** Based on your answers to part a, does $\int_0^\infty e^{-x^2}\, dx$ appear to be convergent or divergent? If convergent, what seems to be its approximate value?

    **c.** Explain why this integral should be convergent by comparing $e^{-x^2}$ with $e^{-x}$ for $x > 1$.

    *Note:* The integral is convergent, with a value of $\sqrt{\pi}/2$.

**41. a.** Use your calculator to approximate $\int_0^b e^{-0.00001x}\, dx$ for $b = 10, 50, 100,$ and $1000$.

    **b.** Based on your answers to part a, does $\int_0^\infty e^{-0.00001x}\, dx$ appear to be convergent or divergent?

    **c.** To what value does the integral actually converge?

**For Exercises 42 and 43 use the integration feature on a graphing calculator and successively larger values of $b$ to estimate $\int_0^\infty f(x)$.**

**42.** $\int_0^b e^{-x} \sin x\, dx$

**43.** $\int_0^b e^{-x} \cos x\, dx$

## LIFE SCIENCE APPLICATIONS

**44. Drug Reaction** The rate of reaction to a drug is given by

$$r'(t) = 2t^2 e^{-t},$$

where $t$ is the number of hours since the drug was administered. Find the total reaction to the drug over all the time since it was administered, assuming this is an infinite time interval. (*Hint:* $\lim_{t \to \infty} t^k e^{-t} = 0$ for all real numbers $k$.)

**45. Drug Epidemic** In an epidemiological model used to study the spread of drug use, a single drug user is introduced into a population of $N$ non-users. Under certain assumptions, the number of people expected to use drugs as a result of direct influence from each drug user is given by

$$S = N \int_0^\infty \frac{a(1 - e^{-kt})}{k} e^{-bt}\, dt,$$

where $a$, $b$, and $k$ are constants. Find the value of $S$. *Source: Mathematical Biology.*

**46. Present Value** When harvesting a population, such as fish, the present value of the resource is given by

$$P = \int_0^\infty e^{-rt} n(t) y(t)\, dt,$$

where $r$ is a discount factor, $n(t)$ is the net revenue at time $t$, and $y(t)$ is the harvesting effort. Suppose $y(t) = K$ and $n(t) = at + b$. Find the present value. *Source: Some Mathematical Questions in Biology.*

## OTHER APPLICATIONS

**Radioactive Waste** Radioactive waste is entering the atmosphere over an area at a decreasing rate. Use the improper integral

$$\int_0^\infty P e^{-kt} dt$$

with $P = 50$ to find the total amount of the waste that will enter the atmosphere for each value of $k$.

**47.** $k = 0.06$

**48.** $k = 0.04$

**YOUR TURN ANSWERS**

**1. a.** Divergent    **b.** 3/2

**2.** 5/2

# 8

# CHAPTER REVIEW

## SUMMARY

In this chapter, we covered several topics related to integration. Numerical integration can be used to find the definite integral when finding an antiderivative is not feasible. The technique known as integration by parts, which is derived from the product rule for derivatives, was then introduced. We also developed definite integral formulas to calculate the volume of a solid of revolution and the average value of a function on some interval. Finally, we learned how to evaluate improper integrals that have upper or lower limits of $\infty$ or $-\infty$.

**Trapezoidal Rule**
$$\int_a^b f(x)\,dx \approx \left(\frac{b-a}{n}\right)\left[\frac{1}{2}f(x_0) + f(x_1) + \cdots + f(x_{n-1}) + \frac{1}{2}f(x_n)\right]$$

**Simpson's Rule**
$$\int_a^b f(x)\,dx \approx \left(\frac{b-a}{3n}\right)[f(x_0) + 4f(x_1) + 2f(x_2) + 4f(x_3) + \cdots$$
$$+ 2f(x_{n-2}) + 4f(x_{n-1}) + f(x_n)]$$

**Integration by Parts** If $u$ and $v$ are differentiable functions, then
$$\int u\,dv = uv - \int v\,du.$$

**Volume of a Solid of Revolution** If $f(x)$ is nonnegative and $R$ is the region between $f(x)$ and the $x$-axis from $x = a$ to $x = b$, the volume of the solid formed by rotating $R$ about the $x$-axis is given by
$$V = \int_a^b \pi[f(x)]^2\,dx.$$

**Average Value of a Function** The average value of a function $f$ on the interval $[a, b]$ is
$$\frac{1}{b-a}\int_a^b f(x)\,dx,$$
provided the indicated definite integral exists.

**Improper Integrals** If $f$ is continuous on the indicated interval and if the indicated limits exist, then
$$\int_a^\infty f(x)\,dx = \lim_{b\to\infty}\int_a^b f(x)\,dx,$$
$$\int_{-\infty}^b f(x)\,dx = \lim_{a\to-\infty}\int_a^b f(x)\,dx,$$
$$\int_{-\infty}^\infty f(x)\,dx = \int_{-\infty}^c f(x)\,dx + \int_c^\infty f(x)\,dx,$$
for real numbers $a$, $b$, and $c$, where $c$ is arbitrarily chosen.

## KEY TERMS

**8.1**
numerical integration
trapezoidal rule
Simpson's rule

**8.2**
integration by parts
column integration

**8.3**
solid of revolution
average value of a function

**8.4**
improper integral
convergent integral
divergent integral

# REVIEW EXERCISES

## CONCEPT CHECK

**Determine whether each of the following statements is true or false, and explain why.**

1. In the trapezoidal rule, the number of subintervals must be even.

2. Simpson's rule usually gives a better approximation than the trapezoidal rule.

3. Integration by parts should be used to evaluate $\int_0^1 \frac{x^2}{x^3 + 1} \, dx$.

4. Integration by parts should be used to evaluate $\int_0^1 xe^{10x} \, dx$.

5. We would need to apply the method of integration by parts twice to determine
$$\int x^3 e^{-x^2} dx.$$

6. Integration by parts should be used to determine $\int \ln(4x) \, dx$.

7. The average value of the function $f(x) = 2x^2 + 3$ on $[1, 4]$ is given by
$$\frac{1}{3} \int_1^4 \pi (2x^2 + 3)^2 \, dx.$$

8. The volume of the solid formed by revolving the function $f(x) = \sqrt{x^2 + 1}$ about the $x$-axis on the interval $[1, 2]$ is given by
$$\int_1^2 \pi \sqrt{x^2 + 1} \, dx.$$

9. The volume of the solid formed by revolving the function $f(x) = x + 4$ about the $x$-axis on the interval $[-4, 5]$ is given by
$$\int_{-4}^5 \pi (x + 4)^2 \, dx.$$

10. $\int_{-\infty}^{\infty} xe^{-2x} \, dx = \lim_{c \to \infty} \int_{-c}^c xe^{-2x} \, dx$

## PRACTICE AND EXPLORATIONS

11. Describe the type of integral for which numerical integration is useful.

12. Describe the type of integral for which integration by parts is useful.

13. Compare finding the average value of a function with finding the average of $n$ numbers.

14. What is an improper integral? Explain why improper integrals must be treated in a special way.

**Use the trapezoidal rule with $n = 4$ to approximate the value of each integral. Then find the exact value and compare the two answers.**

15. $\int_1^3 \frac{\ln x}{x} \, dx$

16. $\int_2^{10} \frac{x \, dx}{x - 1}$

17. $\int_0^1 e^x \sqrt{e^x + 4} \, dx$

18. $\int_0^2 xe^{-x^2} dx$

**Use Simpson's rule with $n = 4$ to approximate the value of each integral. Compare your answers with the answers to Exercises 15–18.**

19. $\int_1^3 \frac{\ln x}{x} \, dx$

20. $\int_2^{10} \frac{x \, dx}{x - 1}$

21. $\int_0^1 e^x \sqrt{e^x + 4} \, dx$

22. $\int_0^2 xe^{-x^2} dx$

23. Find the area of the region between the graphs of $y = \sqrt{x - 1}$ and $2y = x - 1$ from $x = 1$ to $x = 5$ in three ways.
   a. Use antidifferentiation.
   b. Use the trapezoidal rule with $n = 4$.
   c. Use Simpson's rule with $n = 4$.

24. Find the area of the region between the graphs of $y = \frac{1}{x + 1}$ and $y = \frac{x + 2}{2}$ from $x = 0$ to $x = 4$ in three ways.
   a. Use antidifferentiation.
   b. Use the trapezoidal rule with $n = 4$.
   c. Use Simpson's rule with $n = 4$.

25. Let $f(x) = [x(x - 1)(x + 1)(x - 2)(x + 2)]^2$.
   a. Find $\int_{-2}^2 f(x) \, dx$ using the trapezoidal rule with $n = 4$.
   b. Find $\int_{-2}^2 f(x) \, dx$ using Simpson's rule with $n = 4$.
   c. Without evaluating $\int_{-2}^2 f(x) \, dx$, explain why your answers to parts a and b cannot possibly be correct.
   d. Explain why the trapezoidal rule and Simpson's rule with $n = 4$ give incorrect answers for $\int_{-2}^2 f(x) \, dx$ with this function.

26. Suppose that $u$ and $v$ are differentiable functions of $x$ with $\int_0^1 v \, du = 2$ and the following functional values.

| $x$ | $u(x)$ | $v(x)$ |
|-----|--------|--------|
| 0 | $-3$ | 4 |
| 1 | 5 | $-1$ |

Use this information to determine $\int_0^1 u \, dv$.

⊕ **Find each integral, using techniques from this or the previous chapter.**

**27.** $\int x(8 - x)^{3/2} \, dx$

**28.** $\int \dfrac{3x}{\sqrt{x - 2}} \, dx$

**29.** $\int xe^x \, dx$

**30.** $\int (3x + 6)e^{-3x} \, dx$

**31.** $\int \ln|4x + 5| \, dx$

**32.** $\int (x - 1) \ln|x| \, dx$

**33.** $\int \dfrac{x}{25 - 9x^2} \, dx$

**34.** $\int \dfrac{x}{\sqrt{16 + 8x^2}} \, dx$

**35.** $\int_1^e x^3 \ln x \, dx$

**36.** $\int_0^1 x^2 e^{x/2} \, dx$

**37.** $\int (x + 2) \sin x \, dx$

**38.** $\int x^2 \cos 2x \, dx$

**39.** Find the area between $y = (3 + x^2)e^{2x}$ and the $x$-axis from $x = 0$ to $x = 1$.

**40.** Find the area between $y = x^3(x^2 - 1)^{1/3}$ and the $x$-axis from $x = 1$ to $x = 3$.

**Find the volume of the solid of revolution formed by rotating each bounded region about the $x$-axis.**

**41.** $f(x) = 3x - 1$, $\quad y = 0$, $\quad x = 2$

**42.** $f(x) = \sqrt{x - 4}$, $\quad y = 0$, $\quad x = 13$

**43.** $f(x) = e^{-x}$, $\quad y = 0$, $\quad x = -2$, $\quad x = 1$

**44.** $f(x) = \dfrac{1}{\sqrt{x - 1}}$, $\quad y = 0$, $\quad x = 2$, $\quad x = 4$

**45.** $f(x) = 4 - x^2$, $\quad y = 0$, $\quad x = -1$, $\quad x = 1$

**46.** $f(x) = \dfrac{x^2}{4}$, $\quad y = 0$, $\quad x = 4$

**47.** A frustum is what remains of a cone when the top is cut off by a plane parallel to the base. Suppose a right circular frustum (that is, one formed from a right circular cone) has a base with radius $r$, a top with radius $r/2$, and a height $h$. (See the figure below.) Find the volume of this frustum by rotating about the $x$-axis the region below the line segment from $(0, r)$ to $(h, r/2)$.

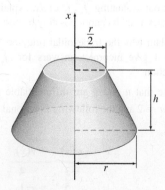

 **48.** How is the average value of a function found?

**49.** Find the average value of $f(x) = \sqrt{x + 1}$ over the interval $[0, 8]$.

**50.** Find the average value of $f(x) = 7x^2(x^3 + 1)^6$ over the interval $[0, 2]$.

**Find the value of each integral that converges.**

**51.** $\int_{10}^{\infty} x^{-1} \, dx$

**52.** $\int_{-\infty}^{-5} x^{-2} \, dx$

**53.** $\int_0^{\infty} \dfrac{dx}{(3x + 1)^2}$

**54.** $\int_1^{\infty} 6e^{-x} \, dx$

**55.** $\int_{-\infty}^{0} \dfrac{x}{x^2 + 3} \, dx$

**56.** $\int_4^{\infty} \ln(5x) \, dx$

⊕ **Find the area between the graph of each function and the $x$-axis over the given interval, if possible.**

**57.** $f(x) = \dfrac{5}{(x - 2)^2}$, $\quad$ for $(-\infty, 1]$

**58.** $f(x) = 3e^{-x}$, $\quad$ for $[0, \infty)$

## LIFE SCIENCE APPLICATIONS

**59. Milk Production** Researchers report that the average amount of milk produced (in kilograms per day) by a 4- to 5-year-old cow weighing 700 kg can be approximated by

$$y = 1.87t^{1.49}e^{-0.189(\ln t)^2},$$

where $t$ is the number of days into lactation. ***Source: Journal of Dairy Science.***

**a.** Approximate the total amount of milk produced from $t = 1$ to $t = 321$ using the trapezoidal rule with $n = 8$.

**b.** Repeat part a using Simpson's rule with $n = 8$.

**c.** Repeat part a using the integration feature of a graphing calculator, and compare your answer with the answers to parts a and b.

**60. Complimentary Error Function** An article on how household chemicals are transported through septic systems used the complimentary error function

$$\text{erfc}(x) = \dfrac{2}{\sqrt{\pi}} \int_x^{\infty} e^{-t^2} \, dt.$$

***Source: Ground Water.***

**a.** To investigate the value of $\text{erfc}(0)$, use Simpson's rule with $n = 20$ to evaluate

$$\dfrac{2}{\sqrt{\pi}} \int_x^k e^{-t^2} \, dt$$

for $k = 2, 3, 4,$ and $5$. Based on these answers, what does $\text{erfc}(0)$ seem to equal?

**b.** The error function is defined as

$$\operatorname{erf}(x) = \frac{2}{\sqrt{\pi}} \int_0^x e^{-t^2}\, dt.$$

Show that $\operatorname{erf}(x) = 1 - \operatorname{erfc}(x)$.

**61. Drug Reaction** The reaction rate to a new drug $t$ hours after the drug is administered is

$$r'(t) = 0.5te^{-t}.$$

Find the total reaction over the first 5 hours.

**62. Oil Leak Pollution** An oil leak from an uncapped well is polluting a bay at a rate of $f(t) = 125e^{-0.025t}$ gallons per year. Use an improper integral to find the total amount of oil that will enter the bay, assuming the well is never capped.

## OTHER APPLICATIONS

**63. Average Temperatures** Suppose the temperature (degrees F) in a river at a point $x$ meters downstream from a factory that is discharging hot water into the river is given by

$$T(x) = 160 - 0.05x^2.$$

Find the average temperature over each interval.

**a.** $[0, 10]$  **b.** $[10, 40]$  **c.** $[0, 40]$

**64. Total Revenue** The rate of change of revenue from the sale of $x$ toaster ovens is

$$R'(x) = x(x - 50)^{1/2}.$$

Find the total revenue from the sale of the 50th to the 75th ovens.

# EXTENDED APPLICATION

# FLOW SYSTEMS

One of the most important biological applications of integration is the determination of properties of flow systems. Among the significant characteristics of a flow system are its flow rate (a heart's output in liters per minute, for example), its volume (for blood, the vascular volume), and the quantity of a substance flowing through it (a pollutant in a stream, perhaps).

A flow system can be anything from a river to an oil pipeline to an artery. We assume that the system has some well-defined volume between two points, and a more or less constant flow rate between them. In arteries the blood pulses, but we shall measure the flow over long enough intervals for the flow rate to be regarded as constant.

The standard method for analyzing flow system behavior is the so-called indicator dilution method, but we shall also discuss the more recent thermodilution technique for determining cardiac output. In the indicator dilution method, a known or unknown quantity of a substance such as a dye, a pollutant, or a radioactive tracer is injected or seeps into a flow system at some entry point. The substance is assumed to mix well with the fluid, and at some downstream measuring point its concentration is sampled either serially or continuously. The concentration function $c(t)$ is then integrated over an appropriate time interval by any of several methods: numerical approximation, a planimeter, electronic calculation, or by curve-fitting and the fundamental theorem of calculus. As we shall see, the integral

$$\int_0^T c(t)\, dt$$

(whose units are mass $\times$ time/volume) is extremely useful for finding unknown flow system properties.

The basis for the analysis of flow systems is the principle of mass conservation, or "what goes in must come out." We are going to assume that the indicator does not leak from the stream (for example, by diffusion into tissue), and that the time interval is over before any of the indicator recirculates (less than a minute in the case of blood). What goes *into* the system is a (known or unknown) quantity $Q_{in}$ of indicator. What comes *out* of the system, or more precisely, what passes by the downstream measuring point at a flow rate $F$, is a volume of fluid with a variable concentration $c(t)$ of the indicator. The product of $F$ and $c(t)$ is called the **mass rate** of the indicator flowing past the measuring point. The total quantity $Q_{out}$ of indicator flowing by during the time interval from 0 to $T$ is the integral of the mass rate from 0 to $T$:

$$Q_{out} = \int_0^T Fc(t)\, dt.$$

By our assumption of mass conservation (no leakage and no recirculation), we have $Q_{in} = Q_{out}$,

$$Q_{in} = \int_0^T Fc(t)dt = F\int_0^T c(t)dt, \qquad \textbf{(1)}$$

provided of course that $F$ is essentially constant.

Equation (1) is the basic relationship of indicator dilution. We shall illustrate its use with an example.

## EXAMPLE 1

Suppose that an industrial plant is discharging toxic kryptonite into the Cahaba River at an unknown variable rate. Downriver, a field station of the Environmental Protection Agency measures the following concentrations:

| Time | Concentration |
|---|---|
| 2 A.M. | $3.0 \text{ mg/m}^3$ |
| 6 A.M. | $3.5 \text{ mg/m}^3$ |
| 10 A.M. | $2.5 \text{ mg/m}^3$ |
| 2 P.M. | $1.0 \text{ mg/m}^3$ |
| 6 P.M. | $0.5 \text{ mg/m}^3$ |
| 10 P.M. | $1.0 \text{ mg/m}^3$ |

The EPA has found the flow rate of the Cahaba in the vicinity to be $1{,}000 \text{ ft}^3/\text{sec} = 1.02 \times 10^5 \text{ m}^3/\text{hr}$. How much kryptonite is being discharged?

**SOLUTION** We use Equation (1) to estimate the total amount of kryptonite discharged during a 24-hour period by approximating the concentration integral. Dividing the day into six 4-hour subintervals (so that $\Delta t = 4$) and using the EPA data as our function values $c(t)$, we compute as follows:

$$\int_0^{24} c(t)\,dt \approx \sum_{i=1}^{6} c(t_i)\Delta t = \Delta t \sum_{i=1}^{6} c(t_i)$$
$$= 4(3.0 + 3.5 + 2.5 + 1.0 + 0.5 + 1.0)$$
$$= 4(11.5) = 46,$$

or $46 \text{ mg} \times \text{hrs/m}^3$. Hence, by Equation (1), the total daily discharge is

$$Q_{\text{in}} = F \int_0^{24} c(t)\,dt$$
$$\approx \left(1.02 \times 10^5 \frac{\text{m}^3}{\text{hour}}\right)\left(46\,\frac{\text{mg} \times \text{hour}}{\text{m}^3}\right)$$
$$= 4.7 \times 10^6 \text{ mg} = 4.7 \text{ kilograms of kryptonite.}$$

## Determining Cardiac Output by Thermodilution

Knowledge of a patient's cardiac output is a valuable aid to the diagnosis and treatment of heart damage. By cardiac output we mean the volume flow rate of venous blood pumped by the right ventricle through the pulmonary artery to the lungs (to be oxygenated). Normal resting cardiac output is 4 to 5 liters per minute, but in critically ill patients it may fall well below 3 liters per minute.

Unfortunately, cardiac output is difficult to determine by dye dilution because of the need to calibrate the dye and to sample the blood repeatedly. There are also problems with dye instability, recirculation, and slow dissipation. With the development in 1970 of a suitable pulmonary artery catheter, it became possible to make rapid and routine bedside measurements of cardiac output by thermodilution, and to do so with very few patient complications.

During a typical application in the coronary intensive care unit, a balloon-tipped catheter is inserted in an arm vein and guided to the superior vena cava or right atrium (the upper chamber of the heart). A temperature-sensing device called a thermistor is moved through the right side of the heart and then positioned a few centimeters into the pulmonary artery. Then, 10 ml of cold D5W (5% dextrose in water) is injected into the vena cava or atrium. As the cold D5W mixes with blood in the right heart, temperature variations in the flow are detected by the thermistor. Figure 16 shows a typical record of temperature change within a minute after injection. Since the cold D5W is warmed by the body before it recirculates, repeated measurements can be made reliably.

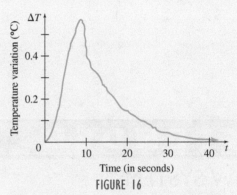

**FIGURE 16**

This curve shows the temperature variation in the pulmonary artery that resulted from injecting a few milliliters of cold dextrose solution into the right atrium of the heart.

We now derive an equation for thermodilution comparable to Equation (1). First, we need to define the thermal equivalent of mass. We define the "quantity of cold" injected to be

$$Q_{\text{in}} = V_i(T_b - T_i),$$

where $V_i$ is the volume of injectate, $T_i$ is its temperature, and $T_b$ is the body temperature. Thus, the "quantity of cold" injected is jointly proportional to how cold the injectate is (below body heat) and to how much is injected.

As with indicator dilution, we shall assume "conservation of cold," that is, no loss and no recirculation. (Actually, there is some loss of cold as the injectate travels up through the catheter to the injection point, but that can be corrected for.) The quantity of cold passing by the thermistor in a short time interval $dt$ can be calculated by multiplying the flow volume $F\,dt$ and the temperature variation $\Delta T$ during the interval. The total quantity of cold flowing by is

$$Q_{\text{out}} = \int_0^{\infty} F\Delta T\,dt$$
$$= F \int_0^{\infty} \Delta T\,dt.$$

Equating $Q_{in}$ to $Q_{out}$ and solving for the flow rate $F$ produces the analog of Equation (1):

$$Q_{out} = Q_{in} = V_i(T_b - T_i)$$

$$F\int_0^\infty \Delta T \, dt = V_i(T_b - T_i)$$

(2)

$$F = \frac{V_i(T_b - T_i)}{\int_0^\infty \Delta T \, dt}$$

If time is measured in seconds and temperature in degrees Celsius, then the denominator of Equation (2) has units of (°C)(sec) and the numerator has units of (ml)(°C). Hence, $F$ has units ml/sec, which is converted to liters/minute by multiplying by $60/1000 = 0.06$.

Although thermistor signals are usually calibrated and integrated electronically, we shall illustrate Equation (2) with an example.

## EXAMPLE 2

Suppose $T_b = 37°C$, $T_i = 0°C$, and $V_i = 10$ ml. Suppose that the change in temperature with respect to time is represented by the curve $\Delta T = 0.1t^2 e^{-0.31t}$, shown in Figure 17. Finally, suppose that the net effect of all the correction factors is to multiply the right side of equation (2) by 0.891. Find the flow rate.

**SOLUTION**   Integration by parts in Equation (2) gives

$$F = \frac{(10)(37-0)(0.891)}{\int_0^\infty (0.1)t^2 e^{-0.31t} dt}$$

$$= \frac{330}{(0.1)\dfrac{2}{(0.31)^3}}$$

$$= 49.16 \text{ ml/sec}$$

$$= 2.95 \text{ liters/minute}.$$

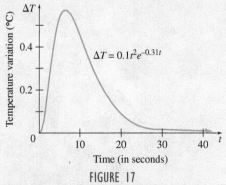

**FIGURE 17**

The illustration of equation (2) uses this smooth curve in place of the thermodilution curve in Figure 16.

*Source: The UMAP Journal.*

## EXERCISES

1. Approximate the integral in Example 1 using **(a)** the trapezoidal rule and **(b)** Simpson's rule. Let $t$ represent the number of hours after 2 A.M., so that 2 A.M. $= 0$, 6 A.M. $= 4$, and so on. Assume the reading after 24 hours is 3.0 mg/m³ again.

2. A dye dilution technique for measuring cardiac output involves injection of a dye into a main vein near the heart and measurement by a catheter in the aorta at regular intervals. Suppose the results from such measurement are as given in the table.

| Time (seconds) | Concentration (ml/liter) |
|---|---|
| 0 | 0 |
| 4 | 0.6 |
| 8 | 2.7 |
| 12 | 4.1 |
| 16 | 2.9 |
| 20 | 0.9 |
| 24 | 0 |

If 8 ml of dye is injected, approximate the cardiac output $F$ in the equation

$$Q_{in} = F\int_0^t c(t) \, dt$$

by **(a)** the trapezoidal rule and **(b)** Simpson's rule.

3. In Example 2 let $T_b = 37.1°C$, $T_i = 5°C$, and $v_i = 8$ ml. Find the cardiac output $F$.

## DIRECTIONS FOR GROUP PROJECT

*Suppose you and three others are employed by a local environmental agency. Unfortunately, instances of hazardous substances being discharged (either by accident or on purpose) are common. Find a real-life example of this situation and estimate the total discharge using Example 1 as a guide. Prepare a presentation for the agency that describes your findings. Presentation software, such as Microsoft PowerPoint, should be used.*

# 9

# Multivariable Calculus

9.1 Functions of Several Variables

9.2 Partial Derivatives

9.3 Maxima and Minima

9.4 Total Differentials and Approximations

9.5 Double Integrals

Chapter 9 Review

Extended Application: Optimization for a Predator

Safe diving requires an understanding of how the increased pressure below the surface affects the body's intake of nitrogen. An exercise in Section 2 of this chapter investigates a formula for nitrogen pressure as a function of two variables, depth and dive time. Partial derivatives tell us how this function behaves when one variable is held constant as the other changes. Dive tables based on the formula help divers to choose a safe time for a given depth, or a safe depth for a given time.

464

We have thus far limited our study of calculus to functions of one variable. There are other phenomena that require more than one variable to adequately model the situation. For example, the number of cases of a disease in a region might depend on the maximum and minimum temperature in the region, the humidity, and the population size and density. In this case, the number of cases of the disease is a function of more than one variable. To analyze and better understand situations like this, we will extend the ideas of calculus, including differentiation and integration, to functions of more than one variable.

# 9.1    Functions of Several Variables

APPLY IT    How is the carbon dioxide released from the lungs related to the total output of blood from the heart?
We will investigate this question in Example 2 of this section.

If a person takes $x$ tablets of vitamin C, for instance, then the total amount of the vitamin that is ingested is given by

$$C(x) = 500x,$$

where each tablet contains 500 mg of vitamin C, and $C(x)$ is total amount (in mg). This is a function of one variable, the number of tablets taken. If the person also drinks $y$ glasses of orange juice, with each glass of juice providing 100 mg of vitamin C, then the total amount of vitamin C ingested is a function of two independent variables, $x$ and $y$. By generalizing $f(x)$ notation, the total amount of vitamin C can be written as $C(x, y)$, where

$$C(x, y) = 500x + 100y.$$

When $x = 2$ and $y = 3$ the total amount of vitamin C ingested is written $C(2, 3)$, with

$$C(2, 3) = 500 \cdot 2 + 100 \cdot 3 = 1300.$$

A general definition follows.

## Function of Two Variables

$z = f(x, y)$ is a **function of two variables** if a unique value of $z$ is obtained from each ordered pair of real numbers $(x, y)$. The variables $x$ and $y$ are **independent variables**, and $z$ is the **dependent variable**. The set of all ordered pairs of real numbers $(x, y)$ such that $f(x, y)$ exists is the **domain** of $f$; the set of all values of $f(x, y)$ is the **range**. Similar definitions could be given for functions of three, four, or more independent variables.

### EXAMPLE 1    Evaluating Functions

Let $f(x, y) = 4x^2 + 2xy + 3/y$ and find each of the following.
**(a)** $f(-1, 3)$

**SOLUTION**  Replace $x$ with $-1$ and $y$ with 3.

$$f(-1, 3) = 4(-1)^2 + 2(-1)(3) + \frac{3}{3} = 4 - 6 + 1 = -1$$

**(b)** $f(2, 0)$

**SOLUTION** Because of the quotient $3/y$, it is not possible to replace $y$ with 0, so $f(2, 0)$ is undefined. By inspection, we see that the domain of the function is the set of all $(x, y)$ such that $y \neq 0$.

**(c)** $\dfrac{f(x + h, y) - f(x, y)}{h}$

**SOLUTION** Calculate as follows:

$$\frac{f(x + h, y) - f(x, y)}{h} = \frac{4(x + h)^2 + 2(x + h)y + 3/y - [4x^2 + 2xy + 3/y]}{h}$$

$$= \frac{4x^2 + 8xh + 4h^2 + 2xy + 2hy + 3/y - 4x^2 - 2xy - 3/y}{h}$$

$$= \frac{8xh + 4h^2 + 2hy}{h} \qquad \text{Simplify the numerator.}$$

$$= \frac{h(8x + 4h + 2y)}{h} \qquad \text{Factor } h \text{ from the numerator.}$$

$$= 8x + 4h + 2y. \qquad \text{TRY YOUR TURN 1}$$

**YOUR TURN 1** For the function in Example 1, find $f(2,3)$.

---

**EXAMPLE 2** Carbon Dioxide

Let $x$ represent the number of milliliters (ml) of carbon dioxide released by the lungs in one minute. Let $y$ be the change in the carbon dioxide content of the blood as it leaves the lungs ($y$ is measured in ml of carbon dioxide per 100 ml of blood). The total output of blood from the heart in one minute (measured in ml) is given by $C$, where $C$ is a function of $x$ and $y$ such that

$$C = C(x, y) = \frac{100x}{y}.$$

Find $C(320, 6)$.

**APPLY IT**    **SOLUTION** Replace $x$ with 320 and $y$ with 6 to get

$$C(320, 6) = \frac{100(320)}{6}$$

$$\approx 5333 \text{ ml of blood per minute.}$$

---

**EXAMPLE 3** Evaluating a Function

**YOUR TURN 2** For the function in Example 3, find $f(1, 2, 3)$.

Let $f(x, y, z) = 4xz - 3x^2y + 2z^2$. Find $f(2, -3, 1)$.

**SOLUTION** Replace $x$ with 2, $y$ with $-3$, and $z$ with 1.

$$f(2, -3, 1) = 4(2)(1) - 3(2)^2(-3) + 2(1)^2 = 8 + 36 + 2 = 46$$

TRY YOUR TURN 2

## Graphing Functions of Two Independent Variables

Functions of one independent variable are graphed by using an $x$-axis and a $y$-axis to locate points in a plane. The plane determined by the $x$- and $y$-axes is called the *xy-plane*. A third axis is needed to graph functions of two independent variables—the $z$-axis, which goes through the origin in the $xy$-plane and is perpendicular to both the $x$-axis and the $y$-axis.

Figure 1 shows one possible way to draw the three axes. In Figure 1, the *yz*-plane is in the plane of the page, with the $x$-axis perpendicular to the plane of the page.

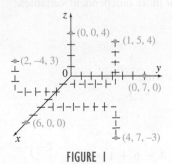

FIGURE 1

FOR REVIEW

Graph the following lines. Refer to Section 1.1 if you need to review.

**1.** $2x + 3y = 6$

**2.** $x = 4$

**3.** $y = 2$

**Answers**

**1.**
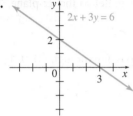
$2x + 3y = 6$

**2.**
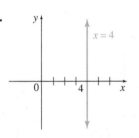
$x = 4$

**3.**
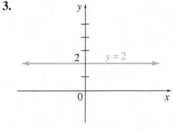
$y = 2$

**YOUR TURN 3**   Graph $x + 2y + 3z = 12$ in the first octant.

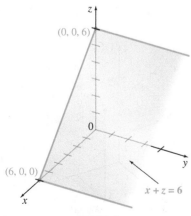

FIGURE 3

Just as we graphed ordered pairs earlier we can now graph **ordered triples** of the form $(x, y, z)$. For example, to locate the point corresponding to the ordered triple $(2, -4, 3)$, start at the origin and go 2 units along the positive $x$-axis. Then go 4 units in a negative direction (to the left) parallel to the $y$-axis. Finally, go up 3 units parallel to the $z$-axis. The point representing $(2, -4, 3)$ is shown in Figure 1, together with several other points. The region of three-dimensional space where all coordinates are positive is called the **first octant**.

In Chapter 1 we saw that the graph of $ax + by = c$ (where $a$ and $b$ are not both 0) is a straight line. This result generalizes to three dimensions.

> **Plane**
> The graph of
> $$ax + by + cz = d$$
> is a **plane** if $a$, $b$, and $c$ are not all 0.

**EXAMPLE 4**   Graphing a Plane

Graph $2x + y + z = 6$.

**SOLUTION**   The graph of this equation is a plane. Earlier, we graphed straight lines by finding $x$- and $y$-intercepts. A similar idea helps in graphing a plane. To find the $x$-intercept, which is the point where the graph crosses the $x$-axis, let $y = 0$ and $z = 0$.

$$2x + 0 + 0 = 6$$
$$x = 3$$

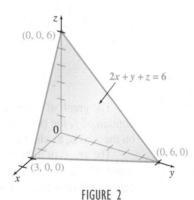

FIGURE 2

The point $(3, 0, 0)$ is on the graph. Letting $x = 0$ and $z = 0$ gives the point $(0, 6, 0)$, while $x = 0$ and $y = 0$ lead to $(0, 0, 6)$. The plane through these three points includes the triangular surface shown in Figure 2. This surface is the first-octant part of the plane that is the graph of $2x + y + z = 6$. The plane does not stop at the axes but extends without bound.

TRY YOUR TURN 3

**EXAMPLE 5**   Graphing a Plane

Graph $x + z = 6$.

**SOLUTION**   To find the $x$-intercept, let $y = 0$ and $z = 0$, giving $(6, 0, 0)$. If $x = 0$ and $y = 0$, we get the point $(0, 0, 6)$. Because there is no $y$ in the equation $x + z = 6$, there can be no $y$-intercept. A plane that has no $y$-intercept is parallel to the $y$-axis. The first-octant portion of the graph of $x + z = 6$ is shown in Figure 3.

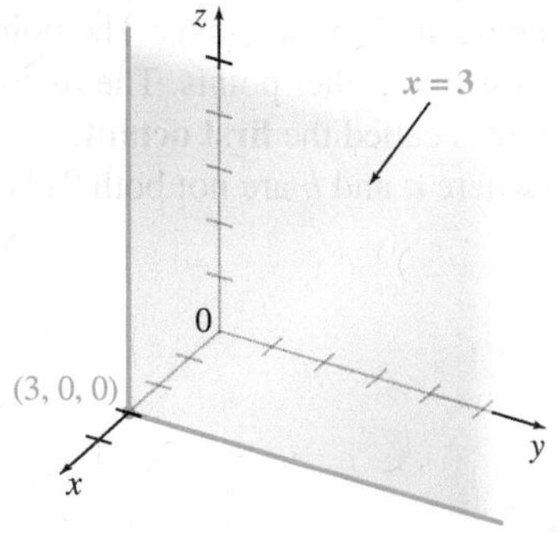

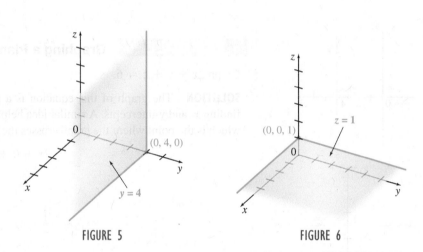

**EXAMPLE 6** Graphing Planes

Graph each equation in the first octant.

**(a)** $x = 3$

**SOLUTION** This graph, which passes through $(3, 0, 0)$, can have no $y$-intercept and no $z$-intercept. It is, therefore, a plane parallel to the $y$-axis and the $z$-axis and, therefore, to the $yz$-plane. The first-octant portion of the graph is shown in Figure 4.

**(b)** $y = 4$

**SOLUTION** This graph passes through $(0, 4, 0)$ and is parallel to the $xz$-plane. The first-octant portion of the graph is shown in Figure 5.

**(c)** $z = 1$

**SOLUTION** The graph is a plane parallel to the $xy$-plane, passing through $(0, 0, 1)$. Its first-octant portion is shown in Figure 6.

FIGURE 4

FIGURE 5

FIGURE 6

The graph of a function of one variable, $y = f(x)$, is a curve in the plane. If $x_0$ is in the domain of $f$, the point $(x_0, f(x_0))$ on the graph lies directly above or below the number $x_0$ on the $x$-axis, as shown in Figure 7.

The graph of a function of two variables, $z = f(x, y)$, is a **surface** in three-dimensional space. If $(x_0, y_0)$ is in the domain of $f$, the point $(x_0, y_0, f(x_0, y_0))$ lies directly above or below the point $(x_0, y_0)$ in the $xy$-plane, as shown in Figure 8.

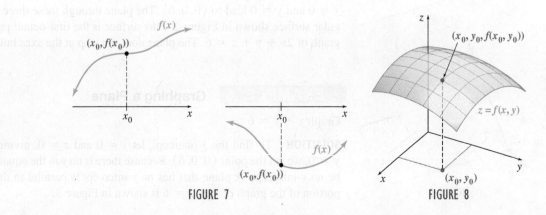

FIGURE 7

FIGURE 8

Although computer software is available for drawing the graphs of functions of two independent variables, you can often get a good picture of the graph without it by finding various **traces**—the curves that result when a surface is cut by a plane. The **xy-trace** is the intersection of the surface with the $xy$-plane. The **yz-trace** and **xz-trace** are defined similarly. You can also determine the intersection of the surface with planes parallel to the $xy$-plane. Such planes are of the form $z = k$, where $k$ is a constant, and the curves that result when they cut the surface are called **level curves**.

### EXAMPLE 7 Graphing a Function

Graph $z = x^2 + y^2$.

**SOLUTION** The $yz$-plane is the plane in which every point has a first coordinate of 0, so its equation is $x = 0$. When $x = 0$, the equation becomes $z = y^2$, which is the equation of a parabola in the $yz$-plane, as shown in Figure 9(a). Similarly, to find the intersection of the surface with the $xz$-plane (whose equation is $y = 0$), let $y = 0$ in the equation. It then becomes $z = x^2$, which is the equation of a parabola in the $xz$-plane, as shown in Figure 9(a). The $xy$-trace (the intersection of the surface with the plane $z = 0$) is the single point $(0, 0, 0)$ because $x^2 + y^2$ is equal to 0 only when $x = 0$ and $y = 0$.

Next, we find the level curves by intersecting the surface with the planes $z = 1, z = 2$, $z = 3$, etc. (all of which are parallel to the $xy$-plane). In each case, the result is a circle:

$$x^2 + y^2 = 1, \qquad x^2 + y^2 = 2, \qquad x^2 + y^2 = 3,$$

and so on, as shown in Figure 9(b). Drawing the traces and level curves on the same set of axes suggests that the graph of $z = x^2 + y^2$ is the bowl-shaped figure, called a **paraboloid**, that is shown in Figure 9(c).

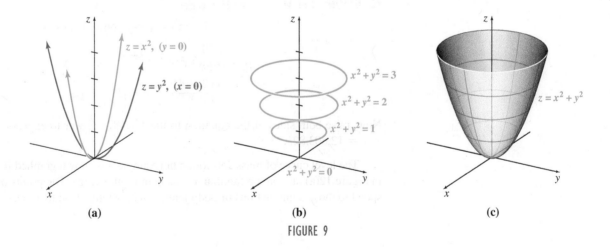

(a)            (b)            (c)

FIGURE 9

Figure 10 on the next page shows the level curves from Example 7 plotted in the $xy$-plane. The picture can be thought of as a topographical map that describes the surface generated by $z = x^2 + y^2$, just as the topographical map in Figure 11 describes the surface of the land in a part of New York state.

One application of level curves occurs in economics with production functions. A **production function** $z = f(x, y)$ is a function that gives the quantity $z$ of an item produced as a function of $x$ and $y$, where $x$ is the amount of labor and $y$ is the amount of capital (in appropriate units) needed to produce $z$ units. For production functions, level curves are used to indicate combinations of the values of $x$ and $y$ that produce the same value of production $z$. Similarly, level curves can be used in the life sciences where various combinations of the independent variables produce the same value of a function.

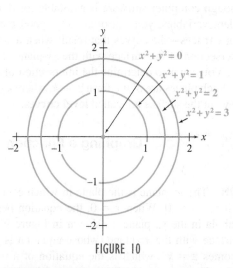

FIGURE 10

FIGURE 11

## EXAMPLE 8  Polar Bears

Researchers in Alaska have developed models that can be used to estimate the mass of a polar bear. For bears that are caught in the spring of the year,

$$M(x, y) = 0.000199x^{1.2823}y^{1.4874},$$

where $M(x, y)$ is the mass of the bear (in kg), $x$ is the straight-line body length (in cm), measured from the tip of the nose to the base of the tail, and $y$ is the girth (in cm), measured directly behind the shoulders. ***Source: Wildlife Society Bulletin.*** Find the level curve for a polar bear with a mass of 250 kg.

**SOLUTION**  Let $M(x, y) = 250$ to get

$$250 = 0.000199x^{1.2823}y^{1.4874}$$

$$\frac{250}{0.000199x^{1.2823}} = y^{1.4874}$$

$$\frac{1,256,281.41}{x^{1.2823}} = y^{1.4874}.$$

Now, raise both sides of the equation to the $1/1.4874$ power to express $y$ as a function of $x$: $y = 12,603.82/x^{0.8622}$.

The level curve of mass 250 found in Example 8 is shown graphed in three dimensions in Figure 12(a) and on the familiar $xy$-plane in Figure 12(b). The points of the graph correspond to those combinations of body length and girth that lead to a mass of 250 kg.

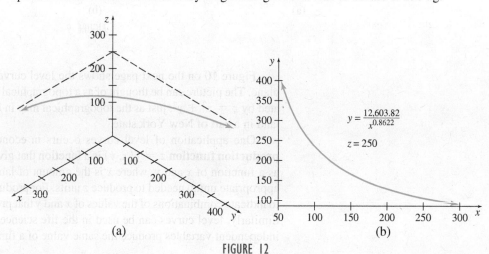

$$y = \frac{12,603.82}{x^{0.8622}}$$

$$z = 250$$

(a)

(b)

FIGURE 12

The curve in Figure 12 is called an *isoquant*, for *iso* (equal) and *quant* (amount). In Example 8, the "amounts" all "equal" 250.

Because of the difficulty of drawing the graphs of more complicated functions, we merely list some common equations and their graphs. We encourage you to explore why these graphs look the way they do by studying their traces, level curves, and axis intercepts. These graphs were drawn by computer, a very useful method of depicting three-dimensional surfaces.

**Paraboloid,** $z = x^2 + y^2$

*xy*-trace: point
*yz*-trace: parabola
*xz*-trace: parabola

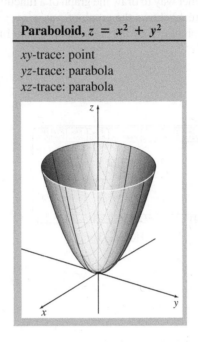

**Ellipsoid,** $\dfrac{x^2}{a^2} + \dfrac{y^2}{b^2} + \dfrac{z^2}{c^2} = 1$

*xy*-trace: ellipse
*yz*-trace: ellipse
*xz*-trace: ellipse

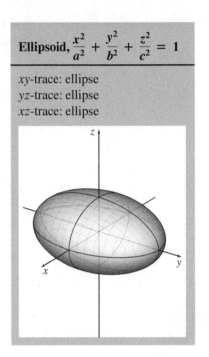

**Hyperbolic Paraboloid,** $z = x^2 - y^2$
(sometimes called a *saddle*)

*xy*-trace: two intersecting lines
*yz*-trace: parabola
*xz*-trace: parabola

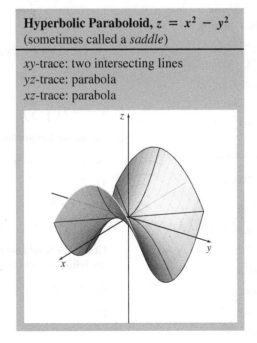

**Hyperboloid of Two Sheets,**
$-x^2 - y^2 + z^2 = 1$

*xy*-trace: none
*yz*-trace: hyperbola
*xz*-trace: hyperbola

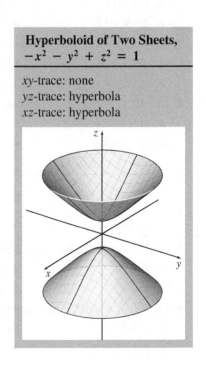

Notice that not all the graphs correspond to functions of two variables. In the ellipsoid, for example, if $x$ and $y$ are both 0, then $z$ can equal $c$ or $-c$, whereas a function can take on only one value. We can, however, interpret the graph as a **level surface** for a function of three variables. Let

$$w(x, y, z) = \frac{x^2}{a^2} + \frac{y^2}{b^2} + \frac{z^2}{c^2}.$$

Then $w = 1$ produces the level surface of the ellipsoid shown, just as $z = c$ gives level curves for the function $z = f(x, y)$.

Another way to draw the graph of a function of two variables is with a graphing calculator. Figure 13 shows the graph of $z = x^2 + y^2$ generated by a TI-89. Figure 14 shows the same graph drawn by the computer program Maple™.

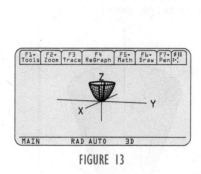

FIGURE 13

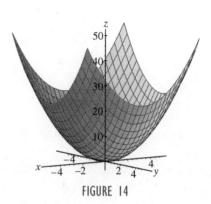

FIGURE 14

# 9.1 EXERCISES

**1.** Let $f(x, y) = 2x - 3y + 5$. Find the following.

   **a.** $f(2, -1)$           **b.** $f(-4, 1)$

   **c.** $f(-2, -3)$        **d.** $f(0, 8)$

**2.** Let $g(x, y) = x^2 - 2xy + y^3$. Find the following.

   **a.** $g(-2, 4)$         **b.** $g(-1, -2)$

   **c.** $g(-2, 3)$         **d.** $g(5, 1)$

**3.** Let $h(x, y) = \sqrt{x^2 + 2y^2}$. Find the following.

   **a.** $h(5, 3)$           **b.** $h(2, 4)$

   **c.** $h(-1, -3)$        **d.** $h(-3, -1)$

**4.** Let $f(x, y) = \dfrac{\sqrt{9x + 5y}}{\log x}$. Find the following.

   **a.** $f(10, 2)$         **b.** $f(100, 1)$

   **c.** $f(1000, 0)$       **d.** $f\left(\dfrac{1}{10}, 5\right)$

**5.** Let $f(x, y) = e^x + \ln(x + y)$. Find the following

   **a.** $f(1, 0)$           **b.** $f(2, -1)$

   **c.** $f(0, e)$           **d.** $f(0, e^2)$

**6.** Let $f(x, y) = xe^{x+y}$. Find the following.

   **a.** $f(1, 0)$           **b.** $f(2, -2)$

   **c.** $f(3, 2)$           **d.** $f(-1, 4)$

**7.** Let $f(x, y) = x \sin(x^2 y)$. Find the following.

   **a.** $f\left(1, \dfrac{\pi}{2}\right)$         **b.** $f\left(\dfrac{1}{2}, \pi\right)$

   **c.** $f\left(\sqrt{\pi}, \dfrac{1}{2}\right)$      **d.** $f\left(-1, -\dfrac{\pi}{2}\right)$

**8.** Let $f(x, y) = y \cos(x + y)$. Find the following.

   **a.** $f(1, -1)$         **b.** $f\left(\dfrac{\pi}{4}, \dfrac{\pi}{4}\right)$

   **c.** $f\left(\dfrac{\pi}{2}, \dfrac{\pi}{2}\right)$      **d.** $f\left(\dfrac{\pi}{2}, \dfrac{\pi}{6}\right)$

**Graph the first-octant portion of each plane.**

   **9.** $x + y + z = 9$         **10.** $x + y + z = 15$

  **11.** $2x + 3y + 4z = 12$     **12.** $4x + 2y + 3z = 24$

  **13.** $x + y = 4$           **14.** $y + z = 5$

  **15.** $x = 5$              **16.** $z = 4$

**Graph the level curves in the first quadrant of the $xy$-plane for the following functions at heights of $z = 0$, $z = 2$, and $z = 4$.**

  **17.** $3x + 2y + z = 24$     **18.** $3x + y + 2z = 8$

  **19.** $y^2 - x = -z$         **20.** $2y - \dfrac{x^2}{3} = z$

**21.** Discuss how a function of three variables in the form $w = f(x, y, z)$ might be graphed.

**22.** Suppose the graph of a plane $ax + by + cz = d$ has a portion in the first octant. What can be said about $a, b, c$, and $d$?

**23.** In Chapter 1 on Functions, the vertical line test was presented, which tells whether a graph is the graph of a function. Does this test apply to functions of two variables? Explain.

**24.** A graph that was not shown in this section is the *hyperboloid of one sheet*, described by the equation $x^2 + y^2 - z^2 = 1$. Describe it as completely as you can.

**Match each equation in Exercises 25–30 with its graph in a–f.**

**25.** $z = x^2 + y^2$

**26.** $z^2 - y^2 - x^2 = 1$

**27.** $x^2 - y^2 = z$

**28.** $z = y^2 - x^2$

**29.** $\dfrac{x^2}{16} + \dfrac{y^2}{25} + \dfrac{z^2}{4} = 1$

**30.** $z = 5(x^2 + y^2)^{-1/2}$

**a.**

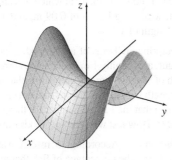

**b.**

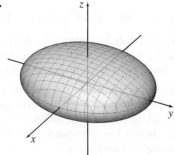

**c.**

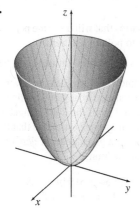

**d.**

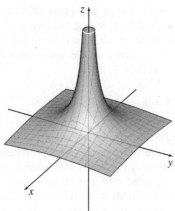

**e.**

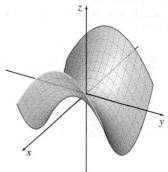

**f.**

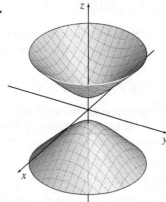

**31.** Let $f(x, y) = 4x^2 - 2y^2$, and find the following.

**a.** $\dfrac{f(x + h, y) - f(x, y)}{h}$

**b.** $\dfrac{f(x, y + h) - f(x, y)}{h}$

**c.** $\lim\limits_{h \to 0} \dfrac{f(x + h, y) - f(x, y)}{h}$

**d.** $\lim\limits_{h \to 0} \dfrac{f(x, y + h) - f(x, y)}{h}$

**32.** Let $f(x, y) = 5x^3 + 3y^2$, and find the following.

**a.** $\dfrac{f(x + h, y) - f(x, y)}{h}$

**b.** $\dfrac{f(x, y + h) - f(x, y)}{h}$

**c.** $\lim\limits_{h \to 0} \dfrac{f(x + h, y) - f(x, y)}{h}$

**d.** $\lim\limits_{h \to 0} \dfrac{f(x, y + h) - f(x, y)}{h}$

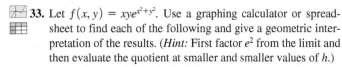

**33.** Let $f(x, y) = xye^{x^2+y^2}$. Use a graphing calculator or spreadsheet to find each of the following and give a geometric interpretation of the results. (*Hint:* First factor $e^2$ from the limit and then evaluate the quotient at smaller and smaller values of $h$.)

**a.** $\lim_{h \to 0} \dfrac{f(1 + h, 1) - f(1, 1)}{h}$

**b.** $\lim_{h \to 0} \dfrac{f(1, 1 + h) - f(1, 1)}{h}$

**34.** The following table provides values of the function $f(x, y)$. However, because of potential errors in measurement, the functional values may be slightly inaccurate. Using the statistical package included with a graphing calculator or spreadsheet and critical thinking skills, find the function $f(x, y) = a + bx + cy$ that best estimates the table where $a$, $b$, and $c$ are integers. (*Hint:* Do a linear regression on each column with the value of $y$ fixed and then use these four regression equations to determine the coefficient $c$.)

| x \ y | 0 | 1 | 2 | 3 |
|---|---|---|---|---|
| **0** | 4.02 | 7.04 | 9.98 | 13.00 |
| **1** | 6.01 | 9.06 | 11.98 | 14.96 |
| **2** | 7.99 | 10.95 | 14.02 | 17.09 |
| **3** | 9.99 | 13.01 | 16.01 | 19.02 |

## LIFE SCIENCE APPLICATIONS

**35. Life Span** Researchers have estimated the maximum life span (in years) for various species of mammals according to the formula

$$L(E, P) = 23E^{0.6}P^{-0.267},$$

where $E$ is the average brain mass and $P$ is the average body mass (both in g). *Source: The Quarterly Review of Biology.* Find $L$ for the following species. Do these values seem reasonable?

**a.** Black rat, with $E = 7.35$ g, $P = 150$ g

**b.** Humans, with $E = 14{,}100$ g, $P = 68{,}700$ g

**36. Heat Loss** The rate of heat loss (in watts) in harbor seal pups has been approximated by

$$H(m, T, A) = \frac{15.2m^{0.67}(T - A)}{10.23 \ln m - 10.74},$$

where $m$ is the body mass of the pup (in kg), and $T$ and $A$ are the body core temperature and ambient water temperature, respectively (in °C). Find the heat loss for the following data. *Source: Functional Ecology.*

**a.** Body mass = 21 kg; body core temperature = 36°C; ambient water temperature = 4°C

**b.** Body mass = 29 kg; body core temperature = 38°C; ambient water temperature = 16°C

**37. Body Surface Area** The surface area of a human (in square meters) has been approximated by

$$A(h, m) = 0.024265h^{0.3964}m^{0.5378},$$

where $h$ is the height (in cm) and $m$ is the mass (in kg). Find $A$ for the following data. *Source: The Journal of Pediatrics.*

**a.** Height, 178 cm; mass, 72 kg

**b.** Height, 140 cm; mass, 65 kg

**c.** Height, 160 cm; mass, 70 kg

**d.** Using your mass and height, find your own surface area.

**38. Dinosaur Running** An article entitled "How Dinosaurs Ran" explains that the locomotion of different sized animals can be compared when they have the same Froude number, defined as

$$F = \frac{v^2}{gl},$$

where $v$ is the velocity, $g$ is the acceleration of gravity (9.81 m per sec²), and $l$ is the leg length (in meters). *Source: Scientific American.*

**a.** One result described in the article is that different animals change from a trot to a gallop at the same Froude number, roughly 2.56. Find the velocity at which this change occurs for a ferret, with a leg length of 0.09 m, and a rhinoceros, with a leg length of 1.2 m.

**b.** Ancient footprints in Texas of a sauropod, a large herbivorous dinosaur, are roughly 1 m in diameter, corresponding to a leg length of roughly 4 m. By comparing the stride divided by the leg length with that of various modern creatures, it can be determined that the Froude number for these dinosaurs is roughly 0.025. How fast were the sauropods traveling?

**39. Pollution Intolerance** According to research at the Great Swamp in New York, the percentage of fish that are intolerant to pollution can be estimated by the function

$$P(W, R, A) = 48 - 2.43W - 1.81R - 1.22A,$$

where $W$ is the percentage of wetland, $R$ is the percentage of residential area, and $A$ is the percentage of agricultural area surrounding the swamp. *Source: Northeastern Naturalist.*

**a.** Use this function to estimate the percentage of fish that will be intolerant to pollution if 5 percent of the land is classified as wetland, 15 percent is classified as residential, and 0 percent is classified as agricultural. (*Note:* The land can also be classified as forest land.)

**b.** What is the maximum percentage of fish that will be intolerant to pollution?

**c.** Develop two scenarios that will drive the percentage of fish that are intolerant to pollution to zero.

**d.** Which variable has the greatest influence on $P$?

**40. Dengue Fever** In tropical regions, dengue fever is a significant health problem that affects nearly 100 million people each year. Using data collected from the 2002 dengue epidemic in Colima, Mexico, researchers have estimated that the incidence $I$ (number of new cases in a given year) of dengue can be predicted by the following function.

$$I(p, a, m, n, e) = (25.54 + 0.04p - 7.92a + 2.62m + 4.46n + 0.15e)^2,$$

where $p$ is the precipitation (mm), $a$ is the mean temperature (°C), $m$ is the maximum temperature (°C), $n$ is the minimum temperature (°C), and $e$ is the evaporation (mm). *Source: Journal of Environmental Health.*

a. Estimate the incidence of a dengue fever outbreak for a region with 80 mm of rainfall, average temperature of 23°C, maximum temperature of 34°C, minimum temperature of 16°C, and evaporation of 50 mm.

b. Which variable has a negative influence on the incidence of dengue? Describe this influence and what can be inferred mathematically about the biology of the fever.

**41. Deer-Vehicle Accidents** Using data collected by the U.S. Forest Service, the annual number of deer-vehicle accidents for any given county in Ohio can be estimated by the function

$$A(L, T, U, C) = 53.02 + 0.383L + 0.0015T + 0.0028U - 0.0003C,$$

where $A$ is the estimated number of accidents, $L$ is the road length (in kilometers), $T$ is the total county land area (in hundred-acres (Ha)), $U$ is the urban land area (in hundred-acres), and $C$ is the number of hundred-acres of crop land. *Source: Ohio Journal of Science.*

a. Use this formula to estimate the number of deer-vehicle accidents for Mahoning County, where $L = 266$ km, $T = 107,484$ Ha, $U = 31,697$ Ha, and $C = 24,870$ Ha. The actual value was 396.

b. Given the magnitude and nature of the input numbers, which of the variables have the greatest potential to influence the number of deer-vehicle accidents? Explain your answer.

**42. Deer Harvest** Using data collected by the U.S. Forest Service, the annual number of deer that are harvested for any given county in Ohio can be estimated by the function

$$N(R, C) = 329.32 + 0.0377R - 0.0171C,$$

where $N$ is the estimated number of harvested deer, $R$ is the rural land area (in hundred-acres), and $C$ is the number of hundred-acres of crop land. *Source: Ohio Journal of Science.*

a. Use this formula to estimate the number of harvested deer for Tuscarawas County, where $R = 141,319$ Ha and $C = 37,960$ Ha. The actual value was 4925 deer harvested.

b. Sketch the graph of this function in the first octant.

**43. Agriculture** Pregnant sows tethered in stalls often show high levels of repetitive behavior, such as bar biting and chain chewing, indicating chronic stress. Researchers from Great Britain have developed a function that estimates the relationship between repetitive behavior, the behavior of sows in adjacent stalls, and food allowances such that

$$\ln(T) = 5.49 - 3.00 \ln(F) + 0.18 \ln(C),$$

where $T$ is the percent of time spent in repetitive behavior, $F$ is the amount of food given to the sow (in kilograms per day), and $C$ is the percent of time that neighboring sows spent bar biting and chain chewing. *Source: Applied Animal Behaviour Science.*

a. Solve the above expression for $T$.

b. Find and interpret $T$ when $F = 2$ and $C = 40$.

**44. Cholesterol** Researchers have developed a mathematical relationship between the concentrations of low-density lipoprotein cholesterol (LDL-C), high-density lipoprotein cholesterol (HDL-C), triglyceride (TG), and fasting total cholesterol (TC) as follows:

$$LDL\text{-}C = TC - (HDL\text{-}C + 0.16TG),$$

where each measurement is in milligrams per liter (mg/l). *Source: JAMA.*

a. Express this mathematical relationship using the functional notation, $L(F, H, T)$, where $L$ is the low-density concentration, $F$ is the total fasting concentration, $H$ is the high-density concentration, and $T$ is the triglyceride concentration.

b. Find $L(2000, 1300, 2500)$.

**45. Total Body Water** Accurate prediction of total body water is critical in determining adequate dialysis doses for patients with renal disease. For white males, total body water can be estimated by the function

$$T(A, W, B) = 23.04 - 0.03A + 0.50W - 0.62B,$$

where $T$ is the total body water (in liters), $A$ is age (in yr), $W$ is weight (in kg), and $B$ is the body mass index. *Source: Kidney International.* Find $T(55, 75, 30)$.

**46. Snake River Salmon** Researchers have determined that the date $D$ (day of the year) in which a Chinook salmon passes the Lower Granite Dam is related to the day of release $R$ (day of the year) of the tagged salmon, the length of the salmon $L$ (in mm), and the mean daily flow of the water $F$ (in m³/s) such that

$$\ln(D) = 4.890179 + 0.005163R - 0.004345L - 0.000019F.$$

*Source: North American Journal of Fisheries Management.*

a. Solve this expression for $D$.

b. Estimate the date of passage for a 75-mm salmon that was released on day 150 with an average flow rate of 3500 m³/s. What day of the year would this be, assuming that the year is not a leap year?

**47. Optimal Foraging** Researchers have developed a function that is used to determine optimal foraging strategies for animals. The function measures the profitability of an animal searching for certain types of prey and is given by

$$P = \frac{\lambda_1 E_1 + \lambda_2 E_2 + \cdots + \lambda_k E_k}{1 + \lambda_1 h_1 + \lambda_2 h_2 + \cdots + \lambda_k h_k},$$

where $E_i$ is the caloric value (in kcal), $\lambda_i$ is the number of prey encountered per minute, and $h_i$ is the handling time (in min) for the $i$th prey type. A hypothetical blue jay may eat worms, moths, and grubs. The values of the variables are given in the table below. *Source: The UMAP Journal.*

| Item | $E$ | $\lambda$ | $h$ |
|------|-----|-----------|-----|
| Worm | 162 | 0.2 | 3.6 |
| Moth | 24 | 3.0 | 0.6 |
| Grub | 40 | 3.0 | 1.6 |

**a.** Find the profitability $P$ for a blue jay to eat worms only.

**b.** Find the profitability for a blue jay to eat worms and moths.

**c.** What is the combination of prey that provides maximum profitability for the blue jay?

## OTHER APPLICATIONS

**48. Postage Rates** Extra postage is charged for parcels sent by U.S. mail that are more than 84 in. in length and girth combined. (Girth is the distance around the parcel perpendicular to its length. See the figure.) Express the combined length and girth as a function of $L$, $W$, and $H$.

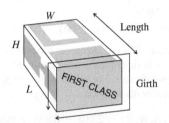

**49. Required Material** Refer to the figure for Exercise 48. Assume $L$, $W$, and $H$ are in feet. Write a function in terms of $L$, $W$, and $H$ that gives the total area of the material required to build the box.

**50. Elliptical Templates** The holes cut in a roof for vent pipes require elliptical templates. A formula for determining the length of the major axis of the ellipse is given by

$$L = f(H, D) = \sqrt{H^2 + D^2},$$

where $D$ is the (outside) diameter of the pipe and $H$ is the "rise" of the roof per $D$ units of "run"; that is, the slope of the roof is $H/D$. (See the figure below.) The width of the ellipse (minor axis) equals $D$. Find the length and width of the ellipse required to produce a hole for a vent pipe with a diameter of 3.75 in. in roofs with the following slopes.

**a.** 3/4          **b.** 2/5

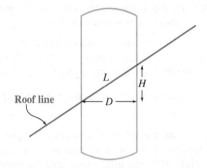

**YOUR TURN ANSWERS**

**1.** 29    **2.** 24    **3.**

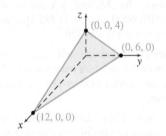

# 9.2 Partial Derivatives

**APPLY IT** What is the relationship between the water pressure on body tissues for various water depths and dive times?

*We will answer this question in Exercise 58 using the concept of partial derivatives.*

In Chapter 3, we found that the derivative $dy/dx$ gives the rate of change of $y$ with respect to $x$. In this section, we show how derivatives are found and interpreted for multivariable functions.

As we saw in an exercise in the previous section, the surface area of a human (in square meters) is a function of both height and mass, and is approximately given by

$$A(h, m) = 0.024265h^{0.3964}m^{0.5378},$$

where $h$ is the height of the person (in cm) and $m$ is the mass (in kg). **Source: The Journal of Pediatrics.** How will a change in $h$ or $m$ affect $A$?

**FOR REVIEW**

You may want to review Chapter 4 on Calculating the Derivative for methods used to find some of the derivatives in this section.

Suppose that the height of baseball player Miguel Cabrera, who is 193 cm tall, is fairly constant, but his weight (which is proportional to his mass) tends to vary. His trainer would like to find the change in surface area with respect to changes in his mass $m$. Here $h$ is fixed at 193 cm. Using this information, we begin by finding a new function $f(m) = A(193, m)$. Let $h = 193$ to get

$$f(m) = A(193, m) = 0.024265(193)^{0.3964}m^{0.5378} = 0.1954m^{0.5378}$$

The function $f(m)$ shows the surface area for a person with a mass of $m$, assuming that $h$ is fixed at 193 cm. Find the derivative $f'(m)$ to get the marginal value with respect to $m$:

$$f'(m) = 0.1954(0.5378)m^{0.5378-1} = 0.1051m^{-0.4622}.$$

In this example, the derivative of the function $f(m)$ was taken with respect to $m$ only; we assume that $h$ was fixed. To generalize, let $z = f(x, y)$. An intuitive definition of the partial derivatives of $f$ with respect to $x$ and $y$ follows.

### Partial Derivatives (Informal Definition)

The **partial derivative of $f$ with respect to $x$** is the derivative of $f$ obtained by treating $x$ as a variable and $y$ as a constant.
The **partial derivative of $f$ with respect to $y$** is the derivative of $f$ obtained by treating $y$ as a variable and $x$ as a constant.

The symbols $f_x(x, y)$ (no prime is used), $\partial z/\partial x$, $z_x$, and $\partial f/\partial x$ are used to represent the partial derivative of $z = f(x, y)$ with respect to $x$, with similar symbols used for the partial derivative with respect to $y$.

Generalizing from the definition of the derivative given earlier, partial derivatives of a function $z = f(x, y)$ are formally defined as follows.

### Partial Derivatives (Formal Definition)

Let $z = f(x, y)$ be a function of two independent variables. Let all indicated limits exist. Then the partial derivative of $f$ with respect to $x$ is

$$f_x(x, y) = \frac{\partial f}{\partial x} = \lim_{h \to 0} \frac{f(x + h, y) - f(x, y)}{h},$$

and the partial derivative of $f$ with respect to $y$ is

$$f_y(x, y) = \frac{\partial f}{\partial y} = \lim_{h \to 0} \frac{f(x, y + h) - f(x, y)}{h}.$$

If the indicated limits do not exist, then the partial derivatives do not exist.

Similar definitions could be given for functions of more than two independent variables.

**EXAMPLE 1** **Partial Derivatives**

Let $f(x, y) = 4x^2 - 9xy + 6y^3$. Find $f_x(x, y)$ and $f_y(x, y)$.

**SOLUTION** To find $f_x(x, y)$, treat $y$ as a constant and $x$ as a variable. The derivative of the first term, $4x^2$, is $8x$. In the second term, $-9xy$, the constant coefficient of $x$ is $-9y$, so the derivative with $x$ as the variable is $-9y$. The derivative of $6y^3$ is zero, since we are treating $y$ as a constant. Thus,

$$f_x(x, y) = 8x - 9y.$$

**YOUR TURN 1** Let
$f(x, y) = 2x^2y^3 + 6x^5y^4$. Find
$f_x(x, y)$ and $f_y(x, y)$.

Now, to find $f_y(x, y)$, treat $y$ as a variable and $x$ as a constant. Since $x$ is a constant, the derivative of $4x^2$ is zero. In the second term, the coefficient of $y$ is $-9x$ and the derivative of $-9xy$ is $-9x$. The derivative of the third term is $18y^2$. Thus,

$$f_y(x, y) = -9x + 18y^2.$$

**TRY YOUR TURN 1**

The next example shows how the chain rule can be used to find partial derivatives.

**EXAMPLE 2** Partial Derivatives

Let $f(x, y) = \ln |x^2 + 3y|$. Find $f_x(x, y)$ and $f_y(x, y)$.

**SOLUTION** Recall the formula for the derivative of a natural logarithm function. If $g(x) = \ln |x|$, then $g'(x) = 1/x$. Using this formula and the chain rule,

$$f_x(x, y) = \frac{1}{x^2 + 3y} \cdot \frac{\partial}{\partial x}(x^2 + 3y) = \frac{1}{x^2 + 3y} \cdot 2x = \frac{2x}{x^2 + 3y},$$

and

**YOUR TURN 2** Let
$f(x, y) = e^{3x^2y}$. Find $f_x(x, y)$ and
$f_y(x, y)$.

$$f_y(x, y) = \frac{1}{x^2 + 3y} \cdot \frac{\partial}{\partial y}(x^2 + 3y) = \frac{1}{x^2 + 3y} \cdot 3 = \frac{3}{x^2 + 3y}.$$

**TRY YOUR TURN 2**

The notation

$$f_x(a, b) \quad \text{or} \quad \frac{\partial f}{\partial y}(a, b)$$

represents the value of the partial derivative when $x = a$ and $y = b$, as shown in the next example.

**EXAMPLE 3** Evaluating Partial Derivatives

Let $f(x, y) = 2x^2 + 9xy^3 + 8y + 5$. Find the following.

**(a)** $f_x(-1, 2)$

**SOLUTION** First, find $f_x(x, y)$ by holding $y$ constant.

$$f_x(x, y) = 4x + 9y^3$$

Now let $x = -1$ and $y = 2$.

$$f_x(-1, 2) = 4(-1) + 9(2)^3 = -4 + 72 = 68$$

**(b)** $\dfrac{\partial f}{\partial y}(-4, -3)$

**SOLUTION** Since $\partial f / \partial y = 27xy^2 + 8$,

$$\frac{\partial f}{\partial y}(-4, -3) = 27(-4)(-3)^2 + 8 = 27(-36) + 8 = -964.$$

**(c)** All values of $x$ and $y$ such that both $f_x(x, y) = 0$ and $f_y(x, y) = 0$. (The importance of such points will be shown in the next section.)

**SOLUTION** From parts (a) and (b),

$$f_x(x, y) = 4x + 9y^3 = 0 \quad \text{and} \quad f_y(x, y) = 27xy^2 + 8 = 0.$$

Solving the first equation for $x$ yields $x = -9y^3/4$. Substituting this into the second equation yields

$$27\left(\frac{-9y^3}{4}\right)y^2 + 8 = 0$$

$$\frac{-243y^5}{4} + 8 = 0$$

$$\frac{-243y^5}{4} = -8$$

$$y^5 = \frac{32}{243}$$

$$y = \frac{2}{3}. \qquad \text{Take the fifth root of both sides.}$$

Substituting $y = 2/3$ yields $x = -9y^3/4 = -9(2/3)^3/4 = -2/3$. Thus, $f_x(x, y) = 0$ and $f_y(x, y) = 0$ when $x = -2/3$ and $y = 2/3$.

**(d)** $f_x(x, y)$ using the formal definition of the partial derivative.

**SOLUTION** Calculate as follows:

$$\frac{f(x + h, y) - f(x, y)}{h} = \frac{2(x + h)^2 + 9(x + h)y^3 + 8y + 5 - (2x^2 + 9xy^3 + 8y + 5)}{h}$$

$$= \frac{2x^2 + 4xh + 2h^2 + 9xy^3 + 9hy^3 + 8y + 5 - 2x^2 - 9xy^3 - 8y - 5}{h}$$

$$= \frac{4xh + 2h^2 + 9hy^3}{h} \qquad \text{Simplify the numerator.}$$

$$= \frac{h(4x + 2h + 9y^3)}{h} \qquad \text{Factor } h \text{ from the numerator.}$$

$$= 4x + 2h + 9y^3.$$

Now take the limit as $h$ goes to 0. Thus,

**YOUR TURN 3** Let $f(x, y) = xye^{x^2+y^3}$. Find $f_x(2, 1)$ and $f_y(2, 1)$.

$$f_x(x, y) = \lim_{h \to 0}\frac{f(x + h, y) - f(x, y)}{h} = \lim_{h \to 0}(4x + 2h + 9y^3)$$

$$= 4x + 9y^3,$$

the same answer we found in part (a). **TRY YOUR TURN 3**

In some cases, the difference quotient may not simplify as easily as it did in Example 3(d). In such cases, the derivative may be approximated by putting a small value for $h$ into $[f(x + h) - f(x)]/h$. In Example 3(d), with $x = -1$ and $y = 2$, the values $h = 10^{-4}$ and $10^{-5}$ give approximations for $f_x(-1, 2)$ as 68.0002 and 68.00002, respectively, compared with the exact value of 68 found in Example 3(a).

The derivative of a function of one variable can be interpreted as the slope of the tangent line to the graph at that point. With some modification, the same is true of partial derivatives of functions of two variables. At a point on the graph of a function of two variables, $z = f(x, y)$, there may be many tangent lines, all of which lie in the same tangent plane, as shown in Figure 15.

In any particular direction, however, there will be only one tangent line. We use partial derivatives to find the slope of the tangent lines in the $x$- and $y$-directions as follows.

Figure 16 on the next page shows a surface $z = f(x, y)$ and a plane that is parallel to the $xz$-plane. The equation of the plane is $y = b$. (This corresponds to holding $y$ fixed.) Since $y = b$ for points on the plane, any point on the curve that represents the intersection of the plane and the surface must have the form $(x, y, z) = (x, b, f(x, b))$. Thus, this curve can be described as $z = f(x, b)$. Since $b$ is constant, $z = f(x, b)$ is a function of one variable. When the derivative of $z = f(x, b)$ is evaluated at $x = a$, it gives the slope of the line

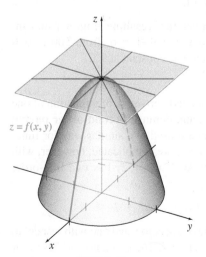

$z = f(x, y)$

**FIGURE 15**

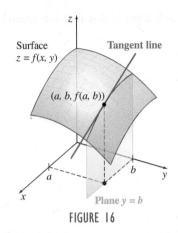

**FIGURE 16**

tangent to this curve at the point $(a, b, f(a, b))$, as shown in Figure 16. Thus, the partial derivative of $f$ with respect to $x$, $f_x(a, b)$, gives the rate of change of the surface $z = f(x, y)$ in the $x$-direction at the point $(a, b, f(a, b))$. In the same way, the partial derivative with respect to $y$ will give the slope of the line tangent to the surface in the $y$-direction at the point $(a, b, f(a, b))$.

## Rate of Change
The derivative of $y = f(x)$ gives the rate of change of $y$ with respect to $x$. In the same way, if $z = f(x, y)$, then $f_x(x, y)$ gives the rate of change of $z$ with respect to $x$, if $y$ is held constant.

---

**EXAMPLE 4  Water Temperature**

Suppose that the temperature of the water at the point on a river where a nuclear power plant discharges its hot waste water is approximated by

$$T(x, y) = 2x + 5y + xy - 40,$$

where $x$ represents the temperature of the river water (in degrees Celsius) before it reaches the power plant and $y$ is the number of megawatts (in hundreds) of electricity being produced by the plant.

**(a)** Find and interpret $T_x(9, 5)$.

**SOLUTION**  First, find the partial derivative $T_x(x, y)$.

$$T_x(x, y) = 2 + y$$

This partial derivative gives the rate of change of $T$ with respect to $x$. Replacing $x$ with 9 and $y$ with 5 gives

$$T_x(9, 5) = 2 + 5 = 7.$$

This result, 7, is the approximate change in temperature of the output water if input water temperature changes by 1 degree, from $x = 9$ to $x = 9 + 1 = 10$, while $y$ remains constant at 5 (500 megawatts of electricity produced).

**(b)** Find and interpret $T_y(9, 5)$.

**SOLUTION**  The partial derivative $T_y(x, y)$ is

$$T_y(x, y) = 5 + x.$$

This partial derivative gives the rate of change of $T$ with respect to $y$ as

$$T_y(9, 5) = 5 + 9 = 14.$$

This result, 14, is the approximate change in temperature resulting from a 1-unit increase in production of electricity from $y = 5$ to $y = 5 + 1 = 6$ (from 500 to 600 megawatts), while the input water temperature $x$ remains constant at 9°C.

## Second-Order Partial Derivatives
The second derivative of a function of one variable is very useful in determining relative maxima and minima. **Second-order partial derivatives** (partial derivatives of a partial derivative) are used in a similar way for functions of two or more variables. The situation is somewhat more complicated, however, with more independent variables. For example, $f(x, y) = 4x + x^2y + 2y$ has two first-order partial derivatives,

$$f_x(x, y) = 4 + 2xy \quad \text{and} \quad f_y(x, y) = x^2 + 2.$$

Since each of these has two partial derivatives, one with respect to $y$ and one with respect to $x$, there are *four* second-order partial derivatives of function $f$. The notations for these four second-order partial derivatives are given on the next page.

## Second-Order Partial Derivatives

For a function $z = f(x, y)$, if the indicated partial derivative exists, then

$$\frac{\partial}{\partial x}\left(\frac{\partial z}{\partial x}\right) = \frac{\partial^2 z}{\partial x^2} = f_{xx}(x, y) = z_{xx} \qquad \frac{\partial}{\partial y}\left(\frac{\partial z}{\partial y}\right) = \frac{\partial^2 z}{\partial y^2} = f_{yy}(x, y) = z_{yy}$$

$$\frac{\partial}{\partial y}\left(\frac{\partial z}{\partial x}\right) = \frac{\partial^2 z}{\partial y \partial x} = f_{xy}(x, y) = z_{xy} \qquad \frac{\partial}{\partial x}\left(\frac{\partial z}{\partial y}\right) = \frac{\partial^2 z}{\partial x \partial y} = f_{yx}(x, y) = z_{yx}$$

**NOTE** For most functions found in applications and for all of the functions in this book, the second-order partial derivatives $f_{xy}(x, y)$ and $f_{yx}(x, y)$ are equal. This is always true when $f_{xy}(x, y)$ and $f_{yx}(x, y)$ are continuous. Therefore, it is not necessary to be particular about the order in which these derivatives are found.

### EXAMPLE 5  Second-Order Partial Derivatives

Find all second-order partial derivatives for

$$f(x, y) = -4x^3 - 3x^2y^3 + 2y^2.$$

**SOLUTION** First find $f_x(x, y)$ and $f_y(x, y)$.

$$f_x(x, y) = -12x^2 - 6xy^3 \qquad \text{and} \qquad f_y(x, y) = -9x^2y^2 + 4y$$

To find $f_{xx}(x, y)$, take the partial derivative of $f_x(x, y)$ with respect to $x$.

$$f_{xx}(x, y) = -24x - 6y^3$$

Take the partial derivative of $f_y(x, y)$ with respect to $y$; this gives $f_{yy}$.

$$f_{yy}(x, y) = -18x^2y + 4$$

Find $f_{xy}(x, y)$ by starting with $f_x(x, y)$, then taking the partial derivative of $f_x(x, y)$ with respect to $y$.

$$f_{xy}(x, y) = -18xy^2$$

Finally, find $f_{yx}(x, y)$ by starting with $f_y(x, y)$; take its partial derivative with respect to $x$.

$$f_{yx}(x, y) = -18xy^2$$

### EXAMPLE 6  Second-Order Partial Derivatives

Let $f(x, y) = 2e^x - 8x^3y^2$. Find all second-order partial derivatives.

**YOUR TURN 4** Let $f(x, y) = x^2e^{7y} + x^4y^5$. Find all second partial derivatives.

**SOLUTION** Here $f_x(x, y) = 2e^x - 24x^2y^2$ and $f_y(x, y) = -16x^3y$. (Recall: If $g(x) = e^x$, then $g'(x) = e^x$.) Now find the second-order partial derivatives.

$$f_{xx}(x, y) = 2e^x - 48xy^2 \qquad f_{xy}(x, y) = -48x^2y$$
$$f_{yy}(x, y) = -16x^3 \qquad f_{yx}(x, y) = -48x^2y \quad \text{TRY YOUR TURN 4}$$

Partial derivatives of functions with more than two independent variables are found in a similar manner. For instance, to find $f_z(x, y, z)$ for $f(x, y, z)$, hold $x$ and $y$ constant and differentiate with respect to $z$.

## EXAMPLE 7    Second-Order Partial Derivatives

Let $f(x, y, z) = 2x^2yz^2 + 3xy^2 - 4yz$. Find $f_x(x, y, z)$, $f_y(x, y, z)$, $f_{xz}(x, y, z)$, and $f_{yz}(x, y, z)$.

**SOLUTION**

$$f_x(x, y, z) = 4xyz^2 + 3y^2$$
$$f_y(x, y, z) = 2x^2z^2 + 6xy - 4z$$

To find $f_{xz}(x, y, z)$, differentiate $f_x(x, y, z)$ with respect to $z$.

$$f_{xz}(x, y, z) = 8xyz$$

Differentiate $f_y(x, y, z)$ with respect to $z$ to get $f_{yz}(x, y, z)$.

$$f_{yz}(x, y, z) = 4x^2z - 4$$

# 9.2    EXERCISES

**1.** Let $z = f(x, y) = 6x^2 - 4xy + 9y^2$. Find the following using the formal definition of the partial derivative.

**a.** $\dfrac{\partial z}{\partial x}$

**b.** $\dfrac{\partial z}{\partial y}$

**c.** $\dfrac{\partial f}{\partial x}(2, 3)$

**d.** $f_y(1, -2)$

**2.** Let $z = g(x, y) = 8x + 6x^2y + 2y^2$. Find the following using the formal definition of the partial derivative.

**a.** $\dfrac{\partial z}{\partial x}$

**b.** $\dfrac{\partial z}{\partial y}$

**c.** $\dfrac{\partial g}{\partial y}(-3, 0)$

**d.** $g_x(2, 1)$

In Exercises 3–22, find $f_x(x, y)$ and $f_y(x, y)$. Then find $f_x(2, -1)$ and $f_y(-4, 3)$. Leave the answers in terms of $e$ in Exercises 7–10, 15–16, and 19–20.

**3.** $f(x, y) = -4xy + 6y^3 + 5$

**4.** $f(x, y) = 9x^2y^2 - 4y^2$

**5.** $f(x, y) = 5x^2y^3$

**6.** $f(x, y) = -3x^4y^3 + 10$

**7.** $f(x, y) = e^{x+y}$

**8.** $f(x, y) = 4e^{3x+2y}$

**9.** $f(x, y) = -6e^{4x-3y}$

**10.** $f(x, y) = 8e^{7x-y}$

**11.** $f(x, y) = \dfrac{x^2 + y^3}{x^3 - y^2}$

**12.** $f(x, y) = \dfrac{3x^2y^3}{x^2 + y^2}$

**13.** $f(x, y) = \ln|1 + 5x^3y^2|$

**14.** $f(x, y) = \ln|4x^4 - 2x^2y^2|$

**15.** $f(x, y) = xe^{x^2y}$

**16.** $f(x, y) = y^2e^{x+3y}$

**17.** $f(x, y) = \sqrt{x^4 + 3xy + y^4 + 10}$

**18.** $f(x, y) = (7x^2 + 18xy^2 + y^3)^{1/3}$

**19.** $f(x, y) = \dfrac{3x^2y}{e^{xy} + 2}$

**20.** $f(x, y) = (7e^{x+2y} + 4)(e^{x^2} + y^2 + 2)$

**21.** $f(x, y) = x\sin(\pi y)$

**22.** $f(x, y) = x\tan(\pi y^2)$

**Find all second-order partial derivatives for the following.**

**23.** $f(x, y) = 4x^2y^2 - 16x^2 + 4y$

**24.** $g(x, y) = 5x^4y^2 + 12y^3 - 9x$

**25.** $R(x, y) = 4x^2 - 5xy^3 + 12y^2x^2$

**26.** $h(x, y) = 30y + 5x^2y + 12xy^2$

**27.** $r(x, y) = \dfrac{6y}{x + y}$

**28.** $k(x, y) = \dfrac{-7x}{2x + 3y}$

**29.** $z = 9ye^x$

**30.** $z = -6xe^y$

**31.** $r = \ln|x + y|$

**32.** $k = \ln|5x - 7y|$

**33.** $z = x\ln|xy|$

**34.** $z = (y + 1)\ln|x^3y|$

**35.** $z = e^x\sin y$

**36.** $z = \tan xy$

**For the functions defined as follows, find all values of $x$ and $y$ such that both $f_x(x, y) = 0$ and $f_y(x, y) = 0$.**

**37.** $f(x, y) = 6x^2 + 6y^2 + 6xy + 36x - 5$

**38.** $f(x, y) = 50 + 4x - 5y + x^2 + y^2 + xy$

**39.** $f(x, y) = 9xy - x^3 - y^3 - 6$

**40.** $f(x, y) = 2200 + 27x^3 + 72xy + 8y^2$

**Find $f_x(x, y, z)$, $f_y(x, y, z)$, $f_z(x, y, z)$, and $f_{yz}(x, y, z)$ for the following.**

**41.** $f(x, y, z) = x^4 + 2yz^2 + z^4$

**42.** $f(x, y, z) = 6x^3 - x^2y^2 + y^5$

**43.** $f(x, y, z) = \dfrac{6x - 5y}{4z + 5}$

**44.** $f(x, y, z) = \dfrac{2x^2 + xy}{yz - 2}$

**45.** $f(x, y, z) = \ln|x^2 - 5xz^2 + y^4|$

**46.** $f(x, y, z) = \ln|8xy + 5yz - x^3|$

In Exercises 47 and 48, approximate the indicated derivative for each function by using the definition of the derivative with small values of $h$.

**47.** $f(x, y) = (x + y/2)^{x+y/2}$

**a.** $f_x(1, 2)$

**b.** $f_y(1, 2)$

**48.** $f(x, y) = (x + y^2)^{2x+y}$

**a.** $f_x(2, 1)$

**b.** $f_y(2, 1)$

# LIFE SCIENCE APPLICATIONS

**49. Calorie Expenditure** The average energy expended for an animal to walk or run 1 km can be estimated by the function

$$f(m, v) = 25.92m^{0.68} + \frac{3.62m^{0.75}}{v},$$

where $f(m, v)$ is the energy used (in kcal per hour), $m$ is the mass (in g), and $v$ is the speed of movement (in km per hour) of the animal. *Source: Wildlife Feeding and Nutrition.*

**a.** Find $f(300, 10)$.

**b.** Find $f_m(300, 10)$ and interpret.

**c.** If a mouse could run at the same speed that an elephant walks, which animal would expend more energy? How can partial derivatives be used to explore this question?

**50. Heat Loss** The rate of heat loss (in watts) in harbor seal pups has been approximated by

$$H(m, T, A) = \frac{15.2m^{0.67}(T - A)}{10.23 \ln m - 10.74},$$

where $m$ is the body mass of the pup (in kg), and $T$ and $A$ are the body core temperature and ambient water temperature, respectively (in °C). Find the approximate change in heat loss under the following conditions. *Source: Functional Ecology.*

**a.** The body core temperature increases from 37°C to 38°, while the ambient water temperature remains at 8°C and the body mass remains at 24 kg.

**b.** The ambient water temperature increases from 10°C to 11°, while the body core temperature remains at 37°C and the body mass remains at 26 kg.

**51. Body Surface Area** The surface area of a human (in square meters) has been approximated by

$$A = 0.024265h^{0.3964}m^{0.5378},$$

where $h$ is the height (in cm) and $m$ is the mass (in kg). *Source: The Journal of Pediatrics.*

**a.** Find the approximate change in surface area when the mass changes from 72 kg to 73 kg, while the height remains at 180 cm.

**b.** Find the approximate change in surface area when the height changes from 160 cm to 161 cm, while the mass remains at 70 kg.

**52. Blood Flow** According to the Fick Principle, the quantity of blood pumped through the lungs depends on the following variables (in milliliters):

$b$ = quantity of oxygen used by the body in one minute

$a$ = quantity of oxygen per liter of blood that has just gone through the lungs

$v$ = quantity of oxygen per liter of blood that is about to enter the lungs

In one minute,

Amount of oxygen used = Amount of oxygen per liter × Liters of blood pumped.

If $C$ is the number of liters of blood pumped through the lungs in one minute, then

$$b = (a - v) \cdot C \quad \text{or} \quad C = \frac{b}{a - v}.$$

*Source: Anaesthesia UK.*

**a.** Find the number of liters of blood pumped through the lungs in one minute if $a = 160$, $b = 200$, and $v = 125$.

**b.** Find the approximate change in $C$ when $a$ changes from 160 to 161, $b = 200$, and $v = 125$.

**c.** Find the approximate change in $C$ when $a = 160$, $b$ changes from 200 to 201, and $v = 125$.

**d.** Find the approximate change in $C$ when $a = 160$, $b = 200$, and $v$ changes from 125 to 126.

**e.** A change of 1 unit in which quantity of oxygen produces the greatest change in the liters of blood pumped?

**53. Health** A weight-loss counselor has prepared a program of diet and exercise for a client. If the client sticks to the program, the weight loss that can be expected (in pounds per week) is given by

$$\text{Weight loss} = f(n, c) = \frac{1}{8}n^2 - \frac{1}{5}c + \frac{1937}{8},$$

where $c$ is the average daily calorie intake for the week and $n$ is the number of 40-minute aerobic workouts per week.

**a.** How many pounds can the client expect to lose by eating an average of 1200 cal per day and participating in four 40-minute workouts in a week?

**b.** Find and interpret $\partial f/\partial n$.

**c.** The client currently averages 1100 cal per day and does three 40-minute workouts each week. What would be the approximate impact on weekly weight loss of adding a fourth workout per week?

**54. Health** The body mass index is a number that can be calculated for any individual as follows: Multiply a person's weight by 703 and divide by the person's height squared. That is,

$$B = \frac{703w}{h^2},$$

where $w$ is in pounds and $h$ is in inches. The National Heart, Lung and Blood Institute uses the body mass index to determine whether a person is "overweight" ($25 \le B < 30$) or "obese" ($B \ge 30$). *Source: The National Institutes of Health.*

**a.** Calculate the body mass index for St. Louis Rams offensive tackle Jake Long, who weighs 317 lb and is 6'7" tall.

**b.** Calculate $\frac{\partial B}{\partial w}$ and $\frac{\partial B}{\partial h}$ and interpret.

**c.** Using the fact that 1 in. = 0.0254 m and 1 lb = 0.4536 kg, transform this formula to handle metric units.

**55. Body Shape Index** An alternative to the body mass index (BMI) for estimating the health risks of obesity is A Body Shape Index (ABSI), given by

$$ABSI = \frac{w}{b^{2/3}h^{1/2}},$$

where $w$ and $h$ are the waist and height, respectively, in meters, and $b$ is the BMI. (See the definition of BMI in the previous exercise.) The National Heart, Lung and Blood Institute uses the BMI to determine whether a person is "overweight" $(25 \leq BMI < 30)$ or "obese" $(BMI \geq 30)$. *Source: PLoS One.*

a. Suppose a man is 1.85 m tall with a waist of 0.864 m and a BMI of 23.1. Find the ABSI for such a person. How does this compare with the average ABSI for a 52-year-old man of 0.0831?

b. For the person described in part a, find the rate that the ABSI is going up with respect to his waist.

c. For the person described in part a, find the rate that the ABSI is going down with respect to his height.

d. Based on your answers to parts b and c, which has a greater effect on the person's ABSI, an increase in waist or an increase in height of the same amount? Explain why this makes sense intuitively.

56. **Growing Degree Days** The number of growing degree days (GDD) is the sum, over all days in a year in which the temperature is above 40°F, of the amount that the temperature exceeds 40°F. It can be approximated by

$$f(x, y) = \frac{365}{2\pi}x - \frac{365}{2}(40 - y) + \frac{365(40 - y)^2}{\pi x},$$

where $x$ is the difference in degrees F between the average temperature in July and the average temperature in January, and $y$ is the average of those two temperatures. *Source: Journal of Ecology.*

a. The average January temperature in New York City is 32°F, and the average July temperature is 77°F. Use the formula above to approximate the number of GDD. Compare this value with the average of 5945 calculated at www.yourweekendview.com/outlook/agriculture/growing-degree-days/USNY0996.

b. For New York City, find the rate of change of the number of GDD with respect to the average of January and July temperatures if the difference between those averages remains constant.

c. For New York City, find the rate of change of the number of GDD with respect to the difference between average January and July temperatures if the average between those remains constant.

d. Find the average January and July temperature for your home town and compute the number of GDD. Then compare with the actual number from the website given in part a.

57. **Drug Reaction** The reaction to $x$ units of a drug $t$ hours after it was administered is given by

$$R(x, t) = x^2(a - x)t^2e^{-t},$$

for $0 \leq x \leq a$ (where $a$ is a constant). Find the following.

a. $\dfrac{\partial R}{\partial x}$

b. $\dfrac{\partial R}{\partial t}$

c. $\dfrac{\partial^2 R}{\partial x^2}$

d. $\dfrac{\partial^2 R}{\partial x \partial t}$

e. Interpret your answers to parts a and b.

58. **APPLY IT Scuba Diving** In 1908, J. Haldane constructed diving tables that provide a relationship between the water pressure on body tissues for various water depths and dive times. The tables were successfully used by divers to virtually eliminate decompression sickness. The pressure in atmospheres for a no-stop dive is given by the following formula:*

$$p(l, t) = 1 + \frac{l}{33}(1 - 2^{-t/5}),$$

where $t$ is in minutes, $l$ is in feet, and $p$ is in atmospheres (atm). *Source: The UMAP Journal.*

a. Find the pressure at 33 ft for a 10-minute dive.

b. Find $p_l(33, 10)$ and $p_t(33, 10)$ and interpret. (*Hint*: $D_t(a^t) = \ln(a)a^t$.)

c. Haldane estimated that decompression sickness for no-stop dives could be avoided if the diver's tissue pressure did not exceed 2.15 atm. Find the maximum amount of time that a diver could stay down (time includes going down and coming back up) if he or she wants to dive to a depth of 66 ft.

59. **Wind Chill** In 1941, explorers Paul Siple and Charles Passel discovered that the amount of heat lost when an object is exposed to cold air depends on both the temperature of the air and the velocity of the wind. They developed the *Wind Chill Index* as a way to measure the danger of frostbite while doing outdoor activities. The wind chill can be calculated as follows:

$$W(V, T) = 91.4 - \frac{(10.45 + 6.69\sqrt{V} - 0.447V)(91.4 - T)}{22}$$

where $V$ is the wind speed in miles per hour and $T$ is the temperature in Fahrenheit for wind speeds between 4 and 45 mph. *Source: The UMAP Journal.*

a. Find the wind chill for a wind speed of 20 mph and 10°F.

b. If a weather report indicates that the wind chill is $-25°F$ and the actual outdoor temperature is 5°F, use a graphing calculator to find the corresponding wind speed to the nearest mile per hour.

c. Find $W_V(20, 10)$ and $W_T(20, 10)$ and interpret.

d. Using the table command on a graphing calculator or a spreadsheet, develop a wind chill chart for various wind speeds and temperatures.

60. **Life Span** As we saw in the previous section, researchers have estimated the maximum life span (in years) for various species of mammals according to the formula

$$L(E, P) = 23E^{0.6}P^{-0.267},$$

where $E$ is the average brain mass and $P$ is the average body mass (both in g). *Source: The Quarterly Review of Biology.* Find the approximate change in life span under the following conditions. Then calculate the actual change in life span and compare.

*These estimates are conservative. Please consult modern dive tables before making a dive.

a. The brain mass stays at 7.35 g, while the body mass goes from 150 g to 151 g.

b. The body mass stays at 68,700 g, while the brain mass goes from 14,100 g to 14,101 g.

**61. Snake River Salmon** As we saw in the last section, researchers have determined that the date $D$ (day of the year) in which a Chinook salmon passes the Lower Granite Dam can be calculated using the function

$$\ln(D) = 4.890179 + 0.005163R - 0.004345L$$
$$- 0.000019F,$$

where $R$ is the release date (day of the year) of the tagged salmon, $L$ is the length of the salmon $L$ (in mm), and $F$ is the mean daily flow of the water (in m³/s). *Source: North American Journal of Fisheries Management.*

a. Use the result of part a of Exercise 46 of the previous section to find $D_F(R, L, F)$. Interpret your answer.

b. Use the partial derivative calculated above to estimate the effect that a 1000 m³/s increase of water flow will have on the date of passage for a 75-mm salmon that was released on day 150 with an average flow rate of 3500 m³/s.

**62. Brown Trout** Researchers from New Zealand have determined that the length of a brown trout depends on both its weight and age, and that the length can be estimated by

$$L(w, t) = (0.00082t + 0.0955)e^{(\ln w + 10.49)/2.842},$$

where $L(w, t)$ is the length of the trout (in cm), $w$ is the weight of the trout (in g), and $t$ is the age of the trout (in yr). *Source: Transactions of the American Fisheries Society.*

a. Find $L(600, 7)$.

b. Find $L_w(600, 7)$ and $L_t(600, 7)$ and interpret.

**63. Zooplankton Growth** The rate that zooplankton consume phytoplankton has been described by the equation

$$C(x, y) = \frac{k}{a + (x - by)^2},$$

where $a$, $b$, and $k$ are constants, and $x$ and $y$ represent the cell size of zooplankton and phytoplankton, respectively. *Source: The American Naturalist.* Calculate $C_x(x, y)$ and $C_y(x, y)$. What do you notice about the signs of $C_x(x, y)$ and $C_y(x, y)$? Explain why this makes sense.

**64. Heat Index** The chart in the next column shows the heat index, which combines the effects of temperature with humidity to give a measure of the apparent temperature, or how hot it feels to the body. *Source: The Weather Channel.* For example, when the outside temperature is 90°F and the relative humidity is 40%, then the apparent temperature is approximately 93°F. Let $I = f(T, H)$ give the heat index, $I$, as a function of the temperature $T$ (in degrees Fahrenheit) and the percent humidity $H$. Estimate the following.

a. $f(90, 30)$

b. $f(90, 75)$

c. $f(80, 75)$

**Heat Index**

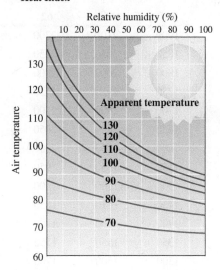

Estimate the following by approximating the partial derivative using a value of $h = 5$ in the difference quotient.

d. $f_T(90, 30)$

e. $f_H(90, 30)$

f. $f_T(90, 75)$

g. $f_H(90, 75)$

h. Describe in words what your answers in parts d–g mean.

**65. Learning** In the exercises of Chapter 4 we saw that for one model of how animals learn, the number of wrong attempts by an animal is given by

$$w = -\frac{1}{a} \ln\left[1 - \frac{(N - 1)b(1 - e^{-kn})}{N}\right],$$

where $n$ is the number of trials an animal makes while attempting to learn a task, $N$ is number of stimuli, and $k$, $a$, and $b$ are constants. *Source: Mathematical Biology of Social Behavior.*

a. Find and interpret $\partial w/\partial n$.

b. Find and interpret $\partial w/\partial N$.

**66. Insect Dispersal** A model for the dispersal of insects due to population pressure is the equation

$$\frac{\partial n}{\partial t} = D_0 \frac{\partial}{\partial x}\left[\left(\frac{n}{n_0}\right)^m \frac{\partial n}{\partial x}\right],$$

where $n$ is the number of insects and $D_0$, $n_0$, and $m$ are positive constants. *Source: Mathematical Biology.* The exact solution of this equation is

$$n(x, t) = n_0[g(t)]^{-1}\left[1 - \left\{\frac{x}{r_0 g(t)}\right\}^2\right]^{1/m},$$

where

$$g(t) = \left(\frac{2D_0(m + 2)t}{r_0^2 m}\right)^{1/(2+m)}$$

and $r_0$ is a positive constant. Show that this solution indeed satisfies the equation for the special case where $n_0 = r_0 = m = 1$ and $D_0 = 1/6$.

**67. Breath Volume** The table on the next page accompanies the Voldyne® 5000 Volumetric Exerciser. The table gives the typical lung capacity (in milliliters) for women of various ages and heights. Based on the chart, it is possible to conclude that the partial derivative of the lung capacity with respect to age and with respect to height has constant values. What are those values?

| Height (in.) | | 58" | 60" | 62" | 64" | 66" | 68" | 70" | 72" | 74" |
|---|---|---|---|---|---|---|---|---|---|---|
| | 20 | 1900 | 2100 | 2300 | 2500 | 2700 | 2900 | 3100 | 3300 | 3500 |
| A | 25 | 1850 | 2050 | 2250 | 2450 | 2650 | 2850 | 3050 | 3250 | 3450 |
| G | 30 | 1800 | 2000 | 2200 | 2400 | 2600 | 2800 | 3000 | 3200 | 3400 |
| E | 35 | 1750 | 1950 | 2150 | 2350 | 2550 | 2750 | 2950 | 3150 | 3350 |
| | 40 | 1700 | 1900 | 2100 | 2300 | 2500 | 2700 | 2900 | 3100 | 3300 |
| I | 45 | 1650 | 1850 | 2050 | 2250 | 2450 | 2650 | 2850 | 3050 | 3250 |
| N | 50 | 1600 | 1800 | 2000 | 2200 | 2400 | 2600 | 2800 | 3000 | 3200 |
| | 55 | 1550 | 1750 | 1950 | 2150 | 2350 | 2550 | 2750 | 2950 | 3150 |
| Y | 60 | 1500 | 1700 | 1900 | 2100 | 2300 | 2500 | 2700 | 2900 | 3100 |
| E | 65 | 1450 | 1650 | 1850 | 2050 | 2250 | 2450 | 2650 | 2850 | 3050 |
| A | 70 | 1400 | 1600 | 1800 | 2000 | 2200 | 2400 | 2600 | 2800 | 3000 |
| R | 75 | 1350 | 1550 | 1750 | 1950 | 2150 | 2350 | 2550 | 2750 | 2950 |
| S | 80 | 1300 | 1500 | 1700 | 1900 | 2100 | 2300 | 2500 | 2700 | 2900 |

## OTHER APPLICATIONS

**68. Gravitational Attraction**  The gravitational attraction $F$ on a body a distance $r$ from the center of Earth, where $r$ is greater than the radius of Earth, is a function of its mass $m$ and the distance $r$ as follows:

$$F = \frac{mgR^2}{r^2},$$

where $R$ is the radius of Earth and $g$ is the force of gravity—about 32 feet per second per second (ft per sec$^2$).

**a.** Find and interpret $F_m$ and $F_r$.

**b.** Show that $F_m > 0$ and $F_r < 0$. Why is this reasonable?

**69. Velocity**  In 1931, Albert Einstein developed the following formula for the sum of two velocities, $x$ and $y$:

$$w(x, y) = \frac{x + y}{1 + \dfrac{xy}{c^2}},$$

where $x$ and $y$ are in miles per second and $c$ represents the speed of light, 186,282 miles per second. *Source: The Mathematics Teacher.*

**a.** Suppose that, relative to a stationary observer, a new super space shuttle is capable of traveling at 50,000 miles per second and that, while traveling at this speed, it launches a rocket that travels at 150,000 miles per second. How fast is the rocket traveling relative to the stationary observer?

**b.** What is the instantaneous rate of change of $w$ with respect to the speed of the space shuttle, $x$, when the space shuttle is traveling at 50,000 miles per second and the rocket is traveling at 150,000 miles per second?

**c.** Hypothetically, if a person is driving at the speed of light, $c$, and she turns on the headlights, what is the velocity of the light coming from the headlights relative to a stationary observer?

**70. Movement Time**  Fitts's law is used to estimate the amount of time it takes for a person, using his or her arm, to pick up a light object, move it, and then place it in a designated target area. Mathematically, Fitts's law for a particular individual is given by

$$T(s, w) = 105 + 265 \log_2\left(\frac{2s}{w}\right),$$

where $s$ is the distance (in feet) the object is moved, $w$ is the width of the area in which the object is being placed, and $T$ is the time (in msec). *Source: Human Factors in Engineering Design.*

**a.** Calculate $T(3, 0.5)$.

**b.** Find $T_s(3, 0.5)$ and $T_w(3, 0.5)$ and interpret these values. (*Hint:* $\log_2 x = \ln x / \ln 2$.)

**71. Ground Temperature**  Mathematical models of ground temperature variation usually involve Fourier series or other sophisticated methods. However, the elementary model

$$u(x, t) = T_0 + A_0 e^{-ax} \cos\left(\frac{\pi}{6}t - ax\right)$$

has been developed for temperature $u(x, t)$ at a given location at a variable time $t$ (in months) and a variable depth $x$ (in centimeters) beneath the Earth's surface. $T_0$ is the annual average surface temperature, and $A_0$ is the amplitude of the seasonal surface temperature variation. *Source: Applications in School Mathematics 1979 Yearbook.*

Assume that $T_0 = 16°C$ and $A_0 = 11°C$ at a certain location. Also assume that $a = 0.00706$ in cgs (centimeter-gram-second) units.

**a.** At what minimum depth $x$ is the amplitude of $u(x, t)$ at most 1°C?

**b.** Suppose we wish to construct a cellar to keep wine at a temperature between 14°C and 18°C. What minimum depth will accomplish this?

**c.** At what minimum depth $x$ does the ground temperature model predict that it will be winter when it is summer at the surface, and vice versa? That is, when will the phase shift correspond to 1/2 year?

**d.** Show that the ground temperature model satisfies the *heat equation*

$$\frac{\partial u}{\partial t} = k\frac{\partial^2 u}{\partial x^2},$$

where $k$ is a constant.

**YOUR TURN ANSWERS**

1. $f_x(x,y) = 4xy^3 + 30x^4y^4$; $f_y(x,y) = 6x^2y^2 + 24x^5y^3$
2. $f_x(x,y) = 6xye^{3x^2y}$; $f_y(x,y) = 3x^2e^{3x^2y}$
3. $f_x(2, 1) = 9e^5$; $f_y(2, 1) = 8e^5$
4. $f_{xx}(x,y) = 2e^{7y} + 12x^2y^5$;
   $f_{yy}(x,y) = 49x^2e^{7y} + 20x^4y^3$;
   $f_{xy}(x,y) = f_{yx}(x, y) = 14xe^{7y} + 20x^3y^4$

# 9.3 Maxima and Minima

**APPLY IT** How many scientists does it take to develop a new antibiotic? What is the minimum cost?

*In this section we will learn how to answer questions such as this one, which is answered in Example 4.*

**FOR REVIEW**

It may be helpful to review Section 5.2 on relative extrema at this point. The concepts presented there are basic to what will be done in this section.

One of the most important applications of calculus is finding maxima and minima of functions. Earlier, we studied this idea extensively for functions of a single independent variable; now we will see that extrema can be found for functions of two variables. In particular, an extension of the second derivative test can be defined and used to identify maxima or minima. We begin with the definitions of relative maxima and minima.

### Relative Maxima and Minima

Let $(a, b)$ be the center of a circular region contained in the $xy$-plane. Then, for a function $z = f(x, y)$ defined for every $(x, y)$ in the region, $f(a, b)$ is a **relative (or local) maximum** if

$$f(a, b) \geq f(x, y)$$

for all points $(x, y)$ in the circular region, and $f(a, b)$ is a **relative (or local) minimum** if

$$f(a, b) \leq f(x, y)$$

for all points $(x, y)$ in the circular region.

As before, the word *extremum* is used for either a relative maximum or a relative minimum. Examples of a relative maximum and a relative minimum are given in Figures 17 and 18.

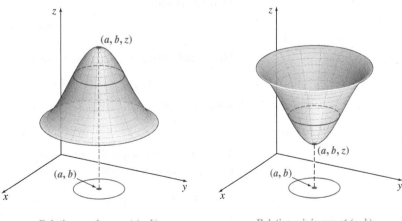

Relative maximum at $(a, b)$

FIGURE 17

Relative minimum at $(a, b)$

FIGURE 18

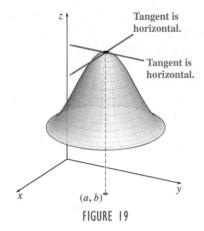

Tangent is horizontal.

Tangent is horizontal.

$x$ $(a, b)$ $y$

**FIGURE 19**

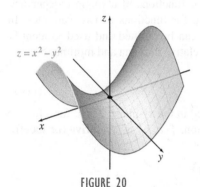

$z = x^2 - y^2$

**FIGURE 20**

**NOTE** When functions of a single variable were discussed, a distinction was made between relative extrema and absolute extrema. The methods for finding absolute extrema are quite involved for functions of two variables, so we will discuss only relative extrema here. In many practical applications the relative extrema coincide with the absolute extrema. In this brief discussion of extrema for multivariable functions, we omit cases where an extremum occurs on a boundary of the domain.

As suggested by Figure 19, at a relative maximum the tangent line parallel to the $xz$-plane has a slope of 0, as does the tangent line parallel to the $yz$-plane. (Notice the similarity to functions of one variable.) That is, if the function $z = f(x, y)$ has a relative extremum at $(a, b)$, then $f_x(a, b) = 0$ and $f_y(a, b) = 0$, as stated in the next theorem.

### Location of Extrema

Let a function $z = f(x, y)$ have a relative maximum or relative minimum at the point $(a, b)$. Let $f_x(a, b)$ and $f_y(a, b)$ both exist. Then

$$f_x(a, b) = 0 \quad \text{and} \quad f_y(a, b) = 0.$$

Just as with functions of one variable, the fact that the slopes of the tangent lines are 0 is no guarantee that a relative extremum has been located. For example, Figure 20 shows the graph of $z = f(x, y) = x^2 - y^2$. Both $f_x(0, 0) = 0$ and $f_y(0, 0) = 0$, and yet $(0, 0)$ leads to neither a relative maximum nor a relative minimum for the function. The point $(0, 0, 0)$ on the graph of this function is called a **saddle point**; it is a minimum when approached from one direction but a maximum when approached from another direction. A saddle point is neither a maximum nor a minimum.

The theorem on location of extrema suggests a useful strategy for finding extrema. First, locate all points $(a, b)$ where $f_x(a, b) = 0$ and $f_y(a, b) = 0$. Then test each of these points separately, using the test given after the next example. For a function $f(x, y)$, the points $(a, b)$ such that $f_x(a, b) = 0$ and $f_y(a, b) = 0$ are called **critical points**.

**NOTE** When we discussed functions of a single variable, we allowed critical points to include points from the domain where the derivative does not exist. For functions of more than one variable, to avoid complications, we will consider only cases in which the function is differentiable.

### EXAMPLE 1   Critical Points

Find all critical points for

$$f(x, y) = 6x^2 + 6y^2 + 6xy + 36x - 5.$$

**SOLUTION**   Find all points $(a, b)$ such that $f_x(a, b) = 0$ and $f_y(a, b) = 0$. Here

$$f_x(x, y) = 12x + 6y + 36 \quad \text{and} \quad f_y(x, y) = 12y + 6x.$$

Set each of these two partial derivatives equal to 0.

$$12x + 6y + 36 = 0 \quad \text{and} \quad 12y + 6x = 0$$

These two equations make up a system of linear equations. We can use the substitution method to solve this system. First, rewrite $12y + 6x = 0$ as follows:

$$12y + 6x = 0$$
$$6x = -12y$$
$$x = -2y.$$

Now substitute $-2y$ for $x$ in the other equation and solve for $y$.

$$12x + 6y + 36 = 0$$
$$12(-2y) + 6y + 36 = 0$$
$$-24y + 6y + 36 = 0$$
$$-18y + 36 = 0$$
$$-18y = -36$$
$$y = 2$$

**YOUR TURN 1** Find all critical points for $f(x,y) = 4x^3 + 3xy + 4y^3$.

From the equation $x = -2y$, $x = -2(2) = -4$. The solution of the system of equations is $(-4, 2)$. Since this is the only solution of the system, $(-4, 2)$ is the only critical point for the given function. By the theorem above, if the function has a relative extremum, it will occur at $(-4, 2)$. TRY YOUR TURN 1

The results of the next theorem can be used to decide whether $(-4, 2)$ in Example 1 leads to a relative maximum, a relative minimum, or neither.

---

### Test for Relative Extrema

For a function $z = f(x, y)$, let $f_{xx}$, $f_{yy}$, and $f_{xy}$ all exist in a circular region contained in the $xy$-plane with center $(a, b)$. Further, let

$$f_x(a, b) = 0 \quad \text{and} \quad f_y(a, b) = 0.$$

Define the number $D$, known as **the discriminant**, by

$$D = f_{xx}(a, b) \cdot f_{yy}(a, b) - [f_{xy}(a, b)]^2.$$

Then

**a.** $f(a, b)$ is a relative maximum if $D > 0$ and $f_{xx}(a, b) < 0$;

**b.** $f(a, b)$ is a relative minimum if $D > 0$ and $f_{xx}(a, b) > 0$;

**c.** $f(a, b)$ is a saddle point (neither a maximum nor a minimum) if $D < 0$;

**d.** if $D = 0$, the test gives no information.

---

This test is comparable to the second derivative test for extrema of functions of one independent variable. The following table summarizes the conclusions of the theorem.

|  | $f_{xx}(a, b) < 0$ | $f_{xx}(a, b) > 0$ |
|---|---|---|
| $D > 0$ | Relative maximum | Relative minimum |
| $D = 0$ | No information | |
| $D < 0$ | Saddle point | |

Notice that in parts a and b of the test for relative extrema, it is necessary to test only the second partial $f_{xx}(a, b)$ and not $f_{yy}(a, b)$. This is because if $D > 0$, $f_{xx}(a, b)$ and $f_{yy}(a, b)$ must have the same sign.

Part of the idea behind the test for relative extrema is that if $f_{xx}(a, b)$ and $f_{yy}(a, b)$ have opposite signs, then the surface is concave down in one direction and concave up in the other, so there is a saddle point. This also causes $D < 0$. If $f_{xx}(a, b)$ and $f_{yy}(a, b)$ are both positive and large enough (compared with $f_{xy}(a, b)$) so that $D > 0$, then the surface is concave up in all directions, and the critical point leads to a relative minimum. Similarly, if $f_{xx}(a, b)$ and $f_{yy}(a, b)$ are both negative and large enough (compared with $f_{xy}(a, b)$) so that $D > 0$, then the surface is concave down in all directions, and the critical point leads to a relative maximum.*

### EXAMPLE 2  Relative Extrema

The previous example showed that the only critical point for the function

$$f(x, y) = 6x^2 + 6y^2 + 6xy + 36x - 5$$

is $(-4, 2)$. Does $(-4, 2)$ lead to a relative maximum, a relative minimum, or neither?

**SOLUTION**  Find out by using the test above. From Example 1,

$$f_x(-4, 2) = 0 \quad \text{and} \quad f_y(-4, 2) = 0.$$

Now find the various second partial derivatives used in finding $D$. From $f_x(x, y) = 12x + 6y + 36$ and $f_y(x, y) = 12y + 6x$,

$$f_{xx}(x, y) = 12, \quad f_{yy}(x, y) = 12, \quad \text{and} \quad f_{xy}(x, y) = 6.$$

(If these second-order partial derivatives had not all been constants, they would have had to be evaluated at the point $(-4, 2)$.) Now

$$D = f_{xx}(-4, 2) \cdot f_{yy}(-4, 2) - [f_{xy}(-4, 2)]^2 = 12 \cdot 12 - 6^2 = 108.$$

Since $D > 0$ and $f_{xx}(-4, 2) = 12 > 0$, part b of the theorem applies, showing that $f(x, y) = 6x^2 + 6y^2 + 6xy + 36x - 5$ has a relative minimum at $(-4, 2)$. This relative minimum is $f(-4, 2) = -77$. A graph of this surface drawn by the computer program Maple™ is shown in Figure 21.

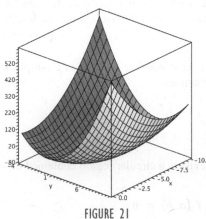

FIGURE 21

### EXAMPLE 3  Relative Extrema and Saddle Point

Find all points where the function

$$f(x, y) = 9xy - x^3 - y^3 - 6$$

has any relative maxima or relative minima.

**SOLUTION**  First find any critical points. Here

$$f_x(x, y) = 9y - 3x^2 \quad \text{and} \quad f_y(x, y) = 9x - 3y^2.$$

Set each of these partial derivatives equal to 0.

$$f_x(x, y) = 0 \qquad f_y(x, y) = 0$$
$$9y - 3x^2 = 0 \qquad 9x - 3y^2 = 0$$
$$9y = 3x^2 \qquad 9x = 3y^2$$
$$3y = x^2 \qquad 3x = y^2$$

The substitution method can be used again to solve the system of equations

$$3y = x^2$$
$$3x = y^2.$$

The first equation, $3y = x^2$, can be rewritten as $y = x^2/3$. Substitute this into the second equation to get

$$3x = y^2 = \left(\frac{x^2}{3}\right)^2 = \frac{x^4}{9}.$$

*For a rigorous proof of the test for relative extrema, see *Thomas' Calculus,* 13th ed., by Thomas, George B. Jr., Maurice D. Weir, and Joel R. Hass, Pearson, 2014.

Solve this equation as follows.

$$27x = x^4 \qquad \text{Multiply both sides by 9.}$$
$$x^4 - 27x = 0$$
$$x(x^3 - 27) = 0 \qquad \text{Factor.}$$
$$x = 0 \quad \text{or} \quad x^3 - 27 = 0 \qquad \text{Set each factor equal to 0.}$$
$$x = 0 \quad \text{or} \qquad x^3 = 27$$
$$x = 0 \quad \text{or} \qquad x = 3 \qquad \text{Take the cube root on both sides.}$$

Use these values of $x$, along with the equation $3y = x^2$, rewritten as $y = x^2/3$, to find $y$. If $x = 0$, $y = 0^2/3 = 0$. If $x = 3$, $y = 3^2/3 = 3$. The critical points are $(0, 0)$ and $(3, 3)$. To identify any extrema, use the test. Here

$$f_{xx}(x, y) = -6x, \qquad f_{yy}(x, y) = -6y, \qquad \text{and} \qquad f_{xy}(x, y) = 9.$$

Test each of the possible critical points.

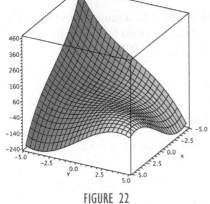

**FIGURE 22**

**YOUR TURN 2** Identify each of the critical points in YOUR TURN 1 as a relative maximum, relative minimum, or saddle point.

For $(0, 0)$:

$$f_{xx}(0, 0) = -6(0) = 0$$
$$f_{yy}(0, 0) = -6(0) = 0$$
$$f_{xy}(0, 0) = 9$$
$$D = 0 \cdot 0 - 9^2 = -81.$$

Since $D < 0$, there is a saddle point at $(0, 0)$.

For $(3, 3)$:

$$f_{xx}(3, 3) = -6(3) = -18$$
$$f_{yy}(3, 3) = -6(3) = -18$$
$$f_{xy}(3, 3) = 9$$
$$D = -18(-18) - 9^2 = 243.$$

Here $D > 0$ and $f_{xx}(3, 3) = -18 < 0$; there is a relative maximum at $(3, 3)$.

Notice that these values are in accordance with the graph generated by the computer program Maple™ shown in Figure 22. **TRY YOUR TURN 2**

---

**EXAMPLE 4** **Scientific Development**

A pharmaceutical company is developing a new antibiotic. The cost in thousands of dollars to develop an antibiotic is approximated by

$$C(x, y) = 2200 + 27x^3 - 72xy + 8y^2,$$

where $x$ is the number of employees working in quality assurance and $y$ is the number of scientists working in the laboratory. Find the number of employees in each area that results in the minimum cost to develop the antibiotic. What is the minimum cost?

**APPLY IT**

**SOLUTION**

**Method 1**
**Calculating by Hand**

Start with the following partial derivatives.

$$C_x(x, y) = 81x^2 - 72y \qquad \text{and} \qquad C_y(x, y) = -72x + 16y$$

Set each of these equal to 0 and solve for $y$.

$$81x^2 - 72y = 0 \qquad\qquad -72x + 16y = 0$$
$$-72y = -81x^2 \qquad\qquad 16y = 72x$$
$$y = \frac{9}{8}x^2 \qquad\qquad\qquad y = \frac{9}{2}x$$

Since $(9/8)x^2$ and $(9/2)x$ both equal $y$, they are equal to each other. Set them equal, and solve the resulting equation for $x$.

$$\frac{9}{8}x^2 = \frac{9}{2}x$$

$$9x^2 = 36x \qquad \text{Multiply both sides by 8.}$$

$$9x^2 - 36x = 0 \qquad \text{Subtract 36x from both sides.}$$

$$9x(x - 4) = 0 \qquad \text{Factor.}$$

$$9x = 0 \quad \text{or} \quad x - 4 = 0 \qquad \text{Set each factor equal to 0.}$$

The equation $9x = 0$ leads to $x = 0$ and $y = 0$, which cannot be a minimizer of $C(x, y)$, since, for example, $C(1, 1) < C(0, 0)$. This fact can also be verified by the test for relative extrema. Substitute $x = 4$, the solution of $x - 4 = 0$, into $y = (9/2)x$ to find $y$.

$$y = \frac{9}{2}x = \frac{9}{2}(4) = 18$$

Now check to see whether the critical point $(4, 18)$ leads to a relative minimum. Here

$$C_{xx}(x, y) = 162x, \quad C_{yy}(x, y) = 16, \quad \text{and} \quad C_{xy}(x, y) = -72.$$

For $(4, 18)$,

$$C_{xx}(4, 18) = 162(4) = 648, \quad C_{yy}(4, 18) = 16, \quad \text{and} \quad C_{xy}(4, 18) = -72,$$

so that

$$D = (648)(16) - (-72)^2 = 5184.$$

Since $D > 0$ and $C_{xx}(4, 18) > 0$, the cost at $(4, 18)$ is a minimum.

To find the minimum cost, go back to the cost function and evaluate $C(4, 18)$.

$$C(x, y) = 2200 + 27x^3 - 72xy + 8y^2$$

$$C(4, 18) = 2200 + 27(4)^3 - 72(4)(18) + 8(18)^2 = 1336$$

The minimum cost for developing a new antibiotic is $1,336,000 when four employees work in quality assurance and 18 scientists work in the laboratory. A graph of this surface drawn by the computer program Maple™ is shown in Figure 23.

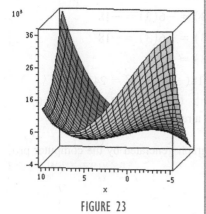

FIGURE 23

---

🖿 **Method 2**
**Spreadsheets**

Finding the maximum or minimum of a function of one or more variables can be done using a spreadsheet. The Solver included with Excel is located in the Tools menu and requires that cells be identified ahead of time for each variable in the problem. (On some versions of Excel, the Solver must be installed from an outside source. For details, see the *Graphing Calculator and Excel Spreadsheet Manual* available with this text.) It also requires that another cell be identified where the function, in terms of the variable cells, is placed. For example, to solve the above problem, we could identify cells A1 and B1 to represent the variables $x$ and $y$, respectively. The Solver requires that we place a guess for the answer in these cells. Thus, our initial value or guess will be to place the number 5 in each of these cells. An expression for the function must be placed in another cell, with $x$ and $y$ replaced by A1 and B1. If we choose cell A3 to represent the function, in cell A3 we would type "= 2200+27*A1^3-72*A1*B1+8*B1^2."

We now click on the Tools menu and choose Solver. This solver will attempt to find a solution that either maximizes or minimizes the value of cell A3. Figure 24 illustrates the Solver box and the items placed in it.

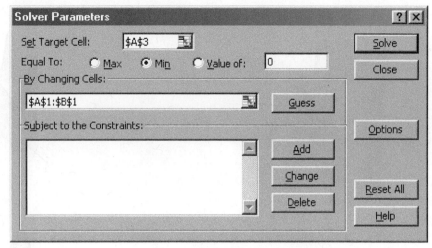

FIGURE 24

To obtain a solution, click on Solve. The rounded solution $x = 4$ and $y = 18$ is located in cells A1 and B1. The minimum cost $C(4, 18) = 1336$ is located in cell A3.

**CAUTION** One must be careful when using Solver because it will not find a maximizer or minimizer of a function if the initial guess is the exact place in which a saddle point occurs. For example, in the problem above, if our initial guess was $(0, 0)$, the Solver would have returned the value of $(0, 0)$ as the place where a minimum occurs. But $(0, 0)$ is a saddle point. Thus, it is always a good idea to run the Solver for two different initial values and compare the solutions.

# 9.3 EXERCISES

**Find all points where the functions have any relative extrema. Identify any saddle points.**

**1.** $f(x, y) = xy + y - 2x$

**2.** $f(x, y) = 3xy + 6y - 5x$

**3.** $f(x, y) = 3x^2 - 4xy + 2y^2 + 6x - 10$

**4.** $f(x, y) = x^2 + xy + y^2 - 6x - 3$

**5.** $f(x, y) = x^2 - xy + y^2 + 2x + 2y + 6$

**6.** $f(x, y) = 2x^2 + 3xy + 2y^2 - 5x + 5y$

**7.** $f(x, y) = x^2 + 3xy + 3y^2 - 6x + 3y$

**8.** $f(x, y) = 5xy - 7x^2 - y^2 + 3x - 6y - 4$

**9.** $f(x, y) = 4xy - 10x^2 - 4y^2 + 8x + 8y + 9$

**10.** $f(x, y) = 4y^2 + 2xy + 6x + 4y - 8$

**11.** $f(x, y) = x^2 + xy - 2x - 2y + 2$

**12.** $f(x, y) = x^2 + xy + y^2 - 3x - 5$

**13.** $f(x, y) = 3x^2 + 2y^3 - 18xy + 42$

**14.** $f(x, y) = 7x^3 + 3y^2 - 126xy - 63$

**15.** $f(x, y) = x^2 + 4y^3 - 6xy - 1$

**16.** $f(x, y) = 3x^2 + 7y^3 - 42xy + 5$

**17.** $f(x, y) = e^{x(y+1)}$     **18.** $f(x, y) = y^2 + 2e^x$

**19.** Describe the procedure for finding critical points of a function in two independent variables.

**20.** How are second-order partial derivatives used in finding extrema?

**Figures a–f on the next page show the graphs of the functions defined in Exercises 21–26. Find all relative extrema for each function, and then match the equation to its graph.**

**21.** $z = -3xy + x^3 - y^3 + \dfrac{1}{8}$

**22.** $z = \dfrac{3}{2}y - \dfrac{1}{2}y^3 - x^2y + \dfrac{1}{16}$

**23.** $z = y^4 - 2y^2 + x^2 - \dfrac{17}{16}$

**24.** $z = -2x^3 - 3y^4 + 6xy^2 + \dfrac{1}{16}$

**25.** $z = -x^4 + y^4 + 2x^2 - 2y^2 + \dfrac{1}{16}$

**26.** $z = -y^4 + 4xy - 2x^2 + \dfrac{1}{16}$

**a.**

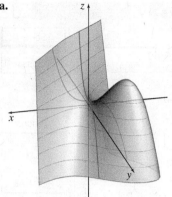

**b.**

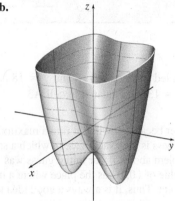

**c.**

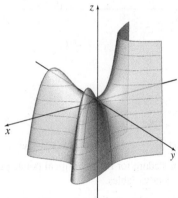

**d.**

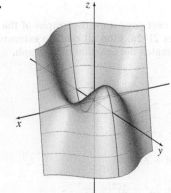

**e.**

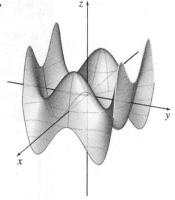

**f.**

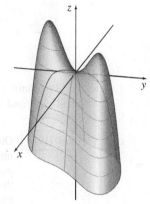

**27.** Show that $f(x, y) = 1 - x^4 - y^4$ has a relative maximum, even though $D$ in the theorem is 0.

**28.** Show that $D = 0$ for $f(x, y) = x^3 + (x - y)^2$ and that the function has no relative extrema.

**29.** A friend taking calculus is puzzled. She remembers that for a function of one variable, if the first derivative is zero at a point and the second derivative is positive, then there must be a relative minimum at the point. She doesn't understand why that isn't true for a function of two variables—that is, why $f_x(x, y) = 0$ and $f_{xx}(x, y) > 0$ doesn't guarantee a relative minimum. Provide an explanation.

**30.** Let $f(x, y) = y^2 - 2x^2y + 4x^3 + 20x^2$. The only critical points are $(-2, 4)$, $(0, 0)$, and $(5, 25)$. Which of the following correctly describes the behavior of $f$ at these points? ***Source: Society of Actuaries.***

**a.** $(-2, 4)$:  local (relative) minimum
$(0, 0)$:  local (relative) minimum
$(5, 25)$:  local (relative) maximum

**b.** $(-2, 4)$:  local (relative) minimum
$(0, 0)$:  local (relative) maximum
$(5, 25)$:  local (relative) maximum

**c.** $(-2, 4)$:  neither a local (relative) minimum nor a local (relative) maximum
$(0, 0)$:  local (relative) maximum
$(5, 25)$:  local (relative) minimum

**d.** $(-2, 4)$:  local (relative) maximum
$(0, 0)$:  neither a local (relative) minimum nor a local (relative) maximum
$(5, 25)$:  local (relative) minimum

**e.** $(-2, 4)$:   neither a local (relative) minimum nor a local (relative) maximum

$(0, 0)$:   local (relative) minimum

$(5, 25)$:   neither a local (relative) minimum nor a local (relative) maximum

**31.** Consider the function $f(x, y) = x^2(y + 1)^2 + k(x + 1)^2 y^2$.

   **a.** For what values of $k$ is the point $(x, y) = (0, 0)$ a critical point?

   **b.** For what values of $k$ is the point $(x, y) = (0, 0)$ a relative minimum of the function?

**32.** In Exercise 9 of Section 1.2, we found the least squares line through a set of $n$ points $(x_1, y_1), (x_2, y_2), \ldots, (x_n, y_n)$ by choosing the slope of the line $m$ and the $y$-intercept $b$ to minimize the quantity

$$S(m, b) = \sum (mx + b - y)^2,$$

where the summation symbol $\Sigma$ means that we sum over all the data points. Minimize $S$ by setting $S_m(m, b) = 0$ and $S_b(m, b) = 0$, and then rearrange the results to derive the equations from Section 1.2

$$\left(\sum x\right)b + \left(\sum x^2\right)m = \sum xy$$
$$nb + \left(\sum x\right)m = \sum y.$$

**33.** Suppose a function $z = f(x, y)$ satisfies the criteria for the test for relative extrema at a point $(a, b)$, and $f_{xx}(a, b) > 0$, while $f_{yy}(a, b) < 0$. What does this tell you about $f(a, b)$? Based on the sign of $f_{xx}(a, b)$ and $f_{yy}(a, b)$, why does this seem intuitively plausible?

## LIFE SCIENCE APPLICATIONS

**34. Satisfaction**   Suppose two animals get satisfaction from making a sound, which is pleasant to the individuals but only to a point. Let $S_1$ and $S_2$ be the satisfaction to the first and second individuals, respectively, and let $x_1$ and $x_2$ be the intensity of the sounds made by the two individuals, respectively. A model of the satisfaction of the two animals is given by

$$S_1 = a_1(x_1 + x_2) - b_1(x_1 + x_2)^2 \quad \text{and}$$
$$S_2 = a_2(x_1 + x_2) - b_2(x_1 + x_2)^2,$$

where $a_1$, $a_2$, $b_1$, and $b_2$ are constants. ***Source: Mathematical Biology of Social Behavior.***

   **a.** Show that if each animal tries to adjust its sound to maximize its own satisfaction, there is no solution unless $a_1/b_1 = a_2/b_2$.

   **b.** If the animals are altruistic, their only interest is the total satisfaction, namely,

$$S = S_1 + S_2.$$

   Show that there are an infinite number of ways to maximize the total satisfaction, namely, all values of $x_1$ and $x_2$ satisfying

$$x_1 + x_2 = \frac{a_1 + a_2}{2(b_1 + b_2)}.$$

## OTHER APPLICATIONS

**35. Political Science**   The probability that a three-person jury will make a correct decision is given by

$$P(\alpha, r, s) = \alpha[3r^2(1 - r) + r^3]$$
$$+ (1 - \alpha)[3s^2(1 - s) + s^3],$$

where $0 < \alpha < 1$ is the probability that the person is guilty of the crime, $r$ is the probability that a given jury member will vote "guilty" when the defendant is indeed guilty of the crime, and $s$ is the probability that a given jury member will vote "innocent" when the defendant is indeed innocent. ***Source: Frontiers of Economics.***

   **a.** Calculate $P(0.9, 0.5, 0.6)$ and $P(0.1, 0.8, 0.4)$ and interpret your answers.

   **b.** Using common sense and without using calculus, what value of $r$ and $s$ would maximize the jury's probability of making the correct verdict? Do these values depend on $\alpha$ in this problem? Should they? What is the maximum probability?

   **c.** Verify your answer for part b using calculus. (*Hint:* There are two critical points. Argue that the maximum value occurs at one of these points.)

**36. Computer Chips**   The table below, which illustrates the dramatic increase in the number of transistors in personal computers since 1985, was given in Section 2.1 on Exponential Functions, Exercise 49.

   **a.** To fit the data to a function of the form $y = ab^t$, where $t$ is the number of years since 1985 and $y$ is the number of transistors (in millions), we could take natural logarithms of both sides of the equation to get $\ln y = \ln a + t \ln b$. We could then let $w = \ln y$, $r = \ln a$, and $s = \ln b$ to form $w = r + st$. Using linear regression, find values for $r$ and $s$ that will fit the data. Then find the function $y = ab^t$. (*Hint:* Take the natural logarithm of the values in the transistors column and then use linear regression to find values of $r$ and $s$ that fit the data. Once you know $r$ and $s$, you can determine the values of $a$ and $b$ by calculating $a = e^r$ and $b = e^s$.)

| Year (since 1985) | Chip | Transistors (in millions) |
|---|---|---|
| 0 | 386 | 0.275 |
| 4 | 486 | 1.2 |
| 8 | Pentium | 3.1 |
| 12 | Pentium II | 7.5 |
| 14 | Pentium III | 9.5 |
| 15 | Pentium 4 | 42 |
| 20 | Pentium D | 291 |
| 22 | Penryn | 820 |
| 24 | Nehalem | 1900 |

**b.** Use the solver capability of a spreadsheet to find a function of the form $y = ab^t$ that fits the data above. (*Hint:* Using the ideas from part a, find values for $a$ and $b$ that minimize the function

$$
\begin{aligned}
f(a, b) = {}& [\ln(0.275) - 0 \ln b - \ln a]^2 \\
& + [\ln(1.2) - 4 \ln b - \ln a]^2 \\
& + [\ln(3.1) - 8 \ln b - \ln a]^2 \\
& + [\ln(7.5) - 12 \ln b - \ln a]^2 \\
& + [\ln(9.5) - 14 \ln b - \ln a]^2 \\
& + [\ln(42) - 15 \ln b - \ln a]^2 \\
& + [\ln(291) - 20 \ln b - \ln a]^2 \\
& + [\ln(820) - 22 \ln b - \ln a]^2 \\
& + [\ln(1900) - 24 \ln b - \ln a]^2.)
\end{aligned}
$$

**c.** Compare your answer to this problem with the one found with a graphing calculator in Section 2.1, Exercise 49.

**37. Food Frying** The process of frying food changes its quality, texture, and color. According to research done at the University of Saskatchewan, the total change in color $E$ (which is measured in the form of energy as kJ/mol) of blanched potato strips can be estimated by the function

$$
\begin{aligned}
E(t, T) = {}& 436.16 - 10.57t - 5.46T - 0.02t^2 \\
& + 0.02T^2 + 0.08Tt,
\end{aligned}
$$

where $T$ is the temperature (in °C) and $t$ is the frying time (in min). *Source: Critical Reviews in Food Science and Nutrition.*

**a.** What is the value of $E$ prior to cooking? (Assume that $T = 0$.)

**b.** Use this function to estimate the total change in color of a potato strip that has been cooked for 10 minutes at 180°C.

**c.** Determine the critical point of this function and determine if a maximum, minimum, or saddle point occurs at that point. Describe what may be happening at this point.

**38. Profit** Suppose that the profit (in hundreds of dollars) of a pharmaceutical company is approximated by

$$
P(x, y) = 1500 + 36x - 1.5x^2 + 120y - 2y^2,
$$

where $x$ is the cost of a unit of labor and $y$ is the cost of a unit of goods. Find values of $x$ and $y$ that maximize profit. Find the maximum profit.

**39. Labor Costs** Suppose the labor cost (in dollars) for manufacturing a medical device can be approximated by

$$
L(x, y) = \frac{3}{2}x^2 + y^2 - 2x - 2y - 2xy + 68,
$$

where $x$ is the number of hours required by a skilled craftsperson and $y$ is the number of hours required by a semiskilled person. Find values of $x$ and $y$ that minimize the labor cost. Find the minimum labor cost.

**40. Cost** The total cost (in dollars) to produce $x$ units of adhesive tape and $y$ units of gauze is given by

$$
C(x, y) = 2x^2 + 2y^2 - 3xy + 4x - 94y + 4200.
$$

Find the number of units of each product that should be produced so that the total cost is a minimum. Find the minimum total cost.

**41. Revenue** The total revenue (in hundreds of dollars) from the sale of $x$ X-ray machines and $y$ bone density scanners is approximated by

$$
R(x, y) = 15 + 169x + 182y - 5x^2 - 7y^2 - 7xy.
$$

Find the number of each that should be sold to produce maximum revenue. Find the maximum revenue.

**YOUR TURN ANSWERS**

**1.** $(0, 0)$ and $(-1/4, -1/4)$

**2.** Saddle point at $(0, 0)$; relative maximum at $(-1/4, -1/4)$

# 9.4 Total Differentials and Approximations

**APPLY IT** How do errors in measuring the length and radius of a blood vessel affect the calculation of its volume?

*In Example 3 in this section, we will see how to answer this question using a total differential.*

In the second section of this chapter we used partial derivatives to find the rate of change of a function with respect to one of the variables while the other variables were held constant. To estimate the change in productivity for a small change in both labor and capital, we can extend the concept of differential, introduced in an earlier chapter for functions of one variable, to the concept of *total differential*.

> ### Total Differential for Two Variables
>
> Let $z = f(x, y)$ be a function of $x$ and $y$. Let $dx$ and $dy$ be real numbers. Then the **total differential** of $z$ is
>
> $$dz = f_x(x, y) \cdot dx + f_y(x, y) \cdot dy.$$
>
> (Sometimes $dz$ is written $df$.)

Recall that the differential for a function of one variable $y = f(x)$ is used to approximate the function by its tangent line. This works because a differentiable function appears very much like a line when viewed closely. Similarly, the differential for a function of two variables $z = f(x, y)$ is used to approximate a function by its tangent plane. A differentiable function of two variables looks like a plane when viewed closely, which is why the earth looks flat when you are standing on it.

**FOR REVIEW**

In Chapter 6 on Applications of the Derivative, we introduced the differential. Recall that the differential of a function defined by $y = f(x)$ is

$$dy = f'(x) \cdot dx,$$

where $dx$, the differential of $x$, is any real number (usually small). We saw that the differential $dy$ is often a good approximation of $\Delta y$, where $\Delta y = f(x + \Delta x) - f(x)$ and $\Delta x = dx$.

**YOUR TURN 1**   For the function $f(x, y) = 3x^2y^4 + 6\sqrt{x^2 - 7y^2}$, find **(a)** $dz$, and **(b)** the value of $dz$ when $x = 4$, $y = 1$, $dx = 0.02$, and $dy = -0.03$.

**EXAMPLE 1**   **Total Differentials**

Consider the function $z = f(x, y) = 9x^3 - 8x^2y + 4y^3$.

**(a)** Find $dz$.

**SOLUTION**   First find $f_x(x, y)$ and $f_y(x, y)$.

$$f_x(x, y) = 27x^2 - 16xy \quad \text{and} \quad f_y(x, y) = -8x^2 + 12y^2$$

By the definition,

$$dz = (27x^2 - 16xy) \, dx + (-8x^2 + 12y^2) \, dy.$$

**(b)** Evaluate $dz$ when $x = 1$, $y = 3$, $dx = 0.01$, and $dy = -0.02$.

**SOLUTION**   Substituting these values into the result from part (a) gives

$$dz = [27(1)^2 - 16(1)(3)](0.01) + [-8(1)^2 + 12(3)^2](-0.02)$$

$$= (-21)(0.01) + (100)(-0.02)$$

$$= -2.21.$$

This result indicates that an increase of 0.01 in $x$ and a decrease of 0.02 in $y$, when $x = 1$ and $y = 3$, will produce an approximate *decrease* of 2.21 in $f(x, y)$.

**TRY YOUR TURN 1**

**Approximations**   Recall that with a function of one variable, $y = f(x)$, the differential $dy$ approximates the change in $y$, $\Delta y$, corresponding to a change in $x$, $\Delta x$ or $dx$. The approximation for a function of two variables is similar.

> ### Approximations
>
> For small values of $dx$ and $dy$,
>
> $$dz \approx \Delta z,$$
>
> where $\Delta z = f(x + dx, y + dy) - f(x, y)$.

EXAMPLE 2 Approximations

Approximate $\sqrt{2.98^2 + 4.01^2}$.

**SOLUTION** Notice that $2.98 \approx 3$ and $4.01 \approx 4$, and we know that $\sqrt{3^2 + 4^2} = \sqrt{25} = 5$. We, therefore, let $f(x, y) = \sqrt{x^2 + y^2}$, $x = 3$, $dx = -0.02$, $y = 4$, and $dy = 0.01$. We then use $dz$ to approximate $\Delta z = \sqrt{2.98^2 + 4.01^2} - \sqrt{3^2 + 4^2}$.

$$dz = f_x(x, y) \cdot dx + f_y(x, y) \cdot dy$$

$$= \left(\frac{1}{2\sqrt{x^2 + y^2}} \cdot 2x\right) dx + \left(\frac{1}{2\sqrt{x^2 + y^2}} \cdot 2y\right) dy$$

$$= \left(\frac{x}{\sqrt{x^2 + y^2}}\right) dx + \left(\frac{y}{\sqrt{x^2 + y^2}}\right) dy$$

$$= \frac{3}{5}(-0.02) + \frac{4}{5}(0.01)$$

$$= -0.004$$

**YOUR TURN 2** Approximate $\sqrt{5.03^2 + 11.99^2}$.

Thus, $\sqrt{2.98^2 + 4.01^2} \approx 5 + (-0.004) = 4.996$. A calculator gives $\sqrt{2.98^2 + 4.01^2} \approx 4.996048$. The error is approximately 0.000048. TRY YOUR TURN 2

For small values of $dx$ and $dy$, the values of $\Delta z$ and $dz$ are approximately equal. Since $\Delta z = f(x + dx, y + dy) - f(x, y)$,

$$f(x + dx, y + dy) = f(x, y) + \Delta z$$

or

$$f(x + dx, y + dy) \approx f(x, y) + dz.$$

Replacing $dz$ with the expression for the total differential gives the following result.

**Approximations by Differentials**

For a function $f$ having all indicated partial derivatives, and for small values of $dx$ and $dy$,

$$f(x + dx, y + dy) \approx f(x, y) + dz,$$

or

$$f(x + dx, y + dy) \approx f(x, y) + f_x(x, y) \cdot dx + f_y(x, y) \cdot dy.$$

The idea of a total differential can be extended to include functions of three or more independent variables.

**Total Differential for Three Variables**

If $w = f(x, y, z)$, then the total differential $dw$ is

$$dw = f_x(x, y, z)\, dx + f_y(x, y, z)\, dy + f_z(x, y, z)\, dz,$$

provided all indicated partial derivatives exist.

### EXAMPLE 3  Blood Vessels

A short length of blood vessel is in the shape of a right circular cylinder (see Figure 25).

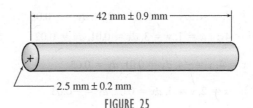

42 mm ± 0.9 mm

2.5 mm ± 0.2 mm

**FIGURE 25**

**APPLY IT**

(a) The length of the vessel is measured as 42 mm, and the radius is measured as 2.5 mm. Suppose the maximum error in the measurement of the length is 0.9 mm, with an error of no more than 0.2 mm in the measurement of the radius. Find the maximum possible error in calculating the volume of the blood vessel.

**SOLUTION**  The volume of a right circular cylinder is given by $V = \pi r^2 h$. To approximate the error in the volume, find the total differential, $dV$.

$$dV = (2\pi rh) \cdot dr + (\pi r^2) \cdot dh$$

Here, $r = 2.5$, $h = 42$, $dr = 0.2$, and $dh = 0.9$. Substitution gives

$$dV = [(2\pi)(2.5)(42)](0.2) + [\pi(2.5)^2](0.9) \approx 149.6.$$

The maximum possible error in calculating the volume is approximately 149.6 mm³.

(b) Suppose that the errors in measuring the radius and length of the vessel are at most 1% and 3%, respectively. Estimate the maximum percent error in calculating the volume.

**SOLUTION**  To find the percent error, calculate $dV/V$.

$$\frac{dV}{V} = \frac{(2\pi rh)dr + (\pi r^2)dh}{\pi r^2 h} = 2\frac{dr}{r} + \frac{dh}{h}$$

Because $dr/r = 0.01$ and $dh/h = 0.03$,

$$\frac{dV}{V} = 2(0.01) + 0.03 = 0.05.$$

The maximum percent error in calculating the volume is approximately 5%.

**TRY YOUR TURN 3**

**YOUR TURN 3**  In Example 3, estimate the maximum percent error in calculating the volume if the errors in measuring the radius and length of the vessel are at most 4% and 2%, respectively.

### EXAMPLE 4  Volume of a Can of Beer

The formula for the volume of a cylinder given in Example 3 also applies to cans of beer, for which $r \approx 1.5$ in. and $h \approx 5$ in. How sensitive is the volume to changes in the radius compared with changes in the height?

**SOLUTION**  Using the formula for $dV$ from the previous example with $r = 1.5$ and $h = 5$ gives

$$dV = (2\pi)(1.5)(5)dr + \pi(1.5)^2 dh = \pi(15dr + 2.25dh).$$

The factor of 15 in front of $dr$ in this equation, compared with the factor of 2.25 in front of $dh$, shows that a small change in the radius has almost 7 times the effect on the volume as a small change in the height. One author argues that this is the reason that beer cans are so tall and thin. ***Source: The College Mathematics Journal.*** The brewers can reduce the radius by a tiny amount and compensate by making the can taller. The resulting can appears larger in volume than the shorter, wider can. (Others have argued that a shorter, wider can does not fit as easily in the hand.)

# 9.4 EXERCISES

**Evaluate $dz$ using the given information.**

1. $z = 2x^2 + 4xy + y^2$; $x = 5, y = -1, dx = 0.03, dy = -0.02$

2. $z = 5x^3 + 2xy^2 - 4y$; $x = 1, y = 3, dx = 0.01, dy = 0.02$

3. $z = \dfrac{y^2 + 3x}{y^2 - x}$; $x = 4, y = -4, dx = 0.01, dy = 0.03$

4. $z = \ln(x^2 + y^2)$; $x = 2, y = 3, dx = 0.02, dy = -0.03$

**Evaluate $dw$ using the given information.**

5. $w = \dfrac{5x^2 + y^2}{z + 1}$; $x = -2, y = 1, z = 1, dx = 0.02,$
$dy = -0.03, dz = 0.02$

6. $w = x \ln(yz) - y \ln\dfrac{x}{z}$; $x = 2, y = 1, z = 4,$
$dx = 0.03, dy = 0.02, dz = -0.01$

**Use the total differential to approximate each quantity. Then use a calculator to approximate the quantity, and give the absolute value of the difference in the two results to 4 decimal places.**

7. $\sqrt{8.05^2 + 5.97^2}$

8. $\sqrt{4.96^2 + 12.06^2}$

9. $(1.92^2 + 2.1^2)^{1/3}$

10. $(2.93^2 - 0.94^2)^{1/3}$

11. $1.03e^{0.04}$

12. $0.98e^{-0.04}$

13. $0.99 \ln 0.98$

14. $2.03 \ln 1.02$

15. $e^{0.02} \cos 0.03$

16. $\ln 1.02 \sec 0.03$

## LIFE SCIENCE APPLICATIONS

17. **Bone Preservative Volume** A piece of bone in the shape of a right circular cylinder is 7 cm long and has a radius of 1.4 cm. It is coated with a layer of preservative 0.09 cm thick. Estimate the volume of preservative used.

18. **Blood Vessel Volume** A portion of a blood vessel is measured as having length 7.9 cm and radius 0.8 cm. If each measurement could be off by as much as 0.15 cm, estimate the maximum possible error in calculating the volume of the vessel.

19. **Blood Volume** In Exercise 52 of Section 2 in this chapter, we found that the number of liters of blood pumped through the lungs in one minute is given by

$$C = \frac{b}{a - v}.$$

Suppose $a = 160, b = 200,$ and $v = 125$. Estimate the change in $C$ if $a$ becomes 145, $b$ becomes 190, and $v$ changes to 130.

20. **Heat Loss** In Exercise 50 of Section 2 of this chapter, we found that the rate of heat loss (in watts) in harbor seal pups could be approximated by

$$H(m, T, A) = \frac{15.2m^{0.67}(T - A)}{10.23 \ln m - 10.74},$$

where $m$ is the body mass of the pup (in kg), and $T$ and $A$ are the body core temperature and ambient water temperature, respectively (in °C). Suppose $m$ is 25 kg, $T$ is 36.0°, and $A$ is 12.0°. Approximate the change in $H$ if $m$ changes to 26 kg, $T$ to 36.5°, and $A$ to 10.0°.

21. **Dialysis** A model that estimates the concentration of urea in the body for a particular dialysis patient, following a dialysis session, is given by

$$C(t, g) = 0.6(0.96)^{(210t/1500)-1}$$
$$+ \frac{gt}{126t - 900}[1 - (0.96)^{(210t/1500)-1}],$$

where $t$ represents the number of minutes of the dialysis session and $g$ represents the rate at which the body generates urea in mg per minute. *Source: Clinical Dialysis.*

a. Find $C(180, 8)$.

b. Using the total differential, estimate the urea concentration if the dialysis session of part a was cut short by 10 minutes and the urea generation rate was 9 mg per minute. Compare this with the actual concentration. (*Hint:* First, replace the variable $g$ with the number 8, thus reducing the function to one variable. Then use your graphing calculator to calculate the partial derivative $C_t(180, 8)$. A similar procedure can be done for $C_g(180, 8)$.)

22. **Horn Volume** The volume of the horns from bighorn sheep was estimated by researchers using the equation

$$V = \frac{h\pi}{3}(r_1^2 + r_1r_2 + r_2^2),$$

where $h$ is the length of a horn segment (in centimeters) and $r_1$ and $r_2$ are the radii of the two ends of the horn segment (in centimeters). *Source: Conservation Biology.*

a. Determine the volume of a segment of horn that is 40 cm long with radii of 5 cm and 3 cm, respectively.

b. Use the total differential to estimate the volume of the segment of horn if the horn segment from part a was actually 42 cm long with radii of 5.1 cm and 2.9 cm, respectively. Compare this with the actual volume.

23. **Eastern Hemlock** Ring shake, which is the separation of the wood between growth rings, is a serious problem in hemlock trees. Researchers have developed the following function that estimates the probability $P$ that a given hemlock tree has ring shake:

$$P(A, B, D) = \frac{1}{1 + e^{3.68 - 0.016A - 0.77B - 0.12D}},$$

where $A$ is the age of the tree (yr), $B$ is 1 if bird pecking is present and 0 otherwise, and $D$ is the diameter (in.) of the tree at breast height. *Source: Forest Products Journal.*

a. Estimate the probability that a 150-year-old tree, with bird pecking present and a breast height diameter of 20 in., will have ring shake.

b. Estimate the probability that a 150-year-old tree, with no presence of bird pecking and a breast height diameter of 20 in., will have ring shake.

c. Develop a statement about what can be said about the influence that the three variables have on the probability of ring shake.

**d.** Using the total differential, estimate the probability if the actual age of the tree was 160 years and the diameter at breast height was 25 in. Assume that no bird pecking was present. Compare your answer to the actual value. (*Hint:* Assume that $B = 0$ and exclude that variable from your calculations.)

**e.** Comment on the practicality of using differentials in part d.

**24. Life Span** As we saw in Exercise 60 of Section 9.2, researchers have estimated the maximum life span (in years) for various species of mammals according to the formula

$$L(E, P) = 23E^{0.6}P^{-0.267},$$

where $E$ is the average brain mass and $P$ is the average body mass (both in g). *Source: The Quarterly Review of Biology.* Consider humans, with $E = 14,100$ g and $P = 68,700$ g. Find the approximate change in life span if the brain mass increases to 14,300 g and the body mass decreases to 68,400 g. Then calculate the actual change in life span and compare.

**25. Hypoxia** The calculation of the differential can be extended to as many variables as necessary. Researchers studying the effect of breathing low oxygen mixtures (hypoxia) in the right lung of humans, and its effect on blood flow distribution to both lungs, used a differential with four variables to calculate the percentage $P$ of a particular error in their study using the formula

$$P = \frac{100a/b}{100a/b + 100c/g}100,$$

where $a$ and $c$ represent the $CO_2$ output of the right and left lungs, respectively, while $b$ and $g$ represent the arteriovenous $CO_2$ content differences in the right and left lungs. *Source: Journal of Applied Physiology.*

**a.** Show that the above expression can be simplified to

$$P = \frac{100ag}{ag + bc}.$$

**b.** The researchers found that $a = 130$ and $c = 167$. They found that $b$ could be as small as 3.6, and $g$ as small as 2.0. Find the corresponding value of $P$.

**c.** The researchers also found that $da/a = 0.07$, $db/b = 0.09$, and $dc/c = dg/g = 0.02$, where $da$, $db$, $dc$, and $dg$ were errors in the values of $a$, $b$, $c$, and $g$. Calculate $dP/P$, the fractional error in $P$, where $dP$ is the differential in $P$, using these values and the numerical values from part b.

**d.** The researchers used a different approximation for $dP$:

$$dP = \sqrt{(P_a\, da)^2 + (P_b\, db)^2 + (P_c\, dc)^2 + (P_g\, dg)^2}.$$

Calculate $dP/P$ using this expression and the values in parts b and c. Compare these with your answers from part c.

## OTHER APPLICATIONS

**26. Swimming** The amount of time in seconds it takes for a swimmer to hear a single, hand-held, starting signal is given by the formula

$$t(x, y, p, C) = \frac{\sqrt{x^2 + (y - p)^2}}{331.45 + 0.6C},$$

where $(x, y)$ is the location of the starter (in meters), $(0, p)$ is the location of the swimmer (in meters), and $C$ is the air temperature (in degrees Celsius). *Source: COMAP.* Assume that the starter is located at the point $(x, y) = (5, -2)$. See the diagram.

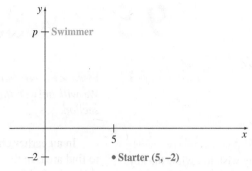

**a.** Calculate $t(5, -2, 20, 20)$ and $t(5, -2, 10, 20)$. Could the difference in time change the outcome of a race?

**b.** Calculate the total differential for $t$ if the starter remains stationary, the swimmer moves from 20 m to 20.5 m away from the starter in the $y$ direction, and the temperature decreases from 20°C to 15°C. Interpret your answer.

**27. Estimating Area** The height of a triangle is measured as 37.5 cm, with the base measured as 15.8 cm. The measurement of the height can be off by as much as 0.8 cm and that of the base by no more than 1.1 cm. Estimate the maximum possible error in calculating the area of the triangle.

**28. Estimating Volume** The height of a cone is measured as 9.3 cm and the radius as 3.2 cm. Each measurement could be off by as much as 0.1 cm. Estimate the maximum possible error in calculating the volume of the cone.

**29. Estimating Volume** Suppose that in measuring the length, width, and height of a box, there is a maximum 1% error in each measurement. Estimate the maximum error in calculating the volume of the box.

**30. Estimating Volume** Suppose there is a maximum error of $a\%$ in measuring the radius of a cone and a maximum error of $b\%$ in measuring the height. Estimate the maximum percent error in calculating the volume of the cone, and compare this value with the maximum percent error in calculating the volume of a cylinder.

**31. Ice Cream Cone** An ice cream cone has a radius of approximately 1 in. and a height of approximately 4 in. By what factor does a change in the radius affect the volume compared with a change in the height?

**32. Hose** A hose has a radius of approximately 0.5 in. and a length of approximately 20 ft. By what factor does a change in the radius affect the volume compared with a change in the length?

**33. Manufacturing** Approximate the volume of aluminum needed for a diet drink can of radius 2.5 cm and height 14 cm. Assume the walls of the can are 0.08 cm thick.

**34. Manufacturing** Approximate the volume of material needed to make a water tumbler of diameter 3 cm and height 9 cm. Assume the walls of the tumbler are 0.2 cm thick.

**35. Volume of a Coating** An industrial coating 0.1 in. thick is applied to all sides of a box of dimensions 10 in. by 9 in. by 18 in. Estimate the volume of the coating used.

### YOUR TURN ANSWERS

**1. (a)** $dz = (6xy^4 + 6x/\sqrt{x^2 - 7y^2})\, dx + (12x^2y^3 - 42y/\sqrt{x^2 - 7y^2})\, dy$  **(b)** $-4.7$

**2.** 13.0023  **3.** 10%

# 9.5 Double Integrals

**How can we find the volume of a bottle with curved sides?**
*We will answer this question in Example 6 using a double integral, the key idea in this section.*

---

**FOR REVIEW**

You may wish to review the key ideas of indefinite and definite integrals from Chapter 7 on Integration before continuing with this section. See the review problems at the end of that chapter.

---

In an earlier chapter, we saw how integrals of functions with one variable may be used to find area. In this section, this idea is extended and used to find volume. We found partial derivatives of functions of two or more variables at the beginning of this chapter by holding constant all variables except one. A similar process is used in this section to find antiderivatives of functions of two or more variables. For example, in

$$\int (5x^3y^4 - 6x^2y + 2)\,dy$$

the notation $dy$ indicates integration with respect to $y$, so we treat $y$ as the variable and $x$ as a constant. Using the rules for antiderivatives gives

$$\int (5x^3y^4 - 6x^2y + 2)\,dy = x^3y^5 - 3x^2y^2 + 2y + C(x).$$

The constant $C$ used earlier must be replaced with $C(x)$ to show that the "constant of integration" here can be any function involving only the variable $x$. Just as before, check this work by taking the derivative (actually the partial derivative) of the answer:

$$\frac{\partial}{\partial y}[x^3y^5 - 3x^2y^2 + 2y + C(x)] = 5x^3y^4 - 6x^2y + 2 + 0,$$

which shows that the antiderivative is correct.

We can use this antiderivative to evaluate a definite integral.

---

**EXAMPLE 1** Definite Integral

Evaluate $\displaystyle\int_1^2 (5x^3y^4 - 6x^2y + 2)\,dy$.

**SOLUTION**

$$\int_1^2 (5x^3y^4 - 6x^2y + 2)\,dy = [x^3y^5 - 3x^2y^2 + 2y + C(x)]\Big|_1^2 \qquad \text{Use the indefinite integral previously found.}$$

$$= x^3 2^5 - 3x^2 2^2 + 2 \cdot 2 + C(x)$$
$$\quad - [x^3 1^5 - 3x^2 1^2 + 2 \cdot 1 + C(x)]$$
$$= 32x^3 - 12x^2 + 4 + C(x)$$
$$\quad - [x^3 - 3x^2 + 2 + C(x)]$$
$$= 31x^3 - 9x^2 + 2 \qquad \text{Simplify.}$$

In the second step, we substituted $y = 2$ and $y = 1$ and subtracted, according to the Fundamental Theorem of Calculus. Notice that $C(x)$ does not appear in the final answer, just as the constant does not appear in a regular definite integral. Therefore, from now on we will not include $C(x)$ when we find the antiderivative for a definite integral with respect to $y$.

**YOUR TURN 1** Evaluate $\displaystyle\int_1^3 (6x^2y^2 + 4xy + 8x^3 + 10y^4 + 3)\,dy$.

**TRY YOUR TURN 1**

By integrating the result from Example 1 with respect to $x$, we can evaluate a double integral.

### EXAMPLE 2   Definite Integral

Evaluate $\displaystyle\int_0^3 \left[ \int_1^2 (5x^3y^4 - 6x^2y + 2)\, dy \right] dx$.

**SOLUTION**

$$\int_0^3 \left[ \int_1^2 (5x^3y^4 - 6x^2y + 2)\, dy \right] dx = \int_0^3 \left[ 31x^3 - 9x^2 + 2 \right] dx \qquad \text{Use the result from Example 1.}$$

$$= \frac{31}{4}x^4 - 3x^3 + 2x \Big|_0^3 \qquad \text{Use the Fundamental Theorem of Calculus.}$$

$$= \frac{31}{4} \cdot 3^4 - 3 \cdot 3^3 + 2 \cdot 3 - \left( \frac{31}{4} \cdot 0^4 - 3 \cdot 0^3 + 2 \cdot 0 \right)$$

$$= \frac{2211}{4}$$

We can integrate the inner integral with respect to $y$ and the outer integral with respect to $x$, as in Example 2, or in the reverse order. The next example shows the same integral done both ways.

### EXAMPLE 3   Definite Integrals

Evaluate each integral.

**(a)** $\displaystyle\int_1^2 \left[ \int_3^5 (6xy^2 + 12x^2y + 4y)\, dx \right] dy$

**SOLUTION**

$$\int_1^2 \left[ \int_3^5 (6xy^2 + 12x^2y + 4y)\, dx \right] dy = \int_1^2 \left[ (3x^2y^2 + 4x^3y + 4xy) \Big|_3^5 \right] dy \qquad \text{Integrate with respect to } x.$$

$$= \int_1^2 \left[ (3 \cdot 5^2 \cdot y^2 + 4 \cdot 5^3 \cdot y + 4 \cdot 5 \cdot y) \right.$$
$$\left. - (3 \cdot 3^2 \cdot y^2 + 4 \cdot 3^3 \cdot y + 4 \cdot 3 \cdot y) \right] dy$$

$$= \int_1^2 \left[ (75y^2 + 500y + 20y) \right.$$
$$\left. - (27y^2 + 108y + 12y) \right] dy$$

$$= \int_1^2 (48y^2 + 400y)\, dy$$

$$= (16y^3 + 200y^2) \Big|_1^2 \qquad \text{Integrate with respect to } y.$$

$$= 16 \cdot 2^3 + 200 \cdot 2^2 - (16 \cdot 1^3 + 200 \cdot 1^2)$$

$$= 128 + 800 - (16 + 200)$$

$$= 712$$

**(b)** $\int_3^5 \left[ \int_1^2 (6xy^2 + 12x^2y + 4y)\, dy \right] dx$

**SOLUTION** (This is the same integrand with the same limits of integration as in part (a), but the order of integration is reversed.)

$$\int_3^5 \left[ \int_1^2 (6xy^2 + 12x^2y + 4y)dy \right] dx = \int_3^5 \left[ (2xy^3 + 6x^2y^2 + 2y^2) \Big|_1^2 \right] dx \qquad \text{Integrate with respect to } y.$$

$$= \int_3^5 [(2x \cdot 2^3 + 6x^2 \cdot 2^2 + 2 \cdot 2^2)$$

$$- (2x \cdot 1^3 + 6x^2 \cdot 1^2 + 2 \cdot 1^2)]\, dx$$

$$= \int_3^5 [(16x + 24x^2 + 8)$$

$$- (2x + 6x^2 + 2)]\, dx$$

$$= \int_3^5 (14x + 18x^2 + 6)\, dx \qquad \text{Integrate with respect to } x.$$

$$= (7x^2 + 6x^3 + 6x) \Big|_3^5$$

$$= 7 \cdot 5^2 + 6 \cdot 5^3 + 6 \cdot 5 - (7 \cdot 3^2 + 6 \cdot 3^3 + 6 \cdot 3)$$

$$= 175 + 750 + 30 - (63 + 162 + 18) = 712$$

**TRY YOUR TURN 2**

**YOUR TURN 2** Evaluate $\int_0^2 \left[ \int_1^3 (6x^2y^2 + 4xy + 8x^3 + 10y^4 + 3)\, dy \right] dx$, and then integrate with the order of integration changed.

**NOTE** In the second step of Example 3 (a), it might help you avoid confusion as to whether to put the limits of 3 and 5 into $x$ or $y$ by writing the integral as

$$\int_1^2 \left[ (3x^2y^2 + 4x^3y + 4xy) \Big|_{x=3}^{x=5} \right] dy.$$

The brackets we have used for the inner integral in Example 3 are not essential because the order of integration is indicated by the order of $dx\, dy$ or $dy\, dx$. For example, if the integral is written as

$$\int_1^2 \int_3^5 (6xy^2 + 12x^2y + 4y)\, dx\, dy,$$

we first integrate with respect to $x$, letting $x$ vary from 3 to 5, and then with respect to $y$, letting $y$ vary from 1 to 2, as in Example 3(a).

The answers in the two parts of Example 3 are equal. It can be proved that for a large class of functions, including most functions that occur in applications, the following equation holds true.

**Fubini's Theorem**

$$\int_a^b \int_c^d f(x, y)\, dy\, dx = \int_c^d \int_a^b f(x, y)\, dx\, dy$$

Either of these integrals is called an **iterated integral**, since it is evaluated by integrating twice, first using one variable and then using the other. The fact that the iterated integrals above are equal makes it possible to define a *double integral*. First, the set of points $(x, y)$,

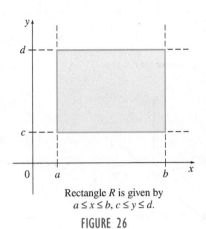

Rectangle $R$ is given by
$a \le x \le b, c \le y \le d.$

**FIGURE 26**

with $a \le x \le b$ and $c \le y \le d$, defines a rectangular region $R$ in the plane, as shown in Figure 26. Then, the *double integral over R* is defined as follows.

### Double Integral

The **double integral** of $f(x, y)$ over a rectangular region $R$ is written

$$\iint_R f(x, y) \, dy \, dx \qquad \text{or} \qquad \iint_R f(x, y) \, dx \, dy,$$

and equals either

$$\int_a^b \int_c^d f(x, y) \, dy \, dx \qquad \text{or} \qquad \int_c^d \int_a^b f(x, y) \, dx \, dy.$$

Extending earlier definitions, $f(x, y)$ is the **integrand** and $R$ is the **region of integration**.

### EXAMPLE 4  Double Integrals

Find $\displaystyle\iint_R \frac{3\sqrt{x} \cdot y}{y^2 + 1} \, dx \, dy$ over the rectangular region $R$ defined by $0 \le x \le 4$ and $0 \le y \le 2$.

**SOLUTION**  Integrate first with respect to $x$; then integrate the result with respect to $y$.

$$\iint_R \frac{3\sqrt{x} \cdot y}{y^2 + 1} \, dx \, dy = \int_0^2 \int_0^4 \frac{3\sqrt{x} \cdot y}{y^2 + 1} \, dx \, dy$$

$$= \int_0^2 \frac{2x^{3/2} \cdot y}{y^2 + 1} \Bigg|_0^4 \, dy \qquad \text{Use the power rule with } x^{1/2}.$$

$$= \int_0^2 \left( \frac{2(4)^{3/2} \cdot y}{y^2 + 1} - \frac{2(0)^{3/2} \cdot y}{y^2 + 1} \right) dy$$

$$= 8 \int_0^2 \frac{2y}{y^2 + 1} \, dy \qquad \text{Factor out } 4^{3/2} = 8.$$

$$= 8 \int_1^5 \frac{du}{u} \qquad \begin{array}{l}\text{Let } u = y^2 + 1. \text{ Change limits}\\ \text{of integration.}\end{array}$$

$$= 8 \ln u \Big|_1^5$$

$$= 8 \ln 5 - 8 \ln 1 = 8 \ln 5$$

As a check, integrate with respect to $y$ first. The answer should be the same.

**TRY YOUR TURN 3**

**YOUR TURN 3**  Find

$\displaystyle\iint_R \frac{1}{\sqrt{x + y + 3}} \, dx \, dy$ over the

rectangular region $R$ defined by $0 \le x \le 5$ and $1 \le y \le 6$.

**Volume**  As shown earlier, the definite integral $\int_a^b f(x) \, dx$ can be used to find the area under a curve. In a similar manner, double integrals are used to find the *volume under a surface*. Figure 27 on the next page shows that portion of a surface $f(x, y)$ directly over a rectangle $R$ in the $xy$-plane. Just as areas were approximated by a large number of small rectangles, volume could be approximated by adding the volumes of a large number of properly drawn small boxes. The height of a typical box would be $f(x, y)$ with the length and width given by $dx$ and $dy$. The formula for the volume of a box would then suggest the following result.

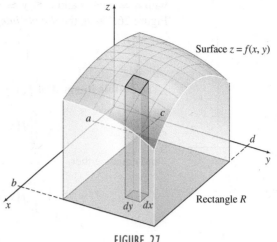

**FIGURE 27**

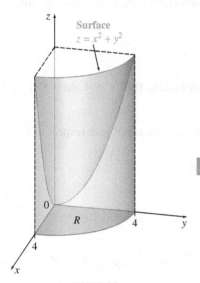

**FIGURE 28**

**YOUR TURN 4** Find the volume under the surface $z = 4 - x^3 - y^3$ over the rectangular region $0 \le x \le 1, 0 \le y \le 1$.

**Volume**

Let $z = f(x, y)$ be a function that is never negative on the rectangular region $R$ defined by $a \le x \le b, c \le y \le d$. The volume of the solid under the graph of $f$ and over the region $R$ is

$$\iint\limits_{R} f(x, y) \, dx \, dy.$$

### EXAMPLE 5   Volume

Find the volume under the surface $z = x^2 + y^2$ shown in Figure 28.

**SOLUTION**   By the equation just given, the volume is

$$\iint\limits_{R} f(x, y) \, dx \, dy,$$

where $f(x, y) = x^2 + y^2$ and $R$ is the region $0 \le x \le 4, 0 \le y \le 4$. By definition,

$$\iint\limits_{R} f(x, y) \, dx \, dy = \int_0^4 \int_0^4 (x^2 + y^2) \, dx \, dy$$

$$= \int_0^4 \left( \frac{1}{3}x^3 + xy^2 \right)\Bigg|_0^4 dy$$

$$= \int_0^4 \left( \frac{64}{3} + 4y^2 \right) dy = \left( \frac{64}{3}y + \frac{4}{3}y^3 \right)\Bigg|_0^4$$

$$= \frac{64}{3} \cdot 4 + \frac{4}{3} \cdot 4^3 - 0 = \frac{512}{3}.$$

**TRY YOUR TURN 4**

### EXAMPLE 6   Perfume Bottle

A product design consultant for a cosmetics company has been asked to design a bottle for the company's newest perfume. The thickness of the glass is to vary so that the outside of the bottle has straight sides and the inside has curved sides, with flat ends shaped like parabolas on the 4-cm sides, as shown in Figure 29. Before presenting the design to management, the consultant needs to make a reasonably accurate estimate of the amount each bottle will hold.

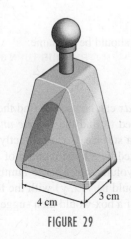

4 cm        3 cm

**FIGURE 29**

If the base of the bottle is to be 4 cm by 3 cm, and if a cross section of its interior is to be a parabola of the form $z = -y^2 + 4y$, what is its internal volume?

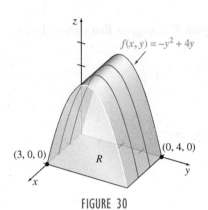

**SOLUTION** The interior of the bottle can be graphed in three-dimensional space, as shown in Figure 30, where $z = 0$ corresponds to the base of the bottle. Its volume is simply the volume above the region $R$ in the $xy$-plane and below the graph of $f(x, y) = -y^2 + 4y$. This volume is given by the double integral

$$\int_0^3 \int_0^4 (-y^2 + 4y)\, dy\, dx = \int_0^3 \left( \frac{-y^3}{3} + \frac{4y^2}{2} \right)\Bigg|_0^4 dx$$

$$= \int_0^3 \left( \frac{-64}{3} + 32 - 0 \right) dx$$

$$= \frac{32}{3} x \Bigg|_0^3$$

$$= 32 - 0 = 32.$$

$f(x,y) = -y^2 + 4y$

$(3, 0, 0)$   $R$   $(0, 4, 0)$

**FIGURE 30**

The bottle holds 32 cm³.

## Double Integrals over Other Regions

In this section, we found double integrals over rectangular regions by evaluating iterated integrals with constant limits of integration. We can also evaluate iterated integrals with *variable* limits of integration. (Notice in the following examples that the variable limits always go on the *inner* integral sign.)

The use of variable limits of integration permits evaluation of double integrals over the types of regions shown in Figure 31. Double integrals over more complicated regions are discussed in more advanced books. Integration over regions such as those in Figure 31 is done with the results of the following theorem.

### Double Integrals over Variable Regions

Let $z = f(x, y)$ be a function of two variables. If $R$ is the region (in Figure 31(a)) defined by $a \le x \le b$ and $g(x) \le y \le h(x)$, then

$$\iint\limits_R f(x, y)\, dy\, dx = \int_a^b \left[ \int_{g(x)}^{h(x)} f(x, y)\, dy \right] dx.$$

If $R$ is the region (in Figure 31(b)) defined by $g(y) \le x \le h(y)$ and $c \le y \le d$, then

$$\iint\limits_R f(x, y)\, dx\, dy = \int_c^d \left[ \int_{g(y)}^{h(y)} f(x, y)\, dx \right] dy.$$

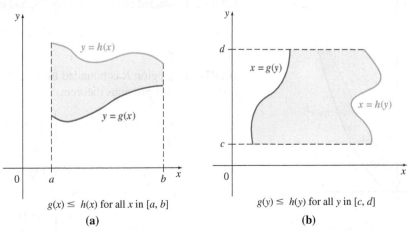

$g(x) \le h(x)$ for all $x$ in $[a, b]$

**(a)**

$g(y) \le h(y)$ for all $y$ in $[c, d]$

**(b)**

FIGURE 31

**EXAMPLE 7** **Double Integrals**

Evaluate $\int_1^2 \int_y^{y^2} xy \, dx \, dy$.

**SOLUTION** The region of integration is shown in Figure 32. Integrate first with respect to $x$, then with respect to $y$.

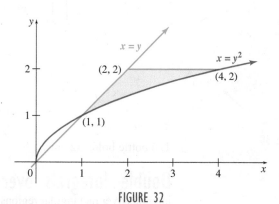

**FIGURE 32**

$$\int_1^2 \int_y^{y^2} xy \, dx \, dy = \int_1^2 \left( \int_y^{y^2} xy \, dx \right) dy = \int_1^2 \left( \frac{1}{2} x^2 y \right) \Big|_y^{y^2} dy$$

Replace $x$ first with $y^2$ and then with $y$, and subtract.

$$\int_1^2 \int_y^{y^2} xy \, dx \, dy = \int_1^2 \left[ \frac{1}{2} (y^2)^2 y - \frac{1}{2} (y)^2 y \right] dy$$

$$= \int_1^2 \left( \frac{1}{2} y^5 - \frac{1}{2} y^3 \right) dy = \left( \frac{1}{12} y^6 - \frac{1}{8} y^4 \right) \Big|_1^2$$

$$= \left( \frac{1}{12} \cdot 2^6 - \frac{1}{8} \cdot 2^4 \right) - \left( \frac{1}{12} \cdot 1^6 - \frac{1}{8} \cdot 1^4 \right)$$

$$= \frac{64}{12} - \frac{16}{8} - \frac{1}{12} + \frac{1}{8} = \frac{27}{8}$$

**YOUR TURN 5** Find
$\iint_R (x^3 + 4y) \, dy \, dx$ over the region
bounded by $y = 4x$ and $y = x^3$ for
$0 \le x \le 2$.

**TRY YOUR TURN 5**

**EXAMPLE 8** **Double Integrals**

Let $R$ be the shaded region in Figure 33, and evaluate

$$\iint_R (x + 2y) \, dy \, dx.$$

**SOLUTION** Region $R$ is bounded by $h(x) = 2x$ and $g(x) = x^2$, with $0 \le x \le 2$. By the first result in the previous theorem,

$$\iint_R (x + 2y) \, dy \, dx = \int_0^2 \int_{x^2}^{2x} (x + 2y) \, dy \, dx$$

$$= \int_0^2 (xy + y^2) \Big|_{x^2}^{2x} dx$$

$$= \int_0^2 [x(2x) + (2x)^2 - [x \cdot x^2 + (x^2)^2]] \, dx$$

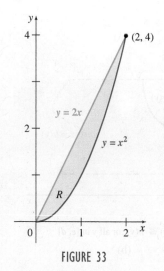

**FIGURE 33**

$$= \int_0^2 [2x^2 + 4x^2 - (x^3 + x^4)]\, dx$$

$$= \int_0^2 (6x^2 - x^3 - x^4)\, dx$$

$$= \left(2x^3 - \frac{1}{4}x^4 - \frac{1}{5}x^5\right)\Big|_0^2$$

$$= 2 \cdot 2^3 - \frac{1}{4} \cdot 2^4 - \frac{1}{5} \cdot 2^5 - 0$$

$$= 16 - 4 - \frac{32}{5} = \frac{28}{5}.$$

## Interchanging Limits of Integration

Sometimes it is easier to integrate first with respect to $x$ and then $y$, while with other integrals the reverse process is easier. The limits of integration can be reversed whenever the region $R$ is like the region in Figure 33, which has the property that it can be viewed as either type of region shown in Figure 31. In practice, this means that all boundaries can be written in terms of $y$ as a function of $x$, or by solving for $x$ as a function of $y$.

For instance, in Example 8, the same result would be found if we evaluated the double integral first with respect to $x$ and then with respect to $y$. In that case, we would need to define the equations of the boundaries in terms of $y$ rather than $x$, so $R$ would be defined by $y/2 \le x \le \sqrt{y}$, $0 \le y \le 4$. The resulting integral is

$$\int_0^4 \int_{y/2}^{\sqrt{y}} (x + 2y)\, dx\, dy = \int_0^4 \left(\frac{x^2}{2} + 2xy\right)\Big|_{y/2}^{\sqrt{y}} dy$$

$$= \int_0^4 \left[\left(\frac{y}{2} + 2y\sqrt{y}\right) - \left(\frac{y^2}{8} + 2\left(\frac{y}{2}\right)y\right)\right] dy$$

$$= \int_0^4 \left(\frac{y}{2} + 2y^{3/2} - \frac{9}{8}y^2\right) dy$$

$$= \left(\frac{y^2}{4} + \frac{4}{5}y^{5/2} - \frac{3}{8}y^3\right)\Big|_0^4$$

$$= 4 + \frac{4}{5} \cdot 4^{5/2} - 24$$

$$= \frac{28}{5}.$$

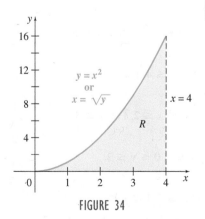

FIGURE 34

**EXAMPLE 9**   **Interchanging Limits of Integration**

Evaluate

$$\int_0^{16} \int_{\sqrt{y}}^4 \sqrt{x^3 + 4}\, dx\, dy.$$

**SOLUTION**   Notice that it is impossible to first integrate this function with respect to $x$. Thus, we attempt to interchange the limits of integration.

For this integral, region $R$ is given by $\sqrt{y} \le x \le 4$, $0 \le y \le 16$. A graph of $R$ is shown in Figure 34.

The same region $R$ can be written in an alternate way. As Figure 34 shows, one boundary of $R$ is $x = \sqrt{y}$. Solving for $y$ gives $y = x^2$. Also, Figure 34 shows that $0 \le x \le 4$. Since $R$ can be written as $0 \le y \le x^2$, $0 \le x \le 4$, the double integral above can be written

$$\int_0^4 \int_0^{x^2} \sqrt{x^3 + 4}\, dy\, dx = \int_0^4 y \sqrt{x^3 + 4}\, \Big|_0^{x^2}\, dx$$

$$= \int_0^4 x^2 \sqrt{x^3 + 4}\, dx$$

$$= \frac{1}{3} \int_0^4 3x^2 \sqrt{x^3 + 4}\, dx \qquad \text{Let } u = x^3 + 4. \text{ Change limits of integration.}$$

$$= \frac{1}{3} \int_4^{68} u^{1/2}\, du$$

$$= \frac{2}{9} u^{3/2} \Big|_4^{68}$$

$$= \frac{2}{9} [68^{3/2} - 4^{3/2}]$$

$$\approx 122.83.$$

**CAUTION** Fubini's Theorem cannot be used to interchange the order of integration when the limits contain variables, as in Example 9. Notice in Example 9 that after the order of integration was changed, the new limits were completely different. It would be a serious error to rewrite the integral in Example 9 as

$$\int_{\sqrt{y}}^4 \int_0^{16} \sqrt{x^3 + 4}\, dy\, dx.$$

# 9.5   EXERCISES

**Evaluate each integral.**

**1.** $\int_0^5 (x^4 y + y)\, dx$

**2.** $\int_1^2 (xy^3 - x)\, dy$

**3.** $\int_4^5 x\sqrt{x^2 + 3y}\, dy$

**4.** $\int_3^6 x\sqrt{x^2 + 3y}\, dx$

**5.** $\int_4^9 \frac{3 + 5y}{\sqrt{x}}\, dx$

**6.** $\int_2^7 \frac{3 + 5y}{\sqrt{x}}\, dy$

**7.** $\int_2^6 e^{2x+3y}\, dx$

**8.** $\int_{-1}^1 e^{2x+3y}\, dy$

**9.** $\int_0^3 ye^{4x+y^2}\, dy$

**10.** $\int_1^5 ye^{4x+y^2}\, dx$

**Evaluate each iterated integral. (Many of these use results from Exercises 1–10.)**

**11.** $\int_1^2 \int_0^5 (x^4 y + y)\, dx\, dy$

**12.** $\int_0^3 \int_1^2 (xy^3 - x)\, dy\, dx$

**13.** $\int_0^1 \int_3^6 x\sqrt{x^2 + 3y}\, dx\, dy$

**14.** $\int_0^3 \int_4^5 x\sqrt{x^2 + 3y}\, dy\, dx$

**15.** $\int_1^2 \int_4^9 \frac{3 + 5y}{\sqrt{x}}\, dx\, dy$

**16.** $\int_{16}^{25} \int_2^7 \frac{3 + 5y}{\sqrt{x}}\, dy\, dx$

**17.** $\int_1^3 \int_1^3 \frac{1}{xy}\, dy\, dx$

**18.** $\int_1^5 \int_2^4 \frac{1}{y}\, dx\, dy$

**19.** $\int_2^4 \int_3^5 \left(\frac{x}{y} + \frac{y}{3}\right)\, dx\, dy$

**20.** $\int_3^4 \int_1^2 \left(\frac{6x}{5} + \frac{y}{x}\right)\, dx\, dy$

**Find each double integral over the rectangular region $R$ with the given boundaries.**

**21.** $\iint_R (3x^2 + 4y)\, dx\, dy; \quad 0 \le x \le 3, 1 \le y \le 4$

**22.** $\iint_R (x^2 + 4y^3)\, dy\, dx; \quad 1 \le x \le 2, 0 \le y \le 3$

**23.** $\iint_R \sqrt{x + y}\, dy\, dx; \quad 1 \le x \le 3, 0 \le y \le 1$

**24.** $\iint_R x^2 \sqrt{x^3 + 2y}\, dx\, dy; \quad 0 \le x \le 2, 0 \le y \le 3$

**25.** $\iint_R \frac{3}{(x + y)^2}\, dy\, dx; \quad 2 \le x \le 4, 1 \le y \le 6$

**26.** $\iint_R \frac{y}{\sqrt{2x + 5y^2}}\, dx\, dy; \quad 0 \le x \le 2, 1 \le y \le 3$

**27.** $\iint_R ye^{x+y^2}\, dx\, dy; \quad 2 \le x \le 3, 0 \le y \le 2$

**28.** $\iint_R x^2 e^{x^3 + 2y}\, dx\, dy; \quad 1 \le x \le 2, 1 \le y \le 3$

**29.** $\iint\limits_{R} x \cos(xy)\, dy\, dx; \quad \dfrac{\pi}{2} \le x \le \pi, 0 \le y \le 1$

**30.** $\iint\limits_{R} x \sec^2 y\, dy\, dx; \quad 0 \le x \le 1, 0 \le y \le \dfrac{\pi}{4}$

**Find the volume under the given surface $z = f(x, y)$ and above the rectangle with the given boundaries.**

**31.** $z = 8x + 4y + 10; \quad -1 \le x \le 1, 0 \le y \le 3$

**32.** $z = 3x + 10y + 20; \quad 0 \le x \le 3, -2 \le y \le 1$

**33.** $z = x^2; \quad 0 \le x \le 2, 0 \le y \le 5$

**34.** $z = \sqrt{y}; \quad 0 \le x \le 4, 0 \le y \le 9$

**35.** $z = x\sqrt{x^2 + y}; \quad 0 \le x \le 1, 0 \le y \le 1$

**36.** $z = yx\sqrt{x^2 + y^2}; \quad 0 \le x \le 4, 0 \le y \le 1$

**37.** $z = \dfrac{xy}{(x^2 + y^2)^2}; \quad 1 \le x \le 2, 1 \le y \le 4$

**38.** $z = e^{x+y}; \quad 0 \le x \le 1, 0 \le y \le 1$

Although it is often true that a double integral can be evaluated by using either $dx$ or $dy$ first, sometimes one choice over the other makes the work easier. Evaluate the double integrals in Exercises 39 and 40 in the easiest way possible.

**39.** $\iint\limits_{R} xe^{xy}\, dx\, dy; \quad 0 \le x \le 2, 0 \le y \le 1$

**40.** $\iint\limits_{R} 2x^3 e^{x^2 y}\, dx\, dy; \quad 0 \le x \le 1, 0 \le y \le 1$

**Evaluate each double integral.**

**41.** $\displaystyle\int_{2}^{4}\int_{2}^{x^2} (x^2 + y^2)\, dy\, dx$

**42.** $\displaystyle\int_{0}^{2}\int_{0}^{3y} (x^2 + y)\, dx\, dy$

**43.** $\displaystyle\int_{0}^{4}\int_{0}^{x} \sqrt{xy}\, dy\, dx$

**44.** $\displaystyle\int_{1}^{4}\int_{0}^{x} \sqrt{x + y}\, dy\, dx$

**45.** $\displaystyle\int_{2}^{6}\int_{2y}^{4y} \dfrac{1}{x}\, dx\, dy$

**46.** $\displaystyle\int_{1}^{4}\int_{x}^{x^2} \dfrac{1}{y}\, dy\, dx$

**47.** $\displaystyle\int_{0}^{4}\int_{1}^{e^x} \dfrac{x}{y}\, dy\, dx$

**48.** $\displaystyle\int_{0}^{1}\int_{2x}^{4x} e^{x+y}\, dy\, dx$

**Use the region $R$ with the indicated boundaries to evaluate each double integral.**

**49.** $\iint\limits_{R} (5x + 8y)\, dy\, dx; \quad 1 \le x \le 3, 0 \le y \le x - 1$

**50.** $\iint\limits_{R} (2x + 6y)\, dy\, dx; \quad 2 \le x \le 4, 2 \le y \le 3x$

**51.** $\iint\limits_{R} (4 - 4x^2)\, dy\, dx; \quad 0 \le x \le 1, 0 \le y \le 2 - 2x$

**52.** $\iint\limits_{R} \dfrac{1}{x}\, dy\, dx; \quad 1 \le x \le 2, 0 \le y \le x - 1$

**53.** $\iint\limits_{R} e^{x/y^2}\, dx\, dy; \quad 1 \le y \le 2, 0 \le x \le y^2$

**54.** $\iint\limits_{R} (x^2 - y)\, dy\, dx; \quad -1 \le x \le 1, -x^2 \le y \le x^2$

**55.** $\iint\limits_{R} x^3 y\, dy\, dx; \quad R$ bounded by $y = x^2, y = 2x$

**56.** $\iint\limits_{R} x^2 y^2\, dx\, dy; \quad R$ bounded by $y = x, y = 2x, x = 1$

**57.** $\iint\limits_{R} \dfrac{1}{y}\, dy\, dx; \quad R$ bounded by $y = x, y = \dfrac{1}{x}, x = 2$

**58.** $\iint\limits_{R} e^{2y/x}\, dy\, dx; \quad R$ bounded by $y = x^2, y = 0, x = 2$

**Evaluate each double integral. If the function seems too difficult to integrate, try interchanging the limits of integration, as in Exercises 39 and 40.**

**59.** $\displaystyle\int_{0}^{\ln 2}\int_{e^y}^{2} \dfrac{1}{\ln x}\, dx\, dy$

**60.** $\displaystyle\int_{0}^{2}\int_{y/2}^{1} e^{x^2}\, dx\, dy$

**61.** Recall from the Volume and Average Value section in the previous chapter that volume could be found with a single integral. In this section volume is found using a double integral. Explain when volume can be found with a single integral and when a double integral is needed.

**62.** Give an example of a region that cannot be expressed by either of the forms shown in Figure 31. (One example is the disk with a hole in the middle between the graphs of $x^2 + y^2 = 1$ and $x^2 + y^2 = 2$ in Figure 10.)

The idea of the average value of a function, discussed earlier for functions of the form $y = f(x)$, can be extended to functions of more than one independent variable. For a function $z = f(x, y)$, the average value of $f$ over a region $R$ is defined as

$$\frac{1}{A}\iint\limits_{R} f(x, y)\, dx\, dy,$$

where $A$ is the area of the region $R$. Find the average value for each function over the regions $R$ having the given boundaries.

**63.** $f(x, y) = 6xy + 2x; \quad 2 \le x \le 5, 1 \le y \le 3$

**64.** $f(x, y) = x^2 + y^2; \quad 0 \le x \le 2, 0 \le y \le 3$

**65.** $f(x, y) = e^{-5y+3x}; \quad 0 \le x \le 2, 0 \le y \le 2$

**66.** $f(x, y) = e^{2x+y}; \quad 1 \le x \le 2, 2 \le y \le 3$

## APPLICATIONS

**67. Packaging** The manufacturer of a fruit juice drink has decided to try innovative packaging in order to revitalize sagging sales. The fruit juice drink is to be packaged in containers in the shape of tetrahedra in which three edges are perpendicular, as shown in the figure on the next page. Two of the perpendicular edges will be 3 in. long, and the third edge will be 6 in. long. Find the volume of the container. (*Hint:* The equation of the plane shown in the figure is $z = f(x, y) = 6 - 2x - 2y$.)

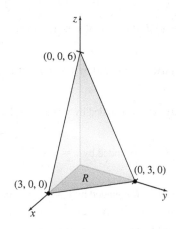

The weekly sales for the first product vary from 100 units to 150 units, and the weekly sales for the second product vary from 40 units to 80 units. Estimate average weekly profit for these two products. (*Hint:* Refer to Exercises 63–66.)

**70. Average Revenue** A company sells two products. The demand functions of the products are given by

$$q_1 = 300 - 2p_1 \quad \text{and} \quad q_2 = 500 - 1.2p_2,$$

where $q_1$ units of the first product are demanded at price $p_1$ and $q_2$ units of the second product are demanded at price $p_2$. The total revenue will be given by

$$R = q_1p_1 + q_2p_2.$$

Find the average revenue if the price $p_1$ varies from \$25 to \$50 and the price $p_2$ varies from \$50 to \$75. (*Hint:* Refer to Exercises 63–66.)

**68. Average Cost** A company's total cost for operating its two warehouses is

$$C(x, y) = \frac{1}{9}x^2 + 2x + y^2 + 5y + 100$$

dollars, where $x$ represents the number of units stored at the first warehouse and $y$ represents the number of units stored at the second. Find the average cost to store a unit if the first warehouse has between 40 and 80 units, and the second has between 30 and 70 units. (*Hint:* Refer to Exercises 63–66.)

**69. Average Profit** The profit (in dollars) from selling $x$ units of one product and $y$ units of a second product is

$$P = -(x - 100)^2 - (y - 50)^2 + 2000.$$

**■ YOUR TURN ANSWERS**

1. $52x^2 + 16x + 16x^3 + 490$
2. $3644/3$
3. $(56\sqrt{14} - 184)/3$
4. $7/2$
5. $5888/105$

# 9 CHAPTER REVIEW

## SUMMARY

In this chapter, we extended our study of calculus to include functions of several variables. We saw that it is possible to produce three-dimensional graphs of functions of two variables and that the process is greatly enhanced using level curves. Level curves are formed by determining the values of $x$ and $y$ that produce a particular functional value. We also saw the graphs of several surfaces, including the

- paraboloid, whose equation is $z = x^2 + y^2$,
- ellipsoid, whose general equation is $\dfrac{x^2}{a^2} + \dfrac{y^2}{b^2} + \dfrac{z^2}{c^2} = 1$,
- hyperbolic paraboloid, whose equation is $z = x^2 - y^2$, and
- hyperboloid of two sheets, whose equation is $-x^2 - y^2 + z^2 = 1$.

Partial derivatives are the extension of the concept of differentiation with respect to one of the variables while the other variables

are held constant. Partial derivatives were used to identify extrema of a function of several variables. In particular, we identified all points where the partial with respect to $x$ and the partial with respect to $y$ are both zero, which we called critical points. We then classified each critical point as a relative maximum, a relative minimum, or a saddle point. Recall that a saddle point is a minimum when approached from one direction but a maximum when approached from another direction. Differentials, introduced earlier for functions of one variable, were generalized to define the total differential. We saw that total differentials can be used to approximate the value of a function using its tangent plane. We concluded the chapter by introducing double integrals, which are simply two iterated integrals, one for each variable. Double integrals were then used to find volume.

**Function of Two Variables**   The expression $z = f(x, y)$ is a function of two variables if a unique value of $z$ is obtained from each ordered pair of real numbers $(x, y)$. The variables $x$ and $y$ are independent variables, and $z$ is the dependent variable. The set of all ordered pairs of real numbers $(x, y)$ such that $f(x, y)$ exists is the domain of $f$; the set of all values of $f(x, y)$ is the range.

**Plane**   The graph of $ax + by + cz = d$ is a plane if $a$, $b$, and $c$ are not all 0.

**Partial Derivatives (Informal Definition)**   The partial derivative of $f$ with respect to $x$ is the derivative of $f$ obtained by treating $x$ as a variable and $y$ as a constant.

The partial derivative of $f$ with respect to $y$ is the derivative of $f$ obtained by treating $y$ as a variable and $x$ as a constant.

**Partial Derivatives (Formal Definition)**   Let $z = f(x, y)$ be a function of two independent variables. Let all indicated limits exist. Then the partial derivative of $f$ with respect to $x$ is

$$f_x(x, y) = \frac{\partial z}{\partial x} = \lim_{h \to 0} \frac{f(x + h, y) - f(x, y)}{h},$$

and the partial derivative of $f$ with respect to $y$ is

$$f_y(x, y) = \frac{\partial z}{\partial y} = \lim_{h \to 0} \frac{f(x, y + h) - f(x, y)}{h}.$$

If the indicated limits do not exist, then the partial derivatives do not exist.

**Second-Order Partial Derivatives**   For a function $z = f(x, y)$, if the partial derivative exists, then

$$\frac{\partial}{\partial x}\left(\frac{\partial z}{\partial x}\right) = \frac{\partial^2 z}{\partial x^2} = f_{xx}(x, y) = z_{xx} \qquad \frac{\partial}{\partial y}\left(\frac{\partial z}{\partial y}\right) = \frac{\partial^2 z}{\partial y^2} = f_{yy}(x, y) = z_{yy}$$

$$\frac{\partial}{\partial y}\left(\frac{\partial z}{\partial x}\right) = \frac{\partial^2 z}{\partial y \partial x} = f_{xy}(x, y) = z_{xy} \qquad \frac{\partial}{\partial x}\left(\frac{\partial z}{\partial y}\right) = \frac{\partial^2 z}{\partial x \partial y} = f_{yx}(x, y) = z_{yx}$$

**Relative Extrema**   Let $(a, b)$ be the center of a circular region contained in the $xy$-plane. Then, for a function $z = f(x, y)$ defined for every $(x, y)$ in the region, $f(a, b)$ is a relative maximum if

$$f(a, b) \geq f(x, y)$$

for all points $(x, y)$ in the circular region, and $f(a, b)$ is a relative minimum if

$$f(a, b) \leq f(x, y)$$

for all points $(x, y)$ in the circular region.

**Location of Extrema**   Let a function $z = f(x, y)$ have a relative maximum or relative minimum at the point $(a, b)$. Let $f_x(a, b)$ and $f_y(a, b)$ both exist. Then

$$f_x(a, b) = 0 \quad \text{and} \quad f_y(a, b) = 0.$$

**Test for Relative Extrema**   For a function $z = f(x, y)$, let $f_{xx}$, $f_{yy}$, and $f_{xy}$ all exist in a circular region contained in the $xy$-plane with center $(a, b)$. Further, let

$$f_x(a, b) = 0 \quad \text{and} \quad f_y(a, b) = 0.$$

Define $D$, known as the discriminant, by

$$D = f_{xx}(a, b) \cdot f_{yy}(a, b) - [f_{xy}(a, b)]^2.$$

Then

**a.** $f(a, b)$ is a relative maximum if $D > 0$ and $f_{xx}(a, b) < 0$;

**b.** $f(a, b)$ is a relative minimum if $D > 0$ and $f_{xx}(a, b) > 0$;

**c.** $f(a, b)$ is a saddle point (neither a maximum nor a minimum) if $D < 0$;

**d.** if $D = 0$, the test gives no information.

**Total Differential for Two Variables**   Let $z = f(x, y)$ be a function of $x$ and $y$. Let $dx$ and $dy$ be real numbers. Then the total differential of $z$ is

$$dz = f_x(x, y) \cdot dx + f_y(x, y) \cdot dy.$$

(Sometimes $dz$ is written $df$.)

**Approximations** For small values of $dx$ and $dy$,

$$dz \approx \Delta z$$

where $\Delta z = f(x + dx, y + dy) - f(x, y)$.

**Approximations by Differentials** For a function $f$ having all indicated partial derivatives, and for small values of $dx$ and $dy$,

$$f(x + dx, y + dy) \approx f(x, y) + dz,$$

or

$$f(x + dx, y + dy) \approx f(x, y) + f_x(x, y) \cdot dx + f_y(x, y) \cdot dy.$$

**Total Differential for Three Variables** If $w = f(x, y, z)$, then the total differential $dw$ is

$$dw = f_x(x, y, z) \cdot dx + f_y(x, y, z) \cdot dy + f_z(x, y, z) \cdot dz,$$

provided all indicated partial derivatives exist.

**Double Integral** The double integral of $f(x, y)$ over a rectangular region $R$ defined by $a \le x \le b, c \le y \le d$ is written

$$\iint_R f(x, y)\, dy\, dx \quad \text{or} \quad \iint_R f(x, y)\, dx\, dy,$$

and equals either

$$\int_a^b \int_c^d f(x, y)\, dy\, dx \quad \text{or} \quad \int_c^d \int_a^b f(x, y)\, dx\, dy.$$

**Volume** Let $z = f(x, y)$ be a function that is never negative on the rectangular region $R$ defined by $a \le x \le b, c \le y \le d$. The volume of the solid under the graph of $f$ and over the region $R$ is

$$\iint_R f(x, y)\, dy\, dx.$$

**Double Integrals over Variable Regions** Let $z = f(x, y)$ be a function of two variables. If $R$ is the region defined by $a \le x \le b$ and $g(x) \le y \le h(x)$, then

$$\iint_R f(x, y)\, dy\, dx = \int_a^b \left[ \int_{g(x)}^{h(x)} f(x, y)\, dy \right] dx.$$

If $R$ is the region defined by $g(y) \le x \le h(y)$ and $c \le y \le d$, then

$$\iint_R f(x, y)\, dx\, dy = \int_c^d \left[ \int_{g(y)}^{h(y)} f(x, y)\, dx \right] dy.$$

# KEY TERMS

**9.1**
function of two variables
independent variable
dependent variable
domain
range
ordered triple
first octant
plane
surface

trace
level curves
paraboloid
production function
ellipsoid
hyperbolic paraboloid
hyperboloid of two sheets
level surface

**9.2**
partial derivative
second-order partial derivative

**9.3**
relative maximum
relative minimum
saddle point
critical point
discriminant

**9.4**
total differential

**9.5**
Fubini's Theorem
iterated integral
double integral
integrand
region of integration

# REVIEW EXERCISES

## CONCEPT CHECK

**Determine whether each of the following statements is true or false, and explain why.**

1. The graph of $6x - 2y + 7z = 14$ is a plane.

2. The graph of $2x + 4y = 10$ is a plane that is parallel to the $z$-axis.

3. A level curve for a paraboloid could be a single point.

4. If the partial derivatives with respect to $x$ and $y$ at some point are both 0, the tangent plane to the function at that point is horizontal.

5. If $f(x, y) = 3x^2 + 2xy + y^2$, then $f(x + h, y) = 3(x + h)^2 + 2xy + h + y^2$.

6. For a function $z = f(x, y)$, suppose that the point $(a, b)$ has been identified such that $f_x(a, b) = f_y(a, b) = 0$. We can conclude that a relative maximum or a relative minimum must exist at $(a, b)$.

7. A saddle point can be a relative maximum or a relative minimum.

8. A function of two variables may have both a relative maximum and an absolute maximum at the same point.

9. $\displaystyle\int_2^4 \int_1^5 (3x + 4y)\, dy\, dx = \int_2^4 \int_1^5 (3x + 4y)\, dx\, dy$

10. $\displaystyle\int_0^1 \int_{-2}^2 xe^y\, dy\, dx = \int_{-2}^2 \int_0^1 xe^y\, dx\, dy$

11. $\displaystyle\int_0^4 \int_1^x (x + xy^2)\, dy\, dx = \int_1^x \int_0^4 (x + xy^2)\, dx\, dy$

## PRACTICE AND EXPLORATIONS

12. Describe in words how to take a partial derivative.

13. Describe what a partial derivative means geometrically.

14. Describe what a total differential is and how it is useful.

**Find $f(-1, 2)$ and $f(6, -3)$ for the following.**

15. $f(x, y) = -4x^2 + 6xy - 3$

16. $f(x, y) = 2x^2y^2 - 7x + 4y$

17. $f(x, y) = \dfrac{x - 2y}{x + 5y}$

18. $f(x, y) = \dfrac{\sqrt{x^2 + y^2}}{x - y}$

**Graph the first-octant portion of each plane.**

19. $x + y + z = 4$          20. $x + 2y + 6z = 6$

21. $5x + 2y = 10$           22. $4x + 3z = 12$

23. $x = 3$                   24. $y = 4$

25. Let $z = f(x, y) = 3x^3 + 4x^2y - 2y^2$. Find the following.

 a. $\dfrac{\partial z}{\partial x}$   b. $\dfrac{\partial f}{\partial y}(-1, 4)$   c. $f_{xy}(2, -1)$

26. Let $z = f(x, y) = \dfrac{x + y^2}{x - y^2}$. Find the following.

 a. $\dfrac{\partial z}{\partial y}$   b. $\dfrac{\partial f}{\partial x}(0, 2)$   c. $f_{xx}(-1, 0)$

**Find $f_x(x, y)$ and $f_y(x, y)$.**

27. $f(x, y) = 6x^2y^3 - 4y$       28. $f(x, y) = 5x^4y^3 - 6x^5y$

29. $f(x, y) = \sqrt{4x^2 + y^2}$   30. $f(x, y) = \dfrac{2x + 5y^2}{3x^2 + y^2}$

31. $f(x, y) = x^3e^{3y}$           32. $f(x, y) = (y - 2)^2 e^{x+2y}$

33. $f(x, y) = \ln|2x^2 + y^2|$    34. $f(x, y) = \ln|2 - x^2y^3|$

35. $f(x, y) = \cos(3x^2 + y^2)$

36. $f(x, y) = x\tan(7x^2 + 4y^2)$

**Find $f_{xx}(x, y)$ and $f_{xy}(x, y)$.**

37. $f(x, y) = 5x^3y - 6xy^2$      38. $f(x, y) = -3x^2y^3 + x^3y$

39. $f(x, y) = \dfrac{3x}{2x - y}$   40. $f(x, y) = \dfrac{3x + y}{x - 1}$

41. $f(x, y) = 4x^2e^{2y}$          42. $f(x, y) = ye^{x^2}$

43. $f(x, y) = \ln|2 - x^2y|$      44. $f(x, y) = \ln|1 + 3xy^2|$

**Find all points where the functions defined below have any relative extrema. Find any saddle points.**

45. $z = 2x^2 - 3y^2 + 12y$

46. $z = x^2 + y^2 + 9x - 8y + 1$

47. $f(x, y) = x^2 + 3xy - 7x + 5y^2 - 16y$

48. $z = x^3 - 8y^2 + 6xy + 4$

49. $z = \dfrac{1}{2}x^2 + \dfrac{1}{2}y^2 + 2xy - 5x - 7y + 10$

50. $f(x, y) = 2x^2 + 4xy + 4y^2 - 3x + 5y - 15$

51. $z = x^3 + y^2 + 2xy - 4x - 3y - 2$

52. $f(x, y) = 7x^2 + y^2 - 3x + 6y - 5xy$

53. Describe the different types of points that might occur when $f_x(x, y) = f_y(x, y) = 0$.

54. Describe how a differential for a function of two variables approximates the function.

**Evaluate $dz$ using the given information.**

55. $z = 6x^2 - 7y^2 + 4xy;\ x = 3,\ y = -1,\ dx = 0.03,\ dy = 0.01$

56. $z = \dfrac{x + 5y}{x - 2y};\quad x = 1,\ y = -2,\ dx = -0.04,\ dy = 0.02$

**Use the total differential to approximate each quantity. Then use a calculator to approximate the quantity, and give the absolute value of the difference in the two results to 4 decimal places.**

57. $\sqrt{5.1^2 + 12.05^2}$          58. $\sqrt{4.06}\, e^{0.04}$

**Evaluate the following.**

**59.** $\displaystyle\int_1^4 \frac{4y-3}{\sqrt{x}}\,dx$

**60.** $\displaystyle\int_1^5 e^{3x+5y}\,dx$

**61.** $\displaystyle\int_0^5 \frac{6x}{\sqrt{4x^2+2y^2}}\,dx$

**62.** $\displaystyle\int_1^3 \frac{y^2}{\sqrt{7x+11y^3}}\,dy$

**Evaluate each iterated integral.**

**63.** $\displaystyle\int_0^2 \int_0^4 (x^2y^2 + 5x)\,dx\,dy$

**64.** $\displaystyle\int_0^3 \int_0^5 (2x + 6y + y^2)\,dy\,dx$

**65.** $\displaystyle\int_3^4 \int_2^5 \sqrt{6x+3y}\,dx\,dy$

**66.** $\displaystyle\int_1^2 \int_3^5 e^{2x-7y}\,dx\,dy$

**67.** $\displaystyle\int_2^4 \int_2^4 \frac{1}{y}\,dx\,dy$

**68.** $\displaystyle\int_1^2 \int_1^2 \frac{1}{x}\,dx\,dy$

**Find each double integral over the region R with boundaries as indicated.**

**69.** $\displaystyle\iint_R (x^2 + 2y^2)\,dx\,dy;\quad 0 \le x \le 5, 0 \le y \le 2$

**70.** $\displaystyle\iint_R \sqrt{2x+y}\,dx\,dy;\quad 1 \le x \le 3, 2 \le y \le 5$

**71.** $\displaystyle\iint_R \sqrt{y+x}\,dx\,dy;\quad 0 \le x \le 7, 1 \le y \le 9$

**72.** $\displaystyle\iint_R ye^{y^2+x}\,dx\,dy;\quad 0 \le x \le 1, 0 \le y \le 1$

**73.** $\displaystyle\iint_R xy\sin(xy^2)\,dy\,dx;\quad 0 \le x \le \frac{\pi}{2}, 0 \le y \le 1$

**74.** $\displaystyle\iint_R ye^x \cos(ye^x)\,dx\,dy;\quad 0 \le x \le 1, 0 \le y \le \pi$

**Find the volume under the given surface $z = f(x, y)$ and above the given rectangle.**

**75.** $z = x + 8y + 4;\quad 0 \le x \le 3, 1 \le y \le 2$

**76.** $z = x^2 + y^2;\quad 3 \le x \le 5, 2 \le y \le 4$

**Evaluate each double integral. If the function seems too difficult to integrate, try interchanging the limits of integration.**

**77.** $\displaystyle\int_0^1 \int_0^{2x} xy\,dy\,dx$

**78.** $\displaystyle\int_1^2 \int_2^{2x^2} y\,dy\,dx$

**79.** $\displaystyle\int_0^1 \int_{x^2}^x x^3y\,dy\,dx$

**80.** $\displaystyle\int_0^1 \int_y^{\sqrt{y}} x\,dx\,dy$

**81.** $\displaystyle\int_0^2 \int_{x/2}^1 \frac{1}{y^2+1}\,dy\,dx$

**82.** $\displaystyle\int_0^8 \int_{x/2}^4 \sqrt{y^2+4}\,dy\,dx$

**Use the region R, with boundaries as indicated, to evaluate the given double integral.**

**83.** $\displaystyle\iint_R (2x + 3y)\,dx\,dy;\quad 0 \le y \le 1, y \le x \le 2 - y$

**84.** $\displaystyle\iint_R (2 - x^2 - y^2)\,dy\,dx;\quad 0 \le x \le 1, x^2 \le y \le x$

## LIFE SCIENCE APPLICATIONS

**85. Blood Vessel Volume** A length of blood vessel is measured as 2.7 cm, with the radius measured as 0.7 cm. If each of these measurements could be off by 0.1 cm, estimate the maximum possible error in the volume of the vessel.

**86. Total Body Water** Accurate prediction of total body water is critical in determining adequate dialysis doses for patients with renal disease. For African American males, total body water can be estimated by the function

$$T(A, M, S) = -18.37 - 0.09A + 0.34M + 0.25S,$$

where $T$ is the total body water (in liters), $A$ is age (in years), $M$ is mass (in kilograms), and $S$ is height (in centimeters). *Source: Kidney International.*

 **a.** Find $T(65, 85, 180)$.

 **b.** Find and interpret $T_A(A, M, S)$, $T_M(A, M, S)$, and $T_S(A, M, S)$.

**87. Brown Trout** Researchers from New Zealand have determined that the length of a brown trout depends on both its mass and age and that the length can be estimated by

$$L(m, t) = (0.00082t + 0.0955)e^{(\ln m + 10.49)/2.842},$$

where $L(m, t)$ is the length of the trout (in centimeters), $m$ is the mass of the trout (in grams), and $t$ is the age of the trout (in years). *Source: Transactions of the American Fisheries Society.*

 **a.** Find $L(450, 4)$.

 **b.** Find $L_m(450, 7)$ and $L_t(450, 7)$ and interpret.

**88. Survival Curves** The figure on the next page shows survival curves (percent surviving as a function of age) for people in the United States in 1900 and 2006. *Source: National Vital Statistics Report.* Let $f(x, y)$ give the proportion surviving at age $x$ in year $y$. Use the graph to estimate the following. Interpret each answer in words.

 **a.** $f(60, 1900)$

 **b.** $f(70, 2006)$

 **c.** $f_x(60, 1900)$

 **d.** $f_x(70, 2006)$

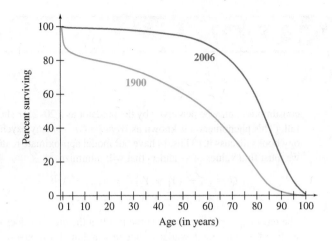

## OTHER APPLICATIONS

**89. Area**  The bottom of a planter is to be made in the shape of an isosceles triangle, with the two equal sides 3 ft long and the third side 2 ft long. The area of an isosceles triangle with two equal sides of length $a$ and third side of length $b$ is

$$f(a, b) = \frac{1}{4}b\sqrt{4a^2 - b^2}.$$

**a.** Find the area of the bottom of the planter.

**b.** The manufacturer is considering changing the shape so that the third side is 2.5 ft long. What would be the approximate effect on the area?

**90. Surface Area**  A closed box with square ends must have a volume of 125 in³. Find the dimensions of such a box that has minimum surface area.

**91. Area**  Find the maximum rectangular area that can be enclosed with 400 ft of fencing, if no fencing is needed along one side.

**92. Charge for Auto Painting**  The charge (in dollars) for painting a sports car is given by

$$C(x, y) = 4x^2 + 5y^2 - 4xy + \sqrt{x},$$

where $x$ is the number of hours of labor needed and $y$ is the number of gallons of paint and sealant used. Find the following.

**a.** The charge for 10 hours and 5 gal of paint and sealant

**b.** The charge for 15 hours and 10 gal of paint and sealant

**c.** The charge for 20 hours and 20 gal of paint and sealant

**93. Manufacturing Costs**  The manufacturing cost (in dollars) for a certain computer is given by

$$c(x, y) = 2x + y^2 + 4xy + 25,$$

where $x$ is the memory capacity of the computer in gigabytes (GB) and $y$ is the number of hours of labor required. For 640 GB and 6 hours of labor, find the following.

**a.** The approximate change in cost for an additional 1 GB of memory

**b.** The approximate change in cost for an additional hour of labor

**94. Cost**  The cost (in dollars) to manufacture $x$ solar cells and $y$ solar collectors is

$$c(x, y) = x^2 + 5y^2 + 4xy - 70x - 164y + 1800.$$

**a.** Find values of $x$ and $y$ that produce minimum total cost.

**b.** Find the minimum total cost.

**95. Cost**  The cost (in dollars) to produce $x$ satellite receiving dishes and $y$ transmitters is given by

$$C(x, y) = 100 \ln(x^2 + y) + e^{xy/20}.$$

Production schedules now call for 15 receiving dishes and 9 transmitters. Use differentials to approximate the change in costs if 1 more dish and 1 fewer transmitter are made.

**96. Production Materials**  Approximate the volume of material needed to manufacture a cone of radius 2 cm, height 8 cm, and wall thickness 0.21 cm.

**97. Production Materials**  A sphere of radius 2 ft is to receive an insulating coating 1 in. thick. Approximate the volume of the coating needed.

**98. Production Error**  The height of a sample cone from a production line is measured as 11.4 cm, while the radius is measured as 2.9 cm. Each of these measurements could be off by 0.2 cm. Approximate the maximum possible error in the volume of the cone.

**99. Profit**  The total profit from 1 acre of a certain crop depends on the amount spent on fertilizer, $x$, and on hybrid seed, $y$, according to the model

$$P(x, y) = 0.01(-x^2 + 3xy + 160x - 5y^2 + 200y + 2600).$$

The budget for fertilizer and seed is limited to $280.

**a.** Use the budget constraint to express one variable in terms of the other. Then substitute into the profit function to get a function with one independent variable. Use the method shown in Chapter 6 on Applications of the Derivative to find the amounts spent on fertilizer and seed that will maximize profit. What is the maximum profit per acre? (*Hint:* Throughout this exercise you may ignore the coefficient of 0.01 until you need to find the maximum profit.)

**b.** Find the amounts spent on fertilizer and seed that will maximize profit using the first method shown in this chapter. (*Hint:* You will not need to use the budget constraint.)

**c.** Look for the relationships between these methods.

## OPTIMIZATION FOR A PREDATOR

A predator is an animal that feeds upon another animal. Foxes, coyotes, wolves, weasels, lions, and tigers are well-known predators, but many other animals also fall into this category. In this case, we set up a mathematical model for predation, and then use partial derivatives to minimize the difference between the desired and the actual levels of food consumption. There are several research studies which show that animals *do* control their activities so as to maximize or minimize variables—lobsters orient their bodies by minimizing the discharge rate from certain organs, for example.

In this case, we assume that the predator has a diet consisting of only two foods, food 1 and food 2. We also assume that the predator will hunt only in two locations, location 1 and location 2. To make this mathematical model meaningful, we would have to gather data on the predators and the prey that we wished to study. For example, suppose that we have gathered the following data:

$u_{11} = 0.4 =$ rate of feeding on food 1 in location 1;
$u_{12} = 0.1 =$ rate of feeding on food 1 in location 2;
$u_{21} = 0.3 =$ rate of feeding on food 2 in location 1;
$u_{22} = 0.3 =$ rate of feeding on food 2 in location 2.

Let $x_1 =$ proportion of time spent feeding in location 1;
$x_2 =$ proportion of time spent feeding in location 2;
$x_3 =$ proportion of time spent on nonfeeding activities.

Using these variables, the total quantity of food 1 consumed is given by $Y_1$, where

$$Y_1 = u_{11}x_1 + u_{12}x_2$$
$$= 0.4x_1 + 0.1x_2. \tag{1}$$

The total quantity of food 2 consumed is given by $Y_2$, where

$$Y_2 = u_{21}x_1 + u_{22}x_2$$
$$= 0.3x_1 + 0.3x_2. \tag{2}$$

The total amount of food consumed is thus given by

$$Y_1 + Y_2 = 0.4x_1 + 0.1x_2 + 0.3x_1 + 0.3x_2$$
$$= 0.7x_1 + 0.4x_2. \tag{3}$$

Again, by gathering experimental data, suppose that

$z_t = 0.4 =$ desired level of total food consumption;

$z_1 = 0.15 =$ desired level of consumption of food 1;

$z_2 = 0.25 =$ desired level of consumption of food 2.

The predator wishes to find values of $x_1$ and $x_2$ such that the difference between the desired level of consumption (the $z$'s) and the actual level of consumption (the $Y$'s) is minimized. However, much experience has shown that the difference between desired food consumption and actual consumption is not perceived by the animal as linear, but rather perhaps as a square. That is, a 5% shortfall in the actual consumption of food, as compared to the desired

consumption, may be perceived by the predator as a 20–25% shortfall. (This phenomenon is known as *Weber's Law*—many psychology books discuss it.) Thus, to have our model approximate reality, we must find values of $x_1$ and $x_2$ that will minimize

$$G = [z_t - (Y_1 + Y_2)]^2 + w_1[z_1 - Y_1]^2$$
$$+ w_2[z_2 - Y_2]^2 + w_3[1 - x_3]^2. \tag{4}$$

The term $w_3[1 - x_3]^2$ represents the fact that the predator does not wish to spend the total available time in searching for food—recall that $x_3$ represents the proportion of available time that is spent on nonfeeding activities. The variables $w_1$, $w_2$, and $w_3$ represent the relative importance, or weights, assigned by the animal to food 1, food 2, and nonfood activities, respectively. Again, it is necessary to gather experimental data; reasonable values of $w_1$, $w_2$, and $w_3$ are as follows:

$$w_1 = 0.5 \quad w_2 = 0.4 \quad w_3 = 0.03.$$

If we substitute 0.4 for $z_t$, 0.15 for $z_1$, 0.25 for $z_2$, 0.5 for $w_1$, 0.4 for $w_2$, and 0.03 for $w_3$, the results of Equation (1) for $Y_1$, (2) for $Y_2$, and (3) for $Y_1 + Y_2$ into Equation (4), we have

$$G = [0.4 - 0.7x_1 - 0.4x_2]^2 + 0.5[0.15 - 0.4x_1 - 0.1x_2]^2$$
$$+ 0.4[0.25 - 0.3x_1 - 0.3x_2]^2 + 0.03[1 - x_3]^2.$$

We want to minimize $G$, subject to the constraint $x_1 + x_2 + x_3 = 1$, or $x_1 + x_2 + x_3 - 1 = 0$. The original article on which this Extended Application is based did this with a mathematical technique known as Lagrange multipliers.* Here, we will instead solve the constraint for $x_3$: $x_3 = 1 - x_1 - x_2$. Substituting this into the equation for $G$ gives

$$G = [0.4 - 0.7x_1 - 0.4x_2]^2 + 0.5[0.15 - 0.4x_1 - 0.1x_2]^2$$
$$+ 0.4[0.25 - 0.3x_1 - 0.3x_2]^2 + 0.03[x_1 + x_2]^2.$$

Now we must find the partial derivatives of $G$ with respect to $x_1$ and $x_2$. Doing this we have

$$\frac{\partial G}{\partial x_1} = 2(-0.7)(0.4 - 0.7x_1 - 0.4x_2)$$
$$+ 2(0.5)(-0.4)(0.15 - 0.4x_1 - 0.1x_2)$$
$$+ 2(0.4)(-0.3)(0.25 - 0.3x_1 - 0.3x_2)$$
$$+ 2(0.03)(x_1 + x_2)$$

$$\frac{\partial G}{\partial x_2} = 2(-0.4)(0.4 - 0.7x_1 - 0.4x_2)$$
$$+ 2(0.5)(-0.1)(0.15 - 0.4x_1 - 0.1x_2)$$
$$+ 2(0.4)(-0.3)(0.25 - 0.3x_1 - 0.3x_2)$$
$$+ 2(0.03)(x_1 + x_2).$$

*For details on Lagrange multipliers, see Lial, Margaret L., Raymond N. Greenwell, and Nathan P. Ritchey, *Calculus with Applications*, 10th ed., Pearson, 2012, pp. 491–499.

Both $\partial G/\partial x_1$ and $\partial G/\partial x_2$ can be simplified, using some rather tedious algebra. Doing this, and placing each partial derivative equal to 0, we get the following system of equations:

$$\frac{\partial G}{\partial x_1} = -0.68 + 1.272x_1 + 0.732x_2 = 0$$

$$\frac{\partial G}{\partial x_2} = -0.395 + 0.732x_1 + 0.462x_2 = 0.$$

Although we shall not go through the details of the solution here, it can be found from the system above that the values of the $x$'s that minimize $G$ are given by

$$x_1 = 0.48 \quad x_2 = 0.09 \quad x_3 = 0.43.$$

(These values have been rounded to the nearest hundredth.) Recall that $x_3 = 1 - x_1 - x_2$. Thus, the predator should spend about 0.48 of the available time searching in location 1, and about 0.09 of the time searching in location 2. This will leave about 0.43 of the time free for nonfeeding activities.

*Source: Ecology*

## EXERCISES

1. Verify the simplification of $\partial G/\partial x_1$ and $\partial G/\partial x_2$.

2. Verify the solution given in the text by solving the system of equations.

3. Suppose $w_1 = 0.4$ and $w_2 = 0.5$ and all other values remain the same. What proportion of time should the predator spend on feeding in each of locations 1 and 2? What proportion of time will be spent on nonfeeding activities?

4. The function $G$ has a graph known as an *elliptic paraboloid*, which is like the paraboloid of Section 9.1 but with an elliptic rather than a circular cross-section. Go to the website WolframAlpha.com and find out what you can about elliptic paraboloids. Then use this website to simplify the function $G$ and to draw its graph. Explain the appearance of the graph.

## DIRECTIONS FOR GROUP PROJECT

*Search through journals or on the Internet for an actual example of a predator-prey system. Use the information you find to estimate reasonable values for $u_{11}$, $u_{12}$, $u_{21}$, $u_{22}$, $z_t$, $z_1$, and $z_2$. Go through an analysis similar to this Extended Application to find the amount of time the predator should spend looking for food in the different locations. Prepare a presentation of your analysis and results, using presentation software such as Microsoft PowerPoint.*

# 10

# Matrices

10.1 Solution of Linear Systems

10.2 Addition and Subtraction of Matrices

10.3 Multiplication of Matrices

10.4 Matrix Inverses

10.5 Eigenvalues and Eigenvectors

Chapter 10 Review

Extended Application: Contagion

When stocking a lake with various species of fish, wildlife biologists need to take into account the nutrient requirements of each fish, as well as the amounts of the various nutrients available in the lake. An exercise in the first section of this chapter solves such a problem using systems of linear equations and matrices.

Many mathematical models require finding the solutions of two or more equations. The solutions must satisfy *all* of the equations in the model. A set of equations related in this way is called a **system of equations**. In this chapter we will discuss systems of equations, introduce the idea of a *matrix*, and then show how matrices are used to solve systems of linear equations. We will also see how matrices provide answers to important questions about population growth.

# 10.1    Solution of Linear Systems

**APPLY IT** How much of each ingredient should be used in an animal feed to meet dietary requirements?

**APPLY IT**     Suppose that an animal feed is made from three ingredients: corn, soybeans, and cottonseed. One gram of each ingredient provides the number of grams of protein, fat, and fiber shown in the table. For example, the entries in the first column, 0.25, 0.4, and 0.3, indicate that one gram of corn provides twenty-five hundredths (one-fourth) of a gram of protein, four-tenths of a gram of fat, and three-tenths of a gram of fiber.

| Nutritional Content of Ingredients | | | |
|---|---|---|---|
| | Corn | Soybeans | Cottonseed |
| Protein | 0.25 | 0.4 | 0.2 |
| Fat | 0.4 | 0.2 | 0.3 |
| Fiber | 0.3 | 0.2 | 0.1 |

Now suppose we need to know the number of grams of each ingredient that should be used to make a feed that contains 22 grams of protein, 28 grams of fat, and 18 grams of fiber. To determine this, we let $x$ represent the required number of grams of corn, $y$ the number of grams of soybeans, and $z$ the number of grams of cottonseed. Since each gram of corn provides 0.25 gram of protein, the amount of protein provided by $x$ grams of corn is $0.25x$. Similarly, the amount of protein provided by $y$ grams of soybeans is $0.4y$ and the amount of protein provided by $z$ grams of cottonseed is $0.2z$. Since the total amount of protein is to be 22 grams,

$$0.25x + 0.4y + 0.2z = 22.$$

The feed must supply 28 grams of fat, so

$$0.4x + 0.2y + 0.3z = 28,$$

and 18 grams of fiber, so

$$0.3x + 0.2y + 0.1z = 18.$$

To solve this problem, we must find values of $x$, $y$, and $z$ that satisfy this system of equations. Verify that $x = 40$ grams, $y = 15$ grams, and $z = 30$ grams is a solution of the system by substituting these numbers into all three equations. In fact, this is the only solution of this system, as we will see in Exercise 45 of this section. Many practical problems lead to such systems of *first-degree equations*, also known as **linear systems**.

A **first-degree equation in $n$ unknowns** is any equation of the form

$$a_1x_1 + a_2x_2 + \cdots + a_nx_n = k,$$

where $a_1, a_2, \ldots, a_n$ and $k$ are all real numbers and $x_1, x_2, \ldots, x_n$ represent variables.* Each of the three equations from the animal feed problem is a first-degree equation. For example, the first equation

$$0.25x + 0.4y + 0.2z = 22$$

is a first-degree equation with $n = 3$ where

$$a_1 = 0.25, \quad a_2 = 0.4, \quad a_3 = 0.2, \quad k = 22,$$

and the variables are $x$, $y$, and $z$. We use $x_1$, $x_2$, etc., rather than $x$, $y$, etc., in the general case because we might have any number of variables. When $n$ is no more than 4, we usually use $x$, $y$, $z$, and $w$ to represent $x_1$, $x_2$, $x_3$, and $x_4$.

A *solution* of the first-degree equation

$$a_1x_1 + a_2x_2 + \cdots + a_nx_n = k$$

is a sequence of numbers $s_1, s_2, \ldots, s_n$ such that

$$a_1s_2 + a_2s_2 + \cdots + a_ns_n = k.$$

A solution of an equation is usually written in parentheses as $(s_1, s_2, \ldots, s_n)$. For example, $(1, 6, 2)$ is a solution of the equation $3x_1 + 2x_2 - 4x_3 = 7$, since $3(1) + 2(6) - 4(2) = 7$. This is an extension of the idea of an ordered pair, which was introduced in Chapter 1. A solution of a first-degree equation in two unknowns is an ordered pair, and the graph of the equation is a straight line. For this reason, all first-degree equations are also called linear equations.

Because the graph of a linear equation in two unknowns is a straight line, there are three possibilities for the solutions of a system of two linear equations in two unknowns.

## Types of Solutions for Two Equations in Two Unknowns

1. The two graphs are lines intersecting at a single point. The system has a **unique solution**, and it is given by the coordinates of this point. See Figure 1(a).

2. The graphs are distinct parallel lines. When this is the case, the system is **inconsistent**; there is no solution common to both equations. See Figure 1(b).

3. The graphs are the same line. In this case, the equations are said to be **dependent**, since any solution of one equation is also a solution of the other. There are infinitely many solutions. See Figure 1(c).

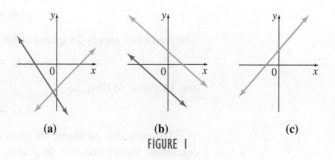

(a)  (b)  (c)

FIGURE 1

*$a_1$ is read "a-sub-one." The notation $a_1, a_2, \ldots a_n$ represents $n$ real-number coefficients (some of which may be equal), and the notation $x_1, x_2, \ldots x_n$ represents $n$ different variables, or unknowns.

In larger systems, with more variables or more equations (or both, as is usually the case), there also may be exactly one solution, no solution, or infinitely many solutions. With more than two variables. the geometrical interpretation becomes. more complicated, so we will consider only the geometry of two variables in two unknowns.

In this section we develop a method for solving a system of linear equations. Although the discussion will be confined to equations with only a few variables, the method of solution can be extended to systems with many variables.

When we solve a system, since the variables are in the same order in each equation, we really need to keep track of just the coefficients and the constants. For example, consider the following system of three equations in three unknowns.

$$2x + y - z = 2$$
$$x + 3y + 2z = 1$$
$$x + y + z = 2$$

This system can be written in an abbreviated form as

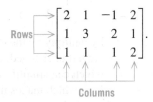

Such a rectangular array of numbers enclosed by brackets is called a **matrix** (plural: **matrices**).* Each number in the array is an **element** or **entry**. To separate the constants in the last column of the matrix from the coefficients of the variables, we use a vertical line, producing the following **augmented matrix**.

$$\begin{bmatrix} 2 & 1 & -1 & 2 \\ 1 & 3 & 2 & 1 \\ 1 & 1 & 1 & 2 \end{bmatrix}$$

Any algebraic operation done to the equations of a linear system is equivalent to performing the same operation on the rows of the augmented matrix, since the matrix is just a shorter way of writing the system. The following **row operations** on the augmented matrix transform the system of equations into an **equivalent system**, that is, one that has the same solutions as the original system.

## Row Operations

For any augmented matrix of a system of equations, the following operations produce the augmented matrix of an equivalent system:

1. interchanging any two rows;
2. multiplying the elements of a row by any nonzero real number;
3. adding a nonzero multiple of the elements of one row to the corresponding elements of a nonzero multiple of some other row.

In steps 2 and 3, we are replacing a row with a new, modified row, which the old row helped to form. This is equivalent to replacing an equation with a new, modified equation.

*The word *matrix*, Latin for "womb," was coined by James Joseph Sylvester (1814–1897) and made popular by his friend Arthur Cayley (1821–1895). Both mathematicians were English, although Sylvester spent much of his life in the United States.

Row operations, like the transformations of systems of equations, are reversible. If they are used to change matrix $A$ to matrix $B$, then it is possible to use row operations to transform $B$ back into $A$.

By the first row operation, we can interchange rows 1 and 2, for example, which we will indicate by the notation $R_2 \rightarrow R_1$ and $R_1 \rightarrow R_2$. ($R$ stands for row.) Row 3 is left unchanged. Thus

$$\begin{bmatrix} 0 & 1 & 2 & | & 3 \\ -2 & -6 & -10 & | & -12 \\ 2 & 1 & -2 & | & -5 \end{bmatrix} \quad \text{becomes} \quad \begin{bmatrix} -2 & -6 & -10 & | & -12 \\ 0 & 1 & 2 & | & 3 \\ 2 & 1 & -2 & | & -5 \end{bmatrix} \quad \begin{matrix} R_2 \rightarrow R_1 \\ R_1 \rightarrow R_2 \end{matrix}$$

by interchanging the first two rows. Row 3 is left unchanged.

The second row operation, multiplying a row by a number, allows us to change

$$\begin{bmatrix} -2 & -6 & -10 & | & -12 \\ 0 & 1 & 2 & | & 3 \\ 2 & 1 & -2 & | & -5 \end{bmatrix} \quad \text{to} \quad \begin{bmatrix} 1 & 3 & 5 & | & 6 \\ 0 & 1 & 2 & | & 3 \\ 2 & 1 & -2 & | & -5 \end{bmatrix} \quad (-1/2)R_1 \rightarrow R_1$$

by multiplying the elements of row 1 of the original matrix by $-1/2$. Note that rows 2 and 3 are left unchanged. This operation is helpful in converting all entries in a matrix to integers, which can simplify the calculations. It is also useful for dividing a common factor out of a row, which makes the numbers smaller and hence easier to work with.

Using the third row operation, adding a multiple of one row to another, we change

$$\begin{bmatrix} 1 & 3 & 5 & | & 6 \\ 0 & 1 & 2 & | & 3 \\ 2 & 1 & -2 & | & -5 \end{bmatrix} \quad \text{to} \quad \begin{bmatrix} 1 & 3 & 5 & | & 6 \\ 0 & 1 & 2 & | & 3 \\ 0 & -5 & -12 & | & -17 \end{bmatrix} \quad -2R_1 + R_3 \rightarrow R_3$$

by first multiplying each element in row 1 of the original matrix by $-2$ and then adding the results to the corresponding elements in the third row of that matrix. Work as follows.

$$\begin{bmatrix} 1 & 3 & 5 & | & 6 \\ 0 & 1 & 2 & | & 3 \\ (-2)1 + 2 & (-2)3 + 1 & (-2)5 - 2 & | & (-2)6 - 5 \end{bmatrix} = \begin{bmatrix} 1 & 3 & 5 & | & 6 \\ 0 & 1 & 2 & | & 3 \\ 0 & -5 & -12 & | & -17 \end{bmatrix}$$

Notice that rows 1 and 2 are left unchanged, *even though the elements of row 1 were used to transform row 3*. This operation is very useful for converting an element in a matrix to 0.

# The Gauss-Jordan Method

The **Gauss-Jordan Method** uses the row operations just described to transform a system of equations into one from which the solution can be immediately read.* Before the Gauss-Jordan method can be used, the system must be in proper form: the terms with variables should be on the left and the constants on the right in each equation, with the variables in the same order in each equation.

The system is then written as an augmented matrix. Using row operations, the goal is to transform the matrix so that it has zeros above and below a diagonal of 1's on the left of the vertical bar. Once this is accomplished, the final solution can be read directly from the last matrix. The following example illustrates the use of the Gauss-Jordan method to solve a system of equations.

*The great German mathematician Carl Friedrich Gauss (1777–1855), sometimes referred to as the "Prince of Mathematicians," originally developed his elimination method for use in finding least squares coefficients. (See Section 1.2.) The German geodesist Wilhelm Jordan (1842–1899) improved his method and used it in surveying problems. Gauss's method had been known to the Chinese at least 1800 years earlier and was described in the *Jiuahang Suanshu (Nine Chapters on the Mathematical Art)*.

EXAMPLE 1    **Gauss-Jordan Method**

Solve the system

$$3x - 4y = \ \ 1 \tag{1}$$
$$5x + 2y = 19. \tag{2}$$

**SOLUTION**

**Method 1**
**1's on Diagonal**

The system is already in the proper form to use the Gauss-Jordan method. Our goal is to transform this matrix, if possible, into the form

$$\begin{bmatrix} 1 & 0 & \bigm| & m \\ 0 & 1 & \bigm| & n \end{bmatrix},$$

where $m$ and $n$ are real numbers. To begin, we change the 3 in the first row to 1 using the second row operation.

$$\begin{bmatrix} 3 & -4 & \bigm| & 1 \\ 5 & 2 & \bigm| & 19 \end{bmatrix} \qquad \text{Augmented matrix}$$

$$\tfrac{1}{3}R_1 \rightarrow R_1 \quad \begin{bmatrix} 1 & -\tfrac{4}{3} & \bigm| & \tfrac{1}{3} \\ 5 & 2 & \bigm| & 19 \end{bmatrix}$$

Using the third row operation, we change the 5 in row 2 to 0.

$$-5R_1 + R_2 \rightarrow R_2 \quad \begin{bmatrix} 1 & -\tfrac{4}{3} & \bigm| & \tfrac{1}{3} \\ 0 & \tfrac{26}{3} & \bigm| & \tfrac{52}{3} \end{bmatrix}$$

We now change 26/3 in row 2 to 1 to complete the diagonal of 1's.

$$\tfrac{3}{26}R_2 \rightarrow R_2 \quad \begin{bmatrix} 1 & -\tfrac{4}{3} & \bigm| & \tfrac{1}{3} \\ 0 & 1 & \bigm| & 2 \end{bmatrix}$$

The final transformation is to change the $-4/3$ in row 1 to 0.

$$\tfrac{4}{3}R_2 + R_1 \rightarrow R_1 \quad \begin{bmatrix} 1 & 0 & \bigm| & 3 \\ 0 & 1 & \bigm| & 2 \end{bmatrix}$$

The last matrix corresponds to the system

$$x = 3$$
$$y = 2,$$

so we can read the solution directly from the last column of the final matrix. Check that $(3, 2)$ is the solution by substitution in the equations of the original system.

**Method 2**
**Fraction-Free**

An alternate form of Gauss-Jordan is to first transform the matrix so that it contains zeros above and below the main diagonal. Then, use the second transformation to get the required 1's. When doing calculations by hand, this second method simplifies the calculations by avoiding fractions and decimals. We will use this method when doing calculations by hand throughout the remainder of this chapter.

To begin, we change the 5 in row 2 to 0.

$$\begin{bmatrix} 3 & -4 & \bigm| & 1 \\ 5 & 2 & \bigm| & 19 \end{bmatrix} \qquad \text{Augmented matrix}$$

$$5R_1 + (-3)R_2 \rightarrow R_2 \quad \begin{bmatrix} 3 & -4 & \bigm| & 1 \\ 0 & -26 & \bigm| & -52 \end{bmatrix}$$

We change the $-4$ in row 1 to 0.

$$-2R_2 + 13R_1 \rightarrow R_1 \quad \begin{bmatrix} 39 & 0 & \bigm| & 117 \\ 0 & -26 & \bigm| & -52 \end{bmatrix}$$

**YOUR TURN 1** Use the Gauss-Jordan method to solve the system

$4x + 5y = 10$

$7x + 8y = 19.$

Then we change the first nonzero number in each row to 1.

$$\begin{array}{c} \frac{1}{39}R_1 \rightarrow R_1 \\ -\frac{1}{26}R_2 \rightarrow R_2 \end{array} \quad \begin{bmatrix} 1 & 0 & | & 3 \\ 0 & 1 & | & 2 \end{bmatrix}$$

The solution is read directly from this last matrix: $x = 3$ and $y = 2$, or $(3, 2)$.

**TRY YOUR TURN 1**

**NOTE** If your solution does not check, the most efficient way to find the error is to substitute back through the equations that correspond to each matrix, starting with the last matrix. When you find a system that is not satisfied by your (incorrect) answers, you have probably reached the matrix just before the error occurred. Look for the error in the transformation to the next matrix. For example, if you erroneously wrote the 2 as $-2$ in the final matrix of the fraction-free method of Example 1, you would find that $(3, -2)$ was not a solution of the system represented by the previous matrix because $-26(-2) \neq -52$, telling you that your error occurred between this matrix and the final one.

When the Gauss-Jordan method is used to solve a system, the final matrix will always have zeros above and below the diagonal of 1's on the left of the vertical bar. To transform the matrix, it is best to work column by column from left to right. Such an orderly method avoids confusion and working in circles. For each column, first perform the steps that give the zeros. When all columns have zeros in place, multiply each row by the reciprocal of the coefficient of the remaining nonzero number in that row to get the required 1's. With dependent equations or inconsistent systems, it will not be possible to get the complete diagonal of 1's.

We will demonstrate the method in an example with three variables and three equations, after which we will summarize the steps.

**EXAMPLE 2** **Gauss-Jordan Method**

Use the Gauss-Jordan method to solve the system

$$x + 5z = -6 + y$$
$$3x + 3y = 10 + z$$
$$x + 3y + 2z = 5.$$

**Method 1**
**Calculating by Hand**

**SOLUTION**

First, rewrite the system in proper form, as follows.

$$x - y + 5z = -6$$
$$3x + 3y - z = 10$$
$$x + 3y + 2z = 5$$

Begin to find the solution by writing the augmented matrix of the linear system.

$$\begin{bmatrix} 1 & -1 & 5 & | & -6 \\ 3 & 3 & -1 & | & 10 \\ 1 & 3 & 2 & | & 5 \end{bmatrix}$$

Row transformations will be used to rewrite this matrix in the form

$$\begin{bmatrix} 1 & 0 & 0 & | & m \\ 0 & 1 & 0 & | & n \\ 0 & 0 & 1 & | & p \end{bmatrix},$$

where $m$, $n$, and $p$ are real numbers (if this form is possible). From this final form of the matrix, the solution can be read: $x = m$, $y = n$, $z = p$, or $(m, n, p)$.

   In the first column, we need zeros in the second and third rows. Multiply the first row by $-3$ and add to the second row to get a zero there. Then multiply the first row by $-1$ and add to the third row to get that zero.

$$
\begin{matrix} -3R_1 + R_2 \to R_2 \\ -1R_1 + R_3 \to R_3 \end{matrix}
\left[\begin{array}{rrr|r} 1 & -1 & 5 & -6 \\ 0 & 6 & -16 & 28 \\ 0 & 4 & -3 & 11 \end{array}\right]
$$

Now get zeros in the second column in a similar way. We want zeros in the first and third rows. Row 2 will not change.

$$
\begin{matrix} R_2 + 6R_1 \to R_1 \\ \\ 2R_2 + (-3)R_3 \to R_3 \end{matrix}
\left[\begin{array}{rrr|r} 6 & 0 & 14 & -8 \\ 0 & 6 & -16 & 28 \\ 0 & 0 & -23 & 23 \end{array}\right]
$$

In transforming the third row, you may have used the operation $4R_2 + (-6)R_3 \to R_3$ instead of $2R_2 + (-3)R_3 \to R_3$. This is perfectly fine; the last row would then have $-46$ and $46$ in place of $-23$ and $23$. To avoid errors, it helps to keep the numbers as small as possible. We observe at this point that all of the numbers can be reduced in size by multiplying each row by an appropriate constant. This next step is not essential, but it simplifies the arithmetic.

$$
\begin{matrix} \frac{1}{2}R_1 \to R_1 \\ \frac{1}{2}R_2 \to R_2 \\ -\frac{1}{23}R_3 \to R_3 \end{matrix}
\left[\begin{array}{rrr|r} 3 & 0 & 7 & -4 \\ 0 & 3 & -8 & 14 \\ 0 & 0 & 1 & -1 \end{array}\right]
$$

Next, we want zeros in the first and second rows of the third column. Row 3 will not change.

$$
\begin{matrix} -7R_3 + R_1 \to R_1 \\ 8R_3 + R_2 \to R_2 \end{matrix}
\left[\begin{array}{rrr|r} 3 & 0 & 0 & 3 \\ 0 & 3 & 0 & 6 \\ 0 & 0 & 1 & -1 \end{array}\right]
$$

Finally, get 1's in each row by multiplying the row by the reciprocal of (or dividing the row by) the number in the diagonal position.

$$
\begin{matrix} \frac{1}{3}R_1 \to R_1 \\ \frac{1}{3}R_2 \to R_2 \end{matrix}
\left[\begin{array}{rrr|r} 1 & 0 & 0 & 1 \\ 0 & 1 & 0 & 2 \\ 0 & 0 & 1 & -1 \end{array}\right]
$$

The linear system associated with the final augmented matrix is

$$
\begin{aligned} x &= 1 \\ y &= 2 \\ z &= -1, \end{aligned}
$$

and the solution is $(1, 2, -1)$. Verify that this is the solution to the original system of equations.

---

| CAUTION | Notice that we have performed two or three operations on the same matrix in one step. This is permissible as long as we do not use a row that we are changing as part of another row operation. For example, when we changed row 2 in the first step, we could not use row 2 to transform row 3 in the same step. To avoid difficulty, use *only* row 1 to get zeros in column 1, row 2 to get zeros in column 2, and so on. |

**Method 2**
**Graphing Calculator**

The row operations of the Gauss-Jordan method can also be done on a graphing calculator. For example, Figure 2 on the next page shows the result when the augmented matrix is entered into a TI-84 Plus. Figures 3 and 4 show how row operations can be used to get zeros in rows 2 and 3 of the first column.

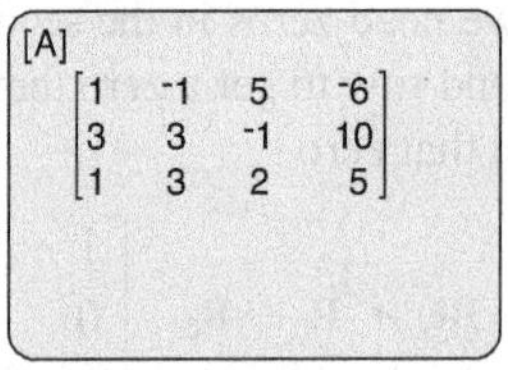

[A]
$$\begin{bmatrix} 1 & \text{-}1 & 5 & \text{-}6 \\ 3 & 3 & \text{-}1 & 10 \\ 1 & 3 & 2 & 5 \end{bmatrix}$$

FIGURE 2

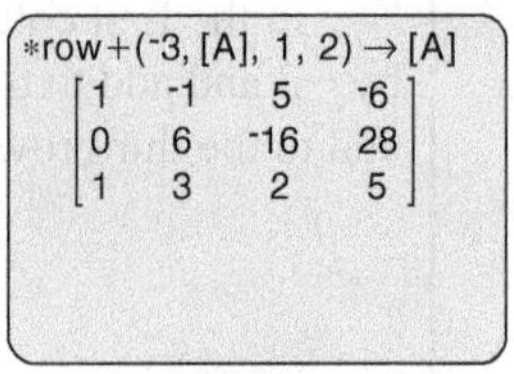

*row+(-3, [A], 1, 2) → [A]
$$\begin{bmatrix} 1 & \text{-}1 & 5 & \text{-}6 \\ 0 & 6 & \text{-}16 & 28 \\ 1 & 3 & 2 & 5 \end{bmatrix}$$

FIGURE 3

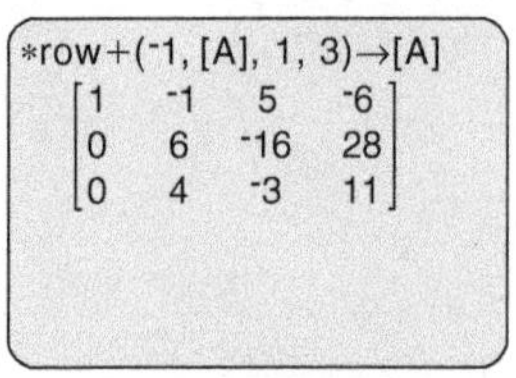

*row+(-1, [A], 1, 3)→[A]
$$\begin{bmatrix} 1 & \text{-}1 & 5 & \text{-}6 \\ 0 & 6 & \text{-}16 & 28 \\ 0 & 4 & \text{-}3 & 11 \end{bmatrix}$$

FIGURE 4

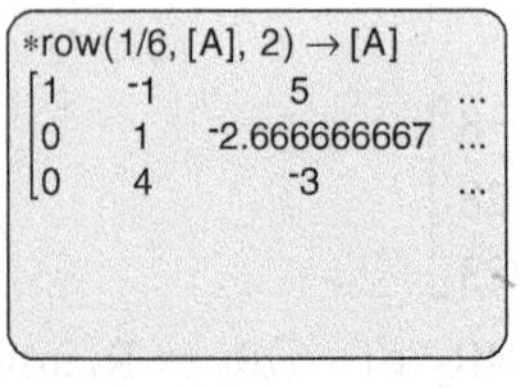

*row(1/6, [A], 2) → [A]
$$\begin{bmatrix} 1 & \text{-}1 & 5 & \cdots \\ 0 & 1 & \text{-}2.666666667 & \cdots \\ 0 & 4 & \text{-}3 & \cdots \end{bmatrix}$$

FIGURE 5

Ans►Frac
$$\begin{bmatrix} 1 & \text{-}1 & 5 & \text{-}6 \\ 0 & 1 & \text{-}\frac{8}{3} & \frac{14}{3} \\ 0 & 4 & \text{-}3 & 11 \end{bmatrix}$$

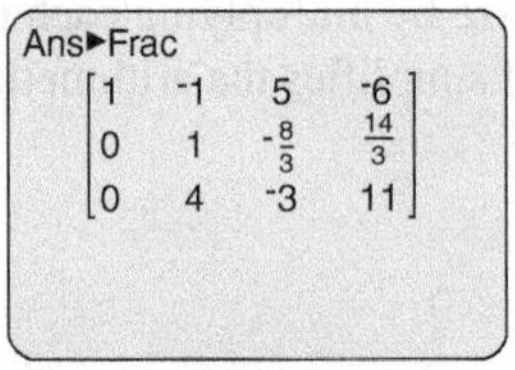

FIGURE 6

Calculators typically do not allow any multiple of a row to be added to any multiple of another row, such as in the operation $2R_2 + 6R_1 \rightarrow R_1$. They normally allow a multiple of a row to be added only to another unmodified row. To get around this restriction, we can convert the diagonal element to a 1 before changing the other elements in the column to 0, as we did in the first method of Example 1. In this example, we change the 6 in row 2, column 2, to a 1 by dividing by 6. The result is shown in Figure 5. (The right side of the matrix is not visible but can be seen by pressing the right arrow key.) Notice that this operation introduces decimals. Converting to fractions is preferable on calculators that have that option; 1/3 is certainly more concise than 0.3333333333. Figure 6 shows such a conversion on the TI-84 Plus.

When performing row operations without a graphing calculator, it is best to avoid fractions and decimals, because these make the operations more difficult and more prone to error. A calculator, on the other hand, encounters no such difficulties.

Continuing in the same manner, the solution $(1, 2, -1)$ is found as shown in Figure 7.

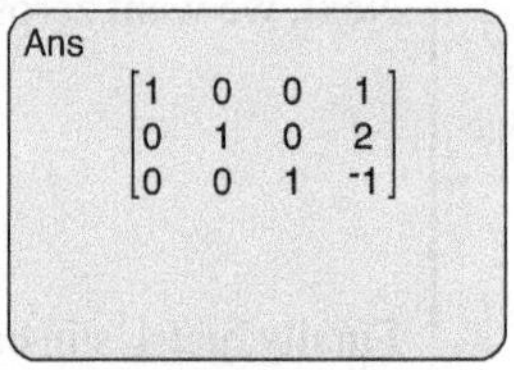

Ans
$$\begin{bmatrix} 1 & 0 & 0 & 1 \\ 0 & 1 & 0 & 2 \\ 0 & 0 & 1 & \text{-}1 \end{bmatrix}$$

FIGURE 7

Some calculators can do the entire Gauss-Jordan process with a single command; on the TI-84 Plus, for example, this is done with the `rref` command. This is very useful in practice, although it does not show any of the intermediate steps. For more details, see the *Graphing Calculator and Excel Spreadsheet Manual* available with this book.

**Method 3 Spreadsheet**

The Gauss-Jordan method can be done using a spreadsheet either by using a macro or by performing row operations using formulas with the copy and paste commands. However, spreadsheets also have built-in methods to solve systems of equations. Although these solvers do not usually employ the Gauss-Jordan method for solving systems of equations, they are, nonetheless, efficient and practical to use.

The Solver included with Excel can solve systems of equations that are both linear and nonlinear. The Solver is located in the Tools menu and requires that cells be identified ahead of time for each variable in the problem. It also requires that the left-hand side of each equation be placed in the spreadsheet as a formula. For example, to solve the above problem, we could identify cells A1, B1, and C1 for the variables $x$, $y$, and $z$, respectively. The Solver requires that we place a guess for the answer in these cells. It is convenient to place a zero in each of these cells. The left-hand side of each equation must be placed in a cell. We could choose A3, A4, and A5 to hold each of these formulas. Thus, in cell A3, we would type "=A1 − B1 + 5*C1" and put the other two equations in cells A4 and A5.

We now click on the Tools menu and choose Solver. (In some versions, it may be necessary to install the Solver. For more details, see the *Graphing Calculator and Excel Spreadsheet Manual* available with this book.) Since this solver attempts to find a solution that is best in some way, we are required to identify a cell with a formula in it that we want to optimize. In this case, it is convenient to use the cell with the left-hand side of the first constraint in it, A3. Figure 8 illustrates the Solver box and the items placed in it.

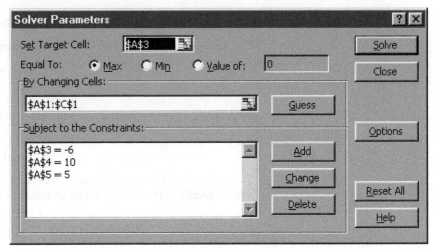

FIGURE 8

YOUR TURN 2 Use the Gauss-Jordan method to solve the system

$$x + 2y + 3z = 2$$
$$2x + 2y - 3z = 27$$
$$3x + 2y + 5z = 10.$$

To obtain a solution, click on Solve. The approximate solution is located in cells A1, B1, and C1, and these correspond to $x$, $y$, and $z$, respectively.

TRY YOUR TURN 2

In summary, the Gauss-Jordan method of solving a linear system requires the following steps.

## Gauss-Jordan Method of Solving a Linear System

1. Write each equation so that variable terms are in the same order on the left side of the equal sign and constants are on the right.
2. Write the augmented matrix that corresponds to the system.
3. Use row operations to transform the first column so that all elements except the element in the first row are zero.
4. Use row operations to transform the second column so that all elements except the element in the second row are zero.
5. Use row operations to transform the third column so that all elements except the element in the third row are zero.
6. Continue in this way, when possible, until the last row is written in the form

$$[0 \quad 0 \quad 0 \quad \cdots \quad 0 \quad j \mid k],$$

   where $j$ and $k$ are constants with $j \neq 0$. When this is not possible, continue until every row has more zeros on the left than the previous row (except possibly for any rows of all zeros at the bottom of the matrix), and the first nonzero entry in each row is the only nonzero entry in its column.
7. Multiply each row by the reciprocal of the nonzero element in that row.

# Systems without a Unique Solution

In the previous examples, we were able to get the last row in the form $[0 \ 0 \ 0 \ \cdots \ 0 \ j \mid k]$, where $j$ and $k$ are constants with $j \neq 0$. We will now look at examples where this is not the case.

## EXAMPLE 3  Solving a System of Equations with No Solution

Use the Gauss-Jordan method to solve the system

$$x - 2y = 2$$
$$3x - 6y = 5.$$

**SOLUTION**  Begin by writing the augmented matrix.

$$\begin{bmatrix} 1 & -2 & \bigm| & 2 \\ 3 & -6 & \bigm| & 5 \end{bmatrix}$$

To get a zero for the second element in column 1, multiply the numbers in row 1 by $-3$ and add the results to the corresponding elements in row 2.

$$-3R_1 + R_2 \to R_2 \quad \begin{bmatrix} 1 & -2 & \bigm| & 2 \\ 0 & 0 & \bigm| & -1 \end{bmatrix}$$

This matrix corresponds to the system

$$x - 2y = 2$$
$$0x + 0y = -1.$$

Since the second equation is $0 = -1$, the system is inconsistent and, therefore, has no solution. The row $[0 \ 0 \mid k]$ for any nonzero $k$ is a signal that the given system is inconsistent.

## EXAMPLE 4  Solving a System of Equations with an Infinite Number of Solutions

Use the Gauss-Jordan method to solve the system

$$x + 2y - z = 0$$
$$3x - y + z = 6$$
$$-2x - 4y + 2z = 0.$$

**SOLUTION**  The augmented matrix is

$$\begin{bmatrix} 1 & 2 & -1 & \bigm| & 0 \\ 3 & -1 & 1 & \bigm| & 6 \\ -2 & -4 & 2 & \bigm| & 0 \end{bmatrix}.$$

We first get zeros in the second and third rows of column 1.

$$\begin{matrix} -3R_1 + R_2 \to R_2 \\ 2R_1 + R_3 \to R_3 \end{matrix} \quad \begin{bmatrix} 1 & 2 & -1 & \bigm| & 0 \\ 0 & -7 & 4 & \bigm| & 6 \\ 0 & 0 & 0 & \bigm| & 0 \end{bmatrix}$$

To continue, we get a zero in the first row of column 2 using the second row, as usual.

$$2R_2 + 7R_1 \to R_1 \quad \begin{bmatrix} 7 & 0 & 1 & \bigm| & 12 \\ 0 & -7 & 4 & \bigm| & 6 \\ 0 & 0 & 0 & \bigm| & 0 \end{bmatrix}$$

We cannot get a zero for the first-row, third-column element without changing the form of the first two columns. We must multiply each of the first two rows by the reciprocal of the first nonzero number.

$$\begin{matrix} \frac{1}{7}R_1 \rightarrow R_1 \\ -\frac{1}{7}R_2 \rightarrow R_2 \end{matrix} \begin{bmatrix} 1 & 0 & \frac{1}{7} & \Big| & \frac{12}{7} \\ 0 & 1 & -\frac{4}{7} & \Big| & -\frac{6}{7} \\ 0 & 0 & 0 & \Big| & 0 \end{bmatrix}$$

To complete the solution, write the equations that correspond to the first two rows of the matrix.

$$x + \frac{1}{7}z = \frac{12}{7}$$
$$y - \frac{4}{7}z = -\frac{6}{7}$$

The third equation of this system, $0 = 0$, gives us no information. Usually the third equation gives us the value of $z$. In this case, $z$ can be any real number. This system is dependent and has an infinite number of solutions, corresponding to the infinite number of values of $z$. The variable $z$ in this case is called a **parameter**. Solve the first equation for $x$ and the second for $y$ to get

$$x = \frac{12 - z}{7} \quad \text{and} \quad y = \frac{4z - 6}{7}.$$

The general solution is written

$$\left( \frac{12 - z}{7}, \frac{4z - 6}{7}, z \right),$$

where $z$ is any real number. For example, $z = 2$ and $z = 12$ lead to the solutions $(10/7, 2/7, 2)$ and $(0, 6, 12)$.

---

**EXAMPLE 5** **Solving a System of Equations with an Infinite Number of Solutions**

Consider the following system of equations.

$$x + 2y + 3z - w = 4$$
$$2x + 3y \quad\quad + w = -3$$
$$3x + 5y + 3z \quad\quad = 1$$

**(a)** Write this system as an augmented matrix, and verify that the result after the Gauss-Jordan method is

$$\begin{bmatrix} 1 & 0 & -9 & 5 & \Big| & -18 \\ 0 & 1 & 6 & -3 & \Big| & 11 \\ 0 & 0 & 0 & 0 & \Big| & 0 \end{bmatrix}.$$

**(b)** Find the solution to this system of equations.

**SOLUTION** To complete the solution, write the equations that correspond to the first two rows of the matrix.

$$x \quad - 9z + 5w = -18$$
$$y + 6z - 3w = 11$$

Because both equations involve both $z$ and $w$, let $z$ and $w$ be parameters. There are an infinite number of solutions, corresponding to the infinite number of values of $z$ and $w$. Solve the first equation for $x$ and the second for $y$ to get

$$x = -18 + 9z - 5w \quad \text{and} \quad y = 11 - 6z + 3w.$$

**YOUR TURN 3** Use the Gauss-Jordan method to solve the system

$2x - 2y + 3z - 4w = 6$

$3x + 2y + 5z - 3w = 7$

$4x + y + 2z - 2w = 8.$

In an analogous manner to problems with a single parameter, the general solution is written

$$(-18 + 9z - 5w, 11 - 6z + 3w, z, w),$$

where $z$ and $w$ are any real numbers. For example, $z = 1$ and $w = -2$ leads to the solution $(1, -1, 1, -2)$.   TRY YOUR TURN 3

Although the examples have used only systems with two equations in two unknowns, three equations in three unknowns, or three equations in four unknowns, the Gauss-Jordan method can be used for any system with $n$ equations and $m$ unknowns. The method becomes tedious with more than three equations in three unknowns; on the other hand, it is very suitable for use with graphing calculators and computers, which can solve fairly large systems quickly. Sophisticated computer programs modify the method to reduce round-off error. Other methods used for special types of large matrices are studied in a course on numerical analysis.

## EXAMPLE 6   Rats

A lab has a group of young rats, adult female rats, and adult male rats to be used for experiments. There are 23 rats in all. The rats consume 376 g of food and 56.9 mL of water each day.

(a) Suppose that a young rat consumes 12 g of food and 2 mL of water each day. The corresponding values for an adult female are 16 and 2.5, and for an adult male are 20 and 2.8. How many young rats, adult females, and adult males are there?

**SOLUTION**   We have organized the information in a table.

| Lab Rats | | | | |
|---|---|---|---|---|
| | Young | Female | Male | Total |
| Number | $x$ | $y$ | $z$ | 23 |
| Food (g) | 12 | 16 | 20 | 376 |
| Water (mL) | 2.0 | 2.5 | 2.8 | 56.9 |

The three rows of the table lead to three equations: one for the total number of rats, one for the food, and one for the water.

$$x + y + z = 23$$
$$12x + 16y + 20z = 376$$
$$2.0x + 2.5y + 2.8z = 56.9$$

Represent this system by an augmented matrix, and verify that the result after the Gauss-Jordan method is

$$\begin{bmatrix} 1 & 0 & 0 & | & 6 \\ 0 & 1 & 0 & | & 9 \\ 0 & 0 & 1 & | & 8 \end{bmatrix}.$$

There are 6 young rats, 9 adult females, and 8 adult males.

(b) Suppose the amounts of water that young rats, adult females, and adult males consume in a new experiment are 2 mL, 4 mL, and 6 mL, respectively, but all other information remains the same. How many young rats, adult females, and adult males are there?

**SOLUTION**   Change the third equation to

$$2x + 4y + 6z = 56.9$$

and use the Gauss-Jordan method again. The result is

$$\begin{bmatrix} 1 & 0 & -1 & 2 \\ 0 & 1 & 2 & 25 \\ 0 & 0 & 0 & -37.1 \end{bmatrix}.$$

(If you do the row operations in a different order in this example, you will have different numbers in the last column.) The last row of this matrix says that $0 = -37.1$, so the system is inconsistent and has no solution. (In retrospect, this is clear, because each rat consumes a whole number of mL of water, and the total amount of water consumed is not a whole number. In general, however, it is not easy to tell whether a system of equations has a solution or not by just looking at it.)

**(c)** Suppose the amounts of water consumed by each rat are the same as in part (b), but the total amount of water consumed is 96 mL. Now how many young rats, adult females, and adult males are there?

**SOLUTION**   The third equation becomes

$$2x + 4y + 6z = 96,$$

and the Gauss-Jordan method leads to

$$\begin{bmatrix} 1 & 0 & -1 & -2 \\ 0 & 1 & 2 & 25 \\ 0 & 0 & 0 & 0 \end{bmatrix}.$$

The system is dependent, similar to Example 4. Let $z$ be the parameter, and solve the first two equations for $x$ and $y$, yielding

$$x = z - 2 \quad \text{and} \quad y = 25 - 2z.$$

Since $x$, $y$, and $z$ represent numbers of rats, they must be nonnegative integers. From the equation for $x$, we must have

$$z - 2 \geq 0, \quad \text{or} \quad z \geq 2,$$

and from the equation for $y$, we must have

$$25 - 2z \geq 0,$$

from which we find

$$z \leq 12.5.$$

We therefore have 11 solutions, corresponding to $z = 2, 3, \ldots, 12$.

**(d)** Give the solutions from part (c) that have the smallest and the largest number of adult males.

**SOLUTION**   For the smallest number of adult males, let $z = 2$, giving $x = 2 - 2 = 0$ and $y = 25 - 2(2) = 21$. There are 0 young rats, 21 adult females, and 2 adult males.

For the largest number of adult males, let $z = 12$, giving $x = 12 - 2 = 10$ and $y = 25 - 2(12) = 1$. There are 10 young rats, 1 adult female, and 12 adult males.

# 10.1 EXERCISES

**Write the augmented matrix for each system. Do not solve.**

1. $3x + y = 6$
   $2x + 5y = 15$

2. $4x - 2y = 8$
   $-7y = -12$

3. $2x + y + z = 3$
   $3x - 4y + 2z = -7$
   $x + y + z = 2$

4. $2x - 5y + 3z = 4$
   $-4x + 2y - 7z = -5$
   $3x - y = 8$

**Write the system of equations associated with each augmented matrix.**

5. $\begin{bmatrix} 1 & 0 & | & 2 \\ 0 & 1 & | & 3 \end{bmatrix}$

6. $\begin{bmatrix} 1 & 0 & | & 5 \\ 0 & 1 & | & -3 \end{bmatrix}$

7. $\begin{bmatrix} 1 & 0 & 0 & | & 4 \\ 0 & 1 & 0 & | & -5 \\ 0 & 0 & 1 & | & 1 \end{bmatrix}$

8. $\begin{bmatrix} 1 & 0 & 0 & | & 4 \\ 0 & 1 & 0 & | & 2 \\ 0 & 0 & 1 & | & 3 \end{bmatrix}$

9. _____ on a matrix correspond to transformations of a system of equations.

10. Describe in your own words what $2R_1 + R_3 \rightarrow R_3$ means.

**Use the indicated row operations to change each matrix.**

11. Replace $R_2$ by $R_1 + (-3)R_2$.

$\begin{bmatrix} 3 & 7 & 4 & | & 10 \\ 1 & 2 & 3 & | & 6 \\ 0 & 4 & 5 & | & 11 \end{bmatrix}$

12. Replace $R_3$ by $(-1)R_1 + 3R_3$.

$\begin{bmatrix} 3 & 2 & 6 & | & 18 \\ 2 & -2 & 5 & | & 7 \\ 1 & 0 & 5 & | & 20 \end{bmatrix}$

13. Replace $R_1$ by $(-2)R_2 + R_1$.

$\begin{bmatrix} 1 & 6 & 4 & | & 7 \\ 0 & 3 & 2 & | & 5 \\ 0 & 5 & 3 & | & 7 \end{bmatrix}$

14. Replace $R_1$ by $R_3 + (-3)R_1$.

$\begin{bmatrix} 1 & 0 & 4 & | & 21 \\ 0 & 6 & 5 & | & 30 \\ 0 & 0 & 12 & | & 15 \end{bmatrix}$

15. Replace $R_1$ by $\frac{1}{3}R_1$.

$\begin{bmatrix} 3 & 0 & 0 & | & 18 \\ 0 & 5 & 0 & | & 9 \\ 0 & 0 & 4 & | & 8 \end{bmatrix}$

16. Replace $R_3$ by $\frac{1}{6}R_3$.

$\begin{bmatrix} 1 & 0 & 0 & | & 30 \\ 0 & 1 & 0 & | & 17 \\ 0 & 0 & 6 & | & 162 \end{bmatrix}$

**Use the Gauss-Jordan method to solve each system of equations.**

17. $x + y = 5$
    $3x + 2y = 12$

18. $x + 2y = 5$
    $2x + y = -2$

19. $x + y = 7$
    $4x + 3y = 22$

20. $4x - 2y = 3$
    $-2x + 3y = 1$

21. $2x - 3y = 2$
    $4x - 6y = 1$

22. $2x + 3y = 9$
    $4x + 6y = 7$

23. $6x - 3y = 1$
    $-12x + 6y = -2$

24. $x - y = 1$
    $-x + y = -1$

25. $y = x - 3$
    $y = 1 + z$
    $z = 4 - x$

26. $x = 1 - y$
    $2x = z$
    $2z = -2 - y$

27. $2x - 2y = -5$
    $2y + z = 0$
    $2x + z = -7$

28. $x - z = -3$
    $y + z = 9$
    $-2x + 3y + 5z = 33$

29. $4x + 4y - 4z = 24$
    $2x - y + z = -9$
    $x - 2y + 3z = 1$

30. $x + 2y - 7z = -2$
    $-2x - 5y + 2z = 1$
    $3x + 5y + 4z = -9$

31. $3x + 5y - z = 0$
    $4x - y + 2z = 1$
    $7x + 4y + z = 1$

32. $3x - 6y + 3z = 11$
    $2x + y - z = 2$
    $5x - 5y + 2z = 6$

33. $5x - 4y + 2z = 6$
    $5x + 3y - z = 11$
    $15x - 5y + 3z = 23$

34. $3x + 2y - z = -16$
    $6x - 4y + 3z = 12$
    $5x - 2y + 2z = 4$

35. $2x + 3y + z = 9$
    $4x + 6y + 2z = 18$
    $-\frac{1}{2}x - \frac{3}{4}y - \frac{1}{4}z = -\frac{9}{4}$

36. $3x - 5y - 2z = -9$
    $-4x + 3y + z = 11$
    $8x - 5y + 4z = 6$

37. $x + 2y - w = 3$
    $2x + 4z + 2w = -6$
    $x + 2y - z = 6$
    $2x - y + z + w = -3$

38. $x + 3y - 2z - w = 9$
    $2x + 4y + 2w = 10$
    $-3x - 5y + 2z - w = -15$
    $x - y - 3z + 2w = 6$

39. $x + y - z + 2w = -20$
    $2x - y + z + w = 11$
    $3x - 2y + z - 2w = 27$

40. $4x - 3y + z + w = 21$
    $-2x - y + 2z + 7w = 2$
    $10x - 5z - 20w = 15$

**41.** $10.47x + 3.52y + 2.58z - 6.42w = 218.65$
$8.62x - 4.93y - 1.75z + 2.83w = 157.03$
$4.92x + 6.83y - 2.97z + 2.65w = 462.3$
$2.86x + 19.10y - 6.24z - 8.73w = 398.4$

**42.** $28.6x + 94.5y + 16.0z - 2.94w = 198.3$
$16.7x + 44.3y - 27.3z + 8.9w = 254.7$
$12.5x - 38.7y + 92.5z + 22.4w = 562.7$
$40.1x - 28.3y + 17.5z - 10.2w = 375.4$

**43.** On National Public Radio, the "Weekend Edition" program on Sunday, July 29, 2001, posed the following puzzle: Draw a three-by-three square (three boxes across by three boxes down). Put the fraction 3/8 in the first square in the first row. Put the fraction 1/4 in the last square in the second row. The object is to put a fraction in each of the remaining boxes, so the three numbers in each row, each column, and each of the long diagonals add up to 1. Solve this puzzle by letting seven variables represent the seven unknown fractions, writing eight equations for the eight sums, and solving by the Gauss-Jordan method.

## LIFE SCIENCE APPLICATIONS

**44. Birds** The date of the first sighting of robins has been occurring earlier each spring over the past 25 years at the Rocky Mountain Biological Laboratory. Scientists from this laboratory have developed two linear equations that estimate the date of the first sighting of robins:

$$y = 759 - 0.338x$$
$$y = 1637 - 0.779x,$$

where $x$ is the year and $y$ is the estimated number of days into the year when a robin can be expected. *Source: Proceedings of the National Academy of Science.*

**a.** Compare the date of first sighting in 2015 for each of these equations.

**b.** Solve this system of equations to find the year in which the two estimates agree.

**45. Animal Feed** Solve the animal feed problem posed at the beginning of this section.

**46. Animal Breeding** An animal breeder can buy four types of food for Vietnamese pot-bellied pigs. Each case of Brand A contains 25 units of fiber, 30 units of protein, and 30 units of fat. Each case of Brand B contains 50 units of fiber, 30 units of protein, and 20 units of fat. Each case of Brand C contains 75 units of fiber, 30 units of protein, and 20 units of fat. Each case of Brand D contains 100 units of fiber, 60 units of protein, and 30 units of fat. How many cases of each should the breeder mix together to obtain a food that provides 1200 units of fiber, 600 units of protein, and 400 units of fat?

**47. Dietetics** A hospital dietician is planning a special diet for a certain patient. The total amount per meal of food groups A, B, and C must equal 400 grams. The diet should include one-third as much of group A as of group B, and the sum of the amounts

of group A and group C should equal twice the amount of group B.

**a.** How many grams of each food group should be included?

**b.** Suppose we drop the requirement that the diet include one-third as much of group A as of group B. Describe the set of all possible solutions.

**c.** Suppose that, in addition to the conditions given in the original problem, foods A and B cost 2 cents per gram and food C costs 3 cents per gram, and that a meal must cost $8. Is a solution possible?

**48. Bacterial Food Requirements** Three species of bacteria are fed three foods, I, II, and III. A bacterium of the first species consumes 1.3 units each of foods I and II and 2.3 units of food III each day. A bacterium of the second species consumes 1.1 units of food I, 2.4 units of food II, and 3.7 units of food III each day. A bacterium of the third species consumes 8.1 units of I, 2.9 units of II, and 5.1 units of III each day. If 16,000 units of I, 28,000 units of II, and 44,000 units of III are supplied each day, how many of each species can be maintained in this environment?

**49. Fish Food Requirements** A lake is stocked each spring with three species of fish, A, B, and C. Three foods, I, II, and III, are available in the lake. Each fish of species A requires an average of 1.32 units of food I, 2.9 units of food II, and 1.75 units of food III each day. Species B fish each require 2.1 units of food I, 0.95 unit of food II, and 0.6 unit of food III daily. Species C fish require 0.86, 1.52, and 2.01 units of I, II, and III per day, respectively. If 490 units of food I, 897 units of food II, and 653 units of food III are available daily, how many of each species should be stocked?

**50. Agriculture** According to data from a Texas agricultural report, the amount of nitrogen (in lb/acre), phosphate (in lb/acre), and labor (in hr/acre) needed to grow honeydews, yellow onions, and lettuce is given by the following table. *Source: The AMATYC Review.*

| | Honeydews | Yellow Onions | Lettuce |
|---|---|---|---|
| *Nitrogen* | 120 | 150 | 180 |
| *Phosphate* | 180 | 80 | 80 |
| *Labor* | 4.97 | 4.45 | 4.65 |

**a.** If the farmer has 220 acres, 29,100 lb of nitrogen, 32,600 lb of phosphate, and 480 hours of labor, is it possible to use all resources completely? If so, how many acres should he allot for each crop?

**b.** Suppose everything is the same as in part a, except that 1061 hours of labor are available. Is it possible to use all resources completely? If so, how many acres should he allot for each crop?

**51. Mixing Plant Foods** Natural Brand plant food is made from three chemicals. The mix must include 10.8% of the first chemical, and the other two chemicals must be in a ratio of 4 to 3 as measured by weight. How much of each chemical is required to make 750 kg of the plant food?

**52. Health** The U.S. National Center for Health Statistics tracks the major causes of death in the United States. After a steady increase, the death rate by cancer has decreased since the early 1990s. The table lists the age-adjusted death rate per 100,000 people for 4 years. *Source: The New York Times 2010 Almanac.*

| Year | Rate |
|------|------|
| 1980 | 183.9 |
| 1990 | 203.3 |
| 2000 | 196.5 |
| 2010 | 186.2 |

**a.** If the relationship between the death rate $R$ and the year $t$ is expressed as $R = at^2 + bt + c$, where $t = 0$ corresponds to 1980, use data from 1980, 1990, and 2000 and a linear system of equations to determine the constants $a$, $b$, and $c$.

**b.** Use the equation from part a to predict the rate in 2010, and compare the result with the actual rate of 186.2. *Source: National Vital Statistics Reports.*

**c.** If the relationship between the death rate $R$ and the year $t$ is expressed as $R = at^3 + bt^2 + ct + d$, where $t = 0$ corresponds to 1980, use all four data points and a linear system of equations to determine the constants $a$, $b$, $c$, and $d$.

**d.** Discuss the appropriateness of the functions used in parts a and c to model this data.

**53. Archimedes' Problem Bovinum** Archimedes is credited with the authorship of a famous problem involving the number of cattle of the sun god. A simplified version of the problem is stated as follows:

> The sun god had a herd of cattle consisting of bulls and cows, one part of which was white, a second black, a third spotted, and a fourth brown.
>
> Among the bulls, the number of white ones was one-half plus one-third the number of the black greater than the brown; the number of the black, one-quarter plus one-fifth the number of the spotted greater than the brown; the number of the spotted, one-sixth and one-seventh the number of the white greater than the brown.
>
> Among the cows, the number of white ones was one-third plus one-quarter of the total black cattle; the number of the black, one-quarter plus one-fifth the total of the spotted cattle; the number of the spotted, one-fifth plus one-sixth the total of the brown cattle; the number of the brown, one-sixth plus one-seventh the total of the white cattle.
>
> What was the composition of the herd?

*Source: 100 Great Problems of Elementary Mathematics.* The problem can be solved by converting the statements into two systems of equations, using $X$, $Y$, $Z$, and $T$ for the number of white, black, spotted, and brown bulls, respectively, and $x$, $y$, $z$, and $t$ for the number of white, black, spotted, and brown cows, respectively. For example, the first statement can be written as $X = (1/2 + 1/3)Y + T$ and then reduced. The result is the following two systems of equations:

$$6X - 5Y = 6T \qquad 12x - 7y = 7Y$$
$$20Y - 9Z = 20T \quad \text{and} \quad 20y - 9z = 9Z$$
$$42Z - 13X = 42T \qquad 30z - 11t = 11T$$
$$-13x + 42t = 13X$$

**a.** Show that these two systems of equations represent Archimedes' Problem Bovinum.

**b.** If it is known that the number of brown bulls, $T$, is 4,149,387, use the Gauss-Jordan method to first find a solution to the $3 \times 3$ system and then use these values and the Gauss-Jordan method to find a solution to the $4 \times 4$ system of equations.

## OTHER APPLICATIONS

**54. Banking** A bank teller has a total of 70 bills in five-, ten-, and twenty-dollar denominations. The total value of the money is $960.

**a.** Find the total number of solutions.

**b.** Find the solution with the smallest number of five-dollar bills.

**c.** Find the solution with the largest number of five-dollar bills.

**55. Surveys** The president of Sam's Supermarkets plans to hire two public relations firms to survey 500 customers by phone, 750 by mail, and 250 by in-person interviews. The Garcia firm has personnel to do 10 phone surveys, 30 mail surveys, and 5 interviews per hour. The Wong firm can handle 20 phone surveys, 10 mail surveys, and 10 interviews per hour. For how many hours should each firm be hired to produce the exact number of surveys needed?

**56. Manufacturing** Fred's Furniture Factory has 1950 machine hours available each week in the cutting department, 1490 hours in the assembly department, and 2160 in the finishing department. Manufacturing a chair requires 0.2 hour of cutting, 0.3 hour of assembly, and 0.1 hour of finishing. A cabinet requires 0.5 hour of cutting, 0.4 hour of assembly, and 0.6 hour of finishing. A buffet requires 0.3 hour of cutting, 0.1 hour of assembly, and 0.4 hour of finishing. How many chairs, cabinets, and buffets should be produced in order to use all the available production capacity?

**57. Manufacturing** Nadir, Inc. produces three models of television sets: deluxe, super-deluxe, and ultra. Each deluxe set requires 2 hours of electronics work, 3 hours of assembly time, and 5 hours of finishing time. Each super-deluxe requires 1, 3, and 2 hours of electronics, assembly, and finishing time, respectively. Each ultra requires 2, 2, and 6 hours of the same work, respectively.

**a.** There are 54 hours available for electronics, 72 hours available for assembly, and 148 hours available for finishing per week. How many of each model should be produced each week if all available time is to be used?

**b.** Suppose everything is the same as in part a, but a super-deluxe set requires 1, rather than 2, hours of finishing time. How many solutions are there now?

**c.** Suppose everything is the same as in part b, but the total hours available for finishing changes from 148 hours to 144 hours. Now how many solutions are there?

**58. Transportation**  An electronics company produces three models of stereo speakers, models A, B, and C, and can deliver them by truck, van, or SUV. A truck holds 2 boxes of model A, 2 of model B, and 3 of model C. A van holds 3 boxes of model A, 4 boxes of model B, and 2 boxes of model C. An SUV holds 3 boxes of model A, 5 boxes of model B, and 1 box of model C.

  **a.** If 25 boxes of model A, 33 boxes of model B, and 22 boxes of model C are to be delivered, how many vehicles of each type should be used so that all operate at full capacity?

  **b.** Model C has been discontinued. If 25 boxes of model A and 33 boxes of model B are to be delivered, how many vehicles of each type should be used so that all operate at full capacity?

**59. Loans**  To get the necessary funds for a planned expansion, a small company took out three loans totaling $25,000. Company owners were able to get interest rates of 8%, 9%, and 10%. They borrowed $1000 more at 9% than they borrowed at 10%. The total annual interest on the loans was $2190.

  **a.** How much did they borrow at each rate?

  **b.** Suppose we drop the condition that they borrowed $1000 more at 9% than at 10%. What can you say about the amount borrowed at 10%? What is the solution if the amount borrowed at 10% is $5000?

  **c.** Suppose the bank sets a maximum of $10,000 at the lowest interest rate of 8%. Is a solution possible that still meets all of the original conditions?

  **d.** Explain why $10,000 at 8%, $8000 at 9%, and $7000 at 10% is not a feasible solution for part c.

**60. Transportation**  A manufacturer purchases a part for use at both of its plants—one at Roseville, California, the other at Akron, Ohio. The part is available in limited quantities from two suppliers. Each supplier has 75 units available. The Roseville plant needs 40 units, and the Akron plant requires 75 units. The first supplier charges $70 per unit delivered to Roseville and $90 per unit delivered to Akron. Corresponding costs from the second supplier are $80 and $120. The manufacturer wants to order a total of 75 units from the first, less expensive supplier, with the remaining 40 units to come from the second supplier. If the company spends $10,750 to purchase the required number of units for the two plants, find the number of units that should be sent from each supplier to each plant.

**61. Broadway Economics**  When Neil Simon opens a new play, he has to decide whether to open the show on Broadway or Off Broadway. For example, in his play *London Suite*, he decided to open it Off Broadway. From information provided by Emanuel Azenberg, his producer, the following equations were developed:

$$43,500x - y = 1,295,000$$
$$27,000x - y = 440,000,$$

where $x$ represents the number of weeks that the show has run and $y$ represents the profit or loss from the show (first equation is for Broadway and second equation is for Off Broadway). ***Source: The Mathematics Teacher.***

  **a.** Solve this system of equations to determine when the profit/loss from the show will be equal for each venue. What is the profit at that point?

  **b.** Discuss which venue is favorable for the show.

**62. Snack Food**  According to the nutrition labels on the package, a single serving of Oreos® has 10 g of fat, 36 g of carbohydrates, 2 g of protein, and 240 calories. The figures for a single serving of Twix® are 14 g, 37 g, 3 g, and 280 calories. The figures for a single serving of trail mix are 20 g, 23 g, 11 g, and 295 calories. How many calories are in each gram of fat, carbohydrate, and protein? ***Source: The Mathematics Teacher.****

**63. Basketball**  Wilt Chamberlain holds the record for the highest number of points scored in a single NBA basketball game. Chamberlain scored 100 points for Philadelphia against the New York Knicks on March 2, 1962. This is an amazing feat, considering he scored all of his points without the help of three-point shots. Chamberlain made a total of 64 baskets, consisting of field goals (worth two points) and foul shots (worth one point). Find the number of field goals and the number of foul shots that Chamberlain made. ***Source: ESPN.***

**64. Basketball**  Kobe Bryant has the second highest single game point total in the NBA. Bryant scored 81 points for the Los Angeles Lakers on January 22, 2006, against the Toronto Raptors. Bryant made a total of 46 baskets, including foul shots (worth one point), field goals (worth two points), and three-point shots (worth three points). The number of field goal shots he made is equal to three times the number of three pointers he made. Find the number of foul shots, field goals, and three pointers Bryant made. ***Source: ESPN.***

**65. Baseball**  Ichiro Suzuki holds the American League record for the most hits in a single baseball season. In 2004, Suzuki had a total of 262 hits for the Seattle Mariners. He hit three fewer triples than home runs, and he hit three times as many doubles as home runs. Suzuki also hit 45 times as many singles as triples. Find the number of singles, doubles, triples, and home runs hit by Suzuki during the season. ***Source: Baseball Almanac.***

**66. Toys**  One hundred toys are to be given out to a group of children. A ball costs $2, a doll costs $3, and a car costs $4. A total of $295 was spent on the toys.

  **a.** A ball weighs 12 oz, a doll 16 oz, and a car 18 oz. The total weight of all the toys is 1542 oz. Find how many of each toy there are.

  **b.** Now suppose the weight of a ball, doll, and car are 11, 15, and 19 oz, respectively. If the total weight is still 1542 oz, how many solutions are there now?

  **c.** Keep the weights as in part b, but change the total weight to 1480 oz. How many solutions are there?

  **d.** Give the solution to part c that has the smallest number of cars.

  **e.** Give the solution to part c that has the largest number of cars.

*For a discussion of some complications that can come up in solving linear systems, read the original article: Szydlik, Stephen D., "The Problem with the Snack Food Problem," *The Mathematics Teacher*, Vol. 103, No. 1, Aug. 2009, pp. 18–28.

**67. Ice Cream** Researchers have determined that the amount of sugar contained in ice cream helps to determine the overall "degree of like" that a consumer has toward that particular flavor. They have also determined that too much or too little sugar will have the same negative affect on the "degree of like" and that this relationship follows a quadratic function. In an experiment conducted at Pennsylvania State University, the following condensed table was obtained. *Source: Journal of Food Science.*

| Percentage of Sugar | Degree of Like |
|---|---|
| 8 | 5.4 |
| 13 | 6.3 |
| 18 | 5.6 |

a. Use this information and the Gauss-Jordan method to determine the coefficients $a$, $b$, $c$, of the quadratic equation

$$y = ax^2 + bx + c,$$

where $y$ is the "degree of like" and $x$ is the percentage of sugar in the ice cream mix.

b. Repeat part a by using the quadratic regression feature on a graphing calculator. Compare your answers.

**68. Modeling War** One of the factors that contribute to the success or failure of a particular army during war is its ability to get new troops ready for service. It is possible to analyze the rate of change in the number of troops of two hypothetical armies with the following simplified model,

Rate of increase (RED ARMY) $= 200{,}000 - 0.5r - 0.3b$

Rate of increase (BLUE ARMY) $= 350{,}000 - 0.5r - 0.7b$,

where $r$ is the number of soldiers in the Red Army at a given time and $b$ is the number of soldiers in the Blue Army at a given time. The factors 0.5 and 0.7 represent each army's efficiency of bringing new soldiers to the fight. *Source: Journal of Peace Research.*

a. Solve this system of equations to determine the number of soldiers in each army when the rate of increase for each is zero.

b. Describe what might be going on in a war when the rate of increase is zero.

**69. Traffic Control** At rush hours, substantial traffic congestion is encountered at the traffic intersections shown in the figure. (The streets are one-way, as shown by the arrows.)

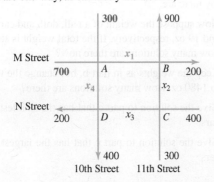

The city wishes to improve the signals at these corners so as to speed the flow of traffic. The traffic engineers first gather data. As the figure shows, 700 cars per hour come down M Street to intersection $A$, and 300 cars per hour come down 10th

Street to intersection $A$. A total of $x_1$ of these cars leave $A$ on M Street, and $x_4$ cars leave $A$ on 10th Street. The number of cars entering $A$ must equal the number leaving, so that

$$x_1 + x_4 = 700 + 300$$

or

$$x_1 + x_4 = 1000.$$

For intersection $B$, $x_1$ cars enter on M Street and $x_2$ on 11th Street. The figure shows that 900 cars leave $B$ on 11th and 200 on M. Thus,

$$x_1 + x_2 = 900 + 200$$
$$x_1 + x_2 = 1100.$$

a. Write two equations representing the traffic entering and leaving intersections $C$ and $D$.

b. Use the four equations to set up an augmented matrix, and solve the system by the Gauss-Jordan method, using $x_4$ as the parameter.

c. Based on your solution to part b, what are the largest and smallest possible values for the number of cars leaving intersection $A$ on 10th Street?

d. Answer the question in part c for the other three variables.

e. Verify that you could have discarded any one of the four original equations without changing the solution. What does this tell you about the original problem?

**70. Lights Out** The Tiger Electronics' game, Lights Out, consists of five rows of five lighted buttons. When a button is pushed, it changes the on/off status of it and the status of all of its vertical and horizontal neighbors. For any given situation where some of the lights are on and some are off, the goal of the game is to push buttons until all of the lights are turned off. It turns out that for any given array of lights, solving a system of equations can be used to develop a strategy for turning the lights out. The following system of equations can be used to solve the problem for a simplified version of the game with 2 rows of 2 buttons where all of the lights are initially turned on:

$$x_{11} + x_{12} + x_{21} = 1$$
$$x_{11} + x_{12} + x_{22} = 1$$
$$x_{11} + x_{21} + x_{22} = 1$$
$$x_{12} + x_{21} + x_{22} = 1,$$

where $x_{ij} = 1$ if the light in row $i$, column $j$, is on and $x_{ij} = 0$ when it is off. The order in which the buttons are pushed does not matter, so we are only seeking which buttons should be pushed. *Source: Mathematics Magazine.*

a. Solve this system of equations and determine a strategy to turn the lights out. (*Hint:* While doing row operations, if an odd number is found, immediately replace this value with a 1; if an even number is found, then immediately replace that number with a zero. This is called modulo 2 arithmetic, and it is necessary in problems dealing with on/off switches.)

b. Resolve the equation with the right side changed to $(0, 1, 1, 0)$.

**YOUR TURN ANSWERS**

1. $(5, -2)$    2. $(7, 2, -3)$
3. $(17/9 + w/3, -4/9 - 2w/3, 4/9 + 2w/3, w)$

# 10.2  Addition and Subtraction of Matrices

APPLY IT
**A pharmaceutical company manufactures different products at its plants in several cities. How might the company keep track of its production most efficiently?**

*In the previous section, matrices were used to store information about systems of linear equations. In this section, we begin to study calculations with matrices, which we will use in Examples 1, 5, and 7 to answer the question posed above.*

The use of matrices has increased in importance in the fields of management, natural science, and social science because matrices provide a convenient way to organize data, as Example 1 demonstrates.

### EXAMPLE 1  Pharmaceutical Manufacturing

The Wellbeing Pharmaceutical Company manufactures decongestants and pain relievers in three forms: capsules, softgels, and tablets. The company has regional manufacturing plants in New York, Chicago, and San Francisco. In August, the New York plant manufactured 10 cases of decongestant capsules, 12 cases of decongestant softgels, and 5 cases of decongestant tablets, as well as 15 cases of pain reliever capsules, 20 cases of pain reliever softgels, and 8 cases of pain reliever tablets. Organize this information in a more readable format.

APPLY IT    **SOLUTION**   To organize this data, we might tabulate the data in a chart.

| Number of Cases of Pharmaceuticals | | | |
| --- | --- | --- | --- |
| | **Capsules** | **Softgels** | **Tablets** |
| *Decongestants* | 10 | 12 | 5 |
| *Pain relievers* | 15 | 20 | 8 |

With the understanding that the numbers in each row refer to the pharmaceutical type (decongestants, pain relievers), and the numbers in each column refer to the form (capsule, softgel, tablet), the same information can be given by a matrix, as follows.

$$M = \begin{bmatrix} 10 & 12 & 5 \\ 15 & 20 & 8 \end{bmatrix}$$

Matrices often are named with capital letters, as in Example 1. Matrices are classified by **size**; that is, by the number of rows and columns they contain. For example, matrix $M$ above has two rows and three columns. This matrix is a $2 \times 3$ (read "2 by 3") matrix. By definition, a matrix with $m$ rows and $n$ columns is an $m \times n$ matrix. The number of rows is always given first.

### EXAMPLE 2  Matrix Size

**(a)** The matrix $\begin{bmatrix} -3 & 5 \\ 2 & 0 \\ 5 & -1 \end{bmatrix}$ is a $3 \times 2$ matrix.  **(b)** $\begin{bmatrix} 0.5 & 8 & 0.9 \\ 0 & 5.1 & -3 \\ -4 & 0 & 5 \end{bmatrix}$ is a $3 \times 3$ matrix.

**(c)** $\begin{bmatrix} 1 & 6 & 5 & -2 & 5 \end{bmatrix}$ is a $1 \times 5$ matrix.  **(d)** $\begin{bmatrix} 3 \\ -5 \\ 0 \\ 2 \end{bmatrix}$ is a $4 \times 1$ matrix.

A matrix with the same number of rows as columns is called a **square matrix**. The matrix in Example 2(b) is a square matrix.

A matrix containing only one row is called a **row matrix** or a **row vector**. The matrix in Example 2(c) is a row matrix, as are

$$[5 \quad 8], \quad [6 \quad -9 \quad 2], \quad \text{and} \quad [-4 \quad 0 \quad 0 \quad 0].$$

A matrix of only one column, as in Example 2(d), is a **column matrix** or a **column vector**.

Equality for matrices is defined as follows.

## Matrix Equality
Two matrices are equal if they are the same size and if each pair of corresponding elements is equal.

By this definition,

$$\begin{bmatrix} 2 & 1 \\ 3 & -5 \end{bmatrix} \quad \text{and} \quad \begin{bmatrix} 1 & 2 \\ -5 & 3 \end{bmatrix}$$

are not equal (even though they contain the same elements and are the same size) since the corresponding elements differ.

## EXAMPLE 3   Matrix Equality

(a) From the definition of matrix equality given above, the only way that the statement

$$\begin{bmatrix} 2 & 1 \\ p & q \end{bmatrix} = \begin{bmatrix} x & y \\ -1 & 0 \end{bmatrix}$$

can be true is if $2 = x$, $1 = y$, $p = -1$, and $q = 0$.

(b) The statement

$$\begin{bmatrix} x \\ y \end{bmatrix} = \begin{bmatrix} 1 \\ -3 \\ 0 \end{bmatrix}$$

can never be true, since the two matrices are different sizes. (One is $2 \times 1$ and the other is $3 \times 1$.)

## Addition
The matrix given in Example 1,

$$M = \begin{bmatrix} 10 & 12 & 5 \\ 15 & 20 & 8 \end{bmatrix},$$

shows the number of cases of decongestants and pain relievers in each of the three forms manufactured by the New York plant of the Wellbeing Pharmaceutical Company in August. If matrix $N$ below gives the amounts manufactured in September, what are the total amounts manufactured over these two months?

$$N = \begin{bmatrix} 45 & 35 & 20 \\ 65 & 40 & 35 \end{bmatrix}$$

If 10 cases of decongestant capsules were manufactured in August and 45 in September, then altogether $10 + 45 = 55$ cases were manufactured in the two months. The other corresponding entries can be added in a similar way to get a new matrix $Q$, which represents the total amounts manufactured over the two months.

$$Q = \begin{bmatrix} 55 & 47 & 25 \\ 80 & 60 & 43 \end{bmatrix}$$

It is convenient to refer to $Q$ as the sum of $M$ and $N$.

The way these two matrices were added illustrates the following definition of addition of matrices.

### Adding Matrices

The sum of two $m \times n$ matrices $X$ and $Y$ is the $m \times n$ matrix $X + Y$ in which each element is the sum of the corresponding elements of $X$ and $Y$.

**CAUTION** It is important to remember that only matrices that are the same size can be added.

### EXAMPLE 4  Adding Matrices

Find each sum, if possible.

**SOLUTION**

**YOUR TURN 1** Find each sum, if possible.

**(a)** $\begin{bmatrix} 3 & 4 & 5 & 6 \\ 1 & 2 & 3 & 4 \end{bmatrix} + \begin{bmatrix} 1 & -2 & 4 \\ -2 & -4 & 8 \end{bmatrix}$

**(b)** $\begin{bmatrix} 3 & 4 & 5 \\ 1 & 2 & 3 \end{bmatrix} + \begin{bmatrix} 1 & -2 & 4 \\ -2 & -4 & 8 \end{bmatrix}$

**(a)** $\begin{bmatrix} 5 & -6 \\ 8 & 9 \end{bmatrix} + \begin{bmatrix} -4 & 6 \\ 8 & -3 \end{bmatrix} = \begin{bmatrix} 5 + (-4) & -6 + 6 \\ 8 + 8 & 9 + (-3) \end{bmatrix} = \begin{bmatrix} 1 & 0 \\ 16 & 6 \end{bmatrix}$

**(b)** The matrices

$$A = \begin{bmatrix} 5 & -8 \\ 6 & 2 \end{bmatrix} \quad \text{and} \quad B = \begin{bmatrix} 3 & -9 & 1 \\ 4 & 2 & -5 \end{bmatrix}$$

are different sizes. Therefore, the sum $A + B$ does not exist.  **TRY YOUR TURN 1**

### EXAMPLE 5  Pharmaceutical Manufacturing

The number of cases manufactured by the Wellbeing Pharmaceutical Company in September at the New York, San Francisco, and Chicago manufacturing plants are given in matrices $N$, $S$, and $C$ below.

$$N = \begin{bmatrix} 45 & 35 & 20 \\ 65 & 40 & 35 \end{bmatrix} \quad S = \begin{bmatrix} 30 & 32 & 28 \\ 43 & 47 & 30 \end{bmatrix} \quad C = \begin{bmatrix} 22 & 25 & 38 \\ 31 & 34 & 35 \end{bmatrix}$$

What was the total amount manufactured at the three plants in September?

**APPLY IT**

**SOLUTION** The total amount manufactured in September is represented by the sum of the three matrices $N$, $S$, and $C$.

$$N + S + C = \begin{bmatrix} 45 & 35 & 20 \\ 65 & 40 & 35 \end{bmatrix} + \begin{bmatrix} 30 & 32 & 28 \\ 43 & 47 & 30 \end{bmatrix} + \begin{bmatrix} 22 & 25 & 38 \\ 31 & 34 & 35 \end{bmatrix}$$
$$= \begin{bmatrix} 97 & 92 & 86 \\ 139 & 121 & 100 \end{bmatrix}$$

For example, this sum shows that the total number of cases of decongestants in tablet form manufactured at the three plants in September was 86.

## Subtraction  Subtraction of matrices is defined similarly to addition.

### Subtracting Matrices

The difference of two $m \times n$ matrices $X$ and $Y$ is the $m \times n$ matrix $X - Y$ in which each element is the difference of the corresponding elements of $X$ and $Y$.

## EXAMPLE 6 Subtracting Matrices

Subtract each pair of matrices, if possible.

**SOLUTION**

(a) $\begin{bmatrix} 8 & 6 & -4 \\ -2 & 7 & 5 \end{bmatrix} - \begin{bmatrix} 3 & 5 & -8 \\ -4 & 2 & 9 \end{bmatrix} = \begin{bmatrix} 8-3 & 6-5 & -4-(-8) \\ -2-(-4) & 7-2 & 5-9 \end{bmatrix} = \begin{bmatrix} 5 & 1 & 4 \\ 2 & 5 & -4 \end{bmatrix}.$

(b) The matrices

$$\begin{bmatrix} -2 & 5 \\ 0 & 1 \end{bmatrix} \quad \text{and} \quad \begin{bmatrix} 3 \\ 5 \end{bmatrix}$$

are different sizes and cannot be subtracted.     **TRY YOUR TURN 2**

**YOUR TURN 2** Calculate
$\begin{bmatrix} 3 & 4 & 5 \\ 1 & 2 & 3 \end{bmatrix} - \begin{bmatrix} 1 & -2 & 4 \\ -2 & -4 & 8 \end{bmatrix}.$

## EXAMPLE 7 Pharmaceutical Manufacturing

**APPLY IT**

During September the Chicago manufacturing plant of the Wellbeing Pharmaceutical Company shipped out the following number of cases of each pharmaceutical.

$$K = \begin{bmatrix} 5 & 10 & 8 \\ 11 & 14 & 15 \end{bmatrix}$$

How many cases were left at the plant on October 1, taking into account only the cases manufactured and sent out during the month?

**Method 1**
**Calculating by Hand**

**SOLUTION** The number of cases of each pharmaceutical manufactured by the Chicago plant during September is given by matrix $C$ from Example 5; the number of cases of each pharmaceutical shipped out during September is given by matrix $K$. The October 1 inventory will be represented by the matrix $C - K$:

$$\begin{bmatrix} 22 & 25 & 38 \\ 31 & 34 & 35 \end{bmatrix} - \begin{bmatrix} 5 & 10 & 8 \\ 11 & 14 & 15 \end{bmatrix} = \begin{bmatrix} 17 & 15 & 30 \\ 20 & 20 & 20 \end{bmatrix}.$$

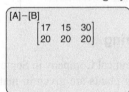
```
[A]-[B]
[17 15 30]
[20 20 20]
```
**FIGURE 9**

**Method 2**
**Graphing Calculator**

Matrix operations are easily performed on a graphing calculator. Figure 9 shows the previous operation; the matrices $A$ and $B$ were already entered into the calculator.

**TECHNOLOGY NOTE**

Spreadsheet programs are designed to effectively organize data that can be represented in rows and columns. Accordingly, matrix operations are also easily performed on spreadsheets. See the *Graphing Calculator and Excel Spreadsheet Manual* available with this book for details.

# 10.2 EXERCISES

Decide whether each statement is true or false. If false, tell why.

1. $\begin{bmatrix} 1 & 3 \\ 5 & 7 \end{bmatrix} = \begin{bmatrix} 1 & 5 \\ 3 & 7 \end{bmatrix}$

2. $\begin{bmatrix} 1 \\ 2 \\ 3 \end{bmatrix} = \begin{bmatrix} 1 & 2 & 3 \end{bmatrix}$

3. $\begin{bmatrix} x \\ y \end{bmatrix} = \begin{bmatrix} -2 \\ 8 \end{bmatrix}$ if $x = -2$ and $y = 8$.

4. $\begin{bmatrix} 3 & 5 & 2 & 8 \\ 1 & -1 & 4 & 0 \end{bmatrix}$ is a 4 × 2 matrix.

5. $\begin{bmatrix} 1 & 9 & -4 \\ 3 & 7 & 2 \\ -1 & 1 & 0 \end{bmatrix}$ is a square matrix.

6. $\begin{bmatrix} 2 & 4 & -1 \\ 3 & 7 & 5 \\ 0 & 0 & 0 \end{bmatrix} = \begin{bmatrix} 2 & 4 & -1 \\ 3 & 7 & 5 \end{bmatrix}$

**Find the size of each matrix. Identify any square, column, or row matrices.**

**7.** $\begin{bmatrix} -4 & 8 \\ 2 & 3 \end{bmatrix}$

**8.** $\begin{bmatrix} 2 & -3 & 7 \\ 1 & 0 & 4 \end{bmatrix}$

**9.** $\begin{bmatrix} -6 & 8 & 0 & 0 \\ 4 & 1 & 9 & 2 \\ 3 & -5 & 7 & 1 \end{bmatrix}$

**10.** $\begin{bmatrix} 8 & -2 & 4 & 6 & 3 \end{bmatrix}$

**11.** $\begin{bmatrix} -7 \\ 5 \end{bmatrix}$

**12.** $\begin{bmatrix} -9 \end{bmatrix}$

**13.** The sum of an $n \times m$ matrix and an $m \times n$ matrix, where $m \neq n$, is _____.

**14.** If $A$ is a $5 \times 2$ matrix and $A + K = A$, what do you know about $K$?

**Find the values of the variables in each equation.**

**15.** $\begin{bmatrix} 3 & 4 \\ -8 & 1 \end{bmatrix} = \begin{bmatrix} 3 & x \\ y & z \end{bmatrix}$

**16.** $\begin{bmatrix} -5 \\ y \end{bmatrix} = \begin{bmatrix} -5 \\ 8 \end{bmatrix}$

**17.** $\begin{bmatrix} s - 4 & t + 2 \\ -5 & 7 \end{bmatrix} = \begin{bmatrix} 6 & 2 \\ -5 & r \end{bmatrix}$

**18.** $\begin{bmatrix} 9 & 7 \\ r & 0 \end{bmatrix} = \begin{bmatrix} m - 3 & n + 5 \\ 8 & 0 \end{bmatrix}$

**19.** $\begin{bmatrix} a + 2 & 3b & 4c \\ d & 7f & 8 \end{bmatrix} + \begin{bmatrix} -7 & 2b & 6 \\ -3d & -6 & -2 \end{bmatrix} = \begin{bmatrix} 15 & 25 & 6 \\ -8 & 1 & 6 \end{bmatrix}$

**20.** $\begin{bmatrix} a + 2 & 3z + 1 & 5m \\ 4k & 0 & 3 \end{bmatrix} + \begin{bmatrix} 3a & 2z & 5m \\ 2k & 5 & 6 \end{bmatrix} = \begin{bmatrix} 10 & -14 & 80 \\ 10 & 5 & 9 \end{bmatrix}$

**Perform the indicated operations, where possible.**

**21.** $\begin{bmatrix} 2 & 4 & 5 & -7 \\ 6 & -3 & 12 & 0 \end{bmatrix} + \begin{bmatrix} 8 & 0 & -10 & 1 \\ -2 & 8 & -9 & 11 \end{bmatrix}$

**22.** $\begin{bmatrix} 1 & 5 \\ 2 & -3 \\ 3 & 7 \end{bmatrix} + \begin{bmatrix} 2 & 3 \\ 8 & 5 \\ -1 & 9 \end{bmatrix}$

**23.** $\begin{bmatrix} 1 & 3 & -2 \\ 4 & 7 & 1 \end{bmatrix} + \begin{bmatrix} 3 & 0 \\ 6 & 4 \\ -5 & 2 \end{bmatrix}$

**24.** $\begin{bmatrix} 8 & 0 & -3 \\ 1 & 19 & -5 \end{bmatrix} - \begin{bmatrix} 1 & -5 & 2 \\ 3 & 9 & -8 \end{bmatrix}$

**25.** $\begin{bmatrix} 2 & 8 & 12 & 0 \\ 7 & 4 & -1 & 5 \\ 1 & 2 & 0 & 10 \end{bmatrix} - \begin{bmatrix} 1 & 3 & 6 & 9 \\ 2 & -3 & -3 & 4 \\ 8 & 0 & -2 & 17 \end{bmatrix}$

**26.** $\begin{bmatrix} 2 & 1 \\ 5 & -3 \\ -7 & 2 \\ 9 & 0 \end{bmatrix} + \begin{bmatrix} 1 & -8 & 0 \\ 5 & 3 & 2 \\ -6 & 7 & -5 \\ 2 & -1 & 0 \end{bmatrix}$

**27.** $\begin{bmatrix} 2 & 3 \\ -2 & 4 \end{bmatrix} + \begin{bmatrix} 4 & 3 \\ 7 & 8 \end{bmatrix} - \begin{bmatrix} 3 & 2 \\ 1 & 4 \end{bmatrix}$

**28.** $\begin{bmatrix} 4 & 3 \\ 1 & 2 \end{bmatrix} - \begin{bmatrix} 1 & 1 \\ 1 & 0 \end{bmatrix} + \begin{bmatrix} 1 & 1 \\ 1 & 4 \end{bmatrix}$

**29.** $\begin{bmatrix} 2 & -1 \\ 0 & 13 \end{bmatrix} - \begin{bmatrix} 4 & 8 \\ -5 & 7 \end{bmatrix} + \begin{bmatrix} 12 & 7 \\ 5 & 3 \end{bmatrix}$

**30.** $\begin{bmatrix} 5 & 8 \\ -3 & 1 \end{bmatrix} + \begin{bmatrix} 0 & 1 \\ -2 & -2 \end{bmatrix} + \begin{bmatrix} -5 & -8 \\ 6 & 1 \end{bmatrix}$

**31.** $\begin{bmatrix} -4x + 2y & -3x + y \\ 6x - 3y & 2x - 5y \end{bmatrix} + \begin{bmatrix} -8x + 6y & 2x \\ 3y - 5x & 6x + 4y \end{bmatrix}$

**32.** $\begin{bmatrix} 4k - 8y \\ 6z - 3x \\ 2k + 5a \\ -4m + 2n \end{bmatrix} - \begin{bmatrix} 5k + 6y \\ 2z + 5x \\ 4k + 6a \\ 4m - 2n \end{bmatrix}$

**33.** For matrices $X = \begin{bmatrix} x & y \\ z & w \end{bmatrix}$ and $0 = \begin{bmatrix} 0 & 0 \\ 0 & 0 \end{bmatrix}$, find the matrix $0 - X$.

Using matrices $O = \begin{bmatrix} 0 & 0 \\ 0 & 0 \end{bmatrix}$, $P = \begin{bmatrix} m & n \\ p & q \end{bmatrix}$, $T = \begin{bmatrix} r & s \\ t & u \end{bmatrix}$, and $X = \begin{bmatrix} x & y \\ z & w \end{bmatrix}$, **verify the statements in Exercises 34–37.**

**34.** $X + T = T + X$ (commutative property of addition of matrices)

**35.** $X + (T + P) = (X + T) + P$ (associative property of addition of matrices)

**36.** $X - X = O$ (inverse property of addition of matrices)

**37.** $P + O = P$ (identity property of addition of matrices)

**38.** Which of the above properties are valid for matrices that are not square?

# LIFE SCIENCE APPLICATIONS

**39. Dietetics** A dietician prepares a diet specifying the amounts a patient should eat of four basic food groups: group I, meats; group II, fruits and vegetables; group III, breads and starches; group IV, milk products. Amounts are given in "exchanges" that represent 1 oz (meat), 1/2 cup (fruits and vegetables), 1 slice (bread), 8 oz (milk), or other suitable measurements.

**a.** The number of "exchanges" for breakfast for each of the four food groups, respectively, are 2, 1, 2, and 1; for lunch, 3, 2, 2, and 1; and for dinner, 4, 3, 2, and 1. Write a $3 \times 4$ matrix using this information.

**b.** The amounts of fat, carbohydrates, and protein (in appropriate units) in each food group, respectively, are as follows.

Fat: 5, 0, 0, 10

Carbohydrates: 0, 10, 15, 12

Protein: 7, 1, 2, 8

Use this information to write a $4 \times 3$ matrix.

**c.** There are 8 calories per exchange of fat, 4 calories per exchange of carbohydrates, and 5 calories per exchange of protein. Summarize this data in a $3 \times 1$ matrix.

40. **Animal Growth** At the beginning of a laboratory experiment, five baby rats measured 5.6, 6.4, 6.9, 7.6, and 6.1 cm in length, and weighed 144, 138, 149, 152, and 146 g, respectively.

   **a.** Write a 2 × 5 matrix using this information.

   **b.** At the end of two weeks, their lengths (in centimeters) were 10.2, 11.4, 11.4, 12.7, and 10.8 and their weights (in grams) were 196, 196, 225, 250, and 230. Write a 2 × 5 matrix with this information.

   **c.** Use matrix subtraction and the matrices found in parts a and b to write a matrix that gives the amount of change in length and weight for each rat.

   **d.** During the third week, the rats grew by the amounts shown in the matrix below.

$$\begin{matrix} \text{Length} \\ \text{Weight} \end{matrix} \begin{bmatrix} 1.8 & 1.5 & 2.3 & 1.8 & 2.0 \\ 25 & 22 & 29 & 33 & 20 \end{bmatrix}$$

   What were their lengths and weights at the end of this week?

41. **Testing Medication** A drug company is testing 200 patients to see if Painfree (a new headache medicine) is effective. Half the patients receive Painfree and half receive a placebo. The data on the first 50 patients is summarized in this matrix:

$$\begin{matrix} \\ \text{Painfree} \\ \text{Placebo} \end{matrix} \overset{\displaystyle \underset{\text{Pain Relief Obtained}}{\begin{matrix} \text{Yes} & \text{No} \end{matrix}}}{\begin{bmatrix} 22 & 3 \\ 8 & 17 \end{bmatrix}}.$$

   **a.** Of those who took the placebo, how many got relief?

   **b.** Of those who took the new medication, how many got no relief?

   **c.** The test was repeated on three more groups of 50 patients each, with the results summarized by these matrices.

$$\begin{bmatrix} 21 & 4 \\ 6 & 19 \end{bmatrix} \quad \begin{bmatrix} 19 & 6 \\ 10 & 15 \end{bmatrix} \quad \begin{bmatrix} 23 & 2 \\ 3 & 22 \end{bmatrix}$$

   Find the total results for all 200 patients.

   **d.** On the basis of these results, does it appear that Painfree is effective?

## OTHER APPLICATIONS

42. **Motorcycle Helmets** The following table shows the percentage of motorcyclists in various regions of the country who used helmets compliant with federal safety regulations and the percentage who used helmets that were noncompliant in two recent years. *Source: NHTSA.*

| 2008 | Compliant | Noncompliant |
|---|---|---|
| Northeast | 45 | 8 |
| Midwest | 67 | 16 |
| South | 61 | 14 |
| West | 71 | 5 |

| 2009 | Compliant | Noncompliant |
|---|---|---|
| Northeast | 61 | 15 |
| Midwest | 67 | 8 |
| South | 65 | 6 |
| West | 83 | 4 |

**a.** Write two matrices for the 2008 and 2009 helmet usage.

**b.** Use the two matrices from part a to write a matrix showing the change in helmet usage from 2008 to 2009.

**c.** Analyze the results from part b and discuss the extent to which changes from 2008 to 2009 differ from one region to another.

43. **Life Expectancy** The following table gives the life expectancy of African American males and females and white American males and females at the beginning of each decade since 1970. *Source: National Center for Health Statistics.*

| | African American | | White American | |
|---|---|---|---|---|
| Year | Male | Female | Male | Female |
| 1970 | 60.0 | 68.3 | 68.0 | 75.6 |
| 1980 | 63.8 | 72.5 | 70.7 | 78.1 |
| 1990 | 64.5 | 73.6 | 72.7 | 79.4 |
| 2000 | 68.2 | 75.1 | 74.7 | 79.9 |
| 2010 | 71.8 | 78.0 | 76.5 | 81.3 |

**a.** Write a matrix for the life expectancy of African Americans.

**b.** Write a matrix for the life expectancy of white Americans.

**c.** Use the matrices from parts a and b to write a matrix showing the difference between the two groups.

**d.** Analyze the results from part c and discuss any noticeable trends.

44. **Educational Attainment** The table below gives the educational attainment of the U.S. population 25 years and older since 1970. *Source: U.S. Census Bureau.*

| | Male | | Female | |
|---|---|---|---|---|
| Year | Percentage with 4 Years of High School or More | Percentage with 4 Years of College or More | Percentage with 4 Years of High School or More | Percentage with 4 Years of College or More |
| 1970 | 55.0 | 14.1 | 55.4 | 8.2 |
| 1980 | 69.2 | 20.9 | 68.1 | 13.6 |
| 1990 | 77.7 | 24.4 | 77.5 | 18.4 |
| 2000 | 84.2 | 27.8 | 84.0 | 23.6 |
| 2010 | 86.6 | 30.3 | 87.6 | 29.6 |

**a.** Write a matrix for the educational attainment of males.

**b.** Write a matrix for the educational attainment of females.

**c.** Use the matrices from parts a and b to write a matrix showing the difference in educational attainment between males and females since 1970.

**45. Educational Attainment** The following table gives the educational attainment of African Americans and Hispanic Americans 25 years and older since 1980. *Source: U.S. Census Bureau.*

| | African American | | Hispanic American | |
|---|---|---|---|---|
| Year | Percentage with 4 Years of High School or More | Percentage with 4 Years of College or More | Percentage with 4 Years of High School or More | Percentage with 4 Years of College or More |
| 1980 | 51.2 | 7.9 | 45.3 | 7.9 |
| 1985 | 59.8 | 11.1 | 47.9 | 8.5 |
| 1990 | 66.2 | 11.3 | 50.8 | 9.2 |
| 1995 | 73.8 | 13.2 | 53.4 | 9.3 |
| 2000 | 78.5 | 16.5 | 57.0 | 10.6 |
| 2010 | 84.2 | 19.8 | 62.9 | 13.9 |

a. Write a matrix for the educational attainment of African Americans.

b. Write a matrix for the educational attainment of Hispanic Americans.

c. Use the matrices from parts a and b to write a matrix showing the difference in educational attainment between African and Hispanic Americans.

**46. Animal Interactions** When two kittens named Cauchy and Cliché were introduced into a household with Jamie (an older cat) and Musk (a dog), the interactions among animals were complicated. The two kittens liked each other and Jamie, but didn't like Musk. Musk liked everybody, but Jamie didn't like any of the other animals.

a. Write a 4 × 4 matrix in which rows (and columns) 1, 2, 3, and 4 refer to Musk, Jamie, Cauchy, and Cliché. Make an element a 1 if the animal for that row likes the animal for that column, and otherwise make the element a 0. Assume every animal likes herself.

b. Within a few days, Cauchy and Cliché decided that they liked Musk after all. Write a 4 × 4 matrix, as you did in part a, representing the new situation.

**47. Car Accidents** The tables in the next column give the death rates, per million person trips, for male and female drivers for various ages and number of passengers. *Source: JAMA.*

| Male Drivers | Number of Passengers | | | |
|---|---|---|---|---|
| Age | 0 | 1 | 2 | ≥ 3 |
| 16 | 2.61 | 4.39 | 6.29 | 9.08 |
| 17 | 1.63 | 2.77 | 4.61 | 6.92 |
| 30–59 | 0.92 | 0.75 | 0.62 | 0.54 |

| Female Drivers | Number of Passengers | | | |
|---|---|---|---|---|
| Age | 0 | 1 | 2 | ≥ 3 |
| 16 | 1.38 | 1.72 | 1.94 | 3.31 |
| 17 | 1.26 | 1.48 | 2.82 | 2.28 |
| 30–59 | 0.41 | 0.33 | 0.27 | 0.40 |

a. Write a matrix for the death rate of male drivers.

b. Write a matrix for the death rate of female drivers.

c. Use the matrices from parts a and b to write a matrix showing the difference between the death rates of males and females.

d. Analyze the results of part c and make some suggestions on how to reduce the rates.

**48. Management** A toy company has plants in Boston, Chicago, and Seattle that manufacture toy phones and calculators. The following matrix gives the production costs (in dollars) for each item at the Boston plant:

$$\begin{array}{cc} & \text{Phones} \quad \text{Calculators} \\ \begin{array}{c} \text{Material} \\ \text{Labor} \end{array} & \begin{bmatrix} 4.27 & 6.94 \\ 3.45 & 3.65 \end{bmatrix} \end{array}$$

a. In Chicago, a phone costs $4.05 for material and $3.27 for labor; a calculator costs $7.01 for material and $3.51 for labor. In Seattle, material costs are $4.40 for a phone and $6.90 for a calculator; labor costs are $3.54 for a phone and $3.76 for a calculator. Write the production cost matrices for Chicago and Seattle.

b. Suppose labor costs increase by $0.11 per item in Chicago and material costs there increase by $0.37 for a phone and $0.42 for a calculator. What is the new production cost matrix for Chicago?

**49. Management** There are three convenience stores in Folsom. This week, store I sold 88 loaves of bread, 48 qt of milk, 16 jars of peanut butter, and 112 lb of cold cuts. Store II sold 105 loaves of bread, 72 qt of milk, 21 jars of peanut butter, and 147 lb of cold cuts. Store III sold 60 loaves of bread, 40 qt of milk, no peanut butter, and 50 lb of cold cuts.

a. Use a 4 × 3 matrix to express the sales information for the three stores.

b. During the following week, sales on these products at store I increased by 25%; sales at store II increased by 1/3; and sales at store III increased by 10%. Write the sales matrix for that week.

c. Write a matrix that represents total sales over the two-week period.

**YOUR TURN ANSWERS**

**1.** (a) Does not exist (b) $\begin{bmatrix} 4 & 2 & 9 \\ -1 & -2 & 11 \end{bmatrix}$ **2.** $\begin{bmatrix} 2 & 6 & 1 \\ 3 & 6 & -5 \end{bmatrix}$

# 10.3 Multiplication of Matrices

APPLY IT What is a wildlife center's total cost of seed for each season of the year for various species of birds?

*Matrix multiplication will be used to answer this question in Example 6.*

We begin by defining the product of a real number and a matrix. In work with matrices, a real number is called a **scalar**.

### Product of a Matrix and a Scalar

The product of a scalar $k$ and a matrix $X$ is the matrix $kX$, in which each element is $k$ times the corresponding element of $X$.

**EXAMPLE 1** Scalar Product of a Matrix

Calculate $-5A$, where $A = \begin{bmatrix} 3 & 4 \\ 0 & -1 \end{bmatrix}$.

**SOLUTION**

$$(-5) \begin{bmatrix} 3 & 4 \\ 0 & -1 \end{bmatrix} = \begin{bmatrix} (-5)(3) & (-5)(4) \\ (-5)(0) & (-5)(-1) \end{bmatrix} = \begin{bmatrix} -15 & -20 \\ 0 & 5 \end{bmatrix}$$

Finding the product of two matrices is more involved, but such multiplication is important in solving practical problems. To understand the reasoning behind matrix multiplication, it may be helpful to consider another example concerning Wellbeing Pharmaceutical Company discussed in the previous section. Suppose Wellbeing makes three antibiotics, which we will label A, B, and C. Matrix $W$ shows the number of cases of each antibiotic manufactured in a month at each plant.

$$
\begin{array}{c}
\phantom{} \\
\text{New York} \\
\text{Chicago} \\
\text{San Francisco}
\end{array}
\begin{array}{c}
\begin{array}{ccc} \text{A} & \text{B} & \text{C} \end{array} \\
\begin{bmatrix} 10 & 7 & 3 \\ 5 & 9 & 6 \\ 4 & 8 & 2 \end{bmatrix} = W
\end{array}
$$

If the selling price for a case of antibiotic A is $800, for a case of antibiotic B is $1000, and for a case of antibiotic C is $1200, the total value of the antibiotics manufactured at the New York plant is found as follows.

| Antibiotic | Number of Cases | | Price of Case | | Total |
|:---:|:---:|:---:|:---:|:---:|:---:|
| | | **Value of Antibiotic** | | | |
| A | 10 | × | $800 | = | $8000 |
| B | 7 | × | $1000 | = | $7000 |
| C | 3 | × | $1200 | = | $3600 |
| | (Total for New York) | | | | $18,600 |

The total value of the three kinds of antibiotics manufactured in New York is $18,600.

The work done in the table is summarized as follows:

$$10(\$800) + 7(\$1000) + 3(\$1200) = \$18,600.$$

In the same way, we find that the Chicago antibiotics have a total value of

$$5(\$800) + 9(\$1000) + 6(\$1200) = \$20{,}200,$$

and in San Francisco, the total value of the antibiotics is

$$4(\$800) + 8(\$1000) + 2(\$1200) = \$13{,}600.$$

The selling prices can be written as a column matrix $P$, and the total value in each location as another column matrix, $V$.

$$\begin{bmatrix} 800 \\ 1000 \\ 1200 \end{bmatrix} = P \qquad \begin{bmatrix} 18{,}600 \\ 20{,}200 \\ 13{,}600 \end{bmatrix} = V$$

Look at the elements of $W$ and $P$ below; multiplying the first, second, and third elements of the first row of $W$ by the first, second, and third elements, respectively, of the column matrix $P$ and then adding these products gives the first element in $V$. Doing the same thing with the second row of $W$ gives the second element of $V$; the third row of $W$ leads to the third element of $V$, suggesting that it is reasonable to write the product of matrices

$$W = \begin{bmatrix} 10 & 7 & 3 \\ 5 & 9 & 6 \\ 4 & 8 & 2 \end{bmatrix} \qquad \text{and} \qquad P = \begin{bmatrix} 800 \\ 1000 \\ 1200 \end{bmatrix}$$

as

$$WP = \begin{bmatrix} 10 & 7 & 3 \\ 5 & 9 & 6 \\ 4 & 8 & 2 \end{bmatrix} \begin{bmatrix} 800 \\ 1000 \\ 1200 \end{bmatrix} = \begin{bmatrix} 18{,}600 \\ 20{,}200 \\ 13{,}600 \end{bmatrix} = V.$$

The product was found by multiplying the elements of *rows* of the matrix on the left and the corresponding elements of the *column* of the matrix on the right, and then finding the sum of these separate products. Notice that the product of a $3 \times 3$ matrix and a $3 \times 1$ matrix is a $3 \times 1$ matrix. Notice also that each element of the product of two matrices is a sum of products. This is exactly what you do when you go to the store and buy 8 candy bars at $0.80 each, 4 bottles of water at $1.25 each, and so forth, and calculate the total as $8 \times 0.80 + 4 \times 1.25 + \cdots$.

The product $AB$ of an $m \times n$ matrix $A$ and an $n \times k$ matrix $B$ is found as follows. Multiply each element of the *first row* of $A$ by the corresponding element of the *first column* of $B$. The sum of these $n$ products is the *first-row, first-column* element of $AB$. Similarly, the sum of the products found by multiplying the elements of the *first row* of $A$ by the corresponding elements of the *second column* of $B$ gives the *first-row, second-column* element of $AB$, and so on.

### Product of Two Matrices

Let $A$ be an $m \times n$ matrix and let $B$ be an $n \times k$ matrix. To find the element in the $i$th row and $j$th column of the **product matrix** $AB$, multiply each element in the $i$th row of $A$ by the corresponding element in the $j$th column of $B$, and then add these products. The product matrix $AB$ is an $m \times k$ matrix.

### EXAMPLE 2  Matrix Product

Find the product $AB$ of matrices

$$A = \begin{bmatrix} 2 & 3 & -1 \\ 4 & 2 & 2 \end{bmatrix} \qquad \text{and} \qquad B = \begin{bmatrix} 1 \\ 8 \\ 6 \end{bmatrix}.$$

**SOLUTION** Since $A$ is $2 \times 3$ and $B$ is $3 \times 1$, we can find the product matrix $AB$.

*Step 1* Multiply the elements of the first row of $A$ and the corresponding elements of the column of $B$.

$$\begin{bmatrix} 2 & 3 & -1 \\ 4 & 2 & 2 \end{bmatrix} \begin{bmatrix} 1 \\ 8 \\ 6 \end{bmatrix} \quad 2 \cdot 1 + 3 \cdot 8 + (-1) \cdot 6 = 20$$

Thus, 20 is the first-row entry of the product matrix $AB$.

*Step 2* Multiply the elements of the second row of $A$ and the corresponding elements of $B$.

$$\begin{bmatrix} 2 & 3 & -1 \\ 4 & 2 & 2 \end{bmatrix} \begin{bmatrix} 1 \\ 8 \\ 6 \end{bmatrix} \quad 4 \cdot 1 + 2 \cdot 8 + 2 \cdot 6 = 32$$

The second-row entry of the product matrix $AB$ is 32.

*Step 3* Write the product as a column matrix using the two entries found above.

$$AB = \begin{bmatrix} 2 & 3 & -1 \\ 4 & 2 & 2 \end{bmatrix} \begin{bmatrix} 1 \\ 8 \\ 6 \end{bmatrix} = \begin{bmatrix} 20 \\ 32 \end{bmatrix}$$

Note that the product of a $2 \times 3$ matrix and a $3 \times 1$ matrix is a $2 \times 1$ matrix.

**EXAMPLE 3** **Matrix Product**

Find the product $CD$ of matrices

$$C = \begin{bmatrix} -3 & 4 & 2 \\ 5 & 0 & 4 \end{bmatrix} \quad \text{and} \quad D = \begin{bmatrix} -6 & 4 \\ 2 & 3 \\ 3 & -2 \end{bmatrix}.$$

**SOLUTION** Since $C$ is $2 \times 3$ and $D$ is $3 \times 2$, we can find the product matrix $CD$.

*Step 1*
$$\begin{bmatrix} -3 & 4 & 2 \\ 5 & 0 & 4 \end{bmatrix} \begin{bmatrix} -6 & 4 \\ 2 & 3 \\ 3 & -2 \end{bmatrix} \quad (-3) \cdot (-6) + 4 \cdot 2 + 2 \cdot 3 = 32$$

*Step 2*
$$\begin{bmatrix} -3 & 4 & 2 \\ 5 & 0 & 4 \end{bmatrix} \begin{bmatrix} -6 & 4 \\ 2 & 3 \\ 3 & -2 \end{bmatrix} \quad (-3) \cdot 4 + 4 \cdot 3 + 2 \cdot (-2) = -4$$

*Step 3*
$$\begin{bmatrix} -3 & 4 & 2 \\ 5 & 0 & 4 \end{bmatrix} \begin{bmatrix} -6 & 4 \\ 2 & 3 \\ 3 & -2 \end{bmatrix} \quad 5 \cdot (-6) + 0 \cdot 2 + 4 \cdot 3 = -18$$

*Step 4*
$$\begin{bmatrix} -3 & 4 & 2 \\ 5 & 0 & 4 \end{bmatrix} \begin{bmatrix} -6 & 4 \\ 2 & 3 \\ 3 & -2 \end{bmatrix} \quad 5 \cdot 4 + 0 \cdot 3 + 4 \cdot (-2) = 12$$

**YOUR TURN 1** Calculate the product $AB$ where $A = \begin{bmatrix} 3 & 4 \\ 1 & 2 \end{bmatrix}$ and $B = \begin{bmatrix} 1 & -2 \\ -2 & -4 \end{bmatrix}$.

***Step 5*** The product is

$$CD = \begin{bmatrix} -3 & 4 & 2 \\ 5 & 0 & 4 \end{bmatrix} \begin{bmatrix} -6 & 4 \\ 2 & 3 \\ 3 & -2 \end{bmatrix} = \begin{bmatrix} 32 & -4 \\ -18 & 12 \end{bmatrix}.$$

Here the product of a $2 \times 3$ matrix and a $3 \times 2$ matrix is a $2 \times 2$ matrix.

**TRY YOUR TURN 1**

**NOTE** One way to avoid errors in matrix multiplication is to lower the first matrix so it is below and to the left of the second matrix, and then write the product in the space between the two matrices. For example, to multiply the matrices in Example 3, we could rewrite the product as shown.

$$\begin{bmatrix} -6 & 4 \\ 2 & 3 \\ 3 & -2 \end{bmatrix}$$
$$\rightarrow \begin{bmatrix} -3 & 4 & 2 \\ 5 & 0 & 4 \end{bmatrix} \begin{bmatrix} \\ * \end{bmatrix}$$

To find the entry where the * is, for example, multiply the row and the column indicated by the arrows: $5 \cdot (-6) + 0 \cdot 2 + 4 \cdot 3 = -18$.

As the definition of matrix multiplication shows,

> the product $AB$ of two matrices $A$ and $B$ can be found only if the number of columns of $A$ is the same as the number of rows of $B$.

The final product will have as many rows as $A$ and as many columns as $B$.

**EXAMPLE 4** **Matrix Product**

Suppose matrix $A$ is $2 \times 2$ and matrix $B$ is $2 \times 4$. Can the products $AB$ and $BA$ be calculated? If so, what is the size of each product?

**SOLUTION** The following diagram helps decide the answers to these questions.

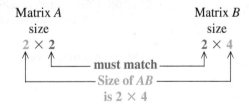

The product of $A$ and $B$ can be found because $A$ has two columns and $B$ has two rows. The size of the product is $2 \times 4$.

Matrix $B$       Matrix $A$
   size           size
  $2 \times 4$         $2 \times 2$
         do not match

The product $BA$ cannot be found because $B$ has 4 columns and $A$ has 2 rows.

| EXAMPLE 5 | Comparing Matrix Products *AB* and *BA* |

Find $AB$ and $BA$, given

$$A = \begin{bmatrix} 1 & -3 \\ 7 & 2 \\ -2 & 5 \end{bmatrix} \quad \text{and} \quad B = \begin{bmatrix} 1 & 0 & -1 \\ 3 & 1 & 4 \end{bmatrix}.$$

**SOLUTION**

**Method 1**
**Calculating by Hand**

$$AB = \begin{bmatrix} 1 & -3 \\ 7 & 2 \\ -2 & 5 \end{bmatrix} \begin{bmatrix} 1 & 0 & -1 \\ 3 & 1 & 4 \end{bmatrix}$$

$$= \begin{bmatrix} -8 & -3 & -13 \\ 13 & 2 & 1 \\ 13 & 5 & 22 \end{bmatrix}$$

**YOUR TURN 2** Calculate each product *AB* and *BA*, if possible, where $A = \begin{bmatrix} 3 & 5 & -1 \\ 2 & 4 & -2 \end{bmatrix}$ and $B = \begin{bmatrix} 3 & -4 \\ -5 & -3 \end{bmatrix}$.

$$BA = \begin{bmatrix} 1 & 0 & -1 \\ 3 & 1 & 4 \end{bmatrix} \begin{bmatrix} 1 & -3 \\ 7 & 2 \\ -2 & 5 \end{bmatrix}$$

$$= \begin{bmatrix} 3 & -8 \\ 2 & 13 \end{bmatrix}$$

**Method 2**
**Graphing Calculator**

Matrix multiplication is easily performed on a graphing calculator. Figure 10 in the margin below shows the results. The matrices *A* and *B* were already entered into the calculator.

**TRY YOUR TURN 2**

**TECHNOLOGY NOTE**

Matrix multiplication can also be easily done with a spreadsheet. See the *Graphing Calculator and Excel Spreadsheet Manual* available with this textbook for details.

```
[A]*[B]
   [-8  -3  -13]
   [13   2    1]
   [13   5   22]
[B]*[A]
   [3  -8]
   [2  13]
```

**FIGURE 10**

Notice in Example 5 that $AB \neq BA$; matrices $AB$ and $BA$ aren't even the same size. In Example 4, we showed that they may not both exist. This means that matrix multiplication is *not* commutative. Even if both *A* and *B* are square matrices, in general, matrices $AB$ and $BA$ are not equal. (See Exercise 31.) Of course, there may be special cases in which they are equal, but this is not true in general.

**CAUTION** Since matrix multiplication is not commutative, always be careful to multiply matrices in the correct order.

Matrix multiplication *is* associative, however. For example, if

$$C = \begin{bmatrix} 3 & 2 \\ 0 & -4 \\ -1 & 1 \end{bmatrix},$$

then $(AB)C = A(BC)$, where *A* and *B* are the matrices given in Example 5. (Verify this.) Also, there is a distributive property of matrices such that, for appropriate matrices *A*, *B*, and *C*,

$$A(B + C) = AB + AC.$$

(See Exercises 32 and 33.) Other properties of matrix multiplication involving scalars are included in the exercises. Multiplicative inverses and multiplicative identities are defined in the next section.

### EXAMPLE 6    Bird Feeding

A wildlife center finds that the number of cardinals and finches that visit a bird feeder on a regular basis varies with the season, as shown in the matrix $P$ below.

$$
\begin{array}{c}
 \\
\text{Spring} \\
\text{Summer} \\
\text{Fall} \\
\text{Winter}
\end{array}
\begin{array}{c}
\text{Cardinals Finches} \\
\begin{bmatrix}
25 & 31 \\
20 & 35 \\
22 & 29 \\
36 & 20
\end{bmatrix} = P
\end{array}
$$

The wildlife center has also found that cardinals and finches eat different amounts of sunflower seed, corn, and millet (in grams) when they visit the feeder. The matrix $Q$ shows how much one bird of each species eats over an entire season.

$$
\begin{array}{c}
 \\
\text{Cardinals} \\
\text{Finches}
\end{array}
\begin{array}{c}
\text{Sunflower Corn Millet} \\
\begin{bmatrix}
32 & 10 & 30 \\
10 & 18 & 27
\end{bmatrix} = Q
\end{array}
$$

The cost (in cents) per gram of each type of seed is shown in the matrix $R$.

$$
\begin{array}{c}
 \\
\text{Sunflower} \\
\text{Corn} \\
\text{Millet}
\end{array}
\begin{array}{c}
\text{Cost} \\
\begin{bmatrix}
3 \\
2 \\
1
\end{bmatrix} = R
\end{array}
$$

**(a)** What is the total cost of seed for each season of the year to feed the cardinals and finches that regularly visit the feeder?

APPLY IT

**SOLUTION**   To find the total cost for each season, first find $PQ$, which shows how much of each type of seed is eaten each season of the year.

$$
PQ = \begin{bmatrix}
25 & 31 \\
20 & 35 \\
22 & 29 \\
36 & 20
\end{bmatrix}
\begin{bmatrix}
32 & 10 & 30 \\
10 & 18 & 27
\end{bmatrix}
$$

$$
= \begin{array}{c}
\text{Sunflower} \quad \text{Corn} \quad \text{Millet} \\
\begin{bmatrix}
1110 & 808 & 1587 \\
990 & 830 & 1545 \\
994 & 742 & 1443 \\
1352 & 720 & 1620
\end{bmatrix}
\end{array}
\begin{array}{c}
\text{Spring} \\
\text{Summer} \\
\text{Fall} \\
\text{Winter}
\end{array}
$$

Now multiply $PQ$ by $R$, the cost matrix, to get the total cost of seed for each season.

$$
\begin{bmatrix}
1110 & 808 & 1587 \\
990 & 830 & 1545 \\
994 & 742 & 1443 \\
1352 & 720 & 1620
\end{bmatrix}
\begin{bmatrix}
3 \\
2 \\
1
\end{bmatrix}
=
\begin{array}{c}
\text{Cost} \\
\begin{bmatrix}
6533 \\
6175 \\
5909 \\
7116
\end{bmatrix}
\end{array}
\begin{array}{c}
\text{Spring} \\
\text{Summer} \\
\text{Fall} \\
\text{Winter}
\end{array}
$$

These costs are in cents, so multiply by 0.01 to convert to dollars. The total cost of seed is $65.33 in spring, $61.75 in summer, $59.09 in fall, and $71.16 in winter.

**(b)** How many grams of each kind of seed are used over an entire year?

**SOLUTION**   The totals of the columns of matrix $PQ$ give a matrix whose elements represent the total number of grams of each seed used over an entire year. Call this matrix $T$, and write it as a row matrix.

$$
T = \begin{bmatrix} 4446 & 3100 & 6195 \end{bmatrix}
$$

Thus, 4446 grams of sunflower seed, 3100 grams of corn, and 6195 grams of millet are used over an entire year.

**(c)** What is the total cost of seed in a year?

**SOLUTION** For the total cost of seed in a year, find the product of the matrix $T$, the matrix showing the total amount of each seed, and matrix $R$, the cost matrix. (To multiply these and get a $1 \times 1$ matrix representing total cost, we need a $1 \times 4$ matrix multiplied by a $4 \times 1$ matrix. This is why $T$ was written as a row matrix in (b) above.)

$$TR = \begin{bmatrix} 4446 & 3100 & 6195 \end{bmatrix} \begin{bmatrix} 3 \\ 2 \\ 1 \end{bmatrix} = \begin{bmatrix} 25{,}733 \end{bmatrix}$$

The total cost is $257.33. Of course, we could have calculated this same answer by adding the amounts found in part (a).

**(d)** Suppose the wildlife center puts bird feeders in four more locations, for a total of five feeders. The locations are very distant from each other, but have similar habitats, so the number of birds and the amount they eat is expected to be the same as at the original feeder. Calculate the total amount of each type of seed expected to be used in each season.

**SOLUTION** Multiply $PQ$ by the scalar 5, as follows.

$$5(PQ) = 5\begin{bmatrix} 1110 & 808 & 1587 \\ 990 & 830 & 1545 \\ 994 & 742 & 1443 \\ 1352 & 720 & 1620 \end{bmatrix} = \begin{bmatrix} 5550 & 4040 & 7935 \\ 4950 & 4150 & 7725 \\ 4970 & 3710 & 7215 \\ 6760 & 3600 & 8100 \end{bmatrix}$$

The total amount of corn used in fall, for example, is 3710 grams.

## Meaning of a Matrix Product

It is helpful to use a notation that keeps track of the quantities a matrix represents. We will use the notation

meaning of the rows/meaning of the columns,

that is, writing the meaning of the rows first, followed by the meaning of the columns. In Example 6, we would use the notation seasons/birds for matrix $P$, birds/seeds for matrix $Q$, and seeds/cost for matrix $R$. In multiplying $PQ$, we are multiplying seasons/birds by birds/seeds. The result is seasons/seeds. Notice that birds, the common quantity in both $P$ and $Q$, was eliminated in the product $PQ$. By this method, the product $(PQ)R$ represents seasons/cost.

In practical problems this notation helps us decide in which order to multiply matrices so that the results are meaningful. In Example 6(c) either $RT$ or $TR$ can be calculated. Since $T$ represents year/seeds and $R$ represents seeds/cost, the product $TR$ gives year/cost, while the product $RT$ is meaningless.

# 10.3 EXERCISES

Let $A = \begin{bmatrix} -2 & 4 \\ 0 & 3 \end{bmatrix}$ and $B = \begin{bmatrix} -6 & 2 \\ 4 & 0 \end{bmatrix}$. Find each value.

**1.** $2A$

**2.** $-3B$

**3.** $-6A$

**4.** $5B$

**5.** $-4A + 5B$

**6.** $7B - 3A$

In Exercises 7–12, the sizes of two matrices A and B are given. Find the sizes of the product AB and the product BA, whenever these products exist.

**7.** $A$ is $2 \times 2$, and $B$ is $2 \times 2$.

**8.** $A$ is $3 \times 3$, and $B$ is $3 \times 3$.

**9.** $A$ is $3 \times 4$, and $B$ is $4 \times 4$.

**10.** $A$ is $4 \times 3$, and $B$ is $3 \times 6$.

**11.** $A$ is $4 \times 2$, and $B$ is $3 \times 4$.

**12.** $A$ is $3 \times 2$, and $B$ is $1 \times 3$.

**13.** To find the product matrix $AB$, the number of _____ of $A$ must be the same as the number of _____ of $B$.

**14.** The product matrix $AB$ has the same number of _____ as $A$ and the same number of _____ as $B$.

**Find each matrix product, if possible.**

**15.** $\begin{bmatrix} 2 & -1 \\ 5 & 8 \end{bmatrix} \begin{bmatrix} 3 \\ -2 \end{bmatrix}$

**16.** $\begin{bmatrix} -1 & 5 \\ 7 & 0 \end{bmatrix} \begin{bmatrix} 6 \\ 2 \end{bmatrix}$

**17.** $\begin{bmatrix} 2 & -1 & 7 \\ -3 & 0 & -4 \end{bmatrix} \begin{bmatrix} 5 \\ 10 \\ 2 \end{bmatrix}$

**18.** $\begin{bmatrix} 5 & 2 \\ 7 & 6 \\ 1 & 0 \end{bmatrix} \begin{bmatrix} 1 & 4 & 0 \\ 2 & -1 & 2 \end{bmatrix}$

**19.** $\begin{bmatrix} 2 & -1 \\ 3 & 6 \end{bmatrix} \begin{bmatrix} -1 & 0 & 4 \\ 5 & -2 & 0 \end{bmatrix}$

**20.** $\begin{bmatrix} 6 & 0 & -4 \\ 1 & 2 & 5 \\ 10 & -1 & 3 \end{bmatrix} \begin{bmatrix} 1 \\ 2 \\ 0 \end{bmatrix}$

**21.** $\begin{bmatrix} 2 & 2 & -1 \\ 3 & 0 & 1 \end{bmatrix} \begin{bmatrix} 0 & 2 \\ -1 & 4 \\ 0 & 2 \end{bmatrix}$

**22.** $\begin{bmatrix} -3 & 1 & 0 \\ 6 & 0 & 8 \end{bmatrix} \begin{bmatrix} 3 \\ -1 \\ -2 \end{bmatrix}$

**23.** $\begin{bmatrix} 1 & 2 \\ 3 & 4 \end{bmatrix} \begin{bmatrix} -1 & 5 \\ 7 & 0 \end{bmatrix}$

**24.** $\begin{bmatrix} 2 & 8 \\ -7 & 5 \end{bmatrix} \begin{bmatrix} 1 & 0 \\ 0 & 1 \end{bmatrix}$

**25.** $\begin{bmatrix} -2 & -3 & 7 \\ 1 & 5 & 6 \end{bmatrix} \begin{bmatrix} 1 \\ 2 \\ 3 \end{bmatrix}$

**26.** $\begin{bmatrix} 2 \\ -9 \\ 12 \end{bmatrix} \begin{bmatrix} 1 & 0 & -1 \end{bmatrix}$

**27.** $\left( \begin{bmatrix} 2 & 1 \\ -3 & -6 \\ 4 & 0 \end{bmatrix} \begin{bmatrix} 1 & -2 \\ 2 & -1 \end{bmatrix} \right) \begin{bmatrix} 3 \\ 1 \end{bmatrix}$

**28.** $\begin{bmatrix} 2 & 1 \\ -3 & -6 \\ 4 & 0 \end{bmatrix} \left( \begin{bmatrix} 1 & -2 \\ 2 & -1 \end{bmatrix} \begin{bmatrix} 3 \\ 1 \end{bmatrix} \right)$

**29.** $\begin{bmatrix} 2 & -2 \\ 1 & -1 \end{bmatrix} \left( \begin{bmatrix} 4 & 3 \\ 1 & 2 \end{bmatrix} + \begin{bmatrix} 7 & 0 \\ -1 & 5 \end{bmatrix} \right)$

**30.** $\begin{bmatrix} 2 & -2 \\ 1 & -1 \end{bmatrix} \begin{bmatrix} 4 & 3 \\ 1 & 2 \end{bmatrix} + \begin{bmatrix} 2 & -2 \\ 1 & -1 \end{bmatrix} \begin{bmatrix} 7 & 0 \\ -1 & 5 \end{bmatrix}$

**31.** Let $A = \begin{bmatrix} -2 & 4 \\ 1 & 3 \end{bmatrix}$ and $B = \begin{bmatrix} -2 & 1 \\ 3 & 6 \end{bmatrix}$.

   **a.** Find $AB$.     **b.** Find $BA$.

   **c.** Did you get the same answer in parts a and b?

   **d.** In general, for matrices $A$ and $B$ such that $AB$ and $BA$ both exist, does $AB$ always equal $BA$?

**Given matrices** $P = \begin{bmatrix} m & n \\ p & q \end{bmatrix}$, $X = \begin{bmatrix} x & y \\ z & w \end{bmatrix}$, **and** $T = \begin{bmatrix} r & s \\ t & u \end{bmatrix}$,
**verify that the statements in Exercises 32–35 are true. The statements are valid for any matrices whenever matrix multiplication and addition can be carried out. This, of course, depends on the size of the matrices.**

**32.** $(PX)T = P(XT)$ (associative property: see Exercises 27 and 28)

**33.** $P(X + T) = PX + PT$ (distributive property: see Exercises 29 and 30)

**34.** $k(X + T) = kX + kT$ for any real number $k$.

**35.** $(k + h)P = kP + hP$ for any real numbers $k$ and $h$.

**36.** Let $I$ be the matrix $I = \begin{bmatrix} 1 & 0 \\ 0 & 1 \end{bmatrix}$, and let matrices $P$, $X$, and $T$ be defined as for Exercises 32–35.

   **a.** Find $IP$, $PI$, and $IX$.

   **b.** Without calculating, guess what the matrix $IT$ might be.

   **c.** Suggest a reason for naming a matrix such as $I$ an *identity matrix*.

**37.** Show that the system of linear equations

$$2x_1 + 3x_2 + x_3 = 5$$
$$x_1 - 4x_2 + 5x_3 = 8$$

can be written as the matrix equation

$$\begin{bmatrix} 2 & 3 & 1 \\ 1 & -4 & 5 \end{bmatrix} \begin{bmatrix} x_1 \\ x_2 \\ x_3 \end{bmatrix} = \begin{bmatrix} 5 \\ 8 \end{bmatrix}.$$

**38.** Let $A = \begin{bmatrix} 1 & 2 \\ -3 & 5 \end{bmatrix}$, $X = \begin{bmatrix} x_1 \\ x_2 \end{bmatrix}$, and $B = \begin{bmatrix} -4 \\ 12 \end{bmatrix}$. Show that the equation $AX = B$ represents a linear system of two equations in two unknowns. Solve the system and substitute into the matrix equation to check your results.

**Use a computer or graphing calculator and the following matrices to find the matrix products and sums in Exercises 39–41.**

$$A = \begin{bmatrix} 2 & 3 & -1 & 5 & 10 \\ 2 & 8 & 7 & 4 & 3 \\ -1 & -4 & -12 & 6 & 8 \\ 2 & 5 & 7 & 1 & 4 \end{bmatrix}$$

$$B = \begin{bmatrix} 9 & 3 & 7 & -6 \\ -1 & 0 & 4 & 2 \\ -10 & -7 & 6 & 9 \\ 8 & 4 & 2 & -1 \\ 2 & -5 & 3 & 7 \end{bmatrix}$$

$$C = \begin{bmatrix} -6 & 8 & 2 & 4 & -3 \\ 1 & 9 & 7 & -12 & 5 \\ 15 & 2 & -8 & 10 & 11 \\ 4 & 7 & 9 & 6 & -2 \\ 1 & 3 & 8 & 23 & 4 \end{bmatrix}$$

$$D = \begin{bmatrix} 5 & -3 & 7 & 9 & 2 \\ 6 & 8 & -5 & 2 & 1 \\ 3 & 7 & -4 & 2 & 11 \\ 5 & -3 & 9 & 4 & -1 \\ 0 & 3 & 2 & 5 & 1 \end{bmatrix}$$

**39. a.** Find $AC$.

   **b.** Find $CA$.

   **c.** Does $AC = CA$?

**40. a.** Find $CD$.

   **b.** Find $DC$.

   **c.** Does $CD = DC$?

**41. a.** Find $C + D$.

   **b.** Find $(C + D)B$.

   **c.** Find $CB$.

   **d.** Find $DB$.

   **e.** Find $CB + DB$.

   **f.** Does $(C + D)B = CB + DB$?

**42.** Which property of matrices does Exercise 41 illustrate?

# LIFE SCIENCE APPLICATIONS

**43. Dietetics** In Exercise 39 from Section 10.2, label the matrices

$$\begin{bmatrix} 2 & 1 & 2 & 1 \\ 3 & 2 & 2 & 1 \\ 4 & 3 & 2 & 1 \end{bmatrix}, \begin{bmatrix} 5 & 0 & 7 \\ 0 & 10 & 1 \\ 0 & 15 & 2 \\ 10 & 12 & 8 \end{bmatrix}, \text{ and } \begin{bmatrix} 8 \\ 4 \\ 5 \end{bmatrix}$$

found in parts a, b, and c, respectively, X, Y, and Z.

**a.** Find the product matrix XY. What do the entries of this matrix represent?

**b.** Find the product matrix YZ. What do the entries represent?

**c.** Find the products $(XY)Z$ and $X(YZ)$ and verify that they are equal. What do the entries represent?

**44. Motorcycle Helmets** In Exercise 42 from Section 10.2, you constructed matrices that represented usage of motorcycle helmets for 2008 and 2009. Use matrix operations to combine these two matrices to form one matrix that represents the average of the two years.

**45. Life Expectancy** In Exercise 43 from Section 10.2, you constructed matrices that represent the life expectancy of African American and white American males and females. Use matrix operations to combine these two matrices to form one matrix that represents the combined life expectancy of both races at the beginning of each decade since 1970. Use the fact that of the combined African and white American population, African Americans are about one-sixth of the total and white Americans about five-sixths. (*Hint:* Multiply the matrix for African Americans by 1/6 and the matrix for the white Americans by 5/6, and then add the results.)

**46. Car Accidents** In Exercise 47 from Section 10.2, you constructed matrices that represent the death rates, per million person trips, for both male and female drivers for various ages and number of passengers. Use matrix operations to combine these two matrices to form one matrix that represents the combined death rates for males and females, per million person trips, of drivers of various ages and number of passengers. (*Hint:* Add the two matrices together and then multiply the resulting matrix by the scalar 1/2.)

**47. Northern Spotted Owl Population** In an attempt to save the endangered northern spotted owl, the U.S. Fish and Wildlife Service imposed strict guidelines for the use of 12 million acres of Pacific Northwest forest. This decision led to a national debate between the logging industry and environmentalists. Mathematical ecologists have created a mathematical model to analyze population dynamics of the northern spotted owl by dividing the female owl population into three categories: juvenile (up to 1 year old), subadult (1 to 2 years), and adult (over 2 years old). By analyzing these three subgroups, it is possible to use the number of females in each subgroup at time $n$ to estimate the number of females in each group at any time $n + 1$ with the following matrix equation:

$$\begin{bmatrix} j_{n+1} \\ s_{n+1} \\ a_{n+1} \end{bmatrix} = \begin{bmatrix} 0 & 0 & 0.33 \\ 0.18 & 0 & 0 \\ 0 & 0.71 & 0.94 \end{bmatrix} \begin{bmatrix} j_n \\ s_n \\ a_n \end{bmatrix},$$

where $j_n$ is the number of juveniles, $s_n$ is the number of subadults, and $a_n$ is the number of adults at time $n$. *Source: Conservation Biology.*

 **a.** If there are currently 4000 female northern spotted owls made up of 900 juveniles, 500 subadults, and 2600 adults, use a graphing calculator or spreadsheet and matrix operations to determine the total number of female owls for each of the next 5 years. (*Hint:* Round each answer to the nearest whole number after each matrix multiplication.)

**b.** With advanced techniques from linear algebra, it is possible to show that in the long run, the following holds.

$$\begin{bmatrix} j_{n+1} \\ s_{n+1} \\ a_{n+1} \end{bmatrix} \approx 0.98359 \begin{bmatrix} j_n \\ s_n \\ a_n \end{bmatrix}$$

What can we conclude about the long-term survival of the northern spotted owl?

**c.** Notice that only 18 percent of the juveniles become subadults. Assuming that, through better habitat management, this number could be increased to 40 percent, rework part a. Discuss possible reasons why only 18 percent of the juveniles become subadults. Under the new assumption, what can you conclude about the long-term survival of the northern spotted owl?

**48. World Population** The 2010 birth and death rates per million for several regions and the world population (in millions) by region are given in the following tables. *Source: U.S. Census Bureau.*

| | Births | Deaths |
|---|---|---|
| *Africa* | 0.0346 | 0.0118 |
| *Asia* | 0.0174 | 0.0073 |
| *Latin America* | 0.0189 | 0.0059 |
| *North America* | 0.0135 | 0.0083 |
| *Europe* | 0.0099 | 0.0103 |

| | Population (millions) | | | | |
|---|---|---|---|---|---|
| Year | Africa | Asia | Latin America | North America | Europe |
| 1970 | 361 | 2038 | 286 | 227 | 460 |
| 1980 | 473 | 2494 | 362 | 252 | 484 |
| 1990 | 627 | 2978 | 443 | 278 | 499 |
| 2000 | 803 | 3435 | 524 | 314 | 511 |
| 2010 | 1013 | 3824 | 591 | 344 | 522 |

a. Write the information in each table as a matrix.

b. Use the matrices from part a to find the total number (in millions) of births and deaths in each year (assuming birth and death rates for all years are close to these in 2010).

c. Using the results of part b, compare the number of births in 1970 and in 2010. Also compare the birth rates from part a. Which gives better information?

d. Using the results of part b, compare the number of deaths in 1980 and in 2010. Discuss how this comparison differs from a comparison of death rates from part a.

**49. Bird Feeding** Consider the matrices $P$, $Q$, and $R$ given in Example 6.

a. Find and interpret the matrix product $QR$.

b. Verify that $P(QR)$ is equal to $(PQ)R$ calculated in Example 6.

## OTHER APPLICATIONS

**50. Cost Analysis** The four departments of Spangler Enterprises need to order the following amounts of the same products.

| | Paper | Tape | Binders | Memo Pads | Pens |
|---|---|---|---|---|---|
| Department 1 | 10 | 4 | 3 | 5 | 6 |
| Department 2 | 7 | 2 | 2 | 3 | 8 |
| Department 3 | 4 | 5 | 1 | 0 | 10 |
| Department 4 | 0 | 3 | 4 | 5 | 5 |

The unit price (in dollars) of each product is given below for two suppliers.

| | Supplier A | Supplier B |
|---|---|---|
| Paper | 2 | 3 |
| Tape | 1 | 1 |
| Binders | 4 | 3 |
| Memo Pads | 3 | 3 |
| Pens | 1 | 2 |

a. Use matrix multiplication to get a matrix showing the comparative costs for each department for the products from the two suppliers.

b. Find the total cost over all departments to buy products from each supplier. From which supplier should the company make the purchase?

**51. Cost Analysis** The Mundo Candy Company makes three types of chocolate candy: Cheery Cherry, Mucho Mocha, and Almond Delight. The company produces its products in San Diego, Mexico City, and Managua using two main ingredients: chocolate and sugar.

a. Each kilogram of Cheery Cherry requires 0.5 kg of sugar and 0.2 kg of chocolate; each kilogram of Mucho Mocha requires 0.4 kg of sugar and 0.3 kg of chocolate; and each kilogram of Almond Delight requires 0.3 kg of sugar and 0.3 kg of chocolate. Put this information into a 2 × 3 matrix called $A$, labeling the rows and columns.

b. The cost of 1 kg of sugar is $4 in San Diego, $2 in Mexico City, and $1 in Managua. The cost of 1 kg of chocolate is $3 in San Diego, $5 in Mexico City, and $7 in Managua. Put this information into a matrix called $C$ in such a way that when you multiply it with your matrix from part a, you get a matrix representing the ingredient cost of producing each type of candy in each city.

c. Only one of the two products $AC$ and $CA$ is meaningful. Determine which one it is, calculate the product, and describe what the entries represent.

d. From your answer to part c, what is the combined sugar-and-chocolate cost to produce 1 kg of Mucho Mocha in Managua?

e. Mundo Candy needs to quickly produce a special shipment of 100 kg of Cheery Cherry, 200 kg of Mucho Mocha, and 500 kg of Almond Delight, and it decides to select one factory to fill the entire order. Use matrix multiplication to determine in which city the total sugar-and-chocolate cost to produce the order is the smallest.

**52. Shoe Sales** Sal's Shoes and Fred's Footwear both have outlets in California and Arizona. Sal's sells shoes for $80, sandals for $40, and boots for $120. Fred's prices are $60, $30, and $150 for shoes, sandals, and boots, respectively. Half of all sales in California stores are shoes, 1/4 are sandals, and 1/4 are boots. In Arizona the fractions are 1/5 shoes, 1/5 sandals, and 3/5 boots.

a. Write a 2 × 3 matrix called $P$ representing prices for the two stores and three types of footwear.

b. Write a 3 × 2 matrix called $F$ representing the fraction of each type of footwear sold in each state.

c. Only one of the two products $PF$ and $FP$ is meaningful. Determine which one it is, calculate the product, and describe what the entries represent.

d. From your answer to part c, what is the average price for a pair of footwear at an outlet of Fred's in Arizona?

**YOUR TURN ANSWERS**

1. $\begin{bmatrix} -5 & -22 \\ -3 & -10 \end{bmatrix}$  2. $AB$ does not exist; $BA = \begin{bmatrix} 1 & -1 & 5 \\ -21 & -37 & 11 \end{bmatrix}$

# 10.4 Matrix Inverses

APPLY IT How can we determine the amount of various kinds of fertilizers to use on a farm, given the requirements for the soil?
*This question is answered in Example 5.*

In this section, we introduce the idea of a matrix inverse, which is comparable to the reciprocal of a real number. This will allow us to solve a matrix equation.

The real number 1 is the *multiplicative* identity for real numbers: for any real number $a$, we have $a \cdot 1 = 1 \cdot a = a$. In this section, we define a *multiplicative identity matrix I* that has properties similar to those of the number 1. We then use the definition of matrix $I$ to find the *multiplicative inverse* of any square matrix that has an inverse.

If $I$ is to be the identity matrix, both of the products $AI$ and $IA$ must equal $A$. This means that an identity matrix exists only for square matrices. The $2 \times 2$ **identity matrix** that satisfies these conditions is

$$I = \begin{bmatrix} 1 & 0 \\ 0 & 1 \end{bmatrix}.$$

To check that $I$, as defined above, is really the $2 \times 2$ identity, let

$$A = \begin{bmatrix} a & b \\ c & d \end{bmatrix}.$$

Then $AI$ and $IA$ should both equal $A$.

$$AI = \begin{bmatrix} a & b \\ c & d \end{bmatrix}\begin{bmatrix} 1 & 0 \\ 0 & 1 \end{bmatrix} = \begin{bmatrix} a(1) + b(0) & a(0) + b(1) \\ c(1) + d(0) & c(0) + d(1) \end{bmatrix} = \begin{bmatrix} a & b \\ c & d \end{bmatrix} = A$$

$$IA = \begin{bmatrix} 1 & 0 \\ 0 & 1 \end{bmatrix}\begin{bmatrix} a & b \\ c & d \end{bmatrix} = \begin{bmatrix} 1(a) + 0(c) & 1(b) + 0(d) \\ 0(a) + 1(c) & 0(b) + 1(d) \end{bmatrix} = \begin{bmatrix} a & b \\ c & d \end{bmatrix} = A$$

This verifies that $I$ has been defined correctly.

It is easy to verify that the identity matrix $I$ is unique. Suppose there is another identity; call it $J$. Then $IJ$ must equal $I$, because $J$ is an identity, and $IJ$ must also equal $J$, because $I$ is an identity. Thus $I = J$.

The identity matrices for $3 \times 3$ matrices and $4 \times 4$ matrices, respectively, are

$$I = \begin{bmatrix} 1 & 0 & 0 \\ 0 & 1 & 0 \\ 0 & 0 & 1 \end{bmatrix} \quad \text{and} \quad I = \begin{bmatrix} 1 & 0 & 0 & 0 \\ 0 & 1 & 0 & 0 \\ 0 & 0 & 1 & 0 \\ 0 & 0 & 0 & 1 \end{bmatrix}.$$

By generalizing, we can find an $n \times n$ identity matrix for any value of $n$.

Recall that the multiplicative inverse of the nonzero real number $a$ is $1/a$. The product of $a$ and its multiplicative inverse $1/a$ is 1. Given a matrix $A$, can a **multiplicative inverse matrix $A^{-1}$** (read "$A$-inverse") that will satisfy both

$$AA^{-1} = I \quad \text{and} \quad A^{-1}A = I$$

be found? For a given matrix, we often can find an inverse matrix by using the row operations of Section 10.1.

**NOTE** $A^{-1}$ does not mean $1/A$; here, $A^{-1}$ is just the notation for the multiplicative inverse of matrix $A$. Also, only square matrices can have inverses because both $A^{-1}A$ and $AA^{-1}$ must exist and be equal to an identity matrix of the same size.

**EXAMPLE 1**  **Inverse Matrices**

Verify that the matrices $A = \begin{bmatrix} 1 & 3 \\ 2 & 5 \end{bmatrix}$ and $B = \begin{bmatrix} -5 & 3 \\ 2 & -1 \end{bmatrix}$ are inverses of each other.

**SOLUTION**  Multiply $A$ times $B$ as in the previous section:

$$AB = \begin{bmatrix} 1 & 3 \\ 2 & 5 \end{bmatrix}\begin{bmatrix} -5 & 3 \\ 2 & -1 \end{bmatrix} = \begin{bmatrix} 1 & 0 \\ 0 & 1 \end{bmatrix}.$$

Similarly,

$$BA = \begin{bmatrix} -5 & 3 \\ 2 & -1 \end{bmatrix}\begin{bmatrix} 1 & 3 \\ 2 & 5 \end{bmatrix} = \begin{bmatrix} 1 & 0 \\ 0 & 1 \end{bmatrix}.$$

Since $AB = BA = I$, $A$ and $B$ are inverses of each other.

If an inverse exists, it is unique. That is, any given square matrix has no more than one inverse. The proof of this is left to Exercise 50 in this section.

As an example, let us find the inverse of

$$A = \begin{bmatrix} 1 & 3 \\ -1 & 2 \end{bmatrix}.$$

Let the unknown inverse matrix be

$$A^{-1} = \begin{bmatrix} x & y \\ z & w \end{bmatrix}.$$

By the definition of matrix inverse, $AA^{-1} = I$, or

$$AA^{-1} = \begin{bmatrix} 1 & 3 \\ -1 & 2 \end{bmatrix}\begin{bmatrix} x & y \\ z & w \end{bmatrix} = \begin{bmatrix} 1 & 0 \\ 0 & 1 \end{bmatrix}.$$

By matrix multiplication,

$$\begin{bmatrix} x + 3z & y + 3w \\ -x + 2z & -y + 2w \end{bmatrix} = \begin{bmatrix} 1 & 0 \\ 0 & 1 \end{bmatrix}.$$

Setting corresponding elements equal gives the system of equations

$$x \quad\quad + 3z \quad\quad\quad = 1 \tag{1}$$
$$y \quad\quad + 3w = 0 \tag{2}$$
$$-x \quad\quad + 2z \quad\quad = 0 \tag{3}$$
$$-y \quad\quad + 2w = 1. \tag{4}$$

Since Equations (1) and (3) involve only $x$ and $z$, while Equations (2) and (4) involve only $y$ and $w$, these four equations lead to two systems of equations,

$$\begin{matrix} x + 3z = 1 \\ -x + 2z = 0 \end{matrix} \quad \text{and} \quad \begin{matrix} y + 3w = 0 \\ -y + 2w = 1. \end{matrix}$$

Writing the two systems as augmented matrices gives

$$\begin{bmatrix} 1 & 3 & | & 1 \\ -1 & 2 & | & 0 \end{bmatrix} \quad \text{and} \quad \begin{bmatrix} 1 & 3 & | & 0 \\ -1 & 2 & | & 1 \end{bmatrix}.$$

Each of these systems can be solved by the Gauss-Jordan method. Notice, however, that the elements to the left of the vertical bar are identical. The two systems can be combined into the single matrix

$$\begin{bmatrix} 1 & 3 & | & 1 & 0 \\ -1 & 2 & | & 0 & 1 \end{bmatrix}.$$

This is of the form $[A \mid I]$. It is solved simultaneously as follows.

$$R_1 + R_2 \rightarrow R_2 \quad \begin{bmatrix} 1 & 3 & | & 1 & 0 \\ 0 & 5 & | & 1 & 1 \end{bmatrix}$$ Get 0 in the second-row, first-column position.

$$-3R_2 + 5R_1 \rightarrow R_1 \quad \begin{bmatrix} 5 & 0 & | & 2 & -3 \\ 0 & 5 & | & 1 & 1 \end{bmatrix}$$ Get 0 in the first-row, second-column position.

$$\begin{matrix} \frac{1}{5}R_1 \rightarrow R_1 \\ \frac{1}{5}R_2 \rightarrow R_2 \end{matrix} \quad \begin{bmatrix} 1 & 0 & | & \frac{2}{5} & -\frac{3}{5} \\ 0 & 1 & | & \frac{1}{5} & \frac{1}{5} \end{bmatrix}$$ Get 1's down the diagonal.

The numbers in the first column to the right of the vertical bar give the values of $x$ and $z$. The second column gives the values of $y$ and $w$. That is,

$$\begin{bmatrix} 1 & 0 & | & x & y \\ 0 & 1 & | & z & w \end{bmatrix} = \begin{bmatrix} 1 & 0 & | & \frac{2}{5} & -\frac{3}{5} \\ 0 & 1 & | & \frac{1}{5} & \frac{1}{5} \end{bmatrix}$$

so that

$$A^{-1} = \begin{bmatrix} x & y \\ z & w \end{bmatrix} = \begin{bmatrix} \frac{2}{5} & -\frac{3}{5} \\ \frac{1}{5} & \frac{1}{5} \end{bmatrix}.$$

To check, multiply $A$ by $A^{-1}$. The result should be $I$.

$$AA^{-1} = \begin{bmatrix} 1 & 3 \\ -1 & 2 \end{bmatrix} \begin{bmatrix} \frac{2}{5} & -\frac{3}{5} \\ \frac{1}{5} & \frac{1}{5} \end{bmatrix} = \begin{bmatrix} \frac{2}{5} + \frac{3}{5} & -\frac{3}{5} + \frac{3}{5} \\ -\frac{2}{5} + \frac{2}{5} & \frac{3}{5} + \frac{2}{5} \end{bmatrix} = \begin{bmatrix} 1 & 0 \\ 0 & 1 \end{bmatrix} = I$$

Verify that $A^{-1}A = I$, also.

## Finding a Multiplicative Inverse Matrix

To obtain $A^{-1}$ for any $n \times n$ matrix $A$ for which $A^{-1}$ exists, follow these steps.

1. Form the augmented matrix $[A|I]$, where $I$ is the $n \times n$ identity matrix.
2. Perform row operations on $[A|I]$ to get a matrix of the form $[I|B]$, if this is possible.
3. Matrix $B$ is $A^{-1}$.

### EXAMPLE 2 Inverse Matrix

Find $A^{-1}$ if $A = \begin{bmatrix} 1 & 0 & 1 \\ 2 & -2 & -1 \\ 3 & 0 & 0 \end{bmatrix}$.

**Method 1**
**Calculating by Hand**

**SOLUTION** Write the augmented matrix $[A \mid I]$.

$$[A \mid I] = \begin{bmatrix} 1 & 0 & 1 & | & 1 & 0 & 0 \\ 2 & -2 & -1 & | & 0 & 1 & 0 \\ 3 & 0 & 0 & | & 0 & 0 & 1 \end{bmatrix}$$

Begin by selecting the row operation that produces a zero for the first element in row 2.

$$\begin{matrix} -2R_1 + R_2 \rightarrow R_2 \\ -3R_1 + R_3 \rightarrow R_3 \end{matrix} \quad \begin{bmatrix} 1 & 0 & 1 & | & 1 & 0 & 0 \\ 0 & -2 & -3 & | & -2 & 1 & 0 \\ 0 & 0 & -3 & | & -3 & 0 & 1 \end{bmatrix}$$ Get 0's in the first column.

Column 2 already has zeros in the required positions, so work on column 3.

$$\begin{matrix} R_3 + 3R_1 \rightarrow R_1 \\ R_3 + (-1)R_2 \rightarrow R_2 \end{matrix} \quad \left[\begin{array}{ccc|ccc} 3 & 0 & 0 & 0 & 0 & 1 \\ 0 & 2 & 0 & -1 & -1 & 1 \\ 0 & 0 & -3 & -3 & 0 & 1 \end{array}\right] \quad \text{Get 0's in the third column.}$$

Now get 1's down the main diagonal.

$$\begin{matrix} \frac{1}{3}R_1 \rightarrow R_1 \\ \frac{1}{2}R_2 \rightarrow R_2 \\ -\frac{1}{3}R_3 \rightarrow R_3 \end{matrix} \quad \left[\begin{array}{ccc|ccc} 1 & 0 & 0 & 0 & 0 & \frac{1}{3} \\ 0 & 1 & 0 & -\frac{1}{2} & -\frac{1}{2} & \frac{1}{2} \\ 0 & 0 & 1 & 1 & 0 & -\frac{1}{3} \end{array}\right] \quad \text{Get 1's down the diagonal.}$$

From the last transformation, the desired inverse is

$$A^{-1} = \begin{bmatrix} 0 & 0 & \frac{1}{3} \\ -\frac{1}{2} & -\frac{1}{2} & \frac{1}{2} \\ 1 & 0 & -\frac{1}{3} \end{bmatrix}.$$

**YOUR TURN 1** Find $A^{-1}$ if
$$A = \begin{bmatrix} 2 & 3 & 1 \\ 1 & -2 & -1 \\ 3 & 3 & 2 \end{bmatrix}.$$

Confirm this by forming the products $A^{-1}A$ and $AA^{-1}$, both of which should equal $I$.

**Method 2** **Graphing Calculator**  The inverse of $A$ can also be found with a graphing calculator, as shown in Figure 11 in the margin below. (The matrix $A$ had previously been entered into the calculator.)

TRY YOUR TURN 1

**TECHNOLOGY NOTE**  Spreadsheets also have the capability of calculating the inverse of a matrix with a simple command. See the *Graphing Calculator and Excel Spreadsheet Manual* available with this book for details.

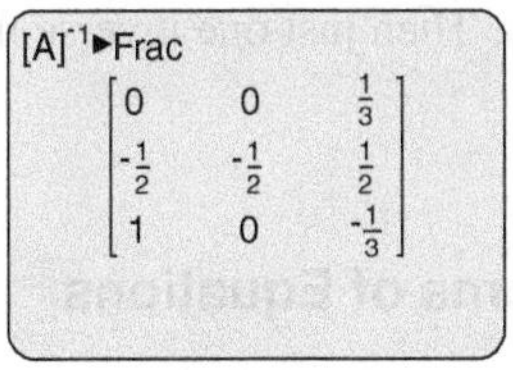

[A]⁻¹▶Frac
$$\begin{bmatrix} 0 & 0 & \frac{1}{3} \\ -\frac{1}{2} & -\frac{1}{2} & \frac{1}{2} \\ 1 & 0 & -\frac{1}{3} \end{bmatrix}$$

**FIGURE 11**

**EXAMPLE 3** **Inverse Matrix**

Find $A^{-1}$ if $A = \begin{bmatrix} 2 & -4 \\ 1 & -2 \end{bmatrix}$.

**SOLUTION** Using row operations to transform the first column of the augmented matrix

$$\left[\begin{array}{cc|cc} 2 & -4 & 1 & 0 \\ 1 & -2 & 0 & 1 \end{array}\right]$$

gives the following results.

$$R_1 + (-2)R_2 \rightarrow R_2 \quad \left[\begin{array}{cc|cc} 2 & -4 & 1 & 0 \\ 0 & 0 & 1 & -2 \end{array}\right]$$

Because the last row has all zeros to the left of the vertical bar, there is no way to complete the process of finding the inverse matrix. What is wrong? Just as the real number 0 has no multiplicative inverse, some matrices do not have inverses. Matrix $A$ is an example of a matrix that has no inverse: there is no matrix $A^{-1}$ such that $AA^{-1} = A^{-1}A = I$.

**NOTE** As is shown in Exercise 46, there is a quick way to find the inverse of a $2 \times 2$ matrix:

When $A = \begin{bmatrix} a & b \\ c & d \end{bmatrix}$ and $ad - bc \neq 0$, $A^{-1} = \dfrac{1}{ad - bc}\begin{bmatrix} d & -b \\ -c & a \end{bmatrix}$.

## Solving Systems of Equations with Inverses

We used matrices to solve systems of linear equations by the Gauss-Jordan method in Section 10.1. Another way to use matrices to solve linear systems is to write the system as a matrix equation $AX = B$, where $A$ is the matrix of the coefficients of the variables of the system, $X$ is the matrix of the variables, and $B$ is the matrix of the constants. Matrix $A$ is called the **coefficient matrix**.

To solve the matrix equation $AX = B$, first see if $A^{-1}$ exists. Assuming $A^{-1}$ exists and using the facts that $A^{-1}A = I$ and $IX = X$ gives

$$AX = B$$
$$A^{-1}(AX) = A^{-1}B \qquad \text{Multiply both sides by } A^{-1}.$$
$$(A^{-1}A)X = A^{-1}B \qquad \text{Associative property}$$
$$IX = A^{-1}B \qquad \text{Multiplicative inverse property}$$
$$X = A^{-1}B. \qquad \text{Identity property}$$

**CAUTION** When multiplying by matrices on both sides of a matrix equation, be careful to multiply in the same order on both sides of the equation, since multiplication of matrices is not commutative (unlike multiplication of real numbers).

The work thus far leads to the following method of solving a system of equations written as a matrix equation.

## Solving a System $AX = B$ Using Matrix Inverses

To solve a system of equations $AX = B$, where $A$ is the square matrix of coefficients and $A^{-1}$ exists, $X$ is the matrix of variables, and $B$ is the matrix of constants, first find $A^{-1}$. Then $X = A^{-1}B$.

This method is most practical in solving several systems that have the same coefficient matrix but different constants, as in Example 5 in this section. Then just one inverse matrix must be found.

## EXAMPLE 4 Inverse Matrices and Systems of Equations

Use the inverse of the coefficient matrix to solve the linear system

$$2x - 3y = 4$$
$$x + 5y = 2.$$

**SOLUTION** To represent the system as a matrix equation, use the coefficient matrix of the system together with the matrix of variables and the matrix of constants:

$$A = \begin{bmatrix} 2 & -3 \\ 1 & 5 \end{bmatrix}, \qquad X = \begin{bmatrix} x \\ y \end{bmatrix}, \qquad \text{and} \qquad B = \begin{bmatrix} 4 \\ 2 \end{bmatrix}.$$

The system can now be written in matrix form as the equation $AX = B$, since

$$AX = \begin{bmatrix} 2 & -3 \\ 1 & 5 \end{bmatrix}\begin{bmatrix} x \\ y \end{bmatrix} = \begin{bmatrix} 2x - 3y \\ x + 5y \end{bmatrix} = \begin{bmatrix} 4 \\ 2 \end{bmatrix} = B.$$

To solve the system, first find $A^{-1}$. Do this by using row operations on matrix $[A \mid I]$ to get

$$\begin{bmatrix} 1 & 0 & \mid & \frac{5}{13} & \frac{3}{13} \\ 0 & 1 & \mid & -\frac{1}{13} & \frac{2}{13} \end{bmatrix}.$$

From this result,

$$A^{-1} = \begin{bmatrix} \frac{5}{13} & \frac{3}{13} \\ -\frac{1}{13} & \frac{2}{13} \end{bmatrix}.$$

Next, find the product $A^{-1}B$.

$$A^{-1}B = \begin{bmatrix} \frac{5}{13} & \frac{3}{13} \\ -\frac{1}{13} & \frac{2}{13} \end{bmatrix} \begin{bmatrix} 4 \\ 2 \end{bmatrix} = \begin{bmatrix} 2 \\ 0 \end{bmatrix}$$

**YOUR TURN 2** Use the inverse of the coefficient matrix to solve the linear system
$$5x + 4y = 23$$
$$4x - 3y = 6.$$

Since $X = A^{-1}B$,

$$X = \begin{bmatrix} x \\ y \end{bmatrix} = \begin{bmatrix} 2 \\ 0 \end{bmatrix}.$$

The solution of the system is $(2, 0)$.  **TRY YOUR TURN 2**

## EXAMPLE 5  Fertilizer

Three brands of fertilizer are available that provide nitrogen, phosphoric acid, and soluble potash to the soil. One bag of each brand provides the following units of each nutrient.

|  |  | Brand | | |
| --- | --- | --- | --- | --- |
|  |  | Fertifun | Big Grow | Soakem |
|  | Nitrogen | 1 | 2 | 3 |
| Nutrient | Phosphoric Acid | 3 | 1 | 2 |
|  | Potash | 2 | 0 | 1 |

For ideal growth, the soil on a Michigan farm needs 18 units of nitrogen, 23 units of phosphoric acid, and 13 units of potash per acre. The corresponding numbers for a California farm are 31, 24, and 11, and for a Kansas farm are 20, 19, and 15. How many bags of each brand of fertilizer should be used per acre for ideal growth on each farm?

**APPLY IT**

**SOLUTION** Rather than solve three separate systems, we consider the single system

$$x + 2y + 3z = a$$
$$3x + y + 2z = b$$
$$2x + z = c,$$

where $a$, $b$, and $c$ represent the units of nitrogen, phosphoric acid, and potash needed for the different farms. The system of equations is then of the form $AX = B$, where

$$A = \begin{bmatrix} 1 & 2 & 3 \\ 3 & 1 & 2 \\ 2 & 0 & 1 \end{bmatrix} \quad \text{and} \quad X = \begin{bmatrix} x \\ y \\ z \end{bmatrix}.$$

$B$ has different values for the different farms. We find $A^{-1}$ first, then use it to solve all three systems.

To find $A^{-1}$, we start with the matrix

$$[A \,|\, I] = \begin{bmatrix} 1 & 2 & 3 & | & 1 & 0 & 0 \\ 3 & 1 & 2 & | & 0 & 1 & 0 \\ 2 & 0 & 1 & | & 0 & 0 & 1 \end{bmatrix}$$

and use row operations to get $[I \,|\, A^{-1}]$. The result is

$$A^{-1} = \begin{bmatrix} -\frac{1}{3} & \frac{2}{3} & -\frac{1}{3} \\ -\frac{1}{3} & \frac{5}{3} & -\frac{7}{3} \\ \frac{2}{3} & -\frac{4}{3} & \frac{5}{3} \end{bmatrix}.$$

Now we can solve each of the three systems by using $X = A^{-1}B$.

For the Michigan farm, $B = \begin{bmatrix} 18 \\ 23 \\ 13 \end{bmatrix}$, and

$$X = \begin{bmatrix} -\frac{1}{3} & \frac{2}{3} & -\frac{1}{3} \\ -\frac{1}{3} & \frac{5}{3} & -\frac{7}{3} \\ \frac{2}{3} & -\frac{4}{3} & \frac{5}{3} \end{bmatrix} \begin{bmatrix} 18 \\ 23 \\ 13 \end{bmatrix} = \begin{bmatrix} 5 \\ 2 \\ 3 \end{bmatrix}.$$

Therefore, $x = 5$, $y = 2$, and $z = 3$. Buy 5 bags of Fertifun, 2 bags of Big Grow, and 3 bags of Soakem.

For the California farm, $B = \begin{bmatrix} 31 \\ 24 \\ 11 \end{bmatrix}$, and

$$X = \begin{bmatrix} -\frac{1}{3} & \frac{2}{3} & -\frac{1}{3} \\ -\frac{1}{3} & \frac{5}{3} & -\frac{7}{3} \\ \frac{2}{3} & -\frac{4}{3} & \frac{5}{3} \end{bmatrix} \begin{bmatrix} 31 \\ 24 \\ 11 \end{bmatrix} = \begin{bmatrix} 2 \\ 4 \\ 7 \end{bmatrix}.$$

Buy 2 bags of Fertifun, 4 bags of Big Grow, and 7 bags of Soakem.

For the Kansas farm, $B = \begin{bmatrix} 20 \\ 19 \\ 15 \end{bmatrix}$. Verify that this leads to $x = 1$, $y = -10$, and $z = 13$.

We cannot have a negative number of bags, so this solution is impossible. In buying enough bags to meet all of the nutrient requirements, the farmer must purchase an excess of some nutrients.

In Example 5, using the matrix inverse method of solving the systems involved considerably less work than using row operations for each of the three systems.

### EXAMPLE 6   Solving an Inconsistent System of Equations

Use the inverse of the coefficient matrix to solve the system

$$2x - 4y = 13$$
$$x - 2y = 1.$$

**SOLUTION**   We saw in Example 3 that the coefficient matrix $\begin{bmatrix} 2 & -4 \\ 1 & -2 \end{bmatrix}$ does not have an inverse. This means that the given system either has no solution or has an infinite number of solutions. Verify that this system is inconsistent and has no solution.

**NOTE**   If a matrix has no inverse, then a corresponding system of equations might have either no solutions or an infinite number of solutions, and we instead use the Gauss-Jordan method. In Example 3, we saw that the matrix $\begin{bmatrix} 2 & -4 \\ 1 & -2 \end{bmatrix}$ had no inverse. The reason is that the first row of this matrix is double the second row. The system

$$2x - 4y = 2$$
$$x - 2y = 1$$

has an infinite number of solutions because the first equation is double the second equation. The system

$$2x - 4y = 13$$
$$x - 2y = \phantom{0}1$$

has no solutions because the left side of the first equation is double the left side of the second equation, but the right side is not.

### EXAMPLE 7 Cryptography

Throughout the Cold War and as the Internet has grown and developed, the need for sophisticated methods of coding and decoding messages has increased. Although there are many methods of encrypting messages, one fairly sophisticated method uses matrix operations. This method first assigns a number to each letter of the alphabet. The simplest way to do this is to assign the number 1 to A, 2 to B, and so on, with the number 27 used to represent a space between words.

For example, the message *math is cool* can be divided into groups of three letters and spaces each and then converted into numbers as follows:

$$\begin{bmatrix} m \\ a \\ t \end{bmatrix} = \begin{bmatrix} 13 \\ 1 \\ 20 \end{bmatrix}.$$

The entire message would then consist of four $3 \times 1$ columns of numbers:

$$\begin{bmatrix} 13 \\ 1 \\ 20 \end{bmatrix}, \quad \begin{bmatrix} 8 \\ 27 \\ 9 \end{bmatrix}, \quad \begin{bmatrix} 19 \\ 27 \\ 3 \end{bmatrix}, \quad \begin{bmatrix} 15 \\ 15 \\ 12 \end{bmatrix}.$$

This code is easy to break, so we further complicate the code by choosing a matrix that has an inverse (in this case a $3 \times 3$ matrix) and calculate the products of the matrix and each of the column vectors above.

If we choose the coding matrix

$$A = \begin{bmatrix} 1 & 3 & 4 \\ 2 & 1 & 3 \\ 4 & 2 & 1 \end{bmatrix},$$

then the products of $A$ with each of the column vectors above produce a new set of vectors

$$\begin{bmatrix} 96 \\ 87 \\ 74 \end{bmatrix}, \quad \begin{bmatrix} 125 \\ 70 \\ 95 \end{bmatrix}, \quad \begin{bmatrix} 112 \\ 74 \\ 133 \end{bmatrix}, \quad \begin{bmatrix} 108 \\ 81 \\ 102 \end{bmatrix}.$$

This set of vectors represents our coded message, and it will be transmitted as 96, 87, 74, 125 and so on.

When the intended person receives the message, it is divided into groups of three numbers, and each group is formed into a column matrix. The message is easily decoded if the receiver knows the inverse of the original matrix. The inverse of matrix $A$ is

$$A^{-1} = \begin{bmatrix} -0.2 & 0.2 & 0.2 \\ 0.4 & -0.6 & 0.2 \\ 0 & 0.4 & -0.2 \end{bmatrix}.$$

Thus, the message is decoded by taking the product of the inverse matrix with each column vector of the received message. For example,

$$A^{-1} \begin{bmatrix} 96 \\ 87 \\ 74 \end{bmatrix} = \begin{bmatrix} 13 \\ 1 \\ 20 \end{bmatrix}.$$

**YOUR TURN 3** Use the matrix $A$ and its inverse $A^{-1}$ in Example 7 to do the following.
**(a)** Encode the message "Behold."
**(b)** Decode the message 96, 87, 74, 141, 117, 114.

Unless the original matrix or its inverse is known, this type of code can be difficult to break. In fact, very large matrices can be used to encrypt data. It is interesting to note that many mathematicians are employed by the National Security Agency to develop encryption methods that are virtually unbreakable. **TRY YOUR TURN 3**

# 10.4 EXERCISES

**Decide whether the given matrices are inverses of each other. (Check to see if their product is the identity matrix $I$.)**

**1.** $\begin{bmatrix} 2 & 1 \\ 5 & 3 \end{bmatrix}$ and $\begin{bmatrix} 3 & -1 \\ -5 & 2 \end{bmatrix}$

**2.** $\begin{bmatrix} 1 & -4 \\ 2 & -7 \end{bmatrix}$ and $\begin{bmatrix} -7 & 4 \\ -2 & 1 \end{bmatrix}$

**3.** $\begin{bmatrix} 2 & 6 \\ 2 & 4 \end{bmatrix}$ and $\begin{bmatrix} -1 & 2 \\ 2 & -4 \end{bmatrix}$

**4.** $\begin{bmatrix} -1 & 2 \\ 3 & -5 \end{bmatrix}$ and $\begin{bmatrix} -5 & -2 \\ -3 & -1 \end{bmatrix}$

**5.** $\begin{bmatrix} 2 & 0 & 1 \\ 1 & 1 & 2 \\ 0 & 1 & 0 \end{bmatrix}$ and $\begin{bmatrix} 1 & 1 & -1 \\ 0 & 1 & 0 \\ -1 & -2 & 2 \end{bmatrix}$

**6.** $\begin{bmatrix} 0 & 1 & 0 \\ 0 & 0 & -2 \\ 1 & -1 & 0 \end{bmatrix}$ and $\begin{bmatrix} 1 & 0 & 1 \\ 1 & 0 & 0 \\ 0 & -1 & 0 \end{bmatrix}$

**7.** $\begin{bmatrix} 1 & 3 & 3 \\ 1 & 4 & 3 \\ 1 & 3 & 4 \end{bmatrix}$ and $\begin{bmatrix} 7 & -3 & -3 \\ -1 & 1 & 0 \\ -1 & 0 & 1 \end{bmatrix}$

**8.** $\begin{bmatrix} 1 & 0 & 0 \\ -1 & -2 & 3 \\ 0 & 1 & 0 \end{bmatrix}$ and $\begin{bmatrix} 1 & 0 & 0 \\ 0 & 0 & 1 \\ \frac{1}{3} & \frac{1}{3} & \frac{2}{3} \end{bmatrix}$

**9.** Does a matrix with a row of all zeros have an inverse? Why?

**10.** Matrix $A$ has $A^{-1}$ as its inverse. What does $(A^{-1})^{-1}$ equal?
(*Hint:* Experiment with a few matrices to see what you get.)

**Find the inverse, if it exists, for each matrix.**

**11.** $\begin{bmatrix} 1 & -1 \\ 2 & 0 \end{bmatrix}$

**12.** $\begin{bmatrix} 1 & 1 \\ 2 & 3 \end{bmatrix}$

**13.** $\begin{bmatrix} 3 & -1 \\ -5 & 2 \end{bmatrix}$

**14.** $\begin{bmatrix} -3 & -8 \\ 1 & 3 \end{bmatrix}$

**15.** $\begin{bmatrix} 1 & -3 \\ -2 & 6 \end{bmatrix}$

**16.** $\begin{bmatrix} 5 & 10 \\ -3 & -6 \end{bmatrix}$

**17.** $\begin{bmatrix} 1 & 0 & 0 \\ 0 & -1 & 0 \\ 1 & 0 & 1 \end{bmatrix}$

**18.** $\begin{bmatrix} 1 & 3 & 0 \\ 0 & 2 & -1 \\ 1 & 0 & 2 \end{bmatrix}$

**19.** $\begin{bmatrix} -1 & -1 & -1 \\ 4 & 5 & 0 \\ 0 & 1 & -3 \end{bmatrix}$

**20.** $\begin{bmatrix} 2 & 1 & 0 \\ 0 & 3 & 1 \\ 4 & -1 & -3 \end{bmatrix}$

**21.** $\begin{bmatrix} 1 & 2 & 3 \\ -3 & -2 & -1 \\ -1 & 0 & 1 \end{bmatrix}$

**22.** $\begin{bmatrix} 2 & 0 & 4 \\ 1 & 0 & -1 \\ 3 & 0 & -2 \end{bmatrix}$

**23.** $\begin{bmatrix} 1 & 3 & -2 \\ 2 & 7 & -3 \\ 3 & 8 & -5 \end{bmatrix}$

**24.** $\begin{bmatrix} 4 & 1 & -4 \\ 2 & 1 & -1 \\ -2 & -4 & 5 \end{bmatrix}$

**25.** $\begin{bmatrix} 1 & -2 & 3 & 0 \\ 0 & 1 & -1 & 1 \\ -2 & 2 & -2 & 4 \\ 0 & 2 & -3 & 1 \end{bmatrix}$

**26.** $\begin{bmatrix} 1 & 1 & 0 & 2 \\ 2 & -1 & 1 & -1 \\ 3 & 3 & 2 & -2 \\ 1 & 2 & 1 & 0 \end{bmatrix}$

**Solve each system of equations by using the inverse of the coefficient matrix if it exists and by the Gauss-Jordan method if the inverse doesn't exist. (The inverses for exercises 35–38 were found in Exercises 19, 20, 23, and 24.)**

**27.** $2x + 5y = 15$
$x + 4y = 9$

**28.** $-x + 2y = 15$
$-2x - y = 20$

**29.** $2x + y = 5$
$5x + 3y = 13$

**30.** $-x - 2y = 8$
$3x + 4y = 24$

**31.** $3x - 2y = 3$
$7x - 5y = 0$

**32.** $3x - 6y = 1$
$-5x + 9y = -1$

**33.** $-x - 8y = 12$
$3x + 24y = -36$

**34.** $2x + 7y = 14$
$4x + 14y = 28$

**35.** $\begin{aligned} -x - y - z &= 1 \\ 4x + 5y &= -2 \\ y - 3z &= 3 \end{aligned}$

**36.** $\begin{aligned} 2x + y &= 1 \\ 3y + z &= 8 \\ 4x - y - 3z &= 8 \end{aligned}$

**37.** $\begin{aligned} x + 3y - 2z &= 4 \\ 2x + 7y - 3z &= 8 \\ 3x + 8y - 5z &= -4 \end{aligned}$

**38.** $\begin{aligned} 4x + y - 4z &= 17 \\ 2x + y - z &= 12 \\ -2x - 4y + 5z &= 17 \end{aligned}$

**39.** $\begin{aligned} 2x - 2y &= 5 \\ 4y + 8z &= 7 \\ x + 2z &= 1 \end{aligned}$

**40.** $\begin{aligned} x + 2z &= -1 \\ y - z &= 5 \\ x + 2y &= 7 \end{aligned}$

**Solve each system of equations by using the inverse of the coefficient matrix. (The inverses were found in Exercises 25 and 26.)**

**41.** $\begin{aligned} x - 2y + 3z &= 4 \\ y - z + w &= -8 \\ -2x + 2y - 2z + 4w &= 12 \\ 2y - 3z + w &= -4 \end{aligned}$

**42.** $\begin{aligned} x + y + 2w &= 3 \\ 2x - y + z - w &= 3 \\ 3x + 3y + 2z - 2w &= 5 \\ x + 2y + z &= 3 \end{aligned}$

Let $A = \begin{bmatrix} a & b \\ c & d \end{bmatrix}$ and $0 = \begin{bmatrix} 0 & 0 \\ 0 & 0 \end{bmatrix}$ in Exercises 43–48.

**43.** Show that $IA = A$.

**44.** Show that $AI = A$.

**45.** Show that $A \cdot O = O$.

**46.** Find $A^{-1}$.
(Assume $ad - bc \neq 0$.)

**47.** Show that $A^{-1}A = I$.

**48.** Show that $AA^{-1} = I$.

**49.** Using the definition and properties listed in this section, show that for square matrices $A$ and $B$ of the same size, if $AB = O$ and if $A^{-1}$ exists, then $B = O$, where $O$ is a matrix whose elements are all zeros.

**50.** Prove that, if it exists, the inverse of a matrix is unique. (*Hint:* Assume there are two inverses $B$ and $C$ for some matrix $A$, so that $AB = BA = I$ and $AC = CA = I$. Multiply the first equation by $C$ and the second by $B$.)

**Use matrices $C$ and $D$ in Exercises 51–55.**

$$C = \begin{bmatrix} -6 & 8 & 2 & 4 & -3 \\ 1 & 9 & 7 & -12 & 5 \\ 15 & 2 & -8 & 10 & 11 \\ 4 & 7 & 9 & 6 & -2 \\ 1 & 3 & 8 & 23 & 4 \end{bmatrix}, \quad D = \begin{bmatrix} 5 & -3 & 7 & 9 & 2 \\ 6 & 8 & -5 & 2 & 1 \\ 3 & 7 & -4 & 2 & 11 \\ 5 & -3 & 9 & 4 & -1 \\ 0 & 3 & 2 & 5 & 1 \end{bmatrix}$$

**51.** Find $C^{-1}$.

**52.** Find $(CD)^{-1}$.

**53.** Find $D^{-1}$.

**54.** Is $C^{-1}D^{-1} = (CD)^{-1}$?

**55.** Is $D^{-1}C^{-1} = (CD)^{-1}$?

**Solve the matrix equation $AX = B$ for $X$ by finding $A^{-1}$, given $A$ and $B$ as follows.**

**56.** $A = \begin{bmatrix} 2 & -5 & 7 \\ 4 & -3 & 2 \\ 15 & 2 & 6 \end{bmatrix}, \quad B = \begin{bmatrix} -2 \\ 5 \\ 8 \end{bmatrix}$

**57.** $A = \begin{bmatrix} 2 & 5 & 7 & 9 \\ 1 & 3 & -4 & 6 \\ -1 & 0 & 5 & 8 \\ 2 & -2 & 4 & 10 \end{bmatrix}, \quad B = \begin{bmatrix} 3 \\ 7 \\ -1 \\ 5 \end{bmatrix}$

**58.** $A = \begin{bmatrix} 3 & 2 & -1 & -2 & 6 \\ -5 & 17 & 4 & 3 & 15 \\ 7 & 9 & -3 & -7 & 12 \\ 9 & -2 & 1 & 4 & 8 \\ 1 & 21 & 9 & -7 & 25 \end{bmatrix}, \quad B = \begin{bmatrix} -2 \\ 5 \\ 3 \\ -8 \\ 25 \end{bmatrix}$

## LIFE SCIENCE APPLICATIONS

**Solve each exercise by using the inverse of the coefficient matrix to solve a system of equations.**

**59. Vitamins** Greg Tobin mixes together three types of vitamin tablets. Each Super Vim tablet contains, among other things, 15 mg of niacin and 12 I.U. of vitamin E. The figures for a Multitab tablet are 20 mg and 15 I.U., and for a Mighty Mix are 25 mg and 35 I.U. How many of each tablet are there if the total number of tablets, total amount of niacin, and total amount of vitamin E are as follows?

   **a.** 225 tablets, 4750 mg of niacin, and 5225 I.U. of vitamin E

   **b.** 185 tablets, 3625 mg of niacin, and 3750 I.U. of vitamin E

   **c.** 230 tablets, 4450 mg of niacin, and 4210 I.U. of vitamin E

**60. Calorie Expenditure** Jennifer Crum has an exercise regimen at the local gym that includes using a treadmill, stair machine, and cycle. According to a website, a 130-pound person like Jennifer will burn, on average, about 600 calories per hour running a moderate pace on the treadmill, about 500 calories per hour on the stair machine, and about 300 calories per hour on the cycle. *Source: Nutristrategy.*

   **a.** Jennifer wants to spend 10 hours a week at the gym, and would like to spend twice as much time on the stair machine as the cycle. She wants to burn 5000 calories a week at the gym. Is this possible? How much time should Jennifer spend on each machine?

   **b.** Jennifer decides to spend 15 hours this week at the gym. She still wants to spend twice as much time on the stair machine as the cycle, but wants to double the total number of calories that she burns. Is this possible?

   **c.** Next week Jennifer can spend only 9 hours at the gym, and still wants to spend twice as much time on the stair machine as the cycle. Can she still burn 5000 total calories? If so, how much time should Jennifer spend on each machine?

## OTHER APPLICATIONS

**61. Analysis of Orders** The Bread Box Bakery sells three types of cakes, each requiring the amounts of the basic ingredients shown in the following matrix.

| | | Type of Cake | | |
|---|---|---|---|---|
| | | I | II | III |
| | Flour (in cups) | 2 | 4 | 2 |
| Ingredient | Sugar (in cups) | 2 | 1 | 2 |
| | Eggs | 2 | 1 | 3 |

To fill its daily orders for these three kinds of cake, the bakery uses 72 cups of flour, 48 cups of sugar, and 60 eggs.

**a.** Write a $3 \times 1$ matrix for the amounts used daily.

**b.** Let the number of daily orders for cakes be a $3 \times 1$ matrix $X$ with entries $x_1$, $x_2$, and $x_3$. Write a matrix equation that can be solved for $X$, using the given matrix and the matrix from part a.

**c.** Solve the equation from part b to find the number of daily orders for each type of cake.

**62. Production Requirements** An electronics company produces transistors, resistors, and computer chips. Each transistor requires 3 units of copper, 1 unit of zinc, and 2 units of glass. Each resistor requires 3, 2, and 1 units of the three materials, and each computer chip requires 2, 1, and 2 units of these materials, respectively. How many of each product can be made with the following amounts of materials?

**a.** 810 units of copper, 410 units of zinc, and 490 units of glass

**b.** 765 units of copper, 385 units of zinc, and 470 units of glass

**c.** 1010 units of copper, 500 units of zinc, and 610 units of glass

**63. Investments** An investment firm recommends that a client invest in AAA-, A-, and B-rated bonds. The average yield on AAA bonds is 6%, on A bonds 6.5%, and on B bonds 8%. The client wants to invest twice as much in AAA bonds as in B bonds. How much should be invested in each type of bond under the following conditions?

**a.** The total investment is $25,000, and the investor wants an annual return of $1650 on the three investments.

**b.** The values in part a are changed to $30,000 and $1985, respectively.

**c.** The values in part a arc changed to $40,000 and $2660, respectively.

**64. Production** Pretzels cost $4 per lb, dried fruit $5 per lb, and nuts $9 per lb. The three ingredients are to be combined in a trail mix containing twice the weight of pretzels as dried fruit. How many pounds of each should be used to produce the following amounts at the given cost?

**a.** 140 lb at $6 per lb

**b.** 100 lb at $7.60 per lb

**c.** 125 lb at $6.20 per lb

**65. Encryption** Use the matrices presented in Example 7 of this section to do the following:

**a.** Encode the message, "All is fair in love and war."

**b.** Decode the message 138, 81, 102, 101, 67, 109, 162, 124, 173, 210, 150, 165.

**66. Encryption** Use the methods presented in Example 7 along with the given matrix $B$ to do the following.

$$B = \begin{bmatrix} 2 & 4 & 6 \\ -1 & -4 & -3 \\ 0 & 1 & -1 \end{bmatrix}$$

**a.** Encode the message, "To be or not to be."

**b.** Find the inverse of $B$.

**c.** Use the inverse of $B$ to decode the message $116, -60, -15, 294, -197, -2, 148, -92, -9, 96, -64, 4, 264, -182, -2$.

**67. Music** During a marching band's half-time show, the band members generally line up in such a way that a common shape is recognized by the fans. For example, as illustrated in the figure, a band might form a letter T, where an x represents a member of the band. As the music is played, the band will either create a new shape or rotate the original shape. In doing this, each member of the band will need to move from one point on the field to another. For larger bands, keeping track of who goes where can be a daunting task. However, it is possible to use matrix inverses to make the process a bit easier. The entire process is calculated by knowing how three band members, all of whom cannot be in a straight line, will move from the current position to a new position. For example, in the figure, we can see that there are band members at $(50, 0)$, $(50, 15)$, and $(45, 20)$. We will assume that these three band members move to $(40, 10)$, $(55, 10)$, and $(60, 15)$, respectively. (Note: The x-coordinate 60 is represented by the right side's 40 yard line in the picture.) *Source: The College Mathematics Journal.*

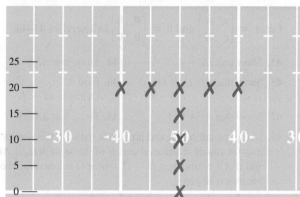

**a.** Find the inverse of $B = \begin{bmatrix} 50 & 50 & 45 \\ 0 & 15 & 20 \\ 1 & 1 & 1 \end{bmatrix}$.

**b.** Find $A = \begin{bmatrix} 40 & 55 & 60 \\ 10 & 10 & 15 \\ 1 & 1 & 1 \end{bmatrix} B^{-1}$.

**c.** Use the result of part b to find the new position of the other band members. What is the shape of the new position? (*Hint:* Multiply the matrix $A$ by a $3 \times 1$ column vector with the first two components equal to the original position of each band member and the third component equal to 1. The new position of the band member is in the first two components of the product.)

**YOUR TURN ANSWERS**

**1.** $\begin{bmatrix} 1/8 & 3/8 & 1/8 \\ 5/8 & -1/8 & -3/8 \\ -9/8 & -3/8 & 7/8 \end{bmatrix}$ **2.** $(3, 2)$ **3. a.** 49, 33, 26, 67, 54, 88 **b.** matrix

# 10.5  Eigenvalues and Eigenvectors

How can we determine a population of insects that grows at a constant rate while keeping the same proportion of adults and juveniles?

*This kind of question is of fundamental interest in mathematical biology, and will be answered in Example 1.*

Suppose there is an insect population consisting of juveniles (insects in their first year of life) and adults (insects in their second year of life). *Source: Population Biology: Concepts and Models.* We will assume that no insects of this species live more than two years.

Let $x_1(t)$ and $x_2(t)$ be the number of juveniles and adults in year $t$. We will assume that a proportion $s$ of the juveniles survive to become adults the following year. We will also assume that a proportion $p$ of the juveniles have offspring that are juveniles the following year, and that a proportion $q$ of the adults have offspring that are juveniles the following year. Then the number of juveniles and adults next year can be calculated from the number this year using the equation

$$x_1(t + 1) = px_1(t) + qx_2(t)$$
$$x_2(t + 1) = sx_1(t).$$

In matrix form, this is

$$\begin{bmatrix} x_1(t + 1) \\ x_2(t + 1) \end{bmatrix} = \begin{bmatrix} p & q \\ s & 0 \end{bmatrix} \begin{bmatrix} x_1(t) \\ x_2(t) \end{bmatrix},$$

which can also be written as

$$X(t + 1) = MX(t),$$

where

$$M = \begin{bmatrix} p & q \\ s & 0 \end{bmatrix}$$

and

$$X(t) = \begin{bmatrix} x_1(t) \\ x_2(t) \end{bmatrix}.$$

The advantage of this last notation is that it can be extended to any number of groups within the population. The matrix $M$ is known as a **Leslie matrix** after P. H. Leslie, who first wrote about them in 1945. *Source: Biometrika.*

## EXAMPLE 1  Insect Population Growth

Suppose

$$M = \begin{bmatrix} 0.6 & 0.8 \\ 0.9 & 0 \end{bmatrix}$$

and that the initial insect population consists of 1000 juveniles and 2000 adults. Determine the population of insects in year 1 and year 2.

**SOLUTION**  Let $X(0)$ be the initial population, so

$$X(0) = \begin{bmatrix} 1000 \\ 2000 \end{bmatrix}.$$

To find the population in year 1, find $X(1) = MX(0)$. Therefore,

$$X(1) = \begin{bmatrix} x_1(1) \\ x_2(1) \end{bmatrix} = \begin{bmatrix} 0.6 & 0.8 \\ 0.9 & 0 \end{bmatrix} \begin{bmatrix} 1000 \\ 2000 \end{bmatrix}$$

$$= \begin{bmatrix} 2200 \\ 900 \end{bmatrix}.$$

Likewise, the population in year 2 is $X(2) = MX(1)$. Thus,

$$X(2) = \begin{bmatrix} x_1(2) \\ x_2(2) \end{bmatrix} = \begin{bmatrix} 0.6 & 0.8 \\ 0.9 & 0 \end{bmatrix} \begin{bmatrix} 2200 \\ 900 \end{bmatrix}$$

$$= \begin{bmatrix} 2040 \\ 1980 \end{bmatrix}.$$

Notice that the population of juveniles has gone from 1000 in year 0 to 2200 in year 1 to 2040 in year 2. Meanwhile, the population of adults has gone from 2000 in year 0 to 900 in year 1 to 1980 in year 2.

In many actual populations, the proportion of the population in each age group stays the same from year to year, even as the population grows. For example, if

$$X(0) = \begin{bmatrix} 400 \\ 300 \end{bmatrix},$$

then

$$X(1) = \begin{bmatrix} 0.6 & 0.8 \\ 0.9 & 0 \end{bmatrix} \begin{bmatrix} 400 \\ 300 \end{bmatrix} = \begin{bmatrix} 480 \\ 360 \end{bmatrix} = 1.2 \begin{bmatrix} 400 \\ 300 \end{bmatrix}.$$

Both the population of juveniles and adults have grown by 20%. If we denote the growth factor by the Greek letter $\lambda$ (pronounced lambda), then

$$MX = \lambda X.$$

A number $\lambda$ that satisfies the above equation is called an **eigenvalue**. (Eigen is the German word meaning "characteristic.") The matrix (or vector) $X$ in the above equation is called an **eigenvector**. Eigenvalues and eigenvectors have a wide variety of applications in mathematics in addition to solving problems about populations. In the next chapter, we will see how they are used in solving systems of differential equations. The important thing to notice here is that multiplying an eigenvector $X$ by the matrix $M$ has the same result as multiplying $X$ by the scalar $\lambda$.

Now we must investigate how to find eigenvalues and eigenvectors. By subtracting $\lambda X$ from both sides of the equation above, we get

$$MX - \lambda X = O,$$

where $O$ represents a square matrix with all zeros. This equation can be rewritten as

$$(M - \lambda I)X = O,$$

where $I$ is the identity matrix. This equation has the trivial solution $X = 0$, but we are interested in nontrivial solutions. In the language of Section 10.1, the above equation would have more than one solution (in fact, an infinite number of solutions) if it describes a dependent system. So we need to know for what values of $\lambda$ the above system is dependent.

We now need to stretch a little to some mathematics that is explained more fully in a text on linear algebra.* It can be shown that the above system has a nontrivial solution if something called the **determinant** of $M - \lambda I$ is zero. (For a proof of this fact in the case where $M$ is a $2 \times 2$ matrix, see Exercise 13.) Computing the determinant of a $2 \times 2$ matrix $A$, denoted $\det(A)$, is easy:

$$\det(A) = \det \begin{bmatrix} a & b \\ c & d \end{bmatrix} = ad - bc.$$

*For example, see Lay, David, *Linear Algebra and Its Applications*, 4th ed., Pearson, 2011.

Recall from the previous section that the expression $ad - bc$ is part of the formula for the inverse of a $2 \times 2$ matrix. You can think of calculating the determinant by starting with the product of the numbers along the diagonal from upper left to lower right, and subtracting the product of the numbers along the diagonal from upper right to lower left:

$$\det \begin{bmatrix} a & b \\ c & d \end{bmatrix} =$$
$$-bc \qquad + \ ad.$$

### EXAMPLE 2   Determinant of 2 × 2 Matrix

Calculate the determinant of the matrix $\begin{bmatrix} 2 & 3 \\ 4 & 5 \end{bmatrix}$.

**SOLUTION**   The determinant is

$$\det \begin{bmatrix} 2 & 3 \\ 4 & 5 \end{bmatrix} =$$
$$-(3 \times 4) \quad + (2 \times 5) = -12 + 10 = -2.$$

**YOUR TURN 1**  Calculate the determinant of $\begin{bmatrix} 4 & 2 \\ 1 & 7 \end{bmatrix}$.

TRY YOUR TURN 1

To find the determinant of a $3 \times 3$ matrix, take the first two columns of the matrix and duplicate them on the right, as we will show in a moment. Then add the products of the numbers along each of the three diagonals going from upper left to lower right, and subtract the products of the numbers along each of the three diagonals from upper right to lower left. Here we demonstrate the process for a general $3 \times 3$ matrix.

$$\det \begin{bmatrix} a & b & c \\ d & e & f \\ g & h & i \end{bmatrix} =$$

$$\begin{matrix} a & b & c & a & b \\ d & e & f & d & e \\ g & h & i & g & h \end{matrix}$$

$$-ceg - afh - bdi \quad + aei + bfg + cdh.$$

### EXAMPLE 3   Determinant of 3 × 3 Matrix

Calculate the determinant of the matrix $\begin{bmatrix} 1 & 2 & 0 \\ -1 & 3 & 4 \\ -2 & 5 & -1 \end{bmatrix}$.

**SOLUTION**   The determinant is

$$\det \begin{bmatrix} 1 & 2 & 0 \\ -1 & 3 & 4 \\ -2 & 5 & -1 \end{bmatrix} =$$

**YOUR TURN 2**  Calculate the determinant of $\begin{bmatrix} 1 & -1 & 4 \\ 2 & 3 & 0 \\ 3 & -2 & -1 \end{bmatrix}$.

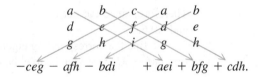

$$-(0 \times 3 \times (-2)) - (1 \times 4 \times 5) - (2 \times (-1) \times (-1)) + (1 \times 3 \times (-1)) + (2 \times 4 \times (-2)) + (0 \times (-1) \times 5)$$
$$= 0 - 20 - 2 - 3 - 16 + 0$$
$$= -41.$$

TRY YOUR TURN 2

CAUTION | The determinant of any square matrix can be calculated, but the procedure is more involved for matrices larger than $3 \times 3$, so we will not discuss those in this book.

Now we return to the problem of finding eigenvalues and their corresponding eigenvectors.

### EXAMPLE 4 Eigenvalues and Eigenvectors

Find the eigenvalues and their corresponding eigenvectors for the matrix

$$M = \begin{bmatrix} 0.6 & 0.8 \\ 0.9 & 0 \end{bmatrix}, \qquad \text{from Example 1.}$$

**SOLUTION** To find the eigenvalues, set $\det(M - \lambda I) = 0$ and solve for $\lambda$. The determinant of $M - \lambda I$ is

$$\det(M - \lambda I) = \det\left(\begin{bmatrix} 0.6 & 0.8 \\ 0.9 & 0 \end{bmatrix} - \lambda \begin{bmatrix} 1 & 0 \\ 0 & 1 \end{bmatrix}\right) = \det\begin{bmatrix} 0.6 - \lambda & 0.8 \\ 0.9 & 0 - \lambda \end{bmatrix}$$

$$= (0.6 - \lambda)(0 - \lambda) - 0.8 \times 0.9$$

$$= \lambda^2 - 0.6\lambda - 0.72.$$

We can solve $\lambda^2 - 0.6\lambda - 0.72 = 0$ with the quadratic formula.

$$\lambda = \frac{0.6 \pm \sqrt{0.6^2 - 4(-0.72)}}{2}$$

$$= \frac{0.6 \pm \sqrt{3.24}}{2}$$

$$= \frac{0.6 \pm 1.8}{2}$$

$$= 1.2, \quad -0.6$$

Therefore, since the $\det(M - \lambda I) = 0$ when $\lambda = 1.2$ and $-0.6$, the eigenvalues are 1.2 and $-0.6$.

To find the corresponding eigenvectors, substitute each eigenvalue back into the equation $(M - \lambda I)X = 0$ and solve the dependent system. First try this with $\lambda = 1.2$.

$$(M - \lambda I)X = \begin{bmatrix} 0.6 - \lambda & 0.8 \\ 0.9 & 0 - \lambda \end{bmatrix}\begin{bmatrix} x_1 \\ x_2 \end{bmatrix} \qquad \text{Let } \lambda = 1.2.$$

$$= \begin{bmatrix} -0.6 & 0.8 \\ 0.9 & -1.2 \end{bmatrix}\begin{bmatrix} x_1 \\ x_2 \end{bmatrix} = \begin{bmatrix} 0 \\ 0 \end{bmatrix}$$

Now create the augmented matrix for this system.

$$\begin{bmatrix} -0.6 & 0.8 & | & 0 \\ 0.9 & -1.2 & | & 0 \end{bmatrix}$$

Before row reducing, let's simplify by getting rid of the decimals, and then dividing out any common factors.

$$\begin{matrix} -10R_1 \to R_1 \\ 10R_2 \to R_2 \end{matrix} \begin{bmatrix} 6 & -8 & | & 0 \\ 9 & -12 & | & 0 \end{bmatrix}$$

$$\begin{matrix} R_1/2 \to R_1 \\ R_2/3 \to R_2 \end{matrix} \begin{bmatrix} 3 & -4 & | & 0 \\ 3 & -4 & | & 0 \end{bmatrix}$$

Next, get a zero in row 2 of column 1.

$$-R_1 + R_2 \to R_2 \begin{bmatrix} 3 & -4 & | & 0 \\ 0 & 0 & | & 0 \end{bmatrix}$$

The zeros in row 2 indicate that we have a dependent system. Since the first row indicates that $3x_1 - 4x_2 = 0$, the simplest solution is $x_1 = 4$ and $x_2 = 3$. There are, of course, an infinite number of solutions, described by letting $x_2$ be an arbitrary parameter and $x_1 = 4x_2/3$, but we only need one solution. An eigenvector corresponding to $\lambda = 1.2$ is therefore

$$\begin{bmatrix} x_1 \\ x_2 \end{bmatrix} = \begin{bmatrix} 4 \\ 3 \end{bmatrix}.$$

Any multiple of this matrix is also an eigenvector.

Now find the eigenvector corresponding to $\lambda = -0.6$.

$$(M - \lambda I)X = \begin{bmatrix} 0.6 - \lambda & 0.8 \\ 0.9 & 0 - \lambda \end{bmatrix}\begin{bmatrix} x_1 \\ x_2 \end{bmatrix} \qquad \text{Let } \lambda = -0.6.$$

$$= \begin{bmatrix} 1.2 & 0.8 \\ 0.9 & 0.6 \end{bmatrix}\begin{bmatrix} x_1 \\ x_2 \end{bmatrix} = \begin{bmatrix} 0 \\ 0 \end{bmatrix}$$

Next create the augmented matrix for this system.

$$\begin{bmatrix} 1.2 & 0.8 & | & 0 \\ 0.9 & 0.6 & | & 0 \end{bmatrix}$$

**YOUR TURN 3** Find the eigenvalues and their corresponding eigenvectors for the matrix
$\begin{bmatrix} 1 & 3 \\ 2 & 0 \end{bmatrix}.$

Verify that one solution to this system is

$$\begin{bmatrix} x_1 \\ x_2 \end{bmatrix} = \begin{bmatrix} 2 \\ -3 \end{bmatrix},$$

which is an eigenvector corresponding to $\lambda = -0.6$.

**TRY YOUR TURN 3**

The matrix $M$ in Example 4 is the Leslie matrix for the insect population first introduced in Example 1. Since a population cannot have a negative value, a population with a stable structure cannot be a multiple of the second eigenvector. This eigenvector can, however, be a component of a population whose age structure is changing with time, as we shall see in Example 6.

Meanwhile, the population in which the proportion of each age group stays the same from one year to the next must be a multiple of the eigenvector

$$\begin{bmatrix} x_1 \\ x_2 \end{bmatrix} = \begin{bmatrix} 4 \\ 3 \end{bmatrix},$$

for which each year's population is 1.2 times the previous year's population. In other words, the population grows 20%, and the proportion of juveniles and adults stays the same.

For example, if we wanted to start in year 0 with a population of size 14,000, then we would take a multiple of the above eigenvector, say,

$$\begin{bmatrix} x_1 \\ x_2 \end{bmatrix} = k\begin{bmatrix} 4 \\ 3 \end{bmatrix} = \begin{bmatrix} 4k \\ 3k \end{bmatrix}.$$

The size of the population is then $4k + 3k$, or $7k$. Setting this equal to 14,000 makes $k = 2000$, so the population in year 0 is

$$\begin{bmatrix} x_1 \\ x_2 \end{bmatrix} = 2000\begin{bmatrix} 4 \\ 3 \end{bmatrix} = \begin{bmatrix} 8000 \\ 6000 \end{bmatrix}.$$

The population the next year is

$$1.2\begin{bmatrix} 8000 \\ 6000 \end{bmatrix} = \begin{bmatrix} 9600 \\ 7200 \end{bmatrix},$$

and the population the following year is

$$1.2\begin{bmatrix} 9600 \\ 7200 \end{bmatrix} = \begin{bmatrix} 11,520 \\ 8640 \end{bmatrix}.$$

Verify that these results are exactly what you would get by multiplying the original population vector by the matrix $M$, and then multiplying that result by $M$. It is, of course, much simpler to multiply a vector by a scalar than by a matrix, which is a nice feature of eigenvectors and eigenvalues.

EXAMPLE 5   **Eigenvalues and Eigenvectors**

Find the eigenvalues and their corresponding eigenvectors for the matrix

$$M = \begin{bmatrix} 5 & 2 & 3 \\ 0 & 8 & 3 \\ 0 & 0 & 4 \end{bmatrix}.$$

**SOLUTION**   Calculate $\det(M - \lambda I)$.

$$\det(M - \lambda I) = \det \begin{bmatrix} 5 - \lambda & 2 & 3 \\ 0 & 8 - \lambda & 3 \\ 0 & 0 & 4 - \lambda \end{bmatrix} =$$

$$-0 \quad -0 \quad -0 + (5 - \lambda)(8 - \lambda)(4 - \lambda) + 0 + 0$$

$$= (5 - \lambda)(8 - \lambda)(4 - \lambda).$$

We now see that $\det(M - \lambda I) = 0$ when $\lambda = 5, 8,$ and $4$. Notice that these are the values along the main diagonal of the original matrix. This will always be the case when all the values above the main diagonal or all the values below the main diagonal are 0.

To find an eigenvector corresponding to the eigenvalue $\lambda = 5$, set $\lambda = 5$ in the augmented matrix corresponding to the system $(M - \lambda I)X = 0$.

$$\begin{bmatrix} 5 - \lambda & 2 & 3 & \bigm| & 0 \\ 0 & 8 - \lambda & 3 & \bigm| & 0 \\ 0 & 0 & 4 - \lambda & \bigm| & 0 \end{bmatrix} \quad \text{Let } \lambda = 5.$$

$$= \begin{bmatrix} 0 & 2 & 3 & \bigm| & 0 \\ 0 & 3 & 3 & \bigm| & 0 \\ 0 & 0 & -1 & \bigm| & 0 \end{bmatrix}$$

Row reduction is simplified by the fact that the first column consists entirely of zeros.

$$\begin{array}{c} R_2/3 \to R_2 \\ -R_3 \to R_3 \end{array} \begin{bmatrix} 0 & 2 & 3 & \bigm| & 0 \\ 0 & 1 & 1 & \bigm| & 0 \\ 0 & 0 & 1 & \bigm| & 0 \end{bmatrix} \quad \text{Simplify the rows.}$$

$$-2R_2 + R_1 \to R_1 \begin{bmatrix} 0 & 0 & 1 & \bigm| & 0 \\ 0 & 1 & 1 & \bigm| & 0 \\ 0 & 0 & 1 & \bigm| & 0 \end{bmatrix} \quad \text{Get zeros in column 2.}$$

$$\begin{array}{c} R_1 - R_3 \to R_1 \\ R_2 - R_3 \to R_2 \end{array} \begin{bmatrix} 0 & 0 & 0 & \bigm| & 0 \\ 0 & 1 & 0 & \bigm| & 0 \\ 0 & 0 & 1 & \bigm| & 0 \end{bmatrix} \quad \text{Get zeros in column 3.}$$

The second row tells us that $x_2 = 0$, and the third row tells us that $x_3 = 0$. Since $x_1$ is arbitrary, we will let $x_1 = 1$. Therefore, an eigenvector corresponding to $\lambda = 5$ is

$$\begin{bmatrix} 1 \\ 0 \\ 0 \end{bmatrix}.$$

Similarly, when $\lambda = 8$, the augmented matrix corresponding to the system $(M - \lambda I)X = 0$ is

$$\left[\begin{array}{ccc|c} -3 & 2 & 3 & 0 \\ 0 & 0 & 3 & 0 \\ 0 & 0 & -4 & 0 \end{array}\right].$$

Verify that this augmented matrix reduces to

$$\left[\begin{array}{ccc|c} -3 & 2 & 0 & 0 \\ 0 & 0 & 1 & 0 \\ 0 & 0 & 0 & 0 \end{array}\right]$$

and that an eigenvector is

$$\begin{bmatrix} 2 \\ 3 \\ 0 \end{bmatrix}.$$

We leave it as an exercise to show that an eigenvector corresponding to $\lambda = 4$ is

$$\begin{bmatrix} 6 \\ 3 \\ -4 \end{bmatrix}.$$

The following result, mentioned in Example 5, is worth noting.

> If all the values above the main diagonal or all the values below the main diagonal of a matrix are 0, the eigenvalues are the values along the main diagonal.

**NOTE**   The eigenvalues of a matrix can be complex numbers. There can also be eigenvalues with a multiplicity greater than 1, in the sense that, rather than having two distinct eigenvalues for a $2 \times 2$ matrix or three distinct eigenvalues for a $3 \times 3$ matrix, some of the eigenvalues may be identical. In the examples and exercises in this text, the eigenvalues will be real and distinct, with the exception of Exercises 14–17 in this section.

Earlier in this section, we showed how a population described by an eigenvector has a stable distribution from one generation to the next. In the next example, we will see how eigenvalues and eigenvectors help us analyze an arbitrary population that is not an eigenvector. The key is to represent the initial population in terms of the eigenvectors.

**EXAMPLE 6**   **Long-term Population**

In the insect population discussed in Examples 1 and 4, recall that the Leslie matrix was $M = \begin{bmatrix} 0.6 & 0.8 \\ 0.9 & 0 \end{bmatrix}$ and that the eigenvectors $\lambda_1 = 1.2$ and $\lambda_2 = -0.6$ corresponded to the eigenvectors $V_1 = \begin{bmatrix} 4 \\ 3 \end{bmatrix}$ and $V_2 = \begin{bmatrix} 2 \\ -3 \end{bmatrix}$, respectively. Recall also that the initial population was $X(0) = \begin{bmatrix} 1000 \\ 2000 \end{bmatrix}$.

**(a)** Find values $a$ and $b$ such that $aV_1 + bV_2 = X(0)$.

**SOLUTION** The equation

$$a\begin{bmatrix} 4 \\ 3 \end{bmatrix} + b\begin{bmatrix} 2 \\ -3 \end{bmatrix} = \begin{bmatrix} 1000 \\ 2000 \end{bmatrix}$$

corresponds to the augmented matrix

$$\begin{bmatrix} 4 & 2 & | & 1000 \\ 3 & -3 & | & 2000 \end{bmatrix}.$$

Use the Gauss-Jordan method to reduce this matrix to

$$\begin{bmatrix} 1 & 0 & | & 3500/9 \\ 0 & 1 & | & -2500/9 \end{bmatrix}.$$

The solution is $a = 3500/9$ and $b = -2500/9$.

**(b)** Use the result from part (a), as well as properties of eigenvalues and eigenvectors, to find the population distribution the following year.

**SOLUTION** The population in the next year is given by

$$X(1) = MX(0) = M\left(\frac{3500}{9}V_1 - \frac{2500}{9}V_2\right) \qquad \text{Use the results of part (a).}$$

$$= \frac{3500}{9}MV_1 - \frac{2500}{9}MV_2$$

$$= \frac{3500}{9}\lambda_1 V_1 - \frac{2500}{9}\lambda_2 V_2 \qquad \text{Use properties of eigenvalues and eigenvectors.}$$

$$= \frac{3500}{9}(1.2)\begin{bmatrix} 4 \\ 3 \end{bmatrix} - \frac{2500}{9}(-0.6)\begin{bmatrix} 2 \\ -3 \end{bmatrix}$$

$$= \begin{bmatrix} 2200 \\ 900 \end{bmatrix}.$$

**(c)** Find a formula for the long-term population distribution.

**SOLUTION** By continuing to multiply by $M$, the population after $n$ years is given by

$$X(n) = M^n X(0) = M^n\left(\frac{3500}{9}V_1 - \frac{2500}{9}V_2\right)$$

$$= \frac{3500}{9}M^n V_1 - \frac{2500}{9}M^n V_2$$

$$= \frac{3500}{9}\lambda_1^n V_1 - \frac{2500}{9}\lambda_2^n V_2$$

$$= \frac{3500}{9}(1.2)^n\begin{bmatrix} 4 \\ 3 \end{bmatrix} - \frac{2500}{9}(-0.6)^n\begin{bmatrix} 2 \\ -3 \end{bmatrix}.$$

Since $\lim\limits_{n \to \infty} (0.6)^n = 0$, the second term becomes smaller and smaller over time. Therefore, when $n$ is large,

$$X(n) \approx \frac{3500}{9}(1.2)^n\begin{bmatrix} 4 \\ 3 \end{bmatrix}.$$

Example 6 demonstrates an important result. Even if the initial population is not an eigenvector, the population over time approaches a multiple of one of the eigenvectors. Since any multiple of an eigenvector is itself an eigenvector, this shows that, at least in the case with two eigenvalues, one greater than 1 in magnitude and one less than 1 in magnitude, the population will approach an eigenvector over time, and that each year's population can be approximately found by multiplying the previous year's population by an eigenvalue. This behavior is typical of a large range of applications in the life sciences.

# $10.5$ EXERCISES

**Find the determinant of the following matrices.**

**1.** $\begin{bmatrix} 3 & -1 \\ 2 & 5 \end{bmatrix}$

**2.** $\begin{bmatrix} 7 & -5 \\ -3 & -1 \end{bmatrix}$

**3.** $\begin{bmatrix} 4 & 1 & 0 \\ -1 & 7 & -2 \\ 2 & 3 & 5 \end{bmatrix}$

**4.** $\begin{bmatrix} -7 & -1 & 4 \\ 9 & 0 & 6 \\ 4 & 2 & 1 \end{bmatrix}$

**For Exercises 5–12, find the eigenvalues and their corresponding eigenvectors.**

**5.** $\begin{bmatrix} 5 & 0 \\ 2 & 1 \end{bmatrix}$

**6.** $\begin{bmatrix} 7 & 4 \\ -3 & -1 \end{bmatrix}$

**7.** $\begin{bmatrix} 3 & 2 \\ 3 & 8 \end{bmatrix}$

**8.** $\begin{bmatrix} 1 & 0 \\ 6 & -1 \end{bmatrix}$

**9.** $\begin{bmatrix} 4 & -3 \\ 2 & -1 \end{bmatrix}$

**10.** $\begin{bmatrix} 2 & 3 \\ 4 & 1 \end{bmatrix}$

**11.** $\begin{bmatrix} 4 & 0 & 0 \\ 3 & -1 & 0 \\ 2 & 5 & -3 \end{bmatrix}$

**12.** $\begin{bmatrix} -5 & 0 & 0 \\ 1 & 2 & 0 \\ -1 & 4 & 4 \end{bmatrix}$

**13.** Show that the system of equations

$$ax + by = 0$$
$$cx + dy = 0$$

is dependent whenever $ad - bc = 0$, that is, when the determinant of the coefficient matrix is 0. (*Hint:* Use the Gauss-Jordan method.)

**In Exercises 14 and 15, each $2 \times 2$ matrix has only one eigenvalue. Find the eigenvalue and the corresponding eigenvector.**

**14.** $\begin{bmatrix} 7 & -2 \\ 2 & 3 \end{bmatrix}$

**15.** $\begin{bmatrix} 10 & -9 \\ 4 & -2 \end{bmatrix}$

**In Exercises 16 and 17, each matrix has complex numbers as eigenvalues. Find the eigenvalues and their corresponding eigenvectors. (Note: These exercises require a knowledge of complex numbers. Complex eigenvalues are important in biology applications.)**

**16.** $\begin{bmatrix} 1 & -2 \\ 1 & 3 \end{bmatrix}$

**17.** $\begin{bmatrix} 1 & 5 \\ -2 & 3 \end{bmatrix}$

## LIFE SCIENCE APPLICATIONS

*Leslie Matrices For each of the following Leslie matrices,*
**(a)** *find a population of size 10,000 for which the proportion of the population in each age group stays the same from one year to the next, and* **(b)** *tell by what factor the population grows or declines each year.*

**18.** $\begin{bmatrix} 0.7 & 0.9 \\ 1.6 & 0 \end{bmatrix}$

**19.** $\begin{bmatrix} 0.5 & 0.9 \\ 1.4 & 0 \end{bmatrix}$

**20.** $\begin{bmatrix} 0.95 & 0.03 \\ 0.05 & 0.97 \end{bmatrix}$

**21.** $\begin{bmatrix} 0.2 & 0.7 \\ 0.7 & 0.2 \end{bmatrix}$

**22.** $\begin{bmatrix} 0.7 & 0.2 \\ 0.2 & 0.4 \end{bmatrix}$

**23.** $\begin{bmatrix} 0.1 & 1.2 \\ 0.4 & 0.3 \end{bmatrix}$

**24. Long-term Population** For the matrix in Exercise 18, find an approximate expression for the population distribution after $n$ years, given that the initial population distribution is given by $X(0) = \begin{bmatrix} 3000 \\ 1000 \end{bmatrix}$.

**25. Long-term Population** Repeat Exercise 24 using the matrix in Exercise 19.

**26. Northern Spotted Owl Population** In Exercise 47 of Section 10.3 on Multiplication of Matrices, we saw that the number of female northern spotted owls at time $n + 1$ could be estimated by the equation

$$\begin{bmatrix} j_{n+1} \\ s_{n+1} \\ a_{n+1} \end{bmatrix} = \begin{bmatrix} 0 & 0 & 0.33 \\ 0.18 & 0 & 0 \\ 0 & 0.71 & 0.94 \end{bmatrix} \begin{bmatrix} j_n \\ s_n \\ a_n \end{bmatrix},$$

where $j_n$ is the number of juveniles, $s_n$ is the number of sub-adults, and $a_n$ is the number of adults at time $n$. In part b of that exercise, it was stated that in the long run, the following approximation holds:

$$\begin{bmatrix} j_{n+1} \\ s_{n+1} \\ a_{n+1} \end{bmatrix} \approx 0.98359 \begin{bmatrix} j_n \\ s_n \\ a_n \end{bmatrix}.$$

Use the concept of eigenvalues to show that this approximation is true if $\begin{bmatrix} j_n \\ s_n \\ a_n \end{bmatrix}$ is an eigenvector. (*Hint:* After calculating $\det(M - \lambda I)$, use a graphing calculator to plot $y = \det(M - \lambda I)$ as a function of $\lambda$, and find where $y = 0$.)

**YOUR TURN ANSWERS**

**1.** 26  **2.** $-57$

**3.** For $\lambda = 3$, the eigenvector is $\begin{bmatrix} 3 \\ 2 \end{bmatrix}$; for $\lambda = -2$, the eigenvector is $\begin{bmatrix} 1 \\ -1 \end{bmatrix}$.

# 10 CHAPTER REVIEW

## SUMMARY

In this chapter we studied systems of linear equations and matrices. The Gauss-Jordan method was developed and used to solve systems of linear equations. We introduced matrices, which are used to store mathematical information. We saw that matrices can be combined using addition, subtraction, scalar multiplication, and matrix multiplication.

We then developed the concept of a multiplicative inverse of a matrix and used such inverses to solve systems of equations. We concluded the chapter by introducing eigenvalues and eigenvectors, which are used to solve certain types of matrix equations that arise in applications.

| | |
|---|---|
| **Row Operations** | For any augmented matrix of a system of equations, the following operations produce the augmented matrix of an equivalent system: |
| | **1.** interchanging any two rows; |
| | **2.** multiplying the elements of a row by a nonzero real number; |
| | **3.** adding a nonzero multiple of the elements of one row to the corresponding elements of a nonzero multiple of some other row. |
| **The Gauss-Jordan Method** | **1.** Write each equation so that variable terms are in the same order on the left side of the equal sign and constants are on the right. |
| | **2.** Write the augmented matrix that corresponds to the system. |
| | **3.** Use row operations to transform the first column so that all elements except the element in the first row are zero. |
| | **4.** Use row operations to transform the second column so that all elements except the element in the second row are zero. |
| | **5.** Use row operations to transform the third column so that all elements except the element in the third row are zero. |
| | **6.** Continue in this way, when possible, until the last row is written in the form |

$$[0 \ 0 \ 0 \ \cdots \ 0 \ j \,|\, k],$$

where $j$ and $k$ are constants with $j \neq 0$. When this is not possible, continue until every row has more zeros on the left than the previous row (except possibly for any rows of all zeros at the bottom of the matrix), and the first nonzero entry in each row is the only nonzero entry in its column.

| | |
|---|---|
| | **7.** Multiply each row by the reciprocal of the nonzero element in that row. |
| **Adding Matrices** | The sum of two $m \times n$ matrices $X$ and $Y$ is the $m \times n$ matrix $X + Y$ in which each element is the sum of the corresponding elements of $X$ and $Y$. |
| **Subtracting Matrices** | The difference of two $m \times n$ matrices $X$ and $Y$ is the $m \times n$ matrix $X - Y$ in which each element is the difference of the corresponding elements of $X$ and $Y$. |
| **Product of a Matrix and a Scalar** | The product of a scalar $k$ and a matrix $X$ is the matrix $kX$, in which each element is $k$ times the corresponding element of $X$. |
| **Product of Two Matrices** | Let $A$ be an $m \times n$ matrix and let $B$ be an $n \times k$ matrix. To find the element in the $i$th row and $j$th column of the product $AB$, multiply each element in the $i$th row of $A$ by the corresponding element in the $j$th column of $B$, and then add these products. The product matrix $AB$ is an $m \times k$ matrix. |
| **Finding a Multiplicative Inverse Matrix** | To obtain $A^{-1}$ for any $n \times n$ matrix $A$ for which $A^{-1}$ exists, follow these steps. |
| | **1.** Form the augmented matrix $[A\,|\,I]$, where $I$ is the $n \times n$ identity matrix. |
| | **2.** Perform row operations on $[A\,|\,I]$ to get a matrix of the form $[I\,|\,B]$, if this is possible. |
| | **3.** Matrix $B$ is $A^{-1}$. |

**Solving a System $AX = B$ Using**    To solve a system of equations $AX = B$, where $A$ is a square matrix of coefficients, $X$ is the matrix
          **Matrix Inverses**    of variables, and $B$ is the matrix of constants, first find $A^{-1}$. Then, $X = A^{-1}B$.

# KEY TERMS

| | | | |
|---|---|---|---|
| system of equations | augmented matrix | **10.3** | **10.5** |
| **10.1** | row operations | scalar | Leslie matrix |
| linear systems | equivalent system | product matrix | eigenvalue |
| first-degree equation in $n$ | Gauss-Jordan method | | eigenvector |
|   unknowns | parameter | **10.4** | determinant |
| unique solution | | identity matrix | |
| inconsistent system | **10.2** | multiplicative inverse matrix | |
| dependent equations | size | coefficient matrix | |
| matrix (matrices) | square matrix | | |
| element (entry) | row matrix (row vector) | | |
| | column matrix (column vector) | | |

# REVIEW EXERCISES

## CONCEPT CHECK

**Determine whether each of the following statements is true or false, and explain why.**

**1.** The only solution to the system of equations

$$2x + 3y = 7$$
$$5x - 4y = 6$$

is $x = 2$ and $y = 1$.

**2.** If a system of equations has three equations and four unknowns, then it could have a unique solution.

**3.** If a system of equations has three equations and three unknowns, then it may have a unique solution, an infinite number of solutions, or no solutions.

**4.** When solving a system of equations by the Gauss-Jordan method, we can add a nonzero multiple of the elements of one column to the corresponding elements of some nonzero multiple of some other column.

**5.** If $A = \begin{bmatrix} 2 & 3 \\ 1 & -1 \end{bmatrix}$ and $B = \begin{bmatrix} 3 & 4 \\ 7 & 4 \\ 1 & 0 \end{bmatrix}$, then $A + B = \begin{bmatrix} 5 & 7 \\ 8 & 3 \\ 1 & 0 \end{bmatrix}$.

**6.** If $A$ is a $2 \times 3$ matrix and $B$ is a $3 \times 4$ matrix, then $A + B$ is a $2 \times 4$ matrix.

**7.** If $A$ is an $n \times k$ matrix and $B$ is a $k \times m$ matrix, then $AB$ is an $n \times m$ matrix.

**8.** If $A$ is a $4 \times 4$ matrix and $B$ is a $4 \times 4$ matrix, then it is always the case that $AB = BA$.

**9.** Every square matrix has an inverse.

**10.** A $3 \times 4$ matrix could have an inverse.

**11.** If $A$, $B$, and $C$ are matrices such that $AB = C$, then $B = \dfrac{C}{A}$.

**12.** If $A$ and $B$ are matrices such that $A = B^{-1}$, then $AB = BA$.

**13.** If $A$, $B$, and $C$ are matrices such that $AB = CB$, then $A = C$.

**14.** The determinant of the matrix $\begin{bmatrix} 1 & 4 \\ 2 & 5 \end{bmatrix}$ is 3.

**15.** The determinant of any square matrix can be calculated.

**16.** The eigenvalues of the matrix $\begin{bmatrix} 1 & 0 \\ 2 & 3 \end{bmatrix}$ are $\lambda = 1$ and $\lambda = 3$.

## PRACTICE AND EXPLORATIONS

**17.** What is true about the number of solutions to a system of $m$ linear equations in $n$ unknowns if $m = n$? If $m < n$? If $m > n$?

**18.** Suppose someone says that a more reasonable way to multiply two matrices than the method presented in the text is to multiply corresponding elements. For example, the result of

$$\begin{bmatrix} 1 & 2 \\ 3 & 4 \end{bmatrix} \cdot \begin{bmatrix} 3 & 5 \\ 7 & 11 \end{bmatrix} \text{ should be } \begin{bmatrix} 3 & 10 \\ 21 & 44 \end{bmatrix},$$

according to this person. How would you respond?

**Solve each system by the Gauss-Jordan method.**

**19.**    $2x + 4y = -6$
       $-3x - 5y = 12$

**20.**    $x - 4y = 10$
       $5x + 3y = 119$

**21.**    $x - y + 3z = 13$
       $4x + y + 2z = 17$
       $3x + 2y + 2z = 1$

**22.**    $x + 2y + 3z = 9$
       $x - 2y = 4$
       $3x + 2z = 12$

**23.**    $3x - 6y + 9z = 12$
       $-x + 2y - 3z = -4$
       $x + y + 2z = 7$

**24.**    $x - 2z = 5$
       $3x + 2y = 8$
       $-x + 2z = 10$

**Find the size of each matrix, find the values of any variables, and identify any square, row, or column matrices.**

**25.** $\begin{bmatrix} 2 & 3 \\ 5 & q \end{bmatrix} = \begin{bmatrix} a & b \\ c & 9 \end{bmatrix}$
**26.** $\begin{bmatrix} 2 & x \\ y & 6 \\ 5 & z \end{bmatrix} = \begin{bmatrix} a & -1 \\ 4 & 6 \\ p & 7 \end{bmatrix}$

**27.** $[2m \quad 4 \quad 3z \quad -12] = [12 \quad k + 1 \quad -9 \quad r - 3]$

**28.** $\begin{bmatrix} a + 5 & 3b & 6 \\ 4c & 2 + d & -3 \\ -1 & 4p & q - 1 \end{bmatrix} = \begin{bmatrix} -7 & b + 2 & 2k - 3 \\ 3 & 2d - 1 & 4l \\ m & 12 & 8 \end{bmatrix}$

**Given the matrices**

$$A = \begin{bmatrix} 4 & 10 \\ -2 & -3 \\ 6 & 9 \end{bmatrix}, \quad B = \begin{bmatrix} 2 & 3 & -2 \\ 2 & 4 & 0 \\ 0 & 1 & 2 \end{bmatrix}, \quad C = \begin{bmatrix} 5 & 0 \\ -1 & 3 \\ 4 & 7 \end{bmatrix},$$

$$D = \begin{bmatrix} 6 \\ 1 \\ 0 \end{bmatrix}, \quad E = [1 \quad 3 \quad -4],$$

$$F = \begin{bmatrix} -1 & 4 \\ 3 & 7 \end{bmatrix}, \quad G = \begin{bmatrix} -2 & 0 \\ 1 & 5 \end{bmatrix},$$

**find each of the following, if it exists.**

**29.** $A + C$

**30.** $2G - 4F$

**31.** $3C + 2A$

**32.** $B - C$

**33.** $2A - 5C$

**34.** $AG$

**35.** $AC$

**36.** $DE$

**37.** $ED$

**38.** $BD$

**39.** $EC$

**40.** $F^{-1}$

**41.** $B^{-1}$

**42.** $(A + C)^{-1}$

**Find the inverse of each matrix that has an inverse.**

**43.** $\begin{bmatrix} 1 & 3 \\ 2 & 7 \end{bmatrix}$

**44.** $\begin{bmatrix} -4 & 2 \\ 0 & 3 \end{bmatrix}$

**45.** $\begin{bmatrix} 3 & -6 \\ -4 & 8 \end{bmatrix}$

**46.** $\begin{bmatrix} 6 & 4 \\ 3 & 2 \end{bmatrix}$

**47.** $\begin{bmatrix} 2 & -1 & 0 \\ 1 & 0 & 1 \\ 1 & -2 & 0 \end{bmatrix}$

**48.** $\begin{bmatrix} 2 & 0 & 4 \\ 1 & -1 & 0 \\ 0 & 1 & -2 \end{bmatrix}$

**49.** $\begin{bmatrix} 1 & 3 & 6 \\ 4 & 0 & 9 \\ 5 & 15 & 30 \end{bmatrix}$

**50.** $\begin{bmatrix} 2 & -3 & 4 \\ 1 & 5 & 7 \\ -4 & 6 & -8 \end{bmatrix}$

**Solve the matrix equation $AX = B$ for $X$ using the given matrices.**

**51.** $A = \begin{bmatrix} 5 & 1 \\ -2 & -2 \end{bmatrix}, \quad B = \begin{bmatrix} -8 \\ 24 \end{bmatrix}$

**52.** $A = \begin{bmatrix} 1 & 2 \\ 2 & 4 \end{bmatrix}, \quad B = \begin{bmatrix} 5 \\ 10 \end{bmatrix}$

**53.** $A = \begin{bmatrix} 1 & 0 & 2 \\ -1 & 1 & 0 \\ 3 & 0 & 4 \end{bmatrix}, \quad B = \begin{bmatrix} 8 \\ 4 \\ -6 \end{bmatrix}$

**54.** $A = \begin{bmatrix} 2 & 4 & 0 \\ 1 & -2 & 0 \\ 0 & 0 & 3 \end{bmatrix}, \quad B = \begin{bmatrix} 72 \\ -24 \\ 48 \end{bmatrix}$

**Solve each system of equations by inverses.**

**55.** $\begin{aligned} x + 2y &= 4 \\ 2x - 3y &= 1 \end{aligned}$

**56.** $\begin{aligned} 5x + 10y &= 80 \\ 3x - 2y &= 120 \end{aligned}$

**57.** $\begin{aligned} x + y + z &= 1 \\ 2x + y &= -2 \\ 3y + z &= 2 \end{aligned}$

**58.** $\begin{aligned} x - 4y + 2z &= -1 \\ -2x + y - 3z &= -9 \\ 3x + 5y - 2z &= 7 \end{aligned}$

**Find the eigenvalues and their corresponding eigenvectors.**

**59.** $\begin{bmatrix} -3 & 12 \\ -2 & 7 \end{bmatrix}$

**60.** $\begin{bmatrix} 5 & 3 \\ 3 & 5 \end{bmatrix}$

**61.** $\begin{bmatrix} 1 & 0 & 0 \\ 2 & 2 & 0 \\ 2 & 1 & -3 \end{bmatrix}$

**62.** $\begin{bmatrix} -2 & 0 & 0 \\ 1 & 3 & 0 \\ -1 & 1 & 2 \end{bmatrix}$

## LIFE SCIENCE APPLICATIONS

**63. Animal Activity** The activities of a grazing animal can be classified roughly into three categories: grazing, moving, and resting. Suppose horses spend 8 hours grazing, 8 moving, and 8 resting; cattle spend 10 grazing, 5 moving, and 9 resting; sheep spend 7 grazing, 10 moving, and 7 resting; and goats spend 8 grazing, 9 moving, and 7 resting. Write this information as a $4 \times 3$ matrix.

**64. CAT Scans** Computer Aided Tomography (CAT) scanners take X-rays of a part of the body from different directions, and put the information together to create a picture of a cross section of the body. The amount by which the energy of the X-ray decreases, measured in linear-attenuation units, tells whether the X-ray has passed through healthy tissue, tumorous tissue, or bone, based on the following table. *Source: The Mathematics Teacher.*

| Type of Tissue | Linear-Attenuation Values |
| --- | --- |
| Healthy tissue | 0.1625–0.2977 |
| Tumorous tissue | 0.2679–0.3930 |
| Bone | 0.3857–0.5108 |

The part of the body to be scanned is divided into cells. If an X-ray passes through more than one cell, the total linear-attenuation value is the sum of the values for the cells. For example, in the figure, let $a$, $b$, and $c$ be the values for cells A, B, and C. The attenuation value for beam 1 is $a + b$ and for beam 2 is $a + c$.

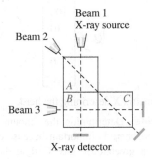

Beam 1
X-ray source

Beam 2

A
B
C

Beam 3

X-ray detector

**a.** Find the attenuation value for beam 3.

**b.** Suppose that the attenuation values are 0.8, 0.55, and 0.65 for beams 1, 2, and 3, respectively. Set up and solve the system of three equations for $a$, $b$, and $c$. What can you conclude about cells A, B, and C?

**c.** Find the inverse of the coefficient matrix from part b to find $a$, $b$, and $c$ for the following three cases, and make conclusions about cells A, B, and C for each.

| | Linear-Attenuation Values | | |
|---|---|---|---|
| Patient | Beam 1 | Beam 2 | Beam 3 |
| X | 0.54 | 0.40 | 0.52 |
| Y | 0.65 | 0.80 | 0.75 |
| Z | 0.51 | 0.49 | 0.44 |

**65. CAT Scans** (Refer to Exercise 64.) Four X-ray beams are aimed at four cells, as shown in the following figure. *Source: The Mathematics Teacher.*

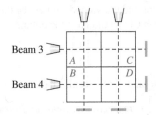

Beam 1  Beam 2

Beam 3

A  C
B  D

Beam 4

**a.** Suppose the attenuation values for beams 1, 2, 3, and 4 are 0.60, 0.75, 0.65, and 0.70, respectively. Do we have enough information to determine the values of $a$, $b$, $c$, and $d$? Explain.

**b.** Suppose we have the data from part a, as well as the following values for $d$. Find the values for $a$, $b$, and $c$, and draw conclusions about cells A, B, C, and D in each case.

**(i)** 0.33    **(ii)** 0.43

**c.** Two X-ray beams are added, as shown in the figure in the next column. In addition to the data in part a, we now have attenuation values for beams 5 and 6 of 0.85 and 0.50. Find the values for $a$, $b$, $c$, and $d$, and make conclusions about cells A, B, C, and D.

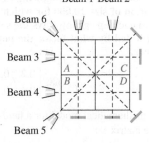

Beam 1  Beam 2

Beam 6

Beam 3

A  C
B  D

Beam 4

Beam 5

**d.** Six X-ray beams are not necessary because four appropriately chosen beams are sufficient. Give two examples of four beams (chosen from beams 1–6 in part c) that will give the solution. (*Note:* There are 12 possible solutions.)

**e.** Discuss what properties the four beams selected in part d must have in order to provide a unique solution.

**66. Hockey** In a recent study, the number of head and neck injuries among hockey players wearing full face shields and half face shields were compared. The following table provides the rates per 1000 athlete-exposures for specific injuries that caused a player wearing either shield to miss one or more events. *Source: JAMA.*

| | Half Shield | Full Shield |
|---|---|---|
| Head and Face Injuries (excluding Concussions) | 3.54 | 1.41 |
| Concussions | 1.53 | 1.57 |
| Neck Injuries | 0.34 | 0.29 |
| Other | 7.53 | 6.21 |

If an equal number of players in a large league wear each type of shield and the total number of athlete-exposures for the league in a season is 8000, use matrix operations to estimate the total number of injuries of each type.

**Leslie Matrices** For each of the following Leslie matrices,
**a.** find a population of size 10,000 for which the proportion of
the population in each age group stays the same from one year
to the next and **b.** tell by what factor the population grows or
declines each year.

**67.** $\begin{bmatrix} 0.3 & 0.2 \\ 0.3 & 0.8 \end{bmatrix}$

**68.** $\begin{bmatrix} 0.2 & 0.3 \\ 0.4 & 0.1 \end{bmatrix}$

**69. Buffalo** A mathematical model for a herd of buffalo is given
by the Leslie matrix

$$\begin{bmatrix} 0.95 & 0 & 0.75 & 0 & 0 & 0 \\ 0 & 0.95 & 0 & 0.75 & 0 & 0 \\ 0 & 0 & 0 & 0 & 0.6 & 0 \\ 0 & 0 & 0 & 0 & 0 & 0.6 \\ 0 & 0.48 & 0 & 0 & 0 & 0 \\ 0 & 0.42 & 0 & 0 & 0 & 0 \end{bmatrix},$$

where the first row corresponds to the number of adult males,
the second to the number of adult females, the third to the number of male yearlings, the fourth to the number of female yearlings, the fifth to the number of male calves, and the sixth to the
number of female calves. *Source: UMAP.* The matrix is used
to give next year's herd as a function of this year's.

**a.** Explain what each row of the matrix means.

**b.** Verify that $\lambda = 0.95$ is an eigenvalue corresponding to the

eigenvector $\begin{bmatrix} k \\ 0 \\ 0 \\ 0 \\ 0 \\ 0 \end{bmatrix}$ for any constant $k$. Explain what this tells

us about the herd.

**c.** The manager of a buffalo herd wants to harvest some of
the herd each year. Suppose that $Q$ is the vector describing
a harvest that leaves the population for next year the same
as this year's. Show that the population $X$ must satisfy the
equation

$$X = (M - I)^{-1}Q.$$

**d.** Suppose the manager in part c wants a harvest given by

$$Q = \begin{bmatrix} 100 \\ 100 \\ 10 \\ 10 \\ 0 \\ 0 \end{bmatrix}.$$

Find the matrix $X$ describing the herd.

**e.** Suppose the manager wants the herd to increase by 10%
each year while still yielding a harvest $Q$. Show that the
population $X$ must satisfy the equation

$$X = (M - 1.1I)^{-1}Q.$$

**f.** Suppose the manager in part e wants a harvest given by

$$Q = \begin{bmatrix} 100 \\ 100 \\ 10 \\ 10 \\ 0 \\ 0 \end{bmatrix}.$$

Find the matrix $X$ describing the herd.

**70. Carbon Dioxide** Determining the amount of carbon dioxide in the atmosphere is important because carbon dioxide is
known to be a greenhouse gas. Carbon dioxide concentrations (in parts per million) have been measured at Mauna Loa,
Hawaii, for more than 40 years. The concentrations have increased quadratically. The table lists readings for 3 years.
*Source: Scripps Institution of Oceanography.*

| Year | CO₂ |
|------|------|
| 1960 | 317 |
| 1980 | 339 |
| 2010 | 390 |

**a.** If the relationship between the carbon dioxide concentration $C$ and the year $t$ is expressed as $C = at^2 + bt + c$,
where $t = 0$ corresponds to 1960, use a linear system of
equations to determine the constants $a$, $b$, and $c$.

**b.** Predict the year when the amount of carbon dioxide in the
atmosphere will double from its 1960 level. (*Hint:* This requires solving a quadratic equation. For review on how to
do this, see Section R.4.)

**71. Chemistry** When carbon monoxide ($CO$) reacts with oxygen
($O_2$), carbon dioxide ($CO_2$) is formed. This can be written as
$CO + (1/2)O_2 = CO_2$ and as a matrix equation. If we form a
$2 \times 1$ column matrix by letting the first element be the number
of carbon atoms and the second element be the number of oxygen atoms, then CO would have the column matrix

$$\begin{bmatrix} 1 \\ 1 \end{bmatrix}$$

Similarly, $O_2$ and $CO_2$ would have the column matrices
$\begin{bmatrix} 0 \\ 2 \end{bmatrix}$ and $\begin{bmatrix} 1 \\ 2 \end{bmatrix}$, respectively. *Source: Journal of Chemical
Education.*

**a.** Use the Gauss-Jordan method to find numbers $x$ and $y$
(known as *stoichiometric numbers*) that solve the system
of equations

$$\begin{bmatrix} 1 \\ 1 \end{bmatrix} x + \begin{bmatrix} 0 \\ 2 \end{bmatrix} y = \begin{bmatrix} 1 \\ 2 \end{bmatrix}.$$

Compare your answers to the equation written above.

**b.** Repeat the process for $xCO_2 + yH_2 + zCO = H_2O$,
where $H_2$ is hydrogen and $H_2O$ is water. In words, what
does this mean?

## OTHER APPLICATIONS

**72. Roof Trusses** Linear systems occur in the design of roof trusses for new homes and buildings. The simplest type of roof truss is a triangle. The truss shown in the figure below is used to frame roofs of small buildings. If a 100-lb force is applied at the peak of the truss, then the forces or weights $W_1$ and $W_2$ exerted parallel to each rafter of the truss are determined by the following linear system of equations.

$$\frac{\sqrt{3}}{2}(W_1 + W_2) = 100$$
$$W_1 - W_2 = 0$$

Solve the system to find $W_1$ and $W_2$. *Source: Structural Analysis.*

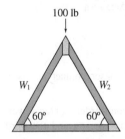

**73. Roof Trusses** (Refer to Exercise 72.) Use the following system of equations to determine the force or weights $W_1$ and $W_2$ exerted on each rafter for the truss shown in the figure.

$$\frac{1}{2}W_1 + \frac{\sqrt{2}}{2}W_2 = 150$$
$$\frac{\sqrt{3}}{2}W_1 - \frac{\sqrt{2}}{2}W_2 = 0$$

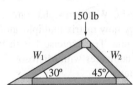

**74. Scheduling Production** An office supply manufacturer makes two kinds of paper clips, standard and extra large. To make 1000 standard paper clips requires 1/4 hour on a cutting machine and 1/2 hour on a machine that shapes the clips. One thousand extra large paper clips require 1/3 hour on each machine. The manager of paper clip production has 4 hours per day available on the cutting machine and 6 hours per day on the shaping machine. How many of each kind of clip can he make?

**75. Production Requirements** The Waputi Indians make woven blankets, rugs, and skirts. Each blanket requires 24 hours for spinning the yarn, 4 hours for dyeing the yarn, and 15 hours for weaving. Rugs require 30, 5, and 18 hours and skirts 12, 3, and 9 hours, respectively. If there are 306, 59, and 201 hours available for spinning, dyeing, and weaving, respectively, how many of each item can be made? (*Hint:* Simplify the equations you write, if possible, before solving the system.)

**76. Distribution** An oil refinery in Tulsa sells 50% of its production to a Chicago distributor, 20% to a Dallas distributor, and 30% to an Atlanta distributor. Another refinery in New Orleans sells 40% of its production to the Chicago distributor, 40% to the Dallas distributor, and 20% to the Atlanta distributor. A third refinery in Ardmore sells the same distributors 30%, 40%, and 30% of its production. The three distributors received 219,000, 192,000, and 144,000 gal of oil, respectively. How many gallons of oil were produced at each of the three plants?

**77. Stock Reports** The New York Stock Exchange reports in daily newspapers give the dividend, price-to-earnings ratio, sales (in hundreds of shares), last price, and change in price for each company. Write the following stock reports as a $4 \times 5$ matrix: American Telephone & Telegraph: 1.33, 17.6, 152,000, 26.75, +1.88; General Electric: 1.00, 20.0, 238,200, 32.36, −1.50; Sara Lee: 0.79, 25.4, 39,110, 16.51, −0.89; Walt Disney Company: 0.27, 21.2, 122,500, 28.60, +0.75.

**78. Filling Orders** A printer has three orders for pamphlets that require three kinds of paper, as shown in the following matrix.

|  | | Order | |
|---|---|---|---|
|  | I | II | III |
| High-grade | 10 | 5 | 8 |
| Paper   Medium-grade | 12 | 0 | 4 |
| Coated | 0 | 10 | 5 |

The printer has on hand 3170 sheets of high-grade paper, 2360 sheets of medium-grade paper, and 1800 sheets of coated paper. All the paper must be used in preparing the order.

**a.** Write a $3 \times 1$ matrix for the amounts of paper on hand.

**b.** Write a matrix of variables to represent the number of pamphlets that must be printed in each of the three orders.

**c.** Write a matrix equation using the given matrix and your matrices from parts a and b.

**d.** Solve the equation from part c.

**79. Students** Suppose 20% of the boys and 30% of the girls in a high school like tennis, and 60% of the boys and 90% of the girls like math. If 500 students like tennis and 1500 like math, how many boys and girls are in the school? Find all possible solutions.

**80. Baseball** In the 2009 Major League Baseball season, slugger Ichiro Suzuki had a total of 225 hits. The number of singles he hit was 11 more than four times the combined total of doubles and home runs. The number of doubles he hit was 1 more than twice the combined total of triples and home runs. The number of singles and home runs together was 15 more than five times the combined total of doubles and triples. Find the number of singles, doubles, triples, and home runs that Suzuki hit during the season. *Source: Baseball-Reference.com.*

**81. Cookies** Regular Nabisco Oreo cookies are made of two chocolate cookie wafers surrounding a single layer of vanilla cream. The claim on the package states that a single serving is 34 g, which is three cookies. Nabisco Double Stuf cookies are made of the same two chocolate cookie wafers surrounding a double layer of vanilla cream. The claim on this package states that a single serving is 29 g, which is two Double Stuf cookies. If the Double Stuf cookies truly have a double layer of vanilla cream, find the weight of a single chocolate wafer and the weight of a single layer of vanilla cream.

# EXTENDED APPLICATION

## CONTAGION

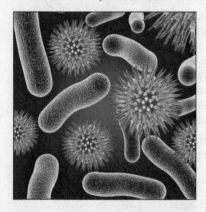

Suppose that three people have contracted a contagious disease. A second group of five people may have been in contact with the three infected persons. A third group of six people may have been in contact with the second group. We can form a $3 \times 5$ matrix $P$ with rows representing the first group of three and columns representing the second group of five. We enter a one in the corresponding position if a person in the first group has contact with a person in the second group. These direct contacts are called *first-order contacts*. Similarly, we form a $5 \times 6$ matrix $Q$ representing the first-order contacts between the second and third group. For example, suppose

$$P = \begin{bmatrix} 1 & 0 & 0 & 1 & 0 \\ 0 & 0 & 1 & 1 & 0 \\ 1 & 1 & 0 & 0 & 0 \end{bmatrix} \text{ and}$$

$$Q = \begin{bmatrix} 1 & 1 & 0 & 1 & 1 & 1 \\ 0 & 0 & 0 & 0 & 1 & 0 \\ 0 & 0 & 0 & 0 & 0 & 0 \\ 0 & 1 & 0 & 1 & 0 & 0 \\ 1 & 0 & 0 & 0 & 1 & 0 \end{bmatrix}.$$

From matrix $P$ we see that the first person in the first group had contact with the first and fourth persons in the second group. Also, none of the first group had contact with the last person in the second group.

A *second-order contact* is an indirect contact between persons in the first and third groups through some person in the second group. The product matrix $PQ$ indicates these contacts. Verify that the second-row, fourth-column entry of $PQ$ is 1. That is, there is one second-order contact between the second person in group one and the fourth person in group three. Let $a_{ij}$ denote the element in the $i$th row and $j$th column of the matrix $PQ$. By looking at the products that form $a_{24}$ below, we see that the common contact was with the fourth individual in group two. (The $p_{ij}$ are entries in $P$, and the $q_{ij}$ are entries in $Q$.)

$$a_{24} = p_{21}q_{14} + p_{22}q_{24} + p_{23}q_{34} + p_{24}q_{44} + p_{25}q_{54}$$
$$= 0 \cdot 1 + 0 \cdot 0 + 1 \cdot 0 + 1 \cdot 1 + 0 \cdot 0$$
$$= 1$$

The second person in group 1 and the fourth person in group 3 both had contact with the fourth person in group 2.

This idea could be extended to third-, fourth-, and larger-order contacts. It indicates a way to use matrices to trace the spread of a contagious disease. It could also pertain to the dispersal of ideas or anything that might pass from one individual to another.

***Source: Finite Mathematics with Applications to Business, Life Sciences, and Social Sciences.***

### EXERCISES

1. Find the second-order contact matrix $PQ$ mentioned in the text.

2. How many second-order contacts were there between the second contagious person and the third person in the third group?

3. Is there anyone in the third group who has had no contacts at all with the first group?

4. The totals of the columns in $PQ$ give the total number of second-order contacts per person, while the column totals in $P$ and $Q$ give the total number of first-order contacts per person. Which person(s) in the third group had the most contacts, counting first- and second-order contacts?

5. Go to the website WolframAlpha.com and enter: "multiply matrices." Study how matrix multiplication can be performed by Wolfram|Alpha. Try Exercise 1 with Wolfram|Alpha and discuss how it compares with Microsoft Excel and with your graphing calculator.

### DIRECTIONS FOR GROUP PROJECT

*Assume that your group (3–5 students) is trying to map the spread of a new disease. Suppose also that the information given above has been obtained from interviews with the first three people that were hospitalized with symptoms of the disease and their contacts. Using the questions above as a guide, prepare a presentation for a public meeting that describes the method of obtaining the data and the data itself, and that addresses the spirit of each question. Formulate a strategy for how to handle the spread of this disease to other people. The presentation should be mathematically sound, grammatically correct, and professionally crafted. Use presentation software, such as Microsoft PowerPoint, to present your findings.*

# Differential Equations

11.1 Solutions of Elementary and Separable Differential Equations

11.2 Linear First-Order Differential Equations

11.3 Euler's Method

11.4 Linear Systems of Differential Equations

11.5 Nonlinear Systems of Differential Equations

11.6 Applications of Differential Equations

Chapter 11 Review

Extended Application: Pollution of the Great Lakes

Mice, like humans, are affected by infectious diseases. Differential equations provide epidemiologists with a useful tool for studying how these diseases cause mouse populations to grow or decline. An exercise in this chapter presents such an equation, which can be solved using a method explained in this chapter.

S uppose a population biologist wants to develop an equation that will forecast the population of an endangered wildcat. By studying data on the population in the past, she hopes to find a relationship between the population and its rate of change. A function giving the rate of change of the population would be the derivative of the function describing the population itself. A **differential equation** is an equation that involves an unknown function $y = f(x)$ and a finite number of its derivatives. Solving the differential equation for $y$ would give the unknown function to be used for forecasting the wildcat population.

Differential equations have been important in the study of physical science and engineering since the 18th century. More recently, differential equations have become useful in the social sciences, life sciences, and economics for solving problems about population growth, ecological balance, and interest rates. In this chapter, we will introduce some methods for solving differential equations and give examples of their applications.

# 11.1 Solutions of Elementary and Separable Differential Equations

APPLY IT **How can we predict the future population of a flock of mountain goats?**
*Using differential equations, we will answer this question in Example 6.*

Usually a solution of an equation is a *number*. A solution of a differential equation, however, is a *function* that satisfies the equation.

### EXAMPLE 1 Solving a Differential Equation

Find all solutions of the differential equation

$$\frac{dy}{dx} = 3x^2 - 2x. \tag{1}$$

**SOLUTION** To say that a function $y(x)$ is a solution of Equation (1) simply means that the derivative of the function $y$ is $3x^2 - 2x$. This is the same as saying that $y$ is an antiderivative of $3x^2 - 2x$, or

$$y = \int (3x^2 - 2x)\, dx = x^3 - x^2 + C. \tag{2}$$

We can verify that the function given by Equation (2) is a solution by taking its derivative. The result is the differential equation (1).

TRY YOUR TURN 1

**YOUR TURN 1** Find all solutions of the differential equation $\frac{dy}{dx} = 12x^5 + \sqrt{x} + e^{5x}$.

**FOR REVIEW**
For review on finding antiderivatives, see the first two sections in Chapter 7 on Integration.

Each different value of $C$ in Equation (2) leads to a different solution of Equation (1), showing that a differential equation can have an infinite number of solutions. Equation (2) is the **general solution** of the differential equation (1). Some of the solutions of Equation (1) are graphed in Figure 1.

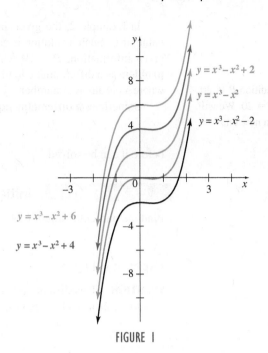

FIGURE 1

The simplest kind of differential equation has the form

$$\frac{dy}{dx} = g(x).$$

Since Equation (1) has this form, the solution of Equation (1) suggests the following generalization.

> **General Solution of $\dfrac{dy}{dx} = g(x)$**
>
> The general solution of the differential equation $\dfrac{dy}{dx} = g(x)$ is
>
> $$y = \int g(x)\, dx.$$

### EXAMPLE 2  Population

The population $P$ of a flock of birds is growing exponentially so that

$$\frac{dP}{dt} = 20e^{0.05t},$$

where $t$ is time in years. Find $P$ in terms of $t$ if there were 20 birds in the flock initially.

**SOLUTION**   To solve the differential equation, first determine the antiderivative of $20e^{0.05t}$, that is,

$$P = \int 20e^{0.05t}\, dt = \frac{20}{0.05}e^{0.05t} + C = 400e^{0.05t} + C.$$

Initially, there were 20 birds, so $P = 20$ at time $t = 0$. We can substitute this information into the equation to determine the value of $C$ that satisfies this condition.

$$20 = 400e^0 + C$$

$$-380 = C$$

Therefore, $P = 400e^{0.05t} - 380$. Verify that $P$ is a solution to the differential equation by taking its derivative. The result should be the original differential equation.

**NOTE**
We can write the condition $P = 20$ when $t = 0$ as $P(0) = 20$. We will use this notation from now on.

In Example 2, the given information was used to produce a solution with a specific value of $C$. Such a solution is called a **particular solution** of the differential equation. The given information, $P = 20$ when $t = 0$, is called an **initial condition**. An **initial value problem** is a differential equation with a value of $y$ given at $x = x_0$ (or $t = t_0$ in this case), where $x_0$ is any real number.

Sometimes a differential equation must be rewritten in the form

$$\frac{dy}{dx} = g(x)$$

before it can be solved.

### EXAMPLE 3 Initial Value Problem

Find the particular solution of

$$\frac{dy}{dx} - 2x = \frac{2x}{\sqrt{x^2 + 3}},$$

given that $y(1) = 8$.

**SOLUTION** This differential equation is not in the proper form, but we can easily fix this by adding $2x$ to both sides of the equation. That is,

$$\frac{dy}{dx} = 2x + \frac{2x}{\sqrt{x^2 + 3}}.$$

To find the general solution, integrate this expression, using the substitution $u = x^2 + 3$ in the second term.

$$y = \int \left(2x + \frac{2x}{\sqrt{x^2 + 3}}\right) dx$$

$$= x^2 + \int \frac{1}{\sqrt{u}}\, du \qquad du = 2x\, dx$$

$$= x^2 + 2\sqrt{u} + C \qquad \text{Use the power rule with } n = -1/2.$$

$$= x^2 + 2\sqrt{x^2 + 3} + C$$

Now use the initial condition to find the value of $C$. Substituting 8 for $y$ and 1 for $x$ gives

$$8 = 1^2 + 2\sqrt{1^2 + 3} + C$$
$$8 = 5 + C$$
$$C = 3.$$

The particular solution is $y = x^2 + 2\sqrt{x^2 + 3} + 3$. Verify that $y(1) = 8$ and that differentiating $y$ leads to the original differential equation.

**TRY YOUR TURN 2**

**YOUR TURN 2** Find the particular solution of
$\frac{dy}{dx} - 12x^3 = 6x^2$, given that $y(2) = 60$.

So far in this section, we have used a method that is essentially the same as that used in the section on antiderivatives, when we first started the topic of integration. But not all differential equations can be solved so easily. For example, if interest on an investment is compounded continuously, then the investment grows at a rate proportional to the amount of money present. If $A$ is the amount in an account at time $t$, then for some constant $k$, the differential equation

$$\frac{dA}{dt} = kA \qquad (3)$$

gives the rate of growth of $A$ with respect to $t$. This differential equation is different from those discussed previously, which had the form

$$\frac{dy}{dx} = g(x).$$

**CAUTION** Since the right-hand side of Equation (3) is a function of $A$, rather than a function of $t$, it would be completely invalid to simply integrate both sides as we did in Examples 1 through 3. That method works only when the side opposite the derivative is simply a function of the independent variable.

Equation (3) is an example of a more general differential equation we will now learn to solve—namely, those that can be written in the form

$$\frac{dy}{dx} = \frac{p(x)}{q(y)}.$$

Suppose we think of $dy/dx$ as a fraction $dy$ over $dx$. This is incorrect, of course; the derivative is actually the limit of a small change in $y$ over a small change in $x$, but the notation is chosen so that this interpretation gives a correct answer, as we shall see. Multiply on both sides by $q(y)\,dx$ to get

$$q(y)\,dy = p(x)\,dx.$$

In this form all terms involving $y$ (including $dy$) are on one side of the equation and all terms involving $x$ (and $dx$) are on the other side. A differential equation that can be put into this form is said to be *separable*, since the variables $x$ and $y$ can be separated. After separation, a **separable differential equation** may be solved by integrating each side with respect to the variable given. This method is known as **separation of variables**.

$$\int q(y)\,dy = \int p(x)\,dx$$

$$Q(y) = P(x) + C,$$

where $P$ and $Q$ are antiderivatives of $p$ and $q$. (We don't need a constant of integration on the left side of the equation; it can be combined with the constant of integration on the right side as $C$.) To show that this answer is correct, differentiate implicitly with respect to $x$.

$$Q'(y)\frac{dy}{dx} = P'(x) \qquad \text{Use the chain rule on the left side.}$$

$$q(y)\frac{dy}{dx} = p(x) \qquad \begin{array}{l} q(y) \text{ is the derivative of } Q(y) \text{ and} \\ p(x) \text{ is the derivative of } P(x). \end{array}$$

$$\frac{dy}{dx} = \frac{p(x)}{q(y)}$$

This last equation is the one we set out to solve.

## EXAMPLE 4 Separation of Variables

Find the general solution of

$$y\frac{dy}{dx} = x^2.$$

**SOLUTION** Begin by separating the variables to get

$$y\,dy = x^2\,dx.$$

The general solution is found by determining the antiderivatives of each side.

$$\int y\,dy = \int x^2\,dx$$

$$\frac{y^2}{2} = \frac{x^3}{3} + C$$

$$y^2 = \frac{2}{3}x^3 + 2C$$

$$y^2 = \frac{2}{3}x^3 + K$$

YOUR TURN 3 Find the general solution of $\dfrac{dy}{dx} = \dfrac{x^2 + 1}{xy^2}$.

The constant $K$ was substituted for $2C$ in the last step. The solution is left in implicit form, not solved explicitly for $y$. In general, we will use $C$ for the arbitrary constant, so that our final answer would be written as $y^2 = (2/3)x^3 + C$. It would also be nice to solve for $y$ by taking the square root of both sides, but since we don't know anything about the sign of $y$, we don't know whether the solution is $y = \sqrt{(2/3)x^3 + C}$ or $y = -\sqrt{(2/3)x^3 + C}$. If $y$ were raised to the third power, we could solve for $y$ by taking the cube root of both sides, since the cube root is unique. **TRY YOUR TURN 3**

### EXAMPLE 5  Separation of Variables

Find the general solution of the differential equation for interest compounded continuously,

$$\frac{dA}{dt} = kA.$$

**SOLUTION** Separating variables leads to

$$\frac{1}{A}\, dA = k\, dt.$$

To solve this equation, determine the antiderivative of each side.

$$\int \frac{1}{A}\, dA = \int k\, dt$$

$$\ln|A| = kt + C$$

(Here $A$ represents a nonnegative quantity, so the absolute value is unnecessary, but we wish to show how to solve equations for which this may not be true.) Use the definition of logarithm to write the equation in exponential form as

$$|A| = e^{kt+C} = e^{kt}e^{C}. \qquad \text{Use the property } e^{m+n} = e^m e^n.$$

Finally, use the definition of absolute value to get

$$A = e^{kt}e^{C} \quad \text{or} \quad A = -e^{kt}e^{C}.$$

Because $e^C$ and $-e^C$ are constants, replace them with the constant $M$, which may be any nonzero real number. (We use $M$ rather than $K$ because we already have a constant $k$ in this example. We could also relabel $M$ as $C$, as we did in Example 4.) The resulting equation,

$$A = Me^{kt},$$

not only describes interest compounded continuously but also defines the exponential growth or decay function that was discussed in Chapter 2 on Exponential, Logarithmic, and Trigonometric Functions.

**FOR REVIEW**

In Chapter 2 on Exponential, Logarithmic, and Trigonometric Functions, we saw that the amount of money in an account with interest compounded continuously is given by

$$A = Pe^{rt},$$

where $P$ is the initial amount in the account, $r$ is the annual interest rate, and $t$ is the time in years. Observe that this is the same as the equation for the amount of money in an account derived here, where $P$ and $r$ have been replaced with $M$ and $k$, respectively. For more applications of exponential growth and decay, see Chapter 2 on Exponential, Logarithmic, and Trigonometric Functions.

**CAUTION** Notice that $A = 0$ is also a solution to the differential equation in Example 5, but after we divide by $A$ (which is not possible if $A = 0$) and integrate, the resulting equation $|A| = e^{kt+C}$ does not allow $y$ to equal 0. In this example, the lost solution can be recovered in the final answer if we allow $M$ to equal 0, a value that was previously excluded. When dividing by an expression in separation of variables, look for solutions that would make this expression 0 and may be lost.

Recall that equations of the form $A = Me^{kt}$ arise in situations where the rate of change of a quantity is proportional to the amount present at time $t$, which is precisely what the differential equation (3) describes. The constant $k$ is called the **growth constant**, while $M$ represents the amount present at time $t = 0$. A positive value of $k$ indicates growth, while a negative value of $k$ indicates decay. The equation was often written in the form $y = y_0 e^{kt}$ in Chapter 2 on Exponential, Logarithmic, and Trigonometric Functions, where we discussed other applications of this equation, such as radioactive decay.

As a model of population growth, the equation $y = Me^{kt}$ is not realistic over the long run for most populations. As shown by graphs of functions of the form $y = Me^{kt}$, with both $M$ and $k$ positive, growth would be unbounded. Additional factors, such as space restrictions or a limited amount of food, tend to inhibit growth of populations as time goes on. In an alternative model that assumes a maximum population of size $N$, the rate of growth of a population is proportional to how close the population is to that maximum, that is, to the difference between $N$ and $y$. These assumptions lead to the differential equation

$$\frac{dy}{dt} = k(N - y),$$

whose solution is the limited growth function mentioned in the third section of Chapter 2 on Exponential, Logarithmic, and Trigonometric Functions.

### EXAMPLE 6    Population

A certain nature reserve can support no more than 4000 mountain goats. Assume that the rate of growth is proportional to how close the population is to this maximum, with a growth rate of 20 percent. There are currently 1000 goats in the area.

**(a)** Solve the general limited growth differential equation, and then write a function describing the goat population at time $t$.

**SOLUTION**    To solve the equation

$$\frac{dy}{dt} = k(N - y),$$

first separate the variables.

$$\frac{dy}{N - y} = k \, dt$$

$$\int \frac{dy}{N - y} = \int k \, dt \qquad \text{Integrate both sides.}$$

$$-\ln(N - y) = kt + C \qquad \begin{array}{l}\text{We assume the population is less} \\ \text{than } N, \text{ so } N - y > 0.\end{array}$$

$$\ln(N - y) = -kt - C$$

$$N - y = e^{-kt-C} \qquad \text{Apply the function } e^x \text{ to both sides.}$$

$$N - y = e^{-kt}e^{-C} \qquad \text{Use the property } e^{m+n} = e^m e^n.$$

$$y = N - e^{-kt}e^{-C} \qquad \text{Solve for } y.$$

$$y = N - Me^{-kt} \qquad \text{Relabel the constant } e^{-C} \text{ as M.}$$

Now apply the initial condition that $y(0) = y_0$.

$$y_0 = N - Me^{-k \cdot 0} = N - M \qquad e^0 = 1$$

$$M = N - y_0 \qquad \text{Solve for } M.$$

Substituting this value of $M$ into the previous solution gives

$$y = N - (N - y_0)e^{-kt}.$$

The graph of this function is shown in Figure 2.

For the goat problem, the maximum population is $N = 4000$, the initial population is $y_0 = 1000$, and the growth rate constant is 20%, or $k = 0.2$. Therefore,

$$y = 4000 - (4000 - 1000)e^{-0.2t} = 4000 - 3000e^{-0.2t}.$$

**FIGURE 2**

**APPLY IT**

**YOUR TURN 4** In Example 6, find the goat population in 5 years if the reserve can support 6000 goats, the growth rate is 15%, and there are currently 1200 goats in the area.

**(b)** What will the goat population be in 5 years?

**SOLUTION** In 5 years, the population will be

$$y = 4000 - 3000e^{-(0.2)(5)} = 4000 - 3000e^{-1}$$
$$\approx 4000 - 1103.6 = 2896.4,$$

or about 2900 goats. **TRY YOUR TURN 4**

## Logistic Growth

Let $y$ be the size of a certain population at time $t$. In the standard model for unlimited growth given by Equation (3), the rate of growth is proportional to the current population size. The constant $k$, the growth rate constant, is the difference between the birth and death rates of the population. The unlimited growth model predicts that the population's growth rate is a constant, $k$.

Growth usually is not unlimited, however, and the population's growth rate is usually not constant because the population is limited by environmental factors to a maximum size $N$, called the **carrying capacity** of the environment for the species. In the limited growth model already given,

$$\frac{dy}{dt} = k(N - y),$$

the rate of growth is proportional to the remaining room for growth, $N - y$.

In the **logistic growth model**

$$\frac{dy}{dt} = k\left(1 - \frac{y}{N}\right)y \tag{4}$$

the rate of growth is proportional to both the current population size $y$ and a factor $(1 - y/N)$ that is equal to the remaining room for growth, $N - y$, divided by $N$. Equation (4) is called the **logistic equation**. Notice that $(1 - y/N) \to 1$ as $y \to 0$, and the differential equation can be approximated as

$$\frac{dy}{dt} = k\left(1 - \frac{y}{N}\right)y \approx k(1)y = ky.$$

In other words, when $y$ is small, the growth of the population behaves as if it were unlimited. On the other hand, $(1 - y/N) \to 0$ as $y \to N$, so

$$\frac{dy}{dt} = k\left(1 - \frac{y}{N}\right)y \approx k(0)y = 0.$$

That is, population growth levels off as $y$ nears the maximum population $N$. Thus, the logistic Equation (4) is the unlimited growth Equation (3) with a factor $(1 - y/N)$ to account for limiting environmental factors when $y$ nears $N$. Let $y_0$ denote the initial population size. Under the assumption $0 < y < N$, the general solution of Equation (4) is

$$y = \frac{N}{1 + be^{-kt}}, \tag{5}$$

where $b = (N - y_0)/y_0$ (see Exercise 39). This solution, called a **logistic curve**, is shown in Figure 3. This function was introduced in Section 4.4 on Derivatives of Exponential Functions in the form

$$G(t) = \frac{m}{1 + \left(\frac{m}{G_0}\right)e^{-kmt}},$$

where $m$ is the limiting value of the population, $G_0$ is the initial number present, and $k$ is a positive constant.

As expected, the logistic curve begins exponentially and subsequently levels off. Another important feature is the point of inflection $((\ln b)/k, N/2)$, where $dy/dx$ is a maximum (see Exercise 41). Notice that the point of inflection is when the population is half of the carrying capacity and that at this point, the population is increasing most rapidly.

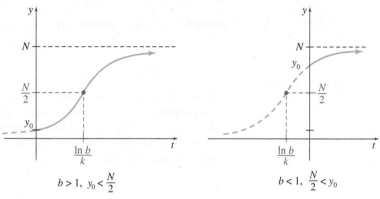

FIGURE 3

Logistic equations arise frequently in the study of populations. In about 1840 the Belgian sociologist P. F. Verhulst fitted a logistic curve to U.S. census figures and made predictions about the population that were subsequently proved to be quite accurate. American biologist Raymond Pearl (circa 1920) found that the growth of a population of fruit flies in a limited space could be modeled by the logistic equation

$$\frac{dy}{dt} = 0.2y - \frac{0.2}{1035}y^2.$$

**TECHNOLOGY NOTE**

Some calculators can fit a logistic curve to a set of data points. For example, the TI-84 Plus has this capability, listed as `Logistic` in the STAT CALC menu, along with other types of regression. See Exercises 48, 49, and 55.

Logistic growth is an example of how a model is modified over time as new insights occur. The model for population growth changed from the early exponential curve $y = Me^{kt}$ to the logistic curve

$$y = \frac{N}{1 + be^{-kt}}.$$

Many other quantities besides population grow logistically. That is, their initial rate of growth is slow, but as time progresses, their rate of growth increases to a maximum value and subsequently begins to decline and to approach zero.

**EXAMPLE 7** **Logistic Curve**

Rapid technological advancements in the last 20 years have made many products obsolete practically overnight. J. C. Fisher and R. H. Pry successfully described the phenomenon of a technically superior new product replacing another product by the logistic equation

$$\frac{dz}{dt} = k(1 - z)z, \tag{6}$$

where $z$ is the market share of the new product and $1 - z$ is the market share of the other product. ***Source: Technological Forecasting and Social Change.*** The new product will initially have little or no market share; that is, $z_0 \approx 0$. Thus, the constant $b$ in Equation (5) will have to be determined in a different way. Let $t_0$ be the time at which $z = 1/2$. Under the assumption $0 < z < 1$, the general solution of Equation (6) is

$$z = \frac{1}{1 + be^{-kt}}, \tag{7}$$

where $b = e^{kt_0}$ (see Exercise 40).

---

**TECHNOLOGY NOTE**

Fisher and Pry applied their model to the fraction of fabric consumed in the United States that was synthetic. Their data are shown in the table below. At the time of the study, natural fabrics in clothing were being replaced with synthetic fabric. Using the logistic regression function on a TI-84 Plus calculator, with $t$ as the number of years since 1930, the best logistic function to fit these data can be shown to be

$$z = \frac{1.293}{1 + 21.80e^{-0.06751t}}.$$

A graph of the data and this function is shown in Figure 4(a). Although the data fit the function well, this function is not of the form studied by Fisher and Pry because it has a numerator of 1.293, rather than 1. This function predicts that the percentage of fabric that is synthetic approaches 129%! A more appropriate function can be found by rewriting Equation (7) as

$$be^{-kt} = \frac{1}{z} - 1,$$

and then finding the best fit exponential function to the points $(t, 1/z - 1)$. This leads to the equation

$$z = \frac{1}{1 + 18.93(0.92703)^t} = \frac{1}{1 + 18.93e^{(\ln 0.92703)t}} = \frac{1}{1 + 18.93e^{-0.07577t}}.$$

As Figure 4(b) shows, this more realistic model also fits the data well. The dangers of extrapolating beyond the data are illustrated by this equation's prediction that 95.8% of fabrics in the United States would be synthetic by 2010. In fact, cotton is still more popular than synthetic fabrics. ***Source: Fabrics Manufacturers.***

| Synthetic Fabric as Percent of U.S. Consumption | | | | | | | | |
|---|---|---|---|---|---|---|---|---|
| *Year* | 1930 | 1935 | 1940 | 1945 | 1950 | 1955 | 1960 | 1965 | 1967 |
| *Fraction synthetic* | 0.044 | 0.079 | 0.10 | 0.14 | 0.22 | 0.28 | 0.29 | 0.43 | 0.47 |

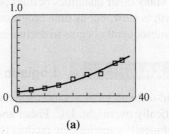

(a)

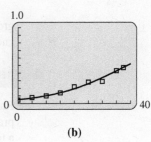

(b)

FIGURE 4

Notice that there are two values of $y$ that make $dy/dt = 0$ in the logistic equation:

$$\frac{dy}{dt} = k\left(1 - \frac{y}{N}\right)y = 0 \quad \text{when} \quad y = 0 \text{ or } y = N.$$

This means that if $y = 0$ or $N$, then $y$ will never change because its rate of change is 0. Such a point is known as an **equilibrium point**. (For more on this topic, see the Extended Application on Managing Renewable Resources in Chapter 4.) For an **autonomous** differential equation—that is, of the form $dy/dt = f(y)$ (so the right side does not depend on the independent variable), finding equilibrium points amounts to solving $f(y) = 0$.

The next thing to determine is whether points near each equilibrium point move closer to the equilibrium point or further away. This is known as finding the **stability** of an equilibrium point. To do this, draw a number line similar to the ones used in Section 5.1 for determining when a function is increasing and decreasing. In Figure 5 we have done this for the logistic equation. Unlike the number lines in Section 5.1, we draw an arrow pointing to the left when $dy/dt < 0$, and to the right when $dy/dt > 0$. (In Figure 5, we consider only $y \geq 0$, since the population $y$ cannot be negative.) Notice from the equation above that if $0 < y < N$, then $dy/dt > 0$, and therefore $y$ increases. Similarly, if $y > N$, then $dy/dt < 0$, so $y$ decreases. This means that $y = N$ is a **stable** equilibrium point, because values of $y$ that are close to $N$ move toward $N$ (increase if they are smaller and decrease if they are larger). Points near $y = 0$ move away from it, so the equilibrium point is said to be **unstable**. Finding equilibrium points and determining their stability is important because it gives us qualitative information about the solution, even when the exact solution is difficult or impossible to find.

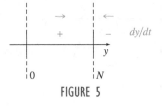

FIGURE 5

### EXAMPLE 8    Equilibrium Points

Find all equilibrium points for the differential equation

$$\frac{dy}{dx} = (1 - e^{2y})(y - 1)^2(y - 2)$$

and determine their stability.

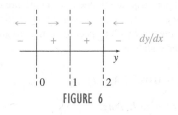

FIGURE 6

**SOLUTION**    The right side of the equation is equal to 0 when $y = 1$ and $y = 2$, as well as when $e^{2y} = 1$, which implies $y = 0$. Figure 6 shows a number line with arrows pointing to the right where $y$ is increasing, and to the left where $y$ is decreasing. Notice that the arrows are pointing away from 0, so 0 is an unstable equilibrium point. The arrows are pointing toward 2, so 2 is a stable equilibrium point. The arrow to the left of 1 points toward it, and the arrow to the right points away from it. We say that 1 is a **semistable** equilibrium point.

## 11.1    EXERCISES

**Find the general solution for each differential equation. Verify that each solution satisfies the original differential equation.**

**1.** $\dfrac{dy}{dx} = -4x + 6x^2$

**2.** $\dfrac{dy}{dx} = 4e^{-3x}$

**3.** $4x^3 - 2\dfrac{dy}{dx} = 0$

**4.** $3x^2 - 3\dfrac{dy}{dx} = 2$

**5.** $y\dfrac{dy}{dx} = x^2$

**6.** $y\dfrac{dy}{dx} = x^2 - x$

**7.** $\dfrac{dy}{dx} = 2xy$

**8.** $\dfrac{dy}{dx} = x^2y$

**9.** $\dfrac{dy}{dx} = 3x^2y - 2xy$

**10.** $(y^2 - y)\dfrac{dy}{dx} = x$

**11.** $\dfrac{dy}{dx} = \dfrac{y}{x}, \ x > 0$

**12.** $\dfrac{dy}{dx} = \dfrac{y}{x^2}$

**13.** $\dfrac{dy}{dx} = \dfrac{y^2 + 6}{2y}$

**14.** $\dfrac{dy}{dx} = \dfrac{e^{y^2}}{y}$

**15.** $\dfrac{dy}{dx} = y^2e^{2x}$

**16.** $\dfrac{dy}{dx} = \dfrac{e^x}{e^y}$

**17.** $\dfrac{dy}{dx} = \dfrac{\cos x}{\sin y}$

**18.** $\dfrac{dy}{dx} = \tan x \cos^2 y$

**Find the particular solution for each initial value problem.**

**19.** $\dfrac{dy}{dx} + 3x^2 = 2x; \ \ y(0) = 5$

**20.** $\dfrac{dy}{dx} = 4x^3 - 3x^2 + x; \ \ y(1) = 0$

**21.** $2\dfrac{dy}{dx} = 4xe^{-x}; \ \ y(0) = 42$

**22.** $x\dfrac{dy}{dx} = x^2 e^{3x}$; $y(0) = \dfrac{8}{9}$

**23.** $\dfrac{dy}{dx} = \dfrac{x^3}{y}$; $y(0) = 5$

**24.** $x^2\dfrac{dy}{dx} - y\sqrt{x} = 0$; $y(1) = e^{-2}$

**25.** $(2x + 3)y = \dfrac{dy}{dx}$; $y(0) = 1$

**26.** $\dfrac{dy}{dx} = \dfrac{x^2 + 5}{2y - 1}$; $y(0) = 11$

**27.** $\dfrac{dy}{dx} = \dfrac{2x + 1}{y - 3}$; $y(0) = 4$

**28.** $x^2\dfrac{dy}{dx} = y$; $y(1) = -1$

**29.** $\dfrac{dy}{dx} = \dfrac{y^2}{x}$; $y(e) = 3$

**30.** $\dfrac{dy}{dx} = x^{1/2}y^2$; $y(4) = 9$

**31.** $\dfrac{dy}{dx} = (y - 1)^2 e^{x-1}$; $y(1) = 2$

**32.** $\dfrac{dy}{dx} = (x + 2)^2 e^y$; $y(1) = 0$

**33.** $\dfrac{dy}{dx} = y \cos x$; $y(0) = 3$

**34.** $\dfrac{dy}{dx} = e^y \sec^2 x$; $y(0) = 0$

**Find all equilibrium points and determine their stability.**

**35.** $\dfrac{dy}{dx} = y(y^2 - 1)$

**36.** $\dfrac{dy}{dx} = (4 - y^2)(y + 1)$

**37.** $\dfrac{dy}{dx} = (e^y - 1)(y - 3)$

**38.** $\dfrac{dy}{dx} = (\ln y - 1)(5 - y)$

**39. a.** Solve the logistic equation (4) in this section by observing that

$$\frac{1}{y} + \frac{1}{N - y} = \frac{N}{(N - y)y}.$$

**b.** Assume $0 < y < N$. Verify that $b = (N - y_0)/y_0$ in Equation (5), where $y_0$ is the initial population size.

**c.** Assume $0 < N < y$ for all $y$. Verify that $b = (y_0 - N)/y_0$.

**40.** Suppose that $0 < z < 1$ for all $z$. Solve the logistic equation (6) as in Exercise 39. Verify that $b = e^{kx_0}$, where $x_0$ is the time at which $z = 1/2$.

**41.** Suppose that $0 < y_0 < N$. Let $b = (N - y_0)/y_0$, and let $y(x) = N/(1 + be^{-kx})$ for all $x$. Show the following.

**a.** $0 < y(x) < N$ for all $x$.

**b.** The lines $y = 0$ and $y = N$ are horizontal asymptotes of the graph.

**c.** $y(x)$ is an increasing function.

**d.** $((\ln b)/k, N/2)$ is a point of inflection of the graph.

**e.** $dy/dx$ is a maximum at $x_0 = (\ln b)/k$.

**42.** Suppose that $0 < N < y_0$. Let $b = (y_0 - N)/y_0$ and let

$$y(x) = \frac{N}{1 - be^{-kx}} \quad \text{for all } x \neq \frac{\ln b}{k}.$$

See the figure. Show the following.

**a.** $0 < b < 1$

**b.** The lines $y = 0$ and $y = N$ are horizontal asymptotes of the graph.

**c.** The line $x = (\ln b)/k$ is a vertical asymptote of the graph.

**d.** $y(x)$ is decreasing on $((\ln b)/k, \infty)$ and on $(-\infty, (\ln b)/k)$.

**e.** $y(x)$ is concave upward on $((\ln b)/k, \infty)$ and concave downward on $(-\infty, (\ln b)/k)$.

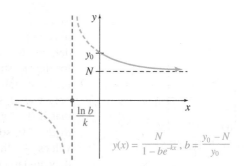

$$y(x) = \frac{N}{1 - be^{-kx}}, \quad b = \frac{y_0 - N}{y_0}$$

## LIFE SCIENCE APPLICATIONS

**43. Tracer Dye** The amount of a tracer dye injected into the bloodstream decreases exponentially, with a decay constant of 3% per minute. If 6 cc are present initially, how many cubic centimeters are present after 10 minutes? (Here $k$ will be negative.)

**44. Soil Moisture** The evapotranspiration index $I$ is a measure of soil moisture. An article on 10- to 14-year-old heath vegetation described the rate of change of $I$ with respect to $W$, the amount of water available, by the equation

$$\frac{dI}{dW} = 0.088(2.4 - I).$$

*Source: Australian Journal of Botany.*

**a.** According to the article, $I$ has a value of 1 when $W = 0$. Solve the initial value problem.

**b.** What happens to $I$ as $W$ becomes larger and larger?

**45. Fish Population** An isolated fish population is limited to 4000 by the amount of food available. If there are now 320 fish and the population is growing with a growth constant of 2% a year, find the expected population at the end of 10 years.

**Dieting** A person's weight depends both on the daily rate of energy intake, say $C$ calories per day, and on the daily rate of energy consumption, typically between 15 and 20 calories per pound per day. Using an average value of 17.5 calories per pound per day, a person weighing $w$ pounds uses $17.5w$ calories per day. If $C = 17.5w$, then weight remains constant, and weight gain or loss occurs according to whether $C$ is greater or less than $17.5w$. *Source: The College Mathematics Journal.*

**46.** To determine how fast a change in weight will occur, the most plausible assumption is that $dw/dt$ is proportional to the net excess (or deficit) $C - 17.5w$ in the number of calories per day.

**a.** Assume $C$ is constant and write a differential equation to express this relationship. Use $k$ to represent the constant of proportionality. What does $C$ being constant imply?

**b.** The units of $dw/dt$ are pounds per day, and the units of $C - 17.5w$ are calories per day. What units must $k$ have?

**c.** Use the fact that 3500 calories is equivalent to 1 lb to rewrite the differential equation in part a.

**d.** Solve the differential equation.

**e.** Let $w_0$ represent the initial weight and use it to express the coefficient of $e^{-0.005t}$ in terms of $w_0$ and $C$.

**47.** (Refer to Exercise 46.) Suppose someone initially weighing 180 lb adopts a diet of 2500 calories per day.

**a.** Write the weight function for this individual.

**b.** Graph the weight function on the window $[0, 300]$ by $[120, 200]$. What is the asymptote? This value of $w$ is the equilibrium weight $w_{eq}$. According to the model, can a person ever achieve this weight?

**c.** How long will it take a dieter to reach a weight just 2 lb more than $w_{eq}$?

**48. H1N1 Virus** The cumulative number of deaths worldwide due to the H1N1 virus, or swine flu, at various days into the epidemic are listed below, where April 21, 2009 was day 1. *Source: BBC.*

| Day | Deaths | Day | Deaths |
|-----|--------|-----|--------|
| 14 | 27 | 148 | 3696 |
| 28 | 74 | 163 | 4334 |
| 43 | 114 | 182 | 4804 |
| 57 | 164 | 206 | 6704 |
| 71 | 315 | 221 | 8450 |
| 84 | 580 | 234 | 9797 |
| 99 | 1049 | 266 | 14,024 |
| 117 | 2074 | 274 | 14,378 |
| 132 | 2967 | | |

Use a calculator with logistic regression capability to complete the following.

**a.** Plot the number of deaths $y$ against the number of days $t$. Discuss the appropriateness of fitting a logistic function to this data.

**b.** Use the logistic regression function on your calculator to determine the logistic equation that best fits the data.

**c.** Plot the logistic regression function from part b on the same graph as the data points. Discuss how well the logistic equation fits the data.

**d.** Assuming the logistic equation found in part b continues to be accurate, what seems to be the limiting size of the number of deaths due to this outbreak of the H1N1 virus?

**e.** Discuss whether a logistic model is more appropriate than an exponential model for estimating the number of deaths due to the H1N1 virus.

**49. Population Growth** The following table gives the historic and projected populations (in millions) of China and India. *Source: United Nations.*

| Year | China | India |
|------|-------|-------|
| 1950 | 544 | 376 |
| 1960 | 651 | 450 |
| 1970 | 814 | 555 |
| 1980 | 984 | 699 |
| 1990 | 1165 | 869 |
| 2000 | 1280 | 1042 |
| 2010 | 1360 | 1206 |
| 2020 | 1386 | 1384 |
| 2030 | 1383 | 1515 |
| 2040 | 1340 | 1609 |
| 2050 | 1268 | 1664 |

Use a calculator with logistic regression capability to complete the following.

**a.** Letting $t$ represent the years since 1950, plot the Chinese population on the $y$-axis against the year on the $t$-axis. Discuss the appropriateness of fitting a logistic function to these data.

**b.** Use the logistic regression function on your calculator to determine the logistic equation that best fits the data. Plot the logistic function on the same graph as the data points. Discuss how well the logistic function fits the data.

**c.** Assuming the logistic equation found in part b continues to be accurate, what seems to be the limiting size of the Chinese population?

**d.** Repeat parts a–c using the population for India.

**50. U.S. Hispanic Population** A recent report by the U.S. Census Bureau predicts that the U.S. Hispanic population will increase from 35.6 million in 2000 to 102.6 million in 2050. *Source: U.S. Census Bureau.* Assuming the unlimited growth model $dy/dt = ky$ fits this population growth, express the population $y$ as a function of the year $t$. Let 2000 correspond to $t = 0$.

**51. U.S. Asian Population** (Refer to Exercise 50.) The report also predicted that the U.S. Asian population would increase from 10.7 million in 2000 to 33.4 million in 2050. *Source: U.S. Census Bureau.* Repeat Exercise 50 using these data.

**52. Guernsey Growth** The growth of Guernsey cows can be approximated by the equation

$$\frac{dW}{dt} = 0.0152(486 - W),$$

where $W$ is the weight in kg after $t$ weeks. *Source: Annales de Zootechnie.*

**a.** Find the solution to the initial value problem with $W(0) = 32$ kg.

**b.** What limit does the weight of a Guernsey cow approach as $t$ goes to infinity?

**53. Flea Beetles** A study of flea beetles found that the change in the rate of flea beetles moving in and out of a patch of beetles could be described by the differential equation

$$\frac{dN}{dt} = mN + i,$$

where $N$ is the number of beetles in a patch, $m$ is the rate at which beetles move out of the patch, and $i$ is the rate at which they move in. *Source: Ecological Monographs.*

**a.** Solve the differential equation above with the initial condition $N(0) = N_0$.

**b.** After the researchers cleared a patch of beetles, so that $N_0 = 0$, they would return 8 hours later and count the number of beetles. Show that the parameter $i$ can then be estimated by the equation

$$i = \frac{mN(8)}{e^{8m} - 1}.$$

**c.** The researchers estimated $m$ using the equation $m = \ln F_{sd}$, where $F_{sd}$ is the fraction of beetles who remained in the patch in which they were released. For the beetles *P. striolata* released in July in the lush interior of patches 5 meters apart, the average values of $F_{sd}$ and $N(8)$ were 0.709 and 4.5, respectively. Find the values of $m$ and $i$.

**54. Plant Growth** Researchers have found that the probability $P$ that a plant will grow to radius $R$ can be described by the differential equation

$$\frac{dP}{dR} = -4\pi DRP^2,$$

where $D$ is the density of the plants in an area. *Source: Ecology.* Given the initial condition $P(0) = 1$, find a formula for $P$ in terms of $R$.

**55. World Population** The following table gives the population of the world at various times over the past two centuries, plus projections for this century. *Source: The New York Times.*

| Year | Population (billions) |
|------|----------------------|
| 1804 | 1 |
| 1927 | 2 |
| 1960 | 3 |
| 1974 | 4 |
| 1987 | 5 |
| 1999 | 6 |
| 2011 | 7 |
| 2025 | 8 |
| 2041 | 9 |
| 2071 | 10 |

Use a calculator with logistic regression capability to complete the following.

**a.** Use the logistic regression function on your calculator to determine the logistic equation that best fits the data.

**b.** Plot the logistic function found in part a and the original data in the same window. Does the logistic function seem to fit the data from 1927 on? Before 1927?

**c.** To get a better fit, subtract 0.99 from each value of the population in the table. (This makes the population in 1804 small, but not 0 or negative.) Find a logistic function that fits the new data.

**d.** Plot the logistic function found in part c and the modified data in the same window. Does the logistic function now seem to be a better fit than in part b?

**e.** Based on the results from parts c and d, predict the limiting value of the world's population as time increases. For comparison, the *New York Times* article predicts a value of 10.73 billion. (*Hint:* After taking the limit, remember to add the 0.99 that was removed earlier.)

**f.** Based on the results from parts c and d, predict the limiting value of the world population as you go further and further back in time. Does that seem reasonable? Explain.

## OTHER APPLICATIONS

**56. Spread of a Rumor** Suppose the rate at which a rumor spreads—that is, the number of people who have heard the rumor over a period of time—increases with the number of people who have heard it. If $y$ is the number of people who have heard the rumor, then

$$\frac{dy}{dt} = ky,$$

where $t$ is the time in days.

**a.** If $y$ is 1 when $t = 0$, and $y$ is 5 when $t = 2$, find $k$.

Using the value of $k$ from part a, find $y$ for each time.

**b.** $t = 3$     **c.** $t = 5$     **d.** $t = 10$

**57. Radioactive Decay** The amount of a radioactive substance decreases exponentially, with a decay constant of 3% per month.

**a.** Write a differential equation to express the rate of change.

**b.** Find a general solution to the differential equation from part a.

**c.** If there are 75 g at the start of the decay process, find a particular solution for the differential equation from part a.

**d.** Find the amount left after 10 months.

**Newton's Law of Cooling** Newton's law of cooling states that the rate of change of temperature of an object is proportional to the difference in temperature between the object and the surrounding medium. Thus, if $T$ is the temperature of the object after $t$ hours and $T_M$ is the (constant) temperature of the surrounding medium, then

$$\frac{dT}{dt} = -k(T - T_M),$$

where $k$ is a constant. Use this equation in Exercises 58–61.

**58.** Show that the solution of this differential equation is

$$T = Ce^{-kt} + T_M,$$

where $C$ is a constant.

**59.** According to the solution in Exercise 58 of the differential equation for Newton's law of cooling, what happens to the temperature of an object after it has been in a surrounding medium with constant temperature for a long period of time? How well does this agree with reality?

**Newton's Law of Cooling** When a dead body is discovered, one of the first steps in the ensuing investigation is for a medical examiner to determine the time of death as closely as possible. Have you ever wondered how this is done? If the temperature of the medium (air, water, or whatever) has been fairly constant and less than 48 hours have passed since the death, Newton's law of cooling can be used. The medical examiner does not actually solve the equation for each case. Instead, a table based on the formula is consulted. Use Newton's law of cooling to work the following exercises. *Source: The College Mathematics Journal.*

**60.** Assume the temperature of a body at death is 98.6°F, the temperature of the surrounding air is 68°F, and at the end of one hour the body temperature is 90°F.

  **a.** What is the temperature of the body after 2 hours?

  **b.** When will the temperature of the body be 75°F?

  **c.** Approximately when will the temperature of the body be within 0.01° of the surrounding air?

**61.** Suppose the air temperature surrounding a body remains at a constant 10°F, $C = 88.6$, and $k = 0.24$.

  **a.** Determine a formula for the temperature at any time $t$.

  **b.** Use a graphing calculator to graph the temperature $T$ as a function of time $t$ on the window $[0, 30]$ by $[0, 100]$.

  **c.** When does the temperature of the body decrease more rapidly: just after death, or much later? How do you know?

  **d.** What will the temperature of the body be after 4 hours?

  **e.** How long will it take for the body temperature to reach 40°F? Use your calculator graph to verify your answer.

**YOUR TURN ANSWERS**

**1.** $y = 2x^6 + (2/3)x^{3/2} + e^{5x}/5 + C$
**2.** $y = 3x^4 + 2x^3 - 4$
**3.** $y = ((3/2)x^2 + 3\ln|x| + C)^{1/3}$
**4.** 3733

# 11.2 Linear First-Order Differential Equations

APPLY IT **What happens over time to the glucose level in a patient's bloodstream?** *The solution to a linear differential equation gives us an answer in Example 4.*

Recall that $f^{(n)}(x)$ represents the $n$th derivative of $f(x)$, and that $f^{(n)}(x)$ is called an $n$th-order derivative. By this definition, the derivative $f'(x)$ is first-order, $f''(x)$ is second-order, and so on. The *order of a differential equation* is that of the highest-order derivative in the equation. In this section only first-order differential equations are discussed.

A **linear first-order differential equation** is an equation of the form

$$\frac{dy}{dx} + P(x)y = Q(x).$$

Notice that a linear differential equation has a $dy/dx$ term, a $y$ term, and a term that is just a function of $x$. It does not have terms involving nonlinear expressions such as $y^2$ or $e^y$, nor does it have terms involving the product or quotient of $y$ and $dy/dx$. Many useful models produce such equations. In this section we develop a general method for solving first-order linear differential equations.

**EXAMPLE 1** **Linear Differential Equation**

Solve the equation

$$x\frac{dy}{dx} + 6y + 2x^4 = 0. \tag{1}$$

**SOLUTION** We need to first get the equation in the form of a linear first-order differential equation. Thus, $dy/dx$ should have a coefficient of 1. To accomplish this, we divide both sides of the equation by $x$ and rearrange the terms to get the linear differential equation

$$\frac{dy}{dx} + \frac{6}{x}y = -2x^3.$$

This equation is not separable and cannot be solved by the methods discussed so far. (Verify this.) Instead, multiply both sides of the equation by $x^6$ (the reason will be explained shortly) to get

$$x^6\frac{dy}{dx} + 6x^5y = -2x^9. \tag{2}$$

On the left, $6x^5$, the coefficient of $y$, is the derivative of $x^6$, the coefficient of $dy/dx$. Recall the product rule for derivatives:

$$D_x(uv) = u\frac{dv}{dx} + \frac{du}{dx}v.$$

If $u = x^6$ and $v = y$, the product rule gives

$$D_x(x^6y) = x^6\frac{dy}{dx} + 6x^5y,$$

which is the left side of Equation (2). Substituting $D_x(x^6y)$ for the left side of Equation (2) gives

$$D_x(x^6y) = -2x^9.$$

Assuming $y = f(x)$, as usual, both sides of this equation can be integrated with respect to $x$ and the result solved for $y$ to get

$$x^6y = \int -2x^9\, dx = -2\left(\frac{x^{10}}{10}\right) + C = -\frac{x^{10}}{5} + C$$

$$y = -\frac{x^4}{5} + \frac{C}{x^6}. \tag{3}$$

Equation (3) is the general solution of Equation (2) and, therefore, of Equation (1).  ▬

The procedure in Example 1 has given us a solution, but what motivated our choice of the multiplier $x^6$? To see where $x^6$ came from, let $I(x)$ represent the multiplier, and multiply both sides of the general equation

$$\frac{dy}{dx} + P(x)y = Q(x)$$

by $I(x)$:

$$I(x)\frac{dy}{dx} + I(x)P(x)y = I(x)Q(x). \tag{4}$$

The method illustrated above will work only if the left side of the equation is the derivative of the product function $I(x) \cdot y$, which is

$$I(x)\frac{dy}{dx} + I'(x)y. \tag{5}$$

Comparing the coefficients of $y$ in Equations (4) and (5) shows that $I(x)$ must satisfy

$$I'(x) = I(x)P(x),$$

or

$$\frac{I'(x)}{I(x)} = P(x).$$

Integrating both sides of this last equation gives

$$\ln |I(x)| = \int P(x)\, dx + C$$

$$|I(x)| = e^{\int P(x)\, dx + C}$$

or

$$I(x) = \pm e^{C} e^{\int P(x)\, dx}.$$

Only one value of $I(x)$ is needed, so let $C = 0$, so that $e^{C} = 1$, and use the positive result, giving

$$I(x) = e^{\int P(x)\, dx}.$$

In summary, choosing $I(x)$ as $e^{\int P(x)\, dx}$ and multiplying both sides of a linear first-order differential equation by $I(x)$ puts the equation in a form that can be solved by integration.

**Integrating Factor**

The function $I(x) = e^{\int P(x)\, dx}$ is called an **integrating factor** for the differential equation

$$\frac{dy}{dx} + P(x)y = Q(x).$$

For Equation (1), written as the linear differential equation

$$\frac{dy}{dx} + \frac{6}{x}y = -2x^3,$$

$P(x) = 6/x$, and the integrating factor is

$$I(x) = e^{\int (6/x)\, dx} = e^{6 \ln |x|} = e^{\ln |x|^6} = e^{\ln x^6} = x^6.$$

This last step used the fact that $e^{\ln a} = a$ for all positive $a$.

In summary, we solve a linear first-order differential equation with the following steps.

**Solving a Linear First-Order Differential Equation**

1. Put the equation in the linear form $\dfrac{dy}{dx} + P(x)y = Q(x)$.
2. Find the integrating factor $I(x) = e^{\int P(x)\, dx}$.
3. Multiply each term of the equation from Step 1 by $I(x)$.
4. Replace the sum of terms on the left with $D_x[I(x)y]$.
5. Integrate both sides of the equation.
6. Solve for $y$.

**EXAMPLE 2**  **Linear Differential Equation**

Give the general solution of $\dfrac{dy}{dx} + 2xy = x$.

**SOLUTION**

*Step 1*  This equation is already in the required form.

*Step 2*  The integrating factor is

$$I(x) = e^{\int 2x\, dx} = e^{x^2}.$$

***Step 3*** Multiplying each term by $e^{x^2}$ gives

$$e^{x^2}\frac{dy}{dx} + 2xe^{x^2}y = xe^{x^2}.$$

***Step 4*** The sum of terms on the left can now be replaced with $D_x(e^{x^2}y)$, to get

$$D_x(e^{x^2}y) = xe^{x^2}.$$

***Step 5*** Integrating on both sides gives

$$e^{x^2}y = \int xe^{x^2}\,dx,$$

or

$$e^{x^2}y = \frac{1}{2}e^{x^2} + C.$$

**YOUR TURN 1** Give the general solution of

$$x\frac{dy}{dx} - y - x^2e^x = 0, x > 0.$$

***Step 6*** Divide both sides by $e^{x^2}$ to get the general solution

$$y = \frac{1}{2} + Ce^{-x^2}.$$

**TRY YOUR TURN 1**

**EXAMPLE 3** **Linear Differential Equation**

Solve the initial value problem $2\left(\dfrac{dy}{dx}\right) - 6y - e^x = 0$ with $y(0) = 5$.

**SOLUTION** Write the equation in the required form by adding $e^x$ to both sides and dividing both sides by 2:

$$\frac{dy}{dx} - 3y = \frac{1}{2}e^x.$$

The integrating factor is

$$I(x) = e^{\int (-3)\,dx} = e^{-3x}.$$

Multiplying each term by $I(x)$ gives

$$e^{-3x}\frac{dy}{dx} - 3e^{-3x}y = \frac{1}{2}e^xe^{-3x},$$

or

$$e^{-3x}\frac{dy}{dx} - 3e^{-3x}y = \frac{1}{2}e^{-2x}.$$

The left side can now be replaced by $D_x(e^{-3x}y)$ to get

$$D_x(e^{-3x}y) = \frac{1}{2}e^{-2x}.$$

Integrating on both sides gives

$$e^{-3x}y = \int \frac{1}{2}e^{-2x}\,dx,$$

$$e^{-3x}y = \frac{1}{2}\left(\frac{e^{-2x}}{-2}\right) + C.$$

Now, multiply both sides by $e^{3x}$ to get

$$y = -\frac{e^x}{4} + Ce^{3x},$$

the general solution. Find the particular solution by substituting 0 for $x$ and 5 for $y$:

$$5 = -\frac{e^0}{4} + Ce^0 = -\frac{1}{4} + C$$

**YOUR TURN 2** Solve the initial value problem

$$\frac{dy}{dx} + 2xy - xe^{-x^2} = 0$$ with $y(0) = 3$.

or

$$\frac{21}{4} = C,$$

which leads to the particular solution

$$y = -\frac{e^x}{4} + \frac{21}{4}e^{3x}.$$

**TRY YOUR TURN 2**

**EXAMPLE 4** **Glucose**

Suppose glucose is infused into a patient's bloodstream at a constant rate of $a$ grams per minute. At the same time, glucose is removed from the bloodstream at a rate proportional to the amount of glucose present. Then the amount of glucose, $G(t)$, present at time $t$ satisfies

$$\frac{dG}{dt} = a - KG$$

for some constant $K$. Solve this equation for $G$. Does the glucose concentration eventually reach a constant? That is, what happens to $G$ as $t \to \infty$? *Source: Ordinary Differential Equations with Applications.*

**APPLY IT**

**SOLUTION** The equation can be written in the form of the linear first-order differential equation

$$\frac{dG}{dt} + KG = a. \tag{6}$$

The integrating factor is

$$I(t) = e^{\int K dt} = e^{Kt}.$$

Multiply both sides of Equation (6) by $I(t) = e^{Kt}$.

$$e^{Kt}\frac{dG}{dt} + Ke^{Kt}G = ae^{Kt}$$

Write the left side as $D_t(e^{Kt}G)$ and solve for $G$ by integrating on both sides.

$$D_t(e^{Kt}G) = ae^{Kt}$$

$$e^{Kt}G = \int ae^{Kt}dt$$

$$e^{Kt}G = \frac{a}{K}e^{Kt} + C$$

Multiply both sides by $e^{-Kt}$ to get

$$G = \frac{a}{K} + Ce^{-Kt}.$$

As $t \to \infty$,

**NOTE**
The equation in Example 4 can also be solved by separation of variables. You are asked to do this in Exercise 24.

$$\lim_{t\to\infty} G = \lim_{t\to\infty}\left(\frac{a}{K} + Ce^{-Kt}\right) = \lim_{t\to\infty}\left(\frac{a}{K} + \frac{C}{e^{Kt}}\right) = \frac{a}{K}.$$

Thus, the glucose concentration stabilizes at $a/K$.

# 11.2 EXERCISES

**Find the general solution for each differential equation.**

**1.** $\dfrac{dy}{dx} + 3y = 6$

**2.** $\dfrac{dy}{dx} + 5y = 12$

**9.** $x\dfrac{dy}{dx} + 2y = x^2 + 6x, \quad x > 0$

**3.** $\dfrac{dy}{dx} + 2xy = 4x$

**4.** $\dfrac{dy}{dx} + 4xy = 4x$

**10.** $x^3\dfrac{dy}{dx} - x^2y = x^4 - 4x^3, \quad x > 0$

**5.** $x\dfrac{dy}{dx} - y - x = 0, \quad x > 0$

**6.** $x\dfrac{dy}{dx} + 2xy - x^2 = 0$

**11.** $y - x\dfrac{dy}{dx} = x^3, \quad x > 0$

**12.** $2xy + x^3 = x\dfrac{dy}{dx}$

**7.** $2\dfrac{dy}{dx} - 2xy - x = 0$

**8.** $3\dfrac{dy}{dx} + 6xy + x = 0$

**13.** $\dfrac{dy}{dx} + y\cot x = x$

**14.** $\dfrac{dy}{dx} + y\tan x = \cos^2 x$

**Solve each differential equation, subject to the given initial condition.**

15. $\dfrac{dy}{dx} + y = 4e^x$;   $y(0) = 50$

16. $\dfrac{dy}{dx} + 4y = 9e^{5x}$;   $y(0) = 25$

17. $\dfrac{dy}{dx} - 2xy - 4x = 0$;   $y(1) = 20$

18. $x\dfrac{dy}{dx} - 3y + 2 = 0$;   $y(1) = 8$

19. $x\dfrac{dy}{dx} + 5y = x^2$;   $y(2) = 12$

20. $2\dfrac{dy}{dx} - 4xy = 5x$;   $y(1) = 10$

21. $x\dfrac{dy}{dx} + (1 + x)y = 3$;   $y(4) = 50$

22. $\dfrac{dy}{dx} + 3x^2y - 2xe^{-x^3} = 0$;   $y(0) = 1000$

## LIFE SCIENCE APPLICATIONS

23. **Population Growth**  The logistic equation introduced in Section 1,

$$\frac{dy}{dx} = k\left(1 - \frac{y}{N}\right)y \qquad (7)$$

can be written as

$$\frac{dy}{dx} = cy - py^2, \qquad (8)$$

where $c$ and $p$ are positive constants. Although this is a nonlinear differential equation, it can be reduced to a linear equation by a suitable substitution for the variable $y$.

    **a.** Letting $y = 1/z$ and $dy/dx = (-1/z^2)dz/dx$, rewrite Equation (8) in terms of $z$. Solve for $z$ and then for $y$.

    **b.** Let $z(0) = 1/y_0$ in part a and find a particular solution for $y$.

    **c.** Find the limit of $y$ as $x \to \infty$. This is the saturation level of the population.

24. **Glucose Level**  Solve the glucose level example (Example 4) using separation of variables.

25. **Excitable Cells**  The Hodgkin-Huxley model for excitable nerve cells is a set of four differential equations, three of the form

$$\frac{dy}{dt} = \alpha(1 - y) - \beta y,$$

where $y$ is a variable between 0 and 1 related to potassium channel activation, sodium channel activation, or sodium channel inactivation in the cells, and $\alpha$ and $\beta$ are rate constants. *Source: The Journal of Physiology.*

    **a.** Solve this equation with the initial condition $y(0) = y_0$.

    **b.** What is the limit of the solution from part a as $t$ goes to infinity?

26. **Drug Use**  The rate of change in the concentration of a drug with respect to time in a user's blood is given by

$$\frac{dC}{dt} = -kC + D(t),$$

where $D(t)$ is dosage at time $t$ and $k$ is the rate that the drug leaves the bloodstream. *Source: Mathematical Biosciences.*

    **a.** Solve this linear equation to show that, if $C(0) = 0$, then

$$C(t) = e^{-kt}\int_0^t e^{ky}D(y)\,dy.$$

    (*Hint:* To integrate both sides of the equation in Step 5 of "Solving a Linear First-Order Differential Equation," integrate from 0 to $t$, and change the variable of integration to $y$.)

    **b.** Show that if $D(y)$ is a constant $D$, then

$$C(t) = \frac{D(1 - e^{-kt})}{k}.$$

27. **Mouse Infection**  A model for the spread of an infectious disease among mice is

$$\frac{dN}{dt} = rN - \frac{\alpha r(\alpha + b + v)}{\beta\left[\alpha - r\left(1 + \dfrac{v}{b + \gamma}\right)\right]},$$

where $N$ is the size of the population of mice, $\alpha$ is the mortality rate due to infection, $b$ is the mortality rate due to natural causes for infected mice, $\beta$ is a transmission coefficient for the rate that infected mice infect susceptible mice, $v$ is the rate the mice recover from infection, and $\gamma$ is the rate that mice lose immunity. *Source: Lectures on Mathematics in the Life Sciences.* Show that the solution to this equation, with the initial condition $N(0) = (\alpha + b + v)/\beta$, can be written as

$$N(t) = \frac{(\alpha + b + v)}{\beta R}[(R - \alpha)e^{rt} + \alpha],$$

where

$$R = \alpha - r\left(1 + \frac{v}{b + \gamma}\right).$$

28. **Arterial Pulse**  During the diastolic phase of a heartbeat, the heart fills with blood, and no blood flows from the heart to the aorta. Assume that the pressure $P$ in the aorta is proportional to the volume $V$, with constant of proportionality $k$. *Source: An Introduction to the Mathematics of Medicine and Biology.* Assume further that the rate of change of the volume with respect to time $t$ is proportional to the pressure, with constant of proportionality $-1/\omega$.

    **a.** Write down the differential equation describing how the pressure $P$ changes with respect to time, and solve this differential equation with the condition that $P = P_0$ at time $t = 0$.

    **b.** During the systolic phase, blood from the heart enters the aorta. The rate that pressure changes with respect to time, described by the differential equation in part a, now has an added term due to the blood entering the aorta. Suppose this term can be described by $A \sin Bt$ for positive constants $A$ and $B$. Solve this differential equation. (*Hint:* Use Formula 37 in the Table of Integrals in the Appendix.)

## OTHER APPLICATIONS

**Immigration and Emigration** If population is changed either by immigration or emigration, the exponential growth model discussed in Section 1 is modified to

$$\frac{dy}{dt} = ky + f(t),$$

where $y$ is the population at time $t$ and $f(t)$ is some (other) function of $t$ that describes the net effect of the emigration/immigration. Assume $k = 0.02$ and $y(0) = 10,000$. Solve this differential equation for $y$, given the following functions $f(t)$.

**29.** $f(t) = e^t$

**30.** $f(t) = e^{-t}$

**31.** $f(t) = -t$

**32.** $f(t) = t$

**33. Newton's Law of Cooling** In Exercises 58–61 in the previous section, we saw that Newton's law of cooling states that the

rate of change of the temperature of an object is proportional to the difference in temperature between the object and the surrounding medium. This leads to the differential equation

$$\frac{dT}{dt} = -k(T - T_M),$$

where $T$ is the temperature of the object after time $t$, $T_M$ is the temperature of the surrounding medium, and $k$ is a constant. In the previous section, we solved this equation by separation of variables. Show that this equation is also linear, and find the solution by the method of this section.

**YOUR TURN ANSWERS**

**1.** $y = xe^x + Cx$

**2.** $y = (x^2 + 6)/(2e^{x^2})$

# 11.3  Euler's Method

**APPLY IT** How can we predict the weight of a goat at 5 weeks, given its weight at birth and a formula for its growth rate?

*This question will be answered in Exercise 36 at the end of this section.*

**FOR REVIEW**

In Section 6.5 on Differentials: Linear Approximation, we defined $\Delta y$ to be the actual change in $y$ as $x$ changed by an amount $\Delta x$:

$$\Delta y = f(x + \Delta x) - f(x).$$

The differential $dy$ is an approximation to $\Delta y$. We find $dy$ by following the tangent line from the point $(x, f(x))$, rather than by following the actual function. Then $dy$ is found by using the formula $dy = (dy/dx)\, dx$ where $dx = \Delta x$. For example, let $f(x) = x^3$, $x = 1$, and $dx = \Delta x = 0.2$. Then $dy = f'(x)\, dx = 3x^2\, dx = 3(1^2)(0.2) = 0.6$. The actual change in $y$ as $x$ changes from 1 to 1.2 is

$$f(x + \Delta x) - f(x)$$
$$= f(1.2) - f(1)$$
$$= 1.2^3 - 1$$
$$= 0.728.$$

Applications sometimes involve differential equations such as

$$\frac{dy}{dx} = \frac{x + y}{y}$$

that cannot be solved by the methods discussed so far, but approximate solutions to these equations often can be found by numerical methods. For many applications, these approximations are quite adequate. In this section we introduce Euler's method, which is only one of numerous mathematical contributions made by Leonhard Euler (1707–1783) of Switzerland. (His name is pronounced "oiler.") He also introduced the $f(x)$ notation used throughout this text. Despite becoming blind during his later years, he was the most prolific mathematician of his era. In fact it took nearly 50 years after his death to publish the works he created in all mathematical fields during his lifetime.

**Euler's method** of solving differential equations gives approximate solutions to differential equations involving $y = f(x)$ where the initial values of $x$ and $y$ are known—that is, equations of the form

$$\frac{dy}{dx} = g(x, y), \quad \text{with} \quad y(x_0) = y_0.$$

Geometrically, Euler's method approximates the graph of the solution $y = f(x)$ with a polygonal line whose first segment is tangent to the curve at the point $(x_0, y_0)$, as shown in Figure 7 on the next page.

To use Euler's method, divide the interval from $x_0$ to another point $x_n$ into $n$ subintervals of equal width (see Figure 7). The width of each subinterval is $h = (x_n - x_0)/n$.

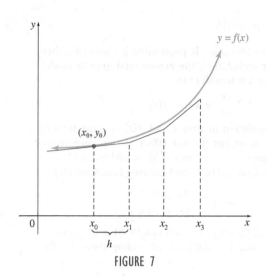

**FIGURE 7**

Recall from Section 6.5 on Differentials: Linear Approximation that if $\Delta x$ is a small change in $x$, then the corresponding small change in $y$, $\Delta y$, is approximated by

$$\Delta y \approx dy = \frac{dy}{dx} \cdot \Delta x.$$

The differential $dy$ is the change in $y$ along the tangent line. On the interval from $x_i$ to $x_{i+1}$, note that $dy$ is just $y_{i+1} - y_i$, where $y_i$ is the approximate solution at $x_i$. We also have $dy/dx = g(x_i, y_i)$ and $\Delta x = h$. Putting these into the previous equation yields

$$y_{i+1} - y_i = g(x_i, y_i)h$$
$$y_{i+1} = y_i + g(x_i, y_i)h.$$

Because $y_0$ is given, we can use the equation just derived with $i = 0$ to get $y_1$. We can then use $y_1$ and the same equation with $i = 1$ to get $y_2$ and continue in this manner until we get $y_n$. A summary of Euler's method follows.

**Euler's Method**

Let $y = f(x)$ be the solution of the differential equation

$$\frac{dy}{dx} = g(x, y), \quad \text{with} \quad y(x_0) = y_0,$$

for $x_0 \leq x \leq x_n$. Let $x_{i+1} = x_i + h$, where $h = (x_n - x_0)/n$ and

$$y_{i+1} = y_i + g(x_i, y_i)h,$$

for $0 \leq i \leq n - 1$. Then

$$f(x_{i+1}) \approx y_{i+1}.$$

As the following examples will show, the accuracy of the approximation varies for different functions. As $h$ gets smaller, however, the approximation improves, although making $h$ too small can make things worse. (See the discussion at the end of this section.) Euler's method is not difficult to program; it then becomes possible to try smaller and smaller values of $h$ to get better and better approximations. Graphing calculator programs for Euler's method are included in the *Graphing Calculator and Excel Spreadsheet Manual* available with this book.

### EXAMPLE 1  Euler's Method

Use Euler's method to approximate the solution of $dy/dx + 2xy = x$, with $y(0) = 1.5$, for $[0, 1]$. Use $h = 0.1$.

**Method 1**
**Calculating by Hand**

**SOLUTION**

The general solution of this equation was found in Example 2 of the last section, so the results using Euler's method can be compared with the actual solution. Begin by writing the differential equation in the required form as

$$\frac{dy}{dx} = x - 2xy, \qquad \text{so that} \qquad g(x, y) = x - 2xy.$$

Since $x_0 = 0$ and $y_0 = 1.5$,

$$g(x_0, y_0) = 0 - 2(0)(1.5) = 0,$$

and

$$y_1 = y_0 + g(x_0, y_0)h = 1.5 + 0(0.1) = 1.5.$$

Now $x_1 = 0.1$, $y_1 = 1.5$, and $g(x_1, y_1) = 0.1 - 2(0.1)(1.5) = -0.2$. Then

$$y_2 = 1.5 + (-0.2)(0.1) = 1.48.$$

The 11 values for $x_i$ and $y_i$ for $0 \le i \le 10$ are shown in the table below, together with the actual values using the result from Example 2 in the last section. (Since the result was only a general solution, replace $x$ with 0 and $y$ with 1.5 to get the particular solution $y = 1/2 + e^{-x^2}$.)

The results in the table look quite good. The graphs in Figure 8 show that the polygonal line follows the actual graph of $f(x)$ quite closely.

| | Approximate Solution Using $h = 0.1$ | | |
|---|---|---|---|
| $x_i$ | Euler's Method $y_i$ | Actual Solution $f(x_i)$ | Difference $y_i - f(x_i)$ |
| 0 | 1.5 | 1.5 | 0 |
| 0.1 | 1.5 | 1.49004983 | 0.0099502 |
| 0.2 | 1.48 | 1.46078944 | 0.0192106 |
| 0.3 | 1.4408 | 1.41393119 | 0.0268688 |
| 0.4 | 1.384352 | 1.35214379 | 0.0322082 |
| 0.5 | 1.31360384 | 1.27880078 | 0.0348031 |
| 0.6 | 1.23224346 | 1.19767633 | 0.0345671 |
| 0.7 | 1.14437424 | 1.11262639 | 0.0317478 |
| 0.8 | 1.05416185 | 1.02729242 | 0.0268694 |
| 0.9 | 0.96549595 | 0.94485807 | 0.0206379 |
| 1.0 | 0.88170668 | 0.86787944 | 0.0138272 |

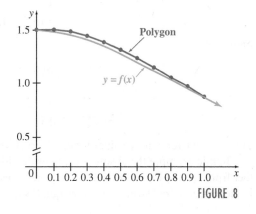
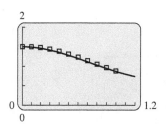

FIGURE 8

**Method 2**
**Graphing Calculator**

A graphing calculator can readily implement Euler's method. To do this example on a TI-84 Plus, start $x$ with the value of $x_0 - h$ by storing $-0.1$ in X, store $y_0$, or 1.5, in Y, and put $X - 2X*Y$ into the function $Y_1$. Then the command $X+.1 \to X:Y+Y_1*.1 \to Y$ gives the next value of $y$, which is still 1.5. Continue to press the ENTER key to get subsequent values of $y$. For more details, see the *Graphing Calculator and Excel Spreadsheet Manual* available with this book.

**Method 3**
**Spreadsheet**

Euler's method can also be performed on a spreadsheet. In Microsoft Excel, for example, store the values of $x$ in column A and the initial value of $y$ in B1. Then put the command "=B1+(A1-2*A1*B1)*.1" into B2 to get the next value of $y$, using the formula for $g(x, y)$ in this example. Copy this formula into the rest of column B to get subsequent values of $y$. For more details, see the *Graphing Calculator and Excel Spreadsheet Manual* available with this book.

**YOUR TURN 1** Use Euler's method to approximate the solution of $dy/dx - x^2y^2 = 1$, with $y(0) = 2$, for $[0,1]$. Use $h = 0.2$.

**TRY YOUR TURN 1**

Euler's method produces a very good approximation for this differential equation because the slope of the solution $f(x)$ is not steep in the interval under investigation. The next example shows that such good results cannot always be expected.

## EXAMPLE 2  Euler's Method

Use Euler's method to solve $dy/dx = 3y + (1/2)e^x$, with $y(0) = 5$, for $[0, 1]$, using 10 subintervals.

**SOLUTION**  This is the differential equation of Example 3 in the last section. The general solution found there, with the initial condition given above, leads to the particular solution

$$y = -\frac{1}{4}e^x + \frac{21}{4}e^{3x}.$$

To solve by Euler's method, start with $g(x, y) = 3y + (1/2)e^x$, $x_0 = 0$, and $y_0 = 5$. For $n = 10$, $h = (1 - 0)/10 = 0.1$ again, and

$$y_{i+1} = y_i + g(x_i, y_i)h = y_i + \left(3y_i + \frac{1}{2}e^{x_i}\right)h.$$

For $y_1$, this gives

$$y_1 = y_0 + \left(3y_0 + \frac{1}{2}e^{x_0}\right)h$$

$$= 5 + \left[3(5) + \frac{1}{2}(e^0)\right](0.1)$$

$$= 5 + (15.5)(0.1) = 6.55.$$

Similarly,

$$y_2 = 6.55 + \left[3(6.55) + \frac{1}{2}e^{0.1}\right](0.1)$$

$$= 6.55 + (19.65 + 0.55258546)(0.1)$$

$$= 8.57025855.$$

These and the remaining values for the interval $[0, 1]$ are shown in the table on the next page.

In this example the absolute value of the differences grows very rapidly as $x_i$ gets farther from $x_0$. See Figure 9 on the next page. These large differences come from the term $e^{3x}$ in the solution; this term grows very quickly as $x$ increases.

| | Approximate Solution Using $h = 0.1$ | | |
|---|---|---|---|
| $x_i$ | Euler's Method $y_i$ | Actual Solution $f(x_i)$ | Difference $y_i - f(x_i)$ |
| 0 | 5 | 5 | 0 |
| 0.1 | 6.55 | 6.810466 | −0.260466 |
| 0.2 | 8.570259 | 9.260773 | −0.690514 |
| 0.3 | 11.202406 | 12.575452 | −1.373045 |
| 0.4 | 14.630621 | 17.057658 | −2.427037 |
| 0.5 | 19.094399 | 23.116687 | −4.022289 |
| 0.6 | 24.905154 | 31.305119 | −6.399965 |
| 0.7 | 32.467806 | 42.368954 | −9.901147 |
| 0.8 | 42.308836 | 57.315291 | −15.006455 |
| 0.9 | 55.112764 | 77.503691 | −22.390927 |
| 1 | 71.769573 | 104.769498 | −32.999925 |

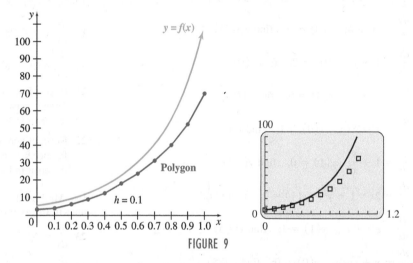

FIGURE 9

As these examples show, numerical methods may produce large errors. The error often can be reduced by using more subintervals of smaller width—letting $n = 100$ or 1000, for example. Approximations for the function in Example 2 with $n = 100$ and $h = (1 - 0)/100 = 0.01$ are shown in the table below. The approximations are considerably improved.

| | Approximate Solution Using $h = 0.01$ | | |
|---|---|---|---|
| $x_i$ | Euler's Method $y_i$ | Actual Solution $f(x_i)$ | Difference $y_i - f(x_i)$ |
| 0 | 5 | 5 | 0 |
| 0.1 | 6.779418 | 6.810466 | −0.031048 |
| 0.2 | 9.177101 | 9.260773 | −0.083672 |
| 0.3 | 12.406341 | 12.575452 | −0.169111 |
| 0.4 | 16.753855 | 17.057658 | −0.303803 |
| 0.5 | 22.605046 | 23.116687 | −0.511642 |
| 0.6 | 30.477945 | 31.305119 | −0.827175 |
| 0.7 | 41.068839 | 42.368954 | −1.300115 |
| 0.8 | 55.313581 | 57.315291 | −2.001710 |
| 0.9 | 74.469995 | 77.503691 | −3.033695 |
| 1 | 100.228621 | 104.769498 | −4.540878 |

We could improve the accuracy of Euler's method by using a smaller $h$, but there are two difficulties. First, this requires more calculations and, consequently, more time. Such calculations are usually done by computer, so the increased time may not matter. But this introduces a second difficulty: The increased number of calculations causes more round-off error, so there is a limit to how small we can make $h$ and still get improvement. The preferred way to get greater accuracy is to use a more sophisticated procedure, such as the Runge-Kutta method. Such methods are beyond the scope of this book but are discussed in numerical analysis and differential equations courses.*

# 11.3 EXERCISES

**Use Euler's method to approximate the indicated function value to 3 decimal places, using $h = 0.1$.**

**1.** $\dfrac{dy}{dx} = x^2 + y^2$; $y(0) = 2$; find $y(0.5)$

**2.** $\dfrac{dy}{dx} = xy + 4$; $y(0) = 0$; find $y(0.5)$

**3.** $\dfrac{dy}{dx} = 1 + y$; $y(0) = 2$; find $y(0.6)$

**4.** $\dfrac{dy}{dx} = x + y^2$; $y(0) = 0$; find $y(0.6)$

**5.** $\dfrac{dy}{dx} = x + \sqrt{y}$; $y(0) = 1$; find $y(0.4)$

**6.** $\dfrac{dy}{dx} = 1 + \dfrac{y}{x}$; $y(1) = 0$; find $y(1.4)$

**7.** $\dfrac{dy}{dx} = 2x\sqrt{1 + y^2}$; $y(1) = 2$; find $y(1.5)$

**8.** $\dfrac{dy}{dx} = e^{-y} + e^x$; $y(1) = 1$; find $y(1.5)$

**9.** $\dfrac{dy}{dx} = y + \cos x$; $y(0) = 0$; find $y(0.5)$

**10.** $\dfrac{dy}{dx} = \sin(x + y)$; $y(1) = 3$; find $y(1.5)$

**Use Euler's method to approximate the indicated function value to 3 decimal places, using $h = 0.1$. Next, solve the differential equation and find the indicated function value to 3 decimal places. Compare the result with the approximation.**

**11.** $\dfrac{dy}{dx} = -4 + x$; $y(0) = 1$; find $y(0.4)$

**12.** $\dfrac{dy}{dx} = 4x + 3$; $y(1) = 0$; find $y(1.5)$

**13.** $\dfrac{dy}{dx} = x^3$; $y(0) = 4$; find $y(0.5)$

**14.** $\dfrac{dy}{dx} = \dfrac{3}{x}$; $y(1) = 2$; find $y(1.4)$

**15.** $\dfrac{dy}{dx} = 2xy$; $y(1) = 1$; find $y(1.6)$

**16.** $\dfrac{dy}{dx} = x^2 y$; $y(0) = 1$; find $y(0.6)$

**17.** $\dfrac{dy}{dx} = ye^x$; $y(0) = 2$; find $y(0.4)$

**18.** $\dfrac{dy}{dx} = \dfrac{2x}{y}$; $y(0) = 3$; find $y(0.6)$

**19.** $\dfrac{dy}{dx} + y = 2e^x$; $y(0) = 100$; find $y(0.3)$

**20.** $\dfrac{dy}{dx} - 2y = e^{2x}$; $y(0) = 10$; find $y(0.4)$

**21.** Repeat Exercise 17 with $h = 0.05$. How does the error (the difference between this answer and the true answer) compare with the error in Exercise 17?

**22.** Repeat Exercise 18 with $h = 0.05$. How does the error (the difference between this answer and the true answer) compare with the error in Exercise 18?

**Use Method 2 or 3 in Example 1 to construct a table like the ones in the examples for $0 \le x \le 1$, with $h = 0.2$.**

**23.** $\dfrac{dy}{dx} = \sqrt[3]{x}$; $y(0) = 0$      **24.** $\dfrac{dy}{dx} = y$; $y(0) = 1$

**25.** $\dfrac{dy}{dx} = 4 - y$; $y(0) = 0$      **26.** $\dfrac{dy}{dx} = x - 2xy$; $y(0) = 1$

**Solve each differential equation and graph the function $y = f(x)$ and the polygonal approximation on the same axes. (The approximations were found in Exercises 23–26.)**

**27.** $\dfrac{dy}{dx} = \sqrt[3]{x}$; $y(0) = 0$      **28.** $\dfrac{dy}{dx} = y$; $y(0) = 1$

**29.** $\dfrac{dy}{dx} = 4 - y$; $y(0) = 0$      **30.** $\dfrac{dy}{dx} = x - 2xy$; $y(0) = 1$

**31. a.** Use Euler's method with $h = 0.2$ to approximate $f(1)$, where $f(x)$ is the solution to the differential equation

$$\frac{dy}{dx} = y^2; \quad y(0) = 1.$$

**b.** Solve the differential equation in part a using separation of variables, and discuss what happens to $f(x)$ as $x$ approaches 1.

**c.** Based on what you learned from parts a and b, discuss what might go wrong when using Euler's method. (More advanced courses on differential equations discuss the question of whether a differential equation has a solution for a given interval in $x$.)

*For example, see Nagle, R. K., E. B. Saff, and A. D. Snider, *Fundamentals of Differential Equations*, 8th ed., Pearson, 2013.

## LIFE SCIENCE APPLICATIONS

**Solve Exercises 32–38 using Euler's method.**

**32. Growth of Algae** The phosphate compounds found in many detergents are highly water soluble and are excellent fertilizers for algae. Assume that there are 5000 algae present at time $t = 0$ and conditions will support at most 500,000 algae. Assume that the rate of growth of algae, in the presence of sufficient phosphates, is proportional both to the number present (in thousands) and to the difference between 500,000 and the number present (in thousands).

  **a.** Write a differential equation using the given information. Use 0.01 for the constant of proportionality.

  **b.** Approximate the number present when $t = 2$, using $h = 0.5$.

**33. Immigration** An island is colonized by immigration from the mainland, where there are 100 species. Let the number of species on the island at time $t$ (in years) equal $y$, where $y = f(t)$. Suppose the rate at which new species immigrate to the island is

$$\frac{dy}{dt} = 0.02(100 - y^{1/2}).$$

Use Euler's method with $h = 0.5$ to approximate $y$ when $t = 5$ if there were 10 species initially.

**34. Insect Population** A population of insects, $y$, living in a circular colony grows at a rate

$$\frac{dy}{dt} = 0.05y - 0.1y^{1/2},$$

where $t$ is time in weeks. If there were 60 insects initially, use Euler's method with $h = 1$ to approximate the number of insects after 6 weeks.

**35. Whale Population** Under certain conditions a population may exhibit a polynomial growth rate function. A population of blue whales is growing according to the function

$$\frac{dy}{dt} = -y + 0.02y^2 + 0.003y^3.$$

Here $y$ is the population in thousands and $t$ is measured in years. Use Euler's method with $h = 1$ to approximate the population in 4 years if the initial population is 15,000.

**36. APPLY IT Goat Growth** The growth of male Saanen goats can be approximated by the equation

$$\frac{dW}{dt} = -0.01189W + 0.92389W^{0.016},$$

where $W$ is the weight (in kilograms) after $t$ weeks. **Source: Annales de Zootechnie.** Find the weight of a goat at 5 weeks, given that the weight at birth is 3.65 kg. Use Euler's method with $h = 1$.

## OTHER APPLICATIONS

**37. Learning** In an early article describing how people learn, the rate of change of the probability that a person performs a task correctly ($p$) with respect to time ($t$) is given by

$$\frac{dp}{dt} = \frac{2k}{\sqrt{m}}(p - p^2)^{3/2},$$

where $k$ and $m$ are constants related to the rate that the person learns the task. **Source: The Journal of General Psychology.** For this exercise, let $m = 4$ and $k = 0.5$.

  **a.** Letting $p = 0.1$ when $t = 0$, use Method 2 or 3 in Example 1 to construct a table for $t_i$ and $p_i$ like the ones in the examples for $0 \le x \le 30$, with $h = 5$.

  **b.** Based on your answer to part a, what does $p$ seem to approach as $t$ increases? Explain why this answer makes sense.

**38. Spread of Rumors** A rumor spreads through a community of 500 people at the rate

$$\frac{dN}{dt} = 0.02(500 - N)N^{1/2},$$

where $N$ is the number of people who have heard the rumor at time $t$ (in hours). Use Euler's method with $h = 0.5$ to find the number who have heard the rumor after 3 hours, if only 2 people heard it initially.

**YOUR TURN ANSWER**

1.

| $x_i$ | $y_i$ |
|---|---|
| 0 | 2 |
| 0.2 | 2.2 |
| 0.4 | 2.43872 |
| 0.6 | 2.82903537 |
| 0.8 | 3.60528313 |
| 1.0 | 5.46903563 |

# 11.4 Linear Systems of Differential Equations

**APPLY IT** How can we describe the way a drug flows from one part of the body to another?

*We will answer this question using a linear system of differential equations.*

In the previous three sections, we have studied differential equations involving one function of one variable. Some of the most important applications of calculus to the life sciences, however, involve more than one function of one variable. In an area of mathematical physiology known as *compartmental analysis*, different organs of the body are considered as different compartments, with material traveling from one compartment to another.* Figure 10 illustrates a simplified version of such a system involving two components.

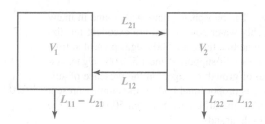

FIGURE 10

For the system in Figure 10, let $V_1$ and $V_2$ be the volumes of the two compartments, and let $x_1$ and $x_2$ be the concentration of the material under consideration (such as a pharmaceutical drug) in each of the two compartments, so that $V_i x_i$ represents the amount of the substance in compartment $i$ for $i = 1$ or 2. The rate at which the substance flows from one compartment to another depends on the amount of the substance in each compartment as well as the fractional rate at which the substance leaves and enters a compartment. To describe these rates, which vary from one compartment to another, we create the matrix

$$L = \begin{bmatrix} -L_{11} & L_{12} \\ L_{21} & -L_{22} \end{bmatrix},$$

where $L_{ij}$ is the fractional rate at which the substance flows from compartment $i$ to compartment $j$ (for $i = 1$ and $j = 2$, or $i = 2$ and $j = 1$), and $L_{ii}$ is the fractional rate at which the substance leaves compartment $i$ (for $i = 1$ or 2). Some of the substance will move to the other compartment (at a rate $L_{ji}$) and the remainder will move outside the system (at a rate $L_{ii} - L_{ji}$). For example, if

$$L = \begin{bmatrix} -2 & 1 \\ 2 & -3 \end{bmatrix},$$

then

$$\frac{d(V_1 x_1)}{dt} = -2(V_1 x_1) + V_2 x_2,$$

where $V_1 x_1$ is the amount of the substance in compartment 1, and $V_2 x_2$ is the amount in compartment 2.

If we define

$$V = \begin{bmatrix} V_1 & 0 \\ 0 & V_2 \end{bmatrix} \quad \text{and} \quad X = \begin{bmatrix} x_1 \\ x_2 \end{bmatrix},$$

then the equations describing how the amount of the drug changes in the compartments over time can be written as

$$\frac{d(VX)}{dt} = LVX.$$

We can simplify further by noting that

$$\frac{d(VX)}{dt} = V \frac{dX}{dt},$$

*For an example of the type of problem solved in compartmental analysis, see Greenwell, Raymond N., "Generalization of Results on Almost Closed Compartment Systems," *Mathematical Biosciences,* Vol. 46, 1979, pp. 125–129.

because $V$ is a matrix of constants. Also, because $V$ is a diagonal matrix, we know $V^{-1}$ exists. Multiplying both sides of the equation above by $V^{-1}$ gives

$$\frac{dX}{dt} = V^{-1}LVX$$

$$= MX,$$

where

$$M = V^{-1}LV.$$

For example, suppose $V_1 = 5$ ml and $V_2 = 10$ ml. Then

$$V = \begin{bmatrix} 5 & 0 \\ 0 & 10 \end{bmatrix}.$$

Verify that

$$V^{-1} = \begin{bmatrix} 1/5 & 0 \\ 0 & 1/10 \end{bmatrix}$$

and that using the matrix $L$ given above yields

$$M = V^{-1}LV$$

$$= \begin{bmatrix} 1/5 & 0 \\ 0 & 1/10 \end{bmatrix} \begin{bmatrix} -2 & 1 \\ 2 & -3 \end{bmatrix} \begin{bmatrix} 5 & 0 \\ 0 & 10 \end{bmatrix}$$

$$= \begin{bmatrix} -2 & 2 \\ 1 & -3 \end{bmatrix}.$$

Our original problem boils down to solving

$$\frac{dX}{dt} = MX.$$

This equation represents the **linear system of differential equations**

$$\frac{dx_1}{dt} = M_{11}x_1 + M_{12}x_2$$

$$\frac{dx_2}{dt} = M_{21}x_1 + M_{22}x_2.$$

This system is a generalization of the linear differential equations of Section 11.2 because it involves more than one differential equation. This particular system is simpler than many because the matrix $M$ is constant, and because we don't have another matrix added to the right-hand side, as in the equation

$$\frac{dX}{dt} = MX + Q,$$

where the entries in the matrix $Q$ might be functions of $t$. (Example 1 shows how to handle the case when the matrix $Q$ is present.) What complicates solving the system is that we can't solve the first equation for $x_1$ without knowing what $x_2$ is, and we can't solve the second equation for $x_2$ without knowing what $x_1$ is. The method for solving this system involves disentangling the equations using ideas from the previous chapter.

To begin, diagonalize the matrix $M$ by finding the eigenvalues and eigenvectors, as we did in the previous chapter. Recall that to find the eigenvalues, set $\det(M - \lambda I) = 0$.

$$M - \lambda I = \begin{bmatrix} -2 - \lambda & 2 \\ 1 & -3 - \lambda \end{bmatrix}$$

$$\det(M - \lambda I) = (-2 - \lambda)(-3 - \lambda) - 2(1)$$

$$= \lambda^2 + 5\lambda + 6 - 2$$

$$= \lambda^2 + 5\lambda + 4$$

Set this expression equal to 0 and solve.

$$0 = \lambda^2 + 5\lambda + 4$$
$$= (\lambda + 4)(\lambda + 1)$$
$$\lambda = -4 \quad \text{and} \quad \lambda = -1$$

We must next find the eigenvectors. For each value of $\lambda$, find a solution to the equation $(M - \lambda I)X = O$, where $O$ is the zero matrix. For $\lambda = -4$,

$$(M - \lambda I)X = \begin{bmatrix} 2 & 2 \\ 1 & 1 \end{bmatrix}\begin{bmatrix} x \\ y \end{bmatrix} = \begin{bmatrix} 0 \\ 0 \end{bmatrix}.$$

This system has an infinite number of solutions; let us arbitrarily choose the solution $\begin{bmatrix} 1 \\ -1 \end{bmatrix}$. For $\lambda = -1$,

$$(M - \lambda I)X = \begin{bmatrix} -1 & 2 \\ 1 & -2 \end{bmatrix}\begin{bmatrix} x \\ y \end{bmatrix} = \begin{bmatrix} 0 \\ 0 \end{bmatrix}.$$

This system also has an infinite number of solutions; let us arbitrarily choose the solution $\begin{bmatrix} 2 \\ 1 \end{bmatrix}$.

We next form the matrix $P$ whose columns are the eigenvectors:

$$P = \begin{bmatrix} 1 & 2 \\ -1 & 1 \end{bmatrix}.$$

Verify that

$$P^{-1} = \begin{bmatrix} 1/3 & -2/3 \\ 1/3 & 1/3 \end{bmatrix} \quad \text{and that} \quad P^{-1}MP = \begin{bmatrix} -4 & 0 \\ 0 & -1 \end{bmatrix}.$$

Notice that $P^{-1}MP$ is the diagonal matrix with the eigenvalues of $M$ on the diagonal.

**NOTE** Since there are an infinite number of choices for the eigenvectors, there are also an infinite number of choices for the matrix $P$.

Now let $X = PY$, so

$$\frac{dX}{dt} = MX$$

becomes

$$\frac{d(PY)}{dt} = MPY.$$

Using the fact that $P$ is a constant, so that $d(PY)/dt = PdY/dt$, and multiplying both sides by $P^{-1}$, we have

$$\frac{dY}{dt} = P^{-1}MPY,$$

or

$$\begin{bmatrix} dy_1/dt \\ dy_2/dt \end{bmatrix} = \begin{bmatrix} -4 & 0 \\ 0 & -1 \end{bmatrix}\begin{bmatrix} y_1 \\ y_2 \end{bmatrix}.$$

This yields two differential equations:

$$\frac{dy_1}{dt} = -4y_1 \quad \text{and} \quad \frac{dy_2}{dt} = -y_2,$$

each of which can be solved by separation of variables or with the use of an integrating factor, yielding

$$\begin{bmatrix} y_1 \\ y_2 \end{bmatrix} = \begin{bmatrix} C_1 e^{-4t} \\ C_2 e^{-t} \end{bmatrix}$$

where $C_1$ and $C_2$ are two arbitrary constants. Finally, calculate $X = PY$.

$$\begin{bmatrix} x_1 \\ x_2 \end{bmatrix} = \begin{bmatrix} 1 & 2 \\ -1 & 1 \end{bmatrix}\begin{bmatrix} C_1 e^{-4t} \\ C_2 e^{-t} \end{bmatrix}$$

$$= \begin{bmatrix} C_1 e^{-4t} + 2C_2 e^{-t} \\ -C_1 e^{-4t} + C_2 e^{-t} \end{bmatrix}$$

The values of $C_1$ and $C_2$ are determined by the initial conditions. For example, if $x_1(0) = 5$ and $x_2(0) = 1$, then

$$C_1 + 2C_2 = 5$$
$$-C_1 + C_2 = 1.$$

This system of equations can be solved by the Gauss-Jordan method, or in this case by simply adding the equations together. The result is $C_1 = 1$ and $C_2 = 2$. Therefore,

$$x_1 = e^{-4t} + 4e^{-t} \qquad \text{and} \qquad x_2 = -e^{-4t} + 2e^{-t}.$$

Let's now look at a case where $Q \neq 0$, and then summarize the steps.

## EXAMPLE 1   Receptor Dynamics

The following linear system of differential equations describes a model of how the body responds when a drug is introduced into the bloodstream:

$$\frac{dx_1}{dt} = -\frac{bc}{k}x_1 + \frac{a}{k}x_2 + g(t)$$

$$\frac{dx_2}{dt} = \frac{bc}{k}x_1 - \frac{a}{k}x_2.$$

Here $x_1$ is the number of free (or active) sites at time $t$; $x_2$ is the number of bound (or inactive) sites; $k$ measures the ratio of the binding rate time scale to the dosage time scale; $bcx_1/k$ gives the binding rate of active sites; $ax_2/k$ gives the rate at which bound sites are freed; and $g(t)$ gives the rate at which new active sites are produced in the body. *Source: Mathematical Biosciences.* The initial conditions are $x_1(0) = p$ and $x_2(0) = 0$. Suppose $a = 3, b = 2, c = 2, k = 1, p = 6$, and $g(t) = e^{-t}$. Solve the initial value problem.

**SOLUTION**   This system of differential equations can be represented in matrix form as

$$\begin{bmatrix} dx_1/dt \\ dx_2/dt \end{bmatrix} = \begin{bmatrix} -4 & 3 \\ 4 & -3 \end{bmatrix}\begin{bmatrix} x_1 \\ x_2 \end{bmatrix} + \begin{bmatrix} e^{-t} \\ 0 \end{bmatrix}.$$

Find the eigenvalues by setting $\det(M - \lambda I) = 0$.

$$M - \lambda I = \begin{bmatrix} -4 - \lambda & 3 \\ 4 & -3 - \lambda \end{bmatrix}$$

$$0 = \det(M - \lambda I) = (-4 - \lambda)(-3 - \lambda) - 3(4)$$

$$0 = \lambda^2 + 7\lambda$$

$$\lambda = 0 \qquad \text{and} \qquad \lambda = -7$$

For $\lambda = 0$,

$$(M - \lambda I)X = \begin{bmatrix} -4 & 3 \\ 4 & -3 \end{bmatrix}\begin{bmatrix} x \\ y \end{bmatrix} = \begin{bmatrix} 0 \\ 0 \end{bmatrix}.$$

One solution of this system is $\begin{bmatrix} 3 \\ 4 \end{bmatrix}$. For $\lambda = -7$,

$$(M - \lambda I)X = \begin{bmatrix} 3 & 3 \\ 4 & 4 \end{bmatrix}\begin{bmatrix} x \\ y \end{bmatrix} = \begin{bmatrix} 0 \\ 0 \end{bmatrix}.$$

One solution of this system is $\begin{bmatrix} 1 \\ -1 \end{bmatrix}$. The matrix of eigenvectors is then

$$P = \begin{bmatrix} 3 & 1 \\ 4 & -1 \end{bmatrix}$$

and we can calculate

$$P^{-1} = \begin{bmatrix} 1/7 & 1/7 \\ 4/7 & -3/7 \end{bmatrix} \quad \text{and} \quad P^{-1}MP = \begin{bmatrix} 0 & 0 \\ 0 & -7 \end{bmatrix}.$$

Now let $X = PY$ and multiply both sides of the differential equation by $P^{-1}$, yielding

$$\frac{dY}{dt} = P^{-1}MPY + P^{-1}\begin{bmatrix} e^{-t} \\ 0 \end{bmatrix},$$

or

$$\begin{bmatrix} dy_1/dt \\ dy_2/dt \end{bmatrix} = \begin{bmatrix} 0 & 0 \\ 0 & -7 \end{bmatrix}\begin{bmatrix} y_1 \\ y_2 \end{bmatrix} + \begin{bmatrix} e^{-t}/7 \\ 4e^{-t}/7 \end{bmatrix}.$$

This yields two differential equations:

$$\frac{dy_1}{dt} = \frac{e^{-t}}{7} \quad \text{and} \quad \frac{dy_2}{dt} = -7y_2 + \frac{4e^{-t}}{7}.$$

The first equation is integrated to yield

$$y_1 = -\frac{e^{-t}}{7} + C_1.$$

The second is a linear equation which can be solved by multiplying by the integrating factor $e^{\int 7\,dt} = e^{7t}$.

$$\frac{dy_2}{dt} + 7y_2 = \frac{4e^{-t}}{7}$$

$$e^{7t}\frac{dy_2}{dt} + 7e^{7t}y_2 = \frac{4e^{-t}}{7}e^{7t}$$

$$D_x\left(e^{7t}y_2\right) = \frac{4e^{6t}}{7}$$

$$e^{7t}y_2 = \int \frac{4e^{6t}}{7}\,dt = \frac{2e^{6t}}{21} + C_2$$

$$y_2 = \frac{2e^{-t}}{21} + C_2 e^{-7t}$$

Finally, calculate $X = PY$.

$$\begin{bmatrix} x_1 \\ x_2 \end{bmatrix} = \begin{bmatrix} 3 & 1 \\ 4 & -1 \end{bmatrix}\begin{bmatrix} -e^{-t}/7 + C_1 \\ 2e^{-t}/21 + C_2 e^{-7t} \end{bmatrix}$$

$$= \begin{bmatrix} -3e^{-t}/7 + 3C_1 + 2e^{-t}/21 + C_2 e^{-7t} \\ -4e^{-t}/7 + 4C_1 - 2e^{-t}/21 - C_2 e^{-7t} \end{bmatrix}$$

Setting $x_1(0) = 6$ and $x_2(0) = 0$ gives

$$-3/7 + 3C_1 + 2/21 + C_2 = 6$$
$$-4/7 + 4C_1 - 2/21 - C_2 = 0.$$

**YOUR TURN 1** Solve the system of differential equations.

$$\frac{dx_1}{dt} = 2x_1 - 3x_2 + 1 + 3e^t$$

$$\frac{dx_2}{dt} = x_1 - 2x_2 + 1 + e^t$$

Adding these two equations eliminates $C_2$, allowing us to solve for $C_1$. Verify that $C_1 = 1$ and $C_2 = 10/3$. Therefore,

$$x_1 = -\frac{3}{7}e^{-t} + 3 + \frac{2}{21}e^{-t} + \frac{10}{3}e^{-7t}$$

and

$$x_2 = -\frac{4}{7}e^{-t} + 4 - \frac{2}{21}e^{-t} - \frac{10}{3}e^{-7t}. \quad \text{TRY YOUR TURN 1}$$

## Solving a Linear System of Differential Equations

$\frac{dX}{dt} = MX + Q$ where $M$ is constant.

1. Find the eigenvalues of the matrix $M$ by solving the equation $\det(M - \lambda I) = 0$.

2. For each eigenvalue, find the corresponding eigenvector by solving the equation $(M - \lambda I)X = O$.

3. Create the matrix $P$ whose columns are the eigenvectors of $M$.

4. Solve the system $dY/dt = (P^{-1}MP)Y + P^{-1}Q$, where $P^{-1}MP$ is the diagonal matrix with the eigenvalues of $M$ on the diagonal. Each equation in this system is linear and involves only one of the functions of $Y$.

5. The solution to the original system is given by $X = PY$.

6. The solution at Step 5 will have arbitrary constants in it. For an initial value problem, substitute the values of the variables and solve the resulting system of linear equations to find the constants.

**NOTE** In this section, we have considered only cases in which the eigenvalues are distinct real numbers. In actual applications, the eigenvalues are often complex numbers. See Exercise 15 for an example of this.

**TECHNOLOGY NOTE**

A graphing calculator can be used to invert the matrix $P$. See Section 10.4 for details. The website WolframAlpha.com can be used to diagonalize a matrix. The command "diagonalize $[[-4, 3], [4, -3]]$" yields the matrices $P$, $P^{-1}$, and $P^{-1}MP$ in Example 1, labeled as $S$, $S^{-1}$, and $J$, and with the rows and columns in a different order.

# 11.4 EXERCISES

**For Exercises 1–8, solve the system of differential equations.**

**1.** $\dfrac{dx_1}{dt} = 3x_1 - 2x_2$

$\dfrac{dx_2}{dt} = x_1$

**2.** $\dfrac{dx_1}{dt} = x_1 + 6x_2$

$\dfrac{dx_2}{dt} = 5x_1 + 2x_2$

**3.** $\dfrac{dx_1}{dt} = 3x_1 + 2x_2$

$\dfrac{dx_2}{dt} = 3x_1 + 8x_2$

**4.** $\dfrac{dx_1}{dt} = 7x_1 + 3x_2$

$\dfrac{dx_2}{dt} = 3x_1 - x_2$

**5.** $\dfrac{dx_1}{dt} = 3x_1$

$\dfrac{dx_2}{dt} = 5x_1 + 2x_2$

$\dfrac{dx_3}{dt} = 2x_1 + 4x_2 - x_3$

**6.** $\dfrac{dx_1}{dt} = 2x_1$

$\dfrac{dx_2}{dt} = 3x_1 + 5x_2$

$\dfrac{dx_3}{dt} = 6x_1 + 5x_2 - 4x_3$

**7.** $\dfrac{dx_1}{dt} = 4x_1 - 2x_2 + e^t$

$\dfrac{dx_2}{dt} = -3x_1 + 9x_2 + e^{2t}$

**8.** $\dfrac{dx_1}{dt} = 2x_1 + 3x_2 + e^{2t}$

$\dfrac{dx_2}{dt} = 3x_1 - 6x_2 + e^{-4t}$

**For Exercises 9–14, find the particular solution for the initial value problem.**

**9.** Exercise 1, $x_1(0) = 7, x_2(0) = 5$

**10.** Exercise 2, $x_1(0) = 12, x_2(0) = 1$

**11.** Exercise 3, $x_1(0) = 3, x_2(0) = 23$

**12.** Exercise 4, $x_1(0) = 9, x_2(0) = 23$

**13.** Exercise 5, $x_1(0) = 6, x_2(0) = 15, x_3(0) = -1$

**14.** Exercise 6, $x_1(0) = -6, x_2(0) = -3, x_3(0) = 0$

## LIFE SCIENCE APPLICATIONS

**15. Population Modeling** An article describing the mathematical modeling of populations considers the example

$$\frac{dx_1}{dt} = x_2 - 2$$

$$\frac{dx_2}{dt} = -x_1 + 2.$$

*Source: Ecology.*

**a.** Show that the characteristic equation is $\lambda^2 + 1 = 0$.

**b.** The equation in part a has no real roots. Show, however, that if we let $i = \sqrt{-1}$, then the equation has two roots, $i$ and $-i$. Notice for later use that $1/i = -i$.

**c.** Show that $\begin{bmatrix} 1 \\ i \end{bmatrix}$ and $\begin{bmatrix} 1 \\ -i \end{bmatrix}$ are eigenvectors corresponding to the eigenvalues found in part b, so that $P = \begin{bmatrix} 1 & 1 \\ i & -i \end{bmatrix}$.

**d.** Verify that $P^{-1} = \begin{bmatrix} 1/2 & -i/2 \\ 1/2 & i/2 \end{bmatrix}$.

**e.** Show that the matrix $P$ in part c leads to the equations

$$\frac{dy_1}{dt} = iy_1 + (-1 - i)$$

$$\frac{dy_2}{dt} = -iy_2 + (-1 + i).$$

**f.** Show that the solutions to the equations in part e are

$$y_1 = C_1 e^{it} + (1 - i)$$

and

$$y_2 = C_2 e^{-it} + (1 + i).$$

**g.** Show that the solutions to the original equations are

$$x_1 = C_1 e^{it} + C_2 e^{-it} + 2$$

and

$$x_2 = iC_1 e^{it} - iC_2 e^{-it} + 2.$$

**h.** The article considered the initial condition $x_1(0) = 2$, $x_2(0) = 3$. Show that $C_1 = -i/2$ and $C_2 = i/2$.

**i.** Use the identity of Euler,

$$e^{it} = \cos t + i \sin t,$$

and its equivalent form

$$e^{-it} = \cos t - i \sin t,$$

to show that the solution to the initial value problem is given by

$$x_1 = \sin t + 2$$

and

$$x_2 = \cos t + 2.$$

Notice that the final answer has no complex numbers.

**j.** Why is the answer to part i biologically reasonable?

**16. Pollution** Compartmental models can be used to describe the dispersal of pollutants. Researchers have studied the model

$$\frac{dX}{dt} = FX,$$

where $x_i$ represents the amount of the pollutant in compartment $i$. The matrix $F$ is the *fate matrix*, in which the negative elements on the diagonal represent the rate at which the pollutant leaves each compartment, and the other terms represent the rate at which the pollutant is transferred from one compartment to another. *Source: Chemosphere.*

**a.** For the simple two-compartment model, let $F = \begin{bmatrix} L_1 & T_{21} \\ T_{12} & L_2 \end{bmatrix}$. Show that the eigenvalues are given by

$$\lambda = (L_1 + L_2 \pm \sqrt{(L_1 - L_2)^2 + 4T_{12}T_{21}})/2,$$

**b.** The researchers studied a seven-compartment model. For simplicity, we have reduced this to four compartments: air, plants, surface soil, and root-zone soil, for which

$$F = \begin{bmatrix} -1.7982 & 0.00191 & 4.8 \times 10^{-5} & 0 \\ 1.740748 & -0.00768 & 4.25 \times 10^{-17} & 5.44 \times 10^{-13} \\ 0.027803 & 0.005767 & -0.00044 & 2.57 \times 10^{-7} \\ 0 & 6.12 \times 10^{-12} & 5.19 \times 10^{-5} & -2.6 \times 10^{-7} \end{bmatrix}.$$

The matrix of eigenvectors is given by the matrix $P$ shown below. Verify that the columns of the matrix $P$ are the eigenvectors corresponding to the eigenvalues $-0.005873653661, -2.260432550 \times 10^{-7}, -1.80005664$, and $-3.897392958 \times 10^{-4}$ by multiplying $F$ by $P$, multiplying each eigenvalue by $P$, and verifying that $F$ times each eigenvector equals the corresponding eigenvalue times that eigenvector.

**c.** Write out the solution to $x_1$ in the linear system of differential equations corresponding to the matrix $F$ in part b.

**d.** According to the answer in part c, what happens to the amount of the pollutant in the air after a long time?

$$P = \begin{bmatrix} -0.001034833155 & -2.300721130 \times 10^{-6} & 0.0995311655 & -3.56625132 \times 10^{-4} \\ -0.9978255773 & -5.218032278 \times 10^{-4} & -0.09671974321 & -0.8520286615 \\ 1.064336125 & -0.06542733470 & -0.001227752248 & -9.966809274 \\ -0.009404958855 & -100.0001234 & 3.548283458 \times 10^{-8} & 1.328125544 \end{bmatrix}$$

**YOUR TURN ANSWER**

**1.** $x_1 = C_1 e^{-t} + 3C_2 e^t + 1 + 3te^t$, $x_2 = C_1 e^{-t} + C_2 e^t + 1 + te^t$

# 11.5 Nonlinear Systems of Differential Equations

**How do the populations of a predator and its prey change over time?**
*We will answer this question in Example 1 using a nonlinear system of differential equations.*

In the previous section we studied linear systems of differential equations. If all systems of differential equations were linear, the world would be a simpler but less interesting place. Most differential equation models in the life sciences are nonlinear. There are no general methods for finding the solution to nonlinear systems, but there are vast amounts of literature on methods for analyzing their behavior. In this section we will touch on a few basic ideas.

Consider the following model of the interaction between baleen whales and Antarctic krill described by a group of researchers in 1979. *Source: Science.* Krill tend to grow according to the logistic model described in Section 11.1, so if $x$ represents the population of krill at time $t$, $k_1$ is the growth rate, and $N$ is the carrying capacity, then

$$\frac{dx_1}{dt} = k_1\left(1 - \frac{x_1}{N}\right)x_1. \tag{1}$$

The population of krill is reduced by the whales, which eat the krill. If the population of whales is denoted by $x_2$, and if the rate at which the whales consume the krill is denoted by $c$, then

$$\frac{dx_1}{dt} = k_1\left(1 - \frac{x_1}{N}\right)x_1 - cx_1x_2. \tag{2}$$

A similar equation describes how the population of whales varies with time, but their carrying capacity is affected by how many krill there are to eat. We will therefore let the carrying capacity of the whales be $qx_1$, where $q$ is a constant describing how many krill are needed to sustain a whale. Thus

$$\frac{dx_2}{dt} = k_2\left(1 - \frac{x_2}{qx_1}\right)x_2. \tag{3}$$

Equations (2) and (3) form a nonlinear system of differential equations. We cannot explicitly solve this system for $x_1$ and $x_2$, even if we knew the values of the constants $k_1$, $k_2$, $c$, and $q$. Nevertheless, we will explore this system to see what conclusions we can make.

One point of interest is the **equilibrium point**, that is, the values of $x_1$ and $x_2$ at which $dx_1/dt = dx_2/dt = 0$. (This generalizes the idea of an equilibrium point for one differential equation, introduced in the first section of this chapter.) This means that if the populations ever reached these values, they would stay there forever, or at least until something changed. Let's start with Equation (3), since it's simpler than Equation (2). Setting $dx_2/dt = 0$ implies either $x_2 = 0$ (that is, there are no whales, an undesirable situation of little interest) or

$$1 - \frac{x_2}{qx_1} = 0,$$

from which we conclude that

$$x_2 = qx_1. \tag{4}$$

Now set $dx_1/dt = 0$ in Equation (2) and substitute the value for $x_2$ from Equation (4).

$$0 = k_1\left(1 - \frac{x_1}{N}\right)x_1 - cx_1(qx_1)$$

$$= x_1\left[k_1\left(1 - \frac{x_1}{N}\right) - cqx_1\right] \qquad \text{Factor out the } x_1.$$

$$= x_1\left[k_1 - \frac{k_1}{N}x_1 - cqx_1\right]$$

$$= x_1\left[k_1 - \left(\frac{k_1}{N} + cq\right)x_1\right]$$

This means that either $x_1 = 0$ (no more whales) or

$$x_1 = \frac{k_1}{\frac{k_1}{N} + cq}$$

$$= \frac{k_1 N}{k_1 + cqN}. \tag{5}$$

From this we find

$$x_2 = qx_1$$

$$= \frac{qk_1 N}{k_1 + cqN}. \tag{6}$$

Together, Equations (5) and (6) give the equilibrium point for the system of differential equations given by Equations (2) and (3).*

Our mathematical model has not yet taken into account the effect of humans harvesting both the whales and the krill for food. Let us denote by $F_1$ and $F_2$ the fishing effort for krill and whales, respectively, scaled so that $F_1 = 1$ and $F_2 = 1$ corresponds to an effort which yields a rate proportional to the natural growth rate. Then the system of differential equations (2) and (3) can be modified to

$$\frac{dx_1}{dt} = k_1\left(1 - \frac{x_1}{N}\right)x_1 - cx_1x_2 - k_1F_1x_1$$

$$\frac{dx_2}{dt} = k_2\left(1 - \frac{x_2}{qx_1}\right)x_2 - k_2F_2x_2. \tag{7}$$

To simplify the system in (7), let $y_1 = x_1/N$ and $y_2 = x_2/(qN)$. Then $y_1 = 1$ corresponds to the krill population being equal to the carrying capacity, and $y_2 = 1$ corresponds to the whale population being equal to the carrying capacity when the krill are at their carrying capacity. Then

$$\frac{d(Ny_1)}{dt} = k_1(1 - y_1)Ny_1 - c(Ny_1)(qNy_2) - k_1F_1Ny_1$$

$$\frac{d(qNy_2)}{dt} = k_2\left(1 - \frac{y_2}{y_1}\right)qNy_2 - k_2F_2qNy_2. \tag{8}$$

---

*The stability of equilibrium points for a system of differential equations can be studied, as we did for a single equation in the first section of this chapter. For more details, see Nagle, R. Kent, Edward B. Saff, and Arthur David Snider, *Fundamentals of Differential Equations*, 8th ed., Pearson, 2013, Sec. 5.4.

Dividing the first equation in (8) by $N$ and the second by $qN$, and factoring $k_1y_1$ from the right-hand side of the first equation and $k_2y_2$ from the right-hand side of the second equation, gives

$$\frac{dy_1}{dt} = k_1y_1\left(1 - y_1 - \frac{cqNy_2}{k_1} - F_1\right)$$

$$\frac{dy_2}{dt} = k_2y_2\left(1 - \frac{y_2}{y_1} - F_2\right). \tag{9}$$

Denoting $cqN/k_1$ by $b$ and rearranging slightly, (9) becomes

$$\frac{dy_1}{dt} = k_1y_1(1 - F_1 - y_1 - by_2)$$

$$\frac{dy_2}{dt} = k_2y_2\left(1 - F_2 - \frac{y_2}{y_1}\right). \tag{10}$$

Note that the constants $c$, $q$, and $N$ have disappeared from the system of differential equations. This means that these constants by themselves do not affect the behavior of the system; it is only their product, which becomes the numerator of the constant $b$, that appears in (9). This is an interesting conclusion that is not at all apparent in the original mathematical model.

### EXAMPLE 1  Predator-Prey

The Austrian mathematician A. J. Lotka (1880–1949) and the Italian mathematician Vito Volterra (1860–1940) proposed the following simple model for the way in which the fluctuations of populations of a predator and its prey affect each other. *Source: Elements of Mathematical Biology.* Let $x_1 = f(t)$ denote the population of the predator and $x_2 = g(t)$ denote the population of the prey at time $t$. The predator might be a wolf and its prey a moose, or the predator might be a ladybug and the prey an aphid.

Assume that if there were no predators present, the population of prey would increase at a rate $px_2$ proportional to their number, but that in the presence of predators, the predators consume the prey at a rate $qx_1x_2$ proportional to the product of the number of prey and the number of predators. The net rate of change $dx_2/dt$ of $x_2$ is the rate of increase of the prey minus the rate at which the prey are eaten, that is,

$$\frac{dx_2}{dt} = px_2 - qx_1x_2, \tag{11}$$

with positive constants $p$ and $q$.

Assume that if there were no prey, the predators would starve and their population would *decrease* at a rate $rx_1$ proportional to their number, but that in the presence of prey, the rate of growth of the population of predators is increased by an amount $sx_1x_2$. These assumptions give a second differential equation,

$$\frac{dx_1}{dt} = -rx_1 + sx_1x_2, \tag{12}$$

with additional positive constants $r$ and $s$.

Equations (11) and (12) form a system of differential equations known as the **Lotka-Volterra equations**.

**(a)** Find the equilibrium point for the Lotka-Volterra equations.

**SOLUTION** Equations (11) and (12) factor as

$$\frac{dx_1}{dt} = -rx_1 + sx_1x_2 = x_1(-r + sx_2) \quad \text{and}$$

$$\frac{dx_2}{dt} = px_2 - qx_1x_2 = x_2(p - qx_1).$$

Assuming that neither $x_1$ nor $x_2$ equals 0 (a trivial equilibrium point),

$$-r + sx_2 = 0 \qquad \text{and} \qquad p - qx_1 = 0,$$

so

$$x_2 = \frac{r}{s} \qquad \text{and} \qquad x_1 = \frac{p}{q}.$$

**(b)** By considering the sign of $dx_1/dt$ and $dx_2/dt$ at points other than the equilibrium point, determine the behavior of any solution to the system of differential equations.

**SOLUTION**   In Figure 11, we have shown the $x_1$- and $x_2$-axes, with the equilibrium point $(p/q, r/s)$ marked. This divides the first quadrant of the plane into four regions as shown.

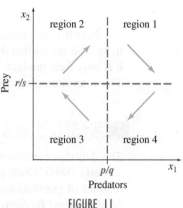

FIGURE 11

In region 1, $x_1 > p/q$ and $x_2 > r/s$. Then $dx_1/dt = x_1(-r + sx_2) > 0$ and $dx_2/dt = x_2(p - qx_1) < 0$. Since $x_1$ is increasing and $x_2$ is decreasing, the graph will go down and to the right, as shown.

In region 2, $x_1 < p/q$ and $x_2 > r/s$. Then $dx_1/dt > 0$ and $dx_2/dt > 0$. Since $x_1$ and $x_2$ are both increasing, the graph will go up and to the right, as shown.

Similarly, in region 3 $dx_1/dt < 0$ and $dx_2/dt > 0$, so the graph will go up and to the left, as shown. Verify the behavior of the solution in region 4.

We can see from this analysis that any solution will move about the equilibrium point in a clockwise direction. The population of prey tends to rise when the population of predators is low. Then, with the resulting abundance of food, the population of predators increases. As a result, the population of prey decreases again, and this forces the population of predators to decline because of the lack of food. The pattern repeats indefinitely. Figure 11 is an example of a **phase plane diagram**.

TRY YOUR TURN 1

**YOUR TURN 1**   Consider the system of differential equations

$$\frac{dx_1}{dt} = x_1x_2 - 4x_1$$

$$\frac{dx_2}{dt} = 3x_1x_2 - 6x_2.$$

**a.** Find all equilibrium points.
**b.** Sketch a phase plane diagram, similar to Figure 11.

We can say more about the graph of the solution by dividing Equation (11) by Equation (12), giving the separable differential equation

$$\frac{dx_2}{dx_1} = \frac{\dfrac{dx_2}{dt}}{\dfrac{dx_1}{dt}} = \frac{px_2 - qx_1x_2}{-rx_1 + sx_1x_2}$$

$$\frac{dx_2}{dx_1} = \frac{x_2(p - qx_1)}{x_1(sx_2 - r)} \tag{13}$$

Equation (13) is solved for specific values of the constants $p$, $q$, $r$, and $s$ in the next example.

### EXAMPLE 2    Predator-Prey

Suppose that $x_1 = f(t)$ (hundreds of predators) and $x_2 = g(t)$ (thousands of prey) satisfy the Lotka-Volterra equations (11) and (12) with $p = 3$, $q = 1$, $r = 4$, and $s = 2$. Suppose that at a time when there are 100 predators $(x_1 = 1)$, there are 1000 prey $(x_2 = 1)$. Find an equation relating $x_1$ and $x_2$.

**SOLUTION**    With the given values of the constants, $p$, $q$, $r$, and $s$, Equation (13) reads

$$\frac{dx_2}{dx_1} = \frac{x_2(3 - x_1)}{x_1(2x_2 - 4)}.$$

Separating the variables yields

$$\frac{2x_2 - 4}{x_2} dx_2 = \frac{3 - x_1}{x_1} dx_1,$$

or

$$\int \left(2 - \frac{4}{x_2}\right) dx_2 = \int \left(\frac{3}{x_1} - 1\right) dx_1.$$

Evaluating the integrals gives

$$2x_2 - 4 \ln x_2 = 3 \ln x_1 - x_1 + C.$$

(It is not necessary to use absolute value for the logarithms, since $x_1$ and $x_2$ are positive.) Use the initial conditions $x_1 = 1$ and $x_2 = 1$ to find $C$.

$$2 - 4(0) = 3(0) - 1 + C$$
$$C = 3$$

The desired equation is

$$x_1 + 2x_2 - 3 \ln x_1 - 4 \ln x_2 = 3. \tag{14}$$

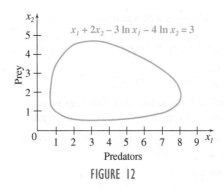

FIGURE 12

**YOUR TURN 2** Letting $p = 4$, $q = 1$, $r = 3$, and $s = 5$ in Example 2, find an equation relating $x$ and $y$, given that there was a time when $x = 1$ and $y = 1$.

A graph of Equation (14), in Figure 12, shows that the variations in the populations of the predator and prey are cyclic. We know from Example 1 that the solution moves about this cycle in a clockwise direction. So even though we cannot solve the nonlinear system of differential equations, we can say quite a bit about the solution.    TRY YOUR TURN 2

Not only have the Lotka-Volterra equations been widely used in biology, but more recently a group of researchers applied these equations to make predictions about where drug gang violence would cluster in Los Angeles. *Source: Criminology.*

# 11.5 EXERCISES

For Exercises 1–4, sketch a phase plane diagram, similar to Figure 11.

**1.** $\dfrac{dx_1}{dt} = 4x_1 - 2x_1 x_2$

$\dfrac{dx_2}{dt} = 3x_2 - x_1 x_2$

**2.** $\dfrac{dx_1}{dt} = 3x_1^2 x_2 - 6x_1^2$

$\dfrac{dx_2}{dt} = x_1 x_2^2 - 2x_2^2$

**3.** $\dfrac{dx_1}{dt} = 3x_1 x_2 - x_1 x_2^2$

$\dfrac{dx_2}{dt} = 2x_1^2 x_2 - 2x_1 x_2$

**4.** $\dfrac{dx_1}{dt} = x_1 x_2^3 - 2x_1 x_2^2$

$\dfrac{dx_2}{dt} = 9x_1^2 x_2 - 3x_1^3 x_2$

In Exercises 5–8, (a) find the nontrivial equilibrium point and (b) sketch a phase plane diagram, similar to Figure 11. Note that in these exercises, the lines where $dx_1/dt = 0$ and $dx_2/dt = 0$ are neither vertical nor horizontal.

**5.** $\dfrac{dx_1}{dt} = x_1^2 + x_1 x_2 - 2x_1$

$\dfrac{dx_2}{dt} = 2x_1 x_2 - x_2^2 - x_2$

**6.** $\dfrac{dx_1}{dt} = x_2^2 - x_1 x_2 - x_2$

$\dfrac{dx_2}{dt} = 2x_1^2 + x_1 x_2 - 7x_1$

**7.** $\dfrac{dx_1}{dt} = x_1^2 x_2 + x_1 x_2^2 - 3x_1 x_2$

$\dfrac{dx_2}{dt} = 2x_1 x_2^2 - x_1^2 x_2 - 3x_1 x_2$

**8.** $\dfrac{dx_1}{dt} = x_1^2 x_2 - x_1^3$

$\dfrac{dx_2}{dt} = 6x_2^2 + x_2^3 - 3x_1 x_2^2$

## LIFE SCIENCE APPLICATIONS

**9. Competing Species** The system of equations

$$\frac{dx_1}{dt} = -2x_1 + 3x_1 x_2$$

$$\frac{dx_2}{dt} = 3x_2 - 2x_1 x_2$$

describes the influence of the populations of two competing species (such as two birds with similar diets) on their growth rates.

**a.** Following Example 2 in the text, find an equation relating $x$ and $y$, assuming $x_2 = 2$ when $x_1 = 1$.

**b.** Find all equilibrium points.

**c.** Determine what happens to the species as time progresses.

**10. Symbiotic Species** When two species, such as the rhinoceros and birds pictured in the next column, coexist in a symbiotic (dependent) relationship, they either increase together or decrease together. Typical equations for the growth rates of two such species might be

$$\frac{dx_1}{dt} = -4x_1 + 4x_1 x_2$$

$$\frac{dx_2}{dt} = -3x_2 + 2x_1 x_2.$$

**a.** Find an equation relating $x_1$ and $x_2$ if $x_1 = 5$ when $x_2 = 1$.

**b.** Find all equilibrium points.

**c.** If both populations are greater than their values at the nonzero equilibrium point, what happens to the populations?

**d.** If both populations are less than their values at the equilibrium point, what happens to the populations?

**e.** A graph of the relationship found in part a is shown in the figure. Based on the differential equations for the growth rate and this graph, what happens to both populations when $x_2 > 1$? When $x_2 < 1$?

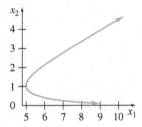

**11. Whales and Krill** For the system of differential equations (10), suppose that $F_1 < 1$ and $F_2 < 1$. Show that the equilibrium point is given by

$$y_1 = \frac{1 - F_1}{1 + b(1 - F_2)} \quad \text{and} \quad y_2 = \frac{(1 - F_1)(1 - F_2)}{1 + b(1 - F_2)}.$$

**12. Whales and Krill** For the system of differential equations (10), consider the sign of $dy_1/dt$ and $dy_2/dt$ to determine what would happen to the whales and krill if

**a.** $F_1 < 1$ and $F_2 > 1$;

**b.** $F_1 > 1$ and $F_2 > 1$.

**13. Mouse Infection** A more detailed model than the one in Exercise 27 of Sec. 11.2 for the spread of an infectious disease among mice is given by the system of differential equations

$$\frac{dx_1}{dt} = a - bx_1 - \beta x_1 x_2 + \gamma x_3$$

$$\frac{dx_2}{dt} = \beta x_1 x_2 - (b + \alpha + v)x_2$$

$$\frac{dx_3}{dt} = vx_2 - (\beta + \gamma)x_3,$$

where $x_1$ is the number of susceptible mice, $x_2$ is the number of infected mice, $x_3$ is the number of immune mice, $\alpha$ is the mortality rate due to infection, $b$ is the mortality rate due to natural causes for infected mice, $\beta$ is a transmission coefficient for the rate that infected mice infect susceptible mice, $\nu$ is the rate at which the mice recover from infection, and $\gamma$ is the rate at which mice lose immunity. ***Source: Lectures on Mathematics in the Life Sciences.*** In the article describing this model, the author states that an equilibrium solution only exists when

$$\frac{a}{b} > \frac{b + \alpha + \nu}{\beta}.$$

Verify this claim by finding the equilibrium solution and determining that the above inequality must be satisfied for the equilibrium values of $x_1$, $x_2$, and $x_3$ to be positive.

**14. Competing Species** A Lotka-Volterra model in which two species compete for the same source of food (as in Exercise 9) can be described by the system of differential equations

$$\frac{dx_1}{dt} = r_1 x_1 \left( 1 - \frac{x_1}{k_1} - b_1 \frac{x_2}{k_1} \right)$$

$$\frac{dx_2}{dt} = r_2 x_2 \left( 1 - \frac{x_2}{k_2} - b_2 \frac{x_1}{k_2} \right),$$

where $r_1$ and $r_2$ are growth rates for the two species, $k_1$ and $k_2$ are the carrying capacity for each species in the absence of the other, and $b_1$ and $b_2$ measure the competitive effect of each species on the other. ***Source: Mathematical Biology.*** Suppose that $k_1 = 6$, $k_2 = 4$, $b_1 = 2$, and $b_2 = 1$.

**a.** Find the equilibrium point.

**b.** Sketch a phase plane diagram, similar to Figure 11.

**c.** Based on your sketch in part b, what do you expect the populations of the two species to do in the long run?

**15. Scarlet Fever** Researchers have modeled scarlet fever epidemics in Liverpool, England, during the 19th century using the system of differential equations

$$\frac{dx}{dt} = \mu - \mu x - N\beta xy (1 + \delta\beta \sin \omega t)$$

$$\frac{dy}{dt} = N\beta xy (1 + \delta\beta \sin \omega t) - (\mu + \nu) y$$

where $N$ is the population size, $x$ is the fraction of the population that is susceptible, $y$ is the fraction of the population that is infected, $\mu$ is the death rate, $\nu$ is the rate of recovery from the disease, $\beta$ is the transmission coefficient, $\delta$ is the fractional variation in the transmission coefficient, and $\omega$ is the angular frequency of oscillations in susceptibility. ***Source: European Journal of Epidemiology.***

**a.** Consider the case when there is no variation in susceptibility, so that $\delta = 0$. Show that the equilibrium point is

$$x_0 = \frac{\mu + \nu}{N\beta} \quad \text{and} \quad y_0 = \frac{\mu(N\beta - \mu - \nu)}{N\beta(\mu + \nu)}$$

**b.** Let $x = x_0 + x_1$ and $y = y_0 + y_1$, where $x_0$ and $y_0$ are the values from part a, and $x_1$ and $y_1$ are small, so the product $x_1 y_1$ is small enough to ignore. Suppose that $\delta$ is also small, so the products $x_1 \delta$ and $y_1 \delta$ are small enough to ignore. Show that the equations for $x_1$ and $y_1$ are approximated by

$$\frac{dx_1}{dt} = -(N\beta y_0 + \mu)x_1 - (\mu + \nu)y_1 - (m\mu + \nu)y_0\delta \sin \omega t$$

$$\frac{dy_1}{dt} = N\beta y_0 x_1 + (\mu + \nu)y_0\delta \sin \omega t.$$

**16. Predator-Prey** Explain in your own words why the solution $(x, y)$ must move clockwise on the curve in Figure 12.

**17. Foraging** The rate of change with respect to time $t$ of a population of a foraging animal $N$, a resource being foraged $R$, and a predator $P$ can be described by the system of equations

$$\frac{dR}{dt} = F - C_0 NRT$$

$$\frac{dN}{dt} = b_1 C_0 RNT - C_1 PNT^2 - D_1 N$$

$$\frac{dP}{dt} = b_2 C_1 PNT^2 - D_2 P$$

where $T$ is the fraction of time that the animal spends foraging $(0 \le T \le 1)$, and $b_1$, $b_2$, $C_0$, $C_1$, $D_1$, $D_2$, and $F$ are constants. ***Source: The American Naturalist.***

**a.** Show that the growth rate of the population of the foraging animal is maximized when $T = b_1 C_0 R/(2C_1 P)$.

**b.** Substitute the value of $T$ from part a into the systems of equations. Then find the equilibrium point. (*Hint*: The first two equations can each be solved for $R^2/P$. Set these two equal and solve the resulting equation for $N$. Put the expressions for $R^2/P$ and $N$ into the third equation and solve for $P$.)

**18. Ecosystem** An ecosystem consisting of populations of nutrients $N$, producers of the nutrient $X$, and decomposers of the nutrient $D$ can be described by the system of equations

$$\frac{dN}{dt} = eQ + kD - aNX - eN$$

$$\frac{dX}{dt} = aNX - bX - eX$$

$$\frac{dD}{dt} = bX - kD - eD$$

where $Q = N + X + D$, the total population, which is assumed to stay constant according to conservation of mass, and $e$, $k$, $a$, and $b$ are constants. ***Source: The American Naturalist.*** Find the equilibrium point, and show that it can be written as

$$N = Q_1, \quad X = \frac{Q - Q_1}{bw}, \quad D = \frac{b}{k + e}X,$$

where $Q_1 = (b + e)/a$ and $w = 1/b + 1/(k + e)$.

**YOUR TURN ANSWERS**

**1. a.** $(0, 0), (2, 4)$      **2.** $x + 5y - 4 \ln x - 3 \ln y = 6$

**b.**

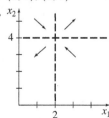

# 11.6 Applications of Differential Equations

During an epidemic, how does the infection rate of a disease vary with time?

**Epidemics** Under certain conditions, the spread of a contagious disease can be described with the logistic growth model, as in the next example.

### EXAMPLE 1 Spread of an Epidemic

APPLY IT

Consider a population of size $N$ that satisfies the following conditions.

1. Initially there is only one infected individual.
2. All uninfected individuals are susceptible, and infection occurs when an uninfected individual contacts an infected individual.
3. Contact between any two individuals is just as likely as contact between any other two individuals.
4. Infected individuals remain infectious.

Let $t$ = the time (in days) and $y$ = the number of individuals infected at time $t$. At any moment there are $(N - y)y$ possible contacts between an uninfected individual and an infected individual. Thus, it is reasonable to assume that the rate of spread of the disease satisfies the following logistic equation (discussed in Section 11.1):

$$\frac{dy}{dt} = k\left(1 - \frac{y}{N}\right)y. \tag{1}$$

As shown in Section 11.1, the general solution of Equation (1) is

$$y = \frac{N}{1 + be^{-kt}}, \tag{2}$$

where $b = (N - y_0)/y_0$. Since just one individual is infected initially, $y_0 = 1$ here. Substituting these values into Equation (2) gives

$$y = \frac{N}{1 + (N - 1)e^{-kt}} \tag{3}$$

as the specific solution of Equation (1).

The infection rate $dy/dt$ will be a maximum when its derivative is 0, that is, when $d^2y/dt^2 = 0$. Since

$$\frac{dy}{dt} = k\left(1 - \frac{y}{N}\right)y,$$

we have

$$\frac{d^2y}{dt^2} = k\left[\left(1 - \frac{y}{N}\right)\left(\frac{dy}{dt}\right) + y\left(-\frac{1}{N}\right)\left(\frac{dy}{dt}\right)\right]$$

$$= k\left(1 - \frac{2y}{N}\right)\left(\frac{dy}{dt}\right)$$

$$= k\left(1 - \frac{2y}{N}\right)k\left(1 - \frac{y}{N}\right)y$$

$$= k^2y\left(1 - \frac{y}{N}\right)\left(1 - \frac{2y}{N}\right).$$

Set $\dfrac{d^2y}{dt^2} = 0$ to get

$$y = 0, \quad 1 - \frac{y}{N} = 0, \quad \text{or} \quad 1 - \frac{2y}{N} = 0.$$

That is, $\dfrac{d^2y}{dt^2} = 0$ when $y = 0$, $y = N$, or $y = \dfrac{N}{2}$.

Notice that since the infection rate $\dfrac{dy}{dt} = 0$ when $y = 0$ or $y = N$, the maximum infection rate does not occur there. Also observe that $\dfrac{d^2y}{dt^2} > 0$ when $0 < y < \dfrac{N}{2}$ and $\dfrac{d^2y}{dt^2} < 0$ when $\dfrac{N}{2} < y < N$. Thus, the maximum infection rate occurs when exactly half the total population is still uninfected and equals

$$\frac{dy}{dt} = k\left(1 - \frac{N/2}{N}\right)\frac{N}{2} = \frac{kN}{4}.$$

Letting $y = N/2$ in Equation (3) and solving for $t$ shows that the maximum infection rate occurs at time

$$t_m = \frac{\ln(N - 1)}{k}.$$

Because $y$ is a function of $t$, the infection rate $dy/dt$ is also a function of $t$. Its graph, shown in Figure 13, is called the *epidemic curve*. It is symmetric about the line $t = t_m$.

**TRY YOUR TURN 1**

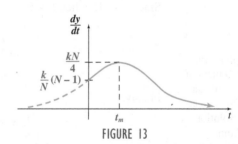

FIGURE 13

## Mixing Problems

The mixing of two solutions can lead to a first-order differential equation, as the next example shows.

**EXAMPLE 2** **Salt Concentration**

Suppose a tank contains 100 gal of a solution of dissolved salt and water, which is kept uniform by stirring. If pure water is allowed to flow into the tank at the rate of 4 gal per minute, and the mixture flows out at the rate of 3 gal per minute (see Figure 14), how much salt will remain in the tank after $t$ minutes if 15 lb of salt are in the mixture initially? *Source: Ordinary Differential Equations with Boundary Value Problems.*

**SOLUTION** Let the amount of salt present in the tank at any specific time be $y = f(t)$. The net rate at which $y$ changes is given by

$$\frac{dy}{dt} = (\text{Rate of salt in}) - (\text{Rate of salt out}).$$

Since pure water is coming in, the rate of salt entering the tank is zero. The rate at which salt is leaving the tank is the product of the amount of salt per gallon (in $V$ gallons) and the number of gallons per minute leaving the tank:

$$\text{Rate of salt out} = \left(\frac{y}{V} \text{ lb per gal}\right)(3 \text{ gal per minute}).$$

**YOUR TURN 1** Suppose that an epidemic in a community of 50,000 starts with 80 people infected, and that 15 days later, 640 are infected. How many are infected 25 days into the epidemic?

4 gal per minute

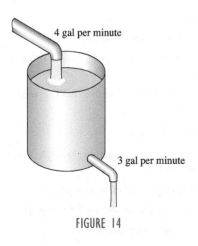

3 gal per minute

FIGURE 14

The differential equation, therefore, can be written as

$$\frac{dy}{dt} = -\frac{3y}{V}; \quad y(0) = 15,$$

where $y(0)$ is the initial amount of salt in the solution. We must take into account the fact that the volume, $V$, of the mixture is not constant but is determined by

$$\frac{dV}{dt} = (\text{Rate of liquid in}) - (\text{Rate of liquid out}) = 4 - 3 = 1,$$

or

$$\frac{dV}{dt} = 1,$$

from which

$$V(t) = t + C_1.$$

Because the volume is known to be 100 at time $t = 0$, we have $C_1 = 100$, and

$$\frac{dy}{dt} = \frac{-3y}{t + 100}; \quad y(0) = 15,$$

a separable equation with solution

$$\frac{dy}{y} = \frac{-3}{t + 100} dt$$

$$\ln y = -3 \ln(t + 100) + C.$$

Since $y = 15$ when $t = 0$,

$$\ln 15 = -3 \ln 100 + C$$

$$\ln 15 + 3 \ln 10^2 = C$$

$$\ln(15 \times 10^6) = C.$$

**YOUR TURN 2** Suppose that a tank initially contains 500 liters of a solution of water and 5 kg of salt. Suppose that pure water flows in at a rate of 6 L/min, and the solution flows out at a rate of 4 L/min. How many kg of salt remain after 20 minutes?

Finally,

$$\ln y = \ln(t + 100)^{-3} + \ln(15 \times 10^6)$$

$$= \ln[(t + 100)^{-3}(15 \times 10^6)]$$

$$y = \frac{15 \times 10^6}{(t + 100)^3}.$$

**TRY YOUR TURN 2**

# 11.6 EXERCISES

## LIFE SCIENCE APPLICATIONS

1. **Spread of an Epidemic** The native Hawaiians lived for centuries in isolation from other peoples. When foreigners finally came to the islands they brought with them diseases such as measles, whooping cough, and smallpox, which decimated the population. Suppose such an island has a native population of 5000, and a sailor from a visiting ship introduces measles, which has an infection rate of $k = 0.25$. Also suppose that the model for spread of an epidemic described in Example 1 applies.

   **a.** Write an equation for the number of natives who remain uninfected. Let $t$ represent time in days.

   **b.** How many are uninfected after 30 days?

   **c.** How many are uninfected after 50 days?

   **d.** When will the maximum infection rate occur?

2. **Spread of an Epidemic** In Example 1, the number of infected individuals is given by Equation (3).

   **a.** Show that the number of uninfected individuals is given by

   $$N - y = \frac{N(N - 1)}{N - 1 + e^{kt}}.$$

   **b.** Graph the equation in part a and Equation (3) on the same axes when $N = 100$ and $k = 1$.

   **c.** Find the common inflection point of the two graphs.

   **d.** What is the significance of the common inflection point?

   **e.** What are the limiting values of $y$ and $N - y$?

3. **Spread of an Epidemic** An influenza epidemic spreads at a rate proportional to the product of the number of people infected and the number not yet infected. Assume that 100 people are infected at the beginning of the epidemic in a community of 20,000 people, and 400 are infected 10 days later.

    **a.** Write an equation for the number of people infected, $y$, after $t$ days.

    **b.** When will half the community be infected?

4. **Spread of an Epidemic** The Gompertz growth law,

$$\frac{dy}{dt} = kye^{-at},$$

for constants $k$ and $a$, is another model used to describe the growth of an epidemic. Repeat Exercise 3, using this differential equation with $a = 0.02$.

5. **Spread of Gonorrhea** Gonorrhea is spread by sexual contact, takes 3 to 7 days to incubate, and can be treated with antibiotics. There is no evidence that a person ever develops immunity. One model proposed for the rate of change in the number of men infected by this disease is

$$\frac{dy}{dt} = -ay + b(f - y)Y,$$

where $y$ is the fraction of men infected, $f$ is the fraction of men who are promiscuous, $Y$ is the fraction of women infected, and $a$ and $b$ are appropriate constants. **Source: An Introduction to Mathematical Modeling.**

    **a.** Assume $a = 1$, $b = 1$, and $f = 0.5$. Choose $Y = 0.01$, and solve for $y$ using $y = 0.02$ when $t$ is 0 as an initial condition. Round your answer to 3 decimal places.

    **b.** A comparable model for women is

$$\frac{dY}{dt} = -AY + B(F - Y)y,$$

where $F$ is the fraction of women who are promiscuous and $A$ and $B$ are constants. Assume $A = 1$, $B = 1$, and $F = 0.03$. Choose $y = 0.1$ and solve for $Y$, using $Y = 0.01$ as an initial condition.

## OTHER APPLICATIONS

**Spread of a Rumor** The equation developed in the text for the spread of an epidemic also can be used to describe diffusion of information. In a population of size $N$, let $y$ be the number who have heard a particular piece of information. Then

$$\frac{dy}{dt} = k\left(1 - \frac{y}{N}\right)y$$

for a positive constant $k$. Use this model in Exercises 6–8.

6. Suppose a rumor starts among 3 people in a certain office building. That is, $y_0 = 3$. Suppose 500 people work in the building and 50 people have heard the rumor in 2 days. Using Equation (2), write an equation for the number who have heard the rumor in $t$ days. How many people will have heard the rumor in 5 days?

7. A rumor spreads at a rate proportional to the product of the number of people who have heard it and the number who have not heard it. Assume that 3 people in an office with 45 employees heard the rumor initially, and 12 people have heard it 3 days later.

    **a.** Write an equation for the number, $y$, of people who have heard the rumor in $t$ days.

    **b.** When will 30 employees have heard the rumor?

8. A news item is heard on the late news by 5 of the 100 people in a small community. By the end of the next day 20 people have heard the news. Using Equation (2), write an equation for the number of people who have heard the news in $t$ days. How many have heard the news after 3 days?

9. Repeat Exercise 7 using the Gompertz growth law,

$$\frac{dy}{dt} = kye^{-at},$$

for constants $k$ and $a$, with $a = 0.1$.

10. **Salt Concentration** A tank holds 100 gal of water that contains 20 lb of dissolved salt. A brine (salt) solution is flowing into the tank at the rate of 2 gal per minute while the solution flows out of the tank at the same rate. The brine solution entering the tank has a salt concentration of 2 lb per gal.

    **a.** Find an expression for the amount of salt in the tank at any time.

    **b.** How much salt is present after 1 hour?

    **c.** As time increases, what happens to the salt concentration?

11. Solve Exercise 10 if the brine solution is introduced at the rate of 3 gal per minute while the rate of outflow remains the same.

12. Solve Exercise 10 if the brine solution is introduced at the rate of 1 gal per minute while the rate of outflow stays the same.

13. Solve Exercise 10 if pure water is added instead of brine.

14. **Chemical in a Solution** Five grams of a chemical is dissolved in 100 liters of alcohol. Pure alcohol is added at the rate of 2 liters per minute and at the same time the solution is being drained at the rate of 1 liter per minute.

    **a.** Find an expression for the amount of the chemical in the mixture at any time.

    **b.** How much of the chemical is present after 30 minutes?

15. Solve Exercise 14 if a 25% solution of the same mixture is added instead of pure alcohol.

16. **Soap Concentration** A prankster puts 4 lb of soap in a fountain that contains 200 gal of water. To clean up the mess a city crew runs clear water into the fountain at the rate of 8 gal per minute, allowing the excess solution to drain off at the same rate. How long will it be before the amount of soap in the mixture is reduced to 1 lb?

**YOUR TURN ANSWERS**

1. 2483   **2.** 4.29 kg

# 11

# CHAPTER REVIEW

## SUMMARY

In this chapter, we studied differential equations, which are equations involving derivatives. Our goal has been to find a function that satisfies the equation. We learned to solve two different types of equations:

- separable equations, using separation of variables, and
- linear equations, using an integrating factor.

For equations that cannot be solved by either of the previous two methods, we introduced a numerical method known as Euler's method.

We then studied linear and nonlinear systems of differential equations. Differential equations have a large number of applications; some of those we studied in this chapter include:

- the logistic equation for populations;
- a predator-prey model;
- and mixing problems.

**General Solution of $\dfrac{dy}{dx} = g(x)$**  The general solution of the differential equation $dy/dx = g(x)$ is

$$y = \int g(x)\,dx.$$

**Separable Differential Equation**  An equation is separable if it can be written in the form

$$\frac{dy}{dx} = \frac{p(x)}{q(y)}.$$

By separating the variables, it is transformed into the equation

$$\int q(y)\,dy = \int p(x)\,dx.$$

**Solving a Linear First-Order Differential Equation**

1. Put the equation in the linear form

$$\frac{dy}{dx} + P(x)y = Q(x).$$

2. Find the integrating factor $I(x) = e^{\int P(x)\,dx}$.
3. Multiply each term of the equation from Step 1 by $I(x)$.
4. Replace the sum of terms on the left with $D_x[I(x)y]$.
5. Integrate both sides of the equation.
6. Solve for $y$.

**Euler's Method**  Let $y = f(x)$ be the solution of the differential equation

$$\frac{dy}{dx} = g(x, y), \text{ with } y(x_0) = y_0,$$

for $x_0 \le x \le x_n$. Let $x_{i+1} = x_i + h$, where $h = (x_n - x_0)/n$ and

$$y_{i+1} = y_i + g(x_i, y_i)h,$$

for $0 \le i \le n - 1$. Then

$$f(x_{i+1}) \approx y_{i+1}.$$

**Solving a Linear System of Differential Equations**  $\dfrac{dX}{dt} = MX + Q$ where $M$ is constant.

1. Find the eigenvalues of the matrix $M$ by solving the equation $\det(M - \lambda I) = 0$.
2. For each eigenvalue, find the corresponding eigenvector by solving the equation $(M - \lambda I)X = O$.

3. Create the matrix $P$ whose columns are the eigenvectors of $M$.

4. Solve the system $dY/dt = (P^{-1}MP)Y + P^{-1}Q$, where $P^{-1}MP$ is the diagonal matrix with the eigenvalues of $M$ on the diagonal. Each equation in this system is linear and involves only one of the functions of $Y$.

5. The solution to the original system is given by $X = PY$.

6. The solution at Step 5 will have arbitrary constants in it. For an initial value problem, substitute the values of the variables and solve the resulting system of linear equations to find the constants.

# KEY TERMS

| | | | |
|---|---|---|---|
| differential equation | carrying capacity | **11.2** | **11.5** |
| **11.1** | logistic growth model | linear first-order differential | equilibrium point |
| general solution | logistic equation | equation | Lotka-Volterra equations |
| particular solution | logistic curve | integrating factor | phase plane diagram |
| initial condition | equilibrium point | | |
| initial value problem | autonomous | **11.3** | |
| separable differential equation | stability | Euler's method | |
| separation of variables | stable | **11.4** | |
| growth constant | unstable | linear system of differential | |
| | semistable | equations | |

# REVIEW EXERCISES

## CONCEPT CHECK

**Determine whether each of the following statements is true or false, and explain why.**

1. To determine a particular solution to a differential equation, you first must find a general solution to the differential equation.

2. The function $y = e^{2x} + 5$ satisfies the differential equation $\dfrac{dy}{dx} = 2y$.

3. The function $y = \dfrac{100}{1 + 99e^{-5t}}$ satisfies the differential equation $\dfrac{dy}{dt} = 5\left(1 - \dfrac{y}{100}\right)y$.

4. The differential equation $y\dfrac{dy}{dx} + xy = 2500e^y$ is a first-order linear differential equation.

5. Every differential equation is either separable or linear.

6. It is possible to solve the following differential equation using the method of separation of variables.
$$x\dfrac{dy}{dx} = (x + 1)(y + 1)$$

7. It is possible to solve the following differential equation using the method of separation of variables.
$$\dfrac{dy}{dx} = x^2 + 4y^2$$

8. The function $I(x) = x^3$ can be used as an integrating factor for the differential equation
$$\dfrac{dy}{dx} + 3\dfrac{y}{x} = \dfrac{1}{x^2}.$$

9. The function $I(x) = e^{5x}$ can be used as an integrating factor for the differential equation
$$x\dfrac{dy}{dx} + 5y = e^{2x}.$$

10. Euler's method can be used to find the general solution to a differential equation.

11. If Euler's method is being used to solve the differential equation $\dfrac{dy}{dx} = x + \sqrt{y + 4}$ with $h = 0.1$, then
$$y_{i+1} = y_i + 0.1(x_i + \sqrt{y_i + 4}).$$

12. The Lotka-Volterra equations are a system of linear differential equations.

## PRACTICE AND EXPLORATIONS

13. What is a differential equation? What is it used for?

14. What is the difference between a particular solution and a general solution to a differential equation?

15. How can you tell that a differential equation is separable? That it is linear?

16. Can a differential equation be both separable and linear? Explain why not, or give an example of an equation that is both.

**Classify each equation as separable, linear, both, or neither.**

17. $y\dfrac{dy}{dx} = 2x + y$

18. $\dfrac{dy}{dx} + y^2 = xy^2$

19. $\sqrt{x}\dfrac{dy}{dx} = \dfrac{1 + \ln x}{y}$

20. $\dfrac{dy}{dx} = xy + e^x$

21. $\dfrac{dy}{dx} + x = xy$

22. $\dfrac{x\,dy}{y\,dx} = 4 + x^{3/2}$

23. $x\dfrac{dy}{dx} + y = e^x(1 + y)$

24. $\dfrac{dy}{dx} = x^2 + y^2$

**Find the general solution for each differential equation.**

25. $\dfrac{dy}{dx} = 3x^2 + 6x$

26. $\dfrac{dy}{dx} = 4x^3 + 6x^5$

27. $\dfrac{dy}{dx} = 4e^{2x}$

28. $\dfrac{dy}{dx} = \dfrac{1}{3x + 2}$

29. $\dfrac{dy}{dx} = \dfrac{3x + 1}{y}$

30. $\dfrac{dy}{dx} = \dfrac{e^x + x}{y - 1}$

31. $\dfrac{dy}{dx} = \dfrac{2y + 1}{x}$

32. $\dfrac{dy}{dx} = \dfrac{3 - y}{e^x}$

33. $\dfrac{dy}{dx} + y = x$

34. $x^4\dfrac{dy}{dx} + 3x^3 y = 1$

35. $x \ln x \dfrac{dy}{dx} + y = 2x^2$

36. $x\dfrac{dy}{dx} + 2y - e^{2x} = 0$

**Find the particular solution for each initial value problem. (Some solutions may give $y$ implicitly.)**

37. $\dfrac{dy}{dx} = x^2 - 6x; \quad y(0) = 3$

38. $\dfrac{dy}{dx} = 5(e^{-x} - 1); \quad y(0) = 17$

39. $\dfrac{dy}{dx} = (x + 2)^3 e^y; \quad y(0) = 0$

40. $\dfrac{dy}{dx} = (3 - 2x)y; \quad y(0) = 5$

41. $\dfrac{dy}{dx} = \dfrac{1 - 2x}{y + 3}; \quad y(0) = 16$    42. $\sqrt{x}\dfrac{dy}{dx} = xy; \quad y(1) = 4$

43. $e^x\dfrac{dy}{dx} - e^x y = x^2 - 1; \quad y(0) = 42$

44. $\dfrac{dy}{dx} + 3x^2 y = x^2; \quad y(0) = 2$

45. $x\dfrac{dy}{dx} - 2x^2 y + 3x^2 = 0; \quad y(0) = 15$

46. $x^2\dfrac{dy}{dx} + 4xy - e^{2x^3} = 0; \quad y(1) = e^2$

47. When is Euler's method useful?

**Use Euler's method to approximate the indicated function value for $y = f(x)$ to 3 decimal places, using $h = 0.2$.**

48. $\dfrac{dy}{dx} = x + y^{-1}; \quad y(0) = 1; \quad$ find $y(1)$

49. $\dfrac{dy}{dx} = e^x + y; \quad y(0) = 1; \quad$ find $y(0.6)$

50. Let $y = f(x)$ and $dy/dx = (x/2) + 4$, with $y(0) = 0$. Use Euler's method with $h = 0.1$ to approximate $y(0.3)$ to 3 decimal places. Then solve the differential equation and find $f(0.3)$ to 3 decimal places. Also, find $y_3 - f(x_3)$.

51. Let $y = f(x)$ and $dy/dx = 3 + \sqrt{y}$, with $y(0) = 0$. Construct a table for $x_i$ and $y_i$ like the one in Section 11.3, Example 2, for $[0, 1]$, with $h = 0.2$. Then graph the polygonal approximation of the graph of $y = f(x)$.

52. What is the logistic equation? Why is it useful?

**Solve each of the following systems of differential equations.**

53. $\dfrac{dx_1}{dt} = 2x_1 + 7x_2 + t$
    $\dfrac{dx_2}{dt} = 7x_1 + 2x_2 + 2t$

54. $\dfrac{dx_1}{dt} = 7x_1 + 2x_2 + 3t$
    $\dfrac{dx_2}{dt} = -4x_1 + x_2 - t$

**In Exercises 55 and 56, (a) find the nontrivial equilibrium point and (b) sketch a phase plane diagram, similar to Figure 11.**

55. $\dfrac{dx_1}{dt} = 4x_1 - x_1^2 - x_1 x_2$
    $\dfrac{dx_2}{dt} = 2x_1^2 - x_1 x_2 - 2x_1$

56. $\dfrac{dx_1}{dt} = x_1 x_2 - x_2^2 - x_2$
    $\dfrac{dx_2}{dt} = 4x_1 x_2 - x_1^2 x_2 - 2x_1 x_2^2$

## LIFE SCIENCE APPLICATIONS

57. **Effect of Insecticide** After use of an experimental insecticide, the rate of decline of an insect population is

$$\frac{dy}{dt} = \frac{-10}{1 + 5t},$$

where $t$ is the number of hours after the insecticide is applied. Assume that there were 50 insects initially.

   **a.** How many are left after 24 hours?

   **b.** How long will it take for the entire population to die?

58. **Growth of a Mite Population** A population of mites grows at a rate proportional to the number present, $y$. If the growth constant is 10% and 120 mites are present at time $t = 0$ (in weeks), find the number present after 6 weeks.

59. **Competing Species** Find an equation relating $x$ to $y$ given the following equations, which describe the interaction of two competing species and their growth rates.

$$\frac{dx}{dt} = 0.2x - 0.5xy$$
$$\frac{dy}{dt} = -0.3y + 0.4xy$$

Find the values of $x$ and $y$ for which both growth rates are 0.

60. **Smoke Content in a Room** The air in a meeting room of 15,000 ft³ has a smoke content of 20 parts per million (ppm). An air conditioner is turned on, which brings fresh air (with no smoke) into the room at a rate of 1200 ft³ per minute and forces the smoky air out at the same rate. How long will it take to reduce the smoke content to 5 ppm?

61. In Exercise 60, how long will it take to reduce the smoke content to 10 ppm if smokers in the room are adding smoke at the rate of 5 ppm per minute?

62. **Spread of Influenza** A small, isolated mountain community with a population of 700 is visited by an outsider who carries influenza. After 6 weeks, 300 people are uninfected.

**a.** Write an equation for the number of people who remain un-infected at time $t$ (in weeks).

**b.** Find the number still uninfected after 7 weeks.

**c.** When will the maximum infection rate occur?

**63. Population Growth** Let

$$y = \frac{N}{1 + be^{-kt}}.$$

If $y$ is $y_1$, $y_2$, and $y_3$ at times $t_1$, $t_2$, and $t_3 = 2t_2 - t_1$ (that is, at three equally spaced times), then prove that

$$N = \frac{1/y_1 + 1/y_3 - 2/y_2}{1/(y_1 y_3) - 1/y_2^2}.$$

**Population Growth** In the following table of U.S. Census figures, $y$ is the population in millions. *Source: U.S. Census Bureau.*

| Year | $y$ | Year | $y$ |
|------|------|------|-------|
| 1790 | 3.9  | 1910 | 92.0  |
| 1800 | 5.3  | 1920 | 105.7 |
| 1810 | 7.2  | 1930 | 122.8 |
| 1820 | 9.6  | 1940 | 131.7 |
| 1830 | 12.9 | 1950 | 150.7 |
| 1840 | 17.1 | 1960 | 179.3 |
| 1850 | 23.2 | 1970 | 203.3 |
| 1860 | 31.4 | 1980 | 226.5 |
| 1870 | 39.8 | 1990 | 248.7 |
| 1880 | 50.2 | 2000 | 281.4 |
| 1890 | 62.9 | 2010 | 308.7 |
| 1900 | 76.0 |      |       |

**64.** Use Exercise 63 and the table to find the following.

**a.** Find $N$ using the years 1800, 1850, and 1900.

**b.** Find $N$ using the years 1850, 1900, and 1950.

**c.** Find $N$ using the years 1870, 1920, and 1970.

**d.** Explain why different values of $N$ were obtained in parts a–c. What does this suggest about the validity of this model and others?

**65.** Let $t = 0$ correspond to 1870, and let every decade correspond to an increase in $t$ of 1. Use the table from Exercise 64.

**a.** Use 1870, 1920, and 1970 to find $N$, 1870 to find $b$, and 1920 to find $k$ in the equation

$$y = \frac{N}{1 + be^{-kt}}.$$

**b.** Estimate the population of the United States in 2010 and compare your estimate to the actual population in 2010.

**c.** Predict the populations of the United States in 2030 and 2050.

**66.** Let $t = 0$ correspond to 1790, and let every decade correspond to an increase in $t$ of 1. Use a calculator with logistic regression capability to complete the following. Use the table from Exercise 64.

**a.** Plot the data points. Do the points suggest that a logistic function is appropriate here?

**b.** Use the logistic regression function on your calculator to determine the logistic equation that best fits the data.

**c.** Plot the logistic equation from part a on the same graph as the data points. How well does the logistic equation seem to fit the data?

**d.** What seems to be the limiting size of the U.S. population?

**67. Growth Functions** In Section 11.1 we studied three models for growth: exponential growth (Example 5), limited growth (Example 6), and logistic growth (before Example 7, and Exercises 39–42). Assume $k > 0$ in all three models, and $N > y_0$ in the limited growth and logistic models.

**a.** For what values of $t$ is each model increasing? Decreasing?

**b.** For what values of $t$ is each model concave up? Concave down?

**c.** Discuss similarities and differences between the models, and under what circumstances each would be appropriate.

## OTHER APPLICATIONS

**68. Education** Researchers have proposed that the amount a full-time student is educated $(x)$ changes with respect to the student's age $t$ according to the differential equation

$$\frac{dx}{dt} = 1 - kx,$$

where $k$ is a constant measuring the rate that education depreciates due to forgetting or technological obsolescence. *Source: Operations Research.*

**a.** Solve the equation using the method of separation of variables.

**b.** Solve the equation using an integrating factor.

**c.** What does $x$ approach over time?

**69. Spread of a Rumor** A rumor spreads through the offices of a company with 200 employees, starting in a meeting with 10 people. After 3 days, 35 people have heard the rumor.

**a.** Write an equation for the number of people who have heard the rumor in $t$ days. (*Hint*: Refer to Exercises 6–8 in Section 11.6.)

**b.** How many people have heard the rumor in 5 days?

**70. Newton's Law of Cooling** A roast at a temperature of 40°F is put in a 300°F oven. After 1 hour the roast has reached a temperature of 150°F. Newton's law of cooling states that

$$\frac{dT}{dt} = k(T - T_M),$$

where $T$ is the temperature of an object, the surrounding medium has temperature $T_M$ at time $t$, and $k$ is a constant. Use Newton's law to find the temperature of the roast after 2 hours.

**71.** In Exercise 70, how long does it take for the roast to reach a temperature of 250°F?

632 CHAPTER 11 Differential Equations

**72. Air Resistance** In Section 7.1 on Antiderivatives, we saw that the acceleration of gravity is a constant if air resistance is ignored. But air resistance cannot always be ignored, or parachutes would be of little use. In the presence of air resistance, the equation for acceleration also contains a term roughly proportional to the velocity squared. Since acceleration forces a falling object downward and air resistance pushes it upward, the air resistance term is opposite in sign to the acceleration of gravity. Thus,

$$a(t) = \frac{dv}{dt} = g - kv^2,$$

where $g$ and $k$ are positive constants. Future calculations will be simpler if we replace $g$ and $k$ by the squared constants $G^2$ and $K^2$, giving

$$\frac{dv}{dt} = G^2 - K^2v^2.$$

**a.** Use separation of variables and the fact that

$$\frac{1}{G^2 - K^2v^2} = \frac{1}{2G}\left(\frac{1}{G - Kv} + \frac{1}{G + Kv}\right)$$

to solve the differential equation above. Assume $v < G/K$, which is certainly true when the object starts falling (with $v = 0$). Write your solution in the form of $v$ as a function of $t$.

**b.** Find $\lim_{t\to\infty} v(t)$, where $v(t)$ is the solution you found in part a. What does this tell you about a falling object in the presence of air resistance?

**c.** According to *Harper's Index*, the terminal velocity of a cat falling from a tall building is 60 mph. ***Source: Harper's.*** Use your answers from part b, plus the fact that 60 mph = 88 ft per second and $g$, the acceleration of gravity, is 32 ft per second$^2$, to find a formula for the velocity of a falling cat (in ft per second) as a function of time (in seconds). (*Hint:* Find $K$ in terms of $G$. Then substitute into the answer from part a.)

# EXTENDED APPLICATION

## POLLUTION OF THE GREAT LAKES

Industrial nations are beginning to face the problems of water pollution. Lakes present a problem, because a polluted lake contains a considerable amount of water that must somehow be cleaned. The main cleanup mechanism is the natural process of gradually replacing the water in the lake. This application deals with pollution in the Great Lakes. The basic idea is to regard the flow in the Great Lakes as a mixing problem.

We make the following assumptions.

1. Rainfall and evaporation balance each other, so the average rates of inflow and outflow are equal.

2. The average rates of inflow and outflow do not vary much seasonally.

3. When water enters the lake, perfect mixing occurs, so that the pollutants are uniformly distributed.

4. Pollutants are not removed from the lake by decay, sedimentation, or in any other way except outflow.

5. Pollutants flow freely out of the lake; they are not retained (as DDT is).

(The first two are valid assumptions; however, the last three are questionable.)

We will use the following variables in the discussion to follow.

$V$ = volume of the lake

$P_L$ = pollution concentration in the lake at time $t$

$P_i$ = pollution concentration in the inflow to the lake at time $t$

$r$ = rate of flow

$t$ = time in years

By the assumptions stated above, the net change in total pollutants during the time interval $\Delta t$ is (approximately)

$$V \cdot \Delta P_L = (P_i - P_L)(r \cdot \Delta t),$$

where $\Delta P_L$ is the change in the pollution concentration. Dividing this equation by $\Delta t$ and by $V$ and taking the limit as $\Delta t \to 0$, we get the differential equation

$$\frac{dP_L}{dt} = \frac{(P_i - P_L)r}{V}.$$

Since we are treating $V$ and $r$ as constants, we replace $r/V$ with $k$, so the equation can be written as the first-order linear equation

$$\frac{dP_L}{dt} + kP_L = kP_i.$$

The solution is

$$P_L(t) = e^{-kt}\left[P_L(0) + k\int_0^t P_i(x)e^{kx}\,dx\right]. \qquad (1)$$

Figure 15 shows values of $1/k$ for each lake (except Huron) measured in years. *If the model is reasonable*, the numbers in the figure can be used in Equation (1) to determine the effect of various pollution abatement schemes. Lake Ontario is excluded from the discussion because about 84% of its inflow comes from Erie and can be controlled only indirectly.

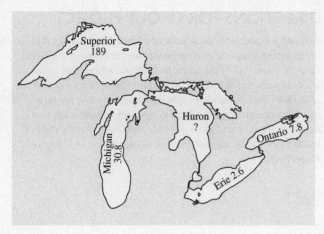

Superior
189

Huron
?

Michigan
30.8

Ontario 7.8

Erie 2.6

FIGURE 15

The fastest possible cleanup will occur if all pollution inflow ceases. This means that $P_i = 0$. In this case, Equation (1) leads to

$$t = \frac{1}{k} \ln\left(\frac{P_L(0)}{P_L(t)}\right).$$

From this we can tell the length of time necessary to reduce pollution to a given percentage of its present level. For example, from the figure, for Lake Superior $1/k = 189$. Thus, to reduce pollution to 50% of its present level, $P_L(0)$, we want

$$\frac{P_L(t)}{P_L(0)} = 0.5 \quad \text{or} \quad \frac{P_L(0)}{P_L(t)} = 2,$$

from which

$$t = 189 \ln 2 \approx 131.$$

The following figures, representing years, were found in this way. Fortunately, the pollution in Lake Superior is quite low at present.

As mentioned before, assumptions 3, 4, and 5 are questionable. For persistent pollutants like DDT, the estimated cleanup times may be too low. For other pollutants, how assumptions 4 and 5 affect cleanup times is unclear. However, the values of $1/k$ given in the figure probably provide rough lower bounds for the cleanup times of persistent pollutants.

| Lake | 50% | 20% | 10% | 5% |
|---|---|---|---|---|
| Erie | 2 | 4 | 6 | 8 |
| Michigan | 21 | 50 | 71 | 92 |
| Superior | 131 | 304 | 435 | 566 |

*Source: An Introduction to Mathematical Modeling.*

## EXERCISES

1. Calculate the number of years to reduce pollution in Lake Erie to each level.

   a. 40%

   b. 30%

2. Repeat Exercise 1 for Lake Michigan.

3. Repeat Exercise 1 for Lake Superior.

4. We claim that Equation (1) is a solution of the differential equation

$$\frac{dP_L}{dt} + kP_L(t) = kP_i(t),$$

where $t$ measures time from the present. The constant $k = r/V$ measures how quickly the water in the lake is replaced through inflow and corresponding outflow. The constant $P_L(0)$ is the current pollution level.

   a. To show that Equation (1) does define a solution of the differential equation, multiply both sides of Equation (1) by $e^{kt}$ and then differentiate both sides with respect to $t$. Remember from the section on the Fundamental Theorem of Calculus that you can differentiate an integral by using the version of the Fundamental Theorem that says

$$\frac{d}{dt} \int_a^t f(x)\, dx = f(t).$$

   b. When you substitute $t = 0$ into the right-hand side of Equation (1), you should get $P_L(0)$. Do you? What happens to the integral? What happens to the factor of $e^{-kt}$?

   c. The map indicates a value of 30.8 for Lake Michigan. What value of $k$ does this correspond to? What percent of the water in Lake Michigan is replaced each year by inflow? Which lake has the biggest annual water turnover?

5. Suppose that instead of assuming that all pollution inflow immediately ceases, we model $P_i(t)$ by a decaying exponential of the form $a \cdot e^{-pt}$, where $p$ is a constant that tells us how fast the inflow is being cleaned of pollution. To simplify things, we'll also assume that initially the inflow and the lake have the same pollution concentration, so $a = P_L(0)$. Now substitute $P_L(0)e^{-px}$ for $P_i(x)$ in Equation (1), and evaluate the integral as a function of $t$.

6. When you simplify the right-hand side of Equation (1) using your new expression for the integral, and then factor out and divide by $P_L(0)$, you'll get the following nice expression for the ratio $P_L(t)/P_L(0)$:

$$\frac{P_L(t)}{P_L(0)} = \frac{1}{k - p}\left(ke^{-pt} - pe^{-kt}\right).$$

   a. Suppose that for Lake Michigan the constant $p$ is equal to 0.02. Use a graph of the ratio $P_L(t)/P_L(0)$ to estimate how long it will take to reduce pollution to 50% of its current value. How does this compare with the time, assuming pollution-free inflow?

   b. If the constant $p$ has the value 0 for Lake Michigan, what does that tell you about the pollution level in the inflow? In this case, what happens to the ratio $P_L(t)/P_L(0)$ over time?

**7.** To solve the initial value problem in this Extended Application using the website WolframAlpha.com, you can enter "$y'(t) = (f(t) - y) * k, \; y(0) = a$", where we have used $y(t)$ to represent $P_L(t)$, $f(t)$ to represent $P_i(t)$, and $a$ to represent $P_L(0)$. Try this, and verify that the solution is equivalent to Equation (1).

**8.** Repeat Exercise 7, but in place of $f(t)$, put $a * e \wedge (-p * t)$, the form of $P_L(t)$ used in Exercises 5 and 6. Verify that the solution is equivalent to the solution given in Exercise 6.

**9.** Repeat Exercise 8, trying other functions of $t$ in place of $f(t)$, such as $t^3$. Find which functions give a recognizable answer, and verify that answer using Equation (1).

## DIRECTIONS FOR GROUP PROJECT

*Suppose you and three others are employed by an agency that is concerned about the environmental health of one of the Great Lakes. Choose one of the lakes, and collect information about levels of pollution in it. Then, using the information you collected along with the information given in this application, prepare a public presentation for a local community organization that describes the lake and gives possible timelines for reducing pollution in the lake. Use presentation software such as Microsoft PowerPoint.*

# 12 Probability

12.1 Sets

12.2 Introduction to Probability

12.3 Conditional Probability; Independent Events; Bayes' Theorem

12.4 Discrete Random Variables; Applications to Decision Making

Chapter 12 Review

Extended Application: Medical Diagnosis

Physicians are required to determine the best treatment strategy for a particular patient with a particular disease. Although it is generally impossible to know beforehand exactly how a patient will respond to certain types of medical procedures or pharmacological agents, probability can help physicians and researchers compare various options. In this chapter, some of the basic ideas from probability theory provide the foundation from which powerful decision tools are constructed.

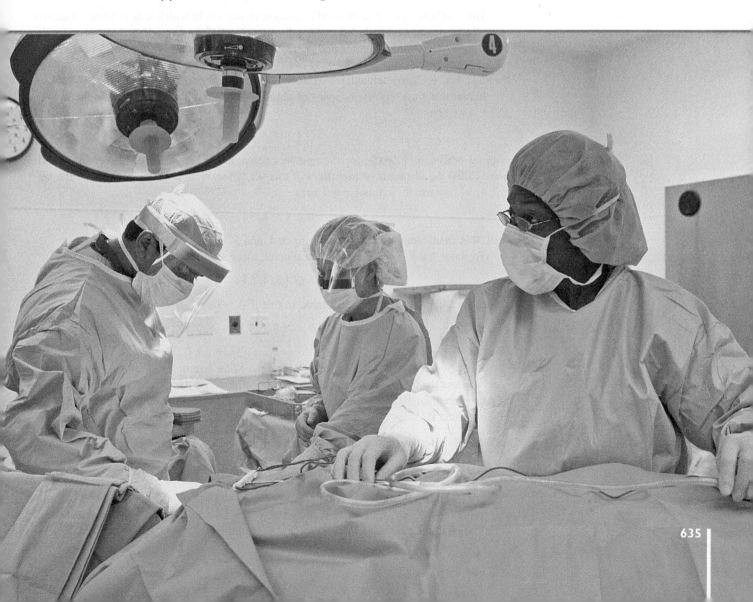

Our lives are bombarded by seemingly chance events — the chance of rain, the risk of an accident, the possibility of catching a cold, the likelihood of winning the lottery — each of whose particular outcome may appear to be quite random. The field of probability attempts to quantify the likelihood of chance events and helps us to prepare for this uncertainty. In short, probability helps us to better understand the world in which we live. In this chapter and the next, we introduce the basic ideas of probability theory and give a sampling of some of its uses. Since the language of sets and set operations is used in the study of probability, we begin there.

# 12.1 Sets

**APPLY IT** How many hockey players sustained head and face injuries while wearing a full shield?
*Using knowledge of sets, we will answer this question in Exercise 78.*

Think of a **set** as a well-defined collection of objects in which it is possible to determine if a given object is included in the collection. A set of bones might include one of each bone found in the human body. Another set might consist of all the students in your biology class. By contrast, a collection of young adults does not constitute a set unless the designation "young adult" is clearly defined. For example, this set might be defined as those aged 18 to 29.

In mathematics, sets often consist of numbers. The set consisting of the numbers 3, 4, and 5 is written

$$\{3, 4, 5\},$$

with set braces, $\{\ \}$, enclosing the numbers belonging to the set. The numbers 3, 4, and 5 are called the **elements** or **members** of this set. To show that 4 is an element of the set $\{3, 4, 5\}$, we use the symbol $\in$ and write

$$4 \in \{3, 4, 5\},$$

read "4 is an element of the set containing 3, 4, and 5." Also, $5 \in \{3, 4, 5\}$.

To show that 8 is *not* an element of this set, place a slash through the symbol:

$$8 \notin \{3, 4, 5\}.$$

Sets often are named with capital letters, so that if

$$B = \{5, 6, 7\},$$

then, for example, $6 \in B$ and $10 \notin B$.

It is possible to have a set with no elements. Some examples are the set of counting numbers less than one, the set of foreign-born presidents of the United States, and the set of men more than 10 feet tall. A set with no elements is called the **empty set** (or **null set**) and is written $\emptyset$.

> **CAUTION** Be careful to distinguish between the symbols 0, $\emptyset$, $\{0\}$, and $\{\emptyset\}$. The symbol 0 represents a *number*; $\emptyset$ represents a *set* with 0 elements; $\{0\}$ represents a set with one element, 0; and $\{\emptyset\}$ represents a set with one element, $\emptyset$.

We use the symbol $n(A)$ to indicate the *number* of unique elements in a finite set $A$. For example, if $A = \{a, b, c, d, e\}$, then $n(A) = 5$. Using this notation, we can write the information in the previous Caution as $n(\emptyset) = 0$ and $n(\{0\}) = n(\{\emptyset\}) = 1$.

Two sets are *equal* if they contain the same elements. The sets $\{5, 6, 7\}$, $\{7, 6, 5\}$, and $\{6, 5, 7\}$ all contain exactly the same elements and are equal. In symbols,

$$\{5, 6, 7\} = \{7, 6, 5\} = \{6, 5, 7\}.$$

This means that the ordering of the elements in a set is unimportant. Note that each element of the set is listed only once. Sets that do not contain exactly the same elements are *not equal*. For example, the sets $\{5, 6, 7\}$ and $\{7, 8, 9\}$ do not contain exactly the same elements and, thus, are not equal. To indicate that these sets are not equal, we write

$$\{5, 6, 7\} \neq \{7, 8, 9\}.$$

Sometimes we are interested in a common property of the elements in a set, rather than a list of the elements. This common property can be expressed by using **set-builder notation**, for example,

$$\{x \,|\, x \text{ has property } P\}$$

(read "the set of all elements $x$ such that $x$ has property $P$") represents the set of all elements $x$ having some stated property $P$.

### EXAMPLE 1 Sets

Write the elements belonging to each set.

**(a)** $\{x \,|\, x \text{ is a natural number less than } 5\}$

**SOLUTION** The natural numbers less than 5 make up the set $\{1, 2, 3, 4\}$.

**(b)** $\{x \,|\, x \text{ is a bone in the middle ear}\}$

**SOLUTION** The bones in the middle ear make up the set {malleus, incus, stapes}.

TRY YOUR TURN 1

**YOUR TURN 1** Write the elements belonging to the set $\{x \mid x \text{ is a state whose name begins with the letter O}\}$.

The **universal set** for a particular discussion is a set that includes all the objects being discussed. In elementary school arithmetic, for instance, the set of whole numbers might be the universal set, while in a college algebra class the universal set might be the set of real numbers. The universal set will be specified when necessary, or it will be clearly understandable from the context of the problem.

## Subsets
Sometimes every element of one set also belongs to another set. For example, if

$$A = \{3, 4, 5, 6\}$$

and

$$B = \{2, 3, 4, 5, 6, 7, 8\},$$

then every element of $A$ is also an element of $B$. This is an example of the following definition.

**Subset**

Set $A$ is a **subset** of set $B$ (written $A \subseteq B$) if every element of $A$ is also an element of $B$. Set $A$ is a *proper subset* (written $A \subset B$) if $A \subseteq B$ and $A \neq B$.

To indicate that $A$ is *not* a subset of $B$, we write $A \nsubseteq B$.

### EXAMPLE 2 Sets

Decide whether the following statements are *true* or *false*.

**(a)** $\{3, 4, 5, 6\} = \{4, 6, 3, 5\}$

**SOLUTION** Both sets contain exactly the same elements, so the sets are equal and the given statement is true. (The fact that the elements are listed in a different order does not matter.)

**(b)** $\{5, 6, 9, 12\} \subseteq \{5, 6, 7, 8, 9, 10, 11\}$

**SOLUTION** The first set is not a subset of the second because it contains an element, 12, that does not belong to the second set. Therefore, the statement is false.

**TRY YOUR TURN 2**

The empty set, Ø, by default, is a subset of every set $A$, since it is impossible to find an element of the empty set (it has no elements) that is not an element of the set $A$. Similarly, $A$ is a subset of itself, since every element of $A$ is also an element of the set $A$.

> **Subset Properties**
> For any set $A$,
> $$\emptyset \subseteq A \quad \text{and} \quad A \subseteq A.$$

## EXAMPLE 3 Subsets

List all possible subsets for each set.

**(a)** $\{7, 8\}$

**SOLUTION** There are 4 subsets of $\{7, 8\}$:
$$\emptyset, \quad \{7\}, \quad \{8\}, \quad \text{and} \quad \{7, 8\}.$$

**(b)** $\{a, b, c\}$

**SOLUTION** There are 8 subsets of $\{a, b, c\}$:
$$\emptyset, \quad \{a\}, \quad \{b\}, \quad \{c\}, \quad \{a, b\}, \quad \{a, c\}, \quad \{b, c\}, \quad \text{and} \quad \{a, b, c\}.$$

A good way to find the subsets of $\{7, 8\}$ and the subsets of $\{a, b, c\}$ in Example 3 is to use a **tree diagram**—a systematic way of listing all the subsets of a given set. Figure 1 shows tree diagrams for finding the subsets of $\{7, 8\}$ and $\{a, b, c\}$.

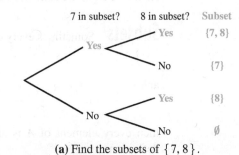

**(a)** Find the subsets of $\{7, 8\}$.

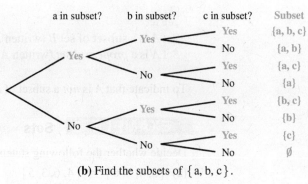

**(b)** Find the subsets of $\{a, b, c\}$.

FIGURE 1

As Figure 1 shows, there are two possibilities for each element (either it's in the subset or it's not), so a set with 2 elements has $2 \cdot 2 = 2^2 = 4$ subsets, and a set with 3 elements has $2^3 = 8$ subsets. This idea can be extended to a set with any finite number of elements, which leads to the following conclusion.

### Number of Subsets

A set of $k$ distinct elements has $2^k$ subsets.

In other words, if $n(A) = k$, then $n(\text{the set of all subsets of } A) = 2^k$.

### EXAMPLE 4  Subsets

Find the number of subsets for each set.

**(a)** $\{3, 4, 5, 6, 7\}$

**SOLUTION**  This set has 5 elements; thus, it has $2^5$ or 32 subsets.

**(b)** $\{x \mid x$ is a day of the week$\}$

**SOLUTION**  This set has 7 elements and therefore has $2^7 = 128$ subsets.

**(c)** $\emptyset$

**SOLUTION**  Since the empty set has 0 elements, it has $2^0 = 1$ subset—itself.

**TRY YOUR TURN 3**

**YOUR TURN 3**  Find the number of subsets for the set $\{x \mid x$ is a season of the year$\}$.

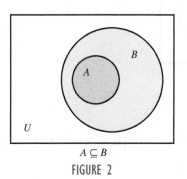

$A \subseteq B$

FIGURE 2

Figure 2 shows a set $A$ that is a subset of set $B$. The rectangle and everything inside it represents the universal set, $U$. Such diagrams, called **Venn diagrams**—after the English logician John Venn (1834–1923), who invented them in 1876—are used to help illustrate relationships among sets.

## Set Operations

It is possible to form new sets by combining or manipulating one or more existing sets. Given a set $A$ and a universal set $U$, the set of all elements of $U$ that do *not* belong to $A$ is called the *complement* of set $A$. For example, if set $A$ is the set of all the female students in a class, and $U$ is the set of all students in the class, then the complement of $A$ would be the set of all students in the class who are not female (that is, who are male). The complement of set $A$ is written $A'$, read "$A$-prime."

### Complement of a Set

Let $A$ be any set, with $U$ representing the universal set. Then the **complement** of $A$, colored pink in the figure, is

$$A' = \{x \mid x \notin A \text{ and } x \in U\}.$$

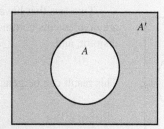

(Recall that the rectangle represents the universal set $U$.)

**EXAMPLE 5**  **Set Operations**

Let $U = \{1, 2, 3, 4, 5, 6, 7, 8, 9, 10, 11\}$, $A = \{1, 2, 4, 5, 7\}$, and $B = \{2, 4, 5, 7, 9, 11\}$. Find each set.

**(a)** $A'$

**SOLUTION**  Set $A'$ contains the elements of $U$ that are not in $A$.

$$A' = \{3, 6, 8, 9, 10, 11\}$$

**(b)** $B' = \{1, 3, 6, 8, 10\}$

**(c)** $\emptyset' = U$ and $U' = \emptyset$

**(d)** $(A')' = A$

Given two sets $A$ and $B$, the set of all elements belonging to *both* set $A$ and set $B$ is called the *intersection* of the two sets, written $A \cap B$. For example, the elements that belong to both set $A = \{1, 2, 4, 5, 7\}$ and set $B = \{2, 4, 5, 7, 9, 11\}$ are 2, 4, 5, and 7, so that

$$A \cap B = \{1, 2, 4, 5, 7\} \cap \{2, 4, 5, 7, 9, 11\} = \{2, 4, 5, 7\}.$$

**Intersection of Two Sets**

The **intersection** of sets $A$ and $B$, shown in green in the figure, is

$$A \cap B = \{x \mid x \in A \text{ and } x \in B\}.$$

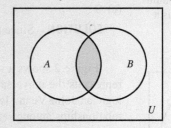

**EXAMPLE 6**  **Set Operations**

Let $A = \{3, 6, 9\}$, $B = \{2, 4, 6, 8\}$, and the universal set $U = \{0, 1, 2, \ldots, 10\}$. Find each set.

**(a)** $A \cap B$

**SOLUTION**

$$A \cap B = \{3, 6, 9\} \cap \{2, 4, 6, 8\} = \{6\}$$

**(b)** $A \cap B'$

**SOLUTION**

$$A \cap B' = \{3, 6, 9\} \cap \{0, 1, 3, 5, 7, 9, 10\} = \{3, 9\}$$

**TRY YOUR TURN 4**

**YOUR TURN 4**  For the sets in Example 6, find $A' \cap B$.

Two sets that have no elements in common are called *disjoint sets*. For example, there are no elements common to both $\{50, 51, 54\}$ and $\{52, 53, 55, 56\}$, so these two sets are disjoint, and

$$\{50, 51, 54\} \cap \{52, 53, 55, 56\} = \emptyset.$$

This result can be generalized as follows.

**Disjoint Sets**

For any sets $A$ and $B$, if $A$ and $B$ are **disjoint sets**, then $A \cap B = \emptyset$.

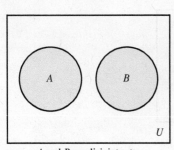

*A and B are disjoint sets.*

**FIGURE 3**

Figure 3 shows a pair of disjoint sets.

The set of all elements belonging to set $A$, to set $B$, or to both sets is called the *union* of the two sets, written $A \cup B$. For example,

$$\{1, 3, 5\} \cup \{3, 5, 7, 9\} = \{1, 3, 5, 7, 9\}.$$

### Union of Two Sets

The **union** of sets $A$ and $B$, shown in blue in the figure, is

$$A \cup B = \{x \mid x \in A \text{ or } x \in B \text{ (or both)}\}.$$

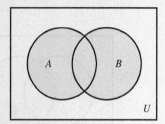

### EXAMPLE 7   Union of Sets

Let $A = \{1, 3, 5, 7, 9, 11\}$, $B = \{3, 6, 9, 12\}$, $C = \{1, 2, 3, 4, 5\}$, and the universal set $U = \{0, 1, 2, \ldots, 12\}$. Find each set.

**(a)** $A \cup B$

   **SOLUTION**   Begin by listing the elements of the first set, $\{1, 3, 5, 7, 9, 11\}$. Then include any elements from the second set *that are not already listed*. Doing this gives

$$A \cup B = \{1, 3, 5, 7, 9, 11\} \cup \{3, 6, 9, 12\} = \{1, 3, 5, 7, 9, 11, 6, 12\}$$
$$= \{1, 3, 5, 6, 7, 9, 11, 12\}.$$

**(b)** $(A \cup B) \cap C'$

   **SOLUTION**   Begin with the expression in parentheses, which we calculated in part (a), and then intersect this with $C'$.

$$(A \cup B) \cap C' = \{1, 3, 5, 6, 7, 9, 11, 12\} \cap \{0, 6, 7, 8, 9, 10, 11, 12\}$$
$$= \{6, 7, 9, 11, 12\}$$

**YOUR TURN 5**   For the sets in Example 7, find $A \cup (B \cap C')$.

TRY YOUR TURN 5

**NOTE**
1. As Example 7 shows, when forming sets, do not list the same element more than once. In our final answer, we listed the elements in numerical order to make it easier to see what elements are in the set, but the set is the same, regardless of the order of the elements.
2. As shown in the definitions, an element is in the *intersection* of sets $A$ and $B$ if it is in $A$ *and* $B$. On the other hand, an element is in the *union* of sets $A$ and $B$ if it is in $A$ *or* $B$ (or both).
3. In mathematics, "$A$ *or* $B$" implies $A$ or $B$, or both. Since "or" includes the possibility of both, we will usually omit the words "or both."

## Applications of Venn Diagrams

Venn diagrams can be used to efficiently represent a variety of complicated sets. The rectangular region of a Venn diagram represents the universal set $U$. Including only a single set $A$ inside the universal set, as in Figure 4, divides $U$ into two regions. Region 1 represents those elements of $U$ outside set $A$ (that is, the elements in $A'$), and region 2 represents those elements belonging to set $A$. (The numbering of these regions is arbitrary.)

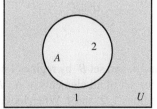

One set leads to 2 regions (numbering is arbitrary).

FIGURE 4

The Venn diagram in Figure 5(a) shows two sets inside $U$. These two sets divide the universal set into four regions. As labeled in Figure 5(a), region 1 represents the set whose elements are outside both set $A$ and set $B$. Region 2 shows the set whose elements belong to $A$ and not to $B$. Region 3 represents the set whose elements belong to both $A$ and $B$. Which set is represented by region 4? (Again, the labeling is arbitrary.)

Two other situations can arise when representing two sets by Venn diagrams. If it is known that $A \cap B = \emptyset$, then the Venn diagram is drawn as in Figure 5(b). If it is known that $A \subseteq B$, then the Venn diagram is drawn as in Figure 5(c). For the material presented throughout this chapter we will refer only to Venn diagrams like the one in Figure 5(a), and note that some of the regions of the Venn diagram may be equal to the empty (or null) set.

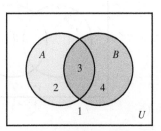

Two sets lead to 4 regions (numbering is arbitrary).

**(a)**

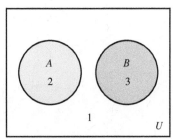

Two sets lead to 3 regions (numbering is arbitrary).

**(b)**

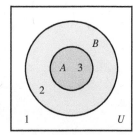

Two sets lead to 3 regions (numbering is arbitrary).

**(c)**

FIGURE 5

### EXAMPLE 8 Venn Diagrams

Draw Venn diagrams similar to Figure 5(a) and shade the regions representing each set.

**(a)** $A' \cap B$

**SOLUTION** Set $A'$ contains all the elements outside set $A$. As labeled in Figure 5(a), $A'$ is represented by regions 1 and 4. Set $B$ is represented by regions 3 and 4. The intersection of sets $A'$ and $B$, the set $A' \cap B$, is given by the region common to both sets. The result is the set represented by region 4, which is blue in Figure 6. When looking for the intersection, remember to choose the area that is in one region *and* the other region.

In addition to the fact that region 4 in Figure 6 is $A' \cap B$, notice that region 1 is $A' \cap B'$, region 2 is $A \cap B'$, and region 3 is $A \cap B$.

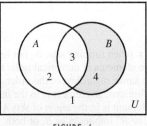

FIGURE 6

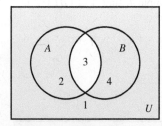

FIGURE 7

**(b)** $A' \cup B'$

**SOLUTION** Again, set $A'$ is represented by regions 1 and 4, and set $B'$ by regions 1 and 2. To find $A' \cup B'$, identify the region that represents the set of all elements in $A'$, $B'$, or both. The result, which is blue in Figure 7, includes regions 1, 2, and 4. When looking for the union, remember to choose the area that is in one region *or* the other region (or both).

**YOUR TURN 6** Draw a Venn diagram and shade the region representing $A \cup B'$.

**TRY YOUR TURN 6**

Venn diagrams also can be drawn with three sets inside $U$. These three sets divide the universal set into eight regions, which can be numbered (arbitrarily) as in Figure 8.

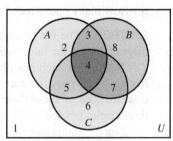

Three sets lead to 8 regions.

FIGURE 8

### EXAMPLE 9  Venn Diagram

In a Venn diagram, shade the region that represents $A' \cup (B \cap C')$.

**SOLUTION**  First find $B \cap C'$. Set $B$ is represented by regions 3, 4, 7, and 8, and set $C'$ by regions 1, 2, 3, and 8. The overlap of these regions (regions 3 and 8) represents the set $B \cap C'$. Set $A'$ is represented by regions 1, 6, 7, and 8. The union of the set represented by regions 3 and 8 and the set represented by regions 1, 6, 7, and 8 is the set represented by regions 1, 3, 6, 7, and 8, which are blue in Figure 9.

**YOUR TURN 7**  Draw a Venn diagram and shade the region representing $A' \cap (B \cup C)$.

TRY YOUR TURN 7

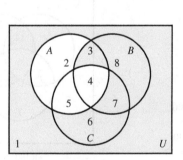

FIGURE 9

## Applications  Venn diagrams can be used to analyze many applications, as illustrated in the following examples.

### EXAMPLE 10  Genetics

After a genetics experiment with 50 pea plants, the number of plants having certain characteristics was tallied, with the following results:

22 were tall;

25 had green peas;

39 had smooth peas;

14 had smooth peas and were tall;

9 were tall and had green peas;

20 had green peas and smooth peas;

6 had all three characteristics.

Use this information to answer the following questions.

**(a)** How many plants had none of the characteristics?

**(b)** How many plants had smooth peas, but were neither tall nor had green peas?

**(c)** How many plants had exactly one of the above characteristics?

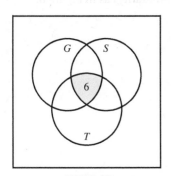

FIGURE 10

**SOLUTION** The list of plant characteristics may seem daunting, but we will use a Venn diagram to sort the information.

Notice that each pea plant is classified into three main categories: being tall (or not), having green peas (or not), and having smooth peas (or not). The rest of the results are combinations of these three characteristics. Therefore, we will use a Venn diagram with three circles labeled $T$ (for tall), $G$ (for green), and $S$ (for smooth), like the one in Figure 10.

Since 6 plants had all three characteristics (tall, green, and smooth), begin by placing 6 in the area that belongs to all three regions, as shown in Figure 10.

Of the 20 plants that had both green and smooth peas, 6 were also tall. Therefore, only $20 - 6 = 14$ plants had just green and smooth peas, and were not tall. Place 14 in the region that is common to green and smooth but not tall, as in Figure 11.

Similarly, $9 - 6 = 3$ plants were tall and had green peas that were not smooth, so we put a 3 in that region. Also, $14 - 6 = 8$ of the plants had smooth peas and were tall, so 8 is placed in that region. See Figure 11.

The data show that 39 of the plants were smooth. However, $8 + 6 + 14 = 28$ of the plants have already been placed in the region representing smooth peas. The balance of this region will contain only $39 - 28 = 11$ plants. These plants had only the characteristic of having smooth peas—they were not tall and did not have green peas. In the same way, 2 plants had only green peas and 5 plants were only tall, as indicated by Figure 12.

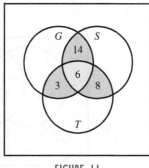

FIGURE 11

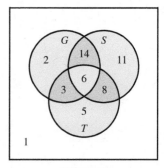

FIGURE 12

**YOUR TURN 8** One hundred students were asked which fast food restaurants they had visited in the past month. The results are as follows:

47 ate at McDonald's;
46 ate at Taco Bell;
44 ate at Wendy's;
17 ate at McDonald's and Taco Bell;
19 ate at Taco Bell and Wendy's;
22 ate at Wendy's and McDonald's;
13 ate at all three.

Determine how many ate only at Taco Bell.

A total of $2 + 14 + 6 + 3 + 11 + 8 + 5 = 49$ plants were placed in the three circles of Figure 12. Since 50 plants were in the experiment, $50 - 49 = 1$ plant had none of the three characteristics, and a 1 is placed outside all three regions.

Now Figure 12 can be used to answer the questions asked above.

**(a)** Only one plant had none of the three characteristics.

**(b)** There were 11 plants with smooth peas that were neither tall nor had green peas.

**(c)** There were 18 plants that had exactly one of the characteristics.

TRY YOUR TURN 8

CAUTION | A common error in solving problems of this type is to make a circle represent one set and another circle represent its complement. In Example 10, with one circle representing those plants that were smooth, we did not draw another for those plants that were not smooth. An additional circle is not only unnecessary (because those not in one set are automatically in the other) but also very confusing, because the region outside or inside both circles must be empty. Similarly, if a problem involves men and women, do not draw one circle for men and another for women. Draw one circle; if you label it "women," for example, then men are automatically those outside the circle.

In the previous example, we started with a piece of information specifying the relationship with all the categories. This is usually the best way to begin solving problems of this type.

As we saw earlier, we use the symbol $n(A)$ to indicate the *number* of elements in a finite set $A$. The following statement about the number of elements in the union of two sets will be used later in our study of probability.

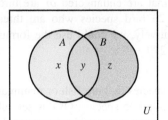

**FIGURE 13**

> ### Union Rule for Sets
>
> $$n(A \cup B) = n(A) + n(B) - n(A \cap B)$$

To prove this statement, let $x + y$ represent $n(A)$, $y$ represent $n(A \cap B)$, and $y + z$ represent $n(B)$, as shown in Figure 13. Then

$$n(A \cup B) = x + y + z,$$
$$n(A) + n(B) - n(A \cap B) = (x + y) + (y + z) - y = x + y + z,$$

so

$$n(A \cup B) = n(A) + n(B) - n(A \cap B).$$

## EXAMPLE 11 School Activities

A group of 10 students meet to plan a school function. All are majoring in biology or chemistry or both. Five of the students are chemistry majors and 7 are majors in biology. How many major in both subjects?

**YOUR TURN 9** A group of students are sitting in the lounge. All are texting or listening to music or both. Eleven are listening to music and 15 are texting. Eight are doing both. How many students are in the lounge?

**SOLUTION** Let $A$ represent the set of chemistry majors and $B$ represent the set of biology majors. Use the union rule, with $n(A) = 5$, $n(B) = 7$, and $n(A \cup B) = 10$. Find $n(A \cap B)$.

$$n(A \cup B) = n(A) + n(B) - n(A \cap B)$$
$$10 = 5 + 7 - n(A \cap B),$$

so

$$n(A \cap B) = 5 + 7 - 10 = 2.$$   **TRY YOUR TURN 9**

When $A$ and $B$ are disjoint, then $n(A \cap B) = 0$, so the union rule simplifies to $n(A \cup B) = n(A) + n(B)$.

> **CAUTION** The rule $n(A \cup B) = n(A) + n(B)$ is only valid when A and B are disjoint. When A and B are *not* disjoint, use the rule $n(A \cup B) = n(A) + n(B) - n(A \cap B)$.

## EXAMPLE 12 Endangered Species

The following table gives the number of threatened and endangered animal species in the world as of January, 2010. *Source: U.S. Fish and Wildlife Service.*

| Endangered and Threatened Species | | | |
|---|---|---|---|
| | Endangered ($E$) | Threatened ($T$) | Totals |
| Amphibians and Reptiles ($A$) | 101 | 52 | 153 |
| Arachnids and Insects ($I$) | 63 | 10 | 73 |
| Birds ($B$) | 256 | 23 | 279 |
| Clams, Crustaceans, Corals, and Snails ($C$) | 108 | 24 | 132 |
| Fishes ($F$) | 85 | 66 | 151 |
| Mammals ($M$) | 325 | 35 | 360 |
| Totals | 938 | 210 | 1148 |

Using the letters given in the table to denote each set, find the number of species in each of the following sets.

**(a)** $E \cap B$

**SOLUTION** The set $E \cap B$ consists of all species that are endangered *and* are birds. From the table, we see that there are 256 such species.

**(b)** $E \cup B$

**SOLUTION** The set $E \cup B$ consists of all species that are endangered *or* are birds. We include all 938 endangered species, plus the 23 bird species who are threatened but not endangered, for a total of 961. Alternatively, we could use the formula $n(E \cup B) = n(E) + n(B) - n(E \cap B) = 938 + 279 - 256 = 961$.

**(c)** $(F \cup M) \cap T'$

**SOLUTION** Begin with the set $F \cup M$, which is all species that are fish or mammals. This consists of the four categories with 85, 66, 325, and 35 species. Of this set, take those that are *not* threatened, for a total of $85 + 325 = 410$ species. This is the number of species of fish and mammals that are not threatened.

# 12.1 EXERCISES

**In Exercises 1–6, write true or false for each statement.**

**1.** $3 \in \{2, 5, 7, 9, 10\}$
**2.** $9 \notin \{2, 1, 5, 8\}$
**3.** $\{2, 5, 8, 9\} = \{2, 5, 9, 8\}$
**4.** $\{x \mid x \text{ is an odd integer}; 6 \le x \le 18\} = \{7, 9, 11, 15, 17\}$
**5.** $0 \in \emptyset$
**6.** $\emptyset \in \{\emptyset\}$

Let $A = \{2, 4, 6, 10, 12\}$, $B = \{2, 4, 8, 10\}$, $C = \{4, 8, 12\}$, $D = \{2, 10\}$, $E = \{6\}$, and $U = \{2, 4, 6, 8, 10, 12, 14\}$. Insert $\subseteq$ or $\not\subseteq$ to make the statement true.

**7.** $A$ ___ $U$
**8.** $E$ ___ $A$
**9.** $A$ ___ $E$
**10.** $B$ ___ $C$
**11.** $\emptyset$ ___ $A$
**12.** $\{0, 2\}$ ___ $D$

**13.** Repeat Exercises 7–12 except insert $\subset$ or $\not\subset$ to make the statement true.

**14.** What is set-builder notation? Give an example.

**Insert a number in each blank to make the statement true, using the sets for Exercises 7–12.**

**15.** There are exactly ___ subsets of $A$.
**16.** There are exactly ___ subsets of $B$.

**Insert $\cap$ or $\cup$ to make each statement true.**

**17.** $\{5, 7, 9, 19\}$ ___ $\{7, 9, 11, 15\} = \{7, 9\}$
**18.** $\{8, 11, 15\}$ ___ $\{8, 11, 19, 20\} = \{8, 11\}$
**19.** $\{2, 1, 7\}$ ___ $\{1, 5, 9\} = \{1, 2, 5, 7, 9\}$
**20.** $\{6, 12, 14, 16\}$ ___ $\{6, 14, 19\} = \{6, 12, 14, 16, 19\}$
**21.** $\{3, 5, 9, 10\}$ ___ $\emptyset = \emptyset$
**22.** $\{3, 5, 9, 10\}$ ___ $\emptyset = \{3, 5, 9, 10\}$
**23.** $\{1, 2, 4\}$ ___ $\{1, 2, 4\} = \{1, 2, 4\}$
**24.** $\{0, 10\}$ ___ $\{10, 0\} = \{0, 10\}$

**25.** Describe the intersection and union of sets. How do they differ?
**26.** Is it possible for two nonempty sets to have the same intersection and union? If so, give an example.

Let $U = \{1, 2, 3, 4, 5, 6, 7, 8, 9\}$, $X = \{2, 4, 6, 8\}$, $Y = \{2, 3, 4, 5, 6\}$, and $Z = \{1, 2, 3, 8, 9\}$. List the members of each set, using set braces.

**27.** $X \cap Y$
**28.** $X \cup Y$
**29.** $X'$
**30.** $X' \cap Y'$
**31.** $Y \cap (X \cup Z)$
**32.** $X' \cap (Y' \cup Z)$
**33.** $(X \cap Y') \cup (Z' \cap Y')$
**34.** $(X \cap Y) \cup (X' \cap Z)$

**35. a.** In Example 6, what set do you get when you calculate $(A \cap B) \cup (A \cap B')$?

**b.** Explain in words why $(A \cap B) \cup (A \cap B') = A$.

**36.** Refer to the sets listed for Exercises 7–12. Which pairs of sets are disjoint?

**37.** Let $A = \{1, 2, 3, \{3\}, \{1, 4, 7\}\}$. Answer each of the following as *true* or *false*.

**a.** $1 \in A$
**b.** $\{3\} \in A$
**c.** $\{2\} \in A$
**d.** $4 \in A$
**e.** $\{\{3\}\} \subset A$
**f.** $\{1, 4, 7\} \in A$
**g.** $\{1, 4, 7\} \subseteq A$

**38.** Let $B = \{a, b, c, \{d\}, \{e, f\}\}$. Answer each of the following as *true* or *false*.

**a.** $a \in B$
**b.** $\{b, c, d\} \subset B$
**c.** $\{d\} \in B$
**d.** $\{d\} \subseteq B$
**e.** $\{e, f\} \in B$
**f.** $\{a, \{e, f\}\} \subset B$
**g.** $\{e, f\} \subset B$

**Sketch a Venn diagram like the one in the figure, and use shading to show each set.**

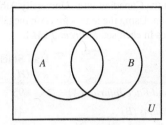

**39.** $B \cap A'$  **40.** $A \cup B'$

**41.** $A' \cup B$  **42.** $A' \cap B'$

**43.** $B' \cup (A' \cap B')$  **44.** $(A \cap B) \cup B'$

**45.** $U'$  **46.** $\emptyset'$

 **47.** Three sets divide the universal set into at most ___ regions.

 **48.** What does the notation $n(A)$ represent?

**Sketch a Venn diagram like the one shown, and use shading to show each set.**

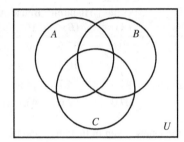

**49.** $(A \cap B) \cap C$  **50.** $(A \cap C') \cup B$

**51.** $A \cap (B \cup C')$  **52.** $A' \cap (B \cap C)$

**53.** $(A' \cap B') \cap C'$  **54.** $(A \cap B') \cap C$

**55.** $A' \cap (B' \cup C)$  **56.** $(A \cup B') \cap C$

**Use the union rule to answer the following questions.**

**57.** If $n(A) = 5$, $n(B) = 12$, and $n(A \cap B) = 4$, what is $n(A \cup B)$?

**58.** If $n(A) = 15$, $n(B) = 30$, and $n(A \cup B) = 33$, what is $n(A \cap B)$?

**59.** Suppose $n(B) = 9$, $n(A \cap B) = 5$, and $n(A \cup B) = 22$. What is $n(A)$?

**60.** Suppose $n(A \cap B) = 5$, $n(A \cup B) = 38$, and $n(A) = 13$. What is $n(B)$?

**Draw a Venn diagram and use the given information to fill in the number of elements for each region.**

**61.** $n(U) = 41, n(A) = 16, n(A \cap B) = 12, n(B') = 20$

**62.** $n(A) = 28, n(B) = 12, n(A \cup B) = 32, n(A') = 19$

**63.** $n(A \cup B) = 24, n(A \cap B) = 6, n(A) = 11,$
  $n(A' \cup B') = 25$

**64.** $n(A') = 31, n(B) = 25, n(A' \cup B') = 46, n(A \cap B) = 12$

**65.** $n(A) = 28, n(B) = 34, n(C) = 25, n(A \cap B) = 14,$
  $n(B \cap C) = 15, n(A \cap C) = 11, n(A \cap B \cap C) = 9,$
  $n(U) = 59$

**66.** $n(A) = 54, n(A \cap B) = 22, n(A \cup B) = 85,$
  $n(A \cap B \cap C) = 4, n(A \cap C) = 15, n(B \cap C) = 16,$
  $n(C) = 44, n(B') = 63$

**67.** $n(A \cap B) = 6, n(A \cap B \cap C) = 4, n(A \cap C) = 7,$
  $n(B \cap C) = 4, n(A \cap C') = 11, n(B \cap C') = 8,$
  $n(C) = 15, n(A' \cap B' \cap C') = 5$

**68.** $n(A) = 13, n(A \cap B \cap C) = 4, n(A \cap C) = 6,$
  $n(A \cap B') = 6, n(B \cap C) = 6, n(B \cap C') = 11,$
  $n(B \cup C) = 22, n(A' \cap B' \cap C') = 5$

**In Exercises 69–72, show that the statement is true by drawing Venn diagrams and shading the regions representing the sets on each side of the equals sign.***

**69.** $(A \cup B)' = A' \cap B'$

**70.** $(A \cap B)' = A' \cup B'$

**71.** $A \cap (B \cup C) = (A \cap B) \cup (A \cap C)$

**72.** $A \cup (B \cap C) = (A \cup B) \cap (A \cup C)$

**73.** Use the union rule of sets to prove that $n(A \cup B \cup C) = n(A) + n(B) + n(C) - n(A \cap B) - n(A \cap C) - n(B \cap C) + n(A \cap B \cap C)$. (*Hint:* Write $A \cup B \cup C$ as $A \cup (B \cup C)$ and use the formula from Exercise 71.)

## LIFE SCIENCE APPLICATIONS

**74. Health** The following table shows some symptoms of an overactive thyroid and an underactive thyroid. *Source: The Merck Manual of Diagnosis and Therapy.*

| Underactive Thyroid | Overactive Thyroid |
|---|---|
| Sleepiness, $s$ | Insomnia, $i$ |
| Dry hands, $d$ | Moist hands, $m$ |
| Intolerance of cold, $c$ | Intolerance of heat, $h$ |
| Goiter, $g$ | Goiter, $g$ |

Let $U$ be the smallest possible set that includes all the symptoms listed, $N$ be the set of symptoms for an underactive thyroid, and O be the set of symptoms for an overactive thyroid. Find each set.

**a.** $O'$  **b.** $N'$

**c.** $N \cap O$  **d.** $N \cup O$

**e.** $N \cap O'$

---

*The statements in Exercises 69 and 70 are known as De Morgan's Laws. They are named for the English mathematician Augustus De Morgan (1806–1871).

**75. Blood Antigens** Human blood can contain the A antigen, the B antigen, both the A and B antigens, or neither antigen. A third antigen, called the Rh antigen, is important in human reproduction, and again may or may not be present in an individual. Blood is called type A-positive if the individual has the A and Rh but not the B antigen. A person having only the A and B antigens is said to have type AB-negative blood. A person having only the Rh antigen has type O-positive blood. Other blood types are defined in a similar manner. Identify the blood types of the individuals in regions (a)–(h) below.

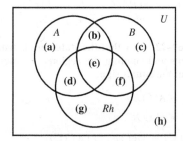

**76. Blood Antigens** (Use the diagram from Exercise 75.) In a certain hospital, the following data were recorded.

25 patients had the A antigen;

8 had the A and not the B antigen;

27 had the B antigen;

22 had the B and Rh antigens;

30 had the Rh antigen;

12 had none of the antigens;

16 had the A and Rh antigens;

15 had all three antigens.

How many patients

a. were represented?     b. had exactly one antigen?

c. had exactly two antigens?     d. had O-positive blood?

e. had AB-positive blood?     f. had B-negative blood?

g. had O-negative blood?     h. had A-positive blood?

**77. Mortality** The table lists the number of deaths in the United States during 2010 according to race and gender. Use this information and the letters given to find the number of people in each set. *Source: National Vital Statistics Reports.*

| | White (W) | Black (B) | American Indian (I) | Asian or Pacific Islander (A) |
|---|---|---|---|---|
| *Female (F)* | 1,063,235 | 141,157 | 7049 | 24,562 |
| *Male (M)* | 1,051,514 | 145,802 | 8516 | 26,600 |

a. $F$     b. $F \cap (I \cup A)$

c. $M \cup B$     d. $W' \cup I' \cup A'$

e. In words, describe the set in part b.

**78. Hockey** The table lists the number of head and neck injuries for 319 ice hockey players wearing either a full shield or half shield in the Canadian Inter-University Athletics Union during one season. Using the letters given in the table, find the number of injuries in each set. *Source: JAMA.*

| | Half Shield (H) | Full Shield (F) |
|---|---|---|
| *Head and Face Injuries (A)* | 95 | 34 |
| *Concussions (B)* | 41 | 38 |
| *Neck Injuries (C)* | 9 | 7 |
| *Other Injuries (D)* | 202 | 150 |

APPLY IT a. $A \cap F$     b. $C \cap (H \cup F)$

c. $D \cup F$     d. $B' \cap C'$

**79. U.S. Population** The projected U.S. population in 2020 (in millions) by age and race or ethnicity is given in the following table. *Source: U.S. Bureau of the Census.*

| | Non-Hispanic White (A) | Hispanic (B) | Black (C) | Asian (D) | American Indian (E) |
|---|---|---|---|---|---|
| *Under 45 (F)* | 110.6 | 37.6 | 30.2 | 13.1 | 2.2 |
| *45–64 (G)* | 55.3 | 10.3 | 9.9 | 4.3 | 0.6 |
| *65 and Over (H)* | 41.4 | 4.7 | 5.0 | 2.2 | 0.3 |
| *Totals* | 207.3 | 52.6 | 45.1 | 19.6 | 3.1 |

Using the letters given in the table, find the number of people in each set.

a. $A \cap F$

b. $G \cup B$

c. $G \cup (C \cap H)$

d. $F \cap (B \cup H)$

e. $H \cup D$

f. $G' \cap (A' \cap C')$

**80. Poultry Analysis** A chicken farmer surveyed his flock with the following results. The farmer had

9 fat red roosters;

13 thin brown hens;

15 red roosters;

11 thin red chickens (hens and roosters);

17 red hens;

56 fat chickens (hens and roosters);

41 roosters;

48 hens.

Assume all chickens are thin or fat, red or brown, and hens (female) or roosters (male). How many chickens were

a. in the flock?     b. red?

c. fat roosters?     d. fat hens?

e. thin and brown?     f. red and fat?

## OTHER APPLICATIONS

**81. Military** The number of female military personnel in 2012 is given in the following table. Use this information and the letters given to find the number of female military personnel in each set. *Source: Department of Defense.*

|  | Army (A) | Air Force (B) | Navy (C) | Marines (D) | Totals |
|---|---|---|---|---|---|
| *Officers (O)* | 16,001 | 12,487 | 8634 | 1348 | 38,470 |
| *Enlisted (E)* | 57,429 | 49,750 | 44,274 | 12,593 | 164,046 |
| *Cadets & Midshipmen (M)* | 752 | 894 | 147 | 0 | 1793 |
| *Totals* | 74,182 | 63,131 | 53,055 | 13,941 | 204,309 |

a. $A \cup B$

b. $E \cup (C \cup D)$

c. $O' \cap M'$

**82. Marital Status** The following table gives the population breakdown (in millions) of the U.S. population in 2010 based on marital status and race or ethnic origin. *Source: U.S. Census Bureau.*

|  | White (W) | Black (B) | Hispanic (H) | Asian or Pacific Islander (A) |
|---|---|---|---|---|
| *Never Married (N)* | 45.1 | 11.7 | 10.9 | 2.7 |
| *Married (M)* | 109.4 | 10.6 | 17.1 | 7.0 |
| *Widowed (I)* | 11.8 | 1.8 | 1.2 | 0.5 |
| *Divorced/ Separated (D)* | 19.4 | 3.2 | 2.6 | 0.5 |

Find the number of people in each set. Describe each set in words.

a. $N \cap (B \cup H)$

b. $(M \cup I) \cap A$

c. $(D \cup W) \cap A'$

d. $M' \cap (B \cup A)$

**83. Chinese New Year** A survey of people attending a Lunar New Year celebration in Chinatown yielded the following results:

120 were women;

150 spoke Cantonese;

170 lit firecrackers;

108 of the men spoke Cantonese;

100 of the men did not light firecrackers;

18 of the non-Cantonese-speaking women lit firecrackers;

78 non-Cantonese-speaking men did not light firecrackers;

30 of the women who spoke Cantonese lit firecrackers.

a. How many attended?

b. How many of those who attended did not speak Cantonese?

c. How many women did not light firecrackers?

d. How many of those who lit firecrackers were Cantonese-speaking men?

**84. Native American Ceremonies** At a pow-wow in Arizona, 75 Native American families from all over the Southwest came to participate in the ceremonies. A coordinator of the pow-wow took a survey and found that

15 families brought food, costumes, and crafts;

25 families brought food and crafts;

42 families brought food;

35 families brought crafts;

14 families brought crafts but not costumes;

10 families brought none of the three items;

18 families brought costumes but not crafts.

a. How many families brought costumes and food?

b. How many families brought costumes?

c. How many families brought food, but not costumes?

d. How many families did not bring crafts?

e. How many families brought food or costumes?

**85. Harvesting Fruit** Toward the middle of the harvesting season, peaches for canning come in three types, early, late, and extra late, depending on the expected date of ripening. During a certain week, the following data were recorded at a fruit delivery station:

34 trucks went out carrying early peaches;

61 carried late peaches;

50 carried extra late;

25 carried early and late;

30 carried late and extra late;

8 carried early and extra late;

6 carried all three;

9 carried only figs (no peaches at all).

a. How many trucks carried only late variety peaches?

b. How many carried only extra late?

c. How many carried only one type of peach?

d. How many trucks (in all) went out during the week?

**86. Cola Consumption** Market research showed that the adult residents of a certain small town in Georgia fit the following categories of cola consumption. (We assume here that no one drinks both regular cola and diet cola.)

| Age | Drink Regular Cola (R) | Drink Diet Cola (D) | Drink No Cola (N) | Totals |
|---|---|---|---|---|
| 21–25 (Y) | 40 | 15 | 15 | 70 |
| 26–35 (M) | 30 | 30 | 20 | 80 |
| Over 35 (O) | 10 | 50 | 10 | 70 |
| Totals | 80 | 95 | 45 | 220 |

Using the letters given in the table, find the number of people in each set.

a. $Y \cap R$

b. $M \cap D$

c. $M \cup (D \cap Y)$

d. $Y' \cap (D \cup N)$

e. $O' \cup N$

f. $M' \cap (R' \cap N')$

g. Describe the set $M \cup (D \cap Y)$ in words.

**87. Living Arrangements** In 2012, there were 73,817 (in thousands) children under the age of 18 living in the United States. Of these children, 50,267 lived with both parents, 2924 lived with their father only, and 2634 lived with neither parent. How many children lived with their mother only? *Source: U.S. Census.*

**88. Sales Calls** Suppose that Kendra Gallegos has appointments with 9 potential customers. Kendra will be ecstatic if all 9 of these potential customers decide to make a purchase from her. Of course, in sales there are no guarantees. How many different sets of customers may place an order with Kendra? (*Hint:* Each set of customers is a subset of the original set of 9 customers.)

**89. Electoral College** U.S. presidential elections are decided by the Electoral College, in which each of the 50 states, plus the District of Columbia, gives all of its votes to a candidate.* Ignoring the number of votes each state has in the Electoral College, but including all possible combinations of states that could be won by either candidate, how many outcomes are possible in the Electoral College if there are two candidates? (*Hint:* The states that can be won by a candidate form a subset of all the states.)

**90. Musicians** A concert featured a cellist, a flutist, a harpist, and a vocalist. Throughout the concert, different subsets of the four musicians performed together, with at least two musicians playing each piece. How many subsets of at least two are possible?

**91. Games** In David Gale's game of Subset Takeaway, the object is for each player, at his or her turn, to pick a nonempty proper subset of a given set subject to the condition that no subset chosen earlier by either player can be a subset of the newly chosen set. The winner is the last person who can make a legal move. Consider the set $A = \{1, 2, 3\}$. Suppose Joe and Dorothy are playing the game and Dorothy goes first. If she chooses the proper subset $\{1\}$, then Joe cannot choose any subset that includes the element 1. Joe can, however, choose $\{2\}$ or $\{3\}$ or $\{2, 3\}$. Develop a strategy for Joe so that he can always win the game if Dorothy goes first. *Source: Scientific American.*

**States** In the following list of states, let $A = \{$states whose name contains the letter $e\}$, let $B = \{$states with a population

*The exceptions are Maine and Nebraska, which allocate their electoral college votes according to the winner in each congressional district.

of more than 4,000,000$\}$, and $C = \{$states with an area greater than 40,000 square miles$\}$. *Source: The World Almanac and Book of Facts 2013.*

| State | Population (1000s) | Area (sq. mi.) |
|---|---|---|
| Alabama | 4803 | 52,419 |
| Alaska | 723 | 663,267 |
| Colorado | 5117 | 104,094 |
| Florida | 19,058 | 65,755 |
| Hawaii | 1375 | 10,931 |
| Indiana | 6517 | 36,418 |
| Kentucky | 4369 | 40,409 |
| Maine | 1328 | 35,385 |
| Nebraska | 1843 | 77,354 |
| New Jersey | 8821 | 8721 |

**92. a.** Describe in words the set $A \cup (B \cap C)'$.
   **b.** List all elements in the set $A \cup (B \cap C)'$.
**93. a.** Describe in words the set $(A \cup B)' \cap C$.
   **b.** List all elements in the set $(A \cup B)' \cap C$.

**YOUR TURN ANSWERS**

1. $\{$Ohio, Oklahoma, Oregon$\}$   2. True   3. $2^4 = 16$ subsets
4. $\{2, 4, 8\}$   5. $\{1, 3, 5, 6, 7, 9, 11, 12\}$
6. 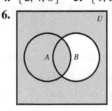   7.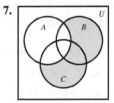
8. 23   9. 18

# 12.2 Introduction to Probability

**APPLY IT** What is the probability that a death in the United States was caused by heart disease or cancer?
*After introducing probability, we will answer this question in Example 13.*

When a physician prescribes an antibiotic, she specifies the *exact* amount of the drug that you should receive. On the other hand, the managing pharmacist is faced with the problem of ordering the medicine. The pharmacist may have a good estimate of the amount of an antibiotic that will be sold during the day, but it is impossible to predict the *exact* amount. The quantity of the antibiotic that customers will purchase during a day is *random*: it cannot be predicted exactly. There are many problems that come up in life science applications of mathematics that involve random phenomena—those for which *exact* prediction is impossible. The best that we can do is to determine the *probability* of the possible outcomes.

## Sample Spaces

In probability, an **experiment** is an activity or occurrence with an observable result. Each repetition of an experiment is called a **trial**. The possible results of each trial are called **outcomes**. The set of all possible outcomes for an experiment is the **sample space** for that experiment. A sample space for the experiment of tossing a coin is made up of the outcomes heads $(h)$ and tails $(t)$. If $S$ represents this sample space, then

$$S = \{h, t\}.$$

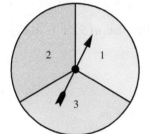

**FIGURE 14**

### EXAMPLE 1  Sample Spaces

Give the sample space for each experiment.

**(a)** A spinner like the one in Figure 14 is spun.

**SOLUTION**  The three outcomes are 1, 2, or 3, so the sample space is

$$S = \{1, 2, 3\}.$$

**(b)** For the purposes of a public opinion poll, respondents are classified as young, middle-aged, or senior, and as male or female.

**SOLUTION**  A sample space for this poll could be written as a set of ordered pairs:

$$S = \{(\text{young, male}), (\text{young, female}), (\text{middle-aged, male}),$$
$$(\text{middle-aged, female}), (\text{senior, male}), (\text{senior, female})\}.$$

**(c)** An experiment consists of studying the numbers of boys and girls in families with exactly 3 children. Let *b* represent *boy* and *g* represent *girl*.

**SOLUTION**  A three-child family can have 3 boys, written *bbb*, 3 girls, *ggg*, or various combinations, such as *bgg*. A sample space with four outcomes (not equally likely) is

$$S_1 = \{3 \text{ boys, 2 boys and 1 girl, 1 boy and 2 girls, 3 girls}\}.$$

Notice that a family with 3 boys or 3 girls can occur in just one way, but a family of 2 boys and 1 girl or 1 boy and 2 girls can occur in more than one way. If the *order* of the births is considered, so that *bgg* is different from *gbg* or *ggb*, for example, another sample space is

$$S_2 = \{bbb, bbg, bgb, gbb, bgg, gbg, ggb, ggg\}.$$

The second sample space, $S_2$, has equally likely outcomes if we assume that boys and girls are equally likely. This assumption, while not quite true, is approximately true, so we will use it throughout this book. The outcomes in $S_1$ are not equally likely, since there is more than one way to get a family with 2 boys and 1 girl (*bbg*, *bgb*, or *gbb*) or a family with 2 girls and 1 boy (*ggb*, *gbg*, or *bgg*), but only one way to get 3 boys (*bbb*) or 3 girls (*ggg*).

TRY YOUR TURN 1

**YOUR TURN 1**  Two coins are tossed, and a head or a tail is recorded for each coin. Give a sample space where each outcome is equally likely.

**CAUTION**  An experiment may have more than one sample space, as shown in Example 1(c). The most convenient sample spaces have equally likely outcomes, but it is not always possible to choose such a sample space.

**Events** An **event** is a subset of a sample space. If the sample space for tossing a coin is $S = \{h, t\}$, then one event is $E = \{h\}$, which represents the outcome "heads."

An ordinary die is a cube whose six different faces show the following numbers of dots: 1, 2, 3, 4, 5, and 6. If the die is fair (not "loaded" to favor certain faces over others), then any one of the faces is equally likely to occur when the die is rolled. The sample space for the experiment of rolling a single fair die is $S = \{1, 2, 3, 4, 5, 6\}$. Some possible events are listed below.

The die shows an even number: $E_1 = \{2, 4, 6\}$.

The die shows a 1: $E_2 = \{1\}$.

The die shows a number less than 5: $E_3 = \{1, 2, 3, 4\}$.

The die shows a multiple of 3: $E_4 = \{3, 6\}$.

Using the notation introduced earlier in this chapter, notice that $n(S) = 6, n(E_1) = 3$, $n(E_2) = 1, n(E_3) = 4$, and $n(E_4) = 2$.

**EXAMPLE 2** Events

For the sample space $S_2$ in Example 1(c), write the following events.

**(a)** Event $H$: The family has exactly two girls.

**SOLUTION** Families with three children can have exactly two girls with either *bgg*, *gbg*, or *ggb*, so event $H$ is

$$H = \{bgg, gbg, ggb\}.$$

**(b)** Event $K$: The three children are the same sex.

**SOLUTION** Two outcomes satisfy this condition: all boys or all girls.

$$K = \{bbb, ggg\}$$

**(c)** Event $J$: The family has three girls.

**SOLUTION** Only *ggg* satisfies this condition, so

$$J = \{ggg\}. \qquad \text{TRY YOUR TURN 2}$$

**YOUR TURN 2** Two coins are tossed, and a head or a tail is recorded for each coin. Write the event $E$: The coins show exactly one head.

In Example 2(c), event $J$ had only one possible outcome, *ggg*. Such an event, with only one possible outcome, is a **simple event**. If event $E$ equals the sample space $S$, then $E$ is called a **certain event**. If event $E = \emptyset$, then $E$ is called an **impossible event**.

**EXAMPLE 3** Events

Suppose a coin is flipped until both a head and a tail appear, or until the coin has been flipped four times, whichever comes first. Write each of the following events in set notation.

**(a)** The coin is flipped exactly three times.

**SOLUTION** This means that the first two flips of the coin did not include both a head and a tail, so they must both be heads or both be tails. Because the third flip is the last one, it must show the side of the coin not yet seen. Thus the event is

$$\{hht, tth\}.$$

**(b)** The coin is flipped at least three times.

**SOLUTION** In addition to the outcomes listed in part (a), there is also the possibility that the coin is flipped four times, which only happens when the first three flips are all heads or all tails. Thus the event is

$$\{hht, tth, hhhh, hhht, tttt, ttth\}.$$

**(c)** The coin is flipped at least two times.

**SOLUTION**    This event consists of the entire sample space:

$$S = \{ ht, th, hht, tth, hhhh, hhht, tttt, ttth \}.$$

This is an example of a certain event.

**(d)** The coin is flipped fewer than two times.

**SOLUTION**    The coin cannot be flipped fewer than two times under the rules described, so the event is the empty set $\emptyset$. This is an example of an impossible event.

Since events are sets, we can use set operations to find unions, intersections, and complements of events. A summary of the set operations for events is given below.

---

**Set Operations for Events**

Let $E$ and $F$ be events for a sample space $S$.

$E \cap F$ occurs when both $E$ **and** $F$ occur;

$E \cup F$ occurs when $E$ **or** $F$ **or both** occur;

$E'$ occurs when $E$ does **not** occur.

---

**EXAMPLE 4**    **Biology Majors**

A study of biology majors grouped such students into various categories, which can be interpreted as events when a student is selected at random. Consider the following events:

$E$: student is under 20;

$F$: student is white;

$G$: student is female.

Describe the following events in words.

**(a)** $E'$

**SOLUTION**    $E'$ is the event that the student is 20 or over.

**(b)** $F \cap G'$

**SOLUTION**    $F \cap G'$ is the event that the student is white and not a female; that is, the student is a white male.

**YOUR TURN 3**    In Example 4, describe the following event in words: $E' \cap F'$.

**(c)** $E \cup G$

**SOLUTION**    $E \cup G$ is the event that the student is under 20 or is female. Note that this event includes all students under 20, both male and female, and all female students of any age.

TRY YOUR TURN 3

Two events that cannot both occur at the same time, such as rolling an even number and an odd number with a single roll of a die, are called *mutually exclusive events*.

---

**Mutually Exclusive Events**

Events $E$ and $F$ are **mutually exclusive events** if $E \cap F = \emptyset$.

---

Any event $E$ and its complement $E'$ are mutually exclusive. By definition, mutually exclusive events are disjoint sets.

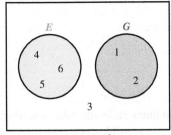

$E \cap G = \emptyset$

**FIGURE 15**

EXAMPLE 5 **Mutually Exclusive Events**

Let $S = \{1, 2, 3, 4, 5, 6\}$, the sample space for tossing a single die. Let $E = \{4, 5, 6\}$, and let $G = \{1, 2\}$. Then $E$ and $G$ are mutually exclusive events since they have no outcomes in common: $E \cap G = \emptyset$. See Figure 15.

## Probability
For sample spaces with *equally likely* outcomes, the probability of an event is defined as follows.

**Basic Probability Principle**

Let $S$ be a sample space of equally likely outcomes, and let event $E$ be a subset of $S$. Then the **probability** that event $E$ occurs is

$$P(E) = \frac{n(E)}{n(S)}.$$

By this definition, the probability of an event is a number that indicates the relative likelihood of the event.

**CAUTION** The basic probability principle applies only when the outcomes are equally likely.

**EXAMPLE 6** **Basic Probabilities**

Suppose a single fair die is rolled. Use the sample space $S = \{1, 2, 3, 4, 5, 6\}$ and give the probability of each event.

**(a)** $E$: The die shows an even number.

**SOLUTION** Here, $E = \{2, 4, 6\}$, a set with three elements. Since $S$ contains six elements,

$$P(E) = \frac{3}{6} = \frac{1}{2}.$$

**(b)** $F$: The die shows a number less than 10.

**SOLUTION** Event $F$ is a certain event, with

$$F = \{1, 2, 3, 4, 5, 6\},$$

so

$$P(F) = \frac{6}{6} = 1.$$

**YOUR TURN 4** In Example 6, find the probability of event $H$: The die shows a number less than 5.

**(c)** $G$: The die shows an 8.

**SOLUTION** This event is impossible, so

$$P(G) = 0.$$ **TRY YOUR TURN 4**

A standard deck of 52 cards has four suits: hearts ($\heartsuit$), clubs ($\clubsuit$), diamonds ($\diamondsuit$), and spades ($\spadesuit$), with 13 cards in each suit. The hearts and diamonds are red, and the spades and clubs are black. Each suit has an ace (A), a king (K), a queen (Q), a jack (J), and cards numbered from 2 to 10. The jack, queen, and king are called *face cards* and for many purposes can be thought of as having values 11, 12, and 13, respectively. The ace can be thought of as the low card (value 1) or the high card (value 14). See Figure 16. We will refer to this standard deck of cards often in our discussion of probability.

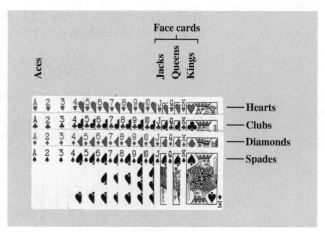

FIGURE 16

### EXAMPLE 7  Playing Cards

If a single playing card is drawn at random from a standard 52-card deck, find the probability of each event.

**(a)** Drawing an ace

**SOLUTION**   There are 4 aces in the deck. The event "drawing an ace" is

$$\{ \text{heart ace, diamond ace, club ace, spade ace} \}.$$

Therefore,

$$P(\text{ace}) = \frac{4}{52} = \frac{1}{13}.$$

**(b)** Drawing a face card

**SOLUTION**   Since there are 12 face cards (three in each of the four suits),

$$P(\text{face card}) = \frac{12}{52} = \frac{3}{13}.$$

**(c)** Drawing a spade

**SOLUTION**   The deck contains 13 spades, so

$$P(\text{spade}) = \frac{13}{52} = \frac{1}{4}.$$

**(d)** Drawing a spade or a heart

**YOUR TURN 5**  Find the probability of drawing a jack or a king.

**SOLUTION**   Besides the 13 spades, the deck contains 13 hearts, so

$$P(\text{spade or heart}) = \frac{26}{52} = \frac{1}{2}. \qquad \text{TRY YOUR TURN 5}$$

In the preceding examples, the probability of each event was a number between 0 and 1. The same thing is true in general. Any event $E$ is a subset of the sample space $S$, so $0 \leq n(E) \leq n(S)$. Since $P(E) = n(E)/n(S)$, it follows that $0 \leq P(E) \leq 1$. Note that a certain event has probability 1 and an impossible event has probability 0, as seen in Example 6.

For any event $E$,     $0 \leq P(E) \leq 1$.

## The Union Rule

To determine the probability of the union of two events $E$ and $F$ in a sample space $S$, use the union rule for sets,

$$n(E \cup F) = n(E) + n(F) - n(E \cap F),$$

which was proved in Section 12.1. Assuming that the events in the sample space $S$ are equally likely, divide both sides by $n(S)$, so that

$$\frac{n(E \cup F)}{n(S)} = \frac{n(E)}{n(S)} + \frac{n(F)}{n(S)} - \frac{n(E \cap F)}{n(S)}$$

$$P(E \cup F) = P(E) + P(F) - P(E \cap F).$$

Although our derivation is valid only for sample spaces with equally likely events, the result is valid for any events $E$ and $F$ from any sample space, and is called the **union rule for probability**.

### Union Rule for Probability

For any events $E$ and $F$ from a sample space $S$,

$$P(E \cup F) = P(E) + P(F) - P(E \cap F).$$

### EXAMPLE 8 Probabilities with Dice

Suppose two fair dice are rolled. Find each probability.

**(a)** The first die shows a 2, or the sum of the results is 6 or 7.

**SOLUTION** The sample space for the throw of two dice is shown in Figure 17, where 1-1 represents the event "the first die shows a 1 and the second die shows a 1," 1-2 represents "the first die shows a 1 and the second die shows a 2," and so on. Let $A$ represent the event "the first die shows a 2," and $B$ represent the event "the sum of the results is 6 or 7." These events are indicated in Figure 17. From the diagram, event $A$ has 6 elements, $B$ has 11 elements, the intersection of $A$ and $B$ has 2 elements, and the sample space has 36 elements. Thus,

$$P(A) = \frac{6}{36}, \quad P(B) = \frac{11}{36}, \quad \text{and} \quad P(A \cap B) = \frac{2}{36}.$$

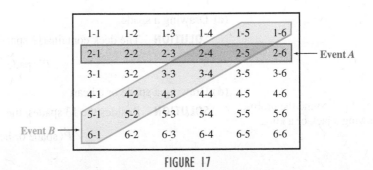

FIGURE 17

By the union rule,

$$P(A \cup B) = P(A) + P(B) - P(A \cap B)$$

$$P(A \cup B) = \frac{6}{36} + \frac{11}{36} - \frac{2}{36} = \frac{15}{36} = \frac{5}{12}.$$

**(b)** The sum of the results is 11, or the second die shows a 5.

**YOUR TURN 6** In Example 8, find the probability that the sum is 8, or both die show the same number.

**SOLUTION** $P(\text{sum is } 11) = 2/36$, $P(\text{second die shows a } 5) = 6/36$, and $P(\text{sum is } 11 \text{ and second die shows a } 5) = 1/36$, so

$$P(\text{sum is } 11 \text{ or second die shows a } 5) = \frac{2}{36} + \frac{6}{36} - \frac{1}{36} = \frac{7}{36}.$$

TRY YOUR TURN 6

CAUTION | You may wonder why we did not use $S = \{2, 3, 4, 5, \ldots, 12\}$ as the sample space in Example 8. Remember, we prefer to use a sample space with equally likely outcomes. The outcomes in set $S$ above are not equally likely—a sum of 2 can occur in just one way, a sum of 3 in two ways, a sum of 4 in three ways, and so on, as shown in Figure 17.

If events $E$ and $F$ are mutually exclusive, then $E \cap F = \emptyset$ by definition; hence, $P(E \cap F) = 0$. In this case the union rule simplifies to $P(E \cup F) = P(E) + P(F)$.

CAUTION | The rule $P(E \cup F) = P(E) + P(F)$ is only valid when $E$ and $F$ are mutually exclusive. When $E$ and $F$ are *not* mutually exclusive, use the rule $P(E \cup F) = P(E) + P(F) - P(E \cap F)$.

# The Complement Rule

By the definition of $E'$, for any event $E$ from a sample space $S$,

$$E \cup E' = S \quad \text{and} \quad E \cap E' = \emptyset.$$

Since $E \cap E' = \emptyset$, events $E$ and $E'$ are mutually exclusive, so that

$$P(E \cup E') = P(E) + P(E').$$

However, $E \cup E' = S$, the sample space, and $P(S) = 1$. Thus

$$P(E \cup E') = P(E) + P(E') = 1.$$

Rearranging these terms gives the following useful rule for complements.

### Complement Rule

$$P(E) = 1 - P(E') \quad \text{and} \quad P(E') = 1 - P(E)$$

### EXAMPLE 9  Complement Rule

If two fair dice are rolled, find the probability that the sum of the numbers rolled is greater than 3. Refer to Figure 17.

**SOLUTION**  To calculate this probability directly, we must find the probabilities that the sum is 4, 5, 6, 7, 8, 9, 10, 11, or 12 and then add them. It is much simpler to first find the probability of the complement, the event that the sum is less than or equal to 3.

$$P(\text{sum} \leq 3) = P(\text{sum is } 2) + P(\text{sum is } 3)$$

$$= \frac{1}{36} + \frac{2}{36}$$

$$= \frac{3}{36} = \frac{1}{12}$$

Now use the fact that $P(E) = 1 - P(E')$ to get

$$P(\text{sum} > 3) = 1 - P(\text{sum} \leq 3)$$

$$= 1 - \frac{1}{12} = \frac{11}{12}. \qquad \text{TRY YOUR TURN 7}$$

**YOUR TURN 7**  Find the probability that when two fair dice are rolled, the sum is less than 11.

# Odds

Sometimes probability statements are given in terms of **odds**, a comparison of $P(E)$ with $P(E')$. For example, suppose $P(E) = 4/5$. Then $P(E') = 1 - 4/5 = 1/5$. These probabilities predict that $E$ will occur 4 out of 5 times and $E'$ will occur 1 out of 5 times. Then we say the *odds in favor* of $E$ are 4 to 1.

## Odds

The **odds in favor** of an event $E$ are defined as the ratio of $P(E)$ to $P(E')$, or

$$\frac{P(E)}{P(E')}, \text{ where } P(E') \neq 0.$$

### EXAMPLE 10 Odds of Flat Feet

About 1 in 7 children in the United States has persistent flat feet, which do not usually require treatment unless they cause pain. Find the odds that a child will have flat feet.

**SOLUTION** Let $E$ be the event "flat feet." Then $E'$ is the event "no flat feet." Since $P(E) = 1/7$, $P(E') = 6/7$. By the definition of odds, the odds in favor of flat feet are

$$\frac{1/7}{6/7} = \frac{1}{6}, \qquad \text{written} \qquad 1 \text{ to } 6, \text{ or } 1:6.$$

**YOUR TURN 8** If the probability that a male is left handed is 1/10, find the odds that a male is left-handed.

On the other hand, the odds that a child will not have flat feet, or the *odds against* flat feet, are

$$\frac{6/7}{1/7} = \frac{6}{1}, \qquad \text{written} \qquad 6 \text{ to } 1, \text{ or } 6:1. \quad \text{TRY YOUR TURN 8}$$

If the odds in favor of an event are, say, 3 to 5, then the probability of the event is 3/8, while the probability of the complement of the event is 5/8. (Odds of 3 to 5 indicate 3 outcomes in favor of the event out of a total of 8 possible outcomes.) This example suggests the following generalization.

If the odds favoring event $E$ are $m$ to $n$, then

$$P(E) = \frac{m}{m + n} \qquad \text{and} \qquad P(E') = \frac{n}{m + n}.$$

### EXAMPLE 11 Motorcycle Accidents

Suppose the odds of dying while riding on a motorcycle are 1 to 1250. What is the probability of death while riding on a motorcycle?

**YOUR TURN 9** If the odds of having twins are 3 to 97, find the probability of not having twins.

**SOLUTION** The odds indicate chances of 1 out of 1251 $(1 + 1250 = 1251)$ that one will die, so

$$P(\text{death}) = \frac{1}{1251}. \qquad \text{TRY YOUR TURN 9}$$

## Empirical Probability
In many real-life problems, it is not possible to establish exact probabilities for events. Instead, useful approximations are often found by drawing on past experience. The next example shows one approach to such **empirical probabilities**.

### EXAMPLE 12 Injuries

The following table lists the estimated number of injuries in the United States associated with recreation equipment. ***Source: National Safety Council.***

| Recreation Equipment Injuries | |
| --- | --- |
| Equipment | Number of Injuries |
| Bicycles | 515,871 |
| Skateboards | 143,682 |
| Trampolines | 107,345 |
| Playground climbing equipment | 77,845 |
| Swings or swing sets | 59,144 |

Find the probability that a randomly selected person whose injury is associated with recreation equipment was hurt on a trampoline.

**YOUR TURN 10** In Example 12, find the probability the randomly selected person was hurt on a skateboard.

**SOLUTION** We first find the total number of injuries. Verify that the amounts in the table sum to 903,887. The probability is then found by dividing the number of people injured on trampolines by the total number of people injured. Thus,

$$P(\text{Trampolines}) = \frac{107{,}345}{903{,}887} \approx 0.1188.$$

TRY YOUR TURN 10

## Probability Distribution
A table listing each possible outcome of an experiment and its corresponding probability is called a **probability distribution**. The assignment of probabilities may be done in any reasonable way (on an empirical basis, as in the next example, or by theoretical reasoning) provided that it satisfies the following conditions.

### Properties of Probability

Let $S$ be a sample space consisting of $n$ distinct outcomes, $s_1, s_2, \ldots, s_n$. An acceptable probability assignment consists of assigning to each outcome $s_i$ a number $p_i$ (the probability of $s_i$) according to these rules.

**1.** The probability of each outcome is a number between 0 and 1, inclusive.

$$0 \le p_1 \le 1, \quad 0 \le p_2 \le 1, \ldots, \quad 0 \le p_n \le 1$$

**2.** The sum of the probabilities of all possible outcomes is 1.

$$p_1 + p_2 + p_3 + \cdots + p_n = 1$$

### EXAMPLE 13  Causes of Death

There were 2,468,435 deaths in the United States in 2010 and they are listed, according to cause, in the following table. *Source: National Vital Statistics Report.*

| Cause | Number of Deaths |
|---|---|
| Heart disease | 597,689 |
| Cancer | 574,743 |
| Cerebrovascular disease | 129,476 |
| Chronic lower respiratory diseases | 138,080 |
| Accidents | 120,859 |
| Pneumonia and influenza | 50,097 |
| Diabetes mellitus | 69,071 |
| All other causes | 788,420 |

**(a)** Construct a probability distribution for the probability that a death was caused by each category.

**SOLUTION** We can find the empirical probability that a death was caused by each category by dividing the number of deaths for each cause by the total number of deaths. Verify that the amounts in the table sum to 2,468,435. The probability that a randomly selected death was caused by an accident, for example, is $P(\text{death by accident}) = 120{,}859/2{,}468{,}435 \approx 0.049$. Similarly, we could divide each amount by

2,468,435, with the results (rounded to three decimal places) shown in the following table.

| Cause | Probabilities |
|---|---|
| Heart disease | 0.242 |
| Cancer | 0.233 |
| Cerebrovascular disease | 0.052 |
| Chronic lower respiratory diseases | 0.056 |
| Accidents | 0.049 |
| Pneumonia and influenza | 0.020 |
| Diabetes mellitus | 0.028 |
| All other causes | 0.319 |

Verify that this distribution satisfies the conditions for probability. Each probability is between 0 and 1, so the first condition holds. The probabilities in this table sum to 0.999. In theory, to satisfy the second condition, they should total 1.000, but this does not always occur when the individual numbers are rounded.

**(b)** Find the probability that a death in the United States was caused by heart disease or cancer.

**SOLUTION**   The categories in the table are mutually exclusive simple events. Thus, to find the probability that a death is caused by heart disease or cancer, we use the union rule to calculate

$$P(\text{heart disease or cancer}) = 0.242 + 0.233 = 0.475$$

We could get this same result by summing the number of deaths by heart disease and cancer and dividing the total by 2,468,435.

Thus, almost half of all deaths were caused by heart disease or cancer.

A Venn diagram is often helpful in determining probabilities, as illustrated in the final example.

**APPLY IT**

## EXAMPLE 14   Health

A run-down college student has determined the probability that she will get a cold this semester is 0.47, the probability she will get the flu is 0.59, and the probability that she will get both is 0.31.

**(a)** Find the probability that she will not get a cold *and* will not get the flu this semester.

**SOLUTION**   Let $C$ represent the event "gets a cold," and $F$ represent "gets the flu." Place the given information on a Venn diagram, starting with 0.31 in the intersection of the regions $C$ and $F$ (see Figure 18). As stated earlier, event $C$ has probability 0.47. Since 0.31 has already been placed inside the intersection of $C$ and $F$,

$$0.47 - 0.31 = 0.16$$

goes inside region $C$, but outside the intersection of $C$ and $F$, that is, in the region $C \cap F'$. In the same way,

$$0.59 - 0.31 = 0.28$$

goes inside the region for $F$, and outside the overlap, that is, in the region $F \cap C'$.

Using regions $C$ and $F$, the event we want is $C' \cap F'$. From the Venn diagram in Figure 18, the labeled regions have a total probability of

$$0.16 + 0.31 + 0.28 = 0.75.$$

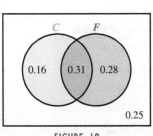

FIGURE 18

Since the entire region of the Venn diagram must have probability 1, the region outside $C$ and $F$, or $C' \cap F'$, has probability

$$1 - 0.75 = 0.25.$$

The probability is 0.25 that she will not get a cold and will not get the flu this semester.

**(b)** Find the probability that she will not get a cold or will not get the flu this semester.

**SOLUTION** The corresponding region, $C' \cup F'$, has probability

$$0.25 + 0.16 + 0.28 = 0.69,$$

or $1 - 0.31 = 0.69$, using the Complement Rule.

# 12.2 EXERCISES

1. What is the sample space for an experiment?

2. Define an event. What is a simple event?

**Write sample spaces for the experiments in Exercises 3–8.**

3. A month of the year is chosen for a wedding.

4. A student is asked how many questions she answered correctly on a recent 80-question test.

5. A patient must decide whether to have surgery, treat the problem with medicine, or wait six months to see if the problem resolves.

6. A record is kept each day for three days about whether a patient's blood pressure goes up or goes down.

7. A coin is tossed, and a die is rolled.

8. A box contains five balls, numbered 1, 2, 3, 4, and 5. A ball is drawn at random, the number on it recorded, and the ball replaced. The box is shaken, a second ball is drawn, and its number is recorded.

**For the experiments in Exercises 9–14, write out the sample space $S$, choosing an $S$ with equally likely outcomes, if possible. Then give the value of $n(S)$ and tell whether the outcomes in $S$ are equally likely. Finally, write the indicated events in set notation.**

9. A committee of 2 people is selected at random from 5 doctors: Alam, Bartolini, Chinn, Dickson, and Ellsberg.

   **a.** Chinn is on the committee.

   **b.** Dickson and Ellsberg are not both on the committee.

   **c.** Both Alam and Chinn are on the committee.

10. Five states are being considered as the location for three new high-energy physics laboratories: California (CA), Colorado (CO), New Jersey (NJ), New York (NY), and Utah (UT). Three states will be chosen at random. Write elements of the sample space in the form (CA, CO, NJ).

   **a.** All three states border an ocean.

   **b.** Exactly two of the three states border an ocean.

   **c.** Exactly one of the three states is west of the Mississippi River.

11. Slips of paper marked with the numbers 1, 2, 3, 4, and 5 are placed in a box. After being mixed, two slips are drawn simultaneously.

   **a.** Both slips are marked with even numbers.

   **b.** One slip is marked with an odd number and the other is marked with an even number.

   **c.** Both slips are marked with the same number.

12. An unprepared student takes a three-question, true/false quiz in which he guesses the answers to all three questions, so each answer is equally likely to be correct or wrong.

   **a.** The student gets three answers wrong.

   **b.** The student gets exactly two answers correct.

   **c.** The student gets only the first answer correct.

13. A coin is flipped until two heads appear, up to a maximum of four flips. (If three tails are flipped, the coin is still tossed a fourth time to complete the experiment.)

   **a.** The coin is tossed four times.

   **b.** Exactly two heads are tossed.

   **c.** No heads are tossed.

14. One jar contains four balls, labeled 1, 2, 3, and 4. A second jar contains five balls, labeled 1, 2, 3, 4, and 5. An experiment consists of taking one ball from the first jar, and then taking a ball from the second jar.

   **a.** The number on the first ball is even.

   **b.** The number on the second ball is even.

   **c.** The sum of the numbers on the two balls is 5.

   **d.** The sum of the numbers on the two balls is 1.

**A single fair die is rolled. Find the probabilities of each event.**

15. Getting a 2

16. Getting an odd number

17. Getting a number less than 5

18. Getting a number greater than 2

19. Getting a 3 or a 4

20. Getting any number except 3

**A card is drawn from a well-shuffled deck of 52 cards. Find the probability of drawing the following.**

**21.** A 9

**22.** A black card

**23.** A black 9

**24.** The 9 of hearts

**25.** A 2 or a queen

**26.** A black 7 or a red 8

**27.** A red card or a 10

**28.** A spade or a king

**A jar contains 3 white, 4 orange, 5 yellow, and 8 black marbles. If a marble is drawn at random, find the probability that it is the following.**

**29.** White

**30.** Yellow

**31.** Not black

**32.** Orange or yellow

**33.** Define mutually exclusive events in your own words.

**34.** Explain the union rule for mutually exclusive events.

**Decide whether the events in Exercises 35–38 are mutually exclusive.**

**35.** Owning a dog and owning a smartphone.

**36.** Being a biology major and being from Texas

**37.** Being a teenager and being 70 years old

**38.** Being one of the ten tallest people in the United States and being under 4 feet tall

**Two dice are rolled. Find the probabilities of rolling the given sums.**

**39. a.** 2    **b.** 4    **c.** 5    **d.** 6

**40. a.** 8    **b.** 9    **c.** 10    **d.** 13

**41. a.** 9 or more    **b.** Less than 7

  **c.** Between 5 and 8 (exclusive)

**42. a.** Not more than 5    **b.** Not less than 8

  **c.** Between 3 and 7 (exclusive)

**Two dice are rolled. Find the probabilities of the following events.**

**43.** The first die is 3 or the sum is 8.

**44.** The second die is 5 or the sum is 10.

**One card is drawn from an ordinary deck of 52 cards. Find the probabilities of drawing the following cards.**

**45. a.** A 9 or 10

  **b.** A red card or a 3

  **c.** A 9 or a black 10

  **d.** A heart or a black card

  **e.** A face card or a diamond

**46. a.** Less than a 4 (count aces as ones)

  **b.** A diamond or a 7

  **c.** A black card or an ace

  **d.** A heart or a jack

  **e.** A red card or a face card

**Kristi Perez invites 13 relatives to a party: her mother, 2 aunts, 3 uncles, 2 brothers, 1 male cousin, and 4 female cousins. If the chances of any one guest arriving first are equally likely, find the probabilities that the first guest to arrive is as follows.**

**47. a.** A brother or an uncle

  **b.** A brother or a cousin

  **c.** A brother or her mother

**48. a.** An uncle or a cousin

  **b.** A male or a cousin

  **c.** A female or a cousin

**The numbers 1, 2, 3, 4, and 5 are written on slips of paper, and 2 slips are drawn at random one at a time without replacement. Find the probabilities in Exercises 49 and 50.**

**49. a.** The sum of the numbers is 9.

  **b.** The sum of the numbers is 5 or less.

  **c.** The first number is 2 or the sum is 6.

**50. a.** Both numbers are even.

  **b.** One of the numbers is even or greater than 3.

  **c.** The sum is 5 or the second number is 2.

**51.** Define what is meant by odds.

**52.** On page 134 of Roger Staubach's autobiography, *First Down, Lifetime to Go*, Staubach makes the following statement regarding his experience in Vietnam: *Source: First Down, Lifetime to Go.*

  "Odds against a direct hit are very low but when your life is in danger, you don't worry too much about the odds."

  Is this wording consistent with our definition of odds, for and against? How could it have been said so as to be technically correct?

**A single fair die is rolled. Find the odds in favor of getting the results in Exercises 53–56.**

**53.** 3

**54.** 4, 5, or 6

**55.** 2, 3, 4, or 5

**56.** Some number less than 6

**57.** A marble is drawn from a box containing 3 yellow, 4 white, and 11 blue marbles. Find the odds in favor of drawing the following.

  **a.** A yellow marble

  **b.** A blue marble

  **c.** A white marble

  **d.** Not drawing a white marble

**58.** Two dice are rolled. Find the odds of rolling the following. (Refer to Figure 17.)

  **a.** A sum of 3

  **b.** A sum of 7 or 11

  **c.** A sum less than 5

  **d.** Not a sum of 6

**Which of Exercises 59–64 are examples of empirical probability?**

**59.** The probability of heads on 5 consecutive tosses of a coin

**60.** The probability that a person is allergic to penicillin

**61.** The probability of drawing an ace from a standard deck of 52 cards

**62.** The probability that a person will get lung cancer from smoking cigarettes

**63.** A surgeon's prediction that a patient has a 90% chance of a full recovery

**64.** A gambler's claim that on a roll of a fair die, $P(\text{even}) = 1/2$

**65.** What is a probability distribution?

**66.** What conditions must hold for a probability distribution to be acceptable?

**An experiment is conducted for which the sample space is $S = \{s_1, s_2, s_3, s_4, s_5\}$. Which of the probability assignments in Exercises 67–72 are possible for this experiment? If an assignment is not possible, tell why.**

**67.**

| Outcomes | $s_1$ | $s_2$ | $s_3$ | $s_4$ | $s_5$ |
|---|---|---|---|---|---|
| Probabilities | 0.09 | 0.32 | 0.21 | 0.25 | 0.13 |

**68.**

| Outcomes | $s_1$ | $s_2$ | $s_3$ | $s_4$ | $s_5$ |
|---|---|---|---|---|---|
| Probabilities | 0.92 | 0.03 | 0 | 0.02 | 0.03 |

**69.**

| Outcomes | $s_1$ | $s_2$ | $s_3$ | $s_4$ | $s_5$ |
|---|---|---|---|---|---|
| Probabilities | 1/3 | 1/4 | 1/6 | 1/8 | 1/10 |

**70.**

| Outcomes | $s_1$ | $s_2$ | $s_3$ | $s_4$ | $s_5$ |
|---|---|---|---|---|---|
| Probabilities | 1/5 | 1/3 | 1/4 | 1/5 | 1/10 |

**71.**

| Outcomes | $s_1$ | $s_2$ | $s_3$ | $s_4$ | $s_5$ |
|---|---|---|---|---|---|
| Probabilities | 0.64 | −0.08 | 0.30 | 0.12 | 0.02 |

**72.**

| Outcomes | $s_1$ | $s_2$ | $s_3$ | $s_4$ | $s_5$ |
|---|---|---|---|---|---|
| Probabilities | 0.05 | 0.35 | 0.5 | 0.2 | −0.3 |

**One way to solve a probability problem is to repeat the experiment many times, keeping track of the results. Then the probability can be approximated using the basic definition of the probability of an event $E$: $P(E) = n(E)/n(S)$, where $E$ occurs $n(E)$ times out of $n(S)$ trials of an experiment. This is called the Monte Carlo method of finding probabilities. If physically repeating the experiment is too tedious, it may be simulated using a random-number generator, available on most computers and scientific or graphing calculators. To simulate a coin toss or the roll of a die on the TI-84 Plus, change the setting to fixed decimal mode with 0 digits displayed, and enter `rand` or `rand*6+.5`, respectively. For a coin toss, interpret 0 as a head and 1 as a tail. In either case, the ENTER key can be pressed repeatedly to perform multiple simulations.**

**73.** Suppose two dice are rolled. Use the Monte Carlo method with at least 50 repetitions to approximate the following probabilities. Compare with the results of Exercise 41.

  **a.** $P(\text{the sum is 9 or more})$

  **b.** $P(\text{the sum is less than 7})$

**74.** Suppose two dice are rolled. Use the Monte Carlo method with at least 50 repetitions to approximate the following probabilities. Compare with the results of Exercise 42.

  **a.** $P(\text{the sum is not more than 5})$

  **b.** $P(\text{the sum is not less than 8})$

**75.** Suppose three dice are rolled. Use the Monte Carlo method with at least 100 repetitions to approximate the following probabilities.

  **a.** $P(\text{the sum is 5 or less})$

  **b.** $P(\text{neither a 1 nor a 6 is rolled})$

**76.** Suppose a coin is tossed 5 times. Use the Monte Carlo method with at least 50 repetitions to approximate the following probabilities.

  **a.** $P(\text{exactly 4 heads})$

  **b.** $P(\text{2 heads and 3 tails})$

**Use Venn diagrams to work Exercises 77 and 78.**

**77.** Suppose $P(E) = 0.26$, $P(F) = 0.41$, and $P(E \cap F) = 0.16$. Find the following.

  **a.** $P(E \cup F)$      **b.** $P(E' \cap F)$

  **c.** $P(E \cap F')$      **d.** $P(E' \cup F')$

**78.** Let $P(Z) = 0.42$, $P(Y) = 0.35$, and $P(Z \cup Y) = 0.59$. Find each probability.

  **a.** $P(Z' \cap Y')$      **b.** $P(Z' \cup Y')$

  **c.** $P(Z' \cup Y)$      **d.** $P(Z \cap Y')$

**79.** The student sitting next to you in class concludes that the probability of the ceiling falling down on both of you before class ends is 1/2, because there are two possible outcomes—the ceiling will fall or not fall. What is wrong with this reasoning?

**80.** The following puzzler was given on the *Car Talk* radio program. ***Source: Car Talk, Feb 24, 2001.***

> "Three different numbers are chosen at random, and one is written on each of three slips of paper. The slips are then placed face down on the table. The objective is to choose the slip upon which is written the largest number. Here are the rules: You can turn over any slip of paper and look at the amount written on it. If for any reason you think this is the largest, you're done; you keep it. Otherwise you discard it and turn over a second slip. Again, if you think this is the one with the biggest number, you keep that one and the game is over. If you don't, you discard that one too. . . . The chance of getting the highest number is one in three. Or is it? Is there a strategy by which you can improve the odds?"

The answer to the puzzler is that you can indeed improve the probability of getting the highest number by the following strategy. Pick one of the slips of paper, and after looking at the number, throw it away. Then pick a second slip; if it has a larger number than the first slip, stop. If not, pick the third slip. Find the probability of winning with this strategy.*

---

*This is a special case of the famous Googol problem. For more details, see "Recognizing the Maximum of a Sequence" by John P. Gilbert and Frederick Mosteller, *Journal of the American Statistical Association*, Vol. 61, No. 313, March 1966, pp. 35–73.

**81.** The following description of the classic "Linda problem" appeared in the *New Yorker*: "In this experiment, subjects are told, 'Linda is thirty-one years old, single, outspoken, and very bright. She majored in philosophy. As a student, she was deeply concerned with issues of discrimination and social justice and also participated in antinuclear demonstrations.' They are then asked to rank the probability of several possible descriptions of Linda today. Two of them are 'bank teller' and 'bank teller and active in the feminist movement.'" Many people rank the second event as more likely. Explain why this violates basic concepts of probability. *Source: New Yorker.*

**82.** Three unusual dice, *A*, *B*, and *C*, are constructed such that die *A* has the numbers 3, 3, 4, 4, 8, 8; die *B* has the numbers 1, 1, 5, 5, 9, 9; and die *C* has the numbers 2, 2, 6, 6, 7, 7.

  **a.** If dice *A* and *B* are rolled, find the probability that *B* beats *A*, that is, the number that appears on die *B* is greater than the number that appears on die *A*.

  **b.** If dice *B* and *C* are rolled, find the probability that *C* beats *B*.

  **c.** If dice *A* and *C* are rolled, find the probability that *A* beats *C*.

  **d.** Which die is better? Explain.

## LIFE SCIENCE APPLICATIONS

**83. Medical Survey** For a medical experiment, people are classified as to whether they smoke, have a family history of heart disease, or are overweight. Define events *E*, *F*, and *G* as follows.

  *E*: person smokes

  *F*: person has a family history of heart disease

  *G*: person is overweight

Describe each event in words.

  **a.** $G'$ **b.** $F \cap G$ **c.** $E \cup G'$

**84. Medical Survey** Refer to Exercise 83. Describe each event in words.

  **a.** $E \cup F$ **b.** $E' \cap F$ **c.** $F' \cup G'$

**85. Body Types** A study on body types gave the following results: 45% were short; 25% were short and overweight; and 24% were tall and not overweight. Find the probabilities that a person is the following.

  **a.** Overweight

  **b.** Short, but not overweight

  **c.** Tall and overweight

**86. Color Blindness** Color blindness is an inherited characteristic that is more common in males than in females. If *M* represents male and *C* represents red-green color blindness, we use the relative frequencies of the incidences of males and red-green color blindness as probabilities to get

$$P(C) = 0.039, P(M \cap C) = 0.035, P(M \cup C) = 0.491.$$

*Source: Parsons' Diseases of the Eye.*

Find the following probabilities.

  **a.** $P(C')$ **b.** $P(M)$ **c.** $P(M')$
  **d.** $P(M' \cap C')$ **e.** $P(C \cap M')$ **f.** $P(C \cup M')$

**87. Genetics** Gregor Mendel, an Austrian monk, was the first to use probability in the study of genetics. In an effort to understand the mechanism of character transmittal from one generation to the next in plants, he counted the number of occurrences of various characteristics. Mendel found that the flower color in certain pea plants obeyed this scheme:

*Pure red crossed with pure white produces red.*

From its parents, the red offspring received genes for both red (*R*) and white (*W*), but in this case red is *dominant* and white *recessive*, so the offspring exhibits the color red. However, the offspring still carries both genes, and when two such offspring are crossed, several things can happen in the third generation. The table below, which is called a *Punnett square*, shows the equally likely outcomes.

| | | **Second Parent** | |
|---|---|---|---|
| | | R | W |
| **First Parent** | R | RR | RW |
| | W | WR | WW |

Use the fact that red is dominant over white to find the following. Assume that there are an equal number of red and white genes in the population.

  **a.** *P*(a flower is red)

  **b.** *P*(a flower is white)

**88. Genetics** Mendel found no dominance in snapdragons, with one red gene and one white gene producing pink-flowered offspring. These second-generation pinks, however, still carry one red and one white gene, and when they are crossed, the next generation still yields the Punnett square from Exercise 87. Find each probability.

  **a.** *P*(red)

  **b.** *P*(pink)

  **c.** *P*(white)

(Mendel verified these probability ratios experimentally and did the same for many characteristics other than flower color. His work, published in 1866, was not recognized until 1890.)

**89. Genetics** In most animals and plants, it is very unusual for the number of main parts of the organism (such as arms, legs, toes, or flower petals) to vary from generation to generation. Some species, however, have *meristic variability*, in which the number of certain body parts varies from generation to generation. One researcher studied the front feet of certain guinea pigs and produced the following probabilities. *Source: Genetics.*

$$P(\text{only four toes, all perfect}) = 0.77$$

$$P(\text{one imperfect toe and four good ones}) = 0.13$$

$$P(\text{exactly five good toes}) = 0.10$$

Find the probability of each event.

**a.** No more than four good toes

**b.** Five toes, whether perfect or not

**90. Doctor Visit** The probability that a visit to a primary care physician's (PCP) office results in neither lab work nor referral to a specialist is 35%. Of those coming to a PCP's office, 30% are referred to specialists and 40% require lab work. Determine the probability that a visit to a PCP's office results in both lab work and referral to a specialist. Choose one of the following. (*Hint:* Use the union rule for probability.) *Source: Society of Actuaries.*

**a.** 0.05 **b.** 0.12 **c.** 0.18

**d.** 0.25 **e.** 0.35

**91. Shoulder Injuries** Among a large group of patients recovering from shoulder injuries, it is found that 22% visit both a physical therapist and a chiropractor, whereas 12% visit neither of these. The probability that a patient visits a chiropractor exceeds by 0.14 the probability that a patient visits a physical therapist. Determine the probability that a randomly chosen member of this group visits a physical therapist. Choose one of the following. (*Hint:* Use the union rule for probability, and let $x = P$(patient visits a physical therapist).) *Source: Society of Actuaries.*

**a.** 0.26 **b.** 0.38 **c.** 0.40

**d.** 0.48 **e.** 0.62

**92. Health Plan** An insurer offers a health plan to the employees of a large company. As part of this plan, the individual employees may choose exactly two of the supplementary coverages A, B, and C, or they may choose no supplementary coverage. The proportions of the company's employees that choose coverages A, B, and C are 1/4, 1/3, and 5/12, respectively. Determine the probability that a randomly chosen employee will choose no supplementary coverage. Choose one of the following. (*Hint:* Draw a Venn diagram with three sets, and let $x = P(A \cap B)$. Use the fact that 4 of the 8 regions in the Venn diagram have a probability of 0.) *Source: Society of Actuaries.*

**a.** 0 **b.** 47/144 **c.** 1/2

**d.** 97/144 **e.** 7/9

**93. U.S. Population** The projected U.S. population (in thousands) by race in 2020 and 2050 is given in the table. *Source: Bureau of the Census.*

| Race | 2020 | 2050 |
|---|---|---|
| White | 207,393 | 207,901 |
| Hispanic | 52,652 | 96,508 |
| Black | 41,538 | 53,555 |
| Asian and Pacific Islander | 18,557 | 32,432 |
| Other | 2602 | 3535 |

Find the probability that a randomly selected person in the given year is of the race specified.

**a.** Hispanic in 2020 **b.** Hispanic in 2050

**c.** Black in 2020 **d.** Black in 2050

## OTHER APPLICATIONS

**94. Employment** The table shows the projected probabilities of a worker employed by different occupational groups in 2018. *Source: U.S. Department of Labor.*

| Occupation | Probability |
|---|---|
| Management and business | 0.1047 |
| Professional | 0.2182 |
| Service | 0.2024 |
| Sales | 0.1015 |
| Office and administrative support | 0.1560 |
| Farming, fishing, forestry | 0.0061 |
| Construction | 0.0531 |
| Production | 0.0585 |
| Other | 0.0995 |

If a worker in 2018 is selected at random, find the following.

**a.** The probability that the worker is in sales or service.

**b.** The probability that the worker is not in construction.

**c.** The odds in favor of the worker being in production.

**95. Labor Force** The 2008 and the 2018 (projected) civilian labor forces by age are given in the following table. *Source: U.S. Department of Labor.*

| Age (in years) | 2008 (in millions) | 2018 (in millions) |
|---|---|---|
| 16 to 24 | 22.0 | 21.1 |
| 25 to 54 | 104.4 | 105.9 |
| 55 and over | 27.9 | 39.8 |
| Total | 154.3 | 166.8 |

**a.** In 2008, find the probability that a member of the civilian labor force is age 55 or older.

**b.** In 2018, find the probability that a member of the civilian labor force is age 55 or over.

**c.** What do these projections imply about the future civilian labor force?

**96. Labor Force** The following table gives the 2018 projected civilian labor force probability distribution by age and gender. *Source: U.S. Department of Labor.*

| Age | Male | Female | Total |
|---|---|---|---|
| 16–24 | 0.066 | 0.061 | 0.127 |
| 25–54 | 0.343 | 0.291 | 0.634 |
| 55 and over | 0.122 | 0.117 | 0.239 |
| Total | 0.531 | 0.469 | 1.000 |

Find the probability that a randomly selected worker is the following.

a. Female and 16 to 24 years old

b. 16 to 54 years old

c. Male or 25 to 54 years old

d. Female or 16 to 24 years old

**97. Book of Odds** The following table gives the probabilities that a particular event will occur. Convert each probability to the odds in favor of the event. *Source: The Book of Odds.*

| Event | Probability for the Event |
|---|---|
| An NFL pass will be intercepted. | 0.03 |
| A U.S. president owned a dog during his term in office. | 0.65 |
| A woman owns a pair of high heels. | 0.61 |
| An adult smokes. | 0.21 |
| A flight is cancelled. | 0.02 |

**98. Book of Odds** The following table gives the odds that a particular event will occur. Convert each odd to the probability that the event will occur. *Source: The Book of Odds.*

| Event | Odds for the Event |
|---|---|
| A Powerball entry will win the jackpot. | 1 to 195,199,999 |
| An adult will be struck by lightning during a year. | 1 to 835,499 |
| An adult will file for personal bankruptcy during a year. | 1 to 157.6 |
| A person collects stamps. | 1 to 59.32 |

**99. Civil War** Estimates of the Union Army's strength and losses for the battle of Gettysburg are given in the following table, where *strength* is the number of soldiers immediately preceding the battle and *loss* indicates a soldier who was killed, wounded, captured, or missing. *Source: Regimental Strengths and Losses of Gettysburg.*

| Unit | Strength | Loss |
|---|---|---|
| I Corps (Reynolds) | 12,222 | 6059 |
| II Corps (Hancock) | 11,347 | 4369 |
| III Corps (Sickles) | 10,675 | 4211 |
| V Corps (Sykes) | 10,907 | 2187 |
| VI Corps (Sedgwick) | 13,596 | 242 |
| XI Corps (Howard) | 9188 | 3801 |
| XII Corps (Slocum) | 9788 | 1082 |
| Cavalry (Pleasonton) | 11,851 | 610 |
| Artillery (Tyler) | 2376 | 242 |
| Total | 91,950 | 22,803 |

a. Find the probability that a randomly selected union soldier was from the XI Corps.

b. Find the probability that a soldier was lost in the battle.

c. Find the probability that a I Corps soldier was lost in the battle.

d. Which group had the highest probability of not being lost in the battle?

e. Which group had the highest probability of loss?

f. Explain why these probabilities vary.

**100. Civil War** Estimates of the Confederate Army's strength and losses for the battle of Gettysburg are given in the following table, where *strength* is the number of soldiers immediately preceding the battle and *loss* indicates a soldier who was killed, wounded, captured, or missing. *Source: Regimental Strengths and Losses at Gettysburg.*

| Unit | Strength | Loss |
|---|---|---|
| I Corps (Longstreet) | 20,706 | 7661 |
| II Corps (Ewell) | 20,666 | 6603 |
| III Corps (Hill) | 22,083 | 8007 |
| Cavalry (Stuart) | 6621 | 286 |
| Total | 70,076 | 22,557 |

a. Find the probability that a randomly selected confederate soldier was from the III Corps.

b. Find the probability that a confederate soldier was lost in the battle.

c. Find the probability that a I Corps soldier was lost in the battle.

d. Which group had the highest probability of not being lost in the battle?

e. Which group had the highest probability of loss?

**101. Perceptions of Threat** Research has been carried out to measure the amount of intolerance that citizens of Russia have for left-wing Communists and right-wing Fascists, as indicated in the table below. Note that the numbers are given as percents and each row sums to 100 (except for rounding). *Source: Political Research Quarterly.*

| Russia | None at All | Don't Know | Not Very Much | Somewhat | Extremely |
|---|---|---|---|---|---|
| Left-Wing Communists | 47.8 | 6.7 | 31.0 | 10.5 | 4.1 |
| Right-Wing Fascists | 3.0 | 3.2 | 7.1 | 27.1 | 59.5 |

**a.** Find the probability that a randomly chosen citizen of Russia would be somewhat or extremely intolerant of right-wing Fascists.

**b.** Find the probability that a randomly chosen citizen of Russia would be completely tolerant of left-wing Communists.

**c.** Compare your answers to parts a and b and provide possible reasons for these numbers.

**102. Perceptions of Threat** Research has been carried out to measure the amount of intolerance that U.S. citizens have for left-wing Communists and right-wing Fascists, as indicated in the table. Note that the numbers are given as percents and each row sums to 100 (except for rounding). *Source: Political Research Quarterly.*

| United States | None at All | Don't Know | Not Very Much | Somewhat | Extremely |
|---|---|---|---|---|---|
| *Left-Wing Communists* | 13.0 | 2.7 | 33.0 | 34.2 | 17.1 |
| *Right-Wing Fascists* | 10.1 | 3.3 | 20.7 | 43.1 | 22.9 |

**a.** Find the probability that a randomly chosen U.S. citizen would have at least some intolerance of right-wing Fascists.

**b.** Find the probability that a randomly chosen U.S. citizen would have at least some intolerance of left-wing Communists.

**c.** Compare your answers to parts a and b and provide possible reasons for these numbers.

**d.** Compare these answers to the answers to Exercise 101.

**103. Earnings** The following data were gathered for 130 adult U.S. workers: 55 were women; 3 women earned more than $40,000; and 62 men earned $40,000 or less. Find the probability that an individual is

**a.** a woman earning $40,000 or less;

**b.** a man earning more than $40,000;

**c.** a man or is earning more than $40,000;

**d.** a woman or is earning $40,000 or less.

**104. Expenditures for Music** A survey of 100 people about their music expenditures gave the following information: 38 bought rock music; 20 were teenagers who bought rock music; and 26 were teenagers. Find the probabilities that a person is

**a.** a teenager who buys nonrock music;

**b.** someone who buys rock music or is a teenager;

**c.** not a teenager;

**d.** not a teenager, but a buyer of rock music.

**105. Refugees** In a refugee camp in southern Mexico, it was found that 90% of the refugees came to escape political oppression, 80% came to escape abject poverty, and 70% came to escape both. What is the probability that a refugee in the camp was not poor nor seeking political asylum?

**106. Community Activities** At the first meeting of a committee to plan a local Lunar New Year celebration, the persons attending are 3 Chinese men, 4 Chinese women, 3 Vietnamese women, 2 Vietnamese men, 4 Korean women, and 2 Korean men. A chairperson is selected at random. Find the probabilities that the chairperson is the following.

**a.** Chinese

**b.** Korean or a woman

**c.** A man or Vietnamese

**d.** Chinese or Vietnamese

**e.** Korean and a woman

**107. Chinese New Year** Exercise 83 of the previous section presented a survey of people attending a Lunar New Year celebration in Chinatown. Use the information given in that exercise to find each of the following probabilities.

**a.** A randomly chosen attendee speaks Cantonese.

**b.** A randomly chosen attendee does not speak Cantonese.

**c.** A randomly chosen attendee was a woman that did not light a firecracker.

**108. Native American Ceremonies** Exercise 84 of the previous section presented a survey of families participating in a pow-wow in Arizona. Use the information given in that exercise to find each probability.

**a.** A randomly chosen family brought costumes and food.

**b.** A randomly chosen family brought crafts, but neither food nor costumes.

**c.** A randomly chosen family brought food or costumes.

**YOUR TURN ANSWERS**

**1.** $S = \{HH, HT, TH, TT\}$

**2.** $E = \{HT, TH\}$

**3.** $E' \cap F'$ is the event that the student is 20 or over and is not white.

**4.** $P(H) = 4/6 = 2/3$.

**5.** $P(\text{jack or a king}) = 8/52 = 2/13$.

**6.** $10/36 = 5/18$

**7.** $33/36 = 11/12$

**8.** 1 to 9 or 1:9

**9.** 97/100

**10.** About 0.1590

# 12.3 Conditional Probability; Independent Events; Bayes' Theorem

APPLY IT   What is the probability that an injury sustained by a player while wearing a full shield was a head and face injury?

*By redefining the sample space, we will determine this probability in Example 1.*

**EXAMPLE 1   Ice Hockey Injuries**

The numbers of head and neck injuries for 319 ice hockey players wearing either a full shield or a half shield in the Canadian Inter-University Athletics Union are listed below. (See Exercise 78 in Section 12.1.) *Source: JAMA.*

| Ice Hockey Head and Neck Injuries | | | |
|---|---|---|---|
| | Half Shield (*H*) | Full Shield (*F*) | Totals |
| Head and Face Injuries (A) | 95 | 34 | 129 |
| Concussions (B) | 41 | 38 | 79 |
| Neck Injuries (C) | 9 | 7 | 16 |
| Other Injuries (D) | 202 | 150 | 352 |
| Totals | 347 | 229 | 576 |

**(a)** If an injured player is selected at random, find the probability that he had a head and face injury as well as the probability that he was wearing a full shield.

**SOLUTION**   Using the letters above, the probability that an injured player had a head and face injury is

$$P(A) = \frac{129}{576} \approx 0.224$$

and the probability that an injured player was wearing a full shield is

$$P(F) = \frac{229}{576} \approx 0.398.$$

**(b)** Find the probability that an injury sustained by a player while wearing a full shield was a head and face injury.

APPLY IT

**SOLUTION**   From the table, of the 229 injuries sustained in players who wore a full shield, 34 injuries were to the head and face, with

$$P(\text{head and face injury while wearing full shield}) = \frac{34}{229} \approx 0.148.$$

The answer to Example 1(b) is a different number from the probability of a head and face injury found in Example 1(a), 0.224, since we have additional information (the player wore a full shield) that has *reduced the sample space*. In other words, we found the probability that an injured player has sustained a head and face injury, *A*, given that the player was wearing a full shield, *F*. This is called the *conditional probability* of event *A*, given that event *F* has occurred, written $P(A|F)$. The probability $P(A|F)$ is read as "the probability of *A* given *F*." In the example above,

$$P(A|F) = \frac{34}{229}.$$

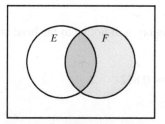

Event $F$ has a total of $n(F)$ elements.

FIGURE 19

To generalize this result, assume that $E$ and $F$ are two events for a particular experiment and that all events in the sample space $S$ are equally likely. We want to find $P(E|F)$, the probability that $E$ occurs given $F$ has occurred. Since we assume that $F$ has occurred, reduce the sample space to $F$: Look only at the elements inside $F$. See Figure 19. Of those $n(F)$ elements, there are $n(E \cap F)$ elements where $E$ also occurs. This makes

$$P(E|F) = \frac{n(E \cap F)}{n(F)}.$$

This equation can also be written as the quotient of two probabilities. Divide numerator and denominator by $n(S)$ to get

$$P(E|F) = \frac{n(E \cap F)/n(S)}{n(F)/n(S)} = \frac{P(E \cap F)}{P(F)}.$$

This last result motivates the definition of conditional probability.

### Conditional Probability

The **conditional probability** of event $E$ given event $F$, written $P(E|F)$, is

$$P(E|F) = \frac{P(E \cap F)}{P(F)}, \quad \text{where } P(F) \neq 0.$$

Although the definition of conditional probability was motivated by an example with equally likely outcomes, it is valid in all cases. However, for *equally likely outcomes*, conditional probability can be found by directly applying the definition, or by first reducing the sample space to event $F$, and then finding the number of outcomes in $F$ that are also in event $E$. Thus,

$$P(E|F) = \frac{n(E \cap F)}{n(F)}.$$

In Example 1(b), the conditional probability could have also been found using the definition of conditional probability:

$$P(A|F) = \frac{P(A \cap F)}{P(F)} = \frac{34/576}{229/576} = \frac{34}{229} \approx 0.148.$$

### EXAMPLE 2  Hockey

Use the information given in the chart in Example 1 to find the following probabilities.

**(a)** $P(F|A)$

**SOLUTION**  This represents the probability that an injured player was wearing a full shield given that a head and face injury occurred. Reduce the sample space to $A$. Then find $n(A \cap F)$ and $n(A)$.

$$P(F|A) = \frac{n(A \cap F)}{n(A)} = \frac{34}{129} \approx 0.264$$

If a head and face injury occurred, then the probability is about 0.264 that the player wore a full shield. This suggests that fewer head and face injuries occur in players who wear a full shield.

**(b)** $P(A'|F)$

**SOLUTION** In words, this is the probability that an injury will not be to the head and face for players who wear full shields.

$$P(A'|F) = \frac{n(A' \cap F)}{n(F)} = \frac{195}{229} \approx 0.852$$

**(c)** $P(F'|A')$

**SOLUTION** Here, we want the probability that an injured player was not wearing a full shield even though the injury was not to the head and face.

$$P(F'|A') = \frac{n(A' \cap F')}{n(A')} = \frac{252}{447} \approx 0.564$$

TRY YOUR TURN 1

**YOUR TURN 1** In Example 2, find $P(A|F')$.

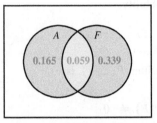

**FIGURE 20**

Venn diagrams are useful for illustrating problems in conditional probability. A Venn diagram for Example 2, in which the probabilities are used to indicate the number in the set defined by each region, is shown in Figure 20. In the diagram, $P(F|A)$ is found by reducing the sample space to just set $A$. Then $P(F|A)$ is the ratio of the number in that part of set $A$ which is also in $F$ to the number in set $A$, or $0.059/0.224 \approx 0.263$. The difference between this value and the one calculated in Example 2(a) is due to rounding.

## EXAMPLE 3  Conditional Probabilities

Given $P(E) = 0.4$, $P(F) = 0.5$, and $P(E \cup F) = 0.7$, find $P(E|F)$.

**SOLUTION** Find $P(E \cap F)$ first. By the union rule,

$$P(E \cup F) = P(E) + P(F) - P(E \cap F)$$
$$0.7 = 0.4 + 0.5 - P(E \cap F)$$
$$P(E \cap F) = 0.2.$$

**YOUR TURN 2** Given $P(E) = 0.56$, $P(F) = 0.64$, and $P(E \cup F) = 0.80$, find $P(E|F)$.

$P(E|F)$ is the ratio of the probability of that part of $E$ that is in $F$ to the probability of $F$, or

$$P(E|F) = \frac{P(E \cap F)}{P(F)} = \frac{0.2}{0.5} = \frac{2}{5}.$$

TRY YOUR TURN 2

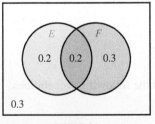

**FIGURE 21**

The Venn diagram in Figure 21 illustrates Example 3.

## EXAMPLE 4  Tossing Coins

Two fair coins were tossed, and it is known that at least one was a head. Find the probability that both were heads.

**SOLUTION** At first glance, the answer may appear to be 1/2, but this is not the case. The sample space has four equally likely outcomes, $S = \{hh, ht, th, tt\}$. Because of the condition that at least one coin was a head, the sample space is reduced to $\{hh, ht, th\}$. Since only one outcome in this reduced sample space is 2 heads,

$$P(2 \text{ heads} | \text{at least 1 head}) = \frac{1}{3}.$$

Alternatively, we could use the conditional probability definition. Define two events:

$$E_1 = \text{at least 1 head} = \{hh, ht, th\}$$

and

$$E_2 = 2 \text{ heads} = \{hh\}.$$

**YOUR TURN 3** In Example 4, find the probability of exactly one head, given that there is at least one tail.

Since there are four equally likely outcomes, $P(E_1) = 3/4$ and $P(E_1 \cap E_2) = 1/4$. Therefore,

$$P(E_2|E_1) = \frac{P(E_2 \cap E_1)}{P(E_1)} = \frac{1/4}{3/4} = \frac{1}{3}.$$ **TRY YOUR TURN 3**

It is important not to confuse $P(A|B)$ with $P(B|A)$. For example, in a criminal trial, a prosecutor may point out to the jury that the probability of the defendant's DNA profile matching that of a sample taken at the scene of the crime, given that the defendant is innocent, $P(D|I)$, is very small. What the jury must decide, however, is the probability that the defendant is innocent, given that the defendant's DNA profile matches the sample, $P(I|D)$. Confusing the two is an error sometimes called "the prosecutor's fallacy," and the 1990 conviction of a rape suspect in England was overturned by a panel of judges, who ordered a retrial, because the fallacy made the original trial unfair. *Source: New Scientist.*

Later in this section, we will see how to compute $P(A|B)$ when we know $P(B|A)$.

# Product Rule
If $P(E) \neq 0$ and $P(F) \neq 0$, then the definition of conditional probability shows that

$$P(E|F) = \frac{P(E \cap F)}{P(F)} \quad \text{and} \quad P(F|E) = \frac{P(F \cap E)}{P(E)}.$$

Using the fact that $P(E \cap F) = P(F \cap E)$, and solving each of these equations for $P(E \cap F)$, we obtain the following rule.

## Product Rule of Probability
If $E$ and $F$ are events, then $P(E \cap F)$ may be found by either of these formulas.

$$P(E \cap F) = P(F) \cdot P(E|F) \quad \text{or} \quad P(E \cap F) = P(E) \cdot P(F|E)$$

The product rule gives a method for finding the probability that events $E$ and $F$ both occur, as illustrated by the next few examples.

## EXAMPLE 5 Biology Majors

In a class with 2/5 women and 3/5 men, 25% of the women are biology majors. Find the probability that a student chosen from the class at random is a female biology major.

**YOUR TURN 4** At a local college, 4/5 of the students live on campus. Of those who live on campus, 25% have cars on campus. Find the probability that a student lives on campus and has a car.

**SOLUTION** Let $B$ and $W$ represent the events "biology major" and "woman," respectively. We want to find $P(B \cap W)$. By the product rule,

$$P(B \cap W) = P(W) \cdot P(B|W).$$

Using the given information, $P(W) = 2/5 = 0.4$ and $P(B|W) = 0.25$. Thus,

$$P(B \cap W) = 0.4(0.25) = 0.10.$$ **TRY YOUR TURN 4**

The next examples show how a tree diagram is used with the product rule to find the probability of a sequence of events.

### EXAMPLE 6 Environmental Inspections

The Environmental Protection Agency is considering inspecting 6 plants for environmental compliance: 3 in Chicago, 2 in Los Angeles, and 1 in New York. Due to a lack of inspectors, they decide to inspect two plants selected at random, one this month and one next month, with each plant equally likely to be selected, but no plant selected twice. What is the probability that 1 Chicago plant and 1 Los Angeles plant are selected?

**SOLUTION** A tree diagram showing the various possible outcomes is given in Figure 22. In this diagram, the events of inspecting a plant in Chicago, Los Angeles, and New York are represented by C, LA, and NY, respectively. For the first inspection, $P(\text{C first}) = 3/6 = 1/2$ because 3 of the 6 plants are in Chicago, and all plants are equally likely to be selected. Likewise, $P(\text{LA first}) = 1/3$ and $P(\text{NY first}) = 1/6$.

For the second inspection, we first note that one plant has been inspected and, therefore, removed from the list, leaving 5 plants. For example, $P(\text{LA second} | \text{C first}) = 2/5$, since 2 of the 5 remaining plants are in Los Angeles. The remaining second inspection probabilities are calculated in the same manner.

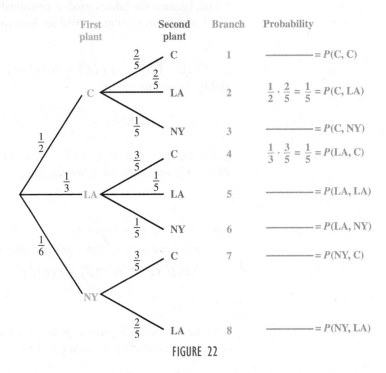

FIGURE 22

We want to find the probability of selecting exactly 1 Chicago plant and 1 Los Angeles plant. This event can occur in two ways: inspecting Chicago this month and Los Angeles next month (branch 2 of the tree diagram), or inspecting Los Angeles this month and Chicago next month (branch 4). For branch 2,

$$P(\text{C first}) \cdot P(\text{LA second} | \text{C first}) = \frac{1}{2} \cdot \frac{2}{5} = \frac{1}{5}.$$

For branch 4, where Los Angeles is inspected first,

$$P(\text{LA first}) \cdot P(\text{C second} | \text{LA first}) = \frac{1}{3} \cdot \frac{3}{5} = \frac{1}{5}.$$

Since the two events are mutually exclusive, the final probability is the sum of these two probabilities.

$$P(1\text{ C}, 1\text{ LA}) = P(\text{C first}) \cdot P(\text{LA second} | \text{C first})$$
$$+ P(\text{LA first}) \cdot P(\text{C second} | \text{LA first})$$
$$= \frac{2}{5}$$

**YOUR TURN 5** In Example 6, what is the probability that 1 New York plant and 1 Chicago plant are selected?

TRY YOUR TURN 5

FOR REVIEW
You may wish to refer to the picture of a deck of cards shown in Figure 16 (Section 12.2) and the description accompanying it.

The product rule is often used with *stochastic processes*, which are mathematical models that evolve over time in a probabilistic manner. For example, selecting factories at random for inspection is such a process, in which the probabilities change with each successive selection.

### EXAMPLE 7  Playing Cards

Two cards are drawn from a standard deck, one after another without replacement.

**(a)** Find the probability that the first card is a heart and the second card is red.

**SOLUTION** Start with the tree diagram in Figure 23. On the first draw, since there are 13 hearts among the 52 cards, the probability of drawing a heart is $13/52 = 1/4$. On the second draw, since a (red) heart has been drawn already, there are 25 red cards in the remaining 51 cards. Thus, the probability of drawing a red card on the second draw, given that the first is a heart, is 25/51. By the product rule of probability,

$$P(\text{heart first and red second})$$
$$= P(\text{heart first}) \cdot P(\text{red second}|\text{heart first})$$
$$= \frac{1}{4} \cdot \frac{25}{51} = \frac{25}{204} \approx 0.123.$$

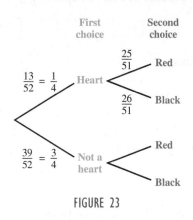

FIGURE 23

**(b)** Find the probability that the second card is red.

**SOLUTION** To solve this, we need to fill out the bottom branch of the tree diagram in Figure 23. Unfortunately, if the first card is not a heart, it is not clear how to find the probability that the second card is red, because it depends upon whether the first card is red or black. One way to solve this problem would be to divide the bottom branch into two separate branches: diamond and black card (club or spade).

There is a simpler way, however, since we don't care whether or not the first card is a heart, as we did in part (a). Instead, we'll consider whether the first card is red or black and then do the same for the second card. The result, with the corresponding probabilities, is in Figure 24. The probability that the second card is red is found by multiplying the probabilities along the two branches and adding.

$$P(\text{red second}) = \frac{1}{2} \cdot \frac{25}{51} + \frac{1}{2} \cdot \frac{26}{51}$$
$$= \frac{1}{2}$$

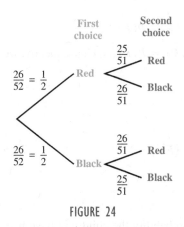

FIGURE 24

The probability is 1/2, exactly the same as the probability that any card is red. If we know nothing about the first card, there is no reason for the probability of the second card to be anything other than 1/2.

## Independent Events
Suppose, in Example 7(a), that we draw the two cards *with* replacement rather than without replacement (that is, we put the first card back before drawing the second card). If the first card is a heart, then the probability of drawing a red card on the second draw is 26/52, rather than 25/51, because there are still 52 cards in the deck, 26 of them red. In this case, $P(\text{red second}|\text{heart first})$ is the same as $P(\text{red second})$. The value of the second card is not affected by the value of the first card. We say that the event that the second card is red is *independent* of the event that the first card is a heart, since the knowledge of the first card does not influence what happens to the second card. On the other hand, when we draw without replacement, the events that the first card is a heart and that the second card is red are *dependent* events. The fact that the first card is a heart means there is one less red card in the deck, influencing the probability that the second card is red.

As another example, consider tossing a fair coin twice. If the first toss shows heads, the probability that the next toss is heads is still 1/2. Coin tosses are independent events, since the outcome of one toss does not influence the outcome of the next toss. Similarly, rolls of a fair die are independent events.

On the other hand, the events "the milk is old" and "the milk is sour" are dependent events; if the milk is old, there is an increased chance that it is sour. Also, in Example 1, the events $A$ (player sustained a head or face injury) and $F$ (player wore a full shield) are dependent events, because information about the use of shields affected the probability of a head or face injury. That is, $P(A|F)$ is different from $P(A)$.

If events $E$ and $F$ are independent, then the knowledge that $E$ has occurred gives no (probability) information about the occurrence or nonoccurrence of event $F$. That is, $P(F)$ is exactly the same as $P(F|E)$, or

$$P(F|E) = P(F).$$

This, in fact, is the formal definition of independent events.

### Independent Events

Events $E$ and $F$ are **independent events** if

$$P(F|E) = P(F) \quad \text{or} \quad P(E|F) = P(E).$$

If the events are not independent, they are **dependent events**.

When $E$ and $F$ are independent events, then $P(F|E) = P(F)$ and the product rule becomes

$$P(E \cap F) = P(E) \cdot P(F|E) = P(E) \cdot P(F).$$

Conversely, if this equation holds, then it follows that $P(F) = P(F|E)$. Consequently, we have this useful fact:

### Product Rule for Independent Events

Events $E$ and $F$ are independent events if and only if

$$P(E \cap F) = P(E) \cdot P(F).$$

### EXAMPLE 8  Product Rule

For a particular couple expecting their first child, the probability the child will have blue eyes is 0.25 and the probability the child will be a boy is 0.5. Find the probability that the child will be a blue-eyed boy.

**SOLUTION**  If the child having blue eyes and the child being a boy are independent events, then

$$P(\text{blue-eyed boy}) = P(\text{child has blue eyes}) \cdot P(\text{child is a boy})$$
$$= (0.25)(0.5) = 0.125.$$

TRY YOUR TURN 6

**YOUR TURN 6**  The probability that you roll a five on a single die is 1/6. Find the probability you roll two fives in a row.

**CAUTION**  It is common for students to confuse the ideas of *mutually exclusive* events and *independent* events. Events $E$ and $F$ are mutually exclusive if $E \cap F = \emptyset$. For example, if a family has exactly one child, the only possible outcomes are $B = \{\text{boy}\}$ and $G = \{\text{girl}\}$. These two events are mutually exclusive. The events are *not* independent, however, since $P(G|B) = 0$ (if a family with only one child has a boy, the probability it has a girl is then 0). Since $P(G|B) \neq P(G)$, the events are not independent.

Of all the families with exactly two children, the events $G_1 = \{\text{first child is a girl}\}$ and $G_2 = \{\text{second child is a girl}\}$ are independent, since $P(G_2|G_1)$ equals $P(G_2)$. However, $G_1$ and $G_2$ are not mutually exclusive, since $G_1 \cap G_2 = \{\text{both children are girls}\} \neq \emptyset$.

To show that two events $E$ and $F$ are independent, show that $P(F|E) = P(F)$ or that $P(E|F) = P(E)$ or that $P(E \cap F) = P(E) \cdot P(F)$. Another way is to observe that knowledge of one outcome does not influence the probability of the other outcome, as we did for coin tosses.

**NOTE** In some cases, it may not be apparent from the physical description of the problem whether two events are independent or not. For example, it is not obvious whether the event that a baseball player gets a hit tomorrow is independent of the event that he got a hit today. In such cases, it is necessary to calculate whether $P(F|E) = P(F)$, or, equivalently, whether $P(E \cap F) = P(E) \cdot P(F)$.

### EXAMPLE 9 Snow in Manhattan

On a typical January day in Manhattan the probability of snow is 0.10, the probability of a traffic jam is 0.80, and the probability of snow or a traffic jam (or both) is 0.82. Are the event "it snows" and the event "a traffic jam occurs" independent?

**SOLUTION** Let $S$ represent the event "it snows" and $T$ represent the event "a traffic jam occurs." We must determine whether

$$P(T|S) = P(T) \qquad \text{or} \qquad P(S|T) = P(S).$$

We know $P(S) = 0.10$, $P(T) = 0.8$, and $P(S \cup T) = 0.82$. We can use the union rule (or a Venn diagram) to find $P(S \cap T) = 0.08$, $P(T|S) = 0.8$, and $P(S|T) = 0.1$. Since

$$P(T|S) = P(T) = 0.8 \qquad \text{and} \qquad P(S|T) = P(S) = 0.1,$$

the events "it snows" and "a traffic jam occurs" are independent. **TRY YOUR TURN 7**

> **YOUR TURN 7** The probability that you do your math homework is 0.8, the probability that you do your history assignment is 0.7, and the probability of you doing your math homework or your history assignment is 0.9. Are the events "do your math homework" and "do your history assignment" independent?

Although we showed $P(T|S) = P(T)$ and $P(S|T) = P(S)$ in Example 9, only one of these results is needed to establish independence. It is also important to note that independence of events does not necessarily follow intuition; it is established from the mathematical definition of independence.

## Bayes' Theorem
Suppose the probability that a person gets lung cancer, given that the person smokes a pack or more of cigarettes daily, is known. For a research project, it might be necessary to know the probability that a person smokes a pack or more of cigarettes daily, given that the person has lung cancer. More generally, if $P(E|F)$ is known for two events $E$ and $F$, then $P(F|E)$ can be found using a tree diagram. Since $P(E|F)$ is known, the first outcome is either $F$ or $F'$. Then for each of these outcomes, either $E$ or $E'$ occurs, as shown in Figure 25.

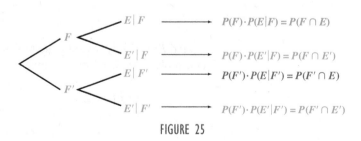

FIGURE 25

The four cases have the probabilities shown on the right. Notice $P(E \cap F)$ is the first case and $P(E)$ is the sum of the first and third cases in the tree diagram. By the definition of conditional probability,

$$P(F|E) = \frac{P(F \cap E)}{P(E)} = \frac{P(F) \cdot P(E|F)}{P(F) \cdot P(E|F) + P(F') \cdot P(E|F')}.$$

This result is a special case of Bayes' theorem, which is generalized later in this section.

**Bayes' Theorem (Special Case)**

$$P(F|E) = \frac{P(F) \cdot P(E|F)}{P(F) \cdot P(E|F) + P(F') \cdot P(E|F')}$$

### EXAMPLE 10   Worker Errors

For a fixed length of time, the probability of a worker error on a certain production line is 0.1, the probability that an accident will occur when there is a worker error is 0.3, and the probability that an accident will occur when there is no worker error is 0.2. Find the probability of a worker error if there is an accident.

**SOLUTION**   Let $E$ represent the event of an accident, and let $F$ represent the event of worker error. From the information given,

$$P(F) = 0.1, \qquad P(E|F) = 0.3, \qquad \text{and} \qquad P(E|F') = 0.2.$$

These probabilities are shown on the tree diagram in Figure 26.

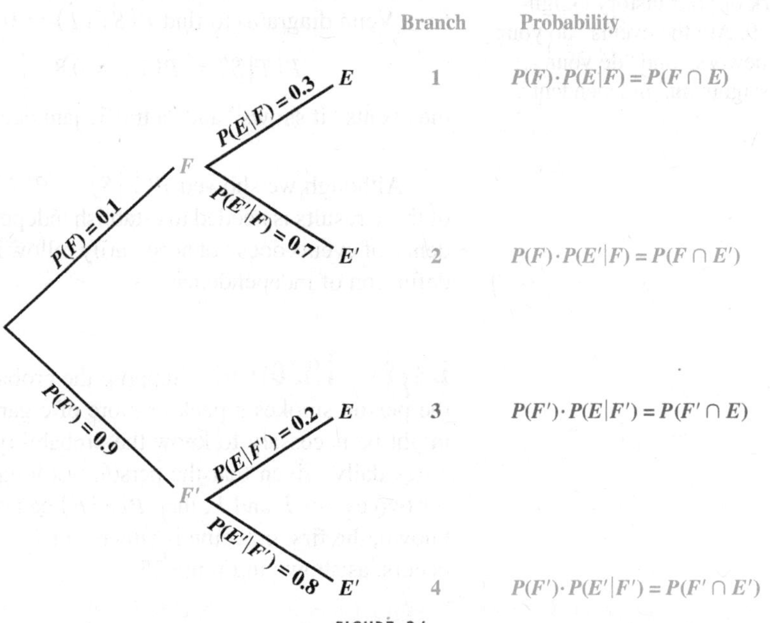

**YOUR TURN 8** The probability that a student will pass a math exam is 0.8 if he or she attends the review session and 0.65 if he or she does not attend the review session. Sixty percent of the students attend the review session. What is the probability that, given a student passed, the student attended the review session?

| | Branch | Probability |
|---|---|---|
| $P(E\|F)=0.3$ — $E$ | 1 | $P(F)\cdot P(E\|F) = P(F \cap E)$ |
| $P(E'\|F)=0.7$ — $E'$ | 2 | $P(F)\cdot P(E'\|F) = P(F \cap E')$ |
| $P(E\|F')=0.2$ — $E$ | 3 | $P(F')\cdot P(E\|F') = P(F' \cap E)$ |
| $P(E'\|F')=0.8$ — $E'$ | 4 | $P(F')\cdot P(E'\|F') = P(F' \cap E')$ |

FIGURE 26

Find $P(F|E)$ by dividing the probability that both $E$ and $F$ occur, given by branch 1, by the probability that $E$ occurs, given by the sum of branches 1 and 3.

$$P(F|E) = \frac{P(F) \cdot P(E|F)}{P(F) \cdot P(E|F) + P(F') \cdot P(E|F')}$$

$$= \frac{(0.1)(0.3)}{(0.1)(0.3) + (0.9)(0.2)} = \frac{0.03}{0.21} = \frac{1}{7} \approx 0.1429$$

TRY YOUR TURN 8

The special case of Bayes' theorem can be generalized to more than two events with the tree diagram in Figure 27. This diagram shows the paths that can produce an event $E$. We assume that the events $F_1, F_2, \ldots, F_n$ are mutually exclusive events (that is, disjoint events) whose union is the sample space, and that $E$ is an event that has occurred. See Figure 28.

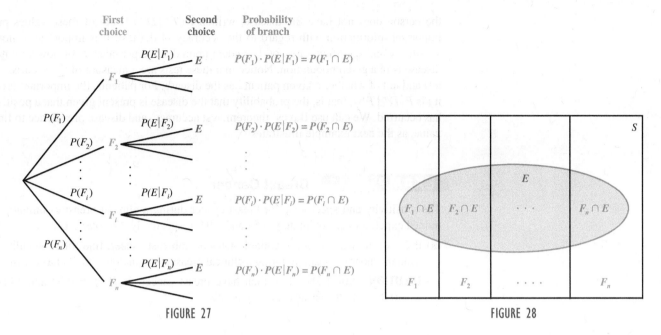

First choice    Second choice    Probability of branch

FIGURE 27

FIGURE 28

The probability $P(F_i \mid E)$, where $1 \le i \le n$, can be found by dividing the probability for the branch containing $P(E \mid F_i)$ by the sum of the probabilities of all the branches producing event $E$.

### Bayes' Theorem

$$P(F_i \mid E) = \frac{P(F_i) \cdot P(E \mid F_i)}{P(F_1) \cdot P(E \mid F_1) + P(F_2) \cdot P(E \mid F_2) + \cdots + P(F_n) \cdot P(E \mid F_n)}$$

This result is known as **Bayes' theorem**, after the Reverend Thomas Bayes (1702–1761), whose paper on probability was published about three years after his death.

The statement of Bayes' theorem can be daunting. Actually, it is easier to remember the formula by thinking of the tree diagram that produced it. Use the following steps.

### Using Bayes' Theorem

1. Start a tree diagram with branches representing $F_1, F_2, \ldots, F_n$. Label each branch with its corresponding probability.

2. From the end of each of these branches, draw a branch for event $E$. Label this branch with the probability of reaching it, $P(E \mid F_i)$.

3. You now have $n$ different paths that result in event $E$. Next to each path, put its probability—the product of the probabilities that the first branch occurs, $P(F_i)$, and that the second branch occurs, $P(E \mid F_i)$; that is, the product $P(F_i) \cdot P(E \mid F_i)$, which equals $P(F_i \cap E)$.

4. $P(F_i \mid E)$ is found by dividing the probability of the branch for $F_i$ by the sum of the probabilities of all the branches producing event $E$.

## Sensitivity and Specificity

The **sensitivity** of a medical test is defined as the probability that a test will be positive given that a person has a disease, written $P(T^+ \mid D^+)$. The **specificity** of a test is defined as the probability that a test will be negative given that the person does not have the disease, written $P(T^-|D^-)$. Both of these values provide important information with regard to the accuracy of the test. It is important to note that sensitivity and specificity are not dependent upon disease prevalence, or how common the disease is in a given population. Notice that these tests are a measure of the accuracy of the test and not of whether a given patient has the disease. For patients, the important probability is $P(D^+|T^+)$, that is, the probability that the disease is present given that a positive test has occurred. We can use Bayes' theorem, test accuracy, and disease prevalence to find this value, as the next example illustrates.

### EXAMPLE 11  Breast Cancer

The sensitivity and specificity for breast cancer during a clinical breast examination by a trained expert are approximately 0.54 and 0.94, respectively. *Source: JAMA.*

**(a)** If 2% of the women in the United States have breast cancer, find the probability that a woman who tests positive during a clinical breast examination actually has breast cancer.

**SOLUTION** Since 2% of women have breast cancer, $P(D^+) = 0.02$ and $P(D^-) = 1 - 0.02 = 0.98$, as shown in Figure 29.

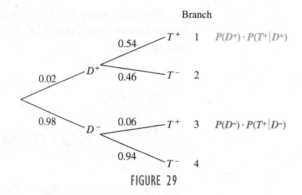

**FIGURE 29**

The sensitivity of the clinical breast examination is 0.54, so the probability that a woman will test positive given that she has the disease is $P(T^+|D^+) = 0.54$. By the complement rule, the probability that she will test negative, given she has the disease, is $P(T^-|D^+) = 1 - 0.54 = 0.46$. See Figure 29.

The specificity of the examination is 0.94, so the probability that a woman will test negative when she does not have the disease is $P(T^-|D^-) = 0.94$. The probability that she will test positive when she does not have the disease is $P(T^+|D^-) = 1 - 0.94 = 0.06$. See Figure 29.

Using Bayes' theorem, we have

$$P(D^+|T^+) = \frac{P(D^+) \cdot P(T^+|D^+)}{P(D^+) \cdot P(T^+|D^+) + P(D^-) \cdot P(T^+|D^-)}$$

$$= \frac{(0.02)(0.54)}{(0.02)(0.54) + (0.98)(0.06)} \approx 0.155.$$

Thus, a positive clinical exam suggests an approximate 16% chance of breast cancer. Notice that, although the probability of breast cancer given a positive test is fairly low, the information gained by the test increases the likelihood of actually having breast cancer eightfold.

**(b)** Suppose that the positive clinical examination has occurred and the woman is directed to receive a mammogram. Suppose that for this particular woman, the sensitivity and specificity for the mammogram are 0.89 and 0.93, respectively. Given that both the clinical exam and the mammogram result in a positive test, what is the probability that the woman has breast cancer? Assume that the mammogram and the clinical examination are independent events.

**SOLUTION**   The patient is having a second test, so we will add another set of branches to the tree diagram in Figure 29. Let $M^+$ and $M^-$ indicate a positive and negative mammogram, respectively.

The sensitivity of the mammogram is 0.89, so the probability the mammogram is positive when she has the disease is 0.89. By the complement rule, the probability is 0.11 that the mammogram will be negative when she has the disease. See Figure 30.

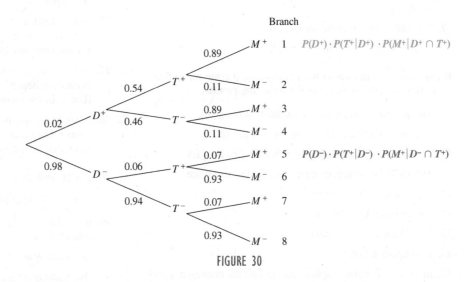

FIGURE 30

The specificity of the mammogram is 0.93, so the probability the mammogram is negative when she does not have the disease is 0.93. The probability the mammogram is positive when she does not have the disease is 0.07. See Figure 30.

We are looking for the probability that the woman has breast cancer given that both the clinical exam and the mammogram are positive $(T^+ \cap M^+)$, so we will use branch 1 and branch 5 in Figure 30. We can calculate this probability using Bayes' theorem. Assuming that events $T$ and $M$ are independent,

$$P(D^+|T^+ \cap M^+) = \frac{P(D^+) \cdot P(T^+|D^+) \cdot P(M^+|D^+ \cap T^+)}{P(D^+) \cdot P(T^+|D^+) \cdot P(M^+|D^+ \cap T^+) + P(D^-) \cdot P(T^+|D^-) \cdot P(M^+|D^- \cap T^+)}$$

$$= \frac{P(D^+) \cdot P(T^+|D^+) \cdot P(M^+|D^+)}{P(D^+) \cdot P(T^+|D^+) \cdot P(M^+|D^+) + P(D^-) \cdot P(T^+|D^-) \cdot P(M^+|D^-)}$$

$$= \frac{(0.02)(0.54)(0.89)}{(0.02)(0.54)(0.89) + (0.98)(0.06)(0.07)} \approx 0.70.$$

Thus, for this woman, who has a positive clinical exam and a positive mammogram, there is a 70% chance that she has breast cancer.

# 12.3 EXERCISES

**If a single fair die is rolled, find the probabilities of the following results.**

1. A 2, given that the number rolled was odd

2. A 4, given that the number rolled was even

3. An even number, given that the number rolled was 6

4. An odd number, given that the number rolled was 6

**If two fair dice are rolled, find the probabilities of the following results.**

5. A sum of 8, given that the sum is greater than 7

6. A sum of 6, given that the roll was a "double" (two identical numbers)

7. A double, given that the sum was 9

8. A double, given that the sum was 8

**If two cards are drawn without replacement from an ordinary deck, find the probabilities of the following results.**

9. The second is a heart, given that the first is a heart.

10. The second is black, given that the first is a spade.

11. The second is a face card, given that the first is a jack.

12. The second is an ace, given that the first is not an ace.

13. A jack and a 10 are drawn.

14. An ace and a 4 are drawn.

15. Two black cards are drawn.

16. Two hearts are drawn.

17. In your own words, explain how to find the conditional probability $P(E|F)$.

18. In your own words, define independent events.

**Decide whether the following pairs of events are dependent or independent.**

19. A red and a green die are rolled. $A$ is the event that the red die comes up even, and $B$ is the event that the green die comes up even.

20. $C$ is the event that it rains more than 10 days in Chicago next June, and $D$ is the event that it rains more than 15 days.

21. $E$ is the event that a resident of Texas lives in Dallas, and $F$ is the event that a resident of Texas lives in either Dallas or Houston.

22. A coin is flipped. $G$ is the event that today is Tuesday, and $H$ is the event that the coin comes up heads.

**In the previous section, we described an experiment in which the numbers 1, 2, 3, 4, and 5 are written on slips of paper, and 2 slips are drawn at random one at a time without replacement. Find each probability in Exercises 23 and 24.**

23. The probability that the first number is 3, given the following.

   a. The sum is 7.

   b. The sum is 8.

24. The probability that the sum is 8, given the following.

   a. The first number is 5.

   b. The first number is 4.

25. Suppose two dice are rolled. Let $A$ be the event that the sum of the two dice is 7. Find an event $B$ related to numbers on the dice such that $A$ and $B$ are

   a. independent;

   b. dependent.

26. Your friend asks you to explain how the product rule for independent events differs from the product rule for dependent events. How would you respond?

27. Another friend asks you to explain how to tell whether two events are dependent or independent. How would you reply? (Use your own words.)

28. A student reasons that the probability in Example 4 of both coins being heads is just the probability that the other coin is a head, that is, 1/2. Explain why this reasoning is wrong.

29. Let $A$ and $B$ be independent events with $P(A) = \frac{1}{4}$ and $P(B) = \frac{1}{5}$. Find $P(A \cap B)$ and $P(A \cup B)$.

30. If $A$ and $B$ are events such that $P(A) = 0.5$ and $P(A \cup B) = 0.7$, find $P(B)$ when

   a. $A$ and $B$ are mutually exclusive;

   b. $A$ and $B$ are independent.

**For two events $M$ and $N$, $P(M) = 0.4, P(N|M) = 0.3,$ and $P(N|M') = 0.4$. Find the following.**

31. $P(M|N)$          32. $P(M'|N)$

**For mutually exclusive events $R_1$, $R_2$, and $R_3$, we have $P(R_1) = 0.15,$ $P(R_2) = 0.55,$ and $P(R_3) = 0.30$. Also, $P(Q|R_1) = 0.40, P(Q|R_2) = 0.20,$ and $P(Q|R_3) = 0.70$. Find the following.**

33. $P(R_1|Q)$          34. $P(R_2|Q)$

35. $P(R_3|Q)$          36. $P(R_1'|Q)$

**Suppose you have three jars with the following contents: 2 black balls and 1 white ball in the first, 1 black ball and 2 white balls in the second, and 1 black ball and 1 white ball in the third. One jar is to be selected, and then 1 ball is to be drawn from the selected jar. If the probabilities of selecting the first, second, or third jar are 1/2, 1/3, and 1/6, respectively, find the probabilities that if a white ball is drawn, it came from the following jars.**

37. The second jar          38. The third jar

**39.** The following problem, submitted by Daniel Hahn of Blairstown, Iowa, appeared in the "Ask Marilyn" column of *Parade* magazine. ***Source: Parade magazine.***

"You discover two booths at a carnival. Each is tended by an honest man with a pair of covered coin shakers. In each shaker is a single coin, and you are allowed to bet upon the chance that both coins in that booth's shakers are heads after the man in the booth shakes them, does an inspection, and can tell you that at least one of the shakers contains a head. The difference is that the man in the first booth always looks inside both of his shakers, whereas the man in the second booth looks inside only one of the shakers. Where will you stand the best chance?"

**40.** The following question was posed in *Chance News* by Craig Fox and Yoval Rotenstrich. You are playing a game in which a fair coin is flipped and a fair die is rolled. You win a prize if both the coin comes up heads and a 6 is rolled on the die. Now suppose the coin is tossed and the die is rolled, but you are not allowed to see either result. You are told, however, that either the head or the 6 occurred. You are then offered the chance to cancel the game and play a new game in which a die is rolled (there is no coin), and you win a prize if a 6 is rolled. ***Source: Chance News.***

   **a.** Is it to your advantage to switch to the new game, or to stick with the original game? Answer this question by calculating your probability of winning in each case.

   **b.** Many people erroneously think that it's better to stick with the original game. Discuss why this answer might seem intuitive, but why it is wrong.

**41.** Suppose a male defendant in a court trial has a mustache, beard, tattoo, and an earring. Suppose, also, that an eyewitness has identified the perpetrator as someone with these characteristics. If the respective probabilities for the male population in this region are 0.35, 0.30, 0.10, and 0.05, is it fair to multiply these probabilities together to conclude that the probability that a person having these characteristics is 0.000525, or 21 in 40,000, and thus decide that the defendant must be guilty?

**42.** In a two-child family, if we assume that the probabilities of a male child and a female child are each 0.5, are the events *all children are the same sex* and *at most one male* independent? Are they independent for a three-child family?

**43.** Laura Johnson, a game show contestant, could win one of two prizes: a shiny new Porsche or a shiny new penny. Laura is given two boxes of marbles. The first box has 50 pink marbles in it and the second box has 50 blue marbles in it. The game show host will pick someone from the audience to be blindfolded and then draw a marble from one of the two boxes. If a pink marble is drawn, she wins the Porsche. Otherwise, Laura wins the penny. Can Laura increase her chances of winning by redistributing some of the marbles from one box to the other? Explain. ***Source: Car Talk.***

**44.** In the "Ask Marilyn" column of *Parade* magazine, a reader wrote about the following game: You and I each roll a die. If your die is higher than mine, you win. Otherwise, I win. The reader thought that the probability that each player wins is 1/2. Is this correct? If not, what is the probability that each player wins? ***Source: Parade magazine.***

## LIFE SCIENCE APPLICATIONS

**45. Genetics** Both of a certain pea plant's parents had a gene for red and a gene for white flowers. (See Exercise 87 in Section 12.2.) If the offspring has red flowers, find the probability that it combined a gene for red and a gene for white (rather than 2 for red).

**46. Medical Experiment** A medical experiment showed that the probability that a new medicine is effective is 0.75, the probability that a patient will have a certain side effect is 0.4, and the probability that both events occur is 0.3. Decide whether these events are dependent or independent.

**47. Genetics** Assuming that boy and girl babies are equally likely, fill in the remaining probabilities on the tree diagram below and use that information to find the probability that a family with three children has all girls, given the following.

   **a.** The first is a girl.

   **b.** The third is a girl.

   **c.** The second is a girl.

   **d.** At least 2 are girls.

   **e.** At least 1 is a girl.

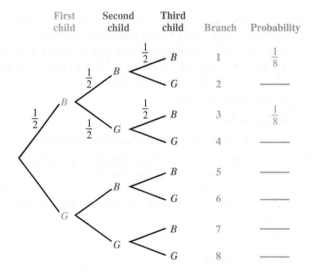

**48. Color Blindness** The following table shows frequencies for red-green color blindness, where M represents "person is male" and C represents "person is color-blind." Use this table to find the following probabilities. (See Exercise 86, Section 12.2.)

|  | **M** | **M′** | **Totals** |
|---|---|---|---|
| **C** | 0.035 | 0.004 | 0.039 |
| **C′** | 0.452 | 0.509 | 0.961 |
| **Totals** | 0.487 | 0.513 | 1.000 |

a. $P(M)$
b. $P(C)$
c. $P(M \cap C)$
d. $P(M \cup C)$
e. $P(M|C)$
f. $P(C|M)$
g. $P(M'|C)$

h. Are the events $C$ and $M$, described above, dependent? What does this mean?

**49. Color Blindness** A scientist wishes to determine whether there is a relationship between color blindness ($C$) and deafness ($D$).

a. Suppose the scientist found the probabilities listed in the table. What should the findings be? (See Exercise 48.)

b. Explain what your answer tells us about color blindness and deafness.

|  | **D** | **D′** | **Totals** |
|---|---|---|---|
| **C** | 0.0008 | 0.0392 | 0.0400 |
| **C′** | 0.0192 | 0.9408 | 0.9600 |
| **Totals** | 0.0200 | 0.9800 | 1.0000 |

**50. Overweight** According to a recent report, 68.3% of men and 64.1% of women in the United States were overweight. Given that 49.3% of Americans are men and 50.7% are women, find the probability that a randomly selected American fits the following description. *Source: JAMA.*

a. An overweight man

b. Overweight

c. Are the events "male" and "overweight" independent?

**51. Blood Pressure** A doctor is studying the relationship between blood pressure and heartbeat abnormalities in her patients. She tests a random sample of her patients and notes their blood pressures (high, low, or normal) and their heartbeats (regular or irregular). She finds that:

(i) 14% have high blood pressure.

(ii) 22% have low blood pressure.

(iii) 15% have an irregular heartbeat.

(iv) Of those with an irregular heartbeat, one-third have high blood pressure.

(v) Of those with normal blood pressure, one-eighth have an irregular heartbeat.

What portion of the patients selected have a regular heartbeat and low blood pressure? Choose one of the following. (*Hint:* Make a table similar to the one for Exercise 48.) *Source: Society of Actuaries.*

a. 2%    b. 5%    c. 8%    d. 9%    e. 20%

**52. Breast Cancer** To explain why the chance of a woman getting breast cancer in the next year goes up each year, while the chance of a woman getting breast cancer in her lifetime goes down, Ruma Falk made the following analogy. Suppose you are looking for a letter that you may have lost. You have 8 drawers in your desk. There is a probability of 0.1 that the letter is in any one of the 8 drawers and a probability of 0.2 that the letter is not in any of the drawers. *Source: Chance News.*

a. What is the probability that the letter is in drawer 1?

b. Given that the letter is not in drawer 1, what is the probability that the letter is in drawer 2?

c. Given that the letter is not in drawer 1 or 2, what is the probability that the letter is in drawer 3?

d. Given that the letter is not in drawers 1–7, what is the probability that the letter is in drawer 8?

e. Based on your answers to parts a–d, what is happening to the probability that the letter is in the next drawer?

f. What is the probability that the letter is in some drawer?

g. Given that the letter is not in drawer 1, what is the probability that the letter is in some drawer?

h. Given that the letter is not in drawer 1 or 2, what is the probability that the letter is in some drawer?

i. Given that the letter is not in drawers 1–7, what is the probability that the letter is in some drawer?

j. Based on your answers to parts f–i, what is happening to the probability that the letter is in some drawer?

**53. Twins** A 1920 study of 17,798 pairs of twins found that 5844 consisted of two males, 5612 consisted of two females, and 6342 consisted of a male and a female. Of course, all of the mixed-gender pairs were not identical twins. The same-gender pairs may or may not have been identical twins. The goal here is to use the data to estimate $p$, the probability that a pair of twins is identical.

a. Use the data to find the proportion of twins who were male.

b. Denoting your answer from part a by $P(B)$, show that the probability that a pair of twins consists of two males is

$$pP(B) + (1 - p)(P(B))^2.$$

(*Hint:* Draw a tree diagram. The first set of branches should be identical and not identical. Then note that if one member of a pair of identical twins is male, the other must also be male.)

c. Using your answers from parts a and b, plus the fact that 5844 of the 17,798 twin pairs consisted of two males, find an estimate for the value of $p$.

d. Find an expression, similar to the one in part b, for the probability that a pair of twins is female.

**e.** Using your answers from parts a and d, plus the fact that 5612 of the 17,798 twin pairs were female, find an estimate for the value of $p$.

**f.** Find an expression, similar to those in parts b and d, for the probability that a pair of twins consists of one male and one female.

**g.** Using your answers from parts a and f, plus the fact that 6342 of the 17,798 twin pairs consisted of one male and one female, find an estimate for the value of $p$.

**54. Cigarette Smokers** The following table gives a recent estimate (in millions) of the smoking status among persons 25 years of age and over and their highest level of education. *Source: National Health Interview Survey.*

| Education | Current Smoker | Former Smoker | Non-smoker | Total |
|---|---|---|---|---|
| Less than a high school diploma | 7.90 | 6.66 | 14.12 | 28.68 |
| High school diploma or GED | 14.38 | 13.09 | 25.70 | 53.17 |
| Some college | 12.41 | 13.55 | 28.65 | 54.61 |
| Bachelor's degree or higher | 4.97 | 12.87 | 38.34 | 56.18 |
| Total | 39.66 | 46.17 | 106.81 | 192.64 |

**a.** Find the probability that a person is a current smoker.

**b.** Find the probability that a person has less than a high school diploma.

**c.** Find the probability that a person is a current smoker and has less than a high school diploma.

**d.** Find the probability that a person is a current smoker, given that the person has less than a high school diploma.

**e.** Are the events "current smoker" and "less than a high school diploma" independent events?*

**55. Colorectal Cancer** Researchers found that only 1 out of 24 physicians could give the correct answer to the following problem: "The probability of colorectal cancer can be given as 0.3%. If a person has colorectal cancer, the probability that the hemoccult test is positive is 50%. If a person does not have colorectal cancer, the probability that he still tests positive is 3%. What is the probability that a person who tests positive actually has colorectal cancer?" What is the correct answer? *Source: Science.*

**56. Hepatitis Blood Test** The probability that a person with certain symptoms has hepatitis is 0.8. The blood test used to confirm this diagnosis gives positive results for 90% of people with the disease and 5% of those without the disease. What is the probability that an individual who has the symptoms and who reacts positively to the test actually has hepatitis?

**57. Sensitivity and Specificity** The sensitivity and specificity for breast cancer during a mammography exam are approximately 79.6% and 90.2%, respectively. *Source: National Cancer Institute.*

**a.** It is estimated that 0.5% of U.S. women under the age of 40 have breast cancer. Find the probability that a woman under 40 who tests positive during a mammography exam actually has breast cancer.

**b.** Given that a woman under 40 tests negative during a mammography exam, find the probability that she does not have breast cancer.

**c.** According to the National Cancer Institute, failure to diagnose breast cancer is the most common cause of medical malpractice litigation. Given that a woman under 40 tests negative for breast cancer, find the probability that she does have breast cancer.

**d.** It is estimated that 1.5% of U.S. women over the age of 50 have breast cancer. Find the probability that a woman over 50 who tests positive actually has breast cancer.

**58. Test for HIV** Clinical studies have demonstrated that rapid HIV tests have a sensitivity of approximately 99.9% and a specificity of approximately 99.8%. *Source: Centers for Disease Control and Prevention.*

**a.** In some HIV clinics, the prevalence of HIV was high, about 5%. Find the probability that a person actually has HIV, given that the test came back positive from one of these clinics.

**b.** In other clinics, like a family planning clinic, the prevalence of HIV was low, about 0.1%. Find the probability that a person actually has HIV, given that the test came back positive from one of these clinics.

**c.** The answers in parts a and b are significantly different. Explain why.

**59. Smokers** A health study tracked a group of persons for five years. At the beginning of the study, 20% were classified as heavy smokers, 30% as light smokers, and 50% as non-smokers. Results of the study showed that light smokers were twice as likely as nonsmokers to die during the five-year study but only half as likely as heavy smokers. A randomly selected participant from the study died over the five-year period. Calculate the probability that the participant was a heavy smoker. Choose one of the following. (*Hint:* Let $x = P$(a nonsmoker dies).) *Source: Society of Actuaries.*

**a.** 0.20

**b.** 0.25

**c.** 0.35

**d.** 0.42

**e.** 0.57

---

*We are assuming here and in other exercises that the events consist entirely of the numbers given in the table. If the numbers are interpreted as a sample of all people fitting the description of the events, then testing for independence is more complicated, requiring a technique from statistics known as a *contingency table*.

**60. Emergency Room** Upon arrival at a hospital's emergency room, patients are categorized according to their condition as critical, serious, or stable. In the past year:

  **(i)** 10% of the emergency room patients were critical;

  **(ii)** 30% of the emergency room patients were serious;

  **(iii)** the rest of the emergency room patients were stable;

  **(iv)** 40% of the critical patients died;

  **(v)** 10% of the serious patients died; and

  **(vi)** 1% of the stable patients died.

  Given that a patient survived, what is the probability that the patient was categorized as serious upon arrival? Choose one of the following. *Source: Society of Actuaries.*

  **a.** 0.06    **b.** 0.29    **c.** 0.30    **d.** 0.39    **e.** 0.64

**61. Blood Test** A blood test indicates the presence of a particular disease 95% of the time when the disease is actually present. The same test indicates the presence of the disease 0.5% of the time when the disease is not present. One percent of the population actually has the disease. Calculate the probability that a person has the disease, given that the test indicates the presence of the disease. Choose one of the following. *Source: Society of Actuaries.*

  **a.** 0.324    **b.** 0.657    **c.** 0.945    **d.** 0.950    **e.** 0.995

**62. Circulation** The probability that a randomly chosen male has a circulation problem is 0.25. Males who have a circulation problem are twice as likely to be smokers as those who do not have a circulation problem. What is the conditional probability that a male has a circulation problem, given that he is a smoker? Choose one of the following. *Source: Society of Actuaries.*

  **a.** 1/4    **b.** 1/3    **c.** 2/5    **d.** 1/2    **e.** 2/3

**63. Seat Belt Effectiveness** A federal study showed that 63.8% of occupants involved in a fatal car crash wore seat belts. Of those in a fatal crash who wore seat belts, 2% were ejected from the vehicle. For those not wearing seat belts, 36% were ejected from the vehicle. *Source: National Highway Traffic Safety Administration.*

  **a.** Find the probability that a randomly selected person in a fatal car crash who was ejected from the vehicle was wearing a seat belt.

  **b.** Find the probability that a randomly selected person in a fatal car crash who was not ejected from the vehicle was not wearing a seat belt.

**Smokers by Age Group** The following table gives the proportion of U.S. adults in each age group in 2008, as well as the proportion in each group who smoke. *Source: Centers for Disease Control and Prevention.*

| Age | Proportion of Population | Proportion That Smoke |
|---|---|---|
| 18–44 years | 0.49 | 0.23 |
| 45–64 years | 0.34 | 0.22 |
| 65–74 years | 0.09 | 0.12 |
| 75 years and over | 0.08 | 0.06 |

**64.** Find the probability that a randomly selected adult who smokes is between 18 and 44 years of age (inclusive).

**65.** Find the probability that a randomly selected adult who does not smoke is between 45 and 64 years of age (inclusive).

## OTHER APPLICATIONS

**66. Working Women** A survey has shown that 52% of the women in a certain community work outside the home. Of these women, 64% are married, while 86% of the women who do not work outside the home are married. Find the probabilities that a woman in that community can be categorized as follows.

  **a.** Married

  **b.** A single woman working outside the home

**67. Rain Forecasts** In a letter to the journal *Nature*, Robert A. J. Matthews gives the following table of outcomes of forecast and weather over 1000 1-hour walks, based on the United Kingdom's Meteorological office's 83% accuracy in 24-hour forecasts. *Source: Nature.*

|  | Rain | No Rain | Totals |
|---|---|---|---|
| *Forecast of Rain* | 66 | 156 | 222 |
| *Forecast of No Rain* | 14 | 764 | 778 |
| *Totals* | 80 | 920 | 1000 |

  **a.** Verify that the probability that the forecast called for rain, given that there was rain, is indeed 83%. Also verify that the probability that the forecast called for no rain, given that there was no rain, is also 83%.

  **b.** Calculate the probability that there was rain, given that the forecast called for rain.

  **c.** Calculate the probability that there was no rain, given that the forecast called for no rain.

  **d.** Observe that your answer to part c is higher than 83% and that your answer to part b is much lower. Discuss which figure best describes the accuracy of the weather forecast in recommending whether or not you should carry an umbrella.

**68. Earthquakes** There are seven geologic faults (and possibly more) capable of generating a magnitude 6.7 earthquake in the region around San Francisco. Their probabilities of rupturing by the year 2032 are 27%, 21%, 11%, 10%, 4%, 3%, and 3%. *Source: Science News.*

  **a.** Calculate the probability that at least one of these faults erupts by the year 2032, assuming that these are independent events.

  **b.** Scientists forecast a 62% chance of an earthquake with magnitude at least 6.7 in the region around San Francisco by the year 2032. Compare this with your answer from part a. Consider the realism of the assumption of independence. Also consider the role of roundoff. For example, the probability of 10% for one of the faults is presumably rounded to the nearest percent, with the actual probability between 9.5% and 10.5%.

**69. *Titanic*** The table at the bottom of the page lists the number of passengers who were on the *Titanic* and the number of passengers who survived, according to class of ticket. Use this information to determine the following (round answers to four decimal places). ***Source: The Mathematics Teacher.***

**a.** What is the probability that a randomly selected passenger was second class?

**b.** What is the overall probability of surviving?

**c.** What is the probability of a first-class passenger surviving?

**d.** What is the probability of a child who was also in the third class surviving?

**e.** Given that the survivor is from first class, what is the probability that she was a woman?

**f.** Given that a male has survived, what is the probability that he was in third class?

**g.** Are the events third-class survival and male survival independent events? What does this imply?

**70. Real Estate** A real estate agent trying to sell you an attractive beachfront house claims that it will not collapse unless it is subjected simultaneously to extremely high winds and extremely high waves. According to weather service records, there is a 0.001 probability of extremely high winds, and the same for extremely high waves. The real estate agent claims, therefore, that the probability of both occurring is $(0.001)(0.001) = 0.000001$. What is wrong with the agent's reasoning?

**71. Age and Loans** Suppose 20% of the population are 65 or over, 26% of those 65 or over have loans, and 53% of those under 65 have loans. Find the probabilities that a person fits into the following categories.

**a.** 65 or over and has a loan

**b.** Has a loan

**72. Women Joggers** In a certain area, 15% of the population are joggers and 40% of the joggers are women. If 55% of those who do not jog are women, find the probabilities that an individual from that community fits the following descriptions.

**a.** A woman jogger

**b.** A man who is not a jogger

**c.** A woman

**d.** Are the events that a person is a woman and a person is a jogger independent? Explain.

**73. Driver's License Test** The Motor Vehicle Department has found that the probability of a person passing the test for a driver's license on the first try is 0.75. The probability that an individual who fails on the first test will pass on the second try is 0.80, and the probability that an individual who fails the first and second tests will pass the third time is 0.70. Find the probabilities that an individual will do the following.

**a.** Fail both the first and second tests

**b.** Fail three times in a row

**c.** Require at least two tries

**74. Speeding Tickets** A smooth-talking young man has a 1/3 probability of talking a policeman out of giving him a speeding ticket. The probability that he is stopped for speeding during a given weekend is 1/2. Find the probabilities of the events in parts a and b.

**a.** He will receive no speeding tickets on a given weekend.

**b.** He will receive no speeding tickets on 3 consecutive weekends.

**c.** We have assumed that what happens on the second or third weekend is the same as what happened on the first weekend. Is this realistic? Will driving habits remain the same after getting a ticket?

**75. Studying** A teacher has found that the probability that a student studies for a test is 0.60, the probability that a student gets a good grade on a test is 0.70, and the probability that both occur is 0.52.

**a.** Are these events independent?

**b.** Given that a student studies, find the probability that the student gets a good grade.

**c.** Given that a student gets a good grade, find the probability that the student studied.

**76. Basketball** A basketball player is fouled and now faces a one-and-one free throw situation. She shoots the first free throw. If she misses it, she scores 0 points. If she makes the first free throw, she gets to shoot a second free throw. If she misses the second free throw, she scores only one point for the first shot. If she makes the second free throw, she scores two points (one for each made shot). ***Source: The Mathematics Teacher.***

**a.** If her free-throwing percentage for this season is 60%, calculate the probability that she scores 0 points, 1 point, or 2 points.

**b.** Determine the free-throwing percentage necessary for the probability of scoring 0 points to be the same as the probability of scoring 2 points.* (*Hint:* Let $p$ be the free-throwing percentage, and then solve $p^2 = 1 - p$ for $p$. Use only the positive value of $p$.)

| | Children | | Women | | Men | | Totals | |
|---|---|---|---|---|---|---|---|---|
| | **On** | **Survived** | **On** | **Survived** | **On** | **Survived** | **On** | **Survived** |
| *First Class* | 6 | 6 | 144 | 140 | 175 | 57 | 325 | 203 |
| *Second Class* | 24 | 24 | 93 | 80 | 168 | 14 | 285 | 118 |
| *Third Class* | 79 | 27 | 165 | 76 | 462 | 75 | 706 | 178 |
| *Totals* | 109 | 57 | 402 | 296 | 805 | 146 | 1316 | 499 |

*The solution is the reciprocal of the number $\dfrac{1 + \sqrt{5}}{2}$, known as the golden ratio or the divine proportion. This number has great significance in architecture, science, and mathematics.

77. **Basketball** The same player from Exercise 76 is now shooting two free throws (that is, she gets to shoot a second free throw whether she makes or misses the first shot). Assume her free-throwing percentage is still 60%. Find the probability that she scores 0 points, 1 point, or 2 points. *Source: The Mathematics Teacher.*

78. **Alcohol Abstinence** The Harvard School of Public Health completed a study on alcohol consumption on college campuses in 2001. They concluded that 20.7% of women attending all-women colleges abstained from alcohol, compared to 18.6% of women attending coeducational colleges. Approximately 4.7% of women college students attend all-women schools. *Source: Harvard School of Public Health.*

   a. What is the probability that a randomly selected female student abstains from alcohol?

   b. If a randomly selected female student abstains from alcohol, what is the probability she attends a coeducational college?

79. **Murder** During the murder trial of O. J. Simpson, Alan Dershowitz, an advisor to the defense team, stated on television that only about 0.1% of men who batter their wives actually murder them. Statistician I. J. Good observed that even if, given that a husband is a batterer, the probability he is guilty of murdering his wife is 0.001, what we really want to know is the probability that the husband is guilty, given that the wife was murdered. Good estimates the probability of a battered wife being murdered, given that her husband is not guilty, as 0.001. The probability that she is murdered if her husband is guilty is 1, of course. Using these numbers and Dershowitz's 0.001 probability of the husband being guilty, find the probability that the husband is guilty, given that the wife was murdered. *Source: Nature.*

**Never-Married Adults by Age Group** The following tables give the proportion of men and of women 18 and older in each age group in 2009, as well as the proportion in each group who have never been married. *Source: U.S Census Bureau.*

| | Men | |
|---|---|---|
| Age | Proportion of Population | Proportion Never Married |
| 18–24 | 0.132 | 0.901 |
| 25–34 | 0.186 | 0.488 |
| 35–44 | 0.186 | 0.204 |
| 45–64 | 0.348 | 0.118 |
| 65 or over | 0.148 | 0.044 |

| | Women | |
|---|---|---|
| Age | Proportion of Population | Proportion Never Married |
| 18–24 | 0.121 | 0.825 |
| 25–34 | 0.172 | 0.366 |
| 35–44 | 0.178 | 0.147 |
| 45–64 | 0.345 | 0.092 |
| 65 or over | 0.184 | 0.040 |

80. Find the probability that a randomly selected man who has never married is between 35 and 44 years old (inclusive).

81. Find the probability that a randomly selected woman who has been married is between 18 and 24 (inclusive).

82. Find the probability that a randomly selected woman who has never been married is between 45 and 64 (inclusive).

**Job Qualifications** **Of all the people applying for a certain job, 75% are qualified and 25% are not. The personnel manager claims that she approves qualified people 85% of the time; she approves an unqualified person 20% of the time. Find each probability.**

83. A person is qualified if he or she was approved by the manager.

84. A person is unqualified if he or she was approved by the manager.

85. **Auto Insurance** An auto insurance company insures drivers of all ages. An actuary compiled the following statistics on the company's insured drivers:

| Age of Driver | Probability of Accident | Portion of Company's Insured Drivers |
|---|---|---|
| 16–20 | 0.06 | 0.08 |
| 21–30 | 0.03 | 0.15 |
| 31–65 | 0.02 | 0.49 |
| 66–99 | 0.04 | 0.28 |

A randomly selected driver that the company insures has an accident. Calculate the probability that the driver was age 16–20. Choose one of the following. *Source: Society of Actuaries.*

   a. 0.13    b. 0.16    c. 0.19    d. 0.23    e. 0.40

86. **Terrorists** John Allen Paulos has pointed out a problem with massive, untargeted wiretaps. To illustrate the problem, he supposes that one out of every million Americans has terrorist ties. Furthermore, he supposes that the terrorist profile is 99% accurate, so that if a person has terrorist ties, the profile will pick them up 99% of the time, and if the person does not have terrorist ties, the profile will accidentally pick them up only 1% of the time. Given that the profile has picked up a person, what is the probability that the person actually has terrorist ties? Discuss how your answer affects your opinion on domestic wiretapping. *Source: Who's Counting.*

87. **Three Prisoners** The famous "problem of three prisoners" is as follows.

   Three men, A, B, and C, were in jail. A knew that one of them was to be set free and the other two were to be executed. But he didn't know who was the one to be spared. To the jailer who did know, A said, "Since two out of the three will be executed, it is certain that either B or C will be, at least. You will give me no information about my own chances if you give me the name of one man, B or C, who is going to be executed." Accepting this argument after some thinking, the jailer said "B will be executed." Thereupon A felt happier because now either he or C would go free, so his chance had increased from 1/3 to 1/2. *Source: Cognition.*

**a.** Assume that initially each of the prisoners is equally likely to be set free. Assume also that if both B and C are to be executed, the jailer is equally likely to name either B or C. Show that A is wrong, and that his probability of being freed, given that the jailer says B will be executed, is still 1/3.

**b.** Now assume that initially the probabilities of A, B, and C being freed are 1/4, 1/4, and 1/2, respectively. As in part a, assume also that if both B and C are to be executed, the jailer is equally likely to name either B or C. Now show that A's probability of being freed, given that the jailer says B will be executed, actually drops to 1/5. Discuss the reasonableness of this answer, and why this result might violate someone's intuition.

# 12.4 Discrete Random Variables; Applications to Decision Making

**APPLY IT**  Should patients with malaria-related anemia receive blood transfusions? *In Example 8, we will calculate the expected values of recovery outcomes and use decision analysis to answer this question.*

We shall see that the *expected value* of a probability distribution is a type of average. Probability distributions were introduced briefly earlier in this chapter. Now we take a more complete look at probability distributions. A probability distribution depends on the idea of a *random variable*.

**Random Variables**  When researchers conduct an experiment, it is necessary to quantify the possible outcomes of the experiment. This process will enable the researcher to recognize individual outcomes and analyze the data. The most common way to record the individual outcomes of the experiment is to assign a numerical value to each of the different possible outcomes of the experiment. For example, if a coin is tossed 2 times, the possible outcomes are: *hh*, *ht*, *th*, and *tt*. For each of these possible outcomes, we could record the number of heads. Then the outcome, which we will label $x$, is one of the numbers 0, 1, or 2. Of course, we could have used other numbers, like 00, 01, 10, and 11, to indicate these same outcomes, but the values of 0, 1, and 2 are simpler and provide an immediate description of the exact outcome of the experiment. Notice that using this random variable also gives us a way to readily know how many tails occurred in the experiment. Thus, in some sense, the values of $x$ are random, so $x$ is called a **random variable**.

### Random Variable
A **random variable** is a function that assigns a real number to each outcome of an experiment.

Random variables are generally categorized as either discrete or continuous. A **discrete random variable** is used in experiments where there is either a finite or countably infinite number of simple events. **Continuous random variables** are used to describe situations where the random variable can be assigned any number in a given interval. We will study continuous random variables in the next chapter.

**Probability Distribution** A table that lists the possible values of a random variable, together with the corresponding probabilities, is called a **probability distribution**. The sum of the probabilities in a probability distribution must always equal 1. (The sum in some distributions may vary slightly from 1 because of rounding.)

### EXAMPLE 1 Hospital Satisfaction

Suppose that 100 patients are asked to rate the services provided by a local hospital. The rating options are: 1 (Outstanding), 2 (Good), 3 (Okay), 4 (Poor), and 5 (Terrible). The results of the survey show that 15 people listed the services as Outstanding, 38 as Good, 25 as Okay, 13 as Poor, and 9 as Terrible. Find the probability distribution for the hospital services rating.

**SOLUTION** Let $x$ represent the discrete random variable "hospital services rating." The possible values of $x$ are 1, 2, 3, 4, and 5. Since 15 of the 100 patients surveyed rated the hospital as Outstanding, the probability of an Outstanding rating is $P(1) = 15/100$.

Likewise, the probability of a Good rating is $P(2) = 38/100$, the probability of an Okay rating is $P(3) = 25/100$, the probability of a Poor rating is $P(4) = 13/100$, and the probability of a Terrible rating is $P(5) = 9/100$.

The results can be put in a table called a probability distribution.

| Probability Distribution of Hospital Survey | | | | | |
|---|---|---|---|---|---|
| $x$ | 1 | 2 | 3 | 4 | 5 |
| $P(x)$ | 15/100 | 38/100 | 25/100 | 13/100 | 9/100 |

Instead of writing the probability distribution as a table, we could write the same information as a set of ordered pairs:

$$\{(1, 15/100), (2, 38/100), (3, 25/100), (4, 13/100), (5, 9/100)\}.$$

There is just one probability for each value of the random variable. Thus, a probability distribution defines a function, called a **probability distribution function**, or simply a **probability function**. We shall use the terms "probability distribution" and "probability function" interchangeably.

The information in a probability distribution is often displayed graphically as a special kind of bar graph called a **histogram**. The bars of a histogram all have the same width, usually 1. (The widths might be different from 1 when the values of the random variable are not consecutive integers.) The heights of the bars are determined by the probabilities. A histogram for the data in Example 1 is given in Figure 31. A histogram shows important characteristics of a distribution that may not be readily apparent in tabular form, such as the relative sizes of the probabilities and any symmetry in the distribution.

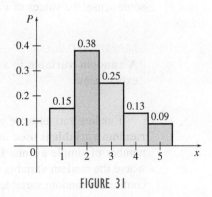

FIGURE 31

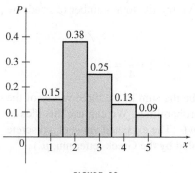

FIGURE 32

The area of the bar above $x = 1$ in Figure 31 is the product of 1 and 15/100, or $1 \cdot 15/100 = 0.15$. Since each bar has a width of 1, its area is equal to the probability that corresponds to that value of $x$. The probability that a particular value will occur is thus given by the area of the appropriate bar of the graph. For example, the probability that a person rated the hospital as Outstanding, Good, or Okay is the sum of the areas for $x = 1$, $x = 2$, and $x = 3$. This area, shown in pink in Figure 32, corresponds to 78/100, which corresponds to 78% of the total area, since

$$P(x \le 3) = P(x = 1) + P(x = 2) + P(x = 3).$$

**EXAMPLE 2** **Probability Distributions**

**(a)** Give the probability distribution for the number of heads showing when two coins are tossed.

**SOLUTION** Let $x$ represent the random variable "number of heads." Then $x$ can take on the values 0, 1, or 2. Now find the probability of each outcome. To find the probability of 0, 1, or 2 heads, notice that there are 4 outcomes in the sample space: $\{hh, ht, th, tt\}$. The results are shown in the table with Figure 33.

| Probability Distribution of Heads | | | |
|---|---|---|---|
| $x$ | 0 | 1 | 2 |
| $P(x)$ | 1/4 | 1/2 | 1/4 |

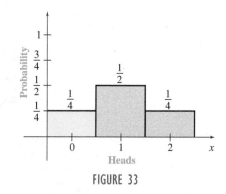

FIGURE 33

**YOUR TURN 1** Find the probability distribution and draw a histogram for the number of tails showing when three coins are tossed.

**(b)** Draw a histogram for the distribution in the table. Find the probability that at least one coin comes up heads.

**SOLUTION** The histogram is shown in Figure 33. The portion in pink represents

$$P(x \ge 1) = P(x = 1) + P(x = 2)$$
$$= \frac{3}{4}.$$

TRY YOUR TURN 1

## Expected Value

In working with probability distributions, it is useful to have a concept of the typical or average value that the random variable takes on. In Example 2, for instance, it seems reasonable that, on the average, one head shows when two coins are tossed. This does not tell what will happen the next time we toss two coins; we may get two heads, or we may get none. If we tossed two coins many times, however, we would expect that, in the long run, we would average about one head for each toss of two coins.

A way to solve such problems in general is to imagine flipping two coins 4 times. Based on the probability distribution in Example 2, we would expect that 1 of the 4 times we would get 0 heads, 2 of the 4 times we would get 1 head, and 1 of the 4 times we would get 2 heads. The total number of heads we would get, then, is

$$0 \cdot 1 + 1 \cdot 2 + 2 \cdot 1 = 4.$$

The expected numbers of heads per toss is found by dividing the total number of heads by the total number of tosses, or

$$\frac{0 \cdot 1 + 1 \cdot 2 + 2 \cdot 1}{4} = 0 \cdot \frac{1}{4} + 1 \cdot \frac{1}{2} + 2 \cdot \frac{1}{4} = 1.$$

Notice that the expected number of heads turns out to be the sum of the three values of the random variable $x$ multiplied by their corresponding probabilities. We can use this idea to define the *expected value* of a random variable as follows. (The expected value of a discrete random variable is also called the mean and is represented by the Greek letter mu, $\mu$.)

## Expected Value (or Mean)

Suppose the discrete random variable $x$ can take on the $n$ values $x_1, x_2, x_3, \ldots, x_n$. Also, suppose the probabilities that these values occur are, respectively, $p_1, p_2, p_3, \ldots, p_n$. Then the **expected value** of the random variable is

$$E(x) = \mu = x_1 p_1 + x_2 p_2 + x_3 p_3 + \cdots + x_n p_n.$$

In summation notation

$$E(X) = \mu = \sum_x x P(X = x) = \sum_x x P(x),$$

where the sum is taken over all possible values of the random variable $X$.

### EXAMPLE 3    Hospital Satisfaction

In Example 1, find the expected satisfaction rating of the hospital.

**SOLUTION**    Using the values in the table from Example 1 and the definition of expected value, we find that

$$E(X) = 1 \cdot P(X = 1) + 2 \cdot P(X = 2) + 3 \cdot P(X = 3) + 4 \cdot P(X = 4) + 5 \cdot P(X = 5)$$

$$= 1 \cdot \frac{15}{100} + 2 \cdot \frac{38}{100} + 3 \cdot \frac{25}{100} + 4 \cdot \frac{13}{100} + 5 \cdot \frac{9}{100} = \frac{263}{100} = 2.63.$$

Thus, the average person will rate the hospital between Good and Okay, but closer to Okay.

Physically, the expected value of a probability distribution represents a balance point. If we think of the histogram in Figure 31 as a series of weights with magnitudes represented by the heights of the bars, then the system would balance if supported at the point corresponding to the expected value.

### EXAMPLE 4    Symphony Orchestra

Suppose a local symphony decides to raise money by raffling an HD television worth $400, a dinner for two worth $80, and 2 CDs worth $20 each. A total of 2000 tickets are sold at $1 each. Find the expected payback for a person who buys one ticket in the raffle.

Method 1
Direct Calculation

**SOLUTION**

Here the random variable represents the possible amounts of payback, where payback = amount won − cost of ticket. The payback of the person winning the television is $400 (amount won) − $1 (cost of ticket) = $399. The payback for each losing ticket is $0 − $1 = −$1.

The paybacks of the various prizes, as well as their respective probabilities, are shown in the table on the next page. The probability of winning $19 is 2/2000 because there are 2 prizes worth $20. We have not reduced the fractions in order to keep all the denominators equal. Because there are 4 winning tickets, there are 1996 losing tickets, so the probability of winning −$1 is 1996/2000.

| Probability Distribution of Prize Winnings | | | | |
|---|---|---|---|---|
| $x$ | $399 | $79 | $19 | $-1 |
| $P(x)$ | 1/2000 | 1/2000 | 2/2000 | 1996/2000 |

The expected payback for a person buying one ticket is

$$399\left(\frac{1}{2000}\right) + 79\left(\frac{1}{2000}\right) + 19\left(\frac{2}{2000}\right) + (-1)\left(\frac{1996}{2000}\right) = -\frac{1480}{2000}$$
$$= -0.74.$$

On average, a person buying one ticket in the raffle will lose $0.74, or 74¢.

It is not possible to lose 74¢ in this raffle: either you lose $1, or you win a prize worth $400, $80, or $20, minus the $1 you pay to play. But if you bought tickets in many such raffles over a long period of time, you would lose 74¢ per ticket on average. It is important to note that the expected value of a random variable may be a number that can never occur in any one trial of the experiment.

**Method 2**
**Alternate Procedure**

**YOUR TURN 2** Suppose that there is a $5 raffle with prizes worth $1000, $500, and $250. Suppose 1000 tickets are sold and you purchase a ticket. Find your expected payback for this raffle.

An alternative way to compute expected value in this and other examples is to calculate the expected amount won and then subtract the cost of the ticket afterward. The amount won is either $400 (with probability 1/2000), $80 (with probability 1/2000), $20 (with probability 2/2000), or $0 (with probability 1996/2000). The expected payback for a person buying one ticket is then

$$400\left(\frac{1}{2000}\right) + 80\left(\frac{1}{2000}\right) + 20\left(\frac{2}{2000}\right) + 0\left(\frac{1996}{2000}\right) - 1 = -\frac{1480}{2000}$$
$$= -0.74.$$

**TRY YOUR TURN 2**

### EXAMPLE 5  Friendly Wager

Each day Donna and Mary toss a coin to see who buys coffee ($1.20 a cup). One tosses and the other calls the outcome. If the person who calls the outcome is correct, the other buys the coffee; otherwise the caller pays. Find Donna's expected payback.

**SOLUTION** Assume that an honest coin is used, that Mary tosses the coin, and that Donna calls the outcome. The possible results and corresponding probabilities are shown below.

| Possible Results | | | | |
|---|---|---|---|---|
| *Result of Toss* | Heads | Heads | Tails | Tails |
| *Call* | Heads | Tails | Heads | Tails |
| *Caller Wins?* | Yes | No | No | Yes |
| *Probability* | 1/4 | 1/4 | 1/4 | 1/4 |

Donna wins a $1.20 cup of coffee whenever the results and calls match, and she loses a $1.20 cup when there is no match. Her expected payback is

$$(1.20)\left(\frac{1}{4}\right) + (-1.20)\left(\frac{1}{4}\right) + (-1.20)\left(\frac{1}{4}\right) + (1.20)\left(\frac{1}{4}\right) = 0.$$

This implies that, over the long run, Donna neither wins nor loses.

A game with an expected value of 0 (such as the one in Example 5) is called a **fair game**. Casinos do not offer fair games. If they did, they would win (on average) $0, and have a hard time paying the help! Casino games have expected winnings for the house that vary from 1.5 cents per dollar to 60 cents per dollar. Exercises 35–37 at the end of the section ask you to find the expected payback for certain games of chance.

The idea of expected value can be very useful in decision making, as shown by the next example.

### EXAMPLE 6 Life Insurance

At age 50, you receive a letter from Mutual of Mauritania Insurance Company. According to the letter, you must tell the company immediately which of the following two options you will choose: take $20,000 at age 60 (if you are alive, $0 otherwise) or $30,000 at age 70 (again, if you are alive, $0 otherwise). Based *only* on the idea of expected value, which should you choose?*

**SOLUTION** Life insurance companies have constructed elaborate tables showing the probability of a person living a given number of years into the future. From a recent such table, the probability of living from age 50 to 60 is 0.88, while the probability of living from age 50 to 70 is 0.64. The expected values of the two options are given below.

First option: $(20,000)(0.88) + (0)(0.12) = 17,600$

Second option: $(30,000)(0.64) + (0)(0.36) = 19,200$

Based strictly on expected values, choose the second option.

## Variance

Although the expected value provides a measure of central tendency, it does not provide a measure of how much the data vary from the expected value. For example, the three sets of numbers in the following table all have an average of 50. But, as you can see, the values in the third distribution vary significantly from the mean.

| List | Values | Mean |
|------|--------|------|
| *Set 1* | 15, 25, 70, 90 | 50 |
| *Set 2* | 45, 46, 47, 48, 49, 51, 52, 53, 54, 55 | 50 |
| *Set 3* | 1, 17, 45, 70, 117 | 50 |

The **variance** (represented by $\sigma^2$ or Var($X$)) of a probability distribution is a measure of the *spread* of the values of a probability distribution. For a discrete distribution, the variance is found by taking the expected value of the squares of the differences of the values of the random variable and the mean. Thus,

$$\sigma^2 = \text{Var}(X) = \sum_x (x - \mu)^2 P(x),$$

where the sum is taken over all possible values of the random variable $X$, and $\mu$ is the expected value (or mean) of $X$.

Think of the variance as the expected value of $(X - \mu)^2$, which measures how far $X$ is from the mean $\mu$. The **standard deviation** (represented by $\sigma$) of $X$ is defined as

$$\sigma = \sqrt{\sigma^2}.$$

*In reality, the value of money now versus money in the future should be considered. For more on this idea see Chapter 5, Mathematics of Finance, in *Finite Mathematics*, 10th edition, by M. Lial, R. Greenwell, and N. Ritchey, Boston, MA: Pearson, 2012.

**YOUR TURN 3** Find the expected value (mean), variance, and standard deviation for the probability distribution in Example 2.

**EXAMPLE 7** Hospital Satisfaction

Find the variance and standard deviation of the hospital satisfaction ratings data.

**SOLUTION**

$$\sigma^2 = (1 - 2.63)^2 \frac{15}{100} + (2 - 2.63)^2 \frac{38}{100} + (3 - 2.63)^2 \frac{25}{100} + (4 - 2.63)^2 \frac{13}{100} + (5 - 2.63)^2 \frac{9}{100}$$
$$= 1.3331.$$

Using the definition of standard deviation we find that $\sigma = \sqrt{1.3331} \approx 1.15$.

TRY YOUR TURN 3

**Decision Analysis** The concept of expected value is used extensively in medicine to help physicians determine diagnostic and treatment strategies that provide the best care for the patients they serve. This area of research is commonly called **decision analysis**. The following example illustrates the power of expected value and decision analysis.

**EXAMPLE 8** Blood Transfusions for Malaria-Related Anemia

Blood transfusions are commonly used to treat childhood malarial anemia in Africa. Although the transfusions are effective in treating the anemia, much of the blood is unscreened and there is a significant risk of transmitting the HIV virus with the transfusion. Given the risk of HIV transmission, what type of transfusion policy would maximize childhood survival? Researchers have used the concept of decision analysis to study this problem. *Source: The American Journal of Tropical Medicine and Hygiene.*

The decision tree in Figure 34 illustrates the two choices of transfusion or no transfusion and the possible effects of both choices. The square □ represents a decision node and the circles ○ represent a chance node. The choice of transfusion includes the possibility of immediate fatal complications. If a fatal complication from the transfusion does not occur, then several outcomes are possible: The blood may or may not be contaminated with the HIV virus, the patient may or may not die of anemia, seroconversion (HIV antigens appearing in the blood) may or may not occur, and HIV may or may not convert to AIDS. If the patient does not receive the transfusion, then he or she will either live or die from the anemia. The values at the end of the tree indicate that the patient has died (0), has fully recovered (1), or has AIDS (0.1) The value of 0.1 indicates that this patient's life expectancy will be one-tenth of that for a patient without AIDS.

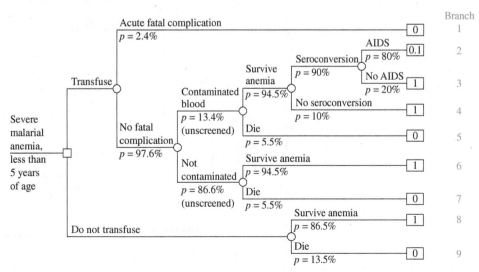

FIGURE 34

To determine the strategy that gives the highest life expectancy, we will calculate the expected value for each of the two choices.

APPLY IT

**Method 1:**
**Probability Distribution Table**

**SOLUTION**

The first seven branches of the decision tree in Figure 34 are related to the choice of giving the transfusion. For each branch, we can calculate the probability that the outcome for a particular patient will follow that branch. For example, the probability that a transfused patient will not have a fatal complication due to the blood transfusion, will receive noncontaminated blood, and will survive the anemia (branch 6) is

$$P = 0.976(0.866)(0.945) = 0.799.$$

We can calculate the probabilities for the other branches in a similar manner, as illustrated in the following table. Verify that the probabilities in the table are correct.

| | **Probabilities for Each Outcome (Branch)** | | | | | | |
|---|---|---|---|---|---|---|---|
| **Branch** | **1** | **2** | **3** | **4** | **5** | **6** | **7** |
| *Outcome* $(x)$ | 0 | 0.1 | 1 | 1 | 0 | 1 | 0 |
| $P(x)$ | 0.024 | 0.089 | 0.022 | 0.012 | 0.007 | 0.799 | 0.046 |

Using the methods discussed earlier, we find that the expected outcome is

$$E_1 = 0(0.024) + 0.1(0.089) + 1(0.022) + 1(0.012) + 0(0.007) + 1(0.799) + 0(0.046)$$
$$\approx 0.842.$$

Similarly, we can calculate the expected outcome of not transfusing.

$$E_2 = 1(0.865) + 0(0.135) = 0.865$$

There are various ways to interpret these expected values. Since 0 is the value that represents death and 1 represents full recovery, the values above indicate that, on average, 84% of the patients who receive a transfusion can expect a full recovery with full life expectancy. Of the patients who do not receive a transfusion, 86.5% can expect a full recovery. Thus, for unscreened blood, it appears that the best strategy is to not provide a transfusion for malaria-related anemia. In Exercise 27 you will be asked to evaluate the same strategies for blood that is screened.

**Method 2:**
**Foldback Procedure**

The expected value of each strategy can also be calculated using a procedure called **folding back**. In this procedure, we start calculating at the end of a branch and by working our way back we compress the entire tree to the decision nodes. For example, in the transfusion branches, we start at the top-most endpoint and calculate inward.

The AIDS/No AIDS branches can be replaced by one endpoint that has the value $0.1(0.80) + 1(0.2) = 0.28$, as illustrated in Figure 35. The Seroconversion/No

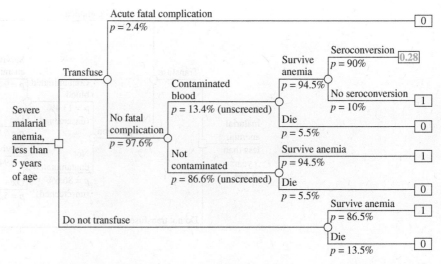

FIGURE 35

Seroconversion branches, with the new value of 0.28, can then be replaced by the value $0.28(0.9) + 1(0.10) = \mathbf{0.352}$, as shown in Figure 36. In a similar manner, the Survive anemia/Die branches, with the newly calculated value of 0.352, can be replaced by the value $0.352(0.945) + 0(0.055) = 0.333$. With this new value, the Contaminated blood/Not contaminated blood branches can be replaced by the single value, $0.333(0.134) + (1(0.945) + 0(0.055))(0.866) = \mathbf{0.863}$, as shown in Figure 37. Continuing the foldback procedure for all branches, the tree is folded back to the original two strategy branches, representing Transfuse/Do not transfuse, with expected values illustrated in Figure 38.

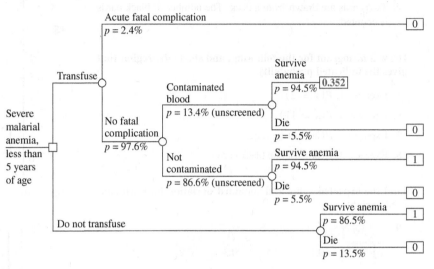

FIGURE 36

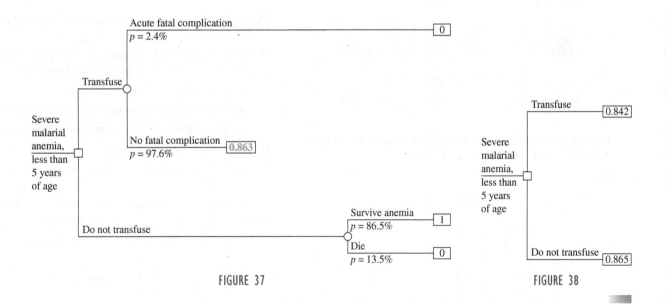

FIGURE 37

FIGURE 38

The random variables representing outcomes that were used in this example provide an illustration of the types of values that are used to analyze various decision strategies. Quality of life indicators for various outcomes are often used in medicine to distinguish between outcomes. Costs are also often used to evaluate strategies. Occasionally, two indicators—cost and risk—are combined to form a cost-benefit decision analysis.

# 12.4 EXERCISES

**For each experiment described below, let *x* determine a random variable, and use your knowledge of probability to prepare a probability distribution.**

1. Four coins are tossed, and the number of heads is noted.

2. Two dice are rolled, and the total number of points is recorded.

3. Three cards are drawn from a deck. The number of aces is counted.

4. Two cards are drawn from a deck. The number of black cards is counted.

**Draw a histogram for the following, and shade the region that gives the indicated probability.**

5. Exercise 1; $P(x \le 2)$

6. Exercise 2; $P(x \ge 11)$

7. Exercise 3; $P(\text{at least one ace})$

8. Exercise 4; $P(\text{at least one black card})$

**Find the expected value and standard deviation for each random variable.**

9.

| x | 2 | 3 | 4 | 5 |
|---|---|---|---|---|
| P(x) | 0.1 | 0.4 | 0.3 | 0.2 |

10.

| y | 4 | 6 | 8 | 10 |
|---|---|---|---|---|
| P(y) | 0.4 | 0.4 | 0.05 | 0.15 |

11.

| z | 9 | 12 | 15 | 18 | 21 |
|---|---|---|---|---|---|
| P(z) | 0.14 | 0.22 | 0.38 | 0.19 | 0.07 |

12.

| x | 30 | 32 | 36 | 38 | 44 |
|---|---|---|---|---|---|
| P(x) | 0.31 | 0.29 | 0.26 | 0.09 | 0.05 |

**Find the expected value for the random variable *x* having the probability function shown in each graph.**

13.

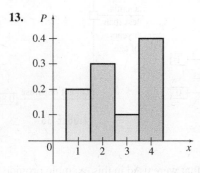

14.

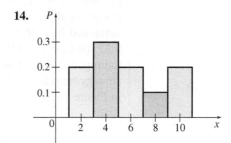

15.

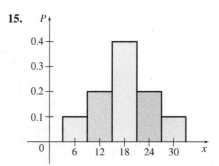

16.

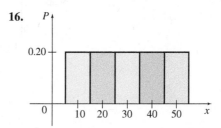

17. For the game in Example 5, find Mary's expected payback. Is it a fair game?

18. Suppose one day Mary brings a 2-headed coin and uses it to toss for the coffee. Since Mary tosses, Donna calls.

   a. Is this still a fair game?

   b. What is Donna's expected payback if she calls heads?

   c. What is Donna's expected payback if she calls tails?

19. Suppose a die is rolled 4 times.

   a. Find the probability distribution for the number of times 1 is rolled.

   b. What is the expected number of times 1 is rolled?

20. Two cards are drawn at one time from a deck of 52 cards.

   a. What is the expected number of diamonds?

   b. Suppose someone offers to pay you $5 if you draw 2 diamonds. He says that you should pay 50 cents for the chance to play. Is this a fair game?

21. Your friend missed class the day probability distributions were discussed. How would you explain probability distribution to him?

22. Explain what expected value means in your own words.

**23.** Four slips of paper numbered 2, 3, 4, and 5 are in a hat. You draw a slip, note the result, and then draw a second slip and note the result (without replacing the first).

  **a.** Find the probability distribution for the sum of the two slips.

  **b.** Draw a histogram for the probability distribution in part a.

  **c.** Find the odds that the sum is even.

  **d.** Find the expected value of the sum.

  **e.** Find the standard deviation of the sum.

## LIFE SCIENCE APPLICATIONS

**24. Animal Offspring** In a certain animal species, the probability that a healthy adult female will have no offspring in a given year is 0.29, while the probabilities of 1, 2, 3, or 4 offspring are, respectively, 0.23, 0.18, 0.16, and 0.14. Find the expected number of offspring.

**25. Ear Infections** Otitis media, or middle ear infection, is initially treated with an antibiotic. Researchers have compared two antibiotics, amoxicillin and cefaclor, for their cost effectiveness. Amoxicillin is inexpensive, safe, and effective. Cefaclor is also safe. However, it is considerably more expensive and it is generally more effective. Use the tree diagram below (where the costs are estimated as the total cost of medication, office visit, ear check, and hours of lost work) to answer the following. *Source: Journal of Pediatric Infectious Disease.*

  **a.** Find the expected cost of using each antibiotic to treat a middle ear infection.

  **b.** To minimize the total expected cost, which antibiotic should be chosen?

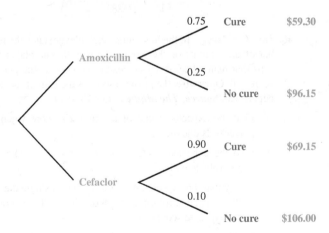

**26. Hospitalization Insurance** An insurance policy pays an individual $100 per day for up to 3 days of hospitalization and $25 per day for each day of hospitalization thereafter. The number of days of hospitalization, $X$, is a discrete random variable with probability function

$$P(X = k) = \begin{cases} \dfrac{6-k}{15} & \text{for } k = 1, 2, 3, 4, 5 \\ 0 & \text{otherwise.} \end{cases}$$

Calculate the expected payment for hospitalization under this policy. Choose one of the following. *Source: Society of Actuaries.*

  **a.** $85   **b.** $163   **c.** $168   **d.** $213   **e.** $255

**27. Blood Transfusions** Suppose in Example 8 that the blood was screened for HIV prior to being used and that the screening process reduces the probability that the blood is contaminated to 1%. Calculate the expected survival rate for each strategy. Should blood transfusions be given after the blood has been screened?

**28. Perinatal Mortality** Preterm birth and low birth weight are consistently the primary causes of perinatal mortality. Researchers have determined that treating infections during pregnancy may be one of the most effective ways to prevent such complications and reduce infant mortality. In the study, 870 women from Germany were screened for bacterial vaginosis. The patients were assigned to one of three different treatment strategies, or practices, according to the clinic which they visited. The decision tree below gives the results of the study. Calculate the expected cost of each strategy. Which strategy yields the lowest expected cost? *Source: The Journal of Reproductive Medicine.*

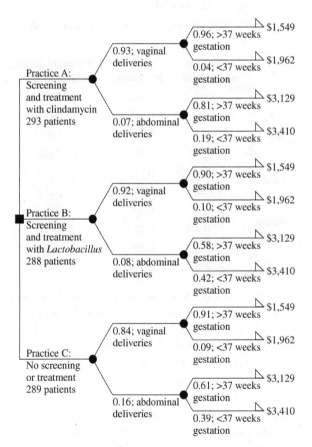

## OTHER APPLICATIONS

**29. Education** A study from Cleveland Clinic reported that 74% of very-low-birth-weight babies graduate from high school by age 20. If 250 very-low-birth-weight babies are followed through high school, how many would you expect to graduate from high school by age 20? *Source: The New England Journal of Medicine.*

**30. Cheating** A poll conducted by *U.S. News and World Report* reported that 84% of college students believe they need to cheat to get ahead in the world today. If 500 college students were surveyed, how many would you expect to say that they need to cheat to get ahead in the world today? *Source: U.S. News and World Report.*

**31. Seeding Storms** One of the few methods that can be used in an attempt to cut the severity of a hurricane is to *seed* the storm. In this process, silver iodide crystals are dropped into the storm. Unfortunately, silver iodide crystals sometimes cause the storm to *increase* its speed. Wind speeds may also increase or decrease even with no seeding. Use the table below to answer the following. *Source: Science.*

a. Find the expected amount of damage under each option, "seed" and "do not seed."

b. To minimize total expected damage, what option should be chosen?

| | Change in wind speed | Probability | Property damage (millions of dollars) |
|---|---|---|---|
| **Seed** | +32% | 0.038 | 335.8 |
| | +16% | 0.143 | 191.1 |
| | 0 | 0.392 | 100.0 |
| | −16% | 0.255 | 46.7 |
| | −34% | 0.172 | 16.3 |
| **Do not seed** | +32% | 0.054 | 335.8 |
| | +16% | 0.206 | 191.1 |
| | 0 | 0.480 | 100.0 |
| | −16% | 0.206 | 46.7 |
| | −34% | 0.054 | 16.3 |

**32. Golf Tournament** At the end of play in a major golf tournament, two players, an "old pro" and a "new kid," are tied. Suppose the first prize is $80,000 and second prize is $20,000. Find the expected winnings for the old pro if

a. both players are of equal ability;

b. the new kid will freeze up, giving the old pro a 3/4 chance of winning.

**33. Raffle** A raffle offers a first prize of $400 and 3 second prizes of $80 each. One ticket costs $2, and 500 tickets are sold. Find the expected payback for a person who buys 1 ticket. Is this a fair game?

**34. Raffle** A raffle offers a first prize of $1000, 2 second prizes of $300 each, and 20 third prizes of $10 each. If 10,000 tickets are sold at 50¢ each, find the expected payback for a person buying 1 ticket. Is this a fair game?

**Find the expected payback for the games of chance described in Exercises 35–37.**

**35. Roulette** In one form of roulette, you bet $1 on "even." If 1 of the 18 even numbers comes up, you get your dollar back, plus another one. If 1 of the 20 noneven (18 odd, 0, and 00) numbers comes up, you lose your dollar.

**36. Roulette** In another form of roulette, there are only 19 non-even numbers (no 00).

**37. Numbers** *Numbers* is a game in which you bet $1 on any three-digit number from 000 to 999. If your number comes up, you get $500.

**38. Contests** A magazine distributor offers a first prize of $100,000, two second prizes of $40,000 each, and two third prizes of $10,000 each. A total of 2,000,000 entries are received in the contest. Find the expected payback if you submit one entry to the contest. If it would cost you $1 in time, paper, and stamps to enter, would it be worth it?

**39. Contests** A contest at a fast-food restaurant offered the following cash prizes and probabilities of winning on one visit. Suppose you spend $1 to buy a bus pass that lets you go to 25 different restaurants in the chain and pick up entry forms. Find your expected value.

| Prize | Probability |
|---|---|
| $100,000 | 1/176,402,500 |
| $25,000 | 1/39,200,556 |
| $5000 | 1/17,640,250 |
| $1000 | 1/1,568,022 |
| $100 | 1/282,244 |
| $5 | 1/7056 |
| $1 | 1/588 |

**40. The Hog Game** In the hog game, each player states the number of dice that he or she would like to roll. The player then rolls that many dice. If a 1 comes up on any die, the player's score is 0. Otherwise, the player's score is the sum of the numbers rolled. *Source: The Mathematics Teacher.*

a. Find the expected value of the player's score when the player rolls one die.

b. Find the expected value of the player's score when the player rolls two dice.

c. Verify that the expected nonzero score of a single die is 4, so that if a player rolls $n$ dice that do not result in a score of 0, the expected score is $4n$.

d. Verify that if a player rolls $n$ dice, there are $5^n$ possible ways to get a nonzero score, and $6^n$ possible ways to roll the dice. Explain why the expected value of the player's score when the player rolls $n$ dice is then

$$\mu = \frac{5^n(4n)}{6^n}.$$

**41. Football** After a team scores a touchdown, it can either attempt to kick an extra point or attempt a two-point conversion. During the 2005 NFL season, two-point conversions were successful 45% of the time and the extra-point kicks were successful 96% of the time. ***Source: NFL.***

    **a.** Calculate the expected value of each strategy.

    **b.** Which strategy, over the long run, will maximize the number of points scored?

    **c.** Using this information, should a team always only use one strategy? Explain.

**42. Complaints** A local used-car dealer gets complaints about his cars as shown in the table below. Let $x$ represent the discrete random variable "number of complaints per day." Find the expected number of complaints per day.

| Number of Complaints per Day | | | | | | |
|---|---|---|---|---|---|---|
| $x$ | 0 | 1 | 2 | 3 | 4 | 5 | 6 |
| Probability | 0.02 | 0.06 | 0.16 | 0.25 | 0.32 | 0.13 | 0.06 |

**43. Payout on Insurance Policies** An insurance company has written 100 policies for $100,000, 500 policies for $50,000, and 1000 policies for $10,000 for people of age 20. If experience shows that the probability that a person will die at age 20 is 0.0012, how much can the company expect to pay out during the year the policies were written?

**44. Device Failure** An insurance policy on an electrical device pays a benefit of $4000 if the device fails during the first year. The amount of the benefit decreases by $1000 each successive year until it reaches 0. If the device has not failed by the beginning of any given year, the probability of failure during that year is 0.4. What is the expected benefit under this policy? Choose one of the following. ***Source: Society of Actuaries.***

    **a.** $2234  **b.** $2400  **c.** $2500  **d.** $2667  **e.** $2694

**YOUR TURN ANSWERS**

**1.**
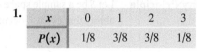
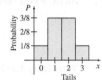

| $x$ | 0 | 1 | 2 | 3 |
|---|---|---|---|---|
| $P(x)$ | 1/8 | 3/8 | 3/8 | 1/8 |

**2.** −$3.25

**3.** 1; 0.5; 0.71

# 12 CHAPTER REVIEW

## SUMMARY

We began this chapter by introducing sets, which are collections of objects. We introduced the following set operations:

- complement ($A'$ is the set of elements not in $A$),
- intersection ($A \cap B$ is the set of elements belonging to both set $A$ and set $B$), and
- union ($A \cup B$ is the set of elements belonging to either set $A$ or set $B$ or both).

We used tree diagrams and Venn diagrams to define and study concepts in set operations as well as in probability. We introduced the following terms:

- experiment (an activity or occurrence with an observable result),
- trial (a repetition of an experiment),
- outcome (a result of a trial),
- sample space (the set of all possible outcomes for an experiment), and
- event (a subset of a sample space).

We investigated how to compute various probabilities and we explored some of the properties of probability. In particular, we studied the following concepts:

- empirical probability (based on how frequently an event actually occurred),
- conditional probability (in which some other event is assumed to have occurred),
- odds (an alternative way of expressing probability),
- independent events (in which the occurrence of one event does not affect the probability of another), and
- Bayes' theorem (used to calculate certain types of conditional probability).

Throughout the chapter, many applications of probability were introduced and analyzed. Finally, the concept of expected value was derived and used to illustrate an important use of decision trees, called decision analysis. In the next chapter we will use calculus to extend the ideas and power of probability.

## Sets Summary

**Number of Subsets** A set of $k$ distinct elements has $2^k$ subsets.

**Disjoint Sets** If sets $A$ and $B$ are disjoint, then

$$A \cap B = \emptyset \quad \text{and} \quad n(A \cap B) = 0.$$

**Union Rule for Sets** For any sets $A$ and $B$,

$$n(A \cup B) = n(A) + n(B) - n(A \cap B).$$

## Probability Summary

**Basic Probability Principle** Let $S$ be a sample space of equally likely outcomes, and let event $E$ be a subset of $S$. Then the probability that event $E$ occurs is

$$P(E) = \frac{n(E)}{n(S)}.$$

**Mutually Exclusive Events** If $E$ and $F$ are mutually exclusive events,

$$E \cap F = \emptyset \quad \text{and} \quad P(E \cap F) = 0.$$

**Union Rule** For any events $E$ and $F$ from a sample space $S$,

$$P(E \cup F) = P(E) + P(F) - P(E \cap F).$$

**Complement Rule** $P(E) = 1 - P(E') \quad$ and $\quad P(E') = 1 - P(E)$

**Odds** The odds in favor of event $E$ are $\dfrac{P(E)}{P(E')}$, where $P(E') \neq 0$.

If the odds favoring event $E$ are $m$ to $n$, then

$$P(E) = \frac{m}{m+n} \quad \text{and} \quad P(E') = \frac{n}{m+n}.$$

**Properties of Probability** 1. For any event $E$ in sample space $S$, $0 \le P(E) \le 1$.

2. The sum of the probabilities of all possible distinct outcomes is 1.

**Conditional Probability** The conditional probability of event $E$, given that event $F$ has occurred, is

$$P(E|F) = \frac{P(E \cap F)}{P(F)}, \quad \text{where } P(F) \neq 0.$$

For equally likely outcomes, conditional probability is found by reducing the sample space to event $F$; then

$$P(E|F) = \frac{n(E \cap F)}{n(F)}.$$

**Product Rule of Probability** If $E$ and $F$ are events, then $P(E \cap F)$ may be found by either of these formulas.

$$P(E \cap F) = P(F) \cdot P(E|F) \quad \text{or} \quad P(E \cap F) = P(E) \cdot P(F|E)$$

**Independent Events** If $E$ and $F$ are independent events,

$$P(E|F) = P(E), \quad P(F|E) = P(F), \quad \text{and} \quad P(E \cap F) = P(E) \cdot P(F).$$

**Bayes' Theorem** $P(F_i|E) = \dfrac{P(F_i) \cdot P(E|F_i)}{P(F_1) \cdot P(E|F_1) + P(F_2) \cdot P(E|F_2) + \cdots + P(F_n) \cdot P(E|F_n)}$

**Expected Value (or Mean)** $E(x) = \mu = x_1 p_1 + x_2 p_2 + x_3 p_3 + \cdots + x_n p_n = \sum_x x P(x)$

where the sum is taken over all possible values of random variable $X$.

**Variance** $\sigma^2 = \sum_x (x - \mu)^2 P(x)$

where the sum is taken over all possible values of random variable $X$.

**Standard Deviation** $\sigma = \sqrt{\sigma^2}$

# KEY TERMS

| 12.1 | 12.2 | 12.3 | histogram |
|------|------|------|-----------|
| set | experiment | conditional probability | expected value |
| element (member) | trial | product rule | fair game |
| empty set (or null set) | outcome | independent events | variance |
| set-builder notation | sample space | dependent events | standard deviation |
| universal set | event | Bayes' theorem | decision analysis |
| subset | simple event | sensitivity | folding back |
| tree diagram | certain event | specificity | |
| Venn diagram | impossible event | | |
| complement | mutually exclusive events | 12.4 | |
| intersection | probability | random variable | |
| disjoint sets | union rule for probability | discrete random variable | |
| union | odds | continuous random variable | |
| union rule for sets | empirical probability | probability distribution | |
| | probability distribution | probability function | |

# REVIEW EXERCISES

## CONCEPT CHECK

**Determine whether each of the following statements is true or false, and explain why.**

1. A set is a subset of itself.

2. A set has more subsets than it has elements.

3. The union of two sets always has more elements than either set.

4. The intersection of two sets always has fewer elements than either set.

5. The number of elements in the union of two sets can be found by adding the number of elements in each set.

6. The probability of an event is always at least 0 and no larger than 1.

7. The probability of the union of two events can be found by adding the probability of each event.

8. The probability of drawing the Queen of Hearts from a deck of cards is an example of empirical probability.

9. If two events are mutually exclusive, then they are independent.

10. The probability of two independent events can be found by multiplying the probabilities of each event.

11. The probability of an event $E$ given an event $F$ is the same as the probability of $F$ given $E$.

12. Bayes' theorem can be useful for calculating conditional probability.

13. A random variable can have negative values.

14. The expected value of a random variable must equal one of the values that the random variable can have.

15. The probabilities in a probability distribution must add up to 1.

16. A fair game can have an expected value that is greater than 0.

## PRACTICE AND EXPLORATIONS

**Write true or false for each statement.**

17. $9 \in \{8, 4, -3, -9, 6\}$     18. $4 \notin \{3, 9, 7\}$

19. $2 \notin \{0, 1, 2, 3, 4\}$     20. $0 \in \{0, 1, 2, 3, 4\}$

21. $\{3, 4, 5\} \subseteq \{2, 3, 4, 5, 6\}$

22. $\{1, 2, 5, 8\} \subseteq \{1, 2, 5, 10, 11\}$

23. $\{3, 6, 9, 10\} \subseteq \{3, 9, 11, 13\}$

24. $\emptyset \subseteq \{1\}$

25. $\{2, 8\} \not\subseteq \{2, 4, 6, 8\}$

26. $0 \subseteq \emptyset$

**In Exercises 27–36, let $U = \{a, b, c, d, e, f, g, h\}$, $K = \{c, d, e, f, h\}$, and $R = \{a, c, d, g\}$. Find the following.**

27. The number of subsets of $K$

28. The number of subsets of $R$

29. $K'$     30. $R'$

31. $K \cap R$     32. $K \cup R$

33. $(K \cap R)'$     34. $(K \cup R)'$

35. $\emptyset'$     36. $U'$

In Exercises 37–42, let

$U = \{\text{all employees of Mercy Hospital}\}$;

$A = \{\text{employees in the pediatrics department}\}$;

$B = \{\text{employees in the maternity department}\}$;

$C = \{\text{female employees}\}$;

$D = \{\text{employees with a nursing degree}\}$.

**Describe each set in words.**

**37.** $A \cap C$        **38.** $B \cap D$

**39.** $A \cup D$        **40.** $A' \cap D$

**41.** $B' \cap C'$       **42.** $(B \cup C)'$

**Draw a Venn diagram and shade each set.**

**43.** $A \cup B'$        **44.** $A' \cap B$

**45.** $(A \cap B) \cup C$      **46.** $(A \cup B)' \cap C$

**Write the sample space $S$ for each experiment, choosing an $S$ with equally likely outcomes, if possible.**

**47.** Rolling a die

**48.** Drawing a card from a deck containing only the 13 spades

**49.** Measuring the weight of a person to the nearest half pound (the scale will not measure more than 300 lb)

**50.** Tossing a coin 4 times

A jar contains 5 balls labeled 3, 5, 7, 9, and 11, respectively, while a second jar contains 4 red and 2 green balls. An experiment consists of pulling 1 ball from each jar, in turn. In Exercises 51–54, write each set using set notation.

**51.** The sample space

**52.** The number on the first ball is greater than 5.

**53.** The second ball is green.

**54.** Are the outcomes in the sample space in Exercise 51 equally likely?

**In Exercises 55–62, find the probability of each event when a single card is drawn from an ordinary deck.**

**55.** A heart         **56.** A red queen

**57.** A face card or a heart     **58.** Black or a face card

**59.** Red, given that it is a queen

**60.** A jack, given that it is a face card

**61.** A face card, given that it is a king

**62.** A king, given that it is not a face card

**63.** Describe what is meant by disjoint sets.

**64.** Describe what is meant by mutually exclusive events.

**65.** How are disjoint sets and mutually exclusive events related?

**66.** Define independent events.

**67.** Are independent events always mutually exclusive? Are they ever mutually exclusive?

**68.** An uproar has raged since September 1990 over the answer to a puzzle published in *Parade* magazine, a supplement of the Sunday newspaper. In the "Ask Marilyn" column, Marilyn vos Savant answered the following question: "Suppose you're on a game show, and you're given the choice of three doors. Behind one door is a car; behind the others, goats. You pick a door, say number 1, and the host, who knows what's behind the other doors, opens another door, say number 3, which has a goat. He then says to you, 'Do you want to pick door number 2?' Is it to your advantage to take the switch?"

Ms. vos Savant estimates that she has since received some 10,000 letters; most of them, including many from mathematicians and statisticians, disagreed with her answer. Her answer has been debated by both professionals and amateurs and tested in classes at all levels from grade school to graduate school. But by performing the experiment repeatedly, it can be shown that vos Savant's answer was correct. Find the probabilities of getting the car if you switch or do not switch, and then answer the question yourself. (*Hint:* Consider the sample space.) ***Source: Parade Magazine.***

**Find the odds in favor of a card drawn from an ordinary deck being the following.**

**69.** A club           **70.** A black jack

**71.** A red face card or a queen    **72.** An ace or a club

**Find the probabilities of getting the following sums when two fair dice are rolled.**

**73.** 8                 **74.** 0

**75.** At least 10

**76.** No more than 5

**77.** An odd number greater than 8

**78.** 12, given that the sum is greater than 10

**79.** 7, given that at least one die shows a 4

**80.** At least 9, given that at least one die shows a 5

**81.** Suppose $P(E) = 0.51$, $P(F) = 0.37$, and $P(E \cap F) = 0.22$. Find the following.
    **a.** $P(E \cup F)$        **b.** $P(E \cap F')$
    **c.** $P(E' \cup F)$       **d.** $P(E' \cap F')$

**82.** An urn contains 10 balls: 4 red and 6 blue. A second urn contains 16 red balls and an unknown number of blue balls. A single ball is drawn from each urn. The probability that both balls are the same color is 0.44. Calculate the number of blue balls in the second urn. Choose one of the following. ***Source: Society of Actuaries.***
    **a.** 4     **b.** 20     **c.** 24     **d.** 44     **e.** 64

**83.** Box A contains 5 red balls and 1 black ball; box B contains 2 red balls and 3 black balls. A box is chosen, and a ball is selected from it. The probability of choosing box A is 3/8. If the selected ball is black, what is the probability that it came from box A?

**84.** Find the probability that the ball in Exercise 83 came from box B, given that it is red.

In Exercises 85 and 86, (a) give a probability distribution, (b) sketch its histogram, (c) find the expected value, and (d) find the standard deviation.

**85.** A coin is tossed 3 times and the number of heads is recorded.

**86.** A pair of dice is rolled and the sum of the results for each roll is recorded.

In Exercises 87 and 88, give the probability that corresponds to the shaded region of each histogram.

**87.**

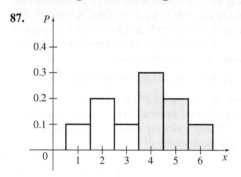

**88.**

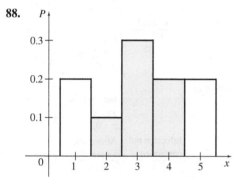

**89.** You pay $6 to play in a game where you will roll a die, with payoffs as follows: $8 for a 6, $7 for a 5, and $4 for any other results. What are your expected winnings? Is the game fair?

**90.** Find the expected number of girls in a family of 3 children.

**91.** Three cards are drawn from a standard deck of 52 cards.

   **a.** What is the expected number of aces?

   **b.** What is the expected number of clubs?

**92.** Suppose someone offers to pay you $100 if you draw 3 cards from a standard deck of 52 cards and all the cards are clubs. What should you pay for the chance to win if it is a fair game?

## LIFE SCIENCE APPLICATIONS

**93. Sickle Cell Anemia** The square at the top of the next column shows the four possible (equally likely) combinations when both parents are carriers of the sickle cell anemia trait. Each carrier parent has normal cells ($N$) and trait cells ($T$).

   **a.** Complete the table.

   **b.** If the disease occurs only when two trait cells combine, find the probability that a child born to these parents will have sickle cell anemia.

   **c.** The child will carry the trait but not have the disease if a normal cell combines with a trait cell. Find this probability.

   **d.** Find the probability that the child neither is a carrier nor has the disease.

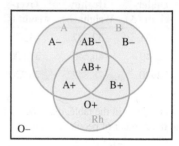

**94. Blood Antigens** In Exercise 75 of Section 12.1, we described the eight types of human blood. The percentage of the population having each type is as follows:

$O^+$: 38%;     $O^-$: 8%;     $A^+$: 32%;     $A^-$: 7%;
$B^+$: 9%;     $B^-$: 2%;     $AB^+$: 3%;     $AB^-$: 1%.

When a person receives a blood transfusion, it is important that the blood be compatible, which means that it introduces no new antigens into the recipient's blood. The following diagram helps illustrate what blood types are compatible. *Source: The Mathematics Teacher.*

The universal blood type is $O^-$, since it has none of the additional antigens. The circles labeled A, B, and Rh contain blood types with the A antigen, B antigen, and Rh antigen, respectively. A person with $O^-$ blood can only be transfused with $O^-$ blood, because any other type would introduce a new antigen. Thus the probability that blood from a random donor is compatible is just 8%. A person with $AB^+$ blood already has all antigens, so the probability that blood from a random donor is compatible is 100%. Find the probability that blood from a random donor is compatible with a person with each blood type.

   **a.** $O^+$     **b.** $A^+$     **c.** $B^+$

   **d.** $A^-$     **e.** $B^-$     **f.** $AB^-$

**95. Risk Factors** An actuary is studying the prevalence of three health risk factors, denoted by A, B, and C, within a population of women. For each of the three factors, the probability is 0.1 that a woman in the population has only this risk factor (and no others). For any two of the three factors, the probability is 0.12 that she has exactly these two risk factors (but not the other). The probability that a woman has all three risk factors, given that she has A and B, is 1/3. What is the probability that a woman has none of the three risk factors, given that she does not have risk factor A? Choose one of the following. *Source: Society of Actuaries.*

   **a.** 0.280     **b.** 0.311     **c.** 0.467

   **d.** 0.484     **e.** 0.700

**96. SIDS** On July 15, 2005, a panel in England ruled that Roy Meadow, a renowned expert on child abuse and co-founder of London's Royal College of Paediatrics and Child Health, should be erased from the register of physicians in Britain for his faulty statistics at the trial of Sally Clark, who was convicted of murdering her first two babies. Meadow testified at the trial that the probability of a baby dying of sudden death syndrome (SIDS) is 1/8543. He then calculated that the probability of two babies in a family dying of SIDS is $(1/8543)^2 \approx 1/73{,}000{,}000$. With such a small probability of both babies dying of SIDS, he concluded that the babies were instead murdered. What assumption did Meadow make in doing this calculation? Discuss reasons why this assumption may be invalid. (*Note:* Clark spent three years in prison before her conviction was reversed.) *Source: Science.*

**97. Body Mass Index** The results of a survey of body mass index among persons 18 years and over in the United States are listed below. (All numbers are in thousands.) *Source: National Center for Health Statistics.*

| | **Body Mass Index** | | | | |
| --- | --- | --- | --- | --- | --- |
| | Under-weight (A) | Healthy Weight (B) | Over-weight (C) | Obese (D) | Totals |
| *Male (M)* | 1071 | 32,207 | 44,970 | 30,868 | 109,116 |
| *Female (F)* | 2849 | 46,640 | 31,800 | 31,157 | 112,446 |
| *Totals* | 3920 | 78,847 | 76,770 | 62,025 | 221,562 |

Using the letters given in the table, describe each set in words and then find the number of people in each set.

**a.** $M \cap C$
**b.** $F \cup B$
**c.** $M' \cap (C \cup D)$
**d.** $F' \cap (C' \cap D')$

**98. Body Mass Index** The results of a survey of body mass index among persons 18 years and over in the United States are given in the table in Exercise 97. (All numbers are in thousands.) *Source: National Center for Health Statistics.* If a person is selected at random, find the probability of the following.

**a.** The person is overweight.

**b.** The person is female and at a healthy weight.

**c.** The person is overweight or obese.

**d.** The person is overweight given that the person is male.

**e.** The person is female given that the person is overweight.

**f.** Are the events "male" and "overweight" independent?

**99. Prostate Cancer** The prostate-specific antigen (PSA) test measures the blood level of PSA, a protein that is produced by the prostate gland. Higher levels of PSA may indicate that a man has prostate cancer. The sensitivity of the PSA test is about 0.205. The specificity of the PSA test is about 0.936. If 0.2% of men in the United States have prostate cancer, find the probability that a man who has a positive PSA test actually has prostate cancer. *Source: National Institutes of Health.*

**100. Infection** A young man has a badly infected compound fracture of his ankle. The infection is spreading and is potentially life threatening. The decision tree below illustrates two choices: The leg can be amputated below the knee immediately or the infection can be treated with antibiotics. If the antibiotics do not work, the patient may require an above-the-knee amputation, or a below-the-knee amputation, or the patient may even die. If the antibiotics do work, the patient may have a full recovery or the patient may have a minor disability of the leg. The values at the end of the tree indicate the quality of life attached to each outcome. A value of 0 is assigned to an outcome of death, and a value of 1 is assigned to an outcome of full recovery with no amputation. *Source: Journal of General Internal Medicine.*

**a.** Find the expected value of life if antibiotics are used.

**b.** Find the expected value of life if a below-the-knee amputation is performed immediately. Compare it with the value of life if antibiotics are used. Which decision gives the higher expected quality of life?

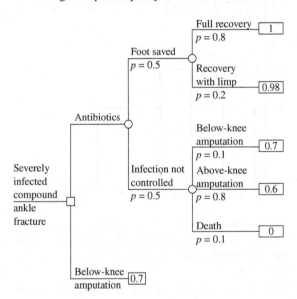

# OTHER APPLICATIONS

**101. Elections** In the 2012 presidential election, over 130 million people voted, of which 47% were male and 53% were female. Of the male voters, 45% voted for Barack Obama and 52% voted for Mitt Romney. Of the female voters, 55% voted for Obama and 44% voted for Romney. *Source: Roper Center.*

**a.** Find the percentage of voters who voted for Obama.

**b.** Find the probability that a randomly selected voter for Obama was male.

**c.** Find the probability that a randomly selected voter for Obama was female.

**102. Television Viewing Habits** A telephone survey of television viewers revealed the following information:

    20 watch situation comedies;
    19 watch game shows;
    27 watch movies;
    19 watch movies but not game shows;
    15 watch situation comedies but not game shows;
    10 watch both situation comedies and movies;
    3 watch all three;
    7 watch none of these.

**a.** How many viewers were interviewed?

**b.** How many viewers watch comedies and movies but not game shows?

**c.** How many viewers watch only movies?

**d.** How many viewers do not watch movies?

**103. Randomized Response Method for Getting Honest Answers to Sensitive Questions** There are many personal questions that most people would rather not answer. In fact, when a person is asked such a question, a common response is to provide a false answer. In 1965, Stanley Warner developed a method to ensure that the identity of an individual who answers a question remains anonymous, thus ensuring an honest answer to the question. For this method, instead of one sensitive question being asked of a person, there are two questions, one sensitive and one nonsensitive, and which question the person answers depends on a randomized procedure. The interviewer, who doesn't know which question a person is answering, simply records the answer given as either "Yes" or "No." In this way, there is no way of knowing to which question the person answered "Yes" or "No." Then, using conditional probability, we can estimate the percentage of people who answer "Yes" to the sensitive question. For example, suppose that the two questions are

    *A*: Does your birth year end in an odd digit? (Nonsensitive)

    *B*: Have you ever intentionally cheated on an examination? (Sensitive)

We already know that $P(\text{"Yes"} \mid \text{Question } A) = 1/2$, since half the years are even and half are odd. The answer we seek is $P(\text{"Yes"} \mid \text{Question } B)$. In this experiment, a student is asked to flip a coin and answer question *A* if the coin comes up heads and otherwise answer question *B*. Note that the interviewer does not know the outcome of the coin flip or which question is being answered. The percentage of students answering "Yes" is used to approximate $P(\text{"Yes"})$, which is then used to estimate the percentage of students who have cheated on an examination. *Source: Journal of the American Statistical Association.*

**a.** Use the fact that the event "Yes" is the union of the event "Yes and Question *A*" with the event "Yes and Question *B*" to prove that

$$P(\text{"Yes"} \mid \text{Question } B)$$
$$= \frac{P(\text{"Yes"}) - P(\text{"Yes"} \mid \text{Question } A) \cdot P(\text{Question } A)}{P(\text{Question } B)}.$$

**b.** If this technique is tried on 100 subjects and 60 answered "Yes," what is the approximated probability that a person randomly selected from the group has intentionally cheated on an examination?

**104. Police Lineup** To illustrate the difficulties with eyewitness identifications from police lineups, John Allen Paulos considers a "lineup" of three pennies, in which we know that two are fair (innocent) and the third (the culprit) has a 75% probability of landing heads. The probability of picking the culprit by chance is, of course, 1/3. Suppose we observe three heads in a row on one of the pennies. If we then guess that this penny is the culprit, what is the probability that we're right? *Source: Who's Counting.*

**105. Earthquake** It has been reported that government scientists have predicted that the odds for a major earthquake occurring in the San Francisco Bay area during the next 30 years are 9 to 1. What is the probability that a major earthquake will occur during the next 30 years in San Francisco? *Source: The San Francisco Chronicle.*

**106. Making a First Down** A first down is desirable in football—it guarantees four more plays by the team making it, assuming no score or turnover occurs in the plays. After getting a first down, a team can get another by advancing the ball at least 10 yards. During the four plays given by a first down, a team's position will be indicated by a phrase such as "third and 4," which means that the team has already had two of its four plays, and that 4 more yards are needed to get 10 yards necessary for another first down. An article in a management journal offers the following results for 189 games for a particular National Football League season. "Trials" represents the number of times a team tried to make a first down, given that it was currently playing either a third or a fourth down. Here, *n* represents the number of yards still needed for a first down. *Source: Management Science.*

| *n* | Trials | Successes | Probability of Making First Down with *n* Yards to Go |
|---|---|---|---|
| 1 | 543 | 388 | |
| 2 | 327 | 186 | |
| 3 | 356 | 146 | |
| 4 | 302 | 97 | |
| 5 | 336 | 91 | |

**a.** Complete the table.

**b.** Why is the sum of the answers in the table not equal to 1?

**107. States** Of the 50 United States, the following is true:

    24 are west of the Mississippi River (western states);*
    22 had populations less than 3.6 million in the 2010 census (small states);
    26 begin with the letters A through M (early states);
    9 are large late (beginning with the letters N through Z) eastern states;
    14 are small western states;
    11 are small early states;
    7 are small early western states.

**a.** How many western states had populations more than 3.6 million in the 2010 census and begin with the letters N through Z?

**b.** How many states east of the Mississippi had populations more than 3.6 million in the 2010 census?

*We count here states such as Minnesota, which has more than half of its area to the west of the Mississippi.

**108. Music** Country-western songs often emphasize three basic themes: love, prison, and trucks. A survey of the local country-western radio station produced the following data:

> 12 songs were about a truckdriver who was in love while in prison;
>
> 13 were about a prisoner in love;
>
> 28 were about a person in love;
>
> 18 were about a truckdriver in love;
>
> 33 were about people not in prison;
>
> 18 were about prisoners;
>
> 15 were about truckdrivers who were in prison;
>
> 16 were about truckdrivers who were not in prison.

**a.** How many songs were surveyed?

Find the number of songs about

**b.** truckdrivers;

**c.** prisoners who are neither truckdrivers nor in love;

**d.** prisoners who are in love but are not truckdrivers;

**e.** prisoners who are not truckdrivers;

**f.** people not in love.

**109. Gambling** The following puzzle was featured on the Puzzler part of the radio program *Car Talk* on February 23, 2002. A con man puts three cards in a bag; one card is green on both sides, one is red on both sides, and the third is green on one side and red on the other. He lets you pick one card out of the bag and put it on a table, so you can see that a red side is face up, but neither of you can see the other side. He offers to bet you even money that the other side is also red. In other words, if you bet $1, you lose if the other side is red but get back $2 if the other side is green. Is this a good bet? What is the probability that the other side is red? *Source: Car Talk.*

**110. Missiles** In his novel *Debt of Honor*, Tom Clancy writes the following:

"There were ten target points—missile silos, the intelligence data said, and it pleased the Colonel [Zacharias] to be eliminating the hateful things, even though the price of that was the lives of other men. There were only three of them [bombers], and his bomber, like the others, carried only eight weapons [smart bombs]. The total number of weapons carried for the mission was only twenty-four, with two designated for each silo, and Zacharias's last four for the last target. Two bombs each. Every bomb had a 95% probability of hitting within four meters of the aim point, pretty good numbers really, except that this sort of mission had precisely no margin for error. Even the paper probability was less than half a percent chance of a double miss, but that number times ten targets meant a 5% chance that [at least] one missile would survive, and that could not be tolerated." *Source: Debt of Honor.* Determine whether the calculations in this quote are correct by the following steps.

**a.** Given that each bomb had a 95% probability of hitting the missile silo on which it was dropped, and that two bombs were dropped on each silo, what is the probability of a double miss?

**b.** What is the probability that a specific silo was destroyed (that is, that at least one bomb of the two bombs struck the silo)?

**c.** What is the probability that all ten silos were destroyed?

**d.** What is the probability that at least one silo survived? Does this agree with the quote?

**e.** What assumptions need to be made for the calculations in parts a through d to be valid? Discuss whether these assumptions seem reasonable.

**111. Viewing Habits** A survey of a group's viewing habits over the last year revealed the following information:

**(i)** 28% watched gymnastics;

**(ii)** 29% watched baseball;

**(iii)** 19% watched soccer;

**(iv)** 14% watched gymnastics and baseball;

**(v)** 12% watched baseball and soccer;

**(vi)** 10% watched gymnastics and soccer;

**(vii)** 8% watched all three sports.

Calculate the percentage of the group that watched none of the three sports during the last year. Choose one of the following. *Source: Society of Actuaries.*

**a.** 24     **b.** 36     **c.** 41     **d.** 52     **e.** 60

**112. Baseball** The number of runs scored in 16,456 half-innings of the 1986 National League Baseball season was analyzed by Hal Stern. Use the table to answer the following questions. *Source: Chance News.*

**a.** What is the probability that a given team scored 5 or more runs in any given half-inning during the 1986 season?

**b.** What is the probability that a given team scored fewer than 2 runs in any given half-inning of the 1986 season?

**c.** What is the expected number of runs that a team scored during any given half-inning of the 1986 season? Interpret this number.

| Runs | Frequency | Probability |
|------|-----------|-------------|
| 0 | 12,087 | 0.7345 |
| 1 | 2451 | 0.1489 |
| 2 | 1075 | 0.0653 |
| 3 | 504 | 0.0306 |
| 4 | 225 | 0.0137 |
| 5 | 66 | 0.0040 |
| 6 | 29 | 0.0018 |
| 7 | 12 | 0.0007 |
| 8 | 5 | 0.0003 |
| 9 | 2 | 0.0001 |

# MEDICAL DIAGNOSIS

When a patient is examined, information (typically incomplete) is obtained about his or her state of health. Probability theory provides a mathematical model appropriate for this situation, as well as a procedure for quantitatively interpreting such partial information to arrive at a reasonable diagnosis.

To develop a model, we list the states of health that can be distinguished in such a way that the patient can be in one and only one state at the time of the examination. For each state of health $H$, we associate a number, $P(H)$, between 0 and 1 such that the sum of all these numbers is 1. This number $P(H)$ represents the probability, before examination, that a patient is in the state of health $H$, and $P(H)$ may be chosen subjectively from medical experience, using any information available prior to the examination. The probability may be most conveniently established from clinical records; that is, a mean probability is established for patients in general, although the number would vary from patient to patient. Of course, the more information that is brought to bear in establishing $P(H)$, the better the diagnosis.

For example, limiting the discussion to the condition of a patient's heart, suppose there are exactly three states of health, with probabilities as follows.

| | State of Health, $H$ | $P(H)$ |
|---|---|---|
| $H_1$ | Patient has a normal heart | 0.8 |
| $H_2$ | Patient has minor heart irregularities | 0.15 |
| $H_3$ | Patient has a severe heart condition | 0.05 |

Having selected $P(H)$, the information from the examination is processed. First, the results of the examination must be classified. The examination itself consists of observing the state of a number of characteristics of the patient. Let us assume that the examination for a heart condition consists of a stethoscope examination and a cardiogram. The outcome of such an examination, $C$, might be one of the following:

$C_1$ = stethoscope shows normal heart
and cardiogram shows normal heart;

$C_2$ = stethoscope shows normal heart
and cardiogram shows minor irregularities;

and so on.

It remains to assess for each state of health $H$ the conditional probability $P(C|H)$ of each examination outcome $C$ using only the knowledge that a patient is in a given state of health. (This may be based on the medical knowledge and clinical experience of the doctor.) The conditional probabilities $P(C|H)$ will not vary from patient to patient (although they should be reviewed periodically), so that they may be built into a diagnostic system.

Suppose the result of the examination is $C_1$. Let us assume the following probabilities:

$$P(C_1|H_1) = 0.9,$$
$$P(C_1|H_2) = 0.4,$$
$$P(C_1|H_3) = 0.1.$$

Now, for a given patient, the appropriate probability associated with each state of health $H$, after examination, is $P(H|C)$, where $C$ is the outcome of the examination. This can be calculated by using Bayes' theorem. For example, to find $P(H_1|C_1)$—that is, the probability that the patient has a normal heart given that the examination showed a normal stethoscope examination and a normal cardiogram—we use Bayes' theorem as follows:

$$P(H_1|C_1)$$
$$= \frac{P(C_1|H_1)P(H_1)}{P(C_1|H_1)P(H_1) + P(C_1|H_2)P(H_2) + P(C_1|H_3)P(H_3)}$$
$$= \frac{(0.9)(0.8)}{(0.9)(0.8) + (0.4)(0.15) + (0.1)(0.05)} \approx 0.92.$$

Hence, the probability is about 0.92 that the patient has a normal heart on the basis of the examination results. This means that in 8 of 100 patients, some abnormality will be present and not be detected by the stethoscope or the cardiogram.
***Source: Some Mathematical Models in Biology.***

## EXERCISES

1. Find $P(H_2|C_1)$.

2. Assuming the following probabilities, find $P(H_1|C_2)$.
   $$P(C_2|H_1) = 0.2 \qquad P(C_2|H_2) = 0.8 \qquad P(C_2|H_3) = 0.3$$

3. Assuming the probabilities of Exercise 2, find $P(H_3|C_2)$.

## DIRECTIONS FOR GROUP PROJECT

*Find an article on medical decision making from a medical journal and develop a doctor–patient scenario for that particular decision. Then create a role-playing activity in which the doctor and nurse present the various options and the mathematics associated with making such a decision to a patient. Make sure to present the mathematics in a manner that the average patient might understand. (Hint: Many leading medical journals include articles on medical decision making. One particular journal that certainly includes such research is* Medical Decision Making.*)*

# 13

# Probability and Calculus

13.1 Continuous Probability Models

13.2 Expected Value and Variance of Continuous Random Variables

13.3 Special Probability Density Functions

Chapter 13 Review

Extended Application: Exponential Waiting Times

Though earthquakes may appear to strike at random, the times between quakes can be modeled with an exponential density function. Such *continuous probability models* have many applications in science, engineering, and medicine. In an exercise in Section 1 of this chapter we'll use an exponential density function to describe the times between major earthquakes in Southern California, and in Section 2 we will compute the mean and standard deviation for this distribution.

n recent years, probability has become increasingly useful in fields ranging from manufacturing to medicine, as well as in all types of research. The foundations of probability were laid in the 17th century by Blaise Pascal (1623–1662) and Pierre de Fermat (1601–1665), who investigated *the problem of the points*. This problem dealt with the fair distribution of winnings in an interrupted game of chance between two equally matched players whose scores were known at the time of the interruption.

Probability has advanced from a study of gambling to a well-developed, deductive mathematical system. In this chapter we give a brief introduction to the use of calculus in probability.

# 13.1    Continuous Probability Models

APPLY IT
**What is the probability that there is a bird's nest within 0.5 kilometer of a given point?**
*In Example 3, we will answer the question posed above.*

In this section, we show how calculus is used to find the probability of certain events. Before discussing probability, however, we need to introduce some new terminology.

Suppose an ornithologist is studying the length of time that sparrows spend feeding at a bird feeder. The lengths of time spent feeding, rounded to the nearest minute, are shown in the following table.

| Time | 1 | 2 | 3 | 4 | 5 | 6 | 7 | 8 | 9 | 10 | |
|------|---|---|---|----|----|----|----|---|---|----|---|
| Frequency | 3 | 5 | 9 | 12 | 15 | 11 | 10 | 6 | 3 | 1 | (Total: 75) |

The table shows, for example, that 9 of the 75 sparrows in the study stayed 3 minutes, 15 sparrows stayed 5 minutes, and 1 sparrow stayed 10 minutes. Because the time for any particular feeding is a random event, the number of minutes for a feeding is a random variable, as defined in the previous chapter. The frequencies can be converted to probabilities by dividing each frequency by the total number of feedings (75) to get the results shown in the next table.*

| Time | 1 | 2 | 3 | 4 | 5 | 6 | 7 | 8 | 9 | 10 |
|------|---|---|---|---|---|---|---|---|---|----|
| Probability | 0.04 | 0.07 | 0.12 | 0.16 | 0.20 | 0.15 | 0.13 | 0.08 | 0.04 | 0.01 |

Because each value of the random variable is associated with just one probability, this table defines a function. As we saw in the last chapter, such a function is called a probability function.

The information in the second table can be displayed graphically with a histogram. Recall that the bars of a histogram have the same width, and their heights are determined by the probabilities of the various values of the random variable. See Figure 1 on the following page.

---

*One definition of the *probability of an event* is the number of outcomes that favor the event divided by the total number of equally likely outcomes in an experiment.

As we learned in the previous chapter, the probability function in the second table is a discrete probability function because it has a finite domain—the integers from 1 to 10, inclusive. A discrete probability function has a finite domain or an infinite domain that can be listed. For example, if we flip a coin until we get heads, and let the random variables be the number of flips, then the domain is 1, 2, 3, 4, . . . . On the other hand, the distribution of heights (in inches) of college women includes infinitely many possible measurements, such as 53, 54.2, 66.5, 72.$\overline{3}$, and so on, *within some real number interval*. Probability functions with such domains are called *continuous probability distributions*.

### Continuous Probability Distribution

A **continuous random variable** can take on any value in some interval of real numbers. The distribution of this random variable is called a **continuous probability distribution**.

Some probability functions are inherently discrete. For example, the number of owls in a forest preserve must be an integer, such as 0, 1, or 2, and could never take on any value in between. But the bird feeder example discussed earlier is different, because you could think of it as a simplification of a continuous distribution. It would have been possible to time the feedings with greater precision—to the nearest tenth of a minute or even to the nearest second (1/60 of a minute), if desired. Theoretically, at least, *t* could take on any positive real-number value between, say, 0 and 11 minutes. The graph of the probabilities $f(t)$ of these feeding times can be thought of as the continuous curve shown in Figure 1. As indicated in Figure 1, the curve was derived from our table by connecting the points at the tops of the bars in the corresponding histogram and smoothing the resulting polygon into a curve.

For a discrete probability function, the area of each bar (or rectangle) gives the probability of a particular feeding time. Thus, by considering the possible feeding times *t* as all the real numbers between 0 and 11, the area under the curve of Figure 2 between any two values of *t* can be interpreted as the probability that a feeding time will be between those two numbers. For example, the shaded region in Figure 2 corresponds to the probability that *t* is between *a* and *b*, written $P(a \leq t \leq b)$.

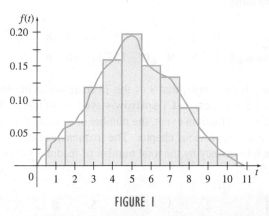

FIGURE 1

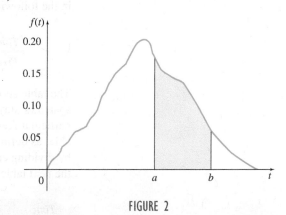

FIGURE 2

It was shown earlier that the definite integral of a continuous function $f$, where $f(x) \geq 0$, gives the area under the graph of $f(x)$ from $x = a$ to $x = b$. If a function $f$ can be found to describe a continuous probability distribution, then the definite integral can be used to find the area under the curve from *a* to *b* that represents the probability that *x* will be between *a* and *b*.

If $X$ is a continuous random variable whose distribution is described by the function $f$ on $[a, b]$, then

$$P(a \leq X \leq b) = \int_a^b f(x) \, dx.$$

## Probability Density Functions

A function $f$ that describes a continuous probability distribution is called a *probability density function*. Such a function must satisfy the following conditions.

---

**Probability Density Function**

The function $f$ is a **probability density function** of a random variable $X$ in the interval $[a, b]$ if

1. $f(x) \geq 0$ for all $x$ in the interval $[a, b]$; and

2. $\displaystyle\int_a^b f(x)\, dx = 1.$

---

Intuitively, Condition 1 says that the probability of a particular event can never be negative. Condition 2 says that the total probability for the interval must be 1; *something* must happen.

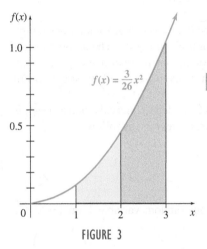

$f(x) = \dfrac{3}{26}x^2$

**FIGURE 3**

**YOUR TURN 1** Repeat Example 1(a) for the function $f(x) = 2/x^2$ on $[1, 2]$. What is the probability that $X$ will be between $3/2$ and 2?

### EXAMPLE 1  Probability Density Function

(a) Show that the function defined by $f(x) = (3/26)x^2$ is a probability density function for the interval $[1, 3]$.

**SOLUTION**   First, note that Condition 1 holds; that is, $f(x) \geq 0$ for the interval $[1, 3]$. Next show that Condition 2 holds.

$$\int_1^3 \frac{3}{26}x^2\,dx = \frac{x^3}{26}\Big|_1^3 = \frac{27-1}{26} = 1$$

Since both conditions hold, $f(x)$ is a probability density function.

(b) Find the probability that $X$ will be between 1 and 2.

**SOLUTION**   The desired probability is given by the area under the graph of $f(x)$ between $x = 1$ and $x = 2$, as shown in blue in Figure 3. The area is found by using a definite integral.

$$P(1 \leq X \leq 2) = \int_1^2 \frac{3}{26}x^2\,dx = \frac{x^3}{26}\Big|_1^2 = \frac{7}{26} \qquad \text{TRY YOUR TURN 1}$$

Earlier, we noted that determining a suitable function is the most difficult part of applying mathematics to actual situations. Sometimes a function appears to model an application well but does not satisfy the requirements for a probability density function. In such cases, we may be able to change the function into a probability density function by multiplying it by a suitable constant, as shown in the next example.

### EXAMPLE 2  Probability Density Function

Is there a constant $k$ such that $f(x) = kx^2$ is a probability density function for the interval $[0, 4]$?

**SOLUTION**   First,

$$\int_0^4 kx^2\,dx = \frac{kx^3}{3}\Big|_0^4 = \frac{64k}{3}.$$

**YOUR TURN 2** Repeat
Example 2 for the function
$f(x) = kx^3$ on the interval $[0, 4]$.

The integral must be equal to 1 for the function to be a probability density function. To convert it to one, let $k = 3/64$. The function defined by $(3/64)x^2$ for $[0, 4]$ will be a probability density function, since $(3/64)x^2 \geq 0$ for all $x$ in $[0, 4]$ and

$$\int_0^4 \frac{3}{64}x^2 = 1.$$

TRY YOUR TURN 2

An important distinction is made between a discrete probability function and a probability density function (which is continuous). In a discrete distribution, the probability that the random variable, $X$, will assume a specific value is given in the distribution for every possible value of $X$. In a probability density function, however, the probability that $X$ equals a specific value, say, $c$ is

$$P(X = c) = \int_c^c f(x) \, dx = 0.$$

For a probability density function, only probabilities of *intervals* can be found. For example, suppose the random variable is the annual rainfall for a given region. The amount of rainfall in one year can take on any value within some continuous interval that depends on the region; however, the probability that the rainfall in a given year will be some specific amount, say 33.25 in., is actually zero.

The definition of a probability density function is extended to intervals such as $(-\infty, b]$, $(-\infty, b)$, $[a, \infty)$, $(a, \infty)$, or $(-\infty, \infty)$ by using improper integrals, as follows.

## Probability Density Functions on $(-\infty, \infty)$

If $f$ is a probability density function for a continuous random variable $X$ on $(-\infty, \infty)$, then

$$P(X \leq b) = P(X < b) = \int_{-\infty}^b f(x) \, dx,$$

$$P(X \geq a) = P(X > a) = \int_a^\infty f(x) \, dx,$$

$$P(-\infty < X < \infty) = \int_{-\infty}^\infty f(x) \, dx = 1.$$

**FOR REVIEW**
Improper integrals, those with
one or two infinite limits, were
discussed in Chapter 8 on
Further Techniques and Applications of Integration. The type of
improper integral we shall need
was defined as

$$\int_a^\infty f(x) \, dx = \lim_{b \to \infty} \int_a^b f(x) \, dx.$$

For example,

$$\int_1^\infty x^{-2} \, dx = \lim_{b \to \infty} \int_1^b x^{-2} \, dx$$

$$= \lim_{b \to \infty} \left( -\frac{1}{x} \Big|_1^b \right)$$

$$= \lim_{b \to \infty} \left( -\frac{1}{b} + \frac{1}{1} \right)$$

$$= 0 + 1 = 1.$$

The total area under the graph of a probability density function of this type must still equal 1.

### EXAMPLE 3   Location of a Bird's Nest

Suppose the random variable $X$ is the distance (in kilometers) from a given point to the nearest bird's nest, with the probability density function of the distribution given by $f(x) = 2xe^{-x^2}$ for $x \geq 0$.

**(a)** Show that $f(x)$ is a probability density function.

**SOLUTION** Since $e^{-x^2} = 1/e^{x^2}$ is always positive, and $x \geq 0$,

$$f(x) = 2xe^{-x^2} \geq 0,$$

and Condition 1 holds.

Use substitution to evaluate the definite integral $\int_0^\infty 2xe^{-x^2}\,dx$. Let $u = -x^2$, so that $du = -2x\,dx$, and

$$\int 2xe^{-x^2}\,dx = -\int e^{-x^2}(-2x\,dx)$$

$$= -\int e^u\,du = -e^u = -e^{-x^2}.$$

Then

$$\int_0^\infty 2xe^{-x^2}\,dx = \lim_{b\to\infty}\int_0^b 2xe^{-x^2}\,dx = \lim_{b\to\infty}(-e^{-x^2})\Big|_0^b$$

$$= \lim_{b\to\infty}\left(-\frac{1}{e^{b^2}} + e^0\right) = 0 + 1 = 1,$$

and Condition 2 holds.

The function defined by $f(x) = 2xe^{-x^2}$ satisfies the two conditions required of a probability density function.

**(b)** Find the probability that there is a bird's nest within 0.5 km of the given point.

**APPLY IT**

**SOLUTION**   Find $P(X \le 0.5)$ where $X \ge 0$. This probability is given by

$$P(0 \le X \le 0.5) = \int_0^{0.5} 2xe^{-x^2}\,dx.$$

Now evaluate the integral. The indefinite integral was found in part (a).

$$P(0 \le X \le 0.5) = \int_0^{0.5} 2xe^{-x^2}\,dx = (-e^{-x^2})\Big|_0^{0.5}$$

$$= -e^{-(0.5)^2} - (-e^0) = -e^{-0.25} + 1$$

$$\approx -0.7788 + 1 = 0.2212$$

**YOUR TURN 3** Using the probability density function of Example 3, find the probability that there is a bird's nest within 1 km of the given point.

The probability that a bird's nest will be found within 0.5 km of the given point is about 0.22.   **TRY YOUR TURN 3**

**TECHNOLOGY**

## EXAMPLE 4   Computing Mortality

According to the National Center for Health Statistics, if we start with 100,000 people who are 50 years old, we can expect a certain number of them to die within each 5-year interval, as indicated by the following table.* *Source: National Vital Statistics Reports.*

| | Life Table | |
|---|---|---|
| **Years from Age 50** | **Midpoint of Interval** | **Number Dying in Each Interval** |
| 0–5 | 2.5 | 2565 |
| 5–10 | 7.5 | 3659 |
| 10–15 | 12.5 | 5441 |
| 15–20 | 17.5 | 7622 |
| 20–25 | 22.5 | 10,498 |
| 25–30 | 27.5 | 13,858 |
| 30–35 | 32.5 | 16,833 |
| 35–40 | 37.5 | 16,720 |
| 40–45 | 42.5 | 13,211 |
| 45–50 | 47.5 | 7068 |
| 50–55 | 52.5 | 2525 |

*For simplicity, we have placed all those who lived past 100 in the class of those who lived from 100 to 105.

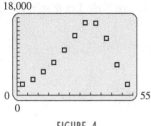

**FIGURE 4**

**(a)** Plot the data.

**SOLUTION**   Figure 4 shows that the plot appears to have the shape of a polynomial.

**(b)** Find a polynomial equation that models the number of deaths, $N(t)$, as a function of the number of years, $t$, since age 50. Use the midpoints and the number of deaths in each interval from the table above.

**SOLUTION**   The highest degree polynomial that the regression feature on a TI-84 Plus calculator can find is fourth degree. As Figure 5(a) shows, this roughly captures the behavior of the data, but it has two drawbacks. For one, it doesn't reach the highest data points. Also, it's decreasing in the beginning when it should be increasing. Higher degree polynomials can be fit using Excel or using the Multiple Regression tool on the Statistics with List Editor application for the TI-89. We were thus able to find that the function

$$N(t) = 5.03958 \times 10^{-5}t^6 - 0.006603t^5 + 0.2992t^4 - 6.0507t^3 +$$
$$67.867t^2 - 110.3t + 2485.1$$

fits the data quite well, as shown in Figure 5(b).

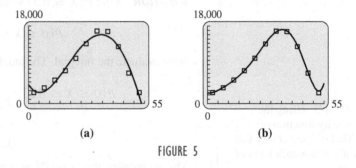

(a)                                      (b)

**FIGURE 5**

**(c)** Use the answer from part (b) to find a probability density function for the random variable $T$ representing the number of additional years that a 50-year-old person lives.

**SOLUTION**   We will construct a density function $S(t) = kN(t)$ by finding a suitable constant $k$, as we did in Example 2. The graph of the function turns upward after $t = 52.5$, which is unlikely for the actual mortality function, so we will restrict the domain of the density function to the interval $[0, 52.5]$, even though this ignores those who live more than 102.5 years. Using the integration feature on our calculator, we find that

$$\int_0^{52.5} S(t)\,dt = k\int_0^{52.5} N(t)\,dt = 497{,}703k.$$

Notice that this number is close to the product of 5 years (interval length) and 100,000 (the total number of people). This is not a coincidence! We set the above integral equal to 1 to get $k = 1/497{,}703$. The function defined by

$$S(t) = \frac{1}{497{,}703}N(t)$$
$$= \frac{1}{497{,}703}\left(5.03958 \times 10^{-5}t^6 - 0.006603t^5 + 0.2992t^4\right.$$
$$\left. - 6.0507t^3 + 67.867t^2 - 110.3t + 2485.1\right)$$

is a probability density function for $[0, 52.5]$ because

$$\int_0^{52.5} S(t)\,dt = 1, \quad \text{and } S(t) \geq 0 \text{ for all } t \text{ in } [0, 52.5].$$

**(d)** Find the probability that a randomly chosen 50-year-old person will live at least until age 70.

**SOLUTION** Again using the integration feature on our calculator,

$$P(T \geq 20) = \int_{20}^{52.5} S(t)\,dt \approx 0.8054.$$

Thus a 50-year-old person has a 80.54% chance of living at least until age 70.

Notice that this value could also be estimated from the table by finding the number of people who have not died by age 70 and then dividing this number by 100,000. Thus, according to our table, there are 80,713 people still alive at age 70, representing 80.7% of the original population. As you can see, our estimate agrees quite well with the actual number.

Another important concept in probability is the *cumulative distribution function*, which gives the probability that a random variable $X$ is less than or equal to an arbitrary value $x$.

**Cumulative Distribution Function**
If $f$ is a probability density function of a random variable in the interval $[a, b]$, then the **cumulative distribution function** is defined as

$$F(x) = P(X \leq x) = \int_a^x f(t)\,dt$$

for $x \geq a$. Also, $F(x) = 0$ for $x < a$.

**NOTE**
1. We integrate with respect to the variable $t$ in the integral, rather than $x$, because we are already using $x$ for the upper limit on the integral. It doesn't matter what variable of integration is used in a definite integral, since that variable doesn't appear in the final answer. We just need to use a variable that's not being used for another purpose.
2. If the random variable is defined on the interval $(-\infty, \infty)$, simply replace $a$ with $-\infty$ in the above definition.

**EXAMPLE 5** **Cumulative Distribution Function**

Consider the random variable $X$ defined in Example 3, giving the distance (in kilometers) from a given point to the nearest bird's nest, with probability density function $f(x) = 2xe^{-x^2}$ for $x \geq 0$.

**(a)** Find the cumulative distribution function for this random variable.

**SOLUTION** The cumulative distribution function is given by

$$F(x) = P(X \leq x) = \int_0^x 2te^{-t^2}dt \quad \text{Use the density function with } t \text{ as the variable.}$$

$$= -e^{-t^2}\Big|_0^x \quad \text{Use the antiderivative found in Example 3.}$$

$$= -e^{-x^2} + 1$$

for $x \geq 0$. The cumulative distribution function can be written as $F(x) = 1 - e^{-x^2}$ for $x \geq 0$. Note that for $x < 0$, $F(x) = 0$.

**YOUR TURN 4** Use part (a) of Example 5 to calculate the probability that there is a bird's nest within 1 km of the given point.

**(b)** Use the solution to part (a) to calculate the probability that there is a bird's nest within 0.5 km of the given point.

**SOLUTION** To find $P(X \leq 0.5)$, calculate $F(0.5) = 1 - e^{-0.5^2} \approx 0.2212$. Notice that this is the same answer that we found in Example 3(b). **TRY YOUR TURN 4**

# 13.1 EXERCISES

**Decide whether the functions defined as follows are probability density functions on the indicated intervals. If not, tell why.**

**1.** $f(x) = \dfrac{1}{9}x - \dfrac{1}{18}$; $[2, 5]$     **2.** $f(x) = \dfrac{1}{3}x - \dfrac{1}{6}$; $[3, 4]$

**3.** $f(x) = \dfrac{x^2}{21}$; $[1, 4]$     **4.** $f(x) = \dfrac{3}{98}x^2$; $[3, 5]$

**5.** $f(x) = 4x^3$; $[0, 3]$     **6.** $f(x) = \dfrac{x^3}{81}$; $[0, 3]$

**7.** $f(x) = \dfrac{x^2}{16}$; $[-2, 2]$     **8.** $f(x) = 2x^2$; $[-1, 1]$

**9.** $f(x) = \dfrac{5}{3}x^2 - \dfrac{5}{90}$; $[-1, 1]$

**10.** $f(x) = \dfrac{3}{13}x^2 - \dfrac{12}{13}x + \dfrac{45}{52}$; $[0, 4]$

**Find a value of $k$ that will make $f$ a probability density function on the indicated interval.**

**11.** $f(x) = kx^{1/2}$; $[1, 4]$     **12.** $f(x) = kx^{3/2}$; $[4, 9]$

**13.** $f(x) = kx^2$; $[0, 5]$     **14.** $f(x) = kx^2$; $[-1, 2]$

**15.** $f(x) = kx$; $[0, 3]$     **16.** $f(x) = kx$; $[2, 3]$

**17.** $f(x) = kx$; $[1, 5]$     **18.** $f(x) = kx^3$; $[2, 4]$

**Find the cumulative distribution function for the probability density function in each of the following exercises.**

**19.** Exercise 1     **20.** Exercise 2

**21.** Exercise 3     **22.** Exercise 4

**23.** Exercise 11     **24.** Exercise 12

**25.** The total area under the graph of a probability density function always equals _____.

**26.** In your own words, define a random variable.

**27.** What is the difference between a discrete probability function and a probability density function?

**28.** Why is $P(X = c) = 0$ for any number $c$ in the domain of a probability density function?

**Show that each function defined as follows is a probability density function on the given interval; then find the indicated probabilities.**

**29.** $f(x) = \dfrac{1}{2}(1 + x)^{-3/2}$; $[0, \infty)$

  **a.** $P(0 \le X \le 2)$

  **b.** $P(1 \le X \le 3)$

  **c.** $P(X \ge 5)$

**30.** $f(x) = e^{-x}$; $[0, \infty)$

  **a.** $P(0 \le X \le 1)$

  **b.** $P(1 \le X \le 2)$

  **c.** $P(X \le 2)$

**31.** $f(x) = (1/2)e^{-x/2}$; $[0, \infty)$

  **a.** $P(0 \le X \le 1)$     **b.** $P(1 \le X \le 3)$

  **c.** $P(X \ge 2)$

**32.** $f(x) = \dfrac{20}{(x + 20)^2}$; $[0, \infty)$

  **a.** $P(0 \le X \le 1)$     **b.** $P(1 \le X \le 5)$

  **c.** $P(X \ge 5)$

**33.** $f(x) = \begin{cases} \dfrac{x^3}{12} & \text{if } 0 \le x \le 2 \\[2mm] \dfrac{16}{3x^3} & \text{if } x > 2 \end{cases}$

  **a.** $P(0 \le X \le 2)$     **b.** $P(X \ge 2)$

  **c.** $P(1 \le X \le 3)$

**34.** $f(x) = \begin{cases} \dfrac{20x^4}{9} & \text{if } 0 \le x \le 1 \\[2mm] \dfrac{20}{9x^5} & \text{if } x > 1 \end{cases}$

  **a.** $P(0 \le X \le 1)$     **b.** $P(X \ge 1)$

  **c.** $P(0 \le X \le 2)$

## LIFE SCIENCE APPLICATIONS

**35. Petal Length**  The length of a petal on a certain flower varies from 1 cm to 4 cm and has a probability density function defined by

$$f(x) = \frac{1}{2\sqrt{x}}.$$

Find the probabilities that the length of a randomly selected petal will be as follows.

  **a.** Greater than or equal to 3 cm

  **b.** Less than or equal to 2 cm

  **c.** Between 2 cm and 3 cm

**36. Clotting Time of Blood**  The clotting time of blood is a random variable $t$ with values from 1 second to 20 seconds and probability density function defined by

$$f(t) = \frac{1}{(\ln 20)t}.$$

Find the following probabilities for a person selected at random.

  **a.** The probability that the clotting time is between 1 and 5 seconds

  **b.** The probability that the clotting time is greater than 10 seconds

**37. Flour Beetles** Researchers who study the abundance of the flour beetle, *Tribolium castaneum*, have developed a probability density function that can be used to estimate the abundance of the beetle in a population. The density function, which is a member of the gamma distribution, is

$$f(x) = 1.185 \times 10^{-9}x^{4.5222}e^{-0.049846x},$$

where $x$ is the size of the population. *Source: Ecology.*

   **a.** Estimate the probability that a randomly selected flour beetle population is between 0 and 150.

   **b.** Estimate the probability that a randomly selected flour beetle population is between 100 and 200.

**38. Flea Beetles** The mobility of an insect is an important part of its survival. Researchers have determined that the probability that a marked flea beetle, *Phyllotreta cruciferae* and *Phyllotreta striolata*, will be recaptured within a certain distance and time after release can be calculated from the probability density function

$$p(x, t) = \frac{e^{-x^2/(4Dt)}}{\int_0^L e^{-u^2/(4Dt)} \, du},$$

where $t$ is the time after release (in hours), $x$ is the distance (in meters) from the release point that recaptures occur, $L$ is the maximum distance from the release point that recaptures can occur, and $D$ is the diffusion coefficient. *Source: Ecology Monographs.*

   **a.** If $t = 12$, $L = 6$, and $D = 38.3$, find the probability that a flea beetle will be recaptured within 3 m of the release point.

   **b.** Using the same values for $t$, $L$, and $D$, find the probability that a flea beetle will be recaptured between 1 and 5 m of the release point.

**39. Time to Learn a Task** The time required for a person to learn a certain task is a random variable with probability density function defined by

$$f(t) = \frac{8}{7(t-2)^2}.$$

The time required to learn the task is between 3 and 10 minutes. Find the probabilities that a randomly selected person will learn the task in the following lengths of time.

   **a.** Less than 4 minutes

   **b.** More than 5 minutes

**40. Drunk Drivers** The frequency of alcohol-related traffic fatalities has dropped in recent years but is still high among young people. Based on data from the National Highway Traffic Safety Administration, the age of a randomly selected, alcohol-impaired driver in a fatal car crash is a random variable with probability density function given by

$$f(t) = \frac{4.045}{t^{1.532}} \quad \text{for } t \text{ in } [16, 80].$$

Find the following probabilities of the age of such a driver. *Source: Traffic Safety Facts.*

   **a.** Less than or equal to 25

   **b.** Greater than or equal to 35

   **c.** Between 21 and 30

   **d.** Find the cumulative distribution function for this random variable.

   **e.** Use the answer to part d to find the probability that a randomly selected alcohol-impaired driver in a fatal car crash is at most 21 years old.

**41. Driving Fatalities** We saw in a review exercise in Chapter 4 on Calculating the Derivative that driver fatality rates were highest for the youngest and oldest drivers. When adjusted for the number of miles driven by people in each age group, the number of drivers in fatal crashes goes down with age, and the age of a randomly selected driver in a fatal car crash is a random variable with probability density function given by

$$f(t) = 0.06049e^{-0.03211t} \quad \text{for } t \text{ in } [16, 84].$$

Find the following probabilities of the age of such a driver. *Source: National Highway Traffic Safety Administration.*

   **a.** Less than or equal to 25

   **b.** Greater than or equal to 35

   **c.** Between 21 and 30

   **d.** Find the cumulative distribution function for this random variable.

   **e.** Use the answer to part d to find the probability that a randomly selected driver in a fatal crash is at most 21 years old.

**42. Time of Traffic Fatality** The National Highway Traffic Safety Administration records the time of day of fatal crashes. The following table gives the time of day (in hours since midnight) and the frequency of fatal crashes. *Source: The National Highway Traffic Safety Administration.*

| Time of Day | Midpoint of Interval (hours) | Frequency |
|---|---|---|
| 0–3 | 1.5 | 4486 |
| 3–6 | 4.5 | 2774 |
| 6–9 | 7.5 | 3236 |
| 9–12 | 10.5 | 3285 |
| 12–15 | 13.5 | 4356 |
| 15–18 | 16.5 | 5325 |
| 18–21 | 19.5 | 5342 |
| 21–24 | 22.5 | 4952 |
| Total | | 33,756 |

   **a.** Plot the data. What type of function appears to best match these data?

   **b.** Use the regression feature on your graphing calculator to find a cubic equation that models the time of day, $t$, and the number of traffic fatalities, $T(t)$. Use the midpoint value to estimate the time in each interval. Graph the function with the plot of the data. Does the graph fit the data?

**c.** By finding an appropriate constant $k$, find a function $S(t) = kT(t)$ that is a probability density function describing the probability of a traffic fatality at a particular time of day.

**d.** For a randomly chosen traffic fatality, find the probabilities that the accident occurred between 12 am and 2 am ($t = 0$ to $t = 2$) and between 4 pm and 5:30 pm ($t = 16$ to $t = 17.5$).

## OTHER APPLICATIONS

**43. Length of a Telephone Call**   The length of a telephone call (in minutes), $t$, for a certain town is a continuous random variable with probability density function defined by

$$f(t) = 3t^{-4}, \quad \text{for } t \text{ in } [1, \infty).$$

Find the probabilities for the following situations.

**a.** The call lasts between 1 and 2 minutes.

**b.** The call lasts between 3 and 5 minutes.

**c.** The call lasts longer than 3 minutes.

**44. Machine Part**   The lifetime of a machine part has a continuous distribution on the interval $(0, 40)$ with probability density function $f$, where $f(x)$ is proportional to $(10 + x)^{-2}$. Calculate the probability that the lifetime of the machine part is less than 6. Choose one of the following. ***Source: Society of Actuaries.***

**a.** 0.04   **b.** 0.15   **c.** 0.47   **d.** 0.53   **e.** 0.94

**45. Insurance**   An insurance policy pays for a random loss $X$ subject to a deductible of $C$, where $0 < C < 1$. The loss amount is modeled as a continuous random variable with density function

$$f(x) = \begin{cases} 2x & \text{for } 0 < x < 1 \\ 0 & \text{otherwise.} \end{cases}$$

Given a random loss $X$, the probability that the insurance payment is less than 0.5 is equal to 0.64. Calculate $C$. Choose one of the following. (*Hint:* The payment is 0 unless the loss is greater than the deductible, in which case the payment is the loss minus the deductible.) ***Source: Society of Actuaries.***

**a.** 0.1   **b.** 0.3   **c.** 0.4   **d.** 0.6   **e.** 0.8

**46. Social Network**   The number of U.S. users (in millions) on Facebook, an online social network, in a recent year is given in the table below. ***Source: Inside Facebook.***

| Age Interval (years) | Midpoint of Interval (year) | Number of Users in Each Interval (millions) |
|---|---|---|
| 13–17 | 15 | 6.049 |
| 18–25 | 21.5 | 19.461 |
| 26–34 | 30 | 13.423 |
| 35–44 | 39.5 | 9.701 |
| 45–54 | 49.5 | 4.582 |
| 55–65 | 60 | 2.849 |
| Total | | 56.065 |

**a.** Plot the data. What type of function appears to best match these data?

**b.** Use the regression feature on your graphing calculator to find a quartic equation that models the number of years, $t$, since birth and the number of Facebook users, $N(t)$. Use the midpoint value to estimate the point in each interval for the age of the Facebook user. Graph the function with the plot of the data. Does the function resemble the data?

**c.** By finding an appropriate constant $k$, find a function $S(t) = kN(t)$ that is a probability density function describing the probability of the age of a Facebook user. (*Hint:* Because the function in part b is negative for values less than 13.4 and greater than 62.0, restrict the domain of the density function to the interval $[13.4, 62.0]$. That is, integrate the function you found in part b from 13.4 to 62.0.)

**d.** For a randomly chosen person who uses Facebook, find the probabilities that the person was at least 35 but less than 45 years old, at least 18 but less than 35 years old, and at least 45 years old. Compare these with the actual probabilities.

**47. Annual Rainfall**   The annual rainfall in a remote Middle Eastern country varies from 0 to 5 in. and is a random variable with probability density function defined by

$$f(x) = \frac{5.5 - x}{15}.$$

Find the following probabilities for the annual rainfall in a randomly selected year.

**a.** The probability that the annual rainfall is greater than 3 in.

**b.** The probability that the annual rainfall is less than 2 in.

**c.** The probability that the annual rainfall is between 1 in. and 4 in.

**48. Earthquakes**   The time between major earthquakes in the Southern California region is a random variable with probability density function

$$f(t) = \frac{1}{960} e^{-t/960},$$

where $t$ is measured in days. ***Source: Journal of Seismology.***

**a.** Find the probability that the time between a major earthquake and the next one is less than 365 days.

**b.** Find the probability that the time between a major earthquake and the next one is more than 960 days.

**49. Earthquakes**   The time between major earthquakes in the Taiwan region is a random variable with probability density function

$$f(t) = \frac{1}{3650.1} e^{-t/3650.1},$$

where $t$ is measured in days. ***Source: Journal of Seismology.***

**a.** Find the probability that the time between a major earthquake and the next one is more than 1 year but less than 3 years.

**b.** Find the probability that the time between a major earthquake and the next one is more than 7300 days.

**■ YOUR TURN ANSWERS**

**1.** $P(3/2 \le X \le 2) = 1/3$   **2.** $k = 1/64$

**3.** 0.6321   **4.** 0.6321

# 13.2 Expected Value and Variance of Continuous Random Variables

APPLY IT   **What is the average age of a drunk driver in a fatal car crash?**
*You will be asked to answer this question in Exercise 31.*

It often is useful to have a single number, a typical or "average" number, that represents a random variable. The **mean** or *expected value* for a discrete random variable is found by multiplying each value of the random variable by its corresponding probability.

As we saw in the previous chapter, if $x$ is a discrete random variable that takes on the $n$ values, $x_1, x_2, \ldots, x_n$, with respective probabilities $p_1, p_2, \ldots, p_n$, then the mean, or expected value, of the random variable is

$$\mu = x_1 p_1 + x_2 p_2 + x_3 p_3 + \cdots + x_n p_n = \sum_{i=1}^{n} x_i p_i.$$

For the bird feeder example in the previous section, the expected value is given by

$$\mu = 1(0.04) + 2(0.07) + 3(0.12) + 4(0.16) + 5(0.20) + 6(0.15) + 7(0.13)$$
$$+ 8(0.08) + 9(0.04) + 10(0.01)$$
$$= 5.09.$$

Thus, the average time a sparrow spends at a bird feeder is 5.09 minutes.

This definition can be extended to continuous random variables by using definite integrals. Suppose a continuous random variable has probability density function $f$ on $[a, b]$. We can divide the interval from $a$ to $b$ into $n$ subintervals of length $\Delta x$, where $\Delta x = (b - a)/n$. In the $i$th subinterval, the probability that the random variable takes a value close to $x_i$ is approximately $f(x_i)\Delta x$, and so

$$\mu \approx \sum_{i=1}^{n} x_i f(x_i) \Delta x.$$

As $n \to \infty$, the limit of this sum gives the expected value

$$\mu = \int_{a}^{b} x f(x) \, dx.$$

As we saw in the previous chapter, the *variance* of a discrete probability distribution is a measure of the *spread* of the values of the distribution. For a discrete distribution, the variance is found by taking the expected value of the squares of the differences of the values of the random variable and the mean. If the random variable $x$ takes the values $x_1, x_2, x_3, \ldots, x_n$, with respective probabilities $p_1, p_2, p_3, \ldots, p_n$ and mean $\mu$, then the variance of $x$ is

$$\text{Var}(x) = \sum_{i=1}^{n} (x_i - \mu)^2 p_i.$$

Think of the variance as the expected value of $(x - \mu)^2$, which measures how far $x$ is from the mean $\mu$. The *standard deviation* of $x$ is defined as

$$\sigma = \sqrt{\text{Var}(x)}.$$

For the bird feeder example in the previous section, the variance and standard deviation are

$$\text{Var}(X) = (1 - 5.09)^2(0.04) + (2 - 5.09)^2(0.07) + (3 - 5.09)^2(0.12)$$
$$+ (4 - 5.09)^2(0.16) + (5 - 5.09)^2(0.20) + (6 - 5.09)^2(0.15)$$
$$+ (7 - 5.09)^2(0.13) + (8 - 5.09)^2(0.08) + (9 - 5.09)^2(0.04)$$
$$+ (10 - 5.09)^2(0.01)$$
$$= 4.1819$$

and,

$$\sigma = \sqrt{\text{Var}(x)} \approx 2.045.$$

Like the mean or expected value, the variance of a continuous random variable is an integral.

$$\text{Var}(x) = \int_a^b (x - \mu)^2 f(x)\,dx$$

To find the standard deviation of a continuous probability distribution, like that of a discrete distribution, we find the square root of the variance. The formulas for the expected value, variance, and standard deviation of a continuous probability distribution are summarized here.

**Expected Value, Variance, and Standard Deviation**
If $x$ is a continuous random variable with probability density function $f$ on $[a, b]$, then the **expected value** of $x$ is

$$E(x) = \mu = \int_a^b x f(x)\,dx.$$

The **variance** of $x$ is

$$\text{Var}(x) = \int_a^b (x - \mu)^2 f(x)\,dx,$$

and the **standard deviation** of $x$ is

$$\sigma = \sqrt{\text{Var}(x)}.$$

**NOTE**  In the formulas for expected value, variance, and standard deviation, and all other formulas in this section, it is possible that $a = -\infty$ or $b = \infty$, in which case the density function $f$ is defined on $[a, \infty)$, $(-\infty, b]$, or $(-\infty, \infty)$. In this case, the integrals in these formulas become improper integrals, which are handled according to the procedure described in Section 8.4 on Improper Integrals. Example 2 will illustrate this procedure.

Geometrically, the expected value (or mean) of a probability distribution represents the balancing point of the distribution. If a fulcrum were placed at $\mu$ on the $x$-axis, the figure would be in balance. See Figure 6.

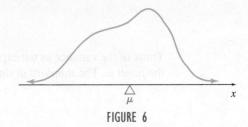

FIGURE 6

The variance and standard deviation of a probability distribution indicate how closely the values of the distribution cluster about the mean. These measures are most useful for comparing different distributions, as in Figure 7.

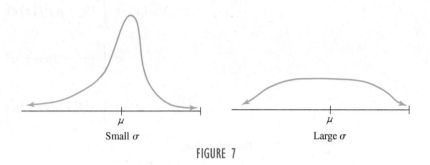

FIGURE 7

## EXAMPLE 1   Expected Value and Variance

Find the expected value and variance of the random variable $x$ with probability density function defined by $f(x) = (3/26)x^2$ on $[1, 3]$.

**SOLUTION**   By the definition of expected value just given,

$$\mu = \int_1^3 xf(x)\,dx$$

$$= \int_1^3 x\left(\frac{3}{26}x^2\right)dx$$

$$= \frac{3}{26}\int_1^3 x^3\,dx$$

$$= \frac{3}{26}\left(\frac{x^4}{4}\right)\bigg|_1^3 = \frac{3}{104}(81 - 1) = \frac{30}{13},$$

or about 2.31.

The variance is

$$\text{Var}(x) = \int_1^3 \left(x - \frac{30}{13}\right)^2\left(\frac{3}{26}x^2\right)dx$$

$$= \int_1^3 \left(x^2 - \frac{60}{13}x + \frac{900}{169}\right)\left(\frac{3}{26}x^2\right)dx \quad \text{Square } \left(x - \tfrac{30}{13}\right).$$

$$= \frac{3}{26}\int_1^3 \left(x^4 - \frac{60}{13}x^3 + \frac{900}{169}x^2\right)dx \quad \text{Multiply.}$$

$$= \frac{3}{26}\left(\frac{x^5}{5} - \frac{60}{13}\cdot\frac{x^4}{4} + \frac{900}{169}\cdot\frac{x^3}{3}\right)\bigg|_1^3 \quad \text{Integrate.}$$

$$= \frac{3}{26}\left[\left(\frac{243}{5} - \frac{60(81)}{52} + \frac{900(27)}{169(3)}\right) - \left(\frac{1}{5} - \frac{60}{52} + \frac{300}{169}\right)\right]$$

$$\approx 0.259.$$

From the variance, the standard deviation is $\sigma \approx \sqrt{0.259} \approx 0.51$. The expected value and standard deviation are shown on the graph of the probability density function in Figure 8.   TRY YOUR TURN 1

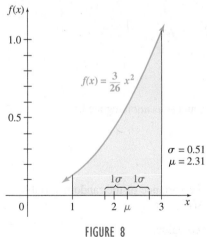

FIGURE 8

**YOUR TURN 1** Repeat Example 1 for the probability density function $f(x) = \dfrac{8}{3x^3}$ on $[1, 2]$.

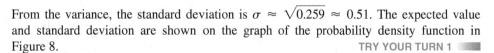

Calculating the variance in the last example was a messy job. An alternative version of the formula for the variance is easier to compute. This alternative formula is derived as follows.

$$\text{Var}(x) = \int_a^b (x - \mu)^2 f(x)\, dx$$

$$= \int_a^b (x^2 - 2\mu x + \mu^2) f(x)\, dx$$

$$= \int_a^b x^2 f(x)\, dx - 2\mu \int_a^b x f(x)\, dx + \mu^2 \int_a^b f(x)\, dx \qquad \textbf{(1)}$$

By definition,

$$\int_a^b x f(x)\, dx = \mu,$$

and, since $f(x)$ is a probability density function,

$$\int_a^b f(x)\, dx = 1.$$

Substitute back into Equation (1) to get the alternative formula,

$$\text{Var}(x) = \int_a^b x^2 f(x)\, dx - 2\mu^2 + \mu^2 = \int_a^b x^2 f(x)\, dx - \mu^2.$$

**Alternative Formula for Variance**

If $x$ is a random variable with probability density function $f$ on $[a, b]$, and if $E(x) = \mu$, then

$$\textbf{Var}(x) = \int_a^b x^2 f(x)\, dx - \mu^2.$$

**CAUTION**  Notice that the term $\mu^2$ comes *after* the $dx$, and so is *not* integrated.

---

**EXAMPLE 2  Variance**

Use the alternative formula for variance to compute the variance of the random variable $x$ with probability density function defined by $f(x) = 3/x^4$ for $x \geq 1$.

**SOLUTION**  To find the variance, first find the expected value:

$$\mu = \int_1^\infty x f(x)\, dx = \int_1^\infty x \cdot \frac{3}{x^4}\, dx = \int_1^\infty \frac{3}{x^3}\, dx$$

$$= \lim_{b \to \infty} \int_1^b \frac{3}{x^3}\, dx = \lim_{b \to \infty} \left( \frac{3}{-2x^2} \right) \Big|_1^b = \frac{3}{2},$$

or 1.5. Now find the variance by the alternative formula for variance:

$$\text{Var}(x) = \int_1^\infty x^2\left(\frac{3}{x^4}\right)dx - \left(\frac{3}{2}\right)^2$$

$$= \int_1^\infty \frac{3}{x^2}dx - \frac{9}{4}$$

$$= \lim_{b\to\infty}\int_1^b \frac{3}{x^2}dx - \frac{9}{4}$$

$$= \lim_{b\to\infty}\left(\frac{-3}{x}\right)\Big|_1^b - \frac{9}{4}$$

$$= 3 - \frac{9}{4} = \frac{3}{4}, \quad \text{or } 0.75. \qquad \textbf{TRY YOUR TURN 2}$$

**YOUR TURN 2** Repeat Example 2 for the probability density function $f(x) = 4/x^5$ for $x \geq 1$.

---

## EXAMPLE 3  Patient Wait

A recent study has shown that the time patients wait at the doctor's office (in hours) after the time of their scheduled appointment is given by the probability density function $f(x) = 6t - 6t^2$ for $0 \leq t \leq 1$.

**(a)** Find and interpret the expected value for this distribution.

**SOLUTION** The expected value is

$$\mu = \int_0^1 t(6t - 6t^2)dt = \int_0^1 (6t^2 - 6t^3)dt$$

$$= \left(2t^3 - \frac{3}{2}t^4\right)\Big|_0^1 = \frac{1}{2},$$

or 0.5. This result indicates that patients wait an average of 1/2 hour past the scheduled office appointment time.

**(b)** Compute the standard deviation.

**SOLUTION** First compute the variance. We use the alternative formula.

$$\text{Var}(t) = \int_0^1 t^2(6t - 6t^2)dt - \left(\frac{1}{2}\right)^2$$

$$= \int_0^1 (6t^3 - 6t^4)dt - \left(\frac{1}{2}\right)^2$$

$$= \left(\frac{3}{2}t^4 - \frac{6}{5}t^5\right)\Big|_0^1 - \frac{1}{4}$$

$$= \frac{3}{10} - \frac{1}{4} = \frac{1}{20} = 0.05$$

The standard deviation is $\sigma = \sqrt{0.05} \approx 0.22$.

**(c)** Calculate the probability that a patient's wait will be within one standard deviation of the mean.

**SOLUTION** Since the mean, or expected value, is 0.5 and the standard deviation is approximately 0.22, we are calculating the probability that a patient will wait between

$$\mu - \sigma = 0.5 - 0.22 = 0.28 \text{ hour}$$

and

$$\mu + \sigma = 0.5 + 0.22 = 0.72 \text{ hour}.$$

The probability is given by

$$P(0.28 \le T \le 0.22) = \int_{0.28}^{0.72} (6t - 6t^2)\, dt$$

$$= (3t^2 - 2t^3)\Big|_{0.28}^{0.72}$$

$$\approx 0.6174.$$

The probability that a patient's wait will be within one standard deviation of the mean is about 0.62.

---

**TECHNOLOGY**

**EXAMPLE 4  Life Expectancy**

In the previous section of this chapter we used statistics compiled by the National Center for Health Statistics to determine a probability density function that can be used to study the proportion of all 50-year-olds who will be alive in $t$ years. The function is given by

$$S(t) = \frac{1}{497{,}703}(5.03958 \times 10^{-5} t^6 - 0.006603 t^5 + 0.2992 t^4$$
$$- 6.0507 t^3 + 67.867 t^2 - 110.3 t + 2485.1)$$

for $0 \le t \le 52.5$.

**(a)** Find the life expectancy of a 50-year-old person.

**SOLUTION**  Since this is a complicated function that is tedious to integrate analytically, we will employ the integration feature on a TI-84 Plus calculator to calculate

$$\mu = \int_0^{52.5} tS(t)\,dt \approx 30.38 \text{ years.}$$

According to life tables, the life expectancy of a person between the ages of 50 and 55 is 30.6 years. Our estimate is remarkably accurate given the limited number of data points and the function used in our original analysis. Life expectancy is generally calculated with techniques from life table analysis. *Source: National Center for Health Statistics.*

**(b)** Find the standard deviation of this probability function.

**SOLUTION**  Using the alternate formula, we first calculate the variance.

$$\text{Var}(T) = \int_0^{52.5} t^2 S(t)\,dt - \mu^2 = 1057.7195 - (30.38)^2 \approx 134.775$$

Thus, $\sigma = \sqrt{\text{Var}(T)} \approx 11.61$ years.

---

As we mentioned earlier, the expected value is also referred to as the mean of the random variable. It is a type of average. There is another type of average, known as the *median*, that is often used. It is the value of the random variable for which there is a 50% probability of being larger and a 50% probability of being smaller. The precise definition is as follows.

**Median**

If $X$ is a random variable with probability density function $f$ on $[a, b]$, then the **median** of $X$ is the number $m$ such that

$$\int_a^m f(x)\,dx = \frac{1}{2}.$$

The median is particularly useful when the random variable is not distributed symmetrically about the mean. An example of this would be a random variable representing the price of homes in a city. There is a small probability that a home will be much more expensive than most of the homes in the city, and this tends to make the mean abnormally high. The median price is a better representation of the average price of a home.

### EXAMPLE 5   Median

Find the median for the random variable described in Example 2, with density function defined by $f(x) = 3/x^4$ for $x \geq 1$.

**SOLUTION**   According to the formula,

$$\int_1^m \frac{3}{x^4}\, dx = \frac{1}{2}.$$

Evaluating the integral on the left, we have

$$-\frac{1}{x^3}\Big|_1^m = -\frac{1}{m^3} + 1.$$

Set this equal to 1/2.

$$-\frac{1}{m^3} + 1 = \frac{1}{2}$$

$$\frac{1}{2} = \frac{1}{m^3} \qquad \text{Subtract 1/2 from both sides, and add } 1/m^3.$$

$$m^3 = 2 \qquad \text{Cross multiply.}$$

$$m = \sqrt[3]{2}$$

**YOUR TURN 3** Repeat Example 5 for the probability density function $f(x) = 4/x^5$ for $x \geq 1$.

The median value is, therefore, $\sqrt[3]{2} \approx 1.2599$. Notice that this is smaller than the mean of 1.5 found in Example 2. This is because the random variable can take on arbitrarily large values, which increases the mean but doesn't affect the median.   **TRY YOUR TURN 3**

Using the notion of cumulative distribution function from the previous section, we can say that the median $m$ is the value for which the cumulative distribution function is 0.5; that is, $F(m) = 0.5$.

# 13.2 EXERCISES

In Exercises 1–8, a probability density function of a random variable is defined. Find the expected value, the variance, and the standard deviation. Round answers to the nearest hundredth.

**1.** $f(x) = \frac{1}{4}$; $[3, 7]$

**2.** $f(x) = \frac{1}{10}$; $[0, 10]$

**3.** $f(x) = \frac{x}{8} - \frac{1}{4}$; $[2, 6]$

**4.** $f(x) = 2(1 - x)$; $[0, 1]$

**5.** $f(x) = 1 - \frac{1}{\sqrt{x}}$; $[1, 4]$

**6.** $f(x) = \frac{1}{11}\left(1 + \frac{3}{\sqrt{x}}\right)$; $[4, 9]$

**7.** $f(x) = 4x^{-5}$; $[1, \infty)$

**8.** $f(x) = 3x^{-4}$; $[1, \infty)$

**9.** What information does the mean (expected value) of a continuous random variable give?

**10.** Suppose two random variables have standard deviations of 0.10 and 0.23, respectively. What does this tell you about their distributions?

**In Exercises 11–14, the probability density function of a random variable is defined.**

a. **Find the expected value to the nearest hundredth.**

b. **Find the variance to the nearest hundredth.**

c. **Find the standard deviation. Round to the nearest hundredth.**

d. **Find the probability that the random variable has a value greater than the mean.**

e. **Find the probability that the value of the random variable is within 1 standard deviation of the mean. Use the value of the standard deviation to the accuracy of your calculator.**

11. $f(x) = \dfrac{\sqrt{x}}{18}$; $[0, 9]$

12. $f(x) = \dfrac{x^{-1/3}}{6}$; $[0, 8]$

13. $f(x) = \dfrac{1}{4}x^3$, $[0, 2]$

14. $f(x) = \dfrac{3}{16}(4 - x^2)$; $[0, 2]$

**For Exercises 15–20, (a) find the median of the random variable with the probability density function given, and (b) find the probability that the random variable is between the expected value (mean) and the median. The expected value for each of these functions was found in Exercises 1–8.**

15. $f(x) = \dfrac{1}{4}$; $[3, 7]$

16. $f(x) = \dfrac{1}{10}$; $[0, 10]$

17. $f(x) = \dfrac{x}{8} - \dfrac{1}{4}$; $[2, 6]$

18. $f(x) = 2(1 - x)$; $[0, 1]$

19. $f(x) = 4x^{-5}$; $[1, \infty)$

20. $f(x) = 3x^{-4}$; $[1, \infty)$

**Find the expected value, the variance, and the standard deviation, when they exist, for each probability density function.**

21. $f(x) = \begin{cases} \dfrac{x^3}{12} & \text{if } 0 \le x \le 2 \\ \dfrac{16}{3x^3} & \text{if } x > 2 \end{cases}$

22. $f(x) = \begin{cases} \dfrac{20x^4}{9} & \text{if } 0 \le x \le 1 \\ \dfrac{20}{9x^5} & \text{if } x > 1 \end{cases}$

23. Let $X$ be a continuous random variable with density function

$$f(x) = \begin{cases} \dfrac{|x|}{10} & \text{for } -2 \le x \le 4 \\ 0 & \text{otherwise.} \end{cases}$$

Calculate the expected value of $X$. **Source: Society of Actuaries.** Choose one of the following.

a. 1/5    b. 3/5

c. 1    d. 28/15

e. 12/5

## LIFE SCIENCE APPLICATIONS

24. **Blood Clotting Time**  The clotting time of blood (in seconds) is a random variable with probability density function defined by

$$f(t) = \dfrac{1}{(\ln 20)t} \quad \text{for } t \text{ in } [1, 20].$$

a. Find the mean clotting time.

b. Find the standard deviation.

c. Find the probability that a person's blood clotting time is within 1 standard deviation of the mean.

d. Find the median clotting time.

25. **Length of a Leaf**  The length of a leaf on a tree is a random variable with probability density function defined by

$$f(x) = \dfrac{3}{32}(4x - x^2) \quad \text{for } x \text{ in } [0, 4].$$

a. What is the expected leaf length?

b. Find $\sigma$ for this distribution.

c. Find the probability that the length of a given leaf is within 1 standard deviation of the expected value.

26. **Petal Length**  The length (in centimeters) of a petal on a certain flower is a random variable with probability density function defined by

$$f(x) = \dfrac{1}{2\sqrt{x}} \quad \text{for } x \text{ in } [1, 4].$$

a. Find the expected petal length.

b. Find the standard deviation.

c. Find the probability that a petal selected at random has a length more than 2 standard deviations above the mean.

d. Find the median petal length.

27. **Flea Beetles**  As we saw in Exercise 38 of the previous section, the probability that a marked flea beetle, *Phyllotreta cruciferae* and *Phyllotreta striolata*, will be recaptured within a certain distance and time after release can be calculated from the probability density function

$$p(x, t) = \dfrac{e^{-x^2/(4Dt)}}{\displaystyle\int_0^L e^{-u^2/(4Dt)}\, du},$$

where $t$ is the time (in hours) after release, $x$ is the distance (in meters) from the release point that recaptures occur, $L$ is the maximum distance from the release point that recaptures can occur, and $D$ is the diffusion coefficient. If $t = 12$, $L = 6$, and $D = 38.3$, find the expected recapture distance. **Source: Ecology Monographs.**

28. **Flour Beetles**  As we saw in Exercise 37 of the previous section, a probability density function has been developed to estimate the abundance of the flour beetle, *Tribolium castaneum*. The density function, which is a member of the gamma distribution, is

$$f(x) = 1.185 \times 10^{-9}x^{4.5222}e^{-0.049846x},$$

where $x$ is the size of the population. Calculate the expected size of a flour beetle population. (*Hint:* Use 1000 as the upper limit of integration.) **Source: Ecology.**

**29. SARS** Researchers investigating the SARS (sever acute respiratory syndrome) epidemic in Hong Kong in 2003 found that the probability distribution of the time from infection to onset of the disease could be described by

$$f(t) = 0.07599t^{1.43}e^{-t/2.62}$$

for $t > 0$, where $t$ is the time in days. ***Source: The Lancet.***

   **a.** Find the mean of this distribution. (Using a calculator to perform the integration, we can use an upper limit of 40, since the probability that the time is greater than 40 days is extremely small.)

   **b.** Find the standard deviation.

   **c.** Find the probability that the time from infection to onset of the disease is between 5 and 10 days.

**30. Time to Learn a Task** In Exercise 39 of the previous section, the probability density function for the time required for a person to learn a certain task was given by

$$f(t) = \frac{8}{7(t-2)^2},$$

for $3 \leq t \leq 10$ minutes. Find the median time for a person to learn the task.

**31. Drunk Drivers** In Exercise 40 of the last section, we saw that the age of a randomly selected, alcohol-impaired driver in a fatal car crash is a random variable with probability density function given by

$$f(t) = \frac{4.045}{t^{1.532}} \quad \text{for } t \text{ in } [16, 80].$$

***Source: Traffic Safety Facts.***

   **a.** APPLY IT Find the expected age of a drunk driver in a fatal car crash.

   **b.** Find the standard deviation of the distribution.

   **c.** Find the probability that such a driver will be younger than 1 standard deviation below the mean.

   **d.** Find the median age of a drunk driver in a fatal car crash.

**32. Driving Fatalities** In Exercise 41 of the last section, we saw that the age of a randomly selected driver in a fatal car crash is a random variable with probability density function given by

$$f(t) = 0.06049e^{-0.03211t} \quad \text{for } t \text{ in } [16, 84].$$

***Source: National Highway Traffic Safety Administration.***

   **a.** Find the expected age of a driver in a fatal car crash.

   **b.** Find the standard deviation of the distribution.

   **c.** Find the probability that such a driver will be younger than 1 standard deviation below the mean.

   **d.** Find the median age of a driver in a fatal car crash.

**33. Time of Traffic Fatality** In Exercise 42 of the previous section, the probability density function for the number of fatal traffic accidents was found to be

$$S(t) = \frac{1}{101,370}(-2.564t^3 + 99.11t^2 - 964.6t + 5631)$$

where $t$ is the number of hours since midnight on $[0, 24]$. Calculate the expected time of day at which a fatal accident will occur. ***Source: The National Highway Traffic Safety Administration.***

**34. Dental Insurance** An insurance policy reimburses dental expense, $X$, up to a maximum benefit of 250. The probability density function for $X$ is

$$f(x) = \begin{cases} ce^{-0.004x} & \text{for } x \geq 0 \\ 0 & \text{otherwise,} \end{cases}$$

where $c$ is a constant. Calculate the median benefit for this policy. Choose one of the following. (*Hint:* As long as the expenses are less than 250, the expenses and the benefit are equal.) ***Source: Society of Actuaries.***

   **a.** 161    **b.** 165    **c.** 173    **d.** 182    **e.** 250

## OTHER APPLICATIONS

**35. Losses After Deductible** A manufacturer's annual losses follow a distribution with density function

$$f(x) = \begin{cases} \dfrac{2.5(0.6)^{2.5}}{x^{3.5}} & \text{for } x > 0.6 \\ 0 & \text{otherwise.} \end{cases}$$

To cover its losses, the manufacturer purchases an insurance policy with an annual deductible of 2. What is the mean of the manufacturer's annual losses not paid by the insurance policy? Choose one of the following. (*Hint:* The loss not paid by the insurance policy will equal the actual loss if the actual loss is less than the deductible. Otherwise it will equal the deductible.) ***Source: Society of Actuaries.***

   **a.** 0.84    **b.** 0.88    **c.** 0.93    **d.** 0.95    **e.** 1.00

**36. Insurance Reimbursement** An insurance policy reimburses a loss up to a benefit limit of 10. The policyholder's loss, $Y$, follows a distribution with density function:

$$f(y) = \begin{cases} \dfrac{2}{y^3} & \text{for } y > 1 \\ 0 & \text{otherwise.} \end{cases}$$

What is the expected value of the benefit paid under the insurance policy? Choose one of the following. (*Hint:* The benefit paid will be equal to the actual loss if the actual loss is less than the limit. Otherwise it will equal the limit.) ***Source: Society of Actuaries.***

   **a.** 1.0    **b.** 1.3    **c.** 1.8    **d.** 1.9    **e.** 2.0

**37. Insurance Claims** An insurance company's monthly claims are modeled by a continuous, positive random variable $X$, whose probability density function is proportional to $(1 + x)^{-4}$, where $0 < x < \infty$. Determine the company's expected monthly claims. Choose one of the following. ***Source: Society of Actuaries.***

   **a.** 1/6    **b.** 1/3
   
   **c.** 1/2    **d.** 1
   
   **e.** 3

**38. Social Network** In Exercise 46 of the previous section, the probability density function for the number of U.S. users of Facebook, an online social network, was found to be

$$S(t) = \frac{1}{466.26}(-0.00007445t^4 + 0.01243t^3 - 0.7419t^2 + 18.18t - 137.5)$$

where $t$ was the number of years since birth on $[13.4, 62.0]$. Calculate the expected age of a Facebook user, as well as the standard deviation. ***Source: Inside Facebook.***

**39. Earthquakes** The time between major earthquakes in the Southern California region is a random variable with probability density function defined by

$$f(t) = \frac{1}{960}e^{-t/960},$$

where $t$ is measured in days. ***Source: Journal of Seismology.*** Find the expected value and the standard deviation of this probability density function.

**40. Annual Rainfall** The annual rainfall in a remote Middle Eastern country is a random variable with probability density function defined by

$$f(x) = \frac{5.5 - x}{15}, \quad \text{for } x \text{ in } [0, 5].$$

**a.** Find the mean annual rainfall.

**b.** Find the standard deviation.

**c.** Find the probability of a year with rainfall less than 1 standard deviation below the mean.

---

**YOUR TURN ANSWERS**

1. 4/3, 0.0706
2. 2/9
3. $\sqrt[4]{2} \approx 1.1892$

---

# 13.3 Special Probability Density Functions

APPLY IT   *What is the probability that the maximum outdoor temperature will be higher than 24°C? What is the probability that a flashlight battery will last longer than 40 hours?*

*These questions, presented in Examples 1 and 2, can be answered if the probability density function for the maximum temperature and for the life of the battery are known.*

In practice, it is not feasible to construct a probability density function for every experiment. Instead, a researcher uses one of several probability density functions that are well known, matching the shape of the experimental distribution to one of the known distributions. In this section we discuss some of the most commonly used probability distributions.

## Uniform Distribution
The simplest probability distribution occurs when the probability density function of a random variable remains constant over the sample space. In this case, the random variable is said to be *uniformly distributed* over the sample space. The probability density function for the **uniform distribution** is defined by

$$f(x) = \frac{1}{b - a} \quad \text{for } x \text{ in } [a, b],$$

where $a$ and $b$ are constant real numbers. The graph of $f(x)$ is shown in Figure 9.

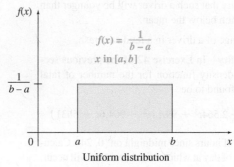

Uniform distribution

**FIGURE 9**

Since $b - a$ is positive, $f(x) \geq 0$, and

$$\int_a^b \frac{1}{b-a}\,dx = \frac{1}{b-a}x\bigg|_a^b = \frac{1}{b-a}(b-a) = 1.$$

Therefore, the function is a probability density function.

The expected value for the uniform distribution is

$$\mu = \int_a^b \left(\frac{1}{b-a}\right)x\,dx = \left(\frac{1}{b-a}\right)\frac{x^2}{2}\bigg|_a^b$$

$$= \frac{1}{2(b-a)}(b^2 - a^2) = \frac{1}{2}(b+a). \quad \text{\small $b^2 - a^2 = (b-a)(b+a)$}$$

The variance is given by

$$\mathrm{Var}(X) = \int_a^b \left(\frac{1}{b-a}\right)x^2\,dx - \left(\frac{b+a}{2}\right)^2$$

$$= \left(\frac{1}{b-a}\right)\frac{x^3}{3}\bigg|_a^b - \frac{(b+a)^2}{4}$$

$$= \frac{1}{3(b-a)}(b^3 - a^3) - \frac{1}{4}(b+a)^2$$

$$= \frac{b^2 + ab + a^2}{3} - \frac{b^2 + 2ab + a^2}{4} \quad \text{\small $b^3 - a^3 = (b-a)(b^2 + ab + a^2)$}$$

$$= \frac{b^2 - 2ab + a^2}{12}. \quad \text{\small Get a common denominator; subtract.}$$

Thus

$$\mathrm{Var}(X) = \frac{1}{12}(b-a)^2, \quad \text{\small Factor.}$$

and

$$\sigma = \frac{1}{\sqrt{12}}(b-a).$$

These properties of the uniform distribution are summarized below.

## Uniform Distribution

If $X$ is a random variable with probability density function

$$f(x) = \frac{1}{b-a} \quad \text{for } x \text{ in } [a,b],$$

then

$$\mu = \frac{1}{2}(b+a) \quad \text{and} \quad \sigma = \frac{1}{\sqrt{12}}(b-a).$$

## EXAMPLE 1 Daily Temperature

A couple is planning to vacation in San Francisco. They have been told that the maximum daily temperature during the time they plan to be there ranges from 15°C to 27°C. Assume that the probability of any temperature between 15°C and 27°C is equally likely for any given day during the specified time period.

(a) What is the probability that the maximum temperature on the day they arrive will be higher than 24°C?

**APPLY IT**

**SOLUTION** If the random variable $T$ represents the maximum temperature on a given day, then the uniform probability density function for $T$ is defined by $f(t) = 1/12$ for the interval $[15, 27]$. By definition,

$$P(T > 24) = \int_{24}^{27} \frac{1}{12}\, dt = \frac{1}{12} t \Big|_{24}^{27} = \frac{1}{4}.$$

**(b)** What average maximum temperature can they expect?

**SOLUTION** The expected maximum temperature is

$$\mu = \frac{1}{2}(27 + 15) = 21,$$

or 21°C.

**(c)** What is the probability that the maximum temperature on a given day will be one standard deviation or more below the mean?

**SOLUTION** First find $\sigma$.

> **YOUR TURN 1** The next vacation for the couple in Example 1 is to a desert with a maximum daily temperature that uniformly ranges from 27°C to 42°C. Find the expected maximum temperature and the probability that the maximum temperature will be within one standard deviation of the mean.

$$\sigma = \frac{1}{\sqrt{12}}(27 - 15) = \frac{12}{\sqrt{12}} = \sqrt{12} = 2\sqrt{3} \approx 3.464.$$

One standard deviation below the mean indicates a temperature of $21 - 3.464 = 17.536$°C.

$$P(T \le 17.536) = \int_{15}^{17.536} \frac{1}{12}\, dt = \frac{1}{12} t \Big|_{15}^{17.536} \approx 0.2113$$

The probability is about 0.21 that the temperature will not exceed 17.5°C.

**TRY YOUR TURN 1**

## Exponential Distribution

The next distribution is very important in reliability and survival analysis. When manufactured items and living things have a constant failure rate over a period of time, the exponential distribution is used to describe their probability of failure. In this case, the random variable is said to be *exponentially distributed* over the sample space. The probability density function for the **exponential distribution** is defined by

$$f(x) = ae^{-ax} \qquad \text{for } x \text{ in } [0, \infty),$$

where $a$ is a positive constant. The graph of $f(x)$ is shown in Figure 10.

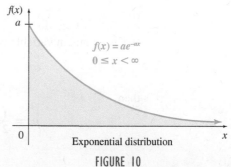

$f(x) = ae^{-ax}$
$0 \le x < \infty$

Exponential distribution

**FIGURE 10**

Here $f(x) \ge 0$, since $e^{-ax}$ and $a$ are both positive for all values of $x$. Also,

$$\int_0^\infty ae^{-ax}\, dx = \lim_{b \to \infty} \int_0^b ae^{-ax}\, dx$$

$$= \lim_{b \to \infty} (-e^{-ax}) \Big|_0^b = \lim_{b \to \infty} \left( \frac{-1}{e^{ab}} + \frac{1}{e^0} \right) = 1,$$

so the function is a probability density function.

The expected value and the standard deviation of the exponential distribution can be found using integration by parts. The results are given below. (See Exercise 28 at the end of this section.)

**Exponential Distribution**

If $X$ is a random variable with probability density function

$$f(x) = ae^{-ax} \quad \text{for } x \text{ in } [0, \infty),$$

then

$$\mu = \frac{1}{a} \quad \text{and} \quad \sigma = \frac{1}{a}.$$

### EXAMPLE 2 Flashlight Battery

Suppose the useful life (in hours) of a flashlight battery is the random variable $T$, with probability density function given by the exponential distribution

$$f(t) = \frac{1}{20}e^{-t/20} \quad \text{for } t \geq 0.$$

**(a)** Find the probability that a particular battery, selected at random, has a useful life of less than 100 hours.

**SOLUTION** The probability is given by

$$P(T \leq 100) = \int_0^{100} \frac{1}{20}e^{-t/20}\, dt = \frac{1}{20}(-20e^{-t/20})\Big|_0^{100}$$
$$= -(e^{-100/20} - e^0) = -(e^{-5} - 1)$$
$$\approx 1 - 0.0067 = 0.9933.$$

**(b)** Find the expected value and standard deviation of the distribution.

**SOLUTION** Use the formulas given above. Both $\mu$ and $\sigma$ equal $1/a$, and since $a = 1/20$ here,

$$\mu = 20 \quad \text{and} \quad \sigma = 20.$$

This means that the average life of a battery is 20 hours, and no battery lasts less than 1 standard deviation below the mean.

**(c)** What is the probability that a battery will last longer than 40 hours?

**SOLUTION** The probability is given by

$$P(T > 40) = \int_{40}^{\infty} \frac{1}{20}e^{-t/20}\, dt = \lim_{b\to\infty}(-e^{-t/20})\Big|_{40}^{b} = \frac{1}{e^2} \approx 0.1353,$$

or about 14%.

TRY YOUR TURN 2

APPLY IT

**YOUR TURN 2** Repeat Example 2 for a flashlight battery with a useful life given by the probability density function $f(t) = \frac{1}{25}e^{-t/25}$ for $t \geq 0$.

There is an interesting connection between the exponential random variable and the discrete random variable known as the *Poisson distribution*, which is used to describe events that occur randomly at a constant rate, such that the probability of two occurrences in a very short time interval is essentially 0. The Poisson distribution is used to describe DNA mutations (per unit length of a strand of DNA, rather than unit time), radioactive decay, defects that occur sporadically at random (as opposed to defects that occur because of something becoming old and worn out), and many other phenomena.

**Poisson Distribution**

If $X$ is a random variable with the **Poisson distribution**, then the probability that $X$ has the value $x$ is given by

$$P(X = x) = f(x) = \frac{a^x e^{-a}}{x!},$$

where $a > 0$ and $x = 0, 1, 2, \ldots$. For the Poisson distribution,

$$\mu = a \quad \text{and} \quad \sigma = \sqrt{a}.$$

**EXAMPLE 3** **Chromosomal Aberrations**

Researchers have used the Poisson distribution to model the number of chromosomal aberrations in cultures of human leukocytes treated with a chemical mutagen. *Source: SIAM Journal on Applied Mathematics.* (The researchers found that a more complicated model based on the Poisson distribution gave a better fit. Interested readers should see the original article as listed in the Sources.) They estimated that $a = 0.5696$ aberration per unit time. Find the probability of the following number of aberrations per unit time.

**(a)** At most 1

**SOLUTION** According to the Poisson distribution, with $a = 0.5696$,

$$P(\text{at most } 1) = P(0) + P(1)$$

$$= \frac{0.5696^0 e^{-0.5696}}{0!} + \frac{0.5696^1 e^{-0.5696}}{1!}$$

$$= 0.56575 + 0.32225$$

$$= 0.8880.$$

There is about a 89% chance of at most 1 aberration.

**(b)** More than 1

**SOLUTION**

$$P(\text{more than } 1) = 1 - P(\text{at most } 1)$$

$$= 1 - 0.8880$$

$$= 0.1120$$

The connection between the Poisson distribution, which is discrete, and the exponential distribution, which is continuous, is that the time until the next occurrence of a Poisson random variable with parameter $a$ is an exponential random variable with parameter $a$.

**EXAMPLE 4** **Chromosomal Aberrations**

For the cultures discussed in Example 3, what is the probability that at least 2 time periods pass before the next aberration?

**SOLUTION** Because the number of aberrations has a Poisson distribution with $a = 0.5696$, the waiting time until the next aberration is distributed exponentially with $a = 0.5696$.

$$P(X > 2) = \int_2^\infty 0.5696 e^{-0.5696t}\, dt$$

$$= \lim_{b \to \infty} \int_2^b 0.5696 e^{-0.5696t}\, dt$$

$$= \lim_{b \to \infty} \left( -e^{-0.5696t} \Big|_2^b \right)$$

$$= \lim_{b \to \infty} \left( \frac{-1}{e^{0.5696b}} + \frac{1}{e^{0.5696(2)}} \right)$$

$$\approx 0.3201$$

## Normal Distribution

The **normal distribution**, with its well-known bell-shaped graph, is undoubtedly the most important probability density function. It is widely used in various applications of statistics. The random variables associated with these applications are said to be normally distributed. The probability density function for the normal distribution has the following characteristics.

---

**Normal Distribution**

If $\mu$ and $\sigma$ are real numbers, $\sigma > 0$, and if $X$ is a random variable with probability density function defined by

$$f(x) = \frac{1}{\sigma\sqrt{2\pi}}e^{-(x-\mu)^2/(2\sigma^2)} \quad \text{for } x \text{ in } (-\infty, \infty),$$

then

$$E(X) = \mu \quad \text{and} \quad \text{Var}(X) = \sigma^2, \quad \text{with standard deviation } \sigma.$$

Notice that the definition of the probability density function includes $\sigma$, which is the standard deviation of the distribution.

---

Advanced techniques can be used to show that

$$\int_{-\infty}^{\infty} \frac{1}{\sigma\sqrt{2\pi}}e^{-(x-\mu)^2/(2\sigma^2)}\,dx = 1.$$

Deriving the expected value and standard deviation for the normal distribution also requires techniques beyond the scope of this text.

Each normal probability distribution has associated with it a bell-shaped curve, called a **normal curve**, such as the one in Figure 11. Each normal curve is symmetric about a vertical line through the mean, $\mu$. Vertical lines at points $+1\sigma$ and $-1\sigma$ from the mean show the inflection points of the graph. (See Exercise 30 at the end of this section.) A normal curve never touches the $x$-axis; it extends indefinitely in both directions.

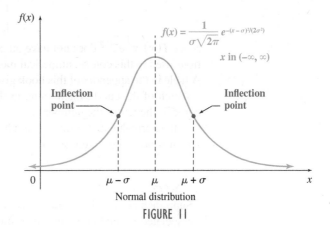

FIGURE 11

The development of the normal curve is credited to the Frenchman Abraham De Moivre (1667–1754). Three of his publications dealt with probability and associated topics: *Annuities upon Lives* (which contributed to the development of actuarial studies), *Doctrine of Chances*, and *Miscellanea Analytica*.

Many different normal curves have the same mean. In such cases, a larger value of $\sigma$ produces a "flatter" normal curve, while smaller values of $\sigma$ produce more values near the mean, resulting in a "taller" normal curve. See Figure 12 on the next page.

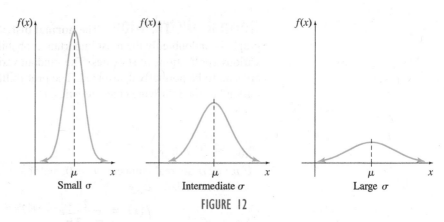

**FIGURE 12**

It would be far too much work to calculate values for the normal probability distribution for various values of $\mu$ and $\sigma$. Instead, values are calculated for the **standard normal distribution**, which has $\mu = 0$ and $\sigma = 1$. The graph of the standard normal distribution is shown in Figure 13.

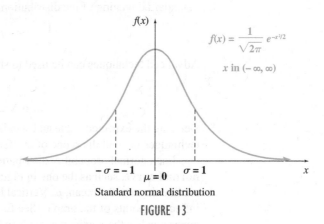

Standard normal distribution

**FIGURE 13**

Probabilities for the standard normal distribution come from the definite integral

$$\int_a^b \frac{1}{\sqrt{2\pi}} e^{-x^2/2}\, dx.$$

Since $f(x) = e^{-x^2/2}$ does not have an antiderivative that can be expressed in terms of functions used in this course, numerical methods are used to find values of this definite integral. A table in the appendix of this book gives areas under the standard normal curve, along with a sketch of the curve. Each value in this table is the total area under the standard normal curve to the left of the number $z$.

If a normal distribution does not have $\mu = 0$ and $\sigma = 1$, we use the following theorem, which is proved in Exercise 29.

**z-Scores Theorem**

Suppose a normal distribution has mean $\mu$ and standard deviation $\sigma$. The area under the associated normal curve that is to the left of the value $x$ is exactly the same as the area to the left of

$$z = \frac{x - \mu}{\sigma}$$

for the standard normal curve.

With this result, the table can be used for *any* normal distribution, regardless of the values of $\mu$ and $\sigma$. The number $z$ in the theorem is called a **z-score**.

### EXAMPLE 5  Life Spans

According to actuarial tables, life spans in the United States are approximately normally distributed with a mean of about 75 years and a standard deviation of about 16 years. By computing the areas under the associated normal curve, find the following probabilities. *Source: Psychological Science.*

**(a)** Find the probability that a randomly selected person lives less than 88 years.

**SOLUTION** Let $T$ represent the life span of a random individual. To find $P(T < 88)$, we calculate the corresponding $z$-score using $t = 88$, $\mu = 75$, and $\sigma = 16$. Round to the nearest hundredths, since this is the extent of our normal curve table.

$$z = \frac{88 - 75}{16} = \frac{13}{16} \approx 0.81$$

Look up 0.81 in the normal curve table in the Appendix. The corresponding area is 0.7910. Thus, the shaded area shown in Figure 14 is 0.7910. This means that the probability of a randomly selected person living less than 88 years is $P(T < 88) = P(Z < 0.81) = 0.7910$, or about 79%.

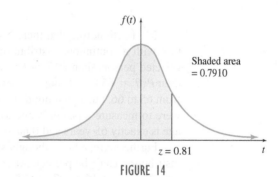

FIGURE 14

**(b)** Find the probability that a randomly selected person lives more than 67 years.

**SOLUTION** To calculate $P(T > 67)$, first find the corresponding $z$-score.

$$z = \frac{67 - 75}{16} = -0.5$$

From the normal curve table, the area to *left* of $z = -0.5$ is 0.3085. Therefore, the area to the *right* is $P(T > 67) = P(Z > -0.5) = 1 - 0.3085 = 0.6915$. See Figure 15. Thus, the probability of a randomly selected person living more than 67 years is 0.6915, or about 69%.

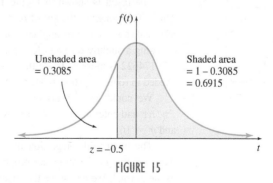

FIGURE 15

**(c)** Find the probability that a randomly selected person lives between 61 and 70 years.

**SOLUTION** Find $z$-scores for both values.

$$z = \frac{61 - 75}{16} = -0.88 \qquad \text{and} \qquad z = \frac{70 - 75}{16} = -0.31$$

Start with the area to the left of $z = -0.31$ and subtract the area to the left of $z = -0.88$. Thus,

$$P(61 \le T \le 70) = P(-0.88 \le Z \le -0.31) = 0.3783 - 0.1894 = 0.1889.$$

The required area is shaded in Figure 16. The probability of a randomly selected person living between 61 and 70 years is about 19%.

**YOUR TURN 3** Using the information provided in Example 5, find the probability that a randomly selected person lives (a) more than 79 years and (b) between 67 and 83 years.

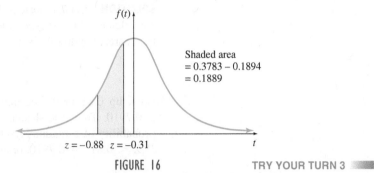

FIGURE 16                    TRY YOUR TURN 3

It's worth noting that there is always some error in approximating a discrete distribution with a continuous distribution. For example, when $T$ is the life span of a randomly selected person, then $P(T = 65)$ is clearly positive, but if $T$ is a normal random variable, then $P(T = 65) = 0$, since it represents no area. The problem is that a person's age jumps from 65 to 66, but a continuous random variable takes on all real numbers in between. If we were to measure a person's age to the nearest nanosecond, the probability that someone's age is exactly 65 years and 0 nanoseconds would be virtually 0.

Furthermore, a bit of thought shows us that the approximation of life spans by a normal distribution can't be perfect. After all, three standard deviations to the left and right of 75 give $75 - 3 \times 16 = 27$ and $75 + 3 \times 16 = 123$. Because of the symmetry of the normal distribution, $P(T < 27)$ and $P(T > 123)$ should be equal. Yet there are people who die before the age of 27, and no human has been verified to live beyond the age of 123.

---

**TECHNOLOGY NOTE**

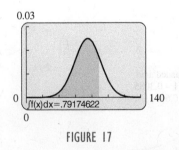

FIGURE 17

As an alternative to using the normal curve table, we can use a graphing calculator. Enter the formula for the normal distribution into the calculator, using $\mu = 75$ and $\sigma = 16$. Plot the function on a window that contains at least four standard deviations to the left and right of $\mu$; for Example 5(a), we will let $0 \le t \le 140$. Then use the integration feature (under CALC on a TI-84 Plus) to find the area under the curve to the left of 88.

The result is shown in Figure 17. In place of $-\infty$, we have used $t = 0$ as the left endpoint. This is far enough to the left of $\mu = 75$ that it can be considered as $-\infty$ for all practical purposes. It also makes sense in this application, since life span can't be a negative number. You can verify that choosing a slightly different lower limit makes little difference in the answer. In fact, the answer of 0.79174622 is more accurate than the answer of 0.7910 that we found in Example 5(a), where we needed to round $13/16 = 0.8125$ to 0.81 in order to use the table.

We could get the answer on a TI-84 Plus without generating a graph using the command `fnInt` and entering $\int_0^{88}(\mathtt{Y_1})\,\mathtt{dx}$, where $\mathtt{Y_1}$ is the formula for the normal distribution with $\mu = 75$ and $\sigma = 16$.

The numerical integration method works with any probability density function. In addition, many graphing calculators are programmed with information about specific density functions, such as the normal. We can solve the first part of Example 5 on the TI-84 Plus by entering `normalcdf` `(-1E99,88,75,16)`. The calculator responds with 0.7917476687. (`-1E99` stands for $-1 \times 10^{99}$, which the calculator uses for $-\infty$.) If you use this method in the exercises, your answers will differ slightly from those in the back of the book, which were generated using the normal curve table in the Appendix.

The z-scores are actually standard deviation multiples; that is, a z-score of 2.5 corresponds to a value 2.5 standard deviations above the mean. For example, looking up $z = 1.00$ and $z = -1.00$ in the table shows that

$$0.8413 - 0.1587 = 0.6826,$$

so that 68.26% of the area under a normal curve is within 1 standard deviation of the mean. Also, using $z = 2.00$ and $z = -2.00$,

$$0.9772 - 0.0228 = 0.9544,$$

meaning 95.44% of the area is within 2 standard deviations of the mean. These results, summarized in Figure 18, can be used to get a quick estimate of results when working with normal curves.

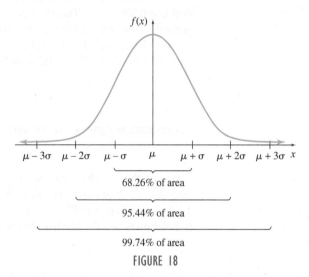

FIGURE 18

Manufacturers make use of the fact that a normal random variable is almost always within 3 standard deviations of the mean to design control charts. When a sample of items produced by a machine has a mean farther than 3 standard deviations from the desired specification, the machine is assumed to be out of control, and adjustments are made to ensure that the items produced meet the tolerance required.

## EXAMPLE 6  Lead Poisoning

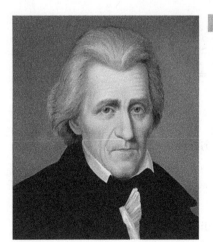

Historians and biographers have collected evidence suggesting that President Andrew Jackson suffered from lead poisoning. Recently, researchers measured the amount of lead in samples of Jackson's hair from 1815. The results of this experiment showed that Jackson had a mean lead level of 130.5 ppm. *Source: JAMA.*

(a) Levels of lead in hair samples from that time period follow a normal distribution with mean 93 and standard deviation 16. *Source: Science.* Find the probability that a randomly selected person from this time period would have a lead level of 130.5 ppm or higher. Does this provide evidence that Jackson suffered from lead poisoning during this time period?

**SOLUTION**  $P(X \geq 130.5) = P\left(Z \geq \dfrac{130.5 - 93}{16}\right) = P(Z \geq 2.34) = 0.0096$

Since this probability is so low, it is likely that Jackson suffered from lead poisoning during this time period.*

*Although this provides evidence that Andrew Jackson had elevated lead levels, the authors of the paper concluded that Andrew Jackson did not die of lead poisoning.

**(b)** Today's normal lead levels follow a normal distribution with approximate mean of 10 ppm and standard deviation of 5 ppm. *Source: Clinical Chemistry.* By today's standards, calculate the probability that a randomly selected person from today would have a lead level of 130.5 ppm or higher. From this, can we conclude that Andrew Jackson had lead poisoning?

**SOLUTION** $P(X \geq 130.5) = P\left(Z \geq \dfrac{130.5 - 10}{5}\right) = P(Z \geq 24.1) \approx 0$

By today's standards, which may not be valid for this experiment, Jackson certainly suffered from lead poisoning.

An amazing result that makes the normal distribution even more important is the **Central Limit Theorem**. This theorem says that averages taken from *any* distribution tend to be normally distributed as the sample size becomes large. Perhaps even more surprising is that the sample size needs to be only 30 or larger for the distribution of the averages to be quite close to normal. To clarify this, we first need to define a *random sample*.

## Random Sample
A **random sample** is a set of independent values of a random variable.

For example, a random sample might consist of the systolic blood pressure of 3 individuals chosen at random from a population, or the bacteria concentration in a pond as measured on 12 days selected at random. The key idea in a random sample is that every member of the population is equally likely to be included in the sample.

## Mean of a Random Sample
Let $X_1, X_2, \ldots, X_n$ be a random sample. The **mean of a random sample** is

$$\overline{X} = \frac{\sum_{i=1}^{n} X_i}{n} = \frac{X_1 + X_2 + \cdots + X_n}{n}.$$

For example, if we chose 3 individuals at random from a population and found their systolic blood pressures to be 120, 145, and 185, then

$$\overline{X} = \frac{120 + 145 + 185}{3} = 150.$$

## Central Limit Theorem
Let $X_1, X_2, \ldots, X_n$ be a random sample from a distribution with mean $\mu$ and standard deviation $\sigma$. Then the distribution of

$$\frac{\overline{X} - \mu}{\sigma/\sqrt{n}}$$

becomes increasingly closer to the standard normal distribution as $n$ gets large.

The power of the Central Limit Theorem comes from the fact that even if we know nothing about the population distribution, we know everything about the standard normal distribution, allowing us to compute probabilities involving the sample mean.

EXAMPLE 7  Central Limit Theorem

In a classic statistical study of plants, R. A. Fisher found that a compound measure of differences between species of irises in terms of their sepal and petal length and width had a mean of 33.816 cm and a standard deviation of 4.781 cm. *Source: Annals of Eugenics.* Fisher examined a sample of size 50. What is the probability that the sample mean of the differences is between 32 and 34 cm?

**SOLUTION**  We wish to find

$$P(32 \leq \overline{X} \leq 34).$$

Using $\mu = 33.816$, $\sigma = 4.781$, and $n = 50$, we calculate

$$P\left(\frac{32 - 33.816}{4.781/\sqrt{50}} \leq \frac{\overline{X} - \mu}{\sigma/\sqrt{n}} \leq \frac{34 - 33.816}{4.781/\sqrt{50}}\right),$$

which simplifies to

$$P\left(-2.69 \leq \frac{\overline{X} - \mu}{\sigma/\sqrt{n}} \leq 0.27\right).$$

Because the sample size is greater than 30, $(\overline{X} - \mu)/(\sigma/\sqrt{n})$ is approximately distributed according to the standard normal distribution, so we can calculate the probability as we did in Example 5.

$$P(-2.69 \leq Z \leq 0.27) = P(Z \leq 0.27) - P(Z \leq -2.69)$$
$$= 0.6064 - 0.0036$$
$$= 0.6028$$

There is a probability of about 60% that the mean of a sample of size 50 would be between 32 and 34 cm. You may want to redo these calculations with larger values of $n$ and verify that the probability approaches 1 as $n$ approaches infinity.

# 13.3 EXERCISES

**Find (a) the mean of the distribution, (b) the standard deviation of the distribution, and (c) the probability that the random variable is between the mean and 1 standard deviation above the mean.**

1. The length (in centimeters) of the leaf of a certain plant is a continuous random variable with probability density function defined by

$$f(x) = \frac{5}{7} \quad \text{for } x \text{ in } [3, 4.4].$$

2. The price of an item (in hundreds of dollars) is a continuous random variable with probability density function defined by

$$f(x) = 4 \quad \text{for } x \text{ in } [2.75, 3].$$

3. The length of time (in years) until a particular radioactive particle decays is a random variable $t$ with probability density function defined by

$$f(t) = 4e^{-4t} \quad \text{for } t \text{ in } [0, \infty).$$

4. The length of time (in years) that a seedling tree survives is a random variable $t$ with probability density function defined by

$$f(t) = 0.05e^{-0.05t} \quad \text{for } t \text{ in } [0, \infty).$$

5. The length of time (in days) required to learn a certain task is a random variable $t$ with probability density function defined by

$$f(t) = \frac{e^{-t/3}}{3} \quad \text{for } t \text{ in } [0, \infty).$$

6. The distance (in meters) that seeds are dispersed from a certain kind of plant is a random variable $x$ with probability density function defined by

$$f(x) = 0.1e^{-0.1x} \quad \text{for } x \text{ in } [0, \infty).$$

**Find the proportion of observations of a standard normal distribution that are between the mean and the given number of standard deviations above the mean.**

7. 3.50                          8. 1.68

**Find the proportion of observations of a standard normal distribution that are between the given z-scores.**

9. 1.28 and 2.05

10. −2.13 and −0.04

**Find a z-score satisfying the conditions given in Exercises 11–14. (Hint: Use the table backwards.)**

11. 10% of the total area is to the left of $z$.

12. 2% of the total area is to the left of $z$.

13. 18% of the total area is to the right of $z$.

14. 22% of the total area is to the right of $z$.

15. Describe the standard normal distribution. What are its characteristics?

16. What is a $z$-score? How is it used?

17. Describe the shape of the graph of each probability distribution.

    **a.** Uniform    **b.** Exponential    **c.** Normal

**Suppose a random variable X has the Poisson distribution with $a = 3.5$. Find each of the following.**

18. $P(2 \le X \le 4)$        19. $P(X = 5)$

20. $P(X \le 3)$        21. $P(X > 3)$

**Find each of the following probabilities for the given sample size, assuming that X is a random variable with a mean of 40 and a standard deviation of 12.**

22. $P(37 \le \overline{X} \le 41)$,   $n = 40$

23. $P(38 \le \overline{X} \le 43)$,   $n = 36$

24. $P(40.5 \le \overline{X} \le 41.5)$,   $n = 48$

25. $P(38.2 \le \overline{X} \le 39.4)$,   $n = 61$

**In the second section of this chapter, we defined the median of a probability distribution as an integral. The median also can be defined as the number $m$ such that $P(X \le m) = P(X \ge m)$.**

26. Find an expression for the median of the uniform distribution.

27. Find an expression for the median of the exponential distribution.

28. Verify the expected value and standard deviation of the exponential distribution given in the text.

29. Prove the z-scores theorem. (*Hint:* Write the formula for the normal distribution with mean $\mu$ and standard deviation $\sigma$, using $t$ instead of $x$ as the variable. Then write the integral representing the area to the left of the value $x$, and make the substitution $u = (t - \mu)/\sigma$.)

30. Show that a normal random variable has inflection points at $x = \mu - \sigma$ and $x = \mu + \sigma$.

31. Use Simpson's rule with $n = 140$, or use the integration feature on a graphing calculator, to approximate the following integrals.

    **a.** $\displaystyle\int_0^{35} 0.5e^{-0.5x}\,dx$

    **b.** $\displaystyle\int_0^{35} 0.5xe^{-0.5x}\,dx$

    **c.** $\displaystyle\int_0^{35} 0.5x^2e^{-0.5x}\,dx$

32. Use your results from Exercise 31 to verify that, for the exponential distribution with $a = 0.5$, the total probability is 1, and both the mean and the standard deviation are equal to $1/a$.

33. Use Simpson's rule with $n = 40$, or the integration feature on a graphing calculator, to approximate the following for the standard normal probability distribution. Use limits of $-6$ and $6$ in place of $-\infty$ and $\infty$.

    **a.** The mean

    **b.** The standard deviation

34. A very important distribution for analyzing the reliability of manufactured goods is the Weibull distribution, whose probability density function is defined by

$$f(x) = abx^{b-1}e^{-ax^b} \quad \text{for } x \text{ in } [0, \infty),$$

where $a$ and $b$ are constants. Notice that when $b = 1$, this reduces to the exponential distribution. The Weibull distribution is more general than the exponential, because it applies even when the failure rate is not constant. Use Simpson's rule with $n = 100$, or the integration feature on a graphing calculator, to approximate the following for the Weibull distribution with $a = 4$ and $b = 1.5$. Use a limit of 3 in place of $\infty$.

    **a.** The mean

    **b.** The standard deviation

35. Determine the cumulative distribution function for the uniform distribution.

36. Determine the cumulative distribution function for the exponential distribution.

## LIFE SCIENCE APPLICATIONS

37. **Insect Life Span** The life span of a certain insect (in days) is uniformly distributed over the interval $[20, 36]$.

    **a.** What is the expected life of this insect?

    **b.** Find the probability that one of these insects, randomly selected, lives longer than 30 days.

38. **Location of a Bee Swarm** A swarm of bees is released from a certain point. The proportion of the swarm located at least 2 m from the point of release after 1 hour is a random variable that is exponentially distributed with $a = 2$.

    **a.** Find the expected proportion under the given conditions.

    **b.** Find the probability that fewer than 1/3 of the bees are located at least 2 m from the release point after 1 hour.

39. **Digestion Time** The digestion time (in hours) of a fixed amount of food is exponentially distributed with $a = 1$.

    **a.** Find the mean digestion time.

    **b.** Find the probability that the digestion time is less than 30 minutes.

40. **Pygmy Heights** The average height of a member of a certain tribe of pygmies is 3.2 ft, with a standard deviation of 0.2 ft. If the heights are normally distributed, what are the largest and smallest heights of the middle 50% of this population?

**41. Finding Prey** H. R. Pulliam found that the time (in minutes) required by a predator to find a prey is a random variable that is exponentially distributed, with $\mu = 25$. *Source: American Naturalist.*

    **a.** According to this distribution, what is the longest time within which the predator will be 90% certain of finding a prey?

    **b.** What is the probability that the predator will have to spend more than 1 hour looking for a prey?

**42. Life Expectancy** According to the National Center for Health Statistics, the life expectancy for a 55-year-old African American female is 26.1 years. Assuming that from age 55, the survival of African American females follows an exponential distribution, determine the following probabilities. *Source: National Vital Statistics Report.*

    **a.** The probability that a randomly selected 55-year-old African American female will live beyond 80 years of age (at least 25 more years)

    **b.** The probability that a randomly selected 55-year-old African American female will live less than 20 more years

**43. Life Expectancy** According to the National Center for Health Statistics, life expectancy for a 70-year-old African American male is 12.3 years. Assuming that from age 70, the survival of African American males follows an exponential distribution, determine the following probabilities. *Source: National Vital Statistics Report.*

    **a.** The probability that a randomly selected 70-year-old African American male will live beyond 90 years of age

    **b.** The probability that a randomly selected 70-year-old African American male will live between 10 and 20 more years

**44. Mercury Poisoning** Historians and biographers have collected evidence that suggests that President Andrew Jackson suffered from mercury poisoning. Recently, researchers measured the amount of mercury in samples of Jackson's hair from 1815. The results of this experiment showed that Jackson had a mean mercury level of 6.0 ppm. *Source: JAMA.*

    **a.** Levels of mercury in hair samples from that time period followed a normal distribution with mean 6.9 and standard deviation 4.6. *Source: Science of the Total Environment.* Find the probability that a randomly selected person from that time period would have a mercury level of 6.0 ppm or higher. Discuss whether this provides evidence that Jackson suffered from mercury poisoning during this time period.

    **b.** Today's accepted normal mercury levels follow a normal distribution with approximate mean 0.6 ppm and standard deviation 0.3 ppm. *Source: Clinical Chemistry.* By today's standards, how likely is it that a randomly selected person from today would have a mercury level of 6.0 ppm or higher? Discuss whether we can conclude from this that Andrew Jackson suffered from mercury poisoning.

**45. Spiders** Researchers have found that 102 spiders scattered over 240 boards were distributed according to the Poisson

distribution with $a = 102/240$. *Source: BioScience.* Find the following probabilities.

    **a.** A board had no spiders.

    **b.** A board had at least one spider.

    **c.** A board had exactly three spiders.

**46. Death Rates** In a recent year, Washington Hospital in Freemont, California, had a death rate of 6.8 of every 100 patients with pneumonia. *Source: San Jose Mercury News.* Assuming these to have a Poisson distribution, find the following probabilities.

    **a.** Exactly 4 of the next 100 patients with pneumonia die.

    **b.** 4 or fewer of the next 100 patients with pneumonia die.

    **c.** More than 4 of the next 100 patients with pneumonia die.

## OTHER APPLICATIONS

**47. Dating a Language** Over time, the number of original basic words in a language tends to decrease as words become obsolete or are replaced with new words. In 1950, C. Feng and M. Swadesh established that of the original 210 basic ancient Chinese words from 950 A.D., 167 were still being used. The proportion of words that remain after $t$ millennia is a random variable that is exponentially distributed with $a = 0.229$. *Source: The UMAP Journal.*

    **a.** Find the life expectancy and standard deviation of a Chinese word.

    **b.** Calculate the probability that a randomly chosen Chinese word will remain after 2000 years.

**48. Rainfall** The rainfall (in inches) in a certain region is uniformly distributed over the interval $[32, 44]$.

    **a.** What is the expected number of inches of rainfall?

    **b.** What is the probability that the rainfall will be between 38 and 40 in.?

**49. Successive Dry Days** Researchers have shown that the number of successive dry days that occur after a rainstorm for particular regions of Catalonia, Spain, is a random variable that is distributed exponentially with a mean of 8 days. *Source: International Journal of Climatology.*

    **a.** Find the probability that 10 or more successive dry days occur after a rainstorm.

    **b.** Find the probability that fewer than 2 dry days occur after a rainstorm.

**50. Earthquakes** The proportion of the times (in days) between major earthquakes in the north-south seismic belt of China is a random variable that is exponentially distributed, with $a = 1/609.5$. *Source: Journal of Seismology.*

    **a.** Find the expected number of days and the standard deviation between major earthquakes for this region.

    **b.** Find the probability that the time between a major earthquake and the next one is more than 1 year.

**51. Soccer** The time between goals (in minutes) for the Wolves soccer team in the English Premier League during a recent season can be approximated by an exponential distribution with $a = 1/90$. *Source: The Mathematical Spectrum.*

    **a.** The Wolves scored their first goal of the season 71 minutes into their first game. Find the probability that the time for a goal is no more than 71 minutes.

    **b.** It was 499 minutes later (in game time) before the Wolves scored their next goal. Find the probability that the time for a goal is 499 minutes or more.

**52. Football** The margin of victory over the point spread (defined as the number of points scored by the favored team minus the number of points scored by the underdog minus the point spread, which is the difference between the previous two, as predicted by oddsmakers) in National Football League games has been found to be normally distributed with mean 0 and standard deviation 13.861. Suppose New England is favored over Miami by 3 points. What is the probability that New England wins? (*Hint:* Calculate the probability that the margin of victory over the point spread is greater than $-3$.) *Source: The American Statistician.*

**53. Customer Expenditures** Customers at a certain pharmacy spend an average of $54.40, with a standard deviation of $13.50. Assuming that the amounts spent are normally distributed, what are the largest and smallest amounts spent by the middle 50% of these customers?

**54. Insured Loss** An insurance policy is written to cover a loss, $X$, where $X$ has a uniform distribution on $[0, 1000]$. At what level must a deductible be set in order for the expected payment to be 25% of what it would be with no deductible? Choose one of the following. (*Hint:* Use a variable, such as $D$, for the deductible. The payment is 0 if the loss is less than $D$, and the loss minus $D$ if the loss is greater than $D$.) *Source: Society of Actuaries.*

    **a.** 250    **b.** 375    **c.** 500    **d.** 625    **e.** 750

**55. High-Risk Drivers** The number of days that elapse between the beginning of a calendar year and the moment a high-risk driver is involved in an accident is exponentially distributed. An insurance company expects that 30% of high-risk drivers will be involved in an accident during the first 50 days of a calendar year. What portion of high-risk drivers are expected to be involved in an accident during the first 80 days of a calendar year? Choose one of the following. *Source: Society of Actuaries.*

    **a.** 0.15    **b.** 0.34    **c.** 0.43    **d.** 0.57    **e.** 0.66

**56. Printer Failure** The lifetime of a printer costing $200 is exponentially distributed with mean of 2 years. The manufacturer agrees to pay a full refund to a buyer if the printer fails during the first year following its purchase, and a one-half refund if it fails during the second year. If the manufacturer sells 100 printers, how much should it expect to pay in refunds? Choose one of the following. *Source: Society of Actuaries.*

    **a.** 6321    **b.** 7358    **c.** 7869    **d.** 10,256    **e.** 12,642

**57. Electronic Device** The time to failure of a component in an electronic device has an exponential distribution with a median of four hours. Calculate the probability that the component will work without failing for at least five hours. Choose one of the following. *Source: Society of Actuaries.*

    **a.** 0.07    **b.** 0.29    **c.** 0.38    **d.** 0.42    **e.** 0.57

**YOUR TURN ANSWERS**

1. 34.5°C, 0.5774
2. (a) 0.9817  (b) 25 and 25  (c) 0.2019
3. (a) 0.4013  (b) 0.3830

# 13   CHAPTER REVIEW

## SUMMARY

In this chapter, we gave a brief introduction to the use of calculus in the study of probability. In particular, the idea of a random variable and its connection to a probability density function and a cumulative distribution function were given. We explored four important concepts:

- expected value (the average value of a random variable that we would expect in the long run),
- variance (a measure of the spread of the values of a distribution),
- standard deviation (the square root of the variance), and
- median (the value of a random variable for which there is a 50% probability of being larger and a 50% probability of being smaller).

Integration techniques were used to determine probabilities, expected value, and variance of continuous random variables. Three probability density functions that have a wide range of applications were studied in detail:

- uniform (when the probability density function remains constant over the sample space),
- exponential (for items that have a constant failure rate over time), and
- normal (for random variables with a bell-shaped distribution).

**Probability Density Function on [a, b]**
1. $f(x) \geq 0$ for all $x$ in the interval $[a, b]$.

2. $\displaystyle\int_a^b f(x)\,dx = 1$.

3. $P(c \leq X \leq d) = \displaystyle\int_c^d f(x)\,dx$ for $c$, $d$ in $[a, b]$.

**Cumulative Distribution Function**   $F(x) = P(X \leq x) = \displaystyle\int_a^x f(t)\,dt$ \quad for $x \geq a$.

**Expected Value for a Density Function on [a, b]**   $E(X) = \mu = \displaystyle\int_a^b xf(x)\,dx$

**Variance for a Density Function on [a, b]**   $\mathrm{Var}(X) = \displaystyle\int_a^b (x - \mu)^2 f(x)\,dx$

**Alternative Formula for Variance**   $\mathrm{Var}(X) = \displaystyle\int_a^b x^2 f(x)\,dx - \mu^2$

**Standard Deviation**   $\sigma = \sqrt{\mathrm{Var}(X)}$

**Median**   The value $m$ such that $\displaystyle\int_a^m f(x)\,dx = \dfrac{1}{2}$

**Uniform Distribution**   $f(x) = \dfrac{1}{b - a}$ on $[a, b]$

$$\mu = \frac{1}{2}(b + a) \qquad \text{and} \qquad \sigma = \frac{1}{\sqrt{12}}(b - a)$$

**Exponential Distribution**   $f(x) = ae^{-ax}$ on $[0, \infty)$

$$\mu = \frac{1}{a} \qquad \text{and} \qquad \sigma = \frac{1}{a}$$

**Poisson Distribution**   $f(x) = \dfrac{a^x e^{-a}}{x!}$ for $x = 0, 1, 2, \ldots$.

$$\mu = a \quad \text{and} \quad \sigma = \sqrt{a}.$$

**Normal Distribution**   $f(x) = \dfrac{1}{\sigma\sqrt{2\pi}} e^{-(x-\mu)^2/(2\sigma^2)}$ on $(-\infty, \infty)$

$$E(X) = \mu \quad \text{and} \quad \mathrm{Var}(X) = \sigma^2$$

**z-Scores Theorem**   For a normal curve with mean $\mu$ and standard deviation $\sigma$, the area to the left of $x$ is the same as the area to the left of

$$z = \frac{x - \mu}{\sigma}$$

for the standard normal curve.

**Central Limit Theorem**   Let $X_1, X_2, \ldots, X_n$ be a random sample from a distribution with mean $\mu$ and standard deviation $\sigma$. Then the distribution of

$$\frac{\overline{X} - \mu}{\sigma/\sqrt{n}}$$

becomes increasingly closer to the standard normal distribution as $n$ gets large.

# KEY TERMS

**13.1**
continuous random variable
continuous probability
  distribution
probability density function
cumulative distribution function

**13.2**
mean
expected value
variance
standard deviation
median

**13.3**
uniform distribution
exponential distribution
Poisson distribution
normal distribution
normal curve

standard normal distribution
$z$-score
random sample
mean of a random sample
Central Limit Theorem

# REVIEW EXERCISES

## CONCEPT CHECK

**Determine whether each of the following statements is true or false, and explain why.**

1. A continuous random variable can take on values greater than 1.

2. A probability density function can take on values greater than 1.

3. A continuous random variable can take on values less than 0.

4. A probability density function can take on values less than 0.

5. The expected value of a random variable must always be at least 0.

6. The variance of a random variable must always be at least 0.

7. The expected value of a uniform random variable is the average of the endpoints of the interval over which the density function is positive.

8. For an exponential random variable, the expected value and standard deviation are always equal.

9. The normal distribution and the exponential distribution have approximately the same shape.

10. In the standard normal distribution, the expected value is 1 and the standard deviation is 0.

## PRACTICE AND EXPLORATIONS

11. In a probability function, the $y$-values (or function values) represent _____.

12. Define a continuous random variable.

13. Give the two conditions that a probability density function for $[a, b]$ must satisfy.

14. In a probability density function, the probability that $X$ equals a specific value, $P(X = c)$, is _____.

**Decide whether each function defined as follows is a probability density function for the given interval.**

15. $f(x) = \sqrt{x}$; $[4, 9]$

16. $f(x) = \dfrac{1}{27}(2x + 4)$; $[1, 4]$

17. $f(x) = 0.7e^{-0.7x}$; $[0, \infty)$

18. $f(x) = 0.4$; $[4, 6.5]$

**In Exercises 19 and 20, find a value of $k$ that will make $f(x)$ define a probability density function for the indicated interval.**

19. $f(x) = kx^2$; $[1, 4]$

20. $f(x) = k\sqrt{x}$; $[4, 9]$

21. The probability density function of a random variable $X$ is defined by

$$f(x) = \frac{1}{10} \quad \text{for } x \text{ in } [10, 20].$$

Find the following probabilities.

a. $P(X \le 12)$

b. $P(X \ge 31/2)$

c. $P(10.8 \le X \le 16.2)$

22. The probability density function of a random variable $X$ is defined by

$$f(x) = 1 - \frac{1}{\sqrt{x - 1}} \quad \text{for } x \text{ in } [2, 5].$$

Find the following probabilities.

a. $P(X \ge 3)$

b. $P(X \le 4)$

c. $P(3 \le X \le 4)$

23. Describe what the expected value or mean of a probability distribution represents geometrically.

24. The probability density functions shown in the following three graphs have the same mean. Which has the smallest standard deviation?

a.

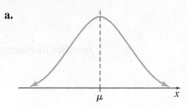

**b.**

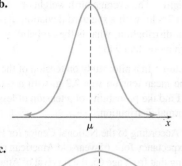

**c.**

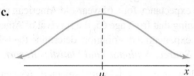

For the probability density functions defined in Exercises 25–28, find (a) the expected value, (b) the variance, (c) the standard deviation, (d) the median, and (e) the cumulative distribution function.

**25.** $f(x) = \frac{2}{9}(x - 2)$; $[2, 5]$    **26.** $f(x) = \frac{1}{5}$; $[4, 9]$

**27.** $f(x) = 5x^{-6}$; $[1, \infty)$

**28.** $f(x) = \frac{1}{20}\left(1 + \frac{3}{\sqrt{x}}\right)$; $[1, 9]$

**29.** The probability density function of a random variable is defined by $f(x) = 4x - 3x^2$ for $x$ in $[0, 1]$. Find the following for the distribution.

   **a.** The mean

   **b.** The standard deviation

   **c.** The probability that the value of the random variable will be less than the mean

   **d.** The probability that the value of the random variable will be within 1 standard deviation of the mean

**30.** Find the median of the random variable of Exercise 29. Then find the probability that the value of the random variable will lie between the median and the mean of the distribution.

For Exercises 31 and 32, find (a) the mean of the distribution, (b) the standard deviation of the distribution, and (c) the probability that the value of the random variable is within 1 standard deviation of the mean.

**31.** $f(x) = 0.01e^{-0.01x}$   for $x$ in $[0, \infty)$

**32.** $f(x) = \frac{5}{112}(1 - x^{-3/2})$   for $x$ in $[1, 25]$

In Exercises 33–40, find the proportion of observations of a standard normal distribution for each region.

**33.** The region to the left of $z = -0.43$

**34.** The region to the right of $z = 1.62$

**35.** The region between $z = -1.17$ and $z = -0.09$

**36.** The region between $z = -1.39$ and $z = 1.28$

**37.** The region that is 1.2 standard deviations or more below the mean

**38.** The region that is up to 2.5 standard deviations above the mean

**39.** Find a $z$-score so that 52% of the area under the normal curve is to the right of $z$.

**40.** Find a $z$-score so that 21% of the area under the normal curve is to the left of $z$.

Suppose a random variable $X$ has the Poisson distribution with $a = 4.2$. Find each of the following.

**41.** $P(3 \leq X \leq 5)$　　　　**42.** $P(X = 7)$

**43.** $P(X \leq 5)$　　　　　　**44.** $P(X > 5)$

Find each of the following probabilities for the given sample size, assuming that $X$ is a random variable with a mean of 60 and a standard deviation of 16.

**45.** $P(58.2 \leq \overline{X} \leq 61.1)$,　$n = 45$

**46.** $P(57.4 \leq \overline{X} \leq 60.6)$,　$n = 52$

**47.** $P(59.3 \leq \overline{X} \leq 61.9)$,　$n = 58$

**48.** $P(56.8 \leq \overline{X} \leq 64.6)$,　$n = 38$

The topics in this short chapter involved much of the material studied earlier in this book, including functions, domain and range, exponential functions, area and integration, improper integrals, integration by parts, and numerical integration. For the following special probability density functions, give

   **a.** the type of distribution;

   **b.** the domain and range;

   **c.** the graph;

   **d.** the mean and standard deviation;

   **e.** $P(\mu - \sigma \leq X \leq \mu + \sigma)$.

**49.** $f(x) = 0.05$   for $x$ in $[10, 30]$

**50.** $f(x) = e^{-x}$   for $x$ in $[0, \infty)$

**51.** $f(x) = \frac{e^{-x^2}}{\sqrt{\pi}}$   for $x$ in $(-\infty, \infty)$ (*Hint:* $\sigma = 1/\sqrt{2}$.)

**52.** The chi-square distribution is important in statistics for testing whether data come from a specified distribution and for testing the independence of two characteristics of a set of data. When a quantity called the *degrees of freedom* is equal to 4, the probability density function is given by

$$f(x) = \frac{xe^{-x/2}}{4} \text{ for } x \text{ in } [0, \infty).$$

   **a.** Verify that this is a probability density function by noting that $f(x) \geq 0$ and by finding $P(0 \leq X < \infty)$.

   **b.** Find $P(0 \leq X \leq 3)$.

**53.** When the degrees of freedom in the chi-square distribution (see the previous exercise) is 1, the probability density function is given by

$$f(x) = \frac{x^{-1/2}e^{-x/2}}{\sqrt{2\pi}} \text{ for } x \text{ in } (0, \infty).$$

Calculating probabilities is now complicated by the fact that the density function cannot be antidifferentiated. Numerical integration is complicated because the density function becomes unbounded as $x$ approaches 0.

**a.** Show that one application of integration by parts (or column integration with just two rows, similar to Example 2 in Section 8.2 on Integration by Parts) allows $P(0 < X \le b)$ to be rewritten as

$$\frac{1}{\sqrt{2\pi}}\left[ 2x^{1/2}e^{-x/2}\Big|_0^b + \int_0^b x^{1/2}e^{-x/2}\,dx \right].$$

**b.** Using Simpson's rule with $n = 12$ in the result from part a, approximate $P(0 < X \le 1)$.

**c.** Using Simpson's rule with $n = 12$ in the result from part a, approximate $P(0 < X \le 10)$.

**d.** What should be the limit as $b \to \infty$ of the expression in part a? Do the results from parts b and c support this?

## LIFE SCIENCE APPLICATIONS

**54. Weight Gain of Rats** The weight gain (in grams) of rats fed a certain vitamin supplement is a continuous random variable with probability density function defined by

$$f(x) = \frac{8}{7}x^{-2} \quad \text{for } x \text{ in } [1, 8].$$

**a.** Find the mean of the distribution.

**b.** Find the standard deviation of the distribution.

**c.** Find the probability that the value of the random variable is within 1 standard deviation of the mean.

**55. Movement of a Released Animal** The distance (in meters) that a certain animal moves away from a release point is exponentially distributed, with a mean of 100 m. Find the probability that the animal will move no farther than 100 m away.

**56. Snowfall** The snowfall (in inches) in a certain area is uniformly distributed over the interval $[2, 30]$.

**a.** What is the expected snowfall?

**b.** What is the probability of getting more than 20 inches of snow?

**57. Body Temperature of a Bird** The body temperature (in degrees Celsius) of a particular species of bird is a continuous random variable with probability density function defined by

$$f(x) = \frac{3}{19{,}696}(x^2 + x) \quad \text{for } x \text{ in } [38, 42].$$

**a.** What is the expected body temperature of this species?

**b.** Find the probability of a body temperature below the mean.

**58. Average Birth Weight** The average birth weight of infants in the United States is 7.8 lb, with a standard deviation of 1.1 lb. Assuming a normal distribution, what is the probability that a newborn will weigh more than 9 lb?

**59. Heart Muscle Tension** In a pilot study on tension of the heart muscle in dogs, the mean tension was 2.2 g, with a standard deviation of 0.4 g. Find the probability of a tension of less than 1.9 g. Assume a normal distribution.

**60. Life Expectancy** According to the National Center for Health Statistics, the life expectancy for a 65-year-old American male is 17.0 years. Assuming that from age 65, the survival of American males follows an exponential distribution, determine the following probabilities. *Source: National Vital Statistics Report.*

**a.** The probability that a randomly selected 65-year-old American male will live beyond 80 years of age (at least 15 more years)

**b.** The probability that a randomly selected 65-year-old American male will live less than 10 more years

**61. Life Expectancy** According to the National Center for Health Statistics, the life expectancy for a 50-year-old American female is 32.5 years. Assuming that from age 50, the survival of American females follows an exponential distribution, determine the following probabilities. *Source: National Vital Statistics Report.*

**a.** The probability that a randomly selected 50-year-old American female will live beyond 90 years of age (at least 40 more years)

**b.** The probability that a randomly selected 50-year-old American female will live between 30 and 50 more years

**62. Assaults** The number of deaths in the United States caused by assault (murder) for each age group is given in the following table. *Source: National Vital Statistics.*

| Age Interval (years) | Midpoint of Interval (year) | Number Dying in Each Interval |
|---|---|---|
| 0–14 | 7 | 957 |
| 15–24 | 19.5 | 4678 |
| 25–34 | 29.5 | 4258 |
| 35–44 | 39.5 | 2473 |
| 45–54 | 49.5 | 1997 |
| 55–64 | 59.5 | 1065 |
| 65–74 | 69.5 | 452 |
| 75–84 | 79.5 | 250 |
| 85+ | 89.5 (est.) | 112 |
| Total | | 16,242 |

**a.** Plot the data. What type of function appears to best match these data?

**b.** Use the regression feature on your graphing calculator to find a quartic equation that models the number of years, $t$, since birth and the number of deaths caused by assault, $N(t)$. Use the midpoint value to estimate the point in each interval when the person died. Graph the function with the plot of the data. Does the function resemble the data?

c. By finding an appropriate constant $k$, find a function $S(t) = kN(t)$ that is a probability density function describing the probability of death by assault. (*Hint:* Because the function in part b is negative for values less than 5.1 and greater than 89.7, restrict the domain of the density function to the interval $[5.1, 89.7]$. That is, integrate the function you found in part b from 5.1 to 89.7.)

d. For a randomly chosen person who was killed by assault, find the probabilities that the person killed was less than 25 years old, at least 45 but less than 65 years old, and at least 75 years old, and compare these with the actual probabilities.

e. Estimate the expected age at which a person will die by assault.

f. Find the standard deviation of this distribution.

63. **Yeast Cells**   The famous statistician William Gosset, who worked for Guinness brewery in Ireland, took measurements of the number of yeast cells per square in a hemocytometer. *Source: Statistical Methods in Biology.* The data can be closely described by a Poisson distribution with $a = 4.68$. Find probabilities of each of the following events.

a. Fewer than 3 yeast cells

b. Exactly 3 yeast cells

c. At least 3 yeast cells

## OTHER APPLICATIONS

64. **Earthquakes**   The time between major earthquakes in the Taiwan region is a random variable with probability density function defined by

$$f(t) = \frac{1}{3650.1}e^{-t/3650.1},$$

where $t$ is measured in days. Find the expected value and standard deviation of this probability density function. *Source: Journal of Seismology.*

65. **State-Run Lotteries**   The average state "take" on lotteries is 40%, with a standard deviation of 13%. Assuming a normal distribution, what is the probability that a state-run lottery will have a "take" of more than 50%?

66. **Equipment Insurance**   A piece of equipment is being insured against early failure. The time from purchase until failure of the equipment is exponentially distributed with mean 10 years. The insurance will pay an amount $x$ if the equipment fails during the first year, and it will pay $0.5x$ if failure occurs during the second or third year. If failure occurs after the first three years, no payment will be made. At what level must $x$ be set if the expected payment made under this insurance is to be 1000? *Source: Society of Actuaries.* Choose one of the following.

a. 3858   b. 4449   c. 5382   d. 5644   e. 7235

# EXTENDED APPLICATION

## EXPONENTIAL WAITING TIMES

We have seen in this chapter how probabilities that are spread out over continuous time intervals can be modeled by continuous probability density functions. The exponential distribution you met in the last section of this chapter is often used to model *waiting times*, the gaps between events that are randomly distributed in time, such as decays of a radioactive nucleus or arrivals of customers in the waiting line at a bank. In this application we investigate some properties of the exponential family of distributions.

Suppose that in a badly run subway system, the times between arrivals of subway trains at your station are exponentially distributed with a mean of 10 minutes. Sometimes trains arrive very close together, sometimes far apart, but if you keep track over many days, you'll find that the *average* time between trains is 10 minutes. According to the last section of this chapter, the exponential distribution with density function $f(t) = ae^{-at}$ has mean $1/a$, so the probability density function for our interarrival times is

$$f(t) = \frac{1}{10}e^{-t/10}.$$

First let's see what these waiting times look like. We have used a random-number generator from a statistical software package to draw 25 waiting times from this distribution. Figure 19 shows cumulative arrival times, which is what you would observe if you recorded the arrival time of each train measured in minutes from an arbitrary 0 point.

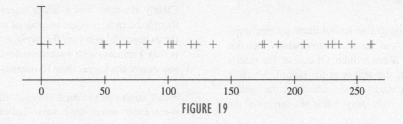

FIGURE 19

You can see that 25 trains arrive in a span of about 260 minutes, so the average interarrival time was indeed close to 10 minutes. You may also notice that there are some large gaps and some cases where trains arrived very close together.

To get a better feeling for the distribution of long and short interarrival times, look at the following list, which gives the 25 interarrival times in minutes, sorted from smallest to largest.

| | | |
|---|---|---|
| 0.016 | 4.398 | 15.659 |
| 0.226 | 4.573 | 15.954 |
| 0.457 | 5.415 | 16.403 |
| 0.989 | 9.570 | 18.978 |
| 1.576 | 10.413 | 20.736 |
| 1.988 | 10.916 | 33.013 |
| 2.738 | 13.109 | 39.073 |
| 3.133 | 13.317 | |
| 3.895 | 14.622 | |

You can see that there were some very short waits. (In fact, the shortest time between trains is only 1 second, which means our model needs to be adjusted somehow to allow for the time trains spend stopped in the station.) The longest time between trains was 39 minutes, almost four times as long as the average! Although the exponential model exaggerates the irregularities of typical subway service, the problem of pile-ups and long gaps is very real for public transportation, especially for bus routes that are subject to unpredictable traffic delays. Anyone who works at a customer service job is also familiar with this behavior: The waiting line at a bank may be empty for minutes at a stretch, and then several customers walk in at nearly the same time. In this case, the customer interarrival times are exponentially distributed.

Planners who are involved with scheduling need to understand this "clumping" behavior. One way to explore it is to find probabilities for ranges of interarrival times. Here integrals are the natural tool. For example, if we want to estimate the fraction of interarrival times that will be less than 2 minutes, we compute

$$\frac{1}{10}\int_0^2 e^{-t/10}\, dt = 1 - e^{-1/5} \approx 0.1813.$$

So, on average, 18% of the interarrival times will be less than 2 minutes, which indicates that clustering of trains will be a problem in our system. (If you have ridden a system like the one in New York City, you may have boarded a train that was ordered to "stand by" for several minutes to spread out a cluster of trains.) We can also compute the probability of a gap of 30 minutes or longer. It will be

$$\frac{1}{10}\int_{30}^{\infty} e^{-t/10}\, dt = e^{-3} \approx 0.0498.$$

So in a random sample of 25 interarrival times we might expect one or two long waits, and our simulation, which includes times of 33 and 39 minutes, is not a fluke. Of course, the rider's experience depends on when she arrives at the station, which is another random input to our model. If she arrives in the middle of a cluster, she'll get a train right away, but if she arrives at the

beginning of a long gap she may have a half-hour wait. So we would also like to model the rider's *waiting time*, the time between the rider's arrival at the station and the arrival of the next train.

A remarkable fact about the exponential distribution is that if our passenger arrives at the station at a random time, the distribution of the rider's waiting times is *the same* as the distribution of interarrival times (that is, exponential with mean 10 minutes). At first this seems paradoxical; since she usually arrives between trains, she should wait less, on average, than the average time between trains. But remember that she's more likely to arrive at the station in one of those long gaps. In our simulation, 72 of 260 minutes is taken up with long gaps, and even if the rider arrives at the middle of such a gap she'll still wait longer than 15 minutes. Because of this feature the exponential distribution is often called *memoryless*: If you dip into the process at random, it is as if you were starting all over. If you arrive at the station just as a train leaves, your waiting time for the next one still has an exponential distribution with mean 10 minutes. The next train doesn't "know" anything about the one that just left.*

Because the riders' waiting times are exponential, the calculations we have already made tell us what riders will experience: A wait of less than 2 minutes has probability about 0.18. The average wait is 10 minutes, but long waits of more than 30 minutes are not all that rare (probability about 0.05).

Customers waiting for service care about the average wait, but they may care even more about the *predictability* of the wait. In this chapter we stated that the standard deviation for an exponential distribution is the same as the mean, so in our model the standard deviation of riders' waiting times will be 10 minutes. This indicates that a wait of twice the average length is not a rare event. (See Exercise 3.)

Let's compare the experience of riders on our exponential subway with the experience of riders of a perfectly regulated service in which trains arrive *exactly* 10 minutes apart. We'll still assume that the passenger arrives at random. But now the waiting time is uniformly distributed on the time interval [0 minutes, 10 minutes]. This uniform distribution has density function

$$f(t) = \begin{cases} \dfrac{1}{10} & \text{for } 0 \le t \le 10 \\[2mm] 0 & \text{otherwise.} \end{cases}$$

The mean waiting time is

$$\int_0^{10} \frac{1}{10} \cdot t\, dt = 5 \text{ minutes}$$

and the standard deviation of the waiting times is

$$\sqrt{\int_0^{10} \frac{(t-5)^2}{10}\, dt} = \sqrt{\frac{25}{3}} \approx 2.89 \text{ minutes.}$$

Clearly the rider has a better experience on this system. Even though the same average number of trains is running per hour as in the exponential subway, the average wait for the uniform subway is only 5 minutes with a standard deviation of 2.89 minutes, and no one ever waits longer than 10 minutes!

*See Chapter 1 in Volume 2 of Feller, William, *An Introduction to Probability Theory and Its Applications*, 2nd ed., New York: Wiley, 1971.

Any subway run is subject to unpredictable accidents and variations, and this random input is always pushing the riders' waiting times toward the exponential model. Indeed, even with uniform scheduling of trains, there will be service bottlenecks because the exponential distribution is also a reasonable model (over a short time period) for interarrival times of *passengers* entering the station. The goal of schedulers is to move passengers efficiently in spite of random train delays and random input of passengers. One proposed solution, the PRT or personal rapid transit system, uses small vehicles holding just a few passengers that can be scheduled to match a fluctuating demand.

The subway scheduling problem is part of a branch of statistics called *queueing theory*, the study of any process in which inputs arrive at a service point and wait in a line or queue to be served. Examples include telephone calls arriving at a customer service center, our passengers entering the subway station, packets of information traveling through the Internet, and even pieces of code waiting for a processor in a multiprocessor computer. The following websites provide a small sampling of work in this very active research area.

- ■ http://web2.uwindsor.ca/math/hlynka/queue.html (A collection of information on queueing theory)
- ■ http://faculty.washington.edu/jbs/itrans/ingsim.htm (an article on scheduling a PRT)

## EXERCISES

1. If $X$ is a continuous random variable, $P(a \leq X \leq b)$ is the same as $P(a < X < b)$. Since these are different events, how can they have the same probability?

2. Someone who rides the subway back and forth to work each weekday makes about 40 trips a month. On the exponential subway, how many times a month can this commuter expect a wait longer than half an hour?

3. Find the probability that a rider of the exponential subway waits more than 20 minutes for a train; that is, find the probability of a wait more than twice as long as the average.

4. On the exponential subway, what is the probability that a randomly arriving passenger has a wait of between 9 and 10 minutes? What is the corresponding probability on the uniform subway?

5. If our system is aiming for an average interarrival time of 10 minutes, we might set a tolerance of plus or minus 2 minutes and try to keep the interarrival times between 8 and 12 minutes. Under the exponential model, what fraction of interarrival times fall in this range? How about under the uniform model?

6. Most mathematical software includes routines for generating "pseudo-random" numbers (that is, numbers that behave randomly even though they are generated by arithmetic). That's what we used to simulate the exponential waiting times for our subway system. But a source on the Internet (http://www.fourmilab.ch/hotbits/) delivers random numbers based on the times between decay events in a sample of Krypton-85. As noted above, the waiting times between decay events have an exponential distribution, so we can see what nature's random numbers look like. Here's a short sample:

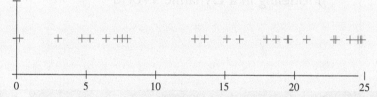

Actually, this source builds its random numbers from random bits, that is, 0's and 1's that occur with equal probability. See if you can think of a way of turning a sequence of exponential waiting times into a random sequence of 0's and 1's.

## DIRECTIONS FOR GROUP PROJECT

*Find a situation in which you and your group can gather actual wait times, such as a bus stop, doctor's office, teller line at a bank, or check-out line at a grocery store. Collect data on interarrival/ service times and determine the mean service time. Using this average, determine whether the data appear to follow an exponential distribution. Develop a table that lists the percentage of the time that particular waiting times occur using both the data and the exponential function. Construct a poster that could be placed near the location where people wait that estimates the waiting time for service.*

# 14

# Discrete Dynamical Systems

14.1 Sequences

14.2 Equilibrium Points

14.3 Determining Stability

Chapter 14 Review

Extended Application: Mathematical Modeling in a Dynamic World

Declining fish stocks throughout the world are a concern not only to those who enjoy eating fish but also to anyone concerned about the health of our planet. Mathematical models help us understand how fishing policies affect the fish population. In this chapter we study several mathematical models for salmon fisheries.

Throughout most of this book we deal with instantaneous rates of change. In the chapter on Differential Equations, for example, we saw how a population described by a differentiable function might be determined if we know the instantaneous rate of change of the population with respect to time. In many situations in biology, however, the population changes only at discrete moments in time, such as each spring when new members of the population are born. Such situations are often described by a **discrete dynamical system**, in which the population at a certain stage is determined by the population at a previous stage. Dynamical systems are an important area of pure mathematical research as well, but in this chapter we will focus on what they tell us about population biology.

# 14.1   Sequences

APPLY IT   If we know the size of a fish population this year, how can we use this information to predict the population for the next four years?
*We will answer this question in Example 4 using recursive sequences.*

A function whose domain is the set of natural numbers, such as

$$a(n) = 2n \quad \text{for } n = 1, 2, 3, 4, \ldots,$$

is a **sequence**. The sequence $a(n) = 2n$ can be written by listing its *terms*, 2, 4, 6, 8, ..., $2n, \ldots$. The letter $n$ is used instead of $x$ as a variable to emphasize the fact that the domain includes only natural numbers. For the same reason, $a$ is used instead of $f$ to name the function. Sequences have many different applications; as we shall see in this chapter, one important example in the life sciences is the prediction of next year's population based on this year's population.

In our definition of sequence we used the example $a(n) = 2n$. The range values of this sequence function,

$$a(1) = 2, \quad a(2) = 4, \quad a(3) = 6, \ldots,$$

are called the **elements** or **terms** of the sequence. Instead of writing $a(4)$ for the fourth term of a sequence, it is customary to write $a_4$; for the sequence above,

$$a_4 = 8.$$

In the same way, for the sequence above, $a_1 = 2, a_2 = 4, a_8 = 16, a_{20} = 40$, and $a_{51} = 102$.

The symbol $a_n$ is used for the **general** or **$n$th term** of a sequence. For example, for the sequence 4, 7, 10, 13, 16, ... the general term might be given by $a_n = 1 + 3n$. This formula for $a_n$ can be used to find any term of the sequence that might be needed. For example, the first three terms of the sequence are

$$a_1 = 1 + 3(1) = 4, \quad a_2 = 1 + 3(2) = 7, \quad \text{and} \quad a_3 = 1 + 3(3) = 10.$$

Also, $a_8 = 25$ and $a_{12} = 37$.

**Sequences**

Find the first four terms of the sequence having general term $a_n = -4n + 2$.

**SOLUTION** Replace $n$, in turn, with 1, 2, 3, and 4.

$$\text{If } n = 1, \quad a_1 = -4(1) + 2 = -4 + 2 = -2.$$
$$\text{If } n = 2, \quad a_2 = -4(2) + 2 = -8 + 2 = -6.$$
$$\text{If } n = 3, \quad a_3 = -4(3) + 2 = -12 + 2 = -10.$$
$$\text{If } n = 4, \quad a_4 = -4(4) + 2 = -16 + 2 = -14.$$

The first four terms of this sequence are $-2, -6, -10$, and $-14$.    **TRY YOUR TURN 1**

**YOUR TURN 1**  Find the first four terms of the sequence having general term $a_n = 6n + 7$.

Often sequences are given in **recursive** form. Rather than giving a formula for $a_n$ as a function of $n$, it is given as a function of $a_{n-1}$, the previous term. In other words, $a_n = f(a_{n-1})$, as in the following example.

**Recursive Sequences**

For the sequence with $a_n = 3a_{n-1} + 2$ and $a_1 = 2$, find the next four terms.

**SOLUTION** Replace $n$, in turn, with 2, 3, 4, and 5.

$$a_2 = 3a_1 + 2 = 3(2) + 2 = 8$$
$$a_3 = 3a_2 + 2 = 3(8) + 2 = 26$$
$$a_4 = 3a_3 + 2 = 3(26) + 2 = 80$$
$$a_5 = 3a_4 + 2 = 3(80) + 2 = 242$$

The first five terms of this sequence are 2, 8, 26, 80, and 242.    **TRY YOUR TURN 2**

**YOUR TURN 2**  Find the next four terms of the sequence $a_n = -2a_{n-1} + 1$ with $a_1 = 3$.

The recursive sequence in the previous example could also be thought of as repeated function composition of the function $f(x) = 3x + 2$. Notice that $a_2 = f(a_1)$, $a_3 = f(a_2) = f(f(a_1))$, and so forth. To simplify the notation, let us denote $f(f(a_1))$ as $f^2(a_1)$ (not to be confused with $f^{(2)}(a_1)$, which represents the second derivative of $f$ evaluated at $a_1$).

$f^n(x)$ represents $f(f(\cdots f(x) \cdots))$, that is, $f$ composed with itself $n$ times. We will say that $f$ is **iterated** $n$ times.

**Iterated Function**

For each of the following functions, find $f^2(x)$, $f^3(x)$, and $f^n(x)$.
**(a)** $f(x) = x + 1$

**SOLUTION**

$$f^2(x) = f(f(x))$$
$$= f(x + 1) \qquad \text{Let } f(x) = x + 1.$$
$$= (x + 1) + 1$$
$$= x + 2$$
$$f^3(x) = f(f^2(x))$$
$$= f(x + 2) \qquad \text{Use the formula for } f^2(x) \text{ just found.}$$
$$= (x + 2) + 1$$
$$= x + 3$$

Continuing in this manner, we see that $f^n(x) = x + n$. This result can be proven more formally with a technique known as *mathematical induction*, which is covered in many college algebra books.[1]

**(b)** $f(x) = 5x$

**SOLUTION**

$$f^2(x) = f(f(x))$$
$$= f(5x) \qquad \text{Let } f(x) = 5x.$$
$$= 5(5x)$$
$$= 5^2 x$$

This last expression could be written as $25x$, but sometimes the factored form makes it easier to find the general term.

$$f^3(x) = f(f^2(x))$$
$$= f(5^2 x) \qquad \text{Use the formula for } f^2(x) \text{ that we just found.}$$
$$= 5(5^2 x)$$
$$= 5^3 x \quad \text{or} \quad 125x.$$

**YOUR TURN 3** For the function $f(x) = x - 1$, find $f^2(x)$, $f^3(x)$, and $f^n(x)$.

Continuing in this manner, we see that $f^n(x) = 5^n x$. **TRY YOUR TURN 3**

Sequences are important in the life sciences for describing the size of populations as well as their long-term trends. For example, suppose a species of fish lays eggs every spring. Putting aside the issue of random variation (a rather large issue to put aside, but necessary if we want to keep things simple for now), it is reasonable to expect the population next year to be a function of the population this year. Label this function $f$, so that $f(x)$ gives the population next year when the population this year is $x$. Letting $a_n$ be the size of the population in year $n$, we have $a_{n+1} = f(a_n)$. If we label as year 1 the first year in which we keep track of the population, then $a_2 = f(a_1)$, $a_3 = f(a_2) = f^2(a_1)$, and $a_4 = f(a_3) = f^3(a_1)$. In general, $a_{n+1} = f(a_n) = f^n(a_1)$.

Let's consider what the function $f$ might look like. If this year's population is too small to lay many eggs, next year's population will also be small. In fact, $f(0) = 0$; if all the fish die, they can't breed to produce any for next year. Also, if the population is too large, overcrowding may cause some fish to starve, so next year's population will be small. Somewhere in between there should be a population that's just the right size, producing a similar, nicely sized population next year.

One function showing such behavior is the logistic growth model described in Section 11.1 on Solutions of Elementary and Separable Differential Equations. In that section we were considering a population that grows continuously, so we used a differential equation. We now consider a population that produces a new generation once a year, so we change the differential equation to a **difference equation**, in which $a_{n+1}$ is given in terms of $a_n$:

$$a_{n+1} = f(a_n) = k\left(1 - \frac{a_n}{N}\right)a_n,$$

where $k$ is a positive constant measuring the growth rate and $N$ is the maximum size of the population. More generally, a difference equation is an equation in which $a_{n+1}$ depends on one or more of the previous terms $a_n$, $a_{n-1}$, etc., and possibly on $n$ itself. An example of such an equation is $a_{n+1} = na_n + 5a_{n-1} + 2a_{n-2}$. For simplicity, in this chapter we will restrict ourselves to difference equations in which $a_{n+1}$ depends only on $a_n$. Such an equation is known as a **first-order difference equation with constant coefficients**.

---

[1]For example, see *College Algebra and Trigonometry*, 11th ed., by Margaret L. Lial, John Hornsby, and David I. Schneider, Pearson, 2013, Sec. 7.5.

FOR REVIEW
To review the graphs of quadratic functions, see Section 1.4. We saw in that section that the graph of a parabola is symmetric with respect to a vertical line through the vertex. Therefore, the x-coordinate of the vertex is halfway between the two x-intercepts.

In our difference equation above, for which $f(x) = k(1 - x/N)x = kx - kx^2/N$, notice that $f(0) = f(N) = 0$. Notice also that $f(x)$ is a quadratic function whose graph is a parabola opening downward, as in Figure 1. The x-coordinate of the vertex is $N/2$ (halfway between 0 and $N$), and the y-coordinate is $f(N/2) = kN/4$.

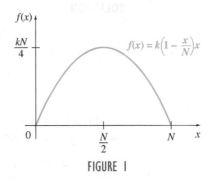

FIGURE 1

It is important to note that both the horizontal and vertical axes in Figure 1 are in terms of population size. We can think of the horizontal axis as representing the size of last year's population, so the vertical axis represents the size of this year's population. The horizontal axis could just as well represent the size of this year's population, so the vertical axis represents the size of next year's population. (Previous chapters have contained many graphs with time on the horizontal axis and population size on the vertical axis, but that is *not* what we are considering here.)

**APPLY IT**   **EXAMPLE 4**   **Logistic Growth**

Suppose a fish population grows according to the logistic growth model, with $N = 1000$ and $k = 3$. Given that the population begins at $a_1 = 200$, find the population in the next four years.

**SOLUTION**   For this problem, $f(x) = 3(1 - x/1000)x$.

$$a_2 = f(a_1) = 3\left(1 - \frac{200}{1000}\right)200 = 480$$

$$a_3 = f(a_2) = 3\left(1 - \frac{480}{1000}\right)480 = 748.8 \approx 749$$

$$a_4 = f(a_3) = 3\left(1 - \frac{749}{1000}\right)749 = 563.997 \approx 564$$

$$a_5 = f(a_4) = 3\left(1 - \frac{564}{1000}\right)564 = 737.712 \approx 738$$

**YOUR TURN 4**   For the logistic growth model with $N = 2500$, $k = 1.6$, and $a_1 = 1200$, find the population in the next four years. Be sure to round the population each year to the nearest integer.

The population in the next four years is 480, 749, 564, and 738. At each step, we have rounded the population to the nearest whole number, since the number of fish must be a whole number.                 TRY YOUR TURN 4

# 14.1   EXERCISES

**List the first 4 terms of the sequence satisfying each of the following conditions.**

1. $a_n = 5n + 2$
2. $a_n = -7n + 12$
3. $a_n = 2(3^n)$
4. $a_n = 3(2^n)$
5. $a_n = \sin n$
6. $a_n = \cos n$

**Find the next 4 terms of the sequence satisfying each of the following conditions.**

7. $a_n = 2a_{n-1} - 3$, $a_1 = 4$
8. $a_n = -3a_{n-1} + 11$, $a_1 = 2$
9. $a_n = -5a_{n-1} + 2$, $a_1 = -1$

**10.** $a_n = 4a_{n-1} - 7, a_1 = 6$

**11.** $a_n = a_{n-1}^2 - 4, a_1 = 3$

**12.** $a_n = (a_{n-1} - 1)^2, a_1 = 3$

**For each function, find $f^2(x), f^3(x)$, and $f^n(x)$.**

**13.** $f(x) = x + 2$       **14.** $f(x) = x - 3$

**15.** $f(x) = 2x$       **16.** $f(x) = -3x$

**17.** $f(x) = 1$       **18.** $f(x) = 2$

**Apply the logistic growth model with the following sets of values to find the population in the next four generations. Be sure to round the population each year to the nearest integer.**

**19.** $N = 3000, k = 1.5, a_1 = 1500$

**20.** $N = 3600, k = 2.8, a_1 = 600$

**21.** $N = 4000, k = 2.5, a_1 = 1800$

**22.** $N = 4800, k = 3.2, a_1 = 1200$

**23.** $N = 6000, k = 3.4, a_1 = 5000$

**24.** $N = 8000, k = 3.8, a_1 = 1000$

## LIFE SCIENCE APPLICATIONS

**25. Ricker Model** Another model of population growth that has been used to model salmon is the Ricker model, given by

$$f(x) = xe^{r(1-x/N)},$$

where $r$ is a positive constant measuring growth rate, and $N$ is a value of the population such that $f(N) = N$. **Source: Journal of the Fisheries Research Board of Canada.** We will see the significance of such points in the next section.

**a.** Find $\lim_{x \to \infty} f(x)$. (*Hint:* See Exercise 72 in Section 3.1.)

**b.** Show that $f'(x) = e^{r(1-x/N)}(1 - rx/N)$.

**c.** Using the result of part b, for what values of $x$ is $f$ increasing? Decreasing? At what value of $x$ does $f$ have an absolute maximum?

**d.** Show that $f''(x) = e^{r(1-x/N)}(rx/N - 2)(r/N)$.

**e.** Using the result of part d, for what values of $x$ is $f$ concave upward? Concave downward? At what value of $x$ does $f$ have an inflection point?

**f.** Using the results from parts a–e, draw a graph of this function.

**g.** In your own words, describe in what ways the graph in part f is similar to the graph of the logistic growth model, and in what ways it is different.

**26. Ricker Model** For the Ricker model of Exercise 25, with $N = 1000$, $r = 3$, and an initial population of 200, find the population for the next four years, where $a_n$ is the population in year $n$ and $a_{n+1} = f(a_n)$. Be sure to round the population each year to the nearest integer.

**27. Ricker Model** For the Ricker model of Exercise 25, with $N = 2000$, $r = 2.5$, and an initial population of 500, find the population for the next four years, where $a_n$ is the population in year $n$ and $a_{n+1} = f(a_n)$. Be sure to round the population each year to the nearest integer.

**28. Beverton-Holt Model** Another model of population growth that has been used to model salmon is the Beverton-Holt model, given by

$$f(x) = \frac{rx}{1 + x/b},$$

where $r$ and $b$ are positive constants. **Source: Nature.**

**a.** Find $\lim_{x \to \infty} f(x)$.

**b.** Show that $f'(x) = \dfrac{r}{(1 + x/b)^2}$.

**c.** Using the result of part b, for what values of $x$ is $f$ increasing? Decreasing? Does $f$ have an absolute maximum?

**d.** Show that $f''(x) = \dfrac{-2r}{b(1 + x/b)^3}$.

**e.** Using the result of part d, for what values of $x$ is $f$ concave upward? Concave downward? Does $f$ have an inflection point?

**f.** Using the results from parts a–e, draw a graph of this function.

**g.** In your own words, describe in what ways the graph in part f is similar to the graph of the logistic growth model and to the Ricker model of Exercise 25, and in what ways it is different.

**29. Beverton-Holt Model** For the Beverton-Holt model of Exercise 28, with $b = 1000$, $r = 3$, and an initial population of 500, find the population for the next four years, where $a_n$ is the population in year $n$ and $a_{n+1} = f(a_n)$. Be sure to round the population each year to the nearest integer.

**30. Beverton-Holt Model** For the Beverton-Holt model of Exercise 28, with $b = 2000$, $r = 4.5$, and an initial population of 1500, find the population for the next four years, where $a_n$ is the population in year $n$ and $a_{n+1} = f(a_n)$. Be sure to round the population each year to the nearest integer.

**31. Shepherd Model** The Shepherd model, a modification of the Beverton-Holt model of Exercise 28, is given by

$$f(x) = \frac{rx}{1 + (x/b)^c},$$

where $r$, $b$, and $c$ are positive constants. **Source: Nature.** For Exercises 31–33, assume $c = 2$.

**a.** Find $\lim_{x \to \infty} f(x)$.

**b.** Show that $f'(x) = \dfrac{r[1 - (x/b)^2]}{[1 + (x/b)^2]^2}$.

**c.** Using the result of part b, for what values of $x$ is $f$ increasing? Decreasing? At what value of $x$ does $f$ have an absolute maximum?

**d.** Show that $f''(x) = \dfrac{-2rx[3 - (x/b)^2]}{b^2(1 + x/b)^3}$.

**e.** Using the result of part d, for what values of $x$ is $f$ concave upward? Concave downward? At what value of $x$ does $f$ have an inflection point?

**f.** Using the results from parts a–e, draw a graph of this function.

**g.** In your own words, describe in what ways the graph in part f is similar to the graph of the logistic growth model and to the Ricker model of Exercise 25, and in what ways it is different.

**32. Shepherd Model** For the Shepherd model of Exercise 31, with $b = 1000$, $r = 3$, and an initial population of 500, find the population for the next four years, where $a_n$ is the population in year $n$ and $a_{n+1} = f(a_n)$. Be sure to round the population each year to the nearest integer.

**33. Shepherd Model** For the Shepherd model of Exercise 31, with $b = 2000$, $r = 4.5$, and an initial population of 1500, find the population for the next four years, where $a_n$ is the population in year $n$ and $a_{n+1} = f(a_n)$. Be sure to round the population each year to the nearest integer.

**YOUR TURN ANSWERS**

**1.** 13, 19, 25, 31
**2.** $-5, 11, -21, 43$
**3.** $x - 2, x - 3, x - n$
**4.** 998, 959, 946, 941

# 14.2 Equilibrium Points

**APPLY IT** If a simple model is used to describe the change in a population from one year to the next, will the behavior predicted by the model be simple?

*As we shall see in Example 2, the answer to this question is "No." Even a simple model can result in very complicated behavior.*

Let's return to the logistic model of the previous section:

$$f(x) = kx(1 - x/N).$$

Recall that $N$ is the maximum size that the population can reach. To simplify matters in this section, we will set $N = 1$, so that $x$ measures a fraction of the maximum possible population. For example, if $x = 0.75$, this means that the population is 75% of its maximum value. One advantage of this change is that we now have only one parameter, $k$, to concern us. Also, we no longer need to concern ourselves with rounding values of $x$ to the nearest integer. For the rest of this section, then, the logistic equation will take the form

$$f(x) = kx(1 - x).$$

One question of interest to biologists is whether there is a population size that never changes, that is, for which $x_{n+1} = x_n$, so the population next year is the same as this year. Such a value of $x$ is called an **equilibrium point**. (The term *fixed point* is also used.) We saw this concept in Sections 11.1 and 11.5 in the chapter on Differential Equations, where we found the equilibrium point by setting the derivative equal to 0. Here, to find any equilibrium points, set $f(x) = x$, since $x$ is this year's population and $f(x)$ is next year's.

$$f(x) = x$$
$$kx(1 - x) = x$$
$$kx - kx^2 = x$$
$$0 = kx^2 - (k - 1)x$$
$$0 = x[kx - (k - 1)]$$
$$x = 0 \quad \text{or} \quad x = (k - 1)/k$$

The first solution is trivial. Of course, if there are no fish this year, there won't be any fish next year, assuming no fish enter from outside. The second solution is only meaningful if $k > 1$.

So what happens if $0 < k \le 1$? It's easy to check with a graphing calculator for a specific value of $k$; we will illustrate using a TI-84 Plus. For example, let $k = 0.8$, and suppose $a_1 = 0.6$. We will keep the current population size in the variable X with the command $.6 \to X$. To update the population size, use the command $.8X(1-X) \to X$. Continue to hit ENTER to repeatedly update the population. You should find that

$$a_2 = 0.192,$$
$$a_3 = 0.1241,$$
$$a_4 = 0.0870,$$
$$a_5 = 0.0635,$$

and so forth.

**NOTE** Note that in the calculations on this page, we have rounded the results to four decimal places. However, all calculations were carried out to the accuracy of the calculator. We will follow this standard throughout the rest of this chapter unless indicated otherwise.

Notice that the population seems to be declining toward 0. We can see graphically why this is so. Figure 2 shows a graph of $y = f(x) = 0.8x(1 - x)$ and $y = x$ on the same axes. Suppose we start at $x = a_1 = 0.6$, as shown in Figure 2. Follow a vertical line to see where it intersects the graph $y = f(x)$. At this point, $y = f(a_1) = a_2 = 0.192$. Now follow a horizontal line until it hits the graph of $y = x$. At this point, $x = a_2 = 0.192$, and we repeat the process. The first several steps of the process are shown in Figure 2, and we can see that $x$ is approaching 0. Such a graph is called a **cobweb diagram**. We will see shortly why it has this name. Cobweb diagrams are useful in economics as well as biology.

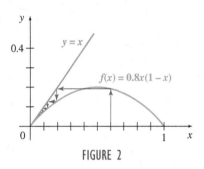

**FIGURE 2**

**TECHNOLOGY**

It's easy to generate graphs such as Figure 2 on a TI-84 Plus. First make sure any StatPlots are off. Then press the MODE button and change the setting from Func to Seq. Press the FORMAT button and change the setting on the first line to Web. Under the Y= menu, set $u(n) = .8u(n-1)(1-u(n-1))$ (or whatever function you want to graph) and set $u(nMin) = \{.6\}$ (or whatever you want the initial value to be). (The "X, T, $\theta$, n" button gives n, and 2nd-7 gives u.) With the window set as in Figure 2 (and making sure PlotStart is set to 1), press GRAPH. After the graph is generated, press TRACE, and then repeatedly press the right arrow key. Press GRAPH at the end to remove the writing from the screen. See the result in Figure 3.

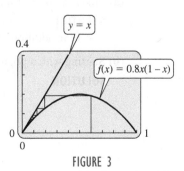

**FIGURE 3**

Now let's see what happens when $k > 1$. As we saw earlier, there are now two fixed points, the trivial one $(x = 0)$ and $x = (k - 1)/k$. For example, consider the case when $k = 2.9$. The fixed point is $(k - 1)/k = (2.9 - 1)/2.9 \approx 0.655$. If the population ever reaches 0.655, it will stay there forever, because $f(0.655) = 0.655$.

As an experiment to see what might happen for the function $f(x) = kx(1 - x)$, with $k = 2.9$ starting with a value of the population other than 0.655, let $a_1 = 0.5$. Then we can generate successive values of the sequence using our calculator as we did before. Verify that the results are as follows:

$$a_2 = 0.725,$$
$$a_3 = 0.5782,$$
$$a_4 = 0.7073,$$
$$a_5 = 0.6004.$$

The values are slowly approaching 0.655, alternating higher and lower. If we follow this sequence long enough, we eventually get as close to 0.655 as we like. For example, $a_{30} = 0.6583$. Once again, the cobweb diagram in Figure 4 shows why this is so. As before, we start with a value of $x$, namely, $a_1 = 0.5$. We follow a vertical line until it intersects the graph of $y = f(x)$. We then follow a horizontal line until it intersects the graph of $y = x$. This gives us a new value whose $x$-coordinate equals the $y$-coordinate of the previous point on the graph. We then repeat the process. Notice that the resulting graph resembles a spider's web, hence the name cobweb diagram.

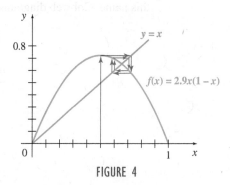

FIGURE 4

For situations in which the population moves toward the equilibrium point, the equilibrium point is said to be **stable**. (Some books use the term *attracting*.)

### EXAMPLE 1  Equilibrium Point

Consider the logistic model with $k = 3$, that is,

$$y = f(x) = 3x(1 - x).$$

(a) Find any equilibrium points.

**SOLUTION**  As always for the logistic model, we have the trivial equilibrium point $x = 0$. In addition, we have $x = (k - 1)/k = (3 - 1)/3 = 2/3 \approx 0.6667$.

(b) Starting with a population $a_1 = 0.5$, find the next four terms in the sequence.

**SOLUTION**

$$a_2 = f(a_1) = 3(0.5)(1 - 0.5) = 0.75$$
$$a_3 = f(a_2) = 0.5625$$
$$a_4 = f(a_3) = 0.7383$$
$$a_5 = f(a_4) = 0.5797$$

(c) Sketch the cobweb diagram for the sequence found in part b.

**SOLUTION**  The cobweb diagram is shown in Figure 5.

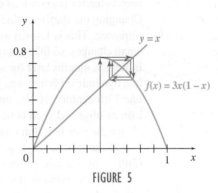

**FIGURE 5**

(d) Determine whether the equilibrium point found in part a is stable or not.

**SOLUTION**  Notice from the results in part b and the cobweb diagram in part c that the population seems to be slowly moving toward the equilibrium point. Generate more points on your own to verify this fact. Thus the equilibrium point appears to be stable.

**APPLY IT**  **EXAMPLE 2**  **Equilibrium Point**

Consider the logistic model with $k = 4$, that is,

$$y = f(x) = 4x(1 - x).$$

(a) Find any equilibrium points.

**SOLUTION**  As always for the logistic model, we have the trivial equilibrium point $x = 0$. In addition, we have $x = (k - 1)/k = (4 - 1)/4 = 3/4 = 0.75$.

(b) Starting with a population $a_1 = 0.6$, find the next five terms in the sequence. Repeat with $a_1 = 0.61$.

**SOLUTION**

$$a_2 = f(a_1) = f(0.6) = 0.96$$
$$a_3 = f(a_2) = 0.1536$$
$$a_4 = f(a_3) = 0.5200$$
$$a_5 = f(a_4) = 0.9984$$
$$a_6 = f(a_5) = 0.0064$$

The equilibrium point in this case is clearly not stable. With $a_1 = 0.61$, which is very close to the previous value of $a_1$, we might expect similar behavior.

$$a_2 = f(a_1) = f(0.61) = 0.9516$$
$$a_3 = f(a_2) = 0.1842$$
$$a_4 = f(a_3) = 0.6012$$
$$a_5 = f(a_4) = 0.9591$$
$$a_6 = f(a_5) = 0.1570$$

**YOUR TURN 1**  Consider the logistic model with $a_1 = 0.7$ and with **a.** $k = 2.8$; **b.** $k = 3.8$.

**i.** Find the nontrivial equilibrium point.

**ii.** Find the next five terms in the sequence.

**iii.** Does the equilibrium point appear to be stable or unstable?

After only five steps, the population is quite different from what it was before. The values of the population no longer approach the equilibrium point, which means that the equilibrium point is **unstable**.  **TRY YOUR TURN 1**

We encourage you to try Example 2 on your calculator and continue to press ENTER, going beyond the values we have generated so far. Any pattern that you might see emerging from the data breaks down as you continue. The numbers appear to be chaotic. In fact, such behavior is known as **chaos**, in which the numbers appear chaotic and without pattern. Changing the starting value very slightly, from 0.6 to 0.61, causes major changes later in the sequence. This is known as the **butterfly effect**, the idea being that weather is sensitive to small changes, so that a butterfly taking off from a flower in Brazil might result in a tornado in Kansas months later (or some other dramatic meteorological event). This type of behavior in a dynamical system was first noticed by the French mathematician Henri Poincaré in the late 19th century. It was only in the 1960s, however, when American meteorologist E. N. Lorenz observed such behavior in a computer program running simple mathematical models for the weather, that chaos began to receive serious study.

In the next section, we will explore more definite ways for determining if an equilibrium point is stable or unstable. We will also see that other types of behavior are possible for a discrete dynamical system. But first, let's look at an example in which $f(x)$ is a third-degree polynomial, to illustrate that equilibrium points can be studied for functions that are not quadratic.

### EXAMPLE 3   Equilibrium Point

Consider the function $f(x) = rx^2(1 - x)$, where $x$ once again represents the fraction of the maximum possible population, so that $0 \le x \le 1$.

**(a)** For what values of $r$ could this function represent next year's population in terms of this year's population?

**SOLUTION**  First, observe that $r > 0$ to ensure that $f(x) > 0$ when $0 < x < 1$. Second, find the maximum value of $f(x)$ to ensure that $f(x) \le 1$. We can do this as we did in Chapter 5 by setting the derivative equal to 0.

$$f(x) = rx^2(1 - x) = rx^2 - rx^3 \qquad \text{Multiply out } f(x).$$
$$f'(x) = 2rx - 3rx^2 = rx(2 - 3x) = 0 \qquad \text{Take the derivative and factor.}$$
$$x = 0 \quad \text{or} \quad x = 2/3$$

Verify that there is a minimum value of 0 at $x = 0$ and a maximum value of $f(2/3) = 4r/27$ at $x = 2/3$. Then

$$\frac{4r}{27} \le 1$$
$$r \le \frac{27}{4}.$$

For this function to be a reasonable model of how the population changes from one year to the next, we require $0 < r \le 27/4$.

**(b)** Find any equilibrium points.

**SOLUTION**  We set $f(x) = x$.

$$rx^2 - rx^3 = x$$
$$0 = rx^3 - rx^2 + x$$
$$= rx\left(x^2 - x + \frac{1}{r}\right)$$

One solution is $x = 0$. Factoring out the $r$ slightly simplifies the process of finding what values of $r$ give real solutions when using the quadratic formula on the quadratic factor, because $r$ appears in only one term.

$$x = \frac{1 \pm \sqrt{1 - 4/r}}{2}$$

A real root requires

$$1 - \frac{4}{r} \geq 0$$

$$1 \geq \frac{4}{r}$$

$$r \geq 4.$$

Therefore, for $4 < r \leq 27/4$ we have three equilibrium points, namely, $x = 0$ and the two values given by the quadratic formula. When $r = 4$, the quadratic formula gives only one root, $x = 1/2$, in addition to $x = 0$. For $0 < r < 4$, the only equilibrium point is $x = 0$.

**(c)** For $r = 5$, use a calculator to explore whether the equilibrium points are stable or unstable.

**SOLUTION**   Using the formula from part b, we calculate that, in addition to $x = 0$, we have equilibrium points at

$$x = \frac{1 \pm \sqrt{1 - 4/5}}{2} \approx 0.2764 \quad \text{or} \quad 0.7236.$$

We pick an arbitrary point in between the equilibrium points 0 and 0.2764, say $a_1 = 0.2$. We calculate

$$a_2 = 5a_1^2(1 - a_1) = 0.16$$
$$a_3 = 0.1075$$
$$a_4 = 0.0516.$$

Verify that for other values of $a_1$ between 0 and 0.2764, the iterations approach 0. It appears that 0 is a stable equilibrium point. We next try a value in the interval $[0.2764, 0.7236]$, say $a_1 = 0.5$.

$$a_2 = 5a_1^2(1 - a_1) = 0.625$$
$$a_3 = 0.7324$$
$$a_4 = 0.7177$$
$$a_5 = 0.7271$$

The points seem to be getting closer and closer to the equilibrium point 0.7236 while oscillating above it and below it. It appears that 0.7236 is a stable equilibrium point. On the other hand, because the iterations move away from 0.2764, it appears to be an unstable equilibrium point. Experiment with other values of $a_1$ to convince yourself that the iterations eventually move either toward $x = 0$ or toward $x = 0.7236$, depending on the initial value $a_1$.

# 14.2   EXERCISES

**Find equilibrium points $x$, $0 \leq x \leq 1$, for each of the following functions.**

**1.** $f(x) = 1 - |2x - 1|$

**2.** $f(x) = \frac{2}{3} - \left|\frac{4}{3}x - \frac{2}{3}\right|$

**3.** $f(x) = 6x^2(1 - x)$

**4.** $f(x) = 6x(1 - x)^2$

**5.** $f(x) = 2x(1 - x^2)$

**6.** $f(x) = \frac{3}{2}x(1 - x^2)$

**For each of the following functions, already studied in Exercises 1–6, find the next six values of the sequence, starting with (a) $x_1 = 0.15$, (b) $x_1 = 0.4$, (c) $x_1 = 0.65$, and (d) $x_1 = 0.85$. Then determine whether each of the equilibrium points found in the corresponding exercise above appears to be stable or unstable.**

**7.** $f(x) = 1 - |2x - 1|$

**8.** $f(x) = \frac{2}{3} - \left|\frac{4}{3}x - \frac{2}{3}\right|$

**9.** $f(x) = 6x^2(1 - x)$

**10.** $f(x) = 6x(1 - x)^2$

**11.** $f(x) = 2x(1 - x^2)$

**12.** $f(x) = \frac{3}{2}x(1 - x^2)$

**For each of the following functions, already studied in Exercises 1–12, draw a cobweb diagram, starting with $x_1 = 0.4$ and iterating four times.**

**13.** $f(x) = 1 - |2x - 1|$     **14.** $f(x) = \frac{2}{3} - \left|\frac{4}{3}x - \frac{2}{3}\right|$

**15.** $f(x) = 6x^2(1 - x)$     **16.** $f(x) = 6x(1 - x)^2$

**17.** $f(x) = 2x(1 - x^2)$     **18.** $f(x) = \frac{3}{2}x(1 - x^2)$

## LIFE SCIENCE APPLICATIONS

**19. Ricker Model** In Exercise 25 of the previous section, we considered the Ricker model, given by

$$f(x) = xe^{r(1-x)},$$

where we have changed $N$ to 1, as in the text.

**a.** Find any equilibrium points.

**b.** For each of the following initial populations and values of $r$, find the next five values of the sequence, and determine whether the equilibrium value is stable or unstable.

   **i.** $a_1 = 0.6, r = 2$     **ii.** $a_1 = 0.6, r = 3$

**c.** Draw a cobweb diagram corresponding to the values in part b.i.

**d.** Draw a cobweb diagram corresponding to the values in part b.ii.

**20. Beverton-Holt Model** In Exercise 28 of the previous section, we considered the Beverton-Holt model, given by

$$f(x) = \frac{rx}{1 + x}.$$

where we have changed $b$ to 1, as in the text.

**a.** Find any equilibrium points.

**b.** For each of the following initial populations and values of $r$, find the next five values of the sequence, and determine whether the nontrivial equilibrium value appears to be stable or unstable.

   **i.** $a_1 = 1.4, r = 2$     **ii.** $a_1 = 2.2, r = 4$

**c.** Draw a cobweb diagram corresponding to the values in part b.i.

**d.** Draw a cobweb diagram corresponding to the values in part b.ii.

**21. Shepherd Model** In Exercise 31 of the previous section, we considered the Shepherd model, given by

$$f(x) = \frac{rx}{1 + x^2},$$

where we have changed $b$ to 1, as in the text, and let $c = 2$ as before.

**a.** Find any equilibrium points.

**b.** For each of the following initial populations and values of $r$, find the next five values of the sequence, and determine whether the nontrivial equilibrium value appears to be stable or unstable.

   **i.** $a_1 = 4, r = 2$     **ii.** $a_1 = 4, r = 10$

**c.** Draw a cobweb diagram corresponding to the values in part b.i.

**d.** Draw a cobweb diagram corresponding to the values in part b.ii.

▬ **YOUR TURN ANSWERS**

**1. a. i.** 0.6429   **ii.** 0.588, 0.6783, 0.6110, 0.6655, 0.6233
**iii.** Stable   **b. i.** 0.7368   **ii.** 0.798, 0.6125, 0.9019, 0.3363,
0.8482   **iii.** Unstable

# 14.3  Determining Stability

APPLY IT  Suppose a population grows according to the logistic model. How can we determine whether any equilibrium points are stable?

*A result in this section will give us an answer to this question.*

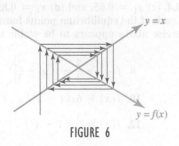

FIGURE 6

In the previous section, we saw that sometimes equilibrium points are stable, so that the population approaches the equilibrium point over time, and sometimes they are unstable, so that the population moves away from the equilibrium point and can even jump around chaotically. In this section, we wish to find methods for determining the nature of any equilibrium points.

As we saw back in Chapter 3, any smooth curve resembles a straight line when viewed closely enough. So let's look closely enough at an equilibrium point so that the graph of $y = f(x)$ appears to be a straight line. In Figure 6, we have shown the neighborhood of the equilibrium point, with $y = f(x)$ having a negative slope that is between $-1$ and $0$. The cobweb diagram starts by going up the longer vertical line on the left, and then spiraling in a clockwise direction toward the equilibrium point. The equilibrium point is stable because the cobweb moves toward it.

In Figure 7, we have a similar graph, but the slope of $f(x)$ is less than $-1$. This time the cobweb starts by going up the longer line near the center of the diagram, and then spirals in a clockwise direction away from the equilibrium point, which must be unstable.

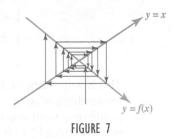

FIGURE 7

In Figure 8, we show a function $y = f(x)$ with a positive slope that is less than 1. Notice how the population moves toward the equilibrium point. In Figure 9, the function $y = f(x)$ has a positive slope that is greater than 1, and the population moves away from the equilibrium point.

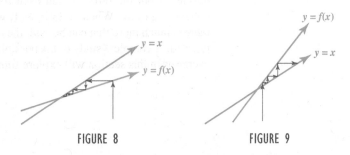

FIGURE 8          FIGURE 9

These results can be summarized by the following theorem.

## Stability of Equilibrium Points

Suppose a function $y = f(x)$ has an equilibrium point at $x = a$. Then the equilibrium point is stable if $\left| f'(a) \right| < 1$ and unstable if $\left| f'(a) \right| > 1$.

The situation is more complicated if $\left| f'(a) \right| = 1$; whether or not the equilibrium point is stable depends upon the nature of the function.

APPLY IT     **EXAMPLE 1**   **Equilibrium Point**

Consider the logistic model

$$y = f(x) = kx(1 - x).$$

Determine the stability of the equilibrium point $x = \dfrac{k - 1}{k}$.

**SOLUTION**   First calculate $f'(x)$:

$$f'(x) = kx(-1) + k(1 - x) = k(1 - 2x).$$

Next, evaluate the derivative at the equilibrium point.

$$f'\left(\frac{k-1}{k}\right) = k\left[1 - 2\left(\frac{k-1}{k}\right)\right] = 2 - k.$$

Now find what values of $k$ result in $\left|f'\left(\frac{k-1}{k}\right)\right| < 1$.

$$|2 - k| < 1$$
$$-1 < 2 - k < 1$$
$$-1 < k - 2 < 1 \qquad \text{Multiply by } -1 \text{ and reverse the order.}$$
$$1 < k < 3$$

**YOUR TURN 1** Find all equilibrium points for $f(x) = x^3 - 3x^2 + 2x$, and determine their stability.

This is why, when we iterated with $k = 2.9$ just before Example 1 of the previous section, the equilibrium point was stable, but with $k = 4$ in Example 2 of the previous section, the equilibrium point was unstable. Notice that with $k = 3$ in Example 1 of the previous section, for which $|f'(x)| = 1$ at the equilibrium point, the equilibrium point was stable.

**TRY YOUR TURN 1**

As we saw in the previous example, the equilibrium point for the logistic model with $k = 3$ has $|f'(x)| = 1$ and is stable. It is *not* the case that an equilibrium point with $|f'(x)| = 1$ is always stable. It can be unstable, or it could be stable from one side and unstable from the other. It can even rotate among a certain number of values, which is known as a **cycle**. When $|f'(x)| > 1$, we know that the equilibrium point is unstable, but there is much more that can be said; the behavior of the dynamical system can be very complicated. A complete study of the possible behavior is beyond the scope of this book, but the exercises in this section will explore some of the possibilities.

## 14.3 EXERCISES

**For each of the following functions studied in exercises in the previous section, use the method of this section to determine the stability of any equilibrium points in the interval [0, 1].**

**1.** $f(x) = 1 - |2x - 1|$

**2.** $f(x) = \frac{2}{3} - \left|\frac{4}{3}x - \frac{2}{3}\right|$

**3.** $f(x) = 6x^2(1 - x)$

**4.** $f(x) = 6x(1 - x)^2$

**5.** $f(x) = 2x(1 - x^2)$

**6.** $f(x) = \frac{3}{2}x(1 - x^2)$

**For each of the following functions, use a graphing calculator to estimate any equilibrium points. Then, using the graphing calculator to estimate the derivative of the function at the equilibrium point, determine whether the equilibrium point is stable or unstable.**

**7.** $f(x) = \cos x$

**8.** $f(x) = \sin 3x$

**9.** $f(x) = 1 - 16\left(x - \frac{1}{2}\right)^4$

**10.** $f(x) = 1 - 64\left(x - \frac{1}{2}\right)^6$

**11.** Consider the function

$$f(x) = 4x^2(1 - x).$$

**a.** Find any equilibrium points where $f(x) = x$.

**b.** Determine the derivative at each of the equilibrium points found in part a.

**c.** What does the theorem on the Stability of Equilibrium Points tell us about each of the equilibrium points found in part a?

**d.** Find the next four iterations of the function for the following starting values.

**i.** $a_1 = 0.4$    **ii.** $a_1 = 0.7$

**e.** Describe the behavior of successive iterations found in part d.

**f.** Discuss how the behavior found in part d relates to the results from part c.

12. Repeat the instructions of Exercise 11 for the function

$$f(x) = 2x^2 + \frac{1}{8}.$$

For part d, use  **i.** $a_1 = 0.2$;  **ii.** $a_1 = 0.3$.

13. Repeat the instructions of Exercise 11 for the function

$$f(x) = x^3 + x.$$

For part d, use  **i.** $a_1 = 0.1$;  **ii.** $a_1 = -0.1$.

14. Repeat the instructions of Exercise 11 for the function

$$f(x) = \frac{1}{x}.$$

For part d, use  **i.** $a_1 = 0.8$;  **ii.** $a_1 = -1.25$.

15. Repeat the instructions of Exercise 11 for the function

$$f(x) = \frac{3}{4} - x^2.$$

For part d, use  **i.** $a_1 = 0.8$;  **ii.** $a_1 = 0.6$.

16. Repeat the instructions of Exercise 11 for the function

$$f(x) = x - 2\ln x.$$

For part d, use  **i.** $a_1 = 0.9$;  **ii.** $a_1 = 1.1$.

When the function has no stable equilibrium points, many outcomes are possible, such as the chaotic behavior described in the previous section. Another outcome is a stable cycle, in which the points get closer and closer to the cycle, where the values rotate among a certain number of values. The following values of *k* lead to a cycle of period 2. Using the logistic function with each of the following values of *k*, and starting with $a_1 = 0.5$, iterate repeatedly until you can determine the two *x*-values in the cycle to three decimal places.

**17.** $k = 3.1$ **18.** $k = 3.2$

The following two exercises are similar to the previous two, except that the values of *k* lead to a cycle of period 4. Starting with $a_1 = 0.5$, iterate repeatedly until you can determine the four *x*-values in the cycle to three decimal places.

**19.** $k = 3.48$ **20.** $k = 3.52$

## LIFE SCIENCE APPLICATIONS

21. **Ricker Model** In Exercise 19 of the previous section, we considered the Ricker model, given by

$$f(x) = xe^{r(1-x)},$$

where *r* is a positive constant. Use the theorem on the Stability of Equilibrium Points to determine the stability of any equilibrium points and how they depend on *r*.

22. **Beverton-Holt Model** In Exercise 20 of the previous section, we considered the Beverton-Holt model, given by

$$f(x) = \frac{rx}{1+x},$$

where *r* is a positive constant. Use the theorem on the Stability of Equilibrium Points to determine the stability of any equilibrium points and how they depend on *r*.

23. **Shepherd Model** In Exercise 21 of the previous section, we considered the Shepherd model, given by

$$f(x) = \frac{rx}{1+x^2},$$

where *r* is a positive constant, and let $c = 2$ as before. Use the theorem on the Stability of Equilibrium Points to determine the stability of any equilibrium points and how they depend on *r*.

24. **Harvesting** None of the models in this chapter involve harvesting. Consider the logistic model in which every year a fixed portion *H* of the maximum population size is harvested.

  **a.** Write a difference equation describing population growth with this harvesting policy.

  **b.** Suppose $H = 0.1$ and $k = 2.9$. Find all equilibrium points, and determine their stability using the theorem from this section. Compare these results with the results from the previous section (before Example 1), when we looked at the case $k = 2.9$ without harvesting.

25. **Harvesting** In the previous exercise, we considered the logistic model with a constant harvest. Suppose instead that each year the size of the harvest is proportional to the population size, where *H* is the constant of proportionality.

  **a.** Write a difference equation describing population growth with this harvesting policy.

  **b.** Suppose $H = 0.1$ and $k = 2.9$. Find all equilibrium points, and determine their stability using the theorem from this section. Compare these results with the results from the previous section (before Example 1), when we looked at the case $k = 2.9$ without harvesting, as well as with the results of the previous exercise.

**YOUR TURN ANSWERS**

1. 0 is unstable; $(3 - \sqrt{5})/2$ is stable; $(3 + \sqrt{5})/2$ is unstable

# 14 CHAPTER REVIEW

## SUMMARY

In this chapter we have looked at linear difference equations, which are a type of discrete dynamical system. We saw how they generate a sequence by iterating a function. We looked at a particular difference equation known as the logistic model, similar to the logistic model described in the chapter on differential equations. We saw

how to find an equilibrium point, and we saw examples of stable and unstable equilibrium points, which we illustrated by means of cobweb diagrams. Finally, we saw how the derivative helps us determine the stability of an equilibrium point.

**Logistic Model**   $a_{n+1} = ka_n(1 - a_n/N)\ (0 \le a_n \le N)$   or   $a_{n+1} = ka_n(1 - a_n)\ (0 \le a_n \le 1)$

**Equilibrium Point**   For the dynamical system $a_{n+1} = f(a_n)$, the equilibrium point is where $f(x) = x$.

**Stability**   Suppose a function $y = f(x)$ has an equilibrium point at $x = a$. Then the equilibrium point is stable if $|f'(a)| < 1$ and unstable if $|f'(a)| > 1$.

## KEY TERMS

discrete dynamical system
**14.1**
sequence
elements (terms)
general (nth) term

recursive
iterated
difference equation
first-order difference equation
   with constant coefficients

**14.2**
equilibrium point
cobweb diagram
stable
unstable

chaos
butterfly effect
**14.3**
cycle

## REVIEW EXERCISES

### CONCEPT CHECK

**For Exercises 1–8 determine whether each of the following statements is true or false, and explain why.**

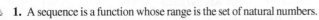

1. A sequence is a function whose range is the set of natural numbers.

2. The equation $a_n = 2^{a_{n-1}}$ is an example of a function defined recursively.

3. The equation $a_{n+1} = 2a_n + 3a_{n-1}$ is an example of a difference equation.

4. The value $x = 0$ is always an equilibrium point for the logistic equation.

5. The logistic equation always has two equilibrium points.

6. For the logistic equation, there is always at least one stable equilibrium point.

7. The value of the derivative always tells us whether or not an equilibrium point is stable.

8. If the equation $x_n = f(x_{n-1})$ has an equilibrium point $x$ with $f'(x) = -1$, the equilibrium point cannot be stable.

**List the first 4 terms of the sequence satisfying each of the following conditions.**

9. $a_n = 6n - 3$

10. $a_n = 8n - 7$

11. $a_n = 3 + (-2)^n$

12. $a_n = \dfrac{4}{2^n}$

**Find the next 4 terms of the sequence satisfying each of the following conditions.**

13. $a_n = -2a_{n-1} - 3, a_1 = 0$

14. $a_n = 3a_{n-1} - 4, a_1 = 1$

15. $a_n = -4a_{n-1} + 2, a_1 = -2$

16. $a_n = 5a_{n-1} - 12, a_1 = 2$

**For each function, find $f^2(x), f^3(x),$ and $f^n(x)$.**

17. $f(x) = 1/x$

18. $f(x) = x$

**Apply the logistic growth model with the following sets of values to find the population in the next four generations. Be sure to round the population each year to the nearest integer.**

**19.** $N = 6000, k = 0.5, a_1 = 3500$

**20.** $N = 7000, k = 1.2, a_1 = 2600$

**21.** $N = 8000, k = 2.3, a_1 = 3200$

**22.** $N = 9000, k = 3.6, a_1 = 2200$

**For each of the following functions, do the following.**

**a.** Find all equilibrium points $x, 0 \le x \le 1$.
**b.** Find the value of $|f'(x)|$ at all equilibrium points.
**c.** Use the theorem in Section 14.3 to determine the stability of all equilibrium points.
**d.** Find the next four values of the sequence, starting with $x_1 = 0.15$.
**e.** Repeat part d, starting with $x_1 = 0.4$.
**f.** Repeat part d, starting with $x_1 = 0.65$.
**g.** Repeat part d, starting with $x_1 = 0.85$.
**h.** Draw a cobweb diagram, starting with $x_1 = 0.4$ and iterating four times.
**i.** Discuss how the results of parts d through g compare with the results of part c.

**23.** $f(x) = \dfrac{1}{4} - \left| \dfrac{x}{2} - \dfrac{1}{4} \right|$  **24.** $f(x) = \dfrac{3}{5} - \left| \dfrac{6}{5}x - \dfrac{3}{5} \right|$

**25.** $f(x) = 4.2x^2(1 - x)$  **26.** $f(x) = 5.2x^2(1 - x)$

## LIFE SCIENCE APPLICATIONS

**27. Population Growth** Suppose a population grows so that each year the population is 10% greater than it was the previous year.

**a.** Write a difference equation for the population in year $n$.

**b.** Find the population in the next four years, given that $a_1 = 100,000$.

**c.** Find a nonrecursive formula for the population in year $n$, given that $a_1 = 100,000$.

**d.** Find any equilibrium points for the equation in part a.

**e.** Find $f'(x)$ for any equilibrium points found in part d.

**f.** Using the theorem from Section 14.3, what can you say about the stability of the equilibrium point found in part c?

**28. Population Growth** Suppose a population grows or declines so that each year the population is a fixed factor $r$ times its value the previous year but, in addition, is increased by a fixed amount $b$ each year. We allow for the possibility that $b < 0$.

**a.** Write a difference equation for the population in year $n$.

**b.** Find any equilibrium points for the equation in part a.

**c.** For what values of $r$ and $b$ does the equation in part a have an equilibrium point that could represent a valid population size?

**d.** For what values of $r$ does the theorem in Section 14.3 guarantee that the equilibrium point found in part b is stable?

**e.** What is the value of $a_n$ if $r = 1$? Express your answer in terms of $a_1$, $n$, and $b$.

**f.** Show that if $r \neq 1$, then the formula

$$a_n = r^n + (1 + r + r^2 + \cdots + r^{n-1})b$$

is a solution to the equation in part a. (*Hint:* What does the formula give for $a_{n-1}$? Multiply this by $r$ and add $b$.)

**29. Malaria** Sir Ronald Ross received the Nobel Prize for Physiology or Medicine for his work on malaria. In this exercise, we explore Ross' mathematical model for the transmission of malaria. *Source: The Prevention of Malaria.*

**a.** Let $p$ be the human population in the region, and let $m_n$ be the fraction infected by malaria in year $n$. The number of people infected in any year equals the number of cases from the previous year, minus the number who have recovered, plus the number of new cases. Let $r$ represent the fraction of those who recover from the disease in any year. Explain why the following equation describes the model.

$$m_n p = m_{n-1} p - r m_{n-1} p + \text{number of new cases}$$

**b.** Let $a$ denote the number of malaria-bearing mosquitoes per human. Let $i$ be the fraction of malaria-infected humans with gametids in their blood (so that a mosquito that bites them can acquire malaria and pass it on to another person). Let $b$ be the fraction of malaria-bearing mosquitoes that succeed in biting a human. Let $s$ be the fraction of malaria-bearing mosquitoes in which the malaria matures. Explain why

$$b^2 s a i m_{n-1} p$$

gives the number of mosquitoes that succeed in infecting humans, and why

$$b^2 s a i m_{n-1} p (1 - m_{n-1})$$

gives the number of new cases of malaria.

**c.** Use the results from parts a and b to show that the equation modeling the number of humans infected by malaria becomes

$$m_n = (1 - r + b^2 sai)m_{n-1} - b^2 sai m_{n-1}^2.$$

**d.** Ross suggested using the values $r = 0.2$ and $b^2 si = 0.005$. In an example he created, he let $m_1 = 0.5$ and $a = 100$. Find the next four terms in the sequence.

**e.** In a second example, Ross let $m_1 = 0.5$ and $a = 60$. Find the next four terms in the sequence.

**f.** In a third example, Ross let $m_1 = 0.5$ and $a = 40$. Find the next four terms in the sequence.

**g.** Using the values $r = 0.2$ and $b^2 si = 0.005$, what are the equilibrium points? One answer will be in terms of $a$; for what values of $a$ is this valid?

**h.** Using the results from part g, what values of $a$ guarantee that the nonzero equilibrium point is stable? Use this answer to explain the results from parts d–f.

**30. Breathing** A simple model for the concentration of a chemical in the lungs is given by

> Amount of the chemical in the lungs
>
> = Amount left over from the previous breath
>
> + Amount breathed in.

*Source: Modeling the Dynamics of Life.*

**a.** Let $c_n$ be the concentration of the chemical after the $n$th breath. Let $a$ be the concentration of the chemical in the ambient air, $0 < a < 1$. Let $V$ be the volume of the lungs, $V_i$ be the volume of a breath, and $V_r$ be the volume remaining in the lungs after breathing out, so that $V = V_i + V_r$, where $V_i > 0$ and $V_r > 0$. Explain why the model can be described by the equation

$$Vc_n = V_i a + V_r c_{n-1}.$$

**b.** Show that

$$c_n = f(c_{n-1}) = pa + (1 - p)c_{n-1},$$

where $p = V_i/V$.

**c.** Suppose $V_i = 1.5 \text{ L}$ and $V_r = 0.5 \text{ L}$. Let oxygen be the chemical, so that $a = 0.2095$, the concentration of oxygen in the atmosphere. Suppose a person has been breathing oxygen-enriched air, so that $c_1 = 0.5$. Find the next four terms in the sequence.

**d.** Show that

$$c_3 = pa[1 + (1 - p)] + (1 - p)^2 c_1,$$
$$c_4 = pa[1 + (1 - p) + (1 - p)^2] + (1 - p)^3 c_1,$$

and, in general,

$$c_n = pa[1 + (1 - p) + \cdots + (1 - p)^{n-2}] + (1 - p)^{n-1}c_1.$$

**e.** From the formula in part b, find the equilibrium point.

**f.** Find $f'(c)$ where $c$ is the equilibrium point from part e, and determine the stability of the equilibrium point.

**g.** Use the formula for the sum of a geometric series

$$1 + r + r^2 + \cdots + r^k = \frac{1 - r^{k+1}}{1 - r}$$

to give a simplified form for $c_n$ from part d.

**h.** Using the result from part g, determine

$$\lim_{n \to \infty} c_n.$$

# EXTENDED APPLICATION

## MATHEMATICAL MODELING IN A DYNAMIC WORLD

You have now been exposed to myriad ways that mathematics is used to better understand the ever-changing world in which we live. To illustrate the complex nature of biological processes, we will focus on the poliomyelitis (polio) epidemic that afflicted many people, including this author's sister.

In 1916, the first large polio epidemic occurred in the United States, with over 27,000 cases. It struck all regions, regardless of economic status, and mostly children. Much like with the more recent AIDS and SARS (severe acute respiratory syndrome) outbreaks, communities reacted with fear. Quarantines were imposed on homes where someone was diagnosed with polio, travel was restricted, and patients with polio were isolated in the hospital. No one was quite sure what was causing this disease. Throughout the first half of the 20th century, the epidemics got worse. *Source: Smithsonian.*

Suppose it is 1952 and that you and a team of people are asked to make predictions about the number of people who would contract polio over the next 20 years. Using the techniques learned in calculus class, you could use existing data to make predictions about the future. Of course, a warning that comes with such futuristic predictions is that the situation may change and all of your predictions could be quite inaccurate.

We begin by examining data from the past. The following table shows the number of new cases of polio, including both paralytic and nonparalytic polio, from 1937 through 1952. *Source: Morbidity and Mortality Weekly Report.*

| Year | Reported New Cases |
|------|--------------------|
| 1937 | 9514 |
| 1938 | 1705 |
| 1939 | 7343 |
| 1940 | 9804 |
| 1941 | 9086 |
| 1942 | 4167 |
| 1943 | 12,450 |
| 1944 | 19,029 |
| 1945 | 13,624 |
| 1946 | 25,698 |
| 1947 | 10,827 |
| 1948 | 27,726 |
| 1949 | 42,033 |
| 1950 | 33,306 |
| 1951 | 28,386 |
| 1952 | 57,879 |

Notice that these data fluctuate from year to year, with some outbreaks of polio larger than other outbreaks. This makes predicting the number of future new cases more difficult. However, as Figure 10 indicates, there is an upward trend in the data.

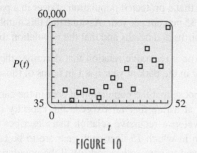

FIGURE 10

Overall, the data appear to be growing exponentially. We enter the data and use the exponential regression feature on a graphing calculator. A model to predict the number of new polio cases, $P$, is found to be

$$P(t) = 12.20(1.172)^t, \qquad \text{(1)}$$

where $t$ is the number of years since 1900. The data with the corresponding curve are shown in Figure 11. The fit is not bad, considering the fluctuations in the data.

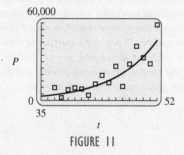

FIGURE 11

According to the model in Equation (1), the number of polio cases is predicted to grow by about 17.2 percent per year. We can therefore predict the number of new cases each year by using the number of cases from the previous year. With the assumption that in 1900 the number of new polio cases was about 12, this relationship is captured by the recursive equation

$$P_{n+1} = 1.172P_n, \qquad \text{(2)}$$

where $P_n$ refers to the number of polio cases in the $n$th year.

Using the regression function (1) to estimate the number of new cases of polio in 1952, we get approximately

$$P(52) = 12.20(1.172)^{52} \approx 46,838.$$

Note that this is an underestimate of the actual number of cases (57,879) in 1952, which is not surprising, given the change in the data. We could use the recursive Equation (2) or the regression Equation (1) to predict that by 1954 we would have about 64,000 new cases of polio. And, by 1956, the equations would predict that there would be nearly 90,000 new cases. At the time, all of the predictive power of mathematics would have suggested that polio in this country was rapidly expanding, with no end in sight.

In 1953, however, things rapidly began to change. Dr. Jonas Salk, an American medical researcher, announced that he had discovered a vaccine that would prevent polio. By 1955, the vaccine was available throughout the United States.

As the table below shows, the number of new cases in 1956 had dropped to 15,140. How could anyone have predicted such a dramatic decrease in cases prior to the introduction of the vaccine? Further, with the up and down fluctuation of the data, how could any mathematical model have made such a prediction?

To place ourselves in the situation of witnessing the effects of the vaccine, suppose that we observed in 1953 the usual decrease in cases from 1952. In 1954 we observed a much smaller increase than we would have anticipated. At this point, would we have been able to see that the tipping point had been reached and the vaccine was making a difference?

| Year | Reported New Cases |
|------|--------------------|
| 1952 | 57,879 |
| 1953 | 35,592 |
| 1954 | 38,476 |
| 1955 | 28,985 |
| 1956 | 15,140 |
| 1957 | 5485 |
| 1958 | 5787 |
| 1959 | 8425 |
| 1960 | 3190 |
| 1961 | 1312 |
| 1962 | 940 |
| 1963 | 449 |
| 1964 | 122 |
| 1965 | 72 |

It is reasonable to think that the percentage of the population that was vaccinated each year would cause the number of new cases to decrease in a declining exponential pattern. In fact, using the data in the preceding table, we can estimate the percentage decrease in new cases of polio after 1953. Using the exponential regression feature on a graphing calculator, we determine a model to predict the number of new polio cases after the vaccine is available, $V$, to be

$$V(t) = 97,894(0.6073)^t, \qquad \text{(3)}$$

where $t$ is the number of years since 1952. The data and the corresponding curve are indicated in Figure 12. Note that the fit in this case is good except for the year 1952.

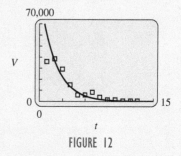

FIGURE 12

As in the earlier data, we can predict that the number of polio cases will decrease by about 40 percent per year after 1952. With the assumption that in 1952, the number of new polio cases was about 98,000 (an overestimate), we can predict the number of cases each year simply by knowing the number of new cases the previous year. This relationship is captured by the recursive equation

$$V_{n+1} = 0.6073V_n, \qquad (4)$$

where $V_n$ refers to the number of polio cases in the $n$th year.

Note that the regression function (4) overestimates the number of new cases in 1952, but soon after, the estimate for the number of new cases becomes more accurate. For example, the number of new cases of polio in 1962 is estimated to be approximately

$$V(10) = 97,894(0.6073)^{10} \approx 668.$$

In just ten years from the peak in the number of new polio cases, the vaccine forced an amazing decline in new cases. This illustrates the amazing power of this vaccine and also gives quite a cautionary note to those who use mathematical modeling to make predictions about the future.

## EXERCISES

1. Use the data from both tables and quadratic, cubic, and quartic regression to find a single function that can be used to accurately describe the polio epidemic. Why is such a single function unlikely to provide an accurate model of this situation?

2. Suppose that a flu epidemic in a community of 1,000,000 is growing by 15 percent per day because one individual was exposed to a virus while traveling. It is also thought that about half of the population has immune systems that will prevent them from getting the flu.

   a. Set up a recursive relation that describes the number of individuals who have the flu on day $n + 1$ in terms of those on day $n$. At this rate, how long will it take until the entire susceptible community has contracted the flu?

   b. Suppose that a flu shot becomes available on day 10 that reduces the rate of growth to 5 percent. How many more days will it take until one-fourth of the population contracts the flu?

3. Suppose that a protected population of deer in a park grows at a rate of 35 percent per year. Assume that this number includes natural births and deaths and that the population today is 500.

   a. Set up a recursive relation that describes the number of deer in the park in year $n + 1$ in terms of those in year $n$.

   b. Suppose that the population of deer in the park are growing at a rate that far exceeds the capacity of the park. Develop a recursive relation that describes a proposed plan in which 15 deer each year are to be captured and moved to some other park. Will the population continue to grow? How many deer would have to be moved so that the population remains at a constant level.

   c. Suppose a second plan is proposed in which $x$ percent of the deer are to be captured and moved to some other park. Find the equilibrium, if one exists.

## DIRECTIONS FOR GROUP PROJECT

*Use the information and the models from this application to develop a mathematical model for some other phenomenon that has experienced drastic change over time. Use data collected from the Internet or other sources to provide future predictions over time.*

*Once your model is developed, create a class presentation that describes the model, your assumptions, results, and future research. Include graphical displays of the results.*

# Appendix A

## Solutions to Prerequisite Skills Diagnostic Test   *(with references to Ch. R)*

For more practice on the material in questions 1–4, see *Beginning and Intermediate Algebra* (5th ed.) by Margaret L. Lial, John Hornsby, and Terry McGinnis, Pearson, 2012.

**1.** $10/50 = 0.20 = 20\%$

**2.**
$$\frac{13}{7} - \frac{2}{5} = \frac{13}{7} \cdot \frac{5}{5} - \frac{2}{5} \cdot \frac{7}{7}$$ 　Get a common denominator.
$$= \frac{65}{35} - \frac{14}{35}$$
$$= \frac{51}{35}$$

**3.** The total number of apples and oranges is $x + y$, so $x + y = 75$.

**4.** The sentence can be rephrased as "The number of students is at least four times the number of professors," or $s \geq 4p$.

**5.**
$$7k + 8 = -4(3 - k)$$
$$7k + 8 = -12 + 4k$$ 　Multiply out.
$$7k - 4k + 8 - 8 = -12 - 8 + 4k - 4k$$ 　Subtract 8 and 4$k$ from both sides.
$$3k = -20$$ 　Simplify.
$$k = -20/3$$ 　Divide both sides by 3.

For more practice, see Sec. R.4.

**6.**
$$\frac{5}{8}x + \frac{1}{16}x = \frac{11}{16} + x$$

$$\frac{5}{8}x + \frac{1}{16}x - x = \frac{11}{16}$$ 　Subtract $x$ from both sides.

$$\frac{5x}{8} \cdot \frac{2}{2} + \frac{x}{16} - x \cdot \frac{16}{16} = \frac{11}{16}$$ 　Get a common denominator.

$$\frac{10x}{16} + \frac{x}{16} - \frac{16x}{16} = \frac{11}{16}$$ 　Simplify.

$$\frac{-5x}{16} = \frac{11}{16}$$ 　Simplify.

$$x = \frac{11}{16} \cdot \frac{16}{-5} = -\frac{11}{5}$$ 　Multiply both sides by the reciprocal of $-5/16$.

For more practice, see Sec. R.4.

**7.** The interval $-2 < x \leq 5$ is written as $(-2, 5]$. For more practice, see Sec. R.5.

**8.** The interval $(-\infty, -3]$ is written as $x \leq -3$. For more practice, see Sec. R.5.

**9.**
$$5(y - 2) + 1 \leq 7y + 8$$
$$5y - 9 \leq 7y + 8$$ 　Multiply out and simplify.
$$5y - 7y - 9 + 9 \leq 7y - 7y + 8 + 9$$ 　Subtract 7$y$ from both sides and add 9.
$$-2y \leq 17$$ 　Simplify.
$$y \geq -17/2$$ 　Divide both sides by $-2$.

**10.**
$$\frac{2}{3}(5p - 3) > \frac{3}{4}(2p + 1)$$

$$\frac{10p}{3} - 2 > \frac{3p}{2} + \frac{3}{4}$$ 　Multiply out and simplify.

$$\frac{10p}{3} - \frac{3p}{2} - 2 + 2 > \frac{3p}{2} - \frac{3p}{2} + \frac{3}{4} + 2$$ 　Subtract 3$p$/2 from both sides, and add 2.

$$\frac{10p}{3} \cdot \frac{2}{2} - \frac{3p}{2} \cdot \frac{3}{3} > \frac{3}{4} + 2 \cdot \frac{4}{4}$$ 　Simplify and get a common denominator.

$$\frac{11p}{6} > \frac{11}{4}$$ 　Simplify.

$$p > \frac{11}{4} \cdot \frac{6}{11}$$ 　Multiply both sides by the reciprocal of 11/6.

$$p > \frac{3}{2}$$ 　Simplify.

For more practice, see Sec. R.5.

**11.** $(5y^2 - 6y - 4) - 2(3y^2 - 5y + 1) = 5y^2 - 6y - 4 - 6y^2 + 10y - 2 = -y^2 + 4y - 6$. For more practice, see Sec. R.1.

**12.** $(x^2 - 2x + 3)(x + 1) = (x^2 - 2x + 3)x + (x^2 - 2x + 3) = x^3 - 2x^2 + 3x + x^2 - 2x + 3 = x^3 - x^2 + x + 3$. For more practice, see Sec. R.1.

**13.** $(a - 2b)^2 = a^2 - 4ab + 4b^2$. For more practice, see Sec. R.1.

**14.** $3pq + 6p^2q + 9pq^2 = 3pq(1 + 2p + 3q)$. For more practice, see Sec. R.2.

**15.** $3x^2 - x - 10 = (3x + 5)(x - 2)$. For more practice, see Sec. R.2.

**16.**
$$\frac{a^2 - 6a}{a^2 - 4} \cdot \frac{a - 2}{a} = \frac{a(a - 6)}{(a - 2)(a + 2)} \cdot \frac{a - 2}{a} \qquad \text{Factor.}$$

$$= \frac{a - 6}{a + 2} \qquad \text{Simplify.}$$

For more practice, see Sec. R.3.

**17.** $\dfrac{x + 3}{x^2 - 1} + \dfrac{2}{x^2 + x} = \dfrac{x + 3}{(x - 1)(x + 1)} + \dfrac{2}{x(x + 1)} \qquad \text{Factor.}$

$$= \frac{x + 3}{(x - 1)(x + 1)} \cdot \frac{x}{x} + \frac{2}{x(x + 1)} \cdot \frac{x - 1}{x - 1} \qquad \text{Get a common denominator.}$$

$$= \frac{x^2 + 3x}{x(x - 1)(x + 1)} + \frac{2x - 2}{x(x + 1)(x - 1)} \qquad \text{Multiply out.}$$

$$= \frac{x^2 + 5x - 2}{x(x - 1)(x + 1)} \qquad \text{Add fractions.}$$

For more practice, see Sec. R.3.

**18.**
$$3x^2 + 4x = 1$$

$$3x^2 + 4x - 1 = 0 \qquad \text{Subtract 1 from both sides.}$$

$$x = \frac{-4 \pm \sqrt{4^2 - 4 \cdot 3(-1)}}{2 \cdot 3} \qquad \text{Use the quadratic formula.}$$

$$= \frac{-4 \pm \sqrt{28}}{6} \qquad \text{Simplify.}$$

$$= \frac{-2 \pm \sqrt{7}}{2} \qquad \text{Simplify.}$$

For more practice, see Sec. R.4.

**19.** First solve the corresponding equation

$$\frac{8z}{z + 3} = 2$$

$$8z = 2(z + 3) \qquad \text{Multiply both sides by } (z + 3).$$
$$8z = 2z + 6 \qquad \text{Multiply out.}$$
$$6z = 6 \qquad \text{Subtract } 2z \text{ from both sides.}$$
$$z = 1 \qquad \text{Divide both sides by 6.}$$

The fraction may also change from being less than 2 to being greater than 2 when the denominator equals 0, namely, at $z = -3$. Testing each of the intervals determined by the numbers $-3$ and 0 shows that the fraction on the left side of the inequality is less than or equal to 2 on $(-\infty, -3) \cup [1, \infty)$. We do not include $x = -3$ in the solution because that would make the denominator 0. For more practice, see Sec. R.5.

**20.**
$$\frac{4^{-1}(x^2y^3)^2}{x^{-2}y^5} = \frac{x^2(x^4y^6)}{4y^5} \qquad \text{Simplify.}$$

$$= \frac{x^6y}{4} \qquad \text{Simplify.}$$

For more practice, see Sec. R.6.

**21.**
$$\frac{4^{1/4}\left(p^{2/3}q^{-1/3}\right)^{-1}}{4^{-1/4}p^{4/3}q^{4/3}} = \frac{4^{1/2}p^{-2/3}q^{1/3}}{p^{4/3}q^{4/3}} \qquad \text{Simplify.}$$

$$= \frac{2}{p^2 q} \qquad \text{Simplify.}$$

For more practice, see Sec. R.6.

**22.**
$$k^{-1} - m^{-1} = \frac{1}{k}\cdot\frac{m}{m} - \frac{1}{m}\cdot\frac{k}{k} \qquad \text{Get a common denominator.}$$

$$= \frac{m - k}{km} \qquad \text{Simplify.}$$

For more practice, see Sec. R.6.

**23.** $(x^2 + 1)^{-1/2}(x + 2) + 3(x^2 + 1)^{1/2} = (x^2 + 1)^{-1/2}[x + 2 + 3(x^2 + 1)] = (x^2 + 1)^{-1/2}(3x^2 + x + 5)$. For more practice, see Sec. R.6.

**24.** $\sqrt[3]{64b^6} = 4b^2$. For more practice, see Sec. R.7.

**25.** $\dfrac{2}{4 - \sqrt{10}}\cdot\dfrac{4 + \sqrt{10}}{4 + \sqrt{10}} = \dfrac{2(4 + \sqrt{10})}{16 - 10} = \dfrac{4 + \sqrt{10}}{3}$. For more practice, see Sec. R.7.

**26.** $\sqrt{y^2 - 10y + 25} = \sqrt{(y - 5)^2} = |y - 5|$. For more practice, see Sec. R.7.

# Appendix B

## Learning Objectives

### CHAPTER R: Algebra Reference

R.1: Polynomials

1. Simplify polynomials

R.2: Factoring

1. Factor polynomials

R.3: Rational Expressions

1. Simplify rational expressions using properties of rational expressions and order of operations

R.4: Equations

1. Solve linear, quadratic, and rational equations

R.5: Inequalities

1. Solve linear, quadratic, and rational inequalities
2. Graph the solutions of linear, quadratic, and rational inequalities
3. Write the solutions of linear, quadratic, and rational inequalities using interval notation

R.6: Exponents

1. Evaluate exponential expressions
2. Simplify exponential expressions

R.7: Radicals

1. Simplify radical expressions
2. Rationalize the denominator in radical expressions
3. Rationalize the numerator in radical expressions

### CHAPTER 1: Functions

1.1: Lines and Linear Functions

1. Find the slope of a line
2. Find the equation of a line using the slope and $y$-intercept
3. Find the equation of a line using a point and the slope
4. Find the equations of parallel and perpendicular lines
5. Graph the equation of a line
6. Evaluate linear functions
7. Write equations for linear models
8. Solve application problems

1.2: The Least Squares Line

1. Draw a scatterplot
2. Calculate the least squares line
3. Compute the response variable in a linear model
4. Calculate the correlation coefficient
5. Interpret the different value meanings for linear correlation

1.3: Properties of Functions

1. Determine if a relation is a function using mappings, tables, graphs, and equations
2. Identify the domains and ranges of functions
3. Use the vertical line test
4. Evaluate nonlinear functions
5. Find the composition of functions
6. Solve application problems

1.4: Quadratic Functions; Translation and Reflection

1. Complete the square for a given quadratic function
2. Obtain the vertex and axis; compute $y$-intercepts and $x$-intercepts
3. Perform translation and reflection rules to graphs of functions
4. Solve application problems

1.5: Polynomial and Rational Functions

1. Perform translation and reflection rules
2. Identify the degree of a polynomial
3. Find vertical and horizontal asymptotes
4. Solve application problems

### CHAPTER 2: Exponential, Logarithmic, and Trigonometric Functions

2.1: Exponential Functions

1. Graph exponential functions
2. Solve exponential equations
3. Solve application problems

2.2: Logarithmic Functions

1. Convert expressions between exponential and logarithmic forms
2. Evaluate logarithmic expressions
3. Simplify expressions using logarithmic properties
4. Change bases of logarithms
5. Solve logarithmic equations
6. Solve exponential equations using logarithms
7. Change bases of exponentials
8. Identify the domain of a logarithmic function
9. Solve application problems

2.3: Applications: Growth and Decay

1. Solve growth and decay application problems

2.4: Trigonometric Functions

1. Convert between degree and radian measurements
2. Evaluate trigonometric functions

3. Find the amplitude and period of trigonometric functions
4. Graph trigonometric functions
5. Solve application problems

# CHAPTER 3: The Derivative

## 3.1: Limits

1. Determine if the limit exists
2. Evaluate limits by completing tables and graphically
3. Use limit rules to evaluate limits
4. Solve application problems

## 3.2: Continuity

1. Identify and apply the conditions for continuity
2. Find points of discontinuity
3. Solve application problems

## 3.3: Rates of Change

1. Find the average rate of change for functions
2. Find the instantaneous rate of change
3. Solve application problems

## 3.4: Definition of the Derivative

1. Obtain the slope of the tangent line
2. Obtain the equation of the secant line
3. Find the derivative of a function (using the definition)
4. Evaluate the derivative
5. Determine where the derivative exists
6. Solve application problems

## 3.5: Graphical Differentiation

1. Interpret the graph of a function and its derivative
2. Sketch the graph of the derivative
3. Solve application problems

# CHAPTER 4: Calculating the Derivative

## 4.1: Techniques for Finding Derivatives

1. Use various notations for the derivative
2. Determine the derivatives of given functions using derivative rules (constant rule, power rule, and sum or difference rule)
3. Solve application problems

## 4.2: Derivatives of Products and Quotients

1. Apply the product rule to find the derivative
2. Apply the quotient rule to find the derivative
3. Solve application problems

## 4.3: The Chain Rule

1. Find the composition of two functions
2. Find the derivative using the chain rule
3. Solve application problems

## 4.4: Derivatives of Exponential Functions

1. Find the derivatives of exponential functions
2. Solve application problems

## 4.5: Derivatives of logarithmic Functions

1. Find the derivatives of logarithmic functions
2. Solve application problems

## 4.6: Derivatives of Trigonometric Functions

1. Find the derivatives of trigonometric functions
2. Solve application problems

# CHAPTER 5: Graphs and the Derivative

## 5.1: Increasing and Decreasing Functions

1. Identify increasing and decreasing intervals for functions
2. Identify critical numbers
3. Solve application problems

## 5.2: Relative Extrema

1. Identify relative extrema points graphically
2. Apply the first derivative test
3. Solve application problems

## 5.3: Higher Derivatives, Concavity, and the Second Derivative Test

1. Find higher derivatives
2. Identify intervals of upward and downward concavity
3. Identify inflection points
4. Apply the second derivative test
5. Solve application problems

## 5.4: Curve Sketching

1. Graph a function using derivative (first and second) information

# CHAPTER 6: Applications of the Derivative

## 6.1: Absolute Extrema

1. Identify absolute extrema locations graphically
2. Find absolute extrema points
3. Solve application problems

## 6.2: Applications of Extrema

1. Solve maxima and minima application problems

## 6.3: Implicit Differentiation

1. Differentiate functions of two variables implicitly
2. Find the equation of the tangent line at a given point .
3. Solve application problems

## 6.4: Related Rates

1. Use related rates to solve application problems

6.5: Differentials: Linear Approximation

1. Obtain differentials of dependent variables
2. Compute approximations using differentials
3. Solve application problems

# CHAPTER 7:    Integration

### 7.1: Antiderivatives

1. Obtain the antiderivative of a function using rules of integration
2. Solve application problems

### 7.2: Substitution

1. Find the indefinite integral using the substitution method
2. Solve application problems

### 7.3: Area and the Definite Integral

1. Understand the difference between indefinite and definite integrals
2. Approximate the area under the graph of a function using sums
3. Find the area under the graph of a function by integration
4. Solve application problems

### 7.4: The Fundamental Theorem of Calculus

1. Evaluate the definite integral of a function
2. Find the area covered by the graph of a function over an interval
3. Solve application problems

### 7.5: The Area Between Two Curves

1. Find the area between two curves
2. Solve application problems

# CHAPTER 8:    Further Techniques and Applications of Integration

### 8.1: Numerical Integration

1. Use approximation rules to evaluate the integral numerically
2. Solve application problems

### 8.2: Integration by Parts

1. Compute the integral using the integration by parts technique
2. Solve application problems

### 8.3: Volume and Average Value

1. Find the volume of a solid formed by rotation about the $x$-axis
2. Find the average value of a function
3. Solve application problems

### 8.4: Improper Integrals

1. Determine the convergence or divergence of improper integrals
2. Find the area between the graph of a function over an infinite interval, when possible
3. Solve application problems

# CHAPTER 9:    Multivariable Calculus

### 9.1: Functions of Several Variables

1. Evaluate functions with several variables
2. Graph functions with several variables
3. Solve application problems

### 9.2: Partial Derivatives

1. Find the partial derivative of functions with several variables
2. Find the second partial derivative of functions with several variables
3. Solve application problems

### 9.3: Maxima and Minima

1. Find relative extrema points
2. Solve application problems

### 9.4: Total Differentials and Approximations

1. Find the total differentials of functions with several variables
2. Find approximations by differentials
3. Solve application problems

### 9.5: Double Integrals

1. Evaluate double integrals
2. Find the volume under a given surface
3. Solve application problems

# CHAPTER 10:    Matrices

### 10.1: Solutions of Linear Systems

1. Perform row operations on matrices
2. Solve linear systems using the Gauss-Jordan method
3. Solve application problems

### 10.2: Addition and Subtraction of Matrices

1. Determine matrix size and equal matrices
2. Add matrices
3. Subtract matrices
4. Solve application problems

### 10.3: Multiplication of Matrices

1. Find scalar products
2. Multiply matrices
3. Solve application problems

### 10.4: Matrix Inverses

1. Find multiplicative inverse matrices
2. Solve linear systems using inverse matrices
3. Solve application problems

### 10.5: Eigenvalues and Eigenvectors

1. Calculate the determinant of a square matrix
2. Find eigenvalues and eigenvectors
3. Solve application problems

# CHAPTER 11:  Differential Equations

## 11.1: Solutions of Elementary and Separable Differential Equations

1. Find the general solution of a differential equation
2. Find the particular solution of a differential equation
3. Find the general solution of a separable differential equation
4. Solve application problems

## 11.2: LinearFirst-Order Differential Equations

1. Find the general solution of a linear differential equation
2. Find the particular solution of a linear differential equation
3. Solve application problems

## 11.3: Euler's Method

1. Use Euler's method to find approximate solutions to differential equations
2. Use Euler's method to find approximate function values
3. Solve application problems

## 11.4: Linear Systems of Differential Equations

1. Solve linear systems of differential equations
2. Solve application problems

## 11.5: Nonlinear Systems of Differential Equations

1. Sketch phase plane diagrams
2. Solve nonlinear system of differential equations
3. Solve application problems

## 11.6: Applications of Differential Equations

1. Solve application problems involving differential equations

# CHAPTER 12:  Probability

## 12.1: Sets

1. Determine elements of sets
2. Determine subsets of sets
3. Use set operations (intersection, union, and complement)
4. Use Venn diagrams
5. Solve application problems

## 12.2: Introduction to Probability

1. Find the sample space of an experiment
2. Identify mutually exclusive events
3. Determine basic probabilities
4. Use probability rules (union and complement)
5. Determine odds in favor of an event
6. Use probability distributions
7. Determine empirical probabilities
8. Solve application problems

## 12.3: Conditional Probability; Independent Events; Bayes' Theorem

1. Determine conditional probabilities
2. Use the product rule of probability

3. Identify independent events
4. Use tree diagrams and Bayes' theorem
5. Solve application problems

## 12.4: Discrete Random Variables; Applications to Decision Making

1. Use random variables
2. Use probability distributions (tables and histograms)
3. Determine expected values
4. Determine standard deviations
5. Solve application problems

# CHAPTER 13:  Probability and Calculus

## 13.1: Continuous Probability Models

1. Identify probability density functions
2. Find the cumulative distribution function of probability density functions
3. Solve application problems

## 13.2: Expected Value and Variance of Continuous Random Variables

1. Find the expected value, variance, and standard deviation of probability density functions
2. Find the median of a random variable with a given probability density function
3. Solve application problems

## 13.3: Special Probability Density Functions

1. Find the mean and standard deviation of a probability distribution
2. Compute the $z$-score
3. Solve application problems

# CHAPTER 14:  Discrete Dynamical Systems

## 14.1: Sequences

1. Find terms of sequences
2. Solve application problems

## 14.2: Equilibrium Points

1. Find equilibrium points of functions
2. Find terms of sequences
3. Determine whether an equilibrium point is stable or unstable
4. Draw cob web diagrams
5. Solve application problems

## 14.3: Determining Stability

1. Determine whether an equilibrium point is stable or unstable
2. Solve application problems

# Appendix C

## MathPrint Operating System for TI-84 and TI-84 Plus Silver Edition

The graphing calculator screens in this text display math in the format of the TI MathPrint operating system. With MathPrint, the math looks more like that seen in a printed book. You can obtain MathPrint and install it by following the instructions in the *Graphing Calculator and Spreadsheet Manual*. Only the TI-84 family of graphing calculators can be updated with the MathPrint operating system. If you own a TI-83 graphing calculator, you can use this brief appendix to help you "translate" what you see in the Classic mode shown on your calculator.

### Translating between MathPrint Mode and Classic Mode
The following table compares displays of several types in MathPrint mode and Classic mode (on a calculator without MathPrint installed).

| Feature | MathPrint | Classic (MathPrint not installed) |
|---|---|---|
| Improper fractions | $\dfrac{2}{5} - \dfrac{1}{3}$     $\dfrac{1}{15}$ <br><br> Press **ALPHA** and F1 and select 1:n/d. | 2/5 − 1/3 ▶ Frac     1/15 <br><br> Enter an expression and press **MATH** and select 1: ▶ Frac. |
| Mixed fractions | $2\dfrac{1}{5} * \left(3\dfrac{2}{3}\right)$     $\dfrac{121}{15}$ <br><br> Press **ALPHA** and F1 and select 2:Un/d. | Not Supported |
| Absolute values | $\lvert 10 - 15 \rvert$     5 <br><br> Press **ALPHA** and F2 and select 1:abs(. | abs(10 − 15)     5 <br><br> Press **MATH** and ▶ and select 1:abs(. |
| Summation | $\sum\limits_{I=1}^{10}(I^2)$     385 <br><br> Press **ALPHA** and F2 and select 2:Σ(. | sum(seq(I²,I,1,10))     385 <br><br> Press **2ND** and LIST and then ▶ twice and select 5:sum( for sum, and press **2ND** and LIST and then ▶ and select 5:seq( for sequence. |
| Numerical derivatives | $\dfrac{d}{dX}(X^2)\Big\vert_{X=3}$     6 <br><br> Press **ALPHA** and F2 and select 3:nDeriv(. | nDeriv(X²,X,3)     6 <br><br> Press **MATH** and select 8:nDeriv(. |
| Numerical values of integrals | $\displaystyle\int_1^5 (X^2)dX$     41.33333333 <br><br> Press **ALPHA** and F2 and select 4:fnInt(. | fnInt(X²,X,1,5)     41.33333333 <br><br> Press **MATH** and select 9:fnInt(. |

*(continued)*

| Feature | MathPrint | Classic (MathPrint not installed) |
|---|---|---|
| Logarithms | $\log_2(32)$<br>    5<br><br>Press **ALPHA** and F2<br>and select 5:logBASE(. | Evaluating logs with bases other than 10 or $e$ cannot be done on a graphing calculator if the MathPrint operating system is not installed. To evaluate $\log_2 32$, use the change-of-base formula:<br><br>$\log(32)/\log(2)$<br>    5 |

# The [Y=] Editor

MathPrint features can be accessed from the [Y=] editor as well as from the home screen. The following table shows examples that illustrate differences between MathPrint in the [Y=] editor and Classic mode.

| Feature | MathPrint | Classic (MathPrint not installed) |
|---|---|---|
| Graphing the derivative of $y = x^2$ | Plot 1  Plot 2  Plot 3<br>$\backslash Y_1 \blacksquare \frac{d}{dX}(X^2)\big|_{X=X}$<br>$\backslash Y_2 =$ | Plot 1  Plot 2  Plot 3<br>$\backslash Y_1 \blacksquare$ nDeriv($X^2$,X,X)<br>$\backslash Y_2 =$ |
| Graphing an antiderivative of $y = x^2$ | Plot 1  Plot 2  Plot 3<br>$\backslash Y_1 \blacksquare \int_0^X (X^2)dX$<br>$\backslash Y_2 =$ | Plot 1  Plot 2  Plot 3<br>$\backslash Y_1 \blacksquare$ fnInt($X^2$,X,0,X)<br>$\backslash Y_2 =$ |

# Appendix D

## Table I — Formulas from Geometry

**PYTHAGOREAN THEOREM**
For a right triangle with legs of lengths $a$ and $b$ and
hypotenuse of length $c$, $a^2 + b^2 = c^2$.

**CIRCLE**
Area: $A = \pi r^2$
Circumference: $C = 2\pi r$

**RECTANGLE**
Area: $A = lw$
Perimeter: $P = 2l + 2w$

**TRIANGLE**
Area: $A = \dfrac{1}{2}bh$

**SPHERE**
Volume: $V = \dfrac{4}{3}\pi r^3$
Surface area: $A = 4\pi r^2$

**CONE**
Volume: $V = \dfrac{1}{3}\pi r^2 h$

**RECTANGULAR BOX**
Volume: $V = lwh$
Surface area: $A = 2lh + 2wh + 2lw$

**CIRCULAR CYLINDER**
Volume: $V = \pi r^2 h$
Surface area: $A = 2\pi r^2 + 2\pi rh$

**TRIANGULAR PRISM**
Volume: $V = \dfrac{1}{2}bhl$

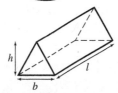

**GENERAL INFORMATION ON SURFACE AREA**
To find the surface area of a figure, break down the total surface area into the
individual components and add up the areas of the components. For example,
a rectangular box has six sides, each of which is a rectangle. A circular cylinder
has two ends, each of which is a circle, plus the side, which forms a rectangle
when opened up.

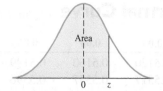

# Table 2 — Area Under a Normal Curve to the Left of $z$, where $z = \dfrac{x - \mu}{\sigma}$

| $z$ | 0.00 | 0.01 | 0.02 | 0.03 | 0.04 | 0.05 | 0.06 | 0.07 | 0.08 | 0.09 |
|------|--------|--------|--------|--------|--------|--------|--------|--------|--------|--------|
| −3.4 | 0.0003 | 0.0003 | 0.0003 | 0.0003 | 0.0003 | 0.0003 | 0.0003 | 0.0003 | 0.0003 | 0.0002 |
| −3.3 | 0.0005 | 0.0005 | 0.0005 | 0.0004 | 0.0004 | 0.0004 | 0.0004 | 0.0004 | 0.0004 | 0.0003 |
| −3.2 | 0.0007 | 0.0007 | 0.0006 | 0.0006 | 0.0006 | 0.0006 | 0.0006 | 0.0005 | 0.0005 | 0.0005 |
| −3.1 | 0.0010 | 0.0009 | 0.0009 | 0.0009 | 0.0008 | 0.0008 | 0.0008 | 0.0008 | 0.0007 | 0.0007 |
| −3.0 | 0.0013 | 0.0013 | 0.0013 | 0.0012 | 0.0012 | 0.0011 | 0.0011 | 0.0011 | 0.0010 | 0.0010 |
| −2.9 | 0.0019 | 0.0018 | 0.0017 | 0.0017 | 0.0016 | 0.0016 | 0.0015 | 0.0015 | 0.0014 | 0.0014 |
| −2.8 | 0.0026 | 0.0025 | 0.0024 | 0.0023 | 0.0023 | 0.0022 | 0.0021 | 0.0021 | 0.0020 | 0.0019 |
| −2.7 | 0.0035 | 0.0034 | 0.0033 | 0.0032 | 0.0031 | 0.0030 | 0.0029 | 0.0028 | 0.0027 | 0.0026 |
| −2.6 | 0.0047 | 0.0045 | 0.0044 | 0.0043 | 0.0041 | 0.0040 | 0.0039 | 0.0038 | 0.0037 | 0.0036 |
| −2.5 | 0.0062 | 0.0060 | 0.0059 | 0.0057 | 0.0055 | 0.0054 | 0.0052 | 0.0051 | 0.0049 | 0.0048 |
| −2.4 | 0.0082 | 0.0080 | 0.0078 | 0.0075 | 0.0073 | 0.0071 | 0.0069 | 0.0068 | 0.0066 | 0.0064 |
| −2.3 | 0.0107 | 0.0104 | 0.0102 | 0.0099 | 0.0096 | 0.0094 | 0.0091 | 0.0089 | 0.0087 | 0.0084 |
| −2.2 | 0.0139 | 0.0136 | 0.0132 | 0.0129 | 0.0125 | 0.0122 | 0.0119 | 0.0116 | 0.0113 | 0.0110 |
| −2.1 | 0.0179 | 0.0174 | 0.0170 | 0.0166 | 0.0162 | 0.0158 | 0.0154 | 0.0150 | 0.0146 | 0.0143 |
| −2.0 | 0.0228 | 0.0222 | 0.0217 | 0.0212 | 0.0207 | 0.0202 | 0.0197 | 0.0192 | 0.0188 | 0.0183 |
| −1.9 | 0.0287 | 0.0281 | 0.0274 | 0.0268 | 0.0262 | 0.0256 | 0.0250 | 0.0244 | 0.0239 | 0.0233 |
| −1.8 | 0.0359 | 0.0352 | 0.0344 | 0.0336 | 0.0329 | 0.0322 | 0.0314 | 0.0307 | 0.0301 | 0.0294 |
| −1.7 | 0.0446 | 0.0436 | 0.0427 | 0.0418 | 0.0409 | 0.0401 | 0.0392 | 0.0384 | 0.0375 | 0.0367 |
| −1.6 | 0.0548 | 0.0537 | 0.0526 | 0.0516 | 0.0505 | 0.0495 | 0.0485 | 0.0475 | 0.0465 | 0.0455 |
| −1.5 | 0.0668 | 0.0655 | 0.0643 | 0.0630 | 0.0618 | 0.0606 | 0.0594 | 0.0582 | 0.0571 | 0.0559 |
| −1.4 | 0.0808 | 0.0793 | 0.0778 | 0.0764 | 0.0749 | 0.0735 | 0.0722 | 0.0708 | 0.0694 | 0.0681 |
| −1.3 | 0.0968 | 0.0951 | 0.0934 | 0.0918 | 0.0901 | 0.0885 | 0.0869 | 0.0853 | 0.0838 | 0.0823 |
| −1.2 | 0.1151 | 0.1131 | 0.1112 | 0.1093 | 0.1075 | 0.1056 | 0.1038 | 0.1020 | 0.1003 | 0.0985 |
| −1.1 | 0.1357 | 0.1335 | 0.1314 | 0.1292 | 0.1271 | 0.1251 | 0.1230 | 0.1210 | 0.1190 | 0.1170 |
| −1.0 | 0.1587 | 0.1562 | 0.1539 | 0.1515 | 0.1492 | 0.1469 | 0.1446 | 0.1423 | 0.1401 | 0.1379 |
| −0.9 | 0.1841 | 0.1814 | 0.1788 | 0.1762 | 0.1736 | 0.1711 | 0.1685 | 0.1660 | 0.1635 | 0.1611 |
| −0.8 | 0.2119 | 0.2090 | 0.2061 | 0.2033 | 0.2005 | 0.1977 | 0.1949 | 0.1922 | 0.1894 | 0.1867 |
| −0.7 | 0.2420 | 0.2389 | 0.2358 | 0.2327 | 0.2296 | 0.2266 | 0.2236 | 0.2206 | 0.2177 | 0.2148 |
| −0.6 | 0.2743 | 0.2709 | 0.2676 | 0.2643 | 0.2611 | 0.2578 | 0.2546 | 0.2514 | 0.2483 | 0.2451 |
| −0.5 | 0.3085 | 0.3050 | 0.3015 | 0.2981 | 0.2946 | 0.2912 | 0.2877 | 0.2843 | 0.2810 | 0.2776 |
| −0.4 | 0.3446 | 0.3409 | 0.3372 | 0.3336 | 0.3300 | 0.3264 | 0.3228 | 0.3192 | 0.3156 | 0.3121 |
| −0.3 | 0.3821 | 0.3783 | 0.3745 | 0.3707 | 0.3669 | 0.3632 | 0.3594 | 0.3557 | 0.3520 | 0.3483 |
| −0.2 | 0.4207 | 0.4168 | 0.4129 | 0.4090 | 0.4052 | 0.4013 | 0.3974 | 0.3936 | 0.3897 | 0.3859 |
| −0.1 | 0.4602 | 0.4562 | 0.4522 | 0.4483 | 0.4443 | 0.4404 | 0.4364 | 0.4325 | 0.4286 | 0.4247 |
| −0.0 | 0.5000 | 0.4960 | 0.4920 | 0.4880 | 0.4840 | 0.4801 | 0.4761 | 0.4721 | 0.4681 | 0.4641 |

(*continued*)

# Table 2 — Area Under a Normal Curve (continued)

| z | 0.00 | 0.01 | 0.02 | 0.03 | 0.04 | 0.05 | 0.06 | 0.07 | 0.08 | 0.09 |
|-----|--------|--------|--------|--------|--------|--------|--------|--------|--------|--------|
| 0.0 | 0.5000 | 0.5040 | 0.5080 | 0.5120 | 0.5160 | 0.5199 | 0.5239 | 0.5279 | 0.5319 | 0.5359 |
| 0.1 | 0.5398 | 0.5438 | 0.5478 | 0.5517 | 0.5557 | 0.5596 | 0.5636 | 0.5675 | 0.5714 | 0.5753 |
| 0.2 | 0.5793 | 0.5832 | 0.5871 | 0.5910 | 0.5948 | 0.5987 | 0.6026 | 0.6064 | 0.6103 | 0.6141 |
| 0.3 | 0.6179 | 0.6217 | 0.6255 | 0.6293 | 0.6331 | 0.6368 | 0.6406 | 0.6443 | 0.6480 | 0.6517 |
| 0.4 | 0.6554 | 0.6591 | 0.6628 | 0.6664 | 0.6700 | 0.6736 | 0.6772 | 0.6808 | 0.6844 | 0.6879 |
| 0.5 | 0.6915 | 0.6950 | 0.6985 | 0.7019 | 0.7054 | 0.7088 | 0.7123 | 0.7157 | 0.7190 | 0.7224 |
| 0.6 | 0.7257 | 0.7291 | 0.7324 | 0.7357 | 0.7389 | 0.7422 | 0.7454 | 0.7486 | 0.7517 | 0.7549 |
| 0.7 | 0.7580 | 0.7611 | 0.7642 | 0.7673 | 0.7704 | 0.7734 | 0.7764 | 0.7794 | 0.7823 | 0.7852 |
| 0.8 | 0.7881 | 0.7910 | 0.7939 | 0.7967 | 0.7995 | 0.8023 | 0.8051 | 0.8078 | 0.8106 | 0.8133 |
| 0.9 | 0.8159 | 0.8186 | 0.8212 | 0.8238 | 0.8264 | 0.8289 | 0.8315 | 0.8340 | 0.8365 | 0.8389 |
| 1.0 | 0.8413 | 0.8438 | 0.8461 | 0.8485 | 0.8508 | 0.8531 | 0.8554 | 0.8577 | 0.8599 | 0.8621 |
| 1.1 | 0.8643 | 0.8665 | 0.8686 | 0.8708 | 0.8729 | 0.8749 | 0.8770 | 0.8790 | 0.8810 | 0.8830 |
| 1.2 | 0.8849 | 0.8869 | 0.8888 | 0.8907 | 0.8925 | 0.8944 | 0.8962 | 0.8980 | 0.8997 | 0.9015 |
| 1.3 | 0.9032 | 0.9049 | 0.9066 | 0.9082 | 0.9099 | 0.9115 | 0.9131 | 0.9147 | 0.9162 | 0.9177 |
| 1.4 | 0.9192 | 0.9207 | 0.9222 | 0.9236 | 0.9251 | 0.9265 | 0.9278 | 0.9292 | 0.9306 | 0.9319 |
| 1.5 | 0.9332 | 0.9345 | 0.9357 | 0.9370 | 0.9382 | 0.9394 | 0.9406 | 0.9418 | 0.9429 | 0.9441 |
| 1.6 | 0.9452 | 0.9463 | 0.9474 | 0.9484 | 0.9495 | 0.9505 | 0.9515 | 0.9525 | 0.9535 | 0.9545 |
| 1.7 | 0.9554 | 0.9564 | 0.9573 | 0.9582 | 0.9591 | 0.9599 | 0.9608 | 0.9616 | 0.9625 | 0.9633 |
| 1.8 | 0.9641 | 0.9649 | 0.9656 | 0.9664 | 0.9671 | 0.9678 | 0.9686 | 0.9693 | 0.9699 | 0.9706 |
| 1.9 | 0.9713 | 0.9719 | 0.9726 | 0.9732 | 0.9738 | 0.9744 | 0.9750 | 0.9756 | 0.9761 | 0.9767 |
| 2.0 | 0.9772 | 0.9778 | 0.9783 | 0.9788 | 0.9793 | 0.9798 | 0.9803 | 0.9808 | 0.9812 | 0.9817 |
| 2.1 | 0.9821 | 0.9826 | 0.9830 | 0.9834 | 0.9838 | 0.9842 | 0.9846 | 0.9850 | 0.9854 | 0.9857 |
| 2.2 | 0.9861 | 0.9864 | 0.9868 | 0.9871 | 0.9875 | 0.9878 | 0.9881 | 0.9884 | 0.9887 | 0.9890 |
| 2.3 | 0.9893 | 0.9896 | 0.9898 | 0.9901 | 0.9904 | 0.9906 | 0.9909 | 0.9911 | 0.9913 | 0.9916 |
| 2.4 | 0.9918 | 0.9920 | 0.9922 | 0.9925 | 0.9927 | 0.9929 | 0.9931 | 0.9932 | 0.9934 | 0.9936 |
| 2.5 | 0.9938 | 0.9940 | 0.9941 | 0.9943 | 0.9945 | 0.9946 | 0.9948 | 0.9949 | 0.9951 | 0.9952 |
| 2.6 | 0.9953 | 0.9955 | 0.9956 | 0.9957 | 0.9959 | 0.9960 | 0.9961 | 0.9962 | 0.9963 | 0.9964 |
| 2.7 | 0.9965 | 0.9966 | 0.9967 | 0.9968 | 0.9969 | 0.9970 | 0.9971 | 0.9972 | 0.9973 | 0.9974 |
| 2.8 | 0.9974 | 0.9975 | 0.9976 | 0.9977 | 0.9977 | 0.9978 | 0.9979 | 0.9979 | 0.9980 | 0.9981 |
| 2.9 | 0.9981 | 0.9982 | 0.9982 | 0.9983 | 0.9984 | 0.9984 | 0.9985 | 0.9985 | 0.9986 | 0.9986 |
| 3.0 | 0.9987 | 0.9987 | 0.9987 | 0.9988 | 0.9988 | 0.9989 | 0.9989 | 0.9989 | 0.9990 | 0.9990 |
| 3.1 | 0.9990 | 0.9991 | 0.9991 | 0.9991 | 0.9992 | 0.9992 | 0.9992 | 0.9992 | 0.9993 | 0.9993 |
| 3.2 | 0.9993 | 0.9993 | 0.9994 | 0.9994 | 0.9994 | 0.9994 | 0.9994 | 0.9995 | 0.9995 | 0.9995 |
| 3.3 | 0.9995 | 0.9995 | 0.9995 | 0.9996 | 0.9996 | 0.9996 | 0.9996 | 0.9996 | 0.9996 | 0.9997 |
| 3.4 | 0.9997 | 0.9997 | 0.9997 | 0.9997 | 0.9997 | 0.9997 | 0.9997 | 0.9997 | 0.9997 | 0.9998 |

# Table 3 — Integrals

(*C* is an arbitrary constant.)

---

**1.** $\displaystyle\int x^n\, dx = \frac{x^{n+1}}{n+1} + C \quad (\text{if } n \neq -1)$

**2.** $\displaystyle\int e^{kx}\, dx = \frac{e^{kx}}{k} + C$

**3.** $\displaystyle\int \frac{a}{x}\, dx = a \ln |x| + C$

**4.** $\displaystyle\int \ln |ax|\, dx = x(\ln |ax| - 1) + C$

**5.** $\displaystyle\int \frac{1}{\sqrt{x^2 + a^2}}\, dx = \ln \left| x + \sqrt{x^2 + a^2} \right| + C$

**6.** $\displaystyle\int \frac{1}{\sqrt{x^2 - a^2}}\, dx = \ln \left| x + \sqrt{x^2 - a^2} \right| + C$

**7.** $\displaystyle\int \frac{1}{a^2 - x^2}\, dx = \frac{1}{2a} \cdot \ln \left| \frac{a + x}{a - x} \right| + C \quad (a \neq 0)$

**8.** $\displaystyle\int \frac{1}{x^2 - a^2}\, dx = \frac{1}{2a} \cdot \ln \left| \frac{x - a}{x + a} \right| + C \quad (a \neq 0)$

**9.** $\displaystyle\int \frac{1}{x\sqrt{a^2 - x^2}}\, dx = -\frac{1}{a} \cdot \ln \left| \frac{a + \sqrt{a^2 - x^2}}{x} \right| + C \quad (a \neq 0)$

**10.** $\displaystyle\int \frac{1}{x\sqrt{a^2 + x^2}}\, dx = -\frac{1}{a} \cdot \ln \left| \frac{a + \sqrt{a^2 + x^2}}{x} \right| + C \quad (a \neq 0)$

**11.** $\displaystyle\int \frac{x}{ax + b}\, dx = \frac{x}{a} - \frac{b}{a^2} \cdot \ln |ax + b| + C \quad (a \neq 0)$

**12.** $\displaystyle\int \frac{x}{(ax + b)^2}\, dx = \frac{b}{a^2(ax + b)} + \frac{1}{a^2} \cdot \ln |ax + b| + C \quad (a \neq 0)$

**13.** $\displaystyle\int \frac{1}{x(ax + b)}\, dx = \frac{1}{b} \cdot \ln \left| \frac{x}{ax + b} \right| + C \quad (b \neq 0)$

**14.** $\displaystyle\int \frac{1}{x(ax + b)^2}\, dx = \frac{1}{b(ax + b)} + \frac{1}{b^2} \cdot \ln \left| \frac{x}{ax + b} \right| + C \quad (b \neq 0)$

**15.** $\displaystyle\int \sqrt{x^2 + a^2}\, dx = \frac{x}{2} \sqrt{x^2 + a^2} + \frac{a^2}{2} \cdot \ln \left| x + \sqrt{x^2 + a^2} \right| + C$

**16.** $\displaystyle\int x^n \cdot \ln |x|\, dx = x^{n+1} \left[ \frac{\ln |x|}{n + 1} - \frac{1}{(n + 1)^2} \right] + C \quad (n \neq -1)$

**17.** $\displaystyle\int x^n e^{ax}\, dx = \frac{x^n e^{ax}}{a} - \frac{n}{a} \cdot \int x^{n-1} e^{ax}\, dx + C \quad (a \neq 0)$

---

# Table 4 — Integrals Involving Trigonometric Functions

**18.** $\displaystyle\int \sin u\, du = -\cos u + C$

**19.** $\displaystyle\int \cos u\, du = \sin u + C$

**20.** $\displaystyle\int \sec^2 u\, du = \tan u + C$

**21.** $\displaystyle\int \csc^2 u\, du = -\cot u + C$

**22.** $\displaystyle\int \sec u \tan u\, du = \sec u + C$

**23.** $\displaystyle\int \csc u \cot u\, du = -\csc u + C$

**24.** $\displaystyle\int \tan u\, du = \ln |\sec u| + C$

**25.** $\displaystyle\int \cot u\, du = \ln |\sin u| + C$

**26.** $\displaystyle\int \sec u\, du = \ln |\sec u + \tan u| + C$

**27.** $\displaystyle\int \csc u\, du = \ln |\csc u - \cot u| + C$

**28.** $\displaystyle\int \sin^n u\, du = -\frac{1}{n} \sin^{n-1} u \cos u + \frac{n-1}{n} \int \sin^{n-2} u\, du \quad (n \neq 0)$

**29.** $\displaystyle\int \cos^n u\, du = \frac{1}{n} \cos^{n-1} u \sin u + \frac{n-1}{n} \int \cos^{n-2} u\, du \quad (n \neq 0)$

**30.** $\displaystyle\int \tan^n u\, du = \frac{1}{n-1} \tan^{n-1} u - \int \tan^{n-2} u\, du \quad (n \neq 1)$

**31.** $\displaystyle\int \sec^n u\, du = \frac{1}{n-1} \tan u \sec^{n-2} u + \frac{n-2}{n-1} \int \sec^{n-2} u\, du \quad (n \neq 1)$

**32.** $\displaystyle\int \sin au \sin bu\, du = \frac{\sin(a-b)u}{2(a-b)} - \frac{\sin(a+b)u}{2(a+b)} + C, \quad |a| \neq |b|$

**33.** $\displaystyle\int \cos au \cos bu\, du = \frac{\sin(a-b)u}{2(a-b)} + \frac{\sin(a+b)u}{2(a+b)} + C, \quad |a| \neq |b|$

**34.** $\displaystyle\int \sin au \cos bu\, du = -\frac{\cos(a-b)u}{2(a-b)} - \frac{\cos(a+b)u}{2(a+b)} + C, \quad |a| \neq |b|$

**35.** $\displaystyle\int u \sin u\, du = \sin u - u \cos u + C$

**36.** $\displaystyle\int u^n \sin u\, du = -u^n \cos u + n \int u^{n-1} \cos u\, du$

**37.** $\displaystyle\int e^{au} \sin bu\, du = \frac{e^{au}}{a^2 + b^2}(a \sin bu - b \cos bu) + C$

**38.** $\displaystyle\int e^{au} \cos bu\, du = \frac{e^{au}}{a^2 + b^2}(a \cos bu + b \sin bu) + C$

# Answers to Selected Exercises

Answers to selected writing exercises are provided.

## Answers to Prerequisite Skills Test

**1.** 20%   **2.** 51/35   **3.** $x + y = 75$   **4.** $s \geq 4p$   **5.** $-20/3$ (Sec. R.4)   **6.** $-11/5$ (Sec. R.4)   **7.** $(-2, 5]$ (Sec. R.5)
**8.** $x \leq -3$ (Sec. R.5)   **9.** $y \geq -17/2$ (Sec. R.5)   **10.** $p > 3/2$ (Sec. R.5)   **11.** $-y^2 + 4y - 6$ (Sec. R.1)
**12.** $x^3 - x^2 + x + 3$ (Sec. R.1)   **13.** $a^2 - 4ab + 4b^2$ (Sec. R.1)   **14.** $3pq(1 + 2p + 3q)$ (Sec. R.2)   **15.** $(3x + 5)(x - 2)$
(Sec. R.2)   **16.** $(a - 6)/(a + 2)$ (Sec. R.3)   **17.** $(x^2 + 5x - 2)/[x(x - 1)(x + 1)]$ (Sec. R.3)   **18.** $(-2 \pm \sqrt{7})/3$ (Sec. R.4)
**19.** $[-\infty, -3) \cup [1, \infty)$ (Sec. R.5)   **20.** $x^6y/4$ (Sec. R.6)   **21.** $2/(p^2q)$ (Sec. R.6)   **22.** $(m - k)/(km)$ (Sec. R.6)
**23.** $(x^2 + 1)^{-1/2}(3x^2 + x + 5)$ (Sec. R.6)   **24.** $4b^2$ (Sec. R.7)   **25.** $(4 + \sqrt{10})/3$ (Sec. R.7)   **26.** $|y - 5|$ (Sec. R.7)

## Chapter R Algebra Reference

### Exercises R.1 (page R-5)

**1.** $-x^2 + x + 9$   **3.** $-16q^2 + 4q + 6$

| For exercises . . . | 1–6 | 7,8,15–22 | 9–14 | 23–26 |
|---|---|---|---|---|
| Refer to example . . . | 2 | 3 | 4 | 5 |

**5.** $-0.327x^2 - 2.805x - 1.458$   **7.** $-18m^3 - 27m^2 + 9m$
**9.** $9t^2 + 9ty - 10y^2$   **11.** $4 - 9x^2$   **13.** $(6/25)y^2 + (11/40)yz + (1/16)z^2$   **15.** $27p^3 - 1$   **17.** $8m^3 + 1$
**19.** $3x^2 + xy + 2xz - 2y^2 - 3yz - z^2$   **21.** $x^3 + 6x^2 + 11x + 6$   **23.** $x^2 + 4x + 4$   **25.** $x^3 - 6x^2y + 12xy^2 - 8y^3$

### Exercises R.2 (page R-7)

**1.** $7a^2(a + 2)$   **3.** $13p^2q(p^2q - 3p + 2q)$   **5.** $(m + 2)(m - 7)$

| For exercises . . . | 1–4 | 5–11 | 12–15 | 16–20 | 21–32 |
|---|---|---|---|---|---|
| Refer to example . . . | 1 | 2 | 3 | 3, 2nd CAUTION | 4 |

**7.** $(z + 4)(z + 5)$   **9.** $(a - 5b)(a - b)$   **11.** $(y - 7z)(y + 3z)$
**13.** $(3a + 7)(a + 1)$   **15.** $(7m + 2n)(3m + n)$   **17.** $3m(m + 3)(m + 1)$   **19.** $2a^2(4a - b)(3a + 2b)$   **21.** $(x + 8)(x - 8)$
**23.** $10(x + 4)(x - 4)$   **25.** $(z + 7y)^2$   **27.** $(3p - 4)^2$   **29.** $(3r - 4s)(9r^2 + 12rs + 16s^2)$   **31.** $(x - y)(x + y)(x^2 + y^2)$

### Exercises R.3 (page R-10)

**1.** $v/7$   **3.** 8/9   **5.** $x - 2$   **7.** $(m - 2)/(m + 3)$

| For exercises . . . | 1–12 | 13–38 |
|---|---|---|
| Refer to example . . . | 1 | 2 |

**9.** $3(x - 1)/(x - 2)$   **11.** $(m^2 + 4)/4$   **13.** $3k/5$   **15.** $9/(5c)$
**17.** 1/4   **19.** $2(a + 4)/(a - 3)$   **21.** $(k - 2)/(k + 3)$   **23.** $(m - 3)/(2m - 3)$   **25.** 1   **27.** $(12 - 15y)/(10y)$
**29.** $(3m - 2)/[m(m - 1)]$   **31.** $14/[3(a - 1)]$   **33.** $(7x + 1)/[(x - 2)(x + 3)(x + 1)]$
**35.** $k(k - 13)/[(2k - 1)(k + 2)(k - 3)]$   **37.** $(4a + 1)/[a(a + 2)]$

### Exercises R.4 (page R-16)

**1.** $-12$   **3.** 12   **5.** $-7/8$   **7.** 4   **9.** $-3, -2$   **11.** 7   **13.** $-1/4, 2/3$

| For exercises . . . | 1–8 | 9–26 | 27–37 |
|---|---|---|---|
| Refer to example . . . | 2 | 3–5 | 6,7 |

**15.** $-3, 3$   **17.** $0, 4$   **19.** $(2 + \sqrt{10})/2 \approx 2.581, (2 - \sqrt{10})/2 \approx -0.581$
**21.** $5 + \sqrt{5} \approx 7.236, 5 - \sqrt{5} \approx 2.764$   **23.** $1, 5/2$   **25.** $(-1 + \sqrt{73})/6 \approx 1.257, (-1 - \sqrt{73})/6 \approx -1.591$
**27.** 3   **29.** $-59/6$   **31.** 3   **33.** $2/3$   **35.** 2   **37.** No solution

### Exercises R.5 (page R-21)

**1.** $(-\infty, 4)$

| For exercises . . . | 1–14 | | 15–26 | 27–38 | 39–42 | 43–54 |
|---|---|---|---|---|---|---|
| Refer to example . . . | Figure 1, Example 2 | 2 | | 3 | 4 | 5–7 |

**3.** $[1, 2)$   **5.** $(-\infty, -9)$

**7.** $-7 \leq x \leq -3$   **9.** $x \leq -1$   **11.** $-2 \leq x < 6$   **13.** $x \leq -4$ or $x \geq 4$

**15.** $(-\infty, 2]$   **17.** $(3, \infty)$   **19.** $(1/5, \infty)$

**21.** $(-4, 6)$   **23.** $[-5, 3)$   **25.** $[-17/7, \infty)$

**27.** $(-5, 3)$   **29.** $(1, 2)$   **31.** $(-\infty, -4) \cup (4, \infty)$

**33.** $(-\infty, -1] \cup [5, \infty)$   **35.** $(-\infty, -1) \cup (1/3, \infty)$

**37.** $(-\infty, -3] \cup [3, \infty)$   **39.** $[-2, 0] \cup [2, \infty)$

**41.** $(-\infty, 0) \cup (1, 6)$ 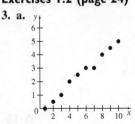 **43.** $(-5, 3]$ **45.** $(-\infty, -2)$ **47.** $[-8, 5)$ **49.** $[2, 3)$

**51.** $(-2, 0] \cup (3, \infty)$ **53.** $[1, 3/2]$

## Exercises R.6 (page R-25)

| For exercises . . . | 1–8 | 9–26 | 27–36 | 37–50 | 51–56 |
|---|---|---|---|---|---|
| Refer to example . . . | 1 | 2 | 3,4 | 5 | 6 |

**1.** $1/64$ **3.** $1$ **5.** $-1/9$ **7.** $36$ **9.** $1/64$ **11.** $1/10^8$ **13.** $x^2$
**15.** $2^3 k^3$ **17.** $x^5/(3y^3)$ **19.** $a^3 b^6$ **21.** $(a + b)/(ab)$
**23.** $2(m - n)/[mn(m + n^2)]$ **25.** $xy/(y - x)$ **27.** $11$ **29.** $4$ **31.** $1/2$ **33.** $1/16$ **35.** $4/3$ **37.** $9$ **39.** $64$ **41.** $x^4/y^4$
**43.** $r$ **45.** $3k^{3/2}/8$ **47.** $a^{2/3} b^2$ **49.** $h^{1/3} t^{1/5}/k^{2/5}$ **51.** $3x(x^2 + 3x)^2(x^2 - 5)$ **53.** $5x(x^2 - 1)^{-1/2}(x^2 + 1)$
**55.** $(2x + 5)(x^2 - 4)^{-1/2}(4x^2 + 5x - 8)$

## Exercises R.7 (page R-28)

| For exercises . . . | 1–22 | 23–26 | 27–40 | 41–44 |
|---|---|---|---|---|
| Refer to example . . . | 1,2 | 3 | 4 | 5 |

**1.** $5$ **3.** $-5$ **5.** $20\sqrt{5}$ **7.** $9$ **9.** $7\sqrt{2}$ **11.** $9\sqrt{7}$ **13.** $5\sqrt[3]{2}$
**15.** $xyz^2\sqrt{2x}$ **17.** $4xy^2 z^3 \sqrt[3]{2y^2}$ **19.** $ab\sqrt{ab}(b - 2a^2 + b^3)$
**21.** $\sqrt[6]{a^5}$ **23.** $|4 - x|$ **25.** Cannot be simplified **27.** $5\sqrt{7}/7$ **29.** $-\sqrt{3}/2$ **31.** $-3(1 + \sqrt{2})$ **33.** $3(2 - \sqrt{2})$
**35.** $(\sqrt{r} + \sqrt{3})/(r - 3)$ **37.** $\sqrt{y} + \sqrt{5}$ **39.** $-2x - 2\sqrt{x(x + 1)} - 1$ **41.** $-1/[2(1 - \sqrt{2})]$ **43.** $-1/[2x - 2\sqrt{x(x + 1)} + 1]$

# Chapter 1 Functions

## Exercises 1.1 (page 13)

| For Exercises . . . | 1–4 | 13, 29, 30 | 14, 31–34 | 15–17, 19–24 | 18, 27 | 25, 26 | 28 | 45–60 |
|---|---|---|---|---|---|---|---|---|
| Refer to Example . . . | 1 | 6 | 7 | 3 | 5 | 2 | 4 | 8, 9 |

| For Exercises . . . | 61–70 | 71–79, 83–85 | 80, 86, 87 | 81, 82 |
|---|---|---|---|---|
| Refer to Example . . . | 10 | 11, 12 | 14 | 13 |

**1.** $3/5$ **3.** Not defined
**5.** $1$ **7.** $5/9$ **9.** Not defined **11.** $0$
**13.** $2$ **15.** $y = -2x + 5$
**17.** $y = -7$ **19.** $y = -(1/3)x + 10/3$
**21.** $y = 6x - 7/2$ **23.** $x = -8$
**25.** $x + 2y = -6$ **27.** $x = -6$ **29.** $3x + 2y = 0$ **31.** $x - y = 7$ **33.** $5x - y = -4$ **35.** No **39.** a **41.** $-4$

**45.** 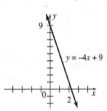 **47.** 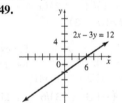 **49.** 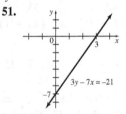 **51.**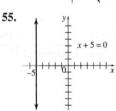

**53.** 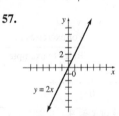 **55.** 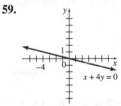 **57.** **59.**

**61.** $-3$ **63.** $22$ **65.** $0$ **67.** $-4$ **69.** $7 - 5t$ **71. a.** $u = 0.85(220 - x) = 187 - 0.85x$, $l = 0.7(220 - x) = 154 - 0.7x$
**b.** 140 to 170 beats per minute **c.** 126 to 153 beats per minute **d.** The women are 16 and 52. Their pulse is 143 beats per minute
**73.** About 86 yr **75. a.** $y = 35 + 2.8t$ **b.** $y = 100 - 2.2t$ **c.** 2003 **77. a.** $y = 2.1t - 2.3$ **b.** $-0.2$, or roughly 0
**c.** 20 years old **79. a.** $T = 0.03t + 15$ **b.** About 2103 **81.** 88 **83. a.** $y = 0.117t + 24.7$ **b.** $y = 0.137t + 22.0$
**c.** Women **d.** 2026 (or 2025, depending on how you round) **e.** 28.3 (or 28.2 if 2025 is used)
**85. a.** $y = -1.55t + 227.5$ **b.** 2021 **87.** $-40°$

## Exercises 1.2 (page 24)

**3. a.**

| For Exercises . . . | 3b, 4, 7a, 10d, 11b, 12a, 13a, 14c, 15d, 16d, 17d, 18b, 20d, 21c, 22a | 3c, 10b, 11c, 12b, 13b, 14b, 15a, 16a,b, 17c, 18c, 20a, 21b, 22b | 3d, 12d, 13e,15b, 16c, 18d, 20b,c, 22c | 5a,b, 6a,b, 8a, 19a,b | 5c, 6c | 15c |
|---|---|---|---|---|---|---|
| Refer to Example . . . | 4 | 1 | 2 | 1, 4 | 5 | 3 |

**b.** 0.993   **c.** $Y = 0.555x - 0.5$

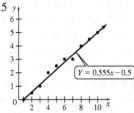

**d.** 5.6   **5. a.** $Y = 0.9783x + 0.0652, 0.9783$   **b.** $Y = 1.5, 0$

**c.**

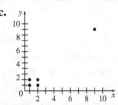

The point $(9, 9)$ is an outlier that has a strong effect on the least squares line and the correlation coefficient.

**7. a.** 0.7746   **b.**

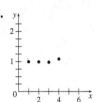

**11. a.**

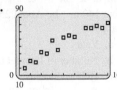

Yes   **b.** 0.959; yes   **c.** $Y = 3.98x + 22.7$

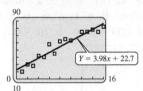

**13. a.** 0.994; very well
**b.** $Y = 1.3525x - 2.51$
**c.**

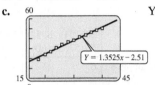

Yes

**15. a.** $Y = 0.212x - 0.309$
**b.** 15.2 chirps per second
**c.** 86.4° F   **d.** 0.835

**17. a.** 4.298 miles per hour
**b.**

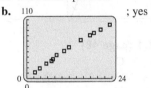

; yes

**c.** $Y = 4.317x + 3.419$
**d.** 0.9971; yes
**e.** 4.317 miles per hour

**d.** 1.3525 cm   **e.** 58.35 cm

**19. a.** $Y = -0.08915x + 74.28, r = -0.1035.$ The taller the student, the shorter is the ideal partner's height.
**b.** Females: $Y = 0.6674x + 27.89, r = 0.9459;$
males: $Y = 0.4348x + 34.04, r = 0.7049$
**c.**

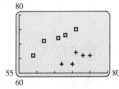

**21. a.**

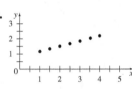

**b.** $Y = 0.366x + 0.803;$ the line seems to fit the data.

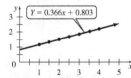

**c.** $r = 0.995$ indicates a good fit, which confirms the conclusion in part b.

**Exercises 1.3 (page 37)**
**1.** Not a function
**3.** Function   **5.** Function
**7.** Not a function

| For Exercises ... | 1–8 | 9–16, 72 | 17–32 | 33–40 | 41–56, 68a, b, 69a, b | 57–62 | 63–66, 68c, 69c | 67, 72 | 70, 71 |
|---|---|---|---|---|---|---|---|---|---|
| Refer to Example ... | 1, 2 | 3b | 3a, c–e | 2d | 4 | 5 | 6 | Basal temp. | 7 |

**9.** $(-2, -1), (-1, 1), (0, 3), (1, 5), (2, 7),$
$(3, 9)$; range: $\{-1, 1, 3, 5, 7, 9\}$

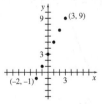

**11.** $(-2, 3/2), (-1, 2), (0, 5/2), (1, 3), (2, 7/2), (3, 4);$
range: $\{3/2, 2, 5/2, 3, 7/2, 4\}$

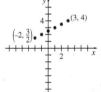

**13.** $(-2, 0), (-1, -1), (0, 0), (1, 3), (2, 8), (3, 15);$
range: $\{-1, 0, 3, 8, 15\}$

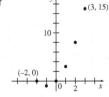

**15.** $(-2, 4), (-1, 1), (0, 0), (1, 1),$
$(2, 4), (3, 9)$; range: $\{0, 1, 4, 9\}$

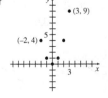

**17.** $(-\infty, \infty)$  **19.** $(-\infty, \infty)$  **21.** $[-2, 2]$  **23.** $[3, \infty)$  **25.** $(-\infty, -1) \cup (-1, 1) \cup (1, \infty)$  **27.** $(-\infty, -4) \cup (4, \infty)$
**29.** $(-\infty, -1] \cup [5, \infty)$  **31.** $(-\infty, -1) \cup (1/3, \infty)$  **33.** Domain: $[-5, 4]$; range: $[-2, 6]$  **35.** Domain: $(-\infty, \infty)$; range: $(-\infty, 12]$
**37.** Domain: $[-2, 4]$; range: $[0, 4]$  **a.** 0  **b.** 4  **c.** 3  **d.** $-1.5, 1.5, 2.5$  **39.** Domain: $[-2, 4]$; range: $[-3, 2]$  **a.** $-3$  **b.** $-2$
**c.** $-1$  **d.** 2.5  **41. a.** 33  **b.** 15/4  **c.** $3a^2 - 4a + 1$  **d.** $12/m^2 - 8/m + 1$ or $(12 - 8m + m^2)/m^2$  **e.** 0, 4/3
**43. a.** 7  **b.** 0  **c.** $(2a + 1)/(a - 4)$ if $a \neq 4$, 7 if $a = 4$  **d.** $(4 + m)/(2 - 4m)$ if $m \neq 1/2$, 7 if $m = 1/2$  **e.** $-5$
**45.** $6t^2 + 12t + 4$  **47.** $r^2 + 2rh + h^2 - 2r - 2h + 5$  **49.** $9/q^2 - 6/q + 5$ or $(9 - 6q + 5q^2)/q^2$  **51. a.** $2x + 2h + 1$
**b.** $2h$  **c.** 2  **53. a.** $2x^2 + 4xh + 2h^2 - 4x - 4h - 5$  **b.** $4xh + 2h^2 - 4h$  **c.** $4x + 2h - 4$  **55. a.** $1/(x + h)$  **b.** $-h/[x(x + h)]$
**c.** $-1/[x(x + h)]$  **57.** Function  **59.** Not a function  **61.** Function  **63. a.** 2  **b.** 8  **c.** $5 - 3x$  **d.** $11 - 3x$
**65. a.** 19  **b.** 23  **c.** $18x^2 + 3x - 2$  **d.** $6x^2 + 15x + 2$  **67. a.** About 140 m  **b.** About 250 m
**69. a. i.** 3.6 kcal/km  **ii.** 61 kcal/km  **b.** $x = g(z) = 1000z$  **c.** $y = 4.4\, z^{0.88}$
**71. a.** $A(w) = (3000 - w)w$  **b.** $[0, 3000]$
**c.**

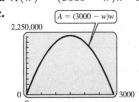

**73. a.** Year
**b.** Number of Internet users
**c.** 1971 million users
**d.** Domain: $[2008, 2012]$; range: $[1574, 2405]$

## Exercises 1.4 (page 48)

**3.** D  **5.** A  **7.** C
**9.** $y = 3(x + 3/2)^2 - 7/4, (-3/2, -7/4)$
**11.** $y = -2(x - 2)^2 - 1, (2, -1)$
**13.** Vertex is $(-5/2, -1/4)$; axis is $x = -5/2$;
x-intercepts are $-3$ and $-2$;
y-intercept is 6.

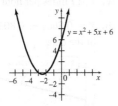

| For Exercises ... | 3–8 | 9–24, 49, 50a, 51c, 58 | 25–38 | 39–46 | 50b, 51a, b, 52, 56, 57, 59, 60 | 53, 55 |
|---|---|---|---|---|---|---|
| Refer to Example ... | 1–3 | 4 | 8 | Before Example 8 | 6 | 7 |

**15.** Vertex is $(-3, 2)$; axis is $x = -3$;
x-intercepts are $-4$ and $-2$;
y-intercept is $-16$.

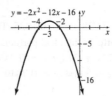

**17.** Vertex is $(-2, -16)$; axis is $x = -2$;
x-intercepts are $-2 \pm 2\sqrt{2} \approx 0.83$ or $-4.83$;
y-intercept is $-8$.

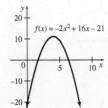

**19.** Vertex is $(1, 3)$; axis is $x = 1$;
no x-intercepts; y-intercept is 5.

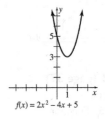

**21.** Vertex is $(4, 11)$; axis is $x = 4$; x-intercepts are
$4 \pm \sqrt{22}/2 \approx 6.35$ or 1.65; y-intercept is $-21$.

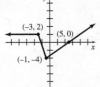

**23.** Vertex is $(4, -5)$; axis is $x = 4$;
x-intercepts are $4 \pm \sqrt{15} = 7.87$ or 0.13; y-intercept is 1/3.

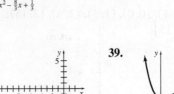

**25.** D  **27.** C  **29.** E
**31.**
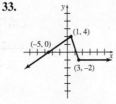
**33.**
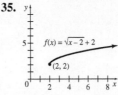
**35.**
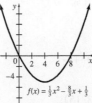
**37.**
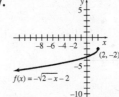
**39.**
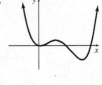

**41.** 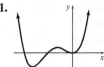  **43.**   **45.**

**47. a.** $r$  **b.** $-r$  **c.** $-r$  **49. a.** 87 yr  **b.** 98 yr

**51. a.** 28.5 weeks  **b.** 0.81  **c.** 0 weeks or 57 weeks of gestation; no

**53. a.**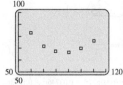

**b.** Quadratic

**c.** $f(x) = 0.021018x^2 - 3.6750x + 227.2$

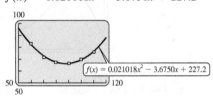

**d.** $f(x) = 0.0245(x - 90)^2 + 67.2$

**e.**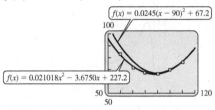

**f.** 72.3, 72.0

**55. a.**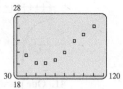

**b.** Quadratic  **c.** $f(x) = 0.002071x^2 - 0.2262x + 26.78$

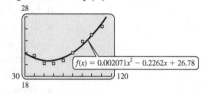

**d.** $f(x) = 0.00248(x - 60)^2 + 20.3$

**e.**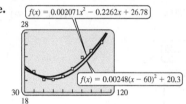

**57. a.** 16 ft  **b.** 2 sec  **59.** 95 ft by 190 ft  **61.** $y = -(1/15)x^2$; $10\sqrt{3}$ m $\approx$ 17.32 m

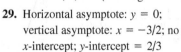

## Exercises 1.5 (page 58)

**3.** 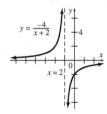  **5.**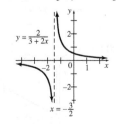

| For Exercises ... | 3–6 | 7–15, 21–26 | 16–20, 27–42, 51–53 | 46–49, 54 | 55, 56 | 57–59 |
|---|---|---|---|---|---|---|
| Refer to Example ... | 1 | 4 | 7 | 2, 3 | 8 | 5 |

**7.** D  **9.** E  **11.** I  **13.** G  **15.** A  **17.** D  **19.** E

**21.** 4, 6, etc. (true degree = 4); +

**23.** 5, 7, etc. (true degree = 5); +

**25.** 7, 9, etc. (true degree = 7); −

**27.** Horizontal asymptote: $y = 0$; vertical asymptote: $x = -2$; no $x$-intercept; $y$-intercept $= -2$

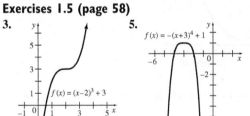

**29.** Horizontal asymptote: $y = 0$; vertical asymptote: $x = -3/2$; no $x$-intercept; $y$-intercept $= 2/3$

**31.** Horizontal asymptote: $y = 2$; vertical asymptote: $x = 3$; $x$-intercept $= 0$; $y$-intercept $= 0$

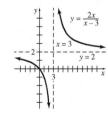

**33.** Horizontal asymptote: $y = 1$; vertical asymptote: $x = 4$; $x$-intercept $= -1$; $y$-intercept $= -1/4$

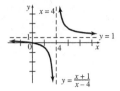

**35.** Horizontal asymptote: $y = -1/2$; vertical asymptote: $x = -5$; $x$-intercept $= 3/2$; $y$-intercept $= 3/20$

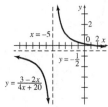

**37.** Horizontal asymptote: $y = -1/3$; vertical asymptote: $x = -2$; $x$-intercept $= -4$; $y$-intercept $= -2/3$

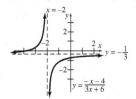

**39.** No asymptotes; hole at $x = -4$; $x$-intercept $= -3$; $y$-intercept $= 3$

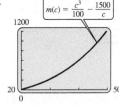

**41.** One possible answer is $y = 2x/(x - 1)$.

**43. a.** 0   **b.** 2, $-3$
 **d.** $(x + 1)(x - 1)(x + 2)$
 **e.** $3(x + 1)(x - 1)(x + 2)$
 **f.** $(x - a)$

**45. a.** Two; one at $x = -1.4$ and one at $x = 1.4$
 **b.** Three; one at $x = -1.414$, one at $x = 1.414$, and one at $x = 1.442$

**47. a.**

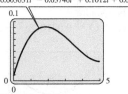

**b.** Close to 2 hours
**c.** About 1.1 to 2.7 hours

**49. a.** 220 g; 602.5 g; 1220 g
 **b.** $c < 19.68$
 **c.**

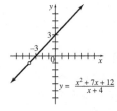

 **d.** 41.9 cm

**51. a.** $[0, \infty)$   **b.**

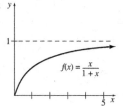

**c.**

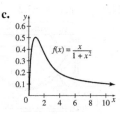

**d.** The population of the next generation, $f(x)$, gets smaller when the current generation, $x$, is larger.

**53. c.** $y = (0.025 + 0.895x)/(0.24 + 0.76x)$
 **d.**    **e.** 0.031

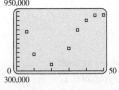

**55. a.** \$6700; \$15,600; \$26,800; \$60,300; \$127,300; \$328,300; \$663,300
 **b.** No   **c.**

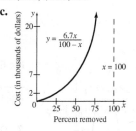

**57. a.**

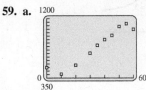

 **b.** $y = 531.27x^2 - 20{,}425x + 712{,}448$;

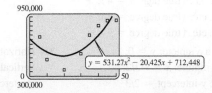

 **c.** $y = -49.713x^3 + 4785.53x^2 - 123{,}025x + 1{,}299{,}118$;

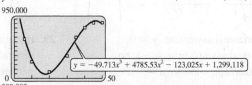

**59. a.**

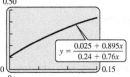

 **b.** $y = 0.06800x^2 + 8.7333x + 400.63$   **c.**

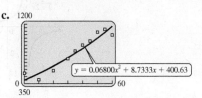

**d.** $y = -0.013270x^3 + 1.2847x^2 - 19.608x + 491.50$   **e.**

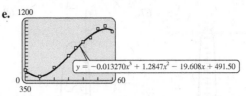

## Chapter 1 Review Exercises (page 65)

**1.** False
**2.** False
**3.** True
**4.** False

| For Exercises... | 1–10, 19, 25–54, 75–77, 88, 89a, b | 11–12, 20, 78, 79, 89c, d, e, 90 | 18, 55–60, 73, 74 | 15–16, 61–64, 80–82, 87 | 13–14, 17, 21–24, 65–72, 83–86, 89f, 91–93 |
|---|---|---|---|---|---|
| Refer to Section... | 1 | 2 | 3 | 4 | 5 |

**5.** True  **6.** False  **7.** True  **8.** False  **9.** False  **10.** False  **11.** False  **12.** True  **13.** True  **14.** False  **15.** True
**16.** True  **17.** False  **18.** False  **20.** $\Sigma x, \Sigma y, \Sigma xy, \Sigma x^2, \Sigma y^2,$ and $n$  **25.** 1  **26.** 2  **27.** $-2/11$  **28.** Undefined
**29.** $-4/3$  **30.** 4  **31.** 0  **32.** 0  **33.** 5  **34.** 1/5  **35.** $y = (2/3)x - 13/3$  **36.** $y = -(1/4)x + 2$  **37.** $y = -x - 3$
**38.** $y = -(7/5)x - 1/5$  **39.** $y = -10$  **40.** $y = 5$  **41.** $2x - y = 10$  **42.** $5x - 8y = -40$
**43.** $x = -1$  **44.** $x = 7$  **45.** $y = -5$  **46.** $x = -3$

**47.**
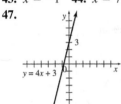

**48.**
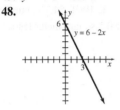

**49.**
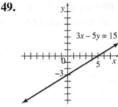

**50.**
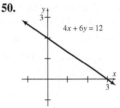

**51.**
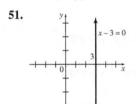

**52.**
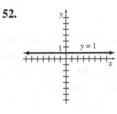

**53.**
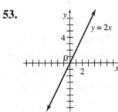

**54.**
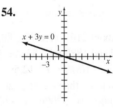

**55.** $(-3, 14), (-2, 5), (-1, 0), (0, -1), (1, 2),$ $(2, 9), (3, 20)$; range: $\{-1, 0, 2, 5, 9, 14, 20\}$

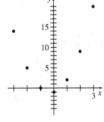

**56.** $(-3, -3/10), (-2, -2/5), (-1, -1/2),$ $(0, 0), (1, 1/2), (2, 2/5), (3, 3/10)$; range: $\{-1/2, -2/5, -3/10, 0, 3/10, 2/5, 1/2\}$

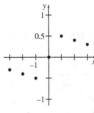

**57. a.** 17  **b.** 4  **c.** $5k^2 - 3$  **d.** $-9m^2 + 12m + 1$  **e.** $5x^2 + 10xh + 5h^2 - 3$  **f.** $-x^2 - 2xh - h^2 + 4x + 4h + 1$
**g.** $10x + 5h$  **h.** $-2x - h + 4$  **58. a.** 23  **b.** 19  **c.** $18m^2 + 5$  **d.** $3k^2 - 4k - 1$  **e.** $2x^2 + 4xh + 2h^2 + 5$
**f.** $3x^2 + 6xh + 3h^2 + 4x + 4h - 1$  **g.** $4x + 2h$  **h.** $6x + 3h + 4$  **59.** $(-\infty, 0) \cup (0, \infty)$  **60.** $[2, \infty)$

**61.**
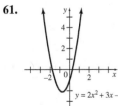

**62.**
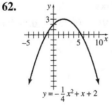

**63.**
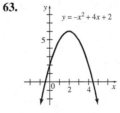

**64.**
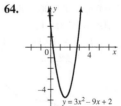

**65.**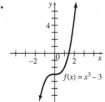

**66.**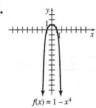

**67.**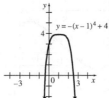

**68.**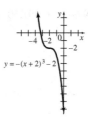

**69.**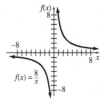

**70.**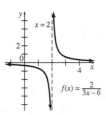

**71.**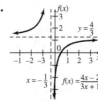

**72.**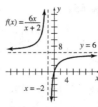

**73.** $(7x^2 + 3x)/(x + 1)^2$; $(4x^2 + 3x)/(4x^2 + 3x + 1)$  **74.** $3x^2 - 18x - 17$; $9x^2 + 6x - 15$

**75. a.** $y = 0.071t + 2.03$  **b.** 3.024 million  **c.** Increasing by 71,000 each year

**76. a.** $b(t) = -0.39t + 64.5$; $p(t) = -0.175t + 47.8$; $c(t) = 0.78t + 42.4$  **b.** Beef: decreasing 0.39 pound per year; pork: decreasing 0.175 pound per year; chicken: increasing 0.78 pound per year  **c.** 1996  **d.** Beef: 54.75 pounds; pork: 43.425 pounds; chicken: 61.9 pounds  **77. a.** 8.06; each additional local species per 0.5 square meter leads to about 8 additional regional species per 0.5 hectare.  **b.** 39.5  **c.** 9.78

**78. a.** 0.8664; yes  **b.** 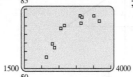 ; yes  **c.** $Y = 0.01423x + 32.19$  **d.** 80.56

**79. a.** $Y = 0.9724x + 31.43$  **b.** About 216  **c.** 0.93  **80. a.** 24 minutes  **b.** 891 g  **c.** $[0, 24]$  **81.** The third day; 104.2°F

**82. b.** 0 and 1.35 square meters; 0 ml and 1889 ml  **83. a.** 1/8; the fraction of radiation let in is 1 over the SPF rating.

**b.**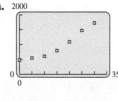
**c.** $UVB = 1 - 1/SPF$  **d.** 12.5%  **e.** 3.3%  **f.** It decreases to 0. 0  **84. a.** 10.915 billion; this is about 4.006 billion more than the estimate of 6.909 billion.  **b.** 26.56 billion; 96.32 billion

**85. a.** $28,000  **b.** $7000  **c.** $63,000  **d.** 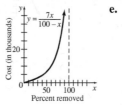  **e.** No

**86. a.** 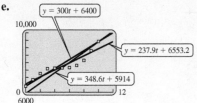  **b.** $Y = 43.11t + 319.5$; $Y = 1.35t^2 + 2.607t + 488.3$; $Y = -0.0362t^3 + 2.98t^2 - 15.5t + 515.5$

**c.** 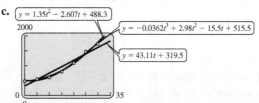  **d.** Linear: 1,828,000; quadratic: 2,233,000; cubic: 2,071,000

**87. a.** The number of acres is increasing from 447,900 to 885,900  **b.** $g(t) = 7.95(t - 2000)^2 - 8.85(t - 2000) + 447.9$

**88.** $y = 1.31t + 48.45$  **89. a.** $y = 300t + 6400$  **b.** $y = 348.6t + 5914$  **c.** $Y = 237.9t + 6553.2$  **d.** 0.9434; yes

**e.**   **f.** $y = 10.11t^3 - 141.87t^2 + 718.26t + 6336.28$;

**90. a.** 0.7485; yes, but the fit is not very good.  **b.** 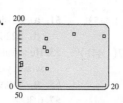  **c.** $Y = 4.179x + 110.6$  **d.** $4179

**91. d.** $w = 0$  **93. a.** $P = 5.48D$; $P = 1.00D^{1.5}$; $P = 0.182D^2$  **b.** 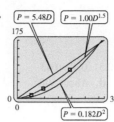  $P = 1.00D^{1.5}$ is the best fit.

**c.** 248.3 yr  **d.** $P = 1.00D^{1.5}$, the same as the function found in part b.

# Chapter 2 Exponential, Logarithmic, and Trigonometric Functions

## Exercises 2.1 (page 80)

| For Exercises . . . | 3–11, 29–32 | 13–20 | 37–39, 42–45, 47 | 40, 41, 46, 48–50 | 51–60 |
|---|---|---|---|---|---|
| Refer to Example . . . | 1, 2 | 3 | 6 | 7 | 4, 5 |

**1.** 2, 4, 8, 16, 32, . . . , 1024; 1.125899907 × 10^{15}

**3.** E  **5.** C  **7.** F  **9.** A  **11.** C  **13.** 5  **15.** $-4$  **17.** $-3$  **19.** 21/4  **21.** $-12/5$  **23.** 2, $-2$  **25.** 0, $-1$  **27.** 0, 1/2

**29.** 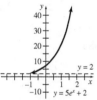  **31.** 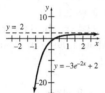  **37. a.** The function gives a population of about 3660 million, which is close to the actual population.  **b.** 6022 million  **c.** 7725 million

**39. a.** 41.93 million  **b.** 12.48 million  **c.** 2.1%, 2.3%; Asian  **d.** 37.99 million
**e.** Hispanic: 2039;      Asian: 2036;      Black: 2079

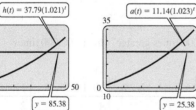

**41. a.** 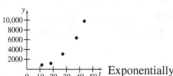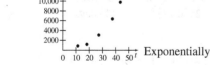 Exponentially
**b.** $f(t) = 487(1.0723)^t$
**c.** 7.2%
**d.** $f(t) = 398.8(1.0762)^t$

**43. a.** 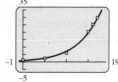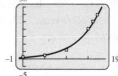 The function fits the data closely.

**b.** About 1.6 percent  **c.** About 6.8 percent
**d.** About 12.6 percent  **e.** About 0.24 percent
**f.** About 1.6 percent  **g.** About 3.1 percent

**45. a.** 0; no cortisone has been administered yet.  **b.** The concentration approaches 0.  **c.** About 0.54 hr
**47. a.** 55 grams  **b.** 10 months

**49. a.** $y = 79.155t + 0.275, y = 3.298t^2 + 0.275, y = 0.275(1.445)^t$
**b.**

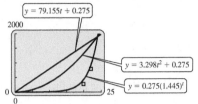
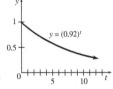

$y = 0.275(1.445)^t$ is the best fit.

**c.** 17,200
**d.** $y = 0.1787(1.441)^t$

**51. a.** $2166.53 **b.** $2189.94 **c.** $2201.90
**d.** $2209.97 **e.** $2214.03
**53.** He should choose the 5.9% investment, which would yield $23.74 additional interest.
**55. a.** $10.94 **b.** $11.27 **c.** $11.62
**57.** 6.30%

**59. a.** 1, 0.92, 0.85, 0.78, 0.72, 0.66, 0.61, 0.56, 0.51, 0.47, 0.43
**b.**

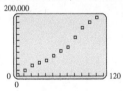

**c.** About $384,000 **d.** About $98

### Exercises 2.2 (page 92)

**1.** $\log_5 125 = 3$ **3.** $\log_3 81 = 4$ **5.** $\log_3(1/9) = -2$
**7.** $2^5 = 32$ **9.** $e^{-1} = 1/e$ **11.** $10^5 = 100,000$ **13.** 2
**15.** 3 **17.** $-4$ **19.** $-2/3$ **21.** 1 **23.** 5/3 **25.** $\log_3 4$
**27.** $\log_5 3 + \log_5 k$ **29.** $1 + \log_3 p - \log_3 5 - \log_3 k$
**31.** $\ln 3 + (1/2) \ln 5 - (1/3) \ln 6$ **33.** $5a$ **35.** $2c + 3a + 1$ **37.** 2.113 **39.** $-0.281$ **41.** $x = 1/6$ **43.** $z = 4/3$
**45.** $r = 25$ **47.** $x = 1$ **49.** No solution **51.** $x = 3$ **53.** $x = 5$ **55.** $x = 1/\sqrt{3e} \approx 0.3502$
**57.** $x = (\ln 6)/(\ln 2) \approx 2.5850$ **59.** $k = 1 + \ln 6 \approx 2.7918$ **61.** $x = (\ln 3)/(\ln(5/3)) \approx 2.1507$
**63.** $x = (\ln 1.25)/(\ln 1.2) \approx 1.2239$ **65.** $e^{(\ln 10)(x+1)}$ **67.** $20.09^x$ **69.** $x < 5$ **75. a.** 23.4 yr **b.** 11.9 yr **c.** 9.0 yr
**77. a.** 2039 **b.** 2036 **79.** 21.3 yr **81. a.** About 0.693 **b.** $\ln 2$ **c.** Yes **83. a.** About 1.099 **b.** About 1.386
**85. a.** 2590 sq cm **b.** 4211 sq cm **c.** 7430 g **87.** About every 7 hr, $T = (3 \ln 5)/\ln 2$
**89. b.** 2015: 0.01207, 2020: 0.01207, 2025: 0.01228
**91. a.**

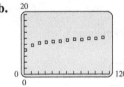

No **b.**

Yes **c.** $Y = 0.02940x + 9.348$
**d.** $Y = 11,471(1.0298)^x$

| For Exercises ... | 1–12 | 13–24 | 27–36 | 37–40 | 41–56 |
|---|---|---|---|---|---|
| Refer to Example ... | 1 | 2 | 3 | 4 | 5 |

| For Exercises ... | 57–64 | 65–68 | 69–70 | | 75–80 | 81–83 |
|---|---|---|---|---|---|---|
| Refer to Example ... | 6 | 7 | First CAUTION | | 8 | 9 |

**93. a.** 530 **b.** 3500 **c.** 6000 **d.** 1800 **e.** $e^{-1/5} \approx 0.8$ **95.** No; 1/10 **97. a.** 1000 times greater **b.** 1,000,000 times greater

### Exercises 2.3 (page 100)

**1.** $y_0$ represents the initial quantity; $k$ represents the rate of growth or decay. **3.** The half-life of a quantity is the time period for the quantity to decay to one-half of the initial amount. **7. a.** $y = y_0 e^{0.05776t}$ **b.** $y = y_0 2^{t/12}$
**c.** 1,048,576; 1,073,741,824 **9. a.** $y = 50,000e^{-0.102t}$ **b.** About 6.8 hours **13.** 0.5% **15.** 87% **17.** About 4100 years ago
**19.** About 1600 years **21. a.** 3.8 g **b.** About 8600 years **23. a.** $y = 25.0 e^{-0.00497t}$ **b.** $y = 25(0.78)^{t/50}$ **c.** 139 days
**25. a.** $y = 10e^{0.0095t}$ **b.** 42.7°C **27.** About 30 minutes

| For Exercises ... | 6 | 6–9, 11, 12, 25 | 10, 14, 26–28 | 13, 15, 17–24 |
|---|---|---|---|---|
| Refer to Example ... | 3 | 1 | 4 | 2 |

### Exercises 2.4 (page 114)

**1.** $\pi/3$ **3.** $5\pi/6$ **5.** $3\pi/2$ **7.** $11\pi/4$
**9.** 225° **11.** $-390°$ **13.** 288° **15.** 105°

*Note: In Exercises 17–24 we give the answers in the following order: sine, cosine, tangent, cotangent, secant, and cosecant.*

| For Exercises ... | 1–16 | 17–20 | 21–24 | 25–48 | 49–54 |
|---|---|---|---|---|---|
| Refer to Example ... | 1 | 2 | 5 | 3, 4, 5 | 7 |

| For Exercises ... | 55–62, 77, 79, 80, 82–84, 88, 89 | 63–74, 85–86 | 78, 81, 87 | 90 |
|---|---|---|---|---|
| Refer to Example ... | 6 | 8 | 9 | 10 |

**17.** $4/5; -3/5; -4/3; -3/4; -5/3; 5/4$ **19.** $-24/25; 7/25; -24/7; -7/24; 25/7; -25/24$ **21.** $+$ $+$ $+$ $+$ $+$
**23.** $-$ $-$ $+$ $+$ $-$ $-$ **25.** $\sqrt{3}/3; \sqrt{3}; 2$ **27.** $\sqrt{3}/2; \sqrt{3}/3; 2\sqrt{3}/3$ **29.** $-1; -1$ **31.** $-\sqrt{3}/2; -2\sqrt{3}/3$ **33.** $\sqrt{3}/2$
**35.** 1 **37.** 2 **39.** $-1$ **41.** $-\sqrt{2}/2$ **43.** $-\sqrt{2}$ **45.** 1 **47.** 1/2 **49.** $\pi/3, 5\pi/3$ **51.** $3\pi/4, 7\pi/4$ **53.** $5\pi/6, 7\pi/6$
**55.** 0.6293 **57.** $-1.5399$ **59.** 0.3558 **61.** 0.3292 **63.** $a = 1; T = 2\pi/3$ **65.** $a = 2; T = 8$

**67.** 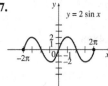   **69.** 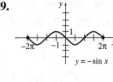   **71.** 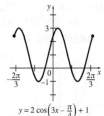   **73.**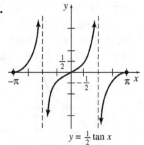

**75. a.** All are 60°.   **b.** 30°, 60°, 90°   **c.** 1, $\sqrt{3}$, 2   **77. a.** $y = 2\sin(2\pi t/0.350)$   **b.** 0.0875 second   **c.** $-1.95°$
**79. a.** 100   **b.** 258   **c.** 122   **d.** 296

**81. a.**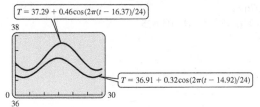

No, the two functions never cross.
**b.** About 2:55 P.M.   **c.** About 4:22 P.M.
**83.** $2.2 \times 10^8$ m per second   **85.** 240°

**87. a.**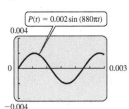

**b.** 0, 1/880, 1/440   **c.** $T = 1/440$; frequency: 440 cycles per second
**89. a.** About 8°F   **b.** About 37°F   **c.** About 45°F   **d.** About 1°F
**e.** 62°F and $-12$°F   **f.** 365
**91.** 60.2 m

## Chapter 2 Review Exercises (page 120)

**1.** False   **2.** True   **3.** False   **4.** False
**5.** False   **6.** False   **7.** True   **8.** False
**9.** False   **10.** True   **11.** True   **12.** False
**13.** True   **14.** False   **15.** False   **16.** True

| For Exercises . . . | 1, 24, 27–30, 35–38, 67, 104a, 106, 107, 109a, b, 110, 111, 117–122 | 2–10, 17, 23, 31–34, 39–66, 68, 69, 104b, 105, 108, 109c, 112, 123, 124 | 11, 18, 125–126 | 12–16, 19–22, 25, 26, 70–103, 113–114, 116, 127 |
|---|---|---|---|---|
| Refer to Example . . . | 1 | 2 | 3 | 4 |

**23.** $(-\infty, -3) \cup (3, \infty)$   **24.** $(-\infty, 0) \cup (0, \infty)$
**25.** $\{x \mid x \neq \pi/2, -3\pi/2, 5\pi/2, -7\pi 2, \ldots\}$   **26.** $\{x \mid x \neq \pm\pi/2, \pm 3\pi/2, \pm 5\pi/2, \ldots\}$
**27.** 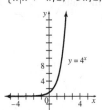   **28.** 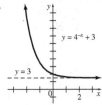   **29.** 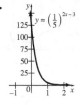   **30.**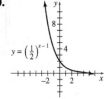

**31.**    **32.** 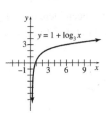   **33.** 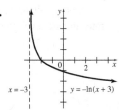   **34.**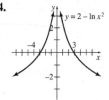

**35.** $-5$   **36.** 1/2   **37.** $-6$   **38.** 1/4   **39.** $\log_3 243 = 5$   **40.** $\log_5 \sqrt{5} = 1/2$   **41.** $\ln 2.22554 = 0.8$   **42.** $\log 12 = 1.07918$
**43.** $2^5 = 32$   **44.** $9^{1/2} = 3$   **45.** $e^{4.41763} = 82.9$   **46.** $10^{0.50651} = 3.21$   **47.** 4   **48.** 4/5   **49.** 3/2   **50.** 3/2
**51.** $\log_5(21k^4)$   **52.** $\log_3(y/4)$   **53.** $\log_3(y^4/x^2)$   **54.** $\log_4(r^4)$   **55.** $p = 1.581$   **56.** $z = 4.183$   **57.** $m = -1.807$
**58.** $k = -0.884$   **59.** $x = -3.305$   **60.** $x = 1.213$   **61.** $m = 2.156$   **62.** $p = 1.830$ or $-6.830$   **63.** $k = 2$
**64.** $x = 119$   **65.** $p = 3/4$   **66.** $m = 14$   **67. a.** $(-\infty, \infty)$   **b.** $(0, \infty)$   **c.** 1   **d.** $y = 0$   **e.** Greater than 1
**f.** Between 0 and 1   **68. a.** $(0, \infty)$   **b.** $(-\infty, \infty)$   **c.** 1   **d.** $x = 0$   **e.** Greater than 1   **f.** Between 0 and 1

**70.** $\pi/2$ **71.** $8\pi/9$ **72.** $5\pi/4$ **73.** $3\pi/2$ **74.** $2\pi$ **75.** $9\pi/4$ **76.** 900° **77.** 135° **78.** 81° **79.** 54° **80.** 117°
**81.** 156° **82.** $\sqrt{3}/2$ **83.** $-\sqrt{3}$ **84.** $\sqrt{2}/2$ **85.** $-2\sqrt{3}/3$ **86.** $2\sqrt{3}/3$ **87.** $-\sqrt{3}/3$ **88.** 1/2 **89.** 1/2 **90.** 2
**91.** $2\sqrt{3}/3$ **92.** 0.7314 **93.** 0.3090 **94.** $-2.1445$ **95.** $-0.8387$ **96.** 0.7058 **97.** 0.6811 **98.** 0.8290 **99.** 3.4868
**100.** 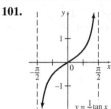 **101.** 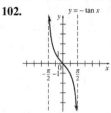 **102.** 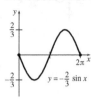 **103.**

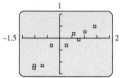 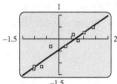

**104. a.** $y = 15{,}000\, e^{0.0313t}$ **b.** About 35 years **105.** About 7.7 m **106.** 0.25; 0.69 minutes
**107. a.** When it is first injected; 0.08 g **b.** Never **c.** It approaches $c/a = 0.0769$ g.
**108. a.** Yes **b.** **c.** $y \approx 0.3565x^{0.7097}$ **d.** 0.9625

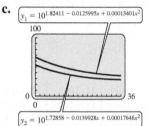 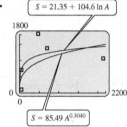

**109. a.** About $4.19 \times 10^{-15}$ cubic meters **b.** $4.19 \times 10^{-15}\, 2^t$ cubic meters **c.** About 27 days **110.** 187.9 cm; 345 kg
**111. a.** $[0, 36]$
 **b.** Decreasing
 **c.**

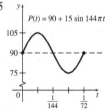

 **d.** 10.7 breaths per minute

**112. a.** $S = 21.35 + 104.6 \ln A$
 **b.** $S = 85.49\, A^{0.3040}$
 **c.**

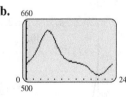

 **d.** 742.2, 694.7

**113.** 105; 75

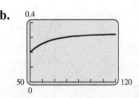

**114. a.** 24 hours **b.**

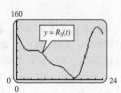

**c.** 511 ng/dl, the value of $C_0$.
**d.** About 634 ng/dl, the value of $C_1$
**e.** No

**115. a.** $g(t) = 0.028t - 0.88$ **b.**

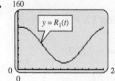

Per capita grain harvests have been increasing, but at a slower rate, and are leveling off at about 0.32 ton per person.

**116. a.**

As more terms are added, the function has greater complexity.
 **b.** $R_1(12) = 20.0$; $R_2(12) = 34.3$; $R_3(12) = 37.2$; $R_3$ gives the most accurate value.

**117.** $4173.68  **118.** $921.95  **119.** $13,701.92  **120.** $15,510.79  **121.** $2574.01  **122.** $17,901.90
**123.** 12 years; 19 years  **124.** 70 quarters or 17.5 years; 111 quarters or 27.75 years  **125. a.** $y = 100,000\,e^{-0.05t}$  **b.** 7.1 years
**126. a.** 0 yr  **b.** $1.85 \times 10^9$ yr  **c.** As $r$ increases, $t$ increases, but at a slower and slower rate. As $r$ decreases, $t$ decreases at a
faster and faster rate.  **127. a.**  **b.** $y = 12.5680 \sin(0.547655t - 2.34990) + 50.2671$
**c.** $y = 12.5680 \sin(0.547655t - 2.34990) + 50.2671$  **d.** About 11.5 months.

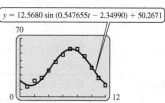

# Chapter 3 The Derivative

## Exercises 3.1 (page 142)
**1.** c  **3.** b  **5. a.** 3  **b.** 1
**7. a.** 0  **b.** Does not exist
**9. a. i.** $-1$  **ii.** $-1/2$
**iii.** Does not exist  **iv.** Does not exist
**b. i.** $-1/2$  **ii.** $-1/2$  **iii.** $-1/2$
**iv.** $-1/2$  **11.** 3  **15.** 4  **17.** 10
**19.** Does not exist  **21.** 1  **23.** $-18$

| For Exercises . . . | 1, 7b, 8b, 9a, 89, 90 | 2, 7a, 10a, 59–62 | 3, 5b, 6a, 17–22 | 4, 95b | 5a, 6b, 8a, 9b, 10b, 15, 16, 95a |
|---|---|---|---|---|---|
| Refer to Example . . . | 5 | 3 | 2 | 4 | 1 |

| For Exercises . . . | 11, 12 | 23–34 | 35–42, 45–48 | 43, 44 | 49–58, 81–87, 91–94, 96–98 | 63–69, 77–80 | 71 |
|---|---|---|---|---|---|---|---|
| Refer to Example . . . | 11 | 6 | 8 | 9 | 12 | 10 | 13 |

**25.** 1/3  **27.** 3  **29.** 512  **31.** 2/3  **33.** 1  **35.** 6  **37.** 3/2  **39.** $-5$  **41.** $-1/9$  **43.** 1/10  **45.** $2x$  **47.** 1  **49.** 3/7
**51.** 3/2  **53.** 0  **55.** $\infty$ (does not exist)  **57.** $-\infty$ (does not exist)  **59.** 1  **61. a.** 2  **b.** Does not exist  **63.** 6  **65.** 1.5
**67. a.** Does not exist  **b.** $x = -2$  **c.** If $x = a$ is a vertical asymptote for the graph of $f(x)$, then $\lim_{x \to a} f(x)$ does not exist.
**71. a.** 0  **b.** $y = 0$  **73. a.** $-\infty$ (does not exist)  **b.** $x = 0$  **77.** 5  **79.** 0.3333 or 1/3  **81. a.** 1.5  **83. a.** $-2$  **85. a.** 8
**89. a.** 2 million units  **b.** Does not exist  **c.** 3.5 million units  **d.** 16 months  **91.** $60; the average cost per test approaches $60.
**93. a.** 65 teeth  **b.** 72 teeth  **95. a.** 0  **b.** $\infty$  **97. a.** 0.572  **b.** 0.526  **c.** 0.503  **d.** 0.5; the numbers in a, b, and c give the
probability that the legislator will vote yes on the second, fourth, and eighth votes. In d, as the number of roll calls increases, the
probability of a yes vote approaches 0.5 but is never less than 0.5.

## Exercises 3.2 (page 152)
**1.** $a = -1$:  **a.** $f(-1)$ does not exist.  **b.** 1/2
**c.** 1/2  **d.** 1/2  **e.** $f(-1)$ does not exist.

| For Exercises . . . | 1–6, 36, 42, 43 | 7–20, 33–35 | 21–30 | 37, 38, 41, 44–46 |
|---|---|---|---|---|
| Refer to Example . . . | 1 | 2 | 3 | 4 |

**3.** $a = 1$:  **a.** 2  **b.** $-2$  **c.** $-2$  **d.** $-2$  **e.** $f(1)$ does not equal the limit.  **5.** $a = -5$:  **a.** $f(-5)$ does not exist
**b.** $\infty$ (does not exist)  **c.** $-\infty$ (does not exist)  **d.** Limit does not exist.  **e.** $f(-5)$ does not exist and the limit does not exist;
$a = 0$:  **a.** $f(0)$ does not exist.  **b.** 0  **c.** 0  **d.** 0  **e.** $f(0)$ does not exist.  **7.** $x = 0$, limit does not exist; $x = 2$, limit does not exist.
**9.** $x = 2$, limit is 4.  **11.** Nowhere  **13.** $x = -2$, limit does not exist.  **15.** $x < 1$, limit does not exist.  **17.** $x = 0, -\infty$ (limit
does not exist); $x = 1, \infty$ (limit does not exist).  **19.** $x = -2$, limit does not exist.

**21. a.**

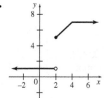

**b.** 2  **c.** 1, 5

**23. a.**

**b.** $-1$  **c.** 11, 3

**25. a.**

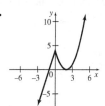

**b.** None

**27.** 2/3  **29.** 4  **33.** Discontinuous at $x = 1.2$  **35.** a

**37. a.** $1.0995 \times 10^{12}$  **b.**

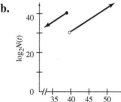

**c.** No, at $t = 40$

**41. a.** About 687 g  **b.** No  **c.**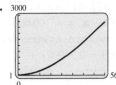

**43. a.** $500  **b.** $1500  **c.** $1000  **d.** Does not exist
**e.** Discontinuous at $x = 10$; a change in shifts  **f.** 15
**45. a.** $100  **b.** $150  **c.** $125  **d.** At $x = 100$

## Exercises 3.3 (page 164)

| For Exercises . . . | 1–12 | 13–22, 31, 35, 36, 46 | 23–28, 34, 38, 47, 48 | 32, 33, 39–43 | 44, 45 |
|---|---|---|---|---|---|
| Refer to Example . . . | 1 | 3, 4 | 5 | 2 | 6 |

**1.** 6  **3.** −15  **5.** 1/3  **7.** −1/3  **9.** 0.4323
**11.** $2\sqrt{2}/\pi$  **13.** 17  **15.** 18  **17.** 5  **19.** 2
**21.** 2  **23.** 1  **25.** 6.773  **27.** 1.121  **31. a.** 6% per day  **b.** 7% per day  **33. a.** 2; from 1 min to 2 min, the population of bacteria increases, on the average, 2 million per min.  **b.** −0.8; from 2 min to 3 min, the population of bacteria decreases, on the average, 0.8 million or 800,000 per min.  **c.** −2.2; from 3 min to 4 min, the population of bacteria decreases, on the average, 2.2 million per min.  **d.** −1: from 4 min to 5 min, the population decreases, on the average, 1 million per min.  **e.** 2 min  **f.** 3 min
**35. a.** 0.288 mm per wk  **c.**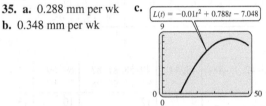
**b.** 0.348 mm per wk

**37. a.** Increased about 873,000 people per year  **b.** Increased about 967,000 people per year  **39. a.** −0.425 percent per year; 0.15 percent per year; −0.1375 percent per year  **b.** −1.3 percent per year; 0.975 percent per year; −0.1625 percent per year.  **c.** −0.775 percent per year; 0.65 percent per year; −0.0625 percent per year  **41. a.** Decreases $23.5 billion per year  **b.** Increases $25.7 billion per year

**43. a.** 4° per 1000 ft  **b.** 1.75° per 1000 ft  **c.** −4/3° per 1000 ft  **d.** 0° per 1000 ft  **e.** 3000 ft; 1000 ft; if 7000 ft is changed to 10,000 ft, the lowest temperature would be at 10,000 ft.  **f.** 9000 ft  **45. a.** 50 mph  **b.** 44 mph  **47. a.** $55.26 per year
**b.** $70.52 per year  **c.** $62.27 per year

## Exercises 3.4 (page 181)

**1. a.** 0  **b.** 1  **c.** −1
**d.** Does not exist  **e.** $m$
**3.** At $x = -2$  **5.** 2  **7.** 1/4  **9.** 0

| For Exercises . . . | 5–10, 53–59 | 11–14 | 15, 16 | 17, 18 | 19, 20 | 21–26 | 27–34, 42–45 | 35–38 | 49–52 |
|---|---|---|---|---|---|---|---|---|---|
| Refer to Example . . . | 3 | 4 | 6 | 7 | 5 | 1, 9 | 2 | 10 | 8 |

**11.** 3; 3; 3; 3  **13.** $-8x + 9$; 25; 9; −15  **15.** $-12/x^2$; −3; does not exist; −4/3  **17.** $1/(2\sqrt{x})$; does not exist; does not exist; $1/(2\sqrt{3})$  **19.** $6x^2$; 24; 0; 54  **21. a.** $y = 10x - 15$  **b.** $y = 8x - 9$  **23. a.** $y = -(1/2)x + 7/2$  **b.** $y = -(5/4)x + 5$
**25. a.** $y = (4/7)x + 48/7$  **b.** $y = (2/3)x + 6$  **27.** −5; −117; 35  **29.** 7.389; 8,886,111; 0.0498  **31.** 1/2; 1/128; 2/9
**33.** $1/(2\sqrt{2})$; 1/8; does not exist  **35.** 0  **37.** −3; −1; 0; 2; 3; 5  **39. a.** $(a, 0)$ and $(b, c)$  **b.** $(0, b)$  **c.** $x = 0$ and $x = b$
**41. a.** Distance  **b.** Velocity  **43.** 56.66  **45.** −0.0158  **49. a.** 8.3 cm  **b.** 0.17 cm of catfish/cm of snake; longer snakes eat bigger catfish at the constant rate of 0.17 cm increase in length of prey to each 1 cm increase in size of snake.  **c.** About 22 cm
**51. a.** 57; the rate of change of the intake of food 5 minutes into a meal is 57 grams per minute.  **c.** $0 \le t \le 24$  **53.** 1000; the population is increasing at a rate of 1000 shellfish per time unit. 570; the population is increasing more slowly at 570 shellfish per time unit. 250; the population is increasing at a much slower rate of 250 shellfish per time unit.  **55.** At 500 ft, the temperature decreases 0.005° per foot. At about 1500 ft, the temperature increases 0.008° per foot. At 5000 ft, the temperature decreases 0.00125° per foot.  **57. a.** 0; about −0.6 mph/oz  **b.** 16 oz  **59.** Answers are in trillions of dollars.  **a.** 2.54; 2.03; 1.02  **b.** 0.061; −0.019; −0.079; −0.120; −0.133

## Exercises 3.5 (page 189)

**3.** $f$: $Y_2$; $f'$: $Y_1$  **5.** $f$: $Y_1$; $f'$: $Y_2$

| For Exercises . . . | 3–6 | 7–25 |
|---|---|---|
| Refer to Example . . . | 4 | 1, 2, 3 |

**7.** 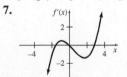  **9.** 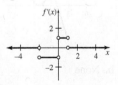  **11.**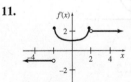

**13.** 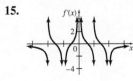  **15.** 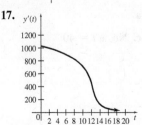  **17.**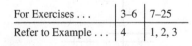

**19.**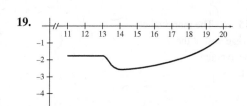

**21.**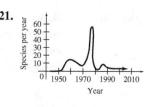

**23.**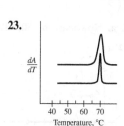

**25.**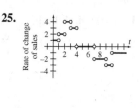

About 9 cm; about 2.6 cm less per year

**Chapter 3 Review Exercises (page 193)**
**1.** True  **2.** True  **3.** True
**4.** False  **5.** False  **6.** False
**7.** False  **8.** True  **9.** True

| For Exercises . . . | 1–3, 17–36, 49–50, 65, 66a-e, 72 | 4–6, 15, 37–48, 66f, 73a | 7–9, 51–54, 67, 74 | 10–12, 55–62, 68, 73b | 63–64, 69–71, 73c |
|---|---|---|---|---|---|
| Refer to Section . . . | 1 | 2 | 3 | 4 | 5 |

**10.** True  **11.** False  **12.** False  **17. a.** 4  **b.** 4  **c.** 4  **d.** 4  **18. a.** −2  **b.** 2  **c.** Does not exist  **d.** −2  **19. a.** ∞
**b.** −∞  **c.** Does not exist  **d.** Does not exist  **20. a.** 1  **b.** 1  **c.** 1  **d.** Does not exist  **21.** ∞  **22.** −3  **23.** 19/9
**24.** Does not exist  **25.** 8  **26.** 7  **27.** −13  **28.** 16  **29.** 1/6  **30.** 1/8  **31.** 2/5  **32.** 0  **33.** 3/8  **34.** −6  **35.** 1  **36.** 1/2
**37.** Discontinuous at $x_2$ and $x_4$  **38.** Discontinuous at $x_1$ and $x_4$  **39.** 0, does not exist, does not exist; −1/3, does not exist, does not
exist  **40.** 1, does not exist, does not exist; −3, does not exist, does not exist  **41.** −5, does not exist, does not exist
**42.** −3, does not exist, −6  **43.** Continuous everywhere  **44.** Continuous everywhere  **45.** 1, does not exist; does not exist
**46.** Continuous everywhere  **47. a.** 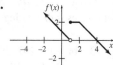  **b.** 1  **c.** 0, 2  **48. a.** 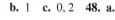  **b.** 2  **c.** 0, 1

**49.** 2  **50.** −13  **51.** 126; 18  **52.** −68; −12  **53.** 9/77; 18/49  **54.** −5/4; −5  **55. a.** $y = 13x − 17$  **b.** $y = 7x − 5$
**56. a.** $y = −(2/3)x + 7/3$  **b.** $y = −4x + 4$  **57. a.** $y = −x + 9$  **b.** $y = −3x + 15$  **58. a.** $y = (2/5)x + 2$
**b.** $y = (1/2)x + 3/2$  **59.** $8x + 3$  **60.** $10x − 6$  **61.** 1.332  **62.** 1.121

**63.** 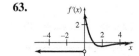  **64.** 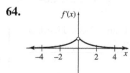  **65.** e  **66. a.** 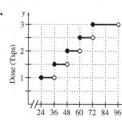  **b.** 1.5 tsp  **c.** 2 tsp
**d.** Does not exist  **e.** 2 tsp
**f.** No, discontinuous at 36, 48, 60, and 72 lb

**67. a.** About 0.13; the number of people aged 65 and over with Alzheimer's disease is going up at a rate of about 0.13 million per
year.  **b.** About 0.34; the number of people aged 65 and over with Alzheimer's disease is going up at a rate of about 0.34 million
per year.  **c.** About 0.16 million people per year

**68. a.** 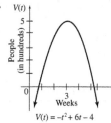  **b.** [0.8, 5.2]
**c.** 3 weeks; 500 cases
**d.** $V'(t) = −2t + 6$
**e.** 0  **f.** +; −

**69. a. i.** About 90 m per minute  **ii.** About 85 m per minute
**b.**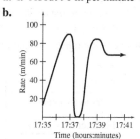

**70. a.** 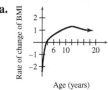  **b.** 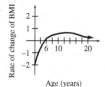  **71.** 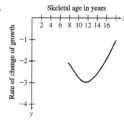  The remaining growth is about 14 cm and the
rate of change is about −2.75 cm per year.

**72. a. i.** $(a/b)^N$ **ii.** 1 **b.** 1    **73. a.** Nowhere    **b.** 50, 130, 230, 770    **74. b.** $x = 7.5$    **c.** The marginal cost equals the average cost at the point where the average cost is smallest.

**c.** $T'(Q)$

[graph with axis marks at 1 and 2, and at 200 400 600 800 $Q$]

# Chapter 4 Calculating the Derivative

## Exercises 4.1 (page 209)

| For Exercises . . . | 1–22, 27–30 | 31–44 | 51–75 |
|---|---|---|---|
| Refer to Example . . . | 1, 2, 3, 5 | 1 (in Section 3.4) | 4, 6, 7 |

**1.** $dy/dx = 36x^2 - 16x + 7$    **3.** $dy/dx = 12x^3 - 18x^2 + (1/4)x$
**5.** $f'(x) = 21x^{2.5} - 5x^{-0.5}$ or $21x^{2.5} - 5/x^{0.5}$
**7.** $dy/dx = 4x^{-1/2} + (9/2)x^{-1/4}$ or $4/x^{1/2} + 9/(2x^{1/4})$    **9.** $dy/dx = -30x^{-4} - 20x^{-5} - 8$ or $-30/x^4 - 20/x^5 - 8$
**11.** $f'(t) = -7t^{-2} + 15t^{-4}$ or $-7/t^2 + 15/t^4$    **13.** $dy/dx = -24x^{-5} + 21x^{-4} - 3x^{-2}$ or $-24/x^5 + 21/x^4 - 3/x^2$
**15.** $p'(x) = 5x^{-3/2} - 12x^{-5/2}$ or $5/x^{3/2} - 12/x^{5/2}$    **17.** $dy/dx = (-3/2)x^{-5/4}$ or $-3/(2x^{5/4})$    **19.** $f'(x) = 2x - 5x^{-2}$ or $2x - 5/x^2$
**21.** $g'(x) = 256x^3 - 192x^2 + 32x$    **23.** b    **27.** $-(9/2)x^{-3/2} - 3x^{-5/2}$ or $-9/(2x^{3/2}) - 3x^{5/2}$    **29.** $-25/3$
**31.** $-28$; $y = -28x + 34$    **33.** 25/6    **35.** (4/9, 20/9)    **37.** $-5, 2$    **39.** $(4 \pm \sqrt{37})/3$    **41.** $(-1/2, -19/2)$
**43.** $(-1, 6), (4, -59)$    **45.** 42    **47. a.** 2    **b.** 1/2    **c.** $[-1, \infty)$    **d.** $(0, \infty)$    **51. a.** 0.4824    **b.** 2.216    **53. a.** 264    **b.** 510
**c.** About 97/4 or 24.25 matings per degree    **55. a.** 1232.62 cm³    **b.** 948.08 cm³/yr    **57.** $R = (2/3)R_0$    **59. a.** $[18, 44]$
**b.** $l'(t) = 0.2356 - 0.005348t$    **c.** 0.1019 cm per wk    **61.** $-38,037$; the number of species is decreasing at a rate of about 38,037 species per gram of body mass, or about 38 for an increase of 1 microgram.

**63. a.**

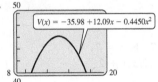

**b.** About 13.58%    **c.** 0
**67. a.** $v(t) = 36t - 13$    **b.** $-13$; 167; 347
**69. a.** $v(t) = -9t^2 + 8t - 10$    **b.** $-10$; $-195$; $-830$
**71. a.** 0 ft/sec; $-32$ ft/sec    **b.** 2 sec    **c.** 64 ft
**73. a.** 18¢; 36¢    **b.** 0.733¢ per year; 1.04¢ per year
**c.** $C(t) = -0.0001493t^3 + 0.02637t^2 - 0.6322t + 3.186$; 0.885¢ per year; 0.865¢ per year    **75. a.** 22.3, 2.64    **b.** 95.8, 0.386

## Exercises 4.2 (page 218)

| For Exercises . . . | 1–10, 29, 34, 39, 43 | 11–26, 30–33, 35, 40, 44, 50–52 | 27, 28, 45 | 41, 42, 46–49 |
|---|---|---|---|---|
| Refer to Example . . . | 1, 2 | 3 | 4 | 5 |

**1.** $dy/dx = 18x^2 - 6x + 4$
**3.** $dy/dx = 8x - 20$
**5.** $k'(t) = 4t^3 - 4t$
**7.** $dy/dx = (3/2)x^{1/2} + (1/2)x^{-1/2} + 2$ or $3x^{1/2}/2 + 1/(2x^{1/2}) + 2$    **9.** $p'(y) = -8y^{-5} + 15y^{-6} + 30y^{-7}$
**11.** $f'(x) = 57/(3x + 10)^2$    **13.** $dy/dt = -17/(4 + t)^2$    **15.** $dy/dx = (x^2 - 2x - 1)/(x - 1)^2$    **17.** $f'(t) = 2t/(t^2 + 3)^2$
**19.** $g'(x) = (4x^2 + 2x - 12)/(x^2 + 3)^2$    **21.** $p'(t) = [-\sqrt{t}/2 - 1/(2\sqrt{t})]/(t - 1)^2$ or $(-t - 1)/[2\sqrt{t}(t - 1)^2]$
**23.** $dy/dx = (5/2)x^{-1/2} - 3x^{-3/2}$ or $(5x - 6)/(2x\sqrt{x})$    **25.** $h'(z) = (-z^{4.4} + 11z^{1.2})/(z^{3.2} + 5)^2$
**27.** $f'(x) = (60x^3 + 57x^2 - 24x + 13)/(5x + 4)^2$    **29.** 77    **31.** In the first step, the numerator should be
$(x^2 - 1)2 - (2x + 5)(2x)$.    **33.** $y = -2x + 9$    **35. a.** $f'(x) = (7x^3 - 4)/x^{5/3}$    **b.** $f'(x) = 7x^{4/3} - 4x^{-5/3}$    **39.** $0, -1.307$,
and 1.307    **41. a.** $s'(x) = m/(m + nx)^2$    **b.** $1/2560 \approx 0.000391$ mm per ml    **43. a.** $N'(t) = 9t^2 - 120t + 300$
**b.** $-84$ million per hr    **c.** 69 million per hr    **d.** The population first declines and then increases.
**45. a.** $dW/dH = (H^2 - 1.86H - 17.7351)/(H - 0.93)^2$    **b.** 5.24 m    **c.** Crows apply optimal foraging techniques.
**47.** $R'(c) = 2ac/(1 + bc^2)^2 - k$; the rate of change of the release of calcium per change in the amount of free calcium
**51. a.** 0.1173    **b.** 2.625

## Exercises 4.3 (page 227)

| For Exercises . . . | 1–6 | 7–14, 53 | 15–20 | 21–28, 57, 58, 63–65 | 29–34, 60 |
|---|---|---|---|---|---|
| Refer to Example . . . | 1 | 2 | 3 | 5, 6 | 7 |

| For Exercises . . . | 35–40, 60 | 55, 61 | 62 | 54, 56 | 45–50 |
|---|---|---|---|---|---|
| Refer to Example . . . | 8 | 9 | 10 | 4 | 1 (in Section 3.4) |

**1.** 1767    **3.** 131    **5.** $320k^2 + 224k + 39$
**7.** $(6x + 55)/8$; $(3x + 164)/4$    **9.** $1/x^2$; $1/x^2$
**11.** $\sqrt{8x^2 - 4}$; $8x + 10$
**13.** $\sqrt{(x - 1)/x}$; $-1/\sqrt{x + 1}$
**15.** If $f(x) = x^{3/5}$ and $g(x) = 5 - x^2$, then $y = f[g(x)]$.    **17.** If $f(x) = -\sqrt{x}$ and $g(x) = 13 + 7x$, then $y = f[g(x)]$.
**19.** If $f(x) = x^{1/3} - 2x^{2/3} + 7$ and $g(x) = x^2 + 5x$, then $y = f[g(x)]$.    **21.** $dy/dx = 4(8x^4 - 5x^2 + 1)^3(32x^3 - 10x)$
**23.** $k'(x) = 288x(12x^2 + 5)^{-7}$    **25.** $s'(t) = (1215/2)t^2(3t^3 - 8)^{1/2}$    **27.** $g'(t) = -63t^2/(2\sqrt{7t^3 - 1})$
**29.** $m'(t) = -6(5t^4 - 1)^3(85t^4 - 1)$    **31.** $dy/dx = 3x^2(3x^4 + 1)^3(19x^4 + 64x + 1)$    **33.** $q'(y) = 2y(y^2 + 1)^{1/4}(9y^2 + 4)$
**35.** $dy/dx = 60x^2/(2x^3 + 1)^3$    **37.** $r'(t) = 2(5t - 6)^3(15t^2 + 18t + 40)/(3t^2 + 4)^2$

**39.** $dy/dx = (-18x^2 + 2x + 1)/(2x - 1)^6$  **43. a.** $-2$  **b.** $-24/7$  **45.** $y = (3/5)x + 16/5$  **47.** $y = x$  **49.** 1, 3
**53.** $P[f(a)] = 18a^2 + 24a + 9$  **55.** About 19.5 mm/wk  **57. a.** 6 million/hr  **b.** 9.75 million/hr  **c.** 19.71 million/hr
**59. a.** $R'(Q) = -Q/[6(C - Q/3)^{1/2}] + (C - Q/3)^{1/2}$  **b.** 2.83  **c.** Increasing  **61.** $p'(t) = Na^N t^{N-1}/(at + 1)^{N+1}$; rate of
change of the probability of a population going extinct with respect to time  **63. a.** $101.22/percent  **b.** $111.86/percent
**c.** $117.59/percent  **65. a.** $8x^7$  **b.** $x^8$; $8x^7$

## Exercises 4.4 (page 235)

**1.** $dy/dx = 4e^{4x}$  **3.** $dy/dx = -24e^{3x}$

| For Exercises ... | 1–12, 25–30 | 13–24, 31–34 | 40, 41, 46–62, 65–69 | 42–45, 63, 64 |
|---|---|---|---|---|
| Refer to Example ... | 1 | 2, 3 | 4 | 5 |

**5.** $dy/dx = -32e^{2x+1}$  **7.** $dy/dx = 2xe^{x^2}$
**9.** $dy/dx = 12xe^{2x^2}$  **11.** $dy/dx = 16xe^{2x^2-4}$  **13.** $dy/dx = xe^x + e^x = e^x(x + 1)$  **15.** $dy/dx = 2(x + 3)(2x + 7)e^{4x}$
**17.** $dy/dx = (2xe^x - x^2e^x)/e^{2x} = x(2 - x)/e^x$  **19.** $dy/dx = [x(e^x - e^{-x}) - (e^x + e^{-x})]/x^2$
**21.** $dp/dt = 8000e^{-0.2t}/(9 + 4e^{-0.2t})^2$  **23.** $f'(z) = 4(2z + e^{-z^2})(1 - ze^{-z^2})$  **25.** $dy/dx = 3(\ln 7)7^{3x+1}$
**27.** $dy/dx = 6x(\ln 4)4^{x^2+2}$  **29.** $ds/dt = (\ln 3)3^{\sqrt{t}}/\sqrt{t}$  **31.** $dy/dt = [(1 - t)e^{3t} - 4e^{2t} + (1 + t)e^t]/(e^{2t} + 1)^2$
**33.** $f'(x) = (9x + 4)e^{x\sqrt{3x+2}}/[2\sqrt{3x + 2}]$  **41. a.** 51,600,000  **b.** 1,070,000 people per year
**43. a.** $G(t) = 5200/(1 + 12e^{-0.52t})$  **b.** 639 clams; 292 clams per year  **c.** 2081 clams; 649 clams per year  **d.** 4877 clams;
157 clams per year  **e.** It increases for a while and then gradually decreases to 0.  **45.** 1.070 hours, 4.251 × 10⁷ per hour
**47. a.** 0.026%, 0.286%, 3.130%  **b.** 0.0025% per year; 0.0274% per year; 0.300% per year  **c.** The percentage of people in
each of the age groups that die in a given year is increasing as indicated by the answers in parts a and b. A person who is 75 years
old has a 3% chance of dying during the year and the rate is increasing by almost 0.3%. The formula implies that everyone will be
dead by age 112.  **49. a.** 2974.15 grams  **b.** 3102 grams  **c.** 124 days  **d.** 2.75 grams per day
**e.** 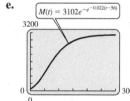  Growth is initially rapid, then tapers off.  **f.**

| Day | Weight | Rate |
|---|---|---|
| 50 | 991 | 24.88 |
| 100 | 2122 | 17.73 |
| 150 | 2734 | 7.60 |
| 200 | 2974 | 2.75 |
| 250 | 3059 | 0.94 |
| 300 | 3088 | 0.32 |

**51. a.** 532 kg  **b.** About 1152 days  **c.** About 0.15 kg per day; on Ayrshire cow will grow about 0.15 kg during day 1000.
**d.** 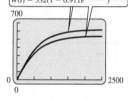  Both approach their maximums asymptotically.
**53. a.** 509.7 kg, 498.4 kg  **b.** 1239 days, 1095 days  **c.** 0.22 kg/day, 0.22 kg/day

**d.** The growth patterns of the two functions
are very similar.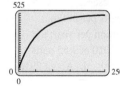

**e.** The graphs of the rates of change of the two functions are also
very similar.

**55. a.** 0.180  **b.** 2024  **c.** 0.023; the marginal increase in the proportion per year in 2010 was approximately 0.023.
**57.** $dI/dS = ae^{-a(S-h)/E}$  **59. a.** 218.5 seconds  **b.** The record is decreasing by 0.034 second per year at the end of 2010.
**c.** 218 seconds. If the estimate is correct, then this is the least amount of time that it will ever take for a human to run 1 mile.
**61. a. (i)** 0.1975, −1.240  **(ii)** 0.009557, −0.01434  **(iii)** 0.001185, −3.296 × 10⁻⁴  **b.** It approaches 0.  **c.** It approaches 0.
**d. (i)** 0.1819, 0.01263, 0.002941  **(ii)** 0.001160, 0.005883, 0.002513  **(iii)** 2.270 × 10⁻¹⁵, 9.988 × 10⁻⁵, 0.001085
**e. (i)** The first  **(ii)** The second  **(iii)** The third  **63. a.** $G(t) = 1/(1 + 270e^{-3.5t})$  **b.** 0.109; 0.341 per century  **c.** 0.802; 0.555
per century  **d.** 0.993; 0.0256 per century  **e.** It increases for a while and then gradually decreases to 0.  **65. a.** 36.8
**b.** 0.00454  **c.** Approximately 0  **d.** $H'(N) = 100e^{-0.1N}$ is always positive, since powers of $e$ are never negative. This means that
repetition always makes a habit stronger.  **67. a.** $I_c = (V/R)e^{-t/RC}$  **b.** 1.35 × 10⁻⁷ amps  **69.** 0.8029 death per 100 students;
0.0540 death per 100 students per year

## Exercises 4.5 (page 244)

1. $dy/dx = 1/x$

| For Exercises ... | 1, 2 | 3, 4, 7–10, 31, 32, 63, 65 | 5, 6, 11–30, 33–44, 57–60 | 61, 62, 64, 66 |
|---|---|---|---|---|
| Refer to Example ... | 1 | 2 | 3 | 4, 5 |

3. $dy/dx = -3/(8 - 3x)$ or $3/(3x - 8)$

5. $dy/dx = (8x - 9)/(4x^2 - 9x)$

7. $dy/dx = 1/[2(x + 5)]$   9. $dy/dx = 3(2x^2 + 5)/[x(x^2 + 5)]$   11. $dy/dx = -15x/(3x + 2) - 5 \ln (3x + 2)$

13. $ds/dt = t + 2t \ln |t|$   15. $dy/dx = [2x - 4(x + 3) \ln (x + 3)]/[x^3(x + 3)]$   17. $dy/dx = (4x + 7 - 4x \ln x)/[x(4x + 7)^2]$

19. $dy/dx = (6x \ln x - 3x)/(\ln x)^2$   21. $dy/dx = 4(\ln |x + 1|)^3/(x + 1)$   23. $dy/dx = 1/(x \ln x)$

25. $dy/dx = e^{x^2}/x + 2xe^{x^2} \ln x$   27. $dy/dx = (xe^x \ln x - e^x)/[x(\ln x)^2]$   29. $g'(z) = 3(e^{2z} + \ln z)^2(2ze^{2z} + 1)/z$

31. $dy/dx = 1/(x \ln 10)$      33. $dy/dx = -1/[(\ln 10)(1 - x)]$ or $1/[(\ln 10)(x - 1)]$   35. $dy/dx = 5/[(2 \ln 5)(5x + 2)]$

37. $dy/dx = 3(x + 1)/[(\ln 3)(x^2 + 2x)]$   39. $dw/dp = (\ln 2)2^p/[(\ln 8)(2^p - 1)]$

41. $f'(x) = e^{\sqrt{x}}\{1/[2\sqrt{x}(\sqrt{x} + 5)] + \ln (\sqrt{x} + 5)/[2\sqrt{x}]\}$

43. $f'(t) = [(t^2 + 1) \ln (t^2 + 1) - t^2 + 2t + 1]/\{(t^2 + 1)[\ln (t^2 + 1) + 1]^2\}$

55. $h'(x) = (x^2 + 1)^{5x}\left[\dfrac{10x^2}{x^2 + 1} + 5 \ln (x^2 + 1)\right]$   57. **a.** About 3 matings   **b.** About 6 matings   **c.** 0.22 mating per degree Celsius

59. **a.** 2590 cm²   **b.** 0.46 g/cm²; when the infant weighs 4000 g, it is gaining 0.46 square centimeter per gram of weight increase.

**c.**
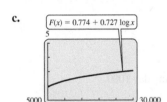
$A(w) = 4.688w^{0.8168-0.0154 \log_{10} w}$

61. **a.** 4 kJ/day   **b.** $1.3 \times 10^{-5}$; when a fawn is 25 kg in size, the rate of change of the energy expenditure of the fawn is about $1.3 \times 10^{-5}$ kJ/day per gram.

**c.**
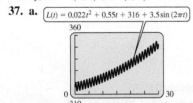
$F(x) = 0.774 + 0.727 \log x$

63. **a.** 0.0306, 1.042   **b.** 0.1073, 1.154

65. **a.** $1.567 \times 10^{11}$ kWh   **b.** 63.4 months

**c.** $4.14 \times 10^{-6}$   **d.** $dM/dE$ decreases and approaches zero.

## Exercises 4.6 (page 253)

1. $dy/dx = 4 \cos 8x$   3. $dy/dx = 108 \sec^2(9x + 1)$

| For Exercises ... | 1–26 | 27–32 | 33–35 | 36 | 37–42 |
|---|---|---|---|---|---|
| Refer to Example ... | 2–6 | 7 | 2–4 | 1 | 8 |

5. $dy/dx = -4 \cos^3 x \sin x$   7. $dy/dx = 8 \tan^7 x \sec^2 x$

9. $dy/dx = -12x \cos 2x - 6 \sin 2x$   11. $dy/dx = -(x \csc x \cot x + \csc x)/x^2$   13. $dy/dx = 4e^{4x} \cos e^{4x}$

15. $dy/dx = (-\sin x)e^{\cos x}$   17. $dy/dx = (4/x) \cos(\ln 3x^4)$   19. $dy/dx = (2x \cos x^2)/\sin x^2$ or $2x \cot x^2$

21. $dy/dx = (6 \cos x)/(3 - 2 \sin x)^2$   23. $dy/dx = \sqrt{\sin 3x} (\sin 3x \cos x - 3 \sin x \cos 3x)/(2\sqrt{\sin x} (\sin^2 3x))$

25. $dy/dx = (3/4) \sec^2 (x/4) - 8 \csc^2 2x + 5 \csc x \cot x - 2e^{-2x}$   27. 1   29. 1/2   31. 1   33. $-\csc^2 x$

37. **a.**
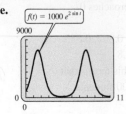
$L(t) = 0.022t^2 + 0.55t + 316 + 3.5 \sin (2\pi t)$

**b.** $L(25) = 343.5$ parts per million; $L(35.5) = 363.25$ parts per million; $L(50.2) = 402.38$ parts per million   **c.** 9.55 parts per million per year; the level of carbon dioxide was increasing at the beginning of 2010 at 9.55 parts per million.   **39. a.** About 1488   **b.** About 5381   **c.** 2000   **d.** About 2916

41. **a.** 1   **b.** $-\sqrt{2}/2 \approx -0.7071$   **c.** 2

**e.**
$f(t) = 1000 \, e^{2 \sin t}$

## Chapter 4 Review Exercises (page 255)

**1.** False  **2.** True  **3.** False
**4.** False  **5.** False  **6.** False
**7.** False  **8.** True  **9.** True
**10.** False  **11.** False  **12.** True

| For Exercises . . . | 1, 2, 13–18, 71, 72, 87–89 | 3, 4, 19–22, 73, 74, 98 | 5, 23–32, 69, 70, 75, 76, 92, 95, 101 | 6, 7, 33–38, 45–48, 77, 78, 90, 91, 93, 94, 96, 97, 99, 100 | 8–10, 39–44, 49–52, 79, 80 | 11, 12, 53–68, 81, 82 |
|---|---|---|---|---|---|---|
| Refer to Section . . . | 1 | 2 | 3 | 4 | 5 | 6 |

**13.** $dy/dx = 15x^2 - 14x - 9$
**14.** $dy/dx = 21x^2 - 8x - 5$  **15.** $dy/dx = 24x^{5/3}$  **16.** $dy/dx = 12x^{-4}$ or $12/x^4$  **17.** $f'(x) = -12x^{-5} + 3x^{-1/2}$ or
$-12/x^5 + 3/x^{1/2}$  **18.** $f'(x) = -19x^{-2} - 4x^{-1/2}$ or $-19/x^2 - 4/x^{1/2}$  **19.** $k'(x) = 21/(4x + 7)^2$  **20.** $r'(x) = -8/(2x + 1)^2$
**21.** $dy/dx = (x^2 - 2x)/(x - 1)^2$  **22.** $dy/dx = (4x^3 + 7x^2 - 20x)/(x + 2)^2$  **23.** $f'(x) = 24x(3x^2 - 2)^3$
**24.** $k'(x) = 90x^2(5x^3 - 1)^5$  **25.** $dy/dt = 7t^6/(2t^7 - 5)^{1/2}$  **26.** $dy/dx = -48t^3(8t^4 - 1)^{-1/2}$ or $-48t^3/(8t^4 - 1)^{1/2}$
**27.** $dy/dx = 3(2x + 1)^2(8x + 1)$  **28.** $dy/dx = 4x(3x - 2)^4(21x - 4)$  **29.** $r'(t) = (-15t^2 + 52t - 7)/(3t + 1)^4$
**30.** $s'(t) = [-4t^3 - 9t^2 + 24t + 6]/(4t - 3)^5$  **31.** $p'(t) = t(t^2 + 1)^{3/2}(7t^2 + 2)$  **32.** $g'(t) = t^2(t^4 + 5)^{5/2}(17t^4 + 15)$
**33.** $dy/dx = -12e^{2x}$  **34.** $dy/dx = 4e^{0.5x}$  **35.** $dy/dx = -6x^2e^{-2x^3}$  **36.** $dy/dx = -8xe^{x^2}$  **37.** $dy/dx = 10xe^{2x} + 5e^{2x} = 5e^{2x}(2x + 1)$
**38.** $dy/dx = 21x^2e^{-3x} - 14xe^{-3x}$ or $7xe^{-3x}(3x - 2)$  **39.** $dy/dx = 2x/(2 + x^2)$  **40.** $dy/dx = 5/(5x + 3)$
**41.** $dy/dx = (x - 3 - x \ln|3x|)/[x(x - 3)^2]$  **42.** $dy/dx = [2(x + 3) - (2x - 1) \ln|2x - 1|]/[(2x - 1)(x + 3)^2]$
**43.** $dy/dx = [e^x(x + 1)(x^2 - 1) \ln(x^2 - 1) - 2x^2e^x]/[(x^2 - 1)[\ln(x^2 - 1)]^2]$
**44.** $dy/dx = \{e^{2x}[2x(\ln x)(x^2 + 1 + x) - (x^2 + 1)]\}/[x(\ln x)^2]$  **45.** $ds/dt = 2(t^2 + e^t)(2t + e^t)$
**46.** $dq/dp = 8e^{2p+1}(e^{2p+1} - 2)^3$  **47.** $dy/dx = -6x(\ln 10) \cdot 10^{-x^2}$  **48.** $dy/dx = [5(\ln 2)2^{\sqrt{x}}]/x^{1/2}$
**49.** $g'(z) = (3z^2 + 1)/[(\ln 2)(z^3 + z + 1)]$  **50.** $h'(z) = e^z/[(\ln 10)(1 + e^z)]$
**51.** $f'(x) = (x + 1)e^{3x}/(xe^x + 1) + 2e^{2x} \ln(xe^x + 1)$
**52.** $f'(x) = e^{\sqrt{x}}[(\sqrt{x} + 1) \ln(\sqrt{x} + 1) - 1]/[2\sqrt{x}(\sqrt{x} + 1)(\ln(\sqrt{x} + 1))^2]$  **53.** $dy/dx = 10 \sec^2 5x$
**54.** $dy/dx = -28 \cos 7x$  **55.** $dy/dx = 6x \csc^2(6 - 3x^2)$  **56.** $dy/dx = 8x \sec^2(4x^2 + 3)$  **57.** $dy/dx = 64x \sin^3(4x^2) \cos(4x^2)$
**58.** $dy/dx = -10 \sin x \cos^4 x$  **59.** $dy/dx = -2x \sin(1 + x^2)$  **60.** $dy/dx = -2x^3 \csc^2(x^4/2)$  **61.** $dy/dx = e^{-2x}(\cos x - 2 \sin x)$
**62.** $dy/dx = -x^2 \csc x \cot x + 2x \csc x$  **63.** $dy/dx = (-2 \cos x \sin x + \cos^2 x \sin x)/(1 - \cos x)^2$
**64.** $dy/dx = 2 \cos x/(\sin x + 1)^2$  **65.** $dy/dx = (\sec^2 x + x \sec^2 x - \tan x)/(1 + x)^2$  **66.** $dy/dx = -[1 + (6 - x) \tan x]/\sec x$ or
$(x - 6) \sin x - \cos x$  **67.** $dy/dx = (\cos x)/(\sin x)$ or $\cot x$  **68.** $dy/dx = -\tan x$  **69. a.** $-3/2$  **b.** $-24/11$
**70. a.** $-36/13$  **b.** $-21/10$  **71.** $-2$; $y = -2x - 4$  **72.** $-2$; $y = -2x + 9$  **73.** $-3/4$; $y = -(3/4)x - 9/4$
**74.** $-5/9$; $y = -(5/9)x + 16/9$  **75.** $3/4$; $y = (3/4)x + 7/4$  **76.** $-4/5$; $y = -(4/5)x - 13/5$  **77.** $1$; $y = x + 1$
**78.** $2e$; $y = 2ex - e$  **79.** $1$; $y = x - 1$  **80.** $2$; $y = 2x - e$  **81.** $1$; $y = x$  **82.** $\pi$; $y = \pi x - \pi$  **83.** No points if $k > 0$;
exactly one point if $k = 0$ or if $k < -1/2$; exactly two points if $-1/2 \leq k \leq 0$.  **85.** 5%; 5.06%  **87. a.** 7.6 fish  **b.** 0.19 caught
walleye/hr/walleye/acre; the catch rate of walleye grows at a constant rate of 0.19 as the density of walleye increases.  **88. a.** 626 kg
**89. a.**

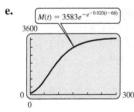

**b.** 201 fish; the processing time begins to decrease.  **c.** 0.015 hr/fish; the rate
at which the processing time is increasing when $n = 40$ is 0.015 hr/fish.
**90.** 50,000 per year  **91. a.** $G(t) = 30,000/(1 + 14e^{-0.15t})$
**b.** 4483; 572 per year  **92. a.** 28.1 cm  **b.** 4.34 cm per year
**c.** 205 g  **d.** 21.2 g per cm  **e.** 92.0 g per year  **93. a.** 3493.76 g
**b.** 3583 g  **c.** 84 days  **d.** 1.76 g/day

**e.** $M(t) = 3583e^{-e^{-0.020(t-66)}}$

Growth is initially rapid, then tapers off.  **f.**

| Day | Weight | Rate |
|---|---|---|
| 50 | 904 | 24.90 |
| 100 | 2159 | 21.87 |
| 150 | 2974 | 11.08 |
| 200 | 3346 | 4.59 |
| 250 | 3494 | 1.76 |
| 300 | 3550 | 0.66 |

**94. a.** 284,000 per year  **b.** 447,000 per year  **95.** 215.15; the balance increases by approximately $215.15 for every 1% increase
in the interest rate when the rate is 5%.  **96.** $218.65; the balance increases by roughly $218.65 for every 1% increase in the interest
rate when the rate is 5%.  **97. a.** 13,769 megawatts/yr  **b.** 42,987 megawatts/yr  **c.** 134,205 megawatts/yr  **98. a.** $-5$ ft/sec
**b.** $-1.7$ ft/sec  **99.** 0.306; the production of corn was increasing at a rate of 0.306 billion bushels a year in 2010.  **100. a.** 169; 167
**b.** 165 words  **c.** $-36$; in the year 2050 the number of words in use will be decreasing by 36 words per millennium.
**101. a.** $-0.4677$ fatality per 1000 licensed drivers per 100 million miles per year; at the age of 20, each extra year results in a
decrease of 0.4677 fatality per 1000 licensed drivers per 100 million miles.  **b.** 0.003672 fatality per 1000 licensed drivers per
100 million miles per year; at the age of 60, each extra year results in an increase of 0.003672 fatality per 1000 licensed drivers
per 100 million miles.  **102. a.** Yes  **b.** $0.18 \leq \alpha \leq 0.41$ (in radians)  **c.** 57.01 ft/radian $\approx 0.995$ ft/degree; the distance will
increase by approximately 1 foot by increasing the angle of the tennis racket by one degree.

# Chapter 5 Graphs and the Derivative

## Exercises 5.1 (page 270)

| For Exercises . . . | 1–8, 41, 51 | 9–20, 25, 26, 29–34, 37–40, 42–45 | 21–24, 35, 36 | 27, 28 | 52–64 |
|---|---|---|---|---|---|
| Refer to Example . . . | 1 | 2 | 4 | 3 | 5, 6 |

**1. a.** $(1, \infty)$ **b.** $(-\infty, 1)$
**3. a.** $(-\infty, -2)$ **b.** $(-2, \infty)$
**5. a.** $(-\infty, -4), (-2, \infty)$ **b.** $(-4, -2)$
**7. a.** $(-7, -4), (-2, \infty)$ **b.** $(-\infty, -7), (-4, -2)$ **9. a.** $(-\infty, -1), (3, \infty)$ **b.** $(-1, 3)$
**11. a.** $(-\infty, -8), (-6, -2.5), (-1.5, \infty)$ **b.** $(-8, -6), (-2.5, -1.5)$ **13. a.** 17/12 **b.** $(-\infty, 17/12)$ **c.** $(17/12, \infty)$
**15. a.** $-3, 4$ **b.** $(-\infty, -3), (4, \infty)$ **c.** $(-3, 4)$ **17. a.** $-3/2, 4$ **b.** $(-\infty, -3/2), (4, \infty)$ **c.** $(-3/2, 4)$ **19. a.** $-2, -1, 0$ -
**b.** $(-2, -1), (0, \infty)$ **c.** $(-\infty, -2), (-1, 0)$ **21. a.** None **b.** None **c.** $(-\infty, \infty)$ **23. a.** None **b.** None
**c.** $(-\infty, -1), (-1, \infty)$ **25. a.** 0 **b.** $(0, \infty)$ **c.** $(-\infty, 0)$ **27. a.** 0 **b.** $(0, \infty)$ **c.** $(-\infty, 0)$ **29. a.** 7 **b.** $(7, \infty)$
**c.** $(3, 7)$ **31. a.** 1/3 **b.** $(-\infty, 1/3)$ **c.** $(1/3, \infty)$ **33. a.** 0, 2/ln 2 **b.** $(0, 2/\ln 2)$ **c.** $(-\infty, 0), (2/\ln 2, \infty)$
**35. a.** 0, 2/5 **b.** $(0, 2/5)$ **c.** $(-\infty, 0), (2/5, \infty)$ **37. a.** $n\pi/2$, where $n$ is an odd integer
**b.** $(n\pi, (n + 1/2)\pi) \cup ((n + 3/2)\pi, (n + 2)\pi)$, where $n$ is an even integer **c.** $((n + 1/2)\pi, (n + 3/2)\pi)$, where $n$ is an
even integer **39. a.** $n\pi$, where $n$ is an integer **b.** $(n\pi, (n + 1/2)\pi) \cup ((n + 1/2)\pi, (n + 1)\pi)$, where $n$ is an even integer
**c.** $((n + 1)\pi, (n + 3/2)\pi) \cup ((n + 3/2)\pi, (n + 2)\pi)$, where $n$ is an even integer **43.** Vertex: $(-b/(2a), (4ac - b^2)/(4a))$;
increasing on $(-\infty, -b/(2a))$; decreasing on $(-b/(2a), \infty)$ **45.** On $(0, \infty)$; nowhere; nowhere **47. a.** About $(567, \infty)$
**b.** About $(0, 567)$ **49. a.** About $(-1.496, 0)$ and $(1.496, \infty)$ **b.** About $(-\infty, -1.496)$ and $(0, 1.496)$ **51. a.** Yes
**b.** April to July; July to November; January to April and November to December **c.** January to April and November to December
**53. a.** $(0, 1.85)$ **b.** $(1.85, 5)$ **55. a.** $(0, 1)$ **b.** $(1, \infty)$ **57. a.** $F'(t) = 175.9e^{-t/1.3}(1 - 0.769t)$ **b.** $(0, 1.3); (1.3, \infty)$
**59. a.** $(0, \infty)$ **b.**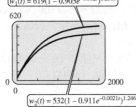
**61.** Increasing on $(4.37, 16.37)$, decreasing on $(0, 4.37) \cup (16.37, 24)$
**63.** $(-\infty, 0); (0, \infty)$

## Exercises 5.2 (page 281)

| For Exercises . . . | 1–8, 39–41 | 13–20, 37, 38 | 21–24 | 25–36 | 46–52 | 45, 54, 55 |
|---|---|---|---|---|---|---|
| Refer to Example . . . | 2 | 3(a) | 3(b) | 3(c) | 4 | 1 |

**1.** Relative minimum of $-4$ at 1
**3.** Relative maximum of 3 at $-2$
**5.** Relative maximum of 3 at $-4$; relative minimum of 1 at $-2$
**7.** Relative maximum of 3 at $-4$; relative minimum of $-2$ at $-7$ and $-2$ **9.** Relative maximum at $-1$; relative minimum at 3
**11.** Relative maxima at $-8$ and $-2.5$; relative minima at $-6$ and $-1.5$ **13.** Relative minimum of 8 at 5 **15.** Relative maximum
of $-8$ at $-3$; relative minimum of $-12$ at $-1$ **17.** Relative maximum of 827/96 at $-1/4$; relative minimum of $-377/6$ at $-5$
**19.** Relative maximum of $-4$ at 0; relative minimum of $-85$ at 3 and $-3$ **21.** Relative maximum of 3 at $-8/3$ **23.** Relative
maximum of 1 at $-1$; relative minimum of 0 at 0 **25.** No relative extrema **27.** Relative maximum of 0 at 1; relative minimum
of 8 at 5 **29.** Relative maximum of $-2.46$ at $-2$; relative minimum of $-3$ at 0 **31.** No relative extrema **33.** Relative
minimum of $e \ln 2$ at $1/\ln 2$ **35.** Relative maximum of 1 at $x = \ldots, -7/2, -3/2, 1/2, 5/2, \ldots$; relative minimum of $-1$ at
$x = \ldots, -5/2, -1/2, 3/2, 7/2, \ldots$ **37.** $(3, 13)$ **39.** Relative maximum of 6.211 at 0.085; relative minimum of $-57.607$ at 2.161
**41.** Relative minimum of 8 at $x = 5$
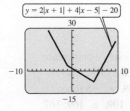
**43.** Relative maximum at $-1$; relative minimum at 3 **45.** 5:04 P.M.; 6:56 A.M.
**47.** 4.96 years; 458.22 kg **49.** $(\ln(n - 1))/n$ **51. c.** $(5/3)^6, (5/2)^6$
**53. a.** $\sin \theta = s/L_2$ **b.** $L_2 = s/\sin \theta$ **c.** $\cot \theta = (L_0 - L_1)/s$ **d.** $L_1 = L_0 - s \cot \theta$
**e.** $R_1 = k \cdot L_1/r_1^4$ **f.** $R_2 = k \cdot L_2/r_2^4$ **g.** $R = k(L_1/r_1^4 + L_2/r_2^4)$
**h.** $R = k(L_0 - s \cot \theta)/r_1^4 + ks/(r_2^4 \sin \theta)$ **i.** $dR/d\theta = ks \csc^2 \theta/r_1^4 - ks \cos \theta/(r_2^4 \sin^2 \theta)$
**j.** $0 = (ks \csc^2 \theta)/r_1^4 - (ks \cos \theta)/(r_2^4 \sin^2 \theta)$ **k.** $k/r_1^4 - k \cos \theta/r_2^4 = 0$
**l.** $\cos \theta = (r_2^4)/(r_1^4)$ **n.** $\cos \theta \approx 0.0039; \theta \approx 90°$ **o.** $84°$ to the nearest degree
**55. a.** 28 ft **b.** 2.57 sec **57. a.** $ds/dt = -2.625 \sin \theta (1 + \cos \theta/\sqrt{15 + \cos^2 \theta})d\theta/dt$
**b.** 21.6 mph

## Exercises 5.3 (page 294)

| For Exercises . . . | 1–32 | 33–56, 60, 71–74, 80, 89, 91, 95 | 65–72, 79, 81, 82, 90, 92 | 78 | 83–86, 88, 90, 93, 94, 100–103 | 96–99 |
|---|---|---|---|---|---|---|
| Refer to Example . . . | 2, 3 | 5 | 6 | 7 | 8 | 4 |

**1.** $f''(x) = 30x - 14; -14; 46$
**3.** $f''(x) = 48x^2 - 18x - 4; -4; 152$
**5.** $f''(x) = 6; 6; 6$

**7.** $f''(x) = 2/(1 + x)^3$; 2; 2/27  **9.** $f''(x) = 4/(x^2 + 4)^{3/2}$; 1/2; $1/(4\sqrt{2})$  **11.** $f''(x) = -6x^{-5/4}$ or $-6/x^{5/4}$; does not exist; $-3/2^{1/4}$
**13.** $f''(x) = 20x^2e^{-x^2} - 10e^{-x^2}$; $-10$; $70e^{-4} \approx 1.282$  **15.** $f''(x) = (-3 + 2\ln x)/(4x^3)$; does not exist; $-0.050$
**17.** $f''(x) = -9x^4 \cos(x^3) - 6x \sin(x^3)$; 0; $-144 \cos 8 - 12 \sin 8$  **19.** $f'''(x) = 168x + 36$; $f^{(4)}(x) = 168$
**21.** $f'''(x) = 300x^2 - 72x + 12$; $f^{(4)}(x) = 600x - 72$  **23.** $f'''(x) = 18(x + 2)^{-4}$ or $18/(x + 2)^4$; $f^{(4)}(x) = -72(x + 2)^{-5}$ or
$-72/(x + 2)^5$  **25.** $f'''(x) = -36(x - 2)^{-4}$ or $-36/(x - 2)^4$; $f^{(4)}(x) = 144(x - 2)^{-5}$ or $144/(x - 2)^5$  **27. a.** $n!$  **b.** 0
**29.** $f''(x) = e^x$; $f'''(x) = e^x$; $f^{(n)}(x) = e^x$  **31.** $f'(x) = \cos x$; $f''(x) = -\sin x$; $f'''(x) = -\cos x$; $f^{(4)}(x) = \sin x$; $f^{(4n)}(x) = \sin x$
**33.** Concave upward on $(2, \infty)$; concave downward on $(-\infty, 2)$; inflection point at $(2, 3)$
**35.** Concave upward on $(-\infty, -1)$ and $(8, \infty)$; concave downward on $(-1, 8)$; inflection points at $(-1, 7)$ and $(8, 6)$
**37.** Concave upward on $(2, \infty)$; concave downward on $(-\infty, 2)$; no inflection points  **39.** Always concave upward; no inflection points
**41.** Concave upward on $(-\infty, 3/2)$; concave downward on $(3/2, \infty)$; inflection point at $(3/2, 525/2)$
**43.** Concave upward on $(5, \infty)$; concave downward on $(-\infty, 5)$; no inflection points  **45.** Concave upward on $(-10/3, \infty)$;
concave downward on $(-\infty, -10/3)$; inflection point at $(-10/3, -250/27)$  **47.** Never concave upward; always concave
downward; no inflection points  **49.** Concave upward on $(-\infty, 0)$ and $(1, \infty)$; concave downward on $(0, 1)$; inflection
points at $(0, 0)$ and $(1, -3)$  **51.** Concave upward on $(-1, 1)$; concave downward on $(-\infty, -1)$ and $(1, \infty)$; inflection
points at $(-1, \ln 2)$ and $(1, \ln 2)$  **53.** Concave upward on $(-\infty, -e^{-3/2})$ and $(e^{-3/2}, \infty)$; concave downward on $(-e^{-3/2}, 0)$
and $(0, e^{-3/2})$; inflection points at $(-e^{-3/2}, -3e^{-3}/(2\ln 10))$ and $(e^{-3/2}, -3e^{-3}/(2\ln 10))$  **55.** Concave upward on
$\ldots \cup (-3\pi/2, -\pi) \cup (-\pi/2, 0) \cup (\pi/2, \pi) \cup \ldots$; concave downward on $\ldots \cup (-\pi, -\pi/2) \cup (0, \pi/2) \cup (\pi, 3\pi/2) \cup \ldots$;
inflection points at $(n\pi/2, 0)$ where $n$ is an integer  **57.** Concave upward on $(-\infty, 0)$ and $(4, \infty)$; concave downward on $(0, 4)$;
inflection points at 0 and 4  **59.** Concave upward on $(-7, 3)$ and $(12, \infty)$; concave downward on $(-\infty, -7)$ and $(3, 12)$; inflection
points at $-7, 3$, and 12  **61.** Choose $f(x) = x^k$ where $1 < k < 2$. For example, $f(x) = x^{4/3}$ has a relative minimum at $x = 0$, and
$f(x) = x^{5/3}$ has an inflection point at $x = 0$.  **63. a.** Close to 0  **b.** Close to 1  **65.** Relative maximum at $-5$
**67.** Relative maximum at 0; relative minimum at 2/3  **69.** Relative minimum at $-3$  **71.** Relative maximum at $-4/7$; relative
minimum at 0  **73. a.** Minimum at about $-0.4$ and 4.0; maximum at about 2.4  **b.** Increasing on about $(-0.4, 2.4)$ and about
$(4.0, \infty)$; decreasing on about $(-\infty, -0.4)$ and $(2.4, 4.0)$  **c.** About 0.7 and 3.3  **d.** Concave upward on about $(-\infty, 0.7)$ and
$(3.3, \infty)$; concave downward on about $(0.7, 3.3)$  **75. a.** Maximum at 1; minimum at $-1$  **b.** Increasing on $(-1, 1)$; decreasing on
$(-\infty, -1)$ and $(1, \infty)$  **c.** About $-1.7, 0$, and about 1.7  **d.** Concave upward on about $(-1.7, 0)$ and $(1.7, \infty)$; concave downward
on about $(-\infty, -1.7)$ and $(0, 1.7)$  **79. a.** 4 hours  **b.** 1160 million  **81. a.** After 2 hours  **b.** 3/4%  **83.** $(38.92, 5000)$
**85.** Inflection point at $t = (\ln c)/k \approx 2.96$ years; this signifies the time when the rate of growth begins to slow down, since $L$ changes
from concave upward to concave downward at this inflection point.  **87.** Always concave downward  **89. a.** $l_1(v)$ is concave upward;
$l_2(v)$ is concave downward.  **91. a.** $(0, 1)$, $(1, \infty)$  **b.** $2(x^3 - 3x - 1)/(x^2 + x + 1)^3$  **c.** $(0, 1.879)$, $(1.879, \infty)$
**95.** $f(t)$ is decreasing and concave upward; $f'(t) < 0$, $f''(t) > 0$.  **97.** $v(t) = 256 - 32t$; $a(t) = -32$; 1024 ft;
16 seconds after being thrown  **99.** $t = 6$  **101.** $(22, 6517.9)$  **103.** $(2.06, 20.7)$

## Exercises 5.4 (page 306)

**1.** 0  **3.**

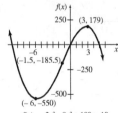

$f(x) = -2x^3 - 9x^2 + 108x - 10$

**5.**

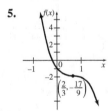

$f(x) = -3x^3 + 6x^2 - 4x - 1$

| For Exercises . . . | 3–10, 44 | 11, 12 | 13–20, 45, 47 | 21–32 |
|---|---|---|---|---|
| Refer to Example . . . | 1 | 2 | 3 | 4 |

**7.**

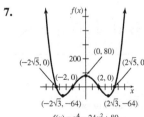

$f(x) = x^4 - 24x^2 + 80$

**9.**

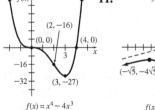

$f(x) = x^4 - 4x^3$

**11.**

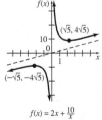

$f(x) = 2x + \dfrac{10}{x}$

**13.**

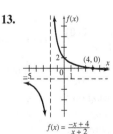

$f(x) = \dfrac{-x + 4}{x + 2}$

**15.**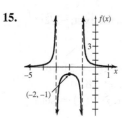

$$f(x) = \frac{1}{x^2 + 4x + 3}$$

**17.**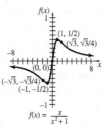

$$f(x) = \frac{x}{x^2 + 1}$$

**19.**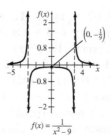

$$f(x) = \frac{1}{x^2 - 9}$$

**21.**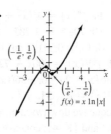

$f(x) = x \ln|x|$

**23.**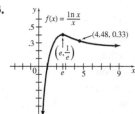

$f(x) = \frac{\ln x}{x}$

**25.**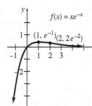

$f(x) = xe^{-x}$

**27.**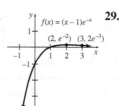

$f(x) = (x-1)e^{-x}$

**29.**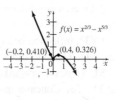

$f(x) = x^{2/3} - x^{5/3}$

**31.**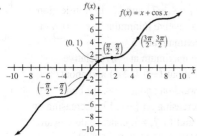

$f(x) = x + \cos x$

**33.** 3, 7, 9, 11, 15  **35.** 17, 19, 23, 25, 27

In Exercises 37–41, other answers are possible.

**37.**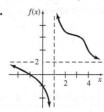

**39.**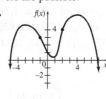

**41.**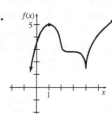

**43.**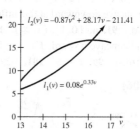

$l_2(v) = -0.87v^2 + 28.17v - 211.41$

$l_1(v) = 0.08e^{0.33v}$

**45.**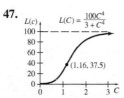

$I(n) = \frac{0.5n^2}{10 + n^2}$

(1.83, 0.125)

**47.**

$L(C) = \frac{100C^4}{3 + C^4}$

(1.16, 37.5)

**Chapter 5 Review Exercises (page 310)**
**1.** True  **2.** False  **3.** False  **4.** True
**5.** False  **6.** True  **7.** False  **8.** False
**9.** False  **10.** False  **11.** True  **12.** False

| For Exercises . . . | 1, 2, 12, 13, 17–26, 72 | 3, 4, 14, 15, 27–36, 71, 73, 79, 80, 82 | 5–9, 16, 37–44, 69, 70, 76, 77, 78, 81 | 10, 11, 45–68, 74, 75 |
|---|---|---|---|---|
| Refer to Example . . . | 1 | 2 | 3 | 4 |

**17.** Increasing on $(-9/2, \infty)$; decreasing on $(-\infty, -9/2)$  **18.** Increasing on $(-\infty, 7/4)$; decreasing on $(7/4, \infty)$
**19.** Increasing on $(-5/3, 3)$; decreasing on $(-\infty, -5/3)$ and $(3, \infty)$  **20.** Increasing on $(-\infty, -2)$ and $(2/3, \infty)$; decreasing on $(-2, 2/3)$  **21.** Never decreasing; increasing on $(-\infty, 3)$ and $(3, \infty)$  **22.** Never increasing; decreasing on $(-\infty, -7/2)$ and $(-7/2, \infty)$  **23.** Decreasing on $(-\infty, -1)$ and $(0, 1)$; increasing on $(-1, 0)$ and $(1, \infty)$  **24.** Increasing on $(-\infty, 1/4)$; decreasing on $(1/4, \infty)$  **25.** Never increasing; decreasing on $(\pi/4 + n\pi/2, 3\pi/4 + n\pi/2)$, where $n$ is an integer
**26.** Increasing on $(\pi/8 + n\pi/2, 3\pi/8 + n\pi/2)$; decreasing on $(-\pi/8 + n\pi/2, \pi/8 + n\pi/2)$, where $n$ is an integer
**27.** Relative maximum of $-4$ at 2  **28.** Relative minimum of $-5$ at 3  **29.** Relative minimum of $-7$ at 2

**30.** Relative maximum of $-14/3$ at $1/3$  **31.** Relative maximum of 101 at $-3$; relative minimum of $-24$ at 2
**32.** Relative maximum of 25 at $-2$; relative minimum of $-2$ at 1  **33.** Relative maximum at $(-0.618, 0.206)$; relative
minimum at $(1.618, 13.203)$  **34.** Relative maximum at $(\sqrt{e}/3, 0.83)$ or $(0.55, 0.83)$
**35.** Relative maximum of 2 at $x = 0, \pm2, \pm4, \ldots$; relative minimum of $-2$ at $x = \pm1, \pm3, \ldots$  **36.** Relative maximum of 5 at
$x = \ldots, -\pi/6, 3\pi/6, 7\pi/6 \ldots$; relative minimum of $-5$ at $x = \ldots, \pi/6, 5\pi/6, 9\pi/6, \ldots$  **37.** $f''(x) = 36x^2 - 10$; 26; 314
**38.** $f''(x) = 54x + 2/x^3$; 56; $-4376/27$  **39.** $f''(x) = 180(3x - 6)^{-3}$ or $180/(3x - 6)^3$; $-20/3$; $-4/75$
**40.** $f''(x) = 112/(4x + 5)^3$; $112/729$; $-16/49$  **41.** $f''(t) = (t^2 + 1)^{-3/2}$ or $1/(t^2 + 1)^{3/2}$; $1/2^{3/2} \approx 0.354$; $1/10^{3/2} \approx 0.032$
**42.** $f''(t) = 5/(5 - t^2)^{3/2}$; 5/8; does not exist  **43.** $f''(x) = 98 \sec^2 7x \tan 7x$; $98 \sec^2 7 \tan 7$; $-98 \sec^2 21 \tan 21$
**44.** $f''(x) = -9x \cos 3x - 6 \sin 3x$; $-9 \cos 3 - 6 \sin 3$; $27 \cos 9 + 6 \sin 9$

**45.**

$f(x) = -2x^3 - \frac{1}{2}x^2 + x - 3$

**46.**

$f(x) = -\frac{4}{3}x^3 + x^2 + 30x - 7$

**47.**

$f(x) = x^4 - \frac{4}{3}x^3 - 4x^2 + 1$

**48.**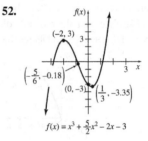

$f(x) = -\frac{2}{3}x^3 + \frac{9}{2}x^2 + 5x + 1$

**49.**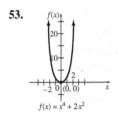

$f(x) = \frac{x - 1}{2x + 1}$

**50.**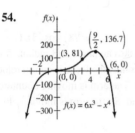

$f(x) = \frac{2x - 5}{x + 3}$

**51.**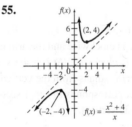

$f(x) = -4x^3 - x^2 + 4x + 5$

**52.**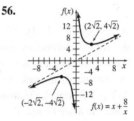

$f(x) = x^3 + \frac{5}{2}x^2 - 2x - 3$

**53.**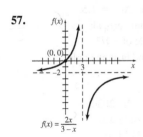

$f(x) = x^4 + 2x^2$

**54.**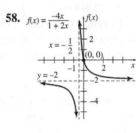

$f(x) = 6x^3 - x^4$

**55.**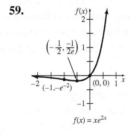

$f(x) = \frac{x^2 + 4}{x}$

**56.**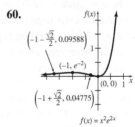

$f(x) = x + \frac{8}{x}$

**57.**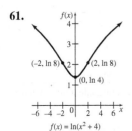

$f(x) = \frac{2x}{3 - x}$

**58.**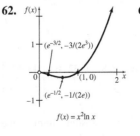
$f(x) = \frac{-4x}{1 + 2x}$

**59.**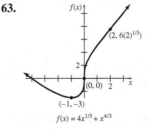

$f(x) = xe^{2x}$

**60.**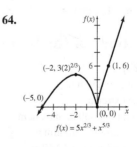

$f(x) = x^2 e^{2x}$

**61.**

$f(x) = \ln(x^2 + 4)$

**62.**

$f(x) = x^2 \ln x$

**63.**

$f(x) = 4x^{1/3} + x^{4/3}$

**64.**

$f(x) = 5x^{2/3} + x^{5/3}$

**65.**
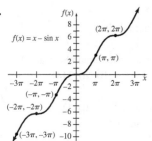

**66.**
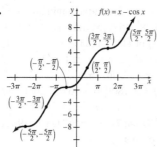

In Exercises 67 and 68, other answers are possible.

**67.**

**68.**
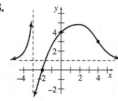

**69. a.** The first derivative has many critical numbers   **b.** The curve is always decreasing except at frequent inflection points.
**70. a.** Metabolic rate and life span are increasing and concave downward. Heartbeat is decreasing and concave upward.

**71. a.** 1486 ml per square meter; for males with 1.88 m² of surface area, the red cell volume increases approximately 1486 ml for each additional square meter of surface area.   **b.** 1.57 m²; 2593 ml (Hurley); 2484 ml (Pearson et al.)   **c.** 1578 ml per m²; for males with 1.57 m² of surface area, the red cell volume increases approximately 1578 ml for each additional square meter of surface area.   **72. d.** $S = 1/(1 + C)$   **i.** Decreasing

**74.**
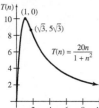

**75. a.**
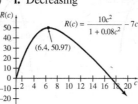

**b.** $0.73 \le c \le 17.13$

**76. a.**
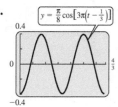

**b.** $v = dy/dt = (-3\pi^2/8) \sin [3\pi(t - 1/3)]$; $a = d^2y/dt^2 = (-9\pi^3/8) \cos [3\pi(t - 1/3)]$
**d.** At $t = 1$ second, the force is clockwise and the arm makes an angle $\pi/8$ radians forward from the vertical. The arm is moving clockwise. At $t = 4/3$ seconds, the force is counterclockwise and the arm makes an angle of $-\pi/8$ radians from the vertical. The arm is moving counterclockwise. At $t = 5/3$ second, the answer corresponds to $t = 1$ second. So the arm is moving clockwise and makes an angle of $\pi/8$ from the vertical.

**77. a.**
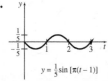

**b.** $v = dy/dt = (\pi/5) \cos [\pi(t - 1)]$; $a = d^2y/dt^2 = (-\pi^2/5)\sin [\pi(t - 1)]$   **d.** At $t = 1.5$, acceleration is negative, arm is moving clockwise and is at an angle of 1/5 radian from vertical; at $t = 2.5$, acceleration is positive, arm is moving counterclockwise and is at an angle of $-1/5$ radian from vertical; at $t = 3.5$, acceleration is negative, arm is moving clockwise and is at an angle of 1/5 radian from vertical.

**78.** 7.6108 yr; the age at which the rate of learning to pass the test begins to slow down   **79. a.** 2010, 380 million   **b.** 2030
**c.** 325 million   **80. a.** At 1965, 1973, 1976, 1983, 1986, and 1988   **b.** Concave upward; this means that the stockpile was increasing at an increasingly rapid rate.   **81. a.** $v(t) = 512 - 32t; a(t) = -32$   **b.** 4096 ft   **c.** After 32 sec; $-512$ ft per sec
**82. a.** 13.55 ft   **c.** 52.39 ft   **d.** $dx/d\alpha = (V^2/16) \cos(2\alpha)$ and $x$ is maximized when $\alpha = \pi/4$.   **e.** 242 ft

# Chapter 6 Applications of the Derivative

## Exercises 6.1 (page 323)
**1.** Absolute maximum at $x_3$; no absolute minimum   **3.** No absolute extrema
**5.** Absolute minimum at $x_1$; no absolute maximum
**7.** Absolute maximum at $x_1$; absolute minimum at $x_2$   **11.** Absolute maximum of 12 at $x = 5$; absolute minimum of $-8$ at $x = 0$ and $x = 3$   **13.** Absolute maximum of 19.67 at $x = -4$; absolute minimum of $-1.17$ at $x = 1$

| For Exercises . . . | 1–6, 33–40 | 7, 8, 11–32, 41 | 43–54 | 55–58 |
|---|---|---|---|---|
| Refer to Example . . . | 2 | 1 | 3 | On Graphical Optimization |

**15.** Absolute maximum of 1 at $x = 0$; absolute minimum of $-80$ at $x = -3$ and $x = 3$   **17.** Absolute maximum of 1/3 at $x = 0$; absolute minimum of $-1/3$ at $x = 3$   **19.** Absolute maximum of 0.21 at $x = 1 + \sqrt{2} \approx 2.4$; absolute minimum of 0 at $x = 1$
**21.** Absolute maximum of 1.710 at $x = 3$; absolute minimum of $-1.587$ at $x = 0$   **23.** Absolute maximum of 7 at $x = 1$; absolute minimum of 0 at $x = 0$   **25.** Absolute maximum of 4.910 at $x = 4$; absolute minimum of $-1.545$ at $x = 2$   **27.** Absolute maximum of 19.09 at $x = -1$; absolute minimum of 0.6995 at $x = (\ln 3)/3$   **29.** Absolute maximum of $\pi/2$ at $x = \pi$; absolute minimum of $\pi/6 - \sqrt{3}/2$ at $x = \pi/3$   **31.** Absolute maximum of 1.356 at $x = 0.6085$; absolute minimum of 0.5 at $x = -1$
**33.** Absolute minimum of 7 at $x = 2$; no absolute maximum   **35.** Absolute maximum of 137 at $x = 3$; no absolute minimum
**37.** Absolute maximum of 0.1 at $x = 4$; absolute minimum of $-0.5$ at $x = -2$   **39.** Absolute maximum of 0.1226 at $x = e^{1/3}$; no absolute minimum   **41. a.** Absolute minimum of $-5$ at $x = -1$; absolute maximum of 0 at $x = 0$   **b.** Absolute maximum of about $-0.76$ at $x = 2$; absolute minimum of $-1$ at $x = 1$   **43.** 6 mo; 6%   **45.** About 7.2 mm
**47. a.**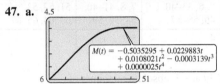
**49. a.**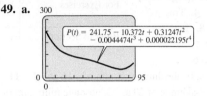

**b.** About 15 days
**b.** Maximum: about 242; minimum: about 44
**51.** The piece formed into a circle should have length $12\pi/(4 + \pi)$ ft, or about 5.28 ft.   **55.** About 11.5 units   **57.** 100 units

## Exercises 6.2 (page 333)

**1. a.** $y = 180 - x$   **b.** $P = x(180 - x)$
**c.** $[0, 180]$   **d.** $dP/dx = 180 - 2x$; $x = 90$
**e.** $P(0) = 0$; $P(180) = 0$; $P(90) = 8100$

| For Exercises . . . | 1–6, 9, 16 | 7, 8, 10–12, 15 | 13, 14, 38, 39, 40, 41 | 17, 32–37 | 18, 19, 43, 44 | 20–25, 26–31, 42 |
|---|---|---|---|---|---|---|
| Refer to Example . . . | 1 | 2 | 6 | 4 | 5 | 3 |

**f.** 8100; 90 and 90   **3. a.** $y = 90 - x$   **b.** $P = x^2(90 - x)$   **c.** $[0, 90]$   **d.** $dP/dx = 180x - 3x^2$; $x = 0, x = 60$
**e.** $P(0) = 0, P(60) = 108{,}000, P(90) = 0$   **f.** 108,000; 60 and 30   **5. a.** 15 days   **b.** 16.875%   **7.** 12.98 thousand
**9. a.** 12 days   **b.** 50 per ml   **c.** 1 day   **d.** 81.365 per ml   **11.** 49.37   **13.** Point $P$ is $3\sqrt{7}/7 \approx 1.134$ mi from Point $A$
**15. a.** Replace $a$ with $e^r$ and $b$ with $r/P$.   **b.** Shepherd: $f'(S) = a[1 + (1 - c)(S/b)^c]/[1 + (S/b)^c]^2$; Ricker: $f'(S) = ae^{-bS}(1 - bS)$;
Beverton–Holt: $f'(S) = a/[1 + (S/b)]^2$   **c.** Shepherd: $a$; Ricker: $a$; Beverton–Holt: $a$; the constant $a$ represents the slope of the
graph of $f(S)$ at $S = 0$.   **d.** 194,000 tons   **e.** 256,000 tons   **17.** 2.18 $\mu$m, 4.36 $\mu$m   **19.** About 40°; about 12 m
**21.** 75 m by 75 m   **23.** 240,000 m$^2$   **25.** \$1600   **27.** 3 ft by 6 ft by 2 ft   **29. a.** Both are 64 square inches.
**b.** Both are 100/9 square feet.   **c.** It appears that the area of the base and the total area of the walls for the box with maximum
volume are equal.   **31.** $3\sqrt{6} + 3$ by $2\sqrt{6} + 2$   **33.** Radius $= 1.08$ ft; height $= 4.34$ ft; cost $= \$44.11$
**35.** Radius $= 5.454$ cm; height $= 10.70$ cm   **37.** Radius $= 10$ cm; height $= 10$ cm   **39.** Point $A$   **41.** 0 mi   **43.** 14.38 ft

## Exercises 6.3 (page 341)

**1.** $dy/dx = -6x/(5y)$   **3.** $dy/dx = (8x - 5y)/(5x - 3y)$
**5.** $dy/dx = 15x^2/(6y + 4)$   **7.** $dy/dx = -3x(2 + y)^2/2$

| For Exercises . . . | 1–18, 42–44 | 19–41 | 45–53 |
|---|---|---|---|
| Refer to Example . . . | 1, 2 | 3 | 4 |

**9.** $dy/dx = \sqrt{y}/[\sqrt{x}(5\sqrt{y} - 2)]$   **11.** $dy/dx = (4x^3y^3 + 6x^{1/2})/(9y^{1/2} - 3x^4y^2)$
**13.** $dy/dx = (5 - 2xye^{x^2y})/(x^2e^{x^2y} - 4)$   **15.** $dy/dx = y(2xy^3 - 1)/(1 - 3x^2y^3)$   **17.** $dy/dx = \sec(xy)/x - y/x$
**19.** $y = (3/4)x + 25/4$   **21.** $y = x + 2$   **23.** $y = x/64 + 7/4$   **25.** $y = (11/12)x - 5/6$   **27.** $y = -(37/11)x + 59/11$
**29.** $y = x/\pi - 1/\pi$   **31.** $y = (5/2)x - 1/2$   **33.** $y = 1$   **35.** $y = -2x + 7$   **37.** $y = -x + 2$   **39.** $y = (8/9)x + 10/9$
**41. a.** $y = -(3/4)x + 25/2$; $y = (3/4)x - 25/2$
**b.**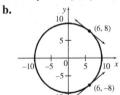
**43. a.** $du/dv = -2u^{1/2}/(2v + 1)^{1/2}$   **b.** $dv/du = -(2v + 1)^{1/2}/(2u^{1/2})$   **c.** They are reciprocals.
**45.** $dy/dx = -x/y$; there is no function $y = f(x)$ that satisfies $x^2 + y^2 + 1 = 0$.
**47.** $1/(3\sqrt{3})$   **49. b.** $\phi'(B) = -p/\lambda^{\alpha - 1}$
**53.** $ds/dt = (-s + 6\sqrt{st})/(8s\sqrt{st} + t)$

## Exercises 6.4 (page 347)

**1.** $-64$   **3.** $-9/7$   **5.** 1/5
**7.** $-3/2$   **9.** $(8 - 4\pi)/[\pi(4 - \pi)]$

| For Exercises . . . | 1–10, 12–20 | 31–34 | 11 | 21, 22, 27, 29, 30 | 23 | 24–26, 28 |
|---|---|---|---|---|---|---|
| Refer to Example . . . | 1 | 5 | 6 | 3 | 2 | 4 |

**11.** 0.067 mm per min   **13.** About 1.9849 g per day   **15. a.** $dm/dt = 46.251w^{-0.46}dw/dt$   **b.** 0.875 kcal/day$^2$
**17.** $-0.0521$ kcal/kg/km/day   **19.** 25.6 crimes per month   **21.** 24/5 ft/min   **23.** $16\pi$ ft$^2$/min   **25.** 2/27 cm/min

**27.** 62.5 ft per min   **29.** $\sqrt{2} \approx 1.41$ ft per sec   **31.** \$384 per month   **33. a.** Revenue is increasing at a rate of \$180 per day.
**b.** Cost is increasing at a rate of \$50 per day.   **c.** Profit is increasing at a rate of \$130 per day.   **35. a.** $200\,\pi$ m/min   **b.** $400\,\pi$ m/min

### Exercises 6.5 (page 354)

**1.** 1.9   **3.** 0.1   **5.** 0.060   **7.** $-0.023$   **9.** 12.0417; 12.0416; 0.0001

| For Exercises . . . | 1–8 | 9–18 | 19–31, 38, 39 | 32–37 |
|---|---|---|---|---|
| Refer to Example . . . | 1 | 2 | 3 | 4 |

**11.** 0.995; 0.9950; 0   **13.** 1.01; 1.0101; 0.0001   **15.** 0.05; 0.0488; 0.0012
**17.** 0.03; 0.0300; 0   **19. a.** 0.007435   **b.** $-0.005105$   **21. a.** 0.347 million
**b.** $-0.022$ million   **23.** $1568\pi$ mm$^3$   **25.** $80\pi$ mm$^2$   **27. a.** About 9.3 kg   **b.** About 9.5 kg   **29.** $-7.2\pi$ cm$^3$   **31.** 0.472 cm$^3$
**33.** 0.00125 cm   **35.** $\pm 1.273$ in$^3$   **37.** $\pm 0.116$ in$^3$   **39.** $21{,}608\pi$ in$^3$

### Chapter 6 Review Exercises (page 357)

**1.** False   **2.** True   **3.** False   **4.** True
**5.** True   **6.** True   **7.** True   **8.** True

| For Exercises . . . | 1–3, 9–18 | 4, 19–32, 47, 48, 50 | 5, 6, 33–40, 49, 55–57 | 7, 8, 41–46, 54, 58, 59 | 51, 52, 53, 60–64, 65–67 |
|---|---|---|---|---|---|
| Refer to Section . . . | 1 | 3 | 4 | 5 | 2 |

**9.** Absolute maximum of 33 at 4;
absolute minimum of 1 at 0 and 6
**10.** Absolute maximum of $-3$ at 0; absolute minimum of $-16$ at $-1$   **11.** Absolute maximum of 39 at $-3$; absolute minimum of
$-319/27$ at 5/3   **12.** Absolute maximum of 29 at $-3$; absolute minimum of $-3$ at $-1$ and 1.   **13.** Absolute maximum of $\pi - 1$ at $\pi$;
absolute minimum of 1 at 0   **14.** Absolute maximum of $1 + \sqrt{2}$ at $3\pi/4$; absolute minimum of 0 at 0
**17. a.** Maximum $= 0.37$; minimum $= 0$   **b.** Maximum $= 0.35$; minimum $= 0.13$   **18. a.** Maximum $= 13.65$;
minimum $= 7.39$   **b.** Maximum $= 44.83$; minimum $= 7.39$   **21.** $dy/dx = (2x - 9x^2y^4)/(8y + 12x^3y^3)$
**22.** $dy/dx = (-2xy^3 - 4y)/(3x^2y^2 + 4x)$   **23.** $dy/dx = 6\sqrt{y-1}/\left[x^{1/3}(1 - \sqrt{y-1})\right]$
**24.** $dy/dx = 9\sqrt{y}/\left[2\sqrt{x}(1 - 12y^{5/2})\right]$   **25.** $dy/dx = -(30 + 50x)/3$   **26.** $dy/dx = (2y - 2y^{1/2})/(4y^{1/2} + 9y - x)$
**27.** $dy/dx = (2xy^4 + 2y^3 - y)/(x - 6x^2y^3 - 6xy^2)$   **28.** $dy/dx = (1 - 2x^2 - 2xy)/(3xy^2 + 3y^3 - 1)$
**29.** $dy/dx = [1 - \sin(x + y)]/[2y + \sin(x + y)]$   **30.** $dy/dx = [y \sec^2(xy) + 3x^2]/[1 - x \sec^2(xy)]$
**31.** $y = (-16/23)x + 94/23$   **32.** $y = (1/12)x + 5/4$   **35.** 272   **36.** 10/3   **37.** $-2$   **38.** $-2/9$   **39.** $-8e^3$
**40.** $-6e/(e + 1)^2$   **43.** 0.00204   **44.** 0.1   **45.** $-0.005$   **46.** $-0.04$   **47. a.** $(2, -5)$ and $(2, 4)$   **b.** $(2, -5)$ is a relative
minimum; $(2, 4)$ is a relative maximum.   **c.** No   **49.** $56\pi$ ft$^2$ per min

**51. a.**
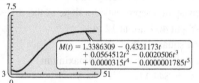

**b.** About the 15th day

**52. a.**
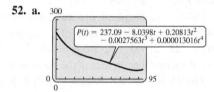

**b.** 237; 44

**53.** About 19.82°C   **54. a.** About 0.04 kg   **b.** About 0.04 kg   **55.** 8/3 ft per min   **56.** 0.0248 ft/min
**57.** $21/16 = 1.3125$ ft per min   **58.** 4.021 in$^3$   **59.** $\pm 0.736$ in$^2$   **60.** 43.1 in.   **61.** $1.25 + 2$ in 1.5   **62.** 225 m by 450 m
**63.** 10 ft; 18.67 sec   **64.** 0 ft; 15.72 sec   **65.** 2 m by 4 m by 4 m   **66.** 3 in.   **67.** 1.684 in.; 4.490 in.   **68.** $\theta = \pi/4$ or $45°$

## Chapter 7 Integration

### Exercises 7.1 (page 373)

**1.** They differ only by a constant.
**5.** $6k + C$   **7.** $z^2 + 3z + C$
**9.** $2t^3 - 4t^2 + 7t + C$

| For Exercises . . . | 5–24, 27, 28 | 25, 26, 29–32, 35, 36, 41, 42, 47, 50, 53 | 33, 34, 37–40, 48 | 43, 44 | 45, 46 | 52–54 | 55–62 |
|---|---|---|---|---|---|---|---|
| Refer to Example . . . | 4, 5 | 7, 8 | 6 | 9 | 12 | 10 | 11 |

**11.** $z^4 + z^3 + z^2 - 6z + C$
**13.** $10z^{3/2}/3 + \sqrt{2}z + C$   **15.** $5x^4/4 - 20x^2 + C$   **17.** $8v^{3/2}/3 - 6v^{5/2}/5 + C$
**19.** $4u^{5/2} - 4u^{7/2} + C$   **21.** $-7/z + C$   **23.** $-\pi^3/(2y^2) - 2\sqrt{\pi y} + C$   **25.** $6t^{-1.5} - 2 \ln |t| + C$   **27.** $-1/(3x) + C$
**29.** $-15e^{-0.2x} + C$   **31.** $-3 \ln |x| - 10e^{-0.4x} + e^{0.1}x + C$   **33.** $(1/4)\ln |t| + t^3/6 + C$   **35.** $e^{2u}/2 + 2u^2 + C$
**37.** $x^3/3 + x^2 + x + C$   **39.** $6x^{7/6}/7 + 3x^{2/3}/2 + C$   **41.** $10^x/(\ln 10) + C$   **43.** $3 \sin x + 4 \cos x + C$
**45.** $f(x) = 3x^{5/3}/5$   **47. a.** $f(t) = -e^{-0.01t} + C$   **b.** 0.095 unit   **49. a.** $c'(t) = (-kA/V)(c_0 - C)e^{-kAt/V}$
**51.** $V(t) = (kP_0/m)e^{-mt} + V_0 - kP_0/m$   **53. a.** $D(t) = 818.9e^{0.03572t} - 118.9$   **b.** About 2740
**55.** $v(t) = 5t^3/3 + 4t + 6$   **57.** $s(t) = -16t^2 + 6400$; 20 sec   **59.** $s(t) = 2t^{5/2} + 3e^{-t} + 1$   **61.** 160 ft/sec, 12 ft

## Exercises 7.2 (page 383)

| For Exercises ... | 2a, 3, 4, 29, 33, 34 | 2c, 5–8, 21, 22 | 9–10, 27, 28–30 | 2d, 11–16, 31, 35, 36 | 17–20, 32 | 2b, 23–26 | 61–67 | 37–50, 55, 57, 58 | 51–54, 56 |
|---|---|---|---|---|---|---|---|---|---|
| Refer to Example ... | 1 | 3 | 2 | 5 | 4 | 6 | 9 | 7 | 8 |

**3.** $2(2x + 3)^5/5 + C$
**5.** $-(2m + 1)^{-2}/2 + C$
**7.** $-(x^2 + 2x - 4)^{-3}/3 + C$
**9.** $(4z^2 - 5)^{3/2}/12 + C$ **11.** $e^{2x^3}/2 + C$ **13.** $e^{2t-t^2}/2 + C$ **15.** $-e^{1/z} + C$ **17.** $[\ln(t^2 + 2)]/2 + C$
**19.** $[\ln(x^4 + 4x^2 + 7)]/4 + C$ **21.** $-1/[2(x^2 + x)^2] + C$ **23.** $(p + 1)^7/7 - (p + 1)^6/6 + C$
**25.** $2(u - 1)^{3/2}/3 + 2(u - 1)^{1/2} + C$ **27.** $(x^2 + 12x)^{3/2}/3 + C$ **29.** $(1 + 3\ln x)^3/9 + C$ **31.** $(1/2)\ln(e^{2x} + 5) + C$
**33.** $(\ln 10)(\log x)^2/2 + C$ **35.** $8^{3x^2+1}/(6 \ln 8) + C$ **37.** $(1/3) \sin 3x + C$ **39.** $(-\cos x^2)/2 + C$ **41.** $-\tan 3x + C$
**43.** $(1/8) \sin^8 x + C$ **45.** $-2 (\cos x)^{3/2} + C$ **47.** $-\ln |1 + \cos x| + C$ **49.** $(1/4) \sin x^8 + C$
**51.** $-3 \ln|\cos (x/3)| + C$ **53.** $(1/6) \ln |\sin x^6| + C$ **55.** $-\cos e^x + C$ **57.** $-\csc e^x + C$
**61. a.** $f(t) = 0.001483[(t - 1980)^{2.75}/2.75 + 1980(t - 1980)^{1.75}/1.75] + 262.951$ **b.** About 1,118,000,000
**63. a.** $N(t) = 50 \ln(t^2 + 2) + 2.343$ **b.** 307 **65.** $\sin k$ ("sink") **67.** $\cos t$ ("cost")

## Exercises 7.3 (page 392)

| For Exercises ... | 5–16, 19–24 | 17, 18, 26–33, 36–38, 42 | 34, 35, 39–41 |
|---|---|---|---|
| Refer to Example ... | 1, 2 | 3 | 4 |

**3. a.** 88 **b.** $\int_0^8 (2x + 5) \, dx$ **5. a.** 21 **b.** 23
**c.** 22 **d.** 22 **7. a.** 10 **b.** 10 **c.** 10 **d.** 11
**9. a.** 8.22 **b.** 15.48 **c.** 11.85 **d.** 10.96 **11. a.** 6.70 **b.** 3.15 **c.** 4.93 **d.** 4.17 **13. a.** 1.896 **b.** 1.896 **c.** 1.896
**d.** 2.052 **15. a.** 4 **b.** 4 **17. a.** 4 **b.** 5 **19.** $4\pi$ **21.** 24 **23. b.** 0.385 **c.** 0.33835 **d.** 0.334334 **e.** 0.333333
**27. a.** Left: about 582,000 cases; right: about 580,000 cases; average: about 581,000 cases **b.** Left: about 146,000 cases; right:
about 144,000 cases; average: about 145,000 cases **29. a.** About 155 cal/cm² **b.** About 564 cal/cm² **31.** Left: 28,514; right:
26,254; average: 27,384 **33.** About 1900 ft; yes **35.** 2451 ft, 2879 ft, 2665 ft **37. a.** About 360 BTU/ft²
**b.** About 170 BTU/ft² **39.** 38 ft; 56 ft **41. a.** 0.0492 mi **b.** 0.0633 mi **c.** 0.0622 mi; b

## Exercises 7.4 (page 407)

| For Exercises ... | 1–6, 9–12, 15–20, 23, 24 | 7, 8, 13, 14, 21, 22, 27–30, 62 | 31–36 | 37–52, 55 | 63–85 | 86 |
|---|---|---|---|---|---|---|
| Refer to Example ... | 1, 2, 3 | 4 | 7 | 5, 6 | 8 | 9 |

**1.** $-18$ **3.** $-3/2$ **5.** $28/3$ **7.** 13 **9.** $-16/3$
**11.** 76 **13.** $4/5$ **15.** $108/25$
**17.** $20e^{0.3} - 20e^{0.2} + 3 \ln 2 - 3 \ln 3 \approx 1.353$
**19.** $e^8/4 - e^4/4 - 1/6 \approx 731.4$ **21.** $91/3$
**23.** $447/7 \approx 63.86$ **25.** $(\ln 2)^2/2 \approx 0.2402$ **27.** 49 **29.** $1/8 - 1/[2(3 + e^2)] \approx 0.07687$ **31.** $1 - \sqrt{2}/2$
**33.** $-\ln(\sqrt{3}/2)$ **35.** $\sqrt{3}/2 - 1$ **37.** 10 **39.** 76 **41.** $41/2$ **43.** $e^2 - 3 + 1/e \approx 4.757$ **45.** $e - 2 + 1/e \approx 1.086$
**47.** 3 **49.** $23/3$ **51.** $e^2 - 2e + 1 \approx 2.952$ **53.** $\int_a^c f(x) \, dx = \int_a^b f(x) \, dx + \int_b^c f(x) \, dx$ **55.** $-8$ **59.** $-12$
**61. a.** $x^5/5 - 1/5$ **c.** $f'(1) \approx 2.746$, and $g(1) = e \approx 2.718$ **63.** No **65. a.** 0.8778 ft **b.** 0.6972 ft
**67. a.** 18.12 **b.** 8.847 **69. b.** $\int_0^{60} n(x) \, dx$ **c.** $2(51^{3/2} - 26^{3/2})/15 \approx 30.89$ million **71. a.** $Q(R) = \pi kR^4/2$
**b.** $0.04k$ mm per min **73. b.** About 505,000 kJ/W$^{0.67}$ **75. a.** About 286 million; the total population aged 0 to 90
**b.** About 64 million **77.** $3 - 2 \ln 2$ **79.** 60,000 **81.** About 32%
**83.** $C(T) = 30(e^{0.04T} - 1)$; 6.64 billion barrels **85.** 4430 hours; this result is relatively close to the actual value.

## Exercises 7.5 (page 416)

| For Exercises ... | 1, 2 | 3–6, 11–18, 20, 23–28 | 7–10, 19, 21, 22, 29, 30, 33 | 31, 32, 35–38 |
|---|---|---|---|---|
| Refer to Example ... | 1 | 3 | 2 | 4 |

**1.** 21 **3.** 20 **5.** $23/3$ **7.** 366.2 **9.** $4/3$
**11.** $2 \ln 2 - \ln 6 + 3/2 \approx 1.095$
**13.** $6 \ln(3/2) - 6 + 2e^{-1} + 2e \approx 2.605$
**15.** $(e^{-2} + e^4)/2 - 2 \approx 25.37$ **17.** $1/2$ **19.** $1/20$ **21.** $3(2^{4/3})/2 - 3(2^{7/3})/7 \approx 1.620$ **23.** $(e^9 + e^6 + 1)/3 \approx 2836$
**25.** $\sqrt{2} - 1$ **27.** $\sqrt{2}/2 - 1 + (\ln 2)/2$ **29.** $-1.9241, -0.4164, 0.6650$ **31. a.** About 71.25 gal **b.** About 25 hr
**c.** About 105 gal **d.** About 47.91 hr **33. a.** 0.019; the lower 10% of the income producers earn 1.9% of the total income of the
population. **b.** 0.184; the lower 40% of the income producers earn 18.4% of the total income of the population.

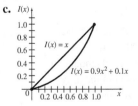

**d.** 0.15 **e.** Income is distributed less equally in 2008 than in 1968.
**35. a.** 8 yr **b.** About $148 **c.** About $771 **37. a.** 39 days
**b.** $3369.18 **c.** $484.02 **d.** $2885.16

## Chapter 7 Review Exercises (page 419)

| For Exercises . . . | 1–4, 15–26 | 5, 13, 14, 27–51, 87, 88 | 6, 8, 12, 52–55, 85, 96 | 9, 10, 56–77, 82–84, 89, 90–92 | 11, 78–81, 93 |
|---|---|---|---|---|---|
| Refer to Section . . . | 1 | 2 | 3 | 4 | 5 |

**1.** True  **2.** False  **3.** False  **4.** True
**5.** True  **6.** False  **7.** False  **8.** True
**9.** True  **10.** False  **11.** True

**15.** $x^2 + 3x + C$  **16.** $5x^2/2 - x + C$  **17.** $x^3/3 - 3x^2/2 + 2x + C$  **18.** $6x - x^3/3 + C$  **19.** $2x^{3/2} + C$  **20.** $x^{3/2}/3 + C$
**21.** $2x^{3/2}/3 + 9x^{1/3} + C$  **22.** $6x^{7/3}/7 + 2x^{1/2} + C$  **23.** $2x^{-2} + C$  **24.** $-5/(3x^3) + C$  **25.** $-3e^{2x}/2 + C$  **26.** $-5e^{-x} + C$
**27.** $e^{3x}/6 + C$  **28.** $e^{x^2} + C$  **29.** $(3\ln|x^2 - 1|)/2 + C$  **30.** $(1/2)\ln|2 - x^2| + C$  **31.** $-(x^3 + 5)^{-3}/9 + C$
**32.** $(x^2 - 5x)^5/5 + C$  **33.** $-e^{-3x^4}/12 + C$  **34.** $e^{3x^2+4}/6 + C$  **35.** $(3\ln x + 2)^5/15 + C$  **36.** $2(5\ln x + 3)^{3/2}/15 + C$
**37.** $(-1/2)\cos 2x + C$  **38.** $(1/5)\sin 5x + C$  **39.** $(-1/7)\ln|\cos 7x| + C$  **40.** $(1/5)\tan 5x + C$  **41.** $8\tan x + C$
**42.** $-4\cot x + C$  **43.** $(-1/12)\cos 4x^3 + C$  **44.** $(5/4)\sec 2x^2 + C$  **45.** $(-2/3)(\cos x)^{3/2} + C$  **46.** $(-1/9)\cos^9 x + C$
**47.** $(-1/22)\ln|\cos 11x^2| + C$  **48.** $(1/24)\ln|\sin 8x^3| + C$  **49.** $(2/5)(\sin x)^{5/2} + C$  **50.** $3(\cos x)^{-1/3} + C$
**51.** $(1/10)\tan^2 5x + C$  **52.** 20  **53. a.** 0  **b.** 4.5  **54.** 24  **55.** 28  **56. a.** $s(T) - s(0)$  **b.** $\int_0^T v(t)\,dt = s(T) - s(0)$ is equivalent to the Fundamental Theorem with $a = 0$ and $b = T$ because $s(t)$ is an antiderivative of $v(t)$.  **58.** 12  **59.** 160.8
**60.** $3\ln 5 + 12/25 \approx 5.308$  **61.** $2\ln 3 + 2/3 \approx 2.864$  **62.** 19/15  **63.** $(25^{4/3} - 1)/12 \approx 6.008$  **64.** $3(1 - e^{-4})/2 \approx 1.473$
**65.** $25(e^2 - e^{0.4})/4 \approx 36.86$  **66.** 1  **67.** 1/2  **68.** $20\pi$  **69.** $\pi - 3/2$  **70.** $\pi/32$  **71.** $2\pi$  **72.** $25\pi/4$  **73.** $9\pi$  **74.** 13/3
**75.** 5504/7  **76.** $(e^4 - 1)/2 \approx 26.80$  **77.** $5 - e^{-4} \approx 4.982$  **78.** 64/3  **79.** 1/6  **80.** 149/3  **81.** 32  **82.** 782
**83.** $50\ln 17 \approx 141.66$; about 142 people  **84. a.** 0.2784  **b.** 0.2784  **85. a.** About 4600 pM  **b.** About 2800 pM
**c.** The area under the curve is about 64% more for the fasting sheep.  **87. a.** 19.736 billion barrels  **b.** 19.736 billion barrels
**c.** $y = 0.001382t^3 - 0.1380t^2 - 0.0206t + 2.1915$; 19.72 billion barrels  **88. a.** $2,464,526  **b.** $2,464,526
**c.** $y = -11,112t + 373,216$; $2,452,352  **89.** $s(t) = t^3/3 - t^2 + 8$.

**90. a.**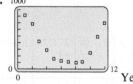
**b.** $C(t) = 648.1\sin(0.3051t + 2.5754) + 828.19$
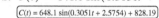
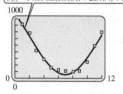
**c.** 6024 trillion BTU; 5933 trillion BTU
**d.** 20.6 months; yes

**91. a.**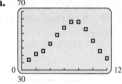
**b.** $y = 12.5680\sin(0.547655t - 2.34990) + 50.2671$
**c.**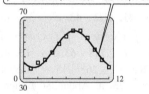
**d.** About 11.5 months; yes

**92. d.** 8.63 in.  **e.** 5.15 in.  **93.** 2.5 yr; about $99,000

# Chapter 8 Further Techniques and Applications of Integration

## Exercises 8.1 (page 434)

| For Exercises . . . | 1a–14a, 17, 23a, 24a, 29 | 1b–14b, 19, 21, 22, 23b, 24b, 29–32, 36, 37 | 25–28, 33–35 |
|---|---|---|---|
| Refer to Example . . . | 1 | 2 | 3 |

**1. a.** 12.25  **b.** 12  **c.** 12  **3. a.** 3.35
**b.** 3.3  **c.** $3\ln 3 \approx 3.296$  **5. a.** 11.34  **b.** 10.5
**c.** 10.5  **7. a.** 0.9436  **b.** 0.8374  **c.** $4/5 = 0.8$
**9. a.** 1.236  **b.** 1.265  **c.** $2 - 2e^{-1} \approx 1.264$  **11. a.** 1.8961  **b.** 2.0046  **c.** 2  **13. a.** 5.991  **b.** 6.167  **c.** 6.283; Simpson's rule  **15.** b is true.  **17. a.** 0.2  **b.** 0.220703, 0.205200, 0.201302, 0.200325, 0.020703, 0.005200, 0.001302, 0.000325
**c.** $p = 2$  **19. a.** 0.2  **b.** 0.2005208, 0.2000326, 0.2000020, 0.2000001, 0.0005208, 0.0000326, 0.0000020, 0.0000001
**c.** $p = 4$  **21.** $M = 0.7355; S = 0.8048$  **23. a.** 1.831  **b.** 1.758  **25.** About 30 mcg(h)/ml; this represents the total amount of drug available to the patient for each ml of blood.  **27.** About 9 mcg(h)/ml; this represents the total effective amount of the drug available to the patient for each ml of blood.  **29. a.** $y = b_0(t/7)^{b_1}e^{-b_2t/7}$  **b.** About 1212 kg; about 1231 kg  **c.** About 1224 kg; about 1250 kg  **31.** 1401 kg

**33. a.** 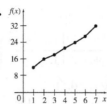 **b.** 6.3  **c.** 6.27  **35. a.** **b.** 128  **c.** 128  **37. a.** 0.6827  **b.** 0.9545  **c.** 0.9973

**Exercises 8.2 (page 443)**

| For Exercises ... | 1–4, 48, 51, 52, 54, 55 | 5–8, 17, 18 | 9, 10, 47, 53, 56 | 11–16 | 19–30 | 31–40 |
|---|---|---|---|---|---|---|
| Refer to Example ... | 1, 3 | 5 | 2 | 4 | 1–4 | 6 |

**1.** $xe^x - e^x + C$

**3.** $-5xe^{-3x}/3 - 5e^{-3x}/9 + 3e^{-3x} + C$ or $-5xe^{-3x}/3 + 22e^{-3x}/9 + C$  **5.** $-5e^{-1} + 3 \approx 1.1606$  **7.** $11 \ln 2 - 3 \approx 4.6246$

**9.** $(x^2 \ln x)/2 - x^2/4 + C$  **11.** $(-6/5) x \sin 5x - (6/25) \cos 5x + C$  **13.** $-8x \cos x + 8 \sin x + C$

**15.** $(-3/4) x^2 \sin 8x - (3/16)x \cos 8x + (3/128) \sin 8x + C$  **17.** $e^4 + e^2 \approx 61.99$  **19.** $x^2 e^{2x}/2 - xe^{2x}/2 + e^{2x}/4 + C$

**21.** $(2/7)(x + 4)^{7/2} - (16/5)(x + 4)^{5/2} + (32/3)(x + 4)^{3/2} + C$ or $(2/3)x^2(x + 4)^{3/2} - (8/15)x(x + 4)^{5/2} + (16/105)(x + 4)^{7/2} + C$

**23.** $(4x^2 + 10x) \ln 5x - 2x^2 - 10x + C$  **25.** $(-e^2/4)(3e^2 + 1) \approx -42.80$  **27.** $2\sqrt{3} - 10/3 \approx 0.1308$  **29.** $\sin e^x + C$

**31.** $16 \ln|x + \sqrt{x^2 + 16}| + C$  **33.** $-(3/11) \ln\left|(11 + \sqrt{121 - x^2})/x\right| + C$  **35.** $-1/(4x + 6) - (1/6) \ln|x/(4x + 6)| + C$

**37.** $(e^{4x}/25)(4 \sin 3x - 3 \cos 3x) + C$  **39.** $\ln|\sec x + \tan x| + C$  **43.** $-18$  **45.** 15

**49. a.** $(2/3) x (x + 1)^{3/2} - (4/15)(x + 1)^{5/2} + C$  **b.** $(2/5)(x + 1)^{5/2} - (2/3)(x + 1)^{3/2} + C$  **51.** 3.431

**53.** $18 \ln 9 - (16/3) \ln 4 - 7619 \approx 23.71$ sq cm  **55. a.** $1 - 1/k + (1/k)e^{-k}$; $k = 1/12$: $12e^{-1/12} - 11 \approx 0.0405$;

$k = 1/24$: $24e^{-1/24} - 23 \approx 0.0205$; $k = 1/48$: $48e^{-1/48} - 47 \approx 0.0103$  **b.** $e^{-6k}/(5k) + (1 - 1/(5k))e^{-k}$; $k = 1/12$:

$(12/5)e^{-1/2} - (7/5)e^{-1/12} \approx 0.1676$; $k = 1/24$: $(24/5)e^{-1/4} - (19/5)e^{-1/24} \approx 0.0933$; $k = 1/48$:$(48/5)e^{-1/8} - (43/5)e^{-1/48} \approx 0.0493$

**Exercises 8.3 (page 451)**

| For Exercises ... | 1–25, 36, 37, 40, 43 | 26–35, 39, 41, 42, 44–47 | 38 |
|---|---|---|---|
| Refer to Example ... | 1–3 | 4 | Derivation of volume formula |

**1.** $9\pi$  **3.** $364\pi/3$  **5.** $386\pi/27$  **7.** $15\pi/2$

**9.** $18\pi$  **11.** $\pi(e^4 - 1)/2 \approx 84.19$

**13.** $4\pi \ln 3 \approx 13.81$  **15.** $3124\pi/5$  **17.** $16\pi/15$  **19.** $\pi$  **21.** $4\pi/3$  **23.** $4\pi r^3/3$  **25.** $\pi r^2 h$  **27.** $13/3 \approx 4.333$  **29.** $38/15 \approx 2.533$

**31.** $e - 1 \approx 1.718$  **33.** $(5e^4 - 1)/8 \approx 34.00$  **35.** $4/\pi$  **37.** 3.758  **39. a.** $110e^{-0.1} - 120e^{-0.2} \approx 1.284$

**b.** $210e^{-1.1} - 220e^{-1.2} \approx 3.640$  **c.** $330e^{-2.3} - 340e^{-2.4} \approx 2.241$  **41. a.** $9(6 \ln 6 - 5) \approx 51.76$ items

**b.** $5(10 \ln 10 - 9) \approx 70.13$ items  **c.** $3(31 \ln 31 - 30)/2 \approx 114.7$ items  **43.** $1.083 \times 10^{21}$ m³  **45.** $42.49  **47.** 200 cases

**Exercises 8.4 (page 456)**

| For Exercises ... | 1–26, 31–34, 37, 42, 43 | 27–30, 35, 36 | 44–48 |
|---|---|---|---|
| Refer to Example ... | 1 | 2 | 3 |

**1.** 1/3  **3.** Divergent  **5.** $-1$  **7.** 10,000  **9.** 1/10

**11.** 3/5  **13.** 1  **15.** 1000  **17.** Divergent  **19.** 1

**21.** Divergent  **23.** Divergent  **25.** Divergent  **27.** 0  **29.** Divergent  **31.** Divergent  **33.** 1  **35.** 0

**39. a.** 2.808, 3.724, 4.417, 6.720, 9.022  **b.** Divergent  **c.** 0.8770, 0.9070, 0.9170, 0.9260, 0.9269  **d.** Convergent

**41. a.** 9.9995, 49.9875, 99.9500, 995.0166  **b.** Divergent  **c.** 100,000  **43.** 1/2  **45.** $Na/[b(b + k)]$  **47.** About 833.3

**Chapter 8 Review Exercises (page 459)**

| For Exercises ... | 1, 2, 11, 15–25, 59, 60 | 3–6, 12, 26–40, 61, 64 | 7–9, 13, 44–50, 63 | 10, 14, 51–58, 62 |
|---|---|---|---|---|
| Refer to Section ... | 1 | 2 | 3 | 4 |

**1.** False  **2.** True  **3.** False  **4.** True

**5.** False  **6.** True  **7.** False  **8.** False

**9.** True  **10.** False  **15.** 0.5833; 0.6035

**16.** 10.46; 10.20  **17.** 4.187; 4.155  **18.** 0.4668; 0.4908  **19.** 0.6011  **20.** 10.28  **21.** 4.156  **22.** 0.4937

**23. a.** 4/3  **b.** 1.146  **c.** 1.252  **24. a.** 6.391  **b.** 6.317  **c.** 6.378  **25. a.** 0  **b.** 0  **26.** 5

**27.** $-(2x/5)(8 - x)^{5/2} - (4/35)(8 - x)^{7/2} + C$  **28.** $6x(x - 2)^{1/2} - 4(x - 2)^{3/2} + C$  **29.** $xe^x - e^x + C$

**30.** $-(x + 2)e^{-3x} - (1/3)e^{-3x} + C$  **31.** $(1/4)(4x + 5)(\ln|4x + 5| - 1) + C$  **32.** $(x^2/2 - x)\ln|x| - x^2/4 + x + C$

**33.** $(-1/18) \ln|25 - 9x^2| + C$  **34.** $(1/8)\sqrt{16 + 8x^2} + C$  **35.** $(3e^4 + 1)/16 \approx 10.30$  **36.** $10e^{1/2} - 16 \approx 0.4872$

**37.** $-(x + 2) \cos x + \sin x + C$  **38.** $(x^2/2) \sin 2x + (x/2) \cos 2x - (1/4) \sin 2x + C$  **39.** $(7/4)(e^2 - 1) \approx 11.18$

**40.** $234/7 \approx 33.43$  **41.** $125\pi/9 \approx 43.63$  **42.** $81\pi/2 \approx 127.2$  **43.** $\pi(e^4 - e^{-2})/2 \approx 85.55$  **44.** $\pi \ln 3 \approx 3.451$

**45.** $406\pi/15 \approx 85.03$  **46.** $64\pi/5 \approx 40.21$  **47.** $7\pi r^2 h/12$  **49.** 13/6  **50.** 2,391,484/3  **51.** Divergent  **52.** 1/5

**53.** 1/3  **54.** $6/e \approx 2.207$  **55.** Divergent  **56.** Divergent  **57.** 5  **58.** 3  **59. a.** About 8208 kg  **b.** About 8430 kg

**c.** About 8558 kg  **60. a.** 0.99532227; 0.99997791; 0.99999998; 1.00000000; erfc$(0) = 1$

**61.** 0.4798  **62.** 5000 gallons  **63. a.** 158.3°  **b.** 125°  **c.** 133.3°  **64.** 16,250/3 $\approx$ $5416.67

# Chapter 9 Multivariable Calculus

## Exercises 9.1 (page 472)

**1. a.** 12 **b.** −6 **c.** 10 **d.** −19
**3. a.** $\sqrt{43}$ **b.** 6 **c.** $\sqrt{19}$ **d.** $\sqrt{11}$
**5. a.** $e$ **b.** $e^2$ **c.** 2 **d.** 3 **7. a.** 1 **b.** $1/(2\sqrt{2})$ **c.** $\sqrt{\pi}$ **d.** 1

| For Exercises . . . | 1–6, 29, 30 | 7–14 | 15–18 | 23–28 | | 33–45, 48 |
|---|---|---|---|---|---|---|
| Refer to Example . . . | 1, 3 | 4–6 | 7 | Material after Example 8 | | 2, 8 |

**9.**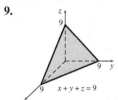
**11.**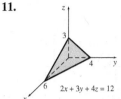
**13.**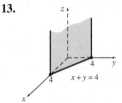
**15.**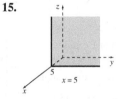
**17.**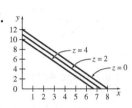

**19.** 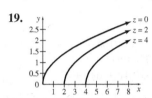 **25.** c **27.** e **29.** b **31. a.** $8x + 4h$ **b.** $-4y - 2h$ **c.** $8x$ **d.** $-4y$
**33. a.** $3e^2$; slope of tangent line in the direction of $x$ at $(1, 1)$ **b.** $3e^2$; slope of tangent line in the direction of $y$ at $(1, 1)$ **35. a.** 20.0 yr **b.** 363 yr **37. a.** 1.89 m² **b.** 1.62 m² **c.** 1.78 m²
**39. a.** 8.7% **b.** 48% **c.** Multiple solutions: $W = 19.75, R = 0, A = 0$ or $W = 10, R = 10, A = 4.59$ **d.** Wetland percentage **41. a.** 397 accidents
**43. a.** $T = 242.257\, C^{0.18}/F^3$ **b.** 58.82; a tethered sow spends nearly 59% of the time doing repetitive behavior when she is fed 2 kg of food a day and neighboring sows spend 40% of the time doing repetitive behavior. **45.** 40.29 L
**47. a.** About 18.8 kcal/min **b.** About 29.7 kcal/min **c.** Worms and moths **49.** $g(L, W, H) = 2LW + 2WH + 2LH$ ft²

## Exercises 9.2 (page 482)

**1. a.** $12x - 4y$ **b.** $-4x + 18y$ **c.** 12 **d.** −40
**3.** $f_x(x, y) = -4y; f_y(x, y) = -4x + 18y^2; 4; 178$
**5.** $f_x(x, y) = 10xy^3; f_y(x, y) = 15x^2y^2; -20; 2160$

| For Exercises . . . | 1–2, 37–40, 47, 48 | 3–22 | 23–36 | 41–46 | 49–71 |
|---|---|---|---|---|---|
| Refer to Example . . . | 3 | 1–3 | 5, 6 | 7 | 4 |

**7.** $f_x(x, y) = e^{x+y}; f_y(x, y) = e^{x+y}; e^1$ or $e; e^{-1}$ or $1/e$ **9.** $f_x(x, y) = -24e^{4x-3y}; f_y(x, y) = 18e^{4x-3y}; -24e^{11}; 18e^{-25}$
**11.** $f_x(x, y) = (-x^4 - 2xy^2 - 3x^2y^3)/(x^3 - y^2)^2; f_y(x, y) = (3x^3y^2 - y^4 + 2x^2y)/(x^3 - y^2)^2; -8/49; -1713/5329$
**13.** $f_x(x, y) = 15x^2y^2/(1 + 5x^3y^2); f_y(x, y) = 10x^3y/(1 + 5x^3y^2); 60/41; 1920/2879$
**15.** $f_x(x, y) = e^{x^2y}(2x^2y + 1); f_y(x, y) = x^3e^{x^2y}; -7e^{-4}; -64e^{48}$ **17.** $f_x(x, y) = (1/2)(4x^3 + 3y)/(x^4 + 3xy + y^4 + 10)^{1/2};$
$f_y(x, y) = (1/2)(3x + 4y^3)/(x^4 + 3xy + y^4 + 10)^{1/2}; 29/(2\sqrt{21}); 48/\sqrt{311}$ **19.** $f_x(x, y) = [6xy(e^{xy} + 2) - 3x^2y^2e^{xy}]/(e^{xy} + 2)^2;$
$f_y(x, y) = [3x^2(e^{xy} + 2) - 3x^3ye^{xy}]/(e^{xy} + 2)^2; -24(e^{-2} + 1)/(e^{-2} + 2)^2; (624e^{-12} + 96)/(e^{-12} + 2)^2$
**21.** $f_x(x, y) = \sin(\pi y); f_y(x, y) = \pi x \cos(\pi y); 0; 4\pi$ **23.** $f_{xx}(x, y) = 8y^2 - 32; f_{yy}(x, y) = 8x^2; f_{xy}(x, y) = f_{yx}(x, y) = 16xy$
**25.** $R_{xx}(x, y) = 8 + 24y^2; R_{yy}(x, y) = -30xy + 24x^2; R_{xy}(x, y) = R_{yx}(x, y) = -15y^2 + 48xy$
**27.** $r_{xx}(x, y) = 12y/(x + y)^3; r_{yy}(x, y) = -12x/(x + y)^3; r_{xy}(x, y) = r_{yx}(x, y) = (6y - 6x)/(x + y)^3$
**29.** $z_{xx} = 9ye^x; z_{yy} = 0; z_{xy} = z_{yx} = 9e^x$ **31.** $r_{xx} = -1/(x + y)^2; r_{yy} = -1/(x + y)^2; r_{xy} = r_{yx} = -1/(x + y)^2$
**33.** $z_{xx} = 1/x; z_{yy} = -x/y^2; z_{xy} = z_{yx} = 1/y$ **35.** $z_{xx} = e^x \sin y; z_{yy} = -e^x \sin y; z_{xy} = e^x \cos y$ **37.** $x = -4, y = 2$
**39.** $x = 0, y = 0$; or $x = 3, y = 3$ **41.** $f_x(x, y, z) = 4x^3; f_y(x, y, z) = 2z^2; f_z(x, y, z) = 4yz + 4z^3; f_{yz}(x, y, z) = 4z$
**43.** $f_x(x, y, z) = 6/(4z + 5); f_y(x, y, z) = -5/(4z + 5); f_z(x, y, z) = -4(6x - 5y)/(4z + 5)^2; f_{yz}(x, y, z) = 20/(4z + 5)^2$
**45.** $f_x(x, y, z) = (2x - 5z^2)/(x^2 - 5xz^2 + y^4); f_y(x, y, z) = 4y^3/(x^2 - 5xz^2 + y^4); f_z(x, y, z) = -10xz/(x^2 - 5xz^2 + y^4);$
$f_{yz}(x, y, z) = 40xy^3z/(x^2 - 5xz^2 + y^4)^2$ **47. a.** 6.773 **b.** 3.386 **49. a.** 1279 kcal per hr **b.** 2.906 kcal per hr per g; the instantaneous rate of change of energy usage for a 300-kg animal traveling at 10 km per hr is about 2.9 kcal per hr per g.
**51. a.** 0.0142 m² **b.** 0.00442 m² **53. a.** 4.125 lb **b.** $\partial f/\partial n = n/4$; the rate of change of weight loss per unit change in workouts
**c.** An additional loss of 3/4 lb **55. a.** 0.0783 **b.** 0.0906 per m **c.** −0.0212 per m **d.** waist **57. a.** $(2ax - 3x^2)t^2e^{-t}$
**b.** $x^2(a - x)(2t - t^2)e^{-t}$ **c.** $(2a - 6x)t^2e^{-t}$ **d.** $(2ax - 3x^2)(2t - t^2)e^{-t}$ **e.** $\partial R/\partial x$ gives the rate of change of the reaction per unit of change in the amount of drug administered. $\partial R/\partial t$ gives the rate of change of the reaction for a 1-hour change in the time after the drug is administered. **59. a.** −24.9°F **b.** 15 mph **c.** $W_V(20, 10) = -1.114$; while holding the temperature fixed at 10°F, the wind chill decreases approximately 1.1°F when the wind velocity increases by 1 mph; $W_T(20, 10) = 1.429$; while holding the wind velocity fixed at 20 mph, the wind chill increases approximately 1.429°F if the actual temperature increases from 10°F to 11°F.
**d.** Sample table

| T\V | 5 | 10 | 15 | 20 |
|---|---|---|---|---|
| 30 | 27 | 16 | 9 | 4 |
| 20 | 16 | 3 | −5 | −11 |
| 10 | 6 | −9 | −18 | −25 |
| 0 | −5 | −21 | −32 | −39 |

**61. a.** $D_F = -0.002527e^{0.005163R - 0.004345L - 0.000019F}$
**b.** About 3.7 days earlier
**63.** $-2k(x - by)/(a + (x - by)^2)^2,$
$2kb(x - by)/(a + (x - by)^2)^2$; they are opposite in sign.

**65. a.** $(N - 1)bke^{-kn}/\{aN[1 - (N - 1)b(1 - e^{-kn})/N]\}$; the rate of change of the number of wrong attempts with respect to the number of trials **b.** $b(1 - e^{-kn})/\{aN^2[1 - (N - 1)b(1 - e^{-kn})/N]\}$; the rate of change of the number of wrong attempts with respect to the number of stimuli **67.** $-10$ ml per year; 100 ml per in. **69. a.** 164,456 m/sec **b.** 0.238 m/sec per m/sec **c.** $c$ **71. a.** About 340 cm **b.** About 242 cm **c.** About 445 cm

## Exercises 9.3 (page 493)

| For Exercises . . . | 1–18, 21–28 | 34, 35, 37–41 |
|---|---|---|
| Refer to Example . . . | 1–3 | 4 |

**1.** Saddle point at $(-1, 2)$ **3.** Relative minimum at $(-3, -3)$
**5.** Relative minimum at $(-2, -2)$ **7.** Relative minimum at $(15, -8)$
**9.** Relative maximum at $(2/3, 4/3)$ **11.** Saddle point at $(2, -2)$ **13.** Saddle point at $(0, 0)$; relative minimum at $(27, 9)$
**15.** Saddle point at $(0, 0)$; relative minimum at $(9/2, 3/2)$ **17.** Saddle point at $(0, -1)$ **21.** Relative maximum of 9/8 at $(-1, 1)$; saddle point at $(0, 0)$; a **23.** Relative minima of $-33/16$ at $(0, 1)$ and at $(0, -1)$; saddle point at $(0, 0)$; b
**25.** Relative maxima of 17/16 at $(1, 0)$ and $(-1, 0)$; relative minima of $-15/16$ at $(0, 1)$ and $(0, -1)$; saddle points at $(0, 0)$, $(-1, 1)$, $(1, -1)$, $(1, 1)$, and $(-1, -1)$; e **31. a.** All values of $k$ **b.** $k \geq 0$ **35. a.** 0.5148; 0.4064; the jury is less likely to make the correct decision in the second situation. **b.** If $r = s = 1$ then $P(\alpha, 1, 1) = 1$ **c.** $P(\alpha, 1, 1) = 1$ is a maximum value. **37. a.** 436.16 kJ/mol **b.** 137.66 kJ/mol **c.** Saddle point at $(1.75, 133°C)$ **39.** Minimum cost of $59 when $x = 4, y = 5$ **41.** Sell 12 X-ray machines and 7 bone density scanners for a maximum revenue of $166,600.

## Exercises 9.4 (page 500)

| For Exercises . . . | 1–6, 19–26 | 7–16 | 17, 18, 27–30, 33–35 | 31, 32 |
|---|---|---|---|---|
| Refer to Example . . . | 1 | 2 | 3 | 4 |

**1.** 0.12 **3.** 0.0311 **5.** $-0.335$ **7.** 10.022; 10.0221; 0.0001
**9.** 2.0067; 2.0080; 0.0013 **11.** 1.07; 1.0720; 0.0020
**13.** $-0.02$; $-0.0200$; 0 **15.** 1.02; 1.0197; 0.0003 **17.** 6.65 cm³ **19.** 2.98 liters **21. a.** 0.2649
**b.** Actual 0.2817; approximation 0.2816 **23. a.** 87% **b.** 75% **d.** 89%, 87% **25. b.** 30.19 **c.** $-0.0140$ **d.** 0.0820
**27.** 26.945 cm² **29.** 3% **31.** 8 **33.** 20.73 cm³ **35.** 86.4 in³

## Exercises 9.5 (page 510)

| For Exercises . . . | 1–10 | 11–20 | 21–30, 63–66, 68–70 | 31–40, 67 | 41–48 | 49–58 | 59, 60 |
|---|---|---|---|---|---|---|---|
| Refer to Example . . . | 1 | 2, 3 | 4 | 5, 6 | 7 | 8 | 9 |

**1.** $630y$
**3.** $(2x/9)[(x^2 + 15)^{3/2} - (x^2 + 12)^{3/2}]$
**5.** $6 + 10y$ **7.** $(1/2)(e^{12+3y} - e^{4+3y})$ **9.** $(1/2)(e^{4x+9} - e^{4x})$ **11.** 945 **13.** $(2/45)(39^{5/2} - 125^{5/2} - 7533)$
**15.** 21 **17.** $(\ln 3)^2$ **19.** $8 \ln 2 + 4$ **21.** 171 **23.** $(4/15)(33 - 25^{5/2} - 35^{5/2})$ **25.** $-3 \ln(3/4)$ or $3 \ln(4/3)$
**27.** $(1/2)(e^7 - e^6 - e^3 + e^2)$ **29.** 1 **31.** 96 **33.** 40/3 **35.** $(2/15)(25^{5/2} - 2)$ **37.** $(1/4) \ln(17/8)$ **39.** $e^2 - 3$
**41.** 97,632/105 **43.** 128/9 **45.** $\ln 16$ or $4 \ln 2$ **47.** 64/3 **49.** 34 **51.** 10/3 **53.** $7(e - 1)/3$
**55.** 16/3 **57.** $4 \ln 2 - 2$ **59.** 1 **63.** 49 **65.** $(e^6 + e^{-10} - e^{-4} - 1)/60$ **67.** 9 in³ **69.** $933.33

## Chapter 9 Review Exercises (page 515)

| For Exercises . . . | 1–5, 15–24, 92 | 6, 12, 13, 25–44, 86–88, 93 | 7–8, 45–53, 90, 91, 94 | 9–11, 59–84, 89 | 14, 54–62, 85, 95–98 |
|---|---|---|---|---|---|
| Refer to Section . . . | 1 | 2 | 3 | 5 | 4 |

**1.** True **2.** True **3.** True **4.** True
**5.** False **6.** False **7.** False **8.** True
**9.** False **10.** True **11.** False
**15.** $-19$; $-255$ **16.** 23; 594 **17.** $-5/9$; $-4/3$ **18.** $-\sqrt{5}/3$; $\sqrt{5}/3$

**19.** 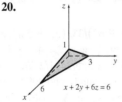 $x + y + z = 4$ **20.** 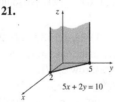 $x + 2y + 6z = 6$ **21.** $5x + 2y = 10$ **22.** 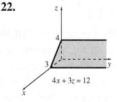 $4x + 3z = 12$ **23.** 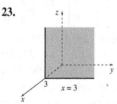 $x = 3$

**24.** 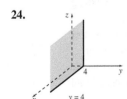 $y = 4$ **25. a.** $9x^2 + 8xy$ **b.** $-12$ **c.** 16 **26. a.** $4xy/(x - y^2)^2$ **b.** $-1/2$ **c.** 0 **27.** $f_x(x, y) = 12xy^3$; $f_y(x, y) = 18x^2y^2 - 4$ **28.** $f_x(x, y) = 20x^3y^3 - 30x^4y$; $f_y(x, y) = 15x^4y^2 - 6x^5$
**29.** $f_x(x, y) = 4x/(4x^2 + y^2)^{1/2}$; $f_y(x, y) = y/(4x^2 + y^2)^{1/2}$
**30.** $f_x(x, y) = (-6x^2 + 2y^2 - 30xy^2)/(3x^2 + y^2)^2$; $f_y(x, y) = (30x^2y - 4xy)/(3x^2 + y^2)^2$
**31.** $f_x(x, y) = 3x^2e^{3y}$; $f_y(x, y) = 3x^3e^{3y}$ **32.** $f_x(x, y) = (y - 2)^2e^{x+2y}$;
$f_y(x, y) = 2(y - 2)(y - 1)e^{x+2y}$ **33.** $f_x(x, y) = 4x/(2x^2 + y^2)$; $f_y(x, y) = 2y/(2x^2 + y^2)$
**34.** $f_x(x, y) = -2xy^3/(2 - x^2y^3)$; $f_y(x, y) = -3x^2y^2/(2 - x^2y^3)$ **35.** $f_x(x, y) = -6x \sin(3x^2 + y^2)$;
$f_y(x, y) = -2y \sin(3x^2 + y^2)$ **36.** $f_x(x, y) = \tan(7x^2 + 4y^2) + 14x^2 \sec^2(7x^2 + 4y^2)$; $f_y(x, y) = 8xy \sec^2(7x^2 + 4y^2)$
**37.** $f_{xx}(x, y) = 30xy$; $f_{xy}(x, y) = 15x^2 - 12y$ **38.** $f_{xx}(x, y) = -6y^3 + 6xy$; $f_{xy}(x, y) = -18xy^2 + 3x^2$
**39.** $f_{xx}(x, y) = 12y/(2x - y)^3$; $f_{xy}(x, y) = (-6x - 3y)/(2x - y)^3$ **40.** $f_{xx}(x, y) = 2(3 + y)/(x - 1)^3$;
$f_{xy}(x, y) = -1/(x - 1)^2$ **41.** $f_{xx}(x, y) = 8e^{2y}$; $f_{xy}(x, y) = 16xe^{2y}$ **42.** $f_{xx}(x, y) = 2ye^{x^2}(2x^2 + 1)$; $f_{xy}(x, y) = 2xe^{x^2}$

**43.** $f_{xx}(x, y) = (-2x^2y^2 - 4y)/(2 - x^2y)^2$; $f_{xy}(x, y) = -4x/(2 - x^2y)^2$   **44.** $f_{xx}(x, y) = -9y^4/(1 + 3xy^2)^2$; $f_{xy}(x, y) = 6y/(1 + 3xy^2)^2$   **45.** Saddle point at $(0, 2)$   **46.** Relative minimum at $(-9/2, 4)$   **47.** Relative minimum at $(2, 1)$

**48.** Relative maximum at $(-3/4, -9/32)$; saddle point at $(0, 0)$   **49.** Saddle point at $(3, 1)$   **50.** Relative minimum at $(11/4, -2)$

**51.** Saddle point at $(-1/3, 11/6)$; relative minimum at $(1, 1/2)$   **52.** Relative minimum at $(-8, -23)$   **55.** 1.22

**56.** −0.0168   **57.** 13.0846; 13.0848; 0.0002   **58.** 2.095; 2.0972; 0.0022   **59.** $8y - 6$   **60.** $(e^{15+5y} - e^{3+5y})/3$

**61.** $(3/2)[(100 + 2y^2)^{1/2} - (2y^2)^{1/2}]$   **62.** $(2/33)[(7x + 297)^{1/2} - (7x + 11)^{1/2}]$   **63.** 1232/9   **64.** 395

**65.** $(2/135)[(42)^{5/2} - (24)^{5/2} - (39)^{5/2} + (21)^{5/2}]$   **66.** $(e^3 + e^{-8} - e^{-4} - e^{-1})/14$   **67.** 2 ln 2 or ln 4   **68.** ln 2

**69.** 110   **70.** $(2/15)(11^{5/2} - 8^{5/2} - 7^{5/2} + 32)$   **71.** $(4/15)(782 - 8^{5/2})$   **72.** $(e^2 - 2e + 1)/2$   **73.** $-1/2 + \pi/4$

**74.** $1/e - 2 - (1/e)\cos(e\pi)$   **75.** 105/2   **76.** 308/3   **77.** 1/2   **78.** 52/5   **79.** 1/48   **80.** 1/12   **81.** ln 2   **82.** $16(5\sqrt{5} - 1)/3$

**83.** 3   **84.** 26/105   **85.** 1.341 cm$^3$   **86. a.** 49.68 liters   **b.** −0.09, the approximate change in total body water if age is increased by 1 yr and mass and height are held constant is −0.09 liter; 0.34, the approximate change in total body water if mass is increased by 1 kg and age and height are held constant is 0.34 liter; 0.25, the approximate change in total body water if height is increased by 1 cm and age and mass are held constant is 0.25 liter.   **87. a.** About 33.98 cm   **b.** About 0.02723 cm per g; about 0.2821 cm/year; the approximate change in the length of a trout if its mass increases from 450 to 451 g while age is held constant at 7 years is 0.027 cm; the approximate change in the length of a trout if its age increases from 7 to 8 years while mass is held constant at 450 g is 0.28 cm.   **88. a.** 50; in 1900, 50% of those born 60 years earlier are still alive.   **b.** 75; in 2006, 75% of those born 70 years earlier are still alive.   **c.** −1.25; in 1900, the percent of those born 60 years earlier who are still alive was dropping at a rate of 1.25 percent per additional year of life.   **d.** −2; in 2006, the percent of those born 70 years earlier who are still alive was dropping at a rate of 2 percent per additional year of life.   **89. a.** 2.828 ft$^2$   **b.** An increase of 0.6187 ft$^2$

**90.** 5 in. by 5 in. by 5 in.   **91.** 20,000 ft$^2$ with dimensions 100 ft by 200 ft   **92. a.** $\$(325 + \sqrt{10}) \approx \$328.16$

**b.** $\$(800 + \sqrt{15}) \approx \$803.87$   **c.** $\$(2000 + \sqrt{20}) \approx \$2004.47$   **93. a.** $26   **b.** $2572

**94. a.** Relative minimum at $(11, 12)$   **b.** $431   **95.** Decrease by $243.82   **96.** 7.92 cm$^3$   **97.** 4.19 ft$^3$   **98.** 15.6 cm$^3$

**99. a.** $200 spent on fertilizer and $80 spent on seed will produce a maximum profit of $266 per acre.   **b.** Same as a.

# Chapter 10 Matrices

**Exercises 10.1 (page 534)**

**1.** $\begin{bmatrix} 3 & 1 & | & 6 \\ 2 & 5 & | & 15 \end{bmatrix}$

| For Exercises ... | 1–8, 11–16 | 17–20, 25, 26, 29, 30, 34, 36–38, 41, 42 | 21, 22, 27, 32 | 23, 24, 28, 31, 33, 35, 39, 40 | 44–70 |
|---|---|---|---|---|---|
| Refer to Example ... | 1–5 | 1, 2 | 3 | 4, 5 | 6 |

**3.** $\begin{bmatrix} 2 & 1 & 1 & | & 3 \\ 3 & -4 & 2 & | & -7 \\ 1 & 1 & 1 & | & 2 \end{bmatrix}$   **5.** $x = 2, y = 3$   **7.** $x = 4, y = -5, z = 1$   **9.** Row operations   **11.** $\begin{bmatrix} 3 & 7 & 4 & | & 10 \\ 0 & 1 & -5 & | & -8 \\ 0 & 4 & 5 & | & 11 \end{bmatrix}$

**13.** $\begin{bmatrix} 1 & 0 & 0 & | & -3 \\ 0 & 3 & 2 & | & 5 \\ 0 & 5 & 3 & | & 7 \end{bmatrix}$   **15.** $\begin{bmatrix} 1 & 0 & 0 & | & 6 \\ 0 & 5 & 0 & | & 9 \\ 0 & 0 & 4 & | & 8 \end{bmatrix}$   **17.** $(2, 3)$   **19.** $(1, 6)$   **21.** No solution   **23.** $((3y + 1)/6, y)$   **25.** $(4, 1, 0)$

**27.** No solution   **29.** $(-1, 23, 16)$   **31.** $((-9z + 5)/23, (10z - 3)/23, z)$   **33.** $((-2z + 62)/35, (3z + 5)/7, z)$

**35.** $((9 - 3y - z)/2, y, z)$   **37.** $(0, 2, -2, 1)$; the answers are given in the order $x, y, z, w$.   **39.** $(-w - 3, -4w - 19, -3w - 2, w)$

**41.** $(28.9436, 36.6326, 9.6390, 37.1036)$   **43.** row 1: 3/8, 1/6, 11/24; row 2: 5/12, 1/3, 1/4; row 3: 5/24, 1/2, 7/24   **45.** 40 units of corn, 15 units of soybeans, and 30 units of cottonseed   **47. a.** 400/9 g of group A, 400/3 g of group B, and 2000/9 g of group C   **b.** For any positive number $z$ of grams of group C, there should be $z$ grams less than 800/3 g of group A and 400/3 g of group B.   **c.** No   **49.** About 244 fish of species A, 39 fish of species B, and 101 fish of species C   **51.** 81 kg of the first chemical, 382.286 kg of the second, and 286.714 kg of the third   **53. b.** 7,206,360 white cows, 4,893,246 black cows, 3,515,820 spotted cows, and 5,439,213 brown cows   **55.** Hire the Garcia firm for 20 hr and the Wong firm for 15 hr.   **57. a.** 10 deluxe, 4 super-deluxe, 15 ultra   **b.** None   **c.** 5   **59. a.** $12,000 at 8%, $7000 at 9%, and $6000 at 10%   **b.** The amount borrowed at 10% must be less than or equal to $9500. If $z = \$5000$, they borrowed $11,000 at 8% and $9000 at 9%.   **c.** No solution   **d.** The total annual interest would be $2220, not $2190, as specified as one of the conditions.   **61. a.** About 51.8 weeks; about $959,091   **63.** 36 field goals, 28 foul shots   **65.** 225 singles, 24 doubles, 5 triples, and 8 home runs   **67. a.** $y = -0.032x^2 + 0.852x + 0.632$

**b.** $y = -0.032x^2 + 0.852x + 0.632$   **69. a.** $x_2 + x_3 = 700, x_3 + x_4 = 600$   **b.** $(1000 - x_4, 100 + x_4, 600 - x_4, x_4)$

**c.** $0 \leq x_4 \leq 600$   **d.** $400 \leq x_1 \leq 1000, 100 \leq x_2 \leq 700, 0 \leq x_3 \leq 600$

## Exercises 10.2 (page 542)

**1.** False; not all corresponding elements are equal.

| For Exercises . . . | 1–6, 15–20 | 7–14 | 21–32 | 41, 48, 49 | 39, 46 | 40 | 42–45, 47 |
|---|---|---|---|---|---|---|---|
| Refer to Example . . . | 3 | 2 | 4, 6 | 5 | 1 | 5, 7 | 7 |

**3.** True  **5.** True  **7.** $2 \times 2$; square
**9.** $3 \times 4$  **11.** $2 \times 1$; column  **13.** Undefined  **15.** $x = 4, y = -8, z = 1$  **17.** $s = 10, t = 0, r = 7$

**19.** $a = 20, b = 5, c = 0, d = 4, f = 1$  **21.** $\begin{bmatrix} 10 & 4 & -5 & -6 \\ 4 & 5 & 3 & 11 \end{bmatrix}$  **23.** Not possible  **25.** $\begin{bmatrix} 1 & 5 & 6 & -9 \\ 5 & 7 & 2 & 1 \\ -7 & 2 & 2 & -7 \end{bmatrix}$  **27.** $\begin{bmatrix} 3 & 4 \\ 4 & 8 \end{bmatrix}$

**29.** $\begin{bmatrix} 10 & -2 \\ 10 & 9 \end{bmatrix}$  **31.** $\begin{bmatrix} -12x + 8y & -x + y \\ x & 8x - y \end{bmatrix}$  **33.** $\begin{bmatrix} -x & -y \\ -z & -w \end{bmatrix}$  **39. a.** $\begin{bmatrix} 2 & 1 & 2 & 1 \\ 3 & 2 & 2 & 1 \\ 4 & 3 & 2 & 1 \end{bmatrix}$  **b.** $\begin{bmatrix} 5 & 0 & 7 \\ 0 & 10 & 1 \\ 0 & 15 & 2 \\ 10 & 12 & 8 \end{bmatrix}$  **c.** $\begin{bmatrix} 8 \\ 4 \\ 5 \end{bmatrix}$

**41. a.** 8  **b.** 3  **c.** $\begin{bmatrix} 85 & 15 \\ 27 & 73 \end{bmatrix}$  **d.** Yes  **43. a.** $\begin{bmatrix} 60.0 & 68.3 \\ 63.8 & 72.5 \\ 64.5 & 73.6 \\ 68.2 & 75.1 \\ 71.8 & 78.0 \end{bmatrix}$  **b.** $\begin{bmatrix} 68.0 & 75.6 \\ 70.7 & 78.1 \\ 72.7 & 79.4 \\ 74.7 & 79.9 \\ 76.5 & 81.3 \end{bmatrix}$  **c.** $\begin{bmatrix} -8.0 & -7.3 \\ -6.9 & -5.6 \\ -8.2 & -5.8 \\ -6.5 & -4.8 \\ -4.7 & -3.3 \end{bmatrix}$

**45. a.** $\begin{bmatrix} 51.2 & 7.9 \\ 59.8 & 11.1 \\ 66.2 & 11.3 \\ 73.8 & 13.2 \\ 78.5 & 16.5 \\ 84.2 & 19.8 \end{bmatrix}$  **b.** $\begin{bmatrix} 45.3 & 7.9 \\ 47.9 & 8.5 \\ 50.8 & 9.2 \\ 53.4 & 9.3 \\ 57.0 & 10.6 \\ 62.9 & 13.9 \end{bmatrix}$  **c.** $\begin{bmatrix} 5.9 & 0 \\ 11.9 & 2.6 \\ 15.4 & 2.1 \\ 20.4 & 3.9 \\ 21.5 & 5.9 \\ 21.3 & 5.9 \end{bmatrix}$  **47. a.** $\begin{bmatrix} 2.61 & 4.39 & 6.29 & 9.08 \\ 1.63 & 2.77 & 4.61 & 6.92 \\ 0.92 & 0.75 & 0.62 & 0.54 \end{bmatrix}$

**b.** $\begin{bmatrix} 1.38 & 1.72 & 1.94 & 3.31 \\ 1.26 & 1.48 & 2.82 & 2.28 \\ 0.41 & 0.33 & 0.27 & 0.40 \end{bmatrix}$  **c.** $\begin{bmatrix} 1.23 & 2.67 & 4.35 & 5.77 \\ 0.37 & 1.29 & 1.79 & 4.64 \\ 0.51 & 0.42 & 0.35 & 0.14 \end{bmatrix}$

**49. a.** $\begin{bmatrix} 88 & 105 & 60 \\ 48 & 72 & 40 \\ 16 & 21 & 0 \\ 112 & 147 & 50 \end{bmatrix}$  **b.** $\begin{bmatrix} 110 & 140 & 66 \\ 60 & 96 & 44 \\ 20 & 28 & 0 \\ 140 & 196 & 55 \end{bmatrix}$  **c.** $\begin{bmatrix} 198 & 245 & 126 \\ 108 & 168 & 84 \\ 36 & 49 & 0 \\ 252 & 343 & 105 \end{bmatrix}$

## Exercises 10.3 (page 552)

| For Exercises . . . | 1–6, 44–46 | 7–12 | 15–31 | 43, 47–52 |
|---|---|---|---|---|
| Refer to Example . . . | 1 | 4 | 2, 3, 5 | 6 |

**1.** $\begin{bmatrix} -4 & 8 \\ 0 & 6 \end{bmatrix}$  **3.** $\begin{bmatrix} 12 & -24 \\ 0 & -18 \end{bmatrix}$  **5.** $\begin{bmatrix} -22 & -6 \\ 20 & -12 \end{bmatrix}$  **7.** $2 \times 2$; $2 \times 2$

**9.** $3 \times 4$; $BA$ does not exist.  **11.** $AB$ does not exist; $3 \times 2$.  **13.** columns; rows  **15.** $\begin{bmatrix} 8 \\ -1 \end{bmatrix}$  **17.** $\begin{bmatrix} 14 \\ -23 \end{bmatrix}$  **19.** $\begin{bmatrix} -7 & 2 & 8 \\ 27 & -12 & 12 \end{bmatrix}$

**21.** $\begin{bmatrix} -2 & 10 \\ 0 & 8 \end{bmatrix}$  **23.** $\begin{bmatrix} 13 & 5 \\ 25 & 15 \end{bmatrix}$  **25.** $\begin{bmatrix} 13 \\ 29 \end{bmatrix}$  **27.** $\begin{bmatrix} 7 \\ -33 \\ 4 \end{bmatrix}$  **29.** $\begin{bmatrix} 22 & -8 \\ 11 & -4 \end{bmatrix}$  **31. a.** $\begin{bmatrix} 16 & 22 \\ 7 & 19 \end{bmatrix}$  **b.** $\begin{bmatrix} 5 & -5 \\ 0 & 30 \end{bmatrix}$  **c.** No  **d.** No

**39. a.** $\begin{bmatrix} 6 & 106 & 158 & 222 & 28 \\ 120 & 139 & 64 & 75 & 115 \\ -146 & -2 & 184 & 144 & -129 \\ 106 & 94 & 24 & 116 & 110 \end{bmatrix}$  **b.** Does not exist  **c.** No

**41. a.** $\begin{bmatrix} -1 & 5 & 9 & 13 & -1 \\ 7 & 17 & 2 & -10 & 6 \\ 18 & 9 & -12 & 12 & 22 \\ 9 & 4 & 18 & 10 & -3 \\ 1 & 6 & 10 & 28 & 5 \end{bmatrix}$  **b.** $\begin{bmatrix} -2 & -9 & 90 & 77 \\ -42 & -63 & 127 & 62 \\ 413 & 76 & 180 & -56 \\ -29 & -44 & 198 & 85 \\ 137 & 20 & 162 & 103 \end{bmatrix}$  **c.** $\begin{bmatrix} -56 & -1 & 1 & 45 \\ -156 & -119 & 76 & 122 \\ 315 & 86 & 118 & -91 \\ -17 & -17 & 116 & 51 \\ 118 & 19 & 125 & 77 \end{bmatrix}$

**d.** $\begin{bmatrix} 54 & -8 & 89 & 32 \\ 114 & 56 & 51 & -60 \\ 98 & -10 & 62 & 35 \\ -12 & -27 & 82 & 34 \\ 19 & 1 & 37 & 26 \end{bmatrix}$ **e.** $\begin{bmatrix} -2 & -9 & 90 & 77 \\ -42 & -63 & 127 & 62 \\ 413 & 76 & 180 & -56 \\ -29 & -44 & 198 & 85 \\ 137 & 20 & 162 & 103 \end{bmatrix}$ **f.** Yes

**43. a.** $\begin{bmatrix} 20 & 52 & 27 \\ 25 & 62 & 35 \\ 30 & 72 & 43 \end{bmatrix}$; the rows give the amounts of fat, carbohydrates, and protein, respectively, in each of the daily meals.

**b.** $\begin{bmatrix} 75 \\ 45 \\ 70 \\ 168 \end{bmatrix}$; the rows give the number of calories in one exchange of each of the food groups. **c.** The rows give the number of calories in each meal.

**45.** $\begin{bmatrix} 66.7 & 74.4 \\ 69.6 & 77.2 \\ 71.3 & 78.4 \\ 73.6 & 79.1 \\ 75.7 & 80.8 \end{bmatrix}$ **47. a.** 3819; 3824; 3763; 3698; 3638 **b.** Extinction **c.** 4017; 4154; 4280; 4399; 4524; does not become extinct

**49. a.** $\begin{bmatrix} 146 \\ 93 \end{bmatrix}$; the rows represent the cost (in cents) for each type of bird.

**51. a.** $\begin{matrix} & \text{CC} & \text{MM} & \text{AD} \\ \text{S} & \\ \text{C} \end{matrix} \begin{bmatrix} 0.5 & 0.4 & 0.3 \\ 0.2 & 0.3 & 0.3 \end{bmatrix}$ **b.** $\begin{matrix} & \text{S} & \text{C} \\ \text{SD} \\ \text{MC} \\ \text{M} \end{matrix} \begin{bmatrix} 4 & 3 \\ 2 & 5 \\ 1 & 7 \end{bmatrix}$ **c.** $CA = \begin{matrix} & \text{CC} & \text{MM} & \text{AD} \\ \text{SD} \\ \text{MC} \\ \text{M} \end{matrix} \begin{bmatrix} 2.6 & 2.5 & 2.1 \\ 2 & 2.3 & 2.1 \\ 1.9 & 2.5 & 2.4 \end{bmatrix}$ **d.** $2.50 **e.** Mexico City

## Exercises 10.4 (page 564)

**1.** Yes **3.** No
**5.** No **7.** Yes

| For Exercises . . . | 1–8 | 11–26, 51–55 | 27–42, 56–58 | 59–64, 67 | 65, 66 |
|---|---|---|---|---|---|
| Refer to Example . . . | 1 | 2, 3 | 4, 6 | 5 | 7 |

**9.** No; the row of all zeros makes it impossible to get all the 1's in the diagonal of the identity matrix, no matter what matrix is used as a candidate for the inverse.

**11.** $\begin{bmatrix} 0 & 1/2 \\ -1 & 1/2 \end{bmatrix}$ **13.** $\begin{bmatrix} 2 & 1 \\ 5 & 3 \end{bmatrix}$ **15.** No inverse **17.** $\begin{bmatrix} 1 & 0 & 0 \\ 0 & -1 & 0 \\ -1 & 0 & 1 \end{bmatrix}$ **19.** $\begin{bmatrix} 15 & 4 & -5 \\ -12 & -3 & 4 \\ -4 & -1 & 1 \end{bmatrix}$ **21.** No inverse

**23.** $\begin{bmatrix} -11/2 & -1/2 & 5/2 \\ 1/2 & 1/2 & -1/2 \\ -5/2 & 1/2 & 1/2 \end{bmatrix}$ **25.** $\begin{bmatrix} 1/2 & 1/2 & -1/4 & 1/2 \\ -1 & 4 & -1/2 & -2 \\ -1/2 & 5/2 & -1/4 & -3/2 \\ 1/2 & -1/2 & 1/4 & 1/2 \end{bmatrix}$ **27.** $(5, 1)$ **29.** $(2, 1)$ **31.** $(15, 21)$

**33.** No inverse, $(-8y - 12, y)$ **35.** $(-8, 6, 1)$ **37.** $(-36, 8, -8)$ **39.** No inverse, no solution for system

**41.** $(-7, -34, -19, 7)$ **51.** $\begin{bmatrix} -0.0447 & -0.0230 & 0.0292 & 0.0895 & -0.0402 \\ 0.0921 & 0.0150 & 0.0321 & 0.0209 & -0.0276 \\ -0.0678 & 0.0315 & -0.0404 & 0.0326 & 0.0373 \\ 0.0171 & -0.0248 & 0.0069 & -0.0003 & 0.0246 \\ -0.0208 & 0.0740 & 0.0096 & -0.1018 & 0.0646 \end{bmatrix}$

**53.** $\begin{bmatrix} 0.0394 & 0.0880 & 0.0033 & 0.0530 & -0.1499 \\ -0.1492 & 0.0289 & 0.0187 & 0.1033 & 0.1668 \\ -0.1330 & -0.0543 & 0.0356 & 0.1768 & 0.1055 \\ 0.1407 & 0.0175 & -0.0453 & -0.1344 & 0.0655 \\ 0.0102 & -0.0653 & 0.0993 & 0.0085 & -0.0388 \end{bmatrix}$ **55.** Yes **57.** $\begin{bmatrix} 1.51482 \\ 0.053479 \\ -0.637242 \\ 0.462629 \end{bmatrix}$

**59. a.** 50 Super Vim, 75 Multitab, and 100 Mighty Mix   **b.** 75 Super Vim, 50 Multitab, and 60 Mighty Mix   **c.** 80 Super Vim, 100 Multitab, and 50 Mighty Mix

**61. a.** $\begin{bmatrix} 72 \\ 48 \\ 60 \end{bmatrix}$  **b.** $\begin{bmatrix} 2 & 4 & 2 \\ 2 & 1 & 2 \\ 2 & 1 & 3 \end{bmatrix} \begin{bmatrix} x_1 \\ x_2 \\ x_3 \end{bmatrix} = \begin{bmatrix} 72 \\ 48 \\ 60 \end{bmatrix}$  **c.** 8 type I, 8 type II, and 12 type III   **63. a.** \$10,000 at 6%, \$10,000 at 6.5%,

and \$5000 at 8%   **b.** \$14,000 at 6%, \$9000 at 6.5%, and \$7000 at 8%   **c.** \$24,000 at 6%, \$4000 at 6.5%, and \$12,000 at 8%

**65. a.** 85, 50, 40, 130, 120, 145, 49, 63, 121, 171, 117, 99, 159, 113, 91, 145, 105, 100, 90, 40, 75, 134, 113, 91, 98, 101, 112

**b.** I love you   **67. a.** $\begin{bmatrix} -1/15 & -1/15 & 13/3 \\ 4/15 & 1/15 & -40/3 \\ -1/5 & 0 & 10 \end{bmatrix}$  **b.** $\begin{bmatrix} 0 & 1 & 40 \\ -1 & 0 & 60 \\ 0 & 0 & 1 \end{bmatrix}$  **c.** The shape is a sideways T whose vertical and horizontal intersection is at the mark $(60, 10)$.

## Exercises 10.5 (page 575)

| For Exercises . . . | 1–4 | 5–23 | 24, 25 |
|---|---|---|---|
| Refer to Example . . . | 2, 3 | 4, 5 | 6 |

**1.** 17   **3.** 165   **5.** 1, 5; $\begin{bmatrix} 0 \\ 1 \end{bmatrix}$, $\begin{bmatrix} 2 \\ 1 \end{bmatrix}$   **7.** 2, 9; $\begin{bmatrix} 2 \\ -1 \end{bmatrix}$, $\begin{bmatrix} 1 \\ 3 \end{bmatrix}$   **9.** 2, 1; $\begin{bmatrix} 3 \\ 2 \end{bmatrix}$, $\begin{bmatrix} 1 \\ 1 \end{bmatrix}$

**11.** 4, -1, -3; $\begin{bmatrix} 35 \\ 21 \\ 25 \end{bmatrix}$, $\begin{bmatrix} 0 \\ 2 \\ 5 \end{bmatrix}$, $\begin{bmatrix} 0 \\ 0 \\ 1 \end{bmatrix}$   **15.** 4; $\begin{bmatrix} 3 \\ 2 \end{bmatrix}$   **17.** $2 + 3i, 2 - 3i$; $\begin{bmatrix} 1 - 3i \\ 2 \end{bmatrix}$, $\begin{bmatrix} 1 + 3i \\ 2 \end{bmatrix}$   **19. a.** $\begin{bmatrix} 5000 \\ 5000 \end{bmatrix}$  **b.** 1.4

**21. a.** $\begin{bmatrix} 5000 \\ 5000 \end{bmatrix}$  **b.** 0.9   **23. a.** $\begin{bmatrix} 6000 \\ 4000 \end{bmatrix}$  **b.** 0.9   **25.** $X(n) \approx (51{,}000/23)\,(1.4)^n \begin{bmatrix} 1 \\ 1 \end{bmatrix}$

## Chapter 10 Review Exercises (page 577)

| For Exercises . . . | 1–4, 19–24, 63–65, 70–76, 79–81 | 5–6, 25–33, 77 | 7–8, 34–39, 66 | 9–13, 40–58, 78 | 14–16, 59–62, 67–69 |
|---|---|---|---|---|---|
| Refer to Section . . . | 1 | 2 | 3 | 4 | 5 |

**1.** True   **2.** False   **3.** True
**4.** False   **5.** False   **6.** False
**7.** True   **8.** False   **9.** False
**10.** False   **11.** False   **12.** True
**13.** False   **14.** False   **15.** True
**16.** True   **19.** $(-9, 3)$   **20.** $(22, 3)$   **21.** $(7, -9, -1)$   **22.** $(2, -1, 3)$   **23.** $(6 - 7z/3, 1 + z/3, z)$   **24.** No solution
**25.** $2 \times 2$ (square); $a = 2, b = 3, c = 5, q = 9$   **26.** $3 \times 2$; $a = 2, x = -1, y = 4, p = 5, z = 7$   **27.** $1 \times 4$ (row);
$m = 6, k = 3, z = -3, r = -9$   **28.** $3 \times 3$ (square); $a = -12, b = 1, k = 9/2, c = 3/4, d = 3, l = -3/4, m = -1, p = 3, q = 9$

**29.** $\begin{bmatrix} 9 & 10 \\ -3 & 0 \\ 10 & 16 \end{bmatrix}$   **30.** $\begin{bmatrix} 0 & -16 \\ -10 & -18 \end{bmatrix}$   **31.** $\begin{bmatrix} 23 & 20 \\ -7 & 3 \\ 24 & 39 \end{bmatrix}$   **32.** Not possible   **33.** $\begin{bmatrix} -17 & 20 \\ 1 & -21 \\ -8 & -17 \end{bmatrix}$   **34.** $\begin{bmatrix} 2 & 50 \\ 1 & -15 \\ -3 & 45 \end{bmatrix}$

**35.** Not possible   **36.** $\begin{bmatrix} 6 & 18 & -24 \\ 1 & 3 & -4 \\ 0 & 0 & 0 \end{bmatrix}$   **37.** [9]   **38.** $\begin{bmatrix} 15 \\ 16 \\ 1 \end{bmatrix}$   **39.** $[-14 \ -19]$   **40.** $\begin{bmatrix} -7/19 & 4/19 \\ 3/19 & 1/19 \end{bmatrix}$   **41.** No inverse

**42.** No inverse   **43.** $\begin{bmatrix} 7 & -3 \\ -2 & 1 \end{bmatrix}$   **44.** $\begin{bmatrix} -1/4 & 1/6 \\ 0 & 1/3 \end{bmatrix}$   **45.** No inverse   **46.** No inverse   **47.** $\begin{bmatrix} 2/3 & 0 & -1/3 \\ 1/3 & 0 & -2/3 \\ -2/3 & 1 & 1/3 \end{bmatrix}$

**48.** $\begin{bmatrix} 1/4 & 1/2 & 1/2 \\ 1/4 & -1/2 & 1/2 \\ 1/8 & -1/4 & -1/4 \end{bmatrix}$   **49.** No inverse   **50.** No inverse   **51.** $X = \begin{bmatrix} 1 \\ -13 \end{bmatrix}$   **52.** Matrix $A$ has no inverse. Solution: $(-2y + 5, y)$

**53.** $X = \begin{bmatrix} -22 \\ -18 \\ 15 \end{bmatrix}$   **54.** $X = \begin{bmatrix} 6 \\ 15 \\ 16 \end{bmatrix}$   **55.** $(2, 1)$   **56.** $(34, -9)$   **57.** $(-1, 0, 2)$   **58.** $(1, 2, 3)$   **59.** 1, 3; $\begin{bmatrix} 3 \\ 1 \end{bmatrix}$, $\begin{bmatrix} 2 \\ 1 \end{bmatrix}$

**60.** 8, 2; $\begin{bmatrix} 1 \\ 1 \end{bmatrix}$, $\begin{bmatrix} 1 \\ -1 \end{bmatrix}$   **61.** 1, 2, -3; $\begin{bmatrix} 1 \\ -2 \\ 0 \end{bmatrix}$, $\begin{bmatrix} 0 \\ 5 \\ 1 \end{bmatrix}$, $\begin{bmatrix} 0 \\ 0 \\ 1 \end{bmatrix}$   **62.** -2, 3, 2; $\begin{bmatrix} 10 \\ -2 \\ 3 \end{bmatrix}$, $\begin{bmatrix} 0 \\ 1 \\ 1 \end{bmatrix}$, $\begin{bmatrix} 0 \\ 0 \\ 1 \end{bmatrix}$   **63.** $\begin{bmatrix} 8 & 8 & 8 \\ 10 & 5 & 9 \\ 7 & 10 & 7 \\ 8 & 9 & 7 \end{bmatrix}$

**64. a.** $b + c$   **b.** A is tumorous, B is bone, and C is healthy.   **c.** For patient X, A and C are healthy; B is tumorous. For patient Y, A and B are tumorous; C is bone. For patient Z, A could be healthy or tumorous; B and C are healthy.

**65. a.** No   **b. (i)** 0.23, 0.37, 0.42; A is healthy; B and D are tumorous; C is bone.   **(ii)** 0.33, 0.27, 0.32; A and C are tumorous, B could be healthy or tumorous, D is bone.   **c.** 0.2, 0.4, 0.45, 0.3; A is healthy; B and C are bone; D is tumorous.   **d.** One example is to choose beams 1, 2, 3, and 6.   **66.** About 20 head and face injuries, 12 concussions, 3 neck injuries, and 55 other injuries

**67.** $\begin{bmatrix} 2500 \\ 7500 \end{bmatrix}$, 0.9   **68.** $\begin{bmatrix} 5000 \\ 5000 \end{bmatrix}$, 0.5   **69. d.** $\begin{bmatrix} 1191 \\ 773 \\ 213 \\ 185 \\ 371 \\ 325 \end{bmatrix}$   **f.** $\begin{bmatrix} 19{,}797 \\ 17{,}233 \\ 4093 \\ 3580 \\ 7520 \\ 6580 \end{bmatrix}$   **70. a.** $C = 0.012t^2 + 0.86t + 317$   **b.** 2091

**71. a.** $x = 1$ and $y = 1/2$   **b.** $x = 1, y = 1$, and $z = -1$   **72.** $W_1 = W_2 = 100\sqrt{3}/3 \approx 58$ lb
**73.** $W_1 \approx 110$ lb and $W_2 \approx 134$ lb   **74.** 8000 standard, 6000 extra large   **75.** 5 blankets, 3 rugs, 8 skirts
**76.** 150,000 gal were produced at Tulsa, 225,000 gal at New Orleans, and 180,000 gal at Ardmore

**77.** $\begin{bmatrix} 1.33 & 17.6 & 152{,}000 & 26.75 & +1.88 \\ 1.00 & 20.0 & 238{,}200 & 32.36 & -1.50 \\ 0.79 & 25.4 & 39{,}110 & 16.51 & -0.89 \\ 0.27 & 21.2 & 122{,}500 & 28.60 & +0.75 \end{bmatrix}$   **78. a.** $\begin{bmatrix} 3170 \\ 2360 \\ 1800 \end{bmatrix}$   **b.** $\begin{bmatrix} x \\ y \\ z \end{bmatrix}$   **c.** $\begin{bmatrix} 10 & 5 & 8 \\ 12 & 0 & 4 \\ 0 & 10 & 5 \end{bmatrix}\begin{bmatrix} x \\ y \\ z \end{bmatrix} = \begin{bmatrix} 3170 \\ 2360 \\ 1800 \end{bmatrix}$   **d.** $\begin{bmatrix} 150 \\ 110 \\ 140 \end{bmatrix}$

**79.** There are $y$ girls and $2500 - 1.5y$ boys, where $y$ is any even integer between 0 and 1666.   **80.** 179 singles, 31 doubles, 4 triples, and 11 home runs   **81.** A chocolate wafer weighs 4.08 g, and a single layer of vanilla cream weighs 3.17 g.

# Chapter 11 Differential Equations

## Exercises 11.1 (page 593)
**1.** $y = -2x^2 + 2x^3 + C$

**3.** $y = x^4/2 + C$

| For Exercises . . . | 1–2 | 3–4,19–22 | 5–18, 23–34 | 35–38 | 43, 50, 51, 56, 57 | 44–47, 52–54, 58-61 | 48, 49, 55 |
|---|---|---|---|---|---|---|---|
| Refer to Example . . . | 1, 2 | 3 | 4, 5 | 8 | 5, 6 | 6 | 7 |

**5.** $y^2 = 2x^3/3 + C$   **7.** $y = ke^{x^2}$   **9.** $y = ke^{x^3 - x^2}$   **11.** $y = Cx$   **13.** $\ln(y^2 + 6) = x + C$   **15.** $y = -1/(e^{2x}/2 + C)$
**17.** $\cos y = -\sin x + C$   **19.** $y = x^2 - x^3 + 5$   **21.** $y = -2xe^{-x} - 2e^{-x} + 44$   **23.** $y^2 = x^4/2 + 25$   **25.** $y = e^{x^2 + 3x}$
**27.** $y^2/2 - 3y = x^2 + x - 4$   **29.** $y = -3/(3\ln|x| - 4)$   **31.** $y = (e^{x-1} - 3)/(e^{x-1} - 2)$   **33.** $y = 3e^{\sin x}$
**35.** $y = -1, 1$: unstable; $y = 0$: stable   **37.** $y = 0$: stable; $y = 3$: unstable   **43.** 4.4 cc   **45.** 987 fish

**47. a.** $w = 143 + 37e^{-0.005t}$
**b.** The asymptote is $w = 143$; 143 will never be attained.

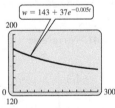

**c.** 584 days

**49. a.**

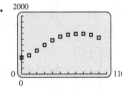

**b.** $y = 1383/(1 + 1.867e^{-0.05615t})$

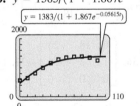

**c.** 1383 million
**d.**

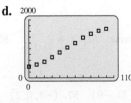

$y = 1959/(1 + 4.800e^{-0.03415t})$; 1959 million

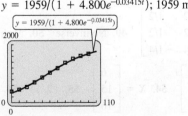

**51.** $y = 10.7e^{0.02277t}$   **53. a.** $N = (N_0 + i/m)e^{mt} - i/m$   **c.** $-0.344, 1.65$

**55. a.** $y = 11.74/[1 + (1.423 \times 10^{22})e^{-0.02554t}]$

**b.** The function seems to fit the data from 1927 on very well; for the year 1804, the function does not fit the data very well.

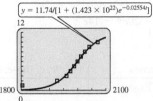

**c.** $y = 9.803/[1 + (2.612 \times 10^{29})e^{-0.03391t}]$

**d.**

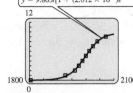

$y = 9.803/[1 + (2.612 \times 10^{29})e^{-0.03391t}]$ Yes

**e.** 10.79 billion  **f.** 0.99 billion; no

**57. a.** $dy/dt = -0.03y$  **b.** $y = Me^{-0.03t}$  **c.** $y = 75e^{-0.03t}$  **d.** 56 g

**59.** The temperature approaches $T_M$, the temperature of the surrounding medium.

**61. a.** $T = 88.6e^{-0.24t} + 10$  **b.**

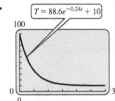

$T = 88.6e^{-0.24t} + 10$

**c.** Just after death—the graph shows that the most rapid decrease occurs in the first few hours.  **d.** About 43.9°F  **e.** About 4.5 hours

## Exercises 11.2 (page 601)

| For Exercises . . . | 1–22, 25, 28–32 |
|---|---|
| Refer to Example . . . | 2–4 |

**1.** $y = 2 + Ce^{-3x}$  **3.** $y = 2 + Ce^{-x^2}$  **5.** $y = x \ln x + Cx$  **7.** $y = -1/2 + Ce^{x^2/2}$
**9.** $y = x^2/4 + 2x + C/x^2$  **11.** $y = -x^3/2 + Cx$  **13.** $y = -x \cot x + 1 + C/\sin x$
**15.** $y = 2e^x + 48e^{-x}$  **17.** $y = -2 + 22e^{x^2-1}$  **19.** $y = x^2/7 + 2560/(7x^5)$  **21.** $y = (3 + 197e^{4-x})/x$
**23. a.** $y = c/(p + Kce^{-cx})$  **b.** $y = cy_0/[py_0 + (c - py_0)e^{-cx}]$  **c.** $c/p$  **25. a.** $y = \alpha/(\alpha + \beta) + [y_0 - \alpha/(\alpha + \beta)]e^{-(\alpha+\beta)t}$
**b.** $\alpha/(\alpha + \beta)$  **29.** $y = 1.02e^t + 9999e^{0.02t}$ (rounded)  **31.** $y = 50t + 2500 + 7500e^{0.02t}$  **33.** $T = Ce^{-kt} + T_M$

## Exercises 11.3 (page 608)

| For Exercises . . . | 1–38 |
|---|---|
| Refer to Example . . . | 1, 2 |

**1.** 8.273  **3.** 4.315  **5.** 1.491  **7.** 6.191  **9.** 0.595  **11.** −0.540; −0.520  **13.** 4.010; 4.016
**15.** 3.806; 4.759  **17.** 3.112; 3.271  **19.** 73.505; 74.691  **21.** 3.186

**23.**

| $x_i$ | $y_i$ | $y(x_i)$ | $y_i - y(x_i)$ |
|---|---|---|---|
| 0 | 0 | 0 | 0 |
| 0.2 | 0 | 0.08772053 | −0.08772053 |
| 0.4 | 0.11696071 | 0.22104189 | −0.10408118 |
| 0.6 | 0.26432197 | 0.37954470 | −0.11522273 |
| 0.8 | 0.43300850 | 0.55699066 | −0.12398216 |
| 1.0 | 0.61867206 | 0.75000000 | −0.13132794 |

**25.**

| $x_i$ | $y_i$ | $y(x_i)$ | $y_i - y(x_i)$ |
|---|---|---|---|
| 0 | 0 | 0 | 0 |
| 0.2 | 0.8 | 0.725077 | 0.07492 |
| 0.4 | 1.44 | 1.3187198 | 0.12128 |
| 0.6 | 1.952 | 1.8047535 | 0.14725 |
| 0.8 | 2.3616 | 2.2026841 | 0.15892 |
| 1.0 | 2.68928 | 2.5284822 | 0.16080 |

**27.**

```
y
0.8
0.6       f(x)
0.4
0.2
0   0.2 0.4 0.6 0.8 1.0   x
```

**29.**

```
y
3.0
2.5
2.0      f(x)
1.5
1.0
0.5
0   0.2 0.4 0.6 0.8 1.0   x
```

**31. a.** 4.109  **b.** $y = 1/(1 - x)$; $y$ approaches $\infty$.
**33.** About 20 species  **35.** About 7000 whales

**37. a.**

| $t_i$ | $p_i$ |
|---|---|
| 0 | 0.1 |
| 5 | 0.1675 |
| 10 | 0.297678 |
| 15 | 0.536660 |
| 20 | 0.846644 |
| 25 | 0.963605 |
| 30 | 0.980024 |

**b.** 1

## Exercises 11.4 (page 615)

**1.** $x_1 = C_1e^t + 2C_2e^{2t}$, $x_2 = C_1e^t + C_2e^{2t}$   **3.** $x_1 = 2C_1e^{2t} + C_2e^{9t}$, $x_2 = -C_1e^{2t} + 3C_2e^{9t}$
**5.** $x_1 = 2C_1e^{3t}$, $x_2 = 10C_1e^{3t} + 3C_2e^{2t}$, $x_3 = 11C_1e^{3t} + 4C_2e^{2t} + C_3e^{-t}$
**7.** $x_1 = 2C_1e^{3t} + C_2e^{10t} - 4e^t/9 - e^{2t}/4$, $x_2 = C_1e^{3t} - 3C_2e^{10t} - e^t/6 - e^{2t}/4$   **9.** $x_1 = 3e^t + 4e^{2t}$, $x_2 = 3e^t + 2e^{2t}$
**11.** $x_1 = -4e^{2t} + 7e^{9t}$, $x_2 = 2e^{2t} + 21e^{9t}$   **13.** $x_1 = 6e^{3t}$, $x_2 = 30e^{3t} - 15e^{2t}$, $x_3 = 33e^{3t} - 20e^{2t} - 14e^{-t}$

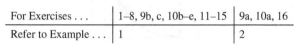

| For Exercises . . . | 1–14 |
|---|---|
| Refer to Example . . . | 1 |

## Exercises 11.5 (page 622)

| For Exercises . . . | 1–8, 9b, c, 10b–e, 11–15 | 9a, 10a, 16 |
|---|---|---|
| Refer to Example . . . | 1 | 2 |

**1.** 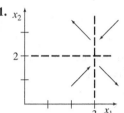   **3.** 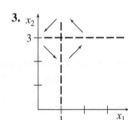   **5. a.** $(1, 1)$   **b.** 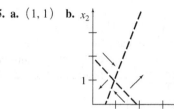   **7. a.** $(1, 2)$   **b.**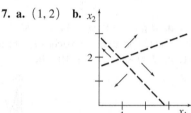

**9. a.** $3 \ln x_1 - 2x_1 + 2 \ln x_2 - 3x_2 = \ln 4 - 8$   **b.** $x_1 = 3/2, x_2 = 2/3$ or $x_1 = 0, x_2 = 0$   **c.** They cycle clockwise in the phase plane, as shown in Figure 11.   **13.** $x_1 = (b + \alpha + v)\beta$, $x_2 = (b + \gamma)[\beta a - b(b + \alpha + v)]/\{\beta[\gamma v + (b + \alpha + v)(b + \gamma)]\}$,
$x_3 = v[\beta a - b(b + \alpha + v)]/\{\beta[\gamma v + (b + \alpha + v)(b + \gamma)]\}$   **17. b.** $R = \sqrt{2C_1 F b_2 D_1/(b_1 D_2 C_0^2)}, N = Fb_1/(2D_1), P = Fb_1 b_2/(2D_2)$

## Exercises 11.6 (page 626)

**1. a.** $y = 24{,}995{,}000/(4999 + e^{0.25t})$   **b.** 3672   **c.** 91   **d.** 34th day
**3. a.** $y = 20{,}000/(1 + 199e^{-0.14t})$ or $20{,}000e^{0.14t}/(e^{0.14t} + 199)$   **b.** About 38 days
**5. a.** $y = 0.005 + 0.015e^{-1.010t}$   **b.** $Y = 0.00727e^{-1.1t} + 0.00273$
**7. a.** $y = 45/(1 + 14e^{-0.54t})$   **b.** About 6 days   **9. a.** $y = 347e^{-4.24e^{-0.1t}}$   **b.** About 5.5 days
**11. a.** $y = [2(t + 100)^3 - 1{,}800{,}000]/(t + 100)^2$   **b.** About 250 lb of salt   **c.** Increases   **13. a.** $y = 20e^{-0.02t}$
**b.** About 6 lb of salt   **c.** Decreases   **15. a.** $y = [0.25(t + 100)^2 - 2000]/(t + 100)$   **b.** About 17.1 g

| For Exercises . . . | 1–9 | 10–16 |
|---|---|---|
| Refer to Example . . . | 1 | 2 |

## Chapter 11 Review Exercises (page 629)

**1.** True   **2.** False   **3.** True   **4.** False
**5.** False   **6.** True   **7.** False   **8.** True
**9.** False   **10.** False   **11.** True   **12.** False
**17.** Neither   **18.** Separable   **19.** Separable
**20.** Linear   **21.** Both   **22.** Both   **23.** Linear   **24.** Neither   **25.** $y = x^3 + 3x^2 + C$   **26.** $y = x^4 + x^6 + C$
**27.** $y = 2e^{2x} + C$   **28.** $y = (1/3) \ln |3x + 2| + C$   **29.** $y^2 = 3x^2 + 2x + C$   **30.** $y^2/2 - y = e^x + x^2/2 + C$
**31.** $y = (Cx^2 - 1)/2$   **32.** $y = 3 + Me^{e^x}$   **33.** $y = x - 1 + Ce^{-x}$   **34.** $y = (\ln |x| + C)/x^3$   **35.** $y = (x^2 + C)/\ln x$
**36.** $y = e^{2x}/(2x) - e^{2x}/(4x^2) + C/x^2$   **37.** $y = x^3/3 - 3x^2 + 3$   **38.** $y = -5e^{-x} - 5x + 22$   **39.** $y = -\ln[5 - (x + 2)^4/4]$
**40.** $y = 5e^{-x^2+3x}$   **41.** $y^2 + 6y = 2x - 2x^2 + 352$   **42.** $y = 2.054e^{(2/3)x^{3/2}}$   **43.** $y = -x^2e^{-x}/2 - xe^{-x}/2 + e^{-x}/4 + 41.75e^x$
**44.** $y = 1/3 + (5/3)e^{-x^3}$   **45.** $y = 3/2 + 27e^{x^2}/2$   **46.** $y = (e^{2x^3} + 5e^2)/(6x^4)$   **48.** 2.138   **49.** 2.608
**50.** 1.215; 1.223; −0.008

| For Exercises . . . | 1–4, 6, 7, 13, 14, 25–32, 37–42, 57, 58, 63–66, 68, 70–72 | 5, 15–24 | 8, 9, 33–36, 43–46 | 10, 11, 47–51 | 12, 52, 55, 56, 59 | 53, 54 | 60–62, 69 |
|---|---|---|---|---|---|---|---|
| Refer to Section . . . | 1 | 1, 2 | 2 | 3 | 5 | 4 | 6 |

**51.**

| $x_i$ | $y_i$ |
|---|---|
| 0 | 0 |
| 0.2 | 0.6 |
| 0.4 | 1.355 |
| 0.6 | 2.188 |
| 0.8 | 3.084 |
| 1.0 | 4.035 |

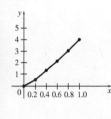

**53.** $x_1 = C_1e^{9t} + C_2e^{-5t} - 4t/15 + 1/675$, $x_2 = C_1e^{9t} - C_2e^{-5t} - t/15 - 26/675$
**54.** $x_1 = C_1e^{3t} + C_2e^{5t} - t/3 + 1/45$, $x_2 = -2C_1e^{3t} - C_2e^{5t} - t/3 - 11/45$
**55. a.** $(2, 2)$   **b.**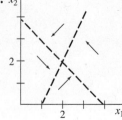

**56. a.** $(2, 1)$  **b.**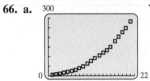

**57. a.** About 40  **b.** About $1.44 \times 10^{10}$ hours  **58.** 219
**59.** $0.2 \ln y - 0.5y + 0.3 \ln x - 0.4x = C$; $x = 3/4$ units, $y = 2/5$ units
**60.** 17.3 min  **61.** It is not possible ($t$ is negative).
**62. a.** $y = 489{,}300/(699 + e^{1.140t})$  **b.** 135 people  **c.** 5.7 wk
**64. a.** 185 million  **b.** 207 million  **c.** 326 million

**65. a.** $N = 326$, $b = 7.20$, $k = 0.248$  **b.** $y \approx 266$ million, which is less than the
table value of 308.7 million.  **c.** About 287 million for 2030; about 301 million for 2050
**66. a.** 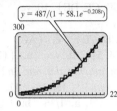  Yes  **b.** $y = 487/(1 + 58.1e^{-0.208t})$  **c.** Well  $y = 487/(1 + 58.1e^{-0.208t})$  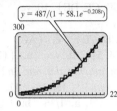  **d.** 487 million

**67. a.** All three models: increasing for all $t$  **b.** Exponential: concave up for all $t$; limited growth: concave down for all $t$; logistic:
concave up for $t < (\ln b)/k$; concave down for $t > (\ln b)/k$  **68. a. and b.** $x = 1/k + Ce^{-kt}$  **c.** $1/k$
**69. a.** $y = 200/(1 + 19e^{-0.4646t})$  **b.** About 70 people  **70.** 213°  **71.** 3 hr  **72. a.** $v = (G/K)(e^{2GKt} - 1)/(e^{2GKt} + 1)$
**b.** $G/K$  **c.** $v = 88(e^{0.727t} - 1)/(e^{0.727t} + 1)$

# Chapter 12 Probability

## Exercises 12.1 (page 646)
**1.** False  **3.** True  **5.** False
**7.** $\subseteq$  **9.** $\not\subseteq$  **11.** $\subseteq$
**13.** $\subset$; $\subset$; $\not\subset$; $\not\subset$; $\subset$; $\not\subset$

| For Exercises ... | 1–13, 37, 38 | 15, 16, 88–90 | 17–34 | 39–46, 61–64 | 49–55, 65–68 | 57–60, 87 | 74, 77–79, 81, 82, 86, 92, 93 | 75, 76, 80, 83–85 |
|---|---|---|---|---|---|---|---|---|
| Refer to Example ... | 2 | 3, 4 | 5, 6, 7 | 8 | 9 | 11 | 12 | 10 |

**15.** 32  **17.** $\cap$  **19.** $\cup$  **21.** $\cap$  **23.** $\cup$ or $\cap$  **27.** $\{2, 4, 6\}$  **29.** $\{1, 3, 5, 7, 9\}$  **31.** $\{2, 3, 4, 6\}$  **33.** $\{7, 8\}$
**35. a.** $\{3, 6, 9\} = A$  **37. a.** True  **b.** True  **c.** False  **d.** False  **e.** True  **f.** True  **g.** False

**39.** 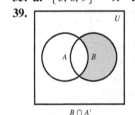  **41.** 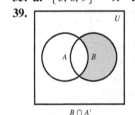  **43.** 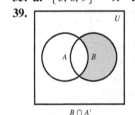  **45.** 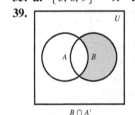  **47.** 8

$B \cap A'$    $A' \cup B$    $B' \cup (A' \cap B')$    $U' = \emptyset$

**49.** 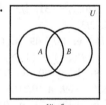  **51.** 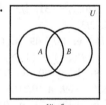  **53.** 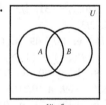  **55.** 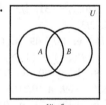  **57.** 13  **59.** 18

$(A \cap B) \cap C$    $A \cap (B \cup C')$    $(A' \cap B') \cap C'$    $A' \cap (B' \cup C)$

**61.** 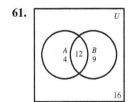  **63.** 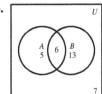  **65.** 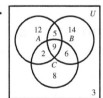  **67.**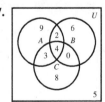

**75. a.** A-negative  **b.** AB-negative  **c.** B-negative  **d.** A-positive  **e.** AB-positive  **f.** B-positive  **g.** O-positive
**h.** O-negative  **77. a.** 1,236,003  **b.** 31,611  **c.** 1,373,589  **d.** 2,468,435  **e.** Females who are either American Indian or
Asian or Pacific Islander  **79. a.** 110.6 million  **b.** 122.7 million  **c.** 85.4 million  **d.** 37.6 million  **e.** 71.0 million

**f.** 60.1 million  **81. a.** 137,313  **b.** 174,175  **c.** 164,046  **83. a.** 342  **b.** 192  **c.** 72  **d.** 86  **85. a.** 12
**b.** 18  **c.** 37  **d.** 97  **87.** 17,992 thousand children  **89.** $2^{51} \approx 2.252 \times 10^{15}$  **93. a.** The set of states who are not among those whose name contains the letter "e" or who are more than 4 million in population, and who also have an area of more than 40,000 square miles,  **b.** { Alaska }

## Exercises 12.2 (page 661)

**3.** { January, February. March, . . . , December }
**5.** { surgery, medicine, wait }
**7.** { $(h, 1)$, $(h, 2)$, $(h, 3)$, $(h, 4)$, $(h, 5)$, $(h, 6)$,
$(t, 1)$, $(t, 2)$, $(t, 3)$, $(t, 4)$, $(t, 5)$, $(t, 6)$ }
**9.** { AB, AC, AD, AE, BC, BD, BE, CD, CE, DE }, 10, yes
**a.** { AC, BC, CD, CE }
**b.** { AB, AC, AD, AE, BC, BD, BE, CD, CE }  **c.** { AC }  **11.** { $(1, 2)$, $(1, 3)$, $(1, 4)$, $(1, 5)$,
$(2, 3)$, $(2, 4)$, $(2, 5)$, $(3, 4)$, $(3, 5)$, $(4, 5)$ }, 10, yes  **a.** { $(2, 4)$ }  **b.** { $(1, 2)$, $(1, 4)$, $(2, 3)$, $(2, 5)$, $(3, 4)$, $(4, 5)$ }
**c.** $\emptyset$  **13.** { *hh, thh, hth, tthh, thth, htth, ttth, tht, thtt, htt, tttt* }, 11, no  **a.** { *tthh, thth, htth, ttth, tht, thtt, httt, tttt* }
**b.** { *hh, thh, hth, tthh, thth, htth* }  **c.** { *tttt* }  **15.** 1/6  **17.** 2/3  **19.** 1/3  **21.** 1/13  **23.** 1/26  **25.** 2/13  **27.** 7/13
**29.** 3/20  **31.** 3/5  **35.** No  **37.** Yes  **39. a.** 1/36  **b.** 1/12  **c.** 1/9  **d.** 5/36  **41. a.** 5/18  **b.** 5/12  **c.** 11/36
**43.** 5/18  **45. a.** 2/13  **b.** 7/13  **c.** 3/26  **d.** 3/4  **e.** 11/26  **47. a.** 5/13  **b.** 7/13  **c.** 3/13  **49. a.** 1/10  **b.** 2/5
**c.** 7/20  **53.** 1 to 5  **55.** 2 to 1  **57. a.** 1 to 5  **b.** 11 to 7  **c.** 2 to 7  **d.** 7 to 2  **59.** Not empirical.  **61.** Not empirical
**63.** Empirical  **67.** Possible  **69.** Not possible; the sum of the probabilities is less than 1.  **71.** Not possible; a probability
cannot be negative.  **73. a.** 0.2778  **b.** 0.4167  **75. a.** 0.0463  **b.** 0.2963  **77. a.** 0.51  **b.** 0.25  **c.** 0.10  **d.** 0.84
**79.** The outcomes are not equally likely.  **83. a.** Person is not overweight.  **b.** Person has a family history of heart disease and is
overweight.  **c.** Person smokes or is not overweight.  **85. a.** 0.56  **b.** 0.20  **c.** 0.31  **87. a.** 3/4  **b.** 1/4  **89. a.** 0.90
**b.** 0.23  **91.** d  **93. a.** 0.1631  **b.** 0.2450  **c.** 0.1287  **d.** 0.1360  **95. a.** 0.1808  **b.** 0.2386
**97.** 3 to 97; 13 to 7; 61 to 39; 21 to 79; 1 to 49  **99. a.** 0.0999  **b.** 0.2480  **c.** 0.4957  **d.** VI Corps  **e.** I Corps
**101. a.** 0.866  **b.** 0.478  **103. a.** 0.4  **b.** 0.1  **c.** 0.6  **d.** 0.9  **105.** 0  **107. a.** 25/57  **b.** 32/57  **c.** 4/19

| For Exercises . . . | 3–8, 9–14 | 9–14, 83, 84 | 15–32, 87, 88, 107, 108 | 33–38 | 39–50, 82, 94–96 |
|---|---|---|---|---|---|
| Refer to Example . . . | 1 | 2, 3, 4 | 6, 7 | 5 | 8, 9 |

| For Exercises . . . | 53–58, 97, 98 | 59–64, 93, 99, 100 | 67–72, 89, 101, 102 | 77, 78, 85, 86, 90–92 |
|---|---|---|---|---|
| Refer to Example . . . | 10, 11 | 12 | 13 | 14 |

## Exercises 12.3 (page 680)

**1.** 0  **3.** 1  **5.** 1/3  **7.** 0  **9.** 4/17  **11.** 11/51  **13.** 8/663
**15.** 25/102  **19.** Independent  **21.** Dependent
**23. a.** 1/4  **b.** 1/2  **25. a.** Many answers are possible.
**b.** Many answers are possible.  **29.** 1/20, 2/5  **31.** 1/3
**33.** 3/19  **35.** 21/38  **37.** 8/17  **45.** 2/3

| For Exercises . . . | 1–12, 45, 48, 49, 51, 54, 67, 69, 75 | 13–16, 23, 24, 47, 50, 53, 66, 71–74, 76, 77 | 29, 30, 46, 48, 68, 70 | 31–38, 59, 60, 62–65, 78–87 | 55–58, 61 |
|---|---|---|---|---|---|
| Refer to Example . . . | 1, 2, 3, 4 | 5, 6, 7 | 8, 9 | 10 | 11 |

**47. a.** 1/4  **b.** 1/4  **c.** 1/4  **d.** 1/4  **e.** 1/7  **49. a.** Color blindness and deafness are independent events.  **51.** e
**53. a.** 0.5065  **c.** 0.2872  **d.** $p(1 - P(B)) + (1 - p)(1 - P(B))^2$  **e.** 0.2872  **f.** $2(1 - p)P(B)(1 - P(B))$
**g.** 0.2872  **55.** 0.0478  **57. a.** 0.039  **b.** 0.999  **c.** 0.001  **d.** 0.110  **59.** d  **61.** b  **63. a.** 0.0892  **b.** 0.2704
**65.** 0.3328  **67. b.** 0.2973  **c.** 0.9820  **69. a.** 0.2166  **b.** 0.3792  **c.** 0.6246  **d.** 0.3418  **e.** 0.6897  **f.** 0.5137
**g.** Not independent  **71. a.** 0.052  **b.** 0.476  **73. a.** 0.05  **b.** 0.015  **c.** 0.25  **75. a.** No  **b.** 0.87  **c.** 0.74
**77.** 0 points: 0.16; 1 point: 0.48; 2 points: 0.36  **79.** 0.500  **81.** 0.0274  **83.** 0.9273  **85.** b

## Exercises 12.4 (page 696)

**1.**

| Number of Heads | 0 | 1 | 2 | 3 | 4 |
|---|---|---|---|---|---|
| Probability | 1/16 | 1/4 | 3/8 | 1/4 | 1/16 |

| For Exercises . . . | 1–8 | 9–20, 23, 24, 26, 29, 30, 32–40, 42–44 | 9–12, 23e | 25, 27, 28, 31, 41 |
|---|---|---|---|---|
| Refer to Example . . . | 1, 2 | 3, 4, 5, 6 | 7 | 8 |

**3.**

| Number of Aces | 0 | 1 | 2 | 3 |
|---|---|---|---|---|
| Probability | 0.7826 | 0.2042 | 0.0130 | 0.0002 |

**5.** 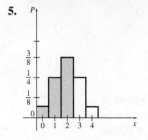  **7.** 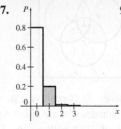  **9.** 3.6; 0.917  **11.** 14.49; 3.315  **13.** 2.7  **15.** 18  **17.** 0; yes

**19 a.**

| x | 0 | 1 | 2 | 3 | 4 |
|---|---|---|---|---|---|
| $P(x)$ | 625/1296 | 125/324 | 25/216 | 5/324 | 1/1296 |

**b.** 2/3

**23. a.**

| Sum | 5 | 6 | 7 | 8 | 9 |
|---|---|---|---|---|---|
| Probability | 1/6 | 1/6 | 1/3 | 1/6 | 1/6 |

**b.**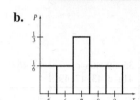

**c.** 1 to 2  **d.** 7  **e.** 1.291  **25. a.** \$68.51; \$72.84  **b.** Amoxicillin
**27.** Transfusion: about 0.92; no transfusion: about 0.87; yes  **29.** 185  **31. a.** Seed: about \$94.0 million; not seed : about \$116.0 million  **b.** Seed  **33.** −\$0.72; no  **35.** −\$0.053  **37.** −\$0.50
**39.** −\$0.878  **41. a.** Two-point conversion: 0.9; extra-point kick: 0.96  **b.** Extra-point kick
**43.** \$54,000

**Chapter 12 Review Exercises (page 701)**
**1.** True  **2.** True  **3.** False  **4.** False  **5.** False
**6.** True  **7.** False  **8.** False  **9.** False  **10.** True
**11.** False  **12.** True  **13.** True  **14.** False
**15.** True  **16.** False  **17.** False  **18.** True

| For Exercises . . . | 1–5, 17–46, 63, 97, 102, 107, 108, 111 | 6–9, 47–58, 64, 65, 69–77, 81, 82, 93, 94, 98abc, 105, 106 | 10–12, 59–62, 66, 67, 78–80, 83, 84, 95, 96, 98def, 99, 101, 104, 109, 110 | 13–16, 85–92, 100, 112 |
|---|---|---|---|---|
| Refer to Section . . . | 1 | 2 | 3 | 4 |

**19.** False  **20.** True  **21.** True  **22.** False  **23.** False  **24.** True  **25.** False  **26.** False  **27.** 32  **28.** 16  **29.** {a, b, g}
**30.** {b, e, f, h}  **31.** {c, d}  **32.** {a, c, d, e, f, g, h}  **33.** {a, b, e, f, g, h}  **34.** {b}  **35.** U  **36.** Ø
**37.** All females employees in the pediatrics department  **38.** All employees in the maternity department with a nursing degree
**39.** All employees in the pediatrics department or with a nursing degree  **40.** All employees with a nursing degrees who are not in the pediatrics department  **41.** All male employees who are not in the maternity department  **42.** All employees who are neither in the maternity department nor female, that is, all male employees not in the maternity department
**43.** **44.** **45.** **46.**

$A \cup B'$     $A' \cap B$     $(A \cap B) \cup C$     $(A \cup B)' \cap C$

**47.** {1, 2, 3, 4, 5, 6}  **48.** {ace, 2, 3, 4, 5, 6, 7, 8, 9, 10, J, Q, K}  **49.** {0, 0.5, 1, 1.5, 2, . . . , 299.5, 300}
**50.** {hhhh, hhht, hhth, hthh, thhh, hhtt, htht, htth, thht, tthh, thth, httt, thtt, ttht, ttth, tttt}  **51.** {(3, r), (3, g), (5, r), (5, g), (7, r), (7, g), (9, r), (9, g), (11, r), (11, g)}  **52.** {(7, r), (7, g), (9, r), (9, g), (11, r), (11, g)}  **53.** {(3, g), (5, g), (7, g), (9, g), (11, g)}  **54.** No  **55.** 1/4  **56.** 1/26  **57.** 11/26  **58.** 8/13  **59.** 1/2  **60.** 1/3  **61.** 1  **62.** 0  **67.** No; yes
**68.** The probability is 2/3 if you switch and 1/3 is if you don't switch. The contestant should switch doors.  **69.** 1 to 3  **70.** 1 to 25
**71.** 2 to 11  **72.** 4 to 9  **73.** 5/36  **74.** 0  **75.** 1/6  **76.** 5/18  **77.** 1/6  **78.** 1/3  **79.** 2/11  **80.** 5/11  **81. a.** 0.66  **b.** 0.29
**c.** 0.71  **d.** 0.34  **82.** a  **83.** 1/7  **84.** 4/9  **85. a.**

| Number of Heads | 0 | 1 | 2 | 3 |
|---|---|---|---|---|
| Probability | 0.125 | 0.375 | 0.375 | 0.125 |

**b.**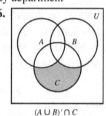

**c.** 1.5  **d.** 0.866  **86. a.**

| Sum | 2 | 3 | 4 | 5 | 6 | 7 | 8 | 9 | 10 | 11 | 12 |
|---|---|---|---|---|---|---|---|---|---|---|---|
| Probability | 1/36 | 1/18 | 1/12 | 1/9 | 5/36 | 1/6 | 5/36 | 1/9 | 1/12 | 1/18 | 1/36 |

**b.**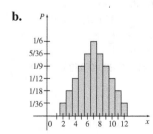

**c.** 7  **d.** 2.415  **87.** 0.6  **88.** 0.6  **89.** −\$0.833; no  **90.** 1.5 girls  **91. a.** 0.231  **b.** 0.75
**92.** \$1.29  **93. a.**

|  | $N_2$ | $T_2$ |
|---|---|---|
| $N_1$ | $N_1N_2$ | $N_1T_2$ |
| $T_1$ | $T_1N_2$ | $T_1T_2$ |

**b.** 1/4  **c.** 1/2  **d.** 1/4  **94. a.** 46%  **b.** 85%
**c.** 57%  **d.** 15%  **e.** 10%  **f.** 18%  **95.** c  **96.** Independence  **97. a.** Overweight males; 44,970 thousand  **b.** Female or a healthy weight; 144,653 thousand  **c.** Overweight or obese females; 62,957 thousand  **d.** Underweight or healthy weight males; 33,278 thousand

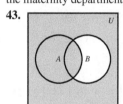

**98. a.** 0.3465 **b.** 0.2105 **c.** 0.6264 **d.** 0.4121 **e.** 0.4142 **f.** No **99.** 0.006378 **100. a.** 0.773 **b.** 0.7; antibiotics
**101. a.** 0.503 **b.** 0.4205 **c.** 0.5795 **102. a.** 53 **b.** 7 **c.** 12 **d.** 26 **103. b.** 7/10 **104.** 0.6279 **105.** 0.90
**106. a.** 0.7145; 0.5688; 0.4101; 0.3212; 0.2708 **107. a.** 4 **b.** 18 **108. a.** 51 **b.** 31 **c.** 2 **d.** 1 **e.** 3 **f.** 23
**109.** No; 2/3 **110. a.** 0.0025 **b.** 0.9975 **c.** 0.9753 **d.** 0.0247; no **e.** Independence **111.** d
**112. a.** 0.0069 **b.** 0.8834 **c.** 0.4651

# Chapter 13 Probability and Calculus

## Exercises 13.1 (page 716)

| For Exercises . . . | 1–10 | 11–18, 29–34, 44 | 19–24, 40d, e, 41d, e | 35–39, 40a–c, 41a–c, 43, 47–49 | 42, 46 |
|---|---|---|---|---|---|
| Refer to Example . . . | 1 | 2 | 5 | 3 | 4 |

**1.** Yes **3.** Yes **5.** No; $\int_0^3 4x^3 dx \neq 1$
**7.** No; $\int_{-2}^2 x^2/16\,dx \neq 1$
**9.** No; $f(x) < 0$ for some $x$ value in $[-1, 1]$ **11.** $k = 3/14$ **13.** $k = 3/125$ **15.** $k = 2/9$ **17.** $k = 1/12$
**19.** $F(x) = (x^2 - x - 2)/18,\ 2 \leq x \leq 5$ **21.** $F(x) = (x^3 - 1)/63,\ 1 \leq x \leq 4$ **23.** $F(x) = (x^{3/2} - 1)/7,\ 1 \leq x \leq 4$
**25.** 1 **29. a.** 0.4226 **b.** 0.2071 **c.** 0.4082 **31. a.** 0.3935 **b.** 0.3834 **c.** 0.3679 **33. a.** 1/3 **b.** 2/3 **c.** 295/432
**35. a.** 0.2679 **b.** 0.4142 **c.** 0.3178 **37. a.** 0.8131 **b.** 0.4901 **39. a.** 0.5714 **b.** 0.2381 **41. a.** 0.2829 **b.** 0.4853
**c.** 0.2409 **d.** $F(t) = 1.8838(0.5982 - e^{-0.03211t}),\ 16 \leq t \leq 84$ **e.** 0.1671 **43. a.** 0.875 **b.** 0.029 **c.** 0.037 **45.** b
**47. a.** 0.2 **b.** 0.6 **c.** 0.6 **49. a.** 0.1640 **b.** 0.1353

## Exercises 13.2 (page 725)

| For Exercises . . . | 1–6, 11a, b, c–14a, b, c, 21–23 | 7, 8 | 11d, e–14d, e, 24a, b, c–26a, b, c, 27–29, 31a–c, 32a–c, 39, 40 | 15–20, 24d, 26d, 30, 31d, 32d | 33, 38 |
|---|---|---|---|---|---|
| Refer to Example . . . | 1 | 2 | 3 | 5 | 4 |

**1.** $\mu = 5$; $\text{Var}(X) \approx 1.33$; $\sigma \approx 1.15$
**3.** $\mu = 14/3 \approx 4.67$; $\text{Var}(X) \approx 0.89$; $\sigma \approx 0.94$
**5.** $\mu = 2.83$; $\text{Var}(X) \approx 0.57$; $\sigma \approx 0.76$
**7.** $\mu = 4/3 \approx 1.33$; $\text{Var}(X) = 2/9 \approx 0.22$; $\sigma \approx 0.47$ **11. a.** 5.40 **b.** 5.55 **c.** 2.36 **d.** 0.5352 **e.** 0.6043
**13. a.** 1.6 **b.** 0.11 **c.** 0.33 **d.** 0.5904 **e.** 0.6967 **15. a.** 5 **b.** 0 **17. a.** 4.828 **b.** 0.0556 **19. a.** $\sqrt[4]{2} \approx 1.189$
**b.** 0.1836 **21.** 16/5; does not exist; does not exist **23.** d **25. a.** 2 **b.** 0.8944 **c.** 0.6261 **27.** 2.990 m
**29. a.** 6.37 days **b.** 4.08 days **c.** 0.39 **31. a.** 35.55 years **b.** 16.98 years **c.** 0.1330 **d.** 30.25 years
**33.** About 1 pm **35.** c **37.** c **39.** 960 days; 960 days

## Exercises 13.3 (page 739)

| For Exercises . . . | 1, 2, 37, 48, 54 | 3–6, 38, 39, 41–43, 47, 49–51, 55–57 | 18–21, 45, 46 | 7–14, 40, 52, 53 | 44 | 22–25 |
|---|---|---|---|---|---|---|
| Refer to Example . . . | 1 | 2 | 3 | 5 | 6 | 7 |

**1. a.** 3.7 cm **b.** 0.4041 cm **c.** 0.2886
**3. a.** 0.25 year **b.** 0.25 year **c.** 0.2325
**5. a.** 3 days **b.** 3 days **c.** 0.2325 **7.** 49.98%
**9.** 8.01% **11.** −1.28 **13.** 0.92 **19.** 0.1322 **21.** 0.4634 **23.** 0.7745 **25.** 0.2273 **27.** $m = (-\ln 0.5)/a$ or $(\ln 2)/a$
**31. a.** 1.00000 **b.** 1.99999 **c.** 8.00000 **33. a.** $\mu \approx 0$ **b.** $\sigma = 0.9999999251 \approx 1$
**35.** $F(x) = (x - a)/(b - a),\ a \leq x \leq b$ **37. a.** 28 days **b.** 0.375 **39. a.** 1 hour **b.** 0.3935 **41. a.** 58 minutes
**b.** 0.0907 **43. a.** 0.1967 **b.** 0.2468 **45. a.** 0.6538 **b.** 0.3462 **c.** 0.0084 **47. a.** 4.37 millennia; 4.37 millennia
**b.** 0.6325 **49. a.** 0.2865 **b.** 0.2212 **51. a.** 0.5457 **b.** 0.0039 **53.** $63.45; $45.36 **55.** c **57.** d

## Chapter 13 Review Exercises (page 744)

| For Exercises . . . | 1–4, 11–22, 62a–d | 5, 6, 23–32, 54, 57, 62e, f, 64 | 7–10, 33–51, 55, 56, 58–61, 63, 65, 66 |
|---|---|---|---|
| Refer to Section . . . | 1 | 2 | 3 |

**1.** True **2.** True **3.** True **4.** False **5.** False **6.** True
**7.** True **8.** True **9.** False **10.** False **11.** probabilities
**13.** 1. $f(x) \geq 0$ for all $x$ in $[a, b]$; 2. $\int_a^b f(x)dx = 1$ **14.** zero
**15.** Not a probability density function **16.** Probability density function **17.** Probability density function
**18.** Probability density function **19.** $k = 1/21$ **20.** $k = 3/38$ **21. a.** $1/5 = 0.2$ **b.** $9/20 = 0.45$ **c.** 0.54 **22. a.** 0.8284
**b.** 0.5359 **c.** 0.3643 **24.** b **25. a.** 4 **b.** 0.5 **c.** 0.7071 **d.** 4.121 **e.** $F(x) = (x - 2)^2/9,\ 2 \leq x \leq 5$ **26. a.** 6.5
**b.** 2.083 **c.** 1.443 **d.** 6.5 **e.** $F(x) = (x - 4)/5,\ 4 \leq x \leq 9$ **27. a.** 5/4 **b.** $5/48 \approx 0.1042$ **c.** 0.3227 **d.** 1.149
**e.** $F(x) = 1 - 1/x^5,\ x \geq 1$ **28. a.** 4.6 **b.** $412/75 \approx 5.493$ **c.** 2.344 **d.** 4.406 **e.** $F(x) = (x + 6\sqrt{x} - 7)/20,\ 1 \leq x \leq 9$
**29. a.** 0.5833 **b.** 0.2444 **c.** 0.4821 **d.** 0.6123 **30.** $m = 0.5970;\ 0.0180$ **31. a.** 100 **b.** 100 **c.** 0.8647
**32. a.** 13.6 **b.** 6.7 **c.** 0.5840 **33.** 33.36% **34.** 5.26% **35.** 34.31% **36.** 81.74% **37.** 11.51% **38.** 99.38%
**39.** −0.05 **40.** −0.81 **41.** 0.5429 **42.** 0.0686 **43.** 0.7531 **44.** 0.2469 **45.** 0.4506 **46.** 0.4854 **47.** 0.4452 **48.** 0.8523

**49. a.** Uniform
**b.** Domain: $[10, 30]$, range: $\{0.05\}$
**c.**

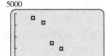

f(x) = 0.05 for [10, 30]

**d.** $\mu = 20$; $\sigma \approx 5.77$   **e.** 0.577

**50. a.** Exponential
**b.** Domain: $[0, \infty)$, range: $(0, 1]$
**c.**

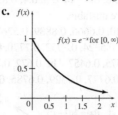

$f(x) = e^{-x}$ for $[0, \infty)$

**d.** $\mu = 1$; $\sigma = 1$   **e.** 0.8647

**51. a.** Normal
**b.** Domain: $(-\infty, \infty)$, range: $(0, 1/\sqrt{\pi}]$
**c.**

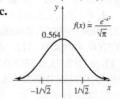

$f(x) \doteq \dfrac{e^{-x^2}}{\sqrt{\pi}}$

**d.** $\mu = 0$; $\sigma = 1/\sqrt{2}$   **e.** 0.6827

**52. a.** 1   **b.** 0.4422   **53. b.** 0.6819   **c.** 0.9716   **d.** l; yes   **54. a.** 2.377 g   **b.** 1.534 g   **c.** 0.8506   **55.** 0.6321   **56. a.** 16 in.
**b.** 0.3571   **57. a.** 40.07°C   **b.** 0.4928   **58.** 0.1379   **59.** 0.2266   **60. a.** 0.4138   **b.** 0.4447   **61. a.** 0.2921   **b.** 0.1826
**62. a.**

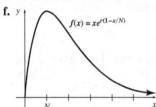

**b.** $N(t) = -0.001703t^4 + 0.39286t^3 - 30.9606t^2 + 887.496t - 3785.25$

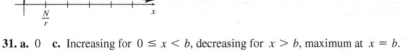

The function fits the model well.

**c.** $S(t) = (-0.001703t^4 + 0.39286t^3 - 30.9606t^2 + 887.496t - 3785.25)/172{,}009$   **d.** Estimates: 0.3812, 0.1449, 0.0306;
actual: 0.3469, 0.1885, 0.0223   **e.** 32.76 years   **f.** 17.00 years   **63. a.** 0.1543   **b.** 0.1585   **c.** 0.8457   **64.** 3650.1 days;
3650.1 days   **65.** 0.2206   **66.** d

# Chapter 14 Discrete Dynamical Systems

## Exercises 14.1 (page 754)

**1.** 7, 12, 17, 22   **3.** 6, 18, 54, 162   **5.** 0.8415, 0.9093, 0.1411, −0.7568
**7.** 5, 7, 11, 19   **9.** 7, −33, 167, −833   **11.** 5, 21, 437, 190, 965
**13.** $x + 4, x + 6, x + 2n$   **15.** $4x, 8x, 2^n x$   **17.** 1, 1, 1
**19.** 1125, 1055, 1026, 1013   **21.** 2475, 2359, 2419, 2390   **23.** 2833, 5084, 2639, 5026

| For Exercises ... | 1–6 | 7–12 | 13–18 | 19–24, 26, 27, 29, 30, 32, 33 |
|---|---|---|---|---|
| Refer to Example ... | 1 | 2 | 3 | 4 |

**25. a.** 0   **c.** Increasing for $0 \le x < N/r$, decreasing for $x > N/r$, maximum at $x = N/r$.
**e.** Concave downward for $0 \le x < 2N/r$, concave upward for $x > 2N/r$, inflection point at $x = 2N/r$.
**f.**

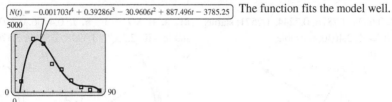

$f(x) = xe^{r(1-x/N)}$

**27.** 3260, 675, 3537, 518   **29.** 1000, 1500, 1800, 1929

**31. a.** 0   **c.** Increasing for $0 \le x < b$, decreasing for $x > b$, maximum at $x = b$.
**e.** Concave downward for $0 \le x < \sqrt{3}b$, concave upward for $x > \sqrt{3}b$, inflection point at $x = \sqrt{3}b$.
**f.**

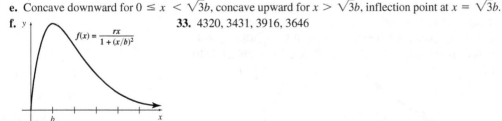

$f(x) = \dfrac{rx}{1 + (x/b)^2}$

**33.** 4320, 3431, 3916, 3646

## Exercises 14.2 (page 761)

| For Exercises . . . | 1–21 |
|---|---|
| Refer to Example . . . | 1, 2 |

**1.** 0, 2/3  **3.** 0, $(3 \pm \sqrt{3})/6$  **5.** 0, $1/\sqrt{2}$  **7. a.** 0.3, 0.6, 0.8, 0.4, 0.8, 0.4  **b.** 0.8, 0.4, 0.8, 0.4, 0.8, 0.4
**c.** 0.7, 0.6, 0.8, 0.4, 0.8, 0.4  **d.** 0.3, 0.6, 0.8, 0.4, 0.8, 0.4; 0 and 2/3 are unstable.
**9. a.** 0.1148, 0.0699, 0.0273, 0.0043, 0.0001, 0.0000  **b.** 0.576, 0.8440, 0.6666, 0.8889, 0.5267, 0.7879
**c.** 0.8873, 0.5325, 0.7954, 0.7766, 0.8084, 0.7512  **d.** 0.6503, 0.8873, 0.5324, 0.7952, 0.7770, 0.8078; 0 is stable; $(3 + \sqrt{3})/6$ and
$(3 - \sqrt{3})/6$ are unstable.  **11. a.** 0.2933, 0.5361, 0.7640, 0.6361, 0.7575, 0.6457  **b.** 0.672, 0.7371, 0.6733, 0.7362, 0.6744, 0.7353
**c.** 0.7508, 0.6552, 0.7479, 0.6592, 0.7455, 0.6623  **d.** 0.4718, 0.7335, 0.6777, 0.7329, 0.6785, 0.7323; 0 is unstable; $1/\sqrt{2}$ is stable.

**13.** 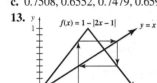  **15.** 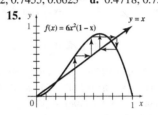  **17.**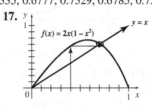

**19. a.** 0, 1  **b. i.** 1.3353, 0.6829, 1.2876, 0.7244, 1.2571; stable  **21. a.** 0, $\sqrt{r-1}$  **b. i.** 0.4706, 0.7705, 0.9670, 0.9994, 1.0000;
**ii.** 1.9921, 0.1016, 1.5042, 0.3314, 2.4630; unstable  stable  **ii.** 2.3529, 3.5998, 2.5789, 3.3708, 2.7267; stable

**c.** 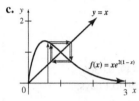  **d.** 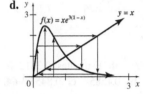  **c.** 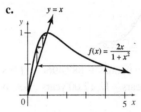  **d.**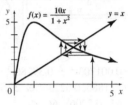

## Exercises 14.3 (page 764)

| For Exercises . . . | 1–25 |
|---|---|
| Refer to Example . . . | 1 |

**1.** 0 and 2/3 are unstable.  **3.** 0 is stable; $(3 + \sqrt{3})/6$ and $(3 - \sqrt{3})/6$ are unstable.
**5.** 0 is unstable; we do not know about $1/\sqrt{2}$.  **7.** 0.7391 is stable.  **9.** 0 is unstable; 0.8239 is unstable.
**11. a.** 0, 1/2  **b.** 0, 1  **c.** 0 is stable; we do not know about 1/2.  **d. i.** 0.384, 0.3633, 0.3362, 0.3001
**ii.** 0.588, 0.5698, 0.5587, 0.5510  **e.** Values of $x$ less than 0.5 move toward 0, while values of $x$ greater than 0.5 move toward 0.5.
**13. a.** 0  **b.** 1  **c.** We do not know.  **d. i.** 0.101, 0.1020, 0.1031, 0.1042  **ii.** $-0.101, -0.1020, -0.1031, -0.1042$
**e.** Values of $x$ move away from 0.  **15. a.** 1/2, $-3/2$  **b.** $-1, 3$  **c.** $-3/2$ is unstable; we do not know about 1/2.
**d. i.** 0.11, 0.7379, 0.2055, 0.7078  **ii.** 0.39, 0.5979, 0.3925, 0.5959  **e.** Values of $x$ move toward 1/2, oscillating from one side
to the other.  **17.** 0.765, 0.558  **19.** 0.487, 0.869, 0.395, 0.832  **21.** $x = 0$ is unstable; $x = 1$ is stable if $0 < r < 2$ and unstable
if $r > 2$.  **23.** $x = 0$ is stable if $0 < r < 1$ and unstable if $r > 1$; $x = \sqrt{r-1}$ only exists if $r > 1$, and it is stable.
**25. a.** $x_{n+1} = kx_n(1 - x_n) - Hx_n$  **b.** 0 is unstable; 0.6207 is stable.

## Chapter 14 Review Exercises (page 766)

| For Exercises . . . | 1–3, 9–22, 23d, e, f, g–26d, e, f, g, 27a–c, 29d–f, 30a–d | 4, 5, 23h–26h, 27d, 28, 29a–c, 30e | 6–8, 23a, b, c–26a, b, c, 27e, f, 29g, h, 30f–h |
|---|---|---|---|
| Refer to Section . . . | 1 | 2 | 3 |

**1.** False  **2.** True  **3.** True  **4.** True  **5.** False
**6.** False  **7.** False  **8.** False  **9.** 3, 9, 15, 21
**10.** 1, 9, 17, 25  **11.** 1, 7, $-5$, 19  **12.** 2, 1, 1/2, 1/4
**13.** $-3, 3, -9, 15$  **14.** $-1, -7, -25, -79$
**15.** 10, $-38$, 154, $-614$  **16.** $-2, -22, -122, -622$  **17.** $x, 1/x, x$ if $n$ is even or $1/x$ if $n$ is odd  **18.** $x, x, x$
**19.** 729, 320, 151, 74  **20.** 1961, 1694, 1541, 1442  **21.** 4416, 4550, 4513, 4524  **22.** 5984, 7219, 5143, 7935
**23. a.** 0  **b.** 1/2  **c.** stable  **24. a.** 0, 6/11  **b.** 6/5, $-6/5$  **c.** Both are unstable.
**d.** 0.075, 0.0375, 0.0188, 0.0094  **d.** 0.18, 0.216, 0.2592, 0.3110  **e.** 0.48, 0.576, 0.5088, 0.5894
**e.** 0.2, 0.1, 0.05, 0.025  **f.** 0.42, 0.504, 0.5952, 0.4858  **g.** 0.18, 0.216, 0.2592, 0.3110
**f.** 0.175, 0.0875, 0.0438, 0.0219  **h.**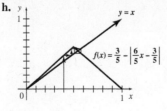
**g.** 0.075, 0.0375, 0.0188, 0.0094
**h.**

**25. a.** 0, 0.3909, 0.6091
**b.** 0, 1.3583, 0.4417
**c.** 0.3909 is unstable; 0 and 0.6091 are stable.
**d.** 0.0803, 0.0249, 0.0025, 0.0000
**e.** 0.4032, 0.4075, 0.4132, 0.4208
**f.** 0.6211, 0.6139, 0.6111, 0.6100
**g.** 0.4552, 0.4741, 0.4965, 0.5213
**h.**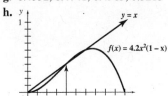

**26. a.** 0, 0.2598, 0.7402
**b.** 0, 1.6490, −0.8490
**c.** 0.2598 is unstable; 0 and 0.7402 are stable.
**d.** 0.0995, 0.0463, 0.0106, 0.0006
**e.** 0.4992, 0.6490, 0.7688, 0.7106
**f.** 0.7690, 0.7104, 0.7600, 0.7209
**g.** 0.5636, 0.7208, 0.7543, 0.7269
**h.**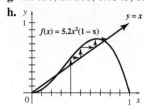

**27. a.** $a_n = 1.10a_{n-1}$   **b.** 110,000, 121,000, 133,100, 146,410   **c.** $a_n = 100,000(1.10)^{n-1}$   **d.** 0   **e.** 1.10   **f.** unstable
**28. a.** $a_n = ra_{n-1} + b$   **b.** $b/(1 - r)$   **c.** $r < 1$ and $b \geq 0$, or $r > 1$ and $b < 0$.   **d.** $-1 < r < 1$   **e.** $a_n = a_1 + (n - 1)b$
**29. d.** 0.525, 0.5447, 0.5598, 0.5710   **e.** 0.475, 0.4548, 0.4382, 0.4244   **f.** 0.45, 0.4095, 0.3760, 0.3477
**g.** $0, 1 - 40/a$ for $a \geq 40$   **h.** $40 < a < 440$   **30. c.** 0.2821, 0.2277, 0.2140, 0.2106   **e.** $a$   **f.** $1 - p$, stable
**g.** $a[1 - (1 - p)^{n-1}] + (1 - p)^{n-1}c_1$   **h.** $a$

# Credits

## Text Credits

**page 25** Exercise 4: Excerpt from the Actuarial Examination on Applied Statistical Methods of the Education and Examination Committee of the Society of Actuaries. Copyright © 1989 by Society of Actuaries. Used by permission of Society of Actuaries.    **page 39** Exercise 67: "Whales Diving" by Peter Tyack © Woods Hole Oceanographic Institution.    **page 40** Exercise 73: From Internet World Stats Usage and Population Statistics. Copyright © Miniwatts Marketing Group. Retrieved from www.internetworldstats.com/emarketing.htm.

**page 195** Exercise 65: Problem 03 from the May 2003 Course 01 Examination of the Education and Examination Committee of the Society of Actuaries. Copyright © by Society of Actuaries. Used by permission of Society of Actuaries.    **page 195** Exercise 69: "Whales Diving" by Peter Tyack. Copyright © Woods Hole Oceanographic Institution.    **page 256** Exercise 83: Excerpt from Japanese University Entrance Examination Problems in Mathematics. Copyright © 2013 Mathematical Association of America. All Rights Reserved.    **page 262** Figure 1: Excerpt from "The Contribution Role of Sleepiness in Highway Vehicle Accidents," by S. Gabarino, L. Nobili, M. Beelke, F. Phy, and F. Ferrillo. Sleep, Vol 24, No. 2, pp. 203–206. Copyright © 2001 by American Academy of Sleep Medicine. Used by permission of American Academy of Sleep Medicine.    **page 323** Exercise 42: Problem 19 from the May 2003 Course 01 Examination of the Education and Examination Committee of the Society of Actuaries. Copyright © by Society of Actuaries. Used by permission of Society of Actuaries.    **page 383** Example10: Excerpt from Chartmasters' Rock 100 by Jim Quirin and Barry Cohen. Copyright © 1992 by Chartmasters. Used by permission of Chartmasters.

**page 461** Extended Application: Excerpt from "Flow System Integrals" by Arthur C. Segal, The UMAP Journal, Vol. 3, No. 1. Copyright © 1982. Published by Consortium for Mathematics and Its Applications (COMAP).    **page 493** Figure 24: "Solver Included with MS Excel" from Microsoft Corporation. Copyright © by Microsoft Corporation. Used by permission of Microsoft Corporation.    **page 494** Exercise 30: Problem 35 from the May 2003 Course 1 Examination of the Education and Examination Committee of the Society of Actuaries. Copyright © by Society of Actuaries. Used by permission of Society of Actuaries.    **page 529** Figure 8: "Solver Included with MS Excel" from Microsoft Corporation. Copyright © by Microsoft Corporation. Used by permission of Microsoft Corporation.    **page 662** Exercise 52: From Roger Staubach, "One Down, Lifetime to Go." W. Publishing Group, 1980.    **page 663** Exercise 80: Puzzler from the Car Talk Radio Program, February 24, 2001. Used by permission of TAPPET BROTHERS LLC.    **page 665** Exercise 90: Problem 2 from the 2005 Sample Exam P of the Education and Examination Committee of the Society of Actuaries. Copyright © by Society of Actuaries. Used by permission of Society of Actuaries.

**page 665** Exercise 91: Problem 8 from the 2005 Sample Exam P of the Education and Examination Committee of the Society of Actuaries. Copyright © by Society of Actuaries. Used by permission of Society of Actuaries.    **page 665** Exercise 92: Problem 15 from the 2005 Sample Exam P of the Education and Examination Committee of the Society of Actuaries. Copyright © by Society of Actuaries. Used by permission of Society of Actuaries.    **page 681** Exercise 39: From "Ask Marilyn." Parade Magazine, June 12, 1994. Copyright © by Marilyn Vos Savant. Used by permission.    **page 681** Exercise 44: From "Ask Marilyn." Parade Magazine, Nov 6, 1994. Copyright © by Marilyn Vos Savant. Used by permission.    **page 682** Exercise 51: Problem 12 from the 2005 Sample Exam P of the Education and Examination Committee of the Society of Actuaries. Copyright © by Society of Actuaries. Used by permission of Society of Actuaries.    **page 683** Exercise 59: Problem 31 from the 2003 Course 1 Examination of the Education and Examination Committee of the Society of Actuaries. Copyright © by Society of Actuaries. Used by permission of Society of Actuaries.    **page 684** Exercise 60: Problem 21 from the 2005 Sample Exam P of the Education and Examination Committee of the Society of Actuaries. Copyright © by Society of Actuaries. Used by permission of Society of Actuaries.

**page 684** Exercise 61: Problem 25 from the 2005 Sample Exam P of the Education and Examination Committee of the Society of Actuaries. Copyright © by Society of Actuaries. Used by permission of Society of Actuaries.    **page 684** Exercise 62: Problem 26 from the 2005 Sample Exam P of the Education and Examination Committee of the Society of Actuaries. Copyright © by Society of Actuaries. Used by permission of Society of Actuaries.    **page 686** Exercise 85: Problem 8 from May 2003 Course 1 Examination of the Education and Examination Committee of the Society of Actuaries. Copyright © by Society of Actuaries. Used by permission of Society of Actuaries.    **page 697** Exercise 26: Problem 36 from May 2003 Course 1 Examination of the Education and Examination Committee of the Society of Actuaries. Copyright © by Society of Actuaries. Used by permission of Society of Actuaries.    **page 699** Exercise 44: Problem 48 from the 2005 Sample Exam P of the Education and Examination Committee of the Society of Actuaries. Copyright © by Society of Actuaries. Used by permission of Society of Actuaries.

**page 702** Exercise 68: From "Ask Marilyn." Parade Magazine, September 9, 1990. Copyright © by Marilyn Vos Savant. Used by permission.    **page 702** Exercise 82: Problem 4 from the 2005 Sample Exam P of the Education and Examination Committee of the Society of Actuaries. Copyright © by Society of Actuaries. Used by permission of Society of Actuaries.    **page 703** Exercise 95: Problem 13 from the 2005 Sample Exam P of the Education and Examination Committee of the Society of Actuaries. Copyright © by Society of Actuaries. Used by permission of Society of Actuaries.    **page 705** Exercise 106: Excerpt from "Optimal Strategies on Fourth Down" by V. Carter and R. Machol, Management Science, Vol. 24, No. 16: 1758-1762. Copyright © 1978. Used by permission of the Institute for Operations Research and the Management Sciences (INFORMS).    **page 706** Exercise 110: "Exercise 110, Missiles" by Tom Clancy from DEBT OF HONOR. Published by G. P. Putnam's Sons, © 1994.    **page 706** Exercise 111: Problem 1 from May 2003 Course 1 Examination of the Education and Examination Committee of the Society of Actuaries. Copyright © by Society of Actuaries. Used by permission of Society of Actuaries.    **page 718** Exercise 44: Problem 34 from May 2003 Course 1 Examination of the Education and Examination Committee of the Society of Actuaries. Copyright © by Society of Actuaries. Used by permission of Society of Actuaries.    **page 718** Exercise 45: Problem 40 from the 2005 Sample Exam P of the Education and Examination Committee of the Society of Actuaries. Copyright © by Society of Actuaries. Used by permission of Society of Actuaries.    **page 726** Exercise 23: Problem 12 from May 2003 Course 1 Examination of the Education and Examination Committee of the Society of Actuaries. Copyright © by Society of Actuaries. Used by permission of Society of Actuaries.    **page 727** Exercise 34: Problem 68 from the 2005 Sample Exam P of the Education and Examination Committee of the Society of Actuaries. Copyright © by Society of Actuaries. Used by permission of Society of Actuaries.    **page 727** Exercise 35: Problem 51 from the 2005 Sample Exam P of the Education and Examination Committee of the Society of Actuaries. Copyright © by Society of Actuaries. Used by permission of Society of Actuaries.

## Photo Credits

Chapter opener photos are repeated at a smaller size on table of contents and preface pp. iii–vii.

# Index of Applications

## ANIMAL SCIENCE

Activity Level, 282
African Wild Dog, 228
Agriculture, 475
Alaskan Moose, 157, 161, 282
Alligator Teeth, 145, 297
Allometric Growth, 94, 347
Animal Activity, 578
Animal Breeding, 535
Animal Feed, 521, 535
Animal Growth, 544
Animal Interactions, 545
Animal Offspring, 697
Ant Colonies, 16
Archimedes' Problem Bovinum, 536
Arctic Foxes, 141, 236, 257
Aryshire Dairy Cattle, 236, 272
Bald Eagle Population, 81, 93
Beagles, 206–207, 409
Beaked Sea Snake, 183
Beef Cattle, 237
Bighorn Sheep, 211, 268–269
Bird Eggs, 25, 212, 452
Bird Feeding, 551–552, 555, 709, 712–713, 719–720
Bird Migration, 334
Bird Population, 585
Birds, 347, 535
Black Crappie, 243–244
Body Temperature of a Bird, 746
Brown Trout, 485, 516
Buffalo, 580
Cactus Wren, 233–234
Calves, 435
Catfish Farming, 293
Cats, 257
Clam Growth, 297
Clam Population, 235
Clam Population Growth, 296–297
Cohesiveness, 283
Crickets Chirping, 27
Cutlassfish, 236–237
Deer Harvest, 475
Deer Population, 61
Deer Ticks, 11–12
Deer-Vehicle Accidents, 475
Dinosaur Running, 474
Dog's Human Age, 212
Elk Energy, 16
Endangered Species, 645–646
Finding Prey, 741
Fish, 257
Fish Food Requirements, 535
Fish Population, 228, 594, 753
Flea Beetles, 596, 717, 726
Flight Speed, 183–184, 190
Flour Beetles, 373, 717, 726
Foot-and-Mouth Epidemic, 394, 435
Foraging, 219, 297, 342, 475–476, 623
Fruit Flies, 246, 308
Goat Growth, 609
Gray Wolves, 355
Growth of a Mite Population, 630
Growth of Algae, 609

Growth Rate, 435, 445
Guernsey Growth, 595
Harvesting Cod, 334
Harvesting Portion of Population, 765
Heat Loss, 474, 483, 500
Holstein Dairy Cattle, 236, 256, 272
Honeycomb, 334
Horn Volume, 500
Immigration, 609
Index of Diversity, 92, 94
Insect Cannibalism, 421
Insect Competition, 229
Insect Dispersal, 485
Insect Growth, 235, 296
Insect Life Span, 740
Insect Mating, 211, 245
Insect Population, 567–568, 571, 573–574, 609
Insect Species, 94, 212
Insulin in Sheep, 421
Life Span, 474, 484–485, 501
Lizards, 348
Location of a Bee Swarm, 740
Location of a Bird's Nest, 709, 712–713
Managing Renewable Resources, 258–260
Marsupials, 348
Mass of Bighorn Yearlings, 161–162, 165
Maximum Sustainable Harvest, 331–333, 358–360
Metabolic Rate, 39
Metabolism Rate, 95
Migratory Animals, 410
Milk Consumption, 280–281, 282, 435
Milk Production, 460
Monkey Eyes, 115
Mountain Goat Population, 589–590
Mouse Infection, 602, 622–623
Movement of a Released Animal, 746
Northern Spotted Owl Population, 554, 575
Optimization for a Predator, 518–519
Oxygen Concentration, 138–140
Pigeon Flight, 333
Pigs, 353, 355, 358
Polar Bear Mass, 122, 470
Pollution Intolerance, 474
Ponies Trotting, 15
Population Growth, 93, 99, 121, 245, 283, 296
Poultry Analysis, 648
Poultry Farming, 154
Predator-Prey, 617–623
Pronghorn Fawns, 245, 405
Rams' Horns, 409
Rats, 532–533
Risk in Fisheries, 421–422
Rumen Fermentation, 445
Salmon Spawning, 324
Sculpin Growth, 99–100
Shad, 256
Shellfish Population, 183, 190
Size of Hunting Parties, 25–26
Snake River Salmon, 475, 485
Spiders, 741
Swimming Energy, 40
Tasmanian Devil, 225–226
Thoroughbred Horses, 297, 307
Velocity of a Marine Organism, 212

Volume, 330–331
Walleye, 256
Water Snakes, 183
Weight Gain, 177–178
Weight Gain of Rats, 746
Whale Population, 609
Whales and Krill, 617–623
Whales Diving, 39, 195–196
Whooping Cranes, 235–236
Zooplankton Growth, 485

## BIOLOGY

Agriculture, 535
Beverton-Holt Model, 755–756, 762, 765
Biochemical Reaction, 342, 410
Botany, 101
Cell Division, 409
Cell Growth, 374
Cell Surface Receptors, 145, 311–312
Cell Traction Force, 219, 312
Cellular Chemistry, 219, 308
Color Blindness, 664, 682
Competing Species, 622–623, 630
Cylindrical Cells, 334
DNA, 191
Eastern Hemlock, 500–501
Ecosystem, 67, 623
Excitable Cells, 602
Extinction, 196, 229, 237
Eye Color, 674
Fertilizer, 561–562
Genetics, 172, 220, 643–644, 664–665, 681
Genotype Fitness, 342
Growing Degree Days, 484
Growth Models, 61, 219
Harvesting Fruit, 649
Jukes-Cantor Distance, 246
Length of a Leaf, 726
Logistic Growth, 257, 358, 754
Mixing Plant Foods, 535
Mutation, 121–122
Organic Farmland, 69
Oscillating System, 410
Petal Length, 716, 726, 739
Phytoplankton Growth, 297, 342
Plant Growth, 272, 297, 596
Population Biology, 61
Ricker Model, 755, 762, 765
Shepherd Model, 755–756, 762, 765
Soil Ingestion, 62
Soil Moisture, 594
Species, 122, 342
Symbiotic Species, 622
Tree Growth, 212, 408
Twins, 682–683
World Grain Harvest, 123
Yeast Cells, 747
Yeast Growth, 236
Yeast Production, 97

## BUSINESS AND ECONOMICS

Age and Loans, 685
Analysis of Orders, 566
Area, 334–335

Automobile Insurance, 686
Average Cost, 145, 197, 512
Average Inventory, 452
Average Price, 452
Average Profit, 512
Average Revenue, 512
Banking, 536
Broadway Economics, 537
Can Design, 335
Charge for Auto Painting, 517
Complaints, 699
Compound Interest, 75–76, 226, 257
Consumer Demand, 145, 191
Container Design, 335
Continuous Compound Interest, 77, 257
Cost, 325, 336, 349, 496, 517
Cost Analysis, 12–13, 16, 152, 155, 555
Cost and Revenue, 349
Cost-Benefit Analysis, 57–58, 62
Cost Function, 29
Cost with Fixed Area, 335
Customer Expenditures, 742
Demand, 381–382
Dental Insurance, 727
Depreciation, 229
Device Failure, 699
Distribution, 581
Doubling Time, 84, 91–92
Earnings, 667
Electronic Device, 742
Employee Training, 220
Employment, 665
Energy Consumption, 423
Equipment Insurance, 747
Error Estimation, 353–354
Filling Orders, 581
Flashlight Battery, 731
Fuel Economy, 417
Governors' Salaries, 69
Hospitalization Insurance, 697
Household Telephones, 157
Housing Starts, 272
Inflation, 84
Injured Loss, 742
Insurance, 718
Insurance Claims, 727
Insurance Reimbursement, 727
Interest, 83–84, 123, 167, 229
Internet Usage, 40
Investments, 566
Job Qualifications, 686
Labor Costs, 496
Labor Force, 665–666
Life Insurance, 692
Loans, 537
Logistic Curve, 591–592
Losses after Deductible, 727
Machine Part, 718
Management, 545
Manufacturing, 501, 536
Manufacturing Cost, 517
Marginal Cost, 162, 415
Material Requirement, 355
Medicare Trust Fund, 166
Money, 213
Net Savings, 417–418, 423
Oil Production, 422

Packaging Cost, 335
Packaging Design, 335–336, 359, 511–512
Payout on Insurance Policies, 699
Perfume Bottle, 506–507
Petroleum Consumption, 411
Pharmaceutical Manufacturing, 539–542, 546–547
Pollution, 68
Postage, 155
Postal Rates, 213
Printer Failure, 742
Production, 154, 566
Production Costs, 491–492
Production Error, 517
Production Materials, 517
Production of Landscape Mulch, 322
Production Requirements, 566, 581
Profit, 325, 418, 496, 517
Rate of Change of Revenue, 445
Real Estate, 685
Revenue, 346, 496
Revenue/Cost/Profit, 349
Revenue from Seasonal Merchandise, 254
Roof Trusses, 581
Sales, 191
Sales Calls, 650
Savings, 415–416
Scheduling Production, 581
Shoe Sales, 555
Social Security Assets, 185
Stock Reports, 581
Surveys, 536
Synthetic Fabric, 592
Total Maintenance Charges, 391
Total Revenue, 436, 461
Total Sales, 436
Training Program, 360–361
Transportation, 385, 537
Use of Materials, 335
Volume of a Coating, 501
Wind Energy Consumption, 394
Worker Errors, 676
Working Women, 684

**GENERAL INTEREST**
Accidental Death Rate, 17–21
Alcohol Abstinence, 686
Area, 36–37, 325, 359, 517
Athletic Records, 27
Baseball, 184, 537, 581, 706
Basketball, 537, 685–686
Book of Odds, 666
Candy, 229
Cola Consumption, 649
Contests, 698
Cookies, 581
Cryptography/Encryption, 563–564, 566
Driver's License Test, 685
Driving Fatalities, 258, 717, 727
Drunk Drivers, 717, 727
Elliptical Templates, 476
Energy Consumption, 40
Estimating Area, 501
Estimating Volume, 501
Expenditures for Music, 667
Food Frying, 496
Food Surplus, 78–79, 257
Football, 29, 698, 742

Friendly Wager, 691–692
Gambling, 706
Games, 650
Golf Tournament, 698
High-Risk Drivers, 742
Hockey, 579, 648
Hog Game, 698
Hose, 501
Ice Cream, 501, 538
Icicle, 345–346
Information Content, 325
Length of a Telephone Call, 718
Lights Out Game, 538
Making a First Down, 705
Maximizing Area, 51
Maximizing Volume, 328–329, 330–331
Measurement Error, 355
Meat Consumption, 67
Mercator's World Map, 423
Minimizing Area, 329–330
Minimizing Time, 327–328
Motorcycle Helmets, 544
Murder, 686
Music, 112, 566, 706
Musicians, 650
Never-Married Adults by Age Group, 686
Numbers Game, 698
Package Dimensions, 359
Parabolic Arch, 51
Parabolic Culvert, 51
Perimeter, 40
Playground, 359
Playing Cards, 655, 673
Popcorn, 212, 297
Popularity Index, 383
Postage Rates, 476
Postal Regulations, 336
Power Functions, 124–126
Probabilities with Dice, 656
Probability, 436
Pursuit, 359
Raffle, 698
Recollection of Facts, 269–270
Required Material, 476
Rotating Camera, 349
Rotating Lighthouse, 349
Roulette, 698
Running, 27, 397
Satisfaction, 324, 495, 688, 690, 693
Seat Belt Effectiveness, 684
Snack Food, 537
Soccer, 742
Speeding Tickets, 685
State-Run Lotteries, 747
States, 650, 705
Street Crossing, 246
Studying, 685
Surface Area, 517
Surfing, 359
Symphony Orchestra, 690–691
Text Messaging, 374
Three Prisoners, 686
Time of Traffic Fatality, 717, 727
*Titanic*, 685
Tolerance, 355
Tossing Coins, 670–671, 689–690
Toys, 537

Track and Field, 211, 237
Traffic Control, 538
Travel Time, 336
Vehicle Waiting Time, 220

**HEALTH SCIENCE**

Alcohol Concentration, 61, 271, 354, 394
Alzheimer's Disease, 115–116, 195, 272
Angioplasty, 49–50
Antibiotics, 650–651
Area of a Bacteria Colony, 355
Arterial Pulse, 602
Average Birth Weight, 746
Bacteria in Sausage, 340
Bacterial Food Requirements, 535
Bacterial Growth, 81, 100, 343–344, 409
Bacterial Population, 81–82, 164–165, 219, 228, 296, 354, 370
Blood Antigens, 648, 703
Blood Clotting Time, 716, 726
Blood Flow, 346–347, 409, 451–452, 483
Blood Level Curves, 435
Blood Pressure, 29, 122, 374, 435, 682, 738
Blood Sugar and Cholesterol Levels, 67–68
Blood Sugar Level, 211
Blood Test, 684
Blood Transfusions, 693–695, 697
Blood Velocity, 347
Blood Vessels, 283, 354, 499, 500, 516
Blood Volume, 68, 311, 500
Body Mass Index, 191, 196, 211, 704
Body Response to Drug in Bloodstream, 613–614
Body Shape Index, 483–484
Body Surface Area, 94, 245, 474, 476–477, 483
Body Temperature, 16
Body Types, 664
Bologna Sausage, 245
Bone Preservation Volume, 500
Brain Mass, 61, 211, 347
Breast Cancer, 236, 297, 678–679, 682, 683
Breath Volume, 485–486
Breathing, 768
Calcium Usage, 228
Calorie Expenditure, 483, 565
Cancer, 49, 122, 210
Cancer Research, 100
Cardiac Output, 60
Cardiology, 209, 271
CAT Scans, 578–579
Causes of Death, 659–660
Cholesterol, 237, 450, 475
Chromosomal Abnormality, 101, 172, 732
Cigarette Smokers, 683, 684
Circadian Testosterone, 122
Circulation, 684
Colorectal Cancer, 683
Contact Lenses, 61
Cortisone Concentration, 82, 123
Death Rates, 741
Decrease in Bacteria, 100
Dengue Fever, 474–475
Dentin Growth, 324, 358
Diabetes, 68–69
Dialysis, 145, 237, 500
Dietetics, 535, 543, 554
Dieting, 594–595
Digestion Time, 740

Disease, 333
Doctor Visit, 665
Drug Concentration, 94, 121, 145, 271, 296, 314–315, 354, 410
Drug Epidemic, 457
Drug Reaction, 228–229, 409, 434, 445, 452, 457, 461, 484
Drug Use, 602
Drugs Administered Intravenously, 197–200, 391
Ear Infections, 697
Eating Behavior, 68, 178, 183
Electrocardiogram, 154
Emergency Room, 684
Epidemics, 282, 384
Exercise Heart Rate, 15
Femoral Angles, 116
Fetal Stature, 26
Fever, 68
Flow Systems, 461–463
Flu Epidemic, 164, 273
Fungal Growth, 324, 358
Giardia, 100
Glucose Concentration, 121
Glucose Level, 601, 602
Health, 483, 536, 647, 660
Health Plan, 665
Heart, 212
Heart Attack Risk, 82
Heart Muscle Tension, 746
Hepatitis Blood Test, 683
HIV Infection, 15
H1N1 virus, 595
Hospital Satisfaction, 688, 690, 693
Human Cough, 211
Human Growth, 190, 196
Human Skin Surface, 324, 358
Human Strength, 166
Hypoxia, 501
Ice Hockey Injuries, 668–670
Immune System Activity, 217, 308
Infection, 704
Infection Rate, 421
Injuries from Recreation Equipment, 658–659
Insecticide, 190
Involutional Psychosis, 237–238
Lead Poisoning, 737–738
Length of Life, 49
Life Expectancy, 15, 70–71, 544, 554, 724, 741, 746
Life Span, 735–736
Lung Cancer, 675
Lyme Disease, 11
Malaria, 767
Medical Diagnosis, 707
Medical Experiment, 681, 687
Medical Literature, 237
Medical School, 26
Medical Survey, 664
Melanoma, 101
Mercury Poisoning, 741
Metabolic Rate, 348
Molars, 165, 324
Mortality, 236, 648, 713–715
Mouse Infection, 602, 622–623
Muscle Reaction, 219
Nervous System, 145, 237, 410
Neuron Communications, 308
Neuron Membrane Potential, 308

Odds of Flat Feet, 658
Outpatient Visits, 384
Overweight, 682
Oxygen Consumption, 78
Oxygen Inhalation, 393
Patient Wait Time, 723–724
Perinatal Mortality, 697
Pharmacology, 154
Physician Demand, 81, 94–95
Plasma Volume, 82
Polio Epidemic, 768–770
Pregnancy, 30–31, 154
Pressure on the Eardrum, 252
Prevalence of Cigarette Smoking, 11
Prostate Cancer, 704
Pygmy Heights, 740
Radioactive Albumin, 238
Radioactive Iron, 238
Recommended Dosage, 146
Registered Nurses, 67
Respiratory Rate, 122, 342
Risk Factors, 703
SARS, 727
Scaling Laws, 311
Scarlet Fever, 623
Shoulder Injuries, 665
Sickle Cell Anemia, 703
SIDS, 704
Silicone Implants, 23–24
Skeletal Maturity, 26
Smoke Content in a Room, 630
Splenic Artery Resistance, 49
Spread of a Virus, 195
Spread of an Epidemic, 624–627
Spread of Gonorrhea, 627
Spread of Infection, 271
Spread of Influenza, 630–631
Sunscreen, 68
Survival Curves, 516–517
Test for HIV, 683
Testing Medication, 544
Thermic Effect of Food, 165, 272, 282, 445
Tobacco Deaths, 15–16
Tooth Length, 49
Total Body Water, 475, 516
Tracer Dye, 594
Transylvania Hypothesis, 115
Tuberculosis in United States, 55
Tumor Growth, 94, 154
TYLENOL Dosage, 195
Vaccinations, 44–45
Vectorial Capacity, 272
Vitamins, 2, 465, 565
Volume of a Tumor, 354
Weight Gain, 190
Weight Loss, 285
Weight of a Young Female, 185–186
Weightlifting, 311
Women Joggers, 685
Work/Rest Cycles, 219
World Health, 67

**PHYSICAL SCIENCES**

Acidity of a Solution, 96
Air Pollution, 115, 271
Area, 348
Area of an Oil Slick, 355

Astronomy, 180–181
Atmospheric Pressure, 83
Automobile Velocity, 396
Average Speed, 69–70, 158–160
Average Temperatures, 123, 461
Batting Power, 62
Biochemical Excretion, 373
Biomechanics, 334
Calcium Kinetics, 219, 312
Carbon Dating, 97–98, 101
Carbon Dioxide, 44, 82, 253, 466, 580
Chemical Dissolution, 102
Chemical Formation, 436
Chemical in a Solution, 627
Chemistry, 580
Cliff Ecology, 394
Coal Consumption, 63
Complementary Error Function, 460–461
Computer Chips, 83, 495–496
Concentration of a Solute, 373
Dating Rocks, 123
Dead Sea, 213
Decay of Radioactivity, 101
Depletion Dates for Minerals, 424–426
Distance, 348, 374, 395–397
Dry Days, 741
Earthquakes, 96, 684, 705, 718, 728, 741, 747
Earth's Volume, 452
Electricity, 238–239
Engine Velocity, 284
Environmental Inspections, 672–673
Flying Gravel, 313
Global Warming, 16
Gravitational Attraction, 486
Ground Temperature, 486
Growth of a Substance, 409
Half-Life, 101
Heat Gain, 396
Heat Index, 239, 485
Height, 284
Height of a Ball, 298
Ice Cube, 348
Intensity of Light, 121
Intensity of Sound, 96
Kite Flying, 349
Ladder, 336
Length of a Pendulum, 28, 62–63
Length of Day, 411
Light Rays, 116
Linear Motion, 422
Maximizing the Height of an Object, 50
Measurement, 117
Motion of a Particle, 253–254
Motion under Gravity, 374
Movement Time, 486
Music Theory, 95
Newton's Law of Cooling, 102, 596–597, 603, 631
Nuclear Energy, 102
Oil Consumption, 410–411
Oil Pollution, 222–223, 228, 461
Oil Production, 123
Oven Temperature, 184
Ozone Depletion, 296
Pedestrian Speed, 243
Piston Velocity, 284
Planets, 70

Pollution, 324, 333, 358, 408, 417, 455–456, 616
Pollution Concentration, 236
Pollution of the Great Lakes, 632–634
Precipitation in Vancouver, Canada, 405–406
Quality Control of Cheese, 183
Radioactive Decay, 83, 99, 101, 232–233, 238, 596
Radioactive Waste, 457
Rain Forecasts, 684
Rainfall, 718, 728, 741
Richter Scale, 246
Rocket, 374
Rocket Science, 374
Salt Concentration, 625–627
Scuba Diving, 484
Sediment, 145, 409
Seeding Storms, 698
Shadow Length, 348
Sliding Ladder, 344–345, 348, 358
Snow in Manhattan, 675
Snowfall, 746
Sound, 116, 253
Spherical Radius, 358
Spread of an Oil Leak, 408
Stopping Distance, 50
Sunrise, 113–114
Sunset, 117
Swimming, 501
Swing of Jogger's Arm, 312
Swing of Runner's Arm, 312
Temperature, 13, 17, 116, 166, 184, 186–187, 196, 423, 729–730
Tennis, 258
Thermal Inversion, 228
Total Distance, 392, 433
Traffic, 397
Velocity, 159–161, 163, 167, 212–213, 349, 374, 486
Velocity and Acceleration, 287–288, 298, 313, 371–372
Voltage, 411
Volume, 348, 355, 359
Volume of a Can of Beer, 499
Water Level, 348, 358
Water Temperature, 480
Wind Chill, 484
Wind Energy, 83, 257
Zenzizenzicube, 229
Zenzizenzizenzic, 229

**SOCIAL SCIENCES**
Accident Rate, 50
Age Distribution, 409
Age of Marriage, 50
AP Examinations, 213
Assaults, 746–747
Attitude Change, 283–284
Automobile Accidents, 394, 422, 545, 554
Bachelor's Degrees, 374
Biology Majors, 653, 671
Centenarians, 208–209
Cheating, 698
Child Mortality Rate, 17
Chinese New Year, 649, 667
Civil War, 666
Community Activities, 667
Crime, 298

Crime Rate, 348
Dating a Language, 257–258, 741
Dead Grandmother Syndrome, 239
Degrees in Dentistry, 374
Drug Use, 165–166, 321
Education, 631, 698
Educational Attainment, 544–545
Educational Psychology, 436
Elections, 704
Electoral College, 650
Emigration, 603
Evolution of Languages, 95–96
Exponential Growth, 257
Gender Ratio, 49
Habit Strength, 238
Head Start, 62
Hispanic Population, 595
Human Mortality, 409
Ideal Partner Height, 28
Immigration, 16, 166, 603
Income Distribution, 410, 417
Learning, 312–313, 485, 609
Legislative Voting, 146
Living Arrangements, 650
Marital Status, 69, 649
Marriage, 16
Memorization Skills, 348
Memory Retention, 220
Military, 649
Minority Population, 81, 93, 165, 235
Missiles, 706
Modeling War, 538
Native American Ceremonies, 649, 667
Nuclear Weapons, 313
Online Learning, 238
Perceptions of Threat, 666–667
Police Lineup, 705
Political Science, 495
Population, 272, 313
Population Growth, 100, 235, 253, 283, 364, 384, 421, 445, 595, 631, 755–756, 767
Population Modeling, 615–616, 756–758, 761–762, 765
Poverty, 28, 69, 246
Present Value of a Population, 457
Production Rate, 452
Randomized Response Method, 705
Refugees, 667
SAT Scores, 28
School Activities, 645
Sleep-Related Accidents, 262
Social Network, 718, 728
Spread of a Rumor, 596, 609, 627, 631
Students, 581
Survival of Manuscripts, 238
Television Viewing Habits, 705
Terrorists, 686
Time to Learn a Task, 717, 727
Toronto's Jewish Population, 95
Typing Speed, 452
U.S. Asian Population, 257, 595
U.S. Population, 648, 665
Viewing Habits, 706
Waiting Times, 747–749
World Population, 68, 80–81, 93–94, 164, 554–555, 596

# Index

Absolute extrema
    explanation of, 317
    graphical optimization, 322
    method for finding, 319–321, 356
Absolute maximum, 317, 319–321
Absolute minimum, 317, 319–321, 330
Absolute value, R-27
Absolute value function, 119, 179
Absolute value sign, 369
Acceleration, 287–288, 371–372
Acrophase, 111
Acute angles, 103
Addition
    of matrices, 539–541, 576
    in order of operations, R-2
    of polynomials, R-2–R-3
    of rational expressions, R-8
Addition property
    of equality, R-11
    of inequality, R-17
Algebra review
    equations, R-11–R-16
    exponents, R-21–R-25
    factoring, R-5–R-7
    importance of algebra, R-1
    inequalities, R-16–R-21
    polynomials, R-2–R-5
    radicals, R-25–R-29
    rational expressions, R-8–R-11
Amplitude, 111
Analysis
    compartmental, 609–610
    decision, 693–695
Angles
    equivalent, 104–105
    explanation of, 102
    radian measure of, 103–105
    special, 106–109
    terminal side of, 102
    trigonometric functions for, 118
    types of, 103
Antiderivatives. *See also* Derivatives
    evaluation of, 401–402
    explanation of, 363
    Fundamental Theorem of Calculus and,
        398–411
    methods for finding, 363–375
Antidifferentiation, 363
Antidifferentiation formulas, 418
Approximations
    of area, 385–391
    of definite integral, 389
    by differentials, 498–499, 514
    explanation of, 497, 514
    linear, 350–355, 356
    Simpson's rule and, 431–433
    trapezoidal rule and, 386
Arc, 103
Archimedes, 385, 387n, 536
Area
    approximation of, 385–391
    under curves, 388–389, 404, 734
    definite integrals and, 385–397, 402–404

    finding of, 404
    minimizing, 329–330
    between two curves, 412–418, 419
Area formula, 37
Associative properties, R-2
Asymptotes
    explanation of, 56, 65
    methods for finding, 137–139, 299
    oblique, 302
Augmented matrices, 523
Autonomous differential equations, 593
Average cost, 58
Average rate of change
    explanation of, 155–157, 192
    formula for, 156
Average value, 449–450, 458
Axes, 2, 41

Basic identities, 247
Bayes, Thomas, 677
Bayes' theorem
    explanation of, 675–677, 700
    use of, 677
Bellard, Fabrice, 387n
Bernoulli, Jakob, 365
Binomial theorem, 204
Binomials, R-4
Brahmagupta, 387n
Break-even point, 13
Briggs, Henry, 87
Butterfly effect, 760

Calculator exercises. *See* Graphing calculator
    exercises
Calculus
    differential, 363
    Fundamental Theorem of, 398–411, 419
    historical background of, 129n
    integral, 363
    multivariable, 464–519
Carbon dating, 97–98
Carrying capacity, 590
Cartesian coordinate system, 2
Cauchy, Augustin-Louis, 129n
Cayley, Arthur, 523n
Celsius, Anders, 13n
Central Limit Theorem, 738–739, 743
Certain event, 652
Chain rule
    alternative form of, 224, 254
    composition of functions, 220–222
    explanation of, 222–223, 254
    use of, 222–227, 231, 249, 587
Change
    average rate of, 155–157, 192
    instantaneous rate of, 158–163
    rate of. *See* Rates of change
    total, 391–392
Change in $x$, 3
Change in $y$, 3
Change-of-base theorem
    for exponentials, 91, 118
    for logarithms, 87–88, 118

Chaos, 760
Circle
    circumference of, 104
    unit, 103
Closed interval, 148, R-17
cobweb diagrams, 757–760
Coefficient matrices, 559
Coefficients
    explanation of, 51, R-2
    leading, 51
Cohen, Barry, 383
Column integration, 438–441
Column matrices, 540
Column vector, 540
Common denominators, R-10
Common logarithms, 87
Commutative properties, R-2
Compartmental analysis, 609–610
Complement of sets, 639
Complement rule, 657, 700
Completing the square, 42
Composite function, 221
Composition, function, 36–37
Compound amount, 75
Compound interest
    chain rule and, 226
    continuous, 77–78
    explanation of, 75
Concave downward, 288–290
Concave upward, 288–289
Concavity
    of graphs, 288–291
    test for, 290, 299–300, 309
Conditional probability, 668–673
    explanation of, 669, 700
    independent events and, 673–675
    product rule and, 671
    Venn diagrams and, 670
Constant
    decay, 97, 314
    derivatives of a, 203–204
    growth, 97, 588
    integration, 364, 366–367
Constant functions, 32, 254
Constant harvesting, 259
Constant multiple rule, 366–367, 418
Constant rule, 203
Constant times a function, 205, 254
Contagion, 582
Continuity, 146–155
    on an open interval, 148
    on a closed interval, 148
    explanation of, 147, 192
    Intermediate Value Theorem and, 152
    from right/left, 148
    at $x = c$, 147
Continuous compounding, 77–78
Continuous functions, 147, 148
Continuous probability distribution, 710
Continuous probability models, 709–718
Continuous random variables, 687, 710,
    719–728
Convergent integrals, 454

Coordinates, 2
Correlation coefficient, 21–24, 64
Cosecant, 105. *See also* Trigonometric functions
Cosine, 105. *See also* Trigonometric functions
Cosine functions, 110–112
Cost analysis, 12–13, 64, 152
Cost-benefit models, 57–58
Cost function, 12
Cotangent, 105. *See also* Trigonometric functions
Critical numbers
    in domain of function, 280
    explanation of, 264
    method for finding, 264–265
    relative extrema at, 278
Critical point theorem, 321, 330
Critical points, 264, 488–489
Cube root, R-23
Cubes
    difference of two, R-7
    sum of two, R-7
Cubic polynomials, 53
Cumulative distribution function, 715, 743
Curve sketching
    explanation of, 299–300, 309
    illustrations of, 300–306
Curves
    area between two, 412–418, 419
    area under, 388–389, 404, 734
    epidemic, 625
    level, 469
    logistic, 590–592
    normal, 733–734
    slope of, 169
Cycle, 764

De Moivre, Abraham, 733
Decay constant, 97, 314
Decision analysis, 693–695
Decreasing functions
    explanation of, 262–273
    test for, 264, 309
Definite integrals
    area and, 385–397, 402–404
    explanation of, 389, 419
    formulas for, 419
    on graphing calculators, 379, 389, 404, 442
    properties of, 400, 419
Degree measure, 103, 118
Demand functions, 381–382
Denominators
    common, R-10
    least common, R-10
    rationalizing, R-27–R-28
Density functions. *See* Probability density
    functions
Dependent equations, 522
Dependent events, 673, 674
Dependent variables, 10, 465
Depletion date estimation, 424–426
Derivative tests
    first, 275–281
    second, 291–294, 309, 489–490
Derivatives, 127–200
    applications of, 316–361
    applications of extrema, 326–336
    calculation of, 201–260

chain rule for, 220–229, 231
concavity of a graph, 288–291
of constant, 203–204
of constant times a function, 205
continuity and, 146–155
curve sketching, 298–308
definition of, 167–185
difference quotient and, 173, 192
existence of, 179–181
explanation of, 167–185
of exponential functions, 230–239
extrema applications and, 317–326
fourth, 285–286
of functions, 172–179, 364
graphical differentiation and, 185–191
on graphing calculators, 174–175, 176, 179,
    187–188, 208
graphs of, 187–188, 261–315
higher, 285–288
implicit differentiation and, 337–342
increasing and decreasing functions and,
    262–273, 299–300, 309
limits and, 128–146
linear approximation and, 350–355
of logarithmic functions, 239–246
notations for, 202, 286
partial, 476–487
power rule and, 204–205
of products and quotients, 214–220
rates of change and, 155–167, 284–285, 480
related rates and, 343–349
relative extrema and, 273–284
rules for, 239–240
second, 285, 286–287
of sum, 207
tangent line and, 168–172
techniques for finding, 202–213
third, 285–286
total cost model and, 360–361
of trigonometric functions, 247–254, 255
Descartes, René, 2n
Determinants
    applications of, 569–570
    explanation of, 568
Deviation, standard. *See* Standard deviation
Difference equations
    with constant coefficients, 753–754
    explanation of, 753–754
Difference quotient
    derivatives and, 173, 192
    explanation of, 159
Differentiable functions, 172
Differential calculus, 363
Differential equations, 583–634
    applications of, 624–627
    autonomous, 593
    elementary, 584–586
    Euler's method and, 603–609, 628
    explanation of, 584
    general solution of, 584–585, 628
    linear first-order, 597–603, 628
    linear system of, 609–616
    nonlinear system of, 617–623
    order of, 597
    separable, 587–590, 628
    solutions to, 584–587, 628

Differentials
    approximations by, 498, 514
    error estimation and, 353–354
    explanation of, 351
    linear approximation and, 350–355, 356
    total, 496–497
Differentiation
    explanation of, 172
    graphical, 185–191
    implicit, 337–342, 356
Discontinuity
    explanation of, 146–147
    removable, 148
Discrete dynamical systems
    determining stability in, 762–765
    equilibrium points and, 756–762
    explanation of, 751
    sequences and, 751–756
Discrete probability functions
    explanation of, 710
    probability density functions vs., 712
Discrete random variables, 687
Discriminant, 489
Disjoint sets
    explanation of, 640, 700
    union rule for, 645
Distribution
    exponential, 730–732, 743, 747–749
    normal, 733–739, 743
    Poisson, 731–732, 743
    probability. *See* Probability distribution
    standard normal, 734
    uniform, 728–730, 743
Distribution function, cumulative, 715, 743
Distributive properties, R-2
Divergent integrals, 454
Division
    in order of operations, R-2
    of rational expressions, R-8
Domain(s)
    agreement on, 33
    explanation of, 30, 64, 465
    of logarithmic functions, 86
    and range, 33–34, 465
    restrictions on, 33
Double integrals, 502–512
    explanation of, 505, 514
    over variable regions, 507–509, 514
    volume and, 505–507
Doubling time, 84
Drugs
    concentration model for orally administered
        medications, 314–315
    intravenous administration of, 197–200
Dynamical systems, discrete. *See* Discrete
    dynamical systems

$e$
    explanation of, 76–77, 118
    exponential functions and, 76–77, 90–91
Eigenvalues
    applications of, 570–574
    explanation of, 568
Eigenvectors
    applications of, 570–574
    explanation of, 568

Einstein, Albert, 486
Elements
    explanation of, 523
    of sequences, 751
    of sets, 636
Ellipsoid, 471
Empirical probabilities, 658–659
Empty sets, 636
Endpoints
    limits at, 330
    of ray, 102
Entry, explanation of, 523
Epidemic curve, 625
Epidemics, 624–625
Equality, properties of, R-11
Equations
    dependent, 522
    differential. *See* Differential equations
    exponential, 74–75, 90–92
    first-degree, 521–522
    with fractions, R-14–R-16
    linear, R-11–R-12, 4
    of lines, 4–7, 64, 169–170
    logarithmic, 88–89
    logistic, 590
    Lotka-Volterra, 619–620
    quadratic, R-12–R-13
    rational, R-14–R-16
    solving system of, 521–538
    systems of. *See* Systems of equations
    of tangent lines, 179
Equilibrium points
    explanation of, 593, 617, 756
    semistable, 593
    stability of, 593, 762–764
    stable, 259, 593, 758
    unstable, 259, 593, 759
    uses for, 756–761
Equivalent system, 523
Error estimation, 353–354
Euler, Leonhard, 76, 603
Euler's method
    accuracy of, 604, 608
    explanation of, 603–604, 628
    use of, 605–608
Events
    certain, 652
    dependent, 673, 674
    explanation of, 652
    impossible, 652
    independent, 673–675, 700
    mutually exclusive, 653–654, 657, 700
    probability of, 709n
    set operations for, 653
    simple, 652
Exhaustion, 385
Expected value
    explanation of, 700, 720–722, 743
    of a probability distribution, 720
    of a random variable, 689–690
Experiments, 651–652
Explicit functions, 337
Exponential distribution
    applications of, 731–732
    explanation of, 730–731, 743
    waiting times and, 747–749

Exponential equations
    explanation of, 74
    solving of, 74–75, 90–92
Exponential functions
    compound interest and, 75–76
    continuity and, 149
    continuous compounding and, 77–78
    derivatives of, 230–239
    $e$ and, 76–77, 90–91
    explanation of, 73, 117, 255
    graphs of, 73–74, 119
    indefinite integrals of, 368
    integration of, 418
Exponential growth and decay functions,
        96–102, 118
Exponentials, change-of-base theorem for, 91, 118
Exponents
    explanation of, R-2, R-21–R-22
    integer, R-21–R-23
    logarithms as, 84–85
    properties of, R-22
    rational, R-23–R-24
    zero and negative, R-22
Extraneous solutions, R-15
Extrapolation, 70–71
Extrema
    absolute, 317–326, 356
    applications of, 326–336
    location of, 488
    relative, 273–284, 488
    solving problems of applied, 317–326, 356
Extreme value theorem, solving problems
        of applied, 318

Factoring
    difference of two cubes, R-7
    difference of two squares, R-7
    explanation of, R-5
    perfect squares, R-7
    of polynomials, R-7
    simplifying by, R-27
    sum of two cubes, R-7
    of trinomials, R-6–R-8
Factoring out, R-5
Factor(s)
    explanation of, R-5
    greatest common, R-5
Fahrenheit, Gabriel, 13n
Fair game, 692
Fate matrix, 616
Fermat, Pierre de, 709
First-degree equation(s)
    explanation of, 521–522
    in $n$, 522
First derivative test
    explanation of, 275–276
    method for, 277–281, 309
First octant, 467
First-order contacts, 582
First-order difference equation with constant
        coefficients, 753–754
Fisher, J. C., 591, 592
Fisher, R. A., 739
Fitts's law, 486
Fixed cost, 12
Flow systems, 461–463

FOIL method, R-4
Folding back, 694–695
Folium of Descartes, 339
For Review features, 4, 6, 33, 41, 73, 74, 78, 140,
        156, 169, 205, 214, 216, 221, 230, 248,
        265, 279, 302, 319, 337, 365, 366, 375,
        388, 437, 442, 455, 467, 476, 487, 497,
        502, 584, 588, 603, 610, 673, 710, 712, 754
Formulas from geometry, A-10
Fourth derivative, 285–286
Fractions
    equations with, R-14–R-16
    inequalities with, R-19–R-20
Frequency, 111
Fubini's Theorem, 504, 510
Function composition, 36–37
Function notation, 10
Functions
    absolute value, 179
    antiderivatives of, 363
    applications for, 29–31
    average value of, 449–450, 458
    composite, 221
    composition of, 220–222
    constant, 32, 254
    constant times $a$, 205, 254
    continuous, 147, 148
    continuous from left, 148
    continuous from right, 148
    continuous on closed interval, 148
    continuous on open interval, 148
    cosine, 110–112
    cumulative distribution, 715, 743
    definite integral of, 389
    definition of, 30
    demand, 380–381
    density, expected value for, 720–722, 743
    density, variance for, 720–723, 743
    derivative of, 172–179, 364
    differentiable, 172
    discontinuous, 146–147
    evaluation of, 34–35
    explanation of, 29–30, 64
    explicit, 337
    exponential. *See* Exponential functions
    exponential growth and decay, 96–102
    graphing of, 65, 466–472
    implicit, 337
    increasing and decreasing, 262–273,
        299–300, 309
    inverse, 86
    iterated, 752–754
    limit of, 128–134, 192
    linear, 2–17, 41
    linear cost, 12–13, 64
    logarithmic. *See* Logarithmic functions
    logistic, 233
    nonlinear, 29–63
    parent-progeny, 332
    periodic, 109, 119
    piecewise, 131, 151
    polynomial. *See* Polynomial functions
    power, 52, 124–126
    probability, 688, 709–710
    probability density. *See* Probability density
        functions

Functions (*continued*)
  properties of, 29–40
  quadratic, 41–51, 64
  rational, 55–58, 65, 149
  relative extrema of, 273–284
  root, 149
  of several variables, 465–476
  sine, 110–112
  spawner-recruit, 332
  translations and reflections of, 46–47
  trigonometric. *See* Trigonometric functions
  of two variables, 465
Fundamental Theorem of Calculus
  applications for, 398–402, 453
  explanation of, 398, 419
  finding area, 402–406
$f(x)$ notation, 10

Gauss, Carl Friedrich, 524n
Gauss-Jordan method
  cautions in using, 527
  explanation of, 524, 576
  on graphing calculators, 527–528
  solution of linear systems by, 524–529
  on spreadsheets, 528–529
General solution to differential equations,
    584–585, 628
General terms, 751
Gosset, William, 747
Graphical differentiation, 185–191
Graphical optimization, 322
Graphing calculator exercises, 25–28, 40, 49–50,
    60–63, 68–70, 81–83, 93–95, 115–117,
    121–123, 144, 153–154, 164–165, 182,
    185, 194–195, 199–200, 210–213, 218,
    235–237, 244–245, 253, 256–258, 271–272,
    282, 284, 295–297, 307, 323–324, 333–335,
    358, 393, 395, 408, 411, 417, 422–423,
    426, 434–436, 451, 457, 460, 474, 482,
    484, 495, 500, 535–536, 538, 553–555,
    565, 580, 595–597, 608–609, 631, 663,
    717–718, 726–728, 740, 746, 765
Graphing calculators
  absolute maximum and, 319
  absolute minimum and, 319, 330
  approximation of area on, 390
  area between two curves and, 414
  cobweb diagrams on, 757, 760
  continuous compounding on, 78–79
  correlation coefficient on, 23–24
  definite integrals on, 379, 389, 404, 442
  degree and radian measure on, 104
  derivatives on, 174–175, 176, 179,
    187–188, 208
  Euler's method on, 606
  exact value of vertex and, 43–44
  exponential equations on, 90
  exponential regression feature on, 79
  extrema on, 278
  functions of two variables on, 472
  Gauss-Jordan method on, 527–528
  graphing polynomials on, 54
  improper integrals on, 455
  increasing and decreasing functions on,
    268
  inflection points and, 291

  instantaneous rate of change on, 163
  intersect feature, 35
  least squares line on, 20–21
  limitations of, 301
  limits on, 131, 133, 137–138
  logarithms on, 88, 306
  logistic curve on, 591, 592
  MathPrint operating system, A-8–A-9
  matrix operations on, 542, 550, 559, 615
  normal curves on, 736
  number $e$ on, 76
  piecewise functions on, 151
  probability density functions on, 713–715
  quadratic functions, 45
  rational function on, 57
  relative extrema on, 278
  tangent lines on, 171
  translations and reflections of graphs, 47
  trapezoidal rule on, 430
  trigonometric functions on, 108, 113–114,
    247, 405–406
  value feature, 35
Graphs
  of absolute value functions, 119
  concavity of, 288–291
  curve sketching and, 298–308
  of derivatives, 187–188, 261–315
  of equations, 2
  explanation of, R-17, 2
  of exponential functions, 73–74, 119
  of functions, 65, 466–472
  of increasing and decreasing functions,
    262–270
  of intervals, R-17
  of linear inequalities, R-18
  of lines, 9
  of logarithmic functions, 85–86, 119,
    305–306
  of planes, 467–469
  of polynomial functions, 53–54, 300–301
  of quadratic functions, 41–48, 64, 119
  of rational functions, 56–57, 119, 301–305
  of sine and cosine functions, 110–112
  translations and reflections of, 46–47
  of trigonometric functions, 109–114, 119
Greatest common factor, R-5
Growth, logistic, 590–593
Growth constant, 97, 588
Growth functions
  exponential, 97
  limited, 99
Gunter, Edmund, 87

Half-life, 97
Half-open interval, R-17
Heraclitus, 317
Histograms, 688, 709
Horizontal asymptotes
  explanation of, 56
  method for finding, 139, 299
Horizontal lines
  equation of, 7
  graphs of, 9
  slope of, 7
Horizontal reflection, 46
Horizontal translation, 42

Hyperbolic paraboloid, 471
Hyperboloid of two sheets, 471

Identities, basic, 247
Identity matrices, 556
Implicit differentiation, 337–342, 356
Impossible event, 652
Improper integrals
  applications of, 453–457, 458
  explanation of, 454
  on graphing calculators, 455
Inconsistent system, 522
Increasing functions
  explanation of, 262–273
  test for, 264, 309
Indefinite integrals
  in definite integral example, 502
  explanation of, 364–365, 368
  of exponential functions, 368
  power rule to find, 365–366
Independent events
  explanation of, 673–675, 700
  product rule for, 674
Independent variables, 10, 465
Indeterminate form, 136
Index, R-26
Index of diversity, 92
Inequalities
  explanation of, R-17
  with fractions, R-19–R-20
  linear, R-17
  polynomial, R-18–R-19
  properties of, R-17
  quadratic, R-18
  rational, R-19–R-20
  symbols for, R-16
Infinity, limits at, 138–141, 192
Inflection points
  explanation of, 288
  graphing calculators and, 291
  method for finding, 299
Initial conditions, 586
Initial side of angle, 102
Initial value problems, 586–587
Instantaneous rate of change
  alternate form of, 160
  explanation of, 158–159, 172–173
  formula for, 159, 160
Integer exponents, R-21–R-23
Integral calculus, 363
Integral sign, 364, 507
Integrals, A-13
  area between two curves and, 412–418, 419
  convergent, 454
  definite, 385–397, 400, 419, 441–443,
    502–504
  divergent, 454
  double, 502–512, 514
  improper, 453–457, 458
  indefinite, 364–365, 368, 502
  iterated, 504
  relationship between sums and, 366
  tables of, 443
  of trigonometric functions, 379–381
Integrand, 364, 505
Integrating factor, 599

Integration, 362–426
　　antiderivatives, 363–375
　　area and the definite integral, 385–397
　　area between two curves, 412–418
　　average value and, 449–450
　　column, 438–441
　　of exponential functions, 418
　　Fundamental Theorem of Calculus,
　　　　398–411
　　improper integrals and, 453–457
　　limits of, 389, 509–510
　　lower limit of, 389
　　numerical, 428–436
　　by parts, 436–445, 458
　　region of, 505
　　rules of, 367–368
　　by substitution, 375–385
　　tabular, 438–441
　　techniques and applications of, 427–463
　　upper limit of, 389
　　variable limits of, 507–509
　　volume and, 446–449
Integration constant, 364, 366–367
Intercepts, 2
Interest
　　compound, 75–76
　　continuously compounded, 77–78
　　explanation of, 75
　　rate of, 75
　　simple, 75
Intermediate Value Theorem, 152
Interpolation, 70
Intersections, 640
Interval notation, R-17
Intervals
　　closed, 148, R-17
　　half-open, R-17
　　open, 148, R-17
　　real number, 710
Intravenous administration of drugs,
　　　197–200
Inverse functions, 86
Irrational numbers, 387n
Isoquant, 471
Itagaki, Koichi, 180
Iterated functions, 752–754
Iterated integrals, 504

Jackson, Andrew, 737–738
Jordan, Wilhelm, 524n

Kepler, Johannes, 447

Law of diminishing returns, 293
Leading coefficients, 51
Least common denominators, R-10
Least squares line
　　calculation of, 19–21, 64
　　correlation and, 21–24
　　explanation of, 17–19
Least squares method, 17
Leibniz, Gottfried Wilhelm, 129n, 202, 365
Leibniz notation, 202
Level curves, 469
Level surface, 472
Like terms, R-2

Limited growth functions, 99
Limits, 128–146
　　at endpoints, 330
　　existence of, 133, 192
　　explanation of, 129–130
　　of function, 128–134, 192
　　on graphing calculators, 131, 133, 137–138
　　at infinity, 138–141, 192
　　of integration, 389, 509–510
　　from left, 129
　　methods for determining, 130–134, 137–138
　　one-sided, 129
　　from right, 129
　　rules for, 134
　　in trigonometric functions, 247–248
　　two-sided, 129
Linear approximation, 350–355, 356
Linear cost function, 12–13, 64
Linear equations
　　explanation of, R-11, 4
　　solving of, R-11–R-12
Linear first-order differential equations
　　explanation of, 597
　　mixing of solutions and, 625–626
　　solving of, 597–601, 628
Linear functions, 2–17
　　cost analysis and, 12–13, 64
　　explanation of, 10, 41
　　least squares method and, 17
　　marginal cost and, 12
　　temperature and, 13
Linear inequalities
　　explanation of, R-17
　　solving of, R-17
Linear systems
　　of differential equations, 609–616, 628
　　explanation of, 521–523
　　Gauss-Jordan method to solve, 524–529, 576
　　inverse matrices and, 559–564
　　transformations of, 524
　　without a unique solution, 530–533
Lines
　　equation of tangent, 179
　　equations of, 4–7, 64, 169–170
　　graphs of, 9
　　horizontal, 6–7, 9
　　parallel, 7–8, 64
　　perpendicular, 8, 64
　　secant, 168
　　slope of, 3–4, 7, 63, 156–157
　　tangent, 168–172, 179
　　vertical, 7
Local extrema. See Relative extrema
Local maximum. See Relative maximum
Local minimum. See Relative minimum
Logarithmic equations
　　explanation of, 88
　　solving of, 88–89
Logarithmic functions, 84–96
　　continuity and, 149
　　derivatives of, 239–246, 255
　　explanation of, 85–86, 118, 255
　　graphs of, 85–86, 119, 305–306
Logarithms
　　change-of-base theorem for, 87–88
　　common, 87

　　evaluation of, 87–88
　　explanation of, 84–85, 118
　　on graphing calculators, 88
　　natural, 87
　　properties of, 86–87, 118
Logistic curve, 590–592
Logistic equations, 590
Logistic function, 233
Logistic growth model
　　application of, 590–593
　　explanation of, 590
Lorenz, E. N., 760
Lotka, A. J., 619
Lotka-Volterra equations, 619–620

Marginal analysis, 162, 353
Marginal cost, 12, 58, 162, 415
Marginal demand, 381
Marginal profit, 162
Marginal revenue, 162
Marginal value, 162
Mathematical induction technique, 753
Mathematical models, 2, 521, 768–770
MathPrint operating system, A-8–A-9
Matrices, 520–582
　　addition of, 539–541, 576
　　applications for, 539–540
　　augmented, 523
　　classification of, 539–540
　　coefficient, 559
　　column, 540
　　eigenvalues and eigenvectors, 556–575
　　equality of, 540
　　explanation of, 523
　　fate, 616
　　on graphing calculators, 542, 550, 559
　　identity, 556
　　inverses, 556–566, 576–577
　　multiplication of, 546–555
　　notation, 552
　　product of, 546–555
　　product of scalar and, 546–547, 576
　　product of two, 547, 576
　　row, 540
　　size of, 539–540
　　solution of linear systems, 521–538
　　square, 540
　　subtraction of, 541–542, 576
Matrix inverses
　　explanation of, 556–564, 576–577
　　finding multiplicative, 556–559
Maximum. See also Extrema
　　absolute, 317, 319–321
　　relative, 274, 276, 487
Maximum sustainable harvest, 258, 331–332
Mean. See Expected value
Median, 724–725, 743
Medical diagnosis, 707
Members of sets, 636
Mendel, Gregor, 664
MESOR, 111
Michaelis-Menten kinetics, 61
Midpoint rule, 386
Minimum. See also Extrema
　　absolute, 317, 319–321
　　relative, 274, 276, 487

Mixing problems, 625–626
Muir, Thomas, 103
Multiplication
    of binomials, R-4
    of matrices, 546–555
    order of operations, R-2
    of polynomials, R-3–R-5
    of rational expressions, R-8
Multiplication property
    of equality, R-11
    of inequality, R-17
Multiplicative inverse matrices, 556–559
Mutually exclusive events, 653–654,
        657, 700

Napier, John, 87
Natural logarithms, 87
Negative exponents, R-22
Newton, Isaac, 129n, 202
Nonlinear functions, 29–63
    explanation of, 29
    exponential functions as, 73–84
    illustrations of, 29–31
    limited growth, 99
    logarithmic functions as, 84–96
    polynomial functions as, 51–55
    properties of, 29–40
    quadratic functions as, 41–51
    rational functions as, 55–58
Nonlinear system of differential equations,
        617–623
Normal curves
    area under, 734
    explanation of, 733–734
Normal distribution
    explanation of, 733–739, 743
    standard, 734
Notation
    for derivatives, 202, 286
    function, 10
    $f(x)$, 10
    interval, R-17
    Leibniz, 202
    matrix, 552
    set-builder, 637
    summation, 17
$n$th term of a sequence, 751
Null sets, 636
Numbers
    critical, 264, 278, 280
    irrational, 387n
    real, R-2
Numerators, rationalizing, R-28
Numerical analysis, 433
Numerical integration, 428–436

Oblique asymptote, 302
Obtuse angles, 103
Odds, 657–658, 700
Open interval, 148, R-17
Operations, order of, R-2
Order of operations, R-2
Ordered pairs, 2
Ordered triples, 467
Origin, explanation of, 2

Outcomes, 651
Outliers, 24

Parabolas
    area of segment of, 431
    explanation of, 41
Paraboloid, 469, 471
Parallel lines, 7–8, 64
Parameter, 531
Parent-progeny function, 332
Parentheses, order of operations, R-2
Partial derivatives
    evaluation of, 478–480
    explanation of, 476–478
    rate of change and, 480
    second-order, 480–482
Particular solutions to differential equations, 586
Pascal, Blaise, 709
Pearl, Raymond, 591
Perfect squares, R-7
Period of function, 109
Periodic functions, 109, 119
Perpendicular lines, 8, 64
Phase plane diagram, 620
Phase shift, 111
Piecewise functions, 131, 151
Plane
    explanation of, 467
    graph of, 467–469
    $xy$-, 466
Plimpton 322, 105n
Poincaré, Henri, 760
Point of diminishing returns, 293–294
Point-slope form, 5, 7
Poisson distribution, 731–732, 743
Pollution of Great Lakes, 632–634
Polynomial functions
    continuity and, 149
    explanation of, 51, 64
    graphs of, 300–301
    properties of, 55, 64
Polynomial inequalities, R-18–R-19
Polynomials
    addition of, R-2–R-3
    cubic, 53
    explanation of, R-2
    factoring of, R-7
    graphing of, 53–54
    identifying degree of, 54
    multiplication of, R-3–R-5
    prime, R-7
    quartic, 53
    subtraction of, R-2–R-3
Popularity Index, 383
Positive root, R-23
Power functions, 52, 124–126
Power rule
    antiderivative and, 365–366
    explanation of, 204–205, 254, 418
Powers, order of operations, R-2
Predator food optimization, 518–519
Predator-Prey model, 617–621
Prime polynomials, R-7
Principal, 75
Principal root, R-23

Probability
    background of, 709
    basic principle of, 654, 700
    Bayes' theorem and, 675–677, 700
    complement rule for, 657, 700
    conditional, 668–675, 700
    empirical, 658–659
    events, 652–654
    experiments and, 651
    explanation of, 654–655
    independent events, 673–675, 700
    mutually exclusive events and, 653–654,
        657, 700
    odds in, 657–658, 700
    probability distribution, 659–661
    product rule of, 671–673, 700
    properties of, 659, 700
    sample spaces, 651
    sensitivity and specificity, 677–679
    union rule for, 655–657, 700
Probability density functions
    discrete probability functions vs., 712
    explanation of, 711–715, 743
    exponential distribution and, 730–732
    on graphing calculators, 713–715
    normal distribution and, 733–739
    special, 728–742
    uniform distribution and, 728–730
Probability distribution
    continuous, 710
    expected value of, 689–690, 720
    explanation of, 659, 688–689
    variance of, 692–693, 700, 719–720
Probability distribution table, 694
Probability functions
    discrete, 710
    explanation of, 688, 709–710
    of random variable, 688
Probability models, continuous, 709–718
Probability of an event, 709n
Problem of the points, 709
Product matrices, 546–555, 576
Product rule
    for derivatives, 214–216, 217, 254
    for independent events, 674
    for probability, 671–673, 700
Production function, 469–470
Proportional, 5
Proportional harvesting, 259
Pry, R. H., 591, 592
Pythagoras, 105n
Pythagorean theorem, 105

Quadrants, 2
Quadratic equations, R-12–R-13
Quadratic formula, R-13–R-14, 372, 413
Quadratic functions
    explanation of, 41, 64
    graphs of, 41–48, 64, 119
    maximum or minimum of, 44
Quadratic inequalities, R-18
Quadratic regression, 44
Quartic polynomials, 53
Quirin, Jim, 383
Quotient rule, explanation of, 216–217, 254

Radian, 103, 104, 118
Radian measure, 103–105
Radical sign, R-26
Radicals
    explanation of, R-25–R-29
    properties of, R-26
Radicand, R-26
Random samples, 738
Random variables
    continuous, 687, 710, 719–728
    discrete, 687
    expected value of, 689–690
    explanation of, 687, 709
    probability function of, 688
Range
    domain and, 33–34, 465
    explanation of, 30, 64
Rates of change
    of derivatives, 172–173, 284–285, 480
    explanation of, 155–157
    formula for average, 156
    formula for instantaneous, 159, 160
Rational equations, R-14–R-16
Rational exponents, R-23–R-24
Rational expressions
    combining of, R-9–R-10
    explanation of, R-8
    properties of, R-8
    reducing of, R-8–R-9
Rational functions
    continuity and, 149
    explanation of, 55, 65
    graphs of, 56–57, 119, 301–305
Rational inequality, R-19–R-20
Rays, 102
Real number interval, 710
Real numbers, R-2
Real zero, 54
Recursive sequences, 752
Reflections of functions, 46–47
Region of integration, 505
Related rates, 343–349, 356
Relative extrema
    explanation of, 273–275
    first derivative test for, 275–276, 309
    on graphing calculators, 278
    methods for finding, 276–281
    for realistic problems, 280–281
    saddle point and, 490–491
    second derivative test for, 292, 309, 489–490
Relative maximum
    explanation of, 274, 276, 487
    in functions of two variables, 487–496
Relative minimum
    explanation of, 274, 276, 487
    in functions of two variables, 487–496
Removable discontinuity, 148
Residuals, 71
Riemann, Georg, 389n
Riemann integral, 389n
Riemann sum, 389n
Right angles, 103
Root functions, 149
Roots
    cube, R-23
    positive, R-23

principal, R-23
square, R-23
Row matrices, 540
Row operations, 523–524, 576
Row vector, 540
Runge-Kutta method, 608

Saddle, 471
Saddle points, 488, 490–491
Sample space, 651, 668
Scalar, 546, 576
Scatterplots, 17, 21
Secant, 105. See also Trigonometric functions
Secant line, 168
Second derivative
    explanation of, 285
    method for finding, 286–287
Second derivative test, 291–294, 309, 489–490
Second-order contacts, 582
Second-order derivatives, 480–482
Semistable equilibrium points, 593
Sensitivity, 677–679
Separable differential equations, 587, 628
Separation of variables, 587–590
Sequences
    explanation of, 751
    general term of, 751
    iterated functions and, 752–754
    nth term of, 751
    recursive, 752
Set-builder notation, 637
Set operations
    for events, 653
    explanation of, 639–641
Sets
    applications of Venn diagrams, 641–646
    complement of, 639
    disjoint, 640, 645, 700
    elements of, 636
    empty, 636
    explanation of, 636
    intersection of, 640
    members of, 636
    set operations, 639–641, 653
    subsets of, 637–639, 700
    union of, 641
    union rule for, 645, 700
    universal, 637
Simple event, 652
Simple interest, 75
Simpson, Thomas, 432
Simpson's rule, 431–433, 458
Sine, 105. See also Trigonometric functions
Sine functions, 110–112
Size of matrices, 539–540
Slope
    of curve, 169
    explanation of, 3
    of line, 3–4, 7, 63
    of tangent line, 168–171, 264, 290
Slope-intercept form, 4–5, 7
Solid of revolution, 446–449, 458
Spawner-recruit function, 332
Special angles, 106–109
Specificity, 677–679

Spreadsheet exercises, 24–29, 67–69, 71, 93, 199–200, 236, 257–258, 361, 395, 426, 474, 484, 496, 535–536, 553–555, 565, 608–609
Spreadsheets
    approximation of area on, 391
    correlation coefficient in, 23
    Euler's method on, 606
    extrema on, 492–493
    Gauss-Jordan method on, 528–529
    least squares line on, 21
    matrix inverses on, 559
    matrix multiplication on, 550
    organizing data on, 542
    trapezoidal rule on, 431
Square matrices, 540
Square root, R-23
Squares
    difference of two, R-7
    perfect, R-7
Stability of equilibrium points, 593, 762–764
Stable equilibrium points, 259, 593, 758
Standard deviation
    explanation of, 692, 700, 719–720, 743
    method for finding, 720
Standard form, R-12
Standard normal distribution, 734
Standard position, 102
Stochastic processes, 673
Straight angles, 103
Subsets
    explanation of, 637
    number of, 639, 700
    properties of, 638
Substitution
    explanation of, 378
    integration by, 375–385
    method of, 381, 401–402, 419
Subtraction
    of matrices, 541–542, 576
    in order of operations, R-2
    of polynomials, R-2–R-3
    of rational expressions, R-8
Sum, derivative of, 207
Sum or difference rule
    explanation of, 207, 254, 418
    indefinite integrals and, 366
Summation notation, 17
Surface
    explanation of, 468
    volume under a, 505
Sylvester, James Joseph, 523n
Systems of equations, 520–582
    addition of matrices, 539–541
    eigenvalues and eigenvectors, 556–575
    explanation of, 521
    inverse matrices and, 556–566
    linear differential equations, 609–616
    multiplication of matrices, 546–555
    nonlinear differential equations, 617–623
    solution of linear systems, 521–538
    subtraction of matrices, 541–542
    transformations of, 524

Tabular integration, 438–441
Tangent, 105. See also Trigonometric functions

Tangent line
    equation of, 179
    explanation of, 168–172
    on graphing calculators, 171
    slope of, 168–171, 290, 339
Technology exercises. *See* Graphing calculator
        exercises and Spreadsheet exercises
Technology notes
    Graphing calculator, 23–24, 35, 45, 47, 53, 79,
        131, 151, 171, 176, 188, 208, 268, 291,
        301, 319, 379, 389, 404, 414, 442, 443,
        448, 591, 592, 615, 713–715, 724, 736
    Spreadsheet, 163, 542, 550, 559
Temperature, 13, 16
Terms
    explanation of, R-2
    like, R-2
    of sequences, 751
    unlike, R-2
Therapeutic window, 314
Thermodilution, 462–463
Third derivative, 285–286
Thomson, James, 103
Three Prisoners problem, 686
Time
    doubling, 84
    explanation of, 75
    minimizing, 327–328
Total change, 391–392
Total cost model, 360–361
Total differentials
    for three variables, 498, 514
    for two variables, 496–497
Traces, 469
Transformations of system, 524
Translations of functions, 46–47
Trapezium, 428n
Trapezoid, 428n
Trapezoidal rule, 386, 428–431, 458
Tree diagrams
    explanation of, 638
    probability applications for, 672–679
Trials, 651
Triangles, right, 107
Trigonometric functions, 102–117
    basic identities of, 247
    for common angles, 118
    definitions of, 105, 118
    derivatives of, 247–254, 255
    estimating limits, 247–248
    on graphing calculators, 108, 113–114, 247,
        405–406
    graphs of, 109–114, 119
    integrals of, 379–381
    values of, 105–109
Trigonometric identities, elementary, 105, 118

Trinomials
    explanation of, R-6
    factoring of, R-6–R-8
Turning points, 53

Uniform distribution, 728–730, 743
Union of sets, 641
Union rule
    for disjoint sets, 645
    for mutually exclusive events, 657
    for probability, 655–657, 700
    for sets, 645, 700
Unique solution
    explanation of, 522
    systems without, 530–533
Unit circle, 103
Universal sets, 637
Unlike terms, R-2
Unstable equilibrium points, 259, 593, 759

Value
    absolute, R-27
    average, 449–450, 458
    expected. *See* Expected value
    marginal, 162
Variables
    dependent, 10, 465
    explanation of, R-2
    functions of several, 465–476
    functions of two or more, 465
    independent, 10, 465
    random. *See* Random variables
    separation of, 587–590
    three, total differential for, 498, 514
    two, total differential for, 496–497
Variance
    alternative formula for, 722, 743
    for density function, 720–723, 743
    explanation of, 719–720, 743
    of a probability distribution, 692–693,
        700, 719–720
Vectors
    column, 540
    row, 540
Velocity
    explanation of, 159, 287–288
    integrals and, 371–372
Venn, John, 639
Venn diagrams
    applications of, 641–646
    conditional probability and, 670
    explanation of, 639
Verhulst, P. F., 591
Vertex, 41, 102
Vertical asymptotes
    explanation of, 56

method for finding, 137–138, 299
Vertical line
    equation of, 7
    slope of, 7
Vertical line test, 35–36, 64
Vertical reflection, 42
Vertical translation, 42
Volterra, Vito, 619
Volume
    double integrals and, 505–507, 514
    maximizing, 328–329, 330–331
    of solid of revolution, 446–449, 458
    under a surface, 505
von Bertalanffy growth curve, 99

Waiting times, exponential, 747–749
Writing exercises, 14–16, 24–25, 27–28, 40,
        48–50, 58, 60, 62–63, 65–70, 80–83,
        93–95, 100, 102, 120, 122–123, 143–144,
        153–154, 164–167, 181, 183–185, 189,
        193, 210–213, 218–219, 227–229,
        235–238, 244–245, 253–254, 255–257,
        270–272, 282, 295, 297–298, 306,
        310–313, 323–325, 334–335, 342,
        357–358, 373, 383–384, 392–393, 397,
        408–409, 411, 417, 419–420, 422–423,
        426, 434, 444–445, 456–457, 459–460,
        473, 475, 483–485, 493–496, 500–501,
        511, 515, 517, 534, 536–538, 542–545,
        553–555, 564, 577, 579–580, 595–597,
        608–609, 616, 622–623, 629–631,
        646–649, 661–667, 680–686, 696, 699,
        701–702, 704–706, 716, 725, 740–741,
        744, 755–756, 764

$x$
    equivalent expressions for change in, 175–176
    function of, 29, 173
$x$-axis, 2
$x$-coordinate
    explanation of, 2
    in exponential functions, 79
$x$-intercept, 2
$xy$-plane, 466
$xy$-trace, 469
$xz$-trace, 469

$y$-axis, 2
$y$-coordinate
    explanation of, 2
    in exponential functions, 79
$y$-intercept, 2
$yz$-trace, 469

$z$-scores, 734–737, 743
Zero-factor property, R-12

# Sources

## Chapter 1

### Section 1.1

1. Example 11 from "Morbidity and Mortality Weekly Report," Centers for Disease Control and Prevention, Vol. 60, No. 35, Sept. 9, 2011, p. 1209.
2. Example 12 from *Science,* Vol. 281, July 17, 1998, pp. 350–351.
3. Exercise 71 from Hockey, Robert V., *Physical Fitness: The Pathway to Healthful Living,* Times Mirror/Mosby College Publishing, 1989, pp. 85–87.
4. Exercise 72 from Alcabes, P., A. Munoz, D. Vlahov, and G. Friedland, "Incubation Period of Human Immunodeficiency Virus," *Epidemiologic Review,* Vol. 15, No. 2, The Johns Hopkins University School of Hygiene and Public Health, 1993, pp. 303–318.
5. Exercise 73 from *Science,* Vol. 254, No. 5034, Nov. 15, 1991, pp. 936–938, and www.cdc.gov/nchs/data.
6. Exercise 74 from *Science,* Vol. 253, No. 5017, July 19, 1991, pp. 306–308.
7. Exercise 75 from *Science,* Vol. 290, Nov. 17, 2000, p. 1291.
8. Exercise 76 from Robbins, Charles T., *Wildlife Feeding and Nutrition,* 2nd ed., Academic Press, 1993, p. 132.
9. Exercise 77 from Wagner, Diane, and Deborah M. Gordon, "Colony Age, Neighborhood Density and Reproductive Potential in Harvester Ants," *Oecologia,* Vol. 119, 1999, pp. 175–182.
10. Exercise 78 from Thompson, L. G., H. H. Brecher, E. Mosley-Thompson, D. R. Hardy, and B. G. Mark, "Glacier Loss on Kilimanjaro Continues Unabated," *Proceedings of the National Academy of Science of the United States of America,* Vol. 106, No. 47, Nov. 24, 2009, pp. 19770–19775.
11. Exercise 79 from *Science News,* June 23, 1990, p. 391.
12. Exercise 80 from *Science News,* Sept. 26, 1992, p. 195. *Science News,* Nov. 7, 1992, p. 399.
13. Exercise 83 from U.S. Census Bureau, www.census.gov/hhes/families/files/ms2.csv.
14. Exercise 84 from *2011 Yearbook of Immigration Statistics,* Office of Immigration Statistics, Sept. 2012, p. 5.
15. Exercise 85 from *Levels and Trends in Child Mortality Report 2011,* World Health Organization, p. 1.

### Section 1.2

1. Page 17 from U.S. Dept. of Health and Human Services, National Center for Health Statistics, found in *New York Times 2010 Almanac,* p. 394, and www.cdc.gov/nchs/faststats/acc-inj.htm.
2. Example 5 from Mena, E. A., et al., "Inflammatory Intermediates Produced by Tissues Encasing Silicone Breast Prostheses," *Journal of Investigative Surgery,* Vol. 8, 1995, p. 33. Copyright © 1995. Reproduced by permission of Taylor & Francis, Inc., www.routledge-ny.com.
3. Exercise 4 from "November 1989 Course 120 Examination Applied Statistical Methods" of the *Education and Examination Committee of The Society of Actuaries.* Reprinted by permission of The Society of Actuaries.
4. Exercise 10 from www.nctm.org/wlme/wlme6/five.htm.
5. Exercise 11 from Stanford, Craig B., "Chimpanzee Hunting Behavior and Human Evolution," *American Scientist,* Vol. 83, May–June 1995, pp. 256–261, and Goetz, Albert, "Using Open-Ended Problems for Assessment," *Mathematics Teacher,* Vol. 99, No. 1, August 2005, pp. 12–17.
6. Exercise 12 from Hensinger, Robert N., *Standards in Pediatric Orthopedics,* Raven Press, 1986, p. 309.
7. Exercise 13 from Shipman, Pat, et al., *The Human Skeleton,* Harvard University Press, 1985, p. 253.
8. Exercise 14 from Association of American Medical Colleges, Table 17: MCAT Scores and GPAs for Applicants and Matriculants to U.S. Medical Schools, 2001–2012, Dec. 17, 2012. www.aamc.org/download/321494/data/2012factstable17.pdf.
9. Exercise 15 from Pierce, George W., *The Songs of Insects,* Cambridge, MA, Harvard University Press, Copyright © 1948 by the President and Fellows of Harvard College.
10. Exercise 16 from Whipp, Brian J., and Susan Ward, "Will Women Soon Outrun Men?" *Nature,* Vol. 355, Jan. 2, 1992, p. 25. The data are from Peter Matthews, *Track and Field Athletics: The Records,* Guinness, 1986, pp. 11, 44; from Robert W. Schultz and Yu-anlong Liu, in *Statistics in Sports,* edited by Bennett, Jay, and Jim Arnold, 1998, p. 189; and from *The World Almanac and Book of Facts 2006,* p. 880.
11. Exercise 17 from www.run100s.com/HR/.
12. Exercise 18 from *Historical Poverty Tables,* U.S. Census Bureau.
13. Exercise 19 from Lee, Grace, Paul Velleman, and Howard Wainer, "Giving the Finger to Dating Services," *Chance,* Vol. 21, No. 3, 2008, pp. 59–61.
14. Exercise 21 from data provided by Gary Rockswold, Mankato State University, MN.
15. Exercise 22 from Carter, Virgil, and Robert E. Machol, *Operations Research,* Vol. 19, 1971, pp. 541–545.

### Section 1.3

1. Exercise 67 from Peter Tyack, © Woods Hole Oceanographic Institution.
2. Exercises 68 and 69 from Robbins, Charles T., *Wildlife Feeding and Nutrition,* 2nd ed., Academic Press, 1993, p. 125.
3. Exercise 72 from U.S. Energy Information Administration, International Energy Outlook 2011, DOE/EIA-0480(2011), Sept. 2011.
4. Exercise 73 from Internet World Stats Usage and Population Statistics, www.internetworldstats.com/emarketing.htm.

### Section 1.4

1. Example 6 from U.S. Environmental Protection Agency, "U.S. Greenhouse Gas Emissions From Human Activities, 1990–2010," *The World Almanac and Book of Facts 2013,* p. 322.
2. Example 7 from National Center for Health Statistics, *Health, United States, 2011: With Special Features on Socioeconomic Status and Health,* Table 86, p. 289.
3. Exercise 49 from Ralph DeMarr, University of New Mexico.
4. Exercise 50 from Harris, Edward F., Joseph D. Hicks, and Betsy D. Barcroft, "Tissue Contributions to Sex and Race; Differences in Tooth Crown Size of Deciduous Molars," *American Journal of Physical Anthropology,* Vol. 115, 2001, pp. 223–237.
5. Exercise 51 from Abuhamad, A. Z., et al., "Doppler Flow Velocimetry of the Splenic Artery in the Human Fetus: Is It a Marker of Chronic Hypoxia?" *American Journal of Obstetrics and Gynecology,* Vol. 172, No. 3, March 1995, pp. 820–825.
6. Exercise 52 from seer.cancer.gov/faststats/selections.php?#output.
7. Exercise 53 from U.S. Census Bureau, Current Population Survey, Annual Social and Economic Supplement, June 2011, and *The New York Times 2010 Almanac,* p. 294.
8. Exercise 54 from Palti-Wasserman, Daphna, et al., "Identifying and Tracking a Guide Wire in the Coronary Arteries During Angioplasty from X-Ray Images," *IEEE Transactions on Biomedical Engineering,* Vol. 44, No. 2, Feb. 1997, pp. 152–163.
9. Exercise 55 from U.S. Census Bureau, *The World Almanac and Book of Facts 2013,* p. 201.
10. Exercise 56 from Ralph DeMarr, University of New Mexico.
11. Exercise 58 from *National Traffic Safety Institute Student Workbook,* 1993, p. 7.

### Section 1.5

1. Example 5 from "TB Incidence in the United States, 1953–2011," Centers for Disease Control and Prevention, www.cdc.gov/tb/statistics/tbcases.ht.
2. Exercises 44 and 45 from Donley, Edward, and Elizabeth Ann George, "Hidden Behavior in Graphs," *Mathematics Teacher,* Vol. 86, No. 6, Sept. 1993.

3. Exercise 47 from Garriott, James C. (ed.), *Medical Aspects of Alcohol Determination in Biological Specimens*, PSG Publishing Company, 1988, p. 57.

4. Exercise 49 from Dobbing, John, and Jean Sands, "Head Circumference, Biparietal Diameter and Brain Growth in Fetal and Postnatal Life," *Early Human Development*, Vol. 2, No. 1, April 1978, pp. 81–87.

5. Exercise 50 data from Bausch & Lomb. The original chart gave all data to 2 decimal places.

6. Exercise 51 from Smith, J. Maynard, *Models in Ecology*, Oxford: Cambridge University Press, 1974.

7. Exercise 52 from Edelstein-Keshet, Leah, *Mathematical Models in Biology*, Random House, 1988.

8. Exercise 53 from Beyer, W. Nelson, et al., "Estimates of Soil Ingestion by Wildlife," *Journal of Wildlife Management*, Vol. 58, No. 2, 1994, pp. 375–382.

9. Exercise 54 from www.rose-hulman.edu /Class/CalculusProbs/Problems/BATTERUP /BATTERUP_3_0_0.html.

10. Exercise 57 from "Head Start Program Fact Sheet Fiscal Year 2011," Head Start, eclkc .ohs.acf.hhs.gov/hslc/mr/factsheets /2011-hs-program-factsheet.html.

11. Exercise 58 from Gary Rockswold, Mankato State University, Mankato, MN.

12. Exercise 59 from *Annual Energy Review*, U.S. Department of Energy, 2011.

**Review Exercises**

1. Exercise 75 from U.S. Department of Labor, Bureau of Labor Statistics, Table 6: The 30 Occupations with the Largest Projected Employment Growth, 2010-20, Feb. 1, 2012, www.bls.gov/news.release/ecopro.t06.htm.

2. Exercise 76 from U.S. Department of Agriculture, Economic Research Service, "Food Consumption, Prices and Expenditures, Annual"; Food Consumption (Per Capita) Data System, www.ers.usda.gov/data/foodconsumption/.

3. Exercise 77 from *Science*, Vol. 286, Nov. 5, 1999, p. 1099.

4. Exercise 78 from Food and Agriculture Organization of the United Nations, FAOSTAT, and UNDESA (2011), "2010 Revision of World Population Prospects."

5. Exercise 80 from Kissileff, H. R., and J. L. Guss, "Microstructure of Eating Behavior in Humans," *Appetite*, Vol. 36, No. 1, Feb. 2001, pp. 70–78.

6. Exercise 82 from Pearson, T. C., et al., "Interpretation of Measured Red Cell Mass and Plasma Volume in Adults," *British Journal of Haematology*, Vol. 89, 1995, pp. 748–756.

7. Exercise 83 from *Family Practice*, May 17, 1993, p. 55.

8. Exercise 84 from Von Foerster, Heinz, Patricia M. Mora, and Lawrence W. Amiot, "Doomsday: Friday, 13 November, A.D. 2026," *Science*, Vol. 132, Nov. 4, 1960, pp. 1291–1295.

9. Exercise 86 from "Diabetes Data and Trends," Centers for Disease Control and Prevention, Department of Health and Human Services, www.cdc.gov/diabetes/statistics/incidence /fig1.htm.

10. Exercise 87 from U.S. Department of Agriculture, Economic Research Service, "Briefing Rooms, Organic Agriculture," www.ers.usda .gov/Briefing/Organic.

11. Exercise 88 from www.census.gov/hhes /families/data/cps2012.html.

12. Exercise 89 from www.census.gov/hhes/ www/poverty/data/historical/families.html.

13. Exercise 90 from www.census.gov/popest /data/state/totals/2012.index.html and *Time Almanac 2013*, pp. 609–613.

14. Exercise 91 is based on a letter by Tom Blazey to *Mathematics Teacher*, Vol. 86, No. 2, Feb. 1993, p. 178.

15. Exercise 92 is based on the article "Problems Whose Solutions Lie on a Hyperbola," by Steven Schwartzman, *AMATYC Review*, Vol. 14, No. 2, Spring 1993, pp. 27–36.

16. Exercise 93 from Ronan, C., *The Natural History of the Universe*, Macmillan, 1991.

**Extended Application**

1. Page 71 from *Health, United States, 2009*, National Center for Health Statistics, U.S. Department of Health and Human Services, Table 24, www.cdc.gov/nchs/data/hus/ hus09.pdf.

## Chapter 2

### Section 2.1

1. Example 7 from Pollan, Michael, "The (Agri)Cultural Contradictions of Obesity," *The New York Times Magazine*, Oct. 12, 2003, p. 41; USDA–National Agriculture Statistics Service, 2006; and *The World Almanac and Book of Facts 2013*, p. 127.

2. Exercise 1 from Thomas, Jamie, "Exponential Functions," *AMATYC Review*, Vol. 18, No. 2, Spring 1997.

3. Exercise 37 from esa.un.org/unpp /index.asp.

4. Exercise 39 from U.S. Census Bureau, "U.S. Interim Projections by Age, Sex, Race, and Hispanic Origin," www.census.gov/ipc/www /usinterimproj/.

5. Exercise 40 from *The Complexities of Physician Supply and Demand: Projections Through 2025*, Association of American Medical Colleges, Nov. 2008, p. 20.

6. Exercise 41 from "Chart and Table of Bald Eagle Breeding Pairs in Lower 48 States,'" U.S. Fish & Wildlife Service, www.fws.gov /midwest/eagle/population/chtofprs.html.

7. Exercise 42 from Zanoni, B., C. Garzardi, S. Anselmi, and G. Rondinini, "Modeling the Growth of *Enterococcus faecium* in Bologna Sausage," *Applied and Environmental Microbiology*, Vol. 59, No. 10, Oct. 1993, pp. 3411–3417.

8. Exercise 43 from *The New York Times*, May 16, 2001, p. A18.

9. Exercise 44 from Hurley, Peter J., "Red Cell and Plasma Volumes in Normal Adults," *Journal of Nuclear Medicine*, Vol. 16, 1975, pp. 46–52; and Pearson, T. C., et al., "Interpretation of Measured Red Cell Mass and Plasma Volume in Adults," *British Journal of Haematology*, Vol. 89, 1995, pp. 748–756.

10. Exercise 45 from Chakraborty, Abhijit, et al., "Mathematical Modeling of Circadian Cortisol Concentrations Using Indirect Response Models: Comparison of Several Methods," *Journal of Pharmocokinetics and Biopharmaceutics*, Vol. 27, No. 1, 1999, pp. 23–43.

11. Exercise 46 from Boden, T., G. Marland, and Bob Andres, "Global $CO_2$ Emissions from Fossil-Fuel Burning, Cement Manufacture, and Gas Flaring: 1751–2008," Carbon Dioxide Information Analysis Center, June 10, 2011.

12. Exercise 48 from Miller, A. and J. Thompson, *Elements of Meteorology*, Charles Merrill, 1975.

13. Exercise 49 from www.intel.com/technology /mooreslaw/.

14. Exercise 50 from www.wwindea.org/.

15. Exercise 60 from Problem 5 in "November 1989 Course 140 Examination, Mathematics of Compound Interest" of the Education and Examination Committee of The Society of Actuaries. Reprinted by permission of The Society of Actuaries.

### Section 2.2

1. Example 9 from Ludwig, John, and James Reynolds, *Statistical Ecology: A Primer on Methods and Computing*, New York: Wiley, 1988, p. 92.

2. Exercise 71 from Lucky Larry #16 by Joan Page, *AMATYC Review*, Vol. 16, No. 1, Fall 1994, p. 67.

3. Exercise 77 from U.S. Census Bureau, "U.S. Interim Projections by Age, Sex, Race, and Hispanic Origin," www.census.gov/ipc/www /usinterimproj/.

4. Exercise 78 from "Chart and Table of Bald Eagle Breeding Pairs in Lower 48 States,'" U.S. Fish & Wildlife Service, www.fws.gov /midwest/eagle/population/chtofprs.html.

5. Exercise 80 from the U.S. Census Bureau.

6. Exercise 84 from Huxley, J. S., *Problems of Relative Growth*, Dover, 1968.

7. Exercise 85 from Sharkey, I., et al., "Body Surface Area Estimation in Children Using Weight Alone: Application in Paediatric Oncology," *British Journal of Cancer*, Vol. 85, No. 1, 2001, pp. 23–28.

8. Exercise 86 from *Science*, Vol. 284, June 18, 1999, p. 1937.

9. Exercise 87 from Horelick, Brindell, and Sinan Koont, "Applications of Calculus to Medicine: Prescribing Safe and Effective Dosage," *UMAP Module 202*, 1977.

10. Exercise 88 from Jansen, Jeroen C., et al., "Estimation of Growth Rate in Patients with Head and Neck Paragangliomas Influences the Treatment Proposal," *Cancer*, Vol. 88, No. 12, June 15, 2000, pp. 2811–2816.
11. Exercise 90 from McNab, Brian K., "Complications Inherent in Scaling the Basal Rate of Metabolism in Mammals," *Quarterly Review of Biology*, Vol. 63, No.1, Mar. 1988, pp. 25–54.
12. Exercise 91 from *The Globe and Mail*, Feb. 17, 1995; and UJA Federation of Greater Toronto. The data were quoted by Ron Lancaster and Charlie Marion, *Mathematics Teacher*, Vol. 90, No. 2, Feb. 1997.
13. Exercise 92 from Tymoczko, Dmitri, "The Geometry of Music Chords," *Science*, Vol. 313, July 7, 2006, pp. 72–74.
14. Exercise 94 from *New York Times*, June 6, 1999, p. 41.
15. Exercise 95 from www.npr.org/templates /rundowns/rundown.php?prgId=2&prgDa te=5-7-2002.
16. Exercise 96c from www.west.net/rperry /PueblaTlaxcala/puebla.html.
17. Exercise 96d from www.history.com /this-day-in-history/earthquake-shakes -mexico-city.
18. Exercise 96g from *New York Times*, Jan. 13, 1995.

## Section 2.3

1. Example 4 from Hoff, Gerald R., "Biology and Ecology of Threaded Sculpin, *Gymnocanthus Pistilliger*, in the Eastern Bering Sea," *Fishery Bulletin*, Vol. 98, No. 4, Oct. 2000, pp. 711–722.
2. Exercise 6 from www.census.gov/ipc/www /idb/worldpopinfo.php.
3. Exercise 7 from Brody, Jane, "Sly Parasite Menaces Pets and Their Owners," *New York Times*, Dec. 21, 1999, p. F7.
4. Exercise 11 from Speroff, Theodore, et al., "A Risk-Benefit Analysis of Elective Bilateral Oophorectomy: Effect of Changes in Compliance with Estrogen Therapy on Outcome," *American Journal of Obstetrics and Gynecology*, Vol. 164, Jan. 1991, pp. 165–174.
5. Exercise 12 from downsyndrome.about.com/ od/diagnosingdownsyndrome/a /Matagechart.htm.
6. Exercise 13 from *Science*, Vol. 277, July 25, 1997, p. 483.
7. Exercise 16 from Seppa, N., "Skin Cancer Makes Unexpected Reappearance," *Science News*, June 21, 1997, p. 383.

## Section 2.4

1. Example 9 from Roederer, Juan, *The Physics and Psychophysics of Music: An Introduction*, New York: Springer-Verlag, 1995.
2. Example 10 from Thomas, Robert, *The Old Farmer's Almanac*, 2000.
3. Exercise 77 from Churchland, M. M., and S. G. J. Lisberger, "Experimental and Computational Analysis of Monkey Smooth Pursuit Eye Movements," *Journal of Neurophysiology*, Vol. 86, No. 2, Aug. 2001, pp. 741–759.
4. Exercise 78 from Neal, R. D., and M. Colledge, "The Effect of the Full Moon on General Practice Consultation Rates," *Family Practice*, Vol. 17, No. 6, Dec. 2000, pp. 472–474.
5. Exercise 81 from Volicer, L., et al., "Sundowning and Circadian Rhythms in Alzheimer's Disease," *American Journal of Psychiatry*, Vol. 158, No. 5, May 2001, pp. 704–711.
6. Exercise 82 from Hensinger, Robert N., *Standards in Pediatric Orthopedics*, Raven Press, 1986, p. 51.
7. Exercise 87 from Roederer, Juan, *The Physics and Psychophysics of Music: An Introduction*, New York: Springer-Verlag, 1995.
8. Exercise 89 from Lando, Barbara, and Clifton Lando, "Is the Graph of Temperature Variation a Sine Curve?" *Mathematics Teacher*, Vol. 70, Sept. 1977, pp. 534–537.
9. Exercise 90 from Thomas, Robert, *The Old Farmer's Almanac*, 2000.

## Review Exercises

1. Exercise 108 from Keightley, Peter D., and Adam Eyre-Walker, "Deleterious Mutations, and the Evolution of Sex," *Science*, Vol. 290, Oct. 13, 2000, p. 331–333.
2. Exercise 110 from Cattet, Marc R. L., et al., "Predicting Body Mass in Polar Bears: Is Morphometry Useful?" *Journal of Wildlife Management*, Vol. 61, No. 4, 1997, pp. 1083–1090.
3. Exercise 111 from Rusconi, Franca, et al., "Reference Values for Respiratory Rate in the First 3 Years of Life," *Pediatrics*, Vol. 94, No. 3, Sept. 1994, pp. 350–355.
4. Exercise 112 from Johnson, Michael P., and Daniel S. Simberloff, "Environmental Determinants of Island Species Numbers in the British Isles," *Journal of Biogeography*, Vol. 1, 1974, pp. 149–154.
5. Exercise 114 from Gupat, Suneel K., et al., "Modeling of Circadian Testosterone in Healthy Men and Hypogonadal Men," *Journal of Clinical Pharmacology*, Vol. 40, No. 7, July 2000, pp. 731–738.
6. Exercise 115 from U.S. Department of Agriculture, "Production, Supply and Distribution," www.fas.usda.gov/psdonline, Aug. 12, 2010; and from *Science*, Vol. 283, Jan. 15, 1990, p. 310.
7. Exercise 116 from Chakraborty, Abhijit, et al., "Mathematical Modeling of Circadian Cortisol Concentrations Using Indirect Response Models: Comparison of Several Methods," *Journal of Pharmocokinetics and Biopharmaceutics*, Vol. 27, No. 1, 1999, pp. 23–43.
8. Exercise 125 from "Monthly Averages for Vancouver, Canada," www.weather.com.

## Extended Application

1. Page 124 from Bornstein, Marc H., and Helen G. Bornstein, "The Pace of Life," *Nature*, Vol. 259, Feb. 19, 1976, pp. 557–559.
2. Page 125 from Johnson, Michael P., and Daniel S. Simberloff, "Environmental Determinants of Island Species Numbers in the British Isles," *Journal of Biogeography*, Vol. 1, 1974, pp. 149–154.
3. Page 125 from Juan Camilo Bohorquez, "Common Ecology Quantifies Human Insurgency," *Nature*, Vol. 462, Dec. 17, 2009, pp. 911–914.
4. Exercise 1 from Gwartney, James D., Richard L. Stroup, and Russell S. Sobel, *Economics: Private and Public Choice*, 9th ed., The Dryden Press, 2000, p. 59.
5. Exercise 2 from White, Craig R., et al., "Phylogenetically Informed Analysis of the Allometry of Mammalian Basal Metabolic Rate Supports Neither Geometric Nor Quarter-Power Scaling," *Evolution*, Vol. 63, Oct. 2009, pp. 2658–2667.

To view the complete source list, visit the Downloadable Student Resources site, www.pearsonhighered.com/mathstatsresources. The complete list is also available to qualified instructors within MyMathLab or through the Pearson Instructor Resource Center, www.pearsonhighered.com/irc.

# KEY DEFINITIONS, THEOREMS, AND FORMULAS

**3.1 Rules for Limits**

Let $a$, $A$, and $B$ be real numbers, and let $f$ and $g$ be functions such that

$$\lim_{x \to a} f(x) = A \quad \text{and} \quad \lim_{x \to a} g(x) = B.$$

1. If $k$ is a constant, then $\lim_{x \to a} k = k$ and $\lim_{x \to a}[k \cdot f(x)] = k \cdot \lim_{x \to a} f(x) = k \cdot A$.

2. $\lim_{x \to a}[f(x) \pm g(x)] = \lim_{x \to a} f(x) \pm \lim_{x \to a} g(x) = A \pm B$
   (The limit of a sum or difference is the sum or difference of the limits.)

3. $\lim_{x \to a}[f(x) \cdot g(x)] = \left[\lim_{x \to a} f(x)\right] \cdot \left[\lim_{x \to a} g(x)\right] = A \cdot B$
   (The limit of a product is the product of the limits.)

4. $\lim_{x \to a} \dfrac{f(x)}{g(x)} = \dfrac{\lim_{x \to a} f(x)}{\lim_{x \to a} g(x)} = \dfrac{A}{B}$ if $B \neq 0$

   (The limit of a quotient is the quotient of the limits, provided the limit of the denominator is not zero.)

5. If $p(x)$ is a polynomial, then $\lim_{x \to a} p(x) = p(a)$.

6. For any real number $k$, $\lim_{x \to a}[f(x)]^k = \left[\lim_{x \to a} f(x)\right]^k = A^k$, provided this limit exists.

7. $\lim_{x \to a} f(x) = \lim_{x \to a} g(x)$ if $f(x) = g(x)$ for all $x \neq a$.

8. For any real number $b > 0$, $\lim_{x \to a} b^{f(x)} = b^{\left[\lim_{x \to a} f(x)\right]} = b^A$.

9. For any real number $b$ such that $0 < b < 1$ or $1 < b$,
   $\lim_{x \to a}[\log_b f(x)] = \log_b[\lim_{x \to a} f(x)] = \log_b A$ if $A > 0$.

**3.1 Limits at Infinity**

For any positive real number $n$,

$$\lim_{x \to \infty} \frac{1}{x^n} = 0 \quad \text{and} \quad \lim_{x \to -\infty} \frac{1}{x^n} = 0.$$

**3.3 Instantaneous Rate of Change**

The instantaneous rate of change for a function $f$ when $x = a$ is

$$\lim_{h \to 0} \frac{f(a + h) - f(a)}{h}, \quad \text{provided this limit exists.}$$

**3.4 Derivative**

The derivative of the function $f$ at $x$, written $f'(x)$, is defined as

$$f'(x) = \lim_{h \to 0} \frac{f(x + h) - f(x)}{h}, \quad \text{provided this limit exists.}$$

**Rules for Derivatives**

The following rules for derivatives are valid when all the indicated derivatives exist.

**4.1**

*Constant Rule* If $f(x) = k$, where $k$ is any real number, then $f'(x) = 0$.

**4.1**

*Power Rule* If $f(x) = x^n$ for any real number $n$, then $f'(x) = nx^{n-1}$.

**4.1**

*Constant Times a Function* Let $k$ be a real number. Then the derivative of $f(x) = k \cdot g(x)$ is

$$f'(x) = k \cdot g'(x).$$

**4.1**     *Sum or Difference Rule*   If $f(x) = u(x) \pm v(x)$, then

$$f'(x) = u'(x) \pm v'(x).$$

**4.2**     *Product Rule*   If $f(x) = u(x) \cdot v(x)$, then

$$f'(x) = u(x) \cdot v'(x) + v(x) \cdot u'(x).$$

**4.2**     *Quotient Rule*   If $f(x) = \dfrac{u(x)}{v(x)}$, and $v(x) \neq 0$, then

$$f'(x) = \frac{v(x) \cdot u'(x) - u(x) \cdot v'(x)}{[v(x)]^2}.$$

**4.3**     *Chain Rule*   If $y$ is a function of $u$, say $y = f(u)$, and if $u$ is a function of $x$, say $u = g(x)$, then $y = f(u) = f[g(x)]$, and

$$\frac{dy}{dx} = \frac{dy}{du} \cdot \frac{du}{dx}.$$

**4.3**     *Chain Rule (Alternate Form)*   If $y = f[g(x)]$, then $dy/dx = f'[g(x)] \cdot g'(x)$.

**4.4**     *Exponential Function*

$$D_x[a^{g(x)}] = (\ln a)a^{g(x)}g'(x)$$
$$D_x[e^{g(x)}] = e^{g(x)}g'(x)$$

**4.5**     *Logarithmic Function*

$$D_x\big(\log_a |g(x)|\big) = \frac{1}{\ln a} \cdot \frac{g'(x)}{g(x)}$$

$$D_x\big(\ln |g(x)|\big) = \frac{g'(x)}{g(x)}$$

**4.6**     *Trigonometric Functions*

$$D_x(\sin x) = \cos x \qquad D_x(\cos x) = -\sin x$$
$$D_x(\tan x) = \sec^2 x \qquad D_x(\cot x) = -\csc^2 x$$
$$D_x(\sec x) = \sec x \tan x \qquad D_x(\csc x) = -\csc x \cot x$$

**5.2 First Derivative Test**     Let $c$ be a critical number for a function $f$. Suppose that $f$ is continuous on $(a, b)$ and differentiable on $(a, b)$ except possibly at $c$, and that $c$ is the only critical number for $f$ in $(a, b)$.

1. $f(c)$ is a relative maximum of $f$ if the derivative $f'(x)$ is positive in the interval $(a, c)$ and negative in the interval $(c, b)$.

2. $f(c)$ is a relative minimum of $f$ if the derivative $f'(x)$ is negative in the interval $(a, c)$ and positive in the interval $(c, b)$.

**5.3 Second Derivative Test**     Let $f''$ exist on some open interval containing $c$ (except possibly at $c$ itself), and let $f'(c) = 0$.

1. If $f''(c) > 0$, then $f(c)$ is a relative minimum.
2. If $f''(c) < 0$, then $f(c)$ is a relative maximum.
3. If $f''(c) = 0$, then the test gives no information about extrema.

**7.1 Basic Trigonometric Integrals**

$$\int \sin x \, dx = -\cos x + C \qquad \int \cos x \, dx = \sin x + C$$

$$\int \sec^2 x \, dx = \tan x + C \qquad \int \csc^2 x \, dx = -\cot x + C$$

$$\int \sec x \tan x \, dx = \sec x + C \qquad \int \csc x \cot x \, dx = -\csc x + C$$

**7.2 General Power Rule for Integrals**

For $u = f(x)$ and $du = f'(x) \, dx$,

$$\int u^n \, du = \frac{u^{n+1}}{n+1} + C.$$

**7.2 Indefinite Integral of $e^u$**

If $u = f(x)$, then $du = f'(x) \, dx$ and

$$\int e^u \, du = e^u + C.$$

**7.2 Indefinite Integral of $u^{-1}$**

If $u = f(x)$, then $du = f'(x) dx$ and

$$\int u^{-1} \, du = \int \frac{du}{u} = \ln|u| + C.$$

**7.4 Fundamental Theorem of Calculus**

Let $f$ be continuous on the interval $[a, b]$, and let $F$ be *any* antiderivative of $f$. Then

$$\int_a^b f(x) \, dx = F(b) - F(a) = F(x)\Big|_a^b.$$

**8.1 Trapezoidal Rule**

Let $f$ be a continuous function on $[a, b]$ and let $[a, b]$ be divided into $n$ equal subintervals by the points $a = x_0, x_1, x_2, \ldots, x_n = b$. Then, by the trapezoidal rule,

$$\int_a^b f(x) dx \approx \left(\frac{b-a}{n}\right)\left[\frac{1}{2}f(x_0) + f(x_1) + \cdots + f(x_{n-1}) + \frac{1}{2}f(x_n)\right].$$

**8.1 Simpson's Rule**

Let $f$ be a continuous function on $[a, b]$ and let $[a, b]$ be divided into an even number $n$ of equal subintervals by the points $a = x_0, x_1, x_2, \ldots, x_n = b$. Then, by Simpson's rule,

$$\int_a^b f(x) \, dx \approx \frac{b-a}{3n}[f(x_0) + 4f(x_1) + 2f(x_2) + 4f(x_3) + \cdots + 2f(x_{n-2}) + 4f(x_{n-1}) + f(x_n)].$$

**8.2 Integration by Parts**

If $u$ and $v$ are differentiable functions, then

$$\int u \, dv = uv - \int v \, du.$$

**8.4 Improper Integrals**

If $f$ is continuous on the indicated interval and if the indicated limits exist, then

$$\int_a^\infty f(x) \, dx = \lim_{b \to \infty} \int_a^b f(x) \, dx,$$

$$\int_{-\infty}^b f(x) \, dx = \lim_{a \to -\infty} \int_a^b f(x) \, dx,$$

$$\int_{-\infty}^\infty f(x) \, dx = \int_{-\infty}^c f(x) \, dx + \int_c^\infty f(x) \, dx,$$

for real numbers $a$, $b$, and $c$, where $c$ is arbitrarily chosen.